Lecture Notes in Computer Science 10185

Commenced Publication in 1973
Founding and Former Series Editors:
Gerhard Goos, Juris Hartmanis, and Jan van Leeuwen

More information about this series at http://www.springer.com/series/7407

T.V. Gopal · Gerhard Jäger
Silvia Steila (Eds.)

Theory and Applications of Models of Computation

14th Annual Conference, TAMC 2017
Bern, Switzerland, April 20–22, 2017
Proceedings

 Springer

Editors
T.V. Gopal
Anna University
Chennai
India

Gerhard Jäger
Universität Bern
Bern
Switzerland

Silvia Steila
Universität Bern
Bern
Switzerland

ISSN 0302-9743   ISSN 1611-3349 (electronic)
Lecture Notes in Computer Science
ISBN 978-3-319-55910-0   ISBN 978-3-319-55911-7 (eBook)
DOI 10.1007/978-3-319-55911-7

Library of Congress Control Number: 2017935403

LNCS Sublibrary: SL1 – Theoretical Computer Science and General Issues

Printed on acid-free paper

This Springer imprint is published by Springer Nature
The registered company is Springer International Publishing AG
The registered company address is: Gewerbestrasse 11, 6330 Cham, Switzerland

Preface

Theory and Applications of Models of Computation (TAMC) is a series of annual conferences that aims at bringing together a wide range of researchers with interest in computational theory and its applications. These conferences have a strong interdisciplinary character; they are distinguished by an appreciation of mathematical depth and scientific rather than heuristic approaches as well as the integration of theory and implementation.

Some of the most important theoretical aspects of a model of computation are its power, generality, simplicity, synthesizability, verifiability, and expressiveness. The TAMC series of conferences explores the algorithmic foundations, computational methods, and computing devices to meet the rapidly emerging challenges of complexity, scalability, sustainability, and interoperability, with wide-ranging impacts on virtually every aspect of human endeavor.

Due to a policy change in China, the 13th such conference had to be canceled. As a consequence, the Steering Committee of TAMC decided to give the authors of those articles that had been accepted for TAMC 2016 the option to present them at TAMC 2017. For TAMC 2016 a total of 24 papers was accepted out of 35 submissions. For TAMC 2017, which was held in Bern during April 20–22, we had 68 submissions and could accept 27. In both cases the reviewing process was rigorous and conducted by international Program Committees. The authors and reviewers for TAMC 2017 were from 29 countries. This volume contains 45 of the 51 accepted submissions of TAMC 2016 and TAMC 2017.

The main themes of TAMC 2017 were computability, computer science logic, complexity, algorithms, models of computation, and systems theory, as reflected also by the choice of the invited speakers.

This volume contains abstracts or full papers of the invited lectures of TAMC 2017 and the written versions of those contributions to TAMC 2016 and TAMC 2017 that were presented at Bern.

> If indeed, as Hilbert asserted, mathematics is a meaningless game played with meaningless marks on paper, the only mathematical experience to which we can refer is the making of marks on paper.
>
> Eric Temple Bell, *The Queen of the Sciences*, 1931

We are very grateful to the Program Committees of TAMC 2016 and TAMC 2017, and the many external reviewers they called on, for the hard work and expertise that they brought to the difficult selection process. We thank all those authors who submitted their work for our consideration. We thank the members of the Editorial Board of *Lecture Notes in Computer Science* and the editors at Springer for their encouragement and cooperation throughout the preparation of this conference.

Last but not least we thank our sponsors for providing the financial and structural basis to have TAMC 2017 in Bern:

- Universität Bern
- Schweizerischer Nationalfonds
- Akademie der Naturwissenschaften Schweiz
- Burgergemeinde Bern
- Kanton Bern
- Engagement Stadt Bern
- Springer
- NASSCOM

April 2017

T.V. Gopal
Gerhard Jäger
Silvia Steila

Organization

Steering Committee

M. Agrawal	Indian Institute of Technology, India
J.Y. Cai	University of Wisconsin, USA
J. Hopcroft	Cornell University, USA
A. Li	Chinese Academy of Sciences, China
Z. Liu	Chinese Academy of Sciences, China

Conference Chair

G. Jäger	University of Bern, Switzerland

Local Organizers

M. Bärtschi	University of Bern, Switzerland
B. Choffat	University of Bern, Switzerland
G. Jäger	University of Bern, Switzerland
L. Jaun	University of Bern, Switzerland
L. Lehmann	University of Bern, Switzerland
T. Rosebrock	University of Bern, Switzerland
K. Sato	University of Bern, Switzerland
N. Savić	University of Bern, Switzerland
S. Steila	University of Bern, Switzerland
T. Strahm	University of Bern, Switzerland
T. Studer	University of Bern, Switzerland
J. Walker	University of Bern, Switzerland

Sponsoring Institutions

Universität Bern
Schweizerischer Nationalfonds
Akademie der Naturwissenschaften Schweiz
Burgergemeinde Bern
Kanton Bern
Engagement Stadt Bern
Springer
NASSCOM

Program Committee TAMC 2016

Chair

J.Y. Cai   University of Wisconsin, USA

Co-chairs

J. Cui   Xidian University, China
X. Sun   Chinese Academy of Sciences, China

Members

A. Abraham   Machine Intelligence Research Labs, USA
A. Bonato   Ryerson University, Canada
Y. Chen   Shanghai Jiao Tong University, China
R.G. Downey   Victoria University of Wellington, New Zealand
H. Fernau   Universität Trier, Germany
D. Fotakis   National Technical University of Athens, Greece
T.V. Gopal   Anna University, India
S. Lempp   University of Wisconsin, USA
J. Liu   Auckland University of Technology, New Zealand
K. Meer   Brandenburgische Technische Universität, Germany
M. Minnes   University of California, USA
P. Moser   Maynooth University, Ireland
M. Ogihara   University of Miami, USA
P. Peng   Technische Universität Dortmund, Germany
H. Wang   Chinese Academy of Sciences, China
W. Wang   Sun Yat-Sen University, China
G. Wu   Nanyang Technological University, Singapore
Y. Yin   Nanjing University, China
M. Ying   University of Technology Sydney, Australia
T. Zeugmann   Hokkaido University, Japan
S. Zhang   Chinese University of Hong Kong, Hong Kong, SAR China
X. Li   Xidian University, China
Z. Fu   University of Wisconsin, USA
D. Zhu   Shandong University, China
S. Lokam   Microsoft Research, India
W. Chen   Microsoft Research, China
X. Chen   Columbia University, USA
Y. Cao   Hong Kong Polytechnic University, Hong Kong, SAR China
K. Yi   Hong Kong University of Science and Technology,
                       Hong Kong, SAR China
J. Zhang   Chinese Academy of Sciences, China
Y. Yu   Shanghai Jiao Tong University, China
M. Li   City University of Hong Kong, Hong Kong, SAR China
Z. Huang   The University of Hong Kong, Hong Kong, SAR China
L. Zhang   Bell Labs Research, USA

X. Bei	Nanyang Technological University, Singapore
M. Xia	Chinese Academy of Sciences, China
A. Fernandez	IMEDIA, Spain
X. Wang	Shandong University, China
Y. Fu	Shanghai Jiao Tong University, China
J. Nelson	Harvard University, USA
J. Xue	Jiangxi Normal University, China
R. Jain	National University of Singapore, Singapore

Program Committee TAMC 2017

Chair

G. Jäger	University of Bern, Switzerland

Co-chair

T.V. Gopal	Anna University, India

Members

S. Artemov	City University of New York, USA
J. Bradfield	The University of Edinburgh, UK
C. Calude	The University of Auckland, New Zealand
V. Chakaravarthy	IBM Research, India
A. Fässler	Bern University of Applied Sciences, Switzerland
H. Fernau	Universität Trier, Germany
D. Fotakis	National Technical University of Athens, Greece
T. Fujito	Toyohashi University of Technology, Japan
C.A. Furia	Chalmers University of Technology, Sweden
A.D. Jaggard	United States Naval Research Laboratory, USA
R. Kuznets	TU Wien, Austria
S. Lempp	University of Wisconsin, USA
J. Liu	Auckland University of Technology, New Zealand
S. Martini	University of Bologna, Italy
K. Meer	Brandenburgische Technische Universität, Germany
M. Minnes	University of California, USA
P. Moser	Maynooth University, Ireland
M. Ogihara	University of Miami, USA
J. Rolim	University of Geneva, Switzerland
H. Schwichtenberg	Ludwig-Maximilians-Universität München, Germany
A. Seth	Indian Institute of Technology, Kanpur, India
R.K. Shyamasundar	Tata Institute of Fundamental Research, India
S. Steila	University of Bern, Switzerland
T. Studer	University of Bern, Switzerland
S. Wainer	University of Leeds, UK
P. Widmayer	ETH Zurich, Switzerland
G. Wu	Nanyang Technological University, Singapore

Y. Yin	Nanjing University, China
M. Ying	University of Technology Sydney, Australia
T. Zeugmann	Hokkaido University, Japan
N. Zhan	Chinese Academy of Sciences, China

Abstracts of Invited Talks

Cognitive Reasoning and Trust in Human-Robot Interactions

Marta Kwiatkowska

Department of Computer Science, University of Oxford, Oxford, UK
marta.kwiatkowska@cs.ox.ac.uk

Abstract. We are witnessing accelerating technological advances in autonomous systems, of which driverless cars and home-assistive robots are prominent examples. As mobile autonomy becomes embedded in our society, we increasingly often depend on decisions made by mobile autonomous robots and interact with them socially. Key questions that need to be asked are how to ensure safety and trust in such interactions. How do we know when to trust a robot? How much should we trust? And how much should the robots trust us? This paper will give an overview of a probabilistic logic for expressing trust between human or robotic agents such as "agent A has 99% trust in agent B's ability or willingness to perform a task" and the role it can play in explaining trust-based decisions and agent's dependence on one another. The logic is founded on a probabilistic notion of belief, supports cognitive reasoning about goals and intentions, and admits quantitative verification via model checking, which can be used to evaluate such trust in human-robot interactions. The paper concludes by summarising recent advances and future challenges for modelling and verification in this important field.

Approximate Counting via Correlation Decay

Pinyan Lu

Shanghai University of Finance and Economics, Yangpu, China
lu.pinyan@mail.shufe.edu.cn

In this talk, I will survey some recent development of approximate counting algorithms based on correlation decay technique. Unlike the previous major approximate counting approach based on sampling such as Markov Chain Monte Carlo (MCMC), correlation decay based approach can give deterministic fully polynomial-time approximation scheme (FPTAS) for a number of counting problems. Based on this new approach, many new approximation schemes are obtained for counting problems on graphs, partition functions in statistical physics and so on.

On Extraction of Programs from Constructive Proofs

Maria Emilia Maietti

Università di Padova, Padua, Italy
maietti@math.unipd.it

A key distinctive feature of constructive mathematics with respect to classical one is the fact that from constructive proofs one can extract computable witnesses of provable existential statements. As a consequence all the constructively definable number theoretic functions are computable.

In this talk we argue that for certain constructive dependent type theories known to satisfy the proofs-as-programs paradigm, the extraction of computable witnesses from existential statements must be done in a stronger proofs-as-programs theory, for example in a realizability model. This is the case both for Coquand's Calculus of Constructions in [1] extended with inductive definitions and implemented in the proof-assistant Coq, and for its predicative version represented by the intensional level of the Minimalist Foundation in [2, 3].

References

1. Coquand, T.: Metamathematical investigation of a calculus of constructions. In: Odifreddi, P. (ed.) Logic in Computer Science, pp. 91–122. Academic Press (1990)
2. Maietti, M.E.: A minimalist two-level foundation for constructive mathematics. Ann. Pure Appl. Logic **160**(3), 319–354 (2009)
3. Maietti, M.E., Sambin, G.: Toward a minimalist foundation for constructive mathematics. In: Crosilla, L., Schuster, P. (eds.) From Sets and Types to Topology and Analysis: Practicable Foundations for Constructive Mathematics. Oxford Logic Guides, vol. 48, pp. 91–114. Oxford University Press (2005)

Computational Complexity for Real Valued Graph Parameters

Johann A. Makowsky

Israel Institute of Technology, Haifa, Israel
janos@cs.technion.ac.il

A *real-valued graph parameter* is a function f which maps (possibly weighted) graphs into the real number $\mathbb{R}$ such that two isomorphic (weighted) graphs receive the same value. Typical examples are graph polynomials $F(G; \bar{X}) \in \mathbb{R}[\bar{X}]$ in k indeterminates, partition functions and Holant functions. The talk is based [1, 2]. We address the following issues:

(i) How to choose the computational model?
the complexity of evaluating $F(G; \bar{a})$ The weighted graphs are best modeled as *metafinite structures*. In our discussion computations over the reals $\mathbb{R}$ are performed in unit cost in the computation model of Blum-Shub-Smale (BSS).

(ii) What are the complexity classes?
In BSS polynomial time $\mathbf{P}_{\mathbb{R}}$ and non-determinsitic time $\mathbf{NP}_{\mathbb{R}}$ are well defined. However, we are looking for an analogue of $\#\mathbf{P}$, which captures the complexity of evaluating a graph polynomial $F(G; \bar{X})$ at real values of $\bar{X}$, for which there are complete problems.

(iii) Are there dichotomy theorems?
For a graph polynomial $F(G; \bar{X})$ we look at the complexity spectrum which describes how the complexity of evaluating $F(G; \bar{a})$ varies for different $\bar{a} \in \mathbb{R}^k$. The analogue of Ladner's Theorem, which states that there are many intermediate complexity classes, unless $\mathbf{P} = \mathbf{NP}$, also holds in BSS. However, in all the cases known in the literature, the complexity of evaluating $F(G; \bar{a})$ is either in $\mathbf{P}_{\mathbb{R}}$ or $\mathbf{NP}_{\mathbb{R}}$-hard (the Difficult Point Dichotomy). We present infinitely many new graph polynomials for which this dichotomy holds. We shall also discuss the difficulties in proving such a dichotomy in general, and formulate various conjectures, which state that such a dichotomy holds for wide classes of graph polynomials.

Johann A. Makowsky—During 2016 partially supported by the Simons Institute, Berkeley, USA.

References

1. Goodall, A., Hermann, M., Kotek, T., Makowsky, J.A., Noble, S.D.: On the complexity of generalized chromatic polynomials (2017). arXiv: 1701.06639
2. Makowsky, J.A., Kotek, T., Ravve, E.V.: A computational framework for the study of partition functions and graph polynomials. In: Proceedings of the 12th Asian Logic Conference 2011, pp. 210–230. World Scientific (2013)

Natural Language Processing, Moving from Rules to Data

Carlos Martín-Vide

Rovira i Virgili University, Tarragona, Spain
carlos.martin@urv.cat

Abstract. During the last decade, we assist to a major change in the direction that theoretical models used in natural language processing follow. We are moving from rule-based systems to corpus-oriented paradigms. In this paper, we analyze several generative formalisms together with newer statistical and data-oriented linguistic methodologies. We review existing methods belonging to deep or shallow learning applied in various subfields of computational linguistics. The continuous, fast improvements obtained by practical, applied machine learning techniques may lead us to new theoretical developments in the classical models as well. We discuss several scenarios for future approaches.

Joint work with Adrian-Horia Dediu and Joana M. Matos.

An All-or-Nothing Flavor
to the Church-Turing Hypothesis

Stefan Wolf[1,2]

[1] Faculty of Informatics, Università della Svizzera italiana (USI),
6900 Lugano, Switzerland
wolfs@usi.ch
[2] Facoltà indipendente di Gandria, Lunga Scala, 6978 Gandria, Switzerland

Abstract. Landauer's principle claims that "Information is Physical." It is not surprising that its conceptual *antithesis*, Wheeler's "It from Bit," has been more popular among computer scientists — in the form of the *Church-Turing hypothesis:* All natural processes can be computed by a universal Turing machine; physical laws then become descriptions of subsets of *observable*, as opposed to merely *possible*, computations. Switching back and forth between the two traditional styles of thought, motivated by quantum-physical Bell correlations and the doubts they raise about fundamental space-time causality, we look for an intrinsic, physical randomness notion and find one around the second law of thermodynamics. Bell correlations combined with *complexity as randomness* tell us that beyond-Turing computations are either physically impossible, or they can be carried out by "devices" as simple as individual photons.

Computing on Streams and Analog Networks

Jeffery Zucker

McMaster University, Hamilton, Canada
zucker@mcmaster.ca

In [5] John Tucker and I defined a general concept of a network of analog processing units or modules connected by analog channels, processing data from a metric space A, and operating with respect to a global continuous clock $\mathbb{T}$, modelled by the set of non-negative reals. The inputs and output of a network are continuous streams $u : \mathbb{T} \to A$, and the input-output behaviour of a network with system parameters from A is modelled by a function of the form

$$\Phi : \mathcal{C}[\mathbb{T}, A]^p \times A^r \to \mathcal{C}[\mathbb{T}, A]^q,$$

where $\mathcal{C}[\mathbb{T}, A]$ is the set of all continuous streams equipped with the compact-open topology. We give an equational specification of the network, and a semantics when some physically motivated conditions on the modules, and a stability condition on the behaviour of the network, are satisfied. This involves solving a fixed point equation over $\mathcal{C}[\mathbb{T}, A]$ using a contraction principle based on the fact that $\mathcal{C}[\mathbb{T}, A]$ can be approximated by metric spaces. We analysed in detail a case study of analogue computation, using a mechanical system involving a mass, spring and damper, in which data are represented by displacements. The curious thing about this solution is that it worked only for certain ranges in the values of the parameters M (mass), K (spring constant) and D (damping constant), namely $M > \max(K, 2D)$, www which has no obvious physical interpretation. (More on this below.)

The fixed points found as above are functions of the parameters of the system, considered as inputs (in our example, the external force applied to the mass, as a function of time). The functionals characterizing the system are then *stream transformations*. Tucker and I showed [6, 7] that these transformations are (respectively) *continuous* (w.r.t. a suitable topology) and *computable*, w.r.t. a suitable notion of *concrete computation* on the reals, where the computable reals, and operations on them, are represented respectively by codes for effective Cauchy sequences, and operations on them – essentially equivalent to Grzegorczyk-Lacombe computability.

The significance of *continuity* of the fixed point function is that it implies *stability* of the fixed point as the solution to the specification. This is related to *Hadamard's principle* which (as (re-)formulated by Courant and Hilbert) states that for a scientific problem to be well posed, the solution must (apart from existing and being unique) depend continuously on the data.

This work was done in collaboration with John Tucker, Nick James and Diogos Poças. Research supported by NSERC (Canada).

Returning to the problem indicated above that the functionals (apparently) converge to fixed points only for certain values in the range of the parameters, this was successfully solved by Nick James [1, 2] who, developing a theory of stream functions on Banach spaces, showed that the important thing is *how* the network is modularized. By re-modularizing our original mass/spring/damper suitably, he obtained an equivalent system in which the (unnatural seeming) limitations on the sizes of the parameters are removed.

We turn to the next development in this theory. Diogo Poças [3] investigated analog networks which were based on Shannon's GPACs (general purpose analog computers). He has made a number of significant developments in this theory, which also impacts on the work described above: (1) He worked in the context of Fréchet spaces, not Banach spaces, which provides a more natural framework for investigating the space $C[\mathbb{T}, A]$. (2) He characterized the class of functions over the reals, computable by GPACS, as the *differential algebraic* functions. (3) He has extended the structure of GPACS to X-GPACS, which allows *two* independent variables: t, ranging over *time*, as before, and also x, ranging over *space*. This to the use of *partial differential equations*.

It turns out that the X-GPAC generable functions are precisely those solvable by partial *differential algebraic systems* [4]. Now certain interesting functions, such as the *gamma function*, and the *Riemann zeta function*, are not differentially algebraic, and (therefore) not X-GPAC computable. (4) This motivates a further extension of the X-GPAC model to the LX-GPAC model, ("L" for limit), which permits the use of limit operations (either discrete, over the naturals, or continuous, over segments of the reals). This facility permits the generation of both the gamma function and Riemann zeta function [3, Ch. 4]

It remains to find a useful characterization, in terms of concrete computation models, of the class of functions generated by LX-GPACs.

References

1. James, N.D.: Existence, continuity and computability of unique fixed points in analog network models. Ph.D. thesis, Department of Computing and Software, McMaster University (2012)
2. James, N.D., Zucker, J.I.: A class of contracting stream operators. Comput. J. **56**, 15–33 (2013)
3. Poças, D.: Analog computability in differential equations. Ph.D. thesis, Department of Mathematics, McMaster University (2017). Draft.
4. Poças, D., Zucker, J.I.: Analog networks on function data streams. Based on Ch. 3 of [3]; submitted for publication (2017)
5. Tucker, J.V., Zucker, J.I.: Computability of analog networks. Theor. Comput. Sci. **371**, 115–146 (2007)
6. Tucker, J.V., Zucker, J.I.: Continuity of operators on continuous and discrete time streams. Theor. Comput. Sci. **412**, 3378–3403 (2011)
7. Tucker, J.V., Zucker, J.I.: Computability of operators on continuous and discrete time streams. Computability **3**, 9–44 (2014)

Contents

Invited Papers

Cognitive Reasoning and Trust in Human-Robot Interactions

Marta Kwiatkowska[✉]

Department of Computer Science, University of Oxford, Oxford, UK
marta.kwiatkowska@cs.ox.ac.uk

Abstract. We are witnessing accelerating technological advances in autonomous systems, of which driverless cars and home-assistive robots are prominent examples. As mobile autonomy becomes embedded in our society, we increasingly often depend on decisions made by mobile autonomous robots and interact with them socially. Key questions that need to be asked are how to ensure safety and trust in such interactions. How do we know when to trust a robot? How much should we trust? And how much should the robots trust us? This paper will give an overview of a probabilistic logic for expressing trust between human or robotic agents such as "agent A has 99% trust in agent B's ability or willingness to perform a task" and the role it can play in explaining trust-based decisions and agent's dependence on one another. The logic is founded on a probabilistic notion of belief, supports cognitive reasoning about goals and intentions, and admits quantitative verification via model checking, which can be used to evaluate trust in human-robot interactions. The paper concludes by summarising future challenges for modelling and verification in this important field.

1 Introduction

Autonomous robotics has made tremendous progress over the past decade, with dramatic advances in areas such as driverless cars, home assistive robots, robot-assisted surgery, and unmanned aerial vehicles. However, high-profile incidents such as the fatal Tesla crash [11] make clear the risks from improper use of this technology. Our decisions whether to rely or not on automation technology are guided by *trust*. Trust is a subjective evaluation made by one agent (the truster) about the ability or willingness of another agent (the trustee) to perform a task [6,14]. A key aspect of a trust-based relationship is that the trustor's decision to trust is made on the expectation of benevolence from the trustee [15], and the trustor is, in fact, vulnerable to the actions of the trustee. Studies of trust in automation [12] have concluded that it is affected by factors such as reliability and predictability: it increases slowly if the system behaves as expected, but drops quickly if we experience failure. However, autonomous robots

This work is supported by EPSRC Mobile Autonomy Programme Grant EP/M019918/1.

T.V. Gopal et al. (Eds.): TAMC 2017, LNCS 10185, pp. 3–11, 2017.
DOI: 10.1007/978-3-319-55911-7_1

are independent decision-makers, and may therefore exhibit unpredictable, and even surprising, behaviour. Further, they need to correctly interpret the social context and form relationships with humans, thus becoming members of society.

Human relationships are built on (social) trust, which is a key influence in decisions whether an autonomous agent, be it a human or a robot, should act or not. Social trust is an expression of a complex cognitive process, informed by the broader context of cultural and social norms. Reasoning with trust and norms is necessary to justify and explain robots' decisions and draw inferences about accountability for failures, and hence induce meaningful communication and relationships with autonomous robots. There are dangers in acting based on inappropriate trust, for example, 'overtrust' in the Tesla crash. We need to program robots so that they can not only be trusted, but also so that they develop human-like trust in humans and other robots, and human-like relationships.

While reliability for computerised systems has been successfully addressed through formal verification techniques such as model checking, trust and ethics for robotics has only recently emerged as an area of study, in response to the rapid technological progress [10]. Elsewhere, for example in management, psychology, philosophy and economics, trust has been studied widely. Digital trust concepts are also prominent in e-commerce, where trust is based on reputation or credentials. However, the notion of trust needed for human-robot partnerships is *social trust*, which has been little studied: it involves cognitive processes (i.e. mental attitude, goals, intentions, emotion) that lead to a decision whether to trust or not, and is influenced through past experience and preferences.

This paper gives an overview of recent progress towards a specification formalism for expressing social trust concepts. The resulting logic, Probabilistic Rational Temporal Logic (PRTL*), is interpreted over stochastic multiagent systems (essentially concurrent stochastic games) extended with goals and intentions, where stochasticity arises from randomness and environmental uncertainty. Trust is defined in terms of (*subjective*) probabilistic belief, which allows one to quantify the amount of trust as a *belief-weighted expectation*, informally understood as a degree of trust. The logic can express, for example, if A is a human rider of an autonomous car B, that "A has 99% trust in B's ability to safely reach the required destination", and "B has 90% trust in A's willingness not to give unwise instructions". The key novelty in the framework is the addition of the *cognitive* dimension, in which (human or robotic) agents carry out their deliberations prior to decision-making; once the decision has been made, the agents act on them, with the actions taking place in the usual *temporal* dimension. The logic, under certain restrictions, admits a model checking procedure, which can be employed in decision-making to evaluate and reason about trust in human-robot relationships, and to assist in establishing accountability. We illustrate the main trust concepts by means of an example, referring the reader to the details in [7].

2   An Illustrative Example

We illustrate the key features of social trust using a variant of a trust game called the Parking Game due to Vincent Conitzer [3], see Fig. 1. Trust games are often used in economics, where it is assumed that players act on the basis of pure self-interest. However, experiments with human subjects consistently show that humans behave differently and are often willing to act on the assumption of the other player's goodwill.

The Parking Game illustrates a situation where cars A and B (let us assume they are autonomous) are waiting for a parking space, with car B behind A. Car A either waits or can move aside to let car B through, on the assumption that B is in a hurry and wants to pass. Car B, however, can steal A's parking space if it becomes available, or pass. Though somewhat artificial, we will also allow an iterated version of this game, where the cars return to compete for the parking space in the same order. The payoffs in this game indicate that the best outcome for both A and B is for A to move aside and B pass. As experience shows, this is a typical situation if the cars were driven by human drivers. However, according to the standard game-theoretic solution the Nash equilibrium is for A to wait, rather than move aside, to avoid the parking space being taken.

In [13] an alternative solution method is proposed that results in the equilibrium of A moving aside and B passing. A similar game is considered in [8], where the computation of the payoff is amended to include trust value. This paper puts forward a different solution, where we explicitly model the evolution of trust starting from some initial value, and update that (subjective) trust based on experience (that is, interactions between agents), preferences and context.

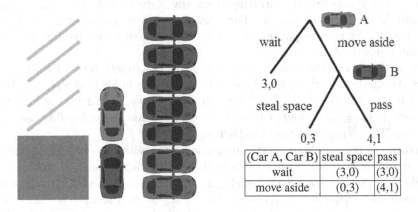

(Car A, Car B)	steal space	pass
wait	(3,0)	(3,0)
move aside	(0,3)	(4,1)

Fig. 1. The 'Parking Game' due to Vincent Conitzer [3] (reproduced with permission).

3   The Model

We work with stochastic multiplayer games as models, to capture both external uncertainty as well as internal probabilistic choices. We sometimes refer to players of the game as agents.

Let $\mathcal{D}(X)$ denote the set of probability distributions on a set X. For simplicity, we present a simplified variant of the model, and remark that partial observability and strategic reasoning can be handled [7].

Definition 1. *A* stochastic multiplayer game (SMG) *is a tuple* $M = (Ags, S, s_{\text{init}}, \{Act_A\}_{A \in Ags}, T)$, *where Ags is a finite set of agents, S is a finite set of states, $s_{\text{init}} \in S$ is an initial state, Act_A is a finite set of actions for agent A, and $T : S \times Act \rightarrow \mathcal{D}(S)$ is a (partial) transition probability function such that $Act = \times_{A \in Ags} Act_A$ and for each state s there exists a unique joint action $a \in Act$ such that a (non-unique) state s' is chosen with probability $T(s, a)(s')$.*

Let a_A be agent A's action in the joint action $a \in Act$. We let $Act(s) = \{a \in Act \mid T(s, a)$ is defined$\}$ and $Act_A(s) = \{a_A \mid a \in Act(s)\}$. For technical reasons, we assume that $Act(s) \neq \emptyset$ for all $s \in S$.

States S are global, and encode agents' local states as well as environment states. In each state s, agents independently (and possibly at random) choose a local action (which may include the silent action $\perp$), the environment performs an update, and the system transitions to a state s' satisfying $T(s, a)(s') > 0$, where a is the *joint* action.

We define a finite, resp. infinite, *path* ρ in the usual way as a sequence of states $s_0 s_1 s_2 \ldots$ such that $T(s_i, -)(s_{i+1}) > 0$ for all $i \geq 0$, and denote the set of finite and infinite paths of M starting in s, respectively, by $\mathrm{FPath}_T^M(s)$ and $\mathrm{IPath}_T^M(s)$, and sets of paths starting from any state by FPath_T^M and IPath_T^M, and omit M if clear from context. For a finite path ρ we write last(ρ) to denote the last state. We refer to paths induced from the transition probability function T as the *temporal dimension*.

For an agent A we define an *action strategy* σ_A as a function $\sigma_A : \mathrm{FPath}_T^M \longrightarrow \mathcal{D}(Act_A)$ such that for all $a_A \in Act_A$ and finite path ρ it holds that $\sigma_A(\rho)(a_A) > 0$ only if $a_A \in Act_A(\text{last}(\rho))$. An *action strategy profile* σ is a vector of action strategies $(\sigma_A)_{A \in Ags}$. Under a fixed σ, one can define a probability measure $\mathrm{Pr}^{M,\sigma}$ on $\mathrm{IPath}_T^M(s_{\text{init}})$ in the standard way.

In order to reason about trust, we endow agents with a *cognitive mechanism* inspired by the BDI framework (beliefs, desires and intentions) in the sense of [2]. We work with *probabilistic* beliefs. A cognitive mechanism includes *goals*, *intentions* and *subjective preferences*. For an agent A, the idea is that, while actions Act_A represent A's actions in the *physical* space, goals and intentions represent the *cognitive* processes that lead to decisions about which action to take. We thus distinguish two *dimensions* of transitions, temporal (behavioural) and cognitive.

Definition 2. *We define a* cognitive mechanism *as a tuple* $\Omega_A = (\{Goal_A\}_{A \in Ags}, \{Int_A\}_{A \in Ags}, \{gp_{A,B}\}_{A,B \in Ags}, \{ip_{A,B}\}_{A,B \in Ags})$, *where* $Goal_A$

is a finite set of goals for agent A; Int_A is a finite set of intentions for agent A; $gp_{A,B} : S \longrightarrow \mathcal{D}(2^{Goal_B})$ assigns to each state, from A's point of view, a distribution over possible goal changes of B; and $ip_{A,B} : S \longrightarrow \mathcal{D}(Int_B)$ assigns to each state, from A's point of view, a distribution over possible intentional changes of B.

An agent can have several goals, not necessarily consistent, but only a single intention. We think of goals as abstract attitudes, for example altruism or risk-taking, whereas intentions are concretely implemented in our (simplified) setting as *action strategies*, thus identifying the next (possibly random) action to be taken in the temporal dimension.

We refer to a stochastic multiplayer game endowed with a cognitive mechanism as an *autonomous* stochastic multiagent system. We extend the set of temporal transitions with cognitive transitions for agent A corresponding to a change of goal (respectively intention) and the transition probability function T in the obvious way. We denote by $\text{FPath}^M(s)$, $\text{IPath}^M(s)$, FPath^M and IPath^M the sets of paths formed by extending the sets $\text{FPath}_T^M(s)$, $\text{IPath}_T^M(s)$, FPath_T^M and IPath_T^M of temporal paths with paths that interleave the cognitive and temporal transitions.

To obtain a probability measure over infinite paths $\text{IPath}^M(s_{\text{init}})$, we need to resolve agents' possible changes to goals or intentions. Similarly to action strategies, we define *cognitive reasoning strategies* g_A and i_A, which are *history dependent* and model *subjective* preferences of A. Formally, we define the cognitive goal strategy as $g_A : \text{FPath} \longrightarrow \mathcal{D}(2^{Goal_A})$, and the intentional strategy as $i_A : \text{FPath} \longrightarrow \mathcal{D}(Int_A)$. We remark that such strategies arise from cognitive architectures, with the subjective view induced by goal and intentional *preference functions*, $gp_{A,B}$ and $ip_{A,B}$, which model *probabilistic prior* knowledge of agent A about goals and intentions of B, informed by prior experience (through observations) and aspects such as personal preferences and social norms. For details see [7].

Example 1. For the Parking Game example, let us consider two possible goals for A, altruism and selfishness. The intention corresponding to altruism is a strategy that always chooses to move aside, whereas for selfishness it is to choose wait. Another goal is absent-mindedness, which is associated with a strategy that chooses between moving aside and waiting at random. A preference function for B could be based on past observations that a Google car is more likely to move aside than, say, a Tesla car.

4   Probabilistic Rational Temporal Logic

We give an overview of the logic PRTL* that combines the probabilistic temporal logic PCTL* with operators for reasoning about agents' beliefs and cognitive trust. The trust operators of the logic are inspired by [4], except we express trust in terms of probabilistic belief, which probabilistically quantifies the degree of trust as a function of subjective certainty, e.g., "I am 99% certain that the autonomous taxi service is trustworthy", or "I trust the autonomous taxi service 99%". The logic captures how the value of 99% can be computed based on the agent's past experience and (social, economic) preferences.

Definition 3. *The syntax of the language PRTL* is:*

$$\phi ::= p \mid \neg\phi \mid \phi \vee \phi \mid \forall\psi \mid \mathrm{P}^{\bowtie q}\psi \mid \mathbb{G}_A\psi \mid \mathbb{I}_A\psi \mid \mathbb{C}_A\psi \mid$$
$$\mathbb{B}_A^{\bowtie q}\psi \mid \mathrm{CT}_{A,B}^{\bowtie q}\psi \mid \mathrm{DT}_{A,B}^{\bowtie q}\psi$$
$$\psi ::= \phi \mid \neg\psi \mid \psi \vee \psi \mid \bigcirc\psi \mid \psi\mathrm{U}\psi \mid \Box\psi$$

where p is an atomic proposition, $A, B \in Ags$, $\bowtie \in \{<, \leq, >, \geq\}$, and $q \in [0,1]$.

In the above, ϕ is a PRTL* formula and ψ an LTL (path) formula. The operator $\forall$ is the path quantifier of CTL* and $\mathrm{P}^{\bowtie d}\psi$ is the probabilistic operator of PCTL [1,5], which denotes the probability of those future infinite paths that satisfy ψ, evaluated in the temporal dimension. We omit the description of standard and derived ($\phi_1 \wedge \phi_2$, $\Diamond\psi$ and $\exists\phi$) operators, and just focus on the added operators.

The *cognitive* operators $\mathbb{G}_A\psi$, $\mathbb{I}_A\psi$ and $\mathbb{C}_A\psi$ consider the task expressed as ψ and respectively quantify, in the cognitive dimension, over possible changes of goals, possible intentions and available intentions. Thus, $\mathbb{G}_A\psi$ expresses that ψ holds in future regardless of agent A changing its goals. Similarly, $\mathbb{I}_A\psi$ states that ψ holds regardless of A changing its (not necessarily available) intention, whereas $\mathbb{C}_A\psi$ quantifies over the available intentions, and thus expresses that agent A can change its intention to achieve ψ.

$\mathbb{B}_A^{\bowtie q}\psi$ is the *belief* operator, which states that agent A believes ψ with probability in relation $\bowtie$ with q. $\mathrm{CT}_{A,B}^{\bowtie q}\psi$ is the *competence trust* operator, meaning that agent A trusts agent B with probability in relation $\bowtie$ with q on its capability of completing the task ψ, where capability is understood to be the existence of a valid intention (in $Int_B(s)$ for s being the current state) to implement the task. $\mathrm{DT}_{A,B}^{\bowtie d}\psi$ is the *disposition trust* operator, which expresses that agent A trusts agent B with probability in relation $\bowtie$ with q on its willingness to do the task ψ, where the state of willingness is interpreted as that the task is unavoidable for all intentions in intentional strategy (i.e., $i_B(\rho)$ for ρ being the path up to the current point in time).

Example 2. For the Parking Game example, the formula

$$\mathrm{DT}_{A,B}^{\geq 0.7}\neg steal_A$$

where $steal_A$ is an atomic proposition, expresses that A's trust in B's willingness not to steal a space is at least 70%, and

$$\mathbb{B}_A^{\geq 0.8}\mathrm{DT}_{B,A}^{\geq 0.7}move_A$$

states that A's belief that B has at least 70% trust in its willingness to move is at least 80%, where $move_A$ is an atomic proposition. Assuming that B has absent-mindedness as its goal, and A has two goals, altruism and selfishness, with the corresponding intentions, as in Example 1, then

$$\mathbb{G}_A \neg \mathbb{DT}_{B,A}^{\geq 0.7} move_A$$

states that, for all goal changes of A, B does not trust in A's willingness to move with probability at least 70%, where $move_A$ is an atomic proposition.

We interpret formulas ϕ in an autonomous stochastic multiagent system M in a state reached after executing a path ρ, in history-dependent fashion. Note that this path ρ may have interleaved cognitive and temporal transitions. The cognitive operators quantify over possible changes of goals and intentions in M in the cognitive dimension only, reflecting the cognitive reasoning processes leading to a decision. The probabilistic operator computes the probability of future paths satisfying ψ (i.e. completing the task ψ) in M in the temporal dimension as for PCTL*, reflecting the physical actions resulting from the cognitive decision, and compares this to the probability bound q. The belief operator corresponds to the belief-weighted expectation of future satisfaction of ψ, which is subjective, as it is influenced by A's prior knowledge about B encoded in the preference function. The competence trust operator reduces to the computation of optimal probability of satisfying ψ in M over possible changes of agent's intention, which is again belief-weighted and compared to the probability bound q. Dispositional trust, on the other hand, computes the optimal probability of satisfying ψ in M over possible states of agent's willingness, weighted by the belief and compared to the probability bound q.

The logic PRTL* can also express strong and weak dependence trust notions of [4]. Strong dependence means that A depends on B to achieve ψ (i.e. ψ can be implemented through intentional change of B), which cannot be achieved otherwise (expressed as a belief in impossibility of ψ in future), and weak dependence that A is better off relying on B compared to doing nothing (meaning intentional changes of B can bring about better outcomes).

Example 3. If B is in a hurry, then

$$\mathbb{DT}_{B,A}^{\geq 0.9} \lozenge leave_B \wedge \neg \mathbb{B}_B^{\geq 0.9} \lozenge leave_B$$

where $leave_B$ is an atomic proposition, expresses that B's leaving the car park strongly depends on A's willingness to cooperate.

Our framework encourages collaboration by allowing agents to update their trust evaluation for other agents and to take into consideration each other's trust when taking decisions. Trust thus evolves dynamically based on agent interactions and the decision to trust can be taken when a specific trust threshold is met. Therefore our notion of social trust helps to explain cases where actual human behaviour is at variance with standard economic and rationality theories.

Example 4. For the Parking Game example, we model the evolution of trust based on interactions and prior knowledge, whereby A's trust in B decreases if

B steals the space, and increases otherwise. A guards its decision whether to move aside by considering the level of trust in B's willingness not to steal, e.g. $\mathbb{DT}_{A,B}^{\geq 0.7} \neg steal_B$.

The precise value of the threshold for trust is context-dependent. The trust value higher than an appropriately calibrated level is known as 'overtrust', which can be expressed using our formalism, see [7].

5   Concluding Remarks

This paper has provided a brief overview of recent advances towards formalisation and quantitative verification of cognitive trust for stochastic multiplayer games based on [7]. Although the full logic is undecidable, we have identified decidable sublogics with reasonable complexity. As the next step we aim to implement the techniques as an extension of the PRISM probabilistic model checker [9] and evaluate them on case studies. To this end, we will define a Bellman operator and integrate with reasoning based on cognitive architectures.

This paper constitutes the first step towards developing design methodologies for capturing the social, trust-based decisions within human-robot partnerships. Pertinent scientific questions arise in the richer and challenging field of ethics and morality. How can we communicate intent in the context of human-robot interactions? How do we incentivise robots to elicit an appropriate response? How do we ensure that robotic assistants will not cause undue harm to others in order to satisfy the desires of their charge? Or that a self-driving car is able to decide between continuing on a path that will cause harm to other road-users, or executing an emergency stop which may harm passengers? These questions call for an in-depth analysis of the role of autonomous robots in society from a variety of perspectives, including philosophical and ethical, in addition to technology development, and for this analysis to inform policy makers, educators and scientists.

References

1. Bianco, A., de Alfaro, L.: Model checking of probabalistic and nondeterministic systems. FSTTCS **1995**, 499–513 (1995)
2. Bratman, M.E.: Intentions, Plans, and Practical Reason. Harvard University Press, Massachusetts (1987)
3. Conitzer, V., Sinnott-Armstrong, W., Borg, J.S., Deng, Y., Kramer, M.: Moral decision making frameworks for Artificial Intelligence. In: AAAI 2017 (2017, to appear)
4. Falcone, R., Castelfranchi, C.: Social trust: a cognitive approach. In: Trust and Deception in Virtual Societies, pp. 55–90. Kluwer (2001)
5. Hansson, H., Jonsson, B.: A logic for reasoning about time and reliability. Form. Aspects Comput. **6**(5), 512–535 (1994)
6. Hardin, R.: Trust and Trustworthiness. Russell Sage Foundation (2002)

7. Huang, X., Kwiatkowska, M.: Reasoning about cognitive trust in stochastic multiagent systems. In: AAAI 2017 (2017, to appear)
8. Kuipers, B.: What is trust and how can my robot get some? (presentation). In: RSS 2016 Workshop on Social Trust in Autonomous Robots (2016)
9. Kwiatkowska, M., Norman, G., Parker, D.: PRISM 4.0: verification of probabilistic real-time systems. In: Gopalakrishnan, G., Qadeer, S. (eds.) CAV 2011. LNCS, vol. 6806, pp. 585–591. Springer, Heidelberg (2011). doi:10.1007/978-3-642-22110-1_47
10. Lahijanian, M., Kwiatkowska, M.: Social trust: a major challenge for the future of autonomous systems. In: AAAI Fall Symposium on Cross-Disciplinary Challenges for Autonomous Systems, AAAI Fall Symposium. AAAI, AAAI Press (2016)
11. Lee, D.: US opens investigation into Tesla after fatal crash. British Broadcasting Corporation (BBC) News, 1 July 2016. http://www.bbc.co.uk/news/technology-36680043
12. Lee, J.D., See, K.A.: Trust in automation: designing for appropriate reliance. Hum. Factors J. Hum. Factors Ergon. Soc. 46(1), 50–80 (2004)
13. Letchford, J., Conitzer, V., Jain, K.: An "ethical" game-theoretic solution concept for two-player perfect-information games. In: Proceedings of the Fourth Workshop on Internet and Network Economics (WINE-08), pp. 696–707 (2008)
14. Mayer, R.C., Davis, J.H., Schoorman, F.D.: An integrative model of organizational trust. Acad. Manage. Rev. 20(3), 709–734 (1995)
15. McKnight, D.H., Choudhury, V., Kacmar, C.: Developing and validating trust measures for e-commerce: an integrative typology. Inf. Syst. Res. 13(3), 334–359 (2002)

On Choice Rules in Dependent Type Theory

Maria Emilia Maietti[✉]

Dipartimento di Matematica, Università di Padova, Padua, Italy
maietti@math.unipd.it

Abstract. In a dependent type theory satisfying the propositions as types correspondence together with the proofs-as-programs paradigm, the validity of the unique choice rule or even more of the choice rule says that the extraction of a computable witness from an existential statement under hypothesis can be performed within the same theory.

Here we show that the unique choice rule, and hence the choice rule, are not valid both in Coquand's Calculus of Constructions with indexed sum types, list types and binary disjoint sums and in its predicative version implemented in the intensional level of the Minimalist Foundation. This means that in these theories the extraction of computational witnesses from existential statements must be performed in a more expressive proofs-as-programs theory.

1 Introduction

Type theory is nowadays both a subfield of mathematical logic and of computer science. A perfect example of type theory studied both by mathematicians and by computer scientists is Martin-Löf's type theory [21], for short **ML**. This is a dependent type theory which can be indeed considered both as a paradigm of a typed functional programming language and as a foundation of constructive mathematics. The reason is that, on one hand, types can be seen to represent data types and typed terms to represent programs. On the other hand, both sets and propositions can be represented as types and both elements of sets and proofs of propositions can be represented as typed terms. These identifications are named in the literature as the propositions-as-types paradigm or the proofs-as-programs correspondence or Curry-Howard correspondence.

An important application of dependent type theory to programming is that one can use a type theory such as **ML** to construct a correct and terminating program as a typed term meeting a certain specification defined as its type. Pushing forward this correspondence one may ask whether from the proof-term $p(x)$ of an existential statement under hypothesis

$$p(x) \in \exists y \in B \ R(x, y) \ [x \in A]$$

one may extract a functional program $f \in A \to B$ whose graph is contained in the graph of $R(x, y)$, namely for which we can prove that there exists a proof-term $q(x)$ such that we can derive

$$q(x) \in R(x, f(x)) \ [x \in A]$$

© Springer International Publishing AG 2017
T.V. Gopal et al. (Eds.): TAMC 2017, LNCS 10185, pp. 12–23, 2017.
DOI: 10.1007/978-3-319-55911-7_2

This property is called *choice rule*. Then, we call *unique choice rule* the corresponding property starting from a proof-term of a unique existential statement

$$p(x) \in \exists! y \in B \ R(x, y) \ [x \in A]$$

from which we may extract a functional term $f \in A \to B$ whose graph is contained in the graph of $R(x, y)$.

It is worth noting that such choice rules *characterizes constructive arithmetics*, i.e. arithmetics within intuitionistic logic, with respect to classical Peano arithmetics (see [25, 26]).

In Martin-Löf's type theory both the unique choice rule and the choice rule are valid given that they follow from the validity of the axiom of choice

$$(AC) \qquad \forall x \in A \ \exists y \in B \ R(x, y) \quad \longrightarrow \quad \exists f \in A \to B \ \forall x \in A \ R(x, f(x))$$

thanks to the identification of the **ML**-existential quantifier with the *the strong indexed sum of a set family*, which characterizes the so called *propositions-as-sets isomorphism*.

However in other dependent type theories proposed as foundations for constructive mathematics the existential quantifier is not identified with the strong indexed sum type whilst it is still a type of its proofs. As a consequence in such theories the validity of the mentioned choice rules is not evident.

A notable example of such a dependent type theory is Coquand's Calculus of Inductive Constructions [6, 7] used as a logical base of the proof-assistant Coq [4, 5] and Matita [2, 3]. Here we consider its fragment **CC**$^+$ extending the original system in [6] with indexed sum types, list types and binary disjoint sums. In **CC**$^+$ propositions are defined primitively by postulating the existence of an impredicative type of propositions. In particular in **CC**$^+$ the identification of the existential quantifier with the strong indexed sum type is not possible because it makes the typed system logically inconsistent (see [6]). In fact in **CC**$^+$ the axiom of choice and even the axiom of unique choice

$$(AC!) \qquad \forall x \in A \ \exists! y \in B \ R(x, y) \quad \longrightarrow \quad \exists f \in A \to B \ \forall x \in A \ R(x, f(x))$$

are not generally provable as shown in [24]. In [24] it was left open whether the choice rule is validated in the original system [6]. Here we show that the choice rule is not validated in **CC**$^+$ by proving that in **CC**$^+$ the unique choice rule implies the axiom of unique choice and hence it is not valid. Of course, from this it follows that also the choice rule is not valid in **CC**$^+$.

Another example of foundation for constructive mathematics based on a dependent type theory where the existential quantifier is given primitively is the Minimalist Foundation, for short **MF**, ideated by the author in joint work with G. Sambin in [16] and completed in [12]. An important feature of **MF**, which is not present in other foundations like **CC**$^+$ or **ML**, is that it constitutes a common core among the most relevant constructive and classical foundations, introduced both in type theory, in category theory and in axiomatic set theory. Moreover it is a two-level system equipped with an *intensional level* suitable for extraction of computational contents from its proofs, an *extensional level* formulated in a language as close as possible to that of ordinary mathematics and an *interpretation of the latter in the former* showing that the extensional level has been obtained by abstraction from the intensional one according to Sambin's forget-restore principle in [23].

The two-level structure of **MF** brings many advantages in comparison to a single level foundation for constructive mathematics as $\mathbf{CC}^+$ or **ML**.

First of all the intensional level of **MF**, called **mTT** in [12] for *Minimalist Type theory*, can be used as a base for computer-aided formalization of proofs done at the extensional level of **MF**. Moreover, we can show the compatibility of **MF** with other constructive foundations at the most appropriate level: the intensional level of **MF** can be easily interpreted in intensional theories such as those formulated in type theory, for example Martin-Löf's type theory [21] or the Calculus of Inductive Constructions, while its extensional level can be easily interpreted in extensional theories such as those formulated in axiomatic set theory, for example Aczel's constructive set theory [1], or those formulated in category theory as topoi [10, 11].

Both intensional and extensional levels of **MF** consist of type systems based on versions of Martin-Löf's type theory with the addition of a primitive notion of propositions: the intensional one is based on [21] and the extensional one is based on [20].

In particular **mTT** constitutes a *predicative* counterpart of $\mathbf{CC}^+$ and to which the argument disproving the validity of the unique choice in $\mathbf{CC}^+$ adapts perfectly well.

As a consequence of our results we get that the extraction of programs computing witnesses of existential statements under hypothesis proved in $\mathbf{CC}^+$ or in **mTT** needs to be performed in a more expressive proofs-as-programs theory. We also believe that the arguments presented here can be adapted to conclude the same statements even for $\mathbf{CC}^+$ with generic inductive definitions.

It is worth noting that we can choose Martin-Löf's type theory as a more expressive theory where to perform the mentioned witness extraction from proofs done in **mTT**. Another option is to perform such a witness extraction in the realizability model of **mTT** extended with the axiom of choice and the formal Church thesis constructed in [9] (note that the axiom of choice and the formal Church thesis say that we can extract computable functions from number-theoretic total relations). Furthermore, in the case we limit ourselves to extract computable witnesses from unique existential statements proven in **mTT** then we can use other realizability models such as that in [17, 18] validating **mTT** extended with the axiom of unique choice and the formal Church thesis.

To perform witness extraction from proofs of existential statements done in $\mathbf{CC}^+$, an impredicative version of Martin-Löf's type theory is not available. We do not even know whether there exists a realizability model proving consistency of $\mathbf{CC}^+$ extended with the axiom of choice and the formal Church thesis.

2   The Dependent Type Theory DT_Σ and the Choice Rules

Here we briefly describe a fragment, called $\mathbf{DT}_\Sigma$, of the intensional type theory $\mathbf{mTT}$ of the Minimalist Foundation in [12] which is sufficient to show that the unique choice rule implies the axiom of unique choice. This is the fragment of $\mathbf{mTT}$ needed to interpret many-sorted predicate intuitionistic logic where sorts are closed under strong indexed sums, dependent products and also comprehension.

$\mathbf{DT}_\Sigma$ is a dependent type theory written in the style of Martin-Löf's type theory [21] by means of the following four kinds of judgements:

$$A \ type \ [\Gamma] \quad A = B \ type \ [\Gamma] \quad a \in A \ [\Gamma] \quad a = b \in A \ [\Gamma]$$

that is the type judgement (expressing that something is a specific type), the type equality judgement (expressing that two types are equal), the term judgement (expressing that something is a term of a certain type) and the term equality judgement (expressing the *definitional equality* between terms of the same type), respectively, all under a context Γ.

The word *type* is used as a meta-variable to indicate two kinds of entities: sets and *small propositions*, namely

$$type \in \{set, props\}$$

Therefore, in $\mathbf{DT}_\Sigma$ types are actually formed by using the following judgements:

$$A \ set \ [\Gamma] \qquad \phi \ props \ [\Gamma]$$

saying that A is a set and that ϕ is a small proposition of $\mathbf{DT}_\Sigma$.

It is worth noting that the adjective *small* is there because in $\mathbf{mTT}$ we defined small propositions as those propositions closed under quantification over sets, while generic propositions may be closed under quantification over collections. In $\mathbf{DT}_\Sigma$ there are no collections and hence all "$\mathbf{DT}_\Sigma$-propositions" are small but we keep the adjective to make $\mathbf{DT}_\Sigma$ a proper fragment of $\mathbf{mTT}$.

As in the intensional version of Martin-Löf's type theory and in $\mathbf{mTT}$, in $\mathbf{DT}_\Sigma$ there are two kinds of equality concerning terms: one is the definitional equality of terms of the same type given by the judgement

$$a = b \in A \ [\Gamma]$$

which is decidable, and the other is the propositional equality written

$$\mathsf{Id}(A, a, b) \ props \ [\Gamma]$$

which is not necessarily decidable.

We now proceed by briefly describing the various kinds of types in $\mathbf{DT}_\Sigma$, starting from small propositions and then passing to sets.

Small Propositions in **mTT** include all the logical constructors of intuitionistic predicate logic with equality and quantifications restricted to sets:

$$\phi \; prop_s \equiv \bot \mid \phi \wedge \psi \mid \phi \vee \psi \mid \phi \rightarrow \psi \mid \forall_{x \in A} \, \phi \mid \exists_{x \in A} \, \phi \mid \mathsf{Id}(A, a, b)$$

provided that A is a set. Their rules are those for the corresponding small propositions in **mTT**.

In order to close sets under comprehension and to define operations on such sets, we need to think of propositions as types of their proofs:

$$\textbf{prop}_s\textbf{-into-set)} \quad \frac{\phi \; prop_s}{\phi \; set}$$

Then *sets* in **DT$_\Sigma$** include the following:

$$A \; set \; \equiv \phi \; prop_s \mid \Sigma_{x \in A} B \mid \Pi_{x \in A} B$$

where the notation $\Sigma_{x \in A} B$ stands for the strong indexed sum of the family of sets $B \; set \; [x \in A]$ indexed on the set A and $\Pi_{x \in A} B$ for the dependent product set of the family of sets $B \; set \; [x \in A]$ indexed on the set A.

Their rules are those for the corresponding sets in **mTT** and we refer to [12] for their precise formulation.

Both **DT$_\Sigma$** as well as **mTT** can be also essentially seen as fragments of a typed system, which we call **CC$^+$**, extending the Calculus of Constructions in [6] with the inductive rules in [4,5,7] defining binary disjoint sums, list types and strong indexed sums (see [12]).

For their crucial role to get the results in this paper, here we just recall the rules of formation, introduction, elimination and conversion of the strong indexed sum as a set:

Strong Indexed Sum

$$\text{F-}\Sigma) \quad \frac{C(x) \; set \; [x \in B]}{\Sigma_{x \in B} C(x) \; set} \qquad \text{I-}\Sigma) \quad \frac{b \in B \quad c \in C(b) \quad C(x) \; set \; [x \in B]}{\langle b, c \rangle \in \Sigma_{x \in B} C(x)}$$

$$\text{E-}\Sigma) \quad \frac{\begin{array}{c} M(z) \; set \; [z \in \Sigma_{x \in B} C(x)] \\ d \in \Sigma_{x \in B} C(x) \quad m(x, y) \in M(\langle x, y \rangle) \; [x \in B, y \in C(x)] \end{array}}{El_\Sigma(d, m) \in M(d)}$$

$$\text{C-}\Sigma) \quad \frac{\begin{array}{c} M(z) \; set \; [z \in \Sigma_{x \in B} C(x)] \\ b \in B \quad c \in C(b) \quad m(x, y) \in M(\langle x, y \rangle) \; [x \in B, y \in C(x)] \end{array}}{El_\Sigma(\langle b, c \rangle, m) = m(b, c) \in M(\langle b, c \rangle)}$$

By using these rules we recall that we can define the following projections

$$\pi_1(z) \equiv El_\Sigma(z, (x, y).x) \in B \; [z \in \Sigma_{x \in B} C(x)]$$
$$\pi_2(z) \equiv El_\Sigma(z, (x, y).y) \in C(\pi_1(z)) \; [z \in \Sigma_{x \in B} C(x)]$$

satisfying

$$\pi_1(\langle b, c \rangle) = b \in B \; [\Gamma] \qquad \qquad \pi_2(\langle b, c \rangle) = c \in C(b) \; [\Gamma]$$

provided that $C(x)$ *set* $[\Gamma, x \in B]$, $b \in B$ $[\Gamma]$ and $c \in C(b)$ $[\Gamma]$ are derivable in **DT**$_\Sigma$.

We also recall the following abbreviations: for set A *set* $[\Gamma]$ and B *set* $[\Gamma]$ we define the set of functions from A to B as

$$A \to B \equiv \Pi_{x \in A} B$$

Now we are ready to define the *choice rule*:

Definition 1. The dependent type theory **DT**$_\Sigma$ satisfies the *choice rule* if for every small proposition $R(x, y)$ *prop$_s$* $[x \in A, y \in B]$ derivable in **DT**$_\Sigma$, for any derivable judgement in **DT**$_\Sigma$ of the form

$$p(x) \in \exists_{y \in B} R(x, y) \; [x \in A]$$

there exists in **DT**$_\Sigma$ a typed term

$$f(x) \in B[x \in A]$$

for which we can find a proof-term $q(x)$ and derive in **DT**$_\Sigma$

$$q(x) \in R(x, f(x)) \; [x \in A]$$

Then we recall the definition of the *axiom of choice* by internalizing the above choice rule as follows:

Definition 2. The *axiom of choice* is the following small proposition

(AC) $\forall x \in A \; \exists y \in B \; R(x, y) \;\; \longrightarrow \;\; \exists f \in A \to B \; \forall x \in A \; R(x, f(x))$

defined for any small proposition $R(x, y)$ *prop$_s$* $[x \in A, y \in B]$.

A special instance of the choice rule is the following *unique choice rule*

Definition 3. The dependent type theory **DT**$_\Sigma$ satisfies the *unique choice rule* if for every small proposition $R(x, y)$ *prop$_s$* $[x \in A, y \in B]$ derivable in **DT**$_\Sigma$, for any derivable judgement in **DT**$_\Sigma$ of the form

$$p(x) \in \exists!_{y \in B} R(x, y) \; [x \in A]$$

there exists a typed term $f(x) \in B[x \in A]$ for which we can find a proof-term $q(x)$ and derive in **DT**$_\Sigma$

$$q(x) \in R(x, f(x)) \; [x \in A]$$

where

$$\exists! y \in B \; R(x, y) \equiv$$
$$\exists y \in B \; R(x, y) \; \wedge \; \forall y_1, y_2 \in B (R(x, y_1) \wedge R(x, y_1) \to \mathsf{Id}(B, y_1, y_2))$$

Then the *axiom of unique choice* is the internal form of the unique choice rule defined as follows:

Definition 4. The *axiom of unique choice* is the following small proposition

(AC!) $\forall x \in A \ \exists! y \in B \ R(x, y) \ \longrightarrow \ \exists f \in A \to B \ \forall x \in A \ R(x, f(x))$

defined for any small proposition $R(x, y) \ prop_s \ [x \in A, y \in B]$.

Observe the following obvious relation between the above rules and the corresponding axioms:

Lemma 1. If $\mathbf{DT}_\Sigma$ satisfies the choice rule then $\mathbf{DT}_\Sigma$ proves the unique choice rule.

Lemma 2. If $\mathbf{DT}_\Sigma$ proves the axiom of choice then $\mathbf{DT}_\Sigma$ proves the axiom of unique choice.

Now we are ready to show the following crucial proposition:

Proposition 1. If $\mathbf{DT}_\Sigma$ satisfies the unique choice rule then $\mathbf{DT}_\Sigma$ proves the axiom of unique choice.

Proof. Suppose that $R(x, y) \ prop_s \ [x \in A, y \in B]$ is derivable in $\mathbf{DT}_\Sigma$. Observe that we can derive in $\mathbf{DT}_\Sigma$

$$\pi_2(z) \in \exists!_{y \in B} \ R(\pi_1(z), y) \ [z \in \Sigma_{x \in A} \ \exists!_{y \in B} \ R(x, y)]$$

Suppose now that the unique choice rule is valid in $\mathbf{DT}_\Sigma$. Then, by using this rule there exists a typed term

$$f(z) \in B \ [z \in \Sigma_{x \in A} \ \exists!_{y \in B} \ R(x, y)]$$

and a proof-term $q(z)$ of $\mathbf{DT}_\Sigma$ for which we can derive

$$q(z) \in R(\pi_1(z), f(z)) \ [z \in \Sigma_{x \in A} \ \exists!_{y \in B} \ R(x, y)]$$

By using these proof-terms we can derive

$$\langle m(w), h(w) \rangle \in \exists_{g \in A \to B} \ \forall x \in A \ R(x, g(x)) \ [w \in \forall_{x \in A} \ \exists!_{y \in B} \ R(x, y)]$$

where

$$m(w) \equiv \lambda x'. f(\langle x', w(x') \rangle)$$

since we can derive

$$w(x') \in \exists!_{y \in B} \ R(x', y) \ [w \in \forall_{x \in A} \ \exists!_{y \in B} \ R(x, y), x' \in A]$$

and

$$\langle x', w(x') \rangle \in \Sigma_{x \in A} \ \exists!_{y \in B} \ R(x, y) \ [w \in \forall_{x \in A} \ \exists!_{y \in B} \ R(x, y), x' \in A]$$

and where

$$h(w) \equiv \lambda x''.q(\langle x'', w(x'')\rangle) \in \forall x \in A \ R(x, m(w)(x))$$

since

$$q(\langle x'', w(x'')\rangle) \in R(\pi_1(\langle x'', w(x'')\rangle), f(\langle x'', w(x'')\rangle)) = R(x'', m(w)(x''))$$

for $x'' \in A$ and $w \in \forall_{x \in A} \ \exists!_{y \in B} \ R(x, y)$.

Finally we conclude that

$$\lambda w.\langle m(w), h(w)\rangle \in$$
$$\forall x \in A \ \exists! y \in B \ R(x, y) \longrightarrow \exists f \in A \to B \ \forall x \in A \ R(x, f(x))$$

i.e. we conclude that the axiom of unique choice is valid in $\mathbf{DT}_\Sigma$.

Observe that the above proof can be adapted to show that also the choice rule implies its axiomatic form by simply replacing $\exists!_{y \in B}$- with $\exists_{y \in B}$- in the proof of Proposition 1 and hence we also get:

Proposition 2. If $\mathbf{DT}_\Sigma$ satisfies the choice rule then $\mathbf{DT}_\Sigma$ satisfies the axiom of choice.

By definition $\mathbf{DT}_\Sigma$ is a fragment of $\mathbf{mTT}$ and therefore by repeating the proofs above we conclude that:

Proposition 3. If $\mathbf{mTT}$ satisfies the unique choice rule then $\mathbf{mTT}$ satisfies the axiom of unique choice.

Proposition 4. If $\mathbf{mTT}$ satisfies the choice rule then $\mathbf{mTT}$ satisfies the axiom of choice.

Now, observe that $\mathbf{DT}_\Sigma$ can be seen essentially as a fragment of $\mathbf{CC}^+$ after interpreting $\mathbf{DT}_\Sigma$-sets as $\mathbf{CC}^+$-sets and $\mathbf{DT}_\Sigma$-small propositions as the corresponding $\mathbf{CC}^+$-propositions. In the same way $\mathbf{mTT}$ can be also viewed essentially as a fragment of $\mathbf{CC}^+$ as first described in [12]. Therefore, we also get the following:

Proposition 5. If $\mathbf{CC}^+$ satisfies the unique choice rule then it satisfies the axiom of unique choice.

Proposition 6. If $\mathbf{CC}^+$ satisfies the choice rule then $\mathbf{CC}^+$ satisfies the axiom of choice.

Note that the above propositions hold also for the extension of $\mathbf{CC}^+$ with inductive definitions [7].

Then, we recall the following result by T. Streicher:

Theorem 1 (T. Streicher). $\mathbf{CC}^+$ does not validate the axiom of unique choice and hence the axiom of choice.

Proof. This is based on [24] and the fact that types are interpreted as assemblies which can be organized into a lextensive regular locally cartesian closed category with a natural numbers object [8,22].

Again since $\mathbf{DT}_\Sigma$ and $\mathbf{mTT}$ can be both viewed essentially as fragments of $\mathbf{CC}^+$, we also get from Theorem 1:

Corollary 1. Both $\mathbf{DT}_\Sigma$ and $\mathbf{mTT}$ do not validate the axiom of unique choice and hence the axiom of choice.

Now from Proposition 1 and Corollary 1 we get:

Theorem 2. $\mathbf{DT}_\Sigma$ does not validate the unique choice rule and hence the choice rule.

Analogously, from Proposition 3 and Corollary 1 we also get:

Theorem 3. $\mathbf{mTT}$ does not validate the unique choice rule and hence the choice rule.

And from Proposition 5 and Theorem 1 we finally get:

Theorem 4. $\mathbf{CC}^+$ does not validate the unique choice rule and hence the choice rule.

We conclude by saying that, as suggested by T. Streicher, it is very plausible that the model in [24] can be extended to interpret generic inductive definitions and hence to show that the axiom of unique choice is not provable in $\mathbf{CC}^+$ extended with generic inductive definitions. Therefore, if this is confirmed, from Proposition 5 we can conclude that the unique choice rule is not valid even in this extension of $\mathbf{CC}^+$.

Remark 1. We believe that in the context of categorical models of dependent type theories we can prove categorical results corresponding to Propositions 1 and 2.

Indeed, the relationship between the choice rules and their axiomatic form in Propositions 1 and 2 was inspired by categorical investigations done in a series of papers [13–15] about setoid models used in dependent type theory to interpret quotient sets. In particular these papers focus their analysis on the quotient model used in [12] to interpret the extensional level of the Minimalist Foundation into its intensional level $\mathbf{mTT}$. As shown in [19], the model used in [12] coincides with the usual exact completion in category theory if and only if the unique choice rule is valid in the completion. We expect to be able to prove also in the context of [19] that the unique choice rule implies the axiom of unique choice, as well as that the choice rule implies the axiom of choice.

3 Conclusion

From the above results it follows that when proving a statement of the form

$$\forall_{x \in A} \exists y \in B \ R(x, y)$$

in the dependent typed theory CC^+ or in mTT we can not always extract a functional term $f \in A \to B$ computing the witness of the existential quantification depending on a $x \in A$, within the theory itself and we need to find it in a more expressive proofs-as-programs theory.

For mTT we can use Martin-Löf's type theory ML as the more expressive theory where to perform the mentioned witness extraction. This is done by first embedding into ML the proof-term

$$p \in \forall_{x \in A} \exists y \in B \ R(x, y)$$

derived in mTT and then using ML-projections to extract f.

Another possibility is to perform this witness extraction in the realizability model in [9] showing consistency of mTT extended with the Formal Church thesis and the axiom of choice. Moreover, in the case we simply want to extract computable witness from unique existential statements proved in mTT under hypothesis we can use also other realizability models such as that in [17,18] showing consistency of mTT with the axiom of unique choice and the formal Church thesis.

For CC^+, and even more for its extension with inductive definitions, it is an open problem whether there is a realizability model showing its consistency with the axiom of choice and the formal Church thesis.

Acknowledgements. We acknowledge very fruitful discussions on this topic with Claudio Sacerdoti Coen, Ferruccio Guidi, Giuseppe Rosolini, Giovanni Sambin, Thomas Streicher. We also thank Tatsuji Kawai and Fabio Pasquali very much for their comments on this paper.

References

1. Aczel, P., Rathjen, M.: Notes on constructive set theory. Mittag-Leffler Technical Report No. 40 (2001)
2. Asperti, A., Ricciotti, W., Sacerdoti Coen, C., Tassi, E.: The Matita interactive theorem prover. In: Bjørner, N., Sofronie-Stokkermans, V. (eds.) CADE 2011. LNCS (LNAI), vol. 6803, pp. 64–69. Springer, Heidelberg (2011). doi:10.1007/978-3-642-22438-6_7
3. Asperti, A., Ricciotti, W., Sacerdoti Coen, C.: Matita tutorial. J. Formalized Reasoning **7**(2), 91–199 (2014)
4. Bertot, Y., Castéran, P.: Interactive Theorem Proving and Program Development. Texts in Theoretical Computer Science. Springer-Verlag, Heidelberg (2004)
5. Coq Development Team: The Coq Proof Assistant Reference Manual: Release 8.4pl6. INRIA, Orsay, France, April 2015

6. Coquand, T.: Metamathematical investigation of a calculus of constructions. In: Odifreddi, P. (ed.) Logic in Computer Science, pp. 91–122. Academic Press (1990)
7. Coquand, T., Paulin, C.: Inductively defined types. In: Martin-Löf, P., Mints, G. (eds.) COLOG 1988. LNCS, vol. 417, pp. 50–66. Springer, Heidelberg (1990). doi:10.1007/3-540-52335-9_47
8. Hyland, J.M.E.: The effective topos. In: The L.E.J. Brouwer Centenary Symposium (Noordwijkerhout 1981). Studies in Logic and the Foundations of Mathematics, vol. 110, pp. 165–216. North-Holland, New York, Amsterdam (1982)
9. Ishihara, H., Maietti, M., Maschio, S., Streicher, T.: Consistency of the Minimalist Foundation with Church's thesis and Axiom of Choice. Submitted
10. Mac Lane, S., Moerdijk, I.: Sheaves in Geometry and Logic: A First Introduction to Topos Theory. Springer Verlag, New York (1992)
11. Maietti, M.E.: Modular correspondence between dependent type theories and categories including pretopoi and topoi. Math. Struct. Comput. Sci. **15**(6), 1089–1149 (2005)
12. Maietti, M.E.: A minimalist two-level foundation for constructive mathematics. Ann. Pure Appl. Logic **160**(3), 319–354 (2009)
13. Maietti, M.E., Rosolini, G.: Elementary quotient completion. Theory Appl. Categories **27**(17), 445–463 (2013)
14. Maietti, M.E., Rosolini, G.: Quotient completion for the foundation of constructive mathematics. Log. Univers. **7**(3), 371–402 (2013)
15. Maietti, M.E., Rosolini, G.: Unifying exact completions. Appl. Categorical Struct. **23**(1), 43–52 (2015)
16. Maietti, M.E., Sambin, G.: Toward a minimalist foundation for constructive mathematics. In: Crosilla, L., Schuster, P. (eds.) From Sets and Types to Topology and Analysis: Practicable Foundations for Constructive Mathematics. Oxford Logic Guides, vol. 48, pp. 91–114. Oxford University Press (2005)
17. Maietti, M.E., Maschio, S.: An extensional Kleene realizability semantics for the Minimalist Foundation. In: Herbelin, H., Letouzey, P., Sozeau, M. (eds.) 20th International Conference on Types for Proofs and Programs (TYPES 2014). Leibniz International Proceedings in Informatics (LIPIcs), vol. 39, pp. 162–186. Schloss Dagstuhl-Leibniz-Zentrum fuer Informatik, Dagstuhl (2015). http://drops.dagstuhl.de/opus/volltexte/2015/5496
18. Maietti, M.E., Maschio, S.: A predicative variant of a realizability tripos for the Minimalist Foundation. IfCoLog J. Logics Appl. (2016). Special Issue Proof Truth Computation
19. Maietti, M., Rosolini, G.: Relating quotient completions via categorical logic. In: Probst, D. (ed.) Concepts of Proof in Mathematics, Philosophy, and Computer Science, pp. 229–250. De Gruyter (2016)
20. Martin-Löf, P.: Intuitionistic Type Theory. Notes by G. Sambin of a series of lectures given in Padua, Bibliopolis, Naples (1984), June 1980
21. Nordström, B., Petersson, K., Smith, J.: Programming in Martin Löf's Type Theory. Clarendon Press, Oxford (1990)
22. van Oosten, J.: Realizability: An Introduction to its Categorical Side. Elsevier, Amsterdam (2008)
23. Sambin, G., Valentini, S.: Building up a toolbox for Martin-Löf's type theory: subset theory. In: Sambin, G., Smith, J. (eds.) Twenty-Five Years of Constructive Type Theory, Proceedings of a Congress held in Venice, October 1995, pp. 221–244. Oxford U. P. (1998)
24. Streicher, T.: Independence of the induction principle and the axiom of choice in the pure calculus of constructions. Theoret. Comput. Sci. **103**(2), 395–408 (1992)

25. Troelstra, A.S., van Dalen, D.: Constructivism in Mathematics: An Introduction. Studies in Logic and the Foundations of Mathematics, vol. I. North-Holland (1988)
26. Troelstra, A.S., van Dalen, D.: Constructivism in Mathematics: An Introduction. Studies in Logic and the Foundations of Mathematics, vol. II. North-Holland (1988)

Natural Language Processing, Moving from Rules to Data

Adrian-Horia Dediu[1]([✉]), Joana M. Matos[2,3], and Carlos Martín-Vide[4]

[1] Superdata, Bucharest, Romania
Adrian-Horia.Dediu@superdata.ro
[2] Centro de Matemática e Aplicações (CMA), FCT, New University of Lisbon,
Caparica, Portugal
jmf.matos@fct.unl.pt
[3] Department of Mathematics, FCT, New University of Lisbon, Caparica, Portugal
[4] Rovira i Virgili University, Tarragona, Spain
carlos.martin@urv.cat

Abstract. During the last decade, we assist to a major change in the direction that theoretical models used in natural language processing follow. We are moving from rule-based systems to corpus-oriented paradigms. In this paper, we analyze several generative formalisms together with newer statistical and data-oriented linguistic methodologies. We review existing methods belonging to deep or shallow learning applied in various subfields of computational linguistics. The continuous, fast improvements obtained by practical, applied machine learning techniques may lead us to new theoretical developments in the classic models as well. We discuss several scenarios for future approaches.

Keywords: Computational models · Computational linguistics · Speech processing methods · Machine translation

1 Introduction

Despite the huge research efforts, models capable to explain even partially two of the most important components of human intelligence, language acquisition and communication, are still missing. According to Winston [86], the search for a comprehensive theory of language understanding will keep linguist busy for many years.

Computational linguistics, also known as *natural language processing* (NLP), aims to learn, understand, and produce human language content (Hirschberg and Manning [36]). According to the 2012 ACM Computing Classification System [2], we find NLP (a subfield of artificial intelligence (AI), which is a branch of computing methodologies) with the following research directions:

This work was partially supported by the Fundação para a Ciência e a Tecnologia (Portuguese Foundation for Science and Technology) through the project UID/MAT/00297/2013 (Centro de Matemática e Aplicações).

© Springer International Publishing AG 2017
T.V. Gopal et al. (Eds.): TAMC 2017, LNCS 10185, pp. 24–38, 2017.
DOI: 10.1007/978-3-319-55911-7_3

- information extraction,
- machine translation,
- discourse, dialogue and pragmatics,
- natural language generation,
- speech recognition,
- lexical semantics,
- phonology/morphology, and
- language resources.

It is generally accepted that a theory of natural language processing should be based on an appropriate model for knowledge representation. Bellegarda and Monz [8] acknowledge that one of the most challenging problems of natural language processing is to find an appropriate model for meaning representations. Many research groups still prefer first-order logic for formal semantic representations (see Montague [59]). To get over the limitation of first-order logic and the lack of expressive power capable to capture all the subtleties of human languages, various extensions represent viable alternatives as we can find for example in Chierchia [12], Dekker [20], Kamp and Reyle [43], Muskens [62], etc.

A new research trend appeared in computational semantics that shifted the interest from formal knowledge representation to distributional semantics, where the meaning of a word depends on the context in which it appears (Rieger [70], Sahlgren [74]). The meaning of larger constituents, such as sentences, phrases, or even paragraphs or a whole document, is then calculated based on the rules of composition that combines semantic representations of constituent elements (see for example Mikolov et al. [55]).

We assist to a continuous improving of NLP performances, moving from symbolic to statistical methods, from hand written rule-based systems to data-driven approaches. This evolution was possible together with a significant qualitative and quantitative development of the following factors:

1. computing power,
2. availability of linguistic data,
3. *machine learning* (ML) methods,
4. understanding the structure of human language.

For a first category of problems that are almost solved, we do not actually need a semantic representation. For example, *part of speech* (POS) tagging (Brill [11], Schmid [76]), currently reaches 97.3% token accuracy, as Manning [51] shows. What do we need to improve the results even furthermore? Manning claims that probably better training data, that is, improved descriptive linguistics could lead us to even better results.

For a second category of problems that should deal directly with the meaning representation, despite the continuous progress from the last decades, we still cannot rely on the automatic tasks performed by computer programs. For example, in *text summarization* tasks, that take input as text document(s) and try to condense them into a summary, the results are still under the expectations, we recall here only a brief overview of the results in this area citing the papers by Hovy and Lin [39], Kastner and Monz [44], Knight and Marcu [46], Zajic et al. [87], etc. The situation is similar for *machine translation* (MT) that involves the use of more than one natural language, with the source language different from the target language. In his article "Machine Translation: The Disappointing Past and Present", Kay [45] says that we cannot view this problem as a mainly linguistic one, but also the alternatives to incorporate in a translation system general knowledge and common sense that humans have is not yet possible either. We mention other works in this domain, Koehn [47] and Lopez [50].

Next we present several computational models for NLP, we could have followed the lines from Cole [13], focusing on classifiers, connectionist techniques, or optimization and search methods, however, we prefer a hierarchical approach. In the next sections we present several models used at morphologic, syntactic and semantic levels. We conclude presenting our vision about future trends and expected results in NLP.

2   Morphologic Level - Finite State Models for NLP

We follow the notations and definitions well established by formal language theory (see for example Hopcroft et al. [37] or Sudkamp [79]), also integrating the practical point of view of Jurafsky and Martin [41], explaining how to use these models in NLP tasks. We assume that the reader is familiar with the basic notions of graph theory, probability theory, and complexity theory. We briefly present an overview of the basic concepts we use in this paper.

We denote by $\mathbb{N}$ the set of natural numbers, that is $\{0, 1, 2, \ldots\}$ and by $\mathbb{R}$ the set of real numbers.

Let f, g be functions defined from $\mathbb{N}$ into $\mathbb{R}$, $n, n_0 \in \mathbb{N}$ and $c \in \mathbb{R}$ (we follow the notations from Rothlauf [72]). We define:

- $f \in O(g) \Leftrightarrow \exists c > 0, \exists n_0 > 0$ such that $|f(n)| \leq c \cdot |g(n)|, \forall n \geq n_0$ (*asymptotic upper bound*).
- $f \in o(g) \Leftrightarrow \forall c > 0, \exists n_0 > 0$ such that $|f(n)| < c \cdot |g(n)|, \forall n \geq n_0$ (*asymptotically negligible*).
- $f \in \Omega(g) \Leftrightarrow g \in O(f)$ (*asymptotic lower bound*).
- $f \in \Theta(g) \Leftrightarrow f \in O(g)$ and $g \in O(f)$ (*asymptotically tight bound*).

Some authors use the "soft-O" notation that ignores also the logarithmic factors, for example, $\tilde{\Theta}(f(m))$ represents $\Theta(f(m) \log^c m)$, for some constant c.

Let Σ be a finite set of symbols, called the *alphabet*. A finite sequence of elements of Σ is called a *string* over Σ. For a given string w, $|w|$ represents the *length* of the string. We denote by ϵ the *empty string*, with $|\epsilon| = 0$. We define a binary operation between strings in the following way. For two strings $w = a_1 \ldots a_n$ and $x = b_1 \ldots b_m$ over Σ, the *concatenation* of the two strings is the string $a_1 \ldots a_n b_1 \ldots b_m$. The concatenation operation is denoted by $w \cdot x$ (or simply wx when there is no confusion).

Let Σ^* be the set of strings over Σ. A *language* L over Σ is a subset of Σ^*. The elements of L are also called *words*. For any given nonnegative integer k, $\Sigma^{\leq k}$ denotes the language of words w with $|w| \leq k$.

Definition 1 (Deterministic finite automata, Freund et al. [29]). *A deterministic finite automaton (DFA[1]) is a tuple $M = (Q, \Sigma, \Gamma, \tau, \lambda, q_0)$, where Q is a finite set of states, Σ is a finite and nonempty alphabet called the input alphabet, Γ is a finite and nonempty alphabet called the output alphabet, τ is a partial function from $Q \times \Sigma$ to Q called the transition function, λ is a mapping from Q to Γ called the output function, and q_0 is a fixed state of Q called the initial state.*

We extend τ to a map from $Q \times \Sigma^*$ to Q in the usual way. We take $\tau(q, \epsilon) = q$ and $\tau(q, \alpha \cdot a) = \tau(\tau(q, \alpha), a)$, for all states q in Q, for all strings α in Σ^*, and for all characters a in Σ, provided that $\tau(q, \alpha)$ and $\tau(\tau(q, \alpha), a)$ are defined. For a state q and a string x over the input alphabet, we denote by qx the state $\tau(q, x)$, and by $q\langle x \rangle$, the sequence of length $|x|+1$ of output labels observed upon executing the transitions from state q dictated by x, that is, the string $\lambda(q)\lambda(qx_1), \ldots, \lambda(qx_1 \ldots x_n)$, where n is the length of the string x and $x_1, \ldots, x_n$ are its characters.

A deterministic *finite acceptor* is a DFA with the output alphabet $\Gamma = \{0, 1\}$; if $\lambda(q) = 1$, then q is an *accepting state*, otherwise, q is a *rejecting state*. A string x is *accepted* or *recognized* by a finite acceptor with the initial state q_0 if $q_0 x$ is an accepting state. The definition of automata with final states (see, for example, Hopcroft and Ullman [38]) is equivalent with our definition of finite acceptors, with the convention that *final states* are the accepting states. For a finite acceptor $A = (Q, \Sigma, \Gamma, \tau, \lambda, q_0)$, we define the *language* $L(A)$ as the set of strings accepted by the acceptor A.

In Chap. 3 of Jurafsky and Martin [41] we find out about using DFA for morphological parsing, that is taking the *surface* or the *input form* of a word and producing a structured output, a *stem* (the "main" morpheme of the word, supplying the main meaning) and an *affix* ("additional" meanings of various kinds, indicating inflectional or derivative forms). Chapter 4 of the same book describes how to use DFA for converting *text-to-speech* (TTS), that is, the output of an input text is the spoken voice. The methodology for using DFA for TTS actually discovers incrementally the DFA by some models of unsupervised machine learning of phonological rules.

We can classify the methods of learning automata as *active* or *passive*. We assume the existence of two entities, a *teacher* who has access to a target language, and a *learner* who acquires the language. In the passive model, the learner has no control over the data received, while in the active model, the learner can experiment with the target language. We mention Gold [33] with *learning in the limit* (an example of passive learning), Valiant [81] with *probably approximately*

---

[1] Angluin et al. [5] use the term "automaton with output". In formal language books, like Hopcroft and Ullman [38], this definition corresponds to a Moore automaton, and the notion of acceptors that we define next, corresponds to a DFA.

correct learning and *query learning* Angluin [3] (an example of active learning), as the initiators of the most important research directions for learning automata. Depending on the way in which the learner has access to the training data, there are two main models, *on-line learning* if the data arrives during the learning process, and *off-line* or *batch learning* if a training data set is available. The data might be positive, if it is examples belonging to the learned language, or negative, if it is examples not in the target language.

An interesting result from Angluin and Becerra-Bonache [4] shows how an extension of DFA can be used as a model of meaning and denotation. A well-known extension of DFA, weighted automata, described by Mohri [58], are successfully used for *automatic speech recognition* (ASR).

3 Syntactic Level

We follow the definitions from Dediu and Tîrnăucă [18]. The trees are finite, their nodes are labeled with symbols from an alphabet Σ, and the branches leaving any given node have a specified order. Starting from the root and going through the leaves, we may associate with every node in a tree an *address* in an inductive way. Note that this is known, especially in linguistics, as *Gorn address*. The root node has the empty address ε or (). For a child node we take the parent address and we add a dot, followed by the child number counted from left to right (the counting always starts from 0). More formally, let U be the set of all finite sequences of nonnegative integers separated by dots, including also the empty sequence ε. The prefix relation $\leq$ in U is: $u \leq v$ iff $u.w = v$ for some $w \in U$. Then $D \subseteq U$ is a (finite) *tree domain* iff: (1) if $u \leq v$ and $v \in D$, then $u \in D$, and (2) if $u.j \in D$, $i, j \in \mathbb{N}$, and $i < j$, then $u.i \in D$. A tree domain D may be viewed as an unlabeled tree with nodes corresponding to the elements of D. The root is ε, and the leaves are the nodes that are maximal in U with respect to $\leq$. A Σ-*tree*, i.e., a labeled finite tree, is a mapping $t \colon D \to \Sigma$, where D is a tree domain and for every $u \in D$, $t(u) \in \Sigma$. An example of a tree with numbered nodes and labeled by symbols in $\Sigma = \{S, V, N, NP, VP, you, see, George\}$ is presented in Fig. 1, where $D = \{(), 0, 1, 0.0, 1.0, 1.1, 0.0.0, 1.0.0, 1.1.0, 1.1.0.0\}$. We denote by T_Σ the set of all finite trees labeled by symbols in Σ. Subsets of T_Σ are called Σ-*tree languages*. We can also speak generally about *trees* and *tree languages* without specifying the alphabet (see Gécseg and Steinby [30,31] for details).

Context-free grammars (CFGs) are a well-known class of language generative devices extensively used for describing syntax of programming languages and some significant number of structures of natural language sentences.

Definition 2 (Hopcroft and Ullman [38]). *A context-free grammar is a tuple* $G = (\Sigma, N, P, S)$, *where:*

- Σ *is a finite alphabet of* terminals,
- N *is a finite alphabet of* nonterminal symbols,
- P *is a finite set of* productions *(or rules) of the form* $A \to \alpha$, *where* A *is a nonterminal and* α *is a string from* $(\Sigma \cup N)^*$,
- $S \in N$ *is a distinguished nonterminal called* start symbol.

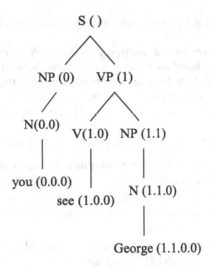

Fig. 1. A tree in which addresses of nodes are marked.

The tree structure of a string, called *derivation tree*, shows how that string can be obtained by applying the rules of a given grammar, and also describes the structure of the sentence. Note that a string can have multiple derivation trees associated. *Parsing* represents the process of analyzing a string (its structural description) and constructing its *derivation tree* (or *parse tree*). A *parser* is an algorithm that takes as input a string and either accepts or rejects it (depending on whether it belongs or not to the language generated by the grammar), and in the case it is accepted, also outputs its derivation trees.

We recall a *yield* mapping denoted by $yd: T_\Sigma \to \Sigma^*$ that extracts a word from each tree. More precisely, for every tree $t \in T_\Sigma$, $yd(t)$ is the string formed by the labels of all its leaves, read from left to right. For example, the yield of the tree depicted in Fig. 1 is *youseeGeorge*.

Even if the usefulness of CFGs in representing the syntax of natural languages is well established, there are still several important linguistics aspects that cannot be covered by this class of grammars. For example, the languages of *multiple agreement* $\{a_1^n a_2^n \ldots a_k^n \mid n \geq 1, k \geq 3\}$, *copy* $\{ww \mid w \in \{a, b\}^*\}$ and *cross agreement* $\{a^n b^m c^n d^m \mid n, m \geq 1\} \subset \{a, b, c, d\}^*$ cannot be generated by any CFG. Thus, a more powerful class of grammars, called *tree adjoining grammars*, was introduced in Joshi [40], yielding interesting mathematical and computational results.

Definition 3 (Joshi [40]). *A tree adjoining grammar (TAG) is a 5-tuple $G = (\Sigma, N, \mathcal{I}, \mathcal{A}, S)$, where:*

- *Σ is a finite alphabet of terminals,*
- *N is a finite alphabet of nonterminal symbols,*
- *$\mathcal{I}$ is a finite set of trees called initial trees, each of them having*
  - *all interior nodes labeled by nonterminal symbols, and*

- *all leaf nodes labeled by terminals, or by nonterminal symbols marked for substitution with ↓,*
- *A is a finite set of trees called* auxiliary trees, *each of them having*
  - *all interior nodes labeled by nonterminal symbols, and*
  - *all leaf nodes labeled by terminals, or by nonterminal symbols marked for substitution with ↓ except for one node, called* foot node, *annotated by "∗" (the label of the foot node must be identical to the label of the root node), and*
- *S ∈ N is a distinguished nonterminal called* start symbol.

The trees with root labeled by the nonterminal A are called A-*type trees*. The elements of $\mathcal{I}$ are usually identified by $\alpha_1, \ldots, \alpha_k$, and the ones of $\mathcal{A}$ by $\beta_1, \ldots, \beta_l$, respectively. Taking Gorn address into consideration, the label of a node of a given tree (auxiliary or initial, i.e., *elementary*) can be uniquely identified by the pair $(treeName, nodeAddress)^2$.

Sikkel [78] gives an overview of several parsing algorithms valid for both, programming languages and NLP, using an interesting formalism suggestively entitled *primordial soup*.

The known parsing algorithms for NLP are ruled-based or probabilistic (see Kakkonen [42] for comparisons, details and further references). Generally, higher grammar costs and computational complexity characterize rule-based algorithms while a higher speed (even linear complexity) and an accuracy lower than 100% are typical for probabilistic parsing algorithms. In such algorithms where probabilities are associated to rules, the idea of supertagging from Bangalore and Joshi [6] (an extension of the notion of "tag" from part of speech to reach syntactic information) followed by a Lightweight Dependency Analysis reached an accuracy of 95 % with a linear complexity. A good overview on such techniques and details about their applications is Bangalore and Joshi [7].

For TAGs, numerous ruled-based parsing algorithms, most of them *Earley-type* (Schabes and Joshi [75]) or *CKY-type* (Vijay-Shanker and Joshi [82]), were developed over the past two decades with the aim to speed up the process, use less resources, offer a better understanding or broad the applications' area. For such algorithms, the complexity in the worst case is known to be $O(n^6)$, where n is the length of the input string.

Grammatical evolution (GE) is a new approach which makes use of the derivation trees generated by CFGs and the searching capabilities of an evolutionary algorithm to perform parsing (O'Neill and Ryan [66], Dempsey et al. [21], Hemberg [35]). Adapting GE for TAGs was done by Dediu and Tîrnăucă [18]. We mention as other works in this domain the papers by Bikel [9] and Collins [14].

4   Word Senses

Word sense disambiguation (WSD) is the capability to assign each occurrence of a word to one or more classes of meaning (*sense*(s)) based on the evidence from

---

[2] Note that infinite derivations are not allowed because input strings are finite. Empty elementary trees can be avoided in the same way as eliminating ε-productions from CFGs.

the context and from external knowledge sources, as Navigli [63] specifies in his survey. Other surveys on WSD are Denkowski [22] and Pal and Saha [67]. A well-known series of evaluations of computational semantic systems is given by SemEval/Senseval [77], the most recognized reference point in WSD, organized as a workshop of the Conference of the Association for Computational Linguistics (ACL [1]).

Understanding word senses is important for quantifying several textual measurements. In fact, Rowcliffe [73] (citing De Beaugrand and Dressler [16]), gives a very nice overview of the seven defining characteristics of text, we present them as Table 1.

Table 1. Seven standards of textuality

Mostly, writer oriented features	Reader interpretation dependent characteristics
1 Cohesion	4 Acceptability
2 Coherence	5 Informativity
3 Intentionality	6 Situationality
7 Intertextuality	

We give further details on the first two characteristics of texts, and briefly explain the rest. *Acceptability* indicates the reader's recognition of a coherent and cohesive text. *Informativity* defines the new knowledge a reader can get. *Situationality* gives the time, space or other contextual information. *Intertextuality* indicates the connectivity of a text with other texts. *Intentionality* associates meanings in the light of some motivation.

Ferenčík [27] explains that *cohesion* is the way in which linguistic items of text constituents are meaningfully interconnected. He describes four types of cohesion:

- *lexical organization* – establishes semantic chains through connections such as repetition, equivalence - synonymy, hyponymy (shows the relationship between more general term (lexical representation) and the more specific instances of it), hyperonymy (hyponymy and hyperonymy are asymmetric relations), paraphrase (alternative way to rephrase some information), collocation (habitual juxtaposition of a particular word with another word or words with a frequency greater than chance).
- *reference* – either points out of the text (to some common knowledge or real world items), or refers to an item within the text. If the references are backward, we call them *anaphoric*, otherwise they are *cataphoric*.
- *ellipsis* is an instance of textual anaphora with the omission or indicating something referred to earlier (e.g., Have some more).
- *conjunction* – Roen [71] describes four kinds of cohesive conjunctions (see Table 2).

Table 2. Cohesive conjunctions

Conjunctive cohesion type	Examples
1 Additive	also, likewise, moreover
2 Adversative	on the other hand, however, and conversely
3 Causal	consequently, as a result, for this reason
4 Temporal	next, finally, and then

Ulbaek [80] (citing De Beaugrand and Dressler [16]) says that a text is *coherent* if it shows a general line of continuity of senses. The connection between the structures of cohesion and coherence of a text was studied by Harabagiu [34].

WSD should rely on the coherence of a text to find the right context and hence the right sense of some word. It is generally accepted that currently WSD faces two major problems (McCarthy [54]). One of the problems is the lack of training data. The other is about the granularity of senses we work with. For example, WordNet (Miller [57] and Fellbaum [26]) which is one of the largest computational lexicon publicly available, has a great many subtle distinctions between various senses of words that may in the end not be required. We would try to briefly explain how the information is organized in WordNet.

Words are organized into synonym sets (*synsets*), each representing one underlying lexical concept. Different relations link the synonym sets. Since the word "word" is commonly used to refer both to an utterance and the associated concept, to reduce ambiguity, we use, "word form" to refer to the utterance and "word meaning" to refer to the semantic concept. We can say that the starting point for lexical semantics is the mapping between forms and meanings (Miller [56]). The same word form F might have different meanings, that is, F is *polysemous*. The same meaning M might have several word forms associated, they are *synonyms* (relative to a context). In fact, some meaning M is represented in WordNet by word forms that can be used to express it: $\{F1, F2, \ldots\}$. The synonym sets do not explain what the concepts are. The only purpose they serve is discriminative, a person knowing English or another entity assumed to have already acquired one concept, should be able to recognize it only from the list of words. As words in different syntactic categories cannot be synonyms, it was necessary to partition words into nouns, verbs, adjectives, and adverbs. Other relations are also stored in WordNet, for example antonyms, hyponyms, meronyms ("has a-" relation), are all present as internal links between words. Morphological relations, that is, inflectional forms of words are incorporated only in the querying interface of WordNet, and not in the internal database.

Although started as a project only for English, WordNet rapidly developed to Global WordNet, sharing and providing WordNets for all languages in the world (Global WordNet [32]). Integrating the lexical concept from WordNet and the encyclopedic knowledge from Wikipedia, we get BabelNet (Navigli and Ponzetto [64]). By using BabelNet, was also possible to create SEW, Semantically Enriched Wikipedia (Raganato et al. [69]). These components are part of a larger project, MultiJEDI, a 5-year ERC (European Research Council) Starting Grant (2011-2016) headed by Roberto Navigli (MultiJEDI [61]).

5   Natural Language Processing Trends

A new paradigm, appeared on our computers, telephones and tablets under the generic name of intelligent personal assistants (IPA). They use ASR with the purpose of operate devices, access information, and manage personal tasks. Weng et al. [84] promote the idea that well-known IPAs like Apple Siri, Google Assistant, Microsoft Cortana, Amazon Echo, etc., marked a significant progress after 2012, mainly due to the recent advances in deep learning technologies [19] — especially deep neural networks (DNNs). Traditionally, ASR was performed based on hidden Markov models (HMMs), however, the biggest limitation of HMM is the assumption that the observations are conditionally independent of all other variables.

IPAs can vary in complexity being able to recognize from several simple commands up to long document writing systems. They can process the information in real time and they are also capable to recognize in context formatting commands, instructions for deleting text or reorganize the sentences. Despite the impressive progress, there are still steps to be done, especially receiving information from the syntactic layer, or from the proper context for a given word. For example, the latest version of the most powerful voice recognition software available today, has problems to distinguish between "suggest" and "suggests" (Branscombe [10]).

We believe that the next research in WSD will try to find an intelligent identification of the context. Starting from the assumption that only some words in the surrounding context are indicative of the correct sense of the target word, an intelligent selection of the right context used in the disambiguation process could potentially lead to much better results and faster solutions than considering all the words in the surrounding context. We can identify *lexical chains* (sequences of semantically related words interconnected via semantic relations, as shown by Erekhinskaya and Moldovan [24]) in the surrounding context, and only include in the context analysis those words that are found in chains containing the target word. For this method to be effective, we need also to adopt a proper size for the surrounding context. The general principle underlying the method of deciding the expansion of a context is that the text must be coherent; however, since coherence is very hard to detect, we must rely on cohesion, which is closely associated to coherence (Fortu and Moldovan [28]). The idea to employ lexical chains to represent text cohesion was inspired by an older work of Morris and Hirst [60]. There are several classes of context types, all initiated by different key words called seeds, then the contexts grow according to the cohesion of sentences (Crossley et al. [15]). The main problem we see with this approach is that the algorithms constructing lexical chains are clearly NP. Closer the word senses for which we construct the lexical chains, easier for the algorithm to construct the chain; for more far away word senses, the algorithm needs more computations.

There are several contingency solutions: one, we can pre-process lists of lexical chains for the most used pairs of words and memorize them; another solution could use some heuristic methods (like evolutionary algorithms) to drop the analysis for non-promising chains.

Despite the fact that there are known results on contexts in AI in general (McCarthy [54]) and in formal languages in particular (Marcus [52], Păun [68], Kudlek et al. [48]), there are fewer results on context taxonomy, boundary detection, and how to use them in language understanding. The representation of contexts for WSD ranges from flat specifications (Mariò et al. [53], Le and Mikolov [49], Fahrenberg et al. [25]) to structured descriptions such as trees or (hyper)graphs (Widdows and Dorow [85], Dorow and Widdows [23], Véronis [83], Navigli and Velardi [65], Dediu et al. [17]).

We estimate that the following NLP processing systems will continue to give an increasing importance of linguistic training data, however, new parallel hybrid system incorporating both symbolic and data driven methods, capable to select the best technique depending on the available processing time, will further improve the current results. At the same time, domain specific applications will be capable to perform better than general applications, mainly due to the reduced size of the external context required by WSD.

Studying WSD problems using for training only written documents it is just a glimpse into the dimension and the quantity of data we could process based on an appropriate semantic model. We face similar problems when processing for example images and trying to interpret them, or even multimedia files. There are so many calling problems in this new world of tremendous data emergence, not only for movies, but also analyzing survey and security data, information retrieval, automatic classification of information.

References

1. Association for Computational Linguistics (ACL). https://www.aclweb.org/. Accessed 19 Jan 2017
2. The 2012 ACM Computing Classification System. http://www.acm.org/publications/class-2012. Accessed 25 Jan 2017
3. Angluin, D.: Learning regular sets from queries and counterexamples. Inf. Comput. **75**(2), 87–106 (1987)
4. Angluin, D., Becerra-Bonache, L.: Learning meaning before syntax. In: Clark, A., Coste, F., Miclet, L. (eds.) ICGI 2008. LNCS (LNAI), vol. 5278, pp. 1–14. Springer, Heidelberg (2008). doi:10.1007/978-3-540-88009-7_1
5. Angluin, D., Becerra-Bonache, L., Dediu, A.H., Reyzin, L.: Learning finite automata using label queries. In: Gavaldà, R., Lugosi, G., Zeugmann, T., Zilles, S. (eds.) ALT 2009. LNCS (LNAI), vol. 5809, pp. 171–185. Springer, Heidelberg (2009). doi:10.1007/978-3-642-04414-4_17
6. Bangalore, S., Joshi, A.K.: Supertagging: an approach to almost parsing. Comput. Linguist. **25**, 237–265 (1999)
7. Bangalore, S., Joshi, A.K. (eds.): Supertagging. A Bradford Book. The MIT Press, Cambridge (2010)

8. Bellegarda, J.R., Monz, C.: State of the art in statistical methods for language and speech processing. Comput. Speech Lang. **35**, 163–184 (2016). http://dx.doi.org/10.1016/j.csl.2015.07.001
9. Bikel, D.M.: Intricacies of Collins' parsing model. Comput. Linguist. **30**(4), 479–511 (2004)
10. Branscombe, M.: Review: Nuance dragon for windows offers strong voice recognition. Computer World, January 2016. http://www.computerworld.com/article/3018071/desktop-apps/review-nuance-dragon-for-windows-offers-strong-voice-recognition.html
11. Brill, E.: Transformation-based error-driven learning and natural language processing: a case study in part-of-speech tagging. Comput. Linguist. **21**(4), 543–565 (1995)
12. Chierchia, G.: Anaphora and dynamic binding. Linguist. Philos. **15**, 111–183 (1992)
13. Cole, R. (ed.): Survey of the State of the Art in Human Language Technology. Cambridge University Press, New York (1997)
14. Collins, M.: Head-Driven Statistical Models for Natural Language Parsing. Ph.D. thesis, University of Pennsylvania, Philadelphia, PA (1999)
15. Crossley, S.A., Kyle, K., McNamara, D.S.: The tool for the automatic analysis of text cohesion (taaco): automatic assessment of local, global, and text cohesion. Behav. Res. Methods **2015**, 1–11 (2015)
16. De Beaugrande, R., Dressler, W.: Introduction to Text Linguistics. Longman Linguistics Library. Routledge, London (2016). https://books.google.pt/books?id=gQrrjwEACAAJ
17. Dediu, A.-H., Klempien-Hinrichs, R., Kreowski, H.-J., Nagy, B.: Contextual hypergraph grammars – a new approach to the generation of hypergraph languages. In: Ibarra, O.H., Dang, Z. (eds.) DLT 2006. LNCS, vol. 4036, pp. 327–338. Springer, Heidelberg (2006). doi:10.1007/11779148_30
18. Dediu, A.H., Tîrnăucă, C.I.: Evolutionary algorithms for parsing tree adjoining grammars. In: Bel-Enguix, G., Jiménez-López, M. (eds.) Bio-Inspired Models for Natural and Formal Languages, pp. 277–304. Cambrige Scholars (2011)
19. Deep Learning. https://en.m.wikipedia.org/wiki/Deep_learning. Accessed 31 Jan 2017
20. Dekker, P.: Coreference and representationalism. In: von Heusinger, K., Egli, U. (eds.) Reference and Anaphorical Relations, pp. 287–310. Kluwer, Dordrecht (2000)
21. Dempsey, I., O'Neill, M., Brabazon, A.: Foundations in Grammatical Evolution for Dynamic Environments. Springer, Heidelberg (2009)
22. Denkowski, M.: A Survey of Techniques for Unsupervised Word Sense Induction. Lang. Stat. II Lit. Rev. (2009)
23. Dorow, B., Widdows, D.: Discovering corpus-specific word senses. In: 82. Proceedings of the 10th Conference of the European Chapter of the Association for Computational Linguistics, Budapest, Hungary (2003)
24. Erekhinskaya, T., Moldovan, D.: Lexical chains on wordnet and extensions. In: Proceedings of the Twenty-Sixth International Florida Artificial Intelligence Research Society Conference, FLAIRS 2013 (2013)
25. Fahrenberg, U., Biondi, F., Corre, K., Jegourel, C., Kongshøj, S., Legay, A.: Measuring global similarity between texts. In: Besacier, L., Dediu, A.-H., Martín-Vide, C. (eds.) SLSP 2014. LNCS (LNAI), vol. 8791, pp. 220–232. Springer, Cham (2014). doi:10.1007/978-3-319-11397-5_17
26. Fellbaum, C. (ed.): WordNet: An Electronic Database. MIT Press, Cambridge (1998)

27. Ferenčík, M.: A Survey of English Stylistics. http://www.pulib.sk/elpub2/FF/ Ferencik/INDEX.HTM. Accessed 22 Jan 2017
28. Fortu, O., Moldovan, D.: Identification of textual contexts. In: Dey, A., Kokinov, B., Leake, D., Turner, R. (eds.) CONTEXT 2005. LNCS (LNAI), vol. 3554, pp. 169–182. Springer, Heidelberg (2005). doi:10.1007/11508373_13
29. Freund, Y., Kearns, M.J., Ron, D., Rubinfeld, R., Schapire, R.E., Sellie, L.: Efficient learning of typical finite automata from random walks. Inf. Comput. **138**(1), 23–48 (1997)
30. Gécseg, F., Steinby, M.: Tree Automata. Akadémiai Kiadó, Budapest (1984)
31. Gécseg, F., Steinby, M.: Tree languages. In: Salomaa, A., Rozenberg, G. (eds.) Handbook of Formal Languages. Beyond Words, vol. 3, pp. 1–68. Springer, New York (1997)
32. Global WordNet Association. http://globalwordnet.org/. Accessed 24 Jan 2017
33. Gold, E.M.: Language identification in the limit. Inf. Control **10**(5), 447–474 (1967)
34. Harabagiu, S.M.: From lexical cohesion to textual coherence: a data driven perspective. Int. J. Patt. Recognit. Artif. Intell. **13**(2), 247–265 (1999)
35. Hemberg, E.A.P.: An Exploration of Grammars in Grammatical Evolution. Ph.D. thesis, University College Dublin, September 2010
36. Hirschberg, J., Manning, C.D.: Advances in natural language processing. Science **349**(6245), 261–266 (2015). http://dx.doi.org/10.1126/science.aaa8685
37. Hopcroft, J.E., Motwani, R., Ullman, J.D.: Introduction to Automata Theory, Languages, and Computation, 3rd edn. Addison-Wesley, Reading (2006)
38. Hopcroft, J.E., Ullman, J.D.: Introduction to Automata Theory, Languages, and Computation. Addison-Wesley, Reading (1979)
39. Hovy, E., Lin, C.Y.: Automated text summarization in SUMMARIST. In: Proceedings of the Intelligent Scalable Text Summarization Workshop, pp. 18–24 (1997)
40. Joshi, A., Levy, L., Takahashi, M.: Tree adjunct grammars. J. Comput. Syst. Sci. **10**(1), 136–163 (1975)
41. Jurafsky, D., Martin, J.H.: Speech and Language Processing: An Introduction to Natural Language Processing, Computational Linguistics, and Speech Recognition, 1st edn. Prentice Hall PTR, Upper Saddle River (2000)
42. Kakkonen, T.: Framework and resources for natural language parser evaluation. Computing Research Repository abs/0712.3705 (2007)
43. Kamp, H., Reyle, U.: From Discourse to Logic. Kluwer, Dordrecht (1993)
44. Kastner, I., Monz, C.: Automatic single-document key fact extraction from newswire articles. In: Proceedings of the 12th Conference on European Chapter of the ACL (EACL 2009), Athens, Greece, pp. 415–423 (2009)
45. Kay, M.: Machine translation: the disappointing past and present. In: Cole, R. (ed.) Survey of the State of the Art in Human Language Technology, pp. 248–250. Cambridge University Press, New York (1997). http://dl.acm.org/citation. cfm?id=278696.278813
46. Knight, K., Marcu, D.: Summarization beyond sentence extraction: a probabilistic approach to sentence compression. Artif. Intell. **13**(1), 91–107 (2001)
47. Koehn, P.: Statistical Machine Translation. Cambridge University Press, Cambridge (2009)
48. Kudlek, M., Martín-Vide, C., Mateescu, A., Mitrana, V.: Contexts and the concept of mild context-sensitivity. Linguist. Philos. **26**, 703–725 (2002)
49. Le, Q., Mikolov, T.: Distributed representations of sentences and documents. In: Proceedings of the 31st International Conference on Machine Learning, pp. 1188–1196 (2014)

50. Lopez, A.: Statistical machine translation. ACM Comput. Surv. **40**(3), 1–8 (2008)
51. Manning, C.D.: Part-of-speech tagging from 97% to 100%: is it time for some linguistics? In: Gelbukh, A.F. (ed.) CICLing 2011. LNCS, vol. 6608, pp. 171–189. Springer, Heidelberg (2011). doi:10.1007/978-3-642-19400-9_14
52. Marcus, S.: Contextual grammars. Rev. Roum. Math. Pures et Appl. **14**(10), 1525–1534 (1969). http://citeseer.ist.psu.edu/marcus69contextual.html
53. Mariòo, J.B., Banchs, R.E., Crego, J.M., de Gispert, A., Lambert, P., Fonollosa, J.A.R., Costa-Jussà, M.R.: N-gram-based machine translation. Comput. Linguist. Arch. **32**(4), 527–549 (2006). MIT Press, Cambridge
54. McCarthy, J.: Notes on formalizing context. In: Proceedings of the 13th International Joint Conference on Artificial Intelligence, IJCAI 1993, vol. 1, pp. 555–560. Morgan Kaufmann Publishers Inc., San Francisco (1993). http://dl.acm.org/citation.cfm?id=1624025.1624103
55. Mikolov, T., Sutskever, I., Chen, K., Corrado, G.S., Dean, J.: Distributed representations of words and phrases and their compositionality. In: Burges, C.J.C., Bottou, L., Welling, M., Ghahramani, Z., Weinberger, K.Q. (eds.) Advances in Neural Information Processing Systems 26, pp. 3111–3119. Curran Associates, Inc. (2013). http://papers.nips.cc/paper/5021-distributed-representations-of-words-and-phrases-and-their-compositionality.pdf
56. Miller, G.A.: Dictionaries of the mind. In: Proceedings of the 23rd Annual Meeting on Association for Computational Linguistics, ACL 1985, pp. 305–314. Association for Computational Linguistics, Stroudsburg (1985). http://dx.doi.org/10.3115/981210.981248
57. Miller, G.A.: WordNet: a lexical database for English. Commun. ACM **38**(11), 39–41 (1995). http://doi.acm.org/10.1145/219717.219748
58. Mohri, M.: Finite-state transducers in language and speech processing. Comput. Linguist. **23**(2), 269–311 (1997). http://dl.acm.org/citation.cfm?id=972695.972698
59. Montague, R.: Universal grammar. Theoria **36**, 373–398 (1970)
60. Morris, J., Hirst, G.: Lexical cohesion computed by thesaural relations as an indicator of the structure of text. Comput. Linguist. **17**(1), 21–48 (1991). http://dl.acm.org/citation.cfm?id=971738.971740
61. MultiJEDI - Multilingual joint word sense disambiguation. http://multijedi.org/. Accessed 28 Jan 2017
62. Muskens, R.: Combining Montague semantics and discourse representation. Linguist. Philos. **19**(2), 143–186 (1996)
63. Navigli, R.: Word sense disambiguation: a survey. ACM Comput. Surv. (CSUR) **41**(2), 1–69 (2009)
64. Navigli, R., Ponzetto, S.P.: Babelnet: the automatic construction, evaluation and application of a wide-coverage multilingual semantic network. Artif. Intell. **193**, 217–250 (2012). http://dx.doi.org/10.1016/j.artint.2012.07.001
65. Navigli, R., Velardi, P.: Structural semantic interconnections: a knowledge-based approach to word sense disambiguation. IEEE Trans. Patt. Anal. Mach. Intell **27**(7), 1075–1088 (2005)
66. O'Neill, M., Ryan, C.: Grammatical Evolution: Evolutionary Automatic Programming in an Arbitrary Language. Kluwer, Dordrecht (2003)
67. Pal, A.R., Saha, D.: Word sense disambiguation: a survey. Int. J. Control Theor. Comput. Model. (IJCTCM) **5**(3), 1–16 (2015)
68. Păun, G.: Marcus Contextual Grammars. Kluwer Academic Publishers, Norwell (1997)

69. Raganato, A., Bovi, C.D., Navigli, R.: Automatic construction and evaluation of a large semantically enriched wikipedia. In: Proceedings of 25th International Joint Conference on Artificial Intelligence (IJCAI 2016), New York, USA, July 2016
70. Rieger, B.B.: On distributed representation in word semantics. Technical report, Forschungsbericht TR-91-012, International Computer Science Institute (ICSI) (1991)
71. Roen, D.H.: The effects of cohesive conjunctions, reference, response rhetorical predicates, and topic on reading rate and written free recall. J. Read. Behav. 16(1), 15–26 (1984)
72. Rothlauf, F.: Design of Modern Heuristics: Principles and Application, 1st edn. Springer, Heidelberg (2011)
73. Rowcliffe, I.C.: Seven Standards of Textuality? http://web.letras.up.pt/icrowcli/textual.html. Accessed 22 Jan 2017
74. Sahlgren, M.: The distributional hypothesis. Ital. J. Linguist. 20(1), 33–54 (2008)
75. Schabes, Y., Joshi, A.K.: An Earley-type parsing algorithm for tree adjoining grammars. In: Proceedings of the 26th Annual Meeting of the Association for Computational Linguistics (ACL 1988), pp. 258–269. Association for Computational Linguistics (1988)
76. Schmid, H.: Probabilistic part-of-speech tagging using decision trees. In: International Conference on New Methods in Language Processing, Manchester, UK, pp. 44–49 (1994)
77. SemEval   Portal.   https://www.aclweb.org/aclwiki/index.php?title=SemEval_Portal. Accessed 19 Jan 2017
78. Sikkel, K.: Parsing Schemata: A Framework for Specification and Analysis of Parsing Algorithms, 1st edn. Springer, Heidelberg (2013)
79. Sudkamp, T.A.: Languages and Machines: An Introduction to the Theory of Computer Science, 3rd edn. Addison-Wesley, Reading (2006)
80. Ulbaek, I.: Second order coherence: a new way of looking at incoherence in texts. Linguist. Beyond and Within 2, 167–179 (2016)
81. Valiant, L.G.: A theory of the learnable. Commun. ACM 27(11), 1134–1142 (1984)
82. Vijay-Shanker, K., Joshi, A.K.: Some computational properties of tree adjoining grammars. In: Proceedings of the 23rd Annual Meeting of the Association for Computational Linguistics (ACL 1985), pp. 82–93. Association for Computational Linguistics (1985)
83. Véronis, J.: Hyperlex: lexical cartography for information retrieval. Comput. Speech Lang. 18(3), 223–252 (2004)
84. Weng, F., Angkititrakul, P., Shriberg, E., Heck, L.P., Peters, S., Hansen, J.H.L.: Conversational in-vehicle dialog systems: the past, present, and future. IEEE Signal Process. Mag. 33(6), 49–60 (2016). http://dx.doi.org/10.1109/MSP.2016.2599201
85. Widdows, D., Dorow, B.: A graph model for unsupervised lexical acquisition. In: Proceedings of the 19th International Conference on Computational Linguistics, COLING, Taipei, Taiwan, pp. 1–7 (2002)
86. Winston, P.H.: Artificial Intelligence, 3rd edn. Addison-Wesley Longman Publishing Co., Inc., Boston (1992)
87. Zajic, D., Dorr, B., Schwartz, R., Monz, C., Lin, J.: A sentence-trimming approach to multi-document summarization. In: Proceedings of EMNLP 2005 Workshop on Text Summarization (2005)

An All-or-Nothing Flavor
to the Church-Turing Hypothesis

Stefan Wolf[1,2]($\boxtimes$)

[1] Faculty of Informatics, Università della Svizzera italiana (USI),
6900 Lugano, Switzerland
wolfs@usi.ch
[2] Facoltà indipendente di Gandria, Lunga Scala, 6978 Gandria, Switzerland

Abstract. Landauer's principle claims that "Information is Physical".
It is not surprising that its conceptual *antithesis*, Wheeler's "It from
Bit", has been more popular among computer scientists — in the form
of the *Church-Turing hypothesis:* All natural processes can be computed
by a universal Turing machine; physical laws then become descriptions
of subsets of *observable*, as opposed to merely *possible*, computations.
Switching back and forth between the two traditional styles of thought,
motivated by quantum-physical Bell correlations and the doubts they
raise about fundamental space-time causality, we look for an intrinsic,
physical randomness notion and find one around the second law of thermodynamics. Bell correlations combined with *complexity as randomness*
tell us that beyond-Turing computations are either physically impossible,
or they can be carried out by "devices" as simple as individual photons.

1 Introduction

1.1 Ice *versus* Fire

According to *Jeanne Hersch* [20], the entire history of philosophy is coined by
an antagonism rooting in the fundamentally opposite world views of the pre-
Socratic philosophers *Parmenides of Elea* (515 B.C.E. – 445 B.C.E.) on the one
hand and *Heraclitus* (520 B.C.E. – 460 B.C.E.) on the other. For Parmenides,
any change, even time itself, is simply an *illusion*, whilst for Heraclitus, *change*
is all there is. The "cold logician" Parmenides has been compared to *ice*, and
Heraclitus' thinking is the *fiery* one [26]. If Hersch is right, and this opposition
between these styles crosses the history of philosophy like a red line, then this
must be true no less for the history of *science*.

A textbook example illustrating the described antagonism is the debate
between *Newton* and *Leibniz* [36]: For Newton, space and time are fundamental
and given *a priori*, just like a stage on which all the play is located. For Leibniz,
on the other hand, space and time are emergent as *relational* properties: The
stage emerges *with* the play and not prior to it, and it is not there without it.
With only a few exceptions — most notably *Ernst Mach* — the course of physical science went for Newton's view; it did so with good success. An important

© Springer International Publishing AG 2017
T.V. Gopal et al. (Eds.): TAMC 2017, LNCS 10185, pp. 39–56, 2017.
DOI: 10.1007/978-3-319-55911-7_4

example here is, of course, Einstein's relativity: Whilst its crystallization point was *Mach's principle*, stating that intertial forces are relational (as opposed to coming from acceleration against an *absolute* space), the resulting theory does *not* follow the principle since there is (the flat) space-time also in a massless universe.

In the present work, we turn our attention to physical phenomena such as the second law of thermodynamics and Bell correlations from quantum theory. We find here again the opposition between Parmenides' and Heraclitus' standpoints, and we directly build on their tension with the goal of obtaining more insight, hereby bridging the gap separating them to some extent.

The *Heraclitean* style can be recognized again in the spirit of *Ferdinand Gonseth*'s "La logique est tout d'abord une science naturelle" — "Logic is, first of all, a natural science". This is a predecessor of Rolf Landauer's famous slogan "Information Is Physical", [24] putting physics at the basis of the concept of information and its treatment.

This is in sharp contrast to *Shannon*'s [31] (very successful) making information *abstract*, independent of the particular physical realization of it (*e.g.*, a specific noisy communication channel). To the *Parmenidean* paradigm belongs also the *Church-Turing hypothesis* [22], stating that all physically possible processes can be simulated by a universal Turing machine. This basing physical reality on information and computation was later summarized by *John Archibald Wheeler* as "It from Bit" [37].

1.2   Non-Locality, Space-Time Causality, and Randomness

After Einstein had made the world mechanistic and "local" (without actions at a distance), he was himself involved in a work [14] paving the (long) way to that locality to fall again. The goal of Einstein *et al.* had, however, been the exact opposite: to *save* locality in view of correlations displayed in the measurement behavior of (potentially physically separated) parts of an *entangled* quantum state. The claim was that quantum theory was an only incomplete description of nature, to be refined by hidden parameters determining the outcomes of all potential, alternative measurements. It took roughly thirty years until that claim was grounded when *John Stewart Bell* [6] showed the impossibility of the program — ironically making the case with the exact same states as "EPR" had introduced. The consequences of Bell's insight are radical: If the values are not predetermined, then there must be *fresh* and at the same time *identical* pieces of classical information popping us spontaneously — this is *non-locality*. The conceptual problem these correlations lead us into is the difficulty of explaining their origin *causally*, *i.e.*, according to *Reichenbach's principle* — which states that a correlation between two space-time events can stem from a *common cause* (in the common past) or a *direct influence* from one event to the other [30]. Bell's result rules out the common cause as an explanation, thus remains the influence. Besides the fact that it is an inelegant overkill to explain a *non*-signaling phenomenon (*not* allowing for transmitting messages from one party to the other) using a *signaling* mechanism, there are further problems: Hidden influences as explanations of Bell correlation require both infinite speed [1,13,33] and precision [39].

In view of this, it appears reasonable to question the (only) assumption made in Reichenbach's principle: The *a priori* causal structure [27].[1] If we turn back the wheel to the *Newton-Leibniz debate*, and choose to follow Leibniz instead, seeing space-time as appearing only *a posteriori*, then there is a first victim to this: *Randomness*: In [12], a piece of information is called *freely random* if it is statistically independent from all other pieces of information except the ones in its *future* light cone. Clearly, when the assumption of an initially given causal structure is dropped, such a definition is not founded any longer.[2] (It is then possible to turn around the affair and base past and future on postulated freeness of bits [5].) In any case, we are now motivated to find an *intrinsic, context-free, physical definition of randomness* and choose to look at: *Watt's Steam Engine.*

2   The Search for an Intrinsic Randomness Notion: From Steam Pipes to the Second Law of Thermodynamics

2.1   The Fragility and the Robustness of the Second Law

The *second law of thermodynamics* has advanced to becoming pop culture.[3] It is, however, much less famous than Einstein's relativity, Heisenberg's uncertainty, or quantum teleportation because it does not have any glamour, fascination, or hope attached to it: The law stands for facts many of us are in denial of or try to escape. We ask whether the attribution of that formalized pessimism to physics has *historical* reasons.

The validity of the second law seems to depend on surprising conditions such as the inexistence of certain life styles (*e.g.*, *Maxwell's demon* or photosynthesizing plants — Kelvin [21] writes: "When light is absorbed *other than in vegetation*, there is dissipation [...]"). To make things worse, there is always a non-zero probability (exponentially small, though) of exceptions where the law fails to hold; we are not used to this from other laws of physics. Can this be taken as an indication that the fundamental way of formulating the law eludes us?

The described *fragility* of the second law is strangely contrasted by its being, in another way, *more robust* than others (such as Bell violations only realizable under extremely precise lab conditions): We certainly do not need to trust experimentalists to be convinced that the second law is acting, everywhere and

---

[1] It has been shown [4] that if causality is dropped but logical consistency maintained, then a rich world opens — comparable to the one between locality and signaling.

[2] Note furthermore that the definition is consistent with full determinism: A random variable with trivial distribution is independent of every other (even itself).

[3] See, *e.g.*, Allen, W., Husbands and Wives (1992): The protagonist Sally is explaining why her marriage did not work out. First she does not know, then she realizes: "It's the *second law of thermodynamics*: sooner or later everything turns to shit. That's my phrasing, not the *Encyclopedia Britannica*"

always. It has even been claimed [35] to hold a "supreme position" among physical laws: It appears easier to imagine a world where relativity or quantum theory do not hold than to figure out a reality lacking the validity of the second law. (Concerning the reasons for this, we can only speculate: Would we be forced to give up the mere *possibility of perception, memory — the arrow of time?*)

2.2 History

This story (see [35]) starts with *Sadi Carnot* (1796–1832) and his study of heat engines such as *James Watt*'s steam pipe. The assumption, in consequence, that the law is closely related to such engines, and to the circular processes involved, is of course not wrong, but it underestimates a fundamental *logical-combinatorial-informational fact*; perhaps steam engines are for the second law what *telescopes* are for Jupiter's moons.

Carnot argued that the maximal efficiency of a heat engine between two heat baths depended only on the two temperatures involved. (The derived formula motivated Lord Kelvin to define the absolute temperature scale.)

Rudolf Clausius' (1822–1888) [11] version of the second law reads: *"Es kann nie Wärme aus einem kälteren in einen wärmeren Körper übergehen, ohne dass eine andere damit zusammenhängende Änderung eintritt"*. — "No process can transport heat from cold to hot and do no further change".

Lord Kelvin (1824–1907) [21] formulated his own version of the second law and concluded — in just the next sentence — that the law may have consequences deeper than what was obvious at first sight: *"Restoration of mechanical energy without dissipation [...] is impossible. Within a finite period of time past, the earth must have been, within a finite time, the earth must again be unfit for the habitation of man"*.

Also for Clausius, it was only a single thinking step from his version of the law to concluding that all temperature differences in the entire universe will vanish (the *Wärmetod*) and that then, no change will be possible anymore. He speaks of *a general tendency of nature for change into a specific direction:* "Wendet man dieses auf das Weltall im Ganzen an, so gelangt man zu einer eigentümlichen Schlussfolgerung, auf welche zuerst W. Thomson [Lord Kelvin] aufmerksam machte, *nachdem er sich meiner Auffassung des zweiten Hauptsatzes angeschlossen hatte*. Wenn [...] im Weltall die Wärme stets das Bestreben zeigt, [...] dass [...] Temperaturdifferenzen ausgeglichen werden, so muss es sich mehr und mehr dem Zustand annähern, wo [...] keine Temperaturdifferenzen mehr existieren." — In short: "He was right after he had realized that I had been right: At the end, no temperature differences will be left in the universe."

Ludwig Boltzmann (1844–1906) brought our understanding of the second law closer to combinatorics and probability theory (in particular, the law of large numbers). His version is based on the fact that it is more likely to end up in a large set (of possible states) than in a small one: The more "microstates" belong to a given "macrostate", the more likely is it that you will find yourself in that macrostate. In other words, if you observe the time evolution of a system (by some reason starting in a very small, "unlikely" macrostate), then the "entropy" of the system — here simply (the logarithm of) the number of corresponding microstates — does not decrease.[4]

The notion of macrostate and its entropy have been much debated. Von Neumann remarked [35]: "No one knows what entropy really is, so in a debate you will always have the advantage". We aim at a version of the second law avoiding this advantage: a view without probabilities or ensembles, but based on intrinsic, one-shot complexity instead. Crucial steps in that direction were made by *Zurek* [40]. We take a Church-Turing view and follow *Landauer* [24] whose role or, more specifically, whose choice of viewpoint around the second law can be compared with *Ernst Specker*'s [32] take on quantum theory: *logical*.

2.3   Reversibility

Landauer investigated the thermodynamic price of logical operations. He was correcting a belief by *John von Neumann* that every bit operation required free energy $kT \ln 2$ (where k is Boltzmann's constant, T the environmental temperature, and $\ln 2$ owed to the fact that 2 is not a natural but a logical constant). According to Landauer — and affirmed by *Fredkin and Toffoli's* "ballistic computer" [18]—, this limitation or condition only concerns (bit) operations which are logically *irreversible*, such as the AND or the OR. On the positive side, it has been observed that every function, bijective or not, can in principle be evaluated in a logically *reversible way, using only "Toffoli gates", i.e., made-reversible and then-universal AND gates*; its computation can be thermodynamically neutral: It does not have to dissipate heat.

Landauer's principle states that erasing (setting the corresponding memory cells to 0) N bits costs $kTN \ln 2$ free energy which must be dissipated as heat to the environment (of temperature T). This *dissipation* is crucial in the argument: Heating up the environment compensates for the *entropy loss* within the memory cell, realized as a physical system (spin, gas molecule, *etc.*).

Let us consider the inverse process: *Work extraction*. Bennett [7] made the key contribution to the resolution of the paradox of *Maxwell's demon*. That demon had been thought of as violating the second law by adaptively handling a frictionless door with the goal of "sorting a gas" in a container. Bennett took the demon's memory (imagined to be in the all-0-state before sorting) into account, which is in the end filled with "random" information, an expression of the original state of the gas. The growth of disorder *inside* the demon compensates for the

---

[4] Boltzmann imagined further that the universe had started in a completely "uniform" state, so the entire, rich reality perceived would be a simple fluctuation. (Note that the fact that this fluctuation is extremely unlikely is irrelevant if we can *condition on our existence*, given our discussing this.) He may have been aware that this way of thinking leads straight into *solipsism:* "My existence alone, simply *imagining* my environment, seems much more likely than the actual existence of all people around me, let alone all the visible galaxies, *etc.*" — he killed himself in a hotel room in Duino, Italy; it has been told that this was also related to "mobbing" by Mach in Vienna. In any case, we choose to comfort us today with the somewhat religious assumption that the universe initiated in a low-entropy state, called the *big bang*.

order she creates *outside* (*i.e.*, in the gas) — the second law is saved. The initial 0-string is the demon's resource allowing for her order creation.

If we break Bennett's argument apart in the middle, we end up with the *converse* of Landauer's principle: The all-0-string has work value, *i.e.*, if we accept the price of the respective memory cells to become "randomized" in the process, we can extract $kTN \ln 2$ free energy from the environment (a heat bath of temperature T). In a *constructivist* manner, we choose to view the work-extraction process as an algorithm which, according to the *Church-Turing hypothesis*, we imagine as being carried out by a universal *Turing machine*. It then follows that the *work value of a string* S is closely related to the possibility of lossless compression of that string: For any concrete data-compression algorithm, we can extract $kT \ln 2$ times the length of S (uncompressed) minus the length of its compression: *Work value is redundancy (in representation) of information.* On the other end of the scale, the upper bound on work extraction is linked to the ultimate compression limit: *Kolmogorov complexity*, *i.e.*, the length of the shortest program for the extraction demon (Turing machine) generating the string in question. This holds because a computation is logically reversible only if it can be carried out in the other direction, step by step.

There is a direct connection between the work value and the *erasure cost* (in the sense of Landauer's principle) of a string. We assume here that for both processes, the extraction demon has access to an additional string X (modeling prior "knowledge" about S) which serves as a catalyst and is to be unchanged at the end of the process. For a string $S \in \{0,1\}^N$, let $\mathrm{WV}(S|X)$ and $\mathrm{EC}(S|X)$ be its work value and erasure costs, respectively, given X. Then[5]

$$\mathrm{WV}(S|X) + \mathrm{EC}(S|X) = N .$$

To see this, consider first the combination extract-then-erase. Since this is *one specific way* of erasing, we have

$$\mathrm{EC}(S|X) \leq N - \mathrm{WV}(S|X) .$$

If, on the other hand, we consider the combination erase-then-extract, this leads to

$$\mathrm{WV}(S|X) \geq N - \mathrm{EC}(S|X) .$$

Given the results on the work value discussed above, as well as this connection between the work value and erasure cost, we obtain the following bounds on the thermodynamic cost of erasing a string S by a demon, modeled as a universal Turing machine $\mathcal{U}$ with initial tape content X.

Landauer's principle, revisited. *Let C be a computable compression function*

$$C : \{0,1\}^* \times \{0,1\}^* \longrightarrow \{0,1\}^*$$

such that $(A, B) \mapsto (C(A, B), B)$ *is injective. Then we have*

$$K_{\mathcal{U}}(S|X) \leq \mathrm{EC}(S|X) \leq \mathrm{len}(C(S, X)) .$$

---

[5] Let $kT \ln 2 = 1$.

Landauer's revised principle puts forward two ideas: First, the erasure cost is an *intrinsic, context-free, physical measure for randomness* (entirely independent of probabilities and counter-factual statements of the form "some value *could* just as well have been *different*", *i.e.*, removing one layer of speculation). The idea that the erasure cost — or the Kolmogorov complexity related to it — is a measure for randomness independent of probabilities can be tested in a context in which randomness has been paramount: *Bell correlations* [6] predicted by quantum theory, see Sect. 3 for details.

The second idea starts from the observation that the price for the *logical* irreversibility of the erasure transformation comes in the form of a *thermodynamic* effort.[6] In an attempt to harmonize this somewhat *hybrid* picture, we invoke Wheeler's [37] *"It from Bit*: Every *it* — every particle, every field of force, even the space-time continuum itself — derives its function, its meaning, its very existence entirely [...] from the apparatus-elicited answers to yes-or-no questions, binary choices, *bits"*. This is an anti-thesis to Landauer's slogan, and we propose the following synthesis of the two: If Wheeler suggests to look at the environment as being *information* as well, then Landauer's principle ends up to be read as: The necessary environmental compensation for the logical irreversibility of the erasure of S is such that *the overall computation, including the environment, is logically reversible: no information ever gets completely lost.*

Second law, Church-Turing view. *If reality is assumed to be computed by a Turing machine, then that computation has the property of being injective: Nature computes with Toffoli, but no AND or OR gates.*

This fact is *a priori* asymmetric in time: The future must uniquely determine the past, not necessarily *vice versa*. (This is identical with *Grete Herrmann's* [19] take on causality.) In case the condition holds also for the reverse time direction, the computation is *deterministic*, and *randomized* otherwise.

2.4   Consequences

If logical reversibility is a simple computational version of a discretized second law, does it have implications resembling the traditional versions of the law?

Logical Reversibility Implies Quasi-Monotonicity

First of all, we find a "Boltzmann-like" form, *i.e.*, the existence of a quantity essentially monotonic in time. More specifically, the logical reversibility of time evolution implies that the Kolmogorov complexity of the global state at time t can be smaller than the one at time 0 only by at most $K(C_t) + O(1)$ if C_t is a string encoding the time span t. The reason is that one possibility of describing the state at time 0 is to give the state at time t, plus t itself; the rest is exhaustive search using only a constant-length program simulating forward time evolution (including possible randomness).

---

[6] Since the amount of the required free energy (and heat dissipation) is proportional to the length of the best compression of the string, the latter can be seen as a *quantification* of the erasure transformation's irreversibility.

Logical Reversibility Implies Clausius-like Law

Similarly, logical reversibility also implies statements resembling the version of the second law due to *Clausius*: "Heat does not spontaneously flow from cold to hot". The rationale here is that if we have a computation — the time evolution — using only (logically reversible) Toffoli gates, then it is *impossible* that this circuit computes a transformation mapping a pair of strings to another pair such that the Hamming-heavier of the two becomes even heavier whilst the lighter gets lighter. A function *accentuating* imbalance, instead of lessening it, is not reversible, as a basic counting argument shows.

Example. Let a circuit consisting of only Toffoli gates map an $N(= 2n)$-bit string to another. We consider the map separately on the first and second halves and assume the computed function to be conservative, *i.e.*, to leave the Hamming weight of the full string unchanged at n (conservativity can be seen as some kind of *first* law, *i.e.*, the preservation of a quantity). We look at the excess of 1's in one of the halves (which equals the deficit of 1's in the other). We observe that the probability (with respect to the uniform distribution over all strings of some Hamming-weight couple $[wn, (1-w)n]$) of the *imbalance substantially growing* is exponentially weak. The key ingredient for the argument is the function's injectivity. Explicitly, the probability that the weight couple changes from $[wn, (1-w)n]$ to $[(w+\Delta)n, (1-w-\Delta)n]$ — or more extremely — , for $1/2 \leq w < 1$ and $0 < \Delta \leq 1-w$, is

$$\frac{\binom{n}{(w+\Delta)n}\binom{n}{(1-w-\Delta)n}}{\binom{n}{wn}\binom{n}{(1-w)n}} = 2^{-\Theta(n)} .$$

Note here that we even get the correct, exponentially weak "error probability" with which the traditional second law can be "violated".

Logical Reversibility Implies Kelvin-like Law

"A single heat bath alone has no work value". This, again, follows from a simple counting argument. There exists no reversible circuit that, for general input environments (with a fixed weight — intuitively: *heat energy*), extracts redundancy, *i.e.*, work value, and concentrates it in some pre-chosen bit positions: *Concentrated* redundancy is *more* of it.

Example. The probability that a fixed circuit maps a "Hamming bath" of length N and Hamming weight w to another such that the first n positions contain all 1's and such that the Hamming weight of the remaining $N - n$ positions is $w - n$ (again, we are assuming conservation here) is

$$\frac{\binom{N-n}{w-n}}{\binom{N}{w}} = 2^{-\Theta(n)} .$$

2.5   Discussion and Questions

We propose a logical view of the second law of thermodynamics: *the injectivity or logical reversibility of time evolution.* This is somewhat ironic as the second law has often been related to its exact opposite: *irreversibility*.[7] It implies, within the Church-Turing view, Clausius-, Kelvin-, and Boltzmann-like statements. We arrive at seeing a law *combinatorial in nature* — and its discovery in the context of steam pipes as a historical incident.

A logically reversible computation can still split up paths [15].[8] This "randomness" may bring in *objective* time asymmetry. What is then the exact mechanism by which randomness implies that a *record* tells more about the past than about the future? (Does it?)

3   Bell Correlations and the Church-Turing Hypothesis

We test the obtained intrinsic notion of randomness, in the form of erasure cost or Kolmogorov complexity, with a physical phenomenon that we have already mentioned above as challenging *a-priori* causality: *"non-local" correlations* from quantum theory. In fact, randomness has been considered crucial in the argument. We put this belief into question in its exclusiveness; at the same time we avoid in our reasoning connecting results of different measurements that, in fact, exclude each other (in other words, we refrain from assuming so-called *counterfactual definiteness*, i.e., that all these measurement outcomes even *exist* altogether).[9] For the sake of comparison, we first review the common, probabilistic, counter-factual reasoning.

3.1   Bell Non-local Correlations

Non-locality, manifested in violations of *Bell inequalities*, expresses the impossibility to prepare parts of an entangled system simultaneously for *all possible mea-*

---

[7] Since new randomness cannot be gotten rid of later, the equation reads: "Logical reversibility plus randomness equals thermodynamic *ir*reversibility". If you *can* go back logically in a random universe, then you certainly *cannot* thermodynamically.

[8] Note that there is no (objective) splitting up, or randomness, if time evolutions are unitary, *e.g.*, come from Schrödinger, heat-propagation, or Maxwell's equations. What is then the origin of the arrow of time? The quantum-physical version of injectivity is *Hugh Everett III's relative-state interpretation*. How do we imagine the bridge from global unitarity to the subjective perception of time asymmetry? When we looked above, with Landauer, at a closed *classical* system of two parts, then the (possible) complexity deficit in one of them must simply be compensated in a corresponding increase in the other. In Everett's view, this means that there can be low-entropy *branches of the wave function* (intuitively, yet too naïvely, called: parallel universes) as long as they are compensated by other, highly complex ones.

[9] The *counter-factual* nature of the reasoning claiming "non-classicality" of quantum theory, that was the main motivation in [38], has already been pointed out by Specker [32]: "In einem gewissen Sinne gehören aber auch die scholastischen Spekulationen über die *Infuturabilien* hieher, das heisst die Frage, ob sich die göttliche Allwissenheit auch auf Ereignisse erstrecke, die eingetreten wären, falls etwas geschehen wäre, was nicht geschehen ist". — "In some sense, this is also related to the scholastic speculations on the *infuturabili*, *i.e.*, the question whether divine omniscience even extends to what would have happened if something had happened that did not happen".

surements. We look at an idealized non-local correlation, the *Popescu-Rohrlich (PR) box* [28]. Let A and B be the respective input bits to the box and X and Y the output bits; the (classical) bits satisfy

$$X \oplus Y = A \cdot B. \tag{1}$$

According to a result by Fine [17], the non-locality of the system (*i.e.*, conditional distribution) $P_{XY|AB}$, which means that it cannot be written as a convex combination of products $P_{X|A} \cdot P_{Y|B}$, is equivalent to the fact that there exists no preparation for all alternative measurement outcomes $P'_{X_0 X_1 Y_0 Y_1}$ such that

$$P'_{X_i Y_j} = P_{XY|A=i, B=j}$$

for all $(i,j) \in \{0,1\}^2$. In this view, non-locality means that the outputs cannot *exist*[10] before the inputs do. Let us make this qualitative statement more precise. We assume a perfect PR box, *i.e.*, a system always satisfying $X \oplus Y = A \cdot B$. Note that this equation alone does not uniquely determine $P_{XY|AB}$ since the marginal of X, for instance, is not determined. If, however, we additionally require *no-signaling*, then the marginals, such as $P_{X|A=0}$ or $P_{Y|B=0}$, must be perfectly unbiased under the assumption that all four (X,Y)-combinations, *i.e.*, $(0,0), (0,1), (1,0)$, and $(1,1)$, are possible. To see this, assume on the contrary that $P_{X|A=0,B=0}(0) > 1/2$. By the PR condition (1), we can conclude the same for Y: $P_{Y|A=0,B=0}(0) > 1/2$. By no-signaling, we also have $P_{X|A=0,B=1}(0) > 1/2$. Using symmetry, and no-signaling again, we obtain both $P_{X|A=1,B=1}(0) > 1/2$ and $P_{Y|A=1,B=1}(0) > 1/2$. This contradicts the PR condition (1) since *two bits which are both biased towards 0 cannot differ with certainty*. Therefore, our original assumption was wrong: The outputs *must* be perfectly unbiased. Altogether, this means that X as well as Y cannot exist (*i.e.*, take a definite value — actually, there cannot even exist a classical value arbitrarily weakly correlated with one of them) *before* the classical bit $f(A,B)$ exists for some nontrivial deterministic function $f : \{0,1\}^2 \to \{0,1\}$. The paradoxical aspect of non-locality — at least if a causal structure is in place — now consists of the fact that *fresh* pieces of information *come to existence* in a *spacelike-separated* manner that are nonetheless *perfectly correlated*.

3.2 Kolmogorov Complexity

We introduce the basic notions required for our alternative, complexity-based view. Let $\mathcal{U}$ be a fixed universal Turing machine (TM).[11] For a finite or infinite

---

[10] What does it mean that a classical bit *exists*? Note first that *classicality* of information implies that it can be measured without disturbance and that the outcome of a "measurement" is always the same; this makes it clear that it is an *idealized* notion requiring the classical bit to be represented in a redundant way over an *infinite* number of degrees of freedom, as a thermodynamic limit. It makes thus sense to say that a *classical bit U exists*, *i.e.*, has taken a definite value.

[11] The introduced asymptotic notions are independent of this choice.

string s, the *Kolmogorov complexity* [23,25] $K(s) = K_{\mathcal{U}}(s)$ is the length of the shortest program for $\mathcal{U}$ such that the machine outputs s. Note that $K(s)$ can be infinite if s is.

Let $a = (a_1, a_2, \ldots)$ be an infinite string. Then

$$a_{[n]} := (a_1, \ldots, a_n, 0, \ldots) .$$

We study the asymptotic behavior of $K(a_{[n]}) : \mathbf{N} \to \mathbf{N}$. For this function, we simply write $K(a)$, similarly $K(a \,|\, b)$ for $K(a_{[n]} \,|\, b_{[n]})$, the latter being the length of the shortest program outputting $a_{[n]}$ upon input $b_{[n]}$. We write

$$K(a) \approx n \ :\Longleftrightarrow\ \lim_{n \to \infty} \frac{K(a_{[n]})}{n} = 1 .$$

We call a string a with this property *incompressible*. We also use $K(a_{[n]}) = \Theta(n)$, as well as

$$K(a) \approx 0 :\Longleftrightarrow \lim_{n \to \infty} \frac{K(a_{[n]})}{n} = 0 \Longleftrightarrow K(a_{[n]}) = o(n) .$$

Note that *computable* strings a satisfy $K(a) \approx 0$, and that incompressibility is, in this sense, the extreme case of uncomputability.

Generally, for functions $f(n)$ and $g(n) \not\approx 0$, we write $f \approx g$ if $f/g \to 1$. *Independence of a and b* is then[12]

$$K(a \,|\, b) \approx K(a)$$

or, equivalently,

$$K(a, b) \approx K(a) + K(b) .$$

If we introduce

$$I_K(x; y) := K(x) - K(x \,|\, y) \approx K(y) - K(y \,|\, x) ,$$

independence of a and b is $I_K(a, b) \approx 0$.

In the same spirit, we can define *conditional independence*: We say that a and b are *independent given c* if

$$K(a, b \,|\, c) \approx K(a \,|\, c) + K(b \,|\, c)$$

or, equivalently,

$$K(a \,|\, b, c) \approx K(a \,|\, c) ,$$

or

$$I_K(a; b \,|\, c) := K(a \,|\, c) - K(a \,|\, b, c) \approx 0 .$$

---

[12] This is inspired by [9] (see also [10]), where (joint) Kolmogorov complexity — or, in practice, any efficient compression method — is used to define a *distance measure* on sets of bit strings (such as literary texts of genetic information of living beings). The resulting structure in that case is a distance measure, and ultimately a clustering as a binary tree.

3.3 Correlations and Computability

We are now ready to discuss non-local correlations with our context-free randomness measure. The mechanism we discover is very similar to what holds probabilistically: If the choices of the measurements are random (uncomputable) and non-signaling holds, then the outputs must be random (uncomputable) as well. We prove the following statement.

Uncomputability from Correlations. *There exist bipartite quantum states with a behavior under measurements such that if the sequences of setting encodings are maximally uncomputable (incompressible), then the sequences of measurement results are uncomputable as well, even given the respective setting sequences.*

Proof. We proceed step by step, starting with the idealized system of the PR box. Let first (a, b, x, y) be infinite binary strings with

$$x_i \oplus y_i = a_i \cdot b_i . \tag{2}$$

Obviously, the intuition is that the strings stand for the inputs and outputs of a PR box. Yet, no dynamic meaning is attached to the strings anymore (or to the "box", for that matter) since there is no *"free choice"* of an input and no generation of an output in function of the input; all we have is a quadruple of strings satisfying the PR condition (2). However, nothing prevents us from defining this (static) situation to be *no-signaling*:

$$K(x \mid a) \approx K(x \mid ab) \text{ and } K(y \mid b) \approx K(y \mid ab) . \tag{3}$$

We argue that if the inputs are incompressible and independent, and no-signaling holds, then the outputs must be uncomputable: To see this, assume now that $(a, b, x, y) \in (\{0, 1\}^{\mathbf{N}})^4$ with $x \oplus y = a \cdot b$ (bit-wisely), no-signaling (3), and

$$K(a, b) \approx 2n ,$$

i.e., the "input" pair is incompressible. We conclude

$$K(a \cdot b \mid b) \approx n/2 .$$

Note first that $b_i = 0$ implies $a_i \cdot b_i = 0$, and second that any further compression of $a \cdot b$, given b, would lead to "structure in (a, b)", *i.e.*, a possibility of describing (programming) a given b in shorter than n and, hence, (a, b) in shorter than $2n$. Observe now

$$K(x \mid b) + K(y \mid b) \geq K(a \cdot b \mid b)$$

which implies

$$K(y \mid b) \geq K(a \cdot b \mid b) - K(x \mid b) \geq n/2 - K(x) . \tag{4}$$

On the other hand,

$$K(y \mid a, b) \approx K(x \mid a, b) \leq K(x) . \tag{5}$$

Now, no-signaling (3) together with (4) and (5) implies

$$n/2 - K(x) \le K(x) ,$$

and

$$K(x) > n/4 = \Theta(n) :$$

(This bound can be improved by a more involved argument [3].) The string x must be uncomputable.

A priori, it is not overly surprising to receive uncomputable outputs upon inputs having the same property. Thus, we now turn our attention to the conditional output complexities given the inputs: We consider the quantities $K(x \mid a)$ and $K(y \mid b)$. Note first

$$K(x \mid a) \approx 0 \Leftrightarrow K(x \mid ab) \approx K(y \mid ab) \approx 0 \Leftrightarrow K(y \mid b) \approx 0 ,$$

i.e., the two expressions vanish simultaneously. We show that, in fact, they both fail to be of order $o(n)$. To see this, assume $K(x \mid a) \approx 0$ and $K(y \mid b) \approx 0$. Hence, there exist programs P_n and Q_n (both of length $o(n)$) for functions f_n and g_n with

$$f_n(a_n) \oplus g_n(b_n) = a_n \cdot b_n . \tag{6}$$

For fixed (families of) functions f_n and g_n, asymptotically how many (a_n, b_n) can at most exist that satisfy (6)? The question boils down to a parallel-repetition analysis of the PR game: A result by Raz [29] implies that the number is of order $(2 - \Theta(1))^{2n}$. Therefore, the two programs P_n and Q_n together with the index, of length

$$(1 - \Theta(1))2n ,$$

of the correct pair (a, b) within the list of length $(2 - \Theta(1))^{2n}$ lead to a program, generating (a, b), that has length

$$o(n) + (1 - \Theta(1))2n ,$$

in contradiction to the assumption of incompressibility of (a, b).

Unfortunately, perfect PR boxes are not predicted by quantum theory. We show that correlations which are achievable in the laboratory [34] allow for the argument to go through; they are based on the chained Bell inequality [2] instead of perfect PR-type non-locality.

To the chained Bell inequality belongs the following idealized system: Let $A, B \in \{1, \ldots, m\}$ be the inputs. We assume the "promise" that B is congruent to A or to $A + 1$ modulo m. Given this promise, the outputs $X, Y \in \{0, 1\}$ must satisfy

$$X \oplus Y = \chi_{A=m, B=1} , \tag{7}$$

where $\chi_{A=m, B=1}$ is the characteristic function of the event $\{A = m, B = 1\}$.

Barrett, Hardy, and Kent [2] showed that if A and B are random, then X and Y must be perfectly unbiased if the system is no-signaling. More precisely, they were even able to show such a statement from the gap between the error probabilities of the best classical — $\Theta(1/m)$ — and quantum — $\Theta(1/m^2)$ — strategies for winning this game.

We assume $(a, b, x, y) \in (\{1, \ldots, m\}^n)^2 \times (\{0, 1\}^n)^2$ to be such that the promise holds, and such that

$$K(a, b) \approx (\log m + 1) \cdot n , \tag{8}$$

i.e., the string $a\|b$ is maximally incompressible given the promise; the system is no-signaling (3); the fraction of quadruples (a_i, b_i, x_i, y_i), $i = 1, \ldots, n$, satisfying (7) is of order $(1 - \Theta(1/m^2))n$. Then $K(x) = \Theta(n)$.

To see this, observe first that $K(a, b)$ being maximal implies

$$K(\chi_{a=m, b=1} \,|\, b) \approx \frac{n}{m} : \tag{9}$$

The fractions of 1's in b must, asymptotically, be $1/m$ due to the string's incompressibility. If we condition on these positions, the string $\chi_{a=m, b=1}$ is incompressible, since otherwise there would be the possibility of compressing (a, b).

Now, we have

$$K(x \,|\, b) + K(y \,|\, b) + h(\Theta(1/m^2))n \geq K(\chi_{a=m, b=1} \,|\, b)$$

since one possibility for "generating" the string $\chi_{a=m, b=1}$, from position 1 to n, is to generate $x_{[n]}$ and $y_{[n]}$ as well as the string indicating the positions where (7) is violated, the complexity of the latter being at most[13]

$$\log \binom{n}{\Theta(1/m^2)n} \approx h(\Theta(1/m^2))n .$$

Let us compare this with $1/m$: Although the binary entropy function has slope ∞ in 0, we have

$$h(\Theta(1/m^2)) < 1/(3m)$$

if m is sufficiently large. To see this, observe first that the dominant term of $h(x)$ for small x is $-x \log x$, and second that

$$c(1/m) \log(m^2/c) < 1/3$$

for m sufficiently large.

Together with (9), we now get

$$K(y \,|\, b) \geq \frac{2n}{3m} - K(x) \tag{10}$$

if m is chosen sufficiently large. On the other hand,

$$K(y \,|\, ab) \leq K(x \,|\, ab) + h(\Theta(1/m^2))n \tag{11}$$

$$\leq K(x) + \frac{n}{3m} . \tag{12}$$

---

[13] Here, h is the binary entropy $h(x) = -p \log_2 p - (1 - p) \log_2(1 - p)$. Usually, p is a probability, but h is invoked here merely as an approximation for binomial coefficients.

Now, (3), (10), and (12) together imply

$$K(x) \leq \frac{n}{6m} = \Theta(n) \; ;$$

in particular, x must be uncomputable. This concludes the proof.   □

3.4 Kolmogorov Amplification and the All-or-Nothing Nature of the Church-Turing Hypothesis

The shown result implies that *if* the experimenters are given access to an incompressible number (such as Ω [8]) for choosing their measurement bases, *then* the measured photon (in a least one of the two labs) is forced to generate an uncomputable number as well, even given the string determining its basis choices.

This is a similar observation as in the probabilistic realm, where certain "free-will theorems" have been formulated in the context. In fact, stronger statements hold there, since non-local correlations allow for *randomness amplification as well as expansion* (see, *e.g.*, [12]): The randomness generated by the photons as their measurement output qualitatively *and* quantitatively *exceeds* what is required for the choices of the measurement settings. This also holds in our complexity-based model: Indeed, it has been shown in [3] that functionalities such as *Kolmogorov-complexity amplification and expansion* are possible using Bell correlations. The consequence is that there is either no incompressibility or uncomputability at all in the world, or it is full of it.

All-or-Nothing Feature of the Church-Turing Hypothesis. *Either no device exists in nature allowing for producing uncomputable sequences, or even a single photon can do it.*

4   Concluding Remarks and Open Questions

The antagonism between the pre-Socratic philosophers *Parmenides* and *Heraclitus* is still vivid in today's thinking traditions: The Parmenidean line puts *logic* is the basis of space-time and dynamics — in the end all of physics. It has inspired researchers such as Leibniz, Mach, or Wheeler. Central here is a doubt about *a priori* absolute space-time causality: Is it possible that these concepts only emerge at a higher level of complexity, along with macroscopic, classical information?

Fundamentally opposed is the Heraclitean style, seeing *physics and its objects* at the center: space, time, causality, and dynamic change is what all rests upon, including logic, computation, or information. To this tradition belong Newton, most physicists including Einstein, the logician Gonseth, certainly Landauer.

According to *Paul Feyerabend* [16], a specific tradition comes with its own criteria for success *etc.*, and it can be judged from the standpoint of another (with those other criteria). In this spirit, it has been the goal of our discourse to build bridges between styles, and to use their tension to serve us. This allowed, for instance, to get more insight into the second law of thermodynamics or the "non-local" correlations from quantum theory. The latter challenge our established views of space and time; they, actually, have us look back to the debate between Newton and Leibniz and to question the path most of science decided to take, at that time.

For the sake of a final thought, assume *à la* Leibniz that space, time, and causality do not exist prior to *classical information* — which we understand as an idealized notion of *macroscopically and highly redundantly represented* information; an ideal classical bit can then be measured without disturbance, copied, and easily recognized as being classical. In this view, classicality is a *thermodynamic* notion. Thus the key to the *quantum measurement process*, and the problems linked to it, may lie within thermodynamics. (Yet, even if this is successful: How come we observe correlations of pieces of *classical* information unexplainable by any reasonable *classical* mechanism? How can quantum correlations and thermodynamic classicality — *Bell & Boltzmann* — be reconciled?)

Acknowledgments. This text is based on a presentation at the *14th Annual Conference on Theory and Applications of Models of Computation (TAMC 2017)* at the Universität Bern. I am grateful to Gerhard Jäger and all the organizers for kindly inviting me to give a talk.

I thank Mateus Araújo, Veronika Baumann, Ämin Baumeler, Charles Bédard, Claus Beisbart, Gilles Brassard, Harvey Brown, Caslav Brukner, Harry Buhrman, Matthias Christandl, Sandro Coretti, Fabio Costa, Bora Dakic, Frédéric Dupuis, Paul Erker, Adrien Feix, Jürg Fröhlich, Manuel Gil, Nicolas Gisin, Esther Hänggi, Arne Hansen, Marcus Huber, Lorenzo Maccone, Alberto Montina, Samuel Ranellucci, Paul Raymond-Robichaud, Louis Salvail, L. Benno Salwey, Martin Schüle, Andreas Winter, and Magdalena Zych for inspiring discussions, and the *Pläfä-Einstein* as well as the *Reitschule* for their inspiring atmosphere.

This research is supported by the Swiss National Science Foundation (SNF), the National Centre of Competence in Research "Quantum Science and Technology" (QSIT), the COST action on Fundamental Problems in Quantum Physics, and by the *Hasler Foundation*.

References

1. Barnea, T.J., Bancal, J.-D., Liang, Y.-C., Gisin, N.: Tripartite quantum state violating the hidden influence constraints. Phys. Rev. A **88**, 022123 (2013)
2. Barrett, J., Hardy, L., Kent, A.: No-signalling and quantum key distribution. Phys. Rev. Lett. **95**, 010503 (2005)
3. Baumeler, Ä., Bédard, C., Brassard, G., Wolf, S.: Kolmogorov amplification from Bell correlation. In: International Symposium in Information Theory (ISIT) 2017 (2017, submitted)
4. Baumeler, Ä., Feix, A., Wolf, S.: Maximal incompatibility of locally classical behavior and global causal order in multi-party scenarios. Phys. Rev. A **90**, 042106 (2014)
5. Baumeler, Ä., Wolf, S.: The space of logically consistent classical processes without causal order. New J. Phys. **18**, 013036 (2016)
6. Bell, J.S.: On the Einstein-Podolsky-Rosen paradox. Physics **1**, 195–200 (1964)

7. Bennett, C.H.: The thermodynamics of computation. Int. J. Theoret. Phys. **21**(12), 905–940 (1982)
8. Chaitin, G.: A theory of program size formally identical to information theory. J. ACM **22**, 329–340 (1975)
9. Cilibrasi, R., Vitányi, P.: Clustering by compression. IEEE Trans. Inf. Theory **51**(4), 1523–1545 (2005)
10. Cilibrasi, R., Vitányi, P.: The Google similarity distance. arXiv:abs/cs/0412098 (2004)
11. Clausius, R.: Über die bewegende Kraft der Wärme und die Gesetze die sich daraus für die Wärmelehre selbst ableiten lassen (1850)
12. Colbeck, R., Renner, R.: Free randomness can be amplified. Nat. Phys. **8**, 450–454 (2012)
13. Coretti, S., Hänggi, E., Wolf, S.: Nonlocality is transitive. Phys. Rev. Lett. **107**, 100402 (2011)
14. Einstein, A., Podolsky, B., Rosen, N.: Can quantum-mechanical description of physical reality be considered complete? Phys. Rev. **47**, 777–780 (1935)
15. Everett, H.: "Relative state" formulation of quantum mechanics. Rev. Mod. Phys. **29**(3), 454–462 (1957)
16. Feyerabend, P.: Wissenschaft als Kunst. Suhrkamp, Berlin (1984)
17. Fine, A.: Hidden variables, joint probability, and the Bell inequalities. Phys. Rev. Lett. **48**, 291–295 (1982)
18. Fredkin, E., Toffoli, T.: Conservative logic. Int. J. Theoret. Phys. **21**(3–4), 219–253 (1982)
19. Hermann, G.: Die naturphilosophischen Grundlagen der Quantenmechanik. Abhandlungen der Fries'schen Schule, Band 6, pp. 69–152 (1935). Die Kausalität in der Physik (1948)
20. Hersch, J.: L'étonnement philosophique (De l'école Milet à Karl Jaspers) (1981)
21. Kelvin, L.: On a universal tendency in nature to the dissipation of mechanical energy (1852)
22. Kleene, S.C.: Mathematical Logic. Wiley, New York (1967)
23. Kolmogorov, A.N.: Three approaches to the quantitative definition of information. Probl. Peredachi Informatsii **1**(1), 3–11 (1965)
24. Landauer, R.: Irreversibility and heat generation in the computing process. IBM J. Res. Dev. **5**, 183–191 (1961)
25. Li, M., Vitányi, P.: An Introduction to Kolmogorov Complexity and Its Applications. Springer, New York (2008)
26. Nietzsche, F.: Die Philosophie im tragischen Zeitalter der Griechen (1872)
27. Oreshkov, O., Costa, F., Brukner, C.: Quantum correlations with no causal order. Nat. Commun. **3**, 1092 (2012)
28. Popescu, S., Rohrlich, D.: Quantum non-locality as an axiom. Found. Phys. **24**, 379–385 (1994)
29. Raz, R.: A parallel repetition theorem. SIAM J. Comput. **27**(3), 763–803 (1998)
30. Reichenbach, H.: The principle of the common cause. In: The Direction of Time, chap. 19, pp. 157–167. California Press, Berkeley (1956)
31. Shannon, C.E., Weaver, W.: The Mathematical Theory of Communication. University of Illinois Press, Urbana (1949)
32. Specker, E.: Die Logik nicht gleichzeitig entscheidbarer Aussagen. Dialectica **14**, 239–246 (1960)
33. Stefanov, A., Zbinden, H., Gisin, N., Suarez, A.: Quantum correlations with space-like separated beam splitters in motion: experimental test of multisimultaneity. Phys. Rev. Lett. **88**, 120404 (2002)

34. Stuart, T.E., Slater, J.A., Colbeck, R., Renner, R., Tittel, W.: An experimental test of all theories with predictive power beyond quantum theory. Phys. Rev. Lett. **109**, 020402 (2012)
35. Uffink, J.: Bluff your way in the second law of thermodynamics (2001)
36. Vailati, E.: Leibniz and Clarke: A Study of Their Correspondence. Oxford University Press, New York (1997)
37. Wheeler, J.A.: Information, physics, quantum: the search for link. In: Proceedings of III International Symposium on Foundations of Quantum Mechanics, pp. 354–368 (1989)
38. Wolf, S.: Non-locality without counterfactual reasoning. Phys. Rev. A **92**(5), 052102 (2015)
39. Wood, C., Spekkens, R.: The lesson of causal discovery algorithms for quantum correlations: causal explanations of Bell-inequality violations require fine-tuning. New J. Phys. **17**, 033002 (2015)
40. Zurek, W.H.: Algorithmic randomness and physical entropy. Phys. Rev. A **40**(8), 4731–4751 (1989)

Contributed Papers

Turbo-Charging Dominating Set with an FPT Subroutine: Further Improvements and Experimental Analysis

Faisal N. Abu-Khzam[1,2], Shaowei Cai[3], Judith Egan[1], Peter Shaw[1], and Kai Wang[1(✉)]

[1] Charles Darwin University, Casuarina, Australia
{judith.egan,peter.shaw,kai.wang}@cdu.edu.au
[2] Department of Computer Science and Mathematics,
Lebanese American University, Beirut, Lebanon
faisal.abukhzam@lau.edu.lb
[3] State Key Lab of Computer Science, Institute of Software,
Chinese Academy of Sciences, Beijing, China
shaoweicai.cs@gmail.com

Abstract. Turbo-charging is a recent algorithmic technique that is based on the fixed-parameter tractability of the dynamic versions of some problems as a way to improve heuristics. We demonstrate the effectiveness of this technique and develop the turbo-charging idea further. A new hybrid heuristic for the DOMINATING SET problem that further improves this method is obtained by combining the turbo-charging technique with other standard heuristic tools including Local Search (LS). We implement both the recently proposed "turbo greedy" algorithm of Downey et al. [8] and a new method presented in this paper. The performance of these new heuristics is assessed on three suites of benchmark datasets, namely DIMACS, BHOSLIB and KONECT. Experiments comparing our algorithm to both heuristic and exact algorithms demonstrate its effectiveness. Our algorithm often produced results that were either exact or better than all the other available algorithms.

1 Introduction

Parameterized problems can be classified as either being fixed-parameter tractable (or FPT) or hard for some other class such as $W[1]$ or $W[2]$. If a problem is FPT, then we often have practical algorithms to solve it exactly when the input parameter is small. It may also be possible to apply efficient kernelization methods. That is, there is a polynomial-time preprocessing algorithm which when applied to an arbitrary problem instance is guaranteed to yield an equivalent instance whose size is bounded by a function of the (relatively small) parameter. For many problems, kernelization reduces the size to a polynomial or even linear function of the parameter. This has been shown to be effective in improving many heuristics for NP-hard problems [4].

© Springer International Publishing AG 2017
T.V. Gopal et al. (Eds.): TAMC 2017, LNCS 10185, pp. 59–70, 2017.
DOI: 10.1007/978-3-319-55911-7_5

When a parameterized problem is not FPT, the seemingly only time-efficient approach is to resort to heuristic algorithms. This is the case of the $W[2]$-Complete DOMINATING SET problem (henceforth DS). Recently, however, a dynamic version of DS was shown to be FPT [8], and it was suggested that a fixed-parameter algorithm for the dynamic version can be used to improve heuristic algorithms. This idea can be used as a general framework to improve heuristic search in some sort of a time-quality trade-off, which is demonstrated in Sect. 2.1.

The notion of a parameterized dynamic problem plays a key role in this paper. A dynamic problem is one whose input is assumed to have changed after some initial solution was found. Thus, a problem instance is accompanied by the initial instance and its solution along with parameters quantifying the changes made and a bound on the difference between the original solution and the one to be found. The DYNAMIC DOMINATING SET (DDS) and GREEDY IMPROVEMENT OF DOMINATING SET (GREEDY-DS) problems were introduced in [8].

Furthermore, the DDS parameterization can be used, in order to improve heuristic search algorithms for DS. This is based on using a "fixed-parameter" DDS algorithm as a subroutine in what was coined in as *turbo-charging* LS problems. In this paper, the "turbo-charging" idea is implemented for the first time. The technique uses a greedy heuristic as the inductive route to then applying a fixed-parameter algorithm for the dynamic version of the problem as a subroutine. Note that the approach is not limited to the heuristic it is applied to. A number of alternative popular greedy and LS heuristics can be used. These heuristics are described in Sect. 1.2.

When it was introduced in [8] the greedy heuristic used was based on choosing the vertex of the highest degree. This was used to illustrate the approach, but it is not the most effective method. Here, this original algorithm is modified to use the classic greedy heuristic (based in the notion of vertex utility) presented by Chvatal as follows: order the vertices from lowest to highest utility and then to each vertex select its highest utility neighbor in the solution. A second improved heuristic (FPT TURBO HYBRID) that instead uses the current-best greedy LS of Sanchis et al. [22] is then presented.

The effectiveness of this approach is then evaluated by comparing its performance using a selection of datasets from DIMACS [17], BHOSLIB [25] (Benchmarks with Hidden Optimum Solutions for Graph Problems) and KONECT (The Koblenz Network Collection) [18].

We implement and test our methods and provide an initial evaluation of the turbo-charging technique. Our experiments show that our improved version outperforms the well-known *Greedy-Vote* heuristic [22] which adopts the greedy randomized adaptive search procedure *GRASP* of [10].

1.1 Terminology

Throughout this paper, we work with simple undirected graphs, i.e., no loops and no multiple edges. Given a graph $G = (V, E)$, V and E (or $V(G)$ and $E(G)$) denote the set of vertices and edges in G, respectively. The *open neighborhood* of a vertex v is the set of all vertices adjacent to v, denoted as $N_G(v)$. The *closed neighborhood* is $N_G[v] = N_G(v) \cup \{v\}$. The subscript G will be dropped when the graph name is known from the context. If $S \subseteq V(G)$, $N_G(S)$ denotes all the vertices of $V(G)\backslash S$ that are adjacent to the vertices in S and $N_G[S] = N_G(S) \cup S$. If $N_G[S] = V(G)$, we say S is a *dominating set* for G. When computing a dominating set we gradually add vertices to a (partial dominating) set S. If $N_G[S] \subsetneq V(G)$, the number of non-dominated neighbors of a vertex v is dubbed the *utility* of v in this paper.

The DS problem and its variants are among the most broadly studied and the most important problems in graph theory and computer science. They find applications in many fields such as message routing with sparse tables [21], replica caching in operation systems and databases [20], server placement in a computer network [3], multicast systems [24], and political science [5]. Among the variants of DS is the classical, naturally parameterized, DS problem which asks whether we can find a dominating set whose size is less than or equal to a parameter k. The problem is known to be NP-hard [13] as well as W[2]-hard [9], and furthermore, it is NP-hard to approximate DS within a ratio of $(1 - \epsilon) \ln n$ for every $\epsilon > 0$ [7].

The current asymptotically-fastest exact algorithm of Fomin et al. [12] finds a dominating set whose size is minimum among all dominating sets with a time complexity in $\mathcal{O}^*(2^{0.61n})$. Despite its exponential time complexity, the algorithm was shown to behave well in practice using the implementation method reported in [1]. On the other hand, current heuristic algorithms for DS are based on either the greedy method with improved objective functions [22] or LS.

The most common greedy approach is based on the assumption that a vertex with high degree is more likely to be included in a dominating set. On the other hand, the common LS approach starts from a feasible solution and tries to improve it iteratively [14]. A common technique used for obtaining an initial feasible solution for LS heuristics is to generate a range of solutions using the random greedy heuristics. These solutions can then be improved using LS heuristics [15]; however, for the DS problem, experimental results suggest that this is not often the case [22]. In [22] an alternative heuristic of GREEDY VOTE was presented, followed by a simple LS to obtain very small reductions in $\mathcal{O}(|V(G)|)$ time.

1.2   Greedy Heuristics

The turbo-charging technique, proposed in [8] will be applied to the DS problem using a variety of greedy heuristics as the inductive route from which a turbo-charging subroutine is called. To do this, we first implement the classic Greedy heuristic of Chvatal [6]. An effective but minor improvement of the objective function for this heuristics is given in Sanchis [22]. We further explore how the original turbo-charging subroutine can be improved using a variety of standard FPT techniques as well as providing some new improvements.

GREEDY CHVATAL: Chvatal's greedy heuristic chooses the vertices of the solution based on the maximum *utility* of a vertex with respect to the remaining vertices. Parekh [19] showed that any solution size obtained using this heuristic is at most $|V| + 1 - \sqrt{2|E| + 1}$. A brief consideration of this bound shows that the heuristic can work relatively well on dense graphs but the guarantee of its performance is not as good on sparse graphs. To avoid ambiguity, we will refer to this heuristic as GREEDY CHVATAL being the most classical one.

GREEDY VOTE: The GREEDY VOTE heuristic of Sanchis et al. [22] is essentially the same as the GREEDY CHVATAL, but instead of using the highest utility to choose the next vertex to add to the solution, it uses a *vote* measure to favor vertices with lower degree neighbors as a tie breaker among vertices with the same utility. These heuristics are further improved using the turbo-technique described below.

GREEDY VOTE GRASP: Sanchis et al. [22] also further improve the GRASP LS of [11] by applying the GREEDY VOTE heuristic as a starting point. For some datasets, this gave a minor improvement.

2   Dynamic FPT Heuristics

The "turbo-charging" technique introduced in [8] uses a simple greedy heuristic to apply the fixed-parameter algorithm for the dynamic version of the DS problem. For simplicity, and as noted earlier, the original algorithm selects vertices based on the degree; however in practice using orders such as maximum utility is more effective. We will refer to this latter algorithm as FPT TURBO I.

Although this algorithm does prove effective, it lacks a number of obvious modifications that were likely to (1) produce better solutions and (2) reduce the running time. We herein abstract these modifications into a general framework.

In general, turbo-charging is expected to perform better when the following techniques are applied:

- Applying reduction rules;
- Using an alternative measure(s) when guessing a solution;
- Checking whether the solution(s) obtained are minimal;
- Using alternative orderings;
- Adding a *heuristic guarantee* (if possible.); and
- Applying an appropriate LS heuristic to refine the solution.

The above framework can be applied to other problems. A remarkable advantage of our use of dynamic problems is the fact that new parameterizations can reveal reduction rules that do not otherwise apply to the non-dynamic versions. Four reduction rules for the DYNAMIC DS problem are given in Sect. 2.2, some of which do not apply to the non-dynamic version.

After presenting the basic FPT TURBO I algorithm, we shall develop a new hybrid algorithm by applying the improvements listed above to produce the FPT TURBO HYBRID algorithm. The motivation and details of this approach are then discussed.

2.1 The Original DS Turbo-Charging Technique: FPT TURBO I

Initially, the FPT TURBO I heuristic was implemented exactly as given in Downey
et al. [8]. Discussion with the authors revealed some minor corrections in the
ordering of vertices. In FPT TURBO I, the vertices are first ordered from the
lowest degree to the highest. Starting from an empty subgraph G_0 with a corresponding empty dominating set S_0, the algorithm proceeds through the list
of vertices one at a time. At each iteration, the vertex v_i is added to G_{i-1} to
form G_i.

Whenever a vertex v_i is found that is not dominated by S_{i-1}, the highest
utility vertex is selected from $N[v_i]$ and added to S_{i-1} (to form S_i) to dominate
v_i. A budget is set for the largest size of $|S_i|$ we will allow. As the algorithm
continues to add vertices to G_i, it may reach a point where it suffers a "moment
of regret", if the number of vertices added to S_i exceeds the allowed budget. In
this paper, the moment of regret is based on the difference in size between S_i and
S_{i-k} (which must be at least two, otherwise the solution cannot be improved).

At this point, the heuristic backs up the process by the last k additions to G_i
and employs the (fixed-parameter) dynamic problem subroutine DDS FPT to
determine if a smaller solution can be found. When calling the DDS FPT algorithm we set $k = d_e(G_i', G_i)$. Here G_i' is a virtually constructed graph from G_i
with (at most k) augmenting edges between S_{i-k} and $V(G_i \backslash G_{i-k})$.

Pseudocode of the FPT TURBO I algorithm is shown below. In the sequel we
shall describe the changes made to obtain the FPT TURBO HYBRID algorithm.

2.2 Reduction Rules

A number of reduction rules are employed in different components of our implementation of the FPT TURBO I and FPT TURBO HYBRID algorithm (described
below). In practice, even for $W[2]$-hard problems, such as DS, the effective use
of reduction rules can significantly reduce the size of many non-synthetic input
datasets [23].

The details of the reduction rules follow. We first implemented the two reduction rules (R1 and R2) that appeared in [2]. Despite being effective on small
graphs, we tested them on many large graphs and noticed they were not efficient
enough for our target data sets. For efficient implementations, it is sometimes
important to apply simple (or fast) reduction rules.

It was noted in [8] that the chief benefit of using these two rules was that
it could disconnect the input graph. We note that in such cases we obtained a
speedup of around a factor of two.

Although Downey et al. [8] showed that the DDS problem has no polynomial-
size kernel, the dynamic formulation of the problem did allow for a number of
reduction rules that do not apply to the DS problem. The soundness of the
following reduction rules can be verified easily by the reader.

Reduction Rule 3 (R3). If there is a non-dominated vertex v of degree 0,
then add v to the graph G_0 and solution set S_0 and update v as dominated.

Algorithm 1. FPT TURBO I

1: **Rank** the list $L = \{v_0, \ldots, v_n\}$ of vertices in G from lowest to highest degree;
2: **Set** $u_0 \leftarrow$ the highest utility vertex in $N_G[v_0]$;
3: $S_0 \leftarrow \{u_0\}$;
4: $i \leftarrow 0$;
5: **do**
6:     $i \leftarrow i + 1$;
7:     **if** v_i is dominated by S_{i-1} **then**
8:         $G_i \leftarrow G_{i-1} \cup \{v_i\}$;
9:         $S_i \leftarrow S_{i-1}$;
10:     **else**
11:         **Set** u_i the highest utility vertex in $N_G[v_i]$;
12:         $S_i \leftarrow S_{i-1} \cup \{u_i\}$;
13:         **Construct** G_i from G_{i-1} with $\{v_i, u_i\}$ and incident edges in G;
14:         **if** is_moment_of_regret(G_i, u_i, S_i) **then**
15:             $r \leftarrow \min(r, |S_i| - |S_{i-k}|)$;
16:             $S_i' \leftarrow$ DDS FPT(G_i', G_i, S_{i-k}, k, r);
17:             **if** $|S_i'| < |S_i|$ **then**
18:                 $S_i \leftarrow S_i'$;   ▷ S_i *becomes the new dominating set for* G_i;
19:             **end if**
20:         **end if**
21:     **end if**
22: **while** (Not all the vertices are dominated);
23: **Return** the final S_i as the dominating solution for G;

Reduction Rule 4 (R4). If there is a non-dominated vertex v of degree 1, then add the neighbor u of v into the solution set S_0, add $N[u]$ into G_0 and update $N[u]$ as dominated.

Reduction Rule 5 (R5). If there is non-dominated vertex v of degree two such that u and w are the neighbors of v and all the non-dominated elements of $N[u]$ are in $N[w]$, then add w into S_0 and add $N[w]$ into G_0.

Reduction Rule 6 (R6). If there is non-dominated vertex v of degree two such that u and w are the neighbors of v, and $N(u) \backslash v$ and $N(w) \backslash v$ are dominated but any, or both, of u and w are not dominated, then add v into S_0 and add $N[v]$ into G_0.

2.3 A Hybrid Turbo-Charged Algorithm

There were a number of obvious key limitations in the way the DDS FPT algorithm was applied by the FPT TURBO I algorithm, for which simple improvements exist. FPT TURBO HYBRID is the result of applying simple and effective improvements to address these issues. The details and reasons for each of these improvements are discussed below.

Using an alternative measure(s) when guessing a solution. As part of the underlying heuristic, the FPT TURBO I first tries to dominate the next non-dominated vertex by considering its closed neighborhood and choosing the vertex with the highest utility to add to the solution. Other measures for choosing the next vertex v_i to add to the dominating set used by the GREEDY VOTE [19] may also be used, instead. These alternative measures yield notably improved results. In fact, this ability to vary the objective function illustrates how turbo-charging can be adapted to other problems and more generally applied to improve heuristics.

Using an alternative vertex-ordering. Varying the ordering of the list L, which controls the order according to which the vertices are included in the graph, obviously affects the quality of the solution. As mentioned earlier, FPT TURBO I orders the vertices from the lowest to the highest degree, as was intended by the authors in [8]. Some variations were considered during our investigation: the FPT TURBO HYBRID algorithm uses the vote measure described in Sanchis et al. [22], which favors vertices with lower degree neighbors in the ordering, whenever the vertices have the same utility. Considering all the orderings improves the solution at the cost of increasing the running time of the heuristic. Our FPT TURBO HYBRID algorithm uses four different orders and selects the best solution.

Checking if the solution is minimal. A key component of the "turbo-charging" technique is to use the dynamic version of the problem as a subroutine; however, Downey et al. [8] proved that the variation of the DYNAMIC DOMINATING SET problems that allowed vertices to be removed from the original solution remains $W[2]$-hard. Thus, the dynamic search FPT algorithm cannot also search for vertices that can be removed from the solution S_i. To address this, we simply check, in each iteration, if the solution found is minimal and remove any unneeded vertices.

Using the GRASP local search heuristic. It has been shown that LS heuristics work best when starting from a quality initial solution than from a number of poorer quality starting points [10]. Thus, the combination of the FPT TURBO I heuristic with a LS seems likely to provide improved results. In addition, the application of LS is an obvious way to address the above mentioned limitation of using dynamic problems, which cannot include removing vertices from existing solution S. Moreover, in some cases, it may be best to apply a LS at each iteration, but for the GRASP LS, we found no benefit in doing this. So, as its application was time-consuming, it was deferred to the end.

Using a heuristic guarantee. The next improvement is to check that the local optimization of the DYNAMIC DOMINATING SET search has not overlooked an obvious global solution. To do this, we employ a simple greedy heuristic to the graph G_i and exchange the solution for S_i if it is smaller. This modification also guarantees that the solution is no worse than the one produced by the Greedy heuristics. In the FPT TURBO HYBRID, this is done by calling GREEDY CHVATAL and GREEDY VOTE in each iteration to get a solution of the graph instance before invoking DDS FPT which can then also be used as an upper bound for r, the number of vertices that can be added by the DDS FPT subroutine to reduce execution time.

3   Experimental Analysis

All algorithms presented in this paper are implemented in the Java programming language. The experiment is run by a computer of the OSX Yosemite operating system with a CPU of 3.5 GHz, 6-Core Intel Xeon E5, and 32 GB memory.

In order to evaluate the performance of the "turbo-charging" technique, the solution sizes as well as execution times obtained were recorded. These results are compared to the results obtained using the two most known DOMINATING SET heuristics; Chvatel [6] and the GREEDY VOTE GRASP improvement given in [22]. Note that the solution sizes for the original FPT TURBO I heuristics is presented in its raw form without improvements such as pre-processing using reduction rules or other techniques described in Sect. 2.3.

The exact algorithm of Fomin et al. [12], implemented using the hybrid method of Abu-Khzam et al. [1], is used to calculate the exact solution and, whenever possible, the solution size of the exact algorithm was used in the comparison.

The DIMACS, BHOSLIB and KONECT datasets were used as benchmark-tests. The DIMACS and BHOSLIB datasets were chosen because they are both commonly-used. Note that due to their constructions – the main purpose of their design was to test Maximum Clique and Maximum Independent Set (MIS) and Minimum Vertex Cover (MVC) algorithms, the DIMACS datasets have a very small and easy-to-obtain dominating sets.

This is due to the relatively dense nature (high number of edges) of this data; however, Parekh [19] showed that the simple greedy heuristics perform very well for dense graphs. For BHOSLIB instances, while the simple Chvatal heuristics still performed well, the exact algorithm was not able to obtain an answer for these datasets in a week's processing.

The KONECT dataset consists of Non-Synthetic data mainly from social networks adding a practical flavor to the data. Unlike the other two datasets, the KONECT graphs are more sparse. The first 11 dataset from KONECT were used. The others were considered to large for the current algorithm and comparison LS heuristic. A detailed description of the graph properties of the KONECT datasets is also available at the KONECT web site [18].

Overall, the performance of GREEDY CHVATAL and GREEDY VOTE GRASP was similar, but the second performed better than the former, on average. FPT TURBO HYBRID performs at least as well as GREEDY CHVATAL and obtains a better solution for around 29.27% of the instances; Similarly, it returns a better solution for about 12.2% of the instances when compared to GREEDY VOTE GRASP. Moreover, the FPT TURBO HYBRID obtained better (smaller) solutions than FPT TURBO I heuristic in almost all the instances. We now focus on the size of solution considering instances from the three sample benchmarks.

(1) **For the KONECT** datasets, the GREEDY VOTE GRASP, FPT TURBO I and FPT TURBO HYBRID were all able to obtain optimum sized solutions. Furthermore, with the additional refinement of the GRASP LS, GREEDY VOTE GRASP was able to obtain smaller solutions in all cases when the obtained

solutions were not already optimum. This mainly applied to the larger non-synthetic instances in the KONECT datasets. In all instances, the FPT TURBO HYBRID either obtained an optimum solution, or a solution that was better (smaller) than that obtained by the other heuristics. Moreover, the improvements of FPT TURBO HYBRID were much more significant on larger sparse instances. The experimental results for the KONECT datasets are presented in Table 1. The first three columns contain the name, order and size of the graph tested. This is then followed by the optimum solution when it was known. **NA** is shown if this could not be obtained. The minimum solution and execution times obtained is shown for each of the algorithms considered.

Table 1. A comparison of the algorithm performance on the KONECT data sets (Time in sec).

DataSet			Opt	GREEDY CHVATAL [6]		GREEDY VOTE GRASP[19]		FPTTURBO I [8]		FPTTURBOHYBRID	
Name	|V|	|E|	Size	Size	Time	Size	Time	Size	Time	Size	Time
Dolphins	62	159	14	17	0.0027	15	0.025	14	1.4	**14**	4.11
Jazz musicians	198	2742	13	14	0.0207	13	0.224	14	0.261	**13**	4.89
PDZBase	212	244	NA	54	0.0115	52	0.584	52	0.189	52	24.9
U.Rovira Virgili	1133	5451	210	229	0.765	213	246	215	6.88	211	607
Euroroad	1174	1417	384	471	0.57	400	271	419	16.1	399	1390
Hamsterster	2426	16631	416	437	4.59	416	1740	417	67.7	**416**	3720

For the FPT TURBO HYBRID algorithm, a box has been placed around the solution when the obtained best result and the results which are known to be optimum are shown in bold. For these experiments, DDS search was restricted to searching for $k = 10$. Instances for which all algorithms performed equally are omitted to save space.

(2) For the DIMACS datasets, the difference in solution size among different heuristics is very small, mostly because the optimum solution for many of these instances have only 2 or 3 vertices. For some instances, the optimum solution could be obtained by the GREEDY CHVATAL heuristic. Nevertheless, the experimental results confirm that the FPT search heuristics did perform well when the optimum solution has a small size. In fact, the FPT TURBO HYBRID algorithm produced optimum results for the 19 DIMACS instances where the optimum was known, and the remaining solutions were at least as good as those found by any of the other techniques. Moreover, even though the solutions were trivial, the FPT TURBO HYBRID still found an improvement on one instance.

(3) For the BHOSLIB datasets, the FPT TURBO HYBRID heuristic was able to outperform all the other heuristics. This is of particular significance as it was not possible to obtain optimum solutions for any of the BHOSLIB datasets. Moreover, the FPT TURBO HYBRID heuristic solutions were never worse than those obtained by the other heuristics. In contrast there are three instances for which the GREEDY CHVATAL heuristic outperformed the GREEDY VOTE

Table 2. A comparison of the algorithm performance on the BHOSLIB data sets.

DataSet			GREEDY CHVATAL [6]		GREEDY VOTE GRASP[19]		FPTTURBO I [8]		FPTTURBOHYBRID	
Name	|V|	|E|	Size	Time	Size	Time	Size	Time	Size	Time
frb35-17-5	595	28143	14	0.252	15	4.24	16	0.532	13	54.5
frb53-24-3	1272	94127	21	1.44	21	34.7	23	2.37	20	264
frb56-25-5	1400	109601	22	1.71	22	43.9	23	2.7	21	323

GRASP heuristic. For the remaining 10 of 32 instances the solution size of GREEDY VOTE GRASP was smaller (better) than that of GREEDY CHVATAL. Table 2 shows highlights from the BOSLIB experimental results. Only instances where the results show significant improvement are provided. The others are not provided due to space requirements.

4   Conclusion

Downey et al. proposed the turbo-charging method as a general framework for improving heuristics [8]. The details for implementing this method were not completely developed. In this paper, we have resolved many of the technical details and the original algorithm has been improved by addressing a limitation to the originally proposed theoretical techniques. Furthermore, using the DOMINATING SET problem as a case study, we demonstrated that the turbo-charging technique, which consists mainly of using a fixed-parameter algorithm for the dynamic FPT problem as a subroutine, can be used to improve the results of heuristics.

We also observed that in some cases the improvement of using some FPT guarantee resulted in a better solution than Chvatel's heuristic [6]. While two FPT heuristics are given, we note that the best results were always obtained by using FPT TURBO HYBRID, the FPT TURBO I heuristic execution time was much faster than both GREEDY VOTE GRASP and FPT TURBO HYBRID heuristics. For this reason, the FPT TURBO I heuristics may also be well suited for use in other heuristics. Future work will explore how to improve the solution using a meta-search based on the order the vertices are processed.

A new algorithm that is based on the inclusion-exclusion technique has been presented in [16] for SET COVERING (SC), parameterized by the size of a small universe. The worst case running time is $O^*(2^k)$. It should be possible to obtain a significantly improved running time for a future FPT turbo-charging algorithm using this Set Cover algorithm.

References

1. Abu-Khzam, F.N., Langston, M.A., Mouawad, A.E., Nolan, C.P.: A hybrid graph representation for recursive backtracking algorithms. In: Lee, D.-T., Chen, D.Z., Ying, S. (eds.) FAW 2010. LNCS, vol. 6213, pp. 136–147. Springer, Heidelberg (2010). doi:10.1007/978-3-642-14553-7_15
2. Alber, J., Fellows, M.R., Niedermeier, R.: Polynomial-time data reduction for dominating set. J. ACM **51**(3), 363–384 (2004)
3. Bar-Ilan, J., Kortsarz, G., Peleg, D.: How to allocate network centers. J. Algorithms **15**(3), 385–415 (1993)
4. Böcker, S., Briesemeister, S., Klau, G.W.: Exact algorithms for cluster editing: evaluation and experiments. Algorithmica **60**(2), 316–334 (2011)
5. Christian, R., Fellows, M., Rosamond, F., Slinko, A.: On complexity of lobbying in multiple referenda. Rev. Econ. Des. **11**(3), 217–224 (2007)
6. Chvatal, V.: A greedy heuristic for the set-covering problem. Math. Oper. Res. **4**(3), 233–235 (1979)
7. Dinur, I., Steurer, D.: Analytical approach to parallel repetition. In: Proceedings of the 46th Annual ACM Symposium on Theory of Computing, STOC 2014, pp. 624–633. ACM, New York (2014)
8. Downey, R.G., Egan, J., Fellows, M.R., Rosamond, F.A., Shaw, P.: Dynamic dominating set and turbo-charging greedy heuristics. Tsinghua Sci. Technol. **19**(4), 329–337 (2014)
9. Downey, R.G., Fellows, M.R.: Parameterized Complexity. Monographs in Computer Science. Springer, New York (1999)
10. Feo, T.A., Resende, M.G.C.: Greedy randomized adaptive search procedures. J. Glob. Optim. **6**(2), 109–133 (1995)
11. Feo, T.A., Resende, M.G.C., Smith, S.H.: A greedy randomized adaptive search procedure for maximum independent set. Oper. Res. **42**(5), 860–878 (1994)
12. Fomin, F.V., Grandoni, F., Kratsch, D.: Measure and conquer: domination – a case study. In: Caires, L., Italiano, G.F., Monteiro, L., Palamidessi, C., Yung, M. (eds.) ICALP 2005. LNCS, vol. 3580, pp. 191–203. Springer, Heidelberg (2005). doi:10. 1007/11523468_16
13. Garey, M.R., Johnson, D.S.: Computers and Intractability: A Guide to the Theory of NP-Completeness. W.H. Freeman, New York (1979)
14. Gaspers, S., Kim, E.J., Ordyniak, S., Saurabh, S., Szeider, S.: Don't be strict in local search! In: Hoffmann, J., Selman, B. (eds.) Proceedings of the Twenty-Sixth AAAI Conference on Artificial Intelligence, Toronto, Ontario, Canada, 22–26 July 2012. AAAI Press (2012)
15. Hoos, H.H., Stützle, T.: Stochastic Local Search: Foundations and Applications. Elsevier/Morgan Kaufmann, Seattle (2004)
16. Hua, Q.-S., Wang, Y., Dongxiao, Y., Lau, F.C.M.: Set multi-covering via inclusion-exclusion. Theor. Comput. Sci. **410**(38–40), 3882–3892 (2009)
17. Johnson, D.S., Trick, M.A.: The Second DIMACS Implementation Challenge (1993). fpt://dimacs.rutgers.edu/pub/challenge
18. Kunegis, J.: KONECT: the Koblenz network collection. In: Carr, L., et al. (eds.) 22nd International World Wide Web Conference, WWW 2013, Rio de Janeiro, Brazil, 13–17 May 2013, pp. 1343–1350. International World Wide Web Conferences Steering Committee/ACM (2013). Companion Volume
19. Parekh, A.K.: Analysis of a greedy heuristic for finding small dominating sets in graphs. Inf. Process. Lett. **39**(5), 237–240 (1991)

20. Peleg, D.: Distributed data structures: a complexity-oriented view. In: Leeuwen, J., Santoro, N. (eds.) WDAG 1990. LNCS, vol. 486, pp. 71–89. Springer, Heidelberg (1991). doi:10.1007/3-540-54099-7_6
21. Peleg, D., Upfal, E.: A trade-off between space and efficiency for routing tables. J. ACM **36**(3), 510–530 (1989)
22. Sanchis, L.A.: Experimental analysis of heuristic algorithms for the dominating set problem. Algorithmica **33**(1), 3–18 (2002)
23. Weihe, K.: Covering trains by stations or the power of data reduction. In: Proceedings of Algorithms and Experiments, pp. 1–8 (1998)
24. Wittmann, R., Zitterbart, M.: Multicast Communication: Protocols, Programming, and Applications. Morgan Kaufmann, San Francisco (2000)
25. Ke, X.: BHOSLIB. http://networkrepository.com/bhoslib.php. Accessed 30 Apr 2015

Multi-interval Pairwise Compatibility Graphs

(Extended Abstract)

Shareef Ahmed[✉] and Md. Saidur Rahman

Graph Drawing and Information Visualization Laboratory,
Department of Computer Science and Engineering,
Bangladesh University of Engineering and Technology, Dhaka, Bangladesh
shareef.tamal@gmail.com, saidurrahman@cse.buet.ac.bd

Abstract. Let T be an edge weighted tree and let d_{min}, d_{max} be two non-negative real numbers where $d_{min} \leq d_{max}$. A pairwise compatibility graph (PCG) of T for d_{min}, d_{max} is a graph G such that each vertex of G corresponds to a distinct leaf of T and two vertices are adjacent in G if and only if the weighted distance between their corresponding leaves lies within the interval $[d_{min}, d_{max}]$. A graph G is a PCG if there exist an edge weighted tree T and suitable d_{min}, d_{max} such that G is a PCG of T. Knowing that all graphs are not PCGs, in this paper we introduce a variant of pairwise compatibility graphs which we call multi-interval PCGs. A graph G is a multi-interval PCG if there exist an edge weighted tree T and some mutually exclusive intervals of nonnegative real numbers such that there is an edge between two vertices in G if and only if the distance between their corresponding leaves in T lies within any such intervals. If the number of intervals is k, then we call the graph a k-interval PCG. We show that every graph is a k-interval PCG for some k. We also prove that wheel graphs and a restricted subclass of series-parallel graphs are 2-interval PCGs.

Keywords: Pairwise compatibility graphs · Phylogenetic trees · Series-parallel graphs

1   Introduction

Let T be an edge weighted tree and let d_{min}, d_{max} be two non-negative real numbers where $d_{min} \leq d_{max}$. A *pairwise compatibility graph* (PCG) of T for d_{min} and d_{max} is a graph $G = (V, E)$ where each vertex of G corresponds to a distinct leaf of T and two vertices are adjacent in G if and only if the weighted distance between their corresponding leaves lies within the interval $[d_{min}, d_{max}]$. The tree T is called a *pairwise compatibility tree* (PCT) of G. We denote a pairwise compatibility graph T for d_{min}, d_{max} by PCG (T, d_{min}, d_{max}). A given graph is a PCG if there exist suitable T, d_{min}, d_{max} such that G is a PCG of T. Figure 1(b) illustrates a pairwise compatibility graph G of the edge weighted tree T in Fig. 1(a) for $d_{min} = 3$ and $d_{max} = 5$. For a pairwise compatibility graph G,

© Springer International Publishing AG 2017
T.V. Gopal et al. (Eds.): TAMC 2017, LNCS 10185, pp. 71–84, 2017.
DOI: 10.1007/978-3-319-55911-7_6

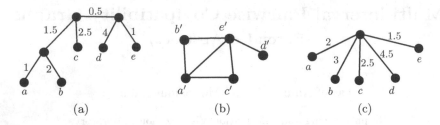

Fig. 1. (a) An edge weighted tree T, (b) a pairwise compatibility graph G of T for $d_{min} = 3$ and $d_{max} = 5$ and (c) another pairwise compatibility tree of G.

pairwise compatibility tree T may not be unique. For example, Fig. 1(c) shows another pairwise compatibility tree of the graph G in Fig. 1(b) for the same d_{min} and d_{max}.

PCGs have their applications in modeling evolutionary relationship among set of organisms from biological data which is also called phylogeny. Phylogenetic relationships are normally represented as a tree called phylogenetic tree. While dealing with a sampling problem from large phylogenetic tree, Kearney *et al.* [9] introduced the concept of PCGs. They also showed that "the clique problem" can be solved in polynomial time for a PCG if a pairwise compatibility tree can be constructed in polynomial time.

Kearney *et al.* [9] conjectured that all graphs are PCGs, but later Yanhaona *et al.* [12] refuted the conjecture by showing a bipartite graph with fifteen vertices which is not a PCG. Later Calamoneri *et al.* proved that every graph with at most seven vertices is a PCG [4]. It is also known that the graphs having cycles as their maximum biconnected components, tree power graphs, Steiner k-power graphs, phylogenetic k-power graphs, some restricted subclasses of bipartite graphs, triangle-free maximum-degree-three outer planar graphs and some superclass of threshold graphs are PCGs [6,11–13]. Calamoneri *et al.* gave some sufficient conditions for split matrogenic graph to be a PCG [5]. Recently a graph with eight vertices and a planar graph with sixteen vertices is proved not to be PCGs [7]. Iqbal *et al.* showed a necessary condition and a sufficient condition for a graph to be PCG [8]. However, the complete characterization of PCGs is not known yet.

As not all graphs are PCGs, some researchers has tried to the relax constraints on PCGs and thus some variants of PCGs are introduced [3,5]. One such variant of PCG is improper PCG which allows multiple leaves corresponding to a vertex of a graph [3]. In this paper we introduce a new variant of PCGs which we call k-interval PCGs. The idea behind a k-interval PCG is to allow k mutually exclusive intervals of nonnegative real numbers instead of one. We call a graph G a *k-interval PCG* of an edge weighted tree T for mutually exclusive intervals $I_1, I_2, \cdots, I_k$ of nonnegative real numbers when each vertex of G corresponds to a leaf of T and there is an edge between two vertices in G if the distance between their corresponding leaves lies in $I_1 \cup I_2 \cup \cdots I_k$. Figure 2(a) illustrates an edge weighted tree T and Fig. 2(b) shows the corresponding 2-interval PCG where $I_1 = [1, 3]$ and $I_2 = [5, 6]$.

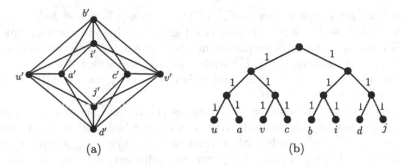

Fig. 2. (a) An edge weighted tree T, (b) a 2-interval PCG G of T where $I_1 = [1,3]$, $I_2 = [5,6]$.

In this paper we show that all graphs are k-interval PCGs for some k. We also show that wheel graphs W_n, which are not yet proved to be PCGs for $n \geq 8$ are 2-interval PCGs. Moreover, we proved that a restricted subclass of series-parallel graphs are 2-interval PCGs and provide an algorithm for constructing 2-interval pairwise compatibility tree for graphs of this subclass.

The remainder of the paper is organized as follows. Section 2 gives some necessary definitions, previous results and preliminary results on k-interval PCGs. In Sect. 3 we give our results on 2-interval PCGs. Finally we conclude in Sect. 4.

2   Preliminaries

In this section we define some terms which will be used throughout this paper and present some preliminary results.

Let, $G = (V, E)$ be a simple, undirected graph with vertex set V and edge set E. An edge between two vertices u and v is denoted by (u, v). If $(u, v) \in E$, then u and v are *adjacent* and the edge (u, v) is incident to u and v. The *degree* of a vertex is the number of edges incident to it. A *path* P_{uv} in G is a sequence of distinct vertices $w_1, w_2, w_3, \cdots, w_n$ in V such that $u = w_1$ and $v = w_n$ and $(w_i, w_{i+1}) \in E$ for $1 \leq i < n$. The vertices u and v are called the *end-vertices* of path P_{uv}. If the end-vertices are the same then the path is called a *cycle*. A *tree* T is a graph with no cycle. A vertex with degree one in a tree is called *leaf* of the tree. All the vertices other than leaves are called *internal nodes*. An *weighted tree* is a tree where each edge is assigned a number as the weight of the edge. The weight of an edge (u, v) is denoted as $w(u, v)$. The distance between two nodes u, v in T is the sum of the weights of the edges on path P_{uv} and denoted by $d_T(u, v)$. A star graph S_n is a tree on n nodes with one node having degree $n-1$ and all other nodes having degree 1. A *caterpillar* is a tree for which deletion of leaves together with their incident edges produces a path. The *spine* of a caterpillar is the longest path to which all other vertices of the caterpillar are adjacent. A *wheel graph* with n vertices, denoted by W_n, is obtained from a cycle graph C_{n-1} with $n-1$ vertices by adding a new vertex p and joining an edge from p to each vertex of C_{n-1}. The vertex p is called *hub*. A graph $G = (V, E)$ is called a *series-parallel* (*SP*) graph with source s and sink t if either G consists of a pair of vertices connected by a single edge or there exists two series-parallel graphs $G_i(V_i, E_i)$ with source s_i and sink t_i for $i = 1, 2$ such that $V = V_1 \cup V_2$, $E = E_1 \cup E_2$ and either $s = s_1$, $t_1 = s_2$ and $t = t_2$ or $s = s_1 = s_2$ and $t = t_1 = t_2$ [10].

We now review a previous result on cycles [11, 13] and show a construction process of a pairwise compatibility tree of a cycle which will be used later in this paper. Let C_n be a cycle with n vertices $v'_1, v'_2, v'_3 \cdots, v'_n$ where (v'_i, v'_{i+1}) are adjacent for $1 \leq i < n$ and (v'_1, v'_n) are also adjacent. We construct an edge weighted caterpillar T as follows. Let $v_1, v_2, v_3 \cdots, v_{n-1}$ be the leaves of T and $u_1, u_2, u_3, \cdots, u_{n-1}$ be the vertices on the spine of T such that u_i is adjacent to v_i for $1 \leq i < n$. We assign weight d to edge (u_i, u_{i+1}) for $1 \leq i < n-1$ and weight w to the edges incident to a leaf where $w > (n+1)\frac{d}{2}$. If n is odd then we put a vertex u_n in the middle of the path $P_{u_1 u_{n-1}}$ as illustrated in Fig. 3(a). If n is even then we use $u_{\frac{n}{2}}$ as u_n which is shown in Fig. 3(b). Then we place the last vertex v_n as a leaf adjacent to u_n. We assign weight $w_n = w - (n-3)\frac{d}{2}$ to the edge (u_n, v_n). This concludes the construction of T and we call this construction process **Algorithm ConstructCyclePCT**. The leaf v_i of T corresponds to the vertex v'_i of C_n. The tree constructed in this way is a PCT of C_n for $d_{min} = 2w + d$ and $d_{max} = 2w + d$. It is easy to observe that $max\{d_T(v_i, v_j)\} = 2w + (n-1)d$.

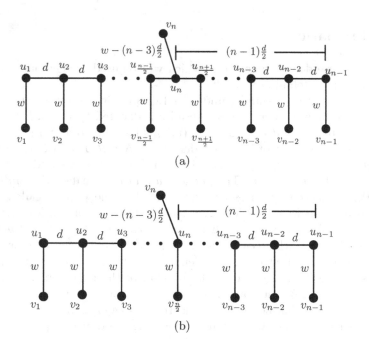

(a)

(b)

Fig. 3. (a) A pairwise compatibility tree of a cycle with odd number of vertices and (b) a pairwise compatibility tree of a cycle with even number of vertices.

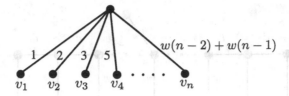

Fig. 4. An $|E|$-interval pairwise compatibility tree for any graph with n vertices.

We now introduce a new concept called k-interval PCG. Let T be an edge weighted tree and $I_1, I_2, I_3, \cdots, I_k$ be k non-negative intervals such that $I_i \cap I_j = \emptyset$ for $i \neq j$. A k-interval PCG of T for $I_1, I_2, I_2, \cdots, I_k$ is a graph $G = (V, E)$ where each vertex $u' \in V$ represent a leaf u in T and there is an edge $(u', v') \in E$ if and only if $d_T(u, v) \in I_1 \cup I_2 \cup I_3 \cup \cdots \cup I_k$. Obviously, a PCG is a k-interval PCG for $k = 1$, but a k-interval PCG may not be a PCG. The graph shown in Fig. 2 is not a PCG [7] but a 2-interval PCG.

The following theorem describes a preliminary result on k-interval PCGs.

Theorem 1. *Every graph is an $|E|$-interval PCG.*

Outline of the Proof: We give a constructive proof. Let $G = (V, E)$ be a graph with n vertices $v'_1, v'_2, v'_3, \cdots, v'_n$. We construct a star T with n leaves $v_1, v_2, v_3, \cdots, v_n$ where v_i corresponds to v'_i of G as illustrated in Fig. 4. Let $w(i)$ be the weight of the edge incident to v_i in T. We take $w(i)$ as follows.

$$w(i) = \begin{cases} 1 & \text{if } i = 1 \\ 2 & \text{if } i = 2 \\ w(i-1) + w(i-2) & \text{if } i > 2 \end{cases}$$

For each edge (v_i, v_j) in E we take an interval $I_{ij} = [d_T(v_i, v_j), d_T(v_i, v_j)]$. Thus we have total $|E|$ number of intervals. Then for every edge $(v_i, v_j) \in E$, $d_T(v_i, v_j) \in I_{ij}$. Similarly, if $(v_i, v_j) \notin E$, then there is no such interval I_{ij} such that $d_T(v_i, v_j) \in I_{ij}$. Thus T is an $|E|$-interval PCT of G. $\qquad \square$

3   2-Interval PCGs

In this section we give some results on 2-interval PCGs.

3.1   Wheel Graphs

In this section we prove that wheel graphs are 2-interval PCGs as in the following theorem.

Theorem 2. *Every wheel graph is a 2-interval PCG.*

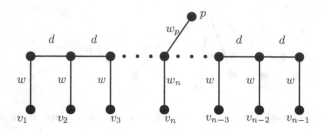

Fig. 5. A 2-interval pairwise compatibility tree of W_n.

Proof. Let W_{n+1} be a wheel graph with $n+1$ vertices $v_1', v_2', v_3' \cdots, v_n', p'$ where p' is the hub and $v_1', v_2', v_3' \cdots, v_n'$ forms the outer cycle C. We first construct a pairwise compatibility tree T for C by **Algorithm ConstructCyclePCT**. Note that the maximum distance between any pair of leaves in T is $2w + (n-1)d$. We then place a vertex p representing the vertex p' in W_{n+1} such that it is adjacent to u_n in T and assign weight w_p to the edge (p, u_n) as illustrated in Fig. 5. We choose w_p such that $w_p > 2w + (n-1)d$.

Clearly $d_T(p, v_i) > 2w + (n-1)d = max\{v_i, v_j\}$ for $i, j \leq n$. Then T is a 2-interval pairwise compatibility tree of W_n for $I_1 = [2w + d, 2w + d]$ and $I_2 = (2w + (n-1)d, \infty)$. $\qquad\square$

3.2   Series-Parallel Graphs

In this section we define a restricted subclass of series-parallel graphs which we call SQQ series-parallel graphs and show that this class of graphs are 2-interval PCGs.

Let $G = (V, E)$ be a series-parallel graph with source s and sink t. A pair of vertices $\{u, v\}$ of a connected graph is a *split pair* if there exist two subgraphs $G_1(V_1, E_1)$ and $G_2(V_2, E_2)$ satisfying following two conditions: 1. $V = V_1 \cup V_2$, $V_1 \cap V_2 = \{u, v\}$; and 2. $E = E_1 \cup E_2$, $E_1 \cap E_2 = \emptyset$, $|E_1| \geq 1$, $|E_2| \geq 1$. The *SPQ-tree* T of a series-parallel graph G with respect to a reference edge (u, v) describes a recursive decomposition of G induced by its split pairs [1,2]. Figure 6(a) illustrates a series-parallel graph G and Fig. 6(b) shows the SPQ-tree of G with respect to s, t. T is a rooted ordered tree and it contains three types of nodes: S, P and Q. Subtrees rooted at each node x of T corresponds to a subgraph of G called its *pertinent graph* $G(x)$. In this paper we use a modified definition of $G(x)$: $G(x)$ contains the leftmost and rightmost children of x in T in order from source to sink if x is a P-node or Q-node; if x is an S-node $G(x)$ does not contain the leftmost and rightmost children. Figure 6(c) illustrates the pertinent graph of the P-node at height 2 in T. Let x be any S-node in T other

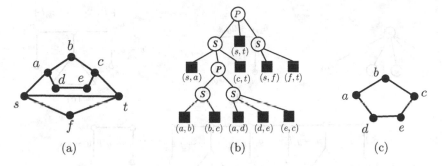

Fig. 6. (a) A series-parallel graph G, (b) An SPQ-tree of G with respect to s and t and (c) pertinent graph of the non-root P-node.

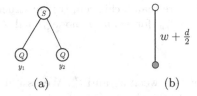

Fig. 7. (a) An S-node x with 2 children and (b) constructed tree T_x for x.

than the root and let $y_1, y_2, y_3, \cdots, y_n$ be the children of x in order from source to sink. If both y_1 and y_n are Q-nodes then we call G an SQQ series-parallel graph. We now give the following theorem.

Theorem 3. *Every SQQ series-parallel graph is a 2-interval PCG.*

Proof. We give a constructive proof. Let $G = (V, E)$ be an SQQ series-parallel graph with source s' and sink t' and $\mathcal{T}$ be an SPQ-tree of G with respect to s' and t'. Note that if $\mathcal{T}$ consists of a single Q-node then G is trivially a 2-interval PCG. We thus assume that $\mathcal{T}$ has at least one S-node or P-node. We construct a 2-interval pairwise compatibility tree of G using a bottom up computation on $\mathcal{T}$. For each internal node x of $\mathcal{T}$ we first compute 2-interval PCT for each of it's child node and then we add additional component and combine them to get a 2-interval PCT T_x of G(x). Let s'_x and t'_x be the source and sink of G(x) and s_x, t_x be the leaves of T_x representing s'_x and t'_x respectively. Depending on the type of the current node we have to consider two cases.

    Case 1: *The current node x is an S-node.* Let $y_1, y_2, y_3, \cdots, y_n$ be the children of x in order from s'_x to t'_x. This is illustrated in Fig. 8(a). According to the property of an SQQ series-parallel graph y_1 and y_n are Q-nodes. If $n = 2$, then we have only one node between s'_x and t'_x in G. In this case we construct a tree T_x with two leaves and one edge between them. One of the two leaves of

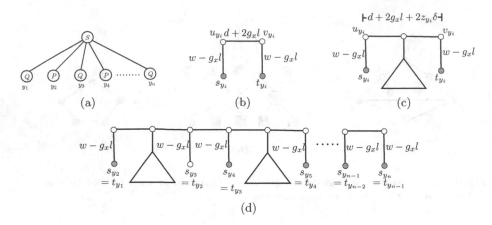

Fig. 8. (a) An S-node with more than 2 children, (b) constructed tree Γ_{y_i} for a child Q-node y_i, (c) constructed tree T_{y_i} for a child P-node y_i and (d) merged tree T_x for S node x.

T_x represents the only node between s'_x and t'_x. We assign weight $w + \frac{d}{2}$ to that edge. This is illustrated in Fig. 7(a) and (b). If x is the root node then we also place two leaves representing s'_x, t'_x and make them adjacent to a leaf in T_x. We then assign weight $w + \frac{d}{2}$ to the newly added edges.

We now consider the case where $n > 2$. In this case we have two subcases.

Case 1(a): y_i *be a* Q-*node.* At first we consider y_i for $i \neq 1, n$. In this case we construct a caterpillar Γ_{y_i} with two leaves s_{y_i}, t_{y_i} and two internal nodes u_{y_i}, v_{y_i} where u_{y_i}, v_{y_i} are adjacent to s_{y_i}, t_{y_i} respectively. Here s_{y_i}, t_{y_i} represent s'_{y_i}, t'_{y_i} of G respectively. Let g_x be an indicator variable which is 1 if depth of x modulo 4 is equal to 0 or 1 in $\mathcal{T}$ and -1 otherwise. We now assign weight $w - g_x l$ to each edge incident to a leaf and weight $d + 2g_x l$ to the edge (u_{y_i}, v_{y_i}) where $l \ll d$ at least as small as $\frac{d}{100|V|}$ as is illustrated in Fig. 8(b). Then for $i = 1, n$ we also construct trees in the way mentioned above if x is the root node of $\mathcal{T}$, otherwise trees will be constructed for y_1 and y_n while processing the parent P-node of x.

Case 1(b): y_i *be a* P-*node.* In this case we have a caterpillar Γ_{y_i} induced by two leaves s_{y_i} and t_{y_i} of T_{y_i} according to the construction process described in case 2 as shown in Fig. 8(c). Let u_{y_i}, v_{y_i} be the vertices on spine of Γ_{y_i} that are adjacent to the leaves s_{y_i} and t_{y_i}.

We thus have a caterpillar Γ_{y_i} for each $i \neq 1, n$. We next merge all this caterpillars such that t_{y_i} and v_{y_i} lie on $s_{y_{i+1}}$ and $u_{y_{i+1}}$ and get a single caterpillar Γ_x with $n - 1$ leaves induced by $s_2, s_3, \cdots, s_n$ as illustrated in Fig. 8(d).

Case 2: *The current node* x *is a* P-*node.* In this case x can have at most one Q-node as its child and if it has one then it represents an (s'_x, t'_x) edge. We first construct a caterpillar Γ_x with two leaves s_x, t_x representing s'_x and t'_x, and two internal nodes u_x, v_x where u_x is adjacent to s_x and v_x is adjacent to t_x. We now assign weight $w + g_{y_i} l$ to each edge incident to a leaf in Γ_x where g_{y_i} is the indicator variable of any child S-node y_i of x in T. If x has a child Q-node in T we assign weight $d - 2g_{y_i}l$ to the edge (u_x, v_x), otherwise we assign $d - 2g_{y_i}l + 2\delta$ where $\delta \ll l$. We now replace the edge (u_x, v_x) by a path u_x, a_x, b_x, v_x where a_x and b_x are two degree 2 vertices. We call a_x, b_x the *port nodes* of Γ_x. Then we reassign weight such that $w(u_x, a_x) = \frac{1}{2}d_{T_x}(u_x, v_x)$ and $w(a_x, b_x) = \delta$. Let z_x be an indicator variable which is 1 if there is a child Q-node of x and 0 otherwise. Then $w(u_x, a_x) = \frac{d}{2} - g_{y_i}l + z_x\delta$, $w(a_x, b_x) = \delta$ and $w(b_x, v_x) = \frac{d}{2} - g_{y_i}l + (z_x - 1)\delta$. See Fig. 9(a).

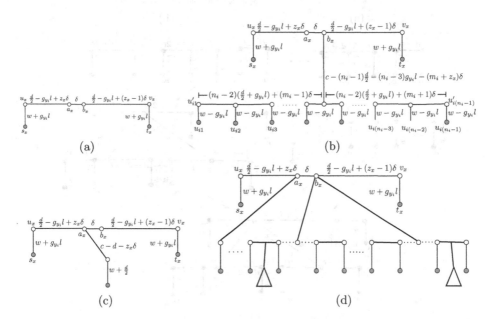

Fig. 9. (a) Constructed tree Γ_x with souce and sink for P-node, (b) merged tree with 2-interval PCT of a children S-node having than 2 children, (c) merged tree with 2-interval PCT of a children S-node having 2 children and (d) final 2-interval PCT T_x of $G(x)$ where x is a P-node.

Let $y_1, y_2, y_3, \cdots, y_n$ be the children of x where y_i is an S-nodes for $1 \le i \le n$. At first we construct 2-interval PCT T_{y_i} of $G(y_i)$ for $1 \le i \le n$ according to case 1. Let y_i be an S-node with n_i children where $n_i > 2$. Then we have a caterpillar Γ_{y_i} with $n_i - 1$ leaves induced by the sources and sinks of some children of y_i in T_{y_i}, which is merged while processing the S-node according to case 1. Let $u_{i1}, u_{i2}, u_{i3}, \cdots, u_{i(n_i-1)}$ be the leaves of Γ_{y_i} and let $u'_{i1}, u'_{i2}, u'_{i3} \cdots, u'_{i(n_i-1)}$ be the vertices on spine where u_{ij} is adjacent to u'_{ij} for $1 \le j \le n$. Note that any edge (u_{ij}, u'_{ij}) has weight w and $(u'_{ij}, u'_{i(j+1)})$ has weight $d + 2g_{y_i}l$ or $d + 2g_{y_i}l + 2\delta$. Let m_i be the number of edges of weight $d + 2g_{y_i}l + 2\delta$ on the spine where $m_i \le (n_i - 2)$. Thus the spine has length of $(n_i - 2)(d + 2g_{y_i}l) + 2m_i\delta$. We now put a vertex v_i on the spine such that $d_{\Gamma_i}(u'_{i1}, v_i) = \frac{(n_i-2)(d+2g_{y_i}l)+2m_i\delta}{2} - \delta$ and

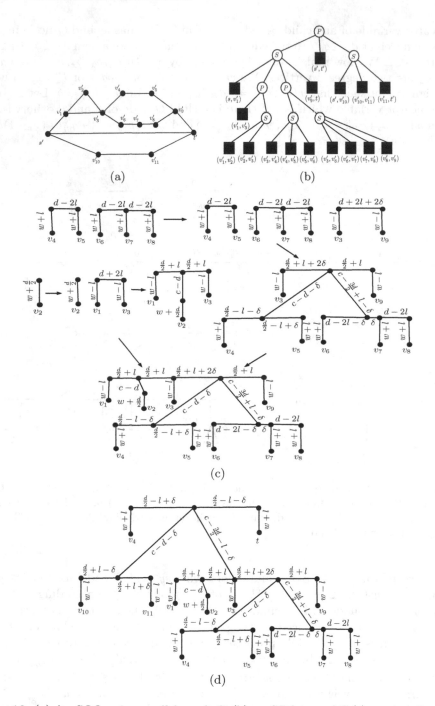

Fig. 10. (a) An SQQ series-parallel graph G, (b) an SPQ-tree of G (c) construction of 2-interval PCT of pertinent graph of the leftmost child of the root which is an S-node and (d) constructed 2-interval PCT of G.

we add an edge between v_i and port node b_x. We assign weight $c - (n_i - 1)\frac{d}{2} - (n_i - 3)g_{y_i}l - (m_i + z_x)\delta$ to the edge (v_i, b) as illustrated in Fig. 9(b). We choose a very large value for c such that $c > 2(d + 2\delta + 2l)|V|$ where $|V|$ is the number of vertices in G.

Let y_j be an S-node with exactly 2 children. Then we have 2-interval PCT T_{y_i} consists of two nodes and the edge between them has weight $w + \frac{d}{2}$. In this case we add an edge between port node a_x and one of the leaves. We assign weight $c - d - z_x\delta$ to the newly added edge as illustrated in Fig. 9(c). We call any edge joining Γ_x with T_{y_i} for $i \leq n$ a *caterpillar-connecting edge*. An example of the construction process is illustrated in Fig. 10.

We now prove that the tree T constructed by the algorithm above is a 2-interval PCT of G for intervals $I_1 = [2w + d, 2w + d]$ and $I_2 = [c + 2w, c + 2w]$. We prove this by an induction on the height $h(T)$ of the SPQ-tree T of G. Let x be the root of T having n children $y_1, y_2, \cdots, y_n$ and n_i be the number of children of y_i.

Assume that G is an SQQ series-parallel graph with $h(T) = 1$. Then T consists of an S-node x as its root and all the children of the root are Q-nodes. In this case the algorithm produces a caterpillar with n leaves where each edge incident to a leaf has weight $w - g_x l$ and each edge on the spine has weight $d + 2g_x l$. Thus if (u, v) is a Q-node in T then $d_T(u, v) = 2w + d$ and otherwise $2w + d < d_T(u, v) < c + 2w$ because of our choice of c being very large. Thus the basis is true.

Assume that $h(T) > 1$ and the claim is true for every SQQ series-parallel graph with $h(T) < h$. Let G be an SQQ series-parallel graph with $h(T) = h$ and let x be the root of T. Let $y_1, y_2, y_3, \cdots, y_n$ be the children of x and pertinent graphs of $y_1, y_2, \cdots, y_n$ are 2-interval PCGs for I_1 and I_2 by the induction hypothesis. Let $T_{y_1}, T_{y_2}, \cdots, T_{y_n}$ be the 2-interval PCTs constructed by the algorithm for $y_1, y_2, \cdots, y_n$.

We first consider the case where x is a P-node. Then according to Case 2 we have $d_T(s_x, t_x) = 2w + d$, if there is an edge (s_x, t_x). Otherwise, we have $2w + d < d_T(s_x, t_x) = 2w + d + 2\delta < c + 2w$. Let y_i be an S-node. If y_i has two children, then there is only one node u'_{i1} between s_x and t_x in $G(y_i)$ and u_{i1} is its corresponding leaf in T_{y_i}. In this case $d_T(s_x, u_{i1}) = d_T(t_x, u_{i2}) = 2w + c$ which lies in interval I_2. If y_i has more than 2 children then the distance $d_T(s_x, u_{i1})$ is computed as follows.

$$d_T(s_x, u_{i1}) = d_{\Gamma_x}(s_x, b) + w(b, v_i) + d_{\Gamma_{y_i}}(v_i, u_{i1})$$

$$= w + g_{y_i}l + \frac{d}{2} - g_{y_i}l + z_x\delta + \delta + c - (n_i - 1)\frac{d}{2} - (n_i - 3)g_{y_i}l$$

$$- (m_i + z_x)\delta + w - g_{y_i}l + (n_i - 2)(\frac{d}{2} + g_{y_i}l) + m_i\delta - \delta$$

$$= 2w + c.$$

Similarly the distance $d_T(s_x, u_{n_i})$ is computed as follows.

$$d_T(s_x, u_{in_i}) = d_{\Gamma_x}(s_x, b) + W(b, v_i) + d_{\Gamma_i}(v_i, u_{in})$$

$$= w + g_{y_i}l + \frac{d}{2} - g_{y_i}l + z_x\delta + \delta + c - (n_i - 1)\frac{d}{2} - (n_i - 3)g_{y_i}l$$

$$- (m_i + z_x)\delta + w - g_{y_i}l + (n_i - 2)(\frac{d}{2} + g_{y_i}l) + m_i\delta + \delta$$

$$= 2w + c + 2\delta.$$

Thus $d_T(s_x, u_{in_i}) > c + 2w$. Now clearly $d_T(s_x, u_{ij}) < 2w + c$ for $j \neq 1, n_i$; as they are at least $d + 2g_{y_i}l$ less than $2w + c + 2\delta$. Again $d_T(s_x, u_{ij}) > 2w + d$ because we choose $c > 2(d + 2\delta + 2l)|V|$. Doing similar calculation for t_x we get, $d_T(t_x, u_{in_i}) = 2w + c$, $d_T(t_x, u_{i1}) = 2w + c - 2\delta$ and $2w + d < d_T(t_x, u_{ij}) < 2w + c$ for $j \neq 1, n_i$. Now the path from u_{ij} and u_{kl} where $i \neq k$ consists of 2 caterpillar-connecting edge, 2 edge from leaf to spine for each leaf and some additional edges on the spines. Thus we get,

$$d_T(u_{ij}, u_{kl}) \geq 2(w - g_{y_i}l) + c - (n_i - 1)\frac{d}{2} - (n_i - 3)g_{y_i}l - (m_i + z_x)\delta$$

$$+ c - (n_k - 1)\frac{d}{2} - (n_k - 3)g_{y_i}l - (m_k + z_x)\delta$$

$$\geq 2w + 2c - (n_i + n_k - 2)\frac{d}{2} - (m_i + m_k + 2z_x)\delta$$

$$- (n_i + n_k - 4)g_{y_i}l$$

$$> 2w + 2c - c$$

$$= 2w + c.$$

The calculation above implies that for any two leaves (u, v) who have more than two caterpillar-connecting edges on path P_{uv} we get $d_T(u, v) > 2w + c$. Thus if x is a P-node then only the distance between s_x, u_{i1} and t_x, u_{in_i} are equal to $2w + c$, distance between s_x, u_{ij} and t_x, u_{ij} are less than $2w + c$ but greater than $2w + d$, any distance between two leaves having two or more caterpillar-connecting edge between them is greater than $2w + c$.

On the other hand if x is an S-node then Γ_x is a caterpillar with $n - 1$ leaves $s_{y_2} = t_{y_1}, s_{y_3} = t_{y_2}, \cdots, s_{y_n} = t_{y_{n-1}}$. If y_i is a child Q-node of x then $d_T(s_{y_i}, t_{y_i}) = 2w + d$ for $i \neq 1, n$. Also $2w + d < d_T(s_{y_i}, s_{y_j})$, $d_T(t_{y_i}, t_{y_j})$, $d_T(s_{y_i}, t_{y_j}) < 2w + c$ for $i \neq j$ as the path between any of the mentioned pair of leaves contains at least two edge with weight $d + 2gl$ or larger and $c > 2(d + 2\delta + 2l)|V|$.

Let y_i be a child P-node of x and r_j be any child S-node of y_i in $\mathcal{T}$. Clearly Γ_x and Γ_{r_j} is connected by a caterpillar connecting edge. Let r_j has n_{r_j} children which implies Γ_{r_j} has $n_{r_j} - 1$ leaves. Let u_1, u_2 be two leaves in Γ_{r_j} where $d_{\Gamma_{r_j}}(u_1, u_2) = max\{d_{\Gamma_{r_j}}(u_i, u_j)\}$. From the proof of processing at P-node we know $d_T(s_{y_i}, u_1) = 2w + c$ and $d_T(t_{y_i}, u_2) = 2w + c$. Let v be a leaf in Γ_{r_j} where $d_{\Gamma_{r_j}}(u_1, v) < d_{\Gamma_{r_j}}(u_2, v)$ and the path P_{u_1v} contains e_{r_j} edges on the spine. We also assume that f_{r_j} edges among those e_{r_j} edges are of weight $d + 2g_{r_j} + 2\delta$.

Thus $d_T(v, s_{y_i}) = 2w + c - e_{r_j}(d + 2g_{r_j}l) - 2f_{r_j}\delta$. Let s_{y_k} be a leaf in Γ_x where $d_{\Gamma_x}(s_{y_k}, s_{y_i}) < d_{\Gamma_x}(s_{y_k}, t_{y_i})$. We also assume that the path $P_{s_{y_i} s_{y_k}}$ contains e_x edges on the spine of Γ_x and f_x edges among them are of weight $d + 2g_x + 2\delta$. Then $d_T(v, s_{y_k}) = 2w + c - e_{r_j}(d + 2g_{r_j}l) - 2f_{r_j}\delta + e_x(d + 2g_xl) + 2f_x\delta = 2w + c + (e_x - e_{r_j})d + 2(e_xg_x - e_{r_j}g_{r_j})l + 2(f_x - f_{r_j})\delta$. Now as r_j is a grandchild of x we get $g_r = -g_{r_j}$. So, $d_T(v, s_{y_k}) > c + 2w$ if $e_x > w_{r_j}$, $2w + d < d_T(v, s_{y_k}) < c + 2w$ if $e_x < w_{r_j}$. On the other hand if $e_x = e_{r_j}$ then $d_T(v, s_{y_k}) > c + 2w$ if $g_x = 1$ and $2w + d < d_T(v, s_{y_k}) < c + 2w$ if $g_x = -1$. Similarly if $d_{\Gamma_x}(s_{y_k}, s_{y_i}) > d_{\Gamma_x}(s_{y_k}, t_{y_i})$, we get $d_T(v, s_{y_k}) = 2w + c + (e_x - e_{r_j})d + 2(e_xg_x - e_{r_j}g_{r_j})l + 2(f_x - f_{r_j} - 2)\delta$. This also implies that $d_T(c, s_{y_k}) \notin I_2$. By doing similar calculation it can be shown that $d_T(v, s_{y_k}) \notin I_2$ if $d_{r_{r_j}}(u_1, v) \geq d_{r_{r_j}}(u_2, v)$. Also the distance between any pair of leaves that have more than two caterpillar-connecting edge in the path between them is greater than $2w + c$. Thus T is a 2-interval PCT of G for $I_1 = [2w + d, 2w + d]$ and $I_2 = [c + 2w, c + 2w]$. □

4 Conclusion

In this paper, we have introduced a new notion named k-interval pairwise compatibility graphs. We have proved that every graph is a k-interval PCGs for some k. We have also showed that wheel graphs and a restricted subclass of series-parallel graphs are 2-interval PCGs. Inception of k-interval PCGs brings in some interesting open problems. It is not known whether some constant number of intervals are sufficient for every graph to be a k-interval PCG. Whether all series-parallel graphs are 2-interval PCGs or not is also unknown.

Acknowledgments. We thank Kazuo Iwama of Kyoto University who pointed out this variant of the problem when the second author discussed the PCG problem with him in 2014.

References

1. Battista, G.D., Tamassia, R.: On-line maintenance of triconnected components with SPQR-trees. Algorithmica **15**(4), 302–318 (1996)
2. Battista, G.D., Tamassia, R., Vismara, L.: Output-sensitive reporting of disjoint paths. Algorithmica **23**(4), 302–340 (1999)
3. Bayzid, M.S.: On Pairwise compatibility graphs. Master's thesis, Bangladesh University of Engineering and Technology, June 2010
4. Calamoneri, T., Frascaria, D., Sinaimeri, B.: All graphs with at most seven vertices are pairwise compatibility graphs. Comput. J. **57**, 882–886 (2013)
5. Calamoneri, T., Petreschi, R., Sinaimeri, B.: On relaxing the constraints in pairwise compatibility graphs. In: Rahman, M.S., Nakano, S. (eds.) WALCOM 2012. LNCS, vol. 7157, pp. 124–135. Springer, Heidelberg (2012). doi:10.1007/978-3-642-28076-4_14
6. Calamoneri, T., Petreschi, R., Sinaimeri, B.: On the pairwise compatibility property of some superclasses of threshold graphs. Discrete Math. Algorithms Appl. **5**(2) (2013)

7. Durocher, S., Mondal, D., Rahman, M.S.: On graphs that are not PCGs. Theor. Comput. Sci. **571**, 78–87 (2015)
8. Hossain, M.I., Salma, S.A., Rahman, M.S.: A necessary condition and a sufficient condition for pairwise compatibility graphs. In: Kaykobad, M., Petreschi, R. (eds.) WALCOM 2016. LNCS, vol. 9627, pp. 107–113. Springer, Cham (2016). doi:10. 1007/978-3-319-30139-6_9
9. Kearney, P., Munro, J.I., Phillips, D.: Efficient generation of uniform samples from phylogenetic trees. In: Benson, G., Page, R.D.M. (eds.) WABI 2003. LNCS, vol. 2812, pp. 177–189. Springer, Heidelberg (2003). doi:10.1007/978-3-540-39763-2_14
10. Rahman, M.S., Egi, N., Nishizeki, T.: No-bend orthogonal drawings of series-parallel graphs. In: Healy, P., Nikolov, N.S. (eds.) GD 2005. LNCS, vol. 3843, pp. 409–420. Springer, Heidelberg (2006). doi:10.1007/11618058_37
11. Salma, S.A., Rahman, M.S., Hossain, M.I.: Triangle-free outerplanar 3-graphs are pairwise compatibility graphs. J. Graph Algorithms Appl. **17**(2), 81–102 (2013)
12. Yanhaona, M.N., Bayzid, M.S., Rahman, M.S.: Discovering pairwise compatibility graphs. Discrete Math. Algorithms Appl. **2**, 607–623 (2010)
13. Yanhaona, M.N., Hossain, K.S.M.T., Rahman, M.S.: Pairwise compatibility graphs. J. Appl. Math. Comput. **30**, 479–503 (2009)

A Note on Effective Categoricity
for Linear Orderings

Nikolay Bazhenov[1,2](✉)

[1] Sobolev Institute of Mathematics, Novosibirsk, Russia
bazhenov@math.nsc.ru
[2] Novosibirsk State University, Novosibirsk, Russia

Abstract. We study effective categoricity for linear orderings. For a computable structure S, the *degree of categoricity* of S is the least Turing degree which is capable of computing isomorphisms among arbitrary computable copies of S.

We build new examples of degrees of categoricity for linear orderings. We show that for an infinite computable ordinal α, every Turing degree c.e. in and above $0^{(2\alpha+2)}$ is the degree of categoricity for some linear ordering. We obtain similar results for linearly ordered abelian groups and decidable linear orderings.

Keywords: Linear ordering · Computable categoricity · Computable structure · Categoricity spectrum · Degree of categoricity · Autostability spectrum · Ordered abelian group · Decidable structure · Autostability relative to strong constructivizations

1 Introduction

The study of computably categorical structures goes back to the works of Fröhlich and Shepherdson [19], and Mal'tsev [25,26]. Since then, the notion of *computable categoricity* and its relativized versions have been the subject of much study.

Definition 1. Let **d** be a Turing degree. A computable structure A is **d**-*computably categorical* if for every computable structure B isomorphic to A, there exists a **d**-computable isomorphism from A onto B. **0**-computably categorical structures are also called *computably categorical*.

The *categoricity spectrum* of a structure A is the set

$$\mathrm{CatSpec}(A) = \{\mathbf{d} : A \text{ is } \mathbf{d}\text{-computably categorical}\}.$$

A Turing degree $\mathbf{d}_0$ is the *degree of categoricity* of A if $\mathbf{d}_0$ is the least degree in the spectrum $\mathrm{CatSpec}(A)$.

Categoricity spectra and degrees of categoricity were introduced by Fokina, Kalimullin, and Miller [18]. Note that some of the literature (see, e.g., [8,9,22])

© Springer International Publishing AG 2017
T.V. Gopal et al. (Eds.): TAMC 2017, LNCS 10185, pp. 85–96, 2017.
DOI: 10.1007/978-3-319-55911-7_7

uses the terms *autostability spectrum* and *degree of autostability* in place of categoricity spectrum and degree of categoricity, respectively.

Suppose that n is a natural number and α is a computable ordinal. Fokina, Kalimullin, and Miller [18] proved that every Turing degree $\mathbf{d}$ that is d.c.e. in and above $\mathbf{0}^{(n)}$ is the degree of categoricity of a computable structure. This result was extended by Csima, Franklin, and Shore [12] to hyperarithmetical degrees. They proved that every degree that is d.c.e. in and above $\mathbf{0}^{(\alpha+1)}$ is a degree of categoricity. They also showed that $\mathbf{0}^{(\alpha)}$ is a degree of cateroricity. Miller [28] constructed the first example of a computable structure with no degree of categoricity. For more results on categoricity spectra, see the survey [17] and the recent papers [1,8,16]

In this paper, we study categoricity spectra for linear orderings. Goncharov and Dzgoev [24], and independently, Remmel [29] proved that a computable linear ordering $\mathcal{L}$ is computably categorical iff its set of successivities $S(\mathcal{L}) = \{(a, b) : (a <_{\mathcal{L}} b)\&\neg\exists x(a <_{\mathcal{L}} x <_{\mathcal{L}} b)\}$ is finite. We summarize known examples of degrees of categoricity for linear orderings in Table 1. The table is read as follows. Frolov [20] proved that every Turing degree d.c.e. in and above $\mathbf{0}''$ is the degree of categoricity for some linear ordering.

Table 1. Known examples of degrees of categoricity for linear orderings.

Conditions for a degree	Reference
d.c.e. in and above $\mathbf{0}''$	[20]
c.e. in and above $\mathbf{0}^{(2k)}$, $2 \leq k < \omega$	[9]
c.e. in and above $\mathbf{0}^{(2\alpha+1)}$, $\omega \leq \alpha < \omega_1^{CK}$	[9]
d.c.e. in and above $\mathbf{0}^{(n)}$, $3 \leq n < \omega$	Frolov announced in his talk at Kazan Federal University (January 2016)

This short note is a companion to [7,9]. Here we prove the following result.

Theorem 1. *Suppose that $1 \leq k < \omega$, α is an infinite computable ordinal, and $\mathbf{d}$ is a Turing degree such that it satisfies one of the following conditions:*

(A) $\mathbf{d} \geq \mathbf{0}^{(2k+1)}$, and $\mathbf{d}$ is c.e. in $\mathbf{0}^{(2k+1)}$;
(B) $\mathbf{d} \geq \mathbf{0}^{(2\alpha+2)}$, and $\mathbf{d}$ is c.e. in $\mathbf{0}^{(2\alpha+2)}$.

Then $\mathbf{d}$ is the degree of categoricity for a computable linear ordering.

The proof is based on the ideas from [7,9]. For simplicity, we give the detailed proof for the case (A). The proof of the case (B) is essentially the same. Theorem 1 and the results from Table 1 yield the following.

Corollary 1. *Suppose that α is a computable successor ordinal, and $\alpha \geq 2$. Every Turing degree c.e. in and above $\mathbf{0}^{(\alpha)}$ is the degree of categoricity for some linear ordering.*

2 Preliminaries

We consider only computable languages, and structures with universe contained in ω. For a structure $\mathcal{S}$, $D(\mathcal{S})$ denotes the atomic diagram of $\mathcal{S}$. For a set X, $card(X)$ is the cardinality of X. We refer the reader to [5,15] for background on computable structures.

We treat linear orderings as structures in the language $L_{LO} = \{\leq^2\}$. Let $\mathcal{B}$ be a substructure of a linear ordering $\mathcal{A}$. The ordering $\mathcal{B}$ is called an *interval* of $\mathcal{A}$ if $b_1, b_2 \in \mathcal{B}$, $a \in \mathcal{A}$, and $b_1 <_{\mathcal{A}} a <_{\mathcal{A}} b_2$ together imply that $a \in \mathcal{B}$. We also assume that the empty set is an interval of $\mathcal{A}$. Suppose that $a, b \in \mathcal{A}$, and $a \leq_{\mathcal{A}} b$. Then $[a, b[_{\mathcal{A}}$ denotes the interval of $\mathcal{A}$ with the universe $\{x : a \leq_{\mathcal{A}} x <_{\mathcal{A}} b\}$. Let $[a, b]_{\mathcal{A}}$ denote the interval with the universe $\{x : a \leq_{\mathcal{A}} x \leq_{\mathcal{A}} b\}$

Suppose that $\{L_n\}_{n \in \omega}$ is a computable sequence of linear orderings, and $\mathcal{A}$ is a computable linear ordering with the universe ω. Then $\sum_{n \in \mathcal{A}} L_n$ denotes the *generalized sum*, i.e. the structure with the universe $\{(x, n) : n \in \omega, x \in L_n\}$ and the ordering defined as follows: $(x, n) \leq (y, m)$ iff $n <_{\mathcal{A}} m$ or $(n = m)\&(x \leq_{L_n} y)$. We identify the ordering $\sum_{n \in \mathcal{A}} L_n$ with its natural computable copy.

Let η denote the linear ordering of rationals, and ζ denote the ordering of integers. For further background on linear orderings, the reader is referred to [13,30].

2.1 Infinitary Formulas

For a language L, *infinitary formulas* of L are formulas of the logic $L_{\omega_1\omega}$. For a countable ordinal α, infinitary Σ_α and Π_α formulas are defined in a standard way (see, e.g., [5, Chap. 6]).

We give a short informal description for the class of *computable infinitary formulas* of L that was introduced in [4]. These formulas allow disjunctions and conjunctions over computably enumerable (c.e.) sets of formulas. Let α be a non-zero computable ordinal.

1. Computable Σ_0 and Π_0 formulas are quantifier-free first-order L-formulas.
2. A computable Σ_α formula is a c.e. disjunction $\bigvee_i \exists \bar{u}_i \psi_i(\bar{x}, \bar{u}_i)$, where ψ_i is a computable Π_{β_i} formula for some $\beta_i < \alpha$.
3. A computable Π_α formula is a c.e. conjunction $\bigwedge_i \forall \bar{u}_i \psi_i(\bar{x}, \bar{u}_i)$, where ψ_i is a computable Σ_{β_i} formula for some $\beta_i < \alpha$.

For the formal definition of computable infinitary formulas and their properties, see Chap. 7 in [5].

For a countable ordinal α, we define infinitary formulas $F_\alpha(x, y)$ and $S_\alpha(x, y)$. These formulas were introduced in [9] and are based on Example 4 from [5, Sect. 6.2]. The formulas satisfy the following: for a well-ordering $\mathcal{A}$ and elements a, b from $\mathcal{A}$, we have:

1. $\mathcal{A} \models F_\alpha(a, b)$ iff $a \leq b$ and the interval $[a, b[_{\mathcal{A}}$ is isomorphic to an ordinal $\beta < \omega^\alpha$;
2. $\mathcal{A} \models S_\alpha(a, b)$ iff $a < b$ and the interval $[a, b[_{\mathcal{A}}$ is isomorphic to ω^α.

Suppose that β is a countable ordinal, and δ is a countable limit ordinal. We set:

$$F_0(x, y) = (x = y),$$
$$S_0(x, y) = (x < y) \ \& \ \neg \exists z \, (x < z < y),$$
$$F_{\beta+1}(x, y) = F_\beta(x, y) \lor \bigvee_{k \in \omega} \exists z_0 \ldots \exists z_{k+1} \big[x = z_0 \ \& \ \underset{i \leq k}{\&} S_\beta(z_i, z_{i+1}) \ \& $$
$$F_\beta(z_{k+1}, y) \big],$$
$$F_\delta(x, y) = \bigvee_{\gamma < \delta} F_\gamma(x, y),$$
$$S_\gamma(x, y) = (x < y) \ \& \ \neg F_\gamma(x, y) \ \& \ \forall z \, (x \leq z < y \to F_\gamma(x, z)),$$
$$\text{where } \gamma \in \{\beta + 1, \delta\}.$$

It is not difficult to verify the following claim.

Lemma 1. *Let α be a computable ordinal. The formula F_α is logically equivalent to a computable $\Sigma_{2\alpha}$ formula, and the formula S_α is equivalent to a computable $\Pi_{2\alpha+1}$ formula.*

2.2 Relative Δ_α^0 Categoricity

Let α be a computable ordinal. A computable structure $\mathcal{A}$ is *relatively Δ_α^0 categorical* if for every $\mathcal{B} \cong \mathcal{A}$, there is a $\Delta_\alpha^0(D(\mathcal{B}))$ isomorphism from $\mathcal{A}$ onto $\mathcal{B}$.

Note that relative $\Delta_{1+\alpha}^0$ categoricity implies $\mathbf{0}^{(\alpha)}$-computable categoricity. Goncharov [23] constructed a computably categorical structure that is not relatively computably categorical. For a computable successor ordinal α, Goncharov, Harizanov, Knight, McCoy, Miller, and Solomon [21] built a $\mathbf{0}^{(\alpha)}$-computably categorical structure that is not relatively $\Delta_{1+\alpha}^0$ categorical. Chisholm, Fokina, Goncharov, Harizanov, Knight, and Quinn [11] extended this result to computable limit ordinals α. Downey, Kach, Lempp, Lewis-Pye, Montalbán, and Turetsky [14] proved that for any computable ordinal α, there exists a computably categorical structure that is not relatively Δ_α^0 categorical.

Let $\mathcal{S}$ be a structure. A *formally Σ_α^0 Scott family* for $\mathcal{S}$ is a c.e. set Φ of computable Σ_α formulas (with a fixed finite tuple of parameters $\bar{c}$ from $\mathcal{S}$) such that

1. every tuple from $\mathcal{S}$ satisfies some $\phi \in \Phi$, and
2. if $\bar{a}$ and $\bar{b}$ are tuples from $\mathcal{S}$ satisfying the same formula $\phi \in \Phi$, then $(\mathcal{S}, \bar{a}) \cong (\mathcal{S}, \bar{b})$.

Theorem 2 ([2, 10]). *Let α be a non-zero computable ordinal. A computable structure $\mathcal{S}$ is relatively Δ_α^0 categorical if and only if it has a formally Σ_α^0 Scott family.*

A computable structure $\mathcal{A}$ is *uniformly relatively Δ^0_α categorical* if, given an X-computable index for a structure $\mathcal{B} \cong \mathcal{A}$, one can effectively find a $\Delta^0_\alpha(X)$-computable index for an isomorphism from $\mathcal{A}$ onto $\mathcal{B}$. We will use the following corollary of Theorem 2.

Corollary 2 ([21, p. 230]). *Suppose that a computable structure $\mathcal{S}$ has a formally Σ^0_α Scott family with no parameters. Then $\mathcal{S}$ is uniformly relatively Δ^0_α categorical.*

Here we give two examples of uniformly relatively Δ^0_α categorical structures. These structures will be used in the proof of Theorem 1.

Proposition 1 ([3, Theorem 8], see also [5, Theorem 17.5]). *Suppose that β is a non-zero computable ordinal. The ordinal ω^β has a formally $\Sigma^0_{2\beta}$ Scott family with no parameters. Thus, the structure ω^β is uniformly relatively $\Delta^0_{2\beta}$ categorical.*

Proposition 2. *Let β be a non-zero computable ordinal. The linear ordering $\mathcal{L} = \omega^\beta \cdot (1 + \eta)$ is a uniformly relatively $\Delta^0_{2\beta+1}$ categorical structure.*

The proof of Proposition 2 appears in Appendix A.

2.3 Pairs of Computable Structures

Suppose that L is a language, $\mathcal{A}$ and $\mathcal{B}$ are L-structures, and α is a countable ordinal. We say that $\mathcal{A} \leq_\alpha \mathcal{B}$ if every infinitary Π_α sentence true in $\mathcal{A}$ is true in $\mathcal{B}$. The relations $\leq_\alpha$ are called *standard back-and-forth relations*.

Let α be a computable ordinal. A family $K = \{\mathcal{A}_i : i \in I\}$ of structures in the language L is called *α-friendly* if the structures $\mathcal{A}_i$ are uniformly computable in $i \in I$, and the relations

$$B_\beta = \left\{ (i, \bar{a}, j, \bar{b}) : i, j \in I, \ \bar{a} \text{ is from } \mathcal{A}_i, \ \bar{b} \text{ is from } \mathcal{A}_j, \ (\mathcal{A}_i, \bar{a}) \leq_\beta (\mathcal{A}_j, \bar{b}) \right\}$$

are computably enumerable uniformly in $\beta < \alpha$.

The proof of Theorem 1 uses the following result on pairs of computable structures.

Theorem 3 ([6, Theorem 3.1]). *Let α be a non-zero computable ordinal. Suppose that $\mathcal{A}$ and $\mathcal{B}$ are computable L-structures such that $\mathcal{B} \leq_\alpha \mathcal{A}$ and the family $\{\mathcal{A}, \mathcal{B}\}$ is α-friendly. Then for any Π^0_α set S, there is a uniformly computable sequence of structures $\{\mathcal{C}_n\}_{n \in \omega}$ such that*

$$\mathcal{C}_n \cong \begin{cases} \mathcal{A}, & \text{if } n \in S, \\ \mathcal{B}, & \text{otherwise.} \end{cases}$$

3 Proof of Theorem 1

Recall that we prove the case (A) of Theorem 1. The proof of the case (B) is essentially the same except that one needs to use the oracle $\emptyset^{(2\alpha+2)}$ in place of $\emptyset^{(2k+1)}$. Assume that k is a non-zero natural number, and a Turing degree $\mathbf{d}$ satisfies the following: $\mathbf{d} \geq \mathbf{0}^{(2k+1)}$, and $\mathbf{d}$ is c.e. in $\mathbf{0}^{(2k+1)}$.

First, we build the computable sequence $\{\mathcal{C}_n\}_{n \in \omega}$ of linear orderings. We give the lemmas which allow us to apply Theorem 3.

Lemma 2 ([6, Proposition 5.4]). *Suppose that α, β, and γ are countable ordinals. Then we have:*

(a) $\omega^\beta \cdot (1+\eta) \leq_\gamma \omega^\alpha$ iff either $(\alpha \leq \beta)\&(\gamma \leq 2\alpha+1)$, or $(\alpha > \beta)\&(\gamma \leq 2\beta+1)$;
(b) $\omega^\alpha \leq_\gamma \omega^\beta \cdot (1+\eta)$ iff either $(\beta < \alpha)\&(\gamma \leq 2\beta+2)$, or $(\beta \geq \alpha)\&(\gamma \leq 2\alpha)$.

Lemma 3 ([6, p. 225]). *Suppose that α and β are computable ordinals, $\beta \neq 0$, and $\kappa = card(\beta)$. There exists an α-friendly family $K = \{\mathcal{A}_i : i \in \omega\} \cup \{\mathcal{B}_j : j < \kappa\}$ with the following properties:*

1. *for an ordinal $\gamma < \omega^\beta$, there is a unique i such that $\mathcal{A}_i \cong \gamma$;*
2. *for an ordinal $\gamma < \beta$, there is a unique j such that $\mathcal{B}_j \cong \omega^\gamma \cdot (1+\eta)$;*
3. *for every $i \in \omega$, there exists an ordinal $\gamma < \omega^\beta$ such that $\mathcal{A}_i \cong \gamma$;*
4. *for every $j < \kappa$, there exists an ordinal $\gamma < \beta$ such that $\mathcal{B}_j \cong \omega^\gamma \cdot (1+\eta)$.*

Lemma 3 implies the following result.

Corollary 3. *Suppose that α, β, and γ are computable ordinals. There exists an α-friendly family $K_\alpha = \{\mathcal{A}, \mathcal{B}\}$ such that $\mathcal{A} \cong \omega^\beta \cdot (1+\eta)$ and $\mathcal{B} \cong \gamma$.*

Assume that S is a Π^0_{2k+2} set such that $S \in \mathbf{d}$. By Corollary 3, there exists a $(2k+2)$-friendly family $K = \{\mathcal{A}, \mathcal{B}\}$ such that $\mathcal{A} \cong \omega^k \cdot (1+\eta)$ and $\mathcal{B} \cong \omega^{k+1}$. By Lemma 2, we have $\omega^{k+1} \leq_{2k+2} \omega^k \cdot (1+\eta)$. Thus, by Theorem 3, there exists a computable sequence $\{\mathcal{C}_n\}_{n \in \omega}$ such that for any $t \in \omega$, we have

$$\mathcal{C}_{4t} \cong \omega^k \cdot (1+\eta), \quad \mathcal{C}_{4t+3} \cong \omega^{k+1},$$

$$\mathcal{C}_{4t+1} \cong \mathcal{C}_{4t+2} \cong \begin{cases} \omega^k \cdot (1+\eta), & \text{if } t \in S, \\ \omega^{k+1}, & \text{if } t \notin S. \end{cases}$$

For a natural number m, the *parity* of m is defined as follows:

$$p(m) = \begin{cases} 0, & \text{if } m \text{ is even,} \\ 1, & \text{if } m \text{ is odd.} \end{cases}$$

For $t, m \in \omega$, set $\mathcal{C}_{[t,m,0]} = \mathcal{C}_{4t+2p(m)}$ and $\mathcal{C}_{[t,m,1]} = \mathcal{C}_{4t+2p(m)+1}$. Fix a computable copy ζ_0 of the ordering ζ with the following properties: ζ_0 has the universe ω, the successor relation $S_0(\zeta_0)$ is a computable set, and $\zeta_0 \models S_0(a,b)$ implies that $p(a) \neq p(b)$.

We define

$$\mathcal{L} = \sum_{t \in \omega} \left(t + 3 + \eta + 1 + \sum_{m \in \zeta_0} (1 + \eta + C_{[t,m,0]} + 2 + \eta + C_{[t,m,1]}) \right).$$

The next two lemmas show that $\mathbf{d}$ is the degree of categoricity for the ordering $\mathcal{L}$.

We define the function $q \colon \zeta_0 \times \mathbb{Z} \to \omega$. For an element $a \in \zeta_0$ and a natural number k, we choose b and c such that $b \leq_{\zeta_0} a \leq_{\zeta_0} c$ and $card([b, a[_{\zeta_0}) = card([a, c[_{\zeta_0}) = k$, and we set $q(a, -k) = b$ and $q(a, k) = c$.

Lemma 4. *Suppose that $\mathcal{S}$ is a computable copy of the structure $\mathcal{L}$. There exists a $\mathbf{d}$-computable isomorphism f from $\mathcal{L}$ onto $\mathcal{S}$.*

Proof. The lemma is similar to Lemma 3.1 in [9]. For the sake of self-completeness, we give a brief sketch of the proof. First, we define auxiliary formulas:

(a) The finitary Π_2 formula $D(x, y)$ says that the interval $[x, y]$ is a dense linear ordering.

$$D(x, y) = (x < y) \ \& \ \forall u \forall v (x \leq u < v \leq y \to \neg S_0(u, v)).$$

(b) The finitary Π_3 formula $D_{\max}(x, y)$ says that $[x, y]$ is a maximal (under inclusion) dense interval.

$$D_{\max}(x, y) = D(x, y) \ \& \ \forall z(z < x \to \neg D(z, y)) \ \& $$
$$\forall w(y < w \to \neg D(x, w)).$$

(c) The finitary Π_3 formula $N(x, y, u, v)$ says that $[x, y]$ and $[u, v]$ are adjacent maximal dense intervals.

$$N(x, y, u, v) = D_{\max}(x, y) \ \& \ (y < u) \ \& \ D_{\max}(u, v) \ \& $$
$$\forall w \forall z (y < w < z < u \to \neg D(w, z)).$$

(d) For $s \in \omega$, the finitary Σ_4 formula $D^s(x, y)$ says that the interval $[x, y]$ is isomorphic to the ordering $(s+1+\eta+1)$, and $[x, y]$ contains a maximal dense interval. Moreover, there is no $z < x$ with the property $[z, y] \cong (s+2+\eta+1)$.

$$D^s(x, y) = \neg \exists z S_0(z, x) \ \& \ \exists z_0 \ldots \exists z_{s+1} [x = z_0 \ \& \ \underset{i \leq s}{\&} S_0(z_i, z_{i+1}) \ \& $$
$$D_{\max}(z_{s+1}, y)].$$

Given a computable copy $\mathcal{S}$ of the ordering $\mathcal{L}$, we describe the construction of the $\mathbf{d}$-computable isomorphism f from $\mathcal{L}$ onto $\mathcal{S}$. Note that $\emptyset^{(3)} \leq_T \emptyset^{(2k+1)} \leq_T \mathcal{S}$. Using the formulas N and D^s, $s \in \omega$, we can effectively in $\emptyset^{(3)}$ find computable indices of the intervals $\mathcal{D}_{t,m,i}$ (where $t, m \in \omega$ and $i \in \{0,1\}$) such that

$$\mathcal{S} = \sum_{t \in \omega} \left(t + 3 + \eta + 1 + \sum_{m \in \zeta_0} (1 + \eta + \mathcal{D}_{t,m,0} + 2 + \eta + \mathcal{D}_{t,m,1}) \right).$$

Note that every interval $\mathcal{D}_{t,m,i}$ is isomorphic either to ω^{k+1}, or to $\omega^k \cdot (1+\eta)$. By Proposition 1, the ordinal ω^{k+1} is uniformly relatively Δ^0_{2k+2} categorical. In addition, by Proposition 2, the structure $\omega^k \cdot (1 + \eta)$ is uniformly relatively Δ^0_{2k+1} categorical. Therefore, now it is sufficient to build a $\mathbf{d}$-computable function $\psi \colon \omega \to \omega$ such that for every $t \in \omega$, we have $\mathcal{C}_{[t,0,0]} \cong \mathcal{D}_{t,\psi(t),0}$ and $\mathcal{C}_{[t,0,1]} \cong \mathcal{D}_{t,\psi(t),1}$.

Indeed, suppose that ψ is such a function. Then, given $t \in \omega$, $z \in \mathbb{Z}$, and $i \in \{0,1\}$, we can effectively in $\mathbf{d}$ find computable indices for the orderings $\mathcal{C}_{[t,q(0,z),i]}$ and $\mathcal{D}_{t,q(\psi(t),z),i}$. It is clear that these orderings are isomorphic. Moreover, since we can use the oracle S, we can determine whether we have $\mathcal{C}_{[t,q(0,z),i]} \cong \omega^{k+1}$, or $\mathcal{C}_{[t,q(0,z),i]} \cong \omega^k \cdot (1 + \eta)$. Hence, uniform relative Δ^0_β categoricity (where $\beta \in \{2k+1, 2k+2\}$) allows us to build (uniformly in t, z, i) a $\mathbf{0}^{(2k+1)}$-computable isomorphism $f_{t,z,i}$ from $\mathcal{C}_{[t,q(0,z),i]}$ onto $\mathcal{D}_{t,q(\psi(t),z),i}$. It is well-known that the ordering η is uniformly relatively computably categorical (see, e.g., Example 3.1 in [13]). Thus, the map $\bigcup_{t,z,i} f_{t,z,i}$ can be extended to a $\mathbf{d}$-computable isomorphism f from $\mathcal{L}$ onto $\mathcal{S}$.

The function ψ is defined as follows. If $t \in S$, then find the least (under the standard ordering of ω) elements a and b such that $\mathcal{S} \models S_k(a,b)$ and $a, b \in \mathcal{D}_{t,m,1}$ for some m. Set $\psi(t) = q(m,1)$. Note that in this case we have $\mathcal{C}_{[t,0,1]} \cong \omega^k \cdot (1+\eta)$ and $\mathcal{D}_{t,m,1} \cong \omega^{k+1}$. Thus, it is easy to note that $\mathcal{C}_{[t,0,0]} \cong \mathcal{C}_{[t,0,1]} \cong \mathcal{D}_{t,\psi(t),0} \cong \mathcal{D}_{t,\psi(t),1} \cong \omega^k \cdot (1 + \eta)$.

If $t \notin S$, then we find the least elements a, b with the property: $\mathcal{S} \models S_k(a,b)$ and $a, b \in \mathcal{D}_{t,m,0}$ for some m. Again, we define $\psi(t) = q(m,1)$. It is not difficult to check that the value $\psi(t)$ satisfies the desired conditions. This concludes the proof of Lemma 4.

Lemma 5. *There exists a computable copy $\mathcal{L}_1$ of the structure $\mathcal{L}$ with the following property: every isomorphism f from $\mathcal{L}$ onto $\mathcal{L}_1$ computes the set S.*

Proof. Using Theorem 3, it is not hard to build a computable sequence $\{\mathcal{D}_n\}_{n \in \omega}$ such that for any $t \in \omega$, we have

$$\mathcal{D}_{4t} \cong \omega^k \cdot (1 + \eta), \quad \mathcal{D}_{4t+1} \cong \omega^{k+1},$$

$$\mathcal{D}_{4t+2} \cong \mathcal{D}_{4t+3} \cong \begin{cases} \omega^k \cdot (1 + \eta), & \text{if } t \in S, \\ \omega^{k+1}, & \text{if } t \notin S. \end{cases}$$

For $t, m \in \omega$, define $\mathcal{D}_{[t,m,0]} = \mathcal{D}_{4t+2p(m)}$ and $\mathcal{D}_{[t,m,1]} = \mathcal{D}_{4t+2p(m)+1}$. We set

$$\mathcal{L}_1 = \sum_{t \in \omega} \left(t + 3 + \eta + 1 + \sum_{m \in \zeta_0} (1 + \eta + \mathcal{D}_{[t,m,0]} + 2 + \eta + \mathcal{D}_{[t,m,1]}) \right).$$

It is straightforward to check that $\mathcal{L}_1$ is a computable copy of the ordering $\mathcal{L}$.

Let a_t be the element in $\mathcal{L}$ such that a_t corresponds to the least element in the ordering $\mathcal{C}_{[t,0,0]}$. Let $b_{t,m}$ be the element in $\mathcal{L}_1$ such that $b_{t,m}$ corresponds to the least element in $\mathcal{D}_{[t,m,0]}$. We may assume that the sequences $\{a_t\}_{t\in\omega}$ and $\{b_{t,m}\}_{t,m\in\omega}$ are computable. For an isomorphism f from $\mathcal{L}$ onto $\mathcal{L}_1$, we have:

a) if $t \in S$, then $f(a_t) = b_{t,2r+1}$ for some r;
b) if $t \notin S$, then $f(a_t) = b_{t,2r}$ for some r.

Therefore, f computes the set S. This concludes the proofs of Lemma 5 and Theorem 1.

4   Corollaries

This section discusses some consequences of Theorem 1. Melnikov [27] constructed the transformation of a linear ordering $\mathcal{L}$ into the linearly ordered abelian group $G(\mathcal{L})$. Using this transformation, it is not hard to prove the following.

Corollary 4. *Suppose that α is an infinite computable ordinal, and $\mathbf{d}$ is a Turing degree c.e. in and above $\mathbf{0}^{(2\alpha+2)}$. Then $\mathbf{d}$ is the degree of categoricity for some computable linearly ordered abelian group.*

Goncharov [22] initiated the systematic study of *spectra of autostability relative to strong constructivizations (SC-autostability spectra)*. We give some basic definitions from this area. Recall that a computable structure $\mathcal{S}$ is *decidable* if, given a first-order formula $\phi(\bar{x})$ and a tuple $\bar{a}$ from $\mathcal{S}$, one can effectively determine whether $\phi(\bar{a})$ is true in $\mathcal{S}$ or not.

Let $\mathbf{d}$ be a Turing degree. A decidable structure $\mathcal{A}$ is *$\mathbf{d}$-SC-autostable* if for every decidable structure $\mathcal{B}$ isomorphic to $\mathcal{A}$, there is a $\mathbf{d}$-computable isomorphism from $\mathcal{A}$ onto $\mathcal{B}$. The *SC-autostability spectrum* of a decidable structure $\mathcal{A}$ is the set

$$\mathrm{SCAutSpec}(\mathcal{A}) = \{\mathbf{d} : \mathcal{A} \text{ is } \mathbf{d}\text{-}SC\text{-autostable}\}.$$

A Turing degree $\mathbf{d}_0$ is the *degree of SC-autostability* for $\mathcal{A}$ if $\mathbf{d}_0$ is the least degree in $\mathrm{SCAutSpec}(\mathcal{A})$.

Informally speaking, the world of decidable structures has its own counterpart to the notion of *computable categoricity*. And this is *SC-autostability*. Theorem 1 and the results from [7] yield the following.

Corollary 5. *Suppose that α is an infinite computable ordinal, and $\mathbf{d}$ is a Turing degree c.e. in and above $\mathbf{0}^{(2\alpha+2)}$. Assume that $\mathcal{L}$ is the linear ordering (from the proof of Theorem 1) such that $\mathbf{d}$ is the degree of categoricity for $\mathcal{L}$. Then the ordering $\zeta \cdot \mathcal{L}$ has a decidable copy, and $\mathbf{d}$ is the degree of SC-autostability for $\zeta \cdot \mathcal{L}$.*

Acknowledgements. The author is grateful to Sergey Goncharov for fruitful discussions on the subject. The reported study was funded by RFBR, according to the research project No. 16-31-60058 mol_a_dk.

A   Appendix: Proof of Proposition 2

For a non-zero countable ordinal α, we define the auxiliary formula $Ord_\alpha(x, y)$. Suppose that the Cantor normal form of α is equal to

$$\omega^{\beta_0} \cdot n_0 + \omega^{\beta_1} \cdot n_1 + \ldots + \omega^{\beta_t} \cdot n_t,$$

where $\beta_0 > \beta_1 > \ldots > \beta_t$ and $0 < n_i < \omega$ for all i. We set:

$$Ord_\alpha(x, y) = \exists z_{0,0} \exists z_{0,1} \ldots \exists z_{0,n_0} \exists z_{1,0} \exists z_{1,1} \ldots \exists z_{1,n_1} \ldots$$
$$\exists z_{t,0} \exists z_{t,1} \ldots \exists z_{t,n_t} \big[x = z_{0,0} \ \& \ \underset{i<n_0}{\&} S_{\beta_0}(z_{0,i}, z_{0,i+1}) \ \& $$
$$z_{0,n_0} = z_{1,0} \ \& \ \underset{i<n_1}{\&} S_{\beta_1}(z_{1,i}, z_{1,i+1}) \ \& \ \ldots \ \& $$
$$z_{t-1,n_{t-1}} = z_{t,0} \ \& \ \underset{i<n_t}{\&} S_{\beta_t}(z_{t,i}, z_{t,i+1}) \ \& \ z_{t,n_t} = y \big].$$

It is not difficult to prove the following claim.

Lemma 6. *Assume that β is a computable ordinal, and $\omega^\beta < \alpha < \omega^{\beta+1}$. For a well-ordering $\mathcal{A}$ and elements $a, b \in \mathcal{A}$, we have $\mathcal{A} \models Ord_\alpha(a, b)$ iff the interval $[a, b[_\mathcal{A}$ is isomorphic to α. Moreover, the formula Ord_α is logically equivalent to a computable $\Sigma_{2\beta+2}$ formula.*

Now suppose that $\mathcal{L}_0$ is a computable copy of the ordinal ω^β with the following property: given a pair of elements $a <_{\mathcal{L}_0} b$, one can effectively find the Cantor normal form of the interval $[a, b[_{\mathcal{L}_0}$. We may assume that $\mathcal{L} = \mathcal{L}_0 \cdot (1+\eta)$.

We describe the formally $\Sigma^0_{2\beta+1}$ Scott family Φ for the structure $\mathcal{L}$. Let

$$\psi(x) = \forall y[y < x \rightarrow \neg F_\beta(y, x)].$$

It is easy to show that ψ is equivalent to a computable $\Pi_{2\beta}$ formula.

First, we define the Scott formula ϕ_a for an element $a \in \mathcal{L}$. If a is the least element in $\mathcal{L}$, then set $\phi_a(x) = \forall y(x \le y)$. If a is not the least element and $\mathcal{L} \models \psi(a)$, then define $\phi_a = \psi$. Now assume that $\mathcal{L} \not\models \psi(a)$. We find the element b such that $\mathcal{L} \models \psi(b) \ \& \ F_\beta(b, a)$. Let γ be the ordinal such that the interval $[b, a[_\mathcal{L}$ is isomorphic to γ. Set

$$\phi_a(x) = \exists y(y < x \ \& \ \phi_b(y) \ \& \ Ord_\gamma(y, x)).$$

Let $\bar{a} = a_0, a_1, \ldots, a_n$ be a tuple from $\mathcal{L}$ such that $a_0 <_\mathcal{L} a_1 <_\mathcal{L} \ldots <_\mathcal{L} a_n$. For $i < n$, we set

$$\psi^i(x, y) = (x < y) \ \&$$
$$\& \begin{cases} \neg F_\beta(x, y), & \text{if } \mathcal{L} \not\models F_\beta(a_i, a_{i+1}), \\ Ord_\gamma(x, y), & \text{if } \mathcal{L} \models F_\beta(a_i, a_{i+1}) \text{ and} \\ & [a_i, a_{i+1}[_\mathcal{L} \cong \gamma; \end{cases}$$
$$\phi_{\bar{a}}(x_0, x_1 \ldots, x_n) = \underset{i \le n}{\&} \phi_{a_i}(x_i) \ \& \ \underset{i<n}{\&} \psi^i(x_i, x_{i+1}).$$

It is straightforward to prove that $\mathcal{L} \models \phi_{\bar{a}}(\bar{a})$. Moreover, for a tuple $\bar{b}$ from $\mathcal{L}$, $\mathcal{L} \models \phi_{\bar{a}}(\bar{b})$ implies that the tuple $\bar{b}$ is automorphic to $\bar{a}$. Furthermore, the formula $\phi_{\bar{a}}$ is logically equivalent to a computable $\Sigma_{2\beta+1}$ formula. We use the effective procedure for calculating Cantor normal forms to construct the formulas $\phi_{\bar{a}}$ and to build the desired c.e. Scott family Φ consisting of computable $\Sigma_{2\beta+1}$ formulas. Note that the formulas have no parameters. Hence, by Corollary 2, the structure $\mathcal{L}$ is uniformly relatively $\Delta^0_{2\beta+1}$ categorical.

References

1. Anderson, B.A., Csima, B.F.: Degrees that are not degrees of categoricity. Notre Dame J. Formal Logic. Advance Publication. doi: 10.1215/00294527-3496154
2. Ash, C., Knight, J., Manasse, M., Slaman, T.: Generic copies of countable structures. Ann. Pure Appl. Logic **42**(3), 195–205 (1989)
3. Ash, C.J.: Recursive labelling systems and stability of recursive structures in hyperarithmetical degrees. Trans. Am. Math. Soc. **298**, 497–514 (1986)
4. Ash, C.J.: Stability of recursive structures in arithmetical degrees. Ann. Pure Appl. Logic **32**, 113–135 (1986)
5. Ash, C.J., Knight, J.F.: Computable Structures and the Hyperarithmetical Hierarchy. Studies in Logic and the Foundations of Mathematics, vol. 144. Elsevier Science B.V, Amsterdam (2000)
6. Ash, C.J., Knight, J.F.: Pairs of recursive structures. Ann. Pure Appl. Logic **46**(3), 211–234 (1990)
7. Bazhenov, N.: Autostability spectra for decidable structures. Math. Struct. Comput. Sci. (accepted)
8. Bazhenov, N.A.: Autostability spectra for Boolean algebras. Algebra Logic **53**(6), 502–505 (2015)
9. Bazhenov, N.A.: Degrees of autostability for linear orderings and linearly ordered abelian groups. Algebra Logic (accepted)
10. Chisholm, J.: Effective model theory vs. recursive model theory. J. Symbolic Logic **55**(3), 1168–1191 (1990)
11. Chisholm, J., Fokina, E.B., Goncharov, S.S., Harizanov, V.S., Knight, J.F., Quinn, S.: Intrinsic bounds on complexity and definability at limit levels. J. Symbolic Logic **74**(3), 1047–1060 (2009)
12. Csima, B.F., Franklin, J.N.Y., Shore, R.A.: Degrees of categoricity and the hyperarithmetic hierarchy. Notre Dame J. Formal Logic **54**(2), 215–231 (2013)
13. Downey, R.G.: Computability theory and linear orderings. In: Ershov, Y., Goncharov, S.S., Nerode, A., Remmel, J.B. (eds.) Handbook of Recursive Mathematics, vol. 2, pp. 823–976. Elsevier Science B.V., Amsterdam (1998). Stud. Logic Found. Math., vol. 139
14. Downey, R.G., Kach, A.M., Lempp, S., Lewis-Pye, A.E.M., Montalbán, A., Turetsky, D.D.: The complexity of computable categoricity. Adv. Math. **268**, 423–466 (2015)
15. Ershov, Y.L., Goncharov, S.S.: Constructive Models. Kluwer Academic/Plenum Publishers, New York (2000)
16. Fokina, E., Frolov, A., Kalimullin, I.: Categoricity spectra for rigid structures. Notre Dame J. Formal Logic **57**(1), 45–57 (2016)

17. Fokina, E.B., Harizanov, V., Melnikov, A.: Computable model theory. In: Downey, R. (ed.) Turing's Legacy: Developments from Turing Ideas in Logic. Lecture Notes in Logic, vol. 42, pp. 124–194. Cambridge University Press, Cambridge (2014)
18. Fokina, E.B., Kalimullin, I., Miller, R.: Degrees of categoricity of computable structures. Arch. Math. Logic **49**(1), 51–67 (2010)
19. Fröhlich, A., Shepherdson, J.C.: Effective procedures in field theory. Philos. Trans. Roy. Soc. London Ser. A **248**, 407–432 (1956)
20. Frolov, A.N.: Effective categoricity of computable linear orderings. Algebra Logic **54**(5), 415–417 (2015)
21. Goncharov, S., Harizanov, V., Knight, J., McCoy, C., Miller, R., Solomon, R.: Enumerations in computable structure theory. Ann. Pure Appl. Logic **136**(3), 219–246 (2005)
22. Goncharov, S.S.: Degrees of autostability relative to strong constructivizations. Proc. Steklov Inst. Math. **274**, 105–115 (2011)
23. Goncharov, S.S.: The quantity of nonautoequivalent constructivizations. Algebra Logic **16**(3), 169–185 (1977)
24. Goncharov, S.S., Dzgoev, V.D.: Autostability of models. Algebra Logic **19**(1), 28–37 (1980)
25. Mal'tsev, A.I.: Constructive algebras I. Russ. Math. Surv. **16**, 77–129 (1961)
26. Mal'tsev, A.I.: On recursive abelian groups. Sov. Math. Dokl. **32**, 1431–1434 (1962)
27. Melnikov, A.G.: Computable ordered abelian groups and fields. In: Ferreira, F., Löwe, B., Mayordomo, E., Mendes Gomes, L. (eds.) CiE 2010. LNCS, vol. 6158, pp. 321–330. Springer, Heidelberg (2010). doi:10.1007/978-3-642-13962-8_36
28. Miller, R.: **d**-computable categoricity for algebraic fields. J. Symb. Log. **74**(4), 1325–1351 (2009)
29. Remmel, J.B.: Recursively categorical linear orderings. Proc. Am. Math. Soc. **83**, 387–391 (1981)
30. Rosenstein, J.G.: Linear Orderings, vol. 98. Academic Press, New York (1982). Pure Appl. Math

On the Shortest Common Superstring
of NGS Reads

Tristan Braquelaire[1]($\boxtimes$), Marie Gasparoux[1,3], Mathieu Raffinot[1,2],
and Raluca Uricaru[1]

[1] LaBRI and CBiB, University of Bordeaux, Bordeaux, France
{`tristan.braquelaire,marie.gasparoux,mathieu.raffinot,`
`raluca.uricaru`}`@labri.fr`
[2] CNRS, LaBRI and CBiB, University of Bordeaux, Bordeaux, France
[3] DIRO, University of Montréal, Montréal, QC, Canada

Abstract. The Shortest Superstring Problem (SSP) consists, for a set of strings $S = \{s_1, \cdots, s_n\}$ (with no s_i substring of s_j), to find a minimum length string that contains all $s_i, 1 \leq i \leq n$, as substrings.

This problem is proved to be NP-Complete and APX-hard. Guaranteed approximation algorithms have been proposed, the current best ratio being $2\frac{11}{30}$, which has been achieved through a long and difficult process. SSP is highly used in practice on Next Generation Sequencing (NGS) data, which plays an increasingly important role in modern biological and medical research. In this note, we show that on NGS data the SSP approximation ratio reached by the classical algorithm of Blum *et al.* [2], is usually below $2\frac{11}{30}$, while assuming specific characteristics of the data that are experimentally verified on a large sampling set. Moreover, we present an efficient linear time test for any input of strings of equal length, which allows to compute the approximation ratio that can be reached using the classical algorithm in [2].

Keywords: Shortest Superstring Problem · Approximation algorithms · Next Generation Sequencing

1 Introduction

The Shortest Superstring Problem (SSP) consists, for a set of strings $S = \{s_1, \cdots, s_n\}$ (with no s_i substring of s_j), in constructing a string s such that any element of S is a substring of s and s is of minimal length. For an arbitrary number of strings n, the problem is known to be NP-Complete [10,11] and APX-hard [2]. Lower bounds for the achievable approximation ratios on a binary alphabet have been given [16,20], and the best approximation ratio so far for the general case is $2\frac{11}{30} \approx 2.3667$ [21], reached after a long series of improvements [1–4,7,15,17–19,22,23,25] leading to increasingly involved algorithms. An SSP greedy algorithm is known to reach good performances in practice but its guaranteed approximation ratio has only been proved to be 3.5 [15] and conjectured 2.

This work was supported by the PEPS INS2I-CNRS project *CompX* and by a *Genotype to Phenotype* project of the Life Sciences Department of University of Bordeaux.

© Springer International Publishing AG 2017
T.V. Gopal et al. (Eds.): TAMC 2017, LNCS 10185, pp. 97–111, 2017.
DOI: 10.1007/978-3-319-55911-7_8

Also, the SSP problem was tackled from the perspective of compression and several compression SSP algorithms have been designed. The idea is to ensure a fixed compression ratio between the sum of the lengths of the strings in the set and the optimal superstring on this set. In this context, the above mentioned greedy algorithm is proved to achieve a compression ratio of at least $\frac{1}{2}$, while the best compression algorithm achieves a ratio of $\frac{38}{63}$ [17].

In this article, we focus on guaranteed algorithms for practical applications of SSP, like assembling reads produced by Next Generation sequencers. Over the past decade, the landscape of sequencing and assembly deeply changed, with the increasing development of Next Generation Sequencing (NGS) devices. These relatively cheap devices produce in a single *run*, millions of randomly read, relatively short, equal length DNA sequences. Such sequences are named *reads* and each read is typically 32 to 1000 bases long, with a low and still decreasing cost per base.

Considering the specificity of read sequences (*i.e.,* equal length, specific period distribution), the question that arises is whether it is possible to propose better approximation algorithms for this type of data. Note that our results can easily be extended to the case where the lengths of the input strings are not necessarily equal but very close, by considering the maximum length. However, for simplicity, we prefer to restrain our presentation to equal length strings.

This research, similar to the one targeting better algorithms for small-world graphs in social networks, aims to better suit the actual data. The greedy SSP algorithm was already studied in the general case of DNA sequences (not necessarily of equal length) in [8] and proved to be efficient.

In this article, we refine the approximation bound of the classical (and one of the simplest) SSP algorithm in [2] taking into account (a) the strings being of equal length and (b) the period distribution of those strings. Then, given any set of equal length strings, we propose a linear-time algorithm to compute the approximation ratio the classical algorithm can reach. If the period distribution fits a specific model (see Sect. 4), this ratio can be better than the best actual ratio of $2\frac{11}{30}$.

Our experiments show that on real NGS datasets, the ratios we obtain are most of the time better than $2\frac{11}{30}$, allowing us to use a better guaranteed and simpler algorithm for its assembling. For instance, on the NGS dataset SRR069579, we reach a 2.0738 approximation ratio (see Table 1). In the appendix we present the result of the ratio computation on 100 sets of reads.

To our knowledge, the only related work where sequences have the same length is [12]. Up to 7 bases, they propose a better approximation ratio, with a De Bruijn graph approach. However, these sequences are much shorter than real-world reads.

Note that some theoretical variations of SSP have also been studied [5,27]. Here we neither dwell on these studies since their focus is far from ours, nor detail the greedy algorithm approximation conjecture, which is a subject in itself [9,15,24].

2 The Classical 3-Approximation Algorithm for SSP

We consider the Shortest Superstring Problem (SSP) on n strings over a finite alphabet Σ, in a set $S = \{s_1, s_2, \ldots, s_n\}$, with no s_i substring of s_j.

For two strings u, v we define the *maximum overlap* of u and v, denoted $ov(u, v)$, as the longest suffix of u that is also a prefix of v. Also, we define the *prefix* of u relatively to v, denoted $\operatorname{pref}(u, v)$, as the string x such that $u = x \cdot ov(u, v)$, *i.e.*, the maximum prefix of u that does not overlap v.

The *prefix graph* (also called the *distance graph*) built on S is a complete directed graph with the vertex set $V = S$ and the edges set $E = \{e_{i,j} = (s_i, s_j) | \forall s_i, s_j \in V\}$, with label $l(e_{i,j}) = \operatorname{pref}(s_i, s_j)$ and weight $w(e_{i,j}) = |\operatorname{pref}(s_i, s_j)|$. Figure 1 shows the prefix graph built on the set of strings $S = \{ACGACG, CGACGA, ATGTAG, TAGATG, GACGAT, ATGATG\}$. For the sake of simplicity, the prefix graph depicted in Fig. 1 is not complete; indeed, edges corresponding to pairs of non overlapping strings are not represented, as they do not impact our example.

Let c be a cycle in the prefix graph and let $w(c)$ denote its weight, that is the sum of the weights of its edges. Now let r be a string corresponding to one of the vertices in this cycle. Then r can be *expressed* by turning around the cycle a certain number of times and concatenating the labels of the edges.

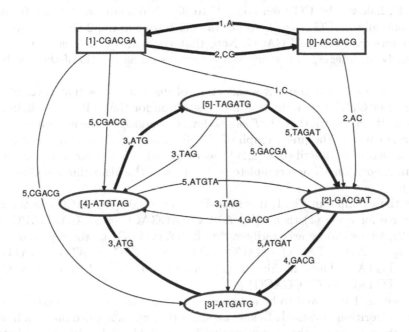

Fig. 1. The prefix graph of $S = \{ACGACG, CGACGA, ATGTAG, TAGATG, GACGAT, ATGATG\}$. Vertices in the graph are numbered from [0] to [5]. We highlight a cycle cover made of two cycles, whose edges are in bold: (1) [0] - [1] and (2) [2] - [3] - [4] - [5].

Algorithm 1. Shortest superstring - factor 3

Data: $S = \{s_1, s_2, \ldots, s_n\}$ a set of n strings

Result: τ a superstring of the strings in S

1 Compute a minimum weight cycle cover $\mathcal{C} = c_1 \ldots c_{nc}$ of the prefix graph

   built on S;

2 For each cycle $c_i = s_{i_1}, \ldots s_{i_l}$ in $\mathcal{C}$, arbitrarily choose one of the strings in

   c_i (let us say s_{i_1}) as the representative string r_i for c_i. Let

   $\sigma_i = \mathrm{pref}(r_i, s_{i_2}) \cdot \ldots \cdot \mathrm{pref}(s_{i_{l-1}}, s_{i_l}) \cdot \mathrm{pref}(s_{i_l}, r_i) \cdot r_i$; Let $S_\sigma = \{\sigma_i\}$ and

   $\|S_\sigma\| = \sum |\sigma_i|$.

3 Compress S_σ using an SSP compression algorithm, like the greedy

   algorithm, and output the result τ.

For instance, on cycle (1) in Fig. 1, ACGACG can be expressed as A from node [0] to [1], followed by CG from node [1] to [0], then again by A from [0] to [1] and eventually by CG from [1] to [0]. In this case, turning 2 times around the cycle allows expressing ACGACG. Note that the number of cycle tours is not necessarily an integer, as a string can be expressed as a ratio of the weight of the cycle.

A *cycle cover* of a graph is a collection of disjoint cycles that cover all the vertices. Let OPT denote the length of a solution for the SSP on the strings in the set S. It was shown that OPT can be lower-bounded by the minimum weight of a cycle cover of the prefix graph of S [26]. Based on these considerations, a classical algorithm described in [2,26] was proposed, whose general framework is given in Algorithm 1 (for a complete description of the algorithm see Chapter 7 in [26]).

For the example in Fig. 1, if we consider vertex [0] (ACGACG) as the representative for cycle (1) then $\sigma_1 = $ A $\cdot$ CG $\cdot$ ACGACG $=$ ACGACGACG. For cycle (2), let us choose arbitrarily vertex [2] (ATGTAG) as the representative. Then $\sigma_2 = $ ATG $\cdot$ TAGAT $\cdot$ GACG $\cdot$ ATG $\cdot$ ATGTAG $=$ ATGTAGATGAC GATGATGTAG. Thus, in Algorithm 1 we obtain $S_\sigma = \{$ATGTAGATGAC GATGATGTAG, ACGACGACG$\}$.

Algorithm 1 is proved to be a 3-approximation when using the greedy compression algorithm of ratio $\frac{1}{2}$. Below, we show that when applied on data having specific characteristics, the approximation factor of Algorithm 1 is in fact better.

3   Algorithm 1 on a Particular Instance of the SSP

In this section we consider a particular instance of the Shortest Superstring Problem, namely when the n strings in S have the same length m with $0 < m \ll n$. We begin this section with several notations, lemmas and corollaries, followed by an analysis of Algorithm 1 on this particular instance of the SSP.

For a string s of length m, an integer $1 \leq p \leq m$ is a *period* of s if $s[i] = s[i+p]$ for all $1 \leq i \leq m - p$. Note that s has at least one period that is its length. The smallest period of s is called *period* of s, and denoted $period(s)$. Let $n(i)$ $(1 \leq i \leq m)$ denote the number of strings of period i.

We denote the *period* of a cycle as being equal to its weight. Note that the period of a cycle is at least 2, as a cycle is composed of at least two vertices thus of at least two edges, and as the weight of an edge is at least 1 (otherwise a string would be a substring of another thus contradicting the definition of SSP).

Lemma 1. *Let $c \in \mathcal{C}$ be a cycle and $s_1 \dots s_k$ the strings in c, then $period(c) \geq \max_{i=1}^{k}\{period(s_i)\}$.*

Proof. Each string in the cycle can be expressed by turning around the cycle. If $period(c) < period(s)$ for a given $s \in \{s_1 \dots s_k\}$, then $period(c)$ is also a period of s, which is smaller than the smallest period of s. Contradiction. Thus $period(c) \geq period(s_i), 1 \leq i \leq k$.

Corollary 1. *Let $1 \leq i \leq m$, the maximal number of cycles of period less or equal to i is bounded by $\frac{1}{2}\sum_{k=1}^{i} n(k)$.*

Proof. By Lemma 1, strings in a cycle c of period i must have a period less than or equal to i. There are $\sum_{k=1}^{i} n(k)$ strings that can compose these cycles. Given that a cycle contains at least two strings, there are at maximum $\frac{1}{2}\sum_{k=1}^{i} n(k)$ cycles of period $\leq i$.

A cycle cover is composed of cycles of different periods. For example, in Fig. 1, the period of cycle (1) is 3, while that of cycle (2) is 16. The period of a cycle determines the number of turns needed in order to express a string on that cycle. Indeed, a string s that is part of a cycle c is expressed by turning around the cycle $\frac{m}{period(c)}$ times. Note that given that all strings have the same length m, the smaller the period of a cycle, the more we need to turn around the cycle in order to express a string (see Fig. 2).

We split the set of cycles in two subsets with respect to their periods: *small* cycles with periods less than or equal to $m\alpha$, and *large* cycles with periods greater than $m\alpha$, where $0 < \alpha \leq 1$ is a parameter that will be discussed later in the paper. For instance for $\alpha = 0.8$, it is straightforward to see that in Fig. 2, cycle (1) is *small* while cycle (2) is *large*.

Corollary 2. *Let $c \in \mathcal{C}$ be a cycle and $s_1 \dots s_k$ the strings in $\mathcal{C}$. If $period(c) \leq m\alpha$, $period(s_i) \leq m\alpha$, where $1 \leq i \leq k$.*

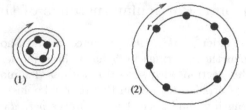

Fig. 2. Expressing a representative r on a small cycle (having a period inferior to $m\alpha$) in (1) versus on a large cycle (with a period superior to $m\alpha$) in (2), where $\alpha = 0.8$.

Proof. By Lemma 1, the periods of the strings in a cycle are smaller than or equal to the period of the cycle.

Based on this corollary, the partition of the set of cycles with respect to $m\alpha$ generates a partition of the set of reads: reads with periods less than or equal to $m\alpha$ that can be part of the *small* cycles, and reads with periods comprised between $m\alpha$ and m that can only be part of the *large* cycles.

Below we will show that $||S_\sigma||$ (see Step 2, Algorithm 1) for equal length strings, approaches OPT when having a maximum number of cycles with large periods.

Lemma 2

$$||S_\sigma|| \leq wt(\mathcal{C}) + wt(\mathcal{C})\frac{1}{\alpha} + \frac{1}{2}\sum_{k=1}^{m\alpha} n(k)(m - \frac{k}{\alpha}) \leq (1 + \frac{1}{\alpha})OPT + \frac{1}{2}\sum_{k=1}^{m\alpha} n(k)(m - \frac{k}{\alpha})$$

Proof. It is straightforward to see that S_σ has the length equal to $\sum_{i=1}^{nc} |\sigma_i|$. A σ_i on a cycle c_i is given by $\text{pref}(r_i, s_{i_2}) \cdot \ldots \cdot \text{pref}(s_{i_{l-1}}, s_{i_l}) \cdot \text{pref}(s_{i_l}, r_i) \cdot r_i$. As $|\text{pref}(r_i, s_{i_2}) \cdot \ldots \cdot \text{pref}(s_{i_{l-1}}, s_{i_l}) \cdot \text{pref}(s_{i_l}, r_i)| = wt(c_i)$, we first sum over these prefixes of each σ_i. This leads to the first global $wt(\mathcal{C})$ in Lemma 2.

We now have to consider $\sum_{i=1}^{nc} |r_i|$. As all r_i have the same length m, this gives $\sum_{i=1}^{nc} |r_i| = m \cdot nc$. However, this is not entirely useful given that it is not possible to finely bound the number of cycles in the cycle cover, nc. Therefore we will have to pursue a different approach, that is further decompose this result on large and small period cycles and exploit the fact that expressing an r_i on a cycle c_i requires $\frac{m}{period(c_i)}$ cycle tours.

We consider the total length of the r_i for the large cycles (with periods $> m\alpha$). From the observation above we get that the expression of such an r_i requires at maximum $\frac{1}{\alpha}$ cycle tours and its length can be formulated as $\frac{1}{\alpha} \cdot wt(c_i)$. Thus their total length is bounded by $\frac{1}{\alpha} \cdot wt(\mathcal{C})$. Note that $wt(\mathcal{C})$ is the sum of the weights of all cycles and not only of the large ones.

We now have to address the total length of the r_i for the small cycles (with periods $\leq m\alpha$). As, by Corollary 1, there are at most $\frac{1}{2}\sum_{k=1}^{m\alpha} n(k)$ such cycles, the sum of the lengths of the corresponding r_i is upper bounded by $\frac{m}{2}\sum_{k=1}^{m\alpha} n(k)$.

This can be further refined given that $\frac{1}{\alpha}$ cycle tours for the small cycles are already taken into account in the computation for the large cycles, $i.e.$, $\frac{k}{\alpha}$ from this upper bound. Note that the worst case for counting the small cycles of period k from 2 to $m\alpha$ is when there is a maximum of small cycles at each step k. By Corollary 1, the number of cycles from step $k-1$ to step k can only be increased by $\frac{n(k)}{2}$. Expressing the representatives of these $\frac{n(k)}{2}$ additional cycles of period k requires $\frac{n(k)}{2}(\frac{m}{k} - \frac{1}{\alpha})$ tours. The number of tours has to be multiplied by k in order to obtain their total length. Finally, we get an upper bound for the sum of the lengths of the r_i for the small cycles of $\frac{1}{2}\sum_{k=1}^{m\alpha} n(k)(m - \frac{k}{\alpha})$. Eventually, as $wt(\mathcal{C}) \le OPT$, the result follows.

Instead of using the greedy algorithm in the last step of Algorithm 1, we $compress$ S_σ using the guaranteed $\frac{38}{63}$ $compression$ algorithm [17], similarly to the classical approaches related to the superstring approximation. We define OPT_σ as the length of an optimal superstring on S_σ and τ as the result of the $\frac{38}{63}$ compression algorithm on S_σ. The next lemma (Lemma 7.7 in [26]) links OPT_σ and OPT.

Lemma 3. $OPT_\sigma \le OPT + wt(\mathcal{C})$

By applying the $\frac{38}{63}$ compression algorithm on S_σ, we thus derive the following result:

Lemma 4

$$|\tau| \le 2OPT + \frac{25}{63}\left(\frac{1-\alpha}{\alpha}\right) OPT + \frac{25}{126}\sum_{k=1}^{m\alpha} n(k)(m - \frac{k}{\alpha})$$

$Proof.$ Let us denote $\Delta = \left(\frac{1-\alpha}{\alpha}\right) OPT + \frac{1}{2}\sum_{k=1}^{m\alpha} n(k)(m-\frac{k}{\alpha})$ (see Fig. 3). Lemma 2 states that $||S_\sigma|| \le 2OPT + \Delta$.

Lemma 3 gives $OPT_\sigma < OPT + wt(\mathcal{C}) \le 2OPT$. As $||S_\sigma|| - |\tau| \ge \frac{38}{63}(||S_\sigma|| - OPT_\sigma)$, the following inequality also holds: $||S_\sigma|| - |\tau| \ge \frac{38}{63}(||S_\sigma|| - 2OPT)$.

Let us define $|\tau_0|$ as the value of $|\tau|$ obtained in the worst case, when $||S_\sigma|| - 2OPT = \Delta$. Then $|\tau_0| \le 2OPT + \frac{25}{63}\Delta$. Assume now that $||S_\sigma||$ is less than the worst case, that is $||S_\sigma|| - 2OPT < \Delta$. Then the $|\tau|$ obtained after compressing $||S_\sigma||$ is less than $|\tau_0|$ (Fig. 3). Thus $|\tau| \le |\tau_0| \le 2OPT + \frac{25}{63}\Delta = 2OPT + \frac{25}{63}\left(\frac{1-\alpha}{\alpha}\right) OPT + \frac{25}{126}\sum_{k=1}^{m\alpha} n(k)(m - \frac{k}{\alpha})$.

An important point is that $OPT \ge n$, since any superstring contains at least one character from each string, as no string is a substring of another. Combined to Lemma 4, this leads to:

Theorem 1

$$\frac{|\tau|}{OPT} \le 2 + \frac{25}{63}\left(\frac{1-\alpha}{\alpha}\right) + \frac{\frac{25}{126}\sum_{k=1}^{m\alpha} n(k)(m - \frac{k}{\alpha})}{OPT} \le 2 + \frac{25}{63}\left(\frac{1-\alpha}{\alpha}\right) + \frac{25}{126n}\sum_{k=1}^{m\alpha} n(k)(m-\frac{k}{\alpha})$$

$Proof.$ The approximation ratio of $|\tau|$ with respect to OPT is computed by dividing $|\tau|$ by OPT. As $OPT \ge n$, $\frac{\frac{25}{126}\sum_{k=1}^{m\alpha} n(k)(m-\frac{k}{\alpha})}{OPT} \le \frac{25}{126n}\sum_{k=1}^{m\alpha} n(k)(m - \frac{k}{\alpha})$.

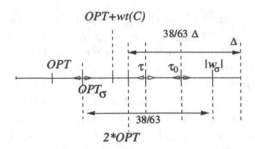

Fig. 3. Compressing S_σ using the $\frac{38}{63}$ algorithm [17].

4   The Shortest Superstring Problem on NGS Reads

In the previous section we have shown that the classical SSP algorithm (Algorithm 1), when applied on strings of equal length, might give a better approximation factor than in the general case, depending on the period distribution of the sequences.

This approximation factor is parametrized by a parameter α, $0 < \alpha \leq 1$, which is inferred from the set of strings given as input. This the novely of our approach. The approximation factor is *ad-hoc*, computed on the data trough a linear time procedure.

The parameter α allows to partition the set of strings in two subsets: strings with periods $\leq m\alpha$ and the others with periods comprised between $m\alpha$ and m, and thanks to Corollary 2 it also roughly partitions the set of cycles (see Sect. 3). Thus the choice of α determines the number of *small* and *large* cycles and as we have seen above (see Lemma 2), the lower the number of *small* cycles with respect to the number of *large* cycles, the better the approximation factor. In this section we describe a procedure that allows to compute the optimal value for the parameter α on a given dataset, and apply this procedure on real NGS datasets composed of equal length reads. We observe that for a large panel of these NGS datasets the optimal value of α is relatively high (between 0.8 and 1), which means that the number of small cycles becomes "negligible" with respect with the number of reads when the period of the reads increases.

As our approach is data driven, we performed extensive tests (see Appendix, Table 5) on 100 sets of reads of differents species, and we took several parameters into account. For instance, in Fig. 4, we show such four sets of reads with lengths of 32, 36, 98 and 200. The x-axis represents the periods, and the y-axis the number of reads on a log10 scale. We plotted three types of curves:

- $n(x)$ (red circles) of the corresponding set of reads;
- random values of $n(x)$ (oblique dotted curve) experimentally generated for the total number of sequences of the set of reads;
- parameters $m\alpha$ (vertical dotted line) computed on the set of reads.

Compared to random periods, this value of $m\alpha$ roughly corresponds to the area where the curves $n(x)$ join the random values. It would be a natural

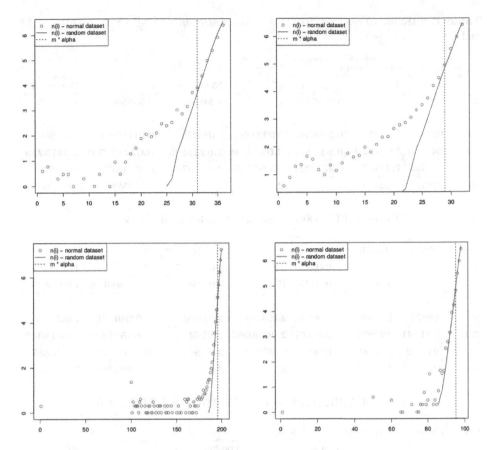

Fig. 4. Four sets of reads from left to right, top to bottom: SRR069579 (human), ERR000009 (yeast), SRR211279 (human), SRR959239 (human). The x-axis is the period and the y-axis is the number of reads on a $\log_{10}$ scale. The circles represent $n(x)$ and the curve represents a distribution of periods on a random dataset. The dash vertical line corresponds to the final $m\alpha$ (computed in Sect. 5). (Color figure online)

approach to fit our experimental curves to the theoretical random distributions under some probabilistic source assumptions (Bernoulli or Markov for instance). However, obtaining these theoretical distributions is an open study [13,14], which is out of the scope of this article.

For a set of equal length strings, the computation of the value of α consists in (a) computing all minimal periods of all input sequences and (b) computing $n(i)$ for all $1 \leq i \leq m$ and (c) computing for each $1 \leq i \leq m$ the parameters α, $\gamma = \frac{25}{126n} \sum_{k=1}^{m\alpha} n(k)(m - \frac{k}{\alpha})$ and $\beta = 2 + \frac{25}{63} \left(\frac{1-\alpha}{\alpha} \right) + \gamma$. Eventually, we keep the lower β as the approximation ratio if it is below $2\frac{11}{30} \approx 2.3667$. Given a string of size m on a fixed alphabet Σ, the computation of its smallest period can be done easily in expected linear time $O(m)$, in worst case $O(m)$ or even in optimal sublinear expected time $O(\sqrt{m.\log_{|\Sigma|} m})$ using a more involved algorithm [6].

Table 1. SRR069579 read set, 3702308 reads of size 36, $\gamma = \frac{25}{126n} \sum_{k=1}^{m\alpha} n(k)(m - \frac{k}{\alpha})$ and $\beta = 2 + \frac{25}{63}\left(\frac{1-\alpha}{\alpha}\right) + \gamma$.

Period	nbseq	cum. nbseq	α	$1 + \frac{1}{\alpha}$	$2 + \frac{25}{63}\left(\frac{1-\alpha}{\alpha}\right)$	γ	β
1	4	4	0.0277778	37	15.888889	0	15.888889
2	6	10	0.0555556	19	8.746032	3.8585968E-6	8.746036
...	...	...	...	...	...	...	...
33	98795	140121	0.9166667	2.090909	2.036075	0.011829347	2.0479045
34	**247451**	**387572**	**0.9444444**	**2.0588236**	**2.0233426**	**0.024179146**	**2.0475218**
35	829535	1217107	0.9722222	2.0285716	2.0113378	0.05718735	2.068525
36	2485201	3702308	1	2	2	0.15358844	2.1535885

Table 2. ERR000009 read set, 4049489 reads of size 32

Period	nbseq	cum. nbseq	α	$1 + \frac{1}{\alpha}$	$2 + \frac{25}{63}\left(\frac{1-\alpha}{\alpha}\right)$	γ	β
1	4	4	0.03125	33	14.301587	0	14.301587
2	101	6895	0.0625	17	7.952381	3.135806E-6	7.9523845
...	...	...	...	...	...	...	...
29	88979	149801	0.90625	2.1034484	2.041051	0.021079484	2.0621305
30	**341091**	**490892**	**0.9375**	**2.0666666**	**2.026455**	**0.03537932**	**2.0618343**
31	953109	1444001	0.96875	2.032258	2.012801	0.073584706	2.0863857
32	2605488	4049489	1	2	2	0.18015388	2.1801538

Table 3. SRR211279 read set, 25103766 reads of size 200

Period	nbseq	cum. nbseq	α	$1 + \frac{1}{\alpha}$	$2 + \frac{25}{63}\left(\frac{1-\alpha}{\alpha}\right)$	γ	β
1	2	2	0.005	201	80.968254	0	80.968254
100	23	25	0.5	3	2.3968253	3.1298662E-6	2.3968284
...	...	...	...	...	...	...	...
195	38013	54574	0.975	2.025641	2.010175	0.002138306	2.0123134
196	**134284**	**188858**	**0.98**	**2.0204082**	**2.0080984**	**0.0028290136**	**2.0109274**
197	473686	662544	0.985	2.0152283	2.006043	0.0050281105	2.011071
198	1685038	2347582	0.99	2.0101008	2.0040083	0.012494448	2.0165029
199	5811666	8159248	0.995	2.0050251	2.0019941	0.038533505	2.0405276
200	16944518	25103766	1	2	2	0.12880057	2.1288006

Performing n such computations and a m-buckets sorting allows to solve the steps (a) and (b) in $O(n\sqrt{m} . \log_{|\Sigma|} m)$ time. Step (c) is then computed in $O(n)$ time.

Table 4. SRR959239 read set, 4143243 reads of size 98

Period	nbseq	cum. nbseq	α	$1+\frac{1}{\alpha}$	$2+\frac{25}{63}\left(\frac{1-\alpha}{\alpha}\right)$	γ	β
1	1	1	0.010204081	99	40.492065	0	40.492065
50	4	5	0.5102041	2.96	2.3809524	4.5991887E-6	2.380957
...	...	...	...	...	...	...	...
94	17083	28491	0.9591837	2.0425532	2.0168862	0.0016021793	2.0184884
95	**65228**	**93719**	**0.96938777**	**2.031579**	**2.0125313**	**0.0038173746**	**2.0163486**
96	302973	396692	0.97959185	2.0208335	2.0082672	0.010567973	2.018835
97	942267	1338959	0.9897959	2.0103092	2.004091	0.036372084	2.0404632
98	2804284	4143243	1	2	2	0.12577005	2.12577

5   Experimental Results

We first present and explain experimental results for the sets of reads SRR069579 (Table 1), ERR000009 (Table 2), SRR211279 (Table 3), and SRR959239 (Table 4).

In each table, for each period i from 1 to m we show : (a) $n(i)$, (b) the cumulative number of sequences, (c) the value of α corresponding to i/m, (d) the value of $1+\frac{1}{\alpha}$, (e) $2+\frac{25}{63}\left(\frac{1-\alpha}{\alpha}\right)$ which corresponds to the term of Theorem 1 due to the large cycles, (f) γ which is the part of the final ratio brought by the small cycles, and eventually (g) β, the final ratio that can be reached by using the value of α from the previous line in the table.

The resulting approximation ratios on the read sets cited above are respectively 2.0475, 2.0618, 2.0109 and 2.0163, which are much lower than 2.3667 using a much simpler algorithm.

We then tested our approach on 100 sets of reads from different organisms (see Appendix, Table 5) in order to compute the ratios with the previous method. One set only (ERR1041316 Tupaia belangeri) is above the best actual bound of 2.3667, all the others being lower and many very close to 2.0 which is the lower bound of our approach.

Appendix

Table 5. Results on 100 sets of reads

Set of reads	β	Set of reads	β
ERR019943 Mus caroli	2.0524175	ERR032202 Martes pennanti	2.0289187
ERR1041207 Chlorocebus	2.1107855	ERR1041268 Otolemur	2.1259987
ERR1041282 Felis catus	2.1143007	ERR1041292 Heterocephalus	2.0799367

Table 5. *(Continued)*

Set of reads	β	Set of reads	β
ERR1041296 Mustela	2.0784738	ERR1041303 Macaca mulatta	2.093892
ERR1041306 Callithrix	2.175384	ERR1041315 Cavia	2.0918944
ERR1041316 Tupaia	2.4811444	ERR1111185 Phoca	2.0267568
ERR1124353 Mus spretus	2.0095918	ERR1197879 Pan paniscus	2.0089042
ERR1197887 Pan troglodytes	2.0089052	ERR1197895 Pongo	2.0088482
ERR1255554 Lynx	2.010125	ERR1353573 Peromyscus	2.00725
ERR1474986 Sarcophilus	2.0161307	ERR264173 Mus cervicolor	2.0531373
ERR266393 Cavia aperea	2.0532582	ERR572143 Lagenorhynchus	2.0864253
ERR576124 Gorilla beringei	2.0059924	ERR650934 Equus hemionus	2.0095165
ERR668455 Lama glama	2.0030081	ERR669551 Equus grevyi	2.009401
ERR866454 Cricetulus	2.0112803	ERR874023 Sus cebifrons	2.0065095
ERR977083 Sus verrucosus	2.0277126	ERR988518 Hippidion	2.0637908
SRR1003027 mouse-rat	2.108472	SRR1036149 Tupaia	2.0358806
SRR109018 Ovis aries	2.0331614	SRR1325023 Microtus	2.1520696
SRR1347548 Physeter	2.020535	SRR1528585 Cercocebus atys	2.0089693
SRR1552610 Vicugna pacos	2.0088954	SRR1653659 Saimiri	2.1446104
SRR1659070 Bison bison	2.0084386	SRR1664251 Piliocolobus	2.0095925
SRR1665158 Odobenus rosmarus	2.0217118	SRR1724110 Miniopterus	2.1718283
SRR1745916 Hipposideros	2.0110621	SRR1758975 Macaca	2.0088358
SRR1802582 Balaenoptera	2.0091696	SRR1929941 Loxodonta	2.0089402
SRR1947236 Camelus	2.0089786	SRR1970869 Panthera	2.0042064
SRR2012205 Elephas	2.0138304	SRR2016453 Phascolarctos	2.0065088

Table 5. *(Continued)*

Set of reads	β	Set of reads	β
SRR2017644 Rhinopithecus bieti	2.0098438	SRR2017692 Rhinopithecus roxellana	2.009081
SRR2017931 Mammuthus primigenius	2.027854	SRR2043984 Nannospalax galili	2.0074072
SRR2058116 Pongo abelii	2.0104375	SRR2137054 Marmota monax	2.060049
SRR2141214 Fukomys damarensis	2.0095904	SRR2154048 Eptesicus fuscus	2.026683
SRR2175612 Mesocricetus auratus	2.0250208	SRR2546897 Tamiasciurus hudsonicus	2.0024154
SRR2547421 Microtus ochrogaster	2.0138535	SRR2737541 Acinonyx jubatus	2.0091531
SRR278622 Sorex araneus	2.1920824	SRR2899805 Hystrix cristata	2.0163338
SRR2925016 Neovison vison	2.015167	SRR2981140 Macaca	2.0073588
SRR2981977 Peromyscus leucopus	2.0187445	SRR2995138 Pteropus alecto	2.0172942
SRR3158616 Papio cynocephalus	2.0101697	SRR3159949 Macaca	2.0033646
SRR317809 Echinops telfairi	2.148989	SRR328420 Chinchilla lanigera	2.0144074
SRR329659 Trichechus manatus latirostris	2.074164	SRR3470780 sus scrofa	2.1215959
SRR3480541 Macaca nemestrina	2.0156894	SRR353137 Condylura cristata	2.0268734
SRR3587085 dog	2.02617	SRR3608910 Monodelphis domestica	2.0083923
SRR361229 Bos mutus	2.02437	SRR3630971 Panthera leo	2.0609856
SRR3634324 Papio anubis	2.0105252	SRR3659132 Capra hircus	2.1219108
SRR3659145 capra	2.1962154	SRR3669996 Rattus norvegicus	2.0878835
SRR3683913 Galeopterus variegatus	2.010047	SRR3709478 Chlorocebus aethiops	2.0311713

Table 5. *(Continued)*

Set of reads	β	Set of reads	β
SRR504904 Ailuropoda melanoleuca	2.0157156	SRR556101 Chlorocebus pygerythrus	2.009286
SRR617095 Myotis brandtii	2.010544	SRR628072 Myotis davidii	2.0363553
SRR709450 Daubentonia madagascariensis	2.0109098	SRR837385 Cynopterus	2.0135381
SRR873443 Equus asinus	2.0088918	SRR942310 Ursus maritimus	2.0092664
SRR955365 Spermophilus dauricus	2.026139	SRR069579	2.0475218
ERR000009	2.0618343	SRR211279	2.0109274
SRR959239	2.0163486	NG-6201 B6Rg1Lg8 apricot tree	2.0427585

References

1. Armen, C., Stein, C.: A $2\frac{2}{3}$ superstring approximation algorithm. Discrete Appl. Math. **88**(1–3), 29–57 (1998). http://dx.doi.org/10.1016/S0166-218X(98)00065-1, http://www.sciencedirect.com/science/article/pii/S0166218X98000651, Computational Molecular Biology DAM - CMB Series
2. Blum, A., Jiang, T., Li, M., Tromp, J., Yannakakis, M.: Linear approximation of shortest superstrings. J. ACM **41**(4), 630–647 (1994). doi:10.1145/179812.179818, http://doi.acm.org/10.1145/179812.179818
3. Breslauer, D., Jiang, T., Jiang, Z.: Rotations of periodic strings and short superstrings. J. Algorithms **24**(2), 340–353 (1997). http://dx.doi.org/10.1006/jagm. 1997.0861, http://www.sciencedirect.com/science/article/pii/S0196677497908610
4. Armen, C., Stein, C.: Improved length bounds for the shortest superstring problem. In: Akl, S.G., Dehne, F., Sack, J.-R., Santoro, N. (eds.) WADS 1995. LNCS, vol. 955, pp. 494–505. Springer, Heidelberg (1995). doi:10.1007/3-540-60220-8_88
5. Crochemore, M., Cygan, M., Iliopoulos, C., Kubica, M., Radoszewski, J., Rytter, W., Waleń, T.: Algorithms for three versions of the shortest common superstring problem. In: Amir, A., Parida, L. (eds.) CPM 2010. LNCS, vol. 6129, pp. 299–309. Springer, Heidelberg (2010). doi:10.1007/978-3-642-13509-5_27
6. Czumaj, A., Gçasieniec, L.: On the complexity of determining the period of a string. In: Giancarlo, R., Sankoff, D. (eds.) CPM 2000. LNCS, vol. 1848, pp. 412–422. Springer, Heidelberg (2000). doi:10.1007/3-540-45123-4_34
7. Czumaj, A., Gasieniec, L., Piotrów, M., Rytter, W.: Sequential and parallel approximation of shortest superstrings. J. Algorithms **23**(1), 74–100 (1997). http://dx.doi.org/10.1006/jagm.1996.0823, http://www.sciencedirect.com/science/article/pii/S0196677496908238
8. Ferragina, P., Landau, G., Ma, B.: Combinatorial pattern matching why greed works for shortest common superstring problem. Theor. Comput. Sci. **410**(51), 5374–5381 (2009). http://dx.doi.org/10.1016/j.tcs.2009.09.014, http://www.sciencedirect.com/science/article/pii/S0304397509006410

9. Fici, G., Kociumaka, T., Radoszewski, J., Rytter, W., Walen, T.: On the greedy algorithm for the shortest common superstring problem with reversals. Inf. Process. Lett. **116**(3), 245–251 (2016). doi:10.1016/j.ipl.2015.11.015

10. Gallant, J., Maier, D., Astorer, J.: On finding minimal length superstrings. J. Comput. Syst. Sci. **20**(1), 50–58 (1980). http://dx.doi.org/10.1016/0022-0000(80)90004-5, http://www.sciencedirect.com/science/article/pii/0022000080900045

11. Garey, M.R., Johnson, D.S.: Computers and Intractability: A Guide to the Theory of NP-Completeness. W.H. Freeman & Co., New York (1990)

12. Golovnev, A., Kulikov, A.S., Mihajlin, I.: Approximating shortest superstring problem using de Bruijn graphs. In: Fischer, J., Sanders, P. (eds.) CPM 2013. LNCS, vol. 7922, pp. 120–129. Springer, Heidelberg (2013). doi:10.1007/978-3-642-38905-4_13

13. Guibas, L.J., Odlyzko, A.M.: Periods in strings. J. Comb. Theor. Ser. A **30**(1), 19–42 (1981). http://dx.doi.org/10.1016/0097-3165(81)90038-8, http://www.sciencedirect.com/science/article/pii/0097316581900388

14. Holub, S., Shallit, J.: Periods and borders of random words. In: STACS, Schloss Dagstuhl - Leibniz-Zentrum fuer Informatik, LIPIcs, vol. 47, pp. 44:1–44:10 (2016)

15. Kaplan, H., Shafrir, N.: The greedy algorithm for shortest superstrings. Inf. Process. Lett. **93**(1), 13–17 (2005). doi:10.1016/j.ipl.2004.09.012

16. Karpinski, M., Schmied, R.: Improved inapproximability results for the shortest superstring and related problems. In: CATS, Australian Computer Society, CRPIT 2013, vol. 141, pp. 27–36 (2013)

17. Kosaraju, S.R., Park, J.K., Stein, C.: Long tours and short superstrings. In: 1994 Proceedings of 35th Annual Symposium on Foundations of Computer Science, pp 166–177 (1994). doi:10.1109/SFCS.1994.365696

18. Li, M.: Towards a DNA sequencing theory (learning a string). In: Proceedings of the 31st Symposium on the Foundations of Computer Science, pp. 125–134. IEEE Computer Society Press, Los Alamitos (1990)

19. Mucha, M.: Lyndon words and short superstrings. In: SODA, pp. 958–972. SIAM (2013)

20. Ott, S.: Lower bounds for approximating shortest superstrings over an alphabet of size 2. In: Widmayer, P., Neyer, G., Eidenbenz, S. (eds.) WG 1999. LNCS, vol. 1665, pp. 55–64. Springer, Heidelberg (1999). doi:10.1007/3-540-46784-X_7

21. Paluch, K.E.: Better approximation algorithms for maximum asymmetric traveling salesman and shortest superstring. CoRR abs/1401.3670 (2014). http://arxiv.org/abs/1401.3670

22. Paluch, K.E., Elbassioni, K.M., van Zuylen, A.: Simpler approximation of the maximum asymmetric traveling salesman problem. In: STACS, Schloss Dagstuhl - Leibniz-Zentrum fuer Informatik, LIPIcs, vol. 14, pp. 501–506 (2012)

23. Sweedyk, Z.: A $2\frac{1}{2}$-approximation algorithm for shortest superstring. SIAM J. Comput. **29**(3), 954–986 (1999). doi:10.1137/S0097539796324661

24. Tarhio, J., Ukkonen, E.: A greedy approximation algorithm for constructing shortest common superstrings. Theor. Comput. Sci. **57**(1), 131–145 (1988). doi:10.1016/0304-3975(88)90167-3, http://www.sciencedirect.com/science/article/pii/0304397588901673

25. Teng, S.H., Yao, F.F.: Approximating shortest superstrings. SIAM J. Comput. **26**(2), 410–417 (1997). doi:10.1137/S0097539794286125

26. Vazirani, V.V.: Approximation Algorithms. Springer-Verlag New York Inc., New York (2001)

27. Yu, Y.W.: Approximation hardness of shortest common superstring variants. CoRR abs/1602.08648 (2016). http://arxiv.org/abs/1602.08648

On the Cost of Simulating a Parallel Boolean Automata Network by a Block-Sequential One

Florian Bridoux[1]([⊠]), Pierre Guillon[2], Kévin Perrot[1], Sylvain Sené[1,3],
and Guillaume Theyssier[2]

[1] Université d'Aix-Marseille, CNRS, LIF, Marseille, France
florian.bridoux@lif.univ-mrs.fr
[2] Université d'Aix-Marseille, CNRS, Centrale Marseille, I2M, Marseille, France
[3] Institut rhône-alpin des systèmes complexes, IXXI, Lyon, France

Abstract. In this article we study the minimum number κ of additional automata that a Boolean automata network (BAN) associated with a given block-sequential update schedule needs in order to simulate a given BAN with a parallel update schedule. We introduce a graph that we call NECC graph built from the BAN and the update schedule. We show the relation between κ and the chromatic number of the NECC graph. Thanks to this NECC graph, we bound κ in the worst case between $n/2$ and $2n/3 + 2$ (n being the size of the BAN simulated) and we conjecture that this number equals $n/2$. We support this conjecture with two results: the clique number of a NECC graph is always less than or equal to $n/2$ and, for the subclass of bijective BANs, κ is always less than or equal to $n/2 + 1$.

Keywords: Boolean automata networks · Intrinsic simulation · Block-sequential update schedules

1 Introduction

In this article, we study Boolean automata networks (BANs). A BAN can be seen as a set of two-states automata interacting with each other and evolving in a discrete time. BANs have been first introduced by McCulloch and Pitts in the 1940[s] [17]. They are common representational models for natural dynamical systems like neural or genetic networks [7,12–14,25], but they are also computational models with which we can study computability or complexity. In this article we are interested in intrinsic simulations between BANs, *i.e.* simulations that focus on the dynamics rather than the computational power. More concretely, given a BAN A we want to find a BAN B which reproduces the dynamics of A while it satisfies some constraints. There have been few studies using intrinsic simulation between BANS before the 2010[s] [2,8,23,24]. More recently, this notion has received a new interest [18–21] and we are convinced that it is essential and deserves to be dealt with. Meanwhile, intrinsic simulation of many other similar objects (cellular automata, tilings, subshifts, self-assembly, etc.) has been really developing since 2000 [3,4,6,11,15,16,22].

© Springer International Publishing AG 2017
T.V. Gopal et al. (Eds.): TAMC 2017, LNCS 10185, pp. 112–128, 2017.
DOI: 10.1007/978-3-319-55911-7_9

A given BAN can be associated with several dynamics, depending on the schedule (*i.e.* the order) chosen to update the automata. In this article, we will consider all block-sequential update schedules: we group automata into blocks, and we update all automata of a block at once, and iterate the blocks sequentially. Among these update schedules are the following classical ones: the parallel one (a unique block composed of n automata) and the $n!$ sequential ones (n blocks of 1 automaton). The pair of a BAN and its update schedule is called a scheduled Boolean automata network (SBAN).

For the last 10 years, people have studied the influence of the update schedules on the dynamics of a BAN [1,5,9,10]. Here, we do the opposite. We take a SBAN, and try to find the smallest SBAN with a constrained update schedule which simulates this dynamics. For example, let N be a parallel SBAN of size 2 with 2 automata that exchange their values. There are no SBANs N' of size 2 with a sequential update schedule which simulates N. Indeed, when we update the first automaton, we necessarily erase its previous value. If we did not previously save it, we cannot use the value of the first automaton to update the second automaton. Thus, N' needs an additional automaton to simulate N under the sequential update schedule constraint. A SBAN N of size n with a parallel update schedule can always be simulated by a SBAN N' of size $2n$ with a given sequential update schedule. Indeed, we just need to add n automata which copy all the information from the original automata and then, we compute sequentially the updates of the originals automata using the saved information. The goal of this article is to establish more precise bounds on the number of required additional automata, function of n, in the worst case.

In Sect. 2, we define BANs and detail the notion of simulation that we use. In Sect. 3, we consider the dynamics of a BAN F with automata set V and the parallel update schedule and we consider a block-sequential update schedule W. We focus on the minimum number $\kappa(F, W)$ of additional automata that a SBAN needs to simulate this dynamics with an update schedule identical to W on V. In Sect. 4, we define a graph which connects configurations depending on a BAN F and a block-sequential update schedule W. We prove that the chromatic number of this graph determines the number $\kappa(F, W)$ defined in the previous section. We also state the following conjecture: $\kappa(F, W)$ is always less than or equal to $n/2$, where n is the size of the BAN F. In Sect. 5, we define another graph constructed from the previous graph where we identify configurations which have the same image. We prove that the chromatic number of this new graph is always greater than that of the previous graph. We deduce an upper bound for $\kappa(F, W)$. In Sect. 6, we try to support our conjecture by finding an upper bound for the clique number of the graph defined in Sect. 4. Finally, in Sect. 7, we study $\kappa(F, W)$ in the case where F is bijective.

2   Definitions and Notations

2.1   BANs and SBANs

In this article, unless otherwise stated, BANs have a size $n \in \mathbb{N}$, which means that they are composed of n automata numbered from 0 to $n - 1$. Usually, we denote this set of automata by $V = \{0, 1, \ldots, n - 1\}$ (which will be abbreviated by $[\![0, n[\![)$. Each automaton can take two states in the Boolean set $\mathbb{B} = \{0, 1\}$. A *configuration* is a Boolean vector of size n, interpreted as the sequence of states of the automata of the BAN. In other words, if x is a configuration, then $x \in \mathbb{B}^n$ and $x = (x_0, \ldots, x_{n-1})$ with x_i the state of automaton i (for all i in V). For all $I \subseteq V$, we denote by x_I the restriction of x to I. In other words, if $I = \{i_1, i_2, \ldots, i_p\}$ with $i_1 < i_2 < \cdots < i_p$ then $x_I = (x_{i_1}, x_{i_2}, \ldots, x_{i_p})$. We also denote by $x_{\overline{I}}$ the restriction of x to $V \setminus I$.

For all $b \in \mathbb{B}$, we denote by $\overline{b}$ the negation of the state of b. In other words, $\overline{0} = 1$ and $\overline{1} = 0$. We also denote by $\overline{x}$ the negation of x, such that $\overline{x} = (\overline{x_0}, \ldots, \overline{x_{n-1}})$. Furthermore, we denote by $\overline{x}^i$ or $\overline{x}^I$ the negation of x respectively restricted to an automaton i or a set I of automata, that is, $\overline{x}_i^I = \overline{x_i}$ if $i \in I$, and $\overline{x}_i^I = x_i$ if $i \in V \setminus I$.

In this article, we only study BANs with block-sequential update schedules. A SBAN $N = (F, W)$ is characterized by:

- a global update function $F : \mathbb{B}^n \to \mathbb{B}^n$ which represents the BAN;
- a block-sequential update schedule W.

The *global update function* of a BAN is the collection of the local update functions of the BAN: we have $F(x) = (f_0(x), \ldots, f_{n-1}(x))$, where for all $i \in V$, $f_i : \mathbb{B}^n \to \mathbb{B}$ is the local update function of automata i. We also use the I-update function F_I, with $I \subseteq V$, which gives a configuration where the states of automata in I are updated and the other ones are not. In other words, $\forall i \in V$, $F_I(x)_i = f_i(x)$ if $i \in I$ and x_i otherwise. And, for singleton, we simply write $F_i(x) = F_{\{i\}}(x)$.

Remark 1. It is important not to confuse $F_I(x)$ and $F(x)_I$. The first one is the I-update function that we have just defined. The second is the configuration $F(x)$ restricted to I.

A *block-sequential update schedule* is an ordered partition of V. The set of ordered partitions of V is denoted by $\overrightarrow{\mathscr{P}}(V)$. Let $W \in \overrightarrow{\mathscr{P}}(V)$ and $p = |W|$ and $W = (W_0, \ldots, W_{p-1})$. We make particular use of F^W defined as $F^W = F_{W_{p-1}} \circ \cdots \circ F_{W_0}$. If $x \in \mathbb{B}^n$ is the configuration of the BAN at some time step, then $F^W(x)$ is the configuration of the BAN at the next step. There are two very particular kinds of block-sequential update schedules:

- the parallel update schedule where all automata are updated at the same time step. So, we have $W = [V]$ (*i.e.* $|W| = 1$ and $W_0 = V$) and $F^W = F$;
- the sequential update schedules where automata are updated one at the time. So, we have $|W| = n$ and $\forall i \in [\![0, n[\![, |W_i| = 1$.

For any $j \in [\![0, p]\!]$, we denote $W_{<j} = \bigcup_{i=0}^{j-1} W_i$. In particular, we have $W_{<0} = \emptyset$ and $W_{<p} = V$. Furthermore, for any $i \in [\![0, p]\!]$, we denote $W^{<i} = (W_0, W_1, \ldots, W_{i-1})$. $W^{<i}$ is an ordered partition of $W_{<i}$. In particular, we have $W^{<0} = [\,]$ (the empty vector) and $W^{<p} = W$.

We will often use the following two notations:

(i) $F^{W^{<j}} = F_{W_{j-1}} \circ \cdots \circ F_{W_0}$ is the function which makes the first j steps of the transition of the SBAN (F, W);

(ii) $F_{W_{<j}} = F_{W_0 \cup \cdots \cup W_{j-1}}$ is the function which updates only the automata in the first j blocks of W.

Let $W \in \vec{\mathscr{P}}(V)$ be an update schedule. We know that each automaton of a block-sequential SBAN is updated only in one step of the update schedule. We denote by $W(i)$ the step at which i is updated. More formally, $\forall i \in V, W(i)$ is the number $j \in [\![0, p[\![$ such that $i \in W_j$.

2.2   Simulation

Here, we define the notion of simulation used in this article. We consider that a SBAN N of size m simulates another SBAN N' of size n if there is a projection from $\mathbb{B}^m$ to $\mathbb{B}^n$ such that the projection of the update in N' equals the update in N of the projection.

Definition 1. *Let $F : \mathbb{B}^n \to \mathbb{B}^n$ and $F' : \mathbb{B}^m \to \mathbb{B}^m$ with $m \geq n$, $V = [\![0, n[\![$ and $V' = [\![0, m[\![$, $W \in \vec{\mathscr{P}}(V)$ and $W' \in \vec{\mathscr{P}}(V')$. Let $h : V \to V'$ be an injective function and $\varphi_h : \mathbb{B}^m \to \mathbb{B}^n$ be defined by $\varphi_h(x) = (x_{h(i)})_{i \in V}$. We say that (F', W') h-simulates (F, W), and note $(F', W') \rhd^h (F, W)$, if $\varphi_h \circ F'^{W'} = F^W \circ \varphi_h$. Moreover, (F', W') simulates (F, W), which is denoted by $(F', W') \rhd (F, W)$ if there is a h such that $(F', W') \rhd^h (F, W)$.*

In this article we often use an *id*-simulation which is a h-simulation with h the identity function ($h(i) = i$).

3   Number of Required Additional Automata

In this section, we define the main object of this article. Given a BAN F with automata V and a block-sequential update schedule $W \in \vec{\mathscr{P}}(V)$, we consider the smallest SBAN (F', W') which simulates the parallel SBAN $(F, [V])$, where W' extends W by preserving its order. We could as well study the problem of finding a block-sequential SBAN (F', W') which simulates another block-sequential SBAN (G, W). However, this problem is in fact the same. Indeed, for any block-sequential SBAN (G, W), the parallel SBAN $(G^W, [V])$ *id*-simulates (G, W).

Let us formalize the notion. From an update schedule W and a BAN of size n, we define the notion of update schedule extending W for a bigger BAN of size m. Let $V' = [\![0, m[\![$. Let $h : V \to V'$ be an injective function. We denote by $\mathscr{E}_h(W, V')$ the set of update schedules W' extending W such that each W' preserves the order of W for the projection by h of the automata of V. That is to say, if one automaton is updated before another one according to W, then the projection of these automata into V' will preserve the same update order in W'. More formally, $\mathscr{E}_h(W, V') = \{W' \in \overrightarrow{\mathscr{P}}(V') \mid \forall i \in V, W(i) \leq W(i') \iff W'(h(i)) \leq W'(h(i'))\}$. In particular, if two automata $i, j \in V$ are updated at the same step $W(i) = W(j)$, then the projections $h(i), h(j)$ of these automata are updated at the same step $W(h(i)) = W(h(j))$ in W'. In other words, h induces a map $\tilde{h} : [\![0, p[\![\to [\![0, p'[\![$ such that $W(h(i)) = \tilde{h}(W(i))$ for all $i \in V$, and $\tilde{h}(W)$ is a subordered partition of W'.

Definition 2. *If F is a BAN over automata $V = [\![0, n[\![$ and $W \in \overrightarrow{\mathscr{P}}(V)$ is an update schedule, we define $\kappa(F, W)$ as the smallest k such that there exist an update schedule $W' \in \mathscr{E}_h(W, V')$ extending W and a BAN $F' : \mathbb{B}^{n+k} \to \mathbb{B}^{n+k}$ such that $(F', W') \rhd (F, [V])$, with $V' = [\![0, n+k[\![$.*

Furthermore, κ_n is the value of $\kappa(F, W)$ in the worst case among all SBANs with automata V. In other words, $\kappa_n = max(\{\kappa(F, W) \mid F : \mathbb{B}^n \to \mathbb{B}^n \text{ and } W \in \overrightarrow{\mathscr{P}}(V)\})$.

4   NECCs Set and NECC Graph

In order to answer the main problem of this article which is is to bound the values of κ_n, we introduce a new concept: the not equivalent and confusable configurations or NECCs and the NECC graph. Theorem 1 will show that the logarithm of the chromatic number of the NECC graph of a SBAN and the κ of this SBAN are equal. NEC (the acronym standing for *non-equivalent configurations*) is the set of pairs of configurations with different images by F. In other words,

$$\text{NEC}_F = \{(x, x') \in \mathbb{B}^n \times \mathbb{B}^n \mid F(x) \neq F(x')\}.$$

We call *confusable configurations* and denote by $\text{CC}_{F,W}$, or simply CC (the acronym standing for *confusable configurations*), the set of pairs of configurations which become identical when we update the first i blocks of W for some $i \in [\![0, p[\![$. Formally,

$$\text{CC} = \{(x, x') \in \mathbb{B}^n \times \mathbb{B}^n \mid \exists i \in [\![0, p]\!], \ F_{W_{<i}}(x) = F_{W_{<i}}(x')\}.$$

Definition 3. $\text{NECC}_{F,W}$, *or simply* NECC *(the acronym standing for* not equivalent and confusable configurations*), is the set of pairs of configurations which are confusable and not equivalent at the same time,* $\text{NECC}_{F,W} = \text{CC}_{F,W} \cap \text{NEC}_F$.

Also, for all $x, x' \in \mathbb{B}^n$, we denote by $\text{CC}_{F,W}(x, x')$ (or just $\text{CC}_{F,W}(x, x')$) the set of time steps i which make them confusable. More formally, $\forall x, x' \in \mathbb{B}^n$, $\text{CC}_{F,W}(x, x') = \{i \in [\![0, p]\!] \mid F_{W_{<i}}(x) = F_{W_{<i}}(x')\}$.

Remark 2. We have $\mathsf{CC}(x, x') = \emptyset$ if and only if $(x, x') \notin \mathsf{CC}$.

Definition 4. *The* NECC *graph, denoted by* $(\mathbb{B}^n, \mathsf{NECC})$, *is the nondirected graph which has the set of configurations* $\mathbb{B}^n$ *as nodes and the set of* NECC *pairs as edges.*

In the sequel, we make a particular use of two concepts of graph theory. A *valid coloring* of G is a coloring of all the nodes of G such that two adjacent nodes do not have the same color. We denote by $\chi(G)$ the *chromatic number* of the graph G, namely the minimum number of colors of a valid coloring of G. Furthermore, the chromatic number of the NECC graph is denoted by $\chi(\mathsf{NECC}) = \chi((\mathbb{B}^n, \mathsf{NECC}))$. We see in Lemma 1 that we can get a valid coloring of the $\mathsf{NECC}_{F,W}$ graph from the SBAN (F', W') which simulates $(F, [V])$. This coloring does not use more than 2^k colors with k the number of additional automata of F'. We color the configuration of the NECC graph using the values of the added automata after the update.

Lemma 1. *For any BAN* $F : \mathbb{B}^n \to \mathbb{B}^n$ *and any block-sequential update schedule* W, $\kappa(F, W) \geq \lceil \log_2(\chi(\mathsf{NECC}_{F,W})) \rceil$.

Proof. Let $h : V \to V'$ injective, $W' \in \mathscr{E}_h(W, V')$, $p = |W|$, $p' = |W'|$ and $F' : \mathbb{B}^{n+k} \to \mathbb{B}^{n+k}$ such that $(F', W') \rhd^h (F, [V])$. We prove that $k \geq \lceil \log_2(\chi(\mathsf{NECC})) \rceil$. Let z, z' be such that $z_{\overline{h(V)}} = z'_{\overline{h(V)}} = [0]^k$ and $(x, x') = (\varphi_h(z), \varphi_h(z')) \in \mathsf{NECC}$, and let us prove that $F'(z)_{\overline{h(V)}} \neq F'(z')_{\overline{h(V)}}$. Suppose the contrary. Since $(x, x') \in \mathsf{NECC}$, we have $F(x) \neq F(x')$ and $\exists j \in [0, p]$, $F_{W_{<j}}(x) = F_{W_{<j}}(x')$. Let $Z = F'^{W'^{<h(j)}}(z) = $ and $Z' = F'^{W'^{<h(j)}}(z')$. By assumption, we have $z_{\overline{h(V)}} = [0]^k = z'_{\overline{h(V)}}$ and $F(z)_{\overline{h(V)}} = F(z')_{\overline{h(V)}}$. Thus, $Z_{\overline{h(V)}} = Z'_{\overline{h(V)}}$. Furthermore, we have $\varphi_h(Z) = F_{W_{<j}}(x) = F_{W_{<j}}(x') = \varphi_h(Z')$. As a result, $Z_{h(V)} = Z'_{h(V)}$ and $Z = Z'$. Consequently, $F'(z) = F_{W'_{p-1}} \circ \cdots \circ F_{W'_{h(j)}}(Z)$ and $F'(z') = F_{W'_{p-1}} \circ \cdots \circ F_{W'_{h(j)}}(Z')$ are equal. However, $(x, x') \in NEC$. Thus, $F'(z)_{h(V)} = F(x) \neq F(x') = F'(z')_{h(V)}$. As a consequence, we have also $F'(z) \neq F'(z')$. There is a contradiction. We have proven that if $(x, x') \in \mathsf{NECC}$ then $F'(z)_{\overline{h(V)}} \neq F'(z')_{\overline{h(V)}}$. In other words, a valid coloring of NECC is obtained by coloring each vertex x by $F(z)_{\overline{h(V)}}$, where $\phi_h(z) = x$ and $x_{\overline{h(V)}} = [0]^k$. Hence $\{F(z)_{\overline{h(V)}} | z_{\overline{h(V)}} = [0]^k\}$ has at least $\chi(\mathsf{NECC})$ different values. To encode these values, we need to have $k = |\overline{h(V)}| \geq \lceil \log_2(\chi(\mathsf{NECC})\rceil$. So $\kappa(F, W) \geq \log_2(\chi(\mathsf{NECC}))$. $\qquad\square$

We see in Lemma 2 that we can get a SBAN (F', W') which simulates $(F, [V])$ from a valid coloring of the $\mathsf{NECC}_{F,W}$ graph.

Lemma 2. *For any BAN* $F : \mathbb{B}^n \to \mathbb{B}^n$ *and any block-sequential update schedule* W, $\kappa(F, W) \leq \lceil \log_2(\chi(\mathsf{NECC}_{F,W})) \rceil$.

Proof. Let $k = \lceil \log_2(\chi(\mathsf{NECC})) \rceil$. We define W' such that we start by updating sequentially the last k nodes, and after this, we update as

$W \colon W' = (\{n\}, \{n+1\}, \ldots, \{n+k-1\}, W_0, W_1, \ldots, W_{p-1})$. Let $\mathsf{color} \colon \mathbb{B}^n \to \mathbb{N}$ be a minimum coloring of the NECC graph. For all $x \in \mathbb{B}^n$, let $\mathrm{COLOR}(x)$ be the number $\mathsf{color}(x)$ encoded with a Boolean vector of size k. It is possible to encode it with k Boolean numbers because with k bits we can encode $2^k \geq \chi(\mathsf{NECC}) = |\mathsf{color}(\mathbb{B}^n)|$ values. Let $x \in \mathbb{B}^n$ and $y \in \mathbb{B}^k$. We define $z = x \| y \in \mathbb{B}^{n+k}$ by $z_{[\![0,n[\![} = x$ and $z_{[\![n,n+k[\![} = y$. For all $j \in [\![0,p]\!]$, let $A_j(x\|y) = \{F(x') \mid x' \in \mathbb{B}^n$ and $\mathrm{COLOR}(x') = y$ and $F_{W_{<j}}(x') = x\}$. We can prove that $|A_j(x\|y)| \leq 1$. For the sake of contradiction, suppose $\exists F(x'), F(x'') \in A_j(x\|y)$, $F(x') \neq F(x'')$. Clearly, $(x', x'') \in \mathsf{NEC}$. Moreover, $F_{W_{<j}}(x') = x = F_{W_{<j}}(x'')$ gives that $(x', x'') \in \mathsf{CC}$. So $(x', x'') \in \mathsf{NECC}$. However, $\mathrm{COLOR}(x') = y = \mathrm{COLOR}(x'')$, which contradicts the construction of the coloring. Let $F' \colon \mathbb{B}^{n+k} \to \mathbb{B}^{n+k}$ be defined for all $x\|y \in \mathbb{B}^{n+k}$ by $F'_{[\![n,n+k[\![}(x\|y) = \mathrm{COLOR}(x)$ and $\forall j \in [\![0,p[\![$, $F'(x\|y)_{W'_{k+j}} = z_{W_j}$ if $A_j(x\|y) = \{z\}$, and $[0]^{|W_j|}$ if $A_j(x\|y)$ is empty. Now, let $z = x\|y \in \mathbb{B}^{n+k}$ and we show that $F'^{W'}(x\|y)_{[\![0,n[\![} = F(x)$. Let us show by induction that $\forall j \in [\![0,p]\!]$, $F'^{W'^{<k+j}}(z)_{[\![0,n[\![} = F_{W_{<j}}(x)$. Let $j = 0$. We have $F'^{W'^{<k+j}}(z)_{[\![0,n[\![} = F'^{W'^{<k}}(z)_{[\![0,n[\![} = x$ (because in the first k steps of W' we only update the automata of $[\![n, n+k[\![)$ and $F_{W_{<j}}(x) = F_{W_{<0}}(x) = x$. So $F'^{W'^{k+j}}(z)_{[\![0,n[\![} = F_{W_{<j}}(x)$. Now let $j \in [\![0,p]\!]$, $z' = F'^{W'^{<k+j}}(z)$, and assume that $z'_{[\![0,n[\![} = F_{W_{<j}}(x)$. We have $F'^{W'^{<k+j+1}}(z)_{[\![0,n[\![} = F'_{W'_{k+j+1}}(z')$. Thus, $F'^{W'^{<k+j}}(z)_{[\![0,n[\![\backslash W'_{k+j+1}} = z'_{[\![0,n[\![\backslash W'_{k+j+1}} = F_{W_{<j}}(x)_{[\![0,n[\![\backslash W_{j+1}} = F_{W_{<j+1}}(x)_{[\![0,n[\![\backslash W_{j+1}}$. Furthermore, $\mathrm{COLOR}(x) = F(z)_{[\![n,n+k[\![} = z'_{[\![n,n+k[\![}$, and by induction hypothesis, $F_{W_{<j}}(x) = z'_{[\![0,n[\![}$. Thus, $F(x) \in A_j(z')$. As a consequence, $F'^{W'^{<k+j+1}}(z)_{W'_{k+j+1}} = F'_{W'_{k+j+1}}(z')_{W'_{k+j+1}}$ was defined as $F(x)_{W'_{k+j+1}} = F(x)_{W_{j+1}}$. As a result, $F'^{W'^{<k+j+1}}(z)_{[\![0,n[\![} = F_{W_{<j+1}}(x)$. Consequently, $\forall z = x\|y \in \mathbb{B}^{n+k}$, $F'^{W'}(z)_{[\![0,n[\![} = F(x)$. Thus, $(F', W') \rhd^{id} (F, [V])$. Finally, $\kappa(F, W) \leq \lceil \log_2(\chi(\mathsf{NECC})) \rceil$. $\qquad\square$

Lemmas 1 and 2 show that there is an equivalence between a coloring of the $\mathsf{NECC}_{F,W}$ graph and a SBAN (F', W') which simulates $(F, [V])$. Moreover, we can see in Lemma 2 that one optimal simulation is always achieved by applying sequentially the additional automata before applying the constrained schedule.

Theorem 1. *For any BAN $F \colon \mathbb{B}^n \to \mathbb{B}^n$ and any block-sequential update schedule W, $\kappa(F, W) = \lceil \log_2(\chi(\mathsf{NECC}_{F,W})) \rceil$.*

In Lemma 3 below, using the example of $n/2$ automata which exchange their values, we find a lower bound for κ_n. We use the fact that if we take the good update schedule W, this $\mathsf{NECC}_{F,W}$ graph has a big clique number.

Lemma 3. *$\forall n \in \mathbb{N}$, $\kappa_n \geq \lfloor n/2 \rfloor$.*

Proof. Let us suppose that n is even (if not, we just have to add a useless automaton and the proof remains valid). Let us consider the BAN F such that:

$$\forall i \in [\![0, n/2[\![, \ f_i(x) = x_{i+n/2} \quad \text{and} \quad \forall i \in [\![n/2, n[\![, \ f_i(x) = x_{i-n/2}.$$

We also consider the simple sequential update schedule $W = (\{0\}, \ldots, \{n\})$. Let $X = \{x \in \mathbb{B}^n \mid x_{[n/2,n[} = [0]^{n/2}\}$, and $x, x' \in X$ such that $x \neq x'$. When we update the first half of the automata, x and x' both become the configuration full of 0. Then, for $i = n/2$, we have $F_{W_{<i}}(x) = [0]^n = F_{W_{<i}}(x')$. Thus, $(x, x') \in \mathrm{CC}$. We also have $x \neq x'$. So $\exists i \in [n/2, n[$ such that $x_i \neq x'_i$ and $f_{i+n/2}(x) = x_i$ and $f_{i+n/2}(x') = x'_i$. Consequently, $f_{i+n/2}(x) \neq f_{i+n/2}(x')$. Then, $F(x) \neq F(x')$ and $(x, x') \in \mathrm{NEC}$. As a result, we have $(x, x') \in \mathrm{NECC}$. We know that X is a clique. Moreover, X is a clique of size $2^{n/2}$. Thus, the chromatic number of the NECC graph is at least $2^{n/2}$ and $\kappa(F, W) \geq n/2$. Hence, $\forall n \in \mathbb{N}$, $\kappa_n \geq n/2$. $\square$

We conjecture that $\lfloor n/2 \rfloor$ is the upper bound as well. This conjecture has not been proven yet, but Theorem 3 supports it by giving an upper bound to the clique number of a NECC graph.

Conjecture 1. $\forall n \in \mathbb{N}$, $\kappa_n \leq \lfloor n/2 \rfloor$.

5   INECC Graph

In this section, we define the INECC graph which is the NECC graph after we quotient its configurations which have the same image. We can prove that the INECC graph has a bigger chromatic number than the NECC graph, find an upper bound of its chromatic number and deduce an upper bound for the NECC graph as well.

Definition 5. *The INECC graph is the graph such that:*

- *the vertex set is $\{F(x) \mid x \in \mathbb{B}^n\}$, i.e. the set of the images of the configurations of the NECC graph;*
- *two vertices y and y' are connected to each other if $\exists x, x' \in \mathbb{B}^n$ such that $F(x) = y$, $F(x') = y'$ and $(x, x') \in \mathrm{NEC}$.*

Let us now prove that we can use a valid coloring of the INECC graph to color the NECC graph.

Lemma 4. $\chi(\mathrm{INECC}) \geq \chi(\mathrm{NECC})$.

Proof. We partition the configurations into sets of equivalent configurations (*i.e.* configurations which have the same image) $E_1, E_2, \ldots, E_k$. We denote by $y^i \in \mathbb{B}^n$ the image of the configurations of E_i for each $i \in [0, k[$. In other words, $\forall i \in [0, k[, \forall x \in E_i$, $F(x) = y^i$. Let color $: [0, k[\to \mathbb{N}^*$ be an optimal coloring of the INECC graph. In the NECC graph, we can color all the configurations of a set E_i by the color of y^i in the INECC graph. Let $x, x' \in \mathbb{B}^n$. If x and x' have the same color:

- either x and x' are in the same set E_i, and then $(x, x') \notin \mathrm{NECC}$ because they are equivalent;

– or they are in two distinct sets E_i and $E_{i'}$. In this case $(x, x') \notin$ NECC otherwise y^i and $y^{i'}$ would be connected in the INECC graph and they would have different colors.

So, the coloring is a valid coloring and does not need more colors than the INECC graph coloring and we conclude that $\chi(\text{INECC}) > \chi(\text{NECC})$.   □

Remark 3. We can see that if we take two SBANs (F, W) and (F, W') with W' a sequentialized version of W (*i.e.* an update schedule that breaks the blocks of W into blocks of size 1), the chromatic number of the NECC graph of (F, W) is always greater than or equal to that of the NECC graph of (F, W'). Indeed, the set of edges of the NECC graph of (F, W) is included in the set of edges of the NECC graph of (F, W'). Thus, the chromatic number of the latter is greater. Furthermore, the same reasoning applies to the INECC graph. *As a result, if we want to find an upper bound to the chromatic number of the NECC or INECC graph, we can restrict our study to SBAN updated sequentially.*

Remark 4. We can see that if we have a SBAN (F, W), with W a sequential update schedule, we can find another SBAN (F', W') with W' the simple sequential update schedule $(\{0\}, \{1\}, \cdots, \{n-1\})$ which will have the same NECC and INECC graphs up to a permutation. As a consequence, their chromatic numbers of their NECC and INECC graphs are equal, respectively. Thus, if we want to find an upper bound to the chromatic number of the NECC or INECC graph, we can restrict our study to the SBAN with the simple sequential update schedule $(\{0\}, \{1\}, \cdots, \{n-1\})$.

Let us find now an upper bound for the chromatic number of the INECC graph, by defining a coloring method of the graph based on a greedy algorithm.

Lemma 5. $\chi(\text{INECC}) \leq 2^{2n/3+2}$.

Proof. Consider the BAN $F : \mathbb{B}^n \to \mathbb{B}^n$ and the simple sequential update schedule $W = (\{0\}, \{1\}, \cdots, \{n-1\})$. We partition the configurations into sets of equivalent configurations $E_1, E_2, \ldots, E_k$. Let us denote by $y^i \in \mathbb{B}^n$ the images of the configurations of E_i for each $i \in [\![1, k]\!]$. In other words, $\forall i \in [\![1, k]\!], \forall x \in E_i, F(x) = y^i$. We denote the neighbors of the ith image by $N(i)$, *i.e.*

$$N(i) = \{i' \mid \exists x \in E_i, x' \in E_{i'}, (x, x') \in \text{NECC}\}.$$

The degree of the ith image is denoted by $D(i) = |N(i)|$. We sort the images by decreasing degree so that $\forall i < i', D(i) \geq D(i')$. To choose the color of y^i, we apply a greedy algorithm. We use the smallest color not already used by a neighbor of y^i: $\text{color}(y^i) = min(\mathbb{N}^* \setminus \{\text{color}(y^{i'}) \mid i' < i \text{ and } i' \in N(i)\})$.

We can see that it is a proper coloring. Let us prove that if $(y^i, y^{i'}) \in$ INECC then $\text{color}(y^i) \neq \text{color}(y^{i'})$. Indeed, let $(y^i, y^{i'}) \in$ INECC. With no loss of generality, let us say that $i' < i$. By definition of INECC, $\exists (x, x') \in$ NECC such that $F(x) = y^i$ and $F(x') = y^{i'}$. So $i' \in N(i)$, and by definition of color, $\text{color}(y^i) \neq \text{color}(y^{i'})$. As a consequence, that is a proper coloring.

Now, let c be the biggest color used and k' the index of (one of) the images which have c as color. By construction, we have $c \leq D(E_{k'}) + 1$ and $c \leq k'$. For all i, we note $\ell_i = \lfloor \log_2(D(E_i) + 1) \rfloor$ and $\ell = \ell_{k'}$. Since $c \leq D(E_{k'}) + 1$, we have $c < 2^{\ell+1}$. Consider $M(i) = \{i' \mid (y^i)_{[\![0,n-\ell_i]\!]} = (y^{i'})_{[\![0,n-\ell_i]\!]}\}$ and $L(i) = N(i) \setminus M(i)$. Clearly, $|M(i)| \leq 2^{\ell_i - 1}$, and $i \in M(i)$. So $L(i) = (N(i) \cup \{i\}) \setminus M(i)$. We also know that $i \notin N(i)$. As a consequence, $|N(i) \cup \{i\}| = D(E_i) + 1 \geq 2^{\ell_i}$. Thus, $|L(i)| \geq 2^{\ell_i} - 2^{\ell_i - 1} = 2^{\ell_i - 1}$.

Moreover, $\forall x \in E_i$, $\{x' \in E_{i'} \mid i' \in L(i) \text{ and } (x, x') \in \text{NECC}\} \subseteq \{x' \mid x_{]\!]n-\ell_i,n[\![} = x'_{]\!]n-\ell_i,n[\![}\}$ because such a pair (x, x') should be confusable at some step $j \leq n - \ell_i$. So $\forall x \in E_i$, $|\{x' \in E_{i'} \mid i' \in L(i) \text{ and } (x, x') \in \text{NECC}\}| \leq 2^{n-\ell_i+1}$. Putting things together, we get:

$$2^{\ell_i - 1} \geq |L(i)|$$
$$\geq |\{(x, x') \in E_i \times E_{i'} \cap \text{NECC} \mid i' \in L(i)\}|$$
$$\geq |E_i| 2^{n-\ell_i+1}.$$

We get $|E_i| \leq 2^{\ell_i - 1} / 2^{n-\ell_i+1} = 2^{2\ell_i - n - 2}$.

Furthermore, $\sum_{i=1}^{k'} |E_i| \leq 2^n$ and $\forall i \leq k'$, $|E_i| \geq 2^{2\ell_i - n - 2} \geq 2^{2\ell - n - 2}$. So $k' 2^{2\ell - n - 2} \leq 2^n$ and $k' \leq 2^{2n+2-2\ell}$. Thus, $c \leq 2^{2n+2-2\ell}$. However, we have also $c \leq 2^{\ell+1}$. An upper bound for c is reached when $2^{\ell+1} = 2^{2n+2-2\ell}$ (see Fig. 1). In other words, when $2^{3\ell} = 2^{2n+1} \iff 2^\ell = 2^{(2n+1)/3}$. So, we have $c \leq 2^{(2n+1)/3+1}$ and $c \leq 2^{2n/3+2}$. Furthermore, $\chi(\text{INECC}) \leq c$. As a result, $\chi(\text{INECC}) \leq 2^{2n/3+2}$. □

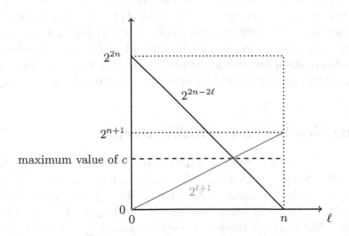

Fig. 1. Upper bound for c.

From Lemmas 4 and 5, we can deduce an upper bound for the chromatic number of a NECC graph. Furthermore, using the relation between the chromatic number of a $\text{NECC}_{F,W}$ graph and $\kappa(F, W)$, we can find an upper bound for κ_n.

Theorem 2. $\forall n \in \mathbb{N}, \kappa_n \leq 2n/3 + 2$.

Proof. Let $F : \mathbb{B}^n \to \mathbb{B}^n$ and $W \in \vec{\mathscr{P}}(V)$. Thanks to Lemmas 4 and 5, we know that $\chi(\text{NECC}_{F,W}) \leq \chi(\text{INECC}_{F,W})$ and $\chi(\text{INECC}_{F,W}) \leq 2^{2n/3+2}$. As a consequence, $\chi(\text{NECC}_{F,W}) \leq 2^{2n/3+2}$, $\log_2(\chi(\text{NECC}_{F,W})) \leq 2n/3 + 2$ and $\kappa(F, W) \leq 2n/3 + 2$. Thus, we have $\forall F : \mathbb{B}^n \to \mathbb{B}^n$ and $W \in \vec{\mathscr{P}}(V)$, $\kappa(F, W) \leq 2n/3 + 2$, which gives by definition, $\kappa_n \leq 2n/3 + 2$. □

(a) INECC graph of (F, W) (b) NECC graph of (F, W)

Fig. 2. INECC and NECC graphs of (F, W).

Remark 5. The chromatic number of the INECC graph gives an upper bound for the NECC graph. However, the NECC graph can have a smaller chromatic number. For instance, let us consider the following BAN. Let $F : \mathbb{B}^4 \to \mathbb{B}^4$ be such that $F((0,0,0,0)) = (0,0,0,0)$, $F((1,1,0,0)) = (0,0,0,0)$, $F((1,0,0,0)) = (0,1,0,0)$, $F((0,1,0,0)) = (0,1,0,1)$, and for all other $x \in \mathbb{B}^4$, $F(x) = (1,1,1,1)$. Let W be the simple sequential schedule $(\{0\}, \{1\}, \{2\}, \{3\})$. Figures 2a and b show that the chromatic number of the INECC and NECC graphs are respectively 3 and 2. So, even if the worst INECC graph had a chromatic number equal to $2^{2n/3}$, it would not disprove the conjecture: we can still hope that the worst NECC graph has a better chromatic number, by coloring some equivalent configurations differently.

6   Clique Number in the NECC Graph

The *clique number* of a graph G, denoted by $\omega(G)$, is the size of the biggest clique of G. We denote by $\omega(\text{NECC})$ the clique number of the NECC graph. In this part, we find the maximum value that $\omega(\text{NECC})$ can get. It is important because we know that the chromatic number is bigger that the clique number. So if in a NECC graph the clique number were bigger than $2^{n/2}$, then the chromatic number would be bigger as well and the conjecture would be wrong. However, if the clique number is smaller than $2^{n/2}$, then we cannot deduce anything about the conjecture. Lemma 6 below proves that the set of steps at which two configurations are confusable is an interval.

Lemma 6. *Let* $(x, x') \in CC$, $I = CC(x, x')$, $a = min(I)$ *and* $b = max(I)$. *Then* $I = [\![a, b]\!]$.

Proof. Since $a = min(I)$ and $b = max(I)$, we have $I \subseteq [\![a, b]\!]$. For the sake of contradiction, let us suppose that there exists $j \in [\![a, b]\!]$ such that $j \notin I$. Let j be the smallest such number. So $F_{W_{<j}}(x) \neq F_{W_{<j}}(x')$, $j \neq a$ because $a \in I$ and $j - 1 \in [\![a, b]\!]$ (because $j \neq a$). Furthermore, $j - 1$ does not valid this propriety, because j is the smallest number which validates it. As a consequence, $F_{W_{<j-1}}(x) = F_{W_{<j-1}}(x')$ and $F_{W_{<j}}(x) \neq F_{W_{<j}}(x')$. So $F(x)_{W_{j-1}} \neq F(x')_{W_{j-1}}$. Furthermore, $F_{W_{<b}}(x)_{W_{j-1}} = F(x)_{W_{j-1}}$ because $j \leq b$ (and then $W_{j-1} \subseteq W_{<b}$) and $F_{W_{<b}}(x')_{W_{j-1}} = F(x')_{W_{j-1}}$. So $F_{W_{<b}}(x)_{W_{j-1}} \neq F_{W_{<b}}(x')_{W_{j-1}}$, and thus $F_{W_{<b}}(x) \neq F_{W_{<b}}(x')$. As a consequence, $b \notin I$ which is a contradiction. This gives $I = [\![a, b]\!]$. □

Lemma 7 shows that if two configurations are confusable with a third one at a given step, then they are also confusable between themselves at this step.

Lemma 7. *Let* $x, x', x'' \in \mathbb{B}^n$. *We have:* $CC(x, x') \cap CC(x, x'') \subseteq CC(x', x'')$.

Proof. Let $i \in CC(x, x') \cap CC(x, x'')$. Thus, $F_{W_{<i}}(x) = F_{W_{<i}}(x')$ and $F_{W_{<i}}(x) = F_{W_{<i}}(x'')$. As a consequence, $F_{W_{<i}}(x') = F_{W_{<i}}(x'')$ and $i \in CC(x', x'')$. Hence, we have $\forall i \in CC(x, x') \cap CC(x, x''), i \in CC(x', x'')$. □

Lemma 8 shows that if two configurations are confusable with a third one, the two former ones are confusable if and only if they are confusable with the third one at some simultaneous step.

Lemma 8. *Let* $x, x', x'' \in \mathbb{B}^n$ *such that* $(x, x') \in CC$ *and* $(x, x'') \in CC$. *Then, we have:* $CC(x, x') \cap CC(x, x'') \neq \emptyset \iff (x', x'') \in CC$.

Proof. Suppose that $CC(x, x') \cap CC(x, x'') \neq \emptyset$. By Lemma 7, we know that $CC(x, x') \cap CC(x, x'') \subseteq CC(x', x'')$. So $CC(x', x'') \neq \emptyset$. As a result, $(x', x'') \in CC$. Now, suppose that we have $(x', x'') \in CC$ and let $[\![a, b]\!] = CC(x, x')$ and $[\![a', b']\!] = CC(x, x'')$. For the sake of contradiction, consider that $CC(x, x') \cap CC(x, x'') = \emptyset$, i.e. $[\![a, b]\!] \cap [\![a', b']\!] = \emptyset$. With no loss of generality, consider that $0 \leq a \leq b < a' \leq b' < p = |W|$. Let $j \in CC(x', x'')$. Thus, $F_{W_{<j}}(x') = F_{W_{<j}}(x'')$. We can show that $j \notin [\![a, b]\!] \cup [\![a', b']\!]$. Indeed, if $j \in [\![a, b]\!]$, then $j \in CC(x, x')$ and $F_{W_{<j}}(x) = F_{W_{<j}}(x')$. So $F_{W_{<j}}(x) = F_{W_{<j}}(x'')$ (because, by definition of j, we have $F_{W_{<j}}(x') = F_{W_{<j}}(x'')$) and, as a consequence, $j \in CC(x, x'')$ and thus $j \in CC(x, x') \cap CC(x, x'')$. As a result, $CC(x, x') \cap CC(x, x'') \neq \emptyset$. There is a contradiction, so $j \notin [\![a, b]\!]$. Similarly, we can prove that $j \notin [\![a', b']\!]$. Now, let us prove that $j \notin [\![0, a[\![$. For the sake of contradiction let us say that $j \in [\![0, a[\![$. Then, $\exists j' \in]\![j, a[\![$, $F(x')_{W_{j'}} \neq F(x'')_{W_{j'}}$. Otherwise, we would have $F_{W_{<a}}(x'') = F_{W_{<a}}(x') = F_{W_{<a}}(x)$ and then $[\![a, b]\!] \cap [\![a', b']\!] \neq \emptyset$. Furthermore, we know that $F_{W_{<a}}(x') = F_{W_{<a}}(x)$ (because $a \in CC(x, x')$) and $W_{j'} \subseteq W_{<a}$ (because $j' < a$) so $F(x')_{W_{j'}} = F(x)_{W_{j'}}$ and thus $F(x'')_{W_{j'}} \neq F(x)_{W_{j'}}$. As a consequence, $F_{W_{<a'}}(x')_{W_{j'}} \neq F_{W_{<a'}}(x)_{W_{j'}}$ (because $W_{j'} \subseteq W_{<a'}$ since $j' < a < a'$). So $a' \notin CC(x, x'')$. This is a contradiction. So $j \notin [\![0, a[\![$. Now, let us prove that $j \notin$

$]b, a'[\cup]b', p[$. If $j \in]b, a'[\cup]b', p[$ then $j > b$. We know that $F(x)_{W_b} \neq F(x')_{W_b}$ (otherwise we would have $F_{W_{<b+1}}(x) \neq F_{W_{<b+1}}(x')$ and then $b + 1 \in \mathsf{CC}(x, x')$). However, we have $F(x)_{W_b} = F(x'')_{W_b}$, because $W_b \subseteq W_{<a'}$ since $b < a'$. So $F(x')_{W_b} \neq F(x'')_{W_b}$. Thus, $F_{W_{<j}}(x') \neq F_{W_{<j}}(x'')$ because $W_{<j}$ because $b < j$, which is a contradiction. As a consequence, $j \notin]b, a'[\cup]b', p[$. As a result, j does not exist. Thus, $\mathsf{CC}(x', x'') = \emptyset$, and finally, $(x', x'') \notin \mathsf{CC}$. □

Lemma 9 shows that all cliques of the NECC graph have at least one step during which all the configurations of the clique are simultaneously confusable.

Lemma 9. *Let X be a clique of the* NECC *graph. Then, we have: $\exists i, \forall x, x' \in X$, $i \in \mathsf{CC}(x, x')$.*

Proof. Let $x \in X = \{x^1, x^2, \ldots, x^k\}$ such that $|X| = k$, and let $I = I_1 \cap I_2 \cap \cdots \cap I_k$ where $I_1 = \mathsf{CC}(x, x^1), \ldots, I_k = \mathsf{CC}(x, x^k)$. We can prove that all the intervals intersect each other two by two. In other words, $\forall i, i' \in [0, k[$, $I_i \cap I_{i'} \neq \emptyset$. For the sake of contradiction, assume that there are disjoint intervals. In this case, we would have $x', x'' \in X$ such that $\mathsf{CC}(x, x') \cap \mathsf{CC}(x, x'') = \emptyset$. By Lemma 8, we would have $(x', x'') \notin \mathsf{CC}$. However, $x', x'' \in X$, so $(x', x'') \in \mathsf{CC}$. There is a contradiction. Consequently, all the intervals intersect each other two by two, and we know that if a set of intervals intersect each other two by two then they have an interval in common. So $I \neq \emptyset$.

Let $i \in I$. Now, let us prove that $\forall x', x'' \in X$, $i \in \mathsf{CC}(x', x'')$. Let $x', x'' \in X$. We have $i \in \mathsf{CC}(x, x')$ and $i \in \mathsf{CC}(x, x'')$. Thus, $F_{W_{<i}}(x) = F_{W_{<i}}(x')$ and $F_{W_{<i}}(x) = F_{W_{<i}}(x'')$, which implies that $F_{W_{<i}}(x') = F_{W_{<i}}(x'')$. As a result, $i \in \mathsf{CC}(x', x'')$ and $\forall x', x'' \in X$, $i \in \mathsf{CC}(x', x'')$. □

Using Lemma 9, Theorem 3 shows that the clique number of any NECC graph is less than or equal to $2^{n/2}$.

Theorem 3. $\omega(\mathsf{NECC}) \leq 2^{\lfloor n/2 \rfloor}$.

Proof. Let X be the biggest clique of the NECC graph, $x \in X$ and i such that $\forall x, x' \in X$, $i \in \mathsf{CC}(x, x')$ (Thanks to Lemma 9, we know there is one). In other words, $\forall x' \in X$, $F_{W_{<i}}(x') = F_{W_{<i}}(x)$. So $\forall x, x' \in X$, $x_{\overline{W_{<i}}} = x'_{\overline{W_{<i}}}$ and $F(x)_{W_{<i}} = F(x')_{W_{<i}}$. Let $x \in X$. There are 2 cases:

- $|W_{<i}| < n/2$. Then, we have $|\overline{W_{<i}}| \geq n/2$. Thus, $|\{x' \mid x'_{\overline{W_{<i}}} = x_{\overline{W_{<i}}}\}| < 2^{n/2}$ and, since $X \subseteq \{x' \mid x'_{W_{<i}} = x_{W_{<i}}\}$, we have $|X| < 2^{n/2}$.
- $|W_{<i}| \geq n/2$. Then, we have $\{F(x') \mid x' \in X\} \subseteq \{x' \mid F(x')_{W_{<i}} = F(x)_{W_{<i}}\}$ and $|\{F(x') \mid F(x')_{W_{<i}} = F(x)_{W_{<i}}\}| \leq 2^{n/2}$. In this case, since all configurations of X are not equivalent, we have $\forall x, x' \in X$, $x \neq x' \implies F(x) \neq F(x')$. Thus, $|X| \leq |\{F(x') \mid x' \in X\}|$. As a consequence, $|X| \leq 2^{n/2}$.

In all cases, we have $|X| \leq 2^{n/2}$. So $\omega(\mathsf{NECC}) \leq 2^{n/2}$. □

This result supports Conjecture 1 because the NECC graphs with the biggest chromatic number that we succeeded to build are graphs with big clique number. It seems we reached the limit of this technique.

7   Class of Bijective BANs

In this part, we study BANs whose global transition functions are bijective, *i.e.* BANs whose dynamics with a parallel update schedule are only composed of recurrent configurations. For this class of BANs, we can prove a result which is really close to the conjecture. We prove this using two intermediate lemmas. The first one is that if two configurations are confusable then either the first parts of the two images are equal or the second parts of the two configurations are.

Lemma 10. *If $W = (0, 1, \ldots, n)$ then $\forall (x, x') \in CC,\ F(x)_{[0,n/2[} = F(x')_{[0,n/2[}$ or $x_{[n/2,n[} = x'_{[n/2,n[}$.*

Proof. Let $(x, x') \in CC$. Then, $\exists i \in [0, n[,\ F_{[0,i[}(x) = F_{[0,i[}(x')$. Let i be the smallest such number. We have: $F(x)_{[0,i[} = F(x')_{[0,i[}$ and $x_{[i,n[} = x'_{[i,n[}$. Then, i can follow the two cases below:

- $i \leq n/2$. Then, $[n/2, n[\subseteq [i, n[$ and $x_{[n/2,n[} = x'_{[n/2,n[}$;
- $i \geq n/2$. Then, $[0, n/2[\subseteq [0, i[$ and $F(x)_{[0,n/2[} = F(x')_{[0,n/2[}$.

And we get the expected result.   □

The next lemma is a simple consequence of Lemma 10: if we take the neighbors of a configuration in a NECC graph and we take the set of images of these configurations when we apply F, then this set has less than $2^{n/2+1} - 2$ elements.

Lemma 11. *If $W' = (0, 1, \ldots, n)$ then $\forall x \in \mathbb{B}^n,\ |\{F(x') \mid (x, x') \in \mathrm{NECC}\}| \leq 2^{n/2+1} - 2$.*

Proof. Let $x \in \mathbb{B}^n$. According to Lemma 10, $\forall x' \in \mathbb{B}^n,\ F(x)_{[0,n/2[} = F(x')_{[0,n/2[}$ or $x_{[n/2,n[} = x'_{[n/2,n[}$. Then, $\{x' \mid (x, x') \in \mathrm{NECC}\} \subseteq \{x' \mid x_{[n/2,n[} = x'_{[n/2,n[}\} \cup \{x' \mid F(x)_{[0,n/2[} = F(x')_{[0,n/2[}\}$. So $\{F(x') \mid (x, x') \in \mathrm{NECC}\} \subseteq \{F(x') \mid x_{[n/2,n[} = x'_{[n/2,n[}\} \cup \{F(x') \mid F(x)_{[0,n/2[} = F(x')_{[0,n/2[}\}$. Thus, $|\{F(x') \mid (x, x') \in \mathrm{NECC}\}| \leq |\{F(x') \mid x_{[n/2,n[} = x'_{[n/2,n[}\}| + |\{F(x') \mid F(x)_{[0,n/2[} = F(x')_{[0,n/2[}\}|$. And we have: $|\{F(x') \mid F(x)_{[0,n/2[} = F(x')_{[0,n/2[}\}| \leq 2^{n/2}$. Furthermore, $|\{x' \mid x_{[n/2,n[} = x'_{[n/2,n[}\}| \leq 2^{n/2}$. As a consequence, $|\{F(x') \mid x_{[n/2,n[} = x'_{[n/2,n[}\}| \leq 2^{n/2}$. So, $|\{F(x') \mid (x, x') \in \mathrm{NECC}\}| \leq 2^{n/2+1}$. Furthermore, $F(x) \in \{F(x') \mid x_{[n/2,n[} = x'_{[n/2,n[}\}$ and $F(x) \in \{F(x') \mid F(x)_{[0,n/2[} = F(x')_{[0,n/2[}\}$ but $F(x) \notin \{F(x') \mid (x, x') \in \mathrm{NECC}\}$. Consequently, $|\{F(x') \mid (x, x') \in \mathrm{NECC}\}| \leq 2^{n/2+1} - 2$, which is the expected result.   □

Using the fact that we are talking about a bijective function, and thanks to Lemma 11, we bound the degree of every configuration in the NECC graph. Then, we deduce a bound for the chromatic number of the NECC and, thus, a bound for κ.

Theorem 4. *If $F : \mathbb{B}^n \to \mathbb{B}^n$ is a bijective function then $\kappa(F, W) \leq n/2 + 1$.*

Proof. Let $F : \mathbb{B}^n \rightarrow \mathbb{B}^n$ be a bijective function. For all $x \in \mathbb{B}^n$, let $d(x)$ be the degree of x in the NECC graph. In other words, $\forall x$, $d(x) = |\{x' \mid (x, x') \in \text{NECC}\}|$. Let $x \in \mathbb{B}^n$ be the configuration with maximal degree. We know by Lemma 11 that $|\{F(x') \mid (x, x') \in \text{NECC}\}| \leq 2^{n/2+1} - 2$. However, since F is a bijective function, we have $|\{F(x') \mid (x, x') \in \text{NECC}\}| = |\{x' \mid (x, x') \in \text{NECC}\}|$ and then, $d(x) \leq 2^{n/2+1} - 2$. So, $\chi(\text{NECC}) \leq 2^{n/2+1} - 1$. Thus, $\log_2(\chi(\text{NECC})) \leq \frac{n}{2} + 1$. As a result, $\kappa(F, W) \leq \frac{n}{2} + 1$. $\qquad\square$

8 Conclusion and Future Research

In this article, we were interested in the minimal number κ of additional automata that a SBAN associated with a block-sequential update schedule needs to simulate another given one with a parallel update schedule, in the worst case. The maximum value that κ can take for all SBANs of size n is denoted by κ_n. To answer this question, we introduced the concept of NECC graph, a graph built from SBANs. We proved that the log of the chromatic number of this graph and the κ of a SBAN are the same quantity. We achieved to bound κ_n in the interval $[n/2, 2n/3 + 2]$ and we conjectured that κ_n is equal to $n/2$. To support this conjecture, we showed that the maximum clique number that a NECC graph can have is equal to $2^{n/2}$. This means that the NECC graph of a SBAN which would have a κ greater than $n/2$ would have a NECC graph with a chromatic number greater than the clique number. Finally, we showed that the conjecture is true (up to one extra automaton) if we restrain to SBANs whose global transition functions are bijective.

More work is needed to close the gap $[n/2, 2n/3 + 2]$ left on κ_n. There is also a related problem where, given a SBAN with a parallel update schedule, we search the number of additional automata needed for a SBAN with any sequential update schedule (*i.e.*, we do not impose any order on the update schedule) to simulate the first SBAN. We can see that for some BANs, this number is really smaller than when we impose an order. We can take the example used in Lemma 3. The BAN has $n/2$ pairs of automata that exchange their values. If the mandatory order is to update one automaton only of every pair of automata and then the other we need $n/2$ additional automata. But if the order is free then we can update all the pairs of automata one at a time and do with only one additional automaton using a parity trick. This is a particular BAN and the problem of finding an upper bound in the worst case better than κ_n is still open.

Furthermore, we could study the issue presented in this article with other kinds of update schedules (which update many times each automata for instance) or other kinds of intrinsic simulations (where many automata can represent one simulated automaton for example).

These results could also help to design new SBANs behaving the same way as a given one, with different update schedule, and as small as possible. Associated with the concept of functional modularity, we could also use them to replace a small functional module with an unexpected behavior in some situations by another module that is more robust to schedule variations.

Acknowledgements. This work has been partially supported by the project PACA APEX FRI.

References

1. Aracena, J.: On the robustness of update schedules in boolean networks. Biosystems **97**, 1–8 (2009)
2. Bruck, J., Goodman, J.W.: A generalized convergence theorem for neural networks. IEEE Trans. Inf. Theor. **34**, 1089–1092 (1988)
3. Delorme, M., Mazoyer, J., Ollinger, N., Theyssier, G.: Bulking I: an abstract theory of bulking. Theor. Comput. Sci. **412**, 3866–3880 (2011)
4. Delorme, M., Mazoyer, J., Ollinger, N., Theyssier, G.: Bulking II: classifications of cellular automata. Theor. Comput. Sci. **412**, 3881–3905 (2011)
5. Demongeot, J., Elena, A., Sené, S.: Robustness in regulatory networks: a multidisciplinary approach. Acta Biotheor. **56**, 27–49 (2008)
6. Doty, D., Lutz, J.H., Patitz, M.J., Schweller, R.T., Summers, S.M., Woods, D.: The tile assembly model is intrinsically universal. In: Proceedings of FOCS 2012, pp. 302–310. IEEE Computer Society (2012)
7. Goles, E., Martínez, S.: Neural and Automata Networks: Dynamical Behavior and Applications. Kluwer Academic Publishers, Dordrecht (1990)
8. Goles, E., Matamala, M.: Computing complexity of symmetric quadratic neural networks. In: Proceedings of ICANN 1993, p. 677 (1993)
9. Goles, E., Noual, M.: Disjunctive networks and update schedules. Adv. Appl. Math. **48**, 646–662 (2012)
10. Goles, E., Salinas, L.: Comparison between parallel and serial dynamics of boolean networks. Theor. Comput. Sci. **396**, 247–253 (2008)
11. Guillon, P.: Projective subdynamics and universal shifts. In: DMTCS Proceedings of AUTOMATA 2011, pp. 123–134 (2011)
12. Hopfield, J.J.: Neural networks and physical systems with emergent collective computational abilities. Proc. Nat. Acad. Sci. USA **79**, 2554–2558 (1982)
13. Kauffman, S.: Gene regulation networks: a theory for their global structures and behaviors. Curr. Top. Dev. Biol. **6**, 145–181 (1971). Springer
14. Kauffman, S.A.: Metabolic stability and epigenesis in randomly constructed genetic nets. J. Theor. Biol. **22**, 437–467 (1969)
15. Lafitte, G., Weiss, M.: Universal tilings. In: Thomas, W., Weil, P. (eds.) STACS 2007. LNCS, vol. 4393, pp. 367–380. Springer, Heidelberg (2007). doi:10.1007/978-3-540-70918-3_32
16. Lafitte, G., Weiss, M.: An almost totally universal tile set. In: Chen, J., Cooper, S.B. (eds.) TAMC 2009. LNCS, vol. 5532, pp. 271–280. Springer, Heidelberg (2009). doi:10.1007/978-3-642-02017-9_30
17. McCulloch, W.S., Pitts, W.: A logical calculus of the ideas immanent in nervous activity. J. Math. Biophys. **5**, 115–133 (1943)
18. Melliti, T., Regnault, D., Richard, A., Sené, S.: On the convergence of boolean automata networks without negative cycles. In: Kari, J., Kutrib, M., Malcher, A. (eds.) AUTOMATA 2013. LNCS, vol. 8155, pp. 124–138. Springer, Heidelberg (2013). doi:10.1007/978-3-642-40867-0_9
19. Melliti, T., Regnault, D., Richard, A., Sené, S.: Asynchronous simulation of boolean networks by monotone boolean networks. In: El Yacoubi, S., Wąs, J., Bandini, S. (eds.) ACRI 2016. LNCS, vol. 9863, pp. 182–191. Springer, Cham (2016). doi:10.1007/978-3-319-44365-2_18

20. Noual, M.: Updating automata networks. Ph.D. thesis, École Normale Supérieure de Lyon (2012)
21. Noual, M., Regnault, D., Sené, S.: About non-monotony in boolean automata networks. Theor. Comput. Sci. **504**, 12–25 (2013)
22. Ollinger, N.: Universalities in cellular automata. In: Rozenberg, G., et al. (eds.) Handbook of Natural Computing, pp. 189–229. Springer, Heidelberg (2012)
23. Orponen, P.: Computing with truly asynchronous threshold logic networks. Theor. Comput. Sci. **174**, 123–136 (1997)
24. Tchuente, M.: Sequential simulation of parallel iterations and applications. Theor. Comput. Sci. **48**, 135–144 (1986)
25. Thomas, R.: Boolean formalization of genetic control circuits. J. Theor. Biol. **42**, 563–585 (1973)

On Resource-Bounded Versions
of the van Lambalgen Theorem

Diptarka Chakraborty[1]([✉]), Satyadev Nandakumar[2], and Himanshu Shukla[2]

[1] Computer Science Institute of Charles University,
Malostranské námesti 25, 118 00 Praha 1, Czech Republic
diptarka@iuuk.mff.cuni.cz
[2] Department of Computer Science and Engineering,
Indian Institute of Technology Kanpur, Kanpur, Uttar Pradesh, India

Abstract. The van Lambalgen theorem is a surprising result in algorithmic information theory concerning the symmetry of relative randomness. It establishes that for any pair of infinite sequences A and B, B is Martin-Löf random and A is Martin-Löf random relative to B if and only if the interleaved sequence $A \uplus B$ is Martin-Löf random. This implies that A is relative random to B if and only if B is random relative to A [1–3]. This paper studies the validity of this phenomenon for different notions of time-bounded relative randomness.

We prove the classical van Lambalgen theorem using martingales and Kolmogorov compressibility. We establish the failure of relative randomness in these settings, for both time-bounded martingales and time-bounded Kolmogorov complexity. We adapt our classical proofs when applicable to the time-bounded setting, and construct counterexamples when they fail. The mode of failure of the theorem may depend on the notion of time-bounded randomness.

1 Introduction

In this paper, we explore the resource-bounded versions of van Lambalgen's theorem in algorithmic information theory. van Lambalgen's theorem deals with the symmetry of relative randomness. The theorem states that an infinite binary sequence B is Martin-Löf random and a sequence A is Martin-Löf random relative to B if and only if the interleaved sequence $A_0B_0A_1B_1\dots$ is Martin-Löf random [1]. It follows that A is Martin-Löf random relative to B if and only if B is Martin-Löf random relative to A.

This result is quite surprising, since it connects the randomness of A with the computational power A possesses [2,3]. Symmetry of relative randomness is desirable for any robust notion of randomness. However, we now know that it fails in several other settings - both Schnorr randomness and computable randomness exhibit a lack of symmetry of relative randomness [4].

D. Chakraborty has been supported by the European Research Council under the European Union's Seventh Framework Programme (FP/2007-2013)/ERC Grant Agreement n. 616787.

© Springer International Publishing AG 2017
T.V. Gopal et al. (Eds.): TAMC 2017, LNCS 10185, pp. 129–143, 2017.
DOI: 10.1007/978-3-319-55911-7_10

We explore whether this symmetry holds when Martin-Löf randomness is replaced with time-bounded randomness. Considering the failure of the analogies of van Lambalgen's theorem in many settings, it is natural to guess that such a resource-bounded version of van Lambalgen's theorem is false. Indeed, the existence of *one-way functions* [5] from strings to strings which are easy to compute but hard to invert, can be expected to have some bearing to the validity of the resource-bounded van Lambalgen's theorem. In the context of polynomial-time compressibility, Longpré and Watanabe [6] establish the connection between polynomial-time symmetry of information and the existence of one-way functions, and analogously, Lee and Romaschenko [7] establish the connection for CD complexity [8].

Modern proofs of van Lambalgen's theorem proceed by defining Solovay tests (see [2,3]). The notion of a resource-bounded Solovay test has not been studied, while the notion of resource-bounded martingales [9] and resource-bounded Kolmogorov complexity have been studied extensively (see Allender et al. [10]). We approach the classical van Lambalgen's theorem using prefix-free incompressibility and martingales, inspired by the Solovay tests. This part may be of independent interest. We then attempt to adapt these proofs to resource-bounded settings.

Our main results are the following. Let t be a superlinear time bound, and t^X denotes t-computable functions with oracle access to the sequence X. Let $A \uplus B$ denotes the interleaving of A and B.

1. Using the notion of t-bounded martingales, we show that there exists a t-nonrandom sequence $A \uplus B$ where B is t-random and A is t^B-random. This result is unconditional, and analogous to the result of Yu [4].
2. There are sequences A and B where A is t^B-nonrandom, but B is t^A-random. However for this pair, $A \uplus B$ is still t-nonrandom. Thus the randomness of the interleaved sequence and mutual relative randomness of the pair are distinct notions for time-bounded martingales. *We establish a sufficient condition under which a t-random B and a t^B-nonrandom A could still create t-random $A \uplus B$. This involves a non-invertibility condition reminiscent of one-way functions.*
3. There are t-compressible $A \uplus B$ such that B is t-incompressible and A is t-incompressible relative to B. This is an unconditional result analogous to 1.
4. If B is t-compressible or A is t-compressible with respect to B, then $A \uplus B$ is t-compressible. This is in contrast to 2.

Thus van Lambalgen's theorem fails in resource-bounded settings. Surprisingly, the manner of failure may depend on the formalism we choose.

The results in the paper also provide indirect evidence that resource-bounded randomness may vary depending on the formalism. In particular, the set of sequences over which resource-bounded martingales fail may not be the same as the set of resource-bounded incompressible sequences. The results in 2 and 4 actually provide us a conditional separation between these two formalisms in case of resource-bounded settings, which were identical in general.

The manner of failure in 2 has to do with the oracle access mechanism, and the proof hinges on a technical obstacle which may be tangential to time-bounded computation. In the final section of the paper, we propose a modified definition which we call t-bounded "lookahead" martingales with which we are able to show that if B is t-lookahead-nonrandom or A is t-lookahead-nonrandom relative to B, then $A \uplus B$ is t-lookahead nonrandom. Here, the van Lambalgen property for t-lookahead martingales fails in precisely the same manner as t-incompressibility. This may be a reasonable model to study resource-bounded martingales.

2 Preliminaries

We assume familiarity with the basic notions of algorithmic randomness at the level of the initial chapters in Downey and Hirschfeldt [3] or Nies [2].

We use the notation $\mathbb{N}$ for the set of natural numbers, $\mathbb{Q}$ for rationals, and $\mathbb{R}$ for reals. We work with the binary alphabet $\Sigma = \{0,1\}$. We denote the set of finite binary strings as Σ^* and the set of infinite binary sequences as Σ^ω. Finite binary strings will be denoted by lower-case Greek letters like σ, ρ etc. and infinite sequences by upper-case Latin symbols like X, Y etc. The length of a string σ is denoted by $|\sigma|$. The letter λ stands for the empty string. For finite strings σ and ρ and any infinite sequence X, $\sigma \preceq \rho$ and $\sigma \preceq X$ denote that σ is a prefix of ρ and X respectively.

The substring of length n starting from the mth position of a finite string σ or an infinite sequence X is denoted by $\sigma[m \ldots m+n-1]$ and $X[m \ldots m+n-1]$, where $m + n - 1 < |\sigma|$. When m is 0, $i.e.$ the first position, we abbreviate the notation as $\sigma \restriction n$ and $X \restriction n$ - e.g. $\sigma \restriction n$ is $\sigma[0 \ldots n-1]$.

The concatenation of σ and τ is written as $\sigma\tau$. The notation $A \uplus B$ stands for the sequence we get by interleaving the bits in A with the bits in B, $i.e.$ $A_0 B_0 A_1 B_1 \ldots$.[1]

A set of finite strings S is said to be *prefix-free* if no string in S can be a proper prefix of another string in S.

Theorem 1 *(van Lambalgen, 1987)* [1]. *For any two infinite sequences A and B, B is Martin-Löf random and A is Martin-Löf random relative to B if and only if $A \uplus B$ is Martin-Löf random.*

3 A Proof Using Incompressibility

We now prove Theorem 1 via incompressibility notions. Throughout the remainder of the paper, we fix a canonical set of prefix-free codes for partial computable functions by $\mathcal{P}$.

Definition 1. *The* self-delimiting Kolmogorov complexity *of $\sigma \in \Sigma^*$ is defined by $K(\sigma) = \min\{|\pi| \mid \pi \in \mathcal{P} \text{ outputs } \sigma\}$.*

---

[1] It is also common to use $\oplus$, but we want to avoid confusion with the bitwise xor operation.

Similarly, the *conditional Kolmogorv complexity* of $\sigma \in \Sigma^*$ given $\tau \in \Sigma^*$ is defined by $K(\sigma \mid \tau) = \min\{|\pi| \mid \pi \in \mathcal{P} \text{ outputs } \sigma \text{ on input } \tau\}$.

Using the notion of incompressibility, it is well-known that we can formulate an equivalent definition of random sequences [2].

Definition 2. *An infinite binary sequence A is said to be* incompressible *if $\exists c \ \forall n \ \ K(A \upharpoonright n) \geq n + c$. The sequence A is* incompressible with respect to *another binary sequence B (or B-incompressible) if $\exists c \ \forall n \ \ K^B(A \upharpoonright n) \geq n+c$.*

The set of Martin-Löf random sequences are precisely the set of incompressible sequences. Relativizing the same result, the set of Martin-Löf random sequences relative to a sequence B is precisely the set of sequences incompressible with respect to B.

We now prove van Lambalgen's theorem using incompressibility. When we consider the issue of resource-bounded van Lambalgen's theorems, we try to either adapt these proofs where applicable, or examine the issues which prevent such an adaptation. As in the martingale proof, we prove the two directions of the van Lambalgen's theorem separately so as to emphasize the issues which arise in the resource-bounded setting.

The proof of the first direction relies on a form of Symmetry of Information, a result first established by Levin and Gács [8]. To this end, we mention basic results from the theory of self-delimiting (prefix-free) Kolmogorov complexity.

Definition 3. *A computably enumerable set $L \subseteq \Sigma^* \times \mathbb{N}$ is said to be a* bounded request set *if $\sum_{(\sigma,n)\in L} \frac{1}{2^n} \leq 1$.*

We may view each element (w, n) as a request to encode w using at most n bits. The boundedness condition is a promise that the requested code lengths satisfy the Kraft inequality. The Machine Existence Theorem states that there is some prefix-free code which can satisfy all requests in a bounded request set.

Theorem 2 *(Machine Existence Theorem) [2]. Let L be a bounded request set. Then there is a prefix-free set of codes $\mathcal{P}$ which, for each $(y, m) \in L$, allocates a prefix-free code $\tau \in \Sigma^m \cap \mathcal{P}$ for y.*

The coding theorem relates the algorithmic probability of a string to its prefix-free Kolmogorov Complexity. We state it here in the form applicable to pairs of strings, but an analogous result holds for strings.

Theorem 3 *(Coding Theorem) [2]. Let τ be a finite string. Let $\mathcal{P}$ be a prefix-free encoding of partial-computable functions outputting pairs of strings. Denote $\mathcal{P}_\tau \subseteq \mathcal{P}$ as the set of prefix-free codes which output pairs (σ, τ) for some arbitrary string σ. Then there is a constant c such that*

$$2^{c-K(\sigma,\tau)} > \sum_{\rho \in \mathcal{P} \text{ outputting } (\sigma,\tau)} 2^{-|\rho|}$$

Using these, we now state and prove the variant of "Symmetry of Information" which we use to establish Lemma 2.

Lemma 1. *Let σ be a finite string with $K(\sigma) > |\sigma| + c$, and τ be a finite string. Then $|\sigma| + K(\tau|\sigma) \leq K(\sigma, \tau) + O(1)$.*

Proof. Let p_i be an arbitrary program in the computable enumeration of $\mathcal{P}$, the set of programs which output string pairs. Consider the program R_{p_i} which can be generated from p_i, defined by the following algorithm.

1. Input σ.
2. Let $U(p_i)$ output the string pair (α, τ).
3. If α is equal to σ, then we output $(\tau, |p_i| - |\sigma| + c)$, where c satisfies the inequality below.

Corresponding to the computable enumeration p_1, p_2, ... of $\mathcal{P}$, we obtain a computable enumeration R_{p_1}, R_{p_2}, ... We now show that this forms a valid enumeration of a *bounded request set* (see, for example, [2] page 78).

Let N_σ be the set of indices $i \in \mathcal{P}$ where $U(p_i)$ outputs a pair of strings of the form (σ, τ) for some τ. First, we have

$$\sum_{i \in N_\sigma} \frac{1}{2^{|p_i| - |\sigma| + c}} < 2^{|\sigma| - c} \sum_{i \in N_\sigma} \frac{1}{2^{|p_i|}} < 2^{|\sigma| - c + c'} \sum_{\tau \in \Sigma^*} \frac{1}{2^{K(\sigma, \tau)}} = \frac{2^{|\sigma| - c + c'}}{2^{K(\sigma)}} < 1,$$

where the second inequality follows from the Coding Theorem (see, for example, Nies [2], Theorem 2.2.25), and the last inequality follows from the assumption.

Hence R_{p_1}, R_{p_2}, ... is a computable enumeration of a bounded request set. By the Machine existence theorem for prefix-free encoding (see for example, [2] Theorem 2.2.17), it follows that for any request $(\tau, |p_i| - |\sigma| + c)$, there is a prefix-free encoding of τ given σ which has length $|p_i| - |\sigma| + c$. Now, consider a shortest prefix-free code p_i for (σ, τ). We have that $|p_i| = K(\sigma, \tau)$. Hence $K(\tau \mid \sigma) \leq K(\sigma, \tau) - |\sigma| + O(1)$. □

Lemma 2. *If B is incompressible and A is B-incompressible, then $A \uplus B$ is incompressible.*

Proof. Suppose that for every n, $K(B \upharpoonright n) \geq n + c$ and $K^B(A \upharpoonright n) \geq n + c'$. This implies that $K(A \upharpoonright n \mid B \upharpoonright (n-1)) \geq n + c'$. By the version of the Symmetry of information in Lemma 1, we have

$$(2n - 1) + c' \quad < \quad (n - 1) + K^B(A \upharpoonright n) \quad \leq \quad K((A \uplus B) \upharpoonright (2n - 1)) + O(1).$$

A similar argument will work for $K((A \uplus B) \upharpoonright 2n)$ and this completes the proof. □

The above proof relied on symmetry of information of prefix-free Kolmogorov Complexity. Since reasonable complexity-theoretic hypotheses imply that this fails in resource-bounded settings, we can foresee that this direction fails in resource-bounded settings, as we show in Sect. 5.

Since the first direction was a consequence of Symmetry of Information, it is reasonable to expect the converse direction to follow from the subadditivity of K: $K((A \uplus B) \upharpoonright 2n) \leq K(B \upharpoonright n) + K^B(A \upharpoonright n) + O(1)$. However, this runs into the following obstacle. If the prefix of B is compressible with complexity, say $n - \log(K(n))$, and the prefix of A is B-incompressible with conditional complexity $n + K(n)$, then we cannot conclude from subadditivity that $K((A \uplus B) \upharpoonright 2n)$ is less than $2n$. Thus concatenating the shortest prefix-codes for $B \upharpoonright n$ and $A \upharpoonright n$ given $B \upharpoonright n$ to obtain a prefix-free code for $(A \uplus B) \upharpoonright 2n$ may be insufficient for our purpose. We now show the converse direction through more succinct prefix-free codes.

Lemma 3. *If B is compressible or A is B-compressible, then $A \uplus B$ is compressible.*

Proof. Let $K(B \upharpoonright n) < n + c$, and let σ be a shortest program from the c.e. set of codes $\mathcal{P}$ which outputs $B \upharpoonright n$. Consider the prefix-free set defined by

$$\mathcal{Q}_n = \{\tau\rho \mid \tau \in \mathcal{P}, |\rho| = n\}. \tag{1}$$

This is a prefix-free c.e. set of codes. Then $\sigma(A \upharpoonright n)$ is a code for $A \uplus B$ for some machine M which first runs $R(\sigma)$ to output $B \upharpoonright n$, then interleaves $A \upharpoonright n$ with $B \upharpoonright n$ to produce $(A \uplus B) \upharpoonright n$. The length of this code is at most $K(B \upharpoonright n) + n + O(1)$, showing that $A \uplus B$ is compressible at length $2n$.

Now, assume that A is B-compressible, and let n and m satisfy

$$K(A \upharpoonright n \mid B \upharpoonright m) \leq n + c.$$

Since we can make redundant queries, without loss of generality, we assume that $m \geq n$. Let $\mathcal{P}$ be the set of prefix-free encodings of one-argument partial computable functions. We construct a prefix-free code to show that $(A \uplus B)$ is compressible at length $2m$. Consider $\mathcal{Q}_{m,n}$ defined by

$$\mathcal{Q}_{m,n} = \{\tau\sigma \mid \tau \in \mathcal{P}, \sigma \in \Sigma^{2m-n}\}. \tag{2}$$

Since $\mathcal{P}$ is a prefix-free set and we append strings of a fixed length to the prefix-free codes, $\mathcal{Q}_{m,n}$ is also a prefix-free set. If $\mathcal{P}$ is computably enumerable, then so is $\mathcal{Q}$. Moreover, there is an encoding of $A_0 B_0 \ldots A_{m-1} B_{m-1}$ in $\mathcal{Q}_{m,n}$ given by $\alpha(B \upharpoonright m)(A[n \ldots m-1])$. This encoding has length $n + c + m + m - n$, which is less than or equal to $2m + c$. $\qquad \square$

We may expect this proof to be easily adapted to resource-bounded settings. Inherent in the above proof is the concept of universality – since there is a universal self-delimiting Turing machine which incurs at most additive loss over any other prefix-free encoding, it suffices to show that there is some prefix-free succinct encoding. We appropriately modify this in resource-bounded settings which lack such universal machines in general.

4   Martingales and van Lambalgen's Theorem

We now approach van Lambalgen's theorem using martingales, adapting the Solovay tests in the literature [2, 3].

Definition 4. *A function* $d : \Sigma^* \to [0, \infty)$ *is said to be a* martingale *if* $d(\lambda) = 1$ *and for every string* w, $d(w) = (d(w0) + d(w1))/2$, *and a* supermartingale *if for every string* w, $d(w) \geq (d(w0) + d(w1))/2$.

A martingale *or a* supermartingale *is said to be* computably enumerable *(c.e.) if there is a Turing Machine* $M : \Sigma^* \times \mathbb{N} \to \mathbb{Q}$ *such that for every string* w, *the sequence* $M(w, n)$ *monotonically converges to* $d(w)$ *from below.*

The rate of convergence in the above definition need not be computable.

Definition 5. *We say that a martingale* d *succeeds on* $X \in \Sigma^\omega$ *if* $\limsup_{n \to \infty} d(X \upharpoonright n) = \infty$, *and write* $X \in S^\infty[d]$. *If no computably enumerable martingale or supermartingale succeeds on* X, *then we say that* X *is Martin-Löf random. We say that* X *is* non-Martin-Löf random *relative to* Y *if there is a computably enumerable oracle martingale* d *such that* $\limsup_{n \to \infty} d^Y (X \upharpoonright n) = \infty$.

Lemma 4. *If* B *is not Martin-Löf random or* A *is not Martin-Löf random relative to* B, *then* $A \uplus B$ *is not Martin-Löf random.*

Proof. Let d_B be a martingale that succeeds on B. Then the martingale d_{AB} defined by setting $d_{AB}(\lambda)$ to 1 and

$$d_{AB}(\sigma_0 \tau_0 \ldots \tau_{n-2} \sigma_{n-1}) = d_{AB}(\sigma_0 \tau_0 \ldots \tau_{n-2}). \qquad (3)$$
$$d_{AB}(\sigma_0 \tau_0 \ldots \sigma_{n-1} \tau_{n-1}) = d_B(\tau_0 \tau_1 \ldots \tau_{n-1}).$$

The above definition is a martingale since for any $n \geq 2$,

$$d_{AB}(\alpha_0 \beta_0 \ldots \beta_{n-2} \alpha_{n-1}) = d_B(\beta_0 \ldots \beta_{n-2}).$$

Clearly, $\limsup_{n \to \infty} d_{AB}(A \uplus B) = \limsup_{n \to \infty} d_B(B)$ and hence d_{AB} succeeds on $A \uplus B$.

Now, suppose d succeeds on A given oracle access to B. Consider martingale m defined by setting $m(\lambda)$ to 1 and setting

$$m(\sigma_0 \tau_0 \ldots \sigma_{n-1})[s] = d^{\tau \upharpoonright s}(\sigma_0 \sigma_1 \ldots \sigma_{n-1})$$
$$m(\sigma_0 \tau_0 \ldots \tau_{n-1})[s] = m(\sigma_0 \tau_o \ldots \sigma_{n-1})[s],$$

where the notation $m(\alpha)[s]$ denotes the value that the computation m assigns to α at stage s and for any string $x \in \Sigma^*$, the value of $m(x) = \limsup_{s \to \infty} m(x)[s]$. Note that in the computation of d in the second step, each fixed initial segment of v can query longer initial segments of w when they become available.

Since d is a c.e. oracle martingale, it follows that m is a c.e. martingale. For every pair of infinite sequences V and W and for every l, there is a number s' computable from $V \upharpoonright l$ and W such that for all large enough stages $s \geq s'$, $d^{W \upharpoonright s}(V \upharpoonright l) = d^W(V \upharpoonright l)$. Thus for each l, the value of $m((V \upharpoonright l) \uplus (W \upharpoonright l))[s]$ is the same as $m((V \upharpoonright l) \uplus (W \upharpoonright l))[s']$ for all $s \geq s'$, for some large enough s'. It follows that m is c.e. martingale. Since d^B succeeds on A, m succeeds on $A \uplus B$ and this completes the proof. □

We mention that the converse also holds. We show later that in time-bounded versions, the analogous results may not hold.

Lemma 5. *If $A \uplus B$ is not Martin-Löf random, then either B is not Martin-Löf random or A is not Martin-Löf random relative to B.*

5   Resource-Bounded Relative Randomness and Incompressibility

We consider time-bounded self-delimiting Kolmogorov complexity in this section. While there are several variants of this notion (see e.g. [6,10]), we deal with the simplest one here.

The time-bound is a function of the lengths of its output[2] as in [6]. We first fix a prefix-free set $\mathcal{P}$ encoding the set of partial-computable functions. We do not insist that $\mathcal{P}$ consist solely of functions which run in t steps, since the results are identical with or without this assumption.

Definition 6. *The t-time-bounded complexity of σ is defined as defined by $K_T(\sigma; t) = \min\{|\pi| \mid \pi \in \mathcal{P}$ outputs σ in $\leq O(t(|\sigma|))$ steps.$\}$ and the conditional t-time-bounded complexity of σ given τ be defined by $K_T(\sigma \mid \tau; t) = \min\{|\pi| \mid \pi \in \mathcal{P}, \pi(\tau)$ outputs σ in $\leq O(t(|\sigma|))$ steps.$\}$.*

For any fixed time bound t, we do not have universal machines within the class of t-bounded machines. However, there are invariance theorems (see e.g. [8] Chap. 7). Hence we can use the definition of time-bounded complexity to define the notion of incompressible infinite sequences.

Definition 7. *An infinite binary sequence X is said to be t-incompressible if $\exists c \ \forall n \ \ K_T(X \upharpoonright n) \geq n + c$ and t^Y-incompressible if $\exists c \ \forall n \ \ K_T(X \upharpoonright n \mid Y \upharpoonright m) \geq n + c$ for some m depending on the value of n.*

If for $t' > t$, a sequence X is t'-incompressible, then it is t-incompressible as well. Moreover, for every X and n, $K_T(X \upharpoonright n) \geq K(X \upharpoonright n)$. Since the set of K-incompressible sequences has measure 1, we know that the set of t-incompressible sequences has measure 1 as well.

We now show that for time-bounded Kolmogorov complexity, only one direction of the implication holds. We first show that it is possible to compress $A \uplus B$ within the time bounds if B can be compressed, or A can be compressed relative to B within the same time bound.

Lemma 6. *If B is t-compressible or A is t^B-compressible, then $A \uplus B$ is t-compressible.*

---

[2] Considering time bound that is dependent on output length is not unnatural for decompressors. To make it input-length dependent it is customary to append 1^l as an additional input where l is the output length.

Proof. Assume that B is t-compressible. Then there is a constant c and infinitely many n such that there is a short program in $\beta \in \mathcal{P}$ with $|\beta| < n + c$ which outputs $B \upharpoonright n$ within $O(t(n))$ steps.

For any such n, consider the prefix-code defined by

$$\mathcal{Q}_n = \{\sigma\alpha \mid \sigma \in \mathcal{P}, |\alpha| = n\}. \tag{4}$$

This forms a prefix encoding, containing a code $[\beta(A \upharpoonright n)]$ for $(A \uplus B) \upharpoonright 2n$. Moreover, it is possible to decode $(A \uplus B) \upharpoonright 2n$ from its code within $O(t(2n))$ steps.

Suppose A is t^B-compressible. Assume that $K_T((A \upharpoonright n) \mid (B \upharpoonright m); t) \le n + c$, witnessed by a code α. Without loss of generality, we may assume $n \le m$. Then consider $\mathcal{Q}_{m,n}$ as defined in (2). We see that $\mathcal{Q}_{m,n}$ is a computably enumerable prefix set. The code $(\alpha(B \upharpoonright m)A[n \ldots m-1]) \in \mathcal{Q}_{m,n}$ of $(A \uplus B) \upharpoonright 2n$ can be decoded in time $O(t(2m))$, and is shorter than $2m + c$.

Hence $K_T((A \uplus B) \upharpoonright 2m) < 2m + c$. $\qquad\square$

The converse of the above lemma is false. We do not appeal to the failure of polynomial-time (in general, resource-bounded) symmetry of information (see for example, [6]), but directly construct a counterexample pair.

Lemma 7. *There are sequences A and B where $A \uplus B$ is t-compressible, but B is t-incompressible and A is t-incompressible relative to B.*

Proof. In $s = 0$, we have $A_s = B_s = \lambda$. Then in stage $s \ge 1$, assume that we have inductively defined prefixes A_{s-1} of A and B_{s-1} of B and additionally $|A_{s-1}| = |B_{s-1}| = l_{s-1}$. We select strings α_s, β_s and γ_s satisfying specific incompressibility properties and then define

$$A_s = A_{s-1}\alpha_s\gamma_s \quad \text{and} \quad B_s = B_{s-1}\beta_s\alpha_s$$

We choose α_s, β_s, γ_s which satisfy following incompressibility requirements.

1. Length requirements: $|\alpha_s| = s$, $|\beta_s| = |\gamma_s| = 2^{t(2l_{s-1})^2}$,
2. Incompressibility requirements for B: $K(\beta_s \mid B_{s-1}) \ge |\beta_s| + c$ for some constant c and $K(\alpha_s \mid B_{s-1}\beta_s) \ge |\alpha_s| + c'$ for some constant c'.
3. Incompressibility requirements for A relative to B: $K(\alpha_s \mid A_{s-1}) \ge |\alpha_s| + c''$ for some constant c'' and $K(\gamma_s \mid A_{s-1}\alpha_s, B_{s-1}\beta_s) \ge |\gamma_s| + c'''$ for some constant c'''.

It suffices to show we can find such strings α_s, β_s and γ_s. We can select the strings in the following order. First, select a string β_s which satisfies $K(\beta_s \mid B_{s-1}) \ge 2^{t(2l_{s-1})^2} + c$ for some constant c. Then select the string α_s to satisfy $K(\alpha_s \mid A_{s-1}, B_{s-1}\beta_s) \ge s + c'$ for some constant c'. Finally, select γ_s satisfying $K(\gamma_s \mid A_{s-1}\alpha_s, B_{s-1}\beta_s) \ge 2^{t(2l_{s-1})^2} + c''$ for some constant c''. Each of these selections is possible because the set of incompressible strings conditioned on any other strings is non-empty (for example, see the Ample Excess Lemma [11]).

By the above construction it is clear that B is incompressible and $A \uplus B$ is compressible for any function $t(n) > n$ due to the shared component α_s for all s between A and B. However by the argument used in the proof of Lemma 8, we can show that A is t^B incompressible and this completes the proof. $\qquad\square$

6   Resource-Bounded Relative Randomness and Martingales

In this section, we show that the symmetry of relative randomness does not hold for resource-bounded martingales. Let $t : \mathbb{N} \to \mathbb{N}$ be a superlinear function. For any input $\sigma \in \Sigma^*$, we henceforth restrict ourselves to martingales computed in time $O(t(|\sigma|))$ and we define t-randomness accordingly.

Lemma 8. *There is a t-random sequence B and a sequence A which is t^B-random, where $A \uplus B$ is t-nonrandom.*

The idea of the construction is that at some positions, substrings in A are copied exactly from regions of B. These regions of B sufficiently far so that it is not possible to consult the relevant region in time $O(t)$. Of course, $A \uplus B$ is nonrandom since a significant suffix of B can be computed directly from the relevant region of A.

Elsewhere, if B is random, and A random relative to B, then we can make B t-random, and A to be t^B-random.

In short, the construction ensures that B has sufficient time to look into the prefix of A, but A does not have time to look into the extension of B.

Proof. We construct two sequences A and B in stages, where at stage $s = 0$, we have $A_s = B_s = \lambda$. At stage $s \geq 1$, let us assume that we have inductively defined prefixes A_{s-1} of A and B_{s-1} of B and additionally $|A_{s-1}| = |B_{s-1}| = l_{s-1}$. We select strings α_s, β_s and γ_s satisfying specific randomness properties and then define

$$A_s = A_{s-1}\alpha_s\gamma_s \quad \text{and} \quad B_s = B_{s-1}\beta_s\alpha_s$$

We choose strings α_s, β_s and γ_s which satisfy all the following randomness requirements.

1. Length requirements: $|\alpha_s| = s$, $|\beta_s| = |\gamma_s| = 2^{t(2l_{s-1})^2}$,
2. Randomness requirements for B: for some universal martingale $d^{B_{s-1}}$ which has an oracle access to the sequence B_{s-1}, for every $v \preceq \beta_s$, $d(v) \leq 1$ and for some universal martingale $d^{B_{s-1}\beta_s}$ which has an oracle access to the sequence $B_{s-1}\beta_s$, for every $v \preceq \alpha_s$, $d(v) \leq 1$.
3. Randomness requirements for A relative to B: for some universal martingale $d^{A_{s-1}}$ which has an oracle access to the sequence A_{s-1}, for every $v \preceq \alpha_s$, $d(v) \leq 1$ and for some universal martingale $d^{A_{s-1}\alpha_s, B_{s-1}\beta_s}$ which has oracle access to the sequence $A_{s-1}\alpha_s$ and $B_{s-1}\beta_s$, for every $v \preceq \gamma_s$, $d(v) \leq 1$.

It suffices to show we can find such strings α_s, β_s and γ_s. We can select the strings in the following order. First, select a string β_s which satisfies the fact that for a universal martingale $d^{B_{s-1}}$ which has an oracle access to the sequence B_{s-1}, for every $v \preceq \beta_s$, $d(v) \leq 1$. Such a string β_s exists because the martingale property together with the Markov inequality allows us to show that the set of strings $w \in \Sigma^j$ such that $\forall v \preceq w, d(v) \leq 1$ has positive probability. By a similar argument we can then select the string α_s such that for a universal martingale $d^{A_{s-1},B_{s-1}\beta_s}$ which has oracle access to the sequence A_{s-1} and $B_{s-1}\beta_s$, for every $v \preceq \alpha_s$, $d(v) \leq 1$. Finally, select γ_s satisfying the fact that for a universal martingale $d^{A_{s-1}\alpha_s,B_{s-1}\beta_s}$ which has oracle access to the sequence $A_{s-1}\alpha_s$ and $B_{s-1}\beta_s$, for every $v \preceq \gamma_s$, $d(v) \leq 1$.

By the above construction it is clear that B is Martin-Löf random and $A \uplus B$ is not t-random for any function $t(n) > n$ due to the shared component α_s for all s between A and B. However we can show that A is t^B-random. By the construction it can be noted that for any martingale d^B with an oracle access to the sequence B, for the sequence A only the part α_s can give value more than one only if it can query the corresponding portion of the sequence B. To calculate the value of $d^B(A \upharpoonright n)$ it needs to query the index bigger than $2^{\omega(t(n))}$ of the sequence B which is not possible if the martingale is only allowed to run within time $O(t(n))$. □

Now, we consider the converse. Before doing that let us state the following simple fact.

Lemma 9. *Let B be an arbitrary t-random sequence. Then there is a sequence t^B-nonrandom sequence A such that B is t^A-random.*

An easy implication of the above stated lemma is the following.

Corollary 1. *There are sequences A and B such that A is t^B nonrandom, $A \uplus B$ is t-nonrandom, and B is t^A random.*

Now let us make one more simple observation.

Lemma 10. *If B is t-nonrandom then for any sequence A, $A \uplus B$ is t-non-random.*

Proof. If d_B be a t-martingale that witnesses the fact that B is t-nonrandom, then the martingale d_{AB} defined in (1) is a t-martingale that succeeds on $A \uplus B$. □

We wish to investigate the question of t-randomness of $A \uplus B$ given that A is t^B-nonrandom. We have weak converses which we now describe. The above corollary suggests that we stipulate "honest" reductions - that a bit at position n in A cannot depend on bits at positions $o(t^{-1}(n))$ in B. With this stipulation, we have the following weak converse to Lemma 8. First, we consider a restricted class of reductions from A to B.

Definition 8. *We say that an infinite sequence A is infinitely often reducible to B in time t via f, written $A \leq_{i.o}^t B$, if $\{n \in \mathbb{N} \mid f(B[n-t(n)\ldots n+t(n)-1]) = A_n\}$ is computable in time $O(t(n))$, i.e., t-computable.*

Note that we have incorporated an honesty requirement into the definition.

Definition 9. *We say that a function $f : \Sigma^* \to \Sigma$ is strongly influenced by the last index if for every $\sigma \in \Sigma^n$, $f(\sigma) \neq f((\sigma \upharpoonright n-1)\overline{\sigma_n})$.*

The function that projects the last bit of its input, and the function computing the parity of all input bits are two examples of such functions.

Lemma 11. *Let B be t-random and $A \leq_{i.o}^t B$ via a function that is strongly influenced by the last index. Then $A \uplus B$ is also t-nonrandom.*

Proof. Consider the t-computable set of positions $S = \{n \mid f(B[n - t(n) \ldots n + t(n) - 1]) = A_n\}$ where A queries B. We define a martingale d with initial capital 1 and which bets evenly on all positions except those in the set T defined by

$$T = \{2(i + t(i)) + 1 \mid i \in S\}.$$

For positions $2(i + t(i)) + 1 \in T$, sets $d(A_0 B_0 \ldots A_{i+t(i)} b)$ to $2d(A_0 \ldots A_{i+t(i)})$ if $f((B \restriction i + t(i) - 1)b) = A_i$, and to 0 otherwise. Then $A \uplus B \in S^\infty[d]$. □

A second weak converse can be obtained by assuming that the t-martingale succeeds on the interleaved sequence in a specific manner.

Definition 10. *We say that a pair of sequences (A, B) is t-resilient if*

1. *For every oracle martingale h runs in time $O(t(n))$, $\limsup_{n \to \infty} h^{B \restriction n-1}(A \restriction n) < \infty$.*
2. *For every oracle martingale g runs in time $O(t(n))$, $\limsup_{n \to \infty} g^{A \restriction n}(B \restriction n) < \infty$.*

We say that a martingale d wins at position i on a sequence X if $d(X \restriction i) > d(X \restriction i - 1)$.

Lemma 12. *$A \uplus B$ is t-random iff (A, B) is a t-resilient pair.*

Proof. Suppose that there exists a martingale d which runs in time $O(t(n))$ and witnesses the fact that $A \uplus B$ is restricted t-nonrandom. Now construct the oracle martingales h and g as follows:

$$h^Y(\sigma) = g(\sigma) = d(\lambda) \text{ if } \sigma = \lambda \text{ or } \sigma \in \Sigma$$

$$h^Y(X \restriction n) = \frac{d(X \uplus Y \restriction 2n - 1)}{d(X \uplus Y \restriction 2n - 2)} \cdot h^Y(X \restriction n - 1)$$

$$g^X(Y \restriction n) = \frac{d(X \uplus Y \restriction 2n)}{d(X \uplus Y \restriction 2n - 1)} \cdot g^X(Y \restriction n - 1)$$

Clearly h is dependent on $B \restriction n - 1$ and g is dependent on $A \restriction n$. Since $\limsup_{n \to \infty} d(A \uplus B \restriction n) \uparrow \infty$, we claim that one of h and g succeeds over A and B given $B \restriction n - 1$ and $A \restriction n$ respectively. We have

$$\limsup_{n \to \infty} h^{B \restriction n-1}(A \restriction n) \cdot g^{A \restriction n}(B \restriction n) = \limsup_{n \to \infty} \frac{d(A \uplus B \restriction 2n)}{c}$$

$$\leq \limsup_{n \to \infty} h^{B \restriction n-1}(A \restriction n) \cdot \limsup_{n \to \infty} g^{A \restriction n}(B \restriction n)$$

for some fixed constant c (independent of n). Note that LHS is ∞ because $d(A \uplus B \restriction n)$ is a sequence which satisfies the property

$$2d(A \uplus B \restriction n - 1) \geq d(A \uplus B \restriction n) \geq 0$$

and $\limsup_{n \to \infty} d(A \uplus B \restriction n) = \infty$. So one of the term involving h or g has to go to ∞. Now we show that h and g are oracle martingales which run in time $O(t(n))$. By construction h and g are oracle functions computable in time $O(t(n))$. Now

$$\sum_{b \in \Sigma} h((A \restriction n)b) = \frac{h(A \restriction n)}{d(A \uplus B \restriction 2n)} \sum_{b \in \Sigma} d((A \uplus B \restriction 2n)b) = 2 \cdot h(A \restriction n)$$

and thus h is a oracle martingale. By a similar argument g will also become a oracle martingale which runs in time $O(t(n))$. Since either $\limsup_{n \to \infty} h^{B \restriction n - 1}(A \restriction n) = \infty$ or $\limsup_{n \to \infty} g^{A \restriction n}(B \restriction n) = \infty$, it follows that (A, B) is a not a t-resilient pair.

Conversely, if (A, B) is not a t-resilient pair then either there is a oracle martingale h runs in time $O(t(n))$ such that $\limsup_{n \to \infty} h^{B \restriction n - 1}(A \restriction n) = \infty$, or a oracle martingale g runs in time $O(t(n))$ such that $\limsup_{n \to \infty} g^{A \restriction n}(B \restriction n) = \infty$. If the first condition holds, then

$$d(A \uplus B \restriction 2n - 1) = h^{B \restriction n - 1}(A \restriction n),$$
$$d(A \uplus B \restriction 2n) = d(A \uplus B \restriction 2n - 1)$$

is a t-martingale witnessing that $A \uplus B$ is t-nonrandom. If the second condition holds then we can define a similar martingale d based on g, witnessing t-nonrandomness of $A \uplus B$.   □

7   A Modified Definition of Resource-Bounded Martingales

In this section, we propose an alternate definition of a time-bounded martingale whose behavior with respect to van Lambalgen's theorem is identical to the definition using time-bounded prefix-complexity. In the light of van Lambalgen's theorem, we may view this as a reasonable variant definition.

Definition 11. *We say that a martingale* $d : \Sigma^* \to [0, \infty)$ *is a* t-*bounded lookahead martingale if the following conditions are satisfied.*

1. $d(\lambda) = 1$ *and* $L_0 = \emptyset$.
2. *For any string* σ, $d(\sigma 0) + d(\sigma 1) = 2d(\sigma)$.
3. *For any string* $\sigma \in \Sigma^{n-1}$, *if* $n - 1 \notin L_{d,\sigma}$ *then to compute* $d(\sigma b)$, $b \in \Sigma$, *the martingale can query a set of positions* $S \subseteq \{0, \ldots, n-2, n, \ldots, O(t(n))\}$ *and* $L_{d,\sigma b}$ *is set to* $L_{d,\sigma} \cup S$; *otherwise set* $d(\sigma b) = d(\sigma)$ *for all* $b \in \Sigma$ *and* $L_{d,\sigma b}$ *is set to* $L_{d,\sigma}$.

Definition 12. *We say that an infinite sequence is t-lookahead-non-random if there is a t-bounded lookahead martingale which succeeds on it.*

To compute $d(X \upharpoonright n)$, the martingale is allowed to wait until an appropriate extension length is available, and base its decision on a few bits ahead. However, we have to be careful not to reveal X_{n-1} itself, and to ensure that positions once revealed can never later be bet on. These restrictions ensure that the betting game is not trivial, and that there are unpredictable or random sequences.

Lemma 13. *There is a t-lookahead random sequence B and a t^B-lookahead random sequence A such that $A \uplus B$ is t-lookahead nonrandom.*

The proof is essentially the same as that of Lemma 8.

With the modified definition, we can now prove result similar to Lemma 6.

Lemma 14. *If B is t-lookahead nonrandom or A is t^B-lookahead nonrandom. Then $A \uplus B$ is t-lookahead nonrandom.*

Proof. Suppose h is a t-lookahead-martingale that succeeds on B. Then define the t-lookahead martingale d by setting $d(\lambda) = 1$ and $L_{d,\lambda} = \emptyset$, and

$$d((X \uplus Y) \upharpoonright 2n + 1) = d((X \uplus Y) \upharpoonright 2n), \quad L_{d,(X \uplus Y) \upharpoonright 2n+1} = L_{d,(X \uplus Y) \upharpoonright 2n}$$
$$d((X \uplus Y) \upharpoonright 2n + 2) = h(Y \upharpoonright n), \quad L_{d,(X \uplus Y) \upharpoonright 2n+2} = \{2i + 1 | i \in L_{h,Y \upharpoonright n}\}.$$

Then clearly $A \uplus B \in S^\infty[d]$ as $B \in S^\infty[h]$.

Now, assume that $A \in S^\infty[g^B]$ for a t-lookahead martingale g. Then we define the t-lookahead martingale d by $d(\lambda) = 1$ with $L_{d,\lambda} = \emptyset$ and

$$d((X \uplus Y) \upharpoonright 2n + 2) = d((X \uplus Y) \upharpoonright 2n + 1), \quad L_{d,(X \uplus Y) \upharpoonright 2n+2} = L_{d,(X \uplus Y) \upharpoonright 2n+1}$$
$$d((X \uplus Y) \upharpoonright 2n + 1) = g^{Y \upharpoonright O(t(n))}(X \upharpoonright n),$$
$$L_{d,(X \uplus Y) \upharpoonright 2n+1} = L_{d,(X \uplus Y) \upharpoonright 2n} \cup \{2i | i \in L_{g,X \upharpoonright n}\} \cup \{2i + 1 | i \in Q_{g,X,Y,n}\},$$

where $Q(g, X, Y, n)$ are the bits in the oracle queried by $g^Y(X \upharpoonright n)$. We know that $A \in S^\infty[g^B]$. Hence $A \uplus B \in S^\infty[d]$. □

Acknowledgments. The authors would like to acknowledge Jack Lutz, Manjul Gupta and Michal Koucký for helpful discussions.

References

1. van Lambalgen, M.: Random Sequences. Academish Proefschri't, Amsterdam (1987)
2. Nies, A.: Computability and Randomness. Oxford University Press, Inc., Oxford (2009)
3. Downey, R., Hirschfeldt, D.: Algorithmic randomness and complexity. Book Draft (2006)
4. Yu, L.: When van Lambalgen's theorem fails. Proc. Am. Math. Soc. **135**(3), 861–864 (2007)

5. Goldreich, O.: The Foundations of Cryptography. Basic Techniques, vol. 1. Cambridge University Press, Cambridge (2001)
6. Longpré, L., Watanabe, O.: On symmetry of information and polynomial-time invertibility. Inf. Comput. **121**, 14–22 (1995)
7. Lee, T., Romaschenko, A.: Resource-bounded symmetry of information revisited. Theor. Comput. Sci. **345**, 386–405 (2005)
8. Li, M., Vitányi, P.M.B.: An Introduction to Kolmogorov Complexity and Its Applications, 3rd edn. Springer, Berlin (2008)
9. Lutz, J.H.: Resource-bounded measure. In: Proceedings of the 13th IEEE Conference on Computational Complexity, pp. 236–248. IEEE Computer Society Press, New York (1998)
10. Allender, E., Buhrman, H., Koucký, M., van Melkebeek, D., Ronneburger, D.: Power from random strings. SIAM J. Comput. **35**(4), 1467–1493 (2006)
11. Miller, J.S., Yu, L.: On initial segment complexity and degrees of randomness. Trans. Am. Math. Soc. **360**(6), 3193–3210 (2008)

Scheduling Fully Parallel Jobs with Integer Parallel Units

Vincent Chau, Minming Li, and Kai Wang$^{(\boxtimes)}$

Department of Computer Science, Hong Kong Baptist University,
Hong Kong SAR, China
minming.li@cityu.edu.hk, kai.wang@my.cityu.edu.hk

Abstract. We consider the following scheduling problem. We have m identical machines, where each machine can accomplish one unit of work at each time unit. We have a set of n jobs, where each job j has s_j units of workload, and each unit workload could be executed on any machine at any time unit. A job is said completed when its whole workload has been executed. The objective is to find a schedule that minimizes the total weighted completion time $\sum w_j C_j$, where w_j is the weight of job j and C_j is the completion time of job j. We first give a PTAS of this problem when m is constant. Then we study the approximation ratio of a greedy algorithm, Largest-Ratio-First algorithm. Any permutation is a possible outcome of this algorithm when $w_j = s_j$ for each job j, and for this special case we show that the approximation ratio depends on the instance size, i.e. n and m. Finally, when jobs have arbitrary weights, we prove that the upper bound of the approximation ratio is $1 + \frac{m-1}{m+2}$.

1 Introduction

In scheduling problems, the total weighted completion time is one of the objective functions and is nowadays well studied [2]. More formally, we have a set of n jobs and a set of m machines. Each job j is defined by a workload s_j and a weight w_j. The goal is to schedule all jobs such that the total weighted completion time is minimized, i.e. $\sum_j w_j C_j$ where C_j is the completion time of job j. In the single machine case, Smith [13] showed that this problem could be solved optimally by a greedy algorithm, the *Largest-Ratio-First* algorithm.

Related Works. For the classical multiple machine case, each job has to be scheduled without preemption, i.e. when a job starts, it has to finish on the same machine before another job is executed on the machine. This problem is proved to be NP-hard in [3] for a fixed number of machines and a dynamic programming algorithm has been proposed in [10]. Kawaguchi and Kyan [6] gave the worst case analysis of LRF schedules. Later, Skutella and Woeginger [12] gave a PTAS for this problem.

The work described in this paper was fully supported by a grant from Research Grants Council of the Hong Kong Special Administrative Region, China (Project No. CityU 11268616).

© Springer International Publishing AG 2017
T.V. Gopal et al. (Eds.): TAMC 2017, LNCS 10185, pp. 144–157, 2017.
DOI: 10.1007/978-3-319-55911-7_11

When jobs have release dates, most of total weighted completion time minimization problems are NP-hard. Afrati et al. [1] gave PTASs for some classes of total weighted completion time minimization problems with release dates. Schulz and Skutella [11] provided a 2-approximation randomized algorithm for the multiple machines case. A variant has also been studied where each job has to be scheduled on a fixed number of machines at the same time. Fishkin et al. [4] gave a PTAS for this problem.

In the *concurrent open shop model* where there are a set of machines, and each job is required to be scheduled on several machines and it can be scheduled at the same time, the problem is proved to be NP-hard [7,9,14], while Garg et al. [5] showed that it is APX-hard. Recently, Mastrolilli et al. [8] proposed a primal-dual 2-approximation algorithm.

In this paper, we study another variant of this problem in which we consider fully parallel environment, aiming to minimize total weighted completion time. This model has been introduced by Zhang et al. in [15]. The difference with the classical problem is that the same job can be scheduled in parallel (at the same time) and always by unit part, i.e. a job cannot be scheduled for a fractional length. They proved that this problem is strongly NP-hard when the number of machines is the input of the problem and proposed a 2-approximation algorithm.

Our Contributions. First, we propose a PTAS for this problem when the number of machines is fixed. Then, we study the worst case of a greedy algorithm, the *Largest-Ratio-First* (LRF) algorithm and give the corresponding approximation ratio, as well as the tight bound with corresponding instance structure. The LRF algorithm may return any order of jobs when jobs have equal density, i.e. $s_j = w_j, \forall j \in J$. For this special case, we prove that the approximation ratio depends on the number of jobs and the number of machines. More specifically, we prove that the approximation ratio is upper bounded by $1 + \frac{2i - 2n/m}{i(i+1)}$ where $i = \lceil \frac{2n}{m} \rceil$. Finally, for the general case in which the jobs have arbitrary weights, we improve the result in [15] and show that the approximation ratio is bounded by $1 + \frac{m-1}{m+2}$.

2   Formulation

Given m as the number of identical machines, an instance J is a set of jobs defined as $J = \{(w_j, s_j) \mid w_j > 0, s_j \in \mathbb{N}\}$, where s_j is the workload and w_j is the weight of job j. We consider fully parallel jobs which are allowed to be executed on any machines at any time unit. Each machine can execute at most one job of one unit workload during one time unit. A schedule S is a table M where $M(i,t)$ defines the job that the i-th machine executes during time unit $[t-1,t)$, where $1 \leq i \leq m, t \in \{1,2,3,...\}$. For job j, the completion time is defined as $C_j = \max_{M(i,t)=j} t$, while the starting time is defined as $R_j = \min_{M(i,t)=j} (t-1)$. All jobs are available at time zero, a schedule is feasible if each job receives required processing resources, i.e. $s_j = |\{ (i,t) \mid M(i,t) = j \}|, \forall j \in J$. The objective is to minimize $T = \sum_{j \in J} w_j C_j$.

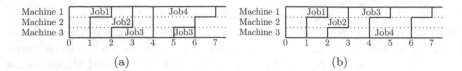

Fig. 1. Fully parallel schedule

First, we assume that the machine is never idle unless there are no more jobs to be executed on that machine. Second, we assume jobs are executed consecutively in any feasible schedule, as the way shown in Fig. 1b. Otherwise, we can always swap the execution to obtain a schedule with better or equal objective value. In Fig. 1a, by swapping the execution of $M(1,5)$ and $M(3,6)$, we get a better schedule, shown in Fig. 1b, where the completion time of job 3 becomes smaller. Therefore, we assume that in a feasible schedule, the execution of one job is consecutive, i.e. no *preemption* is needed to achieve optimality. Hence, a feasible schedule is a permutation of jobs.

We refer to the *length* of job j as the workload s_j, and the *density* of job j as the ratio $\frac{w_j}{s_j}$. Given a feasible schedule S of instance J, we use $i \preceq_S j$ to denote that job i is equal to job j or job i is scheduled before job j in schedule S.

Our Approach. Largest-Ratio-First (LRF) schedule is a schedule that assigns jobs in non-increasing order of $\frac{w}{s}$. It has been proved by Zhang et al. [15] that the upper bound of the approximation ratio of LRF schedule is 2 when jobs have arbitrary weights. However, when the number of machines m and the number of jobs n are fixed, the upper bound of the approximation ratio of LRF schedule can be smaller than 2, and it is related to m and n.

In this paper, we investigate the approximation ratio of the LRF schedule of this problem. Since the LRF schedule may return many possible orders of jobs when some jobs have exactly the same density w/s, we focus on the worst order that an LRF schedule may return, i.e. the order that maximizes the objective value[1]. In the sequel, LRF schedule always refers to the one that maximizes the objective value when there are several possible LRF schedules[2].

In Sect. 3, we present a PTAS. In Sect. 4, we first consider the instance of jobs with equal density, and focus on the worst order of jobs that an LRF schedule could return. Next, we try to locate and explore the properties of the instance (which we call organized instance) that contributes to the maximum approximation ratio of the LRF schedule, when the number of machines m and the number of jobs n are fixed. Then, we give a tight bound of the approximation ratio, which is related to $\left\lceil \frac{2n}{m} \right\rceil$. Finally, we consider the instance of jobs with arbitrary weights, and show that the approximation ratio of the LRF schedule takes maximum value when there are only two jobs.

---

[1] Without preemption and without idle time in order to preserve the order of jobs.

[2] One can always manipulate the order returned by LRF algorithm by changing the weight a little bit.

3   PTAS of Fixed Number of Machines

In this section, we give a PTAS for this problem when the number of machines m is a constant. Let $\epsilon > 0$ and set $q = (m-1)\lceil 1/\epsilon \rceil$.

PTAS: Given job set J, we pick a subset $Q \subseteq J$ such that $|Q| = q$. We make jobs Q as the first q jobs, and try every possibility and every permutation of jobs Q. For the remaining jobs $\overline{Q} = J\backslash Q$, we apply the *Largest-Ratio-First* rule.

Time Complexity: $\binom{n}{q} \cdot (q)! \cdot n = n^{q+1} = O(n^{1+(m-1)(1+1/\epsilon)})$.

Approximation Ratio: Now, we assume the schedule of jobs Q is fixed. Let $S_Q = \sum_{j\in Q} s_j$, then we have $S_Q \geq |Q| = q$. We will compare the total weighted completion time (refer to as cost) of jobs $\overline{Q}$ under two objectives: objective $T_1 = \sum_{j\in\overline{Q}} w_j \cdot \frac{S_Q + \sum_{i\in\overline{Q},i\prec j} s_i}{m}$ and objective $T_2 = \sum_{j\in\overline{Q}} w_j \cdot \left\lceil \frac{S_Q + \sum_{i\in\overline{Q},i\prec j} s_i}{m} \right\rceil$, where the latter is the objective of this problem.

Let σ be the order of jobs $\overline{Q}$ under *Largest-Ratio-First* rule. One should note that under objective T_1, schedule σ is still optimal (adjacent swapping is still effective). Let β be the order of jobs $\overline{Q}$ in the optimal schedule under objective T_2. Therefore the cost of jobs $\overline{Q}$ of our algorithm could be upper bounded:

$$
\begin{aligned}
cost(\overline{Q}) &= \sum_{j\in\overline{Q}} w_j \cdot \left\lceil \frac{S_Q + \sum_{i\in\overline{Q},i\preceq_\sigma j} s_i}{m} \right\rceil \\
&\leq \sum_{j\in\overline{Q}} w_j \cdot \left(\frac{S_Q + \sum_{i\in\overline{Q},i\preceq_\sigma j} s_i}{m} + \frac{m-1}{m} \right) \\
&\leq (1+\epsilon) \sum_{j\in\overline{Q}} w_j \cdot \frac{S_Q + \sum_{i\in\overline{Q},i\preceq_\sigma j} s_i}{m} \\
&\leq (1+\epsilon) \sum_{j\in\overline{Q}} w_j \cdot \frac{S_Q + \sum_{i\in\overline{Q},i\preceq_\beta j} s_i}{m} \\
&\leq (1+\epsilon) \sum_{j\in\overline{Q}} w_j \cdot \left\lceil \frac{S_Q + \sum_{i\in\overline{Q},i\preceq_\beta j} s_i}{m} \right\rceil \\
&\leq (1+\epsilon)OPT(\overline{Q})
\end{aligned}
$$

In the above equation, from line 1 to 2, we relax the ceiling function; from line 2 to 3, we apply $\epsilon \cdot S_Q \geq m-1$; from line 3 to 4, schedule σ is optimal under objective T_1; from line 4 to 5, we simply take ceiling function.

Since we try every possibility of Q, we are able to reach the case where the schedule of jobs Q are the same as the first q jobs scheduled in the optimal schedule. Therefore, we have

$$
cost(J) \leq (1+\epsilon)OPT(J)
$$

4   Worst Case of LRF Algorithm

In this section, we study the worst case of the *Largest-Ratio-First* algorithm. We start from the special case where jobs have equal density, then we extend the result to general case and give the tight upper bound.

Given a feasible schedule S over instance J, we define $L_S(j) = \sum_{i \preceq_S j} s_i$ as the total workload up to job j in schedule S and $L(J) = \sum_{j \in J} s_j$ as the total workload of jobs J. Let C_j^S (resp. R_j^S) be the completion time (resp. starting time) of job j in schedule S, and let $T(S, J) = \sum_{j \in J} w_j C_j^S$ be the corresponding objective value of schedule S. Especially, we denote $T(OPT, J)$ (resp. $T(LRF, J)$) as the objective value of the optimal schedule (resp. LRF schedule) of instance J.

Definition 1. *For any instance J, we define $\alpha(J) = \frac{T(LRF,J)}{T(OPT,J)}$ as the approximation ratio of LRF schedule of instance J. For any set of instances $\mathfrak{J}$, we define $\alpha(\mathfrak{J}) = \max_{J' \in \mathfrak{J}} \alpha(J')$ as the maximum approximation ratio of the instance $J' \in \mathfrak{J}$.*

4.1   Instance of Jobs with Equal Density

When jobs have the same density, i.e. for any pair of jobs i and j, we have $\frac{w_i}{s_i} = \frac{w_j}{s_j}$, any permutation of jobs is an LRF schedule, therefore we focus on the worst case i.e. $T(LRF, J) = \max_{S \in permutation(J)} T(S, J)$. Without loss of generality, we assume that the density of any job j is equal to 1, i.e. $w_j = s_j$, $\forall j \in J$.

Given the number of machines m and the number of jobs n, we aim to find the instance J^* of n jobs such that the corresponding LRF schedule has the maximum approximation ratio. In order to find J^*, we construct three kinds of instance sets and show that one of which always contains the worst instance.

Definition 2. *We refer to the job with $s_i = 1$ as unit job and the job with $s_i = m$ as basic job.*

Definition 3. *Given m as the number of identical machines, we define $\mathfrak{J}^{(m,n)} = \{J \mid \forall j \in J \; w_j = s_j, |J| = n\}$ as the set containing any instance of n jobs in which the density of any job is equal to 1. We define $\mathfrak{J}_{one}^{(m,n)} = \{J \mid \forall j \in J \; 1 \leq s_j \leq m, J \in \mathfrak{J}^{(m,n)}\}$ as the set containing any instance J in which each job has a length no larger than m.*

An organized instance $J = J(y, z, k)$ is an instance composed of y basic jobs, z unit jobs and exactly one job of length k such that $L(J) > m(y + 1)$, where $n = y + z + 1$, $1 < k \leq m$, $0 < z < n$ ($y, z, k \in \mathbb{N}$). We denote $\mathfrak{J}_{org}^{(m,n)}$ as the set containing any organized instance of n jobs.

By Definition 3, we can see that $\mathfrak{J}_{org}^{(m,n)} \subseteq \mathfrak{J}_{one}^{(m,n)} \subseteq \mathfrak{J}^{(m,n)}$, obviously, $\alpha(\mathfrak{J}_{org}^{(m,n)}) \leq \alpha(\mathfrak{J}_{one}^{(m,n)}) \leq \alpha(\mathfrak{J}^{(m,n)})$. However, in the following, we will show that $\alpha(\mathfrak{J}_{org}^{(m,n)}) = \alpha(\mathfrak{J}_{one}^{(m,n)}) = \alpha(\mathfrak{J}^{(m,n)})$. In other words, we show that there always exists an *organized instance* such that it achieves the maximum approximation ratio. Therefore, we focus on the organized instance. One should note that an organized instance contains at most one job which is neither unit job nor basic job. Based on this special structure, we then study the corresponding optimal schedule and LRF schedule and derive the approximation ratio.

In Sects. 4.1.1, 4.1.2 and 4.1.3, we prove that $\alpha(\mathfrak{J}_{org}^{(m,n)}) - \alpha(\mathfrak{J}_{one}^{(m,n)}) = \alpha(\mathfrak{J}^{(m,n)})$. First in Sect. 4.1.1, we introduce the idea of splitting jobs. Then, we show the properties of organized instances in Sect. 4.1.2. We finish the proof in Sect. 4.1.3. In Sect. 4.1.4, we give the tight bound of approximation ratios of LRF schedule on organized instances. More specifically, we divide the organized instances into regions according to the value $\lceil \frac{2n}{m} \rceil$ and give the corresponding tight bound in each region.

4.1.1 Splitting of Jobs

In this part, we first introduce two kinds of jobs in a specific schedule, *free* job, which is executed in one time unit, and *unlucky* job, whose completion time could be reduced by one if the job has a workload just one unit less. Then, we introduce the process of splitting jobs in a specific schedule, that replaces a job by two jobs, keeping the schedule of the remaining jobs unchanged.

Definition 4. *In a schedule S, job j is called **free** job if $C_j^S - R_j^S = 1$, otherwise we call it **non-free** job, job j is called **unlucky** job if $s_j > 1$ and $L_S(j)\%m = 1$.*

Definition 5. *A split process **replaces** a job i with $s_i > 1$ by two jobs i_1 and i_2 $(s_{i_1} > 0, s_{i_2} > 0)$ such that: $s_{i_1} + s_{i_2} = s_i$ and $w_{i_1}/s_{i_1} = w_{i_2}/s_{i_2} = w_i/s_i$*

Symmetrically, a *merge process* which replaces two jobs as one is the reverse of the split process, an example shown in Fig. 2.

Moreover, the process that splits job $j \in J$ into unit jobs means replacing job j by s_j unit jobs, i.e. $L(J) = L(J^{unit})$, where we define the new instance as J^{unit}. We denote $T(J^{unit})$ as the objective value of an arbitrary schedule of J^{unit}, since every job of J^{unit} is a unit job.

Lemma 1. *Given a feasible schedule S of instance J, for any job $i \in J$ with $s_i > 1$, the following properties hold after the split process that splits job i into two jobs i_1 and i_2 (assume job i_2 is scheduled after job i_1), denoting J' (resp. S') as the new instance (resp. new schedule)*

1. $T(S, J) \geq T(S', J')$.
2. *if job i is a free job in schedule S, then $T(S, J) = T(S', J')$.*
3. *if job i is a non-free job with $1 < s_i \leq m$, then $T(S, J) - T(S', J') \leq s_i - 1$.*
4. *if job i is an unlucky job with $1 < s_i \leq m$, then $T(S, J) - T(S', J') = s_{i_1}$.*

Proof. By the definition of *split process* (Definition 5), there is $s_i = s_{i_1} + s_{i_2}$ and $\frac{w_{i_1}}{s_{i_1}} = \frac{w_{i_2}}{s_{i_2}} = \frac{w_i}{s_i} = 1$. After the *split process*, the new schedule S' keeps the execution order of jobs in S, i.e. $R_{i_1}^{S'} = R_i^S$, $C_{i_2}^{S'} = C_i^S$ and $C_{i_1}^{S'} \leq C_i^S$. Thus, $T(S', J') - T(S, J) = w_{i_1}C_{i_1}^{S'} + w_{i_2}C_{i_2}^{S'} - w_iC_i^S = s_{i_1}(C_{i_1}^{S'} - C_i^S)$.

For the first property, since $C_{i_1}^{S'} \le C_i^S$, there is $T(S', J') - T(S, J) \le 0$.

For the second property, since job i is a *free* job in schedule S, that is $C_i^S - R_i^S = 1$, it implies $C_{i_1}^{S'} = C_{i_2}^{S'} = C_i^S$. Consequently, $T(S', J') - T(S, J) = 0$.

For the third property, since $1 < s_i \le m$, we have $C_i^S - 1 \le C_{i_1}^{S'} \le C_i^S$, then $T(S', J') - T(S, J) \ge -s_{i_1}$, i.e. $T(S, J) - T(S', J') \le s_i - 1$.

For the last property, since job i is an *unlucky* job in schedule S and $1 < s_i \le m$, it implies the completion time of job i_1 in S' must be $C_{i_1}^{S'} = C_i^S - 1$, that is $T(S', J') - T(S, J) = -s_{i_1}$.   □

As a consequence of Lemma 1, the process that splits an *unlucky* job i with $1 < s_i \le m$ into unit jobs will decrease the objective value by $s_i - 1$.

Lemma 2. *For any schedule S of instance J, we have $T(S, J) \ge T(J^{unit})$.*

Proof. We apply the *split process* to schedule S and split all jobs into unit jobs. Since the objective value of schedule S does not increase after the *split process* by Lemma 1, we have $T(S, J) \ge T(J^{unit})$.   □

4.1.2   Properties of Organized Instance
In this part, we discuss the optimal schedule and LRF schedule of a given *organized instance*, while the splitting method in the previous section is applied in the proof.

Lemma 3. *For any organized instance $J = J(y, z, k) \in \mathfrak{J}_{org}^{(m,n)}$, schedule $S_1 = (m^y, k, 1^z)$ (which first executes y basic jobs then the job of length k and finally z unit jobs) is an optimal schedule and $T(OPT, J) = T(J^{unit})$.*

Proof. In schedule S_1, we consider the *split process* that splits all jobs into unit jobs, we then obtain the instance J^{unit}. As a matter of fact, all jobs in schedule S_1 are *free* jobs, by Lemma 1 the objective value will not change after the *split process*, i.e. $T(S_1, J) = T(J^{unit})$. By Lemma 2, we have $T(OPT, J) \ge T(J^{unit})$. Therefore, schedule S_1 is an optimal schedule of instance J. Moreover, $T(OPT, J) = T(S_1, J) = T(J^{unit})$.   □

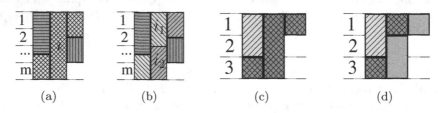

(a)   (b)   (c)   (d)

Fig. 2. A schedule of three jobs is shown in (a) and (b). The second job (job i) is split into two jobs (job i_1 and i_2). In (c) and (d), the LRF schedule of two jobs of length $\{2, 5\}$ with $m = 3$ is shown in (c). Then, the second job is split into a job of length 2 and a basic job, shown in (d).

Lemma 4. *For any organized instance $J = J(y, z, k) \in \mathfrak{J}_{org}^{(m,n)}$, schedule $S_2 = (1, m^y, 1^{m-k}, k, 1^{z+k-1-m})$ is an LRF schedule and $\alpha(J) = 1 + \frac{y(m-1)+(k-1)}{T(OPT,J)}$.*

Proof. We consider the *split process* that splits all jobs of J into unit jobs, then we get the instance J^{unit}. By Lemma 3, we have $T(OPT, J) = T(J^{unit})$.

Since J is an organized instance, it holds that $L(J) = ym + k + z$ and $L(J) > m(y + 1)$, that is, $z + k > m$, also $1 < k \le m$, thus, schedule S_2 is a feasible schedule. As a matter of fact, in schedule S_2 any job that is not a unit job is an *unlucky* job. We then apply the *split process* to schedule S_2. For each *unlucky* job i with length s_i $(1 < s_i \le m)$, after the *split process*, the objective value decreases by $s_i - 1$, according to Lemma 1. Therefore, after the *split process*, we have $T(S_2, J) - T(J^{unit}) = y(m - 1) + (k - 1)$.

Let σ be the LRF schedule of instance J, we then apply the *split process* to schedule σ, the objective value decreases only when splitting the basic jobs and the job of length k. Since J contains y basic jobs and one job of length k, we have $T(\sigma, J) - T(J^{unit}) \le y(m - 1) + (k - 1)$, according to the third property of Lemma 1. Therefore, $T(S_2, J) \ge T(\sigma, J)$, i.e. schedule S_2 is a LRF schedule. Moreover, $\alpha(J) = \frac{T(S_2,J)}{T(OPT,J)} = 1 + \frac{y(m-1)+(k-1)}{T(OPT,J)}$.   □

Lemma 5. *Given $m, n \ge 2$, $\alpha(\mathfrak{J}_{org}^{(m,n)}) > \alpha(\mathfrak{J}_{org}^{(m,n+1)})$.*

Proof. Let instance $J \in \mathfrak{J}_{org}^{(m,n+1)}$ be the instance such that $\alpha(J) = \alpha(\mathfrak{J}_{org}^{(m,n+1)})$. Then, we show that there exists an instance $J' \in \mathfrak{J}_{org}^{(m,n)}$ such that $\alpha(J') > \alpha(J)$.

Suppose instance J is composed of y basic jobs, z unit jobs, and one job of length k with $L(J) > m(y+1)$, where $n+1 = y+z+1$, $0 < z < n+1$, $1 < k \le m$ $(y, z, k \in \mathbb{N})$. Then $\alpha(J) = 1 + \frac{y(m-1)+(k-1)}{T(OPT,J)}$, according to Lemma 4.

Case (i) $1 < k < m$. Since $L(J) = ym + k + z$, we have $k + z > m$, i.e. $z > 1$. Then, we construct an instance J' which is composed of y basic jobs, $(z-1)$ unit jobs, and one job of length $(k+1)$. Since $L(J') = ym + (k+1) + (z-1)$, we have $L(J') = L(J) > m(y + 1)$, which further implies that $J' \in \mathfrak{J}_{org}^{(m,n)}$. Therefore, $\alpha(J') = 1 + \frac{y(m-1)+k}{T(OPT,J')}$, according to Lemma 4. Since $L(J') = L(J)$, we have $T(OPT, J') = T(OPT, J) = T(J^{unit})$, according to Lemma 3. Consequently, we have $\alpha(J) < \alpha(J')$.

Case (ii) $k = m$. If $z > 1$, then, we construct an instance J' which is composed of y basic jobs, $(z-1)$ unit jobs, and one job of length k. Since $k + z > m + 1$, we have $L(J') = ym + k + (z-1) > m(y+1)$, which further implies that $J' \in \mathfrak{J}_{org}^{(m,n)}$. As $L(J') < L(J)$, combined with Lemma 3, we have $T(OPT, J') < T(OPT, J)$. Moreover, by Lemma 4, we have $\alpha(J') = 1 + \frac{y(m-1)+(k-1)}{T(OPT,J')}$. Consequently, we have $\alpha(J) < \alpha(J')$.

Otherwise, $z = 1$. Let $J_{bsc}^{(x)}$ be the instance that contains one unit job and $(x-1)$ basic jobs where $x \ge 2$. It is easy to verify that $J_{bsc}^{(n)} \in \mathfrak{J}_{org}^{(m,n)}$. In this case, $k = m$ and $z = 1$, which means $J = J_{bsc}^{(n+1)}$. Then by Lemma 4, we have

$\alpha(J_{bsc}^{(x)}) = 1 + \frac{(x-1)(m-1)}{m(1+2+...+x-1)+x} = 1 + \frac{(m-1)(x-1)}{mx(x-1)/2+x} = 1 + \frac{1-\frac{1}{m}}{\frac{x}{2}+\frac{1}{m}(1+\frac{1}{x-1})}$. Since

$x - 1 \geq 1 \geq \sqrt{\frac{2}{m}}$, $\alpha(J_{bsc}^{(x)})$ is monotone decreasing on the value of x. That is to

say, $\alpha(J_{bsc}^{(n+1)}) < \alpha(J_{bsc}^{(n)})$. Since $J_{bsc}^{(n)} \in \mathfrak{J}_{org}^{(m,n)}$, we have $\alpha(J) < \alpha(\mathfrak{J}_{org}^{(m,n)})$.   □

4.1.3   Instance Achieving Maximum Approximation Ratio

Lemma 6. *Given $m, n \geq 2$, there is $\alpha(\mathfrak{J}_{org}^{(m,n)}) = \alpha(\mathfrak{J}_{one}^{(m,n)})$.*

Proof. For any instance $J \in \mathfrak{J}_{one}^{(m,n)}$, we construct a series of processes that transform J into J' such that $\alpha(J) \leq \alpha(J')$ where $|J'| \geq n$ and $J' \in \mathfrak{J}_{org}^{(m,|J'|)}$. Since $\alpha(\mathfrak{J}_{org}^{(m,|J'|)}) \leq \alpha(\mathfrak{J}_{org}^{(m,n)})$ by Lemma 5, the proof follows.

Let σ be the LRF schedule of J. If all jobs in σ are *free* jobs, there is $\alpha(J) \leq 1$, by Lemmas 1 and 2. In the following discussion, we assume at least one job in σ is a *non-free* job. Step 1 and 2 show the transformation of instance J in schedule σ, while in step 3 we analyze the approximation ratio of the LRF schedule of the new instance. An example is shown in Fig. 3.

Step 1. For any *free* job in σ, we split it into unit jobs, which will not change the objective value of schedule σ by Lemma 1. For any *non-free* job i in schedule σ, if it is not an *unlucky* job, then we split part of job i (from right side) into unit jobs until $L_\sigma(i)\%m = 1$, i.e. job i is split into an *unlucky* job and several unit jobs. In this process, the completion times of any newly split jobs are the same as job i, i.e. the objective value of the new schedule is the same as σ. We denote J_1 as the instance after all the processes, and denote σ_1 as the corresponding schedule. Then, we have $|J_1| \geq n$ and $T(\sigma_1, J_1) = T(\sigma, J)$.

Step 2. After the first step, it is clear that for any job $i \in J_1$ that is not a unit job, it is an *unlucky* job. We define the set J_b as the set containing any *unlucky* job in J_1, which is not a basic job. More formally, $J_b = \{j \mid j \in J_1, 1 < s_j < m\}$. For any two jobs $i, j \in J_b$, we *split* one unit of job i (from left side) into a unit job and *merge* one unit job with job j (from left side) until either job i turns into a unit job or job j turns into a basic job (keeping j as *unlucky* job, see Fig. 3c and d). Since job j is an *unlucky* job and if $s_j < m$ then there is at least one unit job right before job j, which guarantees that the *merge process* is feasible. The *split process* decreases the objective value by 1, while the *merge process* increases the objective value by 1. Thus, this process neither changes the objective value, nor changes the number of jobs. We repeat this process until the number of jobs in J_b is 0 or 1. Then we convert instance J_1 into instance J_2 (denote σ_2 as the corresponding schedule) with $|J_2| = |J_1|$ and $T(\sigma_2, J_2) = T(\sigma_1, J_1)$ where J_2 contains some unit jobs (at least one since the first job in σ_2 is always an unit job), some basic jobs (maybe zero) and one job c with $1 < s_c \leq m$ (Job c is the remaining job in J_b if $|J_b| = 1$, otherwise we specialize one basic job in J_2 into job c, in which case $s_c = m$). Now we claim that $J_2 \in \mathfrak{J}_{org}^{(m,|J_2|)}$. Suppose there are z unit jobs in J_2. It is clear that only unit jobs and *unlucky* jobs exist in

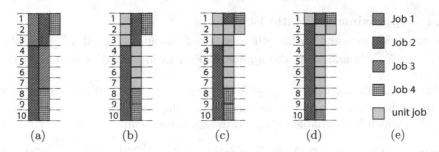

$$(a) \qquad (b) \qquad (c) \qquad (d) \qquad (e)$$

Fig. 3. Four jobs of length $\{3, 10, 4, 5\}$ with $m = 10$. The LRF schedule is shown in (a). In (b), all *free* jobs (Job 1 and Job 3) are split into unit jobs, then in (c) Job 2 and Job 4 are split into two *unlucky* jobs and some unit jobs and finally Job 2 grows while Job 4 is shortened shown in (d).

σ_2. Thus, for job c, we have $s_c + z > m$, which implies that J_2 is an organized instance.

Step 3. We denote the number of jobs in J_2 as n'. After the above two steps, we have $n' \geq n$, $L(J_2) = L(J)$ and $T(\sigma, J) = T(\sigma_2, J_2)$, i.e. the objective value keeps the same. Since $T(\sigma_2, J_2) \leq T(LRF, J_2)$, we have $T(\sigma, J) \leq T(LRF, J_2)$. Since $J_2 \in \mathfrak{J}_{org}^{(m,n')}$, we have $T(OPT, J_2) = T(J^{unit})$ by Lemma 3. By Lemma 2, we have $T(OPT, J) \geq T(J^{unit})$, thus $T(OPT, J) \geq T(OPT, J_2)$. Consequently, we have $\alpha(J) = \frac{T(\sigma, J)}{T(OPT, J)} \leq \frac{T(LRF, J_2)}{T(OPT, J_2)} = \alpha(J_2)$. Since $n' \geq n$, we have $\alpha(\mathfrak{J}_{org}^{(m,n')}) \leq \alpha(\mathfrak{J}_{org}^{(m,n)})$ by Lemma 5. Finally, we conclude that $\alpha(J) \leq \alpha(\mathfrak{J}_{org}^{(m,n)})$. □

Lemma 7. *Given* $n, m \geq 2$, $\alpha(\mathfrak{J}^{(m,n)}) = \alpha(\mathfrak{J}_{org}^{(m,n)})$.

Proof. Suppose instance $J \in \mathfrak{J}^{(m,n)}$ is the instance such that $\alpha(J) = \alpha(\mathfrak{J}^{(m,n)})$. If $J \in \mathfrak{J}_{one}^{(m,n)}$, according to Lemma 6, we have $\alpha(J) = \alpha(\mathfrak{J}^{(m,n)}) = \alpha(\mathfrak{J}_{org}^{(m,n)})$. Otherwise we assume $J \notin \mathfrak{J}_{one}^{(m,n)}$, i.e. there exists a job $j \in J$ such that $s_j > m$. Without loss of generality, we assume $\alpha(J) > 1$.

Consider the *split process* that splits part of job j (from right side) into a basic job. Let σ be a feasible schedule of instance J, we then apply this process to schedule σ. As we can see, in schedule σ, the objective value will decrease by $(s_j - m)$, since the completion time of the left part of job j will decrease by 1 while the completion time of the right part of job j keeps the same (See Fig. 2(c) and (d)). Denoting J^* as the new instance, σ^* as the new schedule, we have $T(\sigma^*, J^*) = T(\sigma, J) - (s_j - m)$, which implies $T(OPT, J^*) \leq T(OPT, J) - (s_j - m)$ and $T(LRF, J^*) \geq T(LRF, J) - (s_j - m)$. Thus, $\alpha(J^*) = \frac{T(LRF, J^*)}{T(OPT, J^*)} \geq \frac{T(LRF, J) - (s_j - m)}{T(OPT, J) - (s_j - m)} > \frac{T(LRF, J)}{T(OPT, J)} = \alpha(J)$. While it is possible, we repeat this process and convert J into instance J' with n' ($n' > n$) jobs and $\alpha(J) < \alpha(J')$ where $J' \in \mathfrak{J}_{one}^{(m,n')}$. By Lemma 5 we have $\alpha(\mathfrak{J}_{org}^{(m,n')}) < \alpha(\mathfrak{J}_{org}^{(m,n)})$ since $n' > n$. By Lemma 6, we have $\alpha(\mathfrak{J}_{one}^{(m,n')}) = \alpha(\mathfrak{J}_{org}^{(m,n')})$. Thus, we have $\alpha(J) < \alpha(\mathfrak{J}_{org}^{(m,n)})$. □

4.1.4 Approximation Ratio Bound

In this section, we aim to find the organized instance $J \in \mathfrak{J}_{org}^{(m,n)}$ such that $\alpha(J) = \alpha(\mathfrak{J}_{org}^{(m,n)})$ and show the approximation ratio of LRF schedule on this instance.

Define function $g(a,b) = \frac{am+b-n}{ma(a+1)/2+b(a+1)}$ $(a > 0, b \geq 0)$. We show that the upper bound is related to the maximum value of this function. We define $i = \lceil \frac{2n}{m} \rceil$, which is a variable only used in this subsection.

Lemma 8. Given $m, n \geq 2$, for any instance $J \in \mathfrak{J}_{org}^{(m,n)}$, $\alpha(J) = 1 + g(a,b)$, where $a = \lfloor \frac{L(J)}{m} \rfloor$, $b = L(J) - am$.

Proof. Since $J \in \mathfrak{J}_{org}^{(m,n)}$, we have $T(LRF, J) = T(OPT, J) + y(m-1) + (k-1)$ by Lemma 4 and $T(OPT, J) = T(J^{unit})$ by Lemma 3. Moreover, $L(J) = ym + k + z$, $n = y + z + 1$. Consequently, $\alpha(J) = 1 + \frac{L(J)-n}{T(J^{unit})}$. Since $L(J) = am + b$ and $0 \leq b < m$, we have $T(J^{unit}) = m(1 + 2 + ... + a) + b(a + 1)$. Hence, $\alpha(J) = 1 + \frac{am+b-n}{ma(a+1)/2+b(a+1)}$. $\square$

Lemma 9. $\forall J_u, J_v \in \mathfrak{J}_{org}^{(m,n)}$, the following two properties hold:

(1) if $L(J_u) \leq L(J_v) \leq im$, then $\alpha(J_u) \leq \alpha(J_v)$.
(2) if $L(J_u) \geq L(J_v) \geq im$, then $\alpha(J_u) \leq \alpha(J_v)$.

Theorem 1 is a corollary of Lemma 9, as the organized instance with total workload im corresponds to the maximum approximation ratio.

Theorem 1. $\forall J \in \mathfrak{J}_{org}^{(m,n)}$, we have $\alpha(J) \leq 1 + \frac{2(im-n)}{i(i+1)m}$.

In the following of this section, we discuss about the tightness of the bound.

Lemma 10. $\forall J \in \mathfrak{J}_{org}^{(m,n)}$, $L(J) = im$ if and only if $i \leq n + 1 - m$.

Proof. Suppose J is composed of y basic jobs, $(n - y - 1)$ unit jobs, and one job of length k, then $L(J) > m(y + 1)$, $0 \leq y < n - 1$, $1 < k \leq m$.

Since $L(J) = im$, we have $y < i - 1$, i.e. $y \leq i - 2$. Since $L(J) = ym + k + (n - y - 1)$ and $k \leq m$, we have $L(J) \leq (i-1)(m-1) + n$, i.e. $i \leq n+1-m$.

Next we show that if $i \leq n+1-m$, there always exists an organized instance $J \in \mathfrak{J}_{org}^{(m,n)}$ such that $L(J) = im$. Take $y = \lfloor \frac{im-n-1}{m-1} \rfloor$, $k = (im - n - 1)\%(m-1)+2$. Obviously, $i = \lceil \frac{2n}{m} \rceil \geq \frac{2n}{m}$, i.e. $im-n-1 > 0$. Since $y(m-1)+(k-2) = im - n - 1$, we have $L(J) = ym + k + (n - y - 1) = im$. If $i \leq n + 1 - m$, then $im - n \leq (i - 1)(m - 1)$, which implies that $y < i - 1$, i.e. $L(J) > m(y + 1)$. Then, J is an organized instance and $J \in \mathfrak{J}_{org}^{(m,n)}$. $\square$

Lemma 10 shows that the bound in Theorem 1 is tight if and only if $i \leq n + 1 - m$. While for the other cases, the bound is not tight, which we show the result in Table 1. We divide the instances with n and m $(n \geq 2, m \geq 2)$ into disjoint but complete regions, according to the value $i = \lceil \frac{2n}{m} \rceil$. (We will put the proof of tightness in each region in the full version).

Table 1. Approximation ratio bound of instances in different regions

Regions	Approximation ratio	Function $g(a,b)$
$B_0 = \{m = 2\}$	$1 + \frac{n-1}{n^2}$ (tight)	$g(n-1, m-1)$
$B_1 = \{m \geq 2n, m \geq 3\}$	$1 + \frac{m-n+1}{m+2}$ (tight)	$g(i, 1)$
$B_2^* - \{m - 2n - 1, m \geq 3\}$	$1 + \frac{m-1}{2m-1}$ (tight)	$g(i-1, n-1)$
$B_2 = \{n \leq m \leq 2n - 2, m \geq 3\}$	$1 + \frac{2m-n+1}{3m+3}$ (tight)	$g(i, 1)$
$B_3^* = \{m = n - 1, m \geq 3, n \neq 4\}$	$1 + \frac{2m-2}{6m-3}$ (tight)	$g(i-1, m-1)$
$B_3 = \{\frac{2n}{3} \leq m \leq n - 2, m \geq 3\}$	$1 + \frac{2(im-n)}{i(i+1)m}$ (tight)	$g(i, 0)$
$B_4 = \{3 < \frac{2n}{m} \leq 4, m \geq 3, n \neq 5\}$		
$B_i = \{i - 1 < \frac{2n}{m} \leq i, m \geq 3\}, \forall i \geq 5$		
$B^* = \{m = 3, 4 \leq n \leq 5\}$	$1 + \frac{2(im-n)}{i(i+1)m}$	

4.2 Instance of Jobs with Arbitrary Weights

Definition 6. *For any job set* $J = \{(w_i, s_i) \mid 1 \leq i \leq n\}$, *let* $J^{(e)} = \{(w_i', s_i) \mid w_i' = s_i, i \in J\}$ *be the corresponding job set of* J *in which the weight of any job equals to its length, i.e.* $w_i' = s_i, \forall i \in J^{(e)}$.

Lemma 11. *For any instance* J, *there always exists a subset* $J_s \subseteq J$ *such that*

$$\alpha(J) \leq \alpha(J_s^{(e)})$$

Proof. We denote σ (resp. γ) as the LRF schedule (optimal schedule) of J. For each job i we denote $\delta_i = \frac{w_i}{s_i}$ as the density of job i. Without loss of generality, we assume jobs are sorted by the order in σ, i.e. $\delta_1 \geq \delta_2 \geq ... \geq \delta_n > 0$.

Given τ, $0 < \tau < min(\{\delta_i - \delta_{i+1} \mid \delta_i - \delta_{i+1} > 0, 1 \leq i < n\} \cup \{\delta_n/2\})$, we define $y > x$ as the meaning of $0 \leq y - x \leq \tau$.

We add a dummy job $n + 1$ into instance J, where $s_{n+1} = m$, $\delta_{n+1} = \tau$. It is clear that, when we take $\tau \to 0$, job $n + 1$ is always the last job in both optimal schedule and LRF schedule of J, and $\lim_{\tau \to 0} \delta_{n+1} s_{n+1} C_{n+1}^{OPT} = \lim_{\tau \to 0} \delta_{n+1} s_{n+1} C_{n+1}^{LRF} = 0$, which implies that the approximation ratio of instance J will not change by adding this dummy job.

If there exists a group of jobs $k \in [i, j]$ with $1 < i \leq j < n + 1$, such that $\delta_{i-1} > \delta_i > ... > \delta_j > \delta_{j+1}$ (in the sense that $\delta_{i-1} > \delta_i + \tau$ and $\delta_j > \delta_{j+1} + \tau$), we create a new instance J' as follows: $\forall k \in [i, j]$, we change δ_k to $\delta_k' := \delta_k + \beta$, i.e. $s_k' := s_k$, $w_k' := s_k(\delta_k + \beta)$, where $\delta_{j+1} - \delta_j + \tau \leq \beta \leq \delta_{i-1} - \delta_i - \tau$.

For simplicity, we denote $a = \sum_{k \in [i,j]} s_k C_k^\sigma$, $c = \sum_{k \in [i,j]} s_k C_k^\gamma$. In J', we have $\delta_i' = \delta_i + \beta < \delta_{i-1}$ and $\delta_j' = \delta_j + \beta > \delta_{j+1}$, which implies new jobs still follow *Largest-Ratio-First* order. However, the LRF schedule of J', in the worst case, may possibly have different order due to the fact that some jobs in J' have exactly equal density. Thus, $T(LRF, J') \geq T(\sigma, J) + \beta a$. Moreover, the optimal schedule of J is a feasible schedule of J', we have $T(OPT, J') \leq T(\gamma, J) + \beta c$. Therefore, $\alpha(J') = \frac{T(LRF, J')}{T(OPT, J')} \geq \frac{T(\sigma, J) + \beta a}{T(\gamma, J) + \beta c}$.

If $\frac{a}{c} > \frac{T(\sigma,J)}{T(\gamma,J)} = \alpha(J)$, then we set $\beta := \delta_{i-1} - \delta_i - \tau > 0$, otherwise, we set $\beta := \delta_{j+1} - \delta_j + \tau < 0$. In either case, we have $\alpha(J') \geq \alpha(J)$.

While it is possible, we repeat the above process until instance J is converted into instance J^* such that $\delta_1^* > \delta_2^* ... > \delta_r^* > \delta_{r+1}^* > ... > \delta_{n+1}^*$, $1 \leq r \leq n$, where $\delta_1^* = \delta_1, \delta_{n+1}^* = \delta_{n+1} = \tau$, and $\alpha(J^*) \geq \alpha(J)$.

Let J_r be the set of first r jobs in J^*. When we take $\tau \to 0$ initially, we have:

(i) $\forall k \in [r+1, n+1]$, $\delta_k^* = 0$, which implies $\alpha(J^*) = \alpha(J_r)$.

(ii) $\forall k \in J_r$, $\delta_k^* = \delta_1$, which implies $T(OPT, J_r) = \delta_1 T(OPT, J_r^{(e)})$ and $T(LRF, J_r) \leq \delta_1 T(LRF, J_r^{(e)})$ (due to the fact that in $J_r^{(e)}$, all jobs have exactly the same density, which is not the case of J_r). Therefore, $\alpha(J_r) \leq \alpha(J_r^{(e)})$.

Finally, we conclude that $\alpha(J) \leq \alpha(J_r^{(e)})$. $\square$

Combine Lemma 11 with the result in the previous section (Lemma 5), we conclude that $\forall J$, $\alpha(J) \leq \alpha(\mathfrak{J}_{org}^{(m,2)}) = 1 + \frac{m-1}{m+2}$, i.e. the approximation ratio is maximum when there are only two jobs, giving the following theorem.

Theorem 2. *For any instance J, there is $\alpha(J) \leq 1 + \frac{m-1}{m+2}$, which is tight.*

This bound is tight since one can always add some dummy jobs (whose weight approaches to 0) to J to make it equivalent to $\mathfrak{J}_{org}^{(m,2)}$.

5   Conclusion and Future Work

In this paper, we first give a PTAS for the problem when the number of machines m is fixed. Then we investigate the approximation ratio of the LRF algorithm and analyze the worst case, including the special case (jobs with equal density) and general case. In the future work, it would be interesting to solve the problem when m is fixed. It remains open that whether the problem is polynomial or not.

Acknowledgement. We gratefully thank Gruia Călinescu for our helpful discussions introducing the idea of the PTAS.

References

1. Afrati, F., Bampis, E., Chekuri, C., Karger, D., Kenyon, C., Khanna, S., Milis, I., Queyranne, M., Skutella, M., Stein, C., Sviridenko, M.: Approximation schemes for minimizing average weighted completion time with release dates. In: 40th Annual Symposium on Foundations of Computer Science, FOCS, pp. 32–44. IEEE Computer Society (1999)
2. Brucker, P.: Scheduling Algorithms, 5th edn. Springer, Heidelberg (2007)
3. Bruno, J., Coffman Jr., E.G., Sethi, R.: Scheduling independent tasks to reduce mean finishing time. Commun. ACM **17**(7), 382–387 (1974)

4. Fishkin, A.V., Jansen, K., Porkolab, L.: On minimizing average weighted completion time: a PTAS for scheduling general multiprocessor tasks. In: Freivalds, R. (ed.) FCT 2001. LNCS, vol. 2138, pp. 495–507. Springer, Heidelberg (2001). doi:10.1007/3-540-44669-9_56
5. Garg, N., Kumar, A., Pandit, V.: Order scheduling models: hardness and algorithms. In: Arvind, V., Prasad, S. (eds.) FSTTCS 2007. LNCS, vol. 4855, pp. 96–107. Springer, Heidelberg (2007). doi:10.1007/978-3-540-77050-3_8
6. Kawaguchi, T., Kyan, S.: Worst case bound of an LRF schedule for the mean weighted flow-time problem. SIAM J. Comput. **15**(4), 1119–1129 (1986)
7. Leung, J.Y.-T., Li, H., Pinedo, M.: Order scheduling in an environment with dedicated resources in parallel. J. Sched. **8**(5), 355–386 (2005)
8. Mastrolilli, M., Queyranne, M., Schulz, A.S., Svensson, O., Uhan, N.A.: Minimizing the sum of weighted completion times in a concurrent open shop. Oper. Res. Lett. **38**(5), 390–395 (2010)
9. Roemer, T.A.: A note on the complexity of the concurrent open shop problem. J. Sched. **9**(4), 389–396 (2006)
10. Sahni, S.: Algorithms for scheduling independent tasks. J. ACM **23**(1), 116–127 (1976)
11. Schulz, A.S., Skutella, M.: Scheduling-LPs bear probabilities randomized approximations for min-sum criteria. In: Burkard, R., Woeginger, G. (eds.) ESA 1997. LNCS, vol. 1284, pp. 416–429. Springer, Heidelberg (1997). doi:10.1007/3-540-63397-9_32
12. Skutella, M., Woeginger, G.J.: A PTAS for minimizing the weighted sum of job completion times on parallel machines. In: Proceedings of the Thirty-First Annual ACM STOC, pp. 400–407. ACM (1999)
13. Smith, W.E.: Various optimizers for single-stage production. Naval Res. Logistics Q. **3**(1–2), 59–66 (1956)
14. Sung, C.S., Yoon, S.H.: Minimizing total weighted completion time at a preassembly stage composed of two feeding machines. Int. J. Prod. Econ. **54**(3), 247–255 (1998)
15. Zhang, Q., Weiwei, W., Li, M.: Minimizing the total weighted completion time of fully parallel jobs with integer parallel units. Theor. Comput. Sci. **507**, 34–40 (2013)

Continuous Firefighting on Infinite Square Grids

Xujin Chen[1], Xiaodong Hu[1], Changjun Wang[2], and Ying Zhang[1(✉)]

[1] Academy of Mathematics and Systems Science, Chinese Academy of Sciences,
Beijing 100190, China
{xchen,xdhu,zhangying}@amss.ac.cn
[2] Beijing Institute for Scientific and Engineering Computing,
Beijing University of Technology, Beijing, China
wcj@bjut.edu.cn

Abstract. The classical firefighter problem, introduced by Bert Hartnell in 1995, is a deterministic discrete-time model of the spread and defence of fire, rumor, or disease. In contrast to the generally "discontinuous" firefighter movements of the classical setting, we propose in the paper the *continuous firefighting model*. Given an undirected graph G, at time 0, all vertices of G are undefended, and fires break out on one or multiple different vertices of G. At each subsequent time step, the fire spreads from each burning vertex to all of its undefended neighbors. A finite number of firefighters are available to be assigned on some vertices of G at time 1, and each firefighter can only move from his current location (vertex) to one of his neighbors or stay still at each time step. A vertex is defended if some firefighter reaches it no later than the fire. We study fire containment on infinite k-dimensional square grids under the continuous firefighting model. We show that the minimum number of firefighters needed is exactly $2k$ for single fire, and 5 for multiple fires when $k = 2$.

Keywords: Firefighter problem · Continuous firefighting · Fire containment · Infinite square grids

1 Introduction

The classical firefighter problem was introduced by Hartnell in 1995 [10]. In the model, given a network G, a fire breaks at some vertex of G at time 0. Subsequently, at each time step, at most a number B of firefighters can be used to defend the fire, and the fire spreads from the burning vertices to all their undefended neighbors. A firefighter at each time step can defend any single unburned vertex of the network. Once a vertex is burned or defended, it remains so from then on.

A mostly studied objective of the firefighter problem is to maximize the number of vertices that are not burned at the end of the process. On the negative side,

---

Research supported in part by NNSF of China under Grant No. 11601022 and 11531014

T.V. Gopal et al. (Eds.): TAMC 2017, LNCS 10185, pp. 158–171, 2017.
DOI: 10.1007/978-3-319-55911-7_12

the maximization problem has been shown to be NP-hard even when the underlying network is a tree with maximum degree three [7] or a cubic graph [11]; the problem on general networks with n vertices does not admit n^c-approximation for any $c < 1$ unless $NP = P$ [1]. On the positive side, MacGillivray and Wang [12] solved the problem to its optimality in polynomial time for some special trees. Cai et al. [2] presented a $(1 - 1/e)$-approximation algorithm for general trees based on LP relaxation and randomized rounding. The approximation ratio was latter improved slightly by Yutaka et al. [18].

Other objectives of the firefighter problem investigated in literature include minimizing the budget B needed per time step in order to protect a given set of vertices [1,3], rumor (or virus) blocking to bring the epidemic threshold above the spreading rate [4,6], and politician's firefighting that exhibits more locality [16]. Among many variants of the classical model, a spreading version of firefighting process, where the firefighting is viewed as a vaccination process that also spreads over the network, was studied by Anshelevich et al. [1]. Once a vertex is defended, at the next time step, all its unburned neighbors will be defended simultaneously. The authors [1] showed that the firefighting spread problem to maximize protection by using targeted vaccinations is reduced to maximizing a submodular function with a matroid constraint. The reader is referred to [8] for an extensive and detailed survey on the firefighter problem.

A large amount of literature on firefighter problems concerns on infinite networks, where the main objective is to determine the smallest number of firefighters that can "contain" the fire in a finite number of steps, and to find a containment with fewest firefighters that minimizes the number of vertices burned. Fogarty [9] gave a main tool, the Hall-like condition, to obtain a lower bound on the number of firefighters needed to contain a fire in an infinite network. The result was generalized by Devlin and Hartke [5].

Most work on firefighter problems in infinite networks focus on various grids [5,9,12,13]. Moeller and Wang [17] proved that a single firefighter can not contain one fire in the infinite 2-dimensional square grid, while two firefighters can contain the fire in eight time units, and this is the best possible. The authors [17] conjectured that $2k - 1$ firefighters per time step are necessary to contain a fire in the k-dimensional square grid for $k \geq 3$. The conjecture was confirmed by Devlin and Hartke [5], who also showed that for any fixed natural number B, there is a outbreak of finitely many fires in which B firefighters per time step are insufficient to contain the outbreak. In addition to square grids, similar problems have been investigated for strong grids [13], triangular grids and hexagonal grids [9]. Variations of the problem that have been considered w.r.t. grids include determining the number of firefighters needed to contain a fire to a given region [13], fractional defending and burning – the firefighters could extinguish some fraction of the fire at a vertex, and then the remaining fraction of the fire is going on to spread [9], and using a varying number of firefighters at each subsequent time step [14,15].

Our contributions. In real-world defending of the fires, firefighters can not "jump" from place to place discontinuously. Instead their movements are usually continuous. In the paper we propose a "continuous" variant of firefighting to model the scenario. In our model, the fire breaks out at time 0 from finitely many source vertices in given network G, and a finite number of firefighters are available to be placed on some vertices of G at time 1. In the defending process, each firefighter can only move from his current location (vertex) to one of his neighbors or stay still at each subsequent time step. Namely, each firefighter defend the fire along a continuous walk step by step. This particularly models the situation that traveling long distance takes long time, and the relative velocity between the fire and firefighters is relatively fixed. It turns out to be more reasonable and realistic than the classical "discontinuous" model, where a long-distance movement and a short-distance one take the same time, and a firefighter may go arbitrarily faster than the fire.

The present paper focus on the fire containment on infinite (square) grids with continuous firefighting. For a single fire on the k-dimensional grid, we show that $2k$ firefighters are necessary and sufficient for containment. For multiple fires on the 2-dimensional grid, we give an efficient way to contain the fire with 5 firefighters regardless of the number and locations of sources. The most difficult part of our work is to establish the lower bound that there is no way for 4 firefighters to contain the fire originating from two adjacent sources. Due to the limitation on pages, we omit some proofs in this extended abstract, and postpone them to the full version of the paper.

2   The Model

The network is modeled by an undirected graph G. At time 0, all vertices of G are *undefended*, and one or multiple fires break out on some *fire source vertices* (*sources*) of G. At each subsequent time step, the fire *spreads* from each burning vertex to all of its undefended neighbor vertices. At time 1, a finite number of firefighters can be placed on some vertices of G. At each subsequent time step each firefighter can only move from his current location (vertex) to one of his neighbors or stay still. A vertex is *defended* if and only if some firefighter reaches it no later than the fire. Once a vertex is burning or defended, it remains so from then on. The burning process stops when there are no more vertices can catch fire. At the end of the process, all defended or unburned vertices are *protected*.

While the general goal of the continuous firefighting problem is to use a minimum number of firefighters to protect a maximum number of vertices, in this paper we focus on the *fire containment* version of the problem when G is an infinite square grid (to be defined in the next paragraph). We study how many firefighters are needed to satisfy the primary requirement that the burning process stops in a finite number of time steps, i.e., only a finite number of vertices are burned finally. Meanwhile, we show how to use a minimum number of firefighters to contain the fire in the grids.

Infinite square grids. Let $k \in \mathbb{N}$ be a positive integer. The *infinite k-dimensional square grid*, denoted as G_k, is the graph whose vertices correspond to the points in $\mathbb{Z}^k$, and two vertices are connected by an edge whenever the corresponding points are at distance 1. Thus in G_k, in each time step, the fire can spread from a burning vertex a distance 1 along *all* the k directions, and the firefighter can either stay still or move a distance 1 along *one* of the k directions. For any two vertices $u, v \in G_k$, we often call a (simple) path between u and v in G_k a *u-v path*. The distance between u and v, denoted as $d(u, v)$, equals the length (i.e., the number of edges) of a shortest u-v path in G_k. Observe that if u is a fire source, then the fire cannot spread to v earlier than time $d(u, v)$.

We establish a k-dimensional coordinate system such that its origin is identical with some vertex of G_k and every vertex $v \in G_k$ has integer coordinates $(v_1, v_2, \cdots, v_k) \in \mathbb{Z}^k$. For convenience, we shall identify a vertex with its coordinate representation.

Lemma 1. $d(u, v) = \sum_{i=1}^{k} |u_i - v_i|$ *for any two vertices* $u, v \in G_k$. □

Fire containment. A major task of the paper is to determine in G_k the minimum number $\Gamma_{k,s}$ of firefighters needed to *contain the fire* starting from s fire sources (i.e., to guarantee that the burning process stops in a finite number of time steps and so with a finite number of burned vertices) *regardless of the locations of fire sources*.

Given any containment $\mathbf{c}$ of the fire, it is specified by its *defending process* – the initial positions and subsequent movements of firefighters. We will use the terms "containment" and "defending process" interchangeably; they share the same symbol $\mathbf{c}$ – the former emphasizing the containing result while the latter laying stress on firefighters' continuous movements.

Let $\mathscr{F}(\mathbf{c})$ denote the set of firefighters used in $\mathbf{c}$, and $\mathcal{D}(\mathbf{c})$ denote the set of vertices defended by the firefighters in $\mathscr{F}(\mathbf{c})$. In the following arguments, when we mention a certain time point at which some vertex $v \in \mathcal{D}(\mathbf{c})$ was defended, we mean the earliest time at which v was defended, and denote it by $\tau_v(\mathbf{c})$. A firefighter in $\mathscr{F}(\mathbf{c})$ who reached (and thus defended) v at time $\tau_v(\mathbf{c})$ is called an *earliest defender* of v. Formally, we aim to determine

$$\Gamma_{k,s} := \max_{\mathfrak{F}} \left\{ \min_{\mathbf{c}} \{ |\mathscr{F}(\mathbf{c})| : \mathbf{c} \text{ is a containment of } \mathfrak{F} \} : \mathfrak{F} \text{ is the fire} \right.$$

$$\left. \text{starting from } s \text{ sources in } G_k \right\}$$

The next lemma states a crucial property implied by the continuity of firefighters' movements.

Lemma 2. *In any containment* $\mathbf{c}$ *of the fire in* G_k, *if* $d(u, v) \geq |\tau_u(\mathbf{c}) - \tau_v(\mathbf{c})| + 1$ *for some vertices* $u, v \in \mathcal{D}(\mathbf{c})$, *then no single firefighter in* $\mathscr{F}(\mathbf{c})$ *could be an earliest defender of both u and v.* □

3 One Fire in G_k

This section studies the single fire case. For convenience of description, we assume without loss of generality that the single fire broke out on the origin at time 0.

Given any containment $\mathbf{c}$ of the fire, for each $i \in [k]$, we take i^+ (resp. i^-) to be the defended vertex on the positive (resp. negative) i-axis that is closest to the origin. Clearly, the set $\mathcal{A}(\mathbf{c}) := \{i^+, i^- : i \in [k]\}$ has cardinality $2k$. For each vertex $v \in \mathcal{A}(\mathbf{c}) \subseteq \mathcal{D}(\mathbf{c})$, let $\phi(v)$ denote the unique non-zero coordinate of v, and F_v denote a (fixed) earliest defender of v.

Lemma 3. $\tau_v(\mathbf{c}) \leq |\phi(v)|$ for each $v \in \mathcal{A}(\mathbf{c})$.

Proof. It is instant from the choice of v – its closeness to the origin (the fire source) that the fire spread to v at time $|\phi(v)|$. Now $v \in \mathcal{D}(\mathbf{c})$ along with the definition of $\tau_v(\mathbf{c})$ enforces the desired inequality. □

Lemma 4. $|\mathscr{F}(\mathbf{c})| \geq 2k$ for any containment $\mathbf{c}$ for a single fire in G_k.

Proof. Note that $\{F_v : v \in \mathcal{A}(\mathbf{c})\}$ is a subset of $\mathscr{F}(\mathbf{c})$. If the lemma failed, then $|\{F_v : v \in \mathcal{A}(\mathbf{c})\}| < 2k$. Since $|\mathcal{A}(\mathbf{c})| = 2k$, there exist distinct $u, v \in A(\mathbf{c})$ who share a common earliest defender $F_* = F_v = F_u$. We may suppose without loss of generality that $1 \leq \tau_u(\mathbf{c}) < \tau_v(\mathbf{c})$. It follows from Lemma 2 that $\tau_v(\mathbf{c}) \geq \tau_u(\mathbf{c}) + d(u, v)$. By Lemma 1, it is easily seen from the choices of u and v that $d(u, v) = |\phi(u)| + |\phi(v)|$. So $\tau_v(\mathbf{c}) \geq \tau_u(\mathbf{c}) + |\phi(u)| + |\phi(v)| > |\phi(v)|$ shows a contradiction to Lemma 3. □

Theorem 1. $\Gamma_{k,1} = 2k$ for any $k \geq 1$.

Proof. Lemma 4 has shown that $\Gamma_{k,1} \geq 2k$. For the reversed inequality, we observe that if $2k$ firefighters are placed at the $2k$ neighbors of the single fire source, respectively at time 1, then the fire stops at time 1, and only the source vertex is burned. □

4   Multiple Fires in G_k

This section investigates the case where multiple fires originate from s different sources. For any $k, k', s, s' \in \mathbb{N}$ with $k \geq k'$ and $s \geq s'$, it is trivial that $\Gamma_{k,s} \geq \Gamma_{k',s}$ and $\Gamma_{k,s} \geq \Gamma_{k,s'}$.

Theorem 2. $\Gamma_{1,s} = 2$ for any $s \in \mathbb{N}$.

Proof. For containment of multiple fires on the line G_1, two firefighters are obviously necessary and sufficient (one on the left of the leftmost fire source and the other on the right of the rightmost fire source). □

The remainder of this section is devoted to verifying the following theorem on the 2-dimensional grid G_2.

Theorem 3. $\Gamma_{2,s} = 5$ for any $s \in \mathbb{N} \backslash \{1\}$. □

In Sect. 4.1 we present a way in which five firefighters fulfill the containing task for any number of fire sources on any locations of G_2 (see Theorem 4). In Sect. 4.2 we show that it is impossible for four firefighters to contain the fire originating from two adjacent sources in G_2 (see Theorem 5). The combination of $\Gamma_{2,s} \leq 5$ and $\Gamma_{2,s} \geq \Gamma_{2,2} \geq 5$ establishes Theorem 3.

4.1   The Upper Bound on $\Gamma_{2,s}$

When finitely many vertices catched on fire at time 0 in G_2 (no matter where they are), we show how five firefighters are placed at time 1 and how they move at each subsequent time step in order to contain the fire.

Let u, v be two vertices in G_2 with the same x- or y-coordinate. We use uv to denote the directional vertical or horizontal line segment from u to v. If a firefighter moves from u to v along uv, we simply say that he moves along uv.

Theorem 4. $\Gamma_{2,s} \leq 5$ for any $s \in \mathbb{N}$.

Proof. Given any set of s fire source vertices in G_2, we observe that the s fire sources are contained in a $W \times H$ rectangle with corners $\alpha, \beta, \gamma, \delta$ that has a minimum area. Suppose without loss generality that the x-axis goes through $\delta\gamma$ and the origin o divides $\delta\gamma$ as evenly as possible. More specially, $\alpha = (-\lfloor W/2 \rfloor, H), \beta = (\lceil W/2 \rceil, H), \gamma = (\lceil W/2 \rceil, 0), \delta = (-\lfloor W/2 \rfloor, 0)$. (See Fig. 1 for an illustration).

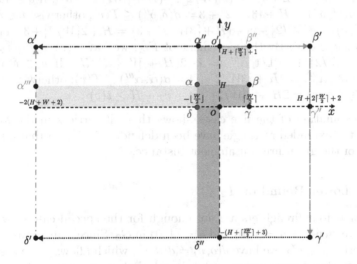

Fig. 1. The proof of $\Gamma_{2,s} \leq 5$.

Based on the rectangle on $\alpha, \beta, \gamma, \delta$, we define an extended rectangle with corners $\alpha' = (-2(H+W+2), H+\lceil W/2 \rceil+1)$, $\beta' = (H+2\lceil W/2 \rceil+2, H+\lceil W/2 \rceil+1)$, $\gamma' = (H+2\lceil W/2 \rceil+2, -(H+\lceil 3W/2 \rceil+3))$ and $\delta' = (-2(H+W+2), -(H+\lceil 3W/2 \rceil+3))$. Let $o' = (0, H+\lceil W/2 \rceil+1)$ be the projection of o onto $\alpha'\beta'$. We consider the defending process $\mathbf{c}$ with five firefighters $F_{o\alpha}, F_{o\beta}, F_{\beta\gamma}, F_{\gamma\delta}, F_{\delta\alpha}$ in which F_{ij} started from i' at time 1 and went through $i'j'$ continuously, for each $ij \in \mathcal{L} := \{o\alpha, o\beta, \beta\gamma, \gamma\delta, \delta\alpha\}$. We treat each element of $\mathcal{L}$ either as an index for firefighter in $\mathscr{F}(\mathbf{c})$ or a side of the extended rectangle.

Given any vertex $v \in \cup_{ij \in \mathcal{L}} ij$, let $t(v)$ and $T(v)$ be the earliest time some firefighter and the fire reached v, respectively. To see that $\mathbf{c}$ is qualified to be a containment of the fire, it suffices to show that $t(v) \leq T(v)$ for all $v \in \cup_{ij \in \mathcal{L}} ij$. We check five possibilities for $v \in ij$ when ij is taken from $\mathcal{L}$.

CASE 1. $v \in o'\alpha'$: By the movement of $F_{o\alpha}$, we have $t(v) \leq 1 + d(o', v) = 1 - v_1$. Let $\alpha'' := (-\lfloor W/2 \rfloor, H + \lceil W/2 \rceil + 1)$ be the projection of α onto $o'\alpha'$. If $v \in o'\alpha''$, then $t(v) \leq 1 + d(o', v) \leq 1 + d(o', \alpha'') = 1 + \lfloor W/2 \rfloor \leq d(\alpha, \alpha'') \leq T(v)$; otherwise $v \in \alpha''\alpha'$ and $T(v) \geq d(\alpha, v) = -\lfloor W/2 \rfloor - v_1 + \lceil W/2 \rceil + 1 \geq t(v)$.

CASE 2. $v \in o'\beta'$: By the movement of $F_{o\beta}$, we have $t(v) \leq 1 + d(o', v) = 1 + v_1$. Let $\beta'' := (\lceil W/2 \rceil, H + \lceil W/2 \rceil + 1)$ be the projection of β onto $o'\beta'$. If $v \in o'\beta''$, then $t(v) \leq 1 + d(o', v) \leq 1 + d(o', \beta'') = 1 + \lceil W/2 \rceil = d(\beta, \beta'') \leq T(v)$; otherwise $v \in \beta''\beta'$ and $T(v) \geq d(\beta, v) = (v_1 - \lceil W/2 \rceil) + (v_2 - H) = (v_1 - \lceil W/2 \rceil) + \lceil W/2 \rceil + 1 = 1 + v_1 \geq t(v)$.

CASE 3. $v \in \beta'\gamma'$: By the movement of $F_{\beta\gamma}$, we have $t(v) \leq 1 + \beta_2' - v_2 = H + \lceil W/2 \rceil + 2 - v_2$. Let $\gamma'' := (H + 2\lceil W/2 \rceil + 2, 0)$. If $v \in \beta'\gamma''$, then $t(v) \leq 1 + d(\beta', \gamma'') = H + \lceil W/2 \rceil + 2 = d(\gamma, \gamma'') \leq T(v)$; otherwise $v \in \gamma''\gamma'$ and $T(v) \geq d(\gamma, v) = v_1 - \lceil W/2 \rceil - v_2 = H + \lceil W/2 \rceil + 2 - v_2 \geq t(v)$.

CASE 4. $v \in \gamma'\delta'$: By the movement of $F_{\gamma\delta}$, we have $t(v) \leq 1 + \gamma_1' - v_1 = H + 2\lceil W/2 \rceil + 3 - v_1$. Let $\delta'' := (-\lfloor W/2 \rfloor, -(H + \lceil 3W/2 \rceil + 3))$. If $v \in \gamma'\delta''$, then $t(v) \leq 1 + d(\gamma', \delta'') = H + \lceil 3W/2 \rceil + 3 = d(\delta, \delta'') \leq T(v)$; otherwise $v \in \delta''\delta'$ and $T(v) \geq d(\delta, v) = -\lfloor W/2 \rfloor - v_1 + (H + \lceil 3W/2 \rceil + 3) = H + 2\lceil W/2 \rceil + 3 - v_1 \geq t(v)$.

CASE 5. $v \in \delta'\alpha'$: By the movement of $F_{\delta\alpha}$, we have $t(v) \leq 1 + v_2 - \delta_2' = v_2 + H + \lceil 3W/2 \rceil + 4$. Let $\alpha''' := (-2(H + W + 2), H)$. If $v \in \delta'\alpha'''$, then $t(v) \leq 1 + d(\delta', \alpha''') = 2H + \lceil 3W/2 \rceil + 4 = d(\alpha, \alpha''') \leq T(v)$; otherwise $v \in \alpha'''\alpha'$ and $T(v) \geq d(\alpha, v) = 2H + \lceil 3W/2 \rceil + 4 + v_2 - H \geq t(v)$.

The combination of the five cases shows that all vertices in the four sides $\cup_{ij \in \mathcal{L}} ij$ of the extended rectangle have been defended, implying that $\mathbf{c}$ is indeed a solution of the given fire containment instance. $\square$

4.2   The Lower Bound on $\Gamma_{2,2}$

We show that four firefighters are not enough for the special case in which the fire source vertices are $o = (0,0)$ and $\sigma = (1,0)$ in G_2. Since o and σ are adjacent, for each vertex $v \in G_2$, we have $d(o, v) \neq d(\sigma, v)$, which allows us to associate v with a *unique* fire source vertex $\varsigma(v) \in \{o, \sigma\}$ that is closer to v. The next fact is guaranteed by the property of shortest paths.

Lemma 5. *Let v be a vertex and Q a shortest $\varsigma(v)$-v path in G_2. Then all vertices of Q have $\varsigma(v)$ as their common closer fire source.* $\square$

The high level idea behind our proof for $\Gamma_{2,2} \geq 5$ goes as follows: We assume on the contrary that the fires originated from the two sources can be contained successfully by four firefighters. Given any such successful defending process, we can always construct a new simpler successful containment with four firefighters defending vertices of a rectangle. Finally, we reach a contradiction by showing that no four firefighters could accomplish such a rectangular containment.

The Counterexample Containment. Suppose on the contrary that there is a containment $\mathbf{c}$ of the fire originating from o and σ such that $\mathscr{F} := \mathscr{F}(\mathbf{c})$ consists of at most four firefighters.

Pioneer vertices. For each defended vertex $v \in \mathcal{D} := \mathcal{D}(\mathbf{c})$, let $F_v \in \mathscr{F}$ denote a fixed earliest defender of v, and write τ_v for $\tau_v(\mathbf{c})$. A defended vertex $v \in \mathcal{D}$ is called a *pioneer (vertex)* if the fire spread to v at time $d(\varsigma(v), v)$, equivalently, there exists a shortest $\varsigma(v)$-v path in which all vertices but v were burned (by the fire originating from $\varsigma(v)$); we call such a path an *evidencing path* of v. Let $\mathcal{P}$ denote the set of all pioneer vertices. The following crucial property will be frequently used in our discussions.

Property 1. *Each pioneer $v \in \mathcal{P}$ was defended at time $\tau_v \leq d(\varsigma(v), v)$.* $\qquad\square$

Clearly, $\mathcal{P} \subseteq \mathcal{D} \neq \emptyset$. The next property particularly implies $\mathcal{P} \neq \emptyset$; its proof follows from the consideration of a defended vertex on the path that is closest to the source.

Property 2. *For each $v \in \mathcal{D}$, if Q is a shortest path from $\varsigma(v)$ to v, then Q contains a pioneer vertex and one of its evidencing paths.* $\qquad\square$

Six special pioneers. Our next step is to prove $|\mathscr{F}| = 4$. To this end, we identify six special pioneers, which are closest to the sources horizontally or vertically from positive or negative direction. Considering the containment of the fire originating from o, we have $\alpha, \beta, \gamma \in \mathcal{P}$ with $\varsigma(\alpha) = \varsigma(\beta) = \varsigma(\gamma) = o$ such that $\alpha = (0, \alpha_2)$, $\beta = (\beta_1, 0)$ and $\gamma = (0, \gamma_2)$ were defended at time

$$\tau_\alpha \leq \alpha_2, \tau_\beta \leq |\beta_1| = -\beta_1 \text{ and } \tau_\gamma \leq |\gamma_2| = -\gamma_2,$$

respectively. Similarly, considering the containment of the fire originating from σ, we have $\kappa, \delta, \eta \in \mathcal{P}$ with $\varsigma(\kappa) = \varsigma(\delta) = \varsigma(\eta) = \sigma$ such that $\kappa = (1, \kappa_2)$, $\delta = (\delta_1, 0)$, and $\eta = (1, \eta_2)$ were defended at time

$$\tau_\kappa \leq \kappa_2, \tau_\delta \leq \delta_1 - 1 \text{ and } \tau_\eta \leq |\eta_2| = -\eta_2,$$

respectively. (See Fig. 2 for an illustration.) For each $i \in \{\alpha, \beta, \gamma, \kappa, \delta, \eta\}$, we use $\mathcal{P}_i := \{v \in \mathcal{P} : F_v = F_i\}$ to denote the set of pioneers defended by F_i.

From Property 1 of pioneers and the continuity hidden in Lemma 2 we derive the following restrictions on the defending areas of firefighters.

Claim 1. *Each of $\mathcal{P}_\alpha \cap \{v : v_2 \leq 0\}$, $\mathcal{P}_\beta \cap \{v : v_1 \geq 0\}$, $\mathcal{P}_\gamma \cap \{v : v_2 \geq 0\}$, $\mathcal{P}_\kappa \cap \{v : v_2 \leq 0\}$, $\mathcal{P}_\delta \cap \{v : v_1 \leq 1\}$, and $\mathcal{P}_\eta \cap \{v : v_2 \geq 0\}$ is empty.* $\qquad\square$

An immediate corollary of the above claim gives $|\mathscr{F}| = 4$ as desired. Let $I := \{\alpha, \beta, \gamma, \delta\}$.

Claim 2. *$\mathscr{F}$ consists of exactly 4 firefighters $F_\alpha(= F_\kappa)$, F_β, $F_\gamma(= F_\eta)$, F_δ, and $\mathcal{P} = \cup_{i \in I} \mathcal{P}_i$.* $\qquad\square$

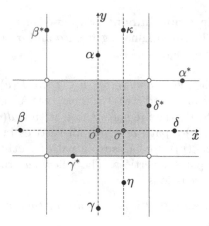

Fig. 2. The counterexample containment **c**.

Four axis-nearest pioneers. We now proceed to identify the rectangular area (for later construction of our new simpler containment), which is enclosed by horizontal and vertical lines through four axis-nearest pioneers. For each $i \in I$, as $i \in \mathcal{P}_i$, we may take $i^* \in \mathcal{P}_i$ such that $|\alpha_2^*|$, $|\beta_1^*|$, $|\gamma_2^*|$ and $|\delta_1^*|$ are minimum, i.e., i^* is the pioneer defended by F_i that is closest to the x-axis (when $i \in \{\alpha, \gamma\}$) or the y-axis (when $i = \beta$) or the vertical line $x = 1$ (when $i = \delta$). Claim 1 and the minimality of $|\alpha_2^*|, |\beta_1^*|, |\gamma_2^*|, |\delta_1^*|$ imply that

Claim 3. *(i)* $\alpha_2^* \geq 1$, $\beta_1^* \leq -1$, $\gamma_2^* \leq -1$, $\delta_1^* \geq 1 + 1$; *and*
(ii) each of $\mathcal{P}_\alpha \cap \{v : v_2 < \alpha_2^*\}$, $\mathcal{P}_\beta \cap \{v : v_1 > \beta_1^*\}$, $\mathcal{P}_\gamma \cap \{v : v_2 > \gamma_2^*\}$ *and* $\mathcal{P}_\delta \cap \{v : v_1 < \delta_1^*\}$ *is empty.* ☐

Then the horizontal lines $y = \alpha_2^*$, $y = \gamma_2^*$ and the vertical lines $x = \beta_1^*$, $x = \delta_1^*$ enclose a rectangle $\mathfrak{R}$ (as shown by the grey area in Fig. 2). Let $int(\mathfrak{R})$ denote the set of vertices in the interior of $\mathfrak{R}$. Combining Claims 2 and 3 we can prove that

Claim 4. $\{o, \sigma\} \subseteq int(\mathfrak{R})$, *and all vertices in* $int(\mathfrak{R})$ *were burned.* ☐

Four sets of blocking pioneers. In view of Claim 4, the fire spread to $\mathfrak{R}$ should be blocked by **c** somewhere on or outside $\mathfrak{R}$. We identify blocking pioneers w.r.t. the four sides of $\mathfrak{R}$. Let $\mathcal{R}_\alpha$ (resp. $\mathcal{R}_\beta$, $\mathcal{R}_\gamma$, $\mathcal{R}_\delta$) denote the top (resp. left, bottom, right) side of $\mathfrak{R}$ with the two end vertices excluded. (These four end vertices are depicted as small circles in Fig. 2.) We abuse notation here by using $\mathcal{R}_i$ ($i \in I$) to denote both the line segment and the set of vertices on it. Notice from Claim 3(i) that $\mathcal{R}_i \neq \emptyset$ for all $i \in I$. For convenience of description, we note that each vertex $i = (i_1, i_2) \in I$ has exactly one nonzero coordinate, whose index we denote as $\xi(i)$. So $\xi(i) \in \{1, 2\}$ with $i_{\xi(i)} \neq 0$ satisfies $\xi(i) := 2$ if $i \in \{\alpha, \gamma\}$ and $\xi(i) := 1$ if $i \in \{\beta, \delta\}$.

For each $i \in I$ and each vertex $v \in \mathcal{R}_i$, we associate v with a fixed *blocking pioneer* $\bar{v} \in \mathcal{P}$ as follows. Let $\mathfrak{A}_v$ denote the ray starting from v that is perpendicular to $\mathcal{R}_i$ and disjoint from $int(\mathfrak{R})$. As only a finite number of vertices were burned, the ray $\mathfrak{A}_v$ must contain a defended vertex, and the union of $\mathfrak{A}_v$ and $int(\mathfrak{R})$ contains a shortest path between this defended vertex and its unique closer fire source. Thus Property 2 asserts that $\mathfrak{A}_v \cup int(\mathfrak{R})$ contains a pioneer vertex, which we set to be $\bar{v}$, and one of its evidencing paths, which we denote as Q_v. Claim 4 along with $\mathcal{P} \subseteq \mathcal{D}$ enforces $\bar{v} \in \mathfrak{A}_v$.

Note that for $i \in I$, if $i^* \in \mathcal{R}_i$, then its blocking pioneer must be i^* itself by the above definition. For technical reasons, even if $i^* \notin \mathcal{R}_i$, we still define $\bar{i}^* := i^*$ to be i^*'s blocking pioneer.

Claim 5. *For each $i \in I$ and each $v \in \mathcal{R}_i$, there holds $d(\varsigma(\bar{v}), \bar{v}) = d(\varsigma(v), v) + |\bar{v}_{\xi(i)} - v_{\xi(i)}|$.*

Proof. We only need to consider the case of $\bar{v} \neq v$. Note that $\bar{v}$'s evidencing path Q_v given above is a shortest $\varsigma(\bar{v})$-$\bar{v}$ path. It is the concatenation of a shortest $\varsigma(\bar{v})$-v path and the (vertical or horizontal) line segment from v to $\bar{v}$ (which has length $|\bar{v}_{\xi(i)} - v_{\xi(i)}|$). Therefore $d(\varsigma(\bar{v}), \bar{v}) = d(\varsigma(\bar{v}), v) + |\bar{v}_{\xi(i)} - v_{\xi(i)}|$. Recalling from Lemma 5 that $\varsigma(\bar{v}) = \varsigma(v)$, the claim follows. □

The next claim specifies simple observations derived from $\bar{v} \in \mathfrak{A}_v$ when v belongs to $\mathcal{R}_\alpha, \mathcal{R}_\beta, \mathcal{R}_\gamma$ or $\mathcal{R}_\delta$.

Claim 6. *For each $i \in I$ and any vertex $v \in \mathcal{R}_i$, exactly one of the following holds:*

(i) $i = \alpha$, $\bar{v}_1 = v_1 \in (\beta_1^*, \delta_1^*)$ and $\bar{v}_2 \geq v_2 = \alpha_2^*$;
(ii) $i = \beta$, $\bar{v}_2 = v_2 \in (\gamma_2^*, \alpha_2^*)$ and $\bar{v}_1 \leq v_1 = \beta_1^*$;
(iii) $i = \gamma$, $\bar{v}_1 = v_1 \in (\beta_1^*, \delta_1^*)$ and $\bar{v}_2 \leq v_2 = \gamma_2^*$;
(iv) $i = \delta$, $\bar{v}_2 = v_2 \in (\gamma_2^*, \alpha_2^*)$ and $\bar{v}_1 \geq v_1 = \delta_1^*$. □

Claim 7. *For each $i \in I$ and any vertices $u, v \in \mathcal{R}_i \cup \{i^*\}$, there holds $d(\bar{u}, \bar{v}) = d(u, v) + |\bar{u}_{\xi(i)} - \bar{v}_{\xi(i)}|$.* □

By the definition of blocking pioneers, it is clear that for each $i \in I$, there is a *one-one correspondence* between $\mathcal{R}_i \cup \{i^*\}$ and $\mathcal{B}_i := \{\bar{v} : v \in \mathcal{R}_i \cup \{i^*\}\} = \{\bar{v} : v \in \mathcal{R}_i\} \cup \{i^*\}$ with each v corresponding to its blocking pioneer $\bar{v}$. The next lemma says that all pioneers in $\mathcal{B}_i$ were defended by the same firefighter F_i, which is a key to our proof.

Claim 8. *For each $i \in I$ and any $\bar{u}, \bar{v} \in \mathcal{B}_i$, there hold $\mathcal{B}_i \subseteq \mathcal{P}_i$ and $d(\bar{u}, \bar{v}) \leq |\tau_{\bar{u}} - \tau_{\bar{v}}|$.*

Proof. First $i^* \in \mathcal{P}_i$ is guaranteed by the choice of i^*. Considering an arbitrary vertex $\bar{v} \in \mathcal{B}_\alpha \setminus \{\alpha^*\}$, we have $\bar{v}_1 = v_1 \in (\beta_1^*, \delta_1^*)$ and $\bar{v}_2 \geq v_2 = \alpha_2^* > \gamma_2^*$ (by Claims 6(i) and 3(i)). It follows from Claim 3(ii) that $\bar{v}$ belongs to none of $\mathcal{P}_\beta$, $\mathcal{P}_\delta$ and $\mathcal{P}_\gamma$, and hence $\bar{v} \in \mathcal{P}$ belongs to $\mathcal{P}_\alpha$ as $\mathcal{P} = \cup_{i \in I} \mathcal{P}_i$ (by Claim 2). The argument and the symmetric counterpart for an arbitrary vertex in $\mathcal{B}_i \setminus \{i^*\}$, $i \in \{\beta, \gamma, \delta\}$ verify $\mathcal{B}_i \subseteq \mathcal{P}_i$ for all $i \in I$.

Since $\{\bar{u}, \bar{v}\} \subseteq \mathcal{P}_i$ says that F_i is an earliest defender of both $\bar{u}$ and $\bar{v}$, it is straightforward from Lemma 2 that $d(\bar{u}, \bar{v}) \leq |\tau_{\bar{u}} - \tau_{\bar{v}}|$. □

A New Defending Process. Given the movements of F_i, $i \in I$ in the defending process $\mathbf{c}$, we aim to define a new containment $\mathbf{c}'$ (possibly $\mathbf{c}'$ is the same as $\mathbf{c}$) which uses four firefighters F_i', $i \in I$ such that F_i' only moves along $\mathcal{R}_i^+$, the shortest elongation of $\mathcal{R}_i$ that covers i^* (Note that $\mathcal{R}_i^+ = \mathcal{R}_i$ in case of $i^* \in \mathcal{R}_i$). By virtue of Claim 8, for each $i \in I$, we can "project" the movement of firefighter F_i for defending the blocking pioneers in $\mathcal{B}_i$ to the movement of F_i' as follows.

Movement projections. For each $i \in I$, we sort the pioneer vertices of $\mathcal{B}_i$ as $b^{(1)}, b^{(2)}, \ldots, b^{(\ell)}$, where $\ell = |\mathcal{B}_i|$, in an increasing order of the time points at which they were defended by (the same) firefighter F_i in $\mathbf{c}$. Accordingly we sort the vertices in $\mathcal{R}_i \cup \{i^*\}$ as $r^{(1)}, r^{(2)}, \ldots, r^\ell$ such that for each $g \in [\ell]$, vertex $b^{(g)}$ is the blocking pioneer $\bar{r}^{(g)}$ of $r^{(g)}$. For any $g, h \in [\ell]$, we refer to $r^{(g)}$ and $b^{(g)}$ as *the g-th vertices* of $\mathcal{R}_i \cup \{i^*\}$ and $\mathcal{B}_i$, respectively; we write $r^{(g)} \prec r^{(h)}$ if and only if $g < h$, and write $r^{(g)} \preceq r^{(h)}$ if and only if $g \le h$. The *movement of F_i' along $\mathcal{R}_i^+$*, as well as *time records $t(v)$*, $v \in \mathcal{R}_i \cup \{i^*\}$ for purpose of technical analysis, is defined in the following recursive way:

– At time 1, F_i' is placed on $r^{(1)}$. Set $t(r^{(1)}) := 1$.
– For each $g = 1, 2, \ldots, |\ell| - 1$, upon reaching $r^{(g)}$, firefighter F_i' moves to the $r^{(g+1)}$ along a shortest $r^{(g)}$-$r^{(g+1)}$ path. (This shortest path is unique and must be on $\mathcal{R}_i^+$, because $r^{(g)}$ and $r^{(g+1)}$ belong to the same (horizontal or vertical) line segment $\mathcal{R}_i^+$.) Set $t(r^{(g+1)}) = t(r^{(g)}) + d(r^{(g)}, r^{(g+1)})$.

Let $\mathbf{c}'$ denote the movements (defending process) of F_i', $i \in I$ defined above. The vertex sorting and the time record setting provide the simple observation that for each $i \in I$ and any distinct $u, v \in \mathcal{R}_i \cup \{i^*\}$,

(i) $u \preceq v$ if and only if $\tau_{\bar{u}} \le \tau_{\bar{v}}$;
(ii) F_i' reached v at time $t(v)$.

Claim 9. $t(v) \le \tau_{\bar{v}}$ for each $v \in \cup_{i \in I}(\mathcal{R}_i \cup \{i^*\})$.

Proof. Suppose that v is the g-th vertex of $\mathcal{R}_i \cup \{i^*\}$. In case of $g = 1$, it is trivial that $t(v) = 1 \le \tau_{\bar{v}}$. Assume that $g \ge 2$, and $t(u) \le \tau_{\bar{u}}$ holds for the $(g-1)$-th vertex u of $\mathcal{R}_i$. Then $\tau_{\bar{u}} \le \tau_{\bar{v}}$. The setting of time records along with Claims 7 and 8 imply $t(v) = t(u) + d(u, v) \le \tau_{\bar{u}} + d(\bar{u}, \bar{v}) \le \tau_{\bar{v}}$. ∎

Claim 10. For each $i \in I$, if $u, v \in \mathcal{R}_i \cup \{i^*\}$ satisfy $u \prec v$, then $\tau_{\bar{v}} - \tau_{\bar{u}} \ge t(v) - t(u) + |\bar{v}_{\xi(i)} - \bar{u}_{\xi(i)}|$. ∎

Successful containment. Next, we show that $\mathbf{c}'$ is indeed a containment of the fire originating from o and σ, given the success of $\mathbf{c}$. For each $v \in \cup_{i \in I} \mathcal{R}_i$, let τ_v' denote the earliest time when some firefighter F_i', $i \in I$ reached it. Obviously, the success of $\mathbf{c}'$ containing the fire is guaranteed by the following lemma.

Lemma 6. For each $i \in I$, firefighter F_i' can defend all vertices in $\mathcal{R}_i$, i.e., $\tau_v' \le d(\varsigma(v), v)$ holds for each vertex $v \in \mathcal{R}_i$. ∎

Proof. We consider an arbitrary $i \in I$ and an arbitrary vertex $v \in \mathcal{R}_i$. Recall that F_i' reached v at time $t(v)$. It is immediate that $\tau_v' \leq t(v)$, which reduces the proof to establishing $t(v) \leq d(\varsigma(v), v)$.

In case of $v \preceq i^*$, i.e., says $\tau_{\bar{v}} \leq \tau_{i^*}$, and Claim 8 gives $\tau_{\bar{v}} + d(\bar{v}, i^*) \leq \tau_{i^*}$. On the other hand, by Property 1, the fact that $i^* \in \mathcal{P}$ gives $\tau_{i^*} \leq d(\varsigma(i^*), i^*) \leq d(\varsigma(v), i^*) \leq d(\varsigma(v), v) + d(v, i^*)$. Therefore

$$d(\varsigma(v), v) \geq \tau_{i^*} - d(v, i^*) \geq \tau_{\bar{v}} + d(\bar{v}, i^*) - d(v, i^*) \geq \tau_{\bar{v}},$$

where the last inequality is guaranteed by the fact that either $v = \bar{v}$ or the triangle on $v, \bar{v}, i^*$ has a right angle at v. Then $t(v) \leq \tau_{\bar{v}}$ in Claim 9 shows $t(v) \leq d(\varsigma(v), v)$ as desired.

It remains to consider the case where $i^* \prec v$. Recalling $\bar{i}^* = i^*$, we have $\tau_{\bar{v}} - \tau_{i^*} \geq t(v) - t(i^*) + |\bar{v}_{\xi(i)} - i^*_{\xi(i)}|$ as stated in Claim 10 and $t(i^*) \leq \tau_{i^*}$ as stated in Claim 9. Combining these two inequalities, we obtain

$$t(v) \leq t(i^*) - \tau_{i^*} + \tau_{\bar{v}} - |\bar{v}_{\xi(i)} - i^*_{\xi(i)}| \leq \tau_{\bar{v}} - |\bar{v}_{\xi(i)} - i^*_{\xi(i)}|.$$

Now by Property 1, $\bar{v} \in \mathcal{P}$ provides $\tau_{\bar{v}} \leq d(\varsigma(\bar{v}), \bar{v})$, and therefore $t(v) \leq d(\varsigma(\bar{v}), \bar{v}) - |\bar{v}_{\xi(i)} - i^*_{\xi(i)}|$. Recalling Claim 6, we deduce from $v \in \mathcal{R}_i$ that $v_{\xi(i)} = i^*_{\xi(i)}$. Hence $t(v) \leq d(\varsigma(\bar{v}), \bar{v}) - |\bar{v}_{\xi(i)} - v_{\xi(i)}| = d(\varsigma(v), v)$, where the last equation is guaranteed by Claim 5. □

Note that in containment (defending process) $\mathbf{c}'$, for each $i \in I$, firefighter F_i' never reached any vertices in $\cup_{j \in I \setminus \{i\}} \mathcal{R}_j$. It follows from Lemma 6 that F_i' is the unique earliest defender of all vertices in $\mathcal{R}_i$. Then Lemma 2 implies the following property.

Property 3. *For each $i \in I$ and each pair of vertices $u, v \in \mathcal{R}_i$, there holds $d(u, v) \leq |\tau_u' - \tau_v'|$.* □

The proof of $\Gamma_{2,2}$'s lower bound. In a counterclockwise visit of the rectangle $\mathfrak{R}$'s sides starting from its upper-right corner, the end vertices of $\mathcal{R}_\alpha$, $\mathcal{R}_\beta$, $\mathcal{R}_\gamma$ and $\mathcal{R}_\delta$ we encounter are $\alpha' := (\delta_1^* - 1, \alpha_2^*)$, $\alpha'' := (\beta_1^* + 1, \alpha_2^*)$, $\beta' := (\beta_1^*, \alpha_2^* - 1)$, $\beta'' := (\beta_1^*, \gamma_2^* + 1)$, $\gamma' := (\beta_1^* + 1, \gamma_2^*)$, $\gamma'' := (\delta_1^* - 1, \gamma_2^*)$, $\delta' := (\delta_1^*, \gamma_2^* + 1)$, and $\delta'' := (\delta_1^*, \alpha_2^* - 1)$. See Fig. 3 for an illustration.

Property 4. *For each $i \in I$, if $\{u, v\} = \{i', i''\}$ and $\tau_u' < \tau_v'$, then $d(\varsigma(v), v) > d(u, v)$.*

Proof. It follows from Property 3 that $\tau_v' \geq \tau_u' + d(u, v) > d(u, v)$. On the other hand, Lemma 6 says $\tau_v' \leq d(\varsigma(v), v)$, implying $d(\varsigma(v), v) > d(u, v)$. □

The contradiction to Lemma 6 establishes the following concluding theorem.

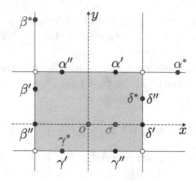

Fig. 3. The proof of $\Gamma_{2,2} \geq 5$.

Theorem 5. $\Gamma_{2,2} \geq 5$.

Proof. If the four firefighters F_i', $i \in I$ all defended i' earlier than i'' or all defended i' later than i'', then by symmetry we may assume that the former case happens. By Property 4, for each $i \in I$, we deduce from $\tau_{i'}' < \tau_{i''}'$ that $d(\varsigma(i''), i'') > d(i', i'')$. Substituting α (resp. δ, γ, β) for i, we obtain

$$d(\varsigma(\alpha''), \alpha'') = d(o, \alpha'') = -\beta_1^* - 1 + \alpha_2^* > d(\alpha', \alpha'') = \delta_1^* - \beta_1^* - 2$$
$$\Rightarrow \alpha_2^* \geq \delta_1^*; \tag{1}$$

$$d(\varsigma(\beta''), \beta'') = d(o, \beta'') = -\beta_1^* + (-\gamma_2^* - 1) > d(\beta', \beta'') = \alpha_2^* - \gamma_2^* - 2$$
$$\Rightarrow -\beta_1^* \geq \alpha_2^*; \tag{2}$$

$$d(\varsigma(\gamma''), \gamma'') = d(\sigma, \gamma'') = \delta_1^* - 2 - \gamma_2^* > d(\gamma', \gamma'') = \delta_1^* - \beta_1^* - 2$$
$$\Rightarrow \beta_1^* \geq \gamma_2^* + 1; \tag{3}$$

$$d(\varsigma(\delta''), \delta'') = d(\sigma, \delta'') = \delta_1^* - 1 + \alpha_2^* - 1 > d(\delta', \delta'') = \alpha_2^* - \gamma_2^* - 2$$
$$\Rightarrow \delta_1^* + \gamma_2^* \geq 1. \tag{4}$$

Adding the four inequalities derived in the above implications (1)–(4), we reach a contradiction that $0 \geq 2$.

It remains to consider the case where there exist $i, j \in I$ such that F_i' defended i' earlier than i'', i.e., $\tau_{i'}' < \tau_{i''}'$, and F_j' defended j' later than j'', i.e., $\tau_{j'}' > \tau_{j''}'$. Then there must exist such i, j such that i' and j'' are adjacent to the same corner vertex. It can be easily seen that such i, j could be taken such that the ordered pair (i, j) is one of (α, δ), (δ, γ), (γ, β) and (β, α). By symmetry, we may assume without loss of generality that $i = \alpha$ and $j = \delta$.

By $\tau_{\alpha'}' < \tau_{\alpha''}'$ and $\tau_{\delta''}' < \tau_{\delta'}'$, it follows from Property 4 that $d(\varsigma(\alpha''), \alpha'') > d(\alpha', \alpha'')$ and $d(\varsigma(\delta'), \delta') > d(\delta', \delta'')$. The former inequality implies $\alpha_2^* \geq \delta_1^*$ as in (1); while the latter gives

$$d(\varsigma(\delta'), \delta') = d(\sigma, \delta') = \delta_1^* - 1 - (\gamma_2^* + 1) > d(\delta', \delta'') = (\alpha_2^* - 1) - (\gamma_2^* + 1) \Rightarrow \delta_1^* > \alpha_2^*.$$

The above contradiction disproves the assumption on the existence of the counterexample containment **c**, establishing $\gamma_{2,2} \geq 5$. $\square$

References

1. Anshelevich, E., Chakrabarty, D., Hate, A., Swamy, C.: Approximation algorithms for the firefighter problem: cuts over time and submodularity. In: Proceedings of the 20th International Symposium on Algorithms and Computation, pp. 974–983 (2009)
2. Cai, L., Verbin, E., Yang, L.: Firefighting on trees: $(1 - 1/e)$-approximation, fixed parameter tractability and a subexponential algorithm. In: Proceedings of the 19th International Symposium on Algorithms and Computation, pp. 258–269 (2008)
3. Chalermsook, P., Chuzhoy, J.: Resource minimization for fire containment. In: Proceedings of the Twenty-First Annual ACM-SIAM Symposium on Discrete Algorithms, pp. 1334–1349 (2010)
4. Comellas, F., Mitjana, M., Peters, J.G.: Epidemics in small world communication networks. Technical report, SFU-CMPT-TR-2002 (2002)
5. Develin, M., Hartke, S.G.: Fire containment in grids of dimension three and higher. Discrete Appl. Math. **155**(17), 2257–2268 (2007)
6. Dezső, Z., Barabási, A.L.: Halting viruses in scale-free networks. Phys. Rev. E **65**(5), 055103 (2002)
7. Finbow, S., King, A., MacGillivray, G., Rizzi, R.: The firefighter problem for graphs of maximum degree three. Discrete Math. **307**(16), 2094–2105 (2007)
8. Finbow, S., MacGillivray, G.: The firefighter problem: a survey of results, directions and questions. Australas. J. Comb. **43**, 57–77 (2009)
9. Fogarty, P.: Catching the fire on grids. Master's thesis, The University of Vermont (2003)
10. Hartnell, B.: Firefighter! an application of domination. presentation. In: 25th Manitoba Conference on Combinatorial Mathematics and Computing, University of Manitoba in Winnipeg, Canada (1995)
11. King, A., MacGillivray, G.: The firefighter problem for cubic graphs. Discrete Math. **310**(3), 614–621 (2010)
12. MacGillivray, G., Wang, P.: On the firefighter problem. J. Comb. Math. Comb. Comput. **47**, 83–96 (2003)
13. Messinger, M.E.: Firefighting on infinite grids. Master's thesis, Department of Mathematics and Statistics, Dalhousie University (2005)
14. Messinger, M.E.: Average firefighting on infinite grids. Australas. J. Comb. **41**, 15 (2008)
15. Ng, K., Raff, P.: Fractional firefighting in the two dimensional grid, pp. 2005–23. Technical report, DIMACS (2005)
16. Scott, A.E., Stege, U., Zeh, N.: Politicians firefighting. In: Proceedings of the 17th International Symposium on Algorithms and Computation, pp. 608–617 (2006)
17. Wang, P., Moeller, S.A.: Fire control on graphs. J. Comb. Math. Comb. Comput. **41**, 19–34 (2002)
18. Yutaka, I., Naoyuki, K., Tomomi, M.: Improved approximation algorithms for firefighter problem on trees. IEICE Trans. Inf. Syst. **94**(2), 196–199 (2013)

Mediated Population Protocols: Leader Election and Applications

Shantanu Das[1], Giuseppe Antonio Di Luna[2(✉)], Paola Flocchini[2],
Nicola Santoro[3], and Giovanni Viglietta[2]

[1] LIF, Aix-Marseille University, and CNRS, Marseille, France
shantanu.das@lif.univ-mrs.fr
[2] University of Ottawa, Ottawa, Canada
{gdiluna,paola.flocchini}@uottawa.ca, gvigliet@uottawa.ca
[3] Carleton University, Ottawa, Canada
santoro@scs.carleton.ca

Abstract. Mediated population protocols are an extension of population protocols in which communication links, as well as agents, have internal states. We study the leader election problem and some applications in constant-state mediated population protocols. Depending on the power of the adversarial scheduler, our algorithms are either stabilizing or allow the agents to explicitly reach a terminal state.

We show how to elect a unique leader if the graph of the possible interactions between agents is complete (as in the traditional population protocol model) or a tree. Moreover, we prove that a leader can be elected in a complete bipartite graph if and only if the two sides have coprime size.

We then describe how to take advantage of the presence of a leader to solve the tasks of token circulation and construction of a shortest-path spanning tree of the network. Finally, we prove that with a leader we can transform any stabilizing protocol into a terminating one that solves the same task.

1 Introduction

Background. The *population protocol* model, introduced in the seminal paper of Angluin et al. [3] has recently received a lot of interest among researchers in distributed computing. The model consists of a set of simple anonymous finite-state agents that interact pairwise, and each interaction changes the state of both agents. Normally each pair of agents is supposed to interact infinitely often in any infinite execution of the protocol; however, these interactions may occur in any arbitrary order. This models the asynchrony and uncertainty in a distributed system. Moreover, as the agents have constant memory independent of the size of the system, this means that the protocol can be scaled to populations of any size. The population protocol model is useful for modeling large-scale networks consisting of small mobile devices, such as sensor networks or swarms of microrobots.

© Springer International Publishing AG 2017
T.V. Gopal et al. (Eds.): TAMC 2017, LNCS 10185, pp. 172–186, 2017.
DOI: 10.1007/978-3-319-55911-7_13

Since the introduction of this model, several variants of population protocols have been studied. For example, there could be restrictions on which only certain pairs of agents are allowed to interact, giving rise to arbitrary *interaction graphs* instead of the complete graph [2]. This paper considers the interaction graph to be an arbitrary connected graph on the set of agents. Another possibility is to consider restrictions on the schedule of interactions, e g., allowing a periodic scheduler, or a k-bounded scheduler, or a probabilistic scheduler [1].

The power of population protocols in terms of what kinds of predicates can be computed by them has been studied extensively. Angluin et al. [5] showed that the class of computable predicates is exactly the class of semilinear predicates (or, equivalently, all predicates that can be defined by first-order logical formulas in Presburger arithmetic). Further studies introduced enhancements in the model to increase its computational power and allow the computation of larger classes of predicates: endowing each agent with non-constant memory [1], assuming the presence of a leader [7], allowing a certain amount of information to be stored on the edges of the interaction graph [12,13]. In the present paper we study the latter category of population protocols, which are called *mediated population protocols*. We assume that the amount of memory per node and per edge of the graph is constant, and we study what can be computed in several restricted classes of interaction graphs and with several types of schedulers.

Our Contributions. In this paper we focus on algorithms to elect a unique *leader* in a mediated population protocol, as well as applications of a leader in several common situations. In Sect. 2, we formally define the mediated population protocol model and related concepts. The types of schedulers we consider are the *recurrent* scheduler, which only implements a bland notion of fairness on interactions, and the *k-bounded* scheduler, which cannot neglect any interaction for too long. We also distinguish between *stabilizing* protocols, which reach a configuration in which no agent changes state any more, and *terminating* protocols, in which the agents "realize" that the configuration is stable (or about to become stable), and explicitly terminate the execution. Typically, when the scheduler is recurrent and the task is not trivial, there exists no terminating protocol to solve it. In these cases, we will give only stabilizing protocols. On the other hand, when the scheduler is k-bounded, we will give terminating protocols.

In Sect. 3, we study the problem of leader election in several network topologies: complete graphs, complete bipartite graphs, and trees. We prove that a unique leader can always be elected in a complete graph and in a tree, and we give a characterization of the complete bipartite graphs in which a leader can be elected under a 1-bounded scheduler: those are the complete bipartite graphs in which the two sides have coprime sizes.

In Sect. 4, we assume that the network contains a unique leader, and we show how to use this feature to accomplish some typical tasks in arbitrary networks. First we show how to solve the *token circulation* problem and how to construct a *shortest-path spanning tree*. Then we show how to convert any stabilizing protocol into an "equivalent" but terminating one. As a byproduct, given any protocol for the 2-bounded scheduler, we can make it work also under the

k-bounded scheduler, for any $k > 2$. By combining these solutions with the leader election protocols of Sect. 3, we can solve the same tasks even if a leader is not given in advance, provided that the network is a complete graph, a complete bipartite graph with coprime sides, or a tree.

Related Work. The task of electing a leader has been extensively studied in the context of (non-mediated) population protocols. The majority of papers focus on self-stabilizing leader election under the assumption that the scheduler is *globally fair*. This is a more powerful scheduler than the recurrent one, and it may or may not be more powerful than the k-bounded one. It has been shown that leader election requires either as many states as agents [9] or the presence of an *oracle* that informs agents about the presence of a leader in the system [15]. Concerning restricted interaction graphs, in [10] self-stabilizing leader election algorithms for trees are given; the case of the ring is studied in [15]. Both papers assume the presence of an oracle. These results have been extended, under the same set of assumptions, to arbitrary graphs in [6]. In [4], a constant-space algorithm for the ring graph is also given. In the context of mediated population protocols, a non-constant-space algorithm for leader election is shown in [16].

Under the globally fair scheduler, the self-stabilizing construction of a spanning tree has been investigated in [4], where an algorithm requiring $\mathcal{O}(\log D)$ states is given, D being the graph's diameter. In the same paper, a self-stabilizing token circulation algorithm for the ring graph is given. In a model similar to population protocols, a token circulation algorithm for arbitrary graphs is discussed in [11], assuming the presence of an oracle.

Several papers assumed a unique leader as a computational tool to enhance the power of population protocols. Counting algorithms are given in [7]; in [8], a self-stabilizing transformer for general protocols has been studied in a slightly different model and under the additional assumption of unbounded memory. In the context of fault tolerance, [14] uses a leader to make any protocol tolerant to omission failures.

Other papers have discussed the computational power of mediated population protocols in terms of the predicates that can be computed [12,13].

In light of the above results, our paper represents a breakthrough in that we show how to elect a leader in some large classes of networks using only a constant number of states per agent and per edge; moreover, we often make weaker assumptions on the scheduler, and we do not resort to querying an oracle. The same holds for the applications of leader election, which in addition apply to all networks in which a leader is present (regardless of the interaction graph's topology).

2   Model and Definitions

Network and Configuration. A *network* is an unoriented connected finite graph G on a set V of at least two vertices, which are called *agents*. Each agent has an *agent state* belonging to a finite set Q. In turn, Q is partitioned into *input states* Q_I, *work states* Q_W, and *terminal states* Q_T.

Each edge of G has a *port* for each of its two endpoints; the set of all ports is denoted by P. If $\{a, b\}$ is an edge of G, we denote by $p(a, b)$ the port on a's side of $\{a, b\}$ and by $p(b, a)$ the port on b's side. Each port has a *port state* belonging to a finite set U; there exists a unique *initial state* $u_0 \in U$.

A *configuration* C is a pair of functions (f, g), where $f \colon V \to Q$ and $g \colon P \to U$. C is said to be *initial* if $f(V) \subseteq Q_I$ and $g(P) = \{u_0\}$.

Interaction and Scheduler. An *interaction* is an ordered pair of agents (a_s, a_r), where a_s is called the *sender* and a_r is called the *receiver*, such that $\{a_s, a_r\}$ is an edge of G. The set of possible interactions of G (i.e., two for each edge) is denoted by I. A *schedule* S is an infinite sequence of interactions, i.e., $S \colon \mathbb{N} \to I$. A *scheduler* is a set of schedules. We define the following schedulers:

- the *recurrent scheduler* is the set of schedules in which each interaction of I appears infinitely often;
- the *k-bounded scheduler*, where k is a positive constant, is the set of schedules that belong to the recurrent scheduler and such that, between two consecutive occurrences of the same interaction within a schedule, none of the other interactions appears more than k times.

Let us observe that all the schedules in the 1-bounded scheduler are periodic.

Transition Function, Execution, and Task. A *transition function* (or *protocol*) is a function $\delta \colon Q \times U \times Q \times U \to Q \times U \times Q \times U$ such that, if $\delta(q_s, u_s, q_r, u_r) = (q'_s, u'_s, q'_r, u'_r)$ and $q_s \in Q_T$ (respectively, $q_r \in Q_T$), then $q_s = q'_s$ (respectively, $q_r = q'_r$). That is, δ leaves terminal states unchanged.

Given a configuration $C = (f, g)$, we say that configuration $C' = (f', g')$ *results* from C by the interaction $i = (a_s, a_r)$ according to transition function δ if f' coincides with f on $V \setminus \{a_s, a_r\}$, g' coincides with g on $P \setminus \{p(a_s, a_r), p(a_r, a_s)\}$, and $(f'(a_s), g'(p(a_s, a_r)), f'(a_r), g'(p(a_r, a_s))) = \delta(f(a_s), g(p(a_s, a_r)), f(a_r), g(p(a_r, a_s)))$. If this is the case, we write $C \xrightarrow{\delta, i} C'$.

The *execution* of a schedule $S = (i_0, i_1, \dots)$ from an initial configuration C_0 according to a transition function δ is the sequence of configurations $(C_0, C_1, \dots)$ such that, for every $j \in \mathbb{N}$, $C_j \xrightarrow{\delta, i_j} C_{j+1}$. Let an execution $E = (C_j)_{j \geq 0}$ be given, with $C_j = (f_j, g_j)$. We say that E is *stable* if there is a $j^* \in \mathbb{N}$ such that, for every $j' > j^*$, $f_{j'} = f_{j'+1}$. We say that E *terminates* if there is a $j^* \in \mathbb{N}$ such that $f_{j^*}(V) \subseteq Q_T$. Note that an execution that terminates is also stable.

A *task* or *problem* is a set of executions. We say that a protocol δ on a set of agent states Q and a set of port states U *solves* a task $\mathcal{T}$ under scheduler $\mathcal{S}$ in a given network if the execution according to δ of every schedule in $\mathcal{S}$ from any initial configuration is in $\mathcal{T}$. If such executions are all stable, the protocol is said to be *stabilizing*. If such executions all terminate, the protocol is said to be *terminating*.

Algorithmic notation. When describing transition functions, we will sometimes use an "algorithmic style" (cf. Figs. 1 and 2). When the interaction (a, b) occurs, the function Transition function is applied to $(a, p(a, b), b, p(b, a))$; note that, with a little abuse of notation, we identify agents and ports with their respective states. In our formalism, a state is seen as a tuple of variables. To refer to variable x of the state of agent a, we use the expression $a.x$.

3   Leader Election

In this section we study the task of electing a leader in several types of networks. Formally, the set of agent states includes some *leader states*, and *leader election* is the task consisting of the executions in which eventually there is a unique agent in the network with a leader state. Note that a protocol solving the leader election problem need not be stabilizing.

3.1   Complete Graphs

If the network is a complete graph, there is a simple leader election protocol that works under the recurrent scheduler.

Theorem 1. *There exists a stabilizing protocol that solves the leader election problem in K_n, for all $n > 1$, under the recurrent scheduler.*

Proof. We use only two agent states: an input state, which is also a leader state, and a work state, which is a non-leader state. There are no terminal states and only one port state. Whenever two agents with the leader state interact, the sender retains the leader state and the receiver takes the non-leader state. In all other cases, the agents retain their states.

As a result, in every execution all agents will initially have the leader state (because it is the only input state) and, whenever two leaders meet, one will be "eliminated". Since all ordered pairs of agents are going to interact infinitely many times (because the network is the complete graph and the scheduler is recurrent), it is obvious that eventually only one agent with the leader state will remain, and its state will never change. Hence the protocol is stabilizing.   □

3.2   Complete Bipartite Graphs

Next we give a characterization of the complete bipartite graphs in which the leader election problem is solvable under the 1-bounded scheduler. When the problem is solvable, we can also give a terminating protocol.

Theorem 2. *There exists a (terminating) protocol that solves the leader election problem in $K_{m,n}$ under the 1-bounded scheduler if and only if m and n are coprime.*

Proof. Suppose that m and n are coprime. Without loss of generality, let $m < n$. The idea of our protocol is to make the m agents in the smaller side of the graph "eliminate" m agents in the larger side. What is left is a smaller complete bipartite graph, on which the same procedure is repeated until only one agent remains: this agent will be the leader.

The protocol uses the fact that the schedule has period $2mn$, and hence the concept of *round* can be defined as a set of $2mn$ consecutive interactions. Whenever an agent a is involved in an interaction for the first time (which is easy to detect, since a still has an input state), and its partner is some agent b, the port $p(a, b)$ is "marked" with a special state that also encodes the role of a in the interaction (i.e., sender or receiver). So, when a sees the marked port again, and its role is the same as the one encoded by the mark, it knows that a new round has started. With this technique, agents can implicitly coordinate their actions and do different things at different rounds.

In the first round, a maximal matching is constructed. Initially all agents are unmatched; whenever two unmatched agents interact, they become matched and change their state accordingly. By the end of the first round, all agents in the smaller side of the graph have been matched. During the second round, all agents discover if they belong to the smaller side or the larger side: the ones in the smaller side will see some unmatched agents, and the ones in the larger side will only see matched agents. Note that each agent can perform this check by having a "flag" in its state that is cleared at the beginning of the round and is set whenever an unmatched agent is encountered. During the third round, the agents in the smaller side revert their state to unmatched, and the matched agents in the larger side become eliminated.

This three-round cycle is then repeated, ignoring the eliminated agents, until only one agent remains unmatched (note that this happens if and only if m and n are coprime). Finally, this situation has to be detected by all agents, so that the protocol can terminate. This is done by adding another flag, which is used in the third round, to check if the encountered unmatched agents are more than one. So, when only one unmatched agent is left, the agents in the opposite side detect it and get a terminal (non-leader) state. As soon as the other agents see some terminated agents, they also get a terminal state (which will be a leader or non-leader state, depending on whether they are unmatched or matched).

Suppose now that m and n are not coprime, and let $d > 1$ be their greatest common divisor. Partition one side of the graph into m/d groups of size d and the other side in n/d groups of size d. Let $A = \{a_0, \ldots, a_{d-1}\}$ and $B = \{b_0, \ldots, b_{d-1}\}$ be two groups of agents on opposite sides, and let $S_{A,B}$ be the following sequence of d^2 interactions: first all the interactions of the form (a_j, b_j), with $0 \leq j < d$, then all the interactions of the form $(a_j, b_{j+1 \mod d})$, then all the interactions of the form $(a_j, b_{j+2 \mod d})$, etc. We then construct a sequence S by concatenating all the sequences $S_{A,B}$ for every ordered pair (A, B) of groups of agents located on opposite sides of the graph. The resulting sequence has length $2mn$ and involves all possible interactions in the network. Finally, we construct a schedule S^* by concatenating infinitely many copies of S.

Suppose that in the initial configuration all agents have the same input state (which is a valid initial configuration, regardless of the protocol), and suppose that the above scheduler S^* is executed. Then, every d interactions, all agents in a same group will have the same state, and in particular there will not be a unique agent with a leader state. This means that no protocol solves the leader election problem under the 1-bounded scheduler (since S^* is 1-bounded).   ⊔

3.3   Tree Graphs

If the network is a tree and the scheduler is recurrent, we can always elect a leader with a stabilizing protocol. Moreover, if the scheduler is k-bounded, we have a terminating protocol.

Theorem 3. *For every* $k \geq 1$, *under the* k-*bounded scheduler (respectively, under the recurrent scheduler), there exists a terminating (respectively, stabilizing) protocol that solves the leader election problem in every tree.*

Proof. First we describe how to elect a stable leader if the scheduler is recurrent: the protocol is summarized in Fig. 1. Then we will show how to make the same protocol terminate under the k-bounded scheduler. The idea of the protocol is to establish a parent-child relation between adjacent vertices of the tree in such a way that eventually the tree becomes rooted: the root will then be the leader. Assuming that $\{a, b\}$ is an edge, the way we represent the fact that a is a parent of b is by setting a *parent flag* in the state of port $p(a, b)$. In the initial configuration, no parent flag is set. Each agent has a *parents variable* too, which counts how many parents the agent has (initially 0, and ranging from 0 to 2). Both agents and ports also have a *busy flag*, initially not set.

Whenever a pair of non-busy agents (a, b) is activated and none of $p(a, b)$ and $p(b, a)$ have their parent flag set, then the parent flag of $p(a, b)$ is set, encoding the fact that b has become a child of a. Then the parents variable of b is incremented; if b has now two parents, both b and $p(b, a)$ become "busy" by setting their respective busy flag. b will then look for its "old parent" c. Note that, while b has its busy flag set, it will accept no more parents or children. When b interacts again with c (which is recognizable because the parent flag of $p(c, b)$ is set and the busy flag of $p(b, c)$ is not set and c is not busy, b becomes a parent of c and c becomes a child of b (i.e., the parent flags of $p(b, c)$ and $p(c, b)$ are switched). Also, the parents variable of c is incremented; if c has two parents, then both c and $p(c, b)$ become busy. At the same time, the parents variable of b is decremented, meaning that b has a unique parent again. So, the next time b interacts with a (which is recognizable because the parent flag of $p(a, b)$ is set), b will clear its own busy flag, as well as the busy flag of $p(b, a)$. In the meantime, if the busy flag of c is set, c looks for its old parent d and does the same operations that b just did; then d will do the same, etc.

Let us see how an execution of this algorithm works globally. Initially, all edges of the network are "unoriented"; as soon as some edge is activated, it gets an "orientation", telling which of the two endpoints is the parent. As the execution continues, a forest of oriented subtrees is constructed, and each tree in this forest has a unique root. When two subtrees meet because an interaction (a, b) occurs, they merge, and the root of a's subtree becomes the root of the new tree. So, the orientations of all the edges in the path from b to the root of b's subtree have to reverse. While edges are being reversed, the agents involved become temporarily busy, so that no other subtrees can merge at those points and interfere with the process. Note that deadlocks are impossible because the network is cycleless. Also, progress will always be made, because the scheduler is recurrent, and therefore all possible interactions will eventually occur. When all the subtrees have finally merged and all edges stop reversing, the entire tree is oriented and has a unique root. The root is also the only agent in the tree whose parent flag is not set. If we define this as a leader state, we have a stabilizing leader election protocol.

```
1:  Agent variables
2:  parents := 0
3:  busy := false
4:
5:  Port variables
6:  parent := false
7:  busy := false
8:
9:  Transition function δ(a, p(a, b), b, p(b, a))
10: if ¬a.busy ∧ ¬b.busy ∧ ¬p(a, b).parent ∧ ¬p(b, a).parent then
11:     p(a, b).parent := true
12:     b.parents := b.parents + 1
13:     if b.parents = 2 then
14:         b.busy := true
15:         p(b, a).busy := true
16: else if a.busy ∧ ¬b.busy ∧ ¬p(a, b).busy ∧ p(b, a).parent then
17:     p(a, b).parent := true
18:     p(b, a).parent := false
19:     a.parents := a.parents − 1
20:     b.parents := b.parents + 1
21:     if b.parents = 2 then
22:         b.busy := true
23:         p(b, a).busy := true
24: else if a.busy ∧ a.parents = 1 ∧ p(b, a).parent then
25:     a.busy := false
26:     p(a, b).busy := false
```

Fig. 1. Stabilizing leader election in a tree

Suppose now that the scheduler is k-bounded, and let us show how to make the above protocol terminate. Observe that a technique similar to the one used in Theorem 2 allows any agent to determine when it has interacted with all of its neighbors in the tree. The first time an agent is involved in an interaction as the sender, it marks the corresponding port. Then it counts how many times that same interaction occurs; at the $k + 1$th occurrence, the agent knows that all the possible interactions have occurred at least once, and therefore it has interacted with all of its neighbors. Since k is fixed, a constant number of agent states is sufficient to implement this counter. Furthermore, an agent can determine whether it is a leaf or an internal vertex of the network: if the agent sees a marked port every time it is activated as the sender of an interaction for $k + 1$ times consecutively, then it is a leaf. Now, if a leaf agent has a parent, it knows that it will never become leader, and therefore it can get a terminal non-leader state. More generally, if an agent with a parent realizes that all its neighboring agents except its parent are in a terminal state, then it can get a terminal non-leader state, as well. Once again, this check can be performed with a flag and a finite counter. It is easy to prove by induction that eventually all agents except the final root of the tree will get a terminal non-leader state. When this happens, the root easily realizes and gets a terminal leader state.   □

4   Applications of a Unique Leader

In this section we will show how the presence of a unique leader can help us solve several different tasks. Formally, these are tasks consisting of executions in which the initial configuration has a unique agent in a leader state.

4.1   Token Circulation

Here we provide a stabilizing solution to the *token circulation* task that works in every network under the recurrent scheduler. Formally, the set of agent states includes some *token states*, and token circulation is the task consisting of the executions in which, if in the initial configuration there is a unique agent with token state, then in every configuration there is a unique agent with token state, and each agent has token state in at least one configuration.

```
 1: Agent variables
 2: token                                    ▷ true for the leader, false for non-leaders
 3: tree := false
 4: summoning := false
 5:
 6: Port variables
 7: parent := false
 8:
 9: Transition function δ(a, p(a, b), b, p(b, a))
10: if a.token ∧ ¬a.tree then
11:     a.tree := true
12: if ¬a.tree ∧ ¬a.summoning ∧ b.tree then
13:     a.summoning := true
14:     p(b, a).parent := true
15: if a.summoning ∧ ¬b.summoning ∧ p(b, a).parent then
16:     b.summoning := true
17:     p(a, b).parent := true
18:     p(b, a).parent := false
19: if a.summoning ∧ b.token ∧ p(a, b).parent then
20:     a.token := true
21:     if ¬a.tree then
22:         a.tree := true
23:         a.summoning := false
24:     b.token := false
25:     b.summoning := false
```

Fig. 2. Token circulation protocol

Protocol Variables. Each agent's state consists of three flags: *token*, *tree*, and *summoning*. Each port's state consists of the single flag *parent*. The token states coincide with the leader states, and are those in which *token* = *true*. All flags of all agents and ports are initially set to *false*, with the exception of the *token* flag of the leader, which is set to *true*.

Protocol Description. Our protocol is given in Fig. 2. The token circulates along the edges of a spanning tree of the network, which is constructed incrementally as the algorithm is executed. Each agent remembers if it has already obtained the token: this is done by setting the flag *tree*. With this flag, the agent also remembers that it belongs to the "partial" spanning tree. The flag *summoning* is used by an agent to remember that the token has to be sent to a new agent that recently joined the spanning tree. The ports of each edge have a *parent* flag that we use to encode a parent-child relationship between the endpoint agents or an orientation of the edge, in the same way as we did in Theorem 3. The resulting oriented edges can point either in the direction of the token (along the spanning tree) or toward an agent that is summoning the token.

The details of the algorithm are as follows. If an agent has the token and is not in the partial spanning tree (i.e., *a.token* and ¬*a.tree*), it sets its own *tree* flag, thus becoming part of the spanning tree. This is an initialization operation that is performed only once in every execution.

If a sender *a* not in the spanning tree and not summoning (i.e., ¬*a.tree* and ¬*a.summoning*) interacts with a receiver *b* in the spanning tree (i.e., *b.tree*), then it sets it own *summoning* flag and orients the edge {*a, b*} toward *b*, setting the *p(b, a).parent* flag.

If a summoning sender *a* (i.e., *a.summoning*) interacts with a non-summoning receiver *b* along an edge of spanning tree that is oriented toward *b* (i.e., ¬*b.summoning* and *p(b, a).parent*), then *b* becomes a summoner as well, and the orientation of the edge {*a, b*} is reversed.

Finally, if a summoning sender *a* interacts with a receiver *b* possessing the token and the edge {*a, b*} is oriented toward *a*, then *a* gets the token, while *b* loses it and ceases to be a summoner (in case it was a summoner). Additionally, if *a* is not in the spanning tree yet, it sets its own *tree* flag and stops being a summoner.

Theorem 4. *The protocol in Fig. 2 solves the token circulation task in any network under the recurrent scheduler, provided that there is a unique leader. Moreover, the protocol is stabilizing and the edges with the parent flag set (on either port) eventually define a spanning tree of the network with edges oriented toward the token.*

Proof. First we shall prove that the edges with the *parent* flag set always form a tree. Indeed, initially no *parent* flag is set. Then, the only line of the algorithm that creates an edge orientation is line 14 (note that lines 16 and 17 only flip an edge that is already oriented). In turn, line 14 is triggered only when *a* is not in the spanning tree and is not summoning. But as line 14 is triggered, *a* becomes summoning. Moreover, when a ceases to be summoning, it also becomes part of the spanning tree (lines 22 and 23). And once a is part of the spanning tree, it never leaves it (because the *tree* flag is never cleared in the algorithm). It follows that a is involved in the execution of line 14 at most once. This shows that the edges with the *parent* flag set never form cycles. Showing that they form a connected sub-network (containing the token) is also easy, because line 14 is executed only when a is a neighbor of an agent b with flag *tree* set, and the *tree* flag is set only by the agent that initially has the token (line 11) and by agents that have incident edges with the *parent* flag set (line 22). This proves that the edges with the *parent* flag set form a tree throughout the execution.

Also note that the agents with the flag *tree* set are vertices of this partial spanning tree: indeed, the flag is first set by the agent with the token (when the tree has no edges yet), and then only by agents that have incident edges with the *parent* flag set. Actually, the only agents that do not have the *tree* flag set and are incident to edges with the *parent* flag set are leaves of the partial spanning tree that have the *summoning* flag set.

Let us now prove the correctness of the protocol. We have to show that eventually all agents in the network set their *tree* flag. Since this only happens when they have the token, this would prove that the token reaches all agents. We will prove by induction that the number of agents with the *tree* flag set is bound to increase. Initially no agents have the *tree* flag set, and nothing happens until the agent with the token is involved in an interaction and sets its own *tree* flag. Note that this must happen sooner or later because the network is connected, there are at least two agents, and the scheduler is recurrent. Then, some agent a whose *tree* flag is not set will interact with an agent b whose *tree* flag is set, triggering lines 12–14. So a will become a summoner and the edge $\{a, b\}$ will be oriented toward b. Of course, this may happen to several different agents, not only to a. What will happen next is that lines 15–18 will be triggered and some edges of the partial spanning tree will start reversing. The idea is that each summoner tries to reach the agent with the token by reversing the edges along the path connecting to it. This path can be easily identified because all the non-summoners that are in the partial spanning tree point toward the token. When an edge is reversed, its non-summoner endpoint becomes a summoner, as well. This prevents lines 15–18 from being triggered more then once on the same agent, and therefore prevents different summoners from interfering with each other. Eventually, an edge reversal will reach the token. When this happens, there is a unique path in the partial spanning tree that is oriented from the token to a summoning leaf. Then lines 19–25 will be triggered, and the token will follow the edges of such an oriented path, until it reaches the summoning leaf. The agents that lose the token will clear their *summoning* flags, but the one with the token will remain a summoner, to avoid triggering lines 15–18 and avoid creating forks in the path. When the summoning leaf obtains the token (the leaf is recognizable because its *tree* flag is not set), it sets its own *tree* flag and clears its own *summoning* flag.

This ends the proof of correctness. Note that the protocol is stabilizing because, as soon as all agents have set their *tree* flag, no new summoners appear and all edges stop reversing.   □

4.2   Construction of Shortest-Path Spanning Trees

Next we show how to solve the task of constructing a spanning tree of the network under a k-bounded scheduler, again assuming that there is a unique leader. As a bonus, the distance of the leader from any agent along the spanning tree coincides with the distance over the whole network. Equivalently, this spanning tree is generated by a breadth-first traversal of the network starting at the leader.

Theorem 5. *For every $k \geq 1$, under the k-bounded scheduler there exists a terminating protocol that constructs a shortest-path spanning tree of any network, provided that there is a unique leader.*

Proof. The tree is created level by level, and the leader coordinates the construction: when a new level of the tree is completed, the leader will be notified and will broadcast a message on the partial spanning tree, ordering the construction of a new level. When a leaf receives a "new level" message, it expands identifying its children among the agents that have not been included in the tree, yet. Each of these agents will be part of the new level. Since the scheduler is k-bounded, a leaf is able to detect when it has seen all its neighbors, as in Theorem 3. Then, the leaf sends a "job done" message to its parent; in the message it also communicates if there is a new level or not. Each upper level collects all termination messages and forwards them, until they reach the leader. Initially, the leader is the only leaf of the tree and it will bootstrap the procedure creating the first level. Note that the leader knows that the task has been completed when it detects that no leaf has been able to add a new level to the spanning tree. At that point it broadcasts a "terminate" message along the spanning tree.   □

4.3   Detection of Stability

Under the k-bounded scheduler, a unique leader can be used to convert any stabilizing protocol into a terminating one, in any network. A similar technique has been used in [17] in the context of detecting stability in message-passing systems. As byproduct, any protocol for the 2-bounded scheduler can be simulated in all k-bounded schedulers, for $k > 2$. First we give some crucial definitions.

Definitions. Let G be a network, and let δ be a transition function for G with agent states Q and port states U. Now let $Q' = Q_S \times Q$ and $U' = U_S \times U$, and let δ' be a transition function for the same network G with agent states Q' and port states U'. Let us refer to the sets Q_S and U_S as the *simulator work states* for agents and ports, while Q and U are the *simulated states*.

A *simulated transition* for δ' is a state transition in which some agents or ports change their simulated states as a result of an interaction according to δ (if, instead, only the simulator work states change, the transition is not considered a simulated one). Given an execution E of δ', its *simulated execution* is the execution of δ that is obtained from E by removing the non-simulated transitions and projecting the agents' and ports' states on Q and U.

We say that δ' under scheduler $\mathcal{S}'$ *simulates* δ under scheduler $\mathcal{S}$ if, for every execution of δ' corresponding to a schedule in $\mathcal{S}'$, its simulated execution is an execution of δ corresponding to a schedule in $\mathcal{S}$. If, additionally, δ is stabilizing under $\mathcal{S}$ and δ' is terminating under $\mathcal{S}'$, we say that δ' *detects the stability* of δ.

Theorem 6. *For every $k > 2$, given a stabilizing protocol δ, there is a protocol δ' that, under the k-bounded scheduler, detects the stability of δ under the 2-bounded scheduler, from any initial configuration with a unique leader.*

Proof. The protocol δ' has an *initialization phase* in which the leader builds a shortest-path spanning tree of the network, as in Theorem 5. Recall that the spanning tree construction is terminating, hence the agents can perform other tasks when they are finished. After the initialization phase, the protocol is structured in two alternating phases: a *reset phase* and a *simulation phase*. When the stability is detected, the leader starts the *termination phase*.

In the reset phase, all the flags used in the simulation by agents and ports are reset. This phase is performed level by level and is coordinated by the leader as in Theorem 5. Once again, note that the agents are able to tell when they have reset all their incident ports, because the scheduler is k-bounded.

Once all flags have been reset, the leader starts the simulation phase. In this phase, one simulated interaction between each ordered pair of neighboring agents is performed, starting from the leader and proceeding to the leaves, following the levels of the spanning tree. Each edge port has a *simulation flag*, which tells if the edge has already been part of a simulated interaction in the direction corresponding to the port. This flag is reset during each reset phase. The simulation at level ℓ proceeds as follows: an agent receiving the order to simulate starts scanning each incident port (whenever the scheduler generates the corresponding interaction) and, if its simulation flag is not set, both endpoints of that edge perform a simulated interaction according to δ and set the port's simulation flag. When all its incident ports have the simulation flag set (which can be verified because the scheduler is k-bounded), the agent sets its own *complete flag* and notifies the leader. The leader waits until it detects that each agent at level ℓ has set the complete flag: this can be done with a convergecast. Then the leader issues another order to simulate, which reaches level $\ell + 1$. The simulation phase ends once the lowest level of the three has finished simulating. Thanks to the complete flag, this phase can be performed in constant space.

During the simulation phase, the agents also perform a "local stability" check on each edge. An edge $\{a, b\}$ is *locally stable* if no (infinite) schedule consisting only of the interactions (a, b) and (b, a) ever causes the simulated state of a or b to change according to δ. Note that the stability of an edge can be verified by its endpoints in a single interaction executing δ'. Each agent has an *unstable flag* that is cleared during the reset phase and is set whenever the agent either changes its simulated state or detects that an incident edge is not locally stable. Then, during the convergecast, agents also communicate the state of their unstable flag to the leader. When the simulation phase is over, the leader knows if the whole network is locally stable. If it is not, it starts the next reset phase; otherwise, it proceeds with the termination phase. The termination phase is simply a broadcast over the spanning tree that orders all agents to get a terminal state.

The correctness of the simulation follows from the fact that, at every phase, each simulation flag is first cleared and then set. This means that all possible simulated interactions occur in some order at each simulation phase. So, the resulting simulated schedule is a sequence of permutations of all the possible interactions in the network. Each permutation contains each interaction exactly once. Therefore, between two occurrences of the same interaction within two consecutive permutations, no other interaction occurs more than twice. In other words, the simulated schedule is 2-bounded.

Let us now show that the stability of δ is correctly detected and that δ' correctly terminates. Of course, when the simulated execution of δ stabilizes, all edges are locally stable and no agents change simulated states, and this is detected by the leader, which then correctly executes the termination phase. We have to prove that the leader cannot start the termination phase "by accident" before the execution of δ has actually stabilized. Equivalently, we have to prove that, if all edges are locally stable at some point during the simulation phase, then the simulated execution of δ has indeed reached a stable simulated configuration. Here the key observation is that, by the way the simulator works, the simulated states of agents and ports change only according to δ. So, if all edges pass the local stability test at some point in the simulation phase (and no agents change their simulated states), it does not matter in what order they are checked, and when. Indeed, in the next simulation phase, the simulated states of the agents will still be the same, and therefore all edges will still be locally stable.   □

If δ' is executed under the 1-bounded scheduler, the simulated execution obtained in Theorem 6 corresponds to a 1-bounded schedule, as well.

Corollary 1. *Given a stabilizing protocol δ, there is a protocol δ' that, under the 1-bounded scheduler, detects the stability of δ under the 1-bounded scheduler, from any initial configuration with a unique leader.*

Proof. Recall from Theorem 6 that the simulated schedule generated by δ' consists of a sequence of permutations of all the possible interactions. Now, if δ' is executed under a 1-bounded scheduler, the schedule will actually be periodic, and so will be the resulting simulated schedule. Therefore, the simulated schedule is a repetition of the same permutation of interactions, which implies that it is 1-bounded, as well.   □

References

1. Aliatarh, D., Gelashvili, R., Vojnovic, M.: Fast and exact majority in population protocols. In: 34th Annual ACM Symposium on Principles of Distributed Computing, PODC, pp. 47–56 (2015)

2. Angluin, D., Aspnes, J., Chan, M., Fischer, M.J., Jiang, H., Peralta, R.: Stably computable properties of network graphs. In: Prasanna, V.K., Iyengar, S.S., Spirakis, P.G., Welsh, M. (eds.) DCOSS 2005. LNCS, vol. 3560, pp. 63–74. Springer, Heidelberg (2005). doi:10.1007/11502593_8

3. Angluin, D., Aspnes, J., Diamadi, Z., Fischer, M.J., Peralta, R.: Computation in networks of passively mobile finite-state sensors. Distrib. Comput. 18(4), 235–253 (2006)

4. Angluin, D., Aspnes, J., Fischer, M.J., Jiang, H.: Self-stabilizing population protocols. ACM Trans. Auton. Adapt. Syst. 3(4), 1–28 (2008)

5. Angluin, D., Aspnes, J., Eisenstat, D.: Stably computable predicates are semilinear. In: 25th Annual ACM Symposium on Principles of Distributed Computing, PODC, pp. 292–299 (2006)

6. Beauquier, J., Blanchard, P., Burman, J.: Self-stabilizing leader election in population protocols over arbitrary communication graphs. In: Baldoni, R., Nisse, N., Steen, M. (eds.) OPODIS 2013. LNCS, vol. 8304, pp. 38–52. Springer, Cham (2013). doi:10.1007/978-3-319-03850-6_4

7. Beauquier, J., Burman, J., Clavière, S., Sohier, D.: Space-optimal counting in population protocols. In: Moses, Y. (ed.) DISC 2015. LNCS, vol. 9363, pp. 631–646. Springer, Heidelberg (2015). doi:10.1007/978-3-662-48653-5_42

8. Beauquier, J., Burman, J., Kutten, S.: A self-stabilizing transformer for population protocols with covering. Theor. Comput. Sci. 412(33), 4247–4259 (2011)

9. Cai, S., Izumi, T., Wada, K.: How to prove impossibility under global fairness: on space complexity of self-stabilizing leader election on a population protocol model. Theor. Comput. Syst. 50(3), 433–445 (2012)

10. Canepa, D., Potop-Butucaru, M.G.: Stabilizing leader election in population protocols, Research Report, inria-00166632 (2007)

11. Canepa, D., Potop-Butucaru, M.G.: Self-stabilizing tiny interaction protocols. In: 3rd International Workshop on Reliability, Availability, and Security, WRAS, pp. 1–6 (2010)

12. Chatzigiannakis, I., Michail, O., Spirakis, P.G.: Stably decidable graph languages by mediated population protocols. In: Dolev, S., Cobb, J., Fischer, M., Yung, M. (eds.) SSS 2010. LNCS, vol. 6366, pp. 252–266. Springer, Heidelberg (2010). doi:10.1007/978-3-642-16023-3_21

13. Chatzigiannakis, I., Michail, O., Spirakis, P.G.: Mediated population protocols. Theor. Comput. Sci. 412(22), 2434–2450 (2011)

14. Di Luna, G.A., Flocchini, P., Izumi, T., Izumi, T., Santoro, N., Viglietta, G.: Population protocols with faulty interactions: the impact of a leader. arXiv:1611.06864 [cs.DC] (2016)

15. Fischer, M., Jiang, H.: Self-stabilizing leader election in networks of finite-state anonymous agents. In: Shvartsman, M.M.A.A. (ed.) OPODIS 2006. LNCS, vol. 4305, pp. 395–409. Springer, Heidelberg (2006). doi:10.1007/11945529_28

16. Mizoguchi, R., Hirotaka, O., Kijima, S., Yamashita, M.: On space complexity of self-stabilizing leader election in mediated population protocol. Distrib. Comput. 25(6), 451–460 (2012)

17. Shavit, N., Francez, N.: A new approach to detection of locally indicative stability. In: Kott, L. (ed.) ICALP 1986. LNCS, vol. 226, pp. 344–358. Springer, Heidelberg (1986). doi:10.1007/3-540-16761-7_84

Learning AC^0 Under k-Dependent Distributions

Ning Ding[1,2]([✉]), Yanli Ren[3], and Dawu Gu[1]

[1] Department of Computer Science and Engineering,
Shanghai Jiao Tong University, Shanghai, China
{dingning,dwgu}@sjtu.edu.cn
[2] State Key Laboratory of Cryptology, P.O. Box 5159, Beijing 100878, China
[3] School of Communication and Information Engineering,
Shanghai University, Shanghai, China
renyanli@shu.edu.cn

Abstract. It is well known that AC^0 circuits can be learned by the Low Degree Algorithm in quasi-polynomial-time under the uniform distribution due to Linial, Mansour and Nisan. Furst *et al.* and Blais *et al.* Then showed that this learnability also holds when the input variables are mutually independent or conform to some product distributions. However, a long-standing question is whether we can learn AC^0 beyond these distributions, e.g. under some non-product distributions.

In this paper we show AC^0 can be non-trivially learned under a sort of distributions, which we call k-dependent distributions. Informally, a k-dependent distribution is one satisfying that for a randomly sampled string (as input to a circuit being learned), some bits of it are mutually independent, of which each other bit is dependent on at most k ones. We note that this sort of distributions contains some natural non-product distributions. We show that with respect to any such distribution, if the dependence relations of all bits of sampled strings are known, AC^0 can be learned in quasi-polynomial-time in the case that k is poly-logarithmic, and otherwise, the learning costs exponential-time but still uses similarly many examples as the former case. We note that in the latter case although the time complexity is exponential, it is significantly smaller than that of the brute-force method (when the size of the circuit being learned is sufficiently large).

1 Introduction

The seminal result of Linial, Mansour and Nisan [16] showed the Fourier spectrum of any function in AC^0 is concentrated on low-degree coefficients and then introduced the Low Degree Algorithm to learn the low-degree coefficients under the uniform distribution and thus generated a function approximately identical to the concept function. Later the Fourier concentration bound for AC^0 has been improved in [5,11,17]. Following [16], some works present various Fourier concentration results for more expressive circuits augmented from AC^0 and thus gain corresponding learning results with the Low Degree Algorithm [3,8,13].

In all above results, the uniform distribution is required. There has been a few successful attempts to extend some of the results to product distributions.

© Springer International Publishing AG 2017
T.V. Gopal et al. (Eds.): TAMC 2017, LNCS 10185, pp. 187–200, 2017.
DOI: 10.1007/978-3-319-55911-7_14

In [7] it is shown how to learn AC^0 circuits under any distribution in which all bits are mutually independent. In [4] it is shown that AC^0 circuits can be learned under any product distribution in which the cardinal number of the probability space of each multiplier distribution is polynomial. [6] points out that if AC^0 could be computed by a polynomial threshold function (PTF) (of arbitrary degree) with weight at most W then under any distribution, some conjunction has correlation at least $1/W$ with some circuit being learned due to discriminator lemma of [9], and thus one can then apply an agnostic learning algorithm for conjunctions such as [14] combined with standard boosting techniques, to PAC learn it in $\max(\exp(\tilde{O}(n^{1/2})), W)$ time. However, currently it is only known that AC^0 can be approximately computed by PTFs (of poly-logarithmic degrees)[1,2,10,18,19].

So a natural and long-standing question is whether we can learn AC^0 under non-product distributions. In this paper we are interested in this question and will have an attempt on it.

1.1 Our Results

We present an attempt on learning AC^0 under non-product distributions. Let x of n bits denote an input to a function to be learned. As shown above, all previous results assume that all input variables of x are mutually independent. Thus we consider a relaxation that some bits of x sampled from a distribution are mutually independent, of which each other bit is dependent on at most k ones. By "each other bit is dependent on at most k ones", we mean that the distribution of each other bit on the condition of all bits (excluding itself) is same as that on the condition of the $\le k$ mutually independent bits. This implies that each other bit is only dependent on these $\le k$ bits essentially. So we call this distribution k-dependent. It can be seen that k-dependent distributions are extensions of individually independent distributions.

There is an example showing that k-dependent distributions can be non-product. Consider the following one, in which $x_1, x_3, \cdots, x_n$ are mutually independent, assuming n is odd; let $f_i, i \in [1, n-2] \cap 2\mathbb{Z} + 1$ denote $\frac{n-1}{2}$ (possibly probabilistic) functions; set $x_2 = f_1(x_1, x_3), x_4 = f_3(x_3, x_5), \cdots, x_{n-1} = f_{n-2}(x_{n-2}, x_n)$. So this distribution of x is two-dependent. Since x_i can be related with x_{i-1} for $\forall i \in [2, n]$, it can be non-product.

Let D be any k-dependent distribution and x is sampled from D. According to whether the dependence relations of all bits of x are known or not to the learning algorithm, we state our results in the following two theorems.

Theorem 1 (informal). *Assuming the dependence relations of all bits of x are known, there exists a learning algorithm that can learn AC^0 to any error and confidence (ϵ, δ) in time $poly(2^{\max(k, \log n)^{O(1)}}, \frac{1}{\epsilon}, \log\frac{1}{\delta})$ with respect to D.*

Theorem 2 (informal). *Assuming the dependence relations of all bits of x are unknown, there exists a learning algorithm that can learn AC^0 to any error and confidence (ϵ, δ) in time $2^{O(kn \log n)} \cdot poly(2^{\max(k, \log n)^{O(1)}}, \frac{1}{\epsilon}, \log\frac{1}{\delta})$ with respect to D, using $poly(2^{\max(k, \log n)^{O(1)}}, \frac{1}{\epsilon}, \log\frac{1}{\delta})$ examples.*

We remark that the running-time and the sample complexity of our first result is basically quasi-polynomial similar to [16] in the case that k is poly-logarithmic. As for the second result, although the time complexity is exponential, the algorithm only uses quasi-polynomial examples as the first one. Even considering the time complexity, in learning an AC^0 circuit of size s, the method of learning by finding a consistent hypothesis, e.g. enumerating all circuits of size s to find the one consistent with the given examples, consumes approximately $O(s2^s)$-time, while the algorithm in the second result uses running-time which is significantly smaller than $O(s2^s)$ when s is a sufficiently large polynomial. Lastly, our results can be extended to learning the classes augmented from AC^0, but for simplicity we only focus on AC^0 in this paper.

Our Techniques. Let us recall one approach in [7] that shows AC^0 can be learned when n bits in input are independent but not uniform. The approach is this. For each input bit x_i, construct a DNF formula X_i which has some boolean variables r_i as input and is such that when r_i is uniform, the output of $X_i(r_i)$ almost conforms to the distribution of x_i. Let X denote $(X_1, \cdots, X_n)$, f' denote the composed function $f \circ X$ that on given input $r = (r_1, \cdots, r_n)$, first computes $x_i \leftarrow X_i(r_i)$ for $1 \leq i \leq n$ (in parallel) and then computes $f(x_1, \cdots, x_n)$. Note that $f \in AC^0$ and each $X_i \in AC^0$. So is f'. Thus the question of learning f can be reduced to the question of learning f'.

We extend this approach to learn AC^0 under any k-dependent distribution D. Our main task is to construct n constant-depth circuits $X = (X_1, \cdots, X_n)$ such that for $x \leftarrow D$, we can efficiently sample an (almost) uniform $r = (r_1, \cdots, r_n)$ such that $X_i(r) = x_i$ for all i with high probability. Still let f' denote the composed function $f \circ X$. Thus when given many examples of form $(x, f(x))$ for $x \leftarrow D$, we can come up with a uniform r satisfying that $X_i(r) = x_i$ for each i, which shows $f(x) = f'(r)$ (with high probability). Thus we obtain many new examples of form $(r, f(x))$. Note that f' is of constant-depth. This shows that we can reduce the question of learning f to that of learning f'.

Since in each new example, there is a small probability that the label is not equal to $f'(r)$, we need a learning algorithm that can agnostically learn f'. Fortunately, as shown in [14], the Low Degree Algorithm is actually a somewhat agnostic learning algorithm. So we can run the Low Degree Algorithm to compute a hypothesis to approximate f', denoted h'.

Once obtaining h', a hypothesis h can be constructed to approximate f as follows. On input $x \leftarrow D$, it first samples a uniform r such that $X_i(r) = x_i$ for all i with high probability, and then outputs $h'(r)$. Since $h'(r) = f'(r)$ and $X_i(r) = x_i$ for all i except for small probability and $f' = f \circ X$, we have $h'(r) = f(x)$ except for small probability. This shows h is indeed an approximation of f.

Organization. The rest of the paper is arranged as follows. Section 2 presents the preliminaries. In Sect. 3 we formalize the notion of k-dependent distributions. In Sect. 4 we show how to identify some conditional probabilities of bits from examples. In Sect. 5 we present our core lemma that shows how to construct the desired circuits $X_1, \cdots, X_n$. In Sect. 6 we present our learning algorithms by combining all obtained results.

2 Preliminaries

This section contains the notations and definitions used throughout this paper.

The **Statistic Difference** between X and Y is defined as the function $\triangle(n) = \frac{1}{2} \cdot \Sigma_\alpha | \Pr[X = \alpha] - \Pr[Y = \alpha]|$. If the statistical difference between X and Y is $\triangle(n)$, we say X is $\triangle(n)$-statistically close to Y.

Chernoff Bound. Let $X_1, \cdots, X_m$ be independent identically distributed random variables such that $X_i \in [0,1]$, $\mathbf{E}[X_i] = p$ and $S_m = \sum_{i=1}^m X_i$. Then $\Pr[S_m/m - p \geq t], \Pr[S_m/m - p \leq -t] \leq e^{-2t^2 m}$.

2.1 Learning Models

Let $\mathcal{C}$ denote a class of functions. In the PAC learning model [20], a labeled example is a pair $(x, f(x))$, where $x \in X$ is an input and $f(x)$ is the value of the target function $f \in \mathcal{C}$ on the input x. A training sample labeled by f is of the form $((x_1, f(x_1)), \cdots, (x_s, f(x_s)))$.

Definition 1 (PAC Learning). An algorithm L is called a learner for $\mathcal{C}$ under distribution D over X, if it is given a training sample in which each x is sampled from D and $\epsilon, \delta \in (0,1)$, with probability at least $1 - \delta$, L outputs a function h (not necessarily in $\mathcal{C}$) such that $\Pr[f(x) \neq h(x)] < \epsilon$ for $x \leftarrow D$.

If L can work under any D, we say L PAC (Probably Approximately Correct) learns $\mathcal{C}$ or simply learns $\mathcal{C}$. We refer to ϵ as the accuracy parameter and δ as the confidence parameter.

The Agnostic learning model [12,15] is an extension of the PAC model, in which each example-label pair is chosen from a distribution D' on $X \times \{0,1\}$.

Definition 2 (Agnostic Learning). Let D' be any distribution on $X \times \{0,1\}$ and let $\mathcal{C}$ be a class of functions. We say that an algorithm L (ϵ, δ)-agnostically learns $\mathcal{C}$ under distribution D' if it is given many random example-label pairs (x, b) according to D', then with probability $1 - \delta$, L outputs a hypothesis h such that $\Pr[h(x) \neq b] < \inf_{f \in \mathcal{C}}(\Pr[f(x) \neq b]) + \epsilon$ where $(x, b) \leftarrow D'$.

2.2 Learning AC^0 with the Low Degree Algorithm

Let AC^0 denote the class of all functions computable by polynomial-size constant-depth unbounded fan-in circuits (of AND, OR, NOT gates and of binary output), $AC_d^0[s]$ denote the class of functions computable by circuits with depth bounded by d and size bounded by s.

[16] shows that for each $f \in AC_d^0[s]$, $\sum_{S:|S|>t} \widehat{f}^2(S) < \epsilon$, where all $\widehat{f}(S)$ denote its Fourier coefficients and $t = (20 \log \frac{s}{\epsilon})^d$. Under the uniform distribution, when obtaining $\frac{1}{2}(\frac{n^t}{\epsilon_0})^{1/2} \cdot \log(\frac{2n^t}{\delta})$ labeled examples of the form $(x, f(x))$, the Low Degree Algorithm can approximate all $\widehat{f}(S)$ for $|S| \leq t$ and with probability $1 - \delta$ recover a function h that is such that $\Pr[h(x) \neq f(x)] < \epsilon + \epsilon_0$.

3 k-Dependent Distributions

Informally, a k-dependent distribution is one such that all bits of x can be divided to two parts, and the bits in the first part are mutually independent and each bit in the second part is dependent on at most k ones of the first part, i.e., the distribution of each such bit on the condition of all other bits is same as that on the condition of the $\leq k$ ones in the first part.

Definition 3 (k-Dependent Distribution). A distribution D over $\{0,1\}^n$ is called a k-dependent distribution if for $x \leftarrow D$, letting $x_1, \cdots, x_n$ denote the n bits of x, there exists a set $V \subset [1,n]$ satisfying the following:

1. All x_i for $i \in V$ are mutually independent.
2. For each $i \notin V$, there exists a set $I_i \subset V$ with $|I_i| \leq k$ such that for any values of $b_w \in \{0,1\}, w \in [1,n]$ satisfying $(b_1, \cdots, b_n)$ has positive probability in D, $\Pr[x_i = b_i | \forall w \in [1,n] - \{i\}, x_w = b_w] = \Pr[x_i = b_i | \forall w \in I_i, x_w = b_w]$.

Recall the example shown in Sect. 1.1 that letting $x_1, x_3, \cdots, x_n$ denote mutually independent random bits for odd n and $f_1, f_3, \cdots, f_{n-2}$ denote $\frac{n-1}{2}$ (possibly probabilistic) functions, $x_2 = f_1(x_1, x_3), x_4 = f_3(x_3, x_5), \cdots, x_{n-1} = f_{n-2}(x_{n-2}, x_n)$. It can be seen that the distribution of x_i on the condition of all other ones equals that on the condition of x_{i-1}, x_{i+1} for even $2 \leq i \leq n-1$. So x conforms to a two-dependent distribution.

4   Approximately Identifying Conditional Probabilities

Let D denote any k-dependent distribution. In this section we present the results of recovering approximately some conditional probabilities when given many examples of $x \leftarrow D$.

First we introduce following notations used throughout this paper. For each $i \notin V$, let S_i denote an $|I_i|$-element set $\{b_w \in \{0,1\} : w \in I_i\}$. For convenience, we will often consider that I_i, S_i are ordered sets and there is a correspondence between I_i and S_i. Note that S_i has $2^{|I_i|}$ different values, since each b_w in it can be 0/1. For each value of S_i, let $I_i = S_i$ denote the formula that $\forall w \in I_i, b_w \in S_i, x_w = b_w$ (with respect to the implicit correspondence).

Let $p_{x_i=b_i}, p_{x_i=b_i | I_i=S_i}$ denote approximations of $\Pr[x_i = b_i], \Pr[x_i = b_i | \forall w \in I_i, b_w \in S_i, x_w = b_w]$ respectively. In each conditional probability formula if the event in the condition occurs with zero probability, the conditional probability is considered as 0. Let c be an integer depending on k which will be determined in Sect. 6.

Claim 1. There exists a poly$(2^k, 2^{\log^c n}, \frac{1}{\epsilon}, \log \frac{1}{\delta})$-time algorithm that when given poly$(2^{\log^c n}, \frac{1}{\epsilon}, \log \frac{1}{\delta})$ examples of $x \leftarrow D$ and $V, I_i, i \notin V$ can output $p_{x_i=b_i}$ for $i \in V$ and all b_i and $p_{x_i=b_i | I_i=S_i}$ for $i \notin V$ and all b_i, S_i such that except for probability $O(\delta n 2^k e^{-2n})$ the following holds:

1. For any $i \in V$ and any b_i, $|p_{x_i=b_i} - \Pr[x_i = b_i]| \leq \epsilon 2^{-\log^c n}$.

2. For any $i \notin V$ and any b_i, S_i, $|p_{x_i=b_i|I_i=S_i} - \Pr[x_i = b_i|I_i = S_i]| \leq \epsilon 2^{-\log^c n}$ if $\Pr[I_i = S_i] \geq \epsilon 2^{-\log^c n}$.

Proof. The desired algorithm runs as follows:

1. Read $q(n) = 2^{3\log^c n+1} \cdot \frac{1}{\epsilon^3} \cdot (n + \log \frac{1}{\delta})$ examples of $x \leftarrow D$. For each $i \in V, b_i$, count the fraction satisfying $x_i = b_i$ in the examples, denoted $p_{x_i=b_i}$.
2. For the $q(n)$ examples, for each $i \notin V$, for each b_i, S_i, if the number of the examples in which $I_i = S_i$ is less than $2^{2\log^c n} \cdot \frac{1}{\epsilon^2} \cdot (n + \log \frac{1}{\delta})$, set $p_{x_i=b_i|I_i=S_i} = 0$. Otherwise, count the fraction of $x_i = b_i$ in the examples where $I_i = S_i$, denoted $p_{x_i=b_i|I_i=S_i}$.

First for each $i \in V$, due to the Chernoff bound, $\Pr[|p_{x_i=b_i} - \Pr[x_i = b_i]| \geq \epsilon 2^{-\log^c n}] \leq O(e^{-\epsilon^2 2^{-2\log^c n+1}q}) = O(\delta e^{-2n})$.

Second, consider the case of $\Pr[I_i = S_i] \geq \epsilon 2^{-\log^c n}$. In this case using the Chernoff bound, the number of the examples in which $I_i = S_i$ is more than $q\Pr[I_i = S_i] - 2^{2\log^c n} \cdot \frac{1}{\epsilon^2} \cdot (n + \log \frac{1}{\delta}) = 2^{2\log^c n} \cdot \frac{1}{\epsilon^2} \cdot (n + \log \frac{1}{\delta})$ with probability $O(e^{-2\epsilon^2 2^{-2\log^c n}q}) = O(\delta e^{-2n})$.

Using the Chernoff bound again, we have $\Pr[|p_{x_i=b_i|I_i=S_i} - \Pr[x_i = b_i|I_i = S_i]| \geq \epsilon 2^{-\log^c n}] \leq O(e^{-2\epsilon^2 2^{-2\log^c n} \cdot 2^{2\log^c n} \cdot \frac{1}{\epsilon^2} \cdot (n+\log \frac{1}{\delta})}) = O(\delta e^{-2n})$.

Considering the union of probability bounds for all i, b_i, S_i, the claim holds. □

5   The Core Lemma

In this section we present our core lemma, which asserts that we can construct n constant-depth circuits $X_1, \cdots, X_n$ and a sampling algorithm such that for $x \leftarrow D$, the sampling algorithm can sample an almost uniform string r satisfying $X_i(r) = x_i$ for all i (for simplicity viewing r as input to all X_i). Let U_l denote the uniform randomness over $\{0,1\}^l$.

Lemma 1. *There is a $poly(2^k, 2^{\log^c n}, 1/\epsilon, \log 1/\delta)$-time algorithm that given V and $p_{x_i=b_i}$ for $i \in V$ and all b_i, and $I_i, i \notin V$ and $p_{x_i=b_i|I_i=S_i}$ for $i \notin V$ and all b_i, S_i, can output n circuits $X_1, \cdots, X_n$ and a sampling algorithm Samp such that:*

1. *For each $i \in [1,n]$, X_i is of constant-depth and of size $O(2^k(\log^c n + \log 1/\epsilon))$. The total input to all $X_1, \cdots, X_n$ is of $O(n(\log^c n + \log 1/\epsilon)^2)$-bit, denoted r, and each X_i has a part of r as input. For convenience of statement, we will sometimes write $X_i(r)$ if there is no need to character explicitly the part of r as input to X_i.*
2. *For $x \leftarrow D$, Samp(x) can sample an input $r \in \{0,1\}^{O(n(\log^c n+\log 1/\epsilon)^2)}$ in time $poly(2^k, 2^{\log^c n}, 1/\epsilon, \log 1/\delta) \cdot 2^{\log^c n}$ such that $x_i = X_i(r)$ for all $i \in [1,n]$ except for probability $O(\epsilon n2^k 2^{-\log^c n})$. Moreover, r is $O(n2^k \epsilon 2^{-\log^c n})$-statistically close to the uniformness.*

Proof. The desired algorithm works as follows:

1. First construct X_i for each $i \in V$. For each such i, if $2\epsilon 2^{-\log^c n} \leq p_{x_i=1} < 1 - 2\epsilon 2^{-\log^c n}$, let X_i denote the $O(\log^c n + \log 1/\epsilon)$-size DNF formula shown in [7] that uses only $\{0,1\}^{O(\log^c n + \log 1/\epsilon)^2}$ input bits, denoted r_i, and is such that $|\Pr[X_i(U_{|r_i|}) = b_i] - p_{x_i=b_i}| \leq \epsilon 2^{\log^c n}$. [1]
   Let Samp_i be the algorithm that given x_i, repeats sampling uniform r_i at most $1/\epsilon \cdot \log 1/\epsilon \cdot 2^{\log^c n} \cdot \log^c n$ times until $X_i(r_i) = x_i$, and if no such r_i can be sampled, simply samples r_i uniformly one more time, and finally outputs r_i.
   Otherwise, if $p_{x_i=1} < 2\epsilon \cdot 2^{-\log^c n}$, let X_i be the 0-constant function (that still has r_i as input and always outputs 0). If $p_{x_i=1} \geq 1 - 2\epsilon \cdot 2^{-\log^c n}$, let X_i be the 1-constant function. In these two cases, let Samp_i be the algorithm that simply samples r_i uniformly and outputs it.

2. Then construct X_i for each $i \notin V$ as follows:
   (a) For each $S_i \in \{0,1\}^{|I_i|}$, if $2\epsilon 2^{-\log^c n} \leq p_{x_i=1|I_i=S_i} < 1 - 2\epsilon 2^{-\log^c n}$, construct a DNF formula, denoted $\varphi_{S_i}(\cdot)$, shown in [7], which has $O(\log^c n + \log 1/\epsilon)^2$ bits as input. Let r_i denote the input. Thus $|\Pr[\varphi_{S_i}(U_{|r_i|}) = b_i] - p_{x_i=b_i|I_i=S_i}| \leq \epsilon 2^{-\log^c n}$.
   Otherwise, if $p_{x_i=1|I_i=S_i} < 2\epsilon 2^{-\log^c n}$, let φ_{S_i} be the 0-constant function. If $p_{x_i=1|I_i=S_i} \geq 1 - 2\epsilon 2^{-\log^c n}$, let φ_{S_i} be the 1-constant function.
   (b) Tentatively, let X_i denote the following boolean function that has $X_w, \forall w \in I_i, r_i$ as input:

   $$X_i \overset{\text{def}}{=} \bigvee_{S_i \in \{0,1\}^{|I_i|}} (\forall w \in I_i, b_w \in S_i, X_w = b_w) \wedge (\varphi_{S_i} = 1)$$

   i.e., X_i outputs the OR of the boolean functions corresponding to all possible values of S_i (viewed as $|I_i|$ bits), in which each one, denoted $(\forall w \in I_i, b_w \in S_i, X_w = b_w) \wedge (\varphi_{S_i} = 1)$, outputs 1 if $\forall w \in I_i, b_w \in S_i, X_w = b_w$ and $\varphi_{S_i}(r_i) = 1$, and outputs 0 otherwise.
   (c) Replace the appearance of all X_w in X_i by their DNF formulas constructed at Step 1. Thus X_i is a function of $r_w, \forall w \in I_i, r_i$ and can be computed by a circuit of constant depth and size $O(2^k(\log^c n + \log 1/\epsilon))$.
   (d) Let Samp_i be the following algorithm that given $x_i, x_w, w \in I_i$ follows the sampling strategy at Step 1 to sample r_i satisfying $\varphi_{S_i}(r_i) = x_i$, in which S_i satisfies $\forall w \in I_i, b_w \in S_i, x_w = b_w$, and finally outputs r_i.

---

[1] We sketch the DNF construction in [7]. Suppose μ is a probability that we wish to approximate. Since now $\epsilon 2^{-\log^c n}$ difference is allowed, we only need to construct a DNF that outputs 1 with probability μ with $O(\log^c n + \log 1/\epsilon)$ bits kept after the binary point. So just assume $\mu = \Sigma_{j=1}^l a_j 2^{-j}$, where $l = O(\log^c n + \log 1/\epsilon)$ and $a_j \in \{0,1\}$. Create one AND for each j satisfying $a_j = 1$ such that the AND on input j uniform bits outputs 1 with probability 2^{-j}. Also insure that at most one AND among all produces 1 on each input. Let the DNF be the OR of all these AND's which totally has $O(l^2)$ bits as input and is of size $O(l)$.

3. Let Samp be the algorithm that given $x \leftarrow D$, runs each Samp_i to generate r_i, and outputs $r = (r_1, \cdots, r_n)$. Finally, output $X_1, \cdots, X_n$ and Samp.

We now show $X_i(r) = x_i$ for all i except for probability $O(\epsilon n 2^{-\log^c n})$, and r is $O(n\epsilon 2^{-\log^c n})$-close to the uniformness. Note that for $i \notin V$, X_i actually has $r_w, w \in I_i$ and r_i as input, where all r_w have been determined prior to the construction of X_i. So for simplicity we will often omit all r_w and just write $X_i(r)$ as $X_i(r_i)$. Moreover, we proceed only considering those values of $S_i, i \notin V$ satisfying for each such value of $S_i, i \notin V$, $\Pr[I_i = S_i] \geq \epsilon 2^{-\log^c n}$ for all $i \notin V$ (which implies $|p_{x_i = b_i | I_i = S_i} - \Pr[x_i = b_i | I_i = S_i]| \leq \epsilon 2^{-\log^c n}$ as Claim 1 shows). We can do this since any of other values of $S_i, i \notin V$ which makes $I_i = S_i$ for some i occurs with probability $n 2^k \epsilon 2^{-\log^c n}$. Thus

$$\Pr[X_1(r_1) = x_1, \cdots, X_n(r_n) = x_n : x \leftarrow D, r \leftarrow \mathsf{Samp}(x)]$$
$$= \prod_{i \in V} \Pr[X_i(r_i) = x_i] \cdot \Pr[\forall i \notin V, X_i(r_i) = x_i | \forall i' \in V, X_{i'} = x_{i'}]$$

In the following we present estimations of $\prod_{i \in V} \Pr[X_i = x_i]$ and $\Pr[\forall i \notin V, X_i = x_i | \forall i' \in V, X_{i'} = x_{i'}]$.

Claim. For each $i \in V$, $\Pr[X_i(r_i) = x_i] = 1 - O(\epsilon 2^{-\log^c n})$. So $\prod_{i \in V} \Pr[X_i = x_i] = 1 - O(n\epsilon 2^{-\log^c n})$.

Proof. For each such X_i, first consider the case of $p_{x_i = 1} > 1 - 2\epsilon 2^{-\log^c n}$. Thus X_i is the 1-constant function and $\Pr[x_i = 1] \geq 1 - 3\epsilon 2^{-\log^c n}$. So any r_i sampled by Samp_i is such that

$$\Pr[X_i(r_i) = x_i] = \Pr[x_i = 1] \cdot \Pr[X_i = 1] + \Pr[x_i = 0] \cdot \Pr[X_i = 0]$$
$$\geq \Pr[x_i = 1] \geq 1 - 3\epsilon 2^{-\log^c n}$$

Second consider the case of $p_{x_i = 1} < 2\epsilon 2^{-\log^c n}$. In this case X_i is the 0-constant function. Thus $\Pr[x_i = 0] \geq 1 - 3\epsilon 2^{-\log^c n}$. With a similar argument, the result also holds.

Finally consider the case of $2\epsilon 2^{-\log^c n} \leq p_{x_i = 1} < 1 - 2\epsilon 2^{-\log^c n}$. In this case in each sampling of Samp_i, Samp_i fails to sample $U_{|r_i|}$ satisfying $X_i(U_{|r_i|}) = x_i$ with probability at most $1 - \epsilon 2^{-\log^c n}$. This is so because if x_i happens to be 1, then $\Pr[X_i(U_{|r_i|}) = 1] \geq p_{x_i = 1} - \epsilon 2^{-\log^c n} \geq \epsilon 2^{-\log^c n}$, and if x_i happens to be 0, then $\Pr[X_i(U_{|r_i|}) = 0] \geq p_{x_i = 0} - \epsilon 2^{-\log^c n} \geq \epsilon 2^{-\log^c n}$.

Thus in the $1/\epsilon \log 1/\epsilon \cdot 2^{\log^c n} \cdot \log^c n$-time repetitions in Samp_i, except for probability $(1 - \epsilon 2^{-\log^c n})^{1/\epsilon \log 1/\epsilon \cdot 2^{\log^c n} \cdot \log^c n} = O(\epsilon 2^{-\log^c n})$, one r_i satisfying $X_i(r_i) = x_i$ can be sampled.

So $\prod_{i \in V} \Pr[X_i = x_i] = (1 - O(\epsilon 2^{-\log^c n}))^{|V|} \geq 1 - O(n\epsilon 2^{-\log^c n})$.   □

Claim. $\Pr[\forall i \notin V, X_i(r_i) = x_i | \forall i' \in V, X_{i'} = x_{i'}] = 1 - O(n 2^k \epsilon 2^{-\log^c n})$.

Proof. Since on the occurrence of variables with order number in V the remainder variables are independent, we have

$$\Pr[\forall i \notin V, X_i(r_i) = x_i | \forall i' \in V, X_{i'} = x_{i'}] = \prod_{i \notin V} \Pr[X_i(r_i) = x_i | \forall i' \in V, X_{i'} = x_{i'}]$$

$$= \prod_{i \notin V} \Pr[X_i(r_i) = x_i | I_i = S_i]$$

Except for probability $n2^k\epsilon 2^{-\log^c n}$ (that the value of S_i is such that $\Pr[I_i = S_i] < \epsilon 2^{-\log^c n}$ for some i), for each $i \notin V$, since $|\Pr[\varphi_{S_i}(U_{|r_i|}) = b_i] - p_{x_i = b_i | I_i = S_i}| \leq \epsilon 2^{-\log^c n}$ and $|\Pr[X_i(r_i) = b_i | I_i = S_i] - p_{x_i = b_i | I_i = S_i}| \leq \epsilon 2^{-\log^c n}$ for $b_i = 0/1$, $|\Pr[\varphi_{S_i}(U_{|r_i|}) = x_i] - \Pr[X_i(r_i) = x_i | I_i = S_i]| \leq O(\epsilon 2^{-\log^c n})$.

So the remainder can be proved with a similar analysis in the previous claim by considering the three possibilities of $p_{x_i = 1 | I_i = S_i}$. Thus $\Pr[X_i(U_{|r_i|}) = x_i | I_i = S_i] = 1 - O(\epsilon 2^{-\log^c n})$. So

$$\prod_{i \notin V} \Pr[X_i(r_i) = x_i | I_i = S_i] = (1 - O(\epsilon 2^{-\log^c n}))^{n - |V|} = 1 - O(n\epsilon 2^{-\log^c n})$$

Lastly, considering the $n2^k\epsilon 2^{-\log^c n}$ probability, the claim holds. $\square$

Thus, combining the above two claims, we have the following.

$$\Pr[X_1(r_1) = x_1, \cdots, X_n(r_n) = x_n : x \leftarrow D, r \leftarrow \mathsf{Samp}(x)]$$

$$= (1 - O(n\epsilon 2^{-\log^c n}))(1 - O(n2^k\epsilon 2^{-\log^c n})) = 1 - O(n2^k\epsilon 2^{-\log^c n})$$

Claim. r is $O(n2^k\epsilon 2^{-\log^c n})$–statistically close to $U_{|r|}$.

Proof. Let Good denote the event that $X_i(r_i) = x_i$ for random x which value makes $\Pr[I_i = S_i] \geq \epsilon 2^{-\log^c n}$ for all $i \notin V$. Thus $\Pr[\mathsf{Good}] = 1 - O(n2^k\epsilon 2^{-\log^c n})$. Let $\mathsf{Good}_{x=b}$ denote the sub-event of Good that $x = b$ for all possible b. Let $E_{i,b}$ denote the set of all values of r_i on the condition of $\mathsf{Good}_{x=b}$.

Let $\Delta_{\mathsf{Good}_{x=b}}(r, U_{|r|})$ (resp. $\Delta_{\mathsf{Good}_{x=b}}(r_i, U_{|r_i|})$) denote the statistic distance between r and $U_{|r|}$ (resp. between r_i and $U_{|r_i|}$) on the condition of $\mathsf{Good}_{x=b}$. On the condition of $\mathsf{Good}_{x=b}$, all r_i are mutually independent. First consider $\Delta_{\mathsf{Good}_{x=b}}(r_i, U_{|r_i|})$ for $i \in V$. Letting $b = (b_1, \cdots, b_n)$, $|E_{i,b}| = |X_i^{-1}(b_i)|$.

$$\Delta_{\mathsf{Good}_{x=b}}(r_i, U_{|r_i|}) = \frac{1}{2}\left(\sum_{a \in E_{i,b}} |\Pr[r_i = a] - \frac{1}{2^{|r_i|}}| + \sum_{a \notin E_{i,b}} |\Pr[r_i = a] - \frac{1}{2^{|r_i|}}| \right)$$

Notice that $\Pr[r_i = a] = \frac{\Pr[x_i = b_i]}{|X_i^{-1}(b_i)|}$ for $a \in E_{i,b}$. Thus the above formula is

$$\Delta_{\mathsf{Good}_{x=b}}(r_i, U_{|r_i|}) = \frac{1}{2} \sum_{a \in E_{i,b}} |\frac{\Pr[x_i = b_i]}{|X_i^{-1}(b_i)|} - \frac{1}{2^{|r_i|}}| + \frac{1}{2} \sum_{a \in \{0,1\}^{|r_i|} - E_{i,b}} \frac{1}{2^{|r_i|}}$$

$$= \frac{1}{2} |\Pr[x_i = b_i] - \frac{|X_i^{-1}(b_i)|}{2^{|r_i|}})| + O(\epsilon 2^{-\log^c n})$$

$$= \frac{1}{2} |\Pr[x_i = b_i] - \Pr[X_i(U_{|r_i|}) = b_i]| + O(\epsilon 2^{-\log^c n}) = O(\epsilon 2^{-\log^c n})$$

For the case of $i \notin V$, we have the same result (changing $X_i^{-1}(b_i)$ to $\varphi_{S_i}^{-1}(b_i)$, $\Pr[x_i = b_i]$ to $\Pr[x_i = b_i | I_i = S_i]$, $\Pr[X_i(U_{|r_i|}) = b_i]$ to $\Pr[X_i(U_{|r_i|}) = b_i | I_i = S_i]$ in the argument). Thus since all r_i are mutually independent, we have

$$\Delta_{\mathsf{Good}_{x=b}}(r, U_{|r|}) = \sum_{i=1}^{n} \Delta_{\mathsf{Good}_{x=b}}(r_i, U_{|r_i|}) = O(n\epsilon 2^{-\log^c n})$$

So finally,

$$\Delta(r, U_{|r|}) \leq \sum_{\mathsf{Good}_{x=b} \text{ for all } b} \Pr[\mathsf{Good}_{x=b}] \cdot \Delta_{\mathsf{Good}_{x=b}}(r, U_{|r|}) + \Pr[\neg\mathsf{Good}]$$
$$= O(n\epsilon 2^{-\log^c n}) + O(n2^k \epsilon 2^{-\log^c n}) = O(n2^k \epsilon 2^{-\log^c n})$$

$\square$

We remark that when replacing the appearance of $X_w, w \in I_i$ in X_i by their DNF formulas, X_i can be computed by a circuit of constant depth (which can be determined) and of size $O(2^k(\log^c n + \log 1/\epsilon))$. $\square$

6   The Learning Algorithms

In this section we present the learning algorithms. The sketch of the algorithms is shown in Sect. 1.1, in which our idea is to reduce the question of learning some unknown $f \in AC^0$ to that of learning some f' and one key step in the idea is to construct $X_1, \cdots, X_n$ and Samp. Notice that Lemma 1 requires that V and all $I_i, i \notin V$ are known (which were called the dependence relations of all bits of x in Sect. 1.1). So in Sect. 6.1 we describe the learning algorithm in detail when the knowledge of V and all $I_i, i \notin V$ is available. In Sect. 6.2 we sketch the algorithm when this knowledge is not available.

6.1   Learning with Knowledge of V and All I_i

We first present the algorithm when it has the knowledge of V and all I_i. Assume f is the function being learned. Let X denote $(X_1, \cdots, X_n)$ shown in Lemma 1, $x \leftarrow X(r)$ denote $(x_1, \cdots, x_n) \leftarrow (X_1(r), \cdots, X_n(r))$. Let f' denote the composed function $f \circ X$ that on input r first computes $x \leftarrow X(r)$ and then computes $f(x)$. Since $f \in AC_d^0[s]$ and X is of constant depth and size $O(n2^k(\log^c n + \log 1/\epsilon))$, f' can be computed by a circuit of depth d' and size $s + O(n2^k(\log^c n + \log 1/\epsilon))$, where $d' - d$ is any upper bound for the depths of all X_i. In the following we present the actual description of the learning algorithm.

Algorithm 1. The learning algorithm for AC^0.
Input: $\epsilon, \delta, s, d, k, V, I_i, \forall i \notin V$, sufficiently many examples of form $(x, f(x))$ where each $x \leftarrow D$.
Output: a function h that is approximately identical to f under D.

1. Choose c satisfying $\log^c n > O(\log n(\log \frac{1}{\epsilon} + k + \log n)^{d'+1})$. Let $s^* = s + O(n2^k(\log^c n + \log 1/\epsilon))$, $m = \frac{1}{2}(\frac{|r|^{t^*}}{\epsilon/3})^{1/2} \cdot \log(\frac{2|r|^{t^*}}{\delta/2})$, where $t^* = O(\log \frac{s^*}{\epsilon})^{d'}$. ($m$ is approximately identical to $n^{O(\log \frac{1}{\epsilon} + k + \log n)^{d'}}$.)

2. Run the algorithm in Claim 1 with necessary input to compute $p_{x_i=b_i}$ for $i \in V$ and all b_i and $p_{x_i=b_i|I_i=S_i}$ for $i \notin V$ and all b_i, S_i.

3. Run the algorithm in Lemma 1 with necessary input to output $X_1, \cdots, X_n$ and algorithm Samp.

4. Generate m labeled examples of form $(r, f(x))$ from the examples of form $(x, f(x))$. Concretely, for each example $(x, f(x))$, run the algorithm Samp on input $X_1, \cdots, X_n$ and x to sample r satisfying $x = X(r)$, and let $(r, f(x))$ be the newly generated example.

5. Run the Low Degree Algorithm with accuracy and confidence $(\epsilon/3, \delta/2)$ and the m newly generated examples to output a hypothesis h'.

6. Output the following hypothesis h:
   input: $x \in \{0, 1\}^n$.
   (a) Run algorithm Samp with $X_1, \cdots, X_n$ hardwired on input x to sample r satisfying $x = X(r)$.
   (b) Output $h'(r)$.

End Algorithm

So what we need to do next is to prove that Algorithm 1 can with probability $1 - \delta$ output a hypothesis h such that $\Pr[h(x) \neq f(x)] < \epsilon$ for $x \leftarrow D$. We decompose the proof into the following claims.

Claim 2. *For each generated example $(r, f(x))$ in Step 4 of Algorithm 1, $\Pr[f'(r) \neq f(x)] \leq O(n2^k \epsilon 2^{-\log^c n})$.*

Proof. It can be seen that $f'(r) \neq f(x)$ only if Samp fails in sampling r, which happens with probability $O(n2^k \epsilon 2^{-\log^c n})$ due to Lemma 1. The claim holds. $\square$

Claim 3. *In Step 5 of Algorithm 1, on input m examples, the Low Degree Algorithm can with probability $1 - \delta/2$ output a hypothesis h' such that $\Pr[f'(r) \neq h'(r)] < 2\epsilon/3$ for r output by Samp.*

Proof. Since $\Pr[f'(r) \neq f(x)] \leq O(n2^k \epsilon 2^{-\log^c n})$ due to Claim 2, there is a small fraction of examples which has wrong labels. So we need the Low Degree Algorithm to be able to agnostically learn f. As shown in [14] (Sect. 2.3, Observation 3), the Low Degree Algorithm is indeed a somewhat agnostic learning algorithm. Applying this result to our setting, we have that the Low Degree algorithm can agnostically learn f' under the uniform distribution to error $8\epsilon^* + \epsilon/3$ in time $n^{O(t^*)}$, where ϵ^* denotes the label error probability.

Moreover, note that f' can be computed by an AC^0 circuit of depth d' and size s^*. Thus m examples are sufficient to learn f' to $(\epsilon/3, \delta/2)$ due to [16].

Assume the sampled r in each example is uniform. Then the Low Degree Algorithm outputs h' which is such that $\Pr[f'(r) \neq h'(r)] \leq O(n2^k \epsilon 2^{-\log^c n}) + \epsilon/3 < 2\epsilon/3$ when r is uniform. Now actually in Step 6, each r in m examples is $O(n2^k\epsilon 2^{-\log^c n})$-statistically close to $U_{|r|}$. So now on input these m examples of form $(r, f(x))$ the Low Degree Algorithm behaves identically except for probability $O(mn2^k\epsilon 2^{-\log^c n})$. Due to the choices of the parameters, $mn2^k 2^{-\log^c n}$ can be sufficiently small. Thus the claim holds.   □

Claim 4. *When Algorithm 1 outputs h, h is such that $\Pr[h(x) \neq f(x)] < \epsilon$ for $x \leftarrow D$ and it runs in $poly(2^k, 2^{\log^c n}, 1/\epsilon, \log 1/\delta)$-time.*

Proof. First, it can be seen that the first step of h runs algorithm Samp, which uses $poly(2^k, 2^{\log^c n}, 1/\epsilon, \log 1/\delta)$-time, and the second step runs h', which running-time is less than the time in the first step. Thus the running-time is as desired. Second, on input $x \leftarrow D$, h can, with probability $1 - O(n2^k\epsilon 2^{-\log^c n})$, sample a desired r which is $O(n2^k\epsilon 2^{-\log^c n})$-statistically close to the uniformness. Due to Claim 3, $\Pr[h'(r) \neq f'(r)] < 2\epsilon/3$ if r is uniform. Considering the statistic difference between r and the uniformness, $\Pr[h'(r) \neq f'(r)] < 2\epsilon/3 + O(n2^k\epsilon 2^{-\log^c n})$ for the sampled r, which means $\Pr[h(x) \neq f(x)] < 2\epsilon/3 + O(n2^k\epsilon 2^{-\log^c n})$. Considering the failure probability of sampling r, the claim holds.   □

Combining all the claims above, we restate Theorem 1 formally as follows.

Theorem 3. *Algorithm 1 is a learning algorithm for AC^0 with respect to any k-dependent distribution D such that on input (ϵ, δ, s, d), $V, I_i, \forall i \notin V$ and given m examples $(x, f(x))$ for some unknown $f \in AC_d^0[s]$ where each x is drawn from D, it can with probability $1 - \delta$ output a hypothesis h in time $poly(2^k, 2^{\log^c n}, \frac{1}{\epsilon}, \log \frac{1}{\delta})$ satisfying $\Pr[h(x) \neq f(x)] < \epsilon$ for $x \leftarrow D$.*

Proof. It can be seen that if Algorithm 1 can finally output h, then due to Claim 4, h is such that $\Pr[h(x) \neq f(x)] < \epsilon$ for $x \leftarrow D$. Note that the learning algorithm needs m examples. Let us then estimate its running-time. Due to Claim 1 and Lemma 1, the learning algorithm needs $poly(2^k, 2^{\log^c n}, \frac{1}{\epsilon}, \log \frac{1}{\delta})$-time. It can be seen that Algorithm 1 fails to output h if any of Steps 2, 5 fails. According to Claim 1 and 3, we conclude the total failure probability is bounded by $O(\delta n2^k e^{-2n}) + \delta/2 < \delta$. The theorem holds.   □

6.2   Learning Without Knowledge of V and All I_i

We then consider the scenario that the learning algorithm has no knowledge of V and all I_i. Our approach to learning is this: enumerate all possible choices of V and all I_i; and for each choice, run Algorithm 1 with the common m examples to gain a hypothesis h; use $poly(\frac{1}{\epsilon}, \log \frac{1}{\delta}, n)$ more examples to choose the one from all h which admits the smallest errors on these examples.

Let us count the number of all choices. First there are 2^n choices for the value of V and for each choice of V, for each variable with order number outside V, there are less than $\sum_{i=1}^{k} \binom{n}{i}$ possibilities of how it is dependent on at most k variables with order number in V. So multiplying all factors we get a upper bound $2^n \cdot (\sum_{i=1}^{k} \binom{n}{i})^n < 2^n (\frac{en}{k})^{kn}$. This means that the learning algorithm in this subsection basically runs in time $2^n (\frac{en}{k})^{kn}$ times that of Algorithm 1.

Notice that in the right enumeration, i.e., the choice of V and all I_i are right, the obtained h, with probability $1 - \delta$, admits error $2\epsilon/3 + O(n2^k \epsilon 2^{-\log^c n}) < \frac{3}{4}\epsilon$ by Claim 4. Thus by using $\text{poly}(\frac{1}{\epsilon}, \log \frac{1}{\delta}, n)$ more examples, we can distinguish the desired h from any function which admits an error ϵ except for probability $O(\delta e^{-\text{poly}(n)})$. So this h can be founded except for probability δ in all enumerations. Thus we can restate Theorem 2 formally as follows.

Theorem 4. *Algorithm 1 is a learning algorithm for AC^0 with respect to any k-dependent distribution D such that on input $(\epsilon, \delta, s, d, k)$ and $m + \text{poly}(\frac{1}{\epsilon}, \log \frac{1}{\delta}, n)$ examples $(x, f(x))$ for some unknown $f \in AC^0_d[s]$ where each x is drawn from D, it can with probability $1 - \delta$ output a h in $2^n (\frac{en}{k})^{kn} \cdot \text{poly}(2^k, 2^{\log^c n}, \frac{1}{\epsilon}, \log \frac{1}{\delta})$ time satisfying $\Pr[h(x) \neq f(x)] < \epsilon$ for $x \leftarrow D$.*

We remark that although the time complexity now is exponential, Algorithm 1 uses similarly many examples as before. Even considering the time complexity, in learning an AC^0 circuit of size s, enumerating all circuits of size s to find the one consistent with the given examples consumes approximately $O(s2^s)$-time, while Algorithm 1 uses roughly $2^n (\frac{en}{k})^{kn} < O(2^{O(kn \log n)})$ times the time of one enumeration which is significantly smaller than $O(s2^s)$ when s is sufficiently large.

Acknowledgments. We are grateful to the reviewers of TAMC 2016 for their useful comments. This work is supported by the National Natural Science Foundation of China (Grant No. 61572309) and Major State Basic Research Development Program (973 Plan) of China (Grant No. 2013CB338004) and Research Fund of Ministry of Education of China and China Mobile (Grant No. MCM20150301).

References

1. Ajtai, M., Ben-Or, M.: A theorem on probabilistic constant depth computations. In: Proceedings of the 16th Annual ACM Symposium on Theory of Computing, Washington, DC, USA, pp. 471–474, 30 April–2 May 1984. http://doi.acm.org/10.1145/800057.808715
2. Aspnes, J., Beigel, R., Furst, M., Rudich, S.: The expressive power of voting polynomials. Combinatorica **14**(2), 1–14 (1994)
3. Beigel, R.: When do extra majority gates help? Polylog (n) majority gates are equivalent to one. Comput. Complex. **4**, 314–324 (1994)
4. Blais, E., O'Donnell, R., Wimmer, K.: Polynomial regression under arbitrary product distributions. Mach. Learn. **80**(2–3), 273–294 (2010). http://dx.doi.org/10.1007/s10994-010-5179-6
5. Boppana, R.B.: The average sensitivity of bounded-depth circuits. Inf. Process. Lett. **63**(5), 257–261 (1997). http://dx.doi.org/10.1016/S0020-0190(97)00131-2
6. Bun, M., Thaler, J.: Hardness amplification and the approximate degree of constant-depth circuits. In: Halldórsson, M.M., Iwama, K., Kobayashi, N., Speckmann, B. (eds.) ICALP 2015. LNCS, vol. 9134, pp. 268–280. Springer, Heidelberg (2015). doi:10.1007/978-3-662-47672-7_22

7. Furst, M.L., Jackson, J.C., Smith, S.W.: Improved learning of AC^0 functions. In: Warmuth, M.K., Valiant, L.G. (eds.) Proceedings of the Fourth Annual Workshop on Computational Learning Theory, COLT 1991, Santa Cruz, California, USA, pp. 317–325. Morgan Kaufmann, 5–7 August 1991. http://dl.acm.org/citation.cfm?id=114866

8. Gopalan, P., Servedio, R.A.: Learning and lower bounds for AC^0 with threshold gates. In: Serna, M., Shaltiel, R., Jansen, K., Rolim, J. (eds.) APPROX/RANDOM - 2010. LNCS, vol. 6302, pp. 588–601. Springer, Heidelberg (2010). doi:10.1007/978-3-642-15369-3_44

9. Hajnal, A., Maass, W., Pudlák, P., Szegedy, M., Turán, G.: Threshold circuits of bounded depth. J. Comput. Syst. Sci. **46**(2), 129–154 (1993). http://dx.doi.org/10.1016/0022-0000(93)90001-D

10. Harsha, P., Srinivasan, S.: On polynomial approximations to AC^0. CoRR abs/1604.08121 (2016). http://arxiv.org/abs/1604.08121

11. Håstad, J.: A slight sharpening of LMN. J. Comput. Syst. Sci. **63**(3), 498–508 (2001). http://dx.doi.org/10.1006/jcss.2001.1803

12. Haussler, D.: Decision theoretic generalizations of the PAC model for neural net and other learning applications. Inf. Comput. **100**(1), 78–150 (1992)

13. Jackson, J.C., Klivans, A., Servedio, R.A.: Learnability beyond AC^0. In: IEEE Conference on Computational Complexity, p. 26. IEEE Computer Society (2002)

14. Kalai, A.T., Klivans, A.R., Mansour, Y., Servedio, R.A.: Agnostically learning halfspaces. SIAM J. Comput. **37**(6), 1777–1805 (2008). http://dx.doi.org/10.1137/060649057

15. Kearns, M.J., Schapire, R.E., Sellie, L.: Toward efficient agnostic learning. Mach. Learn. **17**(2–3), 115–141 (1994)

16. Linial, N., Mansour, Y., Nisan, N.: Constant depth circuits, fourier transform, and learnability. J. ACM **40**(3), 607–620 (1993)

17. Tal, A.: Tight bounds on the fourier spectrum of AC^0. Electron. Colloq. Comput. Complex. (ECCC) **21**, 174 (2014). http://eccc.hpi-web.de/report/2014/174

18. Tarui, J.: Probablistic polynomials, AC^0 functions, and the polynomial-time hierarchy. Theor. Comput. Sci. **113**(1), 167–183 (1993). http://dx.doi.org/10.1016/0304-3975(93)90214-E

19. Toda, S., Ogiwara, M.: Counting classes are at least as hard as the polynomial-time hierarchy. SIAM J. Comput. **21**(2), 316–328 (1992). http://dx.doi.org/10.1137/0221023

20. Valiant, L.G.: A theory of the learnable. Commun. ACM **27**(11), 1134–1142 (1984)

Parikh Images of Matrix Ins-Del Systems

Henning Fernau[1]([✉]) and Lakshmanan Kuppusamy[2]

[1] Fachbereich 4 – Abteilung Informatikwissenschaften,
Universität Trier, 54286 Trier, Germany
fernau@uni-trier.de
[2] School of Computer Science and Engineering,
VIT University, Vellore 632 014, India
klakshma@vit.ac.in

Abstract. Matrix insertion-deletion systems combine the idea of matrix control (as established in regulated rewriting) with that of insertion and deletion (as opposed to replacements). We study families of multisets that can be described as Parikh images of languages generated by this type of systems, focusing on aspects of descriptional complexity. We show that the Parikh images of matrix insertion-deletion systems having length 2 matrices and context-free insertion/deletion contain only semilinear languages and when the matrices length increased to 3, they contain non-semilinear languages. We also characterize the hierarchy of family of languages that is formed with these systems having small sizes. We also introduce a new class, namely, *monotone strict context-free matrix ins-del systems* and analyze the results connecting with families of context-sensitive languages and Parikh images of regular and context-free matrix languages.

Keywords: Parikh images · Semilinearity · Ins-del systems · Matrix grammars

1 Introduction

Classically, computations are defined on words (compare the basic definitions of finite automata or of Turing machines), leading to the description of set of words (i.e., of languages) and hence to the consideration of families of languages (for instance, as in the Chomsky hierarchy REG $\subsetneq$ CF $\subsetneq$ CS $\subsetneq$ RE). Rather recently, the consideration of multiset languages (also known as macrosets [9]) has regained interest, where the basic entities are multisets of letters (as opposed to words), as the sequence of the letters does not matter. This renewed interest is also motivated from computational models such as membrane computing that are inspired by biology [14].

A different and supposedly older approach to multiset languages is via Parikh mappings [12]. Given an alphabet set $\Sigma = \{a_1, \ldots, a_n\}$, the Parikh mapping of a word w can be seen as an n-dimensional vector, listing the number of occurrences of each (alphabet) symbol as it appeared in the word. This naturally

© Springer International Publishing AG 2017
T.V. Gopal et al. (Eds.): TAMC 2017, LNCS 10185, pp. 201–215, 2017.
DOI: 10.1007/978-3-319-55911-7_15

leads to the Parikh image of a language L, denoted by $Ps\,L$. If $\mathcal{L}$ is some word language family, we can associate the *Parikh set family* $Ps\,\mathcal{L}$ to it. The following summarizes what is known about this with respect to the Chomsky hierarchy.

Proposition 1. $Ps\,\mathrm{REG} = Ps\,\mathrm{CF} \subsetneq Ps\,\mathrm{CS} \subsetneq Ps\,\mathrm{RE}$.

This paper is devoted to put Parikh images of matrix insertion-deletion systems into the context of the better known classes of macrosets. This way, we can also ascertain that several classes of matrix insertion-deletion languages are *not* computationally complete, rather are *semilinear* or non-semilinear. This is important, as the question to find matrix insertion-deletion systems of small size that are still computationally complete is one of the major research directions in that area of formal languages. As a further motivation of this study, observe that both multiset computations and insertion-deletion systems draw one of their origins in computational biology, so it is very natural to consider them together. Namely, insertions and deletions are actually a form of mutation and these operations in DNA, especially in connection with DNA computing, are discussed in [16]. In connection with RNA editing, these operations are reported in [1]. We also refer to [14] for various formalizations of DNA computing in general, including multiset computations.

Due to space constraints, several proofs in Sect. 5 have been omitted; we refer to the long version of this paper.

1.1   General Notions and Notations

Let $\mathbb{N}$ denote the set of nonnegative integers, and $[1\ldots k] = \{i \in \mathbb{N}: 1 \leq i \leq k\}$. Let Σ be a finite set of n elements (an alphabet). Σ^* denotes the free monoid generated by Σ. The elements of Σ^* are called *words* or *strings*; λ denotes the empty string, which is the neutral element with respect to the monoid operation on Σ^* called *concatenation*. If $f : X \to Y$ is some mapping, then this can be easily lifted to sets, i.e., $f : 2^X \to 2^Y$ and $f^- : 2^Y \to 2^X$ for the inverse.

We refrain from giving detailed definitions of standard terms in Formal Languages, but rather refer to any textbook from the area, in particular, to [2].

1.2   Semilinear Sets and Beyond

A *Parikh mapping* ψ (or, more precisely, ψ_Σ) from Σ^* into $\mathbb{N}^n$ is a mapping defined by first choosing an enumeration $a_1, \ldots, a_n$ of the elements of Σ and then defining inductively $\psi(\lambda) = (0, \ldots, 0)$, $\psi(a_i) = (\delta_{1,i}, \ldots, \delta_{n,i})$, where $\delta_{j,i} = 0$ if $i \neq j$ and $\delta_{j,i} = 1$ if $i = j$, and $\psi(au) = \psi(a) + \psi(u)$ for all $a \in \Sigma$, $u \in \Sigma^*$. Clearly, $\psi : \Sigma^* \to \mathbb{N}^n$ becomes thus a monoid morphism (where the operation on $\mathbb{N}^n$ is vector addition). Any two Parikh mappings from Σ^* into $\mathbb{N}^n$ differ only by a permutation of the coordinates of $\mathbb{N}^n$. This is inessential when considering sets of multisets of letters, or, in other words, multiset languages. Hence, if $\mathcal{L}$ is some word language family, we can associate the *Parikh set family* $Ps\,\mathcal{L}$ to it. Also, we can express the *commutative closure* of some language $L \subseteq \Sigma^*$ as

$\psi^-(\psi(L))$. In Proposition 1, we stated what is known about the Parikh images of the families in the Chomsky hierarchy. The equality $Ps\,\mathrm{REG} = Ps\,\mathrm{CF}$ is also known as *Parikh's Theorem*, going back to [12], and the inequalities easily follow from separating examples $L_0, L_1 \subsetneq \{a\}^*$ such that $L_0 \in RE \setminus CS$ and $L_1 \in \mathrm{CS} \setminus \mathrm{CF}$; we also refer to [9].

A subset $A \subseteq \mathbb{N}^n$ is said to be *linear* if there are $v, v_1, \ldots, v_m \in \mathbb{N}^n$ such that

$$A = \{v + k_1 v_1 + k_2 v_2 + \cdots + k_m v_m \mid k_1, k_2, \ldots, k_m \in \mathbb{N}\}.$$

A subset $A \subseteq \mathbb{N}^n$ is said to be *semilinear* if it is a finite union of linear sets. It is also well-known that $Ps\,\mathrm{REG}$ coincides with the family of semilinear sets.

Another important classical class of multiset languages is the one that can be seen as reachability sets of Petri nets. We are not going into details of the respective definitions, as there is a different perspective on this that is somehow more appropriate to us, namely matrix grammars with context-free core rules (without appearance checking); see [6]. It might be interesting to know that these macroset classes were also (independently) introduced from the viewpoint of modeling chemical-based computations; see [9,17]. $\mathrm{MAT(CF)}$ ($\mathrm{MAT(CF}-\lambda)$, resp.) denotes the family of languages generated by matrix context-free grammars allowing (disallowing, resp.) erasing rules; see [2].

1.3 Insertion-Deletion Systems

We now give the basic definition of insertion-deletion systems, following [8,14].

Definition 1. *An* insertion-deletion system *is a construct* $\gamma = (V, T, A, R)$, *where* V *is an alphabet,* $T \subseteq V$ *is the terminal alphabet,* A *is a finite language over* V, R *is a finite set of triplets of the form* $(u, \eta, v)_{ins}$ *or* $(u, \delta, v)_{del}$, *where* $(u, v) \in V^* \times V^*$, $\eta, \delta \in V^+$.

The pair (u, v) is called the *context*, η is called the *insertion string*, δ is called the *deletion string* and $x \in A$ is called an *axiom*. An insertion rule of the form $(u, \eta, v)_{ins}$ means that the string η is inserted between u and v. A deletion rule of the form $(u, \delta, v)_{del}$ means that the string δ is deleted between u and v. Applying $(u, \eta, v)_{ins}$ hence corresponds to the rewriting rule $uv \to u\eta v$, and $(u, \delta, v)_{del}$ corresponds to the rewriting rule $u\delta v \to uv$. If $u = v = \lambda$ for a rule, then the corresponding insertion/deletion can be done freely anywhere in the string and is called *context-free insertion/deletion* and is discussed in [11]. If $|u\eta v| = 1$, we speak of *strict context-free insertion*, and similarly of *strict context-free deletion* if $|u\delta v| = 1$. For simplicity, we write $(\eta)_{ins}$ for context-free insertion rules $(\lambda, \eta, \lambda)_{ins}$, and similarly for deletion rules. Consequently, for $x, y \in V^*$ we write $x \Rightarrow y$ if y can be obtained from x by using either an insertion rule or a deletion rule.

For an ins-del system, the descriptional complexity measures are based on the size comprising of (i) the maximal length of the insertion string, denoted by n, (ii) the maximal length of the left context and right context used in insertion rules, denoted by i' and i'', respectively, (iii) the maximal length of the deletion string, denoted by m, (iv) the maximal length of the left context and right context used in deletion rules denoted by j' and j'', respectively. The size of an ins-del system is denoted by $(n, i', i''; m, j', j'')$. For more details, we refer to [19].

1.4   Matrix Insertion-Deletion Systems

A matrix insertion-deletion system [4,10,13] is a construct $\Gamma = (V, T, A, R)$ where V is an alphabet, $T \subseteq V$, A is a finite language over V, R is a finite set of matrices $\{r_1, r_2, \ldots r_l\}$, where each r_i, $1 \le i \le l$, is a matrix of the form

$$r_i = [(u_1, \alpha_1, v_1)_{t_1}, (u_2, \alpha_2, v_2)_{t_2}, \ldots, (u_k, \alpha_k, v_k)_{t_k}]$$

with $t_j \in \{ins, del\}$, $1 \le j \le k$. For $1 \le j \le k$, the triple $(u_j, \alpha_j, v_j)_{t_j}$ is an ins-del rule. Consequently, for $x, y \in V^*$ we write $x \Longrightarrow_{r_i} y$ if y can be obtained from x by applying all the rules of a matrix r_i, $1 \le i \le l$, in order.

By $w \Longrightarrow_* z$, we denote the relation $w \Longrightarrow_{r_{i_1}} w_1 \Longrightarrow_{r_{i_2}} \cdots \Longrightarrow_{r_{i_k}} z$, where for all j, $1 \le j \le k$, we have $1 \le i_j \le l$. The language generated by Γ is defined as $L(\Gamma) = \{w \in T^* \mid x \Longrightarrow_* w, \text{ for some } x \in A\}$. If a matrix ins-del system has at most k rules in a matrix and the size of the underlying ins-del system is $(n, i', i''; m, j', j'')$, then we denote the corresponding class of languages by $\mathrm{MAT}(k; n, i', i''; m, j', j'')$. In the special case when $k = 2$, the system is said to have *binary matrices*. In [4], it is shown that the family of languages $\mathrm{MAT}(k; n, i', i'; m, j', j')$ is closed under reversal (i.e., $\mathrm{MAT}(k; n, i', i''; m, j', j'') = [\mathrm{MAT}(k; n, i'', i'; m, j'', j')]^R$).

Regarding computational completeness of the systems, it is shown in [4,5,13] that the following matrix ins-del systems are computationally complete: $\mathrm{MAT}(3; 1, 1, 0; 1, 1, 0)$, $\mathrm{MAT}(3; 1, 1, 0; 1, 0, 1)$, $\mathrm{MAT}(2; 1, 1, 0; 2, 0, 0)$, $\mathrm{MAT}(2; 2, 0, 0; 1, 1, 0)$, $\mathrm{MAT}(2; 1, 0, 1; 2, 0, 0)$, $\mathrm{MAT}(2; 2, 0, 0; 1, 0, 1)$, $\mathrm{MAT}(3; 1, 0, 1; 1, 0, 1)$, $\mathrm{MAT}(3; 1, 0, 1; 1, 1, 0)$, $\mathrm{MAT}(3; 1, 1, 1; 1, 0, 0)$, $\mathrm{MAT}(2; 1, 1, 1; 1, 0, 1)$ and $\mathrm{MAT}(4; 1, 0, 0; 1, 1, 1)$.

In this paper, we are interested in Parikh images of these language classes, denoted as $Ps\,\mathrm{MAT}(k; n, i', i''; m, j', j'')$. We now discuss a few examples of matrix ins-del system. Later, they are used in the proofs of some theorems.

Example 1. The language $L_1 = \{w \in \{a, b\}^* \mid |w|_a = |w|_b\}$ can be generated by a matrix Ins-del system of size $(2; 1, 0, 0; 0, 0, 0)$ as follows. $\Gamma_1 = (\{a, b\}, \{a, b\}, \{\lambda\}, \{[(a)_{ins}, (b)_{ins}]\})$. It is easy to see that $L(\Gamma_1) = L_1$. Note that L_1 is a non-regular language and cannot be generated by (matrix) ins-del system of size $(1; 1, 0, 0; 1, 0, 0)$. It can be argued that any matrix ins-del system Γ of size $(1; 1, 0, 0; 1, 0, 0)$ must contain an insertion-only matrix, and this can be applied to some axiom $\omega \in L_1$ of Γ to produce a word that is not in L_1.

The commutative closure of L_1 equals L_1 itself. In other words, $\psi^-(\psi(L_1)) = L_1$. However, if we modify Γ_1 a bit by changing the axiom set to $\{ab\}$, then this new system Γ_1' describes a strict subset of L_1 whose commutative closure equals $L_1 \setminus \{\lambda\}$, with $ba \in (L_1 \setminus \{\lambda\}) \setminus L(\Gamma_1')$.   □

Remark 1. $L_1 = \{w \in \{a,b\}^* \mid |w|_a = |w|_b\} \notin \mathrm{MAT}(1;1,0,0;1,0,0)$.

Proof. For the sake of contradiction, let us assume that Γ_1' is a matrix ins-del system of size $(1;1,0,0;1,0,0)$ such that $L(\Gamma_1') = L_1$. Consider a word $w \in L_1$ that is longer than any axiom of Γ_1'. Hence, at some point of the derivation of w, either a terminal symbol a (or a terminal symbol b, resp.) was inserted by using a matrix $[(a)_{ins}]$ (or $[(b)_{ins}]$, resp.). If we skip this mentioned insertion, say, by applying $[(a)_{ins}]$, in the derivation, but keep all other derivation steps, then the finally derived string will have an unequal number of occurrences of a and b, a contradiction to $L(\Gamma_1') = L_1$. More precisely, the derivation of w from an axiom ω can be described by a sequence of matrices $m_1, \ldots, m_\ell$, such that

$$\omega \Rightarrow_{m_1} w_1 \Rightarrow_{m_2} w_2 \Rightarrow \cdots \Rightarrow_{m_\ell} w_\ell = w.$$

Now assume that $m_i = [(a)_{ins}]$ was causing the last insertion of an a in this sequence. Then,

$$\omega \Rightarrow_{m_1} w_1 \Rightarrow_{m_2} w_2 \Rightarrow \cdots \Rightarrow_{m_{i-1}} w_{i-1} \Rightarrow_{m_{i+1}} w_{i+1}' \Rightarrow \cdots \Rightarrow_{m_\ell} w_\ell'$$

is also a valid derivation according to Γ_1' where, for all $j > i$, $w_j = u_j a v_j$ but $w_j' = u_j v_j$. In particular, w_ℓ' satisfies $\#_a w_\ell' = \#_a w_\ell - 1$ and $\#_b w_\ell' = \#_b w_\ell$; hence $\#_a w_\ell' < \#_b w_\ell' = \#_b w = \#_a w$. □

Example 2. Hopcroft and Pansiot [7] described a so-called vector addition system with states that generates the non-semilinear language $L_2 = \{w \in T^* \mid |w|_b + |w|_c \le 2^{|w|_a}\}$. Only a little scrutiny is needed to translate this mechanism into some $MAT(3;1,0,0;1,0,0)$ system Γ_2. The axiom of Γ_2 is Ac. We take the following rules into Γ_2: $[(A)_{del}, (c)_{del}, (A')_{ins}]$, $[(A')_{del}, (b)_{ins}, (A)_{ins}]$, $[(A)_{del}, (B)_{ins}]$, $[(B)_{del}, (b)_{del}, (B')_{ins}]$, $[(B')_{del}, (c)_{ins}, (B'')_{ins}]$, $[(B'')_{del}, (c)_{ins}, (B)_{ins}]$, $[(B)_{del}, (a)_{ins}, (A)_{ins}]$, $[(A)_{del}]$. Lemma 2.8 in [7] shows that, with terminal alphabet $T = \{a,b,c\}$, this system generates L_2. □

2   Preparatory Results

From earlier findings [4] on $\mathrm{MAT}(k;n,i',i'';m,j',j'')$ and its mirror image, we can immediately conclude the following, as the Parikh image of the mirror image of a language L equals the Parikh image of L.

Theorem 1. *For all non-negative integers k,n,i',i'',m,j,j'', we have that*

$$Ps\,\mathrm{MAT}(k;n,i',i'';m,j',j'') = Ps\,\mathrm{MAT}(k;n,i'',i';m,j'',j').$$

This allows us to summarize the known computational completeness results of matrix ins-del systems of small weights [4,13] as follows:

Theorem 2. *(i) $Ps\,RE = Ps\,\mathrm{MAT}(3;1,1,1;1,0,0) = Ps\,\mathrm{MAT}(4;1,0,0;1,1,1)$.*
(ii) Let $i' + i'' \ge 1$ and $j' + j'' \ge 1$. Then, $Ps\,RE = Ps\,\mathrm{MAT}(3;1,i',i'';1,j',j'')$.
(iii) Let $n + m \ge 3$, $\min(n,m) \ge 1$, $n + i' + i'' \ge 2$, $m + j' + j'' \ge 2$, as well as $i' + i'' + j' + j'' \ge 1$. Then, $Ps\,RE = Ps\,\mathrm{MAT}(2;n,i',i'';m,j',j'')$.

Recall that $Ps\,\mathcal{L}$ is a class of macrosets if $\mathcal{L}$ is a class of languages, as a language $L \in \mathcal{L}$ is mapped to a set of vectors $\psi(L)$. Clearly, one can also reverse this process and consider the language $\psi^-(\psi(L)) \supseteq L$. Accordingly, we can form the language class $Ps^-(Ps\,\mathcal{L})$. In general, the relation between $\mathcal{L}$ and $Ps^-(Ps\,\mathcal{L})$ is unclear. However, for matrix ins-del systems, we can show the following.

Theorem 3. *Let $k \geq 2$, $n \geq 1$, $i',i'' \geq 0$, $m \geq 1$, $j',j'' \geq 0$. Then,*

$$Ps^-(Ps\,\mathrm{MAT}(k;n,i',i'';m,j',j'')) \subsetneq \mathrm{MAT}(k;n,i',i'';m,j',j'').$$

In other words, we claim that nearly all the language families that are considered in this paper are closed under commutative closure. It would be interesting to get a complete picture about for which values of n,i',i'',m,j',j'' the language families $\mathrm{MAT}(1;n,i',i'';m,j',j'')$ are closed under commutative closure and for which not. The following proof only works for $k \geq 2$.

Proof. Let $\Gamma = (V,T,A,R)$ be some matrix ins-del system of size $(k;n,i',i'';m,j',j'')$. Let $V' = \{a' \mid a \in V\}$ be the set of primed versions of symbols from V. Consider $\Gamma' = (V \cup V',T,A,R \cup R')$, where $R' = \{[(a)_{del},(a')_{ins}],[(a')_{del},(a)_{ins}]\}$. Then, $L(\Gamma') = \psi_T^-(\psi_T(L(\Gamma)))$. Hence, there is a matrix ins-del system of the required size to describe the commutative closure of $L(\Gamma)$. The claimed strictness of the inclusion follows with Example 1.   □

3 Languages in $Ps\,\mathrm{MAT}(2;1,0,0;1,0,0)$ are semilinear

Consider some matrix ins-del system Γ containing binary matrices only, with context-free insertion rules and context-free deletion rules.

First, we claim that we can assume that, without loss of generality, a derivation of some word $w \in L(\Gamma)$ can be decomposed into three phases that work in the following order:

1. matrices containing two insertion rules are used;
2. matrices containing one insertion and one deletion rule (*mixed type*) are used;
3. only deletion rules are used.

Namely, a derivation of some $w \in L(\Gamma)$ that does not obey this order can be re-ordered by first moving the insertion-only matrices to the beginning (their applicability does not depend on the prior application of any other matrix) and then moving all deletion-only matrices to the end, leaving in particular the relative order of the matrices containing one insertion and one deletion rule.

Based on this observation, we are proving a normal form lemma that is useful to show the main result of this section.

Lemma 1. *Let $L \in \mathrm{MAT}(2;1,0,0;1,0,0)$. Then, there is a matrix ins-del system Γ' of size $(2;1,0,0;1,0,0)$ with $L(\Gamma') = L$ satisfying the following properties.*

1. Γ' contains no matrices of mixed type.
2. $[(a)_{ins},(b)_{ins}]$ is a matrix of Γ' if and only if $[(b)_{ins},(a)_{ins}]$ is a matrix of Γ'.

3. $[(a)_{del}, (b)_{del}]$ *is a matrix of* Γ' *if and only if* $[(b)_{del}, (a)_{del}]$ *is a matrix of* Γ'.
4. Γ' *contains no matrices of the form* $[(a)_{del}]$.
5. If $w \in L$ *can be obtained in* Γ' *by exclusively using deletion-only matrices, then* w *is an axiom of* Γ'.

Proof. Let a matrix ins-del system $\Gamma = (V, T, A, R)$ of size $(2, 1, 0, 0; 1, 0, 0)$ with $L(\Gamma) = L$ be given. We are describing how to transform Γ in order to satisfy the claimed normal form.

Let R_{mixed} collect all matrices of R that are of mixed type. Let $V' = \{a' \mid a \in V\}$ be a collection of new symbols, primed variants of the original alphabet. W.l.o.g., matrices in R_{mixed} contain first a deletion rule and then an insertion rule. Notice that the only case when $[(b)_{ins}, (a)_{del}]$ is applicable but not $[(a)_{del}, (b)_{ins}]$ is when $a = b$ and there is no symbol a in the current sentential form. However, as the matrix $[(b)_{ins}, (a)_{del}]$ has no effect if $a = b$ and there is no symbol a in the current sentential form, we can neglect this in the following. Matrices in R_{mixed} of the form $[(a)_{del}, (b)_{ins}]$ are replaced by the two matrices (i) $[(a')_{ins}, (b)_{ins}]$ (insertion-only matrix) and (ii) $[(a)_{del}, (a')_{del}]$ (deletion-only matrix). We collect all these matrices in R'_{mixed}. Let $V_1 = V \cup V'$, $A_1 = A$, $R_1 = (R \setminus R_{mixed}) \cup R'_{mixed}$ and $\Gamma_1 = (V_1, T, A_1, R_1)$. Γ_1 contains no matrices of mixed type by construction. We will now argue that $L = L(\Gamma_1)$. (i) '$\subseteq$': We only have to bother about the simulation of matrices of mixed type. Now, an application of $[(a)_{del}, (b)_{ins}]$ can clearly be simulated by first applying $[(a')_{ins}, (b)_{ins}]$ and then applying $[(a)_{del}, (a')_{del}]$.

(ii) '$\supseteq$': As argued above, we can assume that the derivation of some word $w \in L(\Gamma_1)$ first applies insertion-only matrices and then deletion-only matrices. Clearly, the sequence of insertion-only matrices that are applied and also the sequence of deletion-only matrices that are applied can be in any order without affecting the fact that w can be generated this way. Let us therefore take the following ordering. First, we apply insertion-only matrices from $R \setminus R_{mixed}$ (in any order). Secondly, we apply insertion-only matrices from R'_{mixed} (in an order that we describe below). Thirdly, we apply deletion-only matrices from R'_{mixed} (in an order that we describe below). Finally, we apply deletion-only matrices from $R \setminus R_{mixed}$ (in any order). Trivially, with Γ we can simulate all matrices from $R \setminus R_{mixed}$. So, we only discuss matrices from R'_{mixed} in the following. Now, observe in order to finally generate $w \in T^*$, for each insertion-only matrix from R'_{mixed} that introduces some symbol a', there must be one deletion-only matrix from R'_{mixed} that deletes such an occurrence of a' again. Let us fix an arbitrary ordering $<$ on V_1. This can be used to define an ordering $<_R$ on R'_{mixed} as follows. First, all insertion-only matrices are situated; these are ordered according to $[(a')_{ins}, (b)_{ins}] <_R [(c')_{ins}, (d)_{ins}]$ if $a' < c'$ or $a' = c'$ and $b < d$. Then, all deletion-only matrices are situated; these are ordered according to $[(a)_{del}, (a')_{del}] <_R [(c)_{del}, (c')_{del}]$ if $c' < a'$. Now observe that this implies that right in the middle, an application of some matrix $[(a')_{ins}, (b)_{ins}]$ is immediately followed by an application of $[(a)_{del}, (a')_{del}]$. These two matrix applications could be easily simulated by the matrix $[(a)_{del}, (b)_{ins}]$ contained in R_{mixed}. Now consider the insertion-only matrix $[(c')_{ins}, (d)_{ins}]$ that was applied before

$[(a')_{ins}, (b)_{ins}]$ (if there is no such matrix, our inductive argument is complete). If $a \neq d$, then we could clearly simulate the sequence of matrix applications of $[(c')_{ins}, (d)_{ins}]$, $[(a')_{ins}, (b)_{ins}]$, $[(a)_{del}, (a')_{del}]$ and $[(c)_{del}, (c')_{del}]$ by first applying $[(a)_{del}, (b)_{ins}]$ and then $[(c)_{del}, (d)_{ins}]$. If $a = d$, the simulation should be carried out by first applying $[(c)_{del}, (a)_{ins}]$ and then $[(a)_{del}, (b)_{ins}]$. By induction, we can see by this argument that we can simulate the whole sequence of matrix applications from R'_{mixed} by using matrices from R_{mixed}. This shows that $w \in L(\Gamma)$.

Let $V_2 = V_1$, $A_2 = A_1$. $R_2 = R_1 \cup \{[(a)_{ins}, (b)_{ins}] \mid [(b)_{ins}, (a)_{ins}] \in R_1\}$. Let $\Gamma_2 = (V_2, T, A_2, R_2)$. Γ_2 satisfies the first two properties. It can be easily seen that $L(\Gamma_1) = L(\Gamma_2)$.

Let $V_3 = V_1$, $A_3 = A_1$. $R_3 = R_2 \cup \{[(a)_{del}, (b)_{del}] \mid [(b)_{del}, (a)_{del}] \in R_1\}$. Let $\Gamma_3 = (V_3, T, A_3, R_3)$. Γ_3 satisfies the first three properties. It can be easily seen that $L(\Gamma_2) = L(\Gamma_3)$.

Let $R_{del} = \{[(a)_{del}] \mid a \in V_1\} \cap R_3$. Let $\Gamma_{del} = (V_3, V_3, A_3, R_{del})$. Let $A_4 = L(\Gamma_{del})$. Notice that A_4 is a finite language, with $A_4 \subseteq L(\Gamma_3)$. Let $R'_{del} = \{[(b)_{ins}] \mid [(a)_{del}] \in R_{del} \wedge [(a)_{ins}, (b)_{ins}] \in R_3\}$. Let $R_4 = (R_3 \setminus R_{del}) \cup R'_{del}$. Let $V_4 = V_1$ and $\Gamma_4 = (V_4, T, A_4, R_4)$. Γ_4 satisfies the first four properties. We now show that $L(\Gamma_4) = L(\Gamma_3)$. Let $w \in L(\Gamma_4)$. This is certified by some derivation, starting from some axiom $\omega \in A_4$. If $\omega \notin A_3$, then we can obtain ω from A_3 by using rules from $R_{del} \subseteq R_3$. By construction, rules $[(b)_{ins}]$ from R'_{del} can be simulated by first applying a matrix $[(a)_{ins}, (b)_{ins}] \in R_3$ and then applying the matrix $[(a)_{del}] \in R_{del}$. As other matrices can be directly carried out within Γ_3, $w \in L(\Gamma_3)$ follows. Conversely, if $w \in L(\Gamma_3)$, then we can assume that all matrices from R_{del} are executed at the end, except those applications that delete symbols from the axiom $\omega \in A_3$; these are carried out first. So, we could simulate the derivation of w in Γ_4 by starting from an axiom ω' that is obtained from ω by rules from R_{del}. The remaining applications of matrices $[(a)_{del}]$ from R_{del} are obviously applied to symbols that have been introduced by insertion-only matrices, say, $[(a)_{ins}, (b)_{ins}]$ followed by $[(a)_{del}]$. This can be simulated by a rule $[(b)_{ins}]$ from R'_{del}. Hence, $w \in L(\Gamma_4)$.

Finally, let R_{del2} collect all deletion-only matrices of R_4 and define $A' = L((V_4, V_4, A_4, R_{del2}))$. Again, A' is a finite subset of $L(\Gamma_4)$. Moreover, with $V' = V_4$, $R' = R_4$, one can see that $\Gamma' = (V', T, A', R')$ satisfies all claims of this lemma.   □

We are going to use this normal form lemma in the proof of the following main theorem of this section.

Theorem 4. $Ps\,\mathrm{MAT}(2; 1, 0, 0; 1, 0, 0) \subseteq Ps\,\mathrm{REG}$.

Proof. Let $\Gamma = (V, T, A, R)$ be a matrix ins-del system of size $(2; 1, 0, 0; 1, 0, 0)$ that satisfies the previous normal form lemma. Notice that we can further assume that $|A| = 1$, because it is known that $Ps\,\mathrm{REG}$ is closed under (finite) union. So, let $A = \{\omega\}$. We are going to define a right-linear grammar $G = (N, T, S, P)$ such that $\psi_T(L(G)) = \psi_T(L(\Gamma))$.

We define
$$N = \{S\} \cup \{S\langle v\rangle \mid v \in V^{\leq \max(|\omega|,2)}\}.$$

The idea is to memorize in the nonterminal $S\langle v\rangle$ those symbols that still have to be inserted, as we already started deletion-only matrices in our simulating derivation, stopping half-way in their simulation.

The axiom ω is dealt with using the rules
$$P_{axiom} = \{S \to wS\langle v\rangle \mid w \in T^* \wedge \exists u \in V^{\leq |\omega|} : \psi_V(uw) = \psi_V(\omega) \wedge uv \Longrightarrow_*^{del} \lambda\}.$$

Here, we use $\Longrightarrow_*^{del}$ to denote the application of deletion-only matrices from Γ (but no insertion-only matrices are applied). In other words, upon executing $S \to wS\langle v\rangle$, we guessed that some symbols (namely those listed in u) have been already deleted from ω by applying $|u|$ many deletion-only matrices half-way. The not yet applied rules are memorized in the string v. These symbols have yet to be generated by insertion-only matrices. Recall that due to our normal form, we can assume that no deletion-only matrix is applied to ω alone.

Then, we enter a phase to simulate the insertion-only matrices.

The simulation of an insertion-only matrix works in a way similar to the construction of the axiom word ω. All the simulation rules are collected in P_{simul}. Let us first consider the simpler case when $[(a)_{ins}]$ has to be simulated. To capture the case that a is not again deleted, which is only possible if $a \in T$, we introduce the rules $S\langle v\rangle \to aS\langle v\rangle$ for all v. If a is going to be deleted by applying some $[(a)_{del}, (b)_{del}] \in R$, we capture this by two types of rules. If b occurs in v, say, $v = xby$, then $S\langle v\rangle \to S\langle xy\rangle$ is added to P_{simul}. If b does not occur in v and if $|v| < \max(|\omega|, 2)$, then $S\langle v\rangle \to S\langle vb\rangle$ is added to P_{simul}.

The simulation of $m = [(a)_{ins}, (a')_{ins}]$ is more complicated yet similar. We have three cases to consider (as also $m' = [(a')_{ins}, (a)_{ins}] \in R$: (i) none of a and a' are later deleted, (ii) only a is later deleted, or (iii) both a and a' are later deleted. Let us consider these cases in details in the following.

In case (i), we know that $aa' \in T^*$. Hence, we can add the rules $S\langle v\rangle \to aa'S\langle v\rangle$ for all v.

In case (ii), $a' \in T$ is known. Assume that a is going to be deleted by applying some $[(a)_{del}, (b)_{del}] \in R$. We have two subcases. If b occurs in v, say, $v = xby$, then $S\langle v\rangle \to a'S\langle xy\rangle$ is added to P_{simul}. If b does not occur in v and if $|v| < \max(|\omega|, 2)$, then $S\langle v\rangle \to a'S\langle vb\rangle$ is added to P_{simul}.

In case (iii), we assume that a is going to be deleted by applying some $[(a)_{del}, (b)_{del}] \in R$ and that a' is going to be deleted by applying some matrix $[(a')_{del}, (b')_{del}] \in R$. We have four subcases. (iii.a) If b and b' occur in v, say, $v = xbyb'z$, then $S\langle v\rangle \to S\langle xyz\rangle$ is added to P_{simul}. (iii.b) If b occurs in v, but b' does not occur in v, say, $v = xby$, then $S\langle v\rangle \to S\langle xyb'\rangle$ is added to P_{simul}. (iii.c) If b' occurs in v, but b does not occur in v, this can be treated symmetrically. (iii.d) If neither b nor b' occur in v and if $|v| + 1 < \max(|\omega|, 2)$, then $S\langle v\rangle \to S\langle vbb'\rangle$ is added to P_{simul}.

The termination is only accomplished via the rule $S\langle\lambda\rangle \to \lambda$. So, in total $P = P_{axiom} \cup P_{simul} \cup \{S\langle\lambda\rangle \to \lambda\}$.

The main point to understand the simulation is that whenever we apply an insertion-only matrix that generates a symbol that will be finally deleted, we can afford storing the possibly existing second symbol that is going to be deleted by the mentioned deletion-only matrix, because we can assume (w.l.o.g.) that the next insertion-only matrix that is applied provides exactly the symbol that has to be deleted next. So, the amount of information that needs to be stored in nonterminals is finite.   □

However, the reverse inclusion of the previous result seems not clear and remains as an open problem.

4   Beyond Semilinearity

In this section, we first define a notation which is often used in stating results in the remainder of this paper.

$$Ps \, \text{MAT}(*; 1, 0, 0; 1, 0, 0) = \bigcup_{k \geq 1} Ps \, \text{MAT}(k; 1, 0, 0; 1, 0, 0).$$

Together with Theorem 4, we can conclude the following hierarchy result.

Corollary 1. *For any $k \geq 3$, we find the following hierarchy of strictly context-free matrix ins-del systems:*

$$Ps \, \text{MAT}(1; 1, 0, 0; 1, 0, 0) \subsetneq Ps \, \text{MAT}(2; 1, 0, 0; 1, 0, 0)$$
$$\subsetneq Ps \, \text{MAT}(3; 1, 0, 0; 1, 0, 0)$$
$$= Ps \, \text{MAT}(k; 1, 0, 0; 1, 0, 0)$$

Proof. The first strict inclusion follows from Example 1, as $\text{MAT}(1; 1, 0, 0; 1, 0, 0)$ can be viewed as ins-del systems of size $(1, 0, 0; 1, 0, 0)$ without any matrix control (see also Remark 1). The second strict inclusion is due to the fact that there are non-semilinear context-free matrix languages, also refer Example 2.   □

Theorem 5. $Ps \, \text{MAT}(3; 1, 0, 0; 1, 0, 0) = Ps \, \text{MAT}(*; 1, 0, 0; 1, 0, 0)$ *and* $Ps \, \text{MAT}(*; 1, 0, 0; 1, 0, 0) = Ps \, \text{MAT}(\text{CF})$.

Proof. First, we are going to simulate a context-free matrix grammar $G = (N, T, S, M)$ in binary normal form, potentially with erasing rules; see [2]. This means that the nonterminal alphabet N is partitioned into $N_1 \cup N_2 \cup \{S\}$, and all matrices $m \in M$ are of the form $m = [p \rightarrow q, X \rightarrow w]$, where $p \in N_1$, $q \in N_1 \cup \{\lambda\}$, $X \in N_2$, $w \in (N_2 \cup T)^*$, except the start rules, which take the form $S \rightarrow pX$ for some $p \in N_1$ and $X \in N_2$, collected in a singleton matrix. Moreover, we can assume that $[p \rightarrow \lambda, X \rightarrow w]$ only occurs when $w = \lambda$. We put all pX with matrices $[S \rightarrow pX] \in M$ into the set of axioms A of the matrix ins-del system $\Gamma = (V, T, A, R)$ that we are going to describe. The alphabet V will contain $T \cup N_1 \cup N_2$ plus several auxiliary new nonterminals whose sole purpose is to guide the application sequence of the matrices. Our description will make clear that $\psi_T(L(G)) = \psi_T(L(\Gamma))$ as required.

For the simulation itself, we consider several cases of matrices $m \in M$.

- If $m = [p \to \lambda, X \to \lambda]$, then we can simply interpret m as a matrix $m' = [(p)_{del}, (X)_{del}]$ containing two deletion rules and put m' into R.
- If $m = [p \to q, X \to \lambda]$ with $p, q \in N_1$, we can simulate this with the following ins-del matrix $m' = [(p)_{del}, (q)_{ins}, (X)_{del}]$: Again, we adjoin m' to R.
- If $m = [p \to q, X \to w]$ with $p, q \in N_1$ and $|w| \geq 1$, say, $w = a_1 \cdots a_n$, $n = |w|$, $a_i \in (N_2 \cup T)$, we can simulate this with the following ins-del matrices $m[r]$, where x is a prefix of w:

$$m[w] = [(p)_{del}, \ (q[w])_{ins}, \ (X)_{del}]$$
$$m[x] = [(q[xa])_{del}, \ (q[x])_{ins}, \ (a)_{ins}]$$

if xa is a prefix of w; we identify $q[\lambda]$ and q. Now, we adjoin all these matrices $m[x]$ to R.
- No other matrices belong to R.

Notice that whenever we can transform a string x to y using some matrix m, then the suggested matrices m' (or $m[x]$) can be used to obtain a string y' from x such that $\psi_V(y) = \psi_V(y')$. By induction, the claim follows from these observations.

The converse direction is detailed in the long version. The basic idea is to introduce a new nonterminal Z and also to work with pseudo-terminals (i.e., nonterminals a' for each terminal a). A deletion rule can be directly interpreted as an erasing rule, while an insertion rule is simulated by a rule of the form $Z \to ZX'$. □

As the reachability problem for Petri nets (without inhibitor arcs) is decidable, the membership problem is decidable for $Ps\,\mathrm{MAT}(*; 1, 0, 0; 1, 0, 0)$. This also implies that the inclusion $Ps\,\mathrm{MAT}(*; 1, 0, 0; 1, 0, 0) \subsetneq Ps\,\mathrm{RE}$ is strict. Recall the many computationally complete matrix ins-del systems that we listed in Theorem 2.

We are now discussing the simulation of context-free matrix via matrix ins-del systems explained above. If we allow the insertion of two symbols at once, we can in fact (easily) reduce the lengths of the matrices by one except for the matrix that simulates an erasing rule. Namely, instead of inserting two symbols one after the other, we can do this now with just one rule. If we add on matrices like $[(X)_{del}, (X)_{ins}]$ for any symbol X, we can also make sure that, whenever two symbols are to be deleted in some matrix of length three, then this can be simulated by deleting two symbols next to each other with one rule. This observation allows us to state:

Corollary 2. $Ps\,\mathrm{MAT(CF)} \subseteq Ps\,\mathrm{MAT}(2; 2, 0, 0; 2, 0, 0)$.

Whether or not the converse inclusion is true is open. The problem is that we would have to check whether two symbols in intermediate sentential forms sit next to each other; this seems to be impossible with context-free matrix grammars without appearance checking.

The previous corollary also shows the strictness of one of the following two inclusions; it is however unclear which is strict.

$$Ps\,\mathrm{MAT}(2; 1, 0, 0; 1, 0, 0) \subseteq Ps\,\mathrm{MAT}(2; 2, 0, 0; 1, 0, 0)$$
$$\subseteq Ps\,\mathrm{MAT}(2; 2, 0, 0; 2, 0, 0)$$

A similar situation arises with:

$$Ps\,\mathrm{MAT}(2; 1, 0, 0; 1, 0, 0) \subseteq Ps\,\mathrm{MAT}(2; 1, 0, 0; 2, 0, 0)$$
$$\subseteq Ps\,\mathrm{MAT}(2; 2, 0, 0; 2, 0, 0)$$

It is known that ins-del systems of size $(2, 0, 0; 2, 0, 0)$ strictly contain context-free languages only [18]. So, $Ps\,\mathrm{MAT}(1, 2, 0, 0; 2, 0, 0) \subseteq Ps\,\mathrm{CF}$. Thus, we see:

Corollary 3. $Ps\,\mathrm{MAT}(1; 2, 0, 0; 2, 0, 0) \subsetneq Ps\,\mathrm{MAT}(3; 1, 0, 0; 1, 0, 0)$.

The following result strengthens the previous argument.

Theorem 6. $\bigcup_{k \geq 1, n \geq 1} Ps\,\mathrm{MAT}(k; n, 0, 0; 1, 0, 0) = Ps\,\mathrm{MAT}(\mathrm{CF})$.

Proof. The inclusion $\supseteq$ follows with Theorem 5. Conversely, notice that we can always replace an insertion rule $(x)_{ins}$, with $x = a_1 \cdots a_\nu$ being a word of length $1 < \nu \leq n$, by the following sequence of insertions (considered as a sub-matrix of the matrix originally containing r): $[(a_1)_{ins}, \ldots, (a_\nu)_{ins}]$. This does not change the Parikh image of the generated language, as deletions are performed only on single symbols. Doing this repeatedly, we can transform any matrix ins-del system Γ of size $(*; n, 0, 0; 1, 0, 0)$ (with terminal alphabet T) into a matrix ins-del system Γ' of size $(*; *, 0, 0; 1, 0, 0)$ such that $\psi_T(\Gamma) = \psi_T(\Gamma')$. This shows that $Ps\,\mathrm{MAT}(*; n, 0, 0; 1, 0, 0) \subseteq Ps\,\mathrm{MAT}(*; 1, 0, 0; 1, 0, 0)$. Applying Theorem 5 once more proves the claim. □

Focusing on binary matrices, it is however unclear if the inclusion

$$\bigcup_{n \geq 1} Ps\,\mathrm{MAT}(2; n, 0, 0; 1, 0, 0) \subseteq Ps\,\mathrm{MAT}(\mathrm{CF})$$

is strict or not. Similar open problems arise with the following chain of inclusions:

$$Ps\,\mathrm{MAT}(2; 1, 0, 0; 1, 0, 0) \subseteq Ps\,\mathrm{MAT}(2; 1, 1, 0; 1, 0, 0)$$
$$= Ps\,\mathrm{MAT}(2; 1, 0, 1; 1, 0, 0)$$
$$\subseteq Ps\,\mathrm{MAT}(2; 1, 1, 1; 1, 0, 0)$$
$$\subseteq Ps\,\mathrm{MAT}(2; 1, 1, 1; 1, 0, 1)$$

As $Ps\,\mathrm{MAT}(2; 1, 1, 1; 1, 0, 1) = Ps\,\mathrm{RE}$ (see [5] for $\mathrm{RE} = \mathrm{MAT}(2; 1, 1, 1; 1, 0, 1)$), one of these inclusions must be strict. For the equality, we use Theorem 1.

Similarly, we can reason about systems with bigger deletion complexity.

Theorem 7. $\bigcup_{k \geq 1, m \geq 1} Ps\,\mathrm{MAT}(k; 1, 0, 0; m, 0, 0) = Ps\,\mathrm{MAT}(\mathrm{CF})$.

Proof. The inclusion $\supseteq$ follows with Theorem 5. Conversely, notice that we can always replace a deletion rule $(x)_{del}$, with $x = a_1 \cdots a_\mu$, $\mu \le m$, being a word of length $\mu > 1$, by the following sequence of deletions (considered as a sub-matrix of the matrix originally containing r): $[(a_1)_{del}, \ldots, (a_\mu)_{del}]$. This does not change the Parikh image of the generated language, as insertions are performed only on single symbols. Doing this repeatedly, we can transform any matrix ins-del system Γ of size $(*; 1, 0, 0; m, 0, 0)$ (with terminal alphabet T) into a matrix ins-del system Γ' of size $(*; 1, 0, 0; 1, 0, 0)$ such that $\psi_T(\Gamma) = \psi_T(\Gamma')$. This shows that $Ps\,\text{MAT}(*; 1, 0, 0; m, 0, 0) \subseteq Ps\,\text{MAT}(*; 1, 0, 0; 1, 0, 0)$. Applying Theorem 5 once more proves the claim. $\qquad\square$

5   Monotonicity Conditions

With matrix ins-del systems, it is at least possible to define a condition that ensures some kind of monotonicity. For simplicity, we define this concept for strict context-free matrix ins-del systems only. So, such a system is called *monotone* if no axiom equals the empty word (this prevents the system from generating the empty word at all, but according to usual convention in formal languages, this does not matter) and each matrix $[(x_1)_{\mu_1}, \ldots, (x_r)_{\mu_r}]$ obeys:

$$\sum_{i=1}^{r} |x_i| \cdot [\mu_i = ins] \ge \sum_{i=1}^{r} |x_i| \cdot [\mu_i = del]$$

where $[cond]$ is the usual interpretation of a logical condition *cond* to yield 1 if it is true and 0 otherwise. We indicate this by adding *mon* as a subscript.

Theorem 8. $CS = \text{MAT}_{mon}(2; 3, 0, 0; 3, 0, 0)$.

Note that possible corollaries for RE are superseded by results from [11].

For the proof of the next characterization theorem, the following normal form comes in handy. A strict context-free matrix ins-del system Γ is said to be in *deletion first normal form* if in every matrix, all deletion rules precede all insertion rules. (This covers, of course, the extreme cases when a matrix contains only deletion rules or only insertion rules, as well.)

Lemma 2. *To every strict context-free matrix ins-del system Γ, there exists a strict context-free matrix ins-del system Γ' in deletion first normal form such that $L(\Gamma) = L(\Gamma')$.*

Theorem 9. *$Ps\,\text{MAT}(CF - \lambda)$ can be characterized by monotone systems. Hence, $Ps\,\text{MAT}(CF-\lambda) = Ps\,\text{MAT}_{mon}(*; 1, 0, 0; 1, 0, 0) = Ps\,\text{MAT}_{mon}(4; 1, 0, 0; 1, 0, 0)$.*

Let us indicate our simulations in the following. A rule $A \to w$, with $w = a_1 \cdots a_j$, in a context-free matrix grammar $G = (N, T, S, P)$ can be simulated by the sequence $(A)_{del}, (a_1)_{ins}, \ldots, (a_j)_{ins}$ of deletion and insertion rules, which is monotone. This idea can be combined with that from Theorem 5 to show the complete simulation. For the reverse direction, we simulate a strict context-free monotone matrix ins-del system $\Gamma = (V, T, A, R)$ in deletion first normal form.

Remark 2. It is sufficient that either all insertion or all deletion rules are strict context-free. The other rules could be just context-free ; compare Theorems 6 and 7 for the non-monotone case.

Corollary 4. *The class of Parikh images of languages generated by monotone strict context-free matrix ins-del systems is a strict subclass of the class of Parikh images of languages generated by monotone context-free matrix ins-del systems.*

Theorem 10. $Ps\,\mathrm{MAT}_{mon}(1; 1, 0, 0; 0, 0, 0) = Ps\,\mathrm{MAT}_{mon}(1; 1, 0, 0; 1, 0, 0) \subsetneq Ps\,\mathrm{MAT}_{mon}(2; 1, 0, 0; 1, 0, 0) \subsetneq Ps\,\mathrm{MAT}_{mon}(3; 1, 0, 0; 1, 0, 0) = Ps\,\mathrm{REG}$.

6   Conclusions

In this paper, we have considered results associated with Parikh images of languages generated by matrix ins-del systems, especially, of small sizes. We show that $Ps\,\mathrm{MAT}(2; 1, 0, 0; 1, 0, 0)$ contains only semilinear languages. Besides, $Ps\,\mathrm{MAT}(3; 1, 0, 0; 1, 0, 0)$ contains non-semilinear languages and we have shown that this family of languages and the family of languages $Ps\,\mathrm{MAT}(\mathrm{CF})$ are the same. We often dealt with binary matrices (matrices of length at most two) with context-free insertion and deletion rules. We have analyzed the characterization of the hierarchy of family of languages that is formed with these small sizes. For many classes of languages, proving the strict inclusion is left open. To the best of our knowledge, the question whether or not the inclusion $Ps\,\mathrm{MAT}(\mathrm{CF} - \lambda) \subseteq Ps\,\mathrm{MAT}(\mathrm{CF})$ is strict is also an (old) open problem, see [2], that obviously relates to the systems that we investigated.

Apart from ins-del systems where no deletions take place at all (as in Example 1), it is unclear if we get somehow context-sensitive languages only. In the context of our paper, we want to finally recall that $\mathrm{MAT}(1; 2, 2, 2; 0, 0, 0)$ contains non-semilinear languages, see [14, Theorem 6.5].

Let us point out once more the many open problems that we listed throughout the paper, concerning whether or not some inclusions between language classes are strict. Notice that in some of these cases, the upper bounds on the inclusion chains were mostly given by computational completeness results for the corresponding classes of word languages. It might well be that a more genuine approach to multiset languages could yield better upper bounds. For the related area of graph-controlled insertion-deletion systems, such an attempt has been undertaken in [3]. It might be also an idea to try to simulate Petri nets with inhibitor arcs in order to prove the strictness of some of the inclusions. The limits of such an approach are discussed in [15].

References

1. Benne, R. (ed.): RNA Editing: The Alteration of Protein Coding Sequences of RNA. Series in Molecular Biology. Ellis Horwood, Chichester (1993)
2. Dassow, J., Păun, G.: Regulated rewriting in formal language theory. In: EATCS Monographs in Theoretical Computer Science, vol. 18. Springer, Heidelberg (1989)

3. Fernau, H.: An essay on general grammars. J. Automata Lang. Comb. **21**, 69–92 (2016)
4. Fernau, H., Kuppusamy, L., Raman, I.: Generative power of matrix insertion-deletion systems with context-free insertion or deletion. In: Amos, M., Condon, A. (eds.) UCNC 2016. LNCS, vol. 9726, pp. 35–48. Springer, Cham (2016). doi:10.1007/978-3-319-41312-9_4
5. Fernau, H., Kuppusamy, L., Raman, I.: Investigations on the power of matrix insertion-deletion systems with small sizes (2016, Submitted to Natural Computing) (to appear)
6. Hauschildt, D., Jantzen, M.: Petri net algorithms in the theory of matrix grammars. Acta Informatica **31**, 719–728 (1994)
7. Hopcroft, J.E., Pansiot, J.-J.: On the reachability problem for 5-dimensional vector addition systems. Theor. Comput. Sci. **8**, 135–159 (1979)
8. Kari, L., Thierrin, G.: Contextual insertions/deletions and computability. Inf. Comput. **131**(1), 47–61 (1996)
9. Kudlek, M., Martín-Vide, C., Păun, G.: Toward a formal macroset theory. In: Calude, C.S., Păun, G., Rozenberg, G., Salomaa, A. (eds.) WMP 2000. LNCS, vol. 2235, pp. 123–133. Springer, Heidelberg (2001). doi:10.1007/3-540-45523-X_7
10. Kuppusamy, L., Mahendran, A.: Modelling DNA and RNA secondary structures using matrix insertion-deletion systems. Intl. J. Appl. Math. Comput. Sci. **26**(1), 245–258 (2016)
11. Margenstern, M., Păun, G., Rogozhin, Y., Verlan, S.: Context-free insertion-deletion systems. Theor. Comput. Sci. **330**(2), 339–348 (2005)
12. Parikh, R.J.: On context-free languages. J. ACM **13**(4), 570–581 (1966)
13. Petre, I., Verlan, S.: Matrix insertion-deletion systems. Theor. Comput. Sci. **456**, 80–88 (2012)
14. Păun, G., Rozenberg, G., Salomaa, A.: DNA Computing: New Computing Paradigms. Springer, Heidelberg (1998)
15. Reinhardt, K.: Counting as Method, Model and Task in Theoretical Computer Science. Habilitationsschrift, Universität Tübingen (2005)
16. Smith, W.D.: DNA computers in vitro and in vivo. In: Lipton, R.J., Baum, E.B. (eds.) DNA Based Computers, Proceedings of a DIMACS Workshop, Princeton, 1995, pp. 121–186. American Mathematical Society (1996)
17. Suzuki, Y., Tanaka, H.: Symbolic chemical system based on abstract rewriting system and its behavior pattern. Artif. Life Robot. **1**(4), 211–219 (1997)
18. Verlan, S.: On minimal context-free insertion-deletion systems. J. Automata Lang. Comb. **12**(1–2), 317–328 (2007)
19. Verlan, S.: Recent developments on insertion-deletion systems. Comput. Sci. J. Moldova **18**(2), 210–245 (2010)

Algorithmic Aspects of the Maximum Colorful Arborescence Problem

Guillaume Fertin, Julien Fradin$^{(\boxtimes)}$, and Géraldine Jean

LS2N UMR CNRS 6004, Université de Nantes, Nantes, France
{guillaume.fertin,julien.fradin,geraldine.jean}@univ-nantes.fr

Abstract. Given a vertex-colored arc-weighted directed acyclic graph G, the MAXIMUM COLORFUL SUBTREE problem (or MCS) aims at finding an arborescence of maximum weight in G, in which no color appears more than once. The problem was originally introduced in [2] in the context of *de novo* identification of metabolites by tandem mass spectrometry. However, a thorough analysis of the initial motivation shows that the formal definition of MCS needs to be amended, since the input graph G actually possesses two extra properties, which are so far unexploited. This leads us to introduce in this paper a more precise model that we call MAXIMUM COLORFUL ARBORESCENCE (MCA), and extensively study it in terms of algorithmic complexity. In particular, we show that exploiting the implied color hierarchy of the input graph can lead to polynomial algorithms. We also develop Fixed-Parameter Tractable (FPT) algorithms for the problem, notably using the "dual parameter" ℓ, defined as the number of vertices of G which are *not* kept in the solution.

1 Introduction

Metabolites are small molecules that are involved in cellular reactions, most of them remaining unknown to this date. Consequently, identifying molecular structures of metabolites is a key problem in biology. To this aim, tandem mass spectrometry is one of the most used technologies: in a tandem mass spectrometry experiment, a metabolite is fragmented into smaller molecules. The mass spectrometer then outputs a fragmentation spectrum that consists of a series of peaks, where ideally each peak corresponds to the mass of a fragment. If we are able to best "explain" the spectrum by finding the molecule which corresponds to each peak it contains, then the input metabolite can be infered as well. Such identification can be achieved by comparison with some reference database. However, the databases at hand are largely incomplete. This is why *de novo* interpretation of the fragments, directly from the spectra, is a promising alternative. In [2], Böcker *et al.* initiated such study, where the problem of *de novo* identifying metabolites from tandem mass spectrometry spectra was formally modeled by the MAXIMUM COLORFUL SUBTREE (or MCS) problem. The main ideas behind MCS are as follows. Let m be an unknown metabolite that we want to infer from a tandem mass spectrum s_m. We then do the following: for

© Springer International Publishing AG 2017
T.V. Gopal et al. (Eds.): TAMC 2017, LNCS 10185, pp. 216–230, 2017.
DOI: 10.1007/978-3-319-55911-7_16

each peak p in s_m (that represents a mass), we generate a set of sub-molecules that lie in the same range of masses as p, and we "connect" two sub-molecules whenever one can be obtained from the other by fragmentation. This situation is represented by a directed acyclic graph (DAG) $G = (V, A)$, in which every node $v \in V$ represents a molecule, two nodes u and v are linked by an arc $\{u, v\}$ if one molecule (represented by vertex v) is possibly the result of the fragmentation of another (represented by vertex u), and each vertex possesses a color corresponding to its mass (or better said, its mass range). We also assign a weight function $w : A \to \mathbb{R}$ to the arcs of G. Informally, weights correspond to a confidence degree concerning the fragmentation of a molecule into its sub-molecule, and because w is logarithmic, such weights may be negative. Note that in such a graph, there exists a unique vertex of indegree 0, whose color is unique: this vertex indeed represents one possible candidate for metabolite m. Now the MCS problem, defined in [2] and further studied in [3,10,11,13], is defined as follows: given a DAG $G = (V, A)$, a set of colors C, a coloring function $\chi : V \to C$ and a weight function $w : A \to \mathbb{R}$, find a subtree T of G such that (1) no two vertices of T carry the same color (we then say that T is *colorful*) and (2) T is of maximum weight. Intuitively, a solution to MCS represents the best possible "fragmentation scenario" for metabolite m with respect to spectrum s_m. However, the formalization of the initial problem via MCS does not completely reflect the precise structure of the input. First, it is easy to see that G is not *any* DAG: more precisely, as discussed above, it has a unique root r (i.e. a vertex having indegree 0). Let us then call such DAGs r-DAGs (stands for "rooted DAGs"). Moreover, the coloring function χ is not *any* function. Indeed, since vertices are colored according to the masses of the molecules they represent, there exists a total order $P(C)$ on the set C of colors: for every pair of colors $c_1, c_2 \in C$ we say that c_1 precedes c_2 if the mass represented by c_1 is smaller than the mass represented by c_2. We thus introduce the following notation: given a total order $P(C)$ and an r-DAG $G = (V, A)$, a coloring function $\chi : V \to C$ is called *P-order-preserving* if there does not exist $u, v \in V$ such that (i) $\chi(u)$ precedes $\chi(v)$ in $P(C)$ and (ii) there exists a (directed) path from v to u in G. Finally, by the nature of the initial problem, the output tree T must necessarily contain the root r. Thus, T is formally an arborescence, i.e. a directed graph $T = (V_T, A_T)$ with a designated root r such that there exists only one path from r to any node $v \in V_T$. This leads us to reformulate the MCS problem into the following MAXIMUM COLORFUL ARBORESCENCE (or MCA) problem, which better reflects the initial motivation.

Maximum Colorful Arborescence (MCA)

Input: An r-DAG $G = (V, A)$ rooted at some vertex r, a set C of colors, a total order $P(C)$ on C, a P-order-preserving coloring function $\chi : V \to C$, a weight function $w : A \to \mathbb{R}$.

Output: A colorful arborescence $T = (V_T, A_T)$ rooted at r and of maximum weight $w(T) = \sum_{a \in A_T} w(a)$.

Because the definition of MCA is more accurate, it is legitimate to initiate a developed analysis of the computational complexity of the problem, as done in this paper. In particular, we will see that the fact that χ is P-order-preserving can be positively exploited in some situations. Moreover, since any instance of MCA is also an instance of MCS, any positive result (such as polynomial-time, approximation or FPT algorithm) for MCS also applies to MCA – with a time complexity and/or approximation ratio that may even be improved for MCA. Besides, a negative result for MCS does not necessarily imply the same result for MCA. Altogether, we believe that, by introducing MCA, we work for a better understanding of the initial biological problem. In this paper, we thus study MCA under an algorithmic viewpoint: a first goal is to distinguish tractable instances from intractable ones (which implies focusing on the case where G is an arborescence); a second one is to provide new polynomial and FPT algorithms for the problem.

The paper is organized as follows. In Sect. 2, we introduce notations that will be used throughout the paper. We then show in Sect. 3 that MCA remains hard even for very constrained input instances. In Sect. 4, we take advantage of the previously unexploited order-preserving nature of the coloring function χ to describe a new range of instances that are polynomial-time solvable, and develop new FPT algorithms. Finally, in Sect. 5, we present FPT algorithms for MCA, essentially focused on parameter ℓ, defined as the number of vertices *not* present in the solution. Due to space constraints, some proofs are omitted from the paper.

2   Preliminaries

Notations. For any integer k, we note $[k] = \{1, 2, .., k\}$. For any vertex-colored and arc-weighted r-DAG $G = (V, A)$ given as input of MCA, we let $n = |V|$, $m = |A|$ and r will always denote the root of G. We denote an arc in G from a vertex x to a vertex y by $\{x, y\}$. For all vertices $v \in V$, $N^+(v)$ denotes the set of outneighbors of v, $d^+(v)$ (resp. $d^-(v)$) the outdegree (resp. indegree) of v and Δ^+ the maximum outdegree of G. Moreover, $G[v]$ denotes the induced r-DAG of G rooted in v. When G is an arborescence, for a vertex $v \in V$ we let $f(v)$ be the father (thus unique inneighbor) of v in G. For any subset V' of V, $\chi(V')$ denotes the multiset of colors assigned to the vertices of V', and $\chi^*(V')$ denotes its underlying set. We say that G is *colorful* when $\chi^*(V) = \chi(V)$. For a vertex $v \in V$, we let $d(r, v)$ denote the distance between r and v and $D_r(G) = \max\{d(r, v) : v \in V\}$. If G is the input graph of MCA and C is the associated set of colors, we note by $CH_G = (C, A')$, the *Color Hierarchy Graph* of G, which is defined as follows: for every two colors $c_i, c_j \in C$, $\{c_i, c_j\} \in A'$ iff there exists $x, y \in V$ such that $\chi(x) = c_i$, $\chi(y) = c_j$ and $\{x, y\} \in A$.

The problem MCA^+ denotes the restriction of MCA to r-DAGs with positive weights, and UMCA the restriction of MCA^+ to instances having uniform arc weights, i.e. $w(a) = w \in \mathbb{R}^+$ for all $a \in A$. Note that MCA^+ is of interest only when w is strictly positive, otherwise a trivial solution is just the root r of G.

The problem MCA-x is the restriction of MCA in which any color $c \in C$ appears at most x times in $\chi(V)$. We can also constrain the input instances of MCA both on the weights and on the maximal number of occurrences of a color, and thus combine the abovementioned variants, leading to e.g. the UMCA-x problem.

We say that a problem is FPT (for Fixed-Parameter Tractable) with respect to a given parameter p if it can be solved in time $\mathcal{O}(f(p) \cdot poly(n))$ for some computable function f, i.e. if the exponential part of its complexity depends only on p. Three parameters will be of importance in this paper: $k = |V_T|$ is the order of the solution output by MCA, $\ell = n - |C|$ is the number of vertices that are *not* part of the solution, and s is the number of arcs that need to be removed from CH_G in order to turn it into an arborescence. Finally, we note that although G, the solution arborescence T and the color hierarchy graph CH_G are by definition directed, we will often, for simplicity and when clear from the context, refer to the underlying undirected graph of some graph H (rather than H) in the rest of the paper. For instance, when we talk about MCA "in trees", we actually mean that the underlying undirected graph of G is a tree.

Previous Results. We summarize here known results about MCS, and also note that actually every result mentioned below concerning MCS also applies to MCA. Indeed, MCA being a particular case of MCS, any positive result for MCS also holds for MCA. Moreover, for all negative results below, the MCS instances that are built in the corresponding proofs turn out to be either "direct" MCA instances, or can be transformed into such instances.

MCS is known to be NP-hard even when every arc weight is equal to 1 [2], and it can be seen that the result also applies to UMCA. MCS is also APX-hard on binary trees (a result discussed in [10], refering to the conference version of [6]), a result that also applies to UMCA-2. The only known FPT result for MCS comes from [2], which is itself very much inspired by [12], and consists in a dynamic programming algorithm that runs in $\mathcal{O}^*(3^{|C|})$ time and uses $\mathcal{O}^*(2^{|C|})$ space. If we now take a close look at Theorem 1 in [10] (which reduces an instance H of MAXIMUM INDEPENDENT SET containing n_H vertices and m_H edges into an instance of MCS), it is easily seen that the constructed instance of MCS is also an instance of MCA-1. Moreover, this MCA-1 instance can also be easily transformed into an MCA-2 instance in which G is a tree: for this, it suffices to duplicate every y-vertex from the instance of MCA-1, and assign the same color to each pair of newly created vertices. This allows us to conclude that MCA-1 (resp. MCA-2 in trees) is W[1]-hard when parameterized by the weight w of the solution. We can also conclude that MCA-1 is W[1]-hard when parameterized by $\ell = n - |C|$. We can finally see that if P $\neq$ NP, there is no polynomial-time approximation algorithm achieving a ratio $\mathcal{O}(n^{\frac{1}{2}-\epsilon})$ with $\epsilon > 0$ for MCA-1 (resp. MCA-2 in trees). As a side note, we point out an error in the inapproximation ratio given in Theorem 1 in [10]: indeed, the instance of MCS constructed in the reduction contains $n = \mathcal{O}(n_H^2)$ for both MCA and MCS instances; thus the correct inapproximation ratio of Theorem 1 should be $\mathcal{O}(n^{\frac{1}{2}-\epsilon})$ instead of $\mathcal{O}(n^{1-\epsilon})$.

Our Results. The main results obtained in this paper are summarized in Table 1. They will be developed in the following sections.

Table 1. Overview of the results presented in this paper for the MCA problem and its variants. Here, $k = |V_T|$, Δ^+ is the maximum outdegree of G, $\ell = n - |C|$ and $s = |A'| - |C| + 1$.

	DAGs	Trees
MCA	$\mathcal{O}^*(k!(\Delta^+)^k)$ (Proposition 2) P when CH_G is a tree (Theorem 4) $\mathcal{O}^*(2^s)$ (Corollary 1)	P in caterpillars (Proposition 1) $\mathcal{O}^*(2^\ell)$ (Proposition 4)
MCA-1	W[1]-hard in ℓ (from [10])	
MCA$^+$	$\mathcal{O}^*(2^\ell)$ (Proposition 3)	$\mathcal{O}^*(1.62^\ell)$ (Proposition 5)
UMCA-2	no $\mathcal{O}^*((2 - \epsilon)^\ell)$ (Theorem 6)	no $2^{log^\delta n}$ approx, $\delta < 1$ (Theorem 1) APX-hard, even in superstars (Theorem 2) no $\mathcal{O}(n^{\frac{1}{3} - \epsilon})$ approx. even in comb-graphs (Theorem 3)

3   MCA in Trees

In this section, we focus on MCA in the case where the input graph G is a tree, aiming at determining which tree structures lead to (in)tractable (resp. (in)approximable) results. We start with the following proposition, that applies to the general case of trees.

Theorem 1. *For any $\delta < 1$, UMCA in trees cannot be approximated within $2^{log^\delta n}$ in polynomial time, unless* NP $\subseteq DTIME[2^{poly\ log\ n}]$.

Proof. Dondi *et al.* introduced the MAXIMUM LEVEL MOTIF (or MLM) problem [6], a maximization variant of the GRAPH MOTIF problem [9] dealing with colorful motifs on trees. Besides, MLM incorporates the notion of a *leveled coloring function* $\chi' : V \to C$ for which two vertices can have the same color only if they are at the same distance from the root. The formal definition of MLM is given below.

Maximum Level Motif
Input: A rooted tree $H = (V, E)$, a color set C, a leveled coloring function $\chi' : V \to C$.
Output: A maximum cardinality subset $S \subseteq V$ such that the induced subgraph $H[S]$ is connected and $\chi'(S) \subseteq C$.

Let I be any instance of MLM. We construct an instance I' of MCA as follows: graph G is built on V, and each edge in H is changed into an arc in G between the same vertices: for this, each arc a is oriented from parent to child. We let

$w(a) = 1$ for any arc, and we also apply the same coloring function χ', given as input of MLM, to color the vertices of G. Since χ' is a leveled coloring function, the colors in C are partially ordered. This partial order can thus be extended into a(n arbitrary) total order $P(C)$ such that χ' is P-order-preserving. Thus, I' is a correct UMCA instance. We now show that there exists a solution S of MLM iff there exists a colorful arborescence $T = (V_T, A_T)$ such that $w(T) \geq k - 1$ in G.

($\Rightarrow$) Suppose there exists a solution S of MLM such that $|S| = k$. Let T be the spanning arborescence of S in G, with $V_T = S$. Trivially, T is colorful and of weight $k - 1$. If $r \notin V_T$, we search for a vertex $x \in V_T$ such that $d(r, x) = \min\{d(r, u) : u \in V_T\}$. Let $V_{r,x}$ (resp. $A_{r,x}$) be the set of vertices (resp. arcs) in the path from r to x in G. We construct a new arborescence $T' = (V_{T'}, A_{T'})$, with $V_{T'} = V_T \cup V_{r,x}$ and $A_{T'} = A_T \cup A_{r,x}$. According to χ', $V_{r,x}$ is colorful and each vertex in $V_{r,x}$ has a different color from any of the vertices in V_T. Thus, $V_{T'}$ is colorful.

($\Leftarrow$) Suppose there exists a colorful arborescence $T = (V_T, A_T)$ of weight $k - 1$ in G. Then, we choose $S = V_T$. Trivially, S is colorful and $|S| = k$.

Dondi et al. proved that, under the condition that $\text{NP} \subseteq DTIME[2^{poly\ log\ n}]$, MLM cannot be approximated within $2^{log^\delta n}$ in polynomial time [6]. By linearity of the above reduction, we reach the same conclusion for UMCA in trees. $\square$

From the above result, it seems natural to further restrict the structure of the input tree, in order to draw the line between tractable and intractable cases. Note that if G is a star (i.e., every arc of G starts from the root r), MCA is clearly in P: discard negatively weighted arcs, and for every color $c \in C$, consider all arcs from r to a vertex of color c, and keep the one with maximum weight. Superstars are a natural extension of stars, since they are defined as trees of height 2. Unfortunately, the MCA problem turns out to be hard in superstars, as shown by the following result.

Theorem 2. UMCA-2 *is* APX-*hard, even if G is a superstar.*

Proof. The proof is by reduction from MAX-2-SAT(3), which is known to be APX-hard [1]. It can be seen as an extension of proof of Theorem 1 in [2].

MAX-2-SAT(3)
Input: A set $X = \{x_1, x_2 \dots x_p\}$ of variables, a CNF-formula ϕ on a set of size-2 clauses $\mathcal{C} = \{C_1, C_2 \dots C_q\}$ built from X, such that each variable occurs in at most 3 clauses.
Output: A boolean assignment β of X that satisfies the maximum number of clauses in $\mathcal{C}$.

For every $j \in [q]$, let $l_{j,1}$ and $l_{j,2}$ be the two literals that appear in clause C_j. The reduction is as follows: for any instance of MAX-2-SAT(3), we create a directed superstar $G - (V, A)$ that we can see as a three-levels graph. The root

r is at level 1, two vertices v_i and $\bar{v}_i$ are created for every $i \in [p]$ at level 2, and two vertices $C_{j,1}, C_{j,2}$ are created for every $j \in [q]$ at level 3. There exists an arc from r to every level 2 vertex. For all $i \in [p]$ and $j \in [q]$, there exists an arc from v_i (resp. $\bar{v}_i$) to $C_{j,1}$ if $l_{j,1} = x_i$ (resp. $l_{j,1} = \bar{x}_i$) or from v_i (resp. $\bar{v}_i$) to $C_{j,2}$ if $l_{j,2} = x_i$ (resp. $l_{j,2} = \bar{x}_i$). The coloring function on $V(G)$ is defined as follows: the root r is assigned a unique color; for all $i \in [p]$, vertices v_i and $\bar{v}_i$ are assigned the same color $c(v_i)$; for all $j \in [q]$, vertices $C_{j,1}$ and $C_{j,2}$ are assigned the same color $c(C_j)$. Clearly, each color occurs at most twice in G, and the coloring function is partially ordered (thus can easily be extended to a total order), because any two vertices having the same color lie at the same level. Finally, every arc in G is assigned a weight of 1.

We now show that there exists a boolean assignment β of X that satisfies k clauses in ϕ iff there exists a colorful arborescence $T = (V_T, A_T)$ of weight $w(T) \geq p + k$ in G.

($\Rightarrow$) Suppose there exists an assignment β of X that satisfies k clauses of ϕ. Let $S_T = \{v_i : i \in [p]$ s.t. $x_i = \text{True in } \beta\}$ and $S_F = \{\bar{v}_i : i \in [p]$ s.t. $x_i = \text{False in } \beta\}$. We let $V_T = \{r\} \cup S_T \cup S_F \cup \{C_{j,1} : j \in [q]$ s.t. $f(C_{j,1}) \in (S_T \cup S_F)\} \cup \{C_{j,2} : j \in [q]$ s.t. $f(C_{j,2}) \in (S_T \cup S_F)$ and $f(C_{j,1}) \notin (S_T \cup S_F)\}$ and we define T as the spanning arborescence of V_T. By construction, there cannot exist $j \in [q], h \in \{1, 2\}$ such that $C_{j,h} \in V_T$ and $f(C_{j,h}) \notin V_T$. Thus, T is connected. Moreover, since β satisfies k clauses, there exists k distinct vertices of type $C_{j,h}$ that belong to T, in addition to the p vertices in $(S_T \cup S_F)$ and the root r. Hence, T is clearly colorful and $w(T) = p + k$.

($\Leftarrow$) Suppose there exists a colorful arborescence $T' = (V_{T'}, A_{T'})$ of weight $w(T') \geq p + k$ in G. If $V_{T'}$ does not contain p vertices from level 2, then it is always possible to extend it to a set V_T such that $V_{T'} \subseteq V_T$, V_T is colorful and it contains p vertices from level 2. Let T be the spanning arborescence of V_T. Note that since T is colorful, for any $1 \leq i \leq p$, either v_i or $\bar{v}_i$ is in V_T. Thus, for every $i \in [p]$, if $v_i \in V_T$ (resp. $\bar{v}_i \in V_T$) then we let $x_i = \text{True}$ (resp. $x_i = \text{False}$) in β. We now claim that β satisfies at least k clauses from $\mathcal{C}$. Indeed, if a vertex $C_{j,h}$, $j \in [q]$ and $h \in \{1, 2\}$ is in V_T, then necessarily $f(C_{j,h}) \in V_T$ and, by construction, C_j is satisfied by β. Moreover, $C_{j,1}$ and $C_{j,2}$ cannot both belong to V_T because T is colorful. Since T has a weight $w(T) \geq p + k$, V_T must contain at least k vertices from level 3, which means that β satisfies at least k clauses.

To conclude the proof, recall that $q \geq k$ since no more than q clauses can be satisfied. Notice also that $2q \leq 3p$ as every variable appears at most three times in ϕ, while every clause is of size 2. This gives us $p \geq \frac{2q}{3} \geq \frac{2k}{3}$ and $p + k \geq \frac{2k}{3} + k \geq \frac{5k}{3}$. Thus, there exists an assignment β that satisfies at least k clauses of ϕ iff there exists a colorful arborescence $T = (V_T, A_T)$ of weight $w(T) \geq \frac{5k}{3}$ in G. The linearity of the reduction combined with the APX-hardness of MAX-2-SAT(3) shows APX-hardness of UMCA-2, even on superstars.   □

The previous result shows that even in trees with height 2, MCA remains APX-hard. Another option consists in constraining the maximum degree of the input tree, which motivates the study of comb-graphs. If $d(v)$ denotes the degree of a vertex v in a graph, a comb-graph is defined as a tree for which $d(v) \leq 3$ for any $v \in V$, and where all vertices of degree 3 lie on a single simple path. Unfortunately, we show in Theorem 3 that UMCA-2 remains APX-hard (with a large inapproximability ratio) even when the input tree is a comb-graph.

Theorem 3. UMCA-2 *cannot be approximated within* $\mathcal{O}(n^{\frac{1}{3}-\epsilon})$, *with* $\epsilon > 0$, *even if G is a comb-graph.*

Similarly to Theorem 1, Theorem 3 is about large inapproximability ratios. It is stronger than Theorem 1 as it applies to UMCA-2 on very specific trees; however its inapproximability ratio is lower, the one of Theorem 1 almost reaching n when δ tends to 1.

Another way of restricting the input tree structure is to consider trees that are "close to paths". When G is a path, it can be easily seen that MCA is in P. The next step is to study caterpillars, which are trees that become paths after removal of their leaves. Notice that a superstar becomes a star, i.e. a special case of caterpillar, after removal of its leaves. Moreover, MCA is APX-hard for superstars (see Theorem 2), and MCA is in P for stars. As shown below, more generally, MCA in caterpillars is in P. Thus, the following theorem allows us to draw a line between intractable and tractable instances for MCA in trees.

Proposition 1. MCA *in caterpillars is in* P.

Proof. The main purpose of the algorithm we present in this proof is to show polynomiality of the problem when G is a caterpillar; no particular effort is made here on optimizing the running time. Let $S = \{v \in V : d^+(v) \neq 0\}$ be the spine of G. Clearly, $G[S]$ is connected according to the definition of a caterpillar. The proposed algorithm works as follows. First, we generate all colorful subsets $S' \in S$ such that $r \in S'$ and $G[S']$ is connected. Second, for each such S', we denote $N^+(S') = \{v \in N^+(u) : u \in S' \text{ and } v \notin S\}$ and we proceed as follows: for all colors $c \in C \setminus \chi(S')$, take $x \in N^+(S')$ of color c with the maximum weighted incoming arc a_x and add x to S' only if $w(a_x) > 0$. From this newly built set S'', we compute the spanning arborescence T'', and finally output the arborescence that reaches the maximum weighted among all S''. Clearly, the algorithm is correct because we generated all possible connected and colorful structures built from the spine (together with r if r does not belong to the spine). From this, extending to the leaves of the caterpillar can be achieved greedily. There exists $\mathcal{O}(n^2)$ subsets of vertices S' and each S' is treated in polynomial time, thus the whole algorithm is polynomial.

4   A Closer Look at the Color Hierarchy Graph CH_G

One major difference between MCS and MCA lies in the P-order-preserving character of the coloring function χ according to some total order $P(C)$ on the set C of colors. In this section, we exploit this fact by focusing on the structure of the Color Hierarchy Graph CH_G. The main result of the section is the following.

Theorem 4. MCA *can be solved in polynomial time whenever* CH_G *is a tree.*

Proof. The algorithm ARBO we designed for the problem is formally described in Algorithm 1. ARBO computes a colorful arborescence $T = (V_T, A_T)$ of maximum weight $w(T)$ for any input $G = (V, A)$ such that CH_G is a tree. In the following, for any $v \in V$, we denote $T_v = (V_v, A_v)$ the colorful arborescence of maximum weight $w(T_v)$ that is rooted in v and computed by ARBO. Moreover, for each such v, we denote by $post(v) = \{u \in N^+(v) : u \in T_v\}$ the set of outneighbors of v that belong to V_T. Finally, let $H(T_v) = D_v(T_v)$ be the height of T_v.

We create two sets of vertices $S = \emptyset$ and $L_S = \{v \in V : v \text{ is a leaf of } G[V \setminus S]\}$ that are initialized in line (1) of Algorithm 1. For each $v \in V$, lines (4–14) describe how to *treat* a vertex $v \in V$: ARBO first computes $post(v)$, then $w(T_v) = \sum_{u \in post(v)} w(T_u) + w(\{v, u\})$. For each such vertex v, lines (2) and (16–17) ensure that v is treated only after all vertices $u \in N^+(v)$ have been treated. Finally, ARBO recursively builds T_r from $post(r)$ in lines (18–25). As each vertex $v \in V$ is treated in polynomial time, the whole algorithm is thus polynomial.

Now, we show that $w(T) = w(T_r)$. For this, we prove by induction that for all $v \in V$ such that $H(T_v) = h, 0 \le h \le H(T_r), T_v$ is *optimal*, i.e. connected, colorful and of maximum weight among all colorful arborescences rooted in v. First, for all vertices $v \in V$ such that $H(T_v) = 0$, T_v is composed of a single vertex v and is thus trivially optimal. Let us now assume that T_v is optimal for all vertices $v \in V$ such that $H(T_v) = h$ with $0 \le h \le H(T_r) - 1$. We are interested in all the vertices $v \in V$ such that $H(T_v) = h+1$. Recall that $V_v = \{r\} \cup \{V_u : u \in N^+(v)\}$ and that $A_v = \{\{v, u\} : u \in post(v)\} \cup \{A_u : u \in post(v)\}$. Thus, T_v is clearly connected as we assumed that T_u is connected for each $u \in N^+(v)$. Moreover, as CH_G is a tree, there cannot exist a colorful set $\{x, y, z\} \subseteq V$ such that $\{x, z\} \in A$ and $\{y, z\} \in A$. In addition to the fact that $post(v)$ is colorful for any $v \in V$, this implies that $\bigcap_{u \in post(v)} V_u = \emptyset$ and thus, recursively, that T_v is a colorful arborescence. Finally, suppose there exists a colorful arborescence $T'_v = (V'_v, A'_v)$ rooted in v such that $w(T'_v) > w(T_v)$. We denote $post'(v)$ the set of outneighbors of v in T'_v. Observe that:

$$w(T_v) = \sum_{c \in \chi(N^+(v))} \max\{0, w(T_u) + w(\{v, u\}) : u \in N^+(v) \text{ and } \chi(u) = c\} \quad (1)$$

This implies that $post'(v) \ne post(v)$ iff there exists at least a vertex $u \in N^+(v)$ such that T_u is not optimal, which contradicts our assumption. As a consequence, T_v is optimal for all $v \in V$ and thus $w(T) = w(T_r)$.   □

We recall that s is defined as the number of arcs that need to be removed from $CH_G = (C, A')$ in order for CH_G to become a tree. Now, let $C' = \{c \in C \text{ s.t. } d^-(c) > 1\}$ be the set of colors in CH_G that have indegree strictly more than one. Clearly, CH_G is not a tree whenever C' is not empty. In the following, for any $C' \ne \emptyset$, we let $p = \min\{d^-(c) : c \in C'\}$.

Algorithm 1. ARBO

Input: A MCA instance G s.t. CH_G is a tree.
Output: A colorful arborescence $T = (V_T, A_T)$ of maximum weight.

```
 1: S ← ∅; L_S ← {v ∈ V s.t. v is a leaf of G[V \ S]};
 2: while (S ≠ V) do                               ▷ Compute post(v) and w(T_v)
 3:     for all (v ∈ L_S) do                       ▷ Ensure N⁺(v) already treated
 4:         for all (c ∈ χ(N⁺(v))) do
 5:             best ← 0; z ← v;
 6:             for all (u ∈ N⁺(v) s.t. χ(u) = c) do
 7:                 if (w(T_u) + w({v, u}) > best) then
 8:                     best ← w(T_u) + w({v, u}); z ← u;
 9:                 end if
10:             end for
11:             if (best > 0) then
12:                 w(T_v) ← w(T_v) + best; post(v) ← post(v) + {z};
13:             end if
14:         end for
15:     end for
16:     S ← S + L_S; L_S ← {v ∈ V s.t. v is a leaf of G[V \ S]};
17: end while
18: S ← {r}; L_S ← ∅; V_T ← {r};                   ▷ Recursively compute T_r
19: while (S ≠ ∅) do
20:     for all (v ∈ S) do
21:         L_S ← L_S + post(v); V_T ← V_T + post(v);
22:     end for
23:     S ← L_S; L_S ← ∅;
24: end while
25: return T                                        ▷ T is the spanning arborescence of V_T
```

Theorem 5. MCA *can be solved in time* $\mathcal{O}^*(p^{\frac{s}{p-1}})$.

Proof. We design a branching algorithm that recursively removes arcs from a set Z, where initially $Z = A(CH_G)$, thus producing a graph CH'_G. For every color $c \in C'$, we recursively branch on the $d^-(c)$ different cases where only one incoming arc of c is not removed from Z. At the end of these branching steps, each color $c \in C$ has an indegree 1 and thus CH'_G is a tree. We then create a graph $G' = (V, A_Z)$ with $A_Z = A \backslash \{\{x, y\} \in A : \{\chi(x), \chi(y)\} \notin Z\}$. Informally, we build G' such that CH'_G is the Color Hierarchy Graph of G'. Hence, by Theorem 4, we know that computing a colorful arborescence of maximum weight in G' is polynomial-time solvable. As a consequence, the above described algorithm is correct.

The size of the induced search tree T_s produced by the algorithm is $|T_s| = \prod_{c \in C'} d^-(c)$. Assuming C' is not empty, we now search for the lowest real X such that the inequality (1) $d^-(c) \leq X^{d^-(c)-1}$ holds for all colors $c \in C'$. From (1), we have $\log(d^-(c)) \leq (d^-(c) - 1) \cdot \log(X)$, thus $X \geq c^{\frac{\log(d^-(c))}{d^-(c)-1}}$, which gives us $X \geq d^-(c)^{\frac{1}{d^-(c)-1}}$. The corresponding function $f(x) = x^{\frac{1}{x-1}}$ is monotonously decreasing for all $x \in [2; +\infty]$. This implies that we can set X to $p^{\frac{1}{p-1}}$. Recall that $|T_s| \leq \prod_{c \in C'} X^{d^-(c)-1}$ and that $s = \sum_{c \in C'} d^-(c) - 1$, which leads to $|T_s| \leq p^{\frac{s}{p-1}}$. $\qquad \square$

For instance, if $p = 3$, then by Theorem 5 we obtain a running time of $\mathcal{O}^*(1.733^s)$ for solving MCA. In general, we always have that $p \geq 2$ for graphs G such that CH_G is not a tree. Thus, by setting p to 2, we obtain the following "universal" corollary.

Corollary 1. MCA *can be solved in time* $\mathcal{O}^*(2^s)$.

5   FPT Results with Respect to Parameters k and ℓ

In this section, we study parameters $k = |V_T|$ and $\ell = n - |C|$, introduced in Sect. 2. Parameter k ("size of the solution") is a classical parameter for FPT studies, and it is natural to ask whether MCA is FPT in k. At this point, we do not have an answer to this question. However, we have the following proposition.

Proposition 2. MCA *is FPT when parameterized by* (k, Δ^+).

Proof. The proof is by a pure combinatorial enumeration algorithm, which works as follows: first, we generate the set X_k of all order k arborescences that are rooted in r and contained in G; second, we output the colorful arborescence of maximum weight in X_k, or otherwise conclude that we have a NO-instance. Since the latter part is polynomial, it suffices to show that $|X_k|$ only depends on k and Δ^+. First, note that we can remove from G all vertices $v \in V$ such that $d(r, v) > k$, as they cannot belong to any arborescence of order k that is rooted in r. Therefore, we can always assume that G is such that $D_r(G) \leq k$. Let $T_x = (V_x, A_x)$ be an arborescence in X_k with $V_x = \{v_i, 1 \leq i \leq k\}$. Informally, we build T_x vertex by vertex, starting from $v_1 = r$. Thus, v_2 is one of the $d^+(v_1) \leq \Delta^+$ outneighbors of v_1. By induction, it can be shown that for any $2 \leq p \leq k$, there are $(\sum_{i=1}^{p-1} d^+(v_i)) - (p - 2) \leq (p - 1)\Delta^+ - (p - 2)$ choices for vertex v_p: indeed, v_p being a child of a previously considered vertex, it cannot be taken among these already considered vertices. The above argument shows that $|X_k| \leq \prod_{i=2}^{k}((i - 1) \cdot \Delta^+ - (i - 2))$ is upper bounded by $\prod_{i=2}^{k}(i - 1) \cdot \Delta^+ = \mathcal{O}(k!(\Delta^+)^k)$. This proves our proposition. $\qquad \square$

Parameter $\ell = n - |C|$ turns out to be of particular interest for the MCA problem. Indeed, in the tests run in [10] on real datasets, the ratio $r = \frac{n}{|C|}$ can be as low as 1.03. Unfortunately, as noted in Sect. 2, MCA-1 parameterized by ℓ is W[1]-hard. However, constraining the input instances allows us to derive several positive results. For instance, we can show that MCA$^+$ is FPT in ℓ.

Proposition 3. *The* MCA$^+$ *problem can be solved in* $\mathcal{O}^*(2^\ell)$.

Proof. We design a recursive branching algorithm based on the colors of the input graph G. We first let $S = V$. If S is not colorful, we consider $u, v \in S$ such that $\chi(u) = \chi(v)$ and recursively branch on two cases: either u or v is removed from S. Recall that r has a unique color, therefore it is never removed. We repeat this branching step until S is colorful. For each set S that we obtain, let G_S be the connected component containing r in the induced r-DAG $G[S]$. We thus look for a maximum weighted spanning arborescence in each such G_S – this leads to the best solution for G_S, as all weights are positive. Such arborescence can be computed in polynomial time [4]. Clearly, the above described algorithm is correct, and its running time is exponential only in the number of nodes of the induced search tree T_S. Since T_S is binary and of height $\ell = n - |C|$, our algorithm runs in $\mathcal{O}^*(2^\ell)$. □

Now, if for MCA we constrain the input r-DAG to be a tree, we obtain the following result.

Proposition 4. MCA *in trees can be solved in* $\mathcal{O}^*(2^\ell)$ *time.*

Proof. We design another recursive branching algorithm with a set $S = V$. While S is not colorful, we consider $u, v \in S$ such that $\chi(u) = \chi(v)$ and recursively branch on two cases: either $G[u]$ or $G[v]$ is removed from S (instead of either u or v in proof of Proposition 3 above). Clearly, for each set S we obtain, $G[S]$ is a colorful tree. Thus, notice that $CH_{G[S]}$ is itself a tree. As a consequence, by Theorem 4, we can compute a maximum weighted arborescence in $G[S]$ in polynomial time. The search tree is binary as in the previous proof and it has a maximum height $\ell = n - |C|$. Hence our algorithm runs in $\mathcal{O}^*(2^\ell)$. □

Finally, we show that Proposition 4 can be improved when all arcs $a \in A$ have positive weights.

Proposition 5. MCA$^+$ *in trees can be solved in* $\mathcal{O}^*(1.62^\ell)$ *time.*

Proof. We improve the branching algorithm discussed in proof of Propositions 3 and 4, by using a different branching procedure. Let $S = V$ and let us apply the following branching rule: if there exist $u, v \in S$ such that (i) $\chi(u) = \chi(v)$ and (ii) $|N^+(u)| > 0$ or $|N^+(v)| > 0$, then we branch on two cases: either $G[u]$ or $G[v]$ is removed from S. We repeat this branching procedure until it can no longer be applied on S. We now denote U_S the set of vertices having a unique color in S. Note that because of condition (ii), two vertices $u, v \in S$ can have the same color only if they are both leaves of $G[S]$, and thus $G[U_S]$ is connected. Recall that G is a tree and that for any arc $a \in A$, its weight $w(a)$ is positive. Thus, $G[U_S]$ is necessarily contained in a maximum colorful arborescence $T = (V_T, A_T)$ built from $G[S]$. We now need to compute T from S: we start by taking in T all vertices from U_S. Then, for every color $c \in \chi(S) \backslash \chi(U_S)$, we add to V_T the vertex $v \in S$ of color c such that $w(\{f(v), v\})$ is maximum – note that $f(v)$ necessarily belongs to U_S. Finally, T is defined as the tree induced by V_T. It can be easily seen that T is connected, colorful and of maximum weight, which ensures correctness of our algorithm. The computational complexity of our algorithm derives from the fact that, at each step, if $|G[u]| = 1$ (resp. $|G[v]| = 1$) then $|G[v]| \geq 2$ (resp. $|G[u]| \geq 2$), and thus the branching vector is $(1, 2)$, which leads to a 1.62^{ℓ} algorithm[1].   □

We now turn to proving a lower bound on the computational complexity of MCA with respect to ℓ. In particular, our result proves that the FPT algorithm given in Proposition 3 is optimal for MCA$^+$.

Theorem 6. *The* UMCA-2 *problem cannot be solved in time* $\mathcal{O}^*((2-\epsilon)^{\ell})$ *unless the Strong Exponential-Time Hypothesis fails.*

Proof. First note that the Strong Exponential-Time Hypothesis (SETH) states that the CNF-SAT problem defined on p variables cannot be solved in time $\mathcal{O}^*((2-\epsilon)^p)$ for any $\epsilon > 0$ [8]. The reduction from CNF-SAT we present here is adapted from proof of Theorem 1 in [7]. We first formally define CNF-SAT.

CNF-SAT
Input: A set $X = \{x_1, x_2 \ldots x_p\}$ of variables, a CNF-formula ϕ on a set $\mathcal{C} = \{C_1, C_2 \ldots C_q\}$ of clauses built from X.
Output: An assignment β of each $x_i \in X$ that satisfies ϕ.

Starting from any instance ϕ of CNF-SAT, we build an instance of UMCA-2 in the form of a three-level graph G. First, let r be the root at level 1. For each variable $x_i \in X$ with $1 \leq i \leq p$, we create two vertices v_i and $\bar{v}_i$ at level 2. For each clause $C_j \in \mathcal{C}$ with $1 \leq j \leq q$, we create a vertex z_j at level 3. We then add an arc from r to v_i and to $\bar{v}_i$ for all $i \in [p]$. There is also an arc from v_i (resp. $\bar{v}_i$) to z_j iff literal x_i (respectively $\bar{x}_i$) appears in clause C_j, for all $i \in [p]$ and for all $j \in [q]$. Every level 1 and level 3 vertex is assigned a unique color. At level 2, for all $i \in [p]$, v_i and $\bar{v}_i$ share the same color c_i. Thus, all colors $c \in C$ can appear at most twice, and these colors can easily be partially ordered (and thus totally ordered) based on their level in the graph. Finally, the weight of every arc is 1, and it can be seen that G is indeed an instance of UMCA-2. We now show that there exists an assignment β that satisfies ϕ iff there exists a colorful arborescence of weight $p + q$ (and thus of order $p + q + 1$) in G.

($\Rightarrow$) Suppose there exists an assignment `True/False` of each $x_i \in X$, say β, that satisfies ϕ. Let I_T (resp. I_F) be the set of indices $i \in [p]$ such that x_i is set to `True` (resp. `False`) by β. Let $S = \{r\} \cup \{v_i$ for all $i \in I_T\} \cup \{\bar{v}_i$ for all $i \in I_F\} \cup \{z_j$ for all $j \in [q]\}$. Necessarily, $G[S]$ is connected: first, r is connected to every level-2 vertex; second, a vertex z_j corresponds to a clause satisfied by some x_i (resp. $\bar{x}_i$), and by definition $G[S]$ contains v_i (resp. $\bar{v}_i$), which is connected to z_j. Now, let $T = (V_T, A_T)$ be a spanning arborescence of $G[S]$. Clearly, T is colorful of total weight $p + q$.
($\Leftarrow$) Suppose there exists a colorful arborescence $T = (V_T, A_T)$ of weight $p+q$, thus of order $p + q + 1$. Note that T contains at most p vertices from level 2,

---

[1] For an introduction to the analysis of branching vectors, see e.g. [5].

and thus at least q vertices from level 3. However, level 3 contains *exactly* q vertices. Thus, V_T must be composed of the root, exactly p vertices at level 2 and exactly q vertices at level 3. Since level 2 is composed of $2p$ vertices where each color appears twice, and since T is colorful, for all $i \in [p]$, either v_i or $\bar{v}_i$ is in V_T. The assignment β is thus the following: if $v_i \in V_T$ (resp. $\bar{v}_i \in V_T$) then x_i is set to True (resp. False). Since T is connected, then for any z_j with $j \in [q]$, there exists $f(z_j) \in T$, which means that every clause in ϕ is satisfied by at least one literal in β.

Hence, since $n = 2p + q + 1$ and $|C| = p + q + 1$, we have that $\ell = n - |C| = p$. As a consequence, every algorithm running in time $\mathcal{O}^*((2 - \epsilon)^\ell)$ for UMCA-2 would imply an algorithm running in time $\mathcal{O}^*((2 - \epsilon)^p)$ for CNF-SAT, which would contradict SETH.   □

6   Conclusion

In this paper, we introduced the MCA problem, a constrained version of the MCS problem, where the input must be an r-DAG, and the coloring function must be P-order-preserving for some total order P on the colors. MCA is designed to better represent the initial motivation of *de novo* inference of metabolites from tandem mass spectra, and leads to better-shaped algorithms. Although we showed that MCA remains APX-hard even for constrained inputs, we also showed that it is possible to take advantage of the order-preserving nature of the coloring function to describe new polynomial-time solvable and FPT cases – notably, MCA is in P when the color hierarchy graph CH_G is a tree. It remains an open problem whether other polynomial-time algorithms based on the structure of CH_G can be designed.

We also introduced parameter $\ell = n - |C|$, that we consider to be promising as experiments show that in real data, n and $|C|$ tend to be close. Although MCA is W[1]-hard when parameterized by ℓ, we gave several positive results, in the form of FPT algorithms, for some variants of MCA. Some of these may be improved, and other results taking ℓ as a parameter are certainly worth investigating too. Finally, we conclude this paper by the two following questions, that concern two other parameters: is MCA FPT in the number k of vertices of the solution? Is it possible to devise an FPT algorithm with parameter $|C|$ running faster than $\mathcal{O}^*(3^{|C|})$?

References

1. Ausiello, G., Crescenzi, P., Gambosi, G., Kann, V., Marchetti-Spaccamela, A., Protasi, M.: Complexity and Approximation: Combinatorial Optimization Problems and Their Approximability Properties. Springer-Verlag, Heidelberg (1999)
2. Böcker, S., Rasche, F.: Towards de novo identification of metabolites by analyzing tandem mass spectra. In: Proceedings of Seventh European Conference on Computational Biology and Bioinformatics, ECCB 2008, vol. 24, no. 16, pp. i49–i55 (2008)

3. Böcker, S., Rasche, F., Steijger, T.: Annotating fragmentation patterns. In: Salzberg, S.L., Warnow, T. (eds.) WABI 2009. LNCS, vol. 5724, pp. 13–24. Springer, Heidelberg (2009). doi:10.1007/978-3-642-04241-6_2

4. Chu, Y.J., Liu, T.H.: On shortest arborescence of a directed graph. Sci. Sinica **14**(10), 1396 (1965)

5. Cygan, M., Fomin, F.V., Kowalik, L., Lokshtanov, D., Marx, D., Pilipczuk, M., Pilipczuk, M., Saurabh, S.: Parameterized Algorithms. Springer, Cham (2015)

6. Dondi, R., Fertin, G., Vialette, S.: Complexity issues in vertex-colored graph pattern matching. J. Discrete Algorithms **9**(1), 82–99 (2011)

7. Fertin, G., Komusiewicz, C.: Graph Motif problems parameterized by dual. In: Grossi, R., Lewenstein, M. (eds.) 27th Annual Symposium on Combinatorial Pattern Matching, CPM 2016. LIPIcs, vol. 54, pp. 7:1–7:12. Schloss Dagstuhl - Leibniz-Zentrum fuer Informatik (2016)

8. Impagliazzo, R., Paturi, R., Zane, F.: Which problems have strongly exponential complexity? J. Comput. Syst. Sci. **63**(4), 512–530 (2001)

9. Lacroix, V., Fernandes, C.G., Sagot, M.: Motif search in graphs: application to metabolic networks. IEEE/ACM Trans. Comput. Biol. Bioinform. **3**(4), 360–368 (2006)

10. Rauf, I., Rasche, F., Nicolas, F., Böcker, S.: Finding maximum colorful subtrees in practice. J. Comput. Biol. **20**(4), 311–321 (2013)

11. Scheubert, K., Hufsky, F., Rasche, F., Böcker, S.: Computing fragmentation trees from metabolite multiple mass spectrometry data. J. Comput. Biol. **18**(11), 1383–1397 (2011)

12. Scott, J., Ideker, T., Karp, R.M., Sharan, R.: Efficient algorithms for detecting signaling pathways in protein interaction networks. J. Comput. Biol. **13**(2), 133–144 (2006)

13. White, W.T.J., Beyer, S., Dührkop, K., Chimani, M., Böcker, S.: Speedy colorful subtrees. In: Xu, D., Du, D., Du, D. (eds.) COCOON 2015. LNCS, vol. 9198, pp. 310–322. Springer, Cham (2015). doi:10.1007/978-3-319-21398-9_25

Incompleteness Theorems, Large Cardinals, and Automata over Finite Words

Olivier Finkel[✉]

Institut de Mathématiques de Jussieu - Paris Rive Gauche,
CNRS et Université Paris 7, Paris, France
Olivier.Finkel@math.univ-paris-diderot.fr

Abstract. We prove that one can construct various kinds of automata over finite words for which some elementary properties are actually independent from strong set theories like $T_n =:$ **ZFC**+ "There exist (at least) n inaccessible cardinals", for integers $n \geq 0$. In particular, we prove independence results for languages of finite words generated by context-free grammars, or accepted by 2-tape or 1-counter automata. Moreover we get some independence results for weighted automata and for some related finitely generated subsemigroups of the set $\mathbb{Z}^{3 \times 3}$ of 3-3 matrices with integer entries. Some of these latter results are independence results from the Peano axiomatic system **PA**.

Keywords: Automata and formal languages · Logic in computer science · Finite words · Context-free grammars · 2-tape automaton · Post correspondence problem · Weighted automaton · Finitely generated matrix subsemigroups of $\mathbb{Z}^{3 \times 3}$ · Models of set theory · Incompleteness theorems · Large cardinals · Inaccessible cardinals · Independence from the axiomatic system "**ZFC** + there exist n inaccessible cardinals" · Independence from Peano Arithmetic

1 Introduction

We pursue in this paper a study of the links between automata theory and set theory we begun in previous papers [Fin09,Fin11,Fin15]

In [Fin09] we proved a surprising result: the topological complexity of an ω-language accepted by a 1-counter Büchi automaton, or of an infinitary rational relation accepted by a 2-tape Büchi automaton, is not determined by the axiomatic system **ZFC**; notice that here the topological complexity refers to the location of an ω-language in hierarchies, like Borel or Wadge hierarchies, in the Cantor space of infinite words over a finite alphabet Σ, and one assume, as usually, that **ZFC** is consistent and thus has a model. In particular, there is a 1-counter Büchi automaton $\mathcal{A}$ (respectively, a 2-tape Büchi automaton $\mathcal{B}$) and two models $\mathbf{V}_1$ and $\mathbf{V}_2$ of **ZFC** such that the ω-language $L(\mathcal{A})$ (respectively, the infinitary rational relation $L(\mathcal{B})$) is Borel in $\mathbf{V}_1$ but not in $\mathbf{V}_2$. We have

© Springer International Publishing AG 2017
T.V. Gopal et al. (Eds.): TAMC 2017, LNCS 10185, pp. 231–246, 2017.
DOI: 10.1007/978-3-319-55911-7_17

proved in [Fin11] other independence results, showing that some basic cardinality questions on automata reading infinite words actually depend on the models of **ZFC**.

The next step in this research project was to determine which properties of automata actually depend on the models of **ZFC**, and to achieve a more complete investigation of these properties.

Recall that a large cardinal in a model of set theory is a cardinal which is in some sense much larger than the smaller ones. This may be seen as a generalization of the fact that ω is much larger than all *finite* cardinals. The inaccessible cardinals are the simplest such large cardinals. Notice that it cannot be proved in **ZFC** that there exists an inaccessible cardinal, but one usually believes that the existence of such cardinals is consistent with the axiomatic theory **ZFC**. The assumed existence of large cardinals have many consequences in Set Theory as well as in many other branches of Mathematics like Algebra, Topology or Analysis, see [Jec02].

In [Fin15], we recently proved that there exist some 1-counter Büchi automata $\mathcal{A}_n$ for which some elementary properties are independent of theories like $T_n =:$ **ZFC** + "There exist (at least) n inaccessible cardinals", for integers $n \geq 1$. We first prove that "$L(\mathcal{A}_n)$ is Borel", "$L(\mathcal{A}_n)$ is arithmetical", "$L(\mathcal{A}_n)$ is ω-regular", "$L(\mathcal{A}_n)$ is deterministic", and "$L(\mathcal{A}_n)$ is unambiguous" are equivalent to the consistency of the theory T_n (denoted $\mathrm{Cons}(T_n)$). This implies that, if T_n is consistent, all these statements are provable from **ZFC** + "There exist (at least) $n + 1$ inaccessible cardinals" but not from **ZFC** + "There exist (at least) n inaccessible cardinals".

We prove in this paper that independence results, even from strong set theories with large cardinals, occur in the theory of various automata over finite words, like 1-counter automata, pushdown automata (equivalent to context-free grammars), 2-tape automata accepting finitary rational relations, weighted automata. We first show that if T is a given recursive theory then there exists an instance of the Post Correspondence Problem (denoted PCP), constituted of two n-tuples $(x_1, x_2, \ldots, x_n)$ and $(y_1, y_2, \ldots, y_n)$ of non-empty words over a finite alphabet Γ, which has no solution if and only if T is consistent. In other words the theory T is consistent if and only if there does not exist any non-empty sequence of indices $i_1, i_2, \ldots, i_k$ such that $x_{i_1} x_{i_2} \cdots x_{i_k} = y_{i_1} y_{i_2} \cdots y_{i_k}$. This allows to find many elementary properties of some pushdown automata, context-free grammars, or 2-tape automata, which are independent from **ZFC** or from some strong theory in the form **ZFC** + "There exist some kind of large cardinals", since many properties of these automata are proved to be undecidable via some effective reductions of the PCP to these properties.

For instance we prove that, for every integer $n \geq 0$, there exist 2-tape automata $\mathcal{A}_n$, $\mathcal{B}_n$, $\mathcal{C}_n$, and $\mathcal{D}_n$, accepting subsets of $A^\star \times B^\star$, for two alphabets A and B, such that $\mathrm{Cons}(T_n)$ is equivalent to each of the following items: (1) $L(\mathcal{A}_n) \cap L(\mathcal{B}_n) = \emptyset$; (2) $L(\mathcal{C}_n) = A^\star \times B^\star$; (3) "$L(\mathcal{D}_n)$ is accepted by a *deterministic* 2-tape automaton"; (4) "$L(\mathcal{D}_n)$ is accepted by a *synchronous* 2-tape automaton". In particular, if **ZFC** + "There exist (at least) n inaccessible cardinals" is consistent, then each of the properties of these 2-tape automata given by Items (1)–(4) is provable from **ZFC** + "There exist (at least) $n + 1$ inaccessible cardinals" but not from **ZFC** + "There exist (at least) n inaccessible cardinals".

We also prove some independence results for weighted automata, via some independence results for finitely generated matrix subsemigroups of $\mathbb{Z}^{3\times 3}$. Notice that in this context we also obtain results of independence from Peano Arithmetic which make sense since in the context of finite words or of integer matrices everything can be formalized in first-order arithmetic. For instance we show that there exists a finite set of matrices $M_1, M_2, \ldots, M_n \in \mathbb{Z}^{3\times 3}$, for some integer $n \geq 1$, such that: (1) "the subsemigroup of $\mathbb{Z}^{3\times 3}$ generated by these matrices does not contain the zero matrix", and (2) "The property (1) is not provable from **PA**".

These results seem of more concrete mathematical nature than the fact that Cons(**PA**) is an arithmetical statement which is true but unprovable from **PA**. Indeed although our results follow from Gödel's Second Incompleteness Theorem, they express some properties about some natural and simple mathematical objects: the finitely generated subsemigroups of the semigroup $\mathbb{Z}^{3\times 3}$ of 3-3-matrices with integer entries.

This could be compared to the fact that if **PA** (respectively, **ZFC**) is consistent then there is a polynomial $P(x_1, \ldots, x_n)$ which has no integer roots, but for which this cannot be proved from **PA** (respectively, **ZFC**); this result can be inferred from Matiyasevich's Theorem, see [EFT94, End of Chap. 10.7]. The above results could also be compared with other independence results obtained by Kanamori and McAloon [KM87].

Notice that we recently discovered that in older papers it had been noted that undecidability and incompleteness in automata theory were intimately related and that one could for instance obtain some results about automata which are true but unprovable in some recursive theory extending Peano Arithmetic like **ZFC**, [Har85, JY81]. However the results presented here, although they are not very difficult to prove, exhibit in our opinion the following novelties:

1. We obtain results of a different kind: we show that a great number of elementary properties of automata over finite words, are actually independent from strong set theories.
2. We show how we can effectively construct some automata, like 1-counter or 2-tape automata, for which many elementary properties *reflect the scale of a hierarchy of large cardinals axioms like* "There exist (at least) n inaccessible cardinals" for integers $n \geq 1$.
3. We show how we can use Post Correspondence Problem to get simple combinatorial statements about finite words which are independent from strong set theories.

Altogether we think that the collection of results presented in this paper will be of interest for computer scientists and also for set theorists.

The paper is organized as follows. We recall some notions and results of set theory in Sect. 2. We prove some independence results for various kinds of automata over finite words in Sect. 3. Concluding remarks are given in Sect. 4.

2   Some Results of Set Theory

We now recall some basic notions of set theory which will be useful in the sequel, and which are exposed in any textbook on set theory, like [Kun80, Jec02].

The usual axiomatic system **ZFC** is Zermelo-Fraenkel system **ZF** plus the axiom of choice **AC**. The axioms of **ZFC** express some natural facts that we consider to hold in the universe of sets. For instance a natural fact is that two sets x and y are equal iff they have the same elements. This is expressed by the *Axiom of Extensionality*:

$$\forall x \forall y \, [\, x = y \leftrightarrow \forall z (z \in x \leftrightarrow z \in y) \,].$$

Another natural axiom is the *Pairing Axiom* which states that for all sets x and y there exists a set $z = \{x, y\}$ whose elements are x and y:

$$\forall x \forall y \, [\, \exists z (\forall w (w \in z \leftrightarrow (w = x \vee w = y)))].$$

Similarly the *Powerset Axiom* states the existence of the set $\mathcal{P}(x)$ of subsets of a set x. Notice that these axioms are first-order sentences in the usual logical language of set theory whose only non logical symbol is the membership binary relation symbol $\in$. We refer the reader to any textbook on set theory for an exposition of the other axioms of **ZFC**.

A model $(\mathbf{V}, \in)$ of an arbitrary set of axioms $\mathbb{A}$ is a collection $\mathbf{V}$ of sets, equipped with the membership relation $\in$, where "$x \in y$" means that the set x is an element of the set y, which satisfies the axioms of $\mathbb{A}$. We often say "the model $\mathbf{V}$" instead of "the model $(\mathbf{V}, \in)$".

We say that two sets A and B have same cardinality iff there is a bijection from A onto B and we denote this by $A \approx B$. The relation $\approx$ is an equivalence relation. Using the axiom of choice **AC**, one can prove that any set A can be well-ordered so there is an ordinal γ such that $A \approx \gamma$. In set theory the cardinal of the set A is then formally defined as the smallest such ordinal γ.

The infinite cardinals are usually denoted by $\aleph_0, \aleph_1, \aleph_2, \ldots, \aleph_\alpha, \ldots$ The cardinal $\aleph_\alpha$ is also denoted by ω_α, when it is considered as an ordinal. The first infinite ordinal is ω and it is the smallest ordinal which is countably infinite so $\aleph_0 = \omega$ (which could be written ω_0). The first uncountable ordinal is ω_1, and formally $\aleph_1 = \omega_1$.

Let **ON** be the class of all ordinals. Recall that an ordinal α is said to be a successor ordinal iff there exists an ordinal β such that $\alpha = \beta + 1$; otherwise the ordinal α is said to be a limit ordinal and in this case $\alpha = \sup\{\beta \in \mathbf{ON} \mid \beta < \alpha\}$.

We recall now the notions of cofinality of an ordinal and of regular cardinal which may be found for instance in [Jec02]. Let α be a limit ordinal, the cofinality of α, denoted $cof(\alpha)$, is the least ordinal β such that there exists a strictly increasing sequence of ordinals $(\alpha_i)_{i<\beta}$, of length β, such that $\forall i < \beta \; \alpha_i < \alpha$ and $\sup_{i<\beta} \alpha_i = \alpha$. This definition is usually extended to 0 and to the successor ordinals: $cof(0) = 0$ and $cof(\alpha + 1) = 1$ for every ordinal α. The cofinality of a limit ordinal is always a limit ordinal satisfying: $\omega \leq cof(\alpha) \leq \alpha$. Moreover $cof(\alpha)$ is in fact a cardinal. A cardinal κ is said to be *regular* iff $cof(\kappa) = \kappa$. Otherwise $cof(\kappa) < \kappa$ and the cardinal κ is said to be *singular*.

A cardinal κ is said to be a *(strongly) inaccessible* cardinal iff $\kappa > \omega$, κ is regular, and for all cardinals $\lambda < \kappa$ it holds that $2^\lambda < \kappa$, where 2^λ is the cardinal of $\mathcal{P}(\lambda)$.

There are many other notions of large cardinals which have been studied in set theory, see [Dra74, Kan97, Jec02]. A remarkable fact is that the strengths of these notions appear to be linearly ordered (and in fact well ordered).

Recall that the class of sets in a model $\mathbf{V}$ of $\mathbf{ZF}$ may be stratified in a transfinite hierarchy, called the *Cumulative Hierarchy*, which is defined by $\mathbf{V} = \bigcup_{\alpha \in \mathbf{ON}} \mathbf{V}_\alpha$, where the sets $\mathbf{V}_\alpha$ are constructed by induction as follows:

(1) $\mathbf{V}_0 = \emptyset$
(2) $\mathbf{V}_{\alpha+1} = \mathcal{P}(\mathbf{V}_\alpha)$ is the set of subsets of $\mathbf{V}_\alpha$, and
(3) $\mathbf{V}_\alpha = \bigcup_{\beta < \alpha} \mathbf{V}_\beta$, for α a limit ordinal.

It is well known that if $\mathbf{V}$ is a model of $\mathbf{ZFC}$ and κ is an inaccessible cardinal in $\mathbf{V}$ then $\mathbf{V}_\kappa$ is also a model of $\mathbf{ZFC}$. If there exist in $\mathbf{V}$ at least n inaccessible cardinals, where $n \geq 1$ is an integer, and if κ is the n-th inaccessible cardinal, then $\mathbf{V}_\kappa$ is also a model of $\mathbf{ZFC}$ + "There exist exactly $n-1$ inaccessible cardinals" (and the same result is true if we replace "inaccessible" by "hyperinaccessible"). This implies that one cannot prove in $\mathbf{ZFC}$ that there exists an inaccessible cardinal, because if κ is the first inaccessible cardinal in $\mathbf{V}$ then $\mathbf{V}_\kappa$ is a model of $\mathbf{ZFC}$ + "There exist no inaccessible cardinals".

We now recall that a (first-order) theory T in the language of set theory is a set of (first-order) sentences, called the axioms of the theory. If T is a theory and φ is a sentence then we write $T \vdash \varphi$ iff there is a formal proof of φ from T; this means that there is a finite sequence of sentences φ_j, $1 \leq j \leq n$, such that $\varphi_1 \vdash \varphi_2 \vdash \ldots \varphi_n$, where φ_n is the sentence φ and for each $j \in [1, n]$, either φ_j is in T or φ_j is a logical axiom or φ_j follows from $\varphi_1, \varphi_2, \ldots \varphi_{j-1}$ by usual rules of inference which can be defined purely syntactically. A theory is said to be consistent iff for no (first-order) sentence φ does $T \vdash \varphi$ and $T \vdash \neg\varphi$. If T is inconsistent, then for every sentence φ it holds that $T \vdash \varphi$. We shall denote Cons(T) the sentence "the theory T is consistent".

Recall that one can code in a recursive manner the sentences in the language of set theory by finite sequences over a finite alphabet, and then simply over the alphabet $\{0, 1\}$, by using a classical Gödel numbering of the sentences. We say that the theory T is recursive iff the set of codes of axioms in T is a recursive set of words over $\{0, 1\}$. In that case one can also code formal proofs from axioms of a recursive theory T and then Cons(T) is an arithmetical statement. The theory $\mathbf{ZFC}$ is recursive and so are the theories $T_n =: \mathbf{ZFC}$ + "There exist (at least) n inaccessible cardinals", for any integer $n \geq 1$.

We now recall Gödel's Second Incompleteness Theorem, [Göd63, Fri11].

Theorem 1 (Gödel 1931 [Göd63]). *Let T be a consistent recursive extension of* **ZF**. *Then $T \nvdash \mathrm{Cons}(T)$.*

We now state the following lemmas.

Lemma 2. *Let T be a recursive theory in the language of set theory. Then there exists a Turing machine $\mathcal{M}_T$, starting on an empty tape, such that $\mathcal{M}_T$ halts iff T is inconsistent.*

Proof. We describe informally the behaviour of the machine $\mathcal{M}_T$. Essentially the machine works as a program which enumerates all the formal proofs from T and enters in an accepting state and then halts iff the last sentence of the proof is the sentence "$\exists x(x \neq x)$". If the theory T is consistent the machine will never enter in an accepting state q_f and never halts. But if the theory is inconsistent then at some point of the computation the machine sees a proof whose last sentence is actually "$\exists x(x \neq x)$" and halts.   □

In [Fin15] we have focused our results on set theories, even if we noticed that some of our results could be extended to weaker arithmetical theories and to other recursive theories. We have shown that some elementary properties of automata may be independent from strong set theories like **ZFC** + "There exist (at least) n inaccessible cardinals". We are going to show in this paper that some similar phenomena still hold for some kinds of automata on finite words. However in the context of automata over finite words, we can notice that automata and their behaviour can be coded by integers and this can be done in Peano arithmetic; this will be often assumed in the sequel. Then we shall also obtain some new independence results from the axiomatic system of Peano Arithmetic **PA**. Indeed while we have first stated Gödel's Second Incompleteness Theorem for consistent recursive extensions of **ZF** in the above Theorem 1, the prooof of this Theorem leads also to the following version, see [Poi00] for a proof.

Theorem 3 (Gödel 1931). *Let* **PA** *be Peano Arithmetic. Then*

$$\mathbf{PA} \nvdash \mathrm{Cons}(\mathbf{PA}).$$

Notice that **PA** is known to be consistent, since the axioms of Peano Arithmetic are satisfied in the standard model of the natural numbers. Thus the above Theorem 3 gives a true arithmetical statement which is not provable from Peano Arithmetic. Notice that Gentzen gave in 1936 a proof of the consistency of Peano Arithmetic which uses only transfinite induction up to the Cantor ordinal ε_0, see [Gen36,Hor14]; this proof can be considered as being finitistic since the ordinal ε_0 can be coded with finite objects, like finite trees.

3   Incompleteness Results for Automata over Finite Words

We assume the reader to be familiar with the theory of formal languages [HMU01]. We recall the usual notations of formal language theory.

If Σ is a finite alphabet, a *non-empty finite word* over Σ is any sequence $x = a_1 \ldots a_k$, where $a_i \in \Sigma$ for $i = 1, \ldots, k$, and k is an integer ≥ 1. The *length* of x is k, denoted by $|x|$. The *empty word* has no letter and is denoted by ε; its length is 0. Σ^* is the *set of finite words* (including the empty word) over Σ.

The usual concatenation product of two finite words u and v is denoted $u.v$ (and sometimes just uv). This product is extended to the product of a finite word u and an ω-word v: the infinite word $u.v$ is then the ω-word such that:

$$(u.v)(k) = u(k) \text{ if } k \leq |u|, \text{ and } (u.v)(k) = v(k - |u|) \text{ if } k > |u|.$$

We now recall the well known Post Correspondence Problem (PCP), see [HMU01, pp. 392–402]. It is one of the famous undecidable problems in Theoretical Computer Science and in Formal Language Theory. The PCP is an abstract problem involving strings, and it has been very useful to prove the undecidability of many other problems by reduction of PCP to those problems. In particular, many problems about context-free languages, those accepted by pushdown automata or generated by context-free grammars, have been shown to be undecidable by this method. For instance it follows from the undecidability of the Post Correspondence Problem that the universality problem, the inclusion and the equivalence problems for context-free languages are also undecidable.

An instance of the Post Correspondence Problem consists of two lists of finite words over some finite alphabet Γ : $(x_1, x_2, \ldots, x_n)$ and $(y_1, y_2, \ldots, y_n)$. Notice that the two lists must have the same length $n \geq 1$. One says that this instance has a solution if there exists a non-empty sequence of indices $i_1, i_2, \ldots, i_k$ such that $x_{i_1} x_{i_2} \cdots x_{i_k} = y_{i_1} y_{i_2} \cdots y_{i_k}$. The Post Correspondence Problem is:

"Given an instance of the PCP, tell whether this instance has a solution".

We now recall Post's result, now well-known as the undecidability of the Post Correspondence Problem.

Theorem 4 *(Post, see [HMU01]). Let Γ be an alphabet having at least two elements. Then it is undecidable to determine, for arbitrary n-tuples $(x_1, x_2, \ldots, x_n)$ and $(y_1, y_2, \ldots, y_n)$ of non-empty words in Γ^*, whether there exists a non-empty sequence of indices $i_1, i_2, \ldots, i_k$ such that $x_{i_1} x_{i_2} \cdots x_{i_k} = y_{i_1} y_{i_2} \cdots y_{i_k}$.*

We now recall the variant of the PCP called the modified Post Correspondence Problem, which is used in the proof of the above Theorem 4.

The MPCP consists, given two n-tuples $(x_1, x_2, \ldots, x_n)$ and $(y_1, y_2, \ldots, y_n)$ of non-empty words in Γ^*, in determining whether there exists a non-empty sequence of indices $i_1, i_2, \ldots, i_k$ such that $x_1 x_{i_1} x_{i_2} \cdots x_{i_k} = y_1 y_{i_1} y_{i_2} \cdots y_{i_k}$.

The proof of Theorem 4 is given in two steps, see [HMU01]. First one can associate in a recursive manner, to each pair $(\mathcal{M}_z, w)$ where $\mathcal{M}_z$ is the Turing machine of index $z \in \mathbb{N}$ and w is an input word for $\mathcal{M}_z$, an instance of the MPCP consisting of two n-tuples $(x_1, x_2, \ldots, x_n)$ and $(y_1, y_2, \ldots, y_n)$ such that there exists a finite sequence of indices $i_1, i_2, \ldots, i_k$, such that $x_1 x_{i_1} x_{i_2} \cdots x_{i_k} = y_1 y_{i_1} y_{i_2} \cdots y_{i_k}$ if and only if the Turing machine $\mathcal{M}_z$ does halt on the input w.

Next we can associate in an effective manner, to each instance of the MPCP consisting of two n-tuples $(x_1, x_2, \ldots, x_n)$ and $(y_1, y_2, \ldots, y_n)$, another instance of the PCP consisting of two $n+2$-tuples $(x'_1, x'_2, \ldots, x'_{n+2})$ and $(y'_1, y'_2, \ldots, y'_{n+2})$, such that $(x_1, x_2, \ldots, x_n)$ and $(y_1, y_2, \ldots, y_n)$ form a solution of the MPCP if and only if $(x'_1, x'_2, \ldots, x'_{n+2})$ and $(y'_1, y'_2, \ldots, y'_{n+2})$ form a solution of the PCP.

We can now state the following result.

Theorem 5. *Let T be a recursive theory in the language of set theory or $T = \mathbf{PA}$. Then there exist two n-tuples $X_T = (x_1, x_2, \ldots, x_n)$ and $Y_T = (y_1, y_2, \ldots, y_n)$ of finite words over a finite alphabet Σ, such that there exists a non-empty sequence of indices $i_1, i_2, \ldots, i_k$ such that $x_{i_1} x_{i_2} \cdots x_{i_k} = y_{i_1} y_{i_2} \cdots y_{i_k}$ iff T is inconsistent.*

Proof. Let T be a recursive theory in the language of set theory or $T = \mathbf{PA}$. Then there exists a Turing machine $\mathcal{M}$, starting with an empty tape, which halts if and only if the theory T is inconsistent. We can now deduce the announced result from the proof of the undecidability of the PCP which is just sketched above. $\square$

Remark 6. *We can easily see that the above theorem is true for the two-letter alphabet $\Sigma = \{a, b\}$. Indeed if $\Sigma = \{a_1, a_2, \ldots, a_p\}$ is an alphabet having more than two letters, we can use the coding given by: $a_j \to b^j a$, where a and b are two letters, which provides the announced claim.*

Corollary 7. *For every integer $n \geq 0$, there exist $p \geq 1$ and two p-tuples $X_{T,n} = (x_{1,n}, x_{2,n}, \ldots, x_{p,n})$ and $Y_{T,n} = (y_{1,n}, y_{2,n}, \ldots, y_{p,n})$ of finite words over $\Sigma = \{a, b\}$, such that: "P_n: there exist no non-empty sequence of indices $i_1, i_2, \ldots, i_k$ such that: $x_{i_1,n} x_{i_2,n} \cdots x_{i_k,n} = y_{i_1,n} y_{i_2,n} \cdots y_{i_k,n}$" iff T_n is consistent.*

In particular, if $\mathbf{ZFC}$ + "There exist (at least) n inaccessible cardinals" is consistent, then P_n is provable from $\mathbf{ZFC}$ + "There exist (at least) $n+1$ inaccessible cardinals" but not from $\mathbf{ZFC}$ + "There exist (at least) n inaccessible cardinals".

Proof. By Theorem 5, for each integer $n \geq 0$, there exist $p \geq 1$ and two p-tuples $X_{T,n} = (x_{1,n}, x_{2,n}, \ldots, x_{p,n})$ and $Y_{T,n} = (y_{1,n}, y_{2,n}, \ldots, y_{p,n})$ of finite words over $\Sigma = \{a, b\}$, such that: "P_n: there exist no non-empty sequence of indices $i_1, i_2, \ldots, i_k$ such that $x_{i_1,n} x_{i_2,n} \cdots x_{i_k,n} = y_{i_1,n} y_{i_2,n} \cdots y_{i_k,n}$" iff T_n is consistent. Recall that one can prove from $\mathbf{ZFC}$ + "There exist (at least) $n+1$ inaccessible cardinals" that if κ is the $n + 1$-th inaccessible cardinal, then the set $\mathbf{V}_\kappa$ of the cumulative hierarchy is also a model of $\mathbf{ZFC}$ + "There exist n inaccessible cardinals". This implies that the theory $\mathbf{ZFC}$ + "There exist n inaccessible cardinals" is consistent and thus this also implies that there exist no non-empty sequence of indices $i_1, i_2, \ldots, i_k$ such that:

$$x_{i_1,n} x_{i_2,n} \cdots x_{i_k,n} = y_{i_1,n} y_{i_2,n} \cdots y_{i_k,n}$$

On the other hand if T_n is consistent, then P_n is not provable from T_n. Indeed T_n is then a consistent recursive extension of $\mathbf{ZFC}$ and thus by Gödel's Second Incompleteness Theorem we know that $T_n \nvdash \mathrm{Cons}(T_n)$. $\square$

Moreover, since **PA** is consistent, we also get the following result.

Corollary 8. *There exist two p-tuples* $X = (x_1, x_2, \ldots, x_p)$ *and* $Y = (y_1, y_2, \ldots, y_p)$ *of finite words over* $\Sigma = \{a, b\}$, *such that:*

(1) there exist no non-empty sequence of indices $i_1, i_2, \ldots, i_k$ *such that:*

$$x_{i_1} x_{i_2} \cdots x_{i_k} = y_{i_1} y_{i_2} \cdots y_{i_k}$$

(2) The property (1) is not provable from **PA**.

We can now infer from Theorem 5 some incompleteness results for context-free languages generated by context-free grammars or equivalently accepted by pushdown automata. We use the reductions of PCP to some problems about context-free grammars ans context-free languages given in [HMU01, pp. 404–408]. We refer the reader to this textbook for background about context-free grammars and context-free languages.

We first state the following result about ambiguity of context-free grammars.

Theorem 9. *Let T be a recursive theory in the language of set theory or* $T =$ **PA**. *Then there exists a context-free grammar* G_T *which is unambiguous iff T is consistent.*

Proof. We refer here to the proof of the undecidability of the unambiguity of a given context-free grammar in [HMU01, pp. 404–406]. From a given instance of the PCP constituted by two n-tuples $(x_1, x_2, \ldots, x_n)$ and $(y_1, y_2, \ldots, y_n)$ of finite words over a finite alphabet Σ, is constructed a context-free grammar G such that G is ambiguous if and only if this instance of PCP has a solution. The result now follows from this construction and from the above Theorem 5.   □

Corollary 10. *For every integer* $n \geq 0$, *there exists a context-free grammar* G_n *such that* G_n *is unambiguous iff* T_n *is consistent.*

In particular, if **ZFC** + *"There exist (at least) n inaccessible cardinals" is consistent, then "* G_n *is unambiguous" is provable from* **ZFC** + *"There exist (at least) n + 1 inaccessible cardinals" but not from* **ZFC** + *"There exist (at least) n inaccessible cardinals".*

We now state some other results about elementary properties of context-free languages.

Theorem 11. *Let T be a recursive theory in the language of set theory or* $T =$ **PA**. *Then there exist context-free grammars* $G_{1,T}$ $G_{2,T}$, $G_{3,T}$, *and* $G_{4,T}$, *such that* $\mathrm{Cons}(T)$ *is equivalent to each of the following items:*

(1) $L(G_{1,T}) \cap L(G_{2,T}) = \emptyset$;
(2) $L(G_{3,T}) = L(G_{4,T})$;
(3) $L(G_{3,T}) = \Gamma^*$, *for some alphabet* Γ.

Corollary 12. *For every integer $n \geq 0$, there exist context-free grammars $G_{1,n}$ $G_{2,n}$, $G_{3,n}$, and $G_{4,n}$, such that $\mathrm{Cons}(T_n)$ is equivalent to each of the following items:*

(1) $L(G_{1,n}) \cap L(G_{2,n}) = \emptyset$;
(2) $L(G_{3,n}) = L(G_{4,n})$;
(3) $L(G_{3,n}) = \Gamma^\star$, for some alphabet Γ.

In particular, if **ZFC** *+ "There exist (at least) n inaccessible cardinals" is consistent, then each of the properties of these context-free languages given by Items (1)–(3) is provable from* **ZFC** *+ "There exist (at least) $n + 1$ inaccessible cardinals" but not from* **ZFC** *+ "There exist (at least) n inaccessible cardinals".*

We are now going to state some similar independence results for other very simple finite machines reading finite words: the class of 2-tape automata (or transducers) accepting finitary rational relations. We shall refer to the book [Ber79] in which some elementary problems about finitary rational relations are proved to be undecidable by reducing the PCP to these problems, see pages 79–82 in this book.

We now state the following results.

Theorem 13. *let T be a recursive theory in the language of set theory or $T =$* **PA**. *Then there exist 2-tape automata $\mathcal{A}$, $\mathcal{B}$, and $\mathcal{C}$, accepting finitary rational relations $X, Y, Z \subseteq A^\star \times B^\star$, for two alphabets A and B having at least two letters, and such that $\mathrm{Cons}(T)$ is equivalent to each of the following items:*

(1) $X \cap Y = \emptyset$;
(2) $Z = A^\star \times B^\star$;
(3) $A^\star \times B^\star \subseteq Z$.

Proof. We refer to the proof of [Ber79, Theorem 8.4, p. 81]. We assume, as in this proof, that A contains exactly two letters and that $A = \{a, b\}$. For two sequences $u_1, u_2, \ldots, u_p$, and $v_1, v_2, \ldots, v_p$, of finite words over the alphabet B, we define $U = \{(ab, u_1), \ldots, (ab^p, u_p)\}$, and $V = \{(ab, v_1), \ldots, (ab^p, v_p)\}$. Then U^+ and V^+ are rational relations and, by [Ber79, Lemma 8.3, p. 80], the relations $\bar{U} = A^\star \times B^\star \setminus U^+$ and $\bar{V} = A^\star \times B^\star \setminus V^+$ are also rational. It is noticed in the proof of Theorem 8.4 in [Ber79] that if we set $X = U^+$ and $Y = V^+$, then it holds that $X \cap Y \neq \emptyset$ iff the instance of the PCP given by $(u_1, u_2, \ldots, u_p)$, and $(v_1, v_2, \ldots, v_p)$ has a solution. Item (1) of the Theorem follows then from the above Theorem 5. Moreover if we set $Z = \bar{U} \cup \bar{V}$, then $Z = A^\star \times B^\star$ iff $X \cap Y = \emptyset$, and this implies Items (2) and (3). $\square$

Using a 2-tape automaton $\mathcal{C}$ accepting the finitary relation Z given by the above theorem, it is easy to construct, with similar methods as in the paper [Fin03] about infinitary rational relations, another 2-tape automaton $\mathcal{D}$ accepting a finitary rational relation $L \subseteq A^\star \times B^\star$ such that L is accepted by a *deterministic* 2-tape automaton iff L is accepted by a *synchronous* 2-tape automaton iff $Z = A^\star \times B^\star$. Thus we can state the following result. The detailed proof is here left to the reader.

Theorem 14. *let T be a recursive theory in the language of set theory or $T = $ **PA***. Then there exists a 2-tape automaton $\mathcal{D}$, accepting a finitary ratio-nal relation $L \subseteq A^\star \times B^\star$, for two alphabets A and B having at least two letters, and such that $\mathrm{Cons}(T)$ is equivalent to each of the following items:*

(1) L is accepted by a deterministic 2-tape automaton;
(2) L is accepted by a synchronous 2-tape automaton.

Corollary 15. *For every integer $n \geq 0$, there exist 2-tape automata $\mathcal{A}_n$, $\mathcal{B}_n$, $\mathcal{C}_n$, and $\mathcal{D}_n$, accepting subsets of $A^\star \times B^\star$, for two alphabets A and B having at least two letters, such that $\mathrm{Cons}(T_n)$ is equivalent to each of the following items:*

(1) $L(\mathcal{A}_n) \cap L(\mathcal{B}_n) = \emptyset$;
(2) $L(\mathcal{C}_n) = A^\star \times B^\star$;
(3) $L(\mathcal{D}_n)$ is accepted by a deterministic 2-tape automaton;
(4) $L(\mathcal{D}_n)$ is accepted by a synchronous 2-tape automaton.

In particular, if **ZFC** *+ "There exist (at least) n inaccessible cardinals" is consistent, then each of the properties of these 2-tape automata given by Items (1)–(4) is provable from* **ZFC** *+ "There exist (at least) $n+1$ inaccessible cardi-nals" but not from* **ZFC** *+ "There exist (at least) n inaccessible cardinals".*

Since **PA** is consistent, we get the following result from Theorems 13 and 14 (where we assume, as we have already said at the beginning of this section, that automata are coded by integers):

Corollary 16. *There exist 2-tape automata $\mathcal{A}$, $\mathcal{B}$, $\mathcal{C}$, and $\mathcal{D}$, accepting subsets of $A^\star \times B^\star$, for two alphabets A and B having at least two letters, such that*

(1) $L(\mathcal{A}) \cap L(\mathcal{B}) = \emptyset$.
(2) $L(\mathcal{C}) = A^\star \times B^\star$.
(3) $L(\mathcal{D})$ is accepted by a deterministic 2-tape automaton.
(4) $L(\mathcal{D})$ is accepted by a synchronous 2-tape automaton.

But none of the items (1)–(4) is provable from **PA***.*

We are now going to state some incompleteness results about weighted automata. We shall also state some incompleteness results about finitely gener-ated semigroups of matrices with integer entries (with the semigroup operation of multiplication of matrices) which can be presented by automata with multi-plicities, see [Har02].

We first recall the notion of an n-state $\mathbb{Z}$-automaton, i.e. a non-deterministic automaton with integer multiplicities, as presented in [Har02].

A non-deterministic $\mathbb{Z}$-automaton is a 5-tuple $\mathcal{A} = (\Sigma, Q, \delta, J, F)$, where: $\Sigma = \{a_1, a_2, \ldots, a_k\}$ is a finite input alphabet and the letter a_i is associated to a matrix $M_i \in \mathbb{Z}^{n \times n}$; $Q = \{1, 2, \ldots, n\}$ is the state set (and i corresponds to the ith row and column of the matrices); J is the set of initial states and $F \subseteq Q$ is the set of final states; δ is the set of transitions that provides the rules

$$r \xrightarrow{\binom{a_i}{m}} s,$$

where $a_i \in \Sigma$, and $m = (M_i)_{rs}$ is the multiplicity of the rule.

A path

$$\pi = s_1 \xrightarrow{\binom{b_1}{m_1}} s_2 \xrightarrow{\binom{b_2}{m_2}} s_3 \longrightarrow \cdots \longrightarrow s_t \xrightarrow{\binom{b_t}{m_t}} s_{t+1}$$

is a computation of the automaton $\mathcal{A}$ reading a word $w = b_1 b_2 \ldots b_t \in \Sigma^*$ and the multiplicity of this path is equal to $\|\pi\| = m_1 m_2 \ldots m_t \in \mathbb{Z}$. For a word $w \in \Sigma^*$ we denote by Π_{rs} the set of the paths of $\mathcal{A}$ reading the word w which go from state r to state s. Then the multiplicity of the word $w = a_{i_1} a_{i_2} \ldots a_{i_t} \in \Sigma^*$ from r to s is the sum

$$\mathcal{A}_{rs}(w) = \sum_{\pi \in \Pi_{rs}} \|\pi\| = (M_{i_1} M_{i_2} \ldots M_{i_t})_{rs}$$

and we get the multiplicity of w in $\mathcal{A}$ from the accepting paths:

$$\mathcal{A}(w) = \sum_{r \in J, s \in F} \mathcal{A}_{rs}(w) = \sum_{r \in J, s \in F} (M_{i_1} M_{i_2} \ldots M_{i_t})_{rs}.$$

We first state the following result.

Theorem 17. *Let T be a recursive theory in the language of set theory or $T =$* **PA**. *Then there exists a finite set of matrices $M_1, M_2, \ldots, M_n \in \mathbb{Z}^{3 \times 3}$, for some integer $n \geq 1$, such that the subsemigroup of $\mathbb{Z}^{3 \times 3}$ generated by these matrices does not contain any matrix M with $M_{13} = 0$ if and only if T is consistent.*

One can easily state corollaries of the above Theorem for strong set theories, as for previous results in this paper. Details are here left to the reader. Moreover, since **PA** is consistent, we also get the following result.

Corollary 18. *There exists a finite set of matrices $M_1, M_2, \ldots, M_n \in \mathbb{Z}^{3 \times 3}$, for some integer $n \geq 1$, such that:*

(1) the subsemigroup of $\mathbb{Z}^{3 \times 3}$ generated by these matrices does not contain any matrix M with $M_{13} = 0$, and
(2) The property (1) is not provable from **PA**.

We also get the following result as a corollary of the above Theorem 17.

Corollary 19. *Let T be a recursive theory in the language of set theory or $T =$* **PA**. *Then there exists a 3-state $\mathbb{Z}$-automaton $\mathcal{A}$ such that $\mathcal{A}$ accepts a word with multiplicity zero iff T is inconsistent.*

Corollary 20. *Let T be a recursive theory in the language of set theory or $T =$* **PA**. *Then there exists two 2-state $\mathbb{N}$-automata $\mathcal{A}$ and $\mathcal{B}$ such that $\mathcal{A}$ and $\mathcal{B}$ accept a word w with the same multiplicity iff T is inconsistent.*

One can easily state corollaries of the above one for strong set theories or for Peano Arithmetic, as for previous results in this paper. Details are here left to the reader.

Following an idea of Paterson, Halava and Harju proved in [HH01] that it is undecidable for finitely generated subsemigroups S of $\mathbb{Z}^{3 \times 3}$ whether S contains a matrix with $M_{11} = 0$. We now prove the following result.

Theorem 21. *Let T be a recursive theory in the language of set theory or $T = $ **PA**. Then there exists a finite set of matrices $M_1, M_2, \ldots, M_n \in \mathbb{Z}^{3 \times 3}$, for some integer $n \geq 1$, such that the subsemigroup of $\mathbb{Z}^{3 \times 3}$ generated by these matrices does not contain any matrix M with $M_{11} = 0$ if and only if T is consistent.*

Recall that Paterson proved in 1970 that the mortality problem for finitely generated subsemigroups S of $\mathbb{Z}^{3 \times 3}$ is undecidable, i.e. that one cannot decide, for a given set of matrices $M_1, M_2, \ldots, M_n \in \mathbb{Z}^{3 \times 3}$, whether the zero matrix (whose all coefficients are equal to zero) belongs to the subsemigroup generated by the matrices $M_1, M_2, \ldots, M_n$, i.e. whether there exists a sequence of integers $i_1, i_2, \ldots i_k$, such that $M_{i_1} M_{i_2} \ldots M_{i_k} = 0$. Halava and Harju gave a proof of this result in [HH01].

We can now state the following result.

Theorem 22. *Let T be a recursive theory in the language of set theory or $T = $ **PA**. Then there exists a finite set of matrices $M_1, M_2, \ldots, M_n \in \mathbb{Z}^{3 \times 3}$, for some integer $n \geq 1$, such that the subsemigroup of $\mathbb{Z}^{3 \times 3}$ generated by these matrices does not contain the zero matrix if and only if T is consistent.*

Corollary 23. *For every integer $p \geq 0$, there exists a finite set of matrices $M_1, M_2, \ldots, M_{n_p} \in \mathbb{Z}^{3 \times 3}$, for some integer $n_p \geq 1$, such that the subsemigroup of $\mathbb{Z}^{3 \times 3}$ generated by these matrices does not contain the zero matrix if and only if T_p is consistent.*

*In particular, if **ZFC** + "There exist (at least) p inaccessible cardinals" is consistent, then the property "The subsemigroup of $\mathbb{Z}^{3 \times 3}$ generated by the matrices $M_1, M_2, \ldots, M_{n_p}$, does not contain the zero matrix" is provable from **ZFC** + "There exist (at least) $p+1$ inaccessible cardinals" but not from **ZFC** + "There exist (at least) p inaccessible cardinals".*

Moreover, since **PA** is consistent, we also get the following result.

Corollary 24. *There exists a finite set of matrices $M_1, M_2, \ldots, M_n \in \mathbb{Z}^{3 \times 3}$, for some integer $n \geq 1$, such that:*

(1) the subsemigroup of $\mathbb{Z}^{3 \times 3}$ generated by these matrices does not contain the zero matrix, and
*(2) The property (1) is not provable from **PA**.*

We have used in the proof of the above results some effective reductions of the PCP to some undecidable problems and an independence result about the solutions of some instances of the PCP. We can also sometimes use directly some effective reductions of the halting problem for Turing machines to some undecidable problems along with the above Lemma 2.

We now give some examples of independence results we can get by using this lemma.

Theorem 25. *Let T be a recursive theory in the language of set theory or $T = $ **PA**. Then there exists a 1-counter automaton $\mathcal{A}$, reading finite words over a finite alphabet Σ, such that $L(\mathcal{A}) = \Sigma^*$ if and only if T is consistent.*

Proof. Recall that Ibarra proved in [Iba79] that the universality problem for languages of 1-counter automata (and actually for some very restricted classes of 1-counter automata) is undecidable. He constructed, for each single-tape Turing machine $\mathcal{M}$, a 1-counter automaton $\mathcal{A}$, reading finite words over a finite alphabet Σ, such that $L(\mathcal{A}) = \Sigma^\star$ iff the machine $\mathcal{M}$ does not halt on the blank tape. The result now follows from the above Lemma 2. $\qquad\square$

We can now prove the following result.

Theorem 26. *Let T be a recursive theory in the language of set theory or $T = $ **PA**. Then there exists a 1-counter automaton $\mathcal{A}$, reading finite words over a finite alphabet Σ, such that $\mathrm{Cons}(T)$ is equivalent to each of the following items:*

(1) $L(\mathcal{A}) = \Sigma^\star$;
(2) $L(\mathcal{A})$ is accepted by a deterministic 1-counter automaton;
(3) $L(\mathcal{A})$ is accepted by an unambiguous 1-counter automaton.

Corollary 27. *For every integer $n \geq 0$, there exists a 1-counter automaton $\mathcal{A}_n$, reading finite words over a finite alphabet Σ, such that $\mathrm{Cons}(T_n)$ is equivalent to each of the following items:*

(1) $L(\mathcal{A}_n) = \Sigma^\star$;
(2) $L(\mathcal{A}_n)$ is accepted by a deterministic 1-counter automaton;
(3) $L(\mathcal{A}_n)$ is accepted by an unambiguous 1-counter automaton.

In particular, if **ZFC** *+ "There exist (at least) n inaccessible cardinals" is consistent, then each of the properties of the 1-counter automaton $\mathcal{A}_n$ given by Items (1)–(3) is provable from* **ZFC** *+ "There exist (at least) $n + 1$ inaccessible cardinals" but not from* **ZFC** *+ "There exist (at least) n inaccessible cardinals".*

Remark 28. *Part of Theorem 26 and of Corollary 27 subsumes Items (2) and (3) of Theorem 11 and of Corollary 12. Indeed we can construct, from a given pushdown automaton (and thus also from a given 1-counter automaton) accepting a finitary language, a context-free grammar generating the same language.*

4   Concluding Remarks

We have shown that some very elementary properties of some automata over finite words are actually independent from strong set theories like **ZFC** + "There exist (at least) n inaccessible cardinals". The results of this paper are true for other large cardinals than inaccessible ones. For instance we can replace inaccessible cardinals by hyperinaccessible, hyperMahlo, measurable, ... and still other ones and obtain similar results.

Some of our results are even more general because they could have been stated for more general recursive theories,

References

[Ber79] Berstel, J.: Transductions and context free languages. Teubner Studienbücher Informatik (1979). http://www-igm.univ-mlv.fr/~berstel/

[Dra74] Drake, F.R.: Set Theory, An Introduction to Large cardinals. Studies in Logic and the Foundations of Mathematics, vol. 76. North-Holland, Amsterdam (1974)

[EFT94] Ebbinghaus, H.-D., Flum, J., Thomas, W.: Mathematical Logic. Undergraduate Texts in Mathematics, 2nd edn. Springer, New York (1994). Translated from the German by Margit Meßmer

[Fin03] Finkel, O.: Undecidability of topological and arithmetical properties of infinitary rational relations. RAIRO-Theoret. Inf. Appl. **37**(2), 115–126 (2003)

[Fin09] Finkel, O.: The complexity of infinite computations in models of set theory. Logical Methods Comput. Sci. **5**(4:4), 1–19 (2009)

[Fin11] Finkel, O.: Some problems in automata theory which depend on the models of set theory. RAIRO-Theoret. Inf. Appl. **45**(4), 383–397 (2011)

[Fin15] Finkel, O.: Incompleteness theorems, large cardinals, and automata over infinite words. In: Halldórsson, M.M., Iwama, K., Kobayashi, N., Speckmann, B. (eds.) ICALP 2015. LNCS, vol. 9135, pp. 222–233. Springer, Heidelberg (2015). doi:10.1007/978-3-662-47666-6_18

[Fri11] Friedman, H.M.: My forty years on his shoulders. In: Gödel, K. (ed.) The Foundations of Mathematics, pp. 399–432. Cambridge Univ. Press, Cambridge (2011)

[Gen36] Gentzen, G.: Die Widerspruchsfreiheit der reinen Zahlentheorie. Math. Ann. **112**(1), 493–565 (1936)

[Göd63] Gödel, K.: On formally undecidable propositions of Principia Mathematica and related systems (1963). Translated by B. Meltzer, with an introduction by R. B. Braithwaite. Basic Books Inc., Publishers, New York

[Har85] Hartmanis, J.: Independence results about context-free languages and lower bounds. Inf. Process. Lett. **20**(5), 241–248 (1985)

[Har02] Harju, T.: Decision questions on integer matrices. In: Kuich, W., Rozenberg, G., Salomaa, A. (eds.) DLT 2001. LNCS, vol. 2295, pp. 57–68. Springer, Heidelberg (2002). doi:10.1007/3-540-46011-X_5

[HH01] Halava, V., Harju, T.: Mortality in matrix semigroups. Am. Math. Mon. **108**(7), 649–653 (2001)

[HMU01] Hopcroft, J.E., Motwani, R., Ullman, J.D.: Introduction to Automata Theory, Languages, and Computation. Addison-Wesley Series in Computer Science. Addison-Wesley Publishing Co., Reading (2001)

[Hor14] Horská, A.: Where is the Gödel-point Hiding: Gentzen's Consistency Proof of 1936 and His Representation of Constructive Ordinals. Springer Briefs in Philosophy. Springer, Cham (2014)

[Iba79] Ibarra, O.H.: Restricted one-counter machines with undecidable universe problems. Math. Syst. Theory **13**, 181–186 (1979)

[Jec02] Jech, T.: Set Theory, 3rd edn. Springer, Heidelberg (2002)

[JY81] Joseph, D., Young, P.: Independence results in computer science? J. Comput. Syst. Sci. **23**(2), 205–222 (1981)

[Kan97] Kanamori, A.: The Higher Infinite. Springer, Heidelberg (1997)

[KM87] Kanamori, A., McAloon, K.: On Gödel incompleteness and finite combinatorics. Ann. Pure Appl. Logic **33**(1), 23–41 (1987)

[Kun80] Kunen, K.: Set Theory. An Introduction to Independence Proofs. Studies in Logic and the Foundations of Mathematics, vol. 102. North-Holland Publishing Co., Amsterdam, New York (1980)

[Poi00] Poizat, B.: A Course in Model Theory: An Introduction to Contemporary Mathematical Logic. Universitext. Springer, New York (2000). Translated from the French by Moses Klein and revised by the Author

Scheduling Tasks to Minimize Active Time on a Processor with Unlimited Capacity

Ken C.K. Fong[1(✉)], Minming Li[1], Yungao Li[3], Sheung-Hung Poon[2],
Weiwei Wu[3], and Yingchao Zhao[4]

[1] Department of Computer Science,
City University of Hong Kong, Hong Kong SAR, China
ken.fong@my.cityu.edu.hk
[2] School of Computing and Informatics,
Universiti Teknologi Burnei, Gadong, Brunei Darussalam
[3] Southeast University, Nanjing, China
[4] Department of Computer Science,
Caritas Institute of Higher Education, Hong Kong SAR, China

Abstract. We study the following scheduling problem on a single processor. We are given n jobs, where each job j_i has an integer release time r_i, processing time p_i as well as deadline d_i. The processor can schedule an unlimited number of jobs at any time t. Our objective is to schedule the jobs together such that the total number of active time slots is minimized. We present an $O(n^3)$ dynamic programming algorithm for the case of agreeable deadlines with $d_i \leq d_j$ whenever $r_i < r_j$ or all jobs are big. In the general case, we present an online algorithm with competitive ratio 4 and show that our analysis is tight.

1 Introduction

Energy efficiency problems have been well studied by researchers in the past decades [1,4,12]. Its objective is to reduce the energy consumption without performance degradation. One way to reduce the energy consumption is to turn off the idle machines when the machines do not have any jobs to process. For instance, the storage cluster in the data centers can be turned off to save energy during low utilization period [2].

"Min-gap" strategy [4,5] is one of the approaches for energy saving. When the machines are idle, they are transited to the suspended state without any energy consumption. However, in practice, a small amount of energy will be consumed in the process of waking up the machines from the suspended state. If the number of idle periods can be minimized, less energy is consumed for waking up the idle machines. Hence, the objective of min-gap strategy is to find a schedule such that the number of idle periods can be minimized. Baptiste [4]

The work described in this paper was partially supported by a grant from Research Grants Council of the Hong Kong Special Administrative Region, China (Project No. UGC/FDS11/E04/15 and Project No. CityU 11268616).

© Springer International Publishing AG 2017
T.V. Gopal et al. (Eds.): TAMC 2017, LNCS 10185, pp. 247–259, 2017.
DOI: 10.1007/978-3-319-55911-7_18

was the first to propose a polynomial-time algorithm with running time of $O(n^7)$. Later, Baptiste et al. [5] improved the complexity to $O(n^5)$. They also presented an $O(n^4)$ algorithm for a special instance of this problem where all jobs have unit length. Angel et al. [3] consider the special setting of this problem, where the jobs have agreeable deadlines with $d_i \le d_j$ whenever $r_i < r_j$. They presented an $O(n^2)$ algorithm for the single processor case, and in the multiprocessor case with m machines, they presented an $O(n^2 m)$ algorithm for unit length jobs. Demaine et al. [8] presented an $O(n^7 m^5)$ algorithm for the general case in the multiprocessor setting where all jobs have unit length.

Besides "min-gap" strategy, the active/busy time minimization problem introduced by Ikura and Gimple is another approach for energy saving [11] which aims to schedule jobs together in order to maximize their idle periods to save the energy. They proposed an algorithm to minimize the completion time for a single batch processing machine with agreeable deadlines.

In the active/busy time minimization problem, the processor can schedule B jobs at any time t. Let J_t be the number of jobs which are scheduled at time t. The objective is to schedule all the jobs in the time slots satisfying $J_t \le B$ to minimize the number of active time slots. In the preemptive setting, Chang et al. [6] proposed a linear time algorithm to schedule unit length jobs. Khandekar et al. [13] proposed a 5-approximation algorithm and Chang et al. [7] proposed a 2-approximation LP rounding algorithm to find a feasible schedule for all jobs where the total active time can be minimized. Koehler and Khuller proposed an $O(n^3)$ algorithm on parallel machines which improves the previously known $O(n^8)$ dynamic programming algorithm [14]. Shalom et al. [15] extends the studies of active/busy time minimization problem to the online setting. They presented a $5 \log len_{\max}$-competitive algorithm where $len_{\max}$ is the length of the longest job. Tian et al. [17] applied this problem in cloud computing for scheduling real-time virtual machines in the cloud data center. They presented a β-competitive algorithm for the general case, where $1 < \beta < B$. Flammini et al. [10] proposed a 4-approximation algorithm to solve the non-preemptive setting of the active/busy time minimization problem. This problem is NP hard when $B > 2$ in preemptive setting [6] and $B = 2$ in non-premmptive setting [10].

Moreover, Khandekar et al. [13] introduced a new variant of the active/busy time minimization problem where a processor can schedule unlimited jobs at any time t without the constraint of B. The recent work of Tavakoli et al. [16] and Fang et al. [9] studied this problem in a different scenario. The scenario they use is the interval data sharing problem which is abstracted from wireless sensor networks. When the base station broadcasts one sampled data, it can be received and used by any number of sensors whose sampling period contains the time point (which justifies the unlimited capacity assumption in the problem). On the condition that each sensor receives a certain amount of data during the sampling period, their objective is to minimize the transmission energy of the base station. In [16], each application only requires discrete data at some time points whereas the work in [9] considers a continuous interval of sampling data.

In this paper, we focus on the active time minimization problem presented in [13] which is illustrated below. We are given a collection of n jobs, where each job j_i has an integer release time r_i, processing time p_i and deadline d_i. We are given a single processor, which can process an unlimited number of jobs at the same time in a non-preemptive way. For any time t, whenever there exists a job in execution, we denote t as "active time". Otherwise we refer to time t as "idle time". The focus of this paper is to schedule all the jobs such that the active time can be minimized.

In the active time minimization problem, Khandekar et al. presented a dynamic programming algorithm with a time complexity of $O(n^4)$ to solve the problem. Later on, a 2-approximation algorithm for the general case is provided by [9] together with an $O(n^2)$ dynamic programming algorithm for the special instance where the lengths of the execution for all the jobs are the same. However, the analysis of the 2-approximation algorithm in [9] is incorrect, and there exists a counter example showing that their algorithm cannot obtain the approximation ratio of 2. The detail of the counter example is presented in Sect. 4. Moreover, we study some particular cases of this problem where jobs have agreeable deadlines or all jobs are big, and obtain a faster algorithm compared to the general case.

Our results. We present an $O(n^3)$ dynamic programming algorithm for the case of agreeable deadlines where $d_i \leq d_k$ if $r_i < r_k$ or all jobs are big. In the general case, we present a counter example to show that the algorithm proposed by [9] cannot obtain the approximation ratio of 2. Moreover, we present an online algorithm with competitive ratio 4 and show that our analysis is tight. To the best of our knowledge, we are the first to study this problem in the online setting.

The remainder of this paper is organized as follows. The problem formulation is given in Sect. 2. We introduce the dynamic programming algorithm for jobs with agreeable deadlines or big jobs in Sect. 3. In Sect. 4, we present the counter example showing that the approximation ratio of the algorithm presented in [9] is not 2 and introduce an online algorithm with competitive ratio 4. Finally, we conclude in Sect. 5.

2   Problem Formulation

We first describe the general case of the active time minimization problem. We are given n jobs, where each job j_i has an integer release time r_i, processing time p_i and deadline d_i. For any job j_i, it can only be scheduled within the $[r_i, d_i]$ interval and we focus on the non-preemptive setting of the problem where the execution starts till it finishes without any interruption in between. For any time t, whenever there exists a job executed in t, we denote t as "active time". Otherwise we refer to time t as "idle time". The objective is to schedule all jobs by deciding the starting time s_i and the finishing time f_i for each job j_i, such that the active time is minimized.

Besides the general case of the problem, we also study the following special cases for this problem with different assumptions on the input jobs.

1. Jobs with agreeable deadlines: for any two jobs j_i and j_k, we have $d_i \leq d_k$ if $r_i < r_k$.
2. Large jobs: for each job j_i, $p_i \geq (d_i - r_i)/2$.

For ease of presentation, we first show some definitions which are used for all the cases throughout the paper.

Definition 1. We define a block to be the maximal interval containing only active time. We say a job belongs to a block if this job is executed within the block's duration.

Definition 2. We define the latest starting time (LST) for j_i to be $LST_i = d_i - p_i$ and $ELST(J) = \min_{j_i \in J}\{LST_i\}$.

Lemma 1. *There exists an optimal schedule such that the starting time of each block is $ELST(\mathcal{Y})$, where $\mathcal{Y}$ is the set of jobs executed in this block.*

Proof. First, the start time of the block cannot be later than $ELST(\mathcal{Y})$ since it will violate some job's deadline. On the other hand, if the starting time of the block is earlier than $ELST(\mathcal{Y})$, then we can push the starting time of the block to $ELST(\mathcal{Y})$, moving every job's starting time forward by the same amount. This gives us a new block with the same length. However, moving every job's execution by the same amount may violate some job's deadline. When that happens, we just keep those jobs' starting time at their LSTs. Because all these LSTs are at least $ELST(\mathcal{Y})$, this will give us a feasible schedule with active time at most the length of the original block.   □

3   Special Jobs

3.1   Jobs with Agreeable Deadlines

If jobs have agreeable deadlines where $d_i \leq d_k$ if $r_i < r_k$, we show that an optimal schedule can be found by grouping the jobs into different blocks sequentially and then picking the combination of blocks where the total length of the blocks is minimized.

Lemma 2. *For jobs with agreeable deadlines, there exists an optimal schedule S such that the jobs are processed in the order of their release times and the starting time of each block is $ELST(\mathcal{Y})$, where $\mathcal{Y}$ is the set of jobs executed in the same block in S.*

Proof. Let J be the set of n jobs, where the jobs are sorted by release times such that $r_1 \leq r_2 \leq \cdots \leq r_n$. For jobs with the same release time, they are sorted by increasing deadlines. Let S be an optimal schedule where the starting time of each block is $ELST(\mathcal{Y})$ which does not follow the processing order of release times. In other words, there exists a pair of jobs, j_i and j_k, with $r_i < r_k$ (or $r_i = r_k$ and $d_i < d_k$) and j_k is scheduled before j_i. We have two cases to discuss, which are $p_i \leq p_k$ and $p_i > p_k$.

In the first case, since $r_i \leq r_k$ *and* $p_i \leq p_k$, it is feasible to start job j_i at s_k without exceeding its deadline d_i. Because $p_i \leq p_k$, if both jobs start at the same time point, the job execution of j_k can cover up the whole execution of job j_i. Therefore, we can always set s_i to be s_k without increasing the number of active time slots which is illustrated in Fig. 1(a). In the second case, since $p_i > p_k$, the execution of job j_i can cover up the whole execution of job j_k. Since $d_i \leq d_k$. We can start job j_k at s_i without exceeding its deadline d_k. Therefore we can schedule both jobs at the same time point s_i without increasing the number of active time slots in an optimal schedule which is illustrated in Fig. 1(b).

Besides, as indicated by Lemma 1, we can push the starting time of the blocks to $ELST(\mathcal{Y})$ by moving every job's starting time forward by the same amount. Let j_e be the job with the earliest latest starting time in a block. After pushing the jobs, the jobs j_y with $r_y \leq r_e$ can always be scheduled at the same time with the job j_e at $ELST(\mathcal{Y})$. Hence, the jobs remain processed in the order of their release times.

Thus, we show that in any case, an optimal schedule can be obtained where the jobs are processed in the order of release times and the starting time of each block is $ELST(\mathcal{Y})$. □

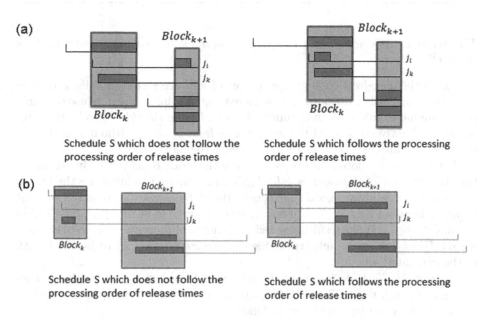

(a)

Schedule S which does not follow the processing order of release times

Schedule S which follows the processing order of release times

(b)

Schedule S which does not follow the processing order of release times

Schedule S which follows the processing order of release times

Fig. 1. Schedule not following the processing order of release times.

Now, we propose Algorithm 1 to compute the optimal schedule for jobs with agreeable deadlines.

Algorithm 1. *SolveAgreeableDeadlines(J)*

Input: $J = \{J_1, J_2, \cdots, J_n\}$, $J_i = \{r_i, p_i, d_i\}$, $LST_i = d_i - p_i$ where $i \geq 1$
Output: The set of blocks $\{i, j\}$ in an optimal schedule

1 Sort the jobs by release times in increasing order, where $r_1 \leq r_2 \leq \cdots \leq r_n$
2 Let $A[i][k]$ be the length of the block that executed jobs from j_i to j_k
3 Initialize $A[i][k] = 0$ for all i, k where $0 \leq i, k \leq n$
4 **for** $(i = 1\ to\ n)$ **do**
5   **for** $(k = 1\ to\ n)$ **do**
6   **if** $(i > k)$ **then**
7             $A[i][k] = A[k][i]$
8   **else if** $(i < k)$ **then**
9   /* We now compute B_e which is the finishing time of the block $Block(i, k - 1)$ */
10             $B_e = ELST(\{j_i, ..., j_k\}) + A[i][k - 1]$
11             $A[i][k] = A[i][k - 1] + p_k - (B_e - r_k - max\{0, B_e - r_k - p_k\})$
12   **else**
13             $A[i][k] = 0$

14 **return** $\min\{OPT(1, k - 1) + A[k, n], k \in [1, n - 1]\}$

Theorem 1. *Algorithm 1 can compute an optimal schedule with running time of $O(n^3)$.*

Proof. In the initialization step, the indices of jobs are sorted by release time in non-decreasing order, breaking ties by increasing deadlines. Then we use dynamic programming to calculate an optimal schedule. Define $Block(i, k)$ to be the minimum total active time used to execute jobs from j_i to j_k without idle time in between. We also use $Block(i, k)$ to represent the block itself.

Let $\mathcal{Y}$ be a set of jobs that are executed in the block $Block(i, k)$. By Lemma 1, the starting time of the block is $ELST(\mathcal{Y})$. Since we aim to minimize the length of the block, it is always good to execute other jobs as early as possible without being earlier than $ELST(\mathcal{Y})$. Therefore, we can set s_m to be $\max\{r_m, ELST(\mathcal{Y})\}$ for all $i \leq m \leq k$. If the resulting schedule consists of more than one blocks, then we set $Block(i, k)$ to be infinity. Otherwise we set $Block(i, k)$ to be the length of the generated schedule.

We use $OPT(i, k)$ to represent the minimum total active time needed to execute jobs from j_i to j_k. Then $OPT(1, n)$, the active time of an optimal schedule for all the jobs can be calculated as follows.

$$OPT(1, n) = min \begin{cases} Block(1, n) \\ OPT(1, k) + Block(k + 1, n) & for\ k = 1, 2, ..., n - 1 \end{cases}$$

where $OPT(1, 1)$ is the processing time of j_1 which is p_1.

The algorithm is conducted in two phases. In the first phase, the algorithm computes $Block(i, k)$ with different i and k in $O(n^3)$ time since each block needs

$O(n)$ time to calculate. In the second phase, the algorithm calculates $OPT(1, k)$ where each needs $O(n)$ time to find the minimum. Therefore the overall running time of the algorithm is $O(n^3)$.   □

3.2   Jobs with Large Sizes

For large jobs where for each job j_i, $p_i \geq (d_i - r_i)/2$, we denote the mid-point of j_i by m_i where $m_i = (d_i + r_i)/2$.

Lemma 3. *Let j_i and j_k with $m_i < m_k$ be scheduled in the same block $Block_j$. Then any job j_x with $m_i < m_x < m_k$ should also be scheduled in $Block_j$.*

Proof. Since $p_i \geq (d_i - r_i)/2$, the execution of the jobs should always contain their mid-points such that $s_i \leq m_i \leq f_i$ for all possible starting time and finishing time. Since m_x is located in $Block_j$ between m_i and m_k, and the execution of the jobs should contain their mid-points, j_x must have a portion of execution in $Block_j$. Hence j_x must be scheduled in $Block_j$.   □

As showed in Lemma 3, for this special case, the processing order in an optimal schedule can be the mid-point order. As the ordering is similar to the agreeable deadlines case, we can use Algorithm 1 to find out an optimal schedule. Instead of using release times as the processing order, the algorithm uses the mid-points as the processing order to group the jobs in different blocks and follows the same principle to find out an optimal schedule.

4   Online Algorithm

In previous sections, we presented algorithms for the problem in the offline setting where all the job information is known initially. However, in practice it may not be possible to know all information about the jobs before they are released. Therefore, we need online algorithms to schedule the real time jobs.

Similar to the offline setting, we are given n jobs, where each job j_i has release time r_i, processing time p_i and deadline d_i. In the online setting, the information of j_i is not available until r_i. Therefore the algorithm is based on the available jobs' information to find the best schedule. The online algorithm maintains an arrival list of jobs denoted as $\mathcal{J}'$. Whenever some job j_k arrives, j_k will be added into the arrival list and j_k will be removed from the list in the completion of its execution.

In the online setting, since the jobs information is unknown before they are released, the result schedule which is outputted by an online algorithm may not be an optimal schedule. Therefore, it is necessary to measure the performance of an online algorithm ALG and it can be measured by its competitive ratio which is defined by $\max_J \frac{ALG(J)}{OPT(J)}$ where $OPT(J)$ is the optimal offline solution, $ALG(J)$ is the solution returned by an online algorithm and J is the set of input jobs.

Recently, a 2-approximation algorithm for the general setting is presented in [9] which is easy to adapt to the online setting with the same performance.

However, we found that there exists a counter example showing that the approximation ratio is not 2, thanks to an anonymous reviewer.

In this section, we will first present the details of the greedy approximation algorithm proposed in [9] and present the counter example. Then we present an online algorithm with competitive ratio of 4.

4.1 Counter Example of Greedy Algorithm Analysis

In the greedy algorithm proposed by [9], the jobs are sorted by deadlines in non-decreasing order in the initialization step. Then the algorithm extracts the job j_x with the earliest deadline and finds the jobs j_y with $r_y < d_x$ to create a single block in order to execute j_x together with all j_y. After that j_x and all j_y are removed from the input job set. The algorithm repeats the extraction step and block creation step for the remaining jobs. The algorithm terminates when the input job set becomes empty.

Consider an example with an interval of length k, there are a unit job with size of 1, and a tiny job with size of ϵ released at each integer point. The parameters of the jobs are illustrated in Fig. 2 below:

1. For job $i \in \{1, \ldots, k\}$ (tiny jobs), its release time and its deadline are respectively i and $i + \varepsilon$.
2. For job $i \in \{k + 1, \ldots, 2k\}$ (unit jobs), its release time and its deadline are respectively $i - k$ and $k + 1$.

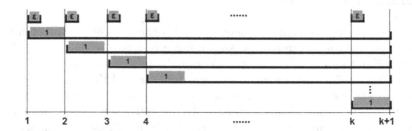

Fig. 2. Counter example.

In the greedy algorithm proposed by [9], since the algorithm extracts the job j_x with the earliest deadline and all jobs j_y with $r_y < d_x$ will be executed together, the tiny job will be executed with the corresponding unit job together. Therefore the length of the schedule is k.

However, in an optimal schedule, all the tiny jobs will be executed separately and all the unit jobs will be executed together at the end. Therefore the length of the schedule is $\epsilon(k - 1) + 1$.

The approximation ratio in this example is $ALG/OPT = k/(k\epsilon + 1) \approx k$. This shows that the algorithm proposed by [9] cannot obtain the ratio of 2.

4.2   Online Algorithm with Competitive Ratio of 4

By Lemma 1, we show that we can always find an optimal schedule with starting time equal to some LST. Let $\mathcal{J}'$ be the set of available jobs. The algorithm initially sorts the jobs in $\mathcal{J}'$ by LST in non-decreasing order. Therefore, the algorithm can wait for more jobs to arrive and schedule the jobs when the current time t reaches the $ELST(\mathcal{J}')$.

When $t = ELST(\mathcal{J}')$, we extract the job j_x from $\mathcal{J}'$ with $LST_x = ELST(\mathcal{J}')$. Then we find all other jobs j_y with $r_y < d_x$ where a portion of job execution can be executed together with j_x in $[LST_x, d_x]$. We denote this portion of the job execution as overlap portion.

For each j_y, if the overlap portion with j_x is at least r, then the algorithm will schedule the job at $max(LST_x, r_y)$. Otherwise the job will be scheduled in later iterations. The algorithm terminates when $\mathcal{J}'$ becomes empty.

Lemma 4. *Algorithm 2 is a 4-competitive algorithm*

Proof. Recall in our algorithm, in each iteration, we first select the job j_k with earliest LST, and fix its execution window as $[d_k - p_k, d_k]$. For all other jobs j_j with $r_j \leq d_k$, we divide these jobs into two cases below:

1. if $r_j > LST_k$, then we allocate its execution window as $[r_j, r_j + p_j]$
2. Otherwise, we allocate its execution window as $[LST_k, LST_k + p_j]$

Algorithm 2. Online algorithm of arbitrary jobs scheduling

Input: $S = \{S_1, S_2, \cdots, S_n\}, S_i = \{r_i, p_i, d_i\}$ where $i \geq 1$
Output: I, the set of jobs interval $\{s_i, f_i\}$ in the near-optimal schedule

```
 1  Initialization: I = ∅, r = 0.5
 2  Jobs arrival. When a new job j_k arrives
 3  j_k is added to the set J'.
 4  Jobs execution. When current time t = ELST(J')
 5  Sort the jobs in J' by LST in non-decreasing order.
 6  while J' ≠ ∅ do
 7      j_x = 1st element in J'
 8      J' = J' \ j_x
 9      s_x = d_x - p_x
10      for j_y ∈ J' with r_y ≤ d_x do
11          if r_y < s_x then
12              s_y = s_x
13          else
14              s_y = r_y
15          if (d_x - s_y)/p_y ≥ r then
16              J' = J' \ j_y
17              I ∪ [s_y, s_y + p_y]
```

Therefore, each job j_j with $r_j \leq d_k$ must contain a portion of the job execution which can be executed together with j_k in $[LST_k, d_k]$. We denote this portion of the job execution as overlap portion. If the ratio between the length of overlap portion and p_j is greater than or equal to r (i.e. $r = 0.5$), the jobs will remain at their current position in the final schedule. Otherwise, the jobs will be postponed to be scheduled in later iterations.

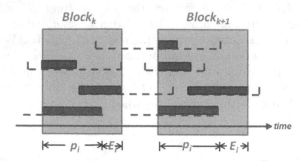

Fig. 3. Blocks creation for the online algorithm.

Observe that in each iteration, we create a block which contains only active time as illustrated in Fig. 3. For each block $Block_k$, the starting time of the block is the $j_i \in Block_k$ with the earliest LST. Let the length of interval between d_i and the finishing time of $Block_k$ be E_i. As we fixed the execution window for j_i, the length of the block is equivalent to the sum of p_i and E_i. Since $r = 0.5$, we have $E_i \leq p_i$. Let ALG be the upper bound of the solution in our online algorithm, and OPT be the lower bound in an optimal schedule. Then,

$$ALG \leq \sum_{i=1}^{m}(p_i + E_i) \leq 2\sum_{i=1}^{m} p_i \qquad (1)$$

where m is the total number of blocks.

In the online algorithm, consider that in the kth iteration, j_i with earliest LST in $Block_k$ is either

1. Not flexible to fit in $Block_{k-1}$ due to $r_i >$ finishing time of $Block_{k-1}$
2. Postponed from the $(k-1)$th iteration due to the r threshold.

Observe that for any two consecutive jobs, j_i and j_k, if $r_k < d_i$, then at most $p_k/2$ will overlap with j_i. Therefore, we require at least $p_k/2$ time to execute j_k and we have the following lower bound.

$$OPT \geq \sum_{i=1}^{m} p_i \geq \sum_{i=1}^{m} \frac{p_i}{2} = 0.5 \sum_{i=1}^{m} p_i \qquad (2)$$

where m is the total number of blocks.

Then we have the competitive ratio $= ALG/OPT \leq 2\sum_{i=1}^{m} p_i / 0.5 \sum_{i=1}^{m} p_i = 4.$ $\square$

Besides the upper bound above, we find an example to show our analysis is tight. In this example, we have u intervals. In each interval, we are given m jobs where their characteristics are illustrated in Fig. 4 below:

1. All jobs have the same release time such that $r_1 = r_2 = \cdots = r_m$
2. In each interval, the processing time of jobs are in the following patterns: $p_1 = 1, p_2 = 2 - \epsilon, p_3 = 2 + \epsilon, p_4 = 4 - \epsilon, p_5 = 4 + 3\epsilon \cdots p_{m-1} = 2^{m-1} + (2^{m-1} - 1)\epsilon, p_m = 2^m$ where ϵ is a small number.
3. In each interval, the deadlines of the first job j_i is $r_i + 2p_i$, the last job is till the end of the schedule, and the deadlines of all intermediate jobs j_j is equal to $d_{j-1} + p_j$ such that $d_1 = r_1 + 2p_1, d_2 = d_1 + p_2, \cdots, d_{m-1} = d_{m-2} + p_{m-1}, d_m = d_{max}$

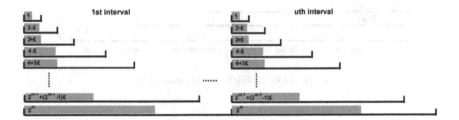

Fig. 4. An example to show the competitive ratio is at least 4.

Consider that in our online algorithm, the algorithm first selects the job j_k with the earliest LST, then for all other jobs, we try to place the jobs as early as $d_k - p_k$ to maximize the overlapping interval between jobs. Observe that in this example, for any two consecutive jobs j_i and j_{i+1} they can be executed together with the duration of $2p_i$ since the overlapping interval require at least r (i.e. $r = 0.5$). Therefore the length of the schedule can be computed by

$$(1 + 2 + \epsilon + 4 + 3\epsilon \cdots + 2^{m-1} + (2^{m-1} - 1)\epsilon) * 2 * u \approx 2^m * 2u = 2^{m-1} * 4u \quad (3)$$

However, since $d_m = d_{max}$ in all u intervals, the last job in each interval can be executed together in the execution window of $[d_{max} - p_m, d_{max}]$. Besides the last job, we only require $2^{m-1} + (2^{m-1} - 1)\epsilon$ to cover all $m - 1$ jobs in each interval. Hence we have

$$(2^{m-1} + (2^{m-1} - 1)\epsilon) * (u - 1) + 2^m \approx 2^{m-1} * (u - 1) + 2^m$$
$$= 2^{m-1} * u + 2^{m-1} = 2^{m-1}(u + 1) \quad (4)$$

which is also illustrated in Fig. 5. Then we have the competitive ratio as follows.

$$ALG/OPT = 2^{m-1} * 4u/2^{m-1}(u + 1) = 4u/(u + 1) \quad (5)$$

This shows that the competitive ratio of Algorithm 2 cannot be better than 4 which shows the tightness of our analysis in Lemma 4.

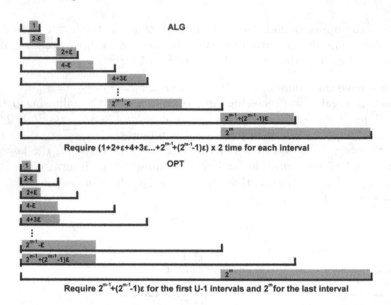

Fig. 5. ALG vs OPT in this example.

5   Conclusion

In this paper, we investigated the active time minimization problem. We present an $O(n^3)$ dynamic programming algorithm for the special case of the problem where all jobs have agreeable deadlines or all jobs are big. In the general case, we present a counter example to show that the algorithm proposed by [9] cannot obtain the approximation ratio of 2. Moreover, we present an online algorithm with competitive ratio of 4 and show that our analysis is tight.

References

1. Albers, S.: Energy-efficient algorithms. Commun. ACM **53**(5), 86–96 (2010)
2. Amur, H., Cipar, J., Gupta, V., Ganger, G.R., Kozuch, M.A., Schwan, K.: Robust and flexible power-proportional storage. In: Proceedings of the 1st ACM SoCC 2010, pp. 217–228. ACM (2010)
3. Angel, E., Bampis, E., Chau, V.: Low complexity scheduling algorithms minimizing the energy for tasks with agreeable deadlines. Discrete Appl. Math. **175**, 1–10 (2014)
4. Baptiste, P.: Scheduling unit tasks to minimize the number of idle periods: a polynomial time algorithm for offline dynamic power management. In: 2006 Proceedings of the 17th Annual ACM-SIAM SODA, pp. 364–367. ACM Press (2006)
5. Baptiste, P., Chrobak, M., Dürr, C.: Polynomial time algorithms for minimum energy scheduling. In: Arge, L., Hoffmann, M., Welzl, E. (eds.) ESA 2007. LNCS, vol. 4698, pp. 136–150. Springer, Heidelberg (2007). doi:10.1007/978-3-540-75520-3_14

6. Chang, J., Gabow, H.N., Khuller, S.: A model for minimizing active processor time. In: Epstein, L., Ferragina, P. (eds.) ESA 2012. LNCS, vol. 7501, pp. 289–300. Springer, Heidelberg (2012). doi:10.1007/978-3-642-33090-2_26

7. Chang, J., Khuller, S., Mukherjee, K.: LP rounding and combinatorial algorithms for minimizing active and busy time. In: 26th ACM SPAA 2014, pp. 118–127. ACM (2014)

8. Demaine, E.D., Ghodsi, M., Hajiaghayi, M.T., Sayedi-Roshkhar, A.S., Zadimoghaddam, M.: Scheduling to minimize gaps and power consumption. J. Sched. **16**(2), 151–160 (2013)

9. Fang, X., Gao, H., Li, J., Li, Y.: Application-aware data collection in wireless sensor networks. In: Proceedings of the IEEE INFOCOM 2013, pp. 1645–1653. IEEE (2013)

10. Flammini, M., Monaco, G., Moscardelli, L., Shachnai, H., Shalom, M., Tamir, T., Zaks, S.: Minimizing total busy time in parallel scheduling with application to optical networks. In: 23rd IEEE IPDPS 2009, pp. 1–12. IEEE (2009)

11. Ikura, Y., Gimple, M.: Efficient scheduling algorithms for a single batch processing machine. Oper. Res. Lett. **5**(2), 61–65 (1986)

12. Irani, S., Pruhs, K.: Algorithmic problems in power management. SIGACT News **36**(2), 63–76 (2005)

13. Khandekar, R., Schieber, B., Shachnai, H., Tamir, T.: Minimizing busy time in multiple machine real-time scheduling. In: IARCS Annual Conference on FSTTCS 2010. LIPIcs, vol. 8, pp. 169–180 (2010)

14. Koehler, F., Khuller, S.: Optimal batch schedules for parallel machines. In: Dehne, F., Solis-Oba, R., Sack, J.-R. (eds.) WADS 2013. LNCS, vol. 8037, pp. 475–486. Springer, Heidelberg (2013). doi:10.1007/978-3-642-40104-6_41

15. Shalom, M., Voloshin, A., Wong, P.W.H., Yung, F.C.C., Zaks, S.: Online optimization of busy time on parallel machines. In: Agrawal, M., Cooper, S.B., Li, A. (eds.) TAMC 2012. LNCS, vol. 7287, pp. 448–460. Springer, Heidelberg (2012). doi:10.1007/978-3-642-29952-0_43

16. Tavakoli, A., Kansal, A., Nath, S.: On-line sensing task optimization for shared sensors. In: Proceedings of the 9th ACM/IEEE International Conference on Information Processing in Sensor Networks, pp. 47–57. ACM (2010)

17. Tian, W., Xue, R., Cao, J., Xiong, Q., Hu, Y.: An energy-efficient online parallel scheduling algorithm for cloud data centers. In: 2013 IEEE Ninth World Congress on Services (SERVICES), pp. 397–402. IEEE (2013)

The Strength of the SCT Criterion

Emanuele Frittaion[1]([⊠]), Silvia Steila[2], and Keita Yokoyama[3]

[1] Mathematical Institute, Tohoku University, Sendai, Japan
emanuelefrittaion@gmail.com
[2] Institute of Computer Science, University of Bern, Bern, Switzerland
steila@inf.unibe.ch
[3] School of Information Science,
Japan Advanced Institute of Science and Technology, Nomi, Japan
y-keita@jaist.ac.jp

Abstract. We undertake the study of size-change analysis in the context of Reverse Mathematics. In particular, we prove that the SCT criterion [9, Theorem 4] is equivalent to $I\Sigma_2^0$ over RCA_0.

Keywords: Ramsey's theorem for pairs · Size-change termination · Reverse Mathematics · Σ_2^0-induction

1 Introduction

Ramsey's theorem for pairs (RT^2) is one of the main characters in Reverse Mathematics. It states that for any natural number k and for any edge coloring of the complete graph with countably many nodes in k-many colors, there exists an infinite homogeneous set, i.e. there exists an infinite subset of nodes whose any two elements are connected with the same color [12].

As highlighted by Gasarch [4], Ramsey's theorem for pairs can be used to prove termination. For instance, Podelski and Rybalchenko characterized the termination of transition based programs as a property of well-founded relations by using Ramsey's theorem for pairs [11]. In [15] we started investigating the termination analysis from the point of view of Reverse Mathematics. We proved the equivalence between the termination theorem of Podelski and Rybalchenko and a corollary of Ramsey's theorem for pairs, which is weaker than Ramsey's theorem for pairs itself.

The termination theorem is not the only result which characterizes the termination of some class of programs. In [9] Lee, Jones and Ben-Amram introduced the notion of size-change termination (SCT) for first order functional programs. Size-change analysis is a general method for *automated termination proofs*. In fact, this method has been applied in the termination analysis of higher-order programs [8], logic programs [2], and term rewrite systems [16].

Informally, a program is size-change terminating (SCT) if every infinite state transition sequence would cause an infinite sequence of data values which is weakly decreasing and strictly decreasing infinitely many times. If the domain of

T.V. Gopal et al. (Eds.): TAMC 2017, LNCS 10185, pp. 260–273, 2017.
DOI: 10.1007/978-3-319-55911-7_19

data values is well-founded, such as the natural numbers, there cannot be such a sequence, thus SCT is a sufficient condition for termination [9, Theorem 1].

Size-change termination is based on the notion of size-change graph (see Subsect. 2.2). Given a first order functional program P we associate to every call $f \to g$ a bipartite graph which describes the relation between source and target parameter values. These graphs are called size-change graphs.

In this paper we start the investigation of size-change termination in the framework of Reverse Mathematics. In particular, we analyse the following criterion for testing SCT [9, Theorem 4]:

Theorem 1 (SCT criterion). *Let $\mathcal{G}$ be a set of size-change graphs for a first order functional program P. Then $\mathcal{G}$ is SCT iff every idempotent $G \in \mathsf{cl}(\mathcal{G})$ has an arc $x \xrightarrow{\downarrow} x$.*

The original proof of the SCT criterion is based on Ramsey's theorem for pairs. In this paper we show that this is far from optimal and pinpoint the exact strength of the SCT criterion from the point of view of Reverse Mathematics. For our analysis we consider the following version, where we consider size-change graphs only.

Theorem 2 (SCT criterion for graphs). *Let $\mathcal{G}$ be a set of size-change graphs. Then $\mathcal{G}$ is SCT iff every idempotent $G \in \mathsf{cl}(\mathcal{G})$ has an arc $x \xrightarrow{\downarrow} x$.*

To the aim of studying the strength of the SCT criterion we introduce and study a corollary of Ramsey's theorem for pairs, called Triangle Ramsey's theorem (Triang). It states that for any natural number k and for any edge coloring of the complete graph with countably many nodes in k-many colors, there is some node which is, for some color $i \in k$, the first node of infinitely many triangles homogeneous in color i. As far as we know this corollary does not appear in the literature.

We show that Triang implies the SCT criterion and that the SCT criterion implies the Strong Pigeonhole Principle (SPP). From these (and some further) results we are able to conclude that both SCT criterion and Triang are equivalent to Σ_2^0-induction ($\mathsf{I}\Sigma_2^0$).

Theorem 3 (RCA$_0$). *The following are equivalent:*

1. $\mathsf{I}\Sigma_2^0$
2. Triang
3. SCT criterion

1.1   Notation

Given a set $X \subseteq \mathbb{N}$, let $[X]^2$ denote the set of 2-element subsets of X. As usual, we identify $[X]^2$ with the set $\{(x,y) \colon x,y \in X \wedge x < y\}$. We also identify a natural number k with the set $\{0, \ldots, k-1\}$. For $k \in \mathbb{N}$, we call a function $c \colon [\mathbb{N}]^2 \to k$ a *coloring* of $[\mathbb{N}]^2$ in k-many colors.

For a set $X \subseteq \mathbb{N}$, $X^{<\mathbb{N}}$ denotes the set of finite sequences of elements in X. Given a set X and a sequence $\sigma \in X^{<\mathbb{N}}$ we denote by $|\sigma|$ the length of the sequence, by $\text{last}(\sigma)$ the last element of the sequence and by $\sigma(i)$ the i-th element of the sequence, if it exists. Note that $k^{<\mathbb{N}}$ is the set of finite sequences of natural numbers less than k.

1.2   Reverse Mathematics

Reverse Mathematics is a program in mathematical logic introduced by Harvey Friedman in [3], which stems from the following question. Given a theorem of ordinary mathematics, what is the weakest subsystem of second order arithmetic in which it is provable?

Amongst the several subsystems of second order arithmetic (see [13] for a detailed description), in this paper we consider only few extensions of RCA_0 (Recursive Comprehension Axiom). RCA_0 is the standard base system of Reverse Mathematics. It consists of the usual axioms of first order arithmetic for $0, 1, +, \times, <$, induction for Σ_1^0-formulas ($\mathsf{I}\Sigma_1^0$) and comprehension for Δ_1^0-formulas.

The infinite pigeonhole principle (RT^1) and Ramsey's theorem for pairs (RT^2) are defined as follows.

(RT_k^1) For any $c \colon \mathbb{N} \to k$ there exists $i < k$ such that $c(x) = i$ for infinitely many x.

(RT^1) $\forall k \in \mathbb{N} \ \mathsf{RT}_k^1$.

(RT_k^2) For any $c \colon [\mathbb{N}]^2 \to k$ there exists an infinite homogeneous set $X \subseteq \mathbb{N}$, that is $c \restriction [X]^2$ is constant.

(RT^2) $\forall k \in \mathbb{N} \ \mathsf{RT}_k^2$.

Let $\mathsf{I}\Sigma_2^0$ be induction for Σ_2^0-formulas. It is known that RT^2 implies the bounding principle for Σ_3^0-formulas ($\mathsf{B}\Sigma_3^0$) over RCA_0 [7], and so in particular $\mathsf{I}\Sigma_2^0$. As a side result here we provide a different proof of the fact that RT^2 implies $\mathsf{I}\Sigma_2^0$. Indeed we introduce an immediate consequence of RT^2, the Triangle Ramsey's theorem (Triang), which turns out to be equivalent to $\mathsf{I}\Sigma_2^0$.

(Triang_k) For any coloring $c \colon [\mathbb{N}]^2 \to k$ there exist $i \in k$ and $t \in \mathbb{N}$ such that $c(t, m) = c(t, l) = c(m, l) = i$ for infinitely many pairs $m < l$.

(Triang) $\forall k \in \mathbb{N} \ \text{Triang}_k$

2   The SCT Framework

In this section we describe the size-change method for first order functional programs as in [9]. All the definitions are made in RCA_0 except for the semantic notion of *safety*.

2.1 Syntax

We consider the following basic first order functional language:

$x \in \text{Par}$ parameter identifier

$f \in \text{Fun}$ function identifier

$o \in \text{Op}$ primitive operator

$a \in \text{AExp}$ arithmetic expression

$$::= x \mid x+1 \mid x-1 \mid o(a,\ldots,a) \mid f(a,\ldots,a)$$

$b \in \text{BExp}$ boolean expression

$$::= x = 0 \mid x = 1 \mid x < y \mid x \leq y \mid b \wedge b \mid b \vee b \mid \neg b$$

$e \in \text{Exp}$ expression

$$::= a \mid \textbf{if } b \textbf{ then } e \textbf{ else } e$$

$d \in \text{Def}$ function definition

$$::= f(x_0,\ldots,x_{n-1}) = e$$

$P \in \text{Prog}$ program

$$::= d_0,\ldots,d_{m-1}$$

Remark 1. This language is Turing complete.

A program P is a list of finitely many defining equations $f(x_0,\ldots,x_{n-1}) = e^f$, where $f \in \text{Fun}$ and e^f is an expression, called the *body* of f. Let $x_0,\ldots,x_{n-1}$ be the *parameters* of f, denoted $\text{Par}(f)$, and let n be the *arity* of f, denoted $\text{ar}(f)$.

By $\text{Fun}(P)$ we denote the set of functions of P. We also assume that a program P specifies an *initial* function $f \in \text{Fun}(P)$. The idea is that P computes the (partial) function $f : \mathbb{N}^{\text{ar}(f)} \to \mathbb{N}$.

In [9] the expression evaluation is based on a *left-to-right call-by-value* strategy given by *denotational semantics*. RCA_0 is not capable to formalize denotational semantics, and hence we need to consider other approaches if we want to study termination over RCA_0 (for instance, by *operational semantics*). Anyway we do not formally discuss semantics. For the sake of exposition, it is enough to say that one evaluates a program function f given an assignment of values $\mathbf{u}$ to its parameters (i.e. an element of $\mathbb{N}^{\text{ar}(f)}$) by evaluating the body of f, that is $f(\mathbf{u}) = e^f(\mathbf{u})$.

Example 1 (Péter-Ackermann).

$$A(x,y) = \textbf{if } x = 0 \textbf{ then } y + 1 \textbf{ else}$$
$$\textbf{if } y = 0 \textbf{ then } A(x-1,1)$$
$$\textbf{else } A(x-1, A(x, y-1))$$

2.2   Size-Change Graphs

In order to express the notion of size-change termination, first of all we need the definition of size-change graph (see [9, Definition 3]).

Definition 1 (size-change graph). *Let P be a program and f, $g \in \mathrm{Fun}(P)$. A size-change graph $G : f \to g$ for P is a bipartite graph on $(\mathrm{Par}(f), \mathrm{Par}(g))$. The set of edges is a subset of $\mathrm{Par}(f) \times \{\downarrow, \Downarrow\} \times \mathrm{Par}(g)$ such that there is at most one edge for any $x \in \mathrm{Par}(f)$ and $y \in \mathrm{Par}(g)$.*

We say that f is the *source* function of G and g is the *target* function of G. We call $(x, \downarrow, y)$ the *decreasing edge* (strict arc), and we denote it by $x \xrightarrow{\downarrow} y$. We call $(x, \Downarrow, y)$ the *weakly-decreasing edge* (non-strict arc), and we denote it by $x \xrightarrow{\Downarrow} y$. We write $x \to y \in G$ as a shorthand for $x \xrightarrow{\downarrow} y \in G \vee x \xrightarrow{\Downarrow} y \in G$.

Note that the absence of edges between two parameters x and y in the size-change graph G indicates either an unknown or an increasing relation in the call $f \to g$.

Informally, a size-change graph is an approximation of the *state transition* relation induced by the program. A size-change graph $G : f \to g$ for a call $\tau : f \to g$ is *safe* if it reflects the relationship between the parameter values in the program call.

In more detail, a *state* of a program P is a pair $(f, \mathbf{u})$, where $f \in \mathrm{Fun}(P)$ and $\mathbf{u}$ is a tuple of length $\mathrm{ar}(f)$. If in the body of $f \in \mathrm{Fun}(P)$ there is a call

$$\ldots \tau : g(e_0, \ldots, e_{m-1})$$

we define a *state transition* $(f, \mathbf{u}) \xrightarrow{\tau} (g, \mathbf{v})$ to be a pair of states such that $\mathbf{v}$ is the sequence of values obtained by the expressions $(e_0, \ldots, e_{m-1})$ when f is evaluated with values $\mathbf{u}$.

Let $\mathrm{Par}(f) = \{x_0, \ldots, x_{n-1}\}$ and $\mathrm{Par}(g) = \{y_0, \ldots y_{m-1}\}$. We say that a size-change graph $G : f \to g$ is *safe* for τ if every edge is safe, where an edge $x_i \xrightarrow{\tau} y_j$ is safe if for any $\mathbf{u} \in \mathbb{N}^n$ and $\mathbf{v} \in \mathbb{N}^m$ such that $(f, \mathbf{u}) \xrightarrow{\tau} (g, \mathbf{v})$, $r = \downarrow$ implies that $\mathbf{u}_i > \mathbf{v}_j$ and $r = \Downarrow$ implies that $\mathbf{u}_i \geq \mathbf{v}_j$.

Note for instance that the size-change graph without edges is always safe.

Example 2 (Péter-Ackermann).

$$
\begin{aligned}
A(x, y) = \ &\mathbf{if}\ \ x = 0\ \ \mathbf{then}\ \ y + 1\ \ \mathbf{else} \\
&\mathbf{if}\ \ y = 0\ \ \mathbf{then}\ \ \tau_0 : A(x - 1, 1) \\
&\mathbf{else}\ \ \tau_1 : A(x - 1, \tau_2 : A(x, y - 1))
\end{aligned}
$$

There are three calls τ_i $(i < 3)$ safely described by the following size-change graphs:

$$x \xrightarrow{\quad \downarrow \quad} x \qquad\qquad x \xrightarrow{\quad \Downarrow \quad} x$$

$$y \qquad\qquad y \qquad\qquad y \xrightarrow{\quad \downarrow \quad} y$$

$$G_{0,1} : A \to A \qquad\qquad G_2 : A \to A$$

The size-change graph $G_{0,1}$ safely describes both calls $\tau_0 : A(x - 1, 1)$ and $\tau_1 : A(x - 1, A(x, y - 1))$. In particular, notice that in the call τ_1 the parameter value x decreases no matter what the value of the expression $A(x, y - 1)$ is. Finally, the size-change graph G_2 safely describes the call $\tau_2 : A(x, y - 1)$.

Note that we could have assumed that for any parameter in the target there is at most one edge, since in every call of the programs we consider any parameter value in the target depends at most from one parameter in the source. However this restriction is not essential. Note also that the SCT framework has been generalized in order to deal with other kinds of *monotonicity constraints* [1], where SCT only deals with two constraints $x > y$ (a strict arc) and $x \geq y$ (a non-strict arc).

Nonetheless we want to emphasize that the notion of size-change graph is clearly independent of that of a program and so we can define it directly. For simplicity we may assume that every function $f \in \text{Fun}$ comes with a set of parameters $\text{Par}(f)$ of size $\text{ar}(f)$.

Definition 2 (size-change graph). *Let $f, g \in \text{Fun}$. A size-change graph $G : f \to g$ is a bipartite graph on $(\text{Par}(f), \text{Par}(g))$. The set of edges is a subset of $\text{Par}(f) \times \{\downarrow, \Downarrow\} \times \text{Par}(g)$ such there is at most one edge for any $x \in \text{Par}(f)$ and $y \in \text{Par}(g)$.*

2.3 SCT Criterion

Definition 3 (composition). *As in [6], given two size-change graphs $G_0 : f \to g$ and $G_1 : g \to h$ we define their composition $G_0; G_1 : f \to h$. The composition of two edges $x \xrightarrow{\Downarrow} y$ and $y \xrightarrow{\Downarrow} z$ is one edge $x \xrightarrow{\Downarrow} z$. In all other cases the composition of two edges from x to y and from y to z is the edge $x \xrightarrow{\downarrow} z$. Formally, $G_0; G_1$ is the size-change graph with the following set of edges:*

$$E = \{x \xrightarrow{\downarrow} z : \exists y \in \text{Par}(g) \ \exists r \in \{\downarrow, \Downarrow\} ((x \xrightarrow{\downarrow} y \in G_0 \wedge y \xrightarrow{r} z \in G_1)$$

$$\vee \, (x \xrightarrow{r} y \in G_0 \wedge y \xrightarrow{\downarrow} z \in G_1))\}$$

$$\cup \{x \xrightarrow{\Downarrow} z : \exists y \in \text{Par}(g)(x \xrightarrow{\Downarrow} y \in G_0 \wedge y \xrightarrow{\Downarrow} z \in G_1) \wedge \forall y \in \text{Par}(g)$$

$$\forall r, r' \in \{\downarrow, \Downarrow\} ((x \xrightarrow{r} y \in G_0 \wedge y \xrightarrow{r'} z \in G_1) \implies r = r' = \Downarrow)\}.$$

Observe that the composition operator ";" is associative. Moreover we say that the size-change graph G is *idempotent* if $G; G = G$.

Given a finite set of size-change graphs $\mathcal{G}$, $\mathsf{cl}(\mathcal{G})$ is the smallest set which contains $\mathcal{G}$ and is closed by composition. Formally $\mathsf{cl}(\mathcal{G})$ is the smallest set such that

- $\mathcal{G} \subseteq \mathsf{cl}(\mathcal{G})$;
- If $G_0 : f \to g$ and $G_1 : g \to h$ are in $\mathsf{cl}(\mathcal{G})$, then $G_0; G_1 \in \mathsf{cl}(\mathcal{G})$.

Definition 4 (multipath). *A multipath $\mathcal{M}$ is a sequence $G_0, \dots, G_n, \dots$ of graphs such that the target function of G_i is the source function of G_{i+1} for all i. A thread is a connected path of edges in $\mathcal{M}$ that starts at some G_t, where $t \in \mathbb{N}$. A multipath $\mathcal{M}$ has* infinite descent *if some thread in $\mathcal{M}$ contains infinitely many decreasing edges.*

Definition 5 (description). *A description $\mathcal{G}$ of P is a finite set of size-change graphs such that to every call $\tau : f \to g$ of P corresponds exactly one $G_\tau \in \mathcal{G}$.*

A description $\mathcal{G}$ of P is *safe* if each graph in $\mathcal{G}$ is safe. Note that there are finitely many descriptions, and in particular finitely many safe descriptions.

Definition 6 (SCT description). *We say that a description $\mathcal{G}$ of P is size-change terminating (SCT) if every infinite multipath $\mathcal{M} = G_0, \dots, G_n, \dots$, where every graph $G_n \in \mathcal{G}$, has an infinite descent.*

It is clear that a program P with a safe SCT description does not have infinite state transition sequences. Thus the existence of a safe SCT description is a sufficient condition for termination.

We now can state the SCT criterion.

Theorem 4 (SCT criterion). *Let $\mathcal{G}$ be a description of P. Then $\mathcal{G}$ is SCT iff every idempotent $G \in \mathsf{cl}(\mathcal{G})$ has an arc $x \xrightarrow{\downarrow} x$.*

To the aim of analysing in Reverse Mathematics it is convenient to state the SCT criterion for arbitrary sets of size-change graphs.

Definition 7 (SCT criterion for graphs). *Let $\mathcal{G}$ be a finite set of size-change graphs. Then $\mathcal{G}$ is SCT iff every idempotent $G \in \mathsf{cl}(\mathcal{G})$ has an arc $x \xrightarrow{\downarrow} x$.*

It is not difficult to see that the two formulations of the SCT criterion are equivalent. In fact, given a finite set $\mathcal{G}$ of size-change graphs, it is straightforward to define a program P such that $\mathcal{G}$ is a description of P. In more detail, let $f_0, \dots, f_m$ be the finite set of source and target functions of $\mathcal{G}$. Without loss of generality, we may assume that all functions have the same arity $n \in \mathbb{N}$. For any i, let $f_{i_0}, \dots, f_{i_{k-1}}$ be the functions (with repetition if there are more graphs with the same source and target functions) which correspond to the target of a graph whose source is f_i. Write the code:

$$f_i(x_0, \ldots x_{n-1}) = \tau_0 : f_{i_0}(e_0^0, \ldots, e_{n-1}^0) \qquad \text{if } x_0 = 0.$$
$$= \ldots$$
$$= \tau_{k-1} : f_{i_{k-1}}(e_0^{k-1}, \ldots, e_{n-1}^{k-1}) \qquad \text{if } x_0 = k - 1.$$

where the expression e_j^h is determined by the source and the kind of the edge to x_j in the corresponding graph, if such an edge exists. Otherwise it is $x_j + 1$.

The union of these codes is a program $P_\mathcal{G}$. Of course, $\mathcal{G}$ is a description of $P_\mathcal{G}$. Therefore:

Proposition 1 (RCA$_0$). *The following are equivalent:*

1. *SCT criterion*
2. *SCT criterion for graphs*

3   Proving the SCT Criterion

The classical proof of the SCT criterion [6] uses Ramsey's theorem for pairs. Actually, what we really need is that there exist infinitely many monochromatic triangles which share a fixed vertex: we need the homogeneous cliques in order to prove that the graph is idempotent and that there are infinitely many strictly decreasing edges in the thread and we need that they share a fixed vertex in order to guarantee the continuity of the path. This is why we introduce the principle Triang.

(Triang$_k$) For any coloring $c : [\mathbb{N}]^2 \to k$ there exist $i \in k$ and $t \in \mathbb{N}$ such that $c(t, m) = c(t, l) = c(m, l) = i$ for infinitely many pairs $\{m, l\}$.
(Triang) $\forall k \in \mathbb{N} \; \text{Triang}_k$.

We also introduce the following strengthening of the infinite pigeonhole principle:

(SPP$_k$) For any coloring $c : \mathbb{N} \to k$ there exists $I \subseteq k$ such that $i \in I$ iff $i < k$ and $c(x) = i$ for infinitely many x.
(SPP) $\forall k \in \mathbb{N} \; \text{SPP}_k$.

For the reversal we use the fact that SPP is equivalent to Σ_2^0-induction.

Lemma 1 (RCA$_0$). *The following are equivalent:*

1. $I\Sigma_2^0$
2. SPP

Proof. It is well-known that $I\Sigma_2^0$ is equivalent over RCA$_0$ to *bounded comprehension* for Π_2^0-formulas, that is the axiom schema

$$\forall k \; \exists X \; \forall i \; (i \in X \leftrightarrow i < k \wedge \varphi(i)),$$

where φ is Π_2^0. It immediately follows that $I\Sigma_2^0$ implies SPP. Let us show that SPP implies bounded Π_2^0-comprehension. Let $\varphi(i) = \forall x \exists y \; \theta(i, x, y)$. We define $c : \mathbb{N} \to k + 1$ by primitive recursion as follows:

1. Let $s = 0$ and $x_i = 0$ for all $i < k$;
2. Suppose we have defined $c(x)$ for every $x < s$. For all $i < k$, if $\exists y < s\, \theta(i, x_i, y)$, let $c(s + i) = i$ and $x_i = x_i + 1$. Otherwise let $c(s + i) = k$;
3. Let $s = s + k$. Return to step 2.

By SPP, the set $I = \{i \leq k \colon \exists^\infty x\ c(x) = i\}$ exists. One can check that $I \setminus k = \{i < k \colon \forall x \exists y\ \theta(i, x, y)\}$.

The following shows that one direction of the SCT criterion is already provable in RCA$_0$.

Proposition 2 (RCA$_0$). *Let $\mathcal{G}$ be a finite set of size-change graphs. If every multipath $M = G_0, \ldots, G_n, \ldots$ has an infinite descent, then every idempotent $G \in \mathsf{cl}(\mathcal{G})$ has an arc $x \overset{\downarrow}{\to} x$.*

Proof. Let G be idempotent. Then $M = G, G, \ldots, G, \ldots$ is a multipath. By hypothesis there exists an infinite descent. Since G is idempotent, one can define an infinite sequence $x_0, x_1, x_2, \ldots$ such that $x_i \overset{\downarrow}{\to} x_{i+1} \in G$. As there are finitely many parameters, by the finite pigeonhole principle, which is provable in RCA$_0$, there exist $i < j$ such that $x = x_i = x_j$. By idempotence of G, $x \overset{\downarrow}{\to} x \in G$.

Theorem 5 (RCA$_0$). Triang *implies the SCT criterion.*

Proof. We prove the SCT criterion for graphs. Let $\mathcal{G}$ be a finite set of size-change graphs and assume that any idempotent graph in $\mathsf{cl}(\mathcal{G})$ has a strict arc $x \overset{\downarrow}{\to} x$ for some parameter x. Let

$$\mathcal{M}_\pi = G_0, \ldots, G_n, \ldots.$$

We aim to prove that $\mathcal{M}_\pi$ has an infinite descent. Define $c : [\mathbb{N}]^2 \to \mathsf{cl}(\mathcal{G})$ as follows:

$$c(i, j) = G_i; \ldots; G_{j-1}.$$

By applying Triang$_{|\mathsf{cl}(\mathcal{G})|}$ to the coloring c, we have:

$$\exists t \exists G \in \mathsf{cl}(\mathcal{G}) \forall n \exists m, l (n < m < l \wedge t < m \wedge c(t, m) = c(t, l) = c(m, l) = G).$$

Then G is idempotent, indeed

$$G; G = c(t, m); c(m, l) = c(t, l) = G.$$

By hypothesis, we have that there exists $x \overset{\downarrow}{\to} x \in G$. By Σ_0^0-comprehension, let $f : \mathbb{N}^3 \to \mathbb{N}$ be such that $f(n, m, l) = 0$ iff $n < m < l$ and $t < m$ and $c(t, m) = c(t, l) = c(m, l) = G$. By minimization (see Simpson [13, Theorem II.3.5]), there exists a function $h : \mathbb{N} \to \mathbb{N}^2$ such that for all n we have that $f(n, h_0(n), h_1(n)) = 0$, where $h(n) = (h_0(n), h_1(n))$. Now define by primitive recursion a Triang witness function $g : \mathbb{N} \to \mathbb{N}$ by letting $g(0) = h(0)$ and $g(n + 1) = h(g_1(n))$, where $g(n) = (g_0(n), g_1(n))$. Therefore, for all n

$- t < g_0(n) < g_1(n) < g_0(n + 1)$ and
$- c(t, g_0(n)) = c(t, g_1(n)) = c(g_0(n), g_1(n)) = G$.

We claim that there exists an infinite descent starting from x in G_t. Since $x \overset{\downarrow}{\to} x \in c(t, g_0(0))$, it is sufficient to show that $x \overset{\downarrow}{\to} x \in c(g_0(n), g_0(n + 1))$ for any n. As $x \overset{\downarrow}{\to} x \in c(t, g_0(n + 1))$, there exists y such that $x \to y \in c(t, g_1(n))$ and $y \to x \in c(g_1(n), g_0(n + 1))$, and at least one of them is strict. Now $c(t, g_1(n)) = c(g_0(n), g_1(n))$, and so $x \to y \in c(g_0(n), g_1(n))$. Therefore we have $x \overset{\downarrow}{\to} x \in c(g_0(n), g_0(n + 1))$, as desired.

Theorem 6 (RCA$_0$). *The SCT criterion implies* SPP.

Proof. We show that the SCT criterion for graphs implies SPP.

We first prove the thesis for $k = 2$. This serves as an illustration of the general case. Note in fact that SPP$_k$ is provable in RCA$_0$ for every standard $k \in \mathbb{N}$.

Given $c : \mathbb{N} \to 2$, we want to show that there exists $I \subseteq 2$ such that $i \in I$ iff $\exists^\infty x\, c(x) = i$. Let us define $\mathcal{G}$ as follows. The set $\mathcal{G}$ consists of three size-change graphs G_0, G_1, G_2 on parameters z_0, z_1, z_2. For $i < 3$, the graph G_i has only one strict arc $z_i \overset{\downarrow}{\to} z_i$ and non-strict arcs $z_j \overset{\Downarrow}{\to} z_j$ for $j > i$. Note that every $G \in \mathrm{cl}(\mathcal{G})$ contains a strict arc $z \overset{\downarrow}{\to} z$. Therefore, by the SCT criterion, every multipath of $\mathcal{G}$ has an infinite descent. Let

$$g(x) = \begin{cases} 0 & \text{if } c(x) = 0 \wedge c(x + 1) = 0 \\ 1 & \text{if } c(x) = 1 \wedge c(x + 1) = 1 \\ 2 & \text{otherwise} \end{cases}$$

Consider the multipath $M = G_{g(0)}, G_{g(1)}, \ldots$. Hence there exists an infinite descent in M. This implies that there exists a parameter z_i that is strictly decreasing infinitely many times, that is $z_i \overset{\downarrow}{\to} z_i \in G_{g(x)}$, viz. $g(x) = i$, for infinitely many x. If $i < 2$, it means that from some point on $c(x) = i$ and so $I = \{i\}$. If $i = 2$, then the color changes infinitely many times and so $I = \{0, 1\}$.

General case. Let $c : \mathbb{N} \to k$ be a given coloring. We want to show that

$$I^\infty = \{i < k : \exists^\infty x\, c(x) = i\}$$

exists.

Let $\mathcal{I}$ be the set of nonempty subset of k and Par$(\mathcal{I})$ consist of parameters z_I for every $I \in \mathcal{I}$. Define size-change graphs G_A on (Par$(\mathcal{I})$, Par$(\mathcal{I})$) for any $A \subseteq \mathcal{I}$ as follows. Let m be the maximum size of an element of A. Then

$- z_I \overset{\downarrow}{\to} z_I \in G$ iff $I \in A$ and $|I| = m$;
$- z_I \overset{\Downarrow}{\to} z_I \in G$ iff $I \notin A$ and $|I| \geq m$.

Let $\mathcal{G} = \{G_A : A \subseteq \mathcal{I}\}$.

Claim. Every idempotent graph $G \in \mathrm{cl}(\mathcal{G})$ has an arc $z_I \overset{\downarrow}{\to} z_I$ for some I.

Proof. We show that every graph $G \in \text{cl}(\mathcal{G})$ has a strict arc $z_I \xrightarrow{\downarrow} z_I$ for some I. Let $G = G_0; G_1; \ldots; G_{l-1}$ with $G_s \in \mathcal{G}$ for all $s < l$. Let $\mathcal{A}_s$ be the $\mathcal{A}$ corresponding to G_s. Choose $I \in \bigcup_{s<l} \mathcal{A}_s$ of maximum size. We claim that $z_I \xrightarrow{\downarrow} z_I \in G$. Let $p < l$ be such that $I \in \mathcal{A}_p$. By definition, $z_I \xrightarrow{\downarrow} z_I \in G_p$. By using the maximality of I it is easy to show that for every $s < l$ either $z_I \xrightarrow{\downarrow} z_I \in G_s$ or $z_I \xrightarrow{\Downarrow} z_I \in G_s$.

We now define a multipath $M = G_0, G_1, \ldots, G_x, \ldots$ as follows.

Let Γ_I be a marker for $I \in \mathcal{I}$. At the beginning every marker Γ_I points to the first color of I (in the standard ordering of the natural numbers). At stage x, if the marker Γ_I points to the color i and $c(x)$ is the color right after i in I (in the standard ordering of the natural numbers), then move the marker to the color $c(x)$. If i is the last color of I and $c(x)$ is the first color of I, move the marker to the first color of I. It is not difficult to see within RCA_0 that every color in I appears infinitely often iff the marker points to the last color of I infinitely often.

Call I a guess at stage x if at the beginning of stage x the marker Γ_I points to the last color of I and $c(x)$ equals the first color of I. The idea is that at stage x we are guessing that $I = I^\infty$. Note that we can have more guesses at the same stage and that I is a guess at infinitely many stages iff $I \subseteq I^\infty$.

Now let $G_x = G_\mathcal{A}$ where $\mathcal{A}$ is the set of guesses at stage x. By the SCT criterion for graphs, we have an infinite descent in M for some parameter z_I starting at some point t. We aim to show that I is the right guess, that is $I = I^\infty$. Now, there exist infinitely many x such that $z_I \xrightarrow{\downarrow} z_I \in G_x$, and in particular I is a guess at stage x for infinitely many x. It follows that $I \subseteq I^\infty$. It is sufficient to show that I is maximal. Suppose not and let $J \supset I$ be such that every color in J appears infinitely often. Therefore there exists $x > t$ such that J is a guess at stage x. By definition, in G_x there is no arc from z_I to z_I, a contradiction.

Therefore we can conclude that Triang $\geq$ SCT criterion $\geq \mathsf{I}\Sigma_2^0$. Actually we can prove that they are all equivalent.

Theorem 7. *Over* RCA_0 *the following are equivalent:*

1. $\mathsf{I}\Sigma_2^0$
2. Triang
3. *SCT criterion*
4. *SCT criterion for graphs*

Proof. We need only to show that $\mathsf{I}\Sigma_2^0$ implies Triang. As shown in [14] RT^2 is Π_1^1-conservative over $\text{B}\Sigma_3^0$, the bounding principle for Σ_3^0-formulas. So, since RT^2 trivially implies Triang (which is a Π_1^1-statement), then also $\text{B}\Sigma_3^0$ does. It is known that $\text{B}\Sigma_3^0$ is $\widetilde{\Pi}_4^0$-conservative over $\mathsf{I}\Sigma_2^0$, where a statement is $\widetilde{\Pi}_4^0$ if it is of the form $\forall X \varphi(X)$ and $\varphi(X) \in \Pi_4^0$. This follows as a particular case from the analogue result in first order arithmetic that $\text{B}\Sigma_{n+1}$ is Π_{n+2}-conservative over $\mathsf{I}\Sigma_n$ for all $n \geq 0$ (see [5, Chapter IV, Sect. 1(f)]). Finally, one can check that Triang is $\widetilde{\Pi}_4^0$, hence the thesis.

Remark 2. One can directly show that the Péter-Ackermann function is SCT in both senses. Indeed let $G_{0,1}, G_2$ be the size change graphs of the Péter-Ackermann function as in Example 2. Let $\mathcal{M} = G'_0, \ldots, G'_n, \ldots$ be an infinite multipath. We have

$$\forall n \, \exists m \geq n \, G'_n = G_{0,1} \vee \exists n \, \forall m \geq n \, G'_n = G_2.$$

In the first case we have an infinite descent for x starting in G'_0. The second case yields an infinite descent for y starting in some G_n, since all graphs in the multipath from n on are G_2. Note that this proof is in classical logic, since it requires the Law of Excluded Middle.

In general, if $\mathcal{G}$ has size k for some standard $k \in \omega$, then RCA_0 proves the SCT criterion for $\mathcal{G}$. This follows from the following:

Proposition 3. *For any standard $k \in \omega$,*

$$\mathsf{RCA}_0 \vdash \mathrm{Triang}_k.$$

Proof. Note that $\mathsf{RCA}_0 \vdash \mathrm{RT}^1_k$ for all standard $k \in \mathbb{N}$. We prove Triang_k by (external) induction on k.

Given a coloring $c : [\mathbb{N}]^2 \to k$, let $c_0 : \mathbb{N} \to k$ such that $c_0(x) = c(0, x)$ and let X be the infinite homogeneous set given by RT^1_k. Let $\{x_n : n \in \mathbb{N}\}$ be the increasing enumeration of X. Suppose $i = c(0, x_0)$. By the law of excluded middle, we have:

$$\forall n \exists m, l(l > m > n \wedge c(x_m, x_l) = i) \vee \exists n \forall m, l(l > m > n \implies c(x_m, x_l) \neq i).$$

In the first case we are done. In the second case let $Y = \{x \in X : x > x_n\}$. Then Y is an infinite homogeneous set in $(k - 1)$-many colors. By the induction hypothesis (on $d : [\mathbb{N}]^2 \to k - 1$ such that $d(a, b) = c(x_{n+a}, x_{n+d})$) we are done again.

4   Conclusion and Further Works

In this paper we addressed the study of size-change analysis in the context of Reverse Mathematics. We determined the exact strength of the SCT criterion by proving that it is equivalent to a weak version of Ramsey's theorem for pairs, which turns out to be equivalent to Σ^0_2-induction over RCA_0. In particular the proof of the SCT criterion does not require full Ramsey's theorem for pairs.

One of the motivations for studying size-change termination in the framework of Reverse Mathematics is that the Péter-Ackermann function is size-change-terminating. Actually, this can be proved in RCA_0, whereas it is well known that the totality of the Péter-Ackermann function is not provable in RCA_0. This arises the question of what is needed in order to show the *soundness* of size-change termination (SCT soundness), that is the statement that every SCT program terminates.

The classical proof is based on the fact that "if a program does not terminate then there exists an infinite state transition sequence". This statement seems to require König's lemma, which is equivalent to Arithmetical Comprehension Axiom ($\mathsf{ACA_0}$) over the base system $\mathsf{RCA_0}$. Roughly, $\mathsf{ACA_0}$ asserts the existence of the jump of every set of natural numbers.

We suspect that a direct proof of the SCT soundness does not require any comprehension (set existence) axiom. In fact, it is known that SCT programs compute exactly the *multiply recursive* functions [1]. On the other hand, the class of multiply recursive functions coincides with the class $\mathcal{M} = \bigcup_{\alpha < \omega^\omega} \mathcal{F}_\alpha$, where $(\mathcal{F}_\alpha)_\alpha$ is the fast growing hierarchy [10]. Since well-foundedness of ω^{ω^ω} implies the totality of every function in $\mathcal{M}$, we thus conjecture that SCT soundness is provable in $\mathsf{RCA_0}$ plus well-foundedness of ω^{ω^ω}.

References

1. Ben-Amram, A.M.: General Size-change termination and lexicographic descent. In: Mogensen, T.Æ., Schmidt, D.A., Sudborough, I.H. (eds.) The Essence of Computation. LNCS, vol. 2566, pp. 3–17. Springer, Heidelberg (2002). doi:10.1007/3-540-36377-7_1
2. Codish, M., Genaim, S.: Proving termination one loop at a time. In: Mesnard, F., Serebrenik, A. (eds.) Proceedings of the 13th International Workshop on Logic Programming Environments, Tata Institute of Fundamental Research, Mumbai, India, 8 December 2003. Report, vol. CW371, pp. 48–59. Katholieke Universiteit Leuven, Department of Computer Science, Celestijnenlaan 200A, B-3001 Heverlee (Belgium) (2003)
3. Friedman, H.: Some systems of second order arithmetic and their use. In: Proceedings of the International Congress of Mathematicians (Vancouver, B.C., 1974), vol. 1, pp. 235–242 (1975)
4. Gasarch, W.: Chapter four - proving programs terminate using well-founded orderings, Ramsey's Theorem, and Matrices. Adv. Comput. **97**, 147–200 (2015)
5. Hájek, P., Pudlák, P.: Metamathematics of First-Order Arithmetic. Perspectives in Mathematical Logic, vol. 3. Springer, Berlin (1998)
6. Heizmann, M., Jones, N.D., Podelski, A.: Size-change termination and transition invariants. In: Cousot, R., Martel, M. (eds.) SAS 2010. LNCS, vol. 6337, pp. 22–50. Springer, Heidelberg (2010). doi:10.1007/978-3-642-15769-1_4
7. Hirst, J.L.: Combinatorics in subsystems of second order arithmetic. Phd thesis, The Pennsylvania State University (1987)
8. Jones, N.D., Bohr, N.: Termination analysis of the untyped λ-Calculus. In: Oostrom, V. (ed.) RTA 2004. LNCS, vol. 3091, pp. 1–23. Springer, Heidelberg (2004). doi:10.1007/978-3-540-25979-4_1
9. Lee, C.S., Jones, N.D., Ben-Amram, A.M.: The size-change principle for program termination. In: Conference Record of POPL 2001: The 28th ACM SIGPLAN-SIGACT Symposium on Principles of Programming Languages, London, UK, 17–19 January 2001, pp. 81–92 (2001)
10. Löb, M.H., Wainer, S.S.: Hierarchies of number-theoretic functions. I. Arch. Math. Logic **13**(1–2), 39–51 (1970)
11. Podelski, A., Rybalchenko, A.: Transition invariants. In: Proceedings of the 19th Annual IEEE Symposium on Logic in Computer Science, LICS 2004, Turku, Finland, 13–17 July, pp. 32–41 (2004)

12. Ramsey, F.P.: On a problem in formal logic. Proc. London Math. Soc. **30**, 264–286 (1930)
13. Simpson, S.G.: Subsystems of Second Order Arithmetic. Perspectives in Mathematical Logic. Springer, Berlin (1999)
14. Slaman, T.A., Yokoyama, K.: The strength of Ramsey's Theorem for pairs and arbitrary many colors (2016) (in preparation)
15. Steila, S., Yokoyama, K.: Reverse mathematical bounds for the Termination Theorem. Ann. Pure Appl. Logic **167**(12), 1213–1241 (2016)
16. Thiemann, R., Giesl, J.: The size-change principle and dependency pairs for termination of term rewriting. Appl. Algebra Eng. Commun. Comput. **16**(4), 229–270 (2005)

Covering Polygons with Rectangles

Roland Glück[✉]

Deutsches Zentrum für Luft- und Raumfahrt,
Am Technologiezentrum 4, 85159 Augsburg, Germany
roland.glueck@dlr.de

Abstract. A well-known and well-investigated family of hard optimization problems deals with nesting, i.e., the non-overlapping placing of polygons to be cut from a rectangle or the plane whilst minimizing the waste. Here we consider the in some sense inverse problem of a subsequent step in production technology: given a set of polygons in the plane and an axis-aligned rectangle (modeling a gripping device), we seek the minimum number of copies of the rectangle such that every polygon is completely covered by at least one copy of the rectangle. As motions of the given rectangle for obtaining the copies we investigate the cases of translation in x-direction, of arbitrary translation and of arbitrary translation combined with rotation. We give a generic algorithm for all three cases which leads to a polynomial-time algorithm for the first case. The other two cases are NP-hard so we introduce a rather straightforward algorithm for the second case and two different approaches to the third one. Finally, we give experimental results and compare them to the theoretical analysis done before.

1 Introduction

The cutting of given polygons out of a rectangle strip whilst minimizing the waste of material is a well-known and long-standing problem class, see e.g. [8] for a survey. In this paper we investigate the subsequent step in production technology: once the pieces are cut out they will be picked off and transported further by a suitable device. Here we use as such a device a rectangular gripper and various degrees of its motion freedom: the easiest case (which admits an efficient solution) is the one of a gripper which can be translated only along the x-axis. Second, we investigate a gripper which can be translated both in x- and in y-direction, and finally, we consider a gripper with an additional degree of rotational freedom. The gripper can grasp only the cut pieces it covers completely. Now the goal is to pick up all cut pieces with as little gripper motions as possible.

Basically, this task corresponds to covering a set of polygons by copies of a rectangle such that every polygon is contained in at least one rectangle. Set cover problems with geometric background can be divided in two classes: one class as e.g. in [1,3] where the geometric objects from which the cover has to be chosen are fixed, and another class as e.g. in [5,9,14] where the covering

© Springer International Publishing AG 2017
T.V. Gopal et al. (Eds.): TAMC 2017, LNCS 10185, pp. 274–288, 2017.
DOI: 10.1007/978-3-319-55911-7_20

objects arise from a given object by translation and/or rotation. Yet another class form the problems investigated e.g. in [13,15]; in contrast to our task the covering objects can have different shapes but are restricted to belong to a certain family, e.g., they are required to be rectangles. A lot of this previous work deals with covering points by disks or rectangles, and only a smaller part with covering connected subsets of the $\mathbb{R}^n$ or polygons in particular. In contrast to work e.g. in [4,12] a rectangle has to cover at least one polygon completely. None of the cited work matches our problem exactly, although one can reduce it to a geometric set cover problem which will be discussed in Sect. 8. We do not propose an approximation algorithm but a family of exact algorithms which works well on practical instances and continues and improves the work from [11].

The paper is organized as follows: Sect. 2 provides basic definitions and states the problem family we consider. In Sect. 3 we collect some important geometric facts while Sect. 4 introduces a generic algorithm for our problem class. The results of these two sections lead to concrete algorithms in Sects. 5, 6 and 7. Finally, we evaluate the algorithms in Sect. 8 and give a conclusion and an outlook to further work in Sect. 9.

2   Basic Definitions and Properties

To formalize our task, we introduce the concept of a packing: a *packing* $\mathbf{P} = \{P_1, P_2, \ldots, P_n\}$ is a set of n possibly overlapping simple polygons $P_1, P_2, \ldots, P_n$ in the plane. We say that a packing $\mathbf{P}$ is *covered* by a set $\mathbf{C} = \{R_1, R_2, \ldots, R_m\}$ of rectangles if each polygon of $\mathbf{P}$ is completely contained in at least one rectangle of $\mathbf{C}$. If a rectangle R' arises from a rectangle R by a translation along the x-axis we call R' a *pushing* of R, if R' arises from R by a translation in possibly both x- and y-direction we call R' a *translation* of R, and if R' arises from R by a translation and a rotation we say that R' is a *general motion* of R (this is equivalent to the term "rigid motion" in [7]). With these namings we can define the main theme of our investigations:

Definition 1. *Let* $\mathbf{P}$ *be a packing and let* R *be an axis-aligned rectangle (the so-called* gripper*). Then we call a set of rectangles* $\mathbf{C} = \{R_1, R_2, \ldots, R_m\}$ *a* pushing *(translational, general) cover of* $\mathbf{P}$ *by* R *if* $\mathbf{C}$ *covers* $\mathbf{P}$ *and all rectangles of* $\mathbf{C}$ *are pushings (translations, general motions) of* R.

Since we are interested in covering a packing with as few rectangles as possible we call a cover of each kind *optimal* if it has minimum cardinality among all covers of that kind. So we define the pushing (translational, general) polygon cover problem or PPC (TPC, GPC) as follows: Given a packing $\mathbf{P}$ and a gripper R, compute an optimal pushing (translational, general) cover of $\mathbf{P}$ by R. The NP-hardness of TPC and GPC can be proved by reductions from 3-SAT along the lines of the proof of the NP-hardness of the BOX-COVER problem given in [9] if one replaces the points by suitable small polygons. However, we will concentrate here more on the algorithmic aspects and hence omit the details.

To ease wording we refer by the term *cover* to a pushing cover as well as to a translational or a general cover. For a packing $\mathbf{P}$ and a rectangle R we denote the set of polygons of $\mathbf{P}$ covered by R by $\mathrm{cov}(R, \mathbf{P})$. Two polygons P_1 and P_2 are called *generally compatible* wrt. to a rectangle R if there is a general motion of R that covers both P_1 and P_2. For running time considerations we introduce the notation $\|\mathbf{P}\|$ for the overall number of vertices in a packing $\mathbf{P}$. We say that p is a point of a packing $\mathbf{P}$ if there is a polygon $P \in \mathbf{P}$ that has p as a vertex.

3   Geometrical Considerations

In this section we state some simple but important observations that help us setting up algorithms to tackle the various kinds of cover problems from Sect. 2. The following three lemmata with proofs in the appendix form a kind of group of "exchange lemmata" since they make similar claims about substitutability of rectangles for each kind of cover. Its first one deals with pushing covers:

Lemma 1. *Let $\mathbf{P}$ be a packing and let $\mathbf{C} = \{R_1, R_2, \ldots, R_m\}$ be an optimal pushing cover of $\mathbf{P}$ by a gripper R. Then there is a point p of $\mathbf{P}$ and an index j together with a pushing R'_j of R with the following properties:*

1. *p has minimum x-coordinate among all points of $\mathbf{P}$,*
2. *p lies on the left side of R'_j, and*
3. *$\mathbf{C} \backslash \{R_j\} \cup \{R'_j\}$ is an optimal pushing cover of $\mathbf{P}$.*

In the next lemma we see an analogous claim for translational covers:

Lemma 2. *Let $\mathbf{P}$ be a packing and $\mathbf{C} = \{R_1, R_2, \ldots, R_m\}$ an optimal translational cover of $\mathbf{P}$ by R. Then there are points p_1 and p_2 of $\mathbf{P}$ and an index j together with an axis-aligned rectangle R'_j fulfilling the following properties:*

1. *p_1 has minimum x-coordinate among all points of $\mathbf{P}$,*
2. *p_1 lies on the left side of R'_j,*
3. *p_2 is a point of maximal y-coordinate for some $P_2 \in \mathbf{P}$ and lies on the upper side of R'_j,*
4. *there are polygons $P_1, P_2 \in \mathbf{P}$ such that for $i \in \{1, 2\}$ p_i is a vertex of P_i and P_i is contained in R'_j, and*
5. *$\mathbf{C} \backslash \{R_j\} \cup \{R'_j\}$ is an optimal translational cover of $\mathbf{P}$.*

Note that we do not require that p_1 and p_2, P_1 and P_2 as well as R_j and R'_j are distinct. Moreover, if p_1 and p_2 are equal then they coincide with the upper left vertex of R'_j.

The normal case of the lemma above is depicted in the left part of Fig. 1: the rectangle in full line has a point of minimal x-coordinate on its left side and another point of a covered polygon on its upper side. Clearly, it can replace the dotted rectangle in an optimal cover. A degenerate example where p_1 and p_2 as well as P_1 and P_2 coincide can be seen in the right part of the same figure.

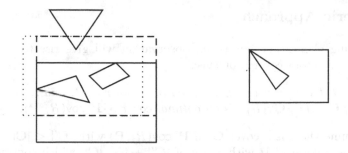

Fig. 1. Translational alignments

Finally, the last lemma of this group concerns general covers:

Lemma 3. *Let* **P** *be a packing and let* $\mathbf{C} = \{R_1, R_2, \ldots, R_m\}$ *be an optimal general cover of* **P** *by R. Then for every polygon* $P_{pi} \in \mathbf{P}$ *there are points* p_1, p_2 *and* p_3 *of* **P** *and an index* j *together with a rectangle* R'_j *fulfilling the following properties:*

1. R'_j *contains* P_{pi},
2. p_1 *and* p_2 *are distinct and lie on different adjacent sides of* R'_j
3. p_1, p_2 *and* p_3 *lie on sides of* R'_j,
4. *there are polygons* $P_1, P_2, P_3 \in \mathbf{P}$ *such that for* $i \in \{1, 2, 3\}$ p_i *is a vertex of* P_i *and* P_i *is contained in* R'_j,
5. $\mathbf{C} \backslash \{R_j\} \cup \{R'_j\}$ *is an optimal general cover of* **P**.

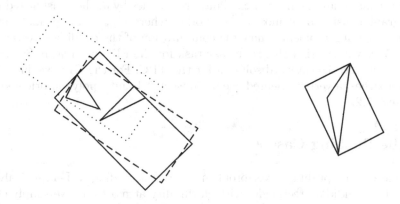

Fig. 2. General motion alignments

In the left part of Fig. 2 we see that the dotted rectangle can be replaced by the one drawn in full line which has three points of a packing on its sides, two of them on adjacent sides. An extremal situation where points coincide with vertices of R'_j is shown in the right part of the same figure.

4   Generic Approach

The following elementary but useful observation will give rise to a family of algorithms for all our cover problems:

Lemma 4. *Let* $\mathbf{P}$ *be a packing,* $\mathbf{C}$ *an optimal cover of* $\mathbf{P}$, *and* R_j *an arbitrary rectangle of* $\mathbf{C}$. *Then* $\mathbf{C}\backslash\{R_j\}$ *is an optimal cover of* $\mathbf{P}\backslash\text{cov}(R_j, \mathbf{P})$.

Proof. Assume there is a cover $\mathbf{C}'$ of $\mathbf{P}\backslash\text{cov}(R_j, \mathbf{P})$ with $|\mathbf{C}'| < |\mathbf{C}| - 1$. Then $\mathbf{C}' \cup \{R_j\}$ is a cover of $\mathbf{P}$ with a size of $|\mathbf{C}'| + 1 < |\mathbf{C}|$, which contradicts the optimality of $\mathbf{C}$.

Let us now assume we had an algorithm candidate_rectangles$(\mathbf{Q}, R)$ which determines for every packing $\mathbf{Q}$ a finite set of pushings, translations or general motions of a gripper R such that for every optimal cover $\mathbf{C}$ of $\mathbf{Q}$ there are an $R_{cov} \in \mathbf{C}$ and an $R_{cand} \in$ candidate_rectangles$(\mathbf{Q})$ with the property $\text{cov}(R_{cov}, \mathbf{Q}) \subseteq \text{cov}(R_{cand}, \mathbf{Q})$. Calling these rectangles *candidate rectangles* we define for a packing $\mathbf{P}$ an edge-labeled graph $G_{search} = ((V, E), r)$ as follows: the nodes of V are subsets of $\mathbf{P}$, and we define $(\mathbf{P}_1, \mathbf{P}_2) \in E$ iff there is a candidate rectangle R for $\mathbf{P}_1$ such that $\mathbf{P}_2 = \mathbf{P}_1\backslash\text{cov}(R, \mathbf{P}_1)$ holds. Moreover, we label each edge $(\mathbf{P}_1, \mathbf{P}_2)$ with an appropriate rectangle $R = r(\mathbf{P}_1, \mathbf{P}_2)$ which fulfills the equality $\mathbf{P}_2 = \mathbf{P}_1\backslash\text{cov}(R, \mathbf{P}_1)$. Clearly, an optimal cover of a packing $\mathbf{P}$ corresponds to a shortest path in G_{search} from $\mathbf{P}$ to $\emptyset$, so we simply perform a BFS in G_{search} starting at $\mathbf{P}$ till we reach the node $\emptyset$. The rectangles of the optimal cover can be recovered from the edge labels of a shortest path in G_{search} from $\mathbf{P}$ to $\emptyset$. In practice we will of course not construct the entire graph G_{search} but generate its nodes step by step on demand till the algorithm terminates. If we denote the maximum number of candidate rectangles by d, the constructed part of the graph consists of at most $d^{|\mathbf{C}_{opt}|}$ nodes, where $\mathbf{C}_{opt}$ is an optimal cover.

The remaining problem is now the computation of the candidate rectangles. In Sect. 5 we will see that this is an easy task for the PPC whereas in Sect. 6 we show a rather straightforward solution for the TPC. For the GPC we introduce two approaches, a vertex-oriented one in Subsect. 7.1 and a polygon-oriented one in Subsect. 7.2.

5   The Pushing Case

In the case of the pushing cover problem, we can according to Lemma 1 always find a set of candidate rectangles with cardinality at most one: we simply chose the unique pushing of R such that a point of $\mathbf{P}$ with minimum x-coordinate lies on its left side. Then the search graph is a simple path from $\mathbf{P}$ to $\emptyset$ of length at most $\mathbf{P} - 1$. Together with an initial sorting of $\mathbf{P}$ according to the minimum x-coordinate of its polygons in ascending order it is straightforward to see that this greedy algorithm can be implemented with an overall running time in $\mathcal{O}\left(\|\mathbf{P}\| + |\mathbf{P}| \cdot \log(|\mathbf{P}|)\right)$.

6   The Translational Case

Confronted with the TPC for a packing $\mathbf{P}$, it is clear that it suffices to consider only the axis-aligned minimal rectangular bounding boxes of each polygon of $\mathbf{P}$. Obviously, this transformation can be done in $\|\mathbf{P}\|$ time and does not affect the asymptotic running time.

Given now a gripper R and a packing $\mathbf{P}$ we can compute a set of candidate rectangles for the translational case using Lemma 2 as follows: first we choose a polygon $P_1 \in \mathbf{P}$ containing a vertex with minimal x-coordinate x_{min} among all points in $\mathbf{P}$. Subsequently, we align R with its left side to x_{min} and push it in negative y-direction, starting at a position where its bottom side has the minimal y-value of all vertices of P_1. We continue this pushing along the y-axis till the top side of R is underneath the top side of P_1. During this process, we register all positions where the upper side contains a point of maximal y-coordinate or its lower side with a point of minimal y-coordinate of some polygon of $\mathbf{P}$. Every such event removes or adds a polygon to the set of covered polygons, and the candidate rectangles are exactly those where a polygon is added the last time before removing one. The cases of simultaneous events are a little bit tedious but not basically different to handle.

Since every polygon causes at most two events there is an overall number of $\mathcal{O}(\mathbf{P})$ events during the described process and hence the number of candidate rectangles is here in $\mathcal{O}(|\mathbf{P}|)$. So the number of nodes in the search graph is in $\mathcal{O}(|\mathbf{P}|^{|\mathbf{C}_{opt}|})$ and the overall running time of the algorithm is in $\mathcal{O}(|\mathbf{P}|^{|\mathbf{C}_{opt}|+1})$ (we need $\mathcal{O}(|\mathbf{P}|)$ time for the construction of a node) if we sort the bounding boxes in a suitable way before starting the algorithm (which takes negligible $\mathcal{O}(|\mathbf{P}| \cdot \log|\mathbf{P}|)$ time). Note that in contrast to the following Sect. 7 we can maintain the set of candidate rectangles by a linked list where we append elements at the end and remove them from the beginning.

7   The General Case

As a general precondition for the general case we assume w.l.o.g. that all polygons of a packing are convex (a polygon can be covered by a rectangle R iff its convex hull can be covered by R). Hence we replace every polygon of the packing by its convex hull. The time required for this is in $\mathcal{O}\left(\sum_{P \in \mathbf{P}} \|\{P\}\| \cdot \log \|\{P\}\| \right)$ (see e.g. [6]) and will have no influence on the overall asymptotic running times.

7.1   A Vertex-Oriented Approach

Using Lemma 3 we can determine the set of candidate rectangles for a gripper R and a packing $\mathbf{P}$ in the general case as follows: first, we choose a pivot polygon $P_{pi} \in \mathbf{P}$. Then we perform a loop over all triples (p_1, p_2, p_3) of points from $\mathbf{P}$ and compute all general motions (if any) of R which fulfill the requirements of Lemma 3. Of course, we discard all triples with three identical points.

The computation of the general motions can be done using elementary geometry. A special challenge is to keep only those rectangles which cover an inclusion-maximal subset of $\mathbf{P}$. We tackled this task by a data structure which maintains a subset $\mathcal{S}$ of the power set of $\{1, 2, \ldots, |\mathbf{P}|\}$ and supports the following operations:

- Given $S \subseteq \{1, 2, \ldots, |\mathbf{P}|\}$, determine whether $\mathcal{S}$ contains a superset of S.
- Given $S \subseteq \{1, 2, \ldots, |\mathbf{P}|\}$, remove all subsets of S from $\mathcal{S}$.
- Given $S \subseteq \{1, 2, \ldots, |\mathbf{P}|\}$, add S to $\mathcal{S}$.

Considering now the rectangles computed above as an input stream, we can for each rectangle determine the polygons it covers, identify it with a subset of $\{1, 2, \ldots, |\mathbf{P}|\}$ and maintain a set of inclusion-maximal subsets of $\{1, 2, \ldots, |\mathbf{P}|\}$, each set associated with a suitable rectangle.

The described data structure was implemented as a linked list of bit vectors corresponding to subsets of $\{1, 2, \ldots, |\mathbf{P}|\}$ in a canonical way. As a solution for the implementation of the bit vectors we used arrays of long integers and operated on them in a bitwise manner in the spirit of word RAM algorithms. In our practical settings, $|\mathbf{P}|$ never exceeded a value of forty, so these arrays consisted all of only one long integer. The worst case running time of the operations above is in $\mathcal{O}\left(|\mathcal{S}| \cdot \frac{|\mathbf{P}|}{w}\right)$ where w denotes the bit length of a long integer.

Per node, we have to iterate over $\|\mathbf{P}\|^3$ triples, and per triple we need $\|\mathbf{P}\|$ time for computing the polygons covered by the rectangle R under consideration by testing whether points of $\mathbf{P}$ are contained in R. So the overall time for processing a node is in $\mathcal{O}\left(\|\mathbf{P}\|^4 \cdot |\mathcal{S}| \cdot \frac{|\mathbf{P}|}{w}\right)$, and because the branching degree of the search graph is also bounded by $|\mathcal{S}|$ we have an overall running time of $\mathcal{O}\left(|\mathcal{S}|^{|\mathbf{C}_{opt}|+1} \cdot \|\mathbf{P}\|^4 \cdot \frac{|\mathbf{P}|}{w}\right)$.

In order to estimate $|\mathcal{S}|$ we denote the maximal number of polygons generally compatible with P_{pi} except P_{pi} by n_{pi}. Then the maximal number of candidate rectangles is bounded by $\sqrt{\frac{2}{\pi}} \frac{2^{n_{pi}}}{\sqrt{n_{pi}}}$; see A.4 for a proof. This means that the overall running time can be bounded by $\mathcal{O}\left((\sqrt{\frac{2}{\pi}} \frac{2^{n_{pi}}}{\sqrt{n_{pi}}})^{|\mathbf{C}_{opt}|+1} \cdot \|\mathbf{P}\|^4 \cdot \frac{|\mathbf{P}|}{w}\right)$. However, this is only a worst case scenario; in practice we will not need to test every node of $\mathbf{P}$ whether it is contained in a rectangle R. In a lot of cases a point and hence the polygon P it belongs to will lie outside of R so the remaining points of P need not to be tested. So the running time will rather be cubic than biquadratic in $\|\mathbf{P}\|$. We will see this also in the results of Sect. 8.

In our implementation we performed a preprocessing step which computes for every polygon P of the initial packing $\mathbf{P}$ the list of polygons $Q_1, Q_2, \ldots, Q_n$ which are generally compatible with P (note that this list also contains P). Furthermore, it obviously suffices to run the outer loop from Q_1 till Q_n, the second loop from the current index of the outer loop till Q_n and the innermost loop from the current index of the second loop till Q_n. Finally, we iterate over sets of two or three different points from the overall set of points of the three (possibly identical) polygons. Unfortunately, this cannot guarantee a number of iterations better than $\theta(\|\mathbf{P}\|^3)$ for the computation of the candidate rectangles.

A special issue is the choice of the pivot polygon P_{pi} which influences the number of candidate rectangles and hence the branching degree of the search graph and eventually the running time. Our experiments showed that a good choice is a polygon whose distance from a vertex of the minimal axis-aligned bounding rectangle of $\mathbf{P}$ is minimal.

7.2   A Polygon-Oriented Approach

The idea of the subsequently described approach is to compute all inclusion-maximal coverable subsets of $\mathbf{P}$ containing a pivot polygon $P_{pi} \in \mathbf{P}$ by a simple backtracking algorithm which enumerates these sets.

If we have a function at our disposal which decides whether a set of polygons (or equivalently, a convex polygon) can be covered by a general motion of a gripper we can enumerate all inclusion-maximal coverable subsets containing P_{pi} by a straightforward backtracking algorithm. Similar to the vertex-oriented approach, we can use preprocessed lists of polygons in order to speed up the computation. However, this will not reduce the worst case number of backtracking iterations below $\mathcal{O}\left(2^{|\mathbf{P}|}\right)$ per computation of a set of candidate packings. As we will see soon one backtracking iteration of the general case can be executed in $\mathcal{O}\left(\|\mathbf{P}\|\right)$ time (including the possible construction of a new node) so the overall running time amounts here to $\mathcal{O}\left(\left(\sqrt{\frac{2}{\pi}} \frac{2^{n_{pi}}}{\sqrt{n_{pi}}}\right)^{|C_{opt}|} \cdot 2^{|\mathbf{P}|} \cdot \|\mathbf{P}\|\right)$.

Packing a Polygon into a Rectangle. The remaining problem is to determine whether for a given gripper there is a general motion such that the gripper covers a given convex polygon, and, if so, to compute such a general motion. Clearly, this is equivalent to the issue of packing such a polygon into a rectangle of given dimensions which is a little bit more intuitive to describe.

The task of packing a convex polygon P given by n vertices $p_1, p_2, \ldots, p_n$ into a rectangle can be solved by an algorithm described in [2]. However, this algorithm is rather complicated to implement (and we are not aware of any implementation), so we developed a special algorithm for our purposes. An algorithm from [7] for maximizing the number of points in a general motion of a convex polygon can be used to solve our problem, leading to a running time of $\mathcal{O}\left(n^3 \log(n)\right)$ in the worst case. Since it is an easy task to determine the translational part of a desired general motion once a suitable rotation angle is known we will concentrate from now on only on the computation of the rotational part.

Our algorithm uses a variant of the rotating calipers (introduced in [18], see also [16]) which were already applied to other problems like maximum or minimum width of a polygon or computing minimal rectangular bounding boxes. The key idea is to determine all pairs of antipodal points of a polygon (i.e., pairs of points which admit two infinite parallel lines through each point which intersect the polygon only in the respective point, see again [16,18] for details) and to compute the solution of the problem using these pairs of points. In our algorithm, we do not only compute pairs of antipodal points but use them to

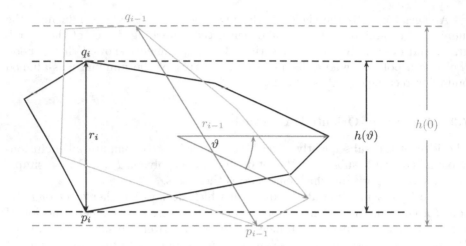

Fig. 3. Height of a polygon for two angles

determine two piecewise defined functions which indicate the height and the width of a rotated polygon depending on the rotation angle.

So for each pair (p_i, q_i) of antipodal points we have a validity interval $[\vartheta_i^l, \vartheta_i^r]$ and constants r_i and φ_i such that for $\vartheta \in [\vartheta_i^l, \vartheta_i^r]$ the height $h(\vartheta)$ of the polygon rotated by ϑ is given by $h(\vartheta) = r_i \cdot cos(\vartheta + \varphi_i)$. Now we determine for each such validity interval its subintervals where $h(\vartheta)$ is less than or equal to the height of the given rectangle. The number of antipodal pairs and hence the number of validity intervals is in $\mathcal{O}(n)$, and each validity interval has at most two subintervals with the above property so we end up with a number of subintervals in $\mathcal{O}(n)$. Since the computation of the pairs of antipodal points, of the validity intervals, the r_i's, the φ_i's and the subintervals can be done in constant time per i, we have an overall time in $\mathcal{O}(n)$ for this part of the computation.

An illustration is given in Fig. 3: the rotation of the original gray polygon around the angle ϑ yields the black polygon. The picture indicates the values of r_{i-1} for the gray polygon and r_i for the black polygon and their associated pairs of antipodal points (p_{i-1}, q_{i-1}) and (p_i, q_i), resp., together with the heights $h(0)$ and $h(\vartheta)$. The values of the offsets are harder to illustrate, however, in this figure, the value of φ_i equals $-\vartheta$ since the height at ϑ equals exactly r_i.

An analogous computation is done for the width function $w(\vartheta)$ (to obtain the width function it suffices to shift the height function by $\frac{\pi}{2}$). Finally, we compute the values of ϑ where $h(\vartheta)$ and $w(\vartheta)$ are less or equal to the height respectively the width of the given rectangle. By a simple sweep-line algorithm this can also be done in $\mathcal{O}(n)$ time, and it yields all rotation angles for which the polygon can be packed into the rectangle.

8   Experimental Results

We tested our algorithms experimentally on instances which are motivated by our practical needs. For the gripper we chose a rectangle of the dimensions 1500 × 2500, and we placed randomly generated convex polygons of maximal diameters between 100 and 500 on a rectangular field of length 3000 and height 2000. The number of polygons runs between 5 and 40, and the number of vertices per polygon between 5 and 55, both in steps of size five. As implementation language we chose Java and executed the programs on an Intel i7-4770 CPU with 3.4 GHz. Since we could solve all translational instances in times under one second we concentrate in the sequel on the general case.

On instances with 40 polygons and more than 30 vertices the vertex-oriented approach takes more than one hour per instance whereas the polygon-oriented algorithm solves instances with 40 polygons and 55 vertices per polygon on average in under nine minutes so we do not take instances with more than 35 polygons into account. For each instance size (i.e., pair of numbers of polygons and numbers of vertices as described above) we generated 30 instances which we processed by the two algorithms under consideration. In order to get rid of statistical outliers we discarded the five fastest and slowest results.

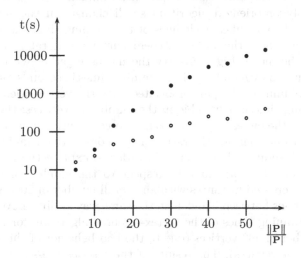

Fig. 4. Running times for 35 polygons

An approach mentioned already in the introduction is to reduce the problem to a (geometric) set cover instance. To make this approach work, we have to compute all inclusion-maximal coverable subsets of the given packing (note that this had to be followed by the actual algorithm for the covering problem). Doing so along the lines of Sects. 7.1 and 7.2 showed that for our test instances this computation takes more time than the execution of our algorithm.

Despite the discouraging running time estimations, the test instances are solvable in reasonable time with at least one of the proposed algorithms. As one could expect, the vertex-oriented algorithm runs faster on instances with a relatively high number of polygons and a low number of vertices per polygon whereas the polygon-oriented version turns out to be favorable if the number of vertices increases. This fact is illustrated in Fig. 4 where the running times of the algorithms on instances with 35 polygons are shown (note the logarithmic scale of the ordinate). The vertex-oriented algorithm, represented by full circles, is faster on instances with five vertices per polygon whereas the polygon-oriented algorithm, depicted in simple circles, is faster on the remaining instances. We made analogous observations for other numbers of polygons, too.

Another phenomenon which can also be observed in Fig. 4 is that the polygon-oriented algorithm is more sensitive against the structure of the input instance and less sensitive against the number of vertices than the vertex-oriented variant. This explains its slower growth rate and the little bulge at instances with 40 vertices per polygon. Moreover, the polygon-oriented algorithm shows a greater variability of running times than the vertex-oriented one. This is not astonishing since the vertex-oriented algorithm iterates rather tenaciously through a set of vertices whose cardinality has a cubic influence on the running time. On the other hand, due to the potentially exponential influence of n_{pi} on the running time of the polygon-oriented algorithm small changes in the structure of an instance can lead to significant changes of the running time. As a measure for the variability we chose the variation coefficient, i.e., the ratio of the standard deviation and the mean. Figure 5 show the means together with the variation coefficients (denoted by μ and c_v) of the running times in seconds for 25 polygons. The last two columns of the part associated with the vertex-oriented algorithm exhibit a running time nearly cubic in the number of vertices: the penultimate column contains the cubes of the ratios of two consecutive numbers of vertices, whereas the last one contains the ratios of two consecutive running times.

We executed additional test series in order to explore the running time of the polygon-oriented algorithm with respect to the number of polygons. For this purpose, we created instances with an overall number of 2000 vertices and a number of polygons between 5 and 40 in the same way as above. Analogously, we measured the running times of the vertex-oriented algorithm for instances with an overall number of 500 vertices (due to the bad behavior of this algorithm for a high number of vertices). The results of these series are shown in Fig. 6. As one can see again, the polygon-oriented algorithm exhibits a strong growth in the number of polygons whereas the vertex-based algorithm is rather indifferent with respect to the number of polygons and shows only a modest growth rate (for random reasons, the running times even decrease at some places).

Contrary to our expectations, we could not confirm a substantial dependency of the running times on the size of an optimal cover (disregarding two outliers in all test series). A possible explanation is that in instances with small optimal covers a lot of polygons are located close to each other. This could lead to a

$\|\mathbf{P}\|$	$\frac{\|\mathbf{P}\|}{\|\mathbf{P}\|}$	polygon-oriented		vertex-oriented			
		μ	c_v	μ	c_v	$\left(\frac{\|\mathbf{P}_n\|}{\|\mathbf{P}_{n-1}\|}\right)^3$	$\frac{\mu_n}{\mu_{n-1}}$
25	5	1.5	0.28	1.3	0.21	-	-
25	10	1.3	0.34	9.0	0.2	8	6.9
25	15	1.1	0.28	32	0.15	3.4	3.6
25	20	1.8	0.44	78	0.21	2.4	2.4
25	25	2.0	0.49	160	0.16	2.0	2.1
25	30	1.3	0.17	290	0.13	1.7	1.8
25	35	1.4	0.34	540	0.26	1.6	1.9
25	40	1.6	0.21	760	0.21	1.5	1.4
25	45	2.5	0.40	960	0.22	1.4	1.3
25	50	3.2	0.60	1300	0.10	1.4	1.4
25	55	2.9	0.29	1700	0.27	1.3	1.3

Fig. 5. Statistical values for 25 polygons

polygon-oriented

$\|\mathbf{P}\|$	$\|\mathbf{P}\|$	μ	$\frac{\mu_n}{\mu_{n-1}}$
2000	5	0.021	-
2000	10	0.064	3.0
2000	15	0.2	3.1
2000	20	1.0	5.0
2000	25	5.2	5.2
2000	30	15	2.9
2000	35	49	3.3
2000	40	400	8.2

vertex-oriented

$\|\mathbf{P}\|$	$\|\mathbf{P}\|$	μ	$\frac{\mu_n}{\mu_{n-1}}$
500	5	61	-
500	10	52	0.9 (!)
500	15	48	0.9 (!)
500	20	70	1.5
500	25	75	1.1
500	30	84	1.1
500	35	110	1.3
500	40	120	1.1

Fig. 6. Running times for fixed overall numbers of vertices

greater number of candidate rectangles and hence increase the branching degree of the search graph which has an effect opposing to the smaller search depth.

9   Further Work

Although we found satisfying solutions for our practical instances the presented algorithms have plenty of room for optimization and further research. So we plan to investigate various strategies for finding a suitable pivot polygon. Another choice than the one described here could be a polygon with a minimum number of polygons in its vicinity. Till now we ran the algorithms only on one single core; an obvious improvement should be to develop a parallelized version of it since parallelizing a breadth first search is a well-investigated topic (see [10,17]).

Theoretical questions concern on the one hand a more precise analysis of the running time of the presented algorithms (the given ones are rather rough and concern only the worst case) and the identification of parameters influencing the computational hardness of an instance. Candidates to investigate are beside the number of polygons and vertices the size of the polygons and their spatial density and distribution. Despite the fact of satisfying results on practical instances, some kind of approximation scheme could be both useful for larger instances and interesting from a theoretical point of view.

Acknowledgments. The author is grateful to Torben Hagerup, Christian Rähtz, Lev Sorokin and the anonymous reviewers for valuable remarks.

A   Deferred Proofs

A.1   Proof of Lemma 1

Proof. Let p be an arbitrary point of $\mathbf{P}$ with minimum x-coordinate. Then there is a rectangle $R_j \in \mathbf{C}$ containing p. Translating this rectangle in positive x-direction till p lies on its left side we obtain a rectangle R'_j with the required properties.

A.2   Proof of Lemma 2

Proof. Let p_1 be a point of $\mathbf{P}$ with minimum x-coordinate and P_1 a polygon of $\mathbf{P}$ which has p_1 as a vertex. Then there is a rectangle $R_j \in \mathbf{C}$ containing P_1. Now we translate R_j in positive x-direction till p_1 lies on the left side of the translated rectangle $\hat{R}_j$. Clearly, we have $\mathsf{cov}(\hat{R}_j, \mathbf{P}) \supseteq \mathsf{cov}(R_j, \mathbf{P})$. Subsequently, we translate $\hat{R}_j$ in negative y-direction till a point p_2 with the following properties lies on the upper side of the translated rectangle R'_j:

1. All polygons of $\mathbf{P}$ in $\hat{R}_j$ with p_2 as a vertex are contained in R'_j, and
2. p_2 is a point with maximum y-coordinate fulfilling the above requirements.

Then we have $\mathsf{cov}(R'_j, \mathbf{P}) \supseteq \mathsf{cov}(\hat{R}_j, \mathbf{P}) \supseteq \mathsf{cov}(R_j, \mathbf{P})$, so $\mathbf{C}\backslash\{R_j\} \cup \{R'_j\}$ is indeed an optimal translational cover of $\mathbf{P}$. Moreover, p_1, p_2 and P_1 meet their requirements by construction, and for P_2 we can chose an arbitrary polygon that has p_2 as a vertex and is contained in R'_j.

In Fig. 1, R_j corresponds to the dotted rectangle, $\hat{R}_j$ to the dashed one, and the final rectangle R'_j is drawn with a full line.

A.3   Proof of Lemma 3

Proof. Let $R_j \in \mathbf{C}$ be a rectangle containing P_{pi}. We apply to R_j similar translations as in Lemma 2 but do not translate in x- and y-direction but in directions parallel to adjacent sides of R_j. Doing so, we end up with a rectangle $\hat{R}_j$ and two (not necessarily distinct!) points p_1 and p_2 with the following properties:

1. $\hat{R}_j$ contains P_{pi},
2. p_1 and p_2 lie on adjacent sides of $\hat{R}_j$,
3. $\text{cov}(\hat{R}_j, \mathbf{P}) \supseteq \text{cov}(R_j, \mathbf{P})$, and
4. there are polygons $P_1, P_2 \in \mathbf{P}$ such that for $i \in \{1, 2\}$, p_i is a vertex of P_i and P_i is contained in $\hat{R}_j$.

Now we perform a general motion of $\hat{R}_j$ combined of a clockwise rotation and suitable translation which keeps p_1 and p_2 on their respective sides. There are two cases:

1. p_1 and p_2 coincide. Then the described general motion is a simple rotation of $\hat{R}_j$ around p_1. This rotation is continued until a point p_3 lies on a side of the resulting rectangle R'_j such that the following properties hold:
   (a) R'_j contains P_{pi},
   (b) there are polygons $P_1, P_3 \in \mathbf{P}$ such that for $i \in \{1, 3\}$ p_i is a vertex of P_i and P_i is contained in R'_j, and
   (c) $\text{cov}(R'_j, \mathbf{P}) \supseteq \text{cov}(\hat{R}_j, \mathbf{P})$.
2. p_1 and p_2 are distinct. Here we continue the general motion till one of the following two cases concerning the resulting rectangle R'_j occurs:
   (a) p_1 or p_2 coincide with a vertex p_3 of R'_j, or
   (b) there is a point p_3 on a side of R'_j such that
       i R'_j contains P_{pi},
       ii there are polygons $P_1, P_2, P_3 \in \mathbf{P}$ such that for $i \in \{1, 2, 3\}$ p_i is a vertex of P_i and P_i is contained in R'_j, and
       iii $\text{cov}(R'_j, \mathbf{P}) \supseteq \text{cov}(\hat{R}_j, \mathbf{P})$.

Now, after possibly necessary renamings, p_1, p_2 and p_3 together with R'_j meet the requirements of the lemma.

An example situation of the proof above is shown in the left part of Fig. 2: the dotted rectangle corresponds to R_j, the dashed one to $\hat{R}_j$ and the fully lined to R'_j. Moreover, P_1 is the left triangle while P_2 and P_3 coincide here in the right triangle. p_1 equals the point of the left triangle lying on the rectangle and p_2 and p_3 are the points of the right triangle lying on the rectangle. An extreme situation where p_1 and p_2 coincide with two opposite vertices of R_j is illustrated by the right part of the same figure (note that two opposite vertices of a rectangle lie on adjacent sides).

A.4   Upper Bound of the Branching Degree for the General Case

Let the candidate rectangles for the general case be computed as described in Subsect. 7.1, and let us replace n_{pi} by n for better readability. According to Sperner's theorem, there are at most $\binom{n}{\lfloor n/2 \rfloor}$ inclusion-maximal coverable sets containing P_{pi}. Ignoring the asymptotically irrelevant Gaussian brackets and expanding the binomial coefficient yields $\frac{n!}{(n/2)!^2}$ which amounts in the asymptotically view to $\frac{\sqrt{2\pi n}\left(\frac{n}{e}\right)^n}{\left(\sqrt{2\pi n/2}\left(\frac{n/2}{e}\right)^{n/2}\right)^2}$ by Stirling's formula. Elementary calculus leads now to the result from Subsect. 7.1.

References

1. Brönnimann, H., Goodrich, M.T.: Almost optimal set covers in finite vc-dimension. Discrete Comput. Geom. **14**(4), 463–479 (1995)
2. Chazelle, B.: The polygon containment problem. Adv. Comput. Res. **1**, 1–33 (1983)
3. Clarkson, K.L., Varadarajan, K.R.: Improved approximation algorithms for geometric set cover. Discrete Comput. Geom. **37**(1), 43–58 (2007)
4. Culberson, J.C., Reckhow, R.A.: Covering polygons is hard. In: 29th Annual Symposium on Foundations of Computer Science, pp. 601–611. IEEE Computer Society (1988)
5. de Berg, M., Cabello, S., Har-Peled, S.: Covering many or few points with unit disks. Theory Comput. Syst. **45**(3), 446–469 (2009)
6. de Berg, M., Cheong, O., van Kreveld, M., Overmars, M.: Computational Geometry, 3rd edn. Springer, Heidelberg (2008)
7. Dickerson, M., Scharstein, D.: Optimal placement of convex polygons to maximize point containment. Comput. Geom. **11**(1), 1–16 (1998)
8. Dowsland, K.A., Dowsland, W.B.: Solution approaches to irregular nesting problems. Eur. J. Oper. Res. **84**(3), 506–521 (1995)
9. Fowler, R.J., Paterson, M., Tanimoto, S.L.: Optimal packing and covering in the plane are np-complete. Inf. Process. Lett. **12**(3), 133–137 (1981)
10. Gibbons, A., Rytter, W.: Efficient Parallel Algorithms. Cambridge University Press, Cambridge (1988)
11. Glück, R.: Covering polygons with rectangles. In: Proceedings of the EuroCG 2016 (2016)
12. Hegedüs, A.: Algorithms for covering polygons by rectangles. Comput.-Aided Des. **14**(5), 257–260 (1982)
13. Heinrich-Litan, L., Lübbecke, M.E.: Rectangle covers revisited computationally. ACM J. Exp. Algorithmics **11** (2006)
14. Hochbaum, D.S., Maass, W.: Approximation schemes for covering and packing problems in image processing and VLSI. J. ACM **32**(1), 130–136 (1985)
15. Kumar, V.S.A., Ramesh, H.: Covering rectilinear polygons with axis-parallel rectangles. In: Vitter, J.S., Larmore, L.L., Leighton, F.T. (eds.) Proceedings of the Thirty-First Annual ACM Symposium on Theory of Computing, pp. 445–454. ACM (1999)
16. Preparata, F.P., Shamos, M.I.: Computational Geometry - An Introduction. Texts and Monographs in Computer Science. Springer, New York (1985)
17. Reif, J.H.: Synthesis of Parallel Algorithms. Morgan Kaufmann Publishers Inc., San Mateo (1993)
18. Shamos, M.I.: Computational Geometry. PhD thesis, Yale University (1978)

The Complexity of Perfect Packings in Dense Graphs

Jie Han[(⊠)]

Instituto de Matemática e Estatística, Universidade de São Paulo,
Rua Do Matão 1010, São Paulo 05508-090, Brazil
jhan@ime.usp.br

Abstract. Let H be a graph of order h and let G be a graph of order n such that $h \mid n$. A perfect H-packing in G is a collection of vertex disjoint copies of H in G that covers all vertices of G. Hell and Kirkpatrick showed that the decision problem whether a graph G has a perfect H-packing is NP-complete if and only if H has a component which contains at least 3 vertices.

We consider the decision problem of containment of a perfect H-packing in graphs G under the additional minimum degree condition. Our main result shows that given any $\gamma > 0$ and any n-vertex graph G with minimum degree at least $(1 - 1/\chi_{cr}(H) + \gamma)n$, the problem of determining whether G has a perfect H-packing can be solved in polynomial time, where $\chi_{cr}(H)$ is the *critical chromatic number* of H. This answers a question of Yuster negatively. Moreover, a hardness result of Kühn and Osthus shows that our main result is essentially best possible and closes a long-standing hardness gap for all complete multi-partite graphs H whose second smallest color class has size at least 2.

Keywords: Perfect packing · Computational complexity · Absorbing method

1 Introduction

1.1 Perfect H-packings

Given two graphs G and H, an H-*packing* (or H-*tiling*) in G is a collection of vertex-disjoint copies of H in G. An H-packing in G is called *perfect* if it covers all vertices of G. When H is a single edge, an H-packing is often called a *matching*. Edmonds' Algorithm [4] determines if a graph has a perfect matching in polynomial time. However, in general, Hell and Kirkpatrick [7] showed that the decision problem whether a graph G has a perfect H-packing is NP-complete if and only if H has a component which contains at least 3 vertices.

Given a graph H, we write $|H|$ for its order and $\chi(H)$ for its chromatic number. For approximating the size of a maximal H-packing, Hurkens and Schrijver [8] gave an $(|H|/2 + \epsilon)$-approximation algorithm (where $\epsilon > 0$ is arbitrary) which runs in polynomial time. On the other hand, Kann [9] proved that the problem

© Springer International Publishing AG 2017
T.V. Gopal et al. (Eds.): TAMC 2017, LNCS 10185, pp. 289–303, 2017.
DOI: 10.1007/978-3-319-55911-7_21

is APX-hard if H has a component which contains at least three vertices. (In other words, it is impossible to approximate the optimum solution within an arbitrary factor unless $P = NP$). In contrast, the results in [7] imply that the remaining cases of the problem can be solved in polynomial time. The following theorem of Alon and Yuster [2] shows that the problem can be solved in polynomial time for instances G which are sufficiently dense. Let $M(n)$ be the time needed to multiply two n by n matrices with $0, 1$ entries over the integer. Determining $M(n)$ is a challenging problem in theoretic computer science, and the up-to-date result $M(n) = O(n^{2.3728639})$ is obtained by Le Gall [18].

Theorem 1 (Alon-Yuster, [2]). *For every $\gamma > 0$ and each graph H there exists an integer $n_0 = n_0(\gamma, H)$ such that every graph G whose order $n \geq n_0$ is divisible by $|H|$ and whose minimum degree is at least $(1-1/\chi(H)+\gamma)n$ contains a perfect H-packing. Moreover, there is an algorithm which finds this H-packing in time $O(M(n))$.*

Note that balanced complete $\chi(H)$-partite graphs show that the minimum degree condition in Theorem 1 is essentially best possible. In [2], they also conjectured that the error term γn in Theorem 1 can be replaced by a constant $C(H) > 0$ depending only on H, which has been verified by Komlós, Sárközy and Szemerédi [15].

Theorem 2 (Komlós-Sárközy-Szemerédi, [15]). *For every graph H there exist integers $C < |H|$ and $n_0 = n_0(H)$ such that every graph G whose order $n \geq n_0$ is divisible by $|H|$ and whose minimum degree is at least $(1-1/\chi(H))n+C$ contains a perfect H-packing. Moreover, there is an algorithm which finds this H-packing in time $O(nM(n))$.*

As observed in [2], there are graphs H for which the constant $C(H)$ cannot be omitted completely. On the other hand, there are graphs H for which the minimum degree condition in Theorem 2 can be improved significantly [3,11], by replacing the chromatic number with the critical chromatic number in Theorem 2. The *critical chromatic number* $\chi_{cr}(H)$ of a graph H is defined as $(\chi(H) - 1)|H|/(|H| - \sigma(H))$, where $\sigma(H)$ denotes the minimum size of the smallest color class in a coloring of H with $\chi(H)$ colors. Note that $\chi(H) - 1 < \chi_{cr}(H) \leq \chi(H)$ and the equality holds if and only if every $\chi(H)$-coloring of H has equal color class sizes. If $\chi_{cr}(H) = \chi(H)$, then we call H *balanced*, otherwise *unbalanced*. Unbalanced complete $\chi(H)$-partite graphs[1] show that for any graph H, one cannot improve Theorem 2 by replacing $\chi(H)$ with a constant smaller than $\chi_{cr}(H)$ [1]. Komlós [14] proved that one can replace $\chi(H)$ with $\chi_{cr}(H)$ in Theorem 2 at the price of obtaining an H-packing covering all but ϵn vertices. He also conjectured that the error term ϵn can be replaced with a constant that only depends on H [14], which was confirmed by Shokoufandeh and Zhao [21] (here we state their result in a slightly weaker form).

---

[1] More precisely, here one should take the complete $\chi(H)$-partite graphs with $\sigma(H)n/|H| - 1$ vertices in one color class, and other color classes of sizes as equal as possible.

Theorem 3 (Shokoufandeh-Zhao, [21]). *For any H there is an $n_0 = n_0(H)$ so that if G is a graph on $n \geq n_0$ vertices and minimum degree at least $(1 - 1/\chi_{cr}(H))n$, then G contains an H-packing that covers all but at most $5|H|^2$ vertices.*

Then the question is, *for which H can we replace $\chi(H)$ with $\chi_{cr}(H)$ in Theorem 2?* Kühn and Osthus [16,17] answered this question completely. To state their result, we need some definitions. Write $k := \chi(H)$. Given a k-coloring c, let $x_1 \leq \cdots \leq x_k$ denote the sizes of the color classes of c and put $D(c) = \{x_{i+1} - x_i \mid i \in [k-1]\}$. Let $D(H)$ be the union of all the sets $D(c)$ taken over all k-colorings c. Denote by $\mathrm{hcf}_\chi(H)$ the highest common factor of all integers in $D(H)$. (If $D(H) = \{0\}$, then set $\mathrm{hcf}_\chi(H) := \infty$.) Write $\mathrm{hcf}_c(H)$ for the highest common factor of all the orders of components of H (for example $\mathrm{hcf}_c(H) = |H|$ if H is connected). If $\chi(H) \neq 2$, then define $\mathrm{hcf}(H) = 1$ if $\mathrm{hcf}_\chi(H) = 1$. If $\chi(H) = 2$, then define $\mathrm{hcf}(H) = 1$ if both $\mathrm{hcf}_c(H) = 1$ and $\mathrm{hcf}_\chi(H) \leq 2$. Then let

$$\chi_*(H) = \begin{cases} \chi_{cr}(H) & \text{if } \mathrm{hcf}(H) = 1, \\ \chi(H) & \text{otherwise.} \end{cases}$$

In particular we have $\chi_{cr}(H) \leq \chi_*(H)$.

Theorem 4 (Kühn-Osthus, [16,17]). *There exist integers $C = C(H)$ and $n_0 = n_0(H)$ such that every graph G whose order $n \geq n_0$ is divisible by $|H|$ and whose minimum degree is at least $(1 - 1/\chi_*(H))n + C$ contains a perfect H-packing.*

Theorem 4 is best possible in the sense that the degree condition cannot be lowered up to the constant C (there are also graphs H such that the constant cannot be omitted entirely). Moreover, this also implies that, *one can replace $\chi(H)$ with $\chi_{cr}(H)$ in Theorem 2 if and only if $\mathrm{hcf}(H) = 1$.*

1.2 The Algorithmic Aspect

Now let us turn to the algorithmic aspect of this problem. Let $\mathbf{Pack}(H, \delta)$ be the decision problem of determining whether a graph G whose minimum degree is at least $\delta|G|$ contains a perfect H-packing. When H contains a component of size at least 3, the result of Hell and Kirkpatrick [7] shows that $\mathbf{Pack}(H, 0)$ is NP-complete. In contrast, Theorem 4 gives that $\mathbf{Pack}(H, \delta)$ is (trivially) in P for any $\delta \in (1 - 1/\chi_*(H), 1]$. In [16], Kühn and Osthus showed that $\mathbf{Pack}(H, \delta)$ is NP-complete for any $\delta \in [0, 1 - 1/\chi_{cr}(H))$ if H is a clique of size at least 3 or a complete k-partite graph such that $k \geq 2$ and the size of the second smallest cluster is at least 2.

Due to lack of knowledge on the range $\delta \in [0, 1 - 1/\chi_*(H))$ for general H, we still do not understand $\mathbf{Pack}(H, \delta)$ well in general. Indeed, even for (unbalanced) complete multi-partite graphs H with $\mathrm{hcf}(H) \neq 1$, there is a substantial hardness gap for $\delta \in [1 - 1/\chi_{cr}(H), 1 - 1/\chi_*(H)]$. In particular, Yuster asked the following question in his survey [23].

Problem 1 (Yuster, [23]). Is it true that **Pack**(H, δ) is NP-complete for all $\delta \in [0, 1 - 1/\chi_*(H))$ and any H which contains a component of size at least 3?

Our main result provides an algorithm showing that **Pack**(H, δ) is in P when $\delta \in (1 - 1/\chi_{cr}(H), 1]$, which confirms a recent conjecture of Treglown [22] and gives a negative answer to Problem 1 (as seen for any H such that $\chi_{cr}(H) < \chi_*(H)$). In fact, this gives the first *nontrivial* polynomial-time algorithm for the decision problem **Pack**(H, δ). In particular, it eliminates the aforementioned hardness gap for unbalanced complete multi-partite graphs H with $\mathrm{hcf}(H) \neq 1$ almost entirely.

Theorem 5. *For any m-vertex k-chromatic graph H and $\delta \in (1 - 1/\chi_{cr}(H), 1]$,* **Pack**$(H, \delta)$ *is in P. That is, for every n-vertex graph G with minimum degree at least δn, there is an algorithm with time $O(n^{\max\{2^{m^{k-1}-1}m+1, m(2m-1)^m - m\}})$, which determines whether G contains a perfect H-packing.*

In view of the aforementioned hardness result of [16], Theorem 5 is asymptotically best possible if H is a complete k-partite graph such that $k \geq 2$ and the size of the second smallest cluster is at least 2 (note that when H is balanced, the result is included in Theorem 1). On the other hand, Theorem 5 complements Theorem 3 in the sense that when the minimum degree condition guarantees an H-packing that covers all but constant number of vertices, we can detect the 'last obstructions' efficiently. A similar phenomenon also appears in the decision problem for perfect matchings in uniform hypergraphs [6, 12].

The rest of the paper is organized as follows. In Sect. 2 we introduce the main ideas of the proof of Theorem 5. We introduce Lemma 1 and Theorem 6 in Sect. 3 and use them to prove Theorem 5. We prove Theorem 6 in Sect. 4. We show the rest of the proofs in the appendix.

Notation. For any graph G, we write $|G|$ for its order, $e(G)$ for its number of edges, $\delta(G)$ for its minimum degree, $\chi(G)$ for its chromatic number and $\chi_{cr}(G)$ for its critical chromatic number defined in Sect. 1. Given $A \subseteq V(G)$, let $G[A]$ be the induced subgraph of G on A; given $A, B \subseteq V(G)$ with $A \cap B = \emptyset$, let $G[A, B]$ be the induced bipartite subgraph of G with parts A and B. Given integers $k \geq 2$ and $a_1, \ldots, a_k$, let $K_{a_1, \ldots, a_k}$ be the complete k-partite graph with color class sizes $a_1, \ldots, a_k$. Throughout this paper, $x \ll y$ means that for any $y > 0$, there exists $x_0 > 0$ such that for any $0 < x \leq x_0$ the following statement holds. Hierarchies of other lengths are defined in the obvious way.

2   Ingredients of the Proof of Theorem 5

Throughout the rest of the paper, we always assume that H is an m-vertex k-chromatic graph. Recall that when H is balanced, the result of Theorem 5 is included in Theorem 1. So we may assume that H is unbalanced, namely, there exists a k-coloring of H, with color class sizes $a_1 \leq a_2 \leq \cdots \leq a_k$, such that $a_1 < a_k$.

There are two main ingredients in our proof of Theorem 5. One is the concepts of *lattices and solubility* introduced by Keevash, Knox and Mycroft [12]; and the other is the *lattice-based absorbing method* developed recently by the author [6].

2.1 Lattices and Solubility

To prove Theorem 5, we need to find a property of G which is both sufficient and necessary for a graph G with appropriate minimum degree to contain a perfect H-packing, and can be tested efficiently. This is done by Theorem 6, the so-called *structural theorem*. However the property is not easy to state – we need the following definitions, which essentially come from [12,13].

Let G be an n-vertex graph. We will work on a partition $\mathcal{P} = \{V_1, \ldots, V_d\}$ of $V(G)$ for some integer $d \geq 1$. In this paper, every partition has an implicit order on its parts. The *index vector* $\mathbf{i}_{\mathcal{P}}(S) \in \mathbb{Z}^d$ of a subset $S \subseteq V(G)$ with respect to $\mathcal{P}$ is the vector whose coordinates are the sizes of the intersections of S with each part of $\mathcal{P}$, namely, $\mathbf{i}_{\mathcal{P}}(S)_X = |S \cap X|$ for $X \in \mathcal{P}$. For any $\mathbf{v} = \{v_1, \ldots, v_d\} \in \mathbb{Z}^d$, let $|\mathbf{v}| = \sum_{i=1}^d v_i$. We say that $\mathbf{v} \in \mathbb{Z}^d$ is an *r-vector* if it has non-negative coordinates and $|\mathbf{v}| = r$.

When Keevash and Mycroft [13] studied perfect matchings in k-uniform hypergraphs, they observed a family of lattice-based constructions, which they named as 'divisibility barriers'. In fact, similar constructions appear in our problem, which we define formally now.

Let $\mathcal{P} = \{V_1, \ldots, V_d\}$ be a partition of $V(G)$ for some integer $d \geq 1$. Let $I_{\mathcal{P}}(G)$ denote the set of all $\mathbf{i} \in \mathbb{Z}^d$ such that G contains at least one copy of H with index vector $\mathbf{i}$ and let $L_{\mathcal{P}}(G)$ denote the lattice in $\mathbb{Z}^d$ generated by $I_{\mathcal{P}}(G)$. A *divisibility barrier* is a graph G which admits a vertex partition $\mathcal{P}$ of $V(G)$ such that $\mathbf{i}_{\mathcal{P}}(V(G)) \notin L_{\mathcal{P}}(G)$. To see that such a G contains no perfect H-packing, let M be an H-packing in G. Then $\mathbf{i}_{\mathcal{P}}(V(M)) = \sum_{F \in M} \mathbf{i}_{\mathcal{P}}(V(F)) \in L_{\mathcal{P}}(G)$. But $\mathbf{i}_{\mathcal{P}}(V(G)) \notin L_{\mathcal{P}}(G)$, so $V(M) \neq V(G)$, namely, M is not perfect.

Note that given a partition $\mathcal{P}$ of bounded number of parts (independent of n), the property $\mathbf{i}_{\mathcal{P}}(V(G)) \in L_{\mathcal{P}}(G)$ can be tested efficiently (it takes time $O(n^m)$ to determine $I_{\mathcal{P}}(G)$, and it takes constant time to generate $L_{\mathcal{P}}(G)$ and check if $\mathbf{i}_{\mathcal{P}}(V(G)) \in L_{\mathcal{P}}(G)$). Moreover, $\mathbf{i}_{\mathcal{P}}(V(G)) \notin L_{\mathcal{P}}(G)$ implies that G contains no perfect H-packing. Unfortunately, the converse is not true. In fact, it is not hard to construct a graph with appropriate minimum degree such that for some m-vector $\mathbf{i}$, all copies of H with index vector $\mathbf{i}$ are intersecting (see [12, Construction 1.6] for an example for matchings in hypergraphs) – but in reality, when we compute $L_{\mathcal{P}}(G)$, we are allowed to use any multiple of $\mathbf{i}$. Thus, it is natural to consider the following robust vectors and robust lattices.

Given $\mu > 0$, let $I_{\mathcal{P}}^\mu(G)$ denote the set of all $\mathbf{i} \in \mathbb{Z}^d$ such that G contains at least μn^m copies of H with index vector $\mathbf{i}$ and let $L_{\mathcal{P}}^\mu(G)$ denote the lattice in $\mathbb{Z}^d$ generated by $I_{\mathcal{P}}^\mu(G)$.

However, $\mathbf{i}_{\mathcal{P}}(V(G)) \notin L_{\mathcal{P}}^\mu(G)$ does not imply that G contains no perfect H-packing. So instead, we consider the following solubility and show that it is the correct property we are looking for. A (possibly empty) H-packing M of size at most $(2m - 1)^d - 1$ is a *solution* for $(\mathcal{P}, L)$ (in G) if $\mathbf{i}_\mathcal{P}(V(G) \setminus V(M)) \in L$; we say that $(\mathcal{P}, L)$ is *soluble* if it has a solution, otherwise *insoluble*.

2.2 Lattice-Based Absorbing Method

We will show that every n-vertex graph G with $\delta(G) \geq (1 - 1/\chi_{cr}(H) + \gamma)n$ admits a partition $\mathcal{P}$ of $V(G)$ (Lemma 1), such that G contains a perfect H-packing if and only if $(\mathcal{P}, L_\mathcal{P}^\mu(G))$ is soluble for some suitable choice of $\mu > 0$ (Theorem 6). The major work is in the proof of the backward implication of Theorem 6. In fact, in view of Theorem 3, it suffices to show that (given that $(\mathcal{P}, L_\mathcal{P}^\mu(G))$ is soluble) we can finish the 'last piece' of the perfect H-packing. The key ingredient of our proof is the so-called lattice-based absorbing method developed recently by the author.

The absorbing technique initiated by Rödl, Ruciński and Szemerédi [20] has been shown to be efficient on finding spanning structures in graphs and hypergraphs. For finding a perfect H-packing, roughly speaking, the goal is to build the *absorbing set* A such that A is a small subset of vertices and for any (much smaller) subset U with $m \mid |U|$, both $G[A]$ and $G[A \cup U]$ contain perfect H-packings. Note that we must have the absorbing property work for *arbitrary* leftover set U, because we cannot foresee the leftover set.

However, for our problem, such an absorbing set cannot be guaranteed in G (because not all G contain perfect H-packings). Roughly speaking, the advance of the lattice-based absorbing method is to utilize the *reachability* information (will be defined formally later) to find a vertex partition $\mathcal{P}$ in polynomial time, for which we can find an absorbing set (not in the usual sense) that works under the lattice structure. If in addition $(\mathcal{P}, L_\mathcal{P}^\mu(G))$ is soluble, then we can find a perfect H-packing by some careful analysis. This method was first used in [6] for solving a complexity problem of Karpiński, Ruciński and Szymańska [10].

Formally, let H be an m-vertex graph and let G be an n-vertex graph. We say that two vertices u and v in $V(G)$ are (H, β, i)-*reachable in* G if there are at least βn^{im-1} $(im - 1)$-sets S such that both $G[S \cup \{u\}]$ and $G[S \cup \{v\}]$ have perfect H-packings. We say a vertex set $U \subseteq V(G)$ is (H, β, i)-*closed in* G if any two vertices $u, v \in U$ are (H, β, i)-reachable in G.

3   Proof of Theorem 5

The following lemma provides a partition $\mathcal{P}$ such that we can utilize the reachability argument and develop the absorbing lemma on $\mathcal{P}$.

Lemma 1. *Let H be an unbalanced m-vertex k-chromatic graph with $k \geq 2$. For any $\gamma > 0$, suppose that $1/n_0 \ll \beta \ll \gamma, 1/m$. Then for each graph G on $n \geq n_0$ vertices with $\delta(G) \geq (1 - 1/\chi_{cr}(H) + \gamma)n$, we find a partition $\mathcal{P} = \{V_1, \ldots, V_d\}$ of $V(G)$ with $d \leq m$ in time $O(n^{2^{m^{k-1}-1}m+1})$ such that for $i \in [d]$, $|V_i| \geq n/m$ and V_i is $(H, \beta, 2^{m^{k-1}-1})$-closed.*

The proof of Lemma 1 is technical and we postpone it to the appendix. Now we are ready to state our main structural theorem.

Theorem 6. *Let H be an unbalanced m-vertex k-chromatic graph with $k \geq 2$. For any $\gamma > 0$, suppose that*

$$1/n_0 \ll \{\beta, \mu\} \ll \gamma, 1/m.$$

Let G be a graph on $n \geq n_0$ vertices such that $\delta(G) \geq (1 - 1/\chi_{cr}(H) + \gamma)n$ with $\mathcal{P}$ found by Lemma 1. Then G contains a perfect H-packing if and only if $(\mathcal{P}, L_{\mathcal{P}}^{\mu}(G))$ is soluble.

3.1   Proof of Theorem 5

By Theorem 6, to determine the existence of an H-packing, it is straightforward to check if $(\mathcal{P}, L_{\mathcal{P}}^{\mu}(G))$ is soluble.

Proof (Proof of Theorem 5). Recall that in view of Theorem 1, it suffices to consider only unbalanced H. Let G be an n-vertex graph with $\delta(G) \geq (1 - 1/\chi_{cr}(H) + \gamma)n$. Note that it is trivial to determine the existence of a perfect H-packing if $n < n_0$ given by Theorem 6. Now assume $n \geq n_0$, we find the partition $\mathcal{P}$ and check if $(\mathcal{P}, L_{\mathcal{P}}^{\mu}(G))$ is soluble. If the answer is 'true', then H contains a perfect H-packing by Theorem 6.

By Lemma 1, we find $\mathcal{P}$ in time $O(n^{2^{m^{k-1}-1}m+1})$. To check the solubility, we check if $\mathbf{i}_{\mathcal{P}}(V(G) \setminus V(M)) \in L_{\mathcal{P}}^{\mu}(G)$ for each H-packing M of size at most $(2m-1)^d - 1$, which can be done in time $O(n^{m((2m-1)^d-1)})$. Since $d \leq m$, the overall running time is $O(n^{\max\{2^{m^{k-1}-1}m+1, \, m(2m-1)^m - m\}})$.

4   Proof of Theorem 6

4.1   Proof of the Forward Implication of Theorem 6

We write $\mathbf{u}_j$ for the 'unit' 1-vector that is 1 in coordinate j and 0 in all other coordinates. Given a partition $\mathcal{P}$ of d parts, we write $L_{\max}^d$ for the lattice generated by all m-vectors. So $L_{\max}^d = \{\mathbf{v} \in \mathbb{Z}^d : m \text{ divides } |\mathbf{v}|\}$.

The following definitions [12] and result are crucial in our proof. Suppose $L \subset L_{\max}^{|\mathcal{P}|}$ is a lattice in $\mathbb{Z}^{|\mathcal{P}|}$, where $\mathcal{P}$ is a partition of a set V. The *coset group* of $(\mathcal{P}, L)$ is $Q = Q(\mathcal{P}, L) = L_{\max}^{|\mathcal{P}|}/L$. For any $\mathbf{i} \in L_{\max}^{|\mathcal{P}|}$, the *residue* of $\mathbf{i}$ in Q is $R_Q(\mathbf{i}) = \mathbf{i} + L$. For any $A \subseteq V$ of size divisible by m, the *residue* of A in Q is $R_Q(A) = R_Q(\mathbf{i}_{\mathcal{P}}(A))$.

Proposition 1. *Let H be an unbalanced m-vertex k-chromatic graph with $k \geq 2$. For any $\gamma > 0$, suppose that*

$$1/n_0 \ll \{\beta, \mu\} \ll \gamma, 1/m.$$

Let G be a graph on $n \geq n_0$ vertices such that $\delta(G) \geq (1 - 1/\chi_{cr}(H) + \gamma)n$ with $\mathcal{P}$ found by Lemma 1. Then $|Q(\mathcal{P}, L_{\mathcal{P}}^{\mu}(G))| < (2m-1)^d$.

We postpone the proof of Proposition 1 to the appendix.

Proof (Proof of the forward implication of Theorem 6). If G contains a perfect H-packing M, then $\mathbf{i}_{\mathcal{P}}(V(G) \setminus V(M)) = \mathbf{0} \in L_{\mathcal{P}}^{\mu}(G)$. We will show that there exists an H-packing $M' \subset M$ of size at most $(2m-1)^d - 1$ such that $\mathbf{i}_{\mathcal{P}}(V(G) \setminus V(M')) \in L_{\mathcal{P}}^{\mu}(G)$ and thus $(\mathcal{P}, L_{\mathcal{P}}^{\mu}(G))$ is soluble. Indeed, suppose $M' \subset M$ is a minimum H-packing such that $\mathbf{i}_{\mathcal{P}}(V(G) \setminus V(M')) \in L_{\mathcal{P}}^{\mu}(G)$ and $|M'| = m' \geq (2m-1)^d$. Let $M' = \{e_1, \dots, e_{m'}\}$ and consider the $m' + 1$ partial sums

$$\sum_{i=1}^{j} \mathbf{i}_{\mathcal{P}}(e_i) + L_{\mathcal{P}}^{\mu}(G) = \sum_{i=1}^{j} R_{Q(\mathcal{P}, L_{\mathcal{P}}^{\mu}(G))}(e_i),$$

for $j = 0, 1, \dots, m'$. Since $|Q(\mathcal{P}, L_{\mathcal{P}}^{\mu}(G))| \leq (2m-1)^d \leq m'$, two of the sums must be equal. That is, there exists $0 \leq j_1 < j_2 \leq m'$ such that

$$\sum_{i=j_1+1}^{j_2} \mathbf{i}_{\mathcal{P}}(e_i) \in L_{\mathcal{P}}^{\mu}(G).$$

So the H-packing $M'' := M' \setminus \{e_{j_1+1}, \dots, e_{j_2}\}$ satisfies that $\mathbf{i}_{\mathcal{P}}(V(G) \setminus V(M'')) \in L_{\mathcal{P}}^{\mu}(G)$ and $|M''| < |M'|$, a contradiction.

4.2   Proof of the Backward Implication of Theorem 6

Suppose I is a set of m-vectors of $\mathbb{Z}^d$ and J is a (finite) set of vectors such that any $\mathbf{i} \in J$ can be written as a linear combination of vectors in I, namely, there exist $a_{\mathbf{v}}(\mathbf{i}) \in \mathbb{Z}$ for all $\mathbf{v} \in I$, such that

$$\mathbf{i} = \sum_{\mathbf{v} \in I} a_{\mathbf{v}}(\mathbf{i})\mathbf{v}.$$

We denote by $C(d, m, I, J)$ as the maximum of $|a_{\mathbf{v}}(\mathbf{i})|, \mathbf{v} \in I$ over all $\mathbf{i} \in J$.

Fix an integer $i > 0$. For an m-vertex set S, we say a set T is an *absorbing i-set for S* if $|T| = i$ and both $G[T]$ and $G[T \cup S]$ contain perfect H-packings. The proof of the following absorbing lemma consists of routine probabilistic arguments and is omitted (a similar proof can be found in [6, Lemma 3.4]).

Lemma 2 (Absorbing Lemma). *Suppose H is an m-vertex graph and*

$$1/n \ll 1/c \ll \{\beta, \mu\} \ll \{1/m, 1/t\}$$

and G is a graph on n vertices. Suppose $\mathcal{P} = \{V_1, \dots, V_d\}$ is a partition of $V(G)$ such that for $i \in [d]$, V_i is (H, β, t)-closed. Then there is a family $\mathcal{F}_{abs}$ of disjoint tm^2-sets with size at most $c \log n$ such that for each $F \in \mathcal{F}_{abs}$, $G[V(F)]$ contains a perfect H-packing and every m-vertex set S with $\mathbf{i}_{\mathcal{P}}(S) \in I_{\mathcal{P}}^{\mu}(G)$ has at least $\sqrt{\log n}$ absorbing tm^2-sets in $\mathcal{F}_{abs}$.

We postpone the proof of the absorbing lemma to the appendix and prove our main goal, the backward implication of Theorem 6, first. Here is an outline of the proof. The proof consists of a few steps. We first fix an H-packing M_1, the solution of $(\mathcal{P}, L_{\mathcal{P}}^{\mu}(G))$. We apply Lemma 2 on G and get a family $\mathcal{F}_{abs}$ of tm^2-sets of size at most $c \log n$. Let $\mathcal{F}_0$ be the subfamily of $\mathcal{F}_{abs}$ that do not intersect $V(M_1)$. Next we find a set M_2 of copies of H, which includes (constantly) many copies of H for each m-vector in $I_{\mathcal{P}}^{\mu}(G)$. Now we are ready to apply Theorem 3 on $G[V \setminus (V(\mathcal{F}_0) \cup V(M_1 \cup M_2))]$ and find an H-packing M_3 covering all but a set U of at most $5 m^2$ vertices. The remaining job is to 'absorb' the vertices in U. Roughly speaking, by the solubility condition, we can release some copies of H in some members of $\mathcal{F}_0$ and M_3, such that the set $D \supseteq U$ of uncovered vertices satisfies that $\mathbf{i}_{\mathcal{P}}(D) \in L_{\mathcal{P}}^{\mu}(G)$. Furthermore, by releasing some copies of H in M_2, we can partition the new set of uncovered vertices as a collection of m-sets S such that $\mathbf{i}_{\mathcal{P}}(S) \in I_{\mathcal{P}}^{\mu}(G)$ for each S. Then we can finish the absorption by the absorbing property of $\mathcal{F}_0$.

Proof (Proof of the backward implication of Theorem 6). Fix $\gamma > 0$. Suppose

$$1/n_0 \ll 1/c \ll \{\beta, \mu\} \ll \gamma, 1/m.$$

Let G be a graph on $n \geq n_0$ vertices such that $\delta(G) \geq (1 - 1/\chi_{cr}(H) + \gamma)n$ with $\mathcal{P} = \{V_1, \ldots, V_d\}$ found by Lemma 1. Moreover, assume that $(\mathcal{P}, L_{\mathcal{P}}^{\mu}(G))$ is soluble. Let $t = 2^{m^{k-1}-1}$. We first apply Lemma 2 on G and get a family $\mathcal{F}_{abs}$ of tm^2-sets of size at most $c \log n$ such that every m-set S of vertices with $\mathbf{i}_{\mathcal{P}}(S) \in I_{\mathcal{P}}^{\mu}(G)$ has at least $\sqrt{\log n}$ absorbing tm^2-sets in $\mathcal{F}_{abs}$.

Since $(\mathcal{P}, L_{\mathcal{P}}^{\mu}(G))$ is soluble, there exists an H-packing M_1 of size at most $(2m - 1)^d - 1$ such that $\mathbf{i}_{\mathcal{P}}(V(G) \setminus V(M_1)) \in L_{\mathcal{P}}^{\mu}(G)$. Note that $V(M_1)$ may intersect $V(\mathcal{F}_{abs})$ in at most $m(2m - 1)^d$ absorbing sets of $\mathcal{F}_{abs}$. Let $\mathcal{F}_0$ be the subfamily of $\mathcal{F}_{abs}$ obtained from removing the tm^2-sets that intersect $V(M_1)$. Let M_0 be the perfect H-packing on $V(\mathcal{F}_0)$ that consists of perfect H-packings on each member of $\mathcal{F}_0$. Note that every m-set S of vertices with $\mathbf{i}_{\mathcal{P}}(S) \in I_{\mathcal{P}}^{\mu}(G)$ has at least $\sqrt{\log n} - m(2m - 1)^d$ absorbing sets in $\mathcal{F}_0$.

Next we want to 'store' some copies of H for each m-vector in $I_{\mathcal{P}}^{\mu}(G)$ for future use. More precisely, let J be the set of all m'-vectors in $L_{\mathcal{P}}^{\mu}(G)$ such that $0 \leq m' \leq 5(2m - 1)^{d+1}$ and $C = C(d, m, I_{\mathcal{P}}^{\mu}(G), J)$. We find an H-packing M_2 in $V(G) \setminus V(M_0 \cup M_1)$ which contains C copies H' of H with $\mathbf{i}_{\mathcal{P}}(H') = \mathbf{i}$ for every $\mathbf{i} \in I_{\mathcal{P}}^{\mu}(G)$. So $|M_2| \leq \binom{m+d-1}{m}C$ and the process is possible because G contains at least μn^m copies of H for each $\mathbf{i} \in I_{\mathcal{P}}^{\mu}(G)$ and $|V(M_0 \cup M_1 \cup M_2)| \leq tm^2c \log n + (2m - 1)^d m + \binom{m+d-1}{m}Cm < \mu n$.

Let $G' := G[V(G) \setminus V(M_0 \cup M_1 \cup M_2)]$. Note that $|V(G')| \geq n - \mu n$. So

$$\delta(G') \geq \delta(G) - \mu n \geq (1 - 1/\chi_{cr}(H))n \geq (1 - 1/\chi_{cr}(H))|G'|.$$

So we can apply Theorem 3 on G' and find an H-packing M_3 covering all but at most $5 m^2$ vertices of G'. Let U be the set of uncovered vertices.

Let $Q = Q(\mathcal{P}, L_{\mathcal{P}}^{\mu}(G))$. Recall that $\mathbf{i}_{\mathcal{P}}(V(G) \setminus V(M_1)) \in L_{\mathcal{P}}^{\mu}(G)$. Note that by definition, the index vectors of all copies in M_2 are in $I_{\mathcal{P}}^{\mu}(G)$. So we have $\mathbf{i}_{\mathcal{P}}(V(G) \setminus V(M_1 \cup M_2)) \in L_{\mathcal{P}}^{\mu}(G)$, namely, $R_Q(V(G) \setminus V(M_1 \cup M_2)) = \mathbf{0} + L_{\mathcal{P}}^{\mu}(G)$. Thus,

$$\sum_{H' \in M_0 \cup M_3} R_Q(V(H')) + R_Q(U) = \mathbf{0} + L_{\mathcal{P}}^{\mu}(G).$$

Suppose $R_Q(U) = \mathbf{v}_0 + L_{\mathcal{P}}^\mu(G)$ for some $\mathbf{v}_0 \in L_{max}^d$ and we get

$$\sum_{H' \in M_0 \cup M_3} R_Q(V(H')) = -\mathbf{v}_0 + L_{\mathcal{P}}^\mu(G).$$

Claim. There exist $H_1, \ldots, H_\ell \in M_0 \cup M_3$ for some $\ell \leq (2m-1)^d - 1$ such that

$$\sum_{i \in [\ell]} R_Q(V(H_i)) = -\mathbf{v}_0 + L_{\mathcal{P}}^\mu(G). \qquad (1)$$

Proof. Assume to the contrary that $H_1, \ldots, H_\ell \in M_0 \cup M_3$ is a minimum set of copies of H such that (1) holds and $\ell \geq (2m-1)^d$. Consider the $\ell+1$ partial sums $\sum_{i \in [j]} R_Q(V(H_i))$ for $j = 0, 1, \ldots, \ell$, where the sum equals $\mathbf{0} + L_{\mathcal{P}}^\mu(H)$ when $j = 0$. By Proposition 1, $|Q| \leq (2m-1)^d$, then two of the partial sums must be equal, that is, there exist $0 \leq \ell_1 < \ell_2 \leq \ell$ such that $\sum_{\ell_1 < i \leq \ell_2} R_Q(V(H_i)) = \mathbf{0} + L_{\mathcal{P}}^\mu(G)$. So we get a smaller set of copies of H such that (1) holds, a contradiction.

So we have $\sum_{i \in [\ell]} \mathbf{i}_{\mathcal{P}}(V(H_i)) + \mathbf{i}_{\mathcal{P}}(U) \in L_{\mathcal{P}}^\mu(G)$. Let $D := \bigcup_{i \in [\ell]} V(H_i) \cup U$ and thus $|D| \leq m\ell + 5m^2 \leq m((2m-1)^d - 1) + 5m^2 \leq 5(2m-1)^{d+1}$. At last, we finish the perfect H-packing by absorption. Since $\mathbf{i}_{\mathcal{P}}(D) \in L_{\mathcal{P}}^\mu(G)$, we have the following equation

$$\mathbf{i}_{\mathcal{P}}(D) = \sum_{\mathbf{v} \in I_{\mathcal{P}}^\mu(G)} a_\mathbf{v} \mathbf{v},$$

where $a_\mathbf{v} \in \mathbb{Z}$ for all $\mathbf{v} \in I_{\mathcal{P}}^\mu(G)$. Since $|D| \leq 5(2m-1)^{d+1}$, by the definition of C, we have $|a_\mathbf{v}| \leq C$ for all $\mathbf{v} \in I_{\mathcal{P}}^\mu(G)$. Noticing that $a_\mathbf{v}$ may be negative, we can assume $a_\mathbf{v} = b_\mathbf{v} - c_\mathbf{v}$ such that one of $b_\mathbf{v}, c_\mathbf{v}$ is $|a_\mathbf{v}|$ and the other is zero for all $\mathbf{v} \in I_{\mathcal{P}}^\mu(G)$. So, we have

$$\sum_{\mathbf{v} \in I_{\mathcal{P}}^\mu(G)} c_\mathbf{v} \mathbf{v} + \mathbf{i}_{\mathcal{P}}(D) = \sum_{\mathbf{v} \in I_{\mathcal{P}}^\mu(G)} b_\mathbf{v} \mathbf{v}.$$

This equation means that given any family $\mathcal{F}$ consisting of disjoint $\sum_\mathbf{v} c_\mathbf{v}$ m-sets $W_1^\mathbf{v}, \ldots, W_{c_\mathbf{v}}^\mathbf{v} \subseteq V(G) \setminus D$ for $\mathbf{v} \in I_{\mathcal{P}}^\mu(G)$ such that $\mathbf{i}_{\mathcal{P}}(W_i^\mathbf{v}) = \mathbf{v}$ for all $i \in [c_\mathbf{v}]$, we can regard $V(\mathcal{F}) \cup D$ as the union of $b_\mathbf{v}$ m-sets $S_1^\mathbf{v}, \ldots, S_{b_\mathbf{v}}^\mathbf{v}$ such that $\mathbf{i}_{\mathcal{P}}(S_j^\mathbf{v}) = \mathbf{v}$, $j \in [b_\mathbf{v}]$ for all $\mathbf{v} \in I_{\mathcal{P}}^\mu(G)$. Since $c_\mathbf{v} \leq C$ for all $\mathbf{v}$ and $V(M_2) \cap D = \emptyset$, we can choose the family $\mathcal{F}$ as a subset of M_2. In summary, starting with the H-packing $M_0 \cup M_1 \cup M_2 \cup M_3$ leaving U uncovered, we delete the copies $H_1, \ldots, H_\ell$ of H from $M_0 \cup M_3$ given by Claim 4.2 and then leave $D = \bigcup_{i \in [\ell]} V(H_i) \cup U$ uncovered. Then we delete the family $\mathcal{F}$ of copies of H from M_2 and leave $V(\mathcal{F}) \cup D$ uncovered. Finally, we regard $V(\mathcal{F}) \cup D$ as the union of at most $\binom{m+d-1}{d} C + 5(2m-1)^{d+1} \leq \sqrt{\log n}/2$ m-sets S with $\mathbf{i}_{\mathcal{P}}(S) \in I_{\mathcal{P}}^\mu(H)$.

Note that by definition, D may intersect at most $5(2m-1)^{d+1}$ absorbing sets in $\mathcal{F}_0$, which cannot be used to absorb those sets we obtained above. Since each m-set S has at least $\sqrt{\log n} - m(2m-1)^d > \sqrt{\log n}/2 + 5(2m-1)^{d+1}$ absorbing tm^2-sets in $\mathcal{F}_0$, we can greedily match each S with a distinct absorbing tm^2-set $F_S \in \mathcal{F}_0$ for S. Replacing the H-packing on $V(F_S)$ in M_0 by the perfect H-packing on $G[F_S \cup S]$ for each S gives a perfect H-packing of G.

Acknowledgments. The author's research is supported by FAPESP Proc. (2013/03447-6, 2014/18641-5, 2015/07869-8). The author would like to thank Andrew Treglown for helpful discussions and thank Andrew Treglown and two anonymous reviewers for valuable comments.

A   Proof of Proposition 1

We need the following simple counting result, which, for example, follows from the result of Erdős [5] on supersaturation.

Proposition 2. *Given $\gamma' > 0$, $\ell_1, \ldots, \ell_k \in \mathbb{N}$, there exists $\mu > 0$ such that the following holds for sufficiently large n. Let T be an n-vertex graph with a vertex partition $V_1 \cup \cdots \cup V_d$. Suppose $i_1, \ldots, i_k \in [d]$ and T contains at least $\gamma' n^k$ copies of K_k with vertex set $\{v_1, \ldots, v_k\}$ such that $v_1 \in V_{i_1}$, $\ldots, v_k \in V_{i_k}$. Then T contains at least $\mu n^{\ell_1 + \cdots + \ell_k}$ copies of $K_{\ell_1, \ldots, \ell_k}$ whose jth part is contained in V_{i_j} for all $j \in [k]$.*

Proof (Proof of Proposition 1). Write $L = L_{\mathcal{P}}^{\mu}(G)$ and $Q = Q(\mathcal{P}, L)$. It suffices to show that for any element $\mathbf{v} \in L_{\max}^d$, there exists $\mathbf{v}' = (v_1', \ldots, v_d')$ such that $-(m-1) \leq v_i' \leq m-1$ for all $i \in [d]$ and $\mathbf{v} - \mathbf{v}' \in L$ – since the number of such $\mathbf{v}'$ is at most $(2m-1)^d$. Recall that since H is unbalanced, there exists a k-coloring with color class sizes $a_1 \leq \cdots \leq a_k$ and $a_1 < a_k$. Let $a = a_k - a_1 < m$.

Define a graph P on the vertex set $[d]$ such that $(i, j) \in P$ if and only if $e(G[V_i, V_j]) \geq \gamma n^2$. We claim that if i and j are connected in P, then $a(\mathbf{u}_i - \mathbf{u}_j) \in L$. Indeed, first assume that $(i, j) \in P$. For each edge uv in (V_i, V_j), by

$$\delta(G) \geq (1 - 1/\chi_{cr}(H) + \gamma)n \geq \left(1 - \frac{m-1}{(k-1)m} + \gamma\right) n,$$

it is easy to see that uv is contained in at least $\frac{1}{m^{k-2}}\binom{n}{k-2}$ copies of K_k. So there are at least $\gamma n^2 \cdot \frac{1}{m^{k-2}}\binom{n}{k-2}/\binom{k}{2}$ copies of K_k in G intersecting both V_i and V_j. By averaging, there exists a k-array $(i_1, \ldots, i_k)$, $i_j \in [d]$ where $i_1 = i$ and $i_k = j$ such that G contains at least

$$\frac{1}{d^{k-2}}\gamma n^2 \cdot \frac{1}{m^{k-2}}\binom{n}{k-2}/\binom{k}{2} \geq \frac{\gamma}{m^{k-2}d^{k-2}k!}n^k$$

copies of K_k with vertex set $\{v_1, \ldots, v_k\}$ such that $v_1 \in V_{i_1}$, $\ldots, v_k \in V_{i_k}$. By applying Proposition 2 with $\ell_i = a_i$, $i \in [k]$, we get that there are at least μn^m copies of $K_{a_1, \ldots, a_k}$ in G whose jth part is contained in V_{i_j} for all $j \in [k]$. We apply Proposition 2 again, this time with $\ell_i = a_i$, for all $2 \leq i \leq k-1$ and $\ell_1 = a_k$, $\ell_k = a_1$ and thus conclude that there are at least μn^m copies of $K_{a_k, a_2, \ldots, a_{k-1}, a_1}$ (with a_1 and a_k exchanged) in G whose jth part is contained in V_{i_j} for all $j \in [k]$. Taking subtraction of index vectors of these two types of copies gives that $a(\mathbf{u}_i - \mathbf{u}_j) \in L$. Furthermore, note that if i and j are connected by a path in P, we can apply the argument above to every edge in the path and conclude that $a(\mathbf{u}_i - \mathbf{u}_j) \in L$, so the claim is proved.

Now we separate two cases.

Case 1: $k \geq 3$. In this case, we first show that P is connected. Indeed, we prove that for any bipartition $A \cup B$ of $[d]$, there exists $i \in A$ and $j \in B$ such that $(i, j) \in P$. Let $V_A = \bigcup_{i \in A} V_i$ and $V_B = \bigcup_{j \in B} V_j$. Without loss of generality, assume that $|V_A| \leq n/2$. Since $\delta(G) \geq \frac{1+(k-2)m}{(k-1)m} n \geq (1/2 + 1/(2m))n$, the number of edges in G that are incident to V_A is at least

$$|V_A| \cdot \left(\frac{1}{2} + \frac{1}{2m} \right) n - \binom{|V_A|}{2} \geq \binom{|V_A|}{2} + \frac{n}{4m}|V_A| \geq \binom{|V_A|}{2} + \gamma n^2 |A||B|,$$

where the last inequality follows from $|A||B| \leq d^2/4$ and $|V_i| \geq n/m$, $i \in [d]$. By averaging, there exists $i \in A$ and $j \in B$ such that $e(G[V_i, V_j]) \geq \gamma n^2$ and thus $(i, j) \in P$.

Now let $\mathbf{v} = (v_1, \ldots, v_d) \in L_{\max}^d$. We fix an arbitrary m-vector $\mathbf{w} \in L$ and let $\mathbf{v}_1 = \mathbf{v} - (|\mathbf{v}|/m)\mathbf{w}$ and thus $|\mathbf{v}_1| = 0$. Since P is connected, the claim above implies that for any $i, j \in [d]$, $a(\mathbf{u}_i - \mathbf{u}_j) \in L$. Thus, we obtain the desired vector $\mathbf{v}'$ by making the difference of any two digits at most a. Since $|\mathbf{v}'| = |\mathbf{v}_1| = 0$, we know that $-(m-1) \leq v_i' \leq m-1$ for all $i \in [d]$ and we are done.

Case 2: $k = 2$. In this case we cannot guarantee that P is connected (we may even have some isolated vertices). First let i be an isolated vertex in P. By the definition of P, we know that $e(G[V_i, V \setminus V_i]) \leq (d-1)\gamma n^2$. Since $\delta(G) \geq n/m$,

$$e(G[V_i]) \geq \frac{1}{2}(|V_i|n/m - (d-1)\gamma n^2) \geq \frac{1}{4m}|V_i|^2.$$

Applying Proposition 2 on V_i shows that there are at least μn^m copies of K_{a_1, a_2} in V_i, i.e., $m\mathbf{u}_i \in L$. Second, if $(i, j) \in P$, then applying Proposition 2 on $[V_i, V_j]$ gives that $a_1 \mathbf{u}_i + a_2 \mathbf{u}_j \in L$. So in both cases, for any component C in P, there exists an m-vector $\mathbf{w} \in L$ such that $\mathbf{w}|_{[d] \setminus C} = \mathbf{0}$.

Now let $\mathbf{v} = (v_1, \ldots, v_d) \in L_{\max}^d$. Consider the connected components C_1, $C_2, \ldots, C_q$ of P, for some $1 \leq q \leq d$. By the conclusion in the last paragraph, there exists $\mathbf{v}_1 \in \mathbb{Z}^d$ such that $\mathbf{v} - \mathbf{v}_1 \in L$ and for each component C_i, $0 \leq |\mathbf{v}_1|_{C_i}| \leq m-1$ (for each component C, we take out a multiple of the vector $\mathbf{w}$ given by the last paragraph). Next, within each nontrivial component C_i, we can use the claim to 'balance' the digits, as in Case 1. At last we obtain a vector $\mathbf{v}'$ with the desired property.

B   Proof of Lemma 1

In this subsection we prove Lemma 1. We will build a partition $\mathcal{P} = \{V_1, \ldots, V_d\}$ of $V(G)$ for some $d \leq m$ such that every V_i is (H, β, t)-closed for some $\beta > 0$ and integer $t \geq 1$ in polynomial time. For any $v \in V(G)$, let $\tilde{N}_{H, \beta, i}(v)$ be the set of vertices in $V(G)$ that are (H, β, i)-reachable to v.

We need the following lemma [19, Lemma 4.2], which was originally stated for k-uniform hypergraphs. We remark that the current form can be easily derived by defining a k-uniform hypergraph G' where each k-set forms a hyperedge if and only if it spans a K_k in G. For any vertex $u \in V(G)$, let $W(u)$ be the collection of $(k-1)$-cliques $S \subseteq N(u)$.

Lemma 3 [19]. *Given H and $\gamma' > 0$, there exists $\alpha > 0$ such that the following holds for sufficiently large n. For any n-vertex graph G, two vertices $x, y \in V(G)$ are $(H, \alpha, 1)$-reachable if the number of $(k-1)$-sets $S \in W(x) \cap W(y)$ with $|N(S)| \geq \gamma' n$ is at least $\gamma' \binom{n}{k-1}$, where $N(S) = \bigcap_{v \in S} N(v)$.*

Proposition 3. *Suppose $0 < 1/n \ll \alpha \ll \gamma \ll 1/m$ and let G be an n-vertex graph with $\delta(G) \geq (1 - 1/\chi_{cr}(H) + \gamma)n$. Then for any $v \in V(G)$, $|\tilde{N}_{H,\alpha,1}(v)| \geq (1/m + \gamma)n$.*

Proof. First note that for each $(k-1)$-clique S, we have $|N(S)| \geq (1/m + k\gamma)n$. Then by Lemma 3, for any distinct $u, v \in V(G)$, $u \in \tilde{N}_{H,\alpha,1}(v)$ if $|W(u) \cap W(v)| \geq \gamma^2 \binom{n}{k-1}$. By double counting, we have

$$\sum_{S \in W(v)} |N(S)| < |\tilde{N}_{H,\alpha,1}(v)| \cdot |W(v)| + n \cdot \gamma^2 \binom{n}{k-1}.$$

Note that any S in the above inequality is a $(k-1)$-clique, thus $|N(S)| \geq (1/m + k\gamma)n$. On the other hand, it is easy to see that $|W(v)| \geq \frac{1}{m^{k-1}} \binom{n}{k-1}$. By $\gamma \ll 1/m$, we have

$$|\tilde{N}_{H,\alpha,1}(v)| > (1/m + k\gamma)n - \frac{\gamma^2 n^k}{|W(v)|} \geq (1/m + \gamma)n.$$
$$\square$$

The following lemma will be used to find the partition $\mathcal{P}$ in Lemma 1. Its proof is almost identical to the one of [6, Lemma 3.8] and thus we omit it.

Lemma 4. *Given $0 < \alpha \ll \{1/c, \delta'\}$, there exists constant $\beta > 0$ such that the following holds for all sufficiently large n. Assume an n-vertex graph T satisfies that $|\tilde{N}_{H,\alpha,1}(v)| \geq \delta' n$ for any $v \in V(T)$ and every set of $c+1$ vertices in $V(T)$ contains two vertices that are $(H, \alpha, 1)$-reachable. Then we can find a partition $\mathcal{P}$ of $V(T)$ into $V_1, \ldots, V_d$ with $d \leq \min\{c, 1/\delta'\}$ such that for any $i \in [d]$, $|V_i| \geq (\delta' - \alpha)n$ and V_i is $(H, \beta, 2^{c-1})$-closed in T, in time $O(n^{2^{c-1}m+1})$.*

Now we are ready to prove Lemma 1.

Proof (Proof of Lemma 1). Fix $\gamma > 0$. Without loss of generality, we may assume that $\gamma \ll 1/m$. We apply Lemma 4 with $c = m^{k-1}$, $\delta' = 1/m + \gamma$ and $\alpha \ll \gamma$ and get $\beta > 0$. Suppose

$$1/n_0 \ll \beta \ll \alpha \ll \gamma \ll \delta'.$$

Let G be a graph on $n \geq n_0$ vertices satisfying $\delta(G) \geq (1 - 1/\chi_{cr}(H) + \gamma)n$. By Proposition 3, for any $v \in V(G)$, $|\tilde{N}_{H,\alpha,1}(v)| \geq \delta' n$. By the degree condition and Lemma 3, for distinct $u, v \in V(G)$, u and v are $(H, \alpha, 1)$-reachable if

$|W(u) \cap W(v)| \geq \gamma^2\binom{n}{k-1}$. So any set of $c+1$ vertices in $V(G)$ contains two vertices that are $(H, \alpha, 1)$-reachable because $|W(u)| \geq \frac{1}{c}\binom{n-1}{k-1}$ for any $u \in V(G)$, and $(c+1)/c - 1 \geq \binom{c+1}{2}\gamma^2$. So we can apply Lemma 4 on G and get a partition $\mathcal{P} = \{V_1, \ldots, V_d\}$ of $V(G)$ in time $O(n^{2^{c-1}m+1})$. Note that $|V_i| \geq (\delta' - \alpha)n \geq n/m$ for all $i \in [d]$. Also $d \leq 1/\delta' \leq m$ and each V_i is $(H, \beta, 2^{c-1})$-closed.

References

1. Alon, N., Fischer, E.: Refining the graph density condition for the existence of almost K-factors. ARS Combin. **52**, 296–308 (1999)
2. Alon, N., Yuster, R.: H-factors in dense graphs. J. Combin. Theory Ser. B **66**(2), 269–282 (1996)
3. Cooley, O., Kühn, D., Osthus, D.: Perfect packings with complete graphs minus an edge. Eur. J. Combin. **28**(8), 2143–2155 (2007)
4. Edmonds, J.: Paths, trees, and flowers. Canad. J. Math. **17**, 449–467 (1965)
5. Erdős, P.: On extremal problems of graphs and generalized graphs. Isr. J. Math. **2**(3), 183–190 (1964)
6. Han, J.: Decision problem for perfect matchings in dense uniform hypergraphs. Trans. Amer. Math. Soc. (accepted)
7. Hell, P., Kirkpatrick, D.G.: On the complexity of general graph factor problems. SIAM J. Comput. **12**(3), 601–609 (1983)
8. Hurkens, C.A.J., Schrijver, A.: On the size of systems of sets every t of which have an SDR, with an application to the worst-case ratio of heuristics for packing problems. SIAM J. Discrete Math. **2**(1), 68–72 (1989)
9. Kann, V.: Maximum bounded H-matching is MAX SNP-complete. Inform. Process. Lett. **49**(6), 309–318 (1994)
10. Karpiński, M., Ruciński, A., Szymańska, E.: Computational complexity of the perfect matching problem in hypergraphs with subcritical density. Internat. J. Found. Comput. Sci. **21**(6), 905–924 (2010)
11. Kawarabayashi, K.: K_4^--factor in a graph. J. Graph Theor. **39**(2), 111–128 (2002)
12. Keevash, P., Knox, F., Mycroft, R.: Polynomial-time perfect matchings in dense hypergraphs. Adv. Math. **269**, 265–334 (2015)
13. Keevash, P., Mycroft, R.: A geometric theory for hypergraph matching. Mem. Am. Math. Soc. **233**(1098) (2014). Monograph
14. Komlós, J.: Tiling Turán theorems. Combinatorica **20**(2), 203–218 (2000)
15. Komlós, J., Sárközy, G., Szemerédi, E.: Proof of the Alon-Yuster conjecture. Discrete Math. **235**(1–3), 255–269 (2001). Combinatorics (prague, 1998)
16. Kühn, D., Osthus, D.: Critical chromatic number and the complexity of perfect packings in graphs. In: Proceedings of the Seventeenth Annual ACM-SIAM Symposium on Discrete Algorithms, pp. 851–859 (2006)
17. Kühn, D., Osthus, D.: The minimum degree threshold for perfect graph packings. Combinatorica **29**(1), 65–107 (2009)
18. Le Gall, F.: Powers of tensors and fast matrix multiplication. In: Proceedings of the 39th International Symposium on Symbolic and Algebraic Computation, ISSAC 2014, pp. 296–303. ACM, New York (2014)
19. Lo, A., Markström, K.: F-factors in hypergraphs via absorption. Graphs Comb. **31**(3), 679–712 (2015)
20. Rödl, V., Ruciński, A., Szemerédi, E.: A Dirac-type theorem for 3-uniform hypergraphs. Comb. Probab. Comput. **15**(1–2), 229–251 (2006)

21. Shokoufandeh, A., Zhao, Y.: Proof of a tiling conjecture of Komlós. Random Struct. Algor. **23**(2), 180–205 (2003)
22. Treglown, A.: Personal communication
23. Yuster, R.: Combinatorial and computational aspects of graph packing and graph decomposition. Comput. Sci. Rev. **1**(1), 12–26 (2007)

On the Maximum Weight Minimal Separator

Tesshu Hanaka[1]([✉]), Hans L. Bodlaender[2,3], Tom C. van der Zanden[2],
and Hirotaka Ono[1]

[1] Department of Economic Engineering, Kyushu University, 6-19-1, Hakozaki,
Higashi-ku, Fukuoka 812-8581, Japan
3EC15004S@s.kyushu-u.ac.jp, hirotaka@econ.kyushu-u.ac.jp
[2] Department of Computer Science, Utrecht University, PO Box 80.089,
3508 TB Utrecht, The Netherlands
{h.l.bodlaender,t.c.vanderzanden}@uu.nl
[3] Department of Mathematics and Computer Science,
Eindhoven University of Technology, PO Box 513,
5600 MB Eindhoven, The Netherlands

Abstract. Given an undirected and connected graph $G = (V, E)$ and
two vertices $s, t \in V$, a vertex subset S that separates s and t is called an
s-t separator, and an *s-t* separator is called *minimal* if no proper subset
of S separates s and t. In this paper, we consider finding a minimal *s-t*
separator with maximum weight on a vertex-weighted graph. We first
prove that this problem is NP-hard. Then, we propose an $\mathbf{tw}^{O(\mathbf{tw})}n$-
time deterministic algorithm based on tree decompositions. Moreover, we
also propose an $O^*(9^{\mathbf{tw}} \cdot W^2)$-time randomized algorithm to determine
whether there exists a minimal *s-t* separator where W is its weight and
$\mathbf{tw}$ is the treewidth of G.

Keywords: Parameterized algorithm · Minimal separator · Treewidth

1 Introduction

Given a connected graph $G = (V, E)$ and two vertices $s, t \in V$, a set $S \subseteq V$
of vertices is called an *s-t separator* if s and t belong to different connected
components in $G \setminus S$, where $G \setminus S = (V \setminus S, E)$. If a set S is an *s-t* separator
for some s and t, it is simply called a *separator*. If an *s-t* separator S is minimal
in terms of set inclusion, that is, no proper subset of S separates s and t, it is
called a *minimal s-t separator*. Similarly, if a separator is minimal in terms of
set inclusion, it is called a *minimal separator*.

Separators and minimal separators have been considered important in several contexts and have been intensively studied indeed. For example, they are
obviously related to the connectivity of graphs, which is an important notion

This study is partially supported by NWO Gravity grant "Networks" (024.002.003),
JSPS KAKENHI Grant Numbers JP26540005, 26241031, and Asahi Glass
Foundation.

© Springer International Publishing AG 2017
T.V. Gopal et al. (Eds.): TAMC 2017, LNCS 10185, pp. 304–318, 2017.
DOI: 10.1007/978-3-319-55911-7_22

in many practical applications, such as network design, supply chain analysis and so on. From a theoretical point of view, minimal separators are related to treewidth or potential maximal cliques, which play key roles in designing fast algorithms [4,6].

In this paper, we consider the problem of finding a maximum weight minimal separator of a given weighted graph. More precisely, the problem is defined as follows: Given a connected graph $G = (V, E)$, vertices $s, t \in V$ and a weight function $w : V \to \mathbb{N}^+$, find a minimal s-t separator whose weight $\sum_{v \in S} w(v)$ is maximum. The decision version of the problem is to decide the existence of minimal s-t separator with weight W. We name the problems MAXIMUM WEIGHT MINIMAL s-t SEPARATOR.

This problem is motivated in the context of supply chain network analysis. When a weighted network represents a supply chain where a vertex represents an industry, s and t are virtual vertices respectively representing source and sink, and the weight of a vertex represents its financial importance, the maximum weight minimal s-t separator is interpreted as the most important set of industries that is influential or vulnerable in the supply chain network.

Unfortunately, the problem is shown to be NP-hard, and we then design an FPT algorithm with respect to treewidth. It should be noted that since the condition of s-t connectivity can be written in Monadic Second Order Logic, it can be solved in $f(\mathbf{tw}) \cdot n$ time by Courcelle's meta-theorem, where f is a computable function and $\mathbf{tw}$ is treewidth of the graph. However, the function f forms a tower of exponentials; the existence of an FPT algorithm with better running time is not obvious.

In this paper, we propose two parameterized algorithms with respect to treewidth. One is a $2^{O(\mathbf{tw} \log \mathbf{tw})}n$-time deterministic algorithm and the other is an $O^*(c^{\mathbf{tw}} \cdot W^2)$-time randomized algorithm for the decision version, where c is a constant and O^* is the order notation omitting the polynomial factor. The former algorithm is based on a standard dynamic programming approach, whereas the latter utilizes two techniques recently developed. The first technique is called *Cut & Count*, and by using this, the running time is bounded by a single exponential of treewidth. Furthermore, by applying the second technique called *fast convolution*, we improve the running time by reducing the base of the exponent from $c = 21$ to $c = 9$; the total running time of the resulting algorithm is $O^*(9^{\mathbf{tw}} \cdot W^2)$, which can be further improved when the graph is unweighted.

1.1 Related Work

The Number of Minimal Separators. Minimal separators have been investigated for a long time in many aspects. As mentioned above, they are related to treewidth or potential maximal cliques, for example [4,6]. In general, a graph has exponentially many minimal separators, and in fact there exists a graph with $\Omega(3^{n/3})$ minimal separators [9]. Recently, this bound was improved to $\Omega(1.4521^n)$ [10]. On the other hand, some graph classes have only polynomially (even linearly) many minimal separators. For example, Bouchitté showed that weakly triangulated (weakly chordal) graphs have a polynomial number of separators [5]. As examples of other graph classes with polynomial minimal separators, there are circular arc graphs [12], and polygon circle graphs, which is a superclass of circle graphs [16,17].

On the other hand, Berry et al. presented an $O(n^3 \cdot R_{sep})$-time algorithm that enumerates all the minimal separators where R_{sep} is the number of these [2]. By combining these results, we know that MAXIMUM WEIGHT MINIMAL s-t SEPARATOR can be solved in polynomial time for the graph classes mentioned above. That is, we just enumerate all the minimal separators and evaluate the weights of these for such graphs.

Proposition 1. MAXIMUM WEIGHT MINIMAL s-t SEPARATOR *can be solved in polynomial time for a graph that has a polynomial number of minimal separators.*

The Relationship Between Minimal Separators and Treewidth. Minimal separators and treewidth are strongly related. As for the number of minimal separators, if a graph has a polynomial minimal separators, we can compute its treewidth in polynomial time [5,6]. Such graph classes include circular-arc $(O(n^2)$ [11,12]), polygon circle $(O(n^2)$ [16]), weakly triangulated $(O(n^2)$ [5]) and so on. On the other hand, computing treewidth is fixed parameter tractable with respect to the maximum size of minimal separators [15]. This parameter corresponds to the solution size of MAXIMUM WEIGHT MINIMAL s-t SEPARATOR on unweighted graphs. In this sense, this paper focuses on the converse relation of two parameters: maximum size of minimal separators and treewidth. For treewidth as the parameter, we consider the fixed parameter tractability of MAXIMUM WEIGHT MINIMAL s-t SEPARATOR.

The remainder of this paper is organized as follows. In Sect. 2, we first give basic terminology, basic notions of algorithm design and NP-hardness for the problem. In Sect. 3, we design a standard dynamic programming algorithm based on tree decompositions. In Sect. 4, we propose randomized algorithms based on the *Cut & Count* technique.

2   Preliminaries

In this section, we give notations, definitions, and some basic concepts. Let $G = (V, E)$ be an undirected and vertex-weighted graph. We assume that G does not have an edge (s, t), that is, $(s, t) \notin E$ because if not then there is no s-t separator. For $V' \subseteq V$, let $G[V']$ denote the subgraph of G induced by V'. Furthermore, we denote the set of neighbors of a vertex v by $N(v)$. We define the function $[p]$ as follows: if p is true, then $[p] = 1$, otherwise $[p] = 0$.

2.1   Tree Decomposition

Our algorithms proposed in Sects. 3 and 4 are based on dynamic programming on tree decompositions. In this subsection, we give the definition of tree decomposition.

Definition 1. *A* tree decomposition *of a graph* $G = (V, E)$ *is defined as a pair* $\langle \mathcal{X}, T \rangle$, *where* $\mathcal{X} = \{X_1, X_2, \ldots, X_N \subseteq V\}$, *and* T *is a tree whose nodes are labeled by* $I \in \{1, 2, \ldots, N\}$, *such that*

1. $\bigcup_{i \in I} X_i = V$.
2. *For all* $\{u, v\} \subset E$, *there exists an* X_i *such that* $\{u, v\} \subseteq X_i$.
3. *For all* $i, j, k \in I$, *if* j *lies on the path from* i *to* k *in* T, *then* $X_i \cap X_k \subseteq X_j$.

In the following, we call T a decomposition tree, and we use term "nodes" (not "vertices") for the elements of T to avoid confusion. Moreover, we call a subset of V corresponding to a node $i \in I$ a *bag* and denote it by X_i. The *width* of a tree decomposition $\langle \mathcal{X}, T \rangle$ is defined by $\max_{i \in I} |X_i| - 1$, and the *treewidth* of G, denoted by $\mathbf{tw}(G)$, is the minimum width over all tree decompositions of G. We sometimes use the notation $\mathbf{tw}$ instead of $\mathbf{tw}(G)$ for simplicity.

In general, computing $\mathbf{tw}(G)$ of a given graph G is NP-hard [1], but fixed-parameter tractable with respect to itself and there exists a linear time algorithm if treewidth is fixed [3]. In the following, we assume that a decomposition tree with the minimum treewidth is given.

Kloks introduced a very useful type of tree decomposition for some algorithms, called *nice tree decomposition* [11]. More precisely, it is a special binary tree decomposition which has four types of nodes, named *leaf, introduce vertex, forget* and *join*. A variant of the notion, using a new type of node named *introduce edge*, was introduced by Cygan et al. [7].

Definition 2. *A tree decomposition* $\langle \mathcal{X}, T \rangle$ *is called* nice tree decomposition *if it satisfies the following:*

1. T *is rooted at a designated node* $X_r \in \mathcal{X}$ *satisfying* $|X_r| = 0$, *called the* root node.
2. *Every node of the tree* T *has at most two children.*
3. *Each node in* T *has one of the following five types:*
   - *A* leaf *node* i *which has no children and its bag* X_i *satisfies* $|X_i| = 0$.
   - *An* introduce vertex *node* i *has one child* j *with* $X_i = X_j \cup \{v\}$ *for a vertex* $v \in V$.
   - *An* introduce edge *node* i *has one child* j *and labeled with an edge* $(u, v) \in E$ *where* $u, v \in X_i = X_j$.
   - *A* forget *node* i *has one child* j *and satisfies* $X_i = X_j \setminus \{v\}$ *for a vertex* $v \in V$.
   - *A* join *node* i *has two children nodes* j_1, j_2 *and satisfies* $X_i = X_{j_1} = X_{j_2}$.

We can transform any tree decomposition to a nice tree decomposition with $O(n)$ bags in linear time [8]. Given a tree decomposition $\langle \mathcal{X}, T \rangle$, we define a subgraph $G_i = (V_i, E_i)$ for each node i where V_i is the union of all bags X_j with $j = i$ or j a descendant of i in T, and $E_i \subseteq E$ is the set of edges with both endpoints in V_i.

2.2 NP-hardness

In this subsection, we mention NP-hardness for MAXIMUM WEIGHT MINIMAL *s-t* SEPARATOR. The proof is omitted from this extended abstract.

Theorem 1. MAXIMUM WEIGHT MINIMAL *s-t* SEPARATOR *is NP-hard even if all the vertex weights are identical.*

3 Dynamic Programming on Tree Decompositions

In this section, we give an FPT algorithm with respect to treewidth. It is a standard dynamic programming algorithm based on tree decompositions, and the running time is $\mathbf{tw}^{O(\mathbf{tw})}$. The running time $\mathbf{tw}^{O(\mathbf{tw})}$ appears in some connectivity problems, for example STEINER TREE, FEEDBACK VERTEX SET and CONNECTED VERTEX COVER [7,8].

We first discuss how MAXIMUM WEIGHT MINIMAL *s-t* SEPARATOR can be viewed as a connectivity problem. To show this, we define connected partitions.

Definition 3. *A* connected partition *of weight W is a partition (S, A, B, Q) of V such that: (1) $s \in A, t \in B$, (2) $G[A]$ is connected, (3) $G[B]$ is connected, (4) $\sum_{v \in S} w(v) = W$, (5) for $\forall v \in S$, there exist vertices $a \in A, b \in B$ such that $(a, v) \in E, (v, b) \in E$ and (6) for sets A, B, Q, there does not exist an edge (u, v) such that u and v are in different sets.*

Note that a connected partition represents a structure of separators. In fact, it corresponds to a minimal separator and we can show the following theorem, which plays a key role of designing dynamic programming algorithms. The proof is omitted from this extended abstract.

Theorem 2. *There exists a minimal s-t separator of weight W if and only if there exists a connected partition (S, A, B, Q) of weight W.*

Using connected partitions, we design an $\mathbf{tw}^{O(\mathbf{tw})}$-time algorithm for MAXIMUM WEIGHT MINIMAL *s-t* SEPARATOR. First, we partition S into $S_\emptyset, S_A, S_B$ and S_{AB}. (See Fig. 1). They are needed for the updating process in the dynamic programming. Set $S_\emptyset$ consists of the vertices in S that have no neighbor in A and B, but may have neighbors in S_A, S_B, S_{AB}, Q. Set S_A (resp., S_B) consists of the vertices in S that has at least one neighbor in A (resp., B), but no neighbor of B (resp., A). They may have neighbors in S_A, S_B, S_{AB}, Q. Set S_{AB} consists of the vertices in S that have neighbors in A and in B and may have neighbors in S_A, S_B, S_{AB}, Q. With these sets, we define a *partial solution* as follows.

Definition 4. *Given a node i of the tree decomposition of G, a* partial solution *for node i is a partition $(S_\emptyset, S_A, S_B, S_{AB}, A, B, Q)$, such that:*

- $S_\emptyset \cup S_A \cup S_A \cup S_{AB} \cup A \cup B \cup Q = V_i$,
- $\forall v \in S_\emptyset, N(v) \cap (A \cup B) = \emptyset$,

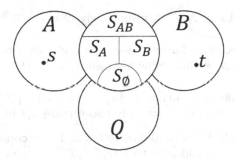

Fig. 1. Connection between vertex sets

- $\forall v \in S_A,\ N(v) \cap B = \emptyset$ and $N(v) \cap A \neq \emptyset$,
- $\forall v \in S_B,\ N(v) \cap B \neq \emptyset$ and $N(v) \cap A = \emptyset$,
- $\forall v \in S_{AB},\ N(v) \cap B \neq \emptyset$ and $N(v) \cap A \neq \emptyset$,
- $s \in V_i \Rightarrow s \in A$ and
- $t \in V_i \Rightarrow t \in B$.

We prepare DP tables for each node. For a representation of the state of v, we define the *coloring* function $c : V \to \{s_\emptyset, s_A, s_B, s_{AB}, a, b, q\}$. Each element of $\{s_\emptyset, s_A, s_B, s_{AB}, a, b, q\}$ is called a *state*. For a coloring c, we denote the state of the coloring of v by $c(v)$. The states of a coloring c represent which set a vertex is in, for example, v is in $S_\emptyset$ if $c(v) = s_\emptyset$.

To consider the connectivity of sets A and B, we use all partitions of these in a bag. That is, we define two partitions $\mathcal{P}^A = \{P_1^A, P_2^A, \ldots, P_\alpha^A\}$ of $X_i \cap A$ and $\mathcal{P}^B = \{P_1^B, P_2^B, \ldots, P_\beta^B\}$ of $X_i \cap B$, where α and β are the number of partitioned sets of $X_i \cap A$ and $X_i \cap B$, respectively, that is, α and β are at most $|X_i \cap A|$ and $|X_i \cap B|$. We call each element of a partition P_ℓ a *block*. They correspond to connected components of $G[A]$(resp., $G[B]$). Note that there are $|X_i|^{O(|X_i|)}$ partitions for each node X_i. Intuitively, just one block $\{\{v\}\}$ is added to $\mathcal{P}$ in each introduce vertex v node; then blocks are merged in the updating process of introduce edge nodes and join nodes. For forget nodes, we have to consider the relationship between connectivity and partitions.

Suppose that introduce vertex nodes, introduce edge nodes and forget nodes have one child node j respectively, and join nodes have two child nodes j_1, j_2. We sometimes denote a coloring in parent node i by c_i and in child node j by c_j to emphasize that we deal with two different nodes. Moreover, we denote the coloring of vertex v by $c_i(v), c_j(v)$, respectively.

Now, we transform a nice tree decomposition by adding $\{s, t\}$ to all bags; thus we can suppose that the root bag X_r contains only two vertices s, t. The width of this tree decomposition is at most $\mathbf{tw} + 2$. We can transform any tree decomposition into such a tree decomposition in polynomial time.

We then define function $f_i(c, \mathcal{P}^A, \mathcal{P}^B)$ as the possible maximum weight of vertices in $S \cap V_i$ under the following conditions. (1) c defines a partial solution $(S_\emptyset, S_A, S_B, S_{AB}, A, B, Q)$ and (2) each block of $\mathcal{P}^A$ and $\mathcal{P}^B$ forms a connected component in $X_i \cap A$ and $X_i \cap B$, respectively. If $c, \mathcal{P}^A, \mathcal{P}^B$ do not satisfy the conditions, let f_i be $-\infty$.

We now define recursive formulas for each node. In a root node, $f_r(\{a\} \times \{b\}, \{\{a\}\}, \{\{b\}\})$ is an optimal value because $X_r = \{s, t\}$.

Leaf node: In a leaf node, we define $f_i(\{a\} \times \{b\}, \{\{a\}\}, \{\{b\}\}) := 0$, otherwise $f_i(c, \mathcal{P}^A, \mathcal{P}^B) := -\infty$ since there are only two vertices s, t in X_i.

Introduce vertex v node: In an introduce vertex node, we consider three cases for colorings. If $c(v) = s_\emptyset$, we add $w(v)$ to $f_j(c, \mathcal{P}^A, \mathcal{P}^B)$ because v is added in S. If $c(v) \in \{a, b, q\}$, the value of f_i does not change since $v \notin S$. Moreover, we add a block $\{\{v\}\}$ to $\mathcal{P}^A$ or $\mathcal{P}^B$ depending on if $c(v) = a$ or $c(v) = b$, respectively. Finally, if $c(v) \in \{s_A, s_B, s_{AB}\}$, a partial solution is invalid by the definition because v has no incident edge and hence no neighbor in A and B. Therefore, we define f_i as follows:

$$
f_i(c, \mathcal{P}^A, \mathcal{P}^B) := \begin{cases}
f_j(c \setminus \{c(v)\}, \mathcal{P}^A, \mathcal{P}^B) + w(v) & \text{if } c(v) = s_\emptyset \\
f_j(c \setminus \{c(v)\}, \mathcal{P}^A \setminus \{\{v\}\}, \mathcal{P}^B) & \text{if } c(v) = a \\
f_j(c \setminus \{c(v)\}, \mathcal{P}^A, \mathcal{P}^B \setminus \{\{v\}\}) & \text{if } c(v) = b \\
f_j(c \setminus \{c(v)\}, \mathcal{P}^A, \mathcal{P}^B) & \text{if } c(v) = q \\
-\infty & \text{otherwise.}
\end{cases}
$$

Introduce edge (u, v) node: In an introduce edge node, we define f_i for the following cases of $c(u), c(v)$.

– If $c(u) = a$ and $c(v) = a$, the vertices u, v are in A. Because edge (u, v) is added, u and v are in the same block of partition $\mathcal{P}^A$. Hence, if not, we set $f_i(c, \mathcal{P}^A, \mathcal{P}^B) := -\infty$. Then, there are two cases: the partitions in $A \cap X_i$ (parent) and $A \cap X_j$ (child) are same or not. In the former case, u and v is in the same block in the partition of $A \cap X_j$, and we then set $f_i(c, \mathcal{P}^A, \mathcal{P}^B) := f_j(c, \mathcal{P}^A, \mathcal{P}^B)$. In the latter case, let $\mathcal{P}'^A$ be a partition of $A \cap X_j$ such that $\mathcal{P}^A \neq \mathcal{P}'^A$ but $\mathcal{P}'^A$ changes to $\mathcal{P}^A$ by merging two blocks of $\mathcal{P}'^A$ including u and v respectively with edge (u, v). Therefore, we take a $\mathcal{P}'^A$ that maximizes $f_i(c, \mathcal{P}'^A, \mathcal{P}^B)$. Then, we set f_i as follows:

$$
f_i(c, \mathcal{P}^A, \mathcal{P}^B) := \max\{f_j(c, \mathcal{P}^A, \mathcal{P}^B), \max_{\mathcal{P}'^A} f_j(c, \mathcal{P}'^A, \mathcal{P}^B)\}.
$$

– The case that $c(u) = b$ and $c(v) = b$ is almost the same as the case that $c(u) = a$ and $c(v) = a$. If u and v are not in the same block of partition $\mathcal{P}^B$, we then set $f_i(c, \mathcal{P}^A, \mathcal{P}^B) := -\infty$. Let $\mathcal{P}'^B$ be a partition of $B \cap X_j$ such that $\mathcal{P}^B \neq \mathcal{P}'^B$ but $\mathcal{P}'^B$ changes to $\mathcal{P}^B$ by merging two blocks of $\mathcal{P}'^B$ including u and v respectively with edge (u, v). Then, we define f_i as follows:

$$
f_i(c, \mathcal{P}^A, \mathcal{P}^B) := \max\{f_j(c, \mathcal{P}^A, \mathcal{P}^B), \max_{\mathcal{P}'^B} f_j(c, \mathcal{P}^A, \mathcal{P}'^B)\}.
$$

- If $c(u), c(v) \in \{s_\emptyset, s_A, s_B, s_{AB}, q\}$, we define f_i as follows:

$$f_i(c, \mathcal{P}^A, \mathcal{P}^B) = f_j(c, \mathcal{P}^A, \mathcal{P}^B).$$

In this case, (u, v) is indifferent to the partitions and the value is not changed because only one edge (u, v) is added.

- If $(c(u), c(v)) = (s_A, a), (a, s_A)$, we consider two cases. One case is that $u \in S_A$ and $v \in A$ in a child node and the other case is that $u \in S_\emptyset$ and $v \in A$ in a child node. In the other case, u is moved from $S_\emptyset$ into S_A by adding (u, v), because u has a neighbor v in A. Thus, we define f_i as follows:

$$f_i(c \times \{s_A\} \times \{a\}, \mathcal{P}^A, \mathcal{P}^B) := \max \{ f_j(c \times \{s_A\} \times \{a\}, \mathcal{P}^A, \mathcal{P}^B),$$
$$f_j(c \times \{s_\emptyset\} \times \{a\}, \mathcal{P}^A, \mathcal{P}^B) \}.$$

- If $(c(u), c(v)) = (s_B, b), (b, s_B)$, we consider almost the same cases as above; that is, $u \in S_B$, $v \in B$ and $u \in S_\emptyset$ and $v \in B$ in a child node.

$$f_i(c \times \{s_B\} \times \{b\}, \mathcal{P}^A, \mathcal{P}^B) := \max \{ f_j(c \times \{s_B\} \times \{b\}, \mathcal{P}^A, \mathcal{P}^B),$$
$$f_j(c \times \{s_\emptyset\} \times \{b\}, \mathcal{P}^A, \mathcal{P}^B) \}.$$

- If $(c(u), c(v)) = (s_{AB}, a), (a, s_{AB})$, there are two cases: (1) $u \in S_{AB}$ and $v \in A$ in a child node and (2) $u \in S_B$ and $v \in A$ in a child node. In the latter case, u is moved from S_B to S_{AB} by adding (u, v), because u has a neighbor v in B. Therefore, we define f_i as follows:

$$f_i(c \times \{s_{AB}\} \times \{a\}, \mathcal{P}^A, \mathcal{P}^B) := \max \{ f_j(c \times \{s_{AB}\} \times \{a\}, \mathcal{P}^A, \mathcal{P}^B),$$
$$f_j(c \times \{s_B\} \times \{a\}, \mathcal{P}^A, \mathcal{P}^B) \}.$$

- If $(c(u), c(v)) = (s_{AB}, b), (b, s_{AB})$, we consider almost the same cases as above; that is, $u \in S_{AB}$, $v \in B$ and $u \in S_A$ and $v \in B$ in a child node.

$$f_i(c \times \{s_{AB}\} \times \{b\}, \mathcal{P}^A, \mathcal{P}^B) := \max \{ f_j(c \times \{s_{AB}\} \times \{b\}, \mathcal{P}^A, \mathcal{P}^B),$$
$$f_j(c \times \{s_A\} \times \{b\}, \mathcal{P}^A, \mathcal{P}^B) \}.$$

- Otherwise, we define $f_i(c, \mathcal{P}^A, \mathcal{P}^B)) := -\infty$ because the rest of cases is invalid. Recall the definition of connected partition and the meaning of states.

Forget v node: In a forget v node, if $c_j(v) \in \{s_\emptyset, s_A, s_B\}$, vertex v never has neighbors both in A and in B and hence such case is invalid because of the definition of connected partition. If $c_j(v) \in \{s_{AB}, q\}$, we need not consider the connectivity of these. In the case that $c_j(v) = a$, we only consider partitions such that there exists at least one vertex u in A included the same block as v. If not, the block including v is never merged. Consequently, $G[A]$ would not be connected in the root node. The case that $c_j(v) = b$ is almost the same. Let $\mathcal{P}'^A, \mathcal{P}'^B$ be a partition satisfying such conditions, then we define f_i as follows:

$$f_i(c, \mathcal{P}^A, \mathcal{P}^B)) := \max \{ f_j(c \times \{s_{AB}\}, \mathcal{P}^A, \mathcal{P}^B)), f_j(c \times \{q\}, \mathcal{P}^A, \mathcal{P}^B),$$
$$\max_{\mathcal{P}'^A} f_j(c \times \{a\}, \mathcal{P}'^A, \mathcal{P}^B)), \max_{\mathcal{P}'^B} f_j(c \times \{b\}, \mathcal{P}^A, \mathcal{P}'^B)) \}.$$

Table 1. This table represents combinations of states of two children nodes j_1, j_2 for each vertex in $X_i = X_{j_1} = X_{j_2}$. The row and column correspond to states of j_1, j_2 respectively and inner elements correspond to states of x. For example, if $c_i(v) = s_A$, there are three combinations such that $(c_{j_1}(v), c_{j_2}(v)) = (s_A, s_\emptyset)$, $(c_{j_1}(v), c_{j_2}(v)) = (s_\emptyset, s_A)$ and $(c_{j_1}(v), c_{j_2}(v)) = (s_A, s_A)$.

	$s_\emptyset$	s_A	s_B	s_{AB}	a	b	q
$s_\emptyset$	$s_\emptyset$	s_A	s_B	s_{AB}			
s_A	s_A	s_A	s_{AB}	s_{AB}			
s_B	s_B	s_{AB}	s_B	s_{AB}			
s_{AB}	s_{AB}	s_{AB}	s_{AB}	s_{AB}			
a					a		
b						b	
q							q

Join node: For a parent node i and two children nodes j_1, j_2, we denote each coloring by c_i, c_{j_1}, c_{j_2} and each partition by $\mathcal{P}_{j_1}^A, \mathcal{P}_{j_1}^B, \mathcal{P}_{j_2}^A, \mathcal{P}_{j_2}^B$. We then define the subset D of tuples of $((c_{j_1}, \mathcal{P}_{j_1}^A, \mathcal{P}_{j_1}^B), (c_{j_2}, \mathcal{P}_{j_2}^A, \mathcal{P}_{j_2}^B))$ such that the combinations of colorings for c_{j_1}, c_{j_2} satisfy the following conditions. (See Table 1):

- $\forall v \in c_i^{-1}(\{s_\emptyset, a, b, q\}), (c_{j_1}(v), c_{j_2}(v)) = (c_i(v), c_i(v))$,
- $\forall v \in c_i^{-1}(\{s_A\}), (c_{j_1}(v), c_{j_2}(v)) = (s_A, s_\emptyset), (s_\emptyset, s_A), (s_A, s_A)$,
- $\forall v \in c_i^{-1}(\{s_B\}), (c_{j_1}(v), c_{j_2}(v)) = (s_B, s_\emptyset), (s_\emptyset, s_B), (s_B, s_B)$, and
- $\forall v \in c_i^{-1}(\{s_{AB}\}), (c_{j_1}(v), c_{j_2}(v)) = (s_{AB}, s_\emptyset), (s_{AB}, s_A), (s_{AB}, s_B),$
  $(s_{AB}, s_{AB}), (s_\emptyset, s_{AB}), (s_A, s_{AB}), (s_B, s_{AB}), (s_A, s_B), (s_B, s_A)$,

and the partition caused by merging $\mathcal{P}_{j_1}^A$ and $\mathcal{P}_{j_2}^A$ equals to $\mathcal{P}^A$ and the partition caused by merging $\mathcal{P}_{j_1}^B, \mathcal{P}_{j_2}^B$ equals to $\mathcal{P}^B$. If $D = \emptyset$ for $c_i, \mathcal{P}^A, \mathcal{P}^B$, we set $f_i(c_i, \mathcal{P}^A, \mathcal{P}^B) := -\infty$. Otherwise, we set $S^* := c_i^{-1}(\{s_\emptyset, s_A, s_B, s_{AB}\})$. Then we define f_i as follows:

$$f_i(c_i, \mathcal{P}^A, \mathcal{P}^B) := \max_{((c_{j_1}, \mathcal{P}_{j_1}^A, \mathcal{P}_{j_1}^B),(c_{j_2}, \mathcal{P}_{j_2}^A, \mathcal{P}_{j_2}^B)) \in D} \{ f_{j_1}(c_{j_1}, \mathcal{P}_{j_1}^A, \mathcal{P}_{j_1}^B)$$
$$+ f_{j_2}(c_{j_2}, \mathcal{P}_{j_2}^A, \mathcal{P}_{j_2}^B) - w(S^*)\}.$$

The subtraction in the right hand side of the equation above is because the weight $w(S^*)$ is counted twice; once in each child node.

We recursively calculate f_i on the decomposition tree. Note that all bags have $|X_i|$ vertices and the number of combinations of colorings and partitions $(c, \mathcal{P}^A, \mathcal{P}^B)$ in each node is $|X_i|^{O(|X_i|)} = \mathbf{tw}^{O(\mathbf{tw})}$. The running time to compute all f_i's in X_i is dominated by join nodes and it is roughly $(\mathbf{tw}^{O(\mathbf{tw})})^3 = \mathbf{tw}^{O(\mathbf{tw})}$ since we scan every coloring and partition in two children nodes X_{j_1} and X_{j_2} for each coloring c_i and each partition $\mathcal{P}^A, \mathcal{P}^B$ and then check all combinations. Therefore, the total running time is $\mathbf{tw}^{O(\mathbf{tw})} n$ and we conclude with the following theorem.

Theorem 3. *For graphs of treewidth at most* **tw**, *there exists an algorithm that solves* MAXIMUM WEIGHT MINIMAL s-t SEPARATOR *in time* $\mathbf{tw}^{O(\mathbf{tw})} n$.

4 Algorithms Using *Cut & Count*

In this section, we give an algorithm that solves the decision version of MAXIMUM WEIGHT MINIMAL s-t SEPARATOR to decide the existence of minimal s-t separator with weight W in time $O^*(9^{\mathbf{tw}} \cdot W^2)$ for graphs of treewidth at most tw. This algorithm is based on the *Cut & Count* technique.

4.1 Isolation Lemma

In this subsection, we explain the Isolation Lemma introduced by Mulmuley et al. [13]. The main idea of the *Cut & Count* technique is to obtain a single solution with high probability; we count modulo 2, and the Isolation Lemma guarantees the existence of such a single solution.

Definition 5 ([13]). *A function* $w' : U \to \mathbb{Z}$ *isolates a set family* $\mathcal{F} \subseteq 2^U$ *if there is a unique* $S' \in \mathcal{F}$ *with* $w'(S') = \min_{S \in \mathcal{F}} w'(S)$ *where* $w'(X) = \sum_{u \in X} w'(u)$.

Lemma 1 (Isolation Lemma [13]**).** *Let* $F \subseteq 2^U$ *be a set family over a universe* U *with* $|F| > 0$. *For each* $u \in U$, *choose a weight* $w'(u) \in \{1, 2, \dots N\}$ *uniformly and independently at random. Then* $Pr[w'$ *isolate* $\mathcal{F}] \geq 1 - |U|/N$.

4.2 Cut & Count

The *Cut & Count* technique was introduced by Cygan et al. for solving connectivity problems [7]. The concept of *Cut & Count* is counting the number of relaxed solutions such that we do not consider whether they are connected or disconnected. Then we compute the number of relaxed solutions modulo 2 and we determine whether there exists a connected solution by cancellation tricks. Now, we define a *consistent cut* to explain the detail of *Cut & Count*.

Definition 6 ([7]). *A cut* (V_1, V_2) *of* $V' \subseteq V$ *such that* $V_1 \cup V_2 = V'$ *and* $V_1 \cap V_2 = \emptyset$ *is consistent if* $v_1 \in V_1$ *and* $v_2 \in V_2$ *implies* $(v_1, v_2) \notin E$.

This means that a consistent cut (V_1, V_2) of V' has no edge between V_1 and V_2. We fix an arbitrary vertex v in V_1. Then, if $G[V]$ is connected, then there only exists one consistent cut, that is, $(V_1, V_2) = (V, \emptyset)$. Therefore, the number of consistent cuts is odd. By this fact, we only compute the number of consistent cuts modulo 2 on decomposition tree and return yes if the number of consistent cuts is odd, otherwise no in a root node. The Isolation Lemma is useful for us as it implies that when the number of solutions is odd, there is a unique solution with high probability; and hence we can use the modulo 2 trick.

Let $\mathcal{S} \subseteq 2^U$ be a set of solutions. According to [7,8], the *Cut & Count* technique is divided into two parts as follows.

- **The Cut part:** Relax the connectivity requirement by considering the set $\mathcal{R} \supseteq \mathcal{S}$ of possibly connected or disconnected candidate solutions. Moreover, consider the set $\mathcal{C}$ of pairs $(X; C)$ where $X \in \mathcal{R}$ and C is a consistent cut of X.
- **The Count part:** Isolate a single solution by sampling weights of all elements in U with high probability by the Isolation Lemma. Then, compute $|\mathcal{C}|$ modulo 2 using a sub-procedure. Disconnected candidate solutions $X \in \mathcal{R} \setminus \mathcal{S}$ cancel since they are consistent with an even number of cuts. If the only connected candidate $x \in \mathcal{S}$ exists, we obtain the odd number of cuts.

Given a set U and a tree decomposition $\langle \mathcal{X}, T \rangle$, the general scheme of *Cut & Count* is as follows:

Step 1. Set the integer weight for every vertex uniformly and independently at random by $w' : U \to \{1, \ldots, 2|U|\}$.

Step 2. For each integer weight $0 \le W' \le 2|U|^2$, compute the number of relaxed solutions of weight W' with consistent cuts modulo 2 on a decomposition tree. Then return yes if it is odd, otherwise no in the root node.

We use the *Cut & Count* technique to determine whether there exists a connected partition (S, A, B, Q) of weight W so that A and B are connected. To apply the above scheme, we newly give the following definition of a *partial solution*. Note that we have to consider two consistent cuts of A and B.

Definition 7. *Given a node i of the tree decomposition of G, a partial solution for that node is a tuple $(S_\emptyset, S_A, S_B, S_{AB}, A_l, A_r, B_l, B_r, Q, w)$, such that:*

- $V_i = S_\emptyset \cup S_A \cup S_A \cup S_{AB} \cup A_l \cup A_r \cup B_l \cup B_r \cup Q$,
- (A_l, A_r) *is a consistent cut: there exists no edge $(u, v) \in E$ such that $u \in A_l$ and $v \in A_r$,*
- (B_l, B_r) *is a consistent cut: there exists no edge $(u, v) \in E$ such that $u \in B_l$ and $v \in B_r$,*
- $w = \Sigma_{v \in S} w(v)$,
- $\forall v \in S_\emptyset$, $N(v) \cap (A_l \cup A_r \cup B_l \cup B_r) = \emptyset$,
- $\forall v \in S_A$, $N(v) \cap (B_l \cup B_r) = \emptyset$ and $N(v) \cap (A_l \cup A_r) \ne \emptyset$,
- $\forall v \in S_B$, $N(v) \cap (B_l \cup B_r) \ne \emptyset$ and $N(v) \cap (A_l \cup A_r) = \emptyset$ and
- $\forall v \in S_{AB}$, $N(v) \cap (B_l \cup B_r) \ne \emptyset$ and $N(v) \cap (A_l \cup A_r) \ne \emptyset$.
- $s \in V_i \Rightarrow s \in A_l$
- $t \in V_i \Rightarrow t \in B_l$

For each vertex v, we set another weight $w'(v)$ by choosing from $\{1, \ldots, 2|V|\}$ and independently at random. We also set the *coloring* $c : V \to \{s_\emptyset, s_A, s_B, s_{AB}, a_l, a_r, b_l, b_r, q\}$. Now, we give a dynamic programming algorithm that computes the number of partial solutions. To count the number of relaxed solutions with consistent cuts, for each c, w and w' we define the counting function $h_i : \{s_\emptyset, s_A, s_B, s_{AB}, a_l, a_r, b_l, b_r, q\}^{|X_i|} \times \mathbb{N} \times \mathbb{N} \to \mathbb{N}$ in each node i on a nice tree decomposition as follows.

Leaf node: In a leaf node, we define $h_i(\emptyset, 0, 0) = 1$, if $S_\emptyset = S_A = S_B = S_{AB} = A_l = A_r = B_l = B_r = \emptyset$ and $w, w' = 0$. Otherwise, $h_i(c, w, w') = 0$.

Introduce vertex v node: The function h_i has five cases in an introduce vertex node. Note that we only add one vertex v without edges. Thus, if $c(v) \in \{s_A, s_B, s_{AB}\}$, a partial solution is invalid by the definition because v has no neighbor. If $c(v) = s_\emptyset$, vertex v is chosen as a vertex of S, and we hence add each weight $w(v)$, $w'(v)$ to w, w', respectively. Moreover, v must not be s, t because s (resp., t) should be in A_l (resp., B_l). If not, it is not a connected partition. Similarly, if $c(v) = a_l$ (resp., b_l), we check whether v is not t (resp., s). As for $c(v) \in \{a_r, b_r, q\}$, we also check whether v is neither s nor t. Therefore, we define h_i in introduce vertex nodes as follows:

$$h_i(c \times \{c(v)\}, w, w') := \begin{cases} [v \neq s, t]h_j(c, w - w(v), w' - w'(v)) & \text{if } c(v) = s_\emptyset \\ [v \neq t]h_j(c, w, w') & \text{if } c(v) = a_l \\ [v \neq s]h_j(c, w, w') & \text{if } c(v) = b_l \\ [v \neq s, t]h_j(c, w, w') & \text{if } c(v) \in \{a_r, b_r, q\} \\ 0 & \text{otherwise.} \end{cases}$$

Introduce edge (u, v) node: In an introduce edge node, we check each state of endpoints of the edge (u, v) and define f_i for some cases.

- If $c(u) = s_\emptyset$, vertex u has no vertices in A, B. Hence, we define the function h_i in this case as follows:

$$h_i(c \times \{c(u)\} \times \{c(v)\}, w, w') := [c(v) \notin \{a_l, a_r, b_l, b_r\}]$$
$$\cdot h_j(c \times \{s_\emptyset\} \times \{c(v)\}, w, w').$$

- If $c(u) = s_A$, vertex u has neighbors of A but no neighbor of B. In this case, we have two cases. The other case is that $u \in S_\emptyset$ and $v \in A$ in a child node, because adding edge (u, v) in the introduce edge (u, v) node, vertex u is moved from $S_\emptyset$ to S_A. The other case is that $u \in S_A$ and $v \notin B$ in a child node. If $v \in B$, vertex u is in S_{AB} in the parent node. We define h_i as follows. Note that only if $c(v) \in \{a_l, a_r\}$, we sum up two cases. If $c(v) \in \{b_l, b_r\}$, $h_i(c \times \{c(u)\} \times \{c(v)\}, w, w') := 0$, otherwise $h_i(c \times \{c(u)\} \times \{c(v)\}, w, w') := h_j(c \times \{s_A\} \times \{c(v)\}, w, w')$.

$$h_i(c \times \{c(u)\} \times \{c(v)\}, w, w') := [c(v) \in \{a_l, a_r\}]h_j(c \times \{s_\emptyset\} \times \{c(v)\}, w, w')$$
$$+ [c(v) \notin \{b_l, b_r\}]h_j(c \times \{s_A\} \times \{c(v)\}, w, w').$$

- If $c(u) = s_B$ is almost the same as above case, that is, we replace A(resp., B) to B(resp., A).

$$h_i(c \times \{c(u)\} \times \{c(v)\}, w, w') := [c(v) \in \{b_l, b_r\}]h_j(c \times \{s_\emptyset\} \times \{c(v)\}, w, w')$$
$$+ [c(v) \notin \{a_l, a_r\}]h_j(c \times \{s_B\} \times \{c(v)\}, w, w').$$

- If $c(u) = s_{AB}$, we consider three cases: $u \in S_A$ and $v \in B$, $u \in S_B$ and $v \in A$, and $u \in S_{AB}$ and v is in arbitrary set in the children node. For first and second cases, vertex u is moved from S_A(resp., S_B) into S_{AB} by adding edge (u, v). If $u \in S_{AB}$, v is allowed to be in any set because a vertex in S_{AB} could connect to all sets. Therefore, we define f_i as follows:

$$h_i(c \times \{c(u)\} \times \{c(v)\}, w, w') := [c(v) \in \{b_l, b_r\}]h_j(c \times \{s_A\} \times \{c(v)\}, w, w')$$
$$+ [c(v) \in \{a_l, a_r\}]h_j(c \times \{s_B\} \times \{c(v)\}, w, w')$$
$$+ h_j(c \times \{s_{AB}\} \times \{c(v)\}, w, w').$$

- If $c(u) \in \{a_l, a_r\}$, then $c(v) \notin \{b_l, b_r, q\}$ since there is no edge between A, B and Q by the definition of connected partition. There is also no edge between A_l and A_r because (A_l, A_r) is a consistent cut. Therefore, if u is in A_l or A_r, then v are in the same set of u or separator sets S_A, S_{AB}. Note that because u is in A, v is not in $S_\emptyset$, S_B.

$$h_i(c \times \{c(u)\} \times \{c(v)\}, w, w') := [c(v) = c(u)]h_j(c \times \{c(u)\} \times \{c(v)\}, w, w')$$
$$+ [c(v) \in \{s_A, s_{AB}\}]h_j(c \times \{c(u)\} \times \{c(v)\}, w, w').$$

- The case that $c(u) \in \{b_l, b_r\}$ is almost the same as above case, that is, we replace A by B.

$$h_i(c \times \{c(u)\} \times \{c(v)\}, w, w') := [c(v) = c(u)]h_j(c \times \{c(u)\} \times \{c(v)\}, w, w')$$
$$+ [c(v) \in \{s_B, s_{AB}\}]h_j(c \times \{c(u)\} \times \{c(v)\}, w, w').$$

- If $c(u) = q$, vertex u is in Q. Hence, v must be in $S_\emptyset, S_A, S_B, S_{AB}$, or Q because a vertex in Q has no neighbor of A and B by the definition of connected partition.

$$h_i(c \times \{c(u)\} \times \{c(v)\}, w, w') := [c(v) \in \{s_\emptyset, s_A, s_B, s_{AB}, q\}]$$
$$\cdot h_j(c \times \{c(u)\} \times \{c(v)\}, w, w').$$

Forget v node: For a forget v node, the state of v would never change forward. Thus, if $c_j(v) \in \{s_\emptyset, s_A, s_B\}$, a partial solution does not satisfy the condition of connected partition because any $v \in S$ must have neighbors of both A and B. For this reason, we only sum up for each state $c_j(v) \in \{s_{AB}, a_l, a_r, b_l, b_r, q\}$. The function h_i in forget nodes is defined as follows:

$$h_i(c, w, w') := \sum_{c_j(v) \in \{s_{AB}, a_l, a_r, b_l, b_r, q\}} h_j(c \times \{c_j(v)\}, w, w').$$

Join node: We denote each coloring and weights of partial solutions in i, j_1, j_2 by c_i, c_{j_1}, c_{j_2} and $w_i, w_{j_1}, w_{j_2}, w_i', w_{j_1}', w_{j_2}'$, respectively. Moreover, for a state subset $L \subseteq \{s_\emptyset, s_A, s_B, s_{AB}, a_l, a_r, b_l, b_r, q\}$, we define $c^{-1}(L)$ as the vertex set such that all vertices satisfy $c(v) \in L$. For a coloring c_i, we also define the subset

D of tuples of (c_{j_1}, c_{j_2}) as the combinations of colorings of c_{j_1}, c_{j_2} like Sect. 3 such that:

- $\forall v \in c_i^{-1}(\{s_\emptyset, a_l, a_r, b_l, b_r, q\})$, $(c_{j_1}(v), c_{j_2}(v)) = (c_i(v), c_i(v))$,
- $\forall v \in c_i^{-1}(\{s_A\})$, $(c_{j_1}(v), c_{j_2}(v)) = (s_A, s_\emptyset), (s_\emptyset, s_A), (s_A, s_A)$,
- $\forall v \in c_i^{-1}(\{s_B\})$, $c_{j_1}(v), c_{j_2}(v)) = (s_B, s_\emptyset), (s_\emptyset, s_B), (s_B, s_B)$, and
- $\forall v \in c_i^{-1}(\{s_{AB}\})$, $(c_{j_1}(v), c_{j_2}(v)) = (s_{AB}, s_\emptyset), (s_{AB}, s_A), (s_{AB}, s_B),$
  $(s_{AB}, s_{AB}), (s_\emptyset, s_{AB}), (s_A, s_{AB}), (s_B, s_{AB}), (s_A, s_B), (s_B, s_A)$.

Let S^* be the vertex subset $c_i^{-1}(\{s_\emptyset, s_A, s_B, s_{AB}\})$. To sum up all combinations of vertex states and weights for counting, we define the function h_i. If $D = \emptyset$, we define $h_i(c_i, w_i, w_i') := 0$. Otherwise,

$$h_i(c_i, w_i, w_i') := \sum_{\substack{w_{j_1} + w_{j_2} \\ = w_i + w(S^*)}} \sum_{\substack{w_{j_1}' + w_{j_2}' \\ = w_i' + w'(S^*)}} \sum_{(c_{j_1}^*, c_{j_2}^*) \in D} h_{j_1}(c_{j_1}^*, w_{j_1}, w_{j_1}') h_{j_2}(c_{j_2}^*, w_{j_2}, w_{j_2}').$$

From now, we analyze the running time of this algorithm. In each leaf, introduce vertex, introduce edge, and forget node, we can compute f_i for each coloring c and weight w, w' in $O(1)$-time because we only use $O(1)$-operations. Therefore, the total running time in them is $O^*(9^{\mathbf{tw}} \cdot W \cdot W')$. However, in a join node, we sum up all weight combinations and coloring combinations satisfying some conditions. There are 21 coloring's combinations for each vertex and $W \cdot W'$ weight's combinations. Therefore, we compute all f_i's in a join node in time $O^*(21^{\mathbf{tw}} \cdot W^2)$. Note that by the definition, $O(W'^2)$ is a polynomial factor.

Theorem 4. *For graphs of treewidth at most* **tw***, there exists a Monte-Carlo algorithm that solves the decision version of* MAXIMUM WEIGHT MINIMAL s-t SEPARATOR *in time* $O^*(21^{\mathbf{tw}} \cdot W^2)$. *It cannot give false positives and may give false negatives with probability at most* $1/2$.

Using the convolution technique [14], we can obtain a faster Monte-Carlo algorithm. The technique helps to speed up the computation for join nodes. The details are omitted from this extended abstract.

Theorem 5. *For graphs of treewidth at most* **tw***, there exists a Monte-Carlo algorithm that solves the decision version of* MAXIMUM WEIGHT MINIMAL s-t SEPARATOR *in time* $O^*(9^{\mathbf{tw}} \cdot W^2)$. *It cannot give false positives and may give false negatives with probability at most* $1/2$. *If the input graph is unweighted, the running time is* $9^{\mathbf{tw}} \cdot |V|^{O(1)}$.

As usual for this type of algorithms, the probability of a false negative can be made arbitrarily small by repeating the algorithm.

Acknowledgments. We are grateful to Dr. Jesper Nederlof for helpful discussions.

References

1. Arnborg, S., Corneil, D.G., Proskurowski, A.: Complexity of finding embeddings in a k-tree. SIAM J. Algebraic Discrete, Methods **8**(2), 277–284 (1987)
2. Berry, A., Bodat, J.P., Cogis, O.: Generating all the minimal separators of a graph. Int. J. Found. Comput. Sci. **11**(3), 397–404 (2000)
3. Bodlaender, H.L.: A linear-time algorithm for finding tree-decompositions of small treewidth. SIAM J. Comput. **25**(6), 1305–1317 (1996)
4. Bodlaender, H.L., Kloks, T., Kratsch, D.: Treewidth and pathwidth of permutation graphs. SIAM J. Discrete Math. **8**(4), 606–616 (1995)
5. Bouchitté, V., Todinca, I.: Treewidth and minimum fill-in: grouping the minimal separators. SIAM J. Comput. **31**(1), 212–232 (2001)
6. Bouchitté, V., Todinca, I.: Listing all potential maximal cliques of a graph. Theoret. Comput. Sci. **276**(1–2), 17–32 (2002)
7. Cygan, M., Nederlof, J., Pilipczuk, M., Pilipczuk, M., van Rooij, J.M.M., Wojtaszczyk, J.O.: Solving connectivity problems parameterized by treewidth in single exponential time. In: Proceedings of the 52nd Annual Symposium on Foundations of Computer Science (FOCS), pp. 150–159 (2011)
8. Cygan, M., Fomin, F.V., Kowalik, Ł., Lokshtanov, D., Marx, D., Pilipczuk, M., Pilipczuk, M., Saurabh, S.: Parameterized Algorithms. Springer International Publishing, Switzerland (2015)
9. Fomin, F.V., Kratsch, D., Todinca, I., Villanger, Y.: Exact algorithms for treewidth and minimum fill-in. SIAM J. Comput. **38**(3), 1058–1079 (2008)
10. Gaspers, S., Mackenzie, S.: On the number of minimal separators in graphs. arXiv:1503.01203v2 (2015)
11. Kloks, T. (ed.): Treewidth: Computations and Approximations. LNCS, vol. 842. Springer, Heidelberg (1994)
12. Kloks, T.: Treewidth of circle graphs. Int. J. Found. Comput. Sci. **7**(2), 111 (1996)
13. Mulmuley, K., Vazirani, U.V., Vazirani, V.V.: Matching is as easy as matrix inversion. Combinatorica **7**(1), 105–113 (1987)
14. Rooij, J.M.M., Bodlaender, H.L., Rossmanith, P.: Dynamic programming on tree decompositions using generalised fast subset convolution. In: Fiat, A., Sanders, P. (eds.) ESA 2009. LNCS, vol. 5757, pp. 566–577. Springer, Heidelberg (2009). doi:10.1007/978-3-642-04128-0_51
15. Skodinis, K.: Efficient analysis of graphs with small minimal separators. In: Widmayer, P., Neyer, G., Eidenbenz, S. (eds.) WG 1999. LNCS, vol. 1665, pp. 155–166. Springer, Heidelberg (1999). doi:10.1007/3-540-46784-X_16
16. Suchan, K.: Minimal separators in intersection graphs. Masters thesis, Akademia Gorniczo- Hutnicza im. Stanislawa Staszica w Krakowie, Cracow (2003)
17. Sundaram, R., Singh, K.S., Rangan, C.P.: Treewidth of circular-arc graphs. SIAM J. Discrete Math. **7**(4), 647–655 (1994)

Pebble Games over Ordered Structural Abstractions

Yuguo He$^{(\boxtimes)}$

School of Computer Science, Beijing Institute of Technology, Beijing 100081, China
hugo274@gmail.com

Abstract. We introduce a new notion called structural abstractions, which is particularly suitable for pebble games over finite ordered graphs. In an example, we show how to apply structural expansions and abstractions in constructions and how to play pebble games over ordered structural abstractions. The proof includes several observations and insights that are fundamental for any games over structural abstractions, which can be used to obtain lower bounds for a number of graph problems with order.

Keywords: Finite model theory · Pebble games · Structural abstraction

1 Introduction

The study of bounded variable fragments of first-order logic (FO, in short) may be traced back to the 19th century [1]. Pebble games [2,4,5] are the main tool in finite model theory for studying expressive power of bounded variable fragments of FO, which are successful for unordered structures. However, taking (linear) orders into account in the study is important since the definitions of many well-known complexity classes depend on an order. It is well known that the presence of orders makes pebble games notoriously difficult to use. The main purpose of this paper is trying to provide a way or paradigm for such games over finite ordered graphs, which is illustrated by an example in depth: defining a triangle in first-order logic needs three variables. The example itself is of very limited interest, which has another simpler proof.[1] Our main contribution is a notion called structural abstractions and a novel way that can be used to obtain lower bounds for number of important graph problems (with order), such as k-Clique, k-Independent-Set, k-Dominating-Set (for fixed k) and subgraph isomorphism (for fixed pattern).

2 Preliminaries

By convention, we use "$\lfloor x \rfloor$" to denote the *floor functions* floor(x), for any real number x. For $n \in \mathbf{N}^+$, we let $[n]$ be the set $\{0, \ldots, n-1\}$ and let $[1, n]$ be the

---

[1] Thanks Stephan Kreutzer for reminding the author in a private communication. The construction is a clever modification of a binary tree.

ⓒ Springer International Publishing AG 2017
T.V. Gopal et al. (Eds.): TAMC 2017, LNCS 10185, pp. 319–332, 2017.
DOI: 10.1007/978-3-319-55911-7_23

set $\{1, \ldots, n\}$. For $n_0, n_1 \in \mathbf{N}^+$ we let $[n_0] \times [n_1]$ be the Cartesian product of $[n_0]$ and $[n_1]$. We use a pair of integers to denote a vertex in a two dimension coordinate plane. For a fixed natural number k (greater than one), the *coordinate congruence number* w.r.t. k (**cc** number, in short) of a vertex (x, y), denoted $\mathbf{cc}(x, y)$, is defined by $(x + y) \bmod k - 1$. In the following sections, we fix k to 3. Hence we can also call **cc** number as coordinate parity or vertex parity.

Let $\sigma = \langle R_1, \ldots, R_m, c_1, \ldots, c_n \rangle$ be a relational signature where R_i is a relation symbol and c_j a constant symbol, a σ-*structure* $\mathfrak{A}$ consists of a universe A together with an interpretation of R_i and c_j. $\mathfrak{A}$ is finite if A is a finite set.

A *linear order* is a binary relation that is transitive, antisymmetric and total. Note that totality implies reflexivity.

A graph $\mathcal{G}' = \langle V', E' \rangle$ is a *subgraph* of a graph $\mathcal{G} = \langle V, E \rangle$ if $V' \subseteq V$ and $E' \subseteq E \cap (V' \times V')$; $\mathcal{G}'$ is an *induced subgraph* of $\mathcal{G}$, denoted by $\mathcal{G}[V']$, if $\mathcal{G}'$ is a subgraph of $\mathcal{G}$ and $E' = E \cap (V' \times V')$. An *ordered graph* is a graph whose vertices are linearly ordered. Two ordered graphs $\mathcal{G}$ and $\mathcal{G}'$ are *isomorphic* if there is a bijective map from V to V' that preserves edge relation as well as order relation, denoted $\mathcal{G} \cong \mathcal{G}'$.

The *quantifier rank* of a formula $\phi \in \mathrm{FO}$ is the maximum depth of nesting of its quantifiers.

Let FO^k be the fragment of FO whose formulae have at most k distinct variables, free or bound.

A game board consists of a pair of structures, e.g. $(\mathfrak{A}, \mathfrak{B})$. An m-round $(k-1)$-pebble game over the game board $(\mathfrak{A}, \mathfrak{B})$, denoted by $\eth_m^{k-1}(\mathfrak{A}, \mathfrak{B})$, is defined as follows. There are two players in the game, called Spoiler and Duplicator. There are $k - 1$ pairs of pebbles, say $(e_1, f_1), \cdots, (e_{k-1}, f_{k-1})$, available for the players, which are off the board at the beginning of the game. In each round, a pair of pebbles, say (e_i, f_i), will be put on the structures wherein e_i is put on an element of $\mathfrak{A}$ and f_i is put on an element of $\mathfrak{B}$. Spoiler first selects a structure and puts a pebble on one element of the selected structure; then Duplicator puts the other pebble in the same pair (matching pebble, for short) on one element of the other structure. If there is no pebble off the board, Spoiler can move a pebble to a new element; then Duplicator should move the matching pebble to some element in the other structure. In the ℓ-th round of the game, let $\overline{c_A} = (a_1, a_2, \ldots, a_n)$, where $n \leq k - 1$, includes all the elements in $\mathfrak{A}$ that have pebbles on them and assume a_i is pebbled before a_{i+1}; let $\overline{c_B} = (b_1, b_2, \ldots, b_n)$ includes all the corresponding pebbled elements in $\mathfrak{B}$. Suppose that, for each i, a_i and b_i are the positions of e_j and f_j for some j in this round. Sometimes we use $((\mathfrak{A}, \overline{c_A}), (\mathfrak{B}, \overline{c_B}))$ to denote the game board in this round. Say that $((\mathfrak{A}, \overline{c_A}), (\mathfrak{B}, \overline{c_B}))$ is in *partial isomorphism* if $\{(a_1, b_1), \ldots, (a_n, b_n)\}$ defines an isomorphism between $\mathfrak{A}[\overline{c_A}]$ and $\mathfrak{B}[\overline{c_B}]$. Spoiler wins the game if the board is not in partial isomorphism in some round; otherwise, Duplicator wins the game. If Duplicator can guarantee a win after m rounds of such $(k - 1)$-pebble game, we say Duplicator has a *winning strategy* in the m-round $(k - 1)$-pebble game, denoted by $\mathfrak{A} \equiv_m^{k-1} \mathfrak{B}$. Without loss of generality, when discussing Duplicator's strategy, we assume *at most* $k - 2$ pairs of pebbles are on the game board *at the start* of each round of a $(k - 1)$-pebble game.

It is well known that pebble games characterise the expressive power of finite variable logics. That is, if for any m we can find a pair of structures, e.g. $(\mathfrak{A}, \mathfrak{B})$, such that $\mathfrak{A}$ satisfies some property while $\mathfrak{B}$ doesn't, and $\mathfrak{A} \equiv_m^{k-1} \mathfrak{B}$, then this property is not expressible in FO^{k-1}. To shorten description, usually we also say that a player *picks* a vertex if the player *puts a pebble on* this vertex.

3 Structural Abstractions

In this section, we construct $\mathfrak{A}_{3,m}$ and $\mathfrak{B}_{3,m}$ for the game board, and use them to prove that three variables are needed to define a triangle on finite ordered graphs. We first construct $\mathfrak{B}'_{3,m}$ in a two dimension coordinate plane, which has some triangles; afterwards, we get $\mathfrak{B}_{3,m}$, which is free of triangle, by forbidding some edges in $\mathfrak{B}'_{3,m}$. Then, we plant a triangle into $\mathfrak{B}_{3,m}$ by adding an edge roughly in its middle, and obtain $\mathfrak{A}_{3,m}$. In this extended abstract, we shall hide some tedious technical details to highlight the main ideas.

Before introducing the main constructions, we use a small graph $\mathcal{B}_3$ to illustrate a notion called "structural expansion", which will be defined formally later. $\mathcal{B}_3$ itself is not useful to the end. But it is small enough for illustration, meanwhile we can get some essence of the notion from studying it. Hopefully it can help the readers understand related notions and intuitions. $\mathcal{B}_3$ is defined in a two dimension coordinate plane. The structure $\mathcal{B}_3$ is an ordered graph over the universe $[2] \times [3]$ and the linear order is defined by the lexicographic ordering on $[3] \times [2]$, i.e. $(x_i, y_i) \le (x_j, y_j)$ if $y_i < y_j$ or $y_i = y_j \wedge x_i \le x_j$. A vertex (x_i, y_i) is adjacent to (x_j, y_j) if and only if

$$y_i \ne y_j \text{ and } \mathbf{cc}(x_i, y_i) \ne \mathbf{cc}(x_j, y_j). \tag{1}$$

Note that $\mathcal{B}_3$ is isomorphic to the graph introduced by Dawar [3] (when $k = 3$), which was used to prove that existential infinitary logic formulae require k variables to define k-Clique on the class of finite ordered graphs.

In Fig. 1, the graph on the left side is $\mathcal{B}_3$. Here, the vertex c is $(0,0)$; b is $(0,1)$; a is $(0,2)$; f is $(1,0)$; e is $(1,1)$ and d is $(1,2)$. The graph on the right side of Fig. 1 is an expansion of $\mathcal{B}_3$. Here, the vertex "0" stand for $(0,0)$; "5" for $(1,0)$; "10" for $(2,0)$; "1" for $(0,1)$, and so on. We call it an expansion because every vertex of $\mathcal{B}_3$ is replaced by two vertices - for any vertex of $\mathcal{B}_3$, we insert a new vertex on its right side and call it the *associated vertex* w.r.t. this expansion. Note that, c of $\mathcal{B}_3$ corresponds to 0 of its expansion; the vertex a corresponds to the vertex 2; d corresponds to 6, and so on. The adjacency of $\mathcal{B}_3$ is *preserved* in the expansion, e.g. 0 is adjacent to 6 while 0 is not adjacent to 2. Moreover, we also use (1) to determine whether a newly inserted vertex is adjacent to another vertex. For instance, the vertex 3 is not adjacent to 5 because $\mathbf{cc}(1,0) = \mathbf{cc}(1,2)$; the vertex 4 is adjacent to 5 because their second coordinates are different and $\mathbf{cc}(1,0) \ne \mathbf{cc}(1,1)$, note that the vertex 6 is $(2,2)$, hence 6 is adjacent to 5 because their second coordinates are different and $\mathbf{cc}(1,0) \ne \mathbf{cc}(2,2)$, and so on.

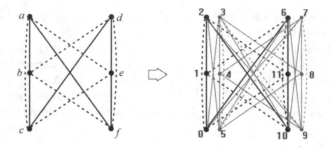

Fig. 1. $\mathcal{B}_3$ (left) and its structural expansion (right).

In the expansion, we can assign a natural number to each vertex as the follows and call this number *vertex index*: we assign "2" to all the vertices corresponding to those of $\mathcal{B}_3$ (i.e. 0, 1, 2, 6, 10 and 11), and assign "1" to all the other vertices. Moreover, we use a set $\mathbb{X}_2^*$ to include all the vertices that have index 2, and use $\mathbb{X}_1^*$ to include all the vertices, i.e. the vertices of index 1 and 2. In short, $\mathbb{X}_i^*$ includes all the vertices that have an index at least i. In other words, the index of a vertex (x, y), denoted $\mathrm{idx}(x, y)$, is the maximum i such that $(x, y) \in \mathbb{X}_i^*$, where $i = 1$ or 2. Note that, $\mathbb{X}_1^*$ subsumes $\mathbb{X}_2^*$.

We call $\mathcal{B}_3$ the *skeleton* of its expansion. If study the expansion carefully, we can see that the subgraph induced by 1, 2 and their associated vertices 4 and 3 is isomorphic to the subgraph induced by 11, 6 and their associated vertices 8 and 7; likewise, the subgraph induced by 0, 2 and their associated vertices 5 and 3 is isomorphic to the subgraph induced by 10, 6 and their associated vertices 9 and 7. In general, the subgraph induced by two vertices of index 2 and their associated vertices is completely determined by these two vertices of index 2. We call such induced subgraphs "elementary bricks". Each pair of these elementary bricks are very similar except that the adjacency between the pair of vertices of index 2 (i.e. those vertices in the skeleton) may be different. Note that, based on the skeleton $\mathcal{B}_3$, we can built its expansion using the elementary bricks.

Observe that the width of $\mathcal{B}_3$ is 2 and the width of the expansion is 4. We use "γ_0^*" to denote the width of $\mathcal{B}_3$ and "γ_1^*" to denote the width of the expansion. Hence the width of the elementary bricks is $\frac{\gamma_1^*}{\gamma_0^*} = 2$. Bigger bricks, whose universes are isomorphic to $[2] \times [3]$, can be built from the elementary ones and they are completely determined by the vertices of index 2. For example, the brick formed by the vertices $0, 1, \ldots, 5$ is isomorphic to $\mathcal{B}_3$; whereas the brick formed by 0, 1, 4, 5, 6 and 7 is akin to $\mathcal{B}_3$ except that the subgraph induced by a, b and c in $\mathcal{B}_3$ is different from that induced by 0, 1 and 6 in the other. In short, the bricks are bound in such a way that their first columns (formed by vertices of index 2) respect the structure (or adjacency) of the graph $\mathcal{B}_3$. Hence, in some sense we can *also* call $\mathcal{B}_3$ an "*abstraction*" of its expansion. Moreover, we can call the expansion the first abstraction, and $\mathcal{B}_3$ the second abstraction. Then, the value $\lfloor x/\frac{\gamma_1^*}{\gamma_0^*} \rfloor$ tells us where (x, y) is in the y-th row of the second abstraction. For example, the vertex 4 is "indistinguishable" from 1 in the second abstraction, therefore this vertex 4 is in the position $(0, 1)$ of the second abstraction. Note that, the universe of the first abstraction is $\mathbb{X}_1^*$ and the universe of the second abstraction is $\mathbb{X}_2^*$. For $i = 1, 2$, we define "$(\!|x|\!)_i$" to be $\lfloor x/\frac{\gamma_1^*}{\gamma_{2-i}^*} \rfloor \times \frac{\gamma_1^*}{\gamma_{2-i}^*}$ so that the following holds: for any $(x, y) \in \mathbb{X}_1^*$ and $(x', y) \in \mathbb{X}_2^*$, if $\lfloor x'/\frac{\gamma_1^*}{\gamma_0^*} \rfloor = \lfloor x/\frac{\gamma_1^*}{\gamma_0^*} \rfloor$, then $(\!|x|\!)_2 = x'$. We call $((\!|x|\!)_i, y)$ the *projection* of the vertex (x, y) in the i-th abstraction. For example, the vertex 2 is the projection of the vertex 3 in the second abstraction; the vertex 11 is the projection of the vertex 8 in the second abstraction; 1 is the projection of 4 in the second abstraction; and any vertex is the projection of itself in the first abstraction. Note that, for any vertex (x, y) of the expansion, (x, y) is in $\mathbb{X}_i^*$ if $x = (\!|x|\!)_i$ for $i = 1, 2$. By definition, $((\!|x|\!)_i, y)$ is in $\mathbb{X}_i^*$, e.g. $((\!|x|\!)_2, y)$ is a vertex of the second abstraction. Note that, for any vertex (x, y) of index i and $1 \leq j \leq i \leq 2$, we have $x = (\!|x|\!)_j$, because $\mathbb{X}_j^*$ subsumes $\mathbb{X}_i^*$.

Note that $\mathrm{idx}((\!|x|\!)_i, y) \geq i$. For example, $\mathrm{idx}((\!|3|\!)_2, 1) = 2$; $\mathrm{idx}((\!|2|\!)_1, 1) > 1$.

We just give a simple example of structural expansion. Such expansion can be performed for many times. Suppose we want to expand $\mathcal{B}_3$ twice. We insert a new vertex on the right side of each vertex of the expansion of $\mathcal{B}_3$. That is, we create an expansion of the expansion of $\mathcal{B}_3$. This time, we assign "3" to the indices of the vertices before the expansions (i.e. those vertices of $\mathcal{B}_3$), and assign "2" to the vertices inserted in the first expansion (the vertices 3, 4, 5 and 7, 8, 9 in Fig. 1). Finally, we assign "1" to the vertices inserted in the second expansion. As before, for each $i \in [1, 3]$, we use a set $\mathbb{X}_i^*$ to include all the vertices that have index at least i. For example, $\mathbb{X}_2^*$ includes all the vertices that have index 2 and index 3. As before, in the second expansion, we also use (1) to determine whether a newly inserted vertex is adjacent to another vertex. The adjacency between the vertices in $\mathbb{X}_2^*$ is preserved in this expansion, just as before.

As have been illustrated in the toy example, the following observations are intuitive, whose proofs are simple. Suppose the structure is produced by $m - 1$ times of structural expansion of $\mathcal{B}_3$ ($\therefore \mathbb{X}_m^*$ is the universe of $\mathcal{B}_3$). Suppose $(x, y) \in \mathbb{X}_1^*$, i.e. it is in the universe of the structure. Let $1 \leq i, j \leq m$.

- For any vertex (x, y) of index i and $j \leq i$, we have $x = (\!|x|\!)_j$.
- For any $(x, y) \in \mathbb{X}_1^*$ and $i \leq j$, we have
  - $\lfloor (\!|x|\!)_i / \frac{\gamma_{m-1}^*}{\gamma_{m-j}^*} \rfloor = \lfloor x / \frac{\gamma_{m-1}^*}{\gamma_{m-j}^*} \rfloor$.
  - $((\!|x|\!)_i)_j = (\!|x|\!)_j$.
- $\mathrm{idx}((\!|x|\!)_i, y) \geq i$.

The structure $\mathcal{B}_{3,m}'$ can be built similarly. It is built from the following structure by $m - 1$ times expansions (we reuse those notations): its universe is $\mathbb{X}_m^* = [4m] \times [3]$; the order is defined by the lexicographic ordering on $[3] \times [4m]$; the edges are defined by (1) as before. We still use γ_0^* to denote its width, i.e. $4m$. In general, we use $\mathbb{X}_{m-i}^*$ to denote the universe of the i-th expansion, and use γ_i^* to denote its width. A definition of structural abstraction can be found in Definition 3. We intend to use $\mathcal{B}_{3,m}'[\mathbb{X}_i^*]$ to denote the i-th abstraction of $\mathcal{B}_{3,m}'$ and we just give the definition of $\mathcal{B}_{3,m}'[\mathbb{X}_m^*]$ - the start point of the expansions. As before, we can regard $\mathcal{B}_{3,m}'[\mathbb{X}_i^*]$ as an "abstraction" of $\mathcal{B}_{3,m}'[\mathbb{X}_{i-1}^*]$.

Hence $\mathfrak{B}'_{3,m}[\mathbb{X}^*_m]$ is the highest abstraction, the maximum element of a chain of abstractions; whereas $\mathfrak{B}'_{3,m}[\mathbb{X}^*_1]$ is the lowest one, the minimum element of this chain. The index of a vertex tells us at which stage it is created in the structural expansions. For example, the vertices of the structure before expansions have index m and the vertices inserted in the first expansion have index $m - 1$, and so on. When $i < j$, we use β^{m-i}_{m-j}, which equals $\frac{\gamma^*_{m-i}}{\gamma^*_{m-j}}$, to tell us the width of the "bricks" we will use to build the i-th abstraction based on the "skeleton" (or the synonym "abstraction") $\mathfrak{B}'_{3,m}[\mathbb{X}^*_j]$, i.e. the number of vertices in $\mathbb{X}^*_i$ that surround (and include) a vertex in $\mathbb{X}^*_j$. It implies that we insert $\beta^{m-j+1}_{m-j} - 1$ vertices of index $j - 1$ surrounding each of the vertices in $\mathbb{X}^*_j$ in the process of the $(m - j + 1)$-th expansion. For example, when $j = m$, it means that we insert $\beta^1_0 - 1$ vertices of index $m - 1$ in the process of the first expansion, i.e. each vertex of index m is replaced by β^1_0 vertices.

As before, we use "$(\!(x)\!)_i, y)$" to denote the projection of (x, y) in the i-th abstraction. Recall that $\mathbb{X}^*_i$ is supposed to include all the vertices that have an index at least i. Intuitively, $\mathbb{X}^*_i$ should include exactly those vertex (x, y) such that $x = (\!(x)\!)_i$ (recall the example in Fig. 1). As before, the value $\lfloor x/\beta^{m-1}_{m-i} \rfloor$ tells us where (x, y) is in the y-th row of the i-th abstraction.

In the following we define $\mathfrak{B}'_{3,m}$ formally.

For any $m, i \in \mathbf{N}^+$, where $m \geq 3$ and $0 < i < m$, let

$$\gamma^*_0 := 4m; \quad \gamma^*_i := 4(m - i)\gamma^*_{i-1} \tag{2}$$

According to (2), we have that, for any i where $0 \leq i < m$,

$$\gamma^*_i = 4^{i+1} \frac{m!}{(m - i - 1)!}. \tag{3}$$

For $x \in [\gamma^*_{m-1}]$ and $1 \leq i \leq j \leq m$, let

$$\beta^{m-i}_{m-j} := \frac{\gamma^*_{m-i}}{\gamma^*_{m-j}} \tag{4}$$

$$[x]_i := \lfloor x/\beta^{m-1}_{m-i} \rfloor \tag{5}$$

$$(\!(x)\!)_i := [x]_i \beta^{m-1}_{m-i} + \frac{1}{2} \sum_{1 < \ell \leq i} \beta^{m-1}_{m-\ell} \tag{6}$$

Let $\mathbb{X}^*_1 := [\gamma^*_{m-1}] \times [3]$. For $1 < i \leq m$, let

$$\mathbb{X}^*_i := \{(x, y) \in \mathbb{X}^*_1 \mid x = (\!(x)\!)_i\}. \tag{7}$$

According to (6) and (7), for any $(x, y) \in \mathbb{X}^*_1$, and any i where $1 < i \leq m$,

$$(x, y) \in \mathbb{X}^*_i \text{ iff } x \equiv \frac{1}{2} \sum_{1 < \ell \leq i} \beta^{m-1}_{m-\ell} \pmod{\beta^{m-1}_{m-i}}. \tag{8}$$

In the following we shall omit some proofs to avoid distracting the readers with the technical details, which are tedious and simple.

Lemma 1. *If $1 \leq j < i$ and $(x, y) \in \mathbb{X}_i^*$, then $[x]_j \equiv 0$ (mod 2).*

It implies that $\mathbf{cc}([x]_j, y) = y$ mod 2, on condition that the premise holds.

We can prove the following lemma which says that $\mathbb{X}_i^*$ subsumes $\mathbb{X}_j^*$ if $i \leq j$.

Lemma 2. *For any i where $1 \leq i \leq m$, if $(x, y) \in \mathbb{X}_i^*$, then $(x, y) \in \mathbb{X}_j^*$ for any $1 \leq j \leq i$.*

Definition 1. *The index of $(x, y) \in \mathbb{X}_1^*$, written $\mathrm{idx}(x, y)$, is the maximum i in $[1, m]$ such that $(x, y) \in \mathbb{X}_i^*$.* ✳

For any (u, v), call $\mathbf{cc}([u]_\ell, v)$ the **cc** number of (u, v) in $\mathbb{X}_\ell^*$.

$\mathfrak{B}'_{3,m}$ is an ordered graph over the universe $\mathbb{X}_1^*$, wherein the linear order is defined as the lexicographic ordering over $[3] \times [\gamma_{m-1}^*]$. For any pair of vertices (x_i, y_i) and (x_j, y_j), if $\mathrm{idx}(x_i, y_i) = \ell \leq \mathrm{idx}(x_j, y_j)$, then (x_i, y_i) is adjacent to (x_j, y_j) iff $y_i \neq y_j$ and $\mathbf{cc}([x_i]_\ell, y_i) \neq \mathbf{cc}([x_j]_\ell, y_j)$.

Imagine that we are looking at $\mathfrak{B}'_{3,m}$ from far away. Assume that a vertex in $\mathbb{X}_i^*$ is "bigger" (so is easier to be observed) than a vertex in $\mathbb{X}_j^*$, if $i > j$. Suppose at the beginning we could only see, or observe, the vertices in $\mathbb{X}_m^*$ clearly, and all the other vertices are indistinguishable. Let us move forward to look a bit closer at the graph, the vertices in $\mathbb{X}_{m-1}^*$ might be observable now, but none of other vertices could. Moving on in this way, we can observe more and more vertices, and finally all the vertices of $\mathfrak{B}'_{3,m}$, i.e., the vertices in $\mathbb{X}_1^*$. In other words, the collection of $\mathfrak{B}'_{3,m}[\mathbb{X}_i^*]$ stand for a sort of "multiresolution" hierarchical structure of $\mathfrak{B}'_{3,m}$.

Definition 2. *Suppose $\mathrm{idx}(x, 1) = \ell < m$. Let $\Omega_x := \{(u, v) \in \mathbb{X}_{\ell+1}^* \mid (u, v)$ is not adjacent to $(\langle x \rangle_{\ell+1}, 1)\}$. The main structure $\mathfrak{B}_{3,m}$ is constructed from $\mathfrak{B}'_{3,m}$ by removing a set of edges: for any vertex $(x, 1)$ where $\mathrm{idx}(x, 1) = \ell < m$ and $[x]_\ell$ is even, we delete any edge between $(x, 1)$ and any vertex in Ω_x.* ✳

In Fig. 2, the black filled circles, except for $(u, 2)$, are vertices in the first row (the bottom row is the 0-th row). The skew dotted line segments indicate that there are no edge between $(u, 2)$ and the other endpoints. The vertex $(x, 1)$ is surrounded by $\beta_{m-\ell}^{m-1} - 1$ vertices of index up to $\ell - 1$ (indicated by the rightmost horizontal dashed line segment in the middle row that is delimited by two smallest grey filled circles in Fig. 2). It is roughly in the middle of this interval, which is indicated by the vertical dotted line. Assume that c is even, i.e. $[x]_\ell$ is even ($\because (x^\star, 1) \in \mathbb{X}_{\ell+1}^*$, by Lemma 1, $[x^\star]_\ell$ is even). Then, by the definition of Ω_x, $(x, 1)$ is not adjacent to $(u, 2)$ since $(x^\star, 1)$ is not adjacent to $(u, 2)$. Similarly, $(x', 1)$ is not adjacent to $(u, 2)$ since $(x, 1)$ is not adjacent to $(u, 2)$ and d is even. Hence, the missing of an edge in the higher abstraction (e.g. there is no edge between $(x^\star, 1)$ and $(u, 2)$) will propagate to the lower abstractions (e.g. there is no edge between $(x', 1)$ and $(u, 2)$), due to the removing of edges according to Ω_x (cf. Definition 2).

Let $mid := 2m\beta_0^{m-1} + \frac{1}{2} \sum_{1 < j \leq m} \beta_{m-j}^{m-1}$. By definition, we know that $mid = \langle mid \rangle_m$, thereby $(mid, y) \in \mathbb{X}_m^*$ for any y. Note that mid is roughly half of γ_{m-1}^*. The structure $\mathfrak{A}_{3,m}$ is built from $\mathfrak{B}_{3,m}$ by adding an edge between $(mid, 0)$ and $(mid, 2)$.

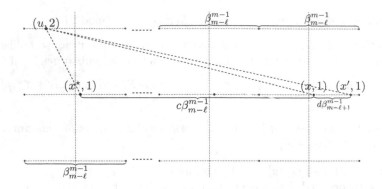

Fig. 2. From $\mathfrak{B}'_{3,m}$ to $\mathfrak{B}_{3,m}$: some edges are forbidden. Suppose $\mathrm{idx}(x,1) = \ell$, $\mathrm{idx}(x',1) = \ell - 1$, $(u,2), (x^*,1) \in \mathbb{X}^*_{\ell+1}$, and $x^* = (\!|x|\!)_{\ell+1}$. Assume c and d are even.

Clearly, $\mathfrak{A}_{3,m}$ has triangles. In particular, it has a triangle formed by the set of vertices $\{(mid,0),(mid,1),(mid,2)\}$, because all the vertices have index m, which implies that $\Omega_{mid} = \emptyset$, and $\mathbf{cc}([mid]_m,i) = i \bmod 2$, which implies that both $(mid,0)$ and $(mid,2)$ are adjacent to $(mid,1)$. In contrast, we have the following observation.

Lemma 3. $\mathfrak{B}_{3,m}$ *has no triangle.*

Proof. We prove it by contradiction. Assume that there are triangles in $\mathfrak{B}_{3,m}$. We can index these triangles such that the index of a triangle is the smallest index of its vertices. Suppose C_3 is such a triangle that has the maximum index, say t. Note that t cannot be m, for otherwise there are two vertices that have the same **cc** number in $\mathbb{X}^*_m$, by the pigeonhole principle. Similarly, C_3 must contain both vertices in $\mathbb{X}^*_t - \mathbb{X}^*_{t+1}$ and vertices in $\mathbb{X}^*_{t+1}$, due to the pigeonhole principle. Let $|C_3| = \{(a,0),(b,1),(c,2)\}$, inasmuch as the second coordinates of the vertices of C_3 must be different. Let $P = \{(x,y) \in \mathbb{X}^*_t - \mathbb{X}^*_{t+1} \mid (x,y) \in |C_3|\}$. Let $Q = \{(x,y) \in \mathbb{X}^*_{t+1} \mid (x,y) \in |C_3|\}$. Note that $P \cap Q = \emptyset$. By Lemma 2, the set of vertices of C_k is exactly $P \cup Q$.

Let $cC_3 := \{\mathbf{cc}([x]_t,y) \mid (x,y) \in |C_3|\}$. Since there are 3 elements in C_3 and $|cC_3| \leq 2$, by pigeonhole principle, there are two vertices such that their **cc** numbers in $\mathbb{X}^*_t$ are the same. If one of them is in P, then there is no edge between them, by definition. Therefore, to have a triangle, both of them should be in Q. Recall that, by Lemma 1, $\mathbf{cc}([x]_t,y) = y \bmod 2$ for any $(x,y) \in \mathbb{X}^*_{t+1}$. Therefore, their **cc** numbers in $\mathbb{X}^*_t$ should be 0. In other words, these two vertices are $(a,0)$ and $(c,2)$. Note that $(b,1) \in P$ since $P \neq \emptyset$ and $\mathbf{cc}([b]_t,1)$ should be 1, for otherwise $(b,1)$ is not adjacent to both $(a,0)$ and $(c,2)$. In other words, $[b]_t$ is even. Note that $((\!|b|\!)_{t+1},1)$ is either not adjacent to $(a,0)$ or not adjacent to $(c,2)$, for otherwise there is a triangle whose index is greater than t. That is, either $(a,0)$ or $(c,2)$ is in Ω_b. Therefore, either $(a,0)$ or $(c,2)$ is not adjacent to $(b,1)$. A contradiction occurs.

Now we formally define the notion of structural abstraction as well as its dual concept structural expansion. Suppose $\mathcal{G}$ and $\mathcal{H}$ are two finite ordered graphs. Let G, H be the vertex set of $\mathcal{G}$, $\mathcal{H}$ respectively. Let $E^{\mathcal{G}}$ be the edge set of $\mathcal{G}$.

Definition 3. *Say that $\mathcal{G}$ is a structural abstraction of $\mathcal{H}$, denoted $\mathcal{H} \preceq_a \mathcal{G}$, if there is a partition P_V of H, a bijection f from G to P_V, a set $\mathfrak{E}$ of finite ordered graphs, a partition P_E of $G \times G$, and a bijection g from P_E to $\mathfrak{E}$ s.t.*

1. *f preserves strict total order: $c <^{\mathcal{H}} d$ if $a <^{\mathcal{G}} b$, for any $c \in f(a)$ and $d \in f(b)$;*
2. *for any $a, b \in G$, $\mathcal{H}[f(a) \cup f(b)] \cong \mathcal{E}$ if $(a, b) \in \mathcal{P} \in P_E$ and $g(\mathcal{P}) = \mathcal{E}$.*

$\mathcal{H}$ is a structural expansion of $\mathcal{G}$ if $\mathcal{G}$ is a structural abstraction of $\mathcal{H}$.   ✳

Lemma 4. *For any finite ordered graph $\mathcal{G}$, $\mathcal{G} \preceq_a \mathcal{G}$.*

Proof. $\mathcal{G} \preceq_a \mathcal{G}$ because we can let (1) $\mathfrak{E} = \{\mathcal{E}^+, \mathcal{E}^-\}$ where $\mathcal{E}^+$ is an edge and $\mathcal{E}^-$ is two isolated vertices; (2) $P_E = \{E^{\mathcal{G}}, \overline{E^{\mathcal{G}}}\}$; (3) $g(E^{\mathcal{G}}) = \mathcal{E}^+$ and $g(\overline{E^{\mathcal{G}}}) = \mathcal{E}^-$; (4) $f(a) = \{a\}$ for $\forall a \in G$; (5) $P_V = \{f(a) \mid a \in G\}$.

In the definition, if $\mathfrak{E}$ is fixed, we can define another version of the notions of structural abstraction and expansion with this set in the similar way. So $\mathcal{G}$ is an abstraction of $\mathcal{H}$ under $\mathfrak{E}$ if the expansions (under $\mathfrak{E}$) of isomorphic subgraphs (w.r.t. P_E) of $\mathcal{G}$ are isomorphic. Here, "isomorphic (w.r.t. P_E)" refers to the usual meaning when P_E (which includes a set of binary relations) is taken into account. For example, suppose $P_E = \{E^{\mathcal{G}}, E^-, E^0\}$ where $E^- \cup E^0 = \overline{E^{\mathcal{G}}}$. Suppose $(a, b) \in E^-$, $(c, d) \in E^0$, and let h be a bijection s.t. $h(a) = c$, $h(b) = d$. Although both $(a, b), (c, d) \notin E^{\mathcal{G}}$, h doesn't define an isomorphism from (a, b) to (c, d) w.r.t. P_E since $E^- \cap E^0 = \emptyset$.

In general, the set of finite ordered graphs ordered by abstractions is not a poset. For example, antisymmetry doesn't hold. To see this, just note that a clique is an abstraction of the corresponding independent set, and vice versa.[2] Nevertheless, we can show that there is a finite chain in the set of subgraphs of $\mathfrak{A}_{3,m}$ ($\mathfrak{B}_{3,m}$ resp.) ordered by abstractions, in which $\mathfrak{A}_{3,m}$ ($\mathfrak{B}_{3,m}$ resp.) is the first element. The length of the chain is m, which coincides with the maximum round of the game.

Definition 4. *Suppose $\mathrm{idx}(x, y) > i$. The set of vertices of index at least i that surround (x, y) in the same row, written $ex_i(x, y)$, is $\{(u, y) \in \mathbb{X}_1^* \mid i \leq \mathrm{idx}(u, y) < \mathrm{idx}(x, y)$ and $(u)_{\mathrm{idx}(x,y)} = x\} \cup \{(x, y)\}$. Say that (x', y) is in the expansion of (x, y) if $(x', y) \in ex_i(x, y)$ for some i.*   ✳

Note that $(u)_{\mathrm{idx}(x,y)} = x$ if, and only if, $[u]_{\mathrm{idx}(x,y)} = [x]_{\mathrm{idx}(x,y)}$.

---

[2] To avoid it, we could also use an alternative definition of structural abstraction wherein an expansion cannot map an edge to isolated vertices. Hence, by this definition, any graph is embeddable to its structural expansion.

Lemma 5. *For* $1 \leq j \leq i \leq m$, $\mathfrak{A}_{3,m}[\mathbb{X}_i^*] \preccurlyeq_a \mathfrak{A}_{3,m}[\mathbb{X}_j^*]$.

Proof. By Lemma 4, it holds when $j = i$. Henceforth, assume $j < i$. For convenience, we let the coordinates of vertices of $\mathfrak{A}_{3,m}[\mathbb{X}_i^*]$ coincide with those of $\mathfrak{A}_{3,m}$. Hence, the first vertex of $\mathfrak{A}_{3,m}[\mathbb{X}_i^*]$ is not $(0,0)$. Instead, let $(c_0, 0)$ be the coordinates of its first vertex (the first vertex in the bottom row) and $(c_1, 0)$ be the second vertex in the bottom row of $\mathfrak{A}_{3,m}[\mathbb{X}_i^*]$. By definition, $(c_0, 0)$ is adjacent to $(c_0, 1)$ and $(c_0, 1)$ is not adjacent to $(c_1, 0)$. We can let $\mathcal{E}_{ij}^+$ ($\mathcal{E}_{ij}^-$ resp.) be isomorphic to $\mathfrak{A}_{3,m}[Y_{ij}^+]$ ($\mathfrak{A}_{3,m}[Y_{ij}^-]$ resp.), where $Y_{ij}^+ = ex_j(c_0, 0) \cup ex_j(c_0, 1)$ and $Y_{ij}^- = ex_j(c_0, 1) \cup ex_j(c_1, 0)$. Note that $|Y_{ij}^+| = |Y_{ij}^-|$. Let $\mathcal{E}_{ij}^0$ be a set of $|Y_{ij}^+|$ isolated vertices, and let $\mathcal{G}_i = (G_i, E^{\mathcal{G}_i}) = \mathfrak{A}_{3,m}[\mathbb{X}_i^*]$.

Now, we let (1) $\mathfrak{E}_{ij} = \{\mathcal{E}_{ij}^+, \mathcal{E}_{ij}^-, \mathcal{E}_{ij}^0\}$; (2) $P_E^i = \{E^{\mathcal{G}_i}, E_i^-, E_i^0\}$ where E_i^- includes all the pairs in $G_i \times G_i$ that are not adjacent and not in the same row, whereas E_i^0 includes all the pairs that are not adjacent and in the same row; (3) $g_i(E^{\mathcal{G}_i}) = \mathcal{E}_{ij}^+$, $g_i(E^-) = \mathcal{E}_{ij}^-$ and $g_i(E^0) = \mathcal{E}_{ij}^0$; (4) $f_{ij}(x, y) = ex_j(x, y)$ for any $(x, y) \in G_i$; (5) $P_V^{ij} = \{f_{ij}(x, y) \mid (x, y) \in G_i\}$.

It's easy to verify that P_E^i and P_V^{ij} are partitions and that f_{ij} preserves strict total order. To prove Lemma 5, it remains to show that every edge of $\mathfrak{A}_{3,m}[\mathbb{X}_i^*]$ is replaced by an isomorphic copy of $\mathcal{E}_{ij}^+$; every pair of vertices that are not adjacent and not in the same row is replaced by an isomorphic copy of $\mathcal{E}_{ij}^-$; and every pair of vertices that are not adjacent and in the same row is replaced by an isomorphic copy of $\mathcal{E}_{ij}^0$ (this case is clear). To this end, we introduce a claim in the following, which is not difficult to verify.

We can regard $\mathfrak{A}_{3,m}[\mathbb{X}_m^*]$ be the starting point of $m-1$ iterative expansions. Then the vertices of index i is the outcome of the $(m-i)$-th expansion. They are replaced by the same number of vertices of index $i-1$ in the next expansion, where they are in the middle w.r.t. the order. Therefore, the vertices of different indices in their expansions should be ordered in the same way after a number of expansions. This intuition is formalized by the following claim.

Claim. For any $1 < \xi \leq m$ and $a, a' \in [\gamma_{m-1}^*]$, if $a - (\!|a|\!)_\xi = a' - (\!|a'|\!)_\xi$, then

(1) $(\!|a|\!)_\xi - (\!|a|\!)_{\xi-1} = (\!|a'|\!)_\xi - (\!|a'|\!)_{\xi-1}$;
(2) $a - (\!|a|\!)_{\xi-1} = a' - (\!|a'|\!)_{\xi-1}$.

Suppose (x, y) is a vertex of index i. We can expect that the index of a vertex (x', y) in the expansion of (x, y) should be determined by their "distance", i.e. $x - x'$, which is implied by the Claim.

Let (x_0, y_0), (x_1, y_1), (x_0', y_0) and (x_1', y_1) be vertices of index i in $\mathfrak{A}_{3,m}[\mathbb{X}_i^*]$, where $y_0 \neq y_1$. Suppose (x_0, y_0) is adjacent to (x_1, y_1) iff (x_0', y_0) is adjacent to (x_1', y_1). Let $(a, y_0) \in ex_j(x_0, y_0)$ and $(a', y_0) \in ex_j(x_0', y_0)$. Assume $a - x_0 = a' - x_0'$. By the Claim, index of a vertex (a, y_0) is completely determined by $a - x_0$. Hence (a, y_0) and (a', y_0) have the same index, say ℓ. Assume $\ell < i$, i.e. $a - x_0 \neq 0$. Note that $\mathbf{cc}([a]_\ell, y_0)$ is completely determined by $a - x_0$, because $\mathbf{cc}([(\!|a|\!)_{\ell+1}]_\ell, y_0) = y_0 \bmod 2$ ($\because$ Lemma 1), and $a - (\!|a|\!)_{\ell+1}$ is completely determined by $a - x_0$ ($\because$ the Claim). Similarly, $\mathbf{cc}([a']_\ell, y_0)$ is completely determined by $a' - x_0' (= a - x_0)$. Hence, $\mathbf{cc}([a]_\ell, y_0) = \mathbf{cc}([a']_\ell, y_0)$.

Note that the adjacency between (a, y_0) and (x_1, y_1) is *completely* determined by two things (they are adjacent iff both hold, provided that $y_0 \neq y_1$):

(i*) whether $(x_1, y_1) \notin \Omega_a$; (cf. Definition 2)
(ii*) whether $\mathbf{cc}([a]_\ell, y_0) \neq \mathbf{cc}([x_1]_\ell, y_1)$.

By Lemma 1, $\mathbf{cc}([x_1]_\ell, y_1) = \mathbf{cc}([x_1']_\ell, y_1) = y_1 \bmod 2$. Therefore, (a, y_0) is adjacent to (x_1, y_1) iff (a', y_0) is adjacent to (x_1', y_1) if ignoring (i*). On the other hand, we can also show that $(x_1, y_1) \notin \Omega_a$ iff $(x_1', y_1) \notin \Omega_{a'}$. Note that $(x_1, y_1), (x_1', y_1) \in \mathbb{X}_{\ell+1}^*$ because $i > \ell$; $[a]_\ell$ is even iff $[a']_\ell$ is even since it has been shown that $\mathbf{cc}([a]_\ell, y_0) = \mathbf{cc}([a']_\ell, y_0)$. Therefore, edges of $\mathfrak{A}_{3,m}[\mathbb{X}_i^*]$ are replaced by graphs isomorphic to $\mathcal{E}_{ij}^+$. Similarly, the disconnected pairs (not in the same row) in $\mathfrak{A}_{3,m}[\mathbb{X}_i^*]$ are also replaced by graphs isomorhpic to $\mathcal{E}_{ij}^-$. This concludes the proof of Lemma 5.

Corollary 1. *The set of graphs $\mathfrak{A}_{3,m}[\mathbb{X}_i^*]$ ($1 \leq i \leq m$) is a chain ordered by structural abstractions.*

We call the i-th element of this finite chain the i-th abstraction of the main structures. In other words, $\mathfrak{A}_{3,m}[\mathbb{X}_i^*]$ is the i-th abstraction of the structure $\mathfrak{A}_{3,m}$. Note that $\mathfrak{A}_{3,m}[\mathbb{X}_1^*]$ is just $\mathfrak{A}_{3,m}$. Henceforth, we call $\mathfrak{A}_{3,m}[\mathbb{X}_i^*]$ a higher abstraction relative to $\mathfrak{A}_{3,m}[\mathbb{X}_{i-1}^*]$.

4 Games over Ordered Abstractions

In the following we show how to play pebble games over ordered abstractions. The strategy introduced here can also be used to prove lower bounds for other graph problems with order.

Lemma 6. *For any $m \geq 3$, $\mathfrak{A}_{3,m} \equiv_m^2 \mathfrak{B}_{3,m}$.*

From here on, we always assume that Spoiler picks (x, y) in some structure and Duplicator responds with (x', y) in the other structure in the current round. Say that *the board is in partial isomorphism over the i-th abstraction* if the projections of pebbled vertices in the i-th abstraction define a partial isomorphism between the structures. Say that *the board is over the i-th abstraction*, if $u - (\!u\!)_i = u' - (\!u'\!)_i$ for any pair of pebbled vertices (u, v) and (u', v), and the board is in partial isomorphism over the i-th abstraction. We have the following important **observation**: *For any $i \in [2, m]$, if the game board is over the i-th abstraction, then it is also over the $(i-1)$-th abstraction.* Therefore, in such case the board is also over the first abstraction, i.e. a partial isomorphism is presented over the original board. The argument is simple. By the Claim in page 10, we have (1) $(\!u\!)_i - (\!u\!)_{i-1} = (\!u'\!)_i - (\!u'\!)_{i-1}$ and (2) $u - (\!u\!)_{i-1} = u' - (\!u'\!)_{i-1}$, for any pair of pebbled vertices (u, v) and (u', v). Because the i-th abstraction of the structures is a structural abstraction of the $(i-1)$-th abstraction (under $\mathfrak{E}_{i(i-1)} = \{\mathcal{E}_{i(i-1)}^+, \mathcal{E}_{i(i-1)}^-, \mathcal{E}_{i(i-1)}^0\}$, cf. the proof of Lemma 5, with j replaced by $i - 1$), (1) implies that the projections of pebbled vertices in the $(i-1)$-th abstraction define a partial isomorphism w.r.t. edge and order, since the expansions (under $\mathfrak{E}_{i(i-1)}$) of isomorphic subgraphs of $\mathfrak{A}_{3,m}[\mathbb{X}_i^*]$ and $\mathfrak{B}_{3,m}[\mathbb{X}_i^*]$ are isomorphic.

This observation is important because it allows Duplicator to resort to lower abstractions for a good pick only when she cannot have a good pick over the higher abstraction; and if Duplicator can ensure a win over the i-th abstraction (i.e. she can ensure that the board is in partial isomorphism over the i-th abstraction), she can also ensure a win over the first abstraction, i.e. over the original board. Let ξ be the maximum i such that the game board is over the i-th abstraction *at the start of the current round*. At the beginning of the game, $\xi = m$. In each round Duplicator uses an auxiliary game over the ξ-th abstraction to help her decide her pick in the original game. A *game over the ξ-th abstraction* is an auxiliary game of the original game: it has the same game board as the original game; in the current round, Spoiler picks $(\langle x \rangle_\xi, y)$ and Duplicator replies by a vertex (a, y) in this round for some a. If the board is over the ξ-th abstraction at the end of this round, then Duplicator wins this round. Otherwise, Spoiler wins. If Duplicator wins this round in the auxiliary game, she pickes (x', y) in the original game s.t. $\langle x' \rangle_\xi = a$ and $x' - \langle x' \rangle_\xi = x - \langle x \rangle_\xi$. It implies that the board is over the ξ-th abstraction. By the previous observation, the original board is in partial isomorphism. Soon we shall see that if Duplicator cannot win this round in the auxiliary game over the ξ-th abstraction, she can *always* win this round in the auxiliary game over the $(\xi - 1)$-th abstraction. We use θ to denote how many rounds are left at the start of the current round. At the beginning, $\theta = m$.

This concludes the outline of the proof. It remains to show that Duplicator can ensure a win if she makes a pick in the $(\xi - 1)$-th abstraction.

Proof. This proof is by induction, wherein we show 1°~4° are preserved at the end of the current round.

1° $x - \langle x \rangle_\xi = x' - \langle x' \rangle_\xi$.
2° The board is in partial isomorphism over the ξ-th abstraction w.r.t. the linear order.
3° The board is in partial isomorphism over the ξ-th abstraction w.r.t. edges.
4° $\theta < \xi$ after the first round.
5° The game board is in partial isomorphism.

Note that, 1°~3° altogether are equivalent to say that "the board is over the ξ-th abstraction" (w.r.t. both the linear order and edges), which implies 5° according to the "observation". Moreover, although all of the conditions should be ensured simultaneously, in the game Duplicator will first try to ensure 2°, then 3°, and then 1° and 4°.

In any round, Duplicator will first try to pick (x', y) such that 1°~3° hold. If she cannot find such a vertex, she *resorts to the $(\xi - 1)$-th abstraction* for a solution. That is, Duplicator tries to ensure 1°~3°, wherein "ξ" is replaced by "$\xi - 1$" in these requirements; and $\xi := \xi - 1$ *at the end of this round*. Recall that in this case Duplicator plays the auxiliary game over the $(\xi - 1)$-th abstraction. Suppose the current round is the ℓ-th round.

In the first round Duplicator is a copycat. It is straightforward to verify that the board is over the m-th abstraction after this round. This forms the **basis** of the proof. In the **induction step**, we first assume that Duplicator can win the first $\ell - 1$ rounds where $1 < \ell \le m$, and $1°\sim4°$ hold, then we prove that she can also win the ℓ-th round. Recall that at the start of the ℓ-th round, i.e. the current round, the board is over the ξ-th abstraction. In the following we show that Duplicator can ensure that the board is over the $(\xi-1)$-th abstraction at the end of the ℓ-th round, provided that the board cannot be over the ξ-th abstraction. Suppose that (a, b), (a', b) are a pair of pebbles on the board at the start of the current round and that Duplicator resorts to the $(\xi - 1)$-th abstraction. By definition, $\theta + \ell = m + 1$. Hence $\xi > \theta = m - \ell + 1$.

Firstly, to ensure $2°$, Duplicator only need to preserve the following, which is called "abstraction-order-condition": (1) If $[x]_\xi \le m - \ell$ or $\gamma^*_{m-\xi} - [x]_\xi \le m - \ell$, then $[x']_\xi = [x]_\xi$; (2) if $m - \ell < [x]_\xi < \gamma^*_{m-\xi} - m + \ell$, then $m - \ell < [x']_\xi < \gamma^*_{m-\xi} - m + \ell$. Roughly speaking, it means Duplicator will pick a vertex close to the boundary of the ξ-th abstraction if and only if Spoiler does so. We can show that the following claim holds: On condition that the abstraction-order-condition and $4°$ hold at the start of the ℓ-th round, the abstraction-order-condition can be preserved after this round, at the price of decreasing ξ by at most one. To prove it, it suffices to show that the abstraction-order-condition holds if x is substituted with a and x' is substituted with a', and ξ is substituted by $\xi - 1$. Here is a brief argument. By induction hypothesis, $a - (\!(a)\!)_\xi = a' - (\!(a')\!)_\xi$. If $[a]_\xi = [a']_\xi$, then $(\!(a)\!)_\xi = (\!(a')\!)_\xi$. Therefore, $a = a'$. It follows that $[a]_{\xi-1} = [a']_{\xi-1}$, which satisfies the condition (with the substitution). Now, assume $[a]_\xi \ne [a']_\xi$, which implies that $((\!(a)\!)_\xi, b)$ and $((\!(a')\!)_\xi, b)$ are not boundary vertices in the b-th row of the ξ-th abstraction. Observe that $((\!(a)\!)_{\xi-1}, b)$ and $((\!(a')\!)_{\xi-1}, b)$ satisfy (2) of the condition: the number of vertices in the $(\xi - 1)$-th abstraction, which surround a vertex of index ξ, is $\beta^{m-\xi+1}_{m-\xi} = 4(\xi - 1) > 4(m - \ell)$, with this vertex of index ξ in the middle of them. Therefore, there are more than $2(m - \ell)$ vertices of index $\xi - 1$ that are on the left side (right side, resp.) of the leftmost (rightmost, resp.) vertex of index ξ in any row of the structure. Therefore, (2) of the condition (with the substitution) holds.

Moreover, the abstraction-order-condition implies that Duplicator can ensure either (L1) $[x]_\xi \le [a]_\xi \Leftrightarrow [x']_\xi \le [a']_\xi$ or (L2) $[x]_{\xi-1} \le [a]_{\xi-1} \Leftrightarrow [x']_{\xi-1} \le [a']_{\xi-1}$. Suppose $y = b$ (the other case is trivial). On the one hand, Duplicator will mimic Spoiler if (x, y) is close to the boundary of the ξ-th abstraction, which implies (L1). On the other hand, if (x, y) is "far away" (i.e., more than $m - \ell$ vertices away) from both boundary vertices of the y-th row of the ξ-th abstraction, and Duplicator cannot do it without violating $2°$, then she resorts to the $(\xi - 1)$-th abstraction such that (L2) holds - now intervals in the ξ-th abstraction will turn into sufficiently large intervals in the $(\xi - 1)$-th abstraction.

Duplicator can also ensure $3°$. Note that, we need to take care of $3°$ only when two pebbles are in different rows. Observe that, any vertex in any abstraction is neither isolated nor adjacent to all the other vertices in the same abstraction. In other words, Duplicator usually doesn't have to resort to the $(\xi - 1)$-th abstraction to ensure 3°. She might have to resort to the $(\xi - 1)$-th abstraction only when Spoiler picks a boundary vertex (i.e. the first or last vertex in a row) in the ξ-th abstraction. The point is that Spoiler cannot distinguish the difference between the interval formed by the boundary in the first abstraction (i.e. the boundary of the board) and the picked vertex in respective structures via the linear order, even using all the remaining rounds: Duplicator has a winning strategy if two linear orders are large enough.

So far, we assume Spoiler picks in the ξ-th abstraction. As explained in the proof outline, if Spoiler tries to pick below the ξ-th abstraction, Duplicator regards it as if $(\llparenthesis x \rrparenthesis_\xi, y)$ were picked and responds by playing the auxiliary game. Then she needs to ensure that the board is still over the ξ-th abstraction, i.e. $x' - \llparenthesis x' \rrparenthesis_\xi = x - \llparenthesis x \rrparenthesis_\xi$, or over the $(\xi - 1)$-th abstraction if she has to.

Recall that, Duplicator simply mimics in the first round. Hence ξ remains to be m and θ be $m - 1$ at the end of this round. In addition, ξ decreases by at most one in each round. Consequently, 4° can be ensured. Finally, 1° can be ensured, because Duplicator resorts to the auxiliary game over abstractions to determine her picks. For example, in the case where Spoiler picks in the ξ-th abstraction and Duplicator replies in the $(\xi - 1)$-th abstraction, 1° holds since both vertices are in the $(\xi - 1)$-th abstraction, i.e. $x' - \llparenthesis x' \rrparenthesis_{\xi-1} = x - \llparenthesis x \rrparenthesis_{\xi-1} = 0$.

By Lemma 6, it is easy to see that *defining a triangle needs three variables in FO on finite ordered graphs*. Suppose there is a $\mathrm{FO}^{k'}$ formula, where $k' < 3$, to describe a triangle, and assume its quantifier rank is m. Suppose $k' < m$. If it is not true, we can define another logically equivalent formula by artificially increasing the quantifier rank of the formula. Consequently, Spoiler has a winning strategy in $\eth^2_m(\mathfrak{A}_{3,m}, \mathfrak{B}_{3,m})$, a contradiction to Lemma 6.

References

1. Andréka, H., Németi, I., Benthem, J.V.: Modal languages and bounded fragments of predicate logic. J. Philos. Logic **27**(3), 217–274 (1998)
2. Barwise, J.: On Moschovakis closure ordinals. J. Symbolic Logic **42**, 292–296 (1977)
3. Dawar, A.: How many first-order variables are needed on finite ordered structures? In: We Will Show Them: Essays in Honour of Dov Gabbay, vol. 1, pp. 489–520. College Publications (2005)
4. Immerman, N.: Upper and lower bounds for first order expressibility. J. Comput. Syst. Sci. **25**(1), 76–98 (1982)
5. Poizat, B.: Deux ou trois choses que je sais de L_n. J. Symbolic Logic **47**(3), 641–658 (1982)

Counting Minimal Dominating Sets

Mamadou Moustapha Kanté[1]([✉]) and Takeaki Uno[2]

[1] Université Clermont Auvergne, LIMOS, CNRS, Aubière, France
mamadou.kante@uca.fr
[2] National Institute of Informatics, Tokyo, Japan

Abstract. A dominating set D in a graph is a subset of its vertex set such that each vertex is either in D or has a neighbour in D. From [M.M. Kanté, V. Limouzy, A. Mary and L. Nourine, On the Enumeration of Minimal Dominating Sets and Related Notions, SIDMA 28(4):1916–1929 (2014)] we know that the counting (resp. enumeration) of (inclusions-wise) minimal dominating sets is equivalent to the counting (resp. enumeration) of (inclusion-wise) minimal transversals in hypergraphs. The existence of an output-polynomial time algorithm for the enumeration of minimal dominating sets in graphs is open for a while, but by now for several graph classes it was shown that there is indeed an output-polynomial time algorithm. Since whenever we can count, we can enumerate in output-polynomial time, it is interesting to know for which graph classes one can count the set of minimal dominating sets in polynomial time (it is known that the problem is already $\#P$-complete in general graphs). In this manuscript we show that for many known graph classes with an output-polynomial time algorithm for the enumeration of minimal dominating sets, the counting version is $\#P$-complete, and for some of them a sub-exponential lower bound is also given (under $\#$ETH).

1 Introduction

Enumeration is a central area in algorithms, for instance in game theory, database theory, artificial intelligence. An *enumeration algorithm* is an algorithm that lists the element of a set without repetitions. As the set to be output is usually exponential in the size of the input, for instance the input is a graph and the goal is to output all matchings of the given graph, the time complexity is often measured by taking into account the cumulated sizes of the input and of the output. Therefore, we can consider an enumeration algorithm as "efficient" if its time complexity is bounded by a polynomial on the cumulated sizes of its input and output, and such enumeration algorithms are called *output-polynomial algorithms*. Indeed, the enumeration of several vertex/edge subsets, satisfying some given property, in (hyper)graphs have been considered and for many of them output-polynomial algorithms have been proposed [9,17], and for others it was proved that under $P \neq NP$ no output-polynomial algorithms exist [17].

M.M. Kanté—Supported by French Agency for Research under the GraphEN project (ANR-15-CE-0009).

T.V. Gopal et al. (Eds.): TAMC 2017, LNCS 10185, pp. 333–347, 2017.
DOI: 10.1007/978-3-319-55911-7_24

One of the central problems in enumeration area is the HYPERGRAPH DUAL-
ISATION problem due to its numerous applications in several areas such as game
theory, artificial intelligence, database theory and integer linear programming to
cite some (see the survey [9]). The HYPERGRAPH DUALISATION problem asks
for the listing of the (inclusion-wise) minimal *hitting sets* of a hypergraph, *i.e.*,
a collection of subsets of a ground set. However, despite the long interest on it
during the last fifty years, it is still open whether it admits an output-polynomial
algorithm, and the best known algorithm is the quasi-polynomial time one by
Fredman and Khachiyan [10].

The MINIMUM DOMINATING SET problem is a classic and well-studied graph
optimisation problem, and has applications in many areas such as networks and
graph theory [12]. A *dominating set* in a graph G is a subset D of its vertex set such that each vertex outside D has a neighbour in D. Recently, the
enumeration of all (inclusion-wise) minimal dominating sets of a graph (DOM-
ENUM for short) have been investigated and it's shown in [13] that this problem
admits an output-polynomial algorithm if and only if the HYPERGRAPH DUAL-
ISATION problem admits one. This is interesting in this area because it brings
graph structural theory and one may explain the lack of success in settling the
HYPERGRAPH DUALISATION problem as somehow a lack of hypergraph structural theory. Indeed, the structure of graphs have been used to propose output-
polynomial algorithms for several graph classes, *e.g.*, [11,14], cases that do not
fit under the already known tractable cases of hypergraph classes.

One of the natural ways to obtain an output-polynomial algorithm for
HYPERGRAPH DUALISATION is a branching algorithm coupled with a count-
ing algorithm. Let $\mathcal{H} := (V, \mathcal{E})$ be a hypergraph, and let $tr(\mathcal{H})$ be the set of
minimal hitting sets of $\mathcal{H}$. From [1] we know that for each $x \in V$, there is a minimal hitting set containing x. Now, take some vertex x, and compute $\#tr(\mathcal{H})$
and $\#tr(\mathcal{H} \setminus x)$ where $\mathcal{H} \setminus x := (V \setminus \{x\}, \{E \setminus \{x\} \mid E \in \mathcal{E}\})$. If the two
values are different, then branch on $\mathcal{H} \setminus x$ and $\mathcal{H}/x$ where $\mathcal{H}/x$ is the hypergraph where we have chosen x to be in all solutions, otherwise branch only on
$\mathcal{H}/x$. If computing $\#tr(\mathcal{H})$ can be done in polynomial time for any hypergraph
$\mathcal{H}$, then this algorithm is clearly output-polynomial. One can therefore expect
$\#$HYPERGRAPH DUALISATION, the problem of computing $\#tr(\mathcal{H})$, to be $\#P$-
complete, and this is indeed the case [8]. However, because the HYPERGRAPH
DUALISATION problem and its counting version are central in database theory
or artificial intelligence, many researchers have investigated the counting in subclasses of hypergraphs, *e.g.*, those arising from conjunctive queries, and many
tractable cases as well as finer $\#P$-completeness results are obtained, see for
instance [4,8].

Our Results. In this paper we investigate the counting of minimal dominating
sets in the following classes of graphs: split graphs, planar bipartite graphs, line
graphs, chordal bipartite graphs and unit-disk graphs (Sect. 3). It is worth noticing that for all of them, except unit-disk graphs, DOM-ENUM have been proved
to admit an output-polynomial algorithm. In all considered cases, we establish
$\#P$-completeness results. While DOM-ENUM is central due to its equivalence with

the HYPERGRAPH DUALISATION problem, its counting version have been considered, to our knowledge, only in [3,11]. Observe however that the counting of dominating sets have been considered in the past, e.g., [15,16].

Our reduction technique is the standard one of polynomial interpolation [21] and uses Vandermonde matrices [18]. We use the following classic #P-complete problems in our reductions: #SET-COVER, #VERTEX-COVER, #INDEPENDENT-SET and #PERFECT-MATCHING. A desirable property is that it rules out the existence of sub-exponential time algorithms, unless #ETH fails. The hypothesis #ETH, the analogue of ETH in counting problems, is the hypothesis ruling out a $2^{o(n)}$ time algorithm for counting the number of satisfiable assignments for a SAT instance with n variables. While the non-existence of FPRAS algorithms for counting problems are widely investigated, the non-existence of sub-exponential time algorithms have been investigated only recently [5], and only few counting problems are known to admit such sub-exponential lower bounds. We additionally show that the Vandermonde reduction fit very well, modulo small changes, in the framework introduced in [5] and obtain besides sub-exponential lower bounds for all considered graph classes, except the unit-disk graphs (Sect. 4).

2   Preliminaries

The power set of a set V is denoted by 2^V and we write $A \setminus B$ to denote the set difference of A and B. If A is a subset of a ground set V, we write $\overline{A}$ to denote the complementary set $V \setminus A$ of A. For a collection $\mathcal{E}$ of subsets of V, we denote by $\min(\mathcal{E})$ the (inclusion-wise) minimal elements of $\mathcal{E}$.

(Hyper)graph Terminology. Our graph terminology is standard, see for instance [7]. The vertex set of a graph G is denoted by $V(G)$ and its edge set by $E(G)$. An edge between x and y is denoted by xy (equivalently yx). For a vertex x, we denote by $N(x)$ the set of neighbours of x, and we let $N[x]$ be $N(x) \cup \{x\}$. For $X \subseteq V(G)$, $N(X) := (\bigcup_{x \in X} N[x]) \setminus X$.

Let G be a graph. For a subset X of $V(G)$ we denote by $G[X]$ the subgraph of G induced by X and write $G \setminus X$ for $G[V(G) \setminus X]$. A subset I of $V(G)$ is an *irredundant set* if for every vertex x in I, there is a vertex y such that $N[y] \cap I = \{x\}$ (such a vertex y is called a *private neighbour* of x). A subset D of $V(G)$ is a *dominating set* of G if every vertex of G is dominated, *i.e.*, is either in D or has a neighbour in D. It is well-known that every (inclusion-wise) minimal dominating set is also an (inclusion-wise) maximal irredundant set, but not all maximal irredundant sets are minimal dominating sets. For instance, if G is the path $a - b - c - d - e$, then $\{b, c\}$ is a maximal irredundant set, but not a minimal dominating set as e is not dominated. The set of minimal dominating sets of a graph G is denoted by $\mathcal{D}(G)$.

A hypergraph is a pair $(V, \mathcal{E} \subseteq 2^V)$ with V its vertex set and $\mathcal{E}$ its hyperedge set. For a hypergraph $\mathcal{H}$, we let $\min(\mathcal{H})$ be the hypergraph with hyperedge set $\min(\mathcal{E})$ and vertex set $\bigcup_{E \in \min(\mathcal{E})} E$. A hypergraph $\mathcal{H}$ is said *simple* if $\min(\mathcal{H}) = \mathcal{H}$ and it is common to denote a simple hypergraph $\mathcal{H}$ by its set of hyperedges.

A *transversal* of $\mathcal{H}$ is a subset T of V that intersects every element of $\mathcal{E}$. The set of (inclusion-wise) minimal transversals of a simple hypergraph $\mathcal{H}$ is denoted by $tr(\mathcal{H})$, and since $tr(tr(\mathcal{H})) = \mathcal{H}$ we can consider tr as a function. It is straightforward to check that $\mathcal{D}(G) = tr(\mathcal{N}(G))$ for every graph G where $\mathcal{N}(G)$ is the simple hypergraph $(V(G), \{N[x] \mid x \in V(G)\})$ called *closed neighbourhood hypergraph*.

Enumeration and Counting Problems. Let two alphabets Σ and Γ. Given a *witness function* $w : \Sigma^* \to 2^{\Gamma^*}$, its associated *enumeration problem* is the problem of listing $w(x)$ for each given $x \in \Sigma^*$ and an *enumeration algorithm* for w lists the elements of $w(x)$ without repetitions. The running time of an enumeration algorithm $\mathcal{A}$ is said to be *output polynomial* if there is a polynomial $p(x, y)$ such that all the elements of $w(x)$ are listed in time bounded by $p(|x|, |w(x)|)$ (this set of problems is denoted by **TotalP**). The associated *counting problem* asks for $|w(x)|$, and a *counting algorithm* is an algorithm that outputs $|w(x)|$, and if it runs in time polynomial in $|x|$, then it is called a *polynomial time algorithm*. The counting class $\#P$ is the set of counting problems with witness functions w such that (1) whenever $y \in w(x)$, then $|y|$ is bounded by a polynomial on $|x|$, and (2) given (x, y) we can check in time polynomial in $|x|$ whether $y \in w(x)$ [22]. As for the complexity class NP, there is a set of counting problems known to be complete for the class $\#P$, under Turing reductions, and an example is the computation of the permanent of a matrix [22]. We recall that the complexity class FP is the class of binary relations $P(x, y)$ for which there exist polynomial time algorithms computing for each x some y such that $P(x, y)$ holds. Observe that the counting problems for which there exist polynomial time counting algorithms are in FP. It was proved that the counting problems of several sets in graphs are $\#P$-complete [21]. Notice however that until now there is no known similar notion of complete problems for **TotalP** as we still fail to define the right notion of reduction between enumeration problems.

It is a fifty-year open problem whether, given a hypergraph $\mathcal{H} = (V, \mathcal{E})$, the enumeration problem for $tr(\mathcal{H})$ (known in the literature as TRANS-ENUM, see for instance the survey [9]) is in **TotalP**. Recently, Kanté et al. proved in [13] that this problem can be polynomially reduced to the enumeration problem for $\mathcal{D}(G)$, given a graph G (a problem called DOM-ENUM). Therefore, the two problems are polynomially equivalent. It is easy to check that, if the following counting problems can be solved in polynomial time, then the two problems DOM-ENUM and TRANS-ENUM can be solved by a backtracking algorithm.

Problem. #DOM-ENUM
Input. A graph G.
Output. The cardinal of $\mathcal{D}(G)$.

Problem. #TRANS-ENUM
Input. A hypergraph $\mathcal{H}$.
Output. The cardinal of $tr(\mathcal{H})$.

Nevertheless, as expected both problems are #P-complete.

Theorem 1 ([8]). #TRANS-ENUM *is #P-complete.*

From [13] one can derive the #P-completeness of #DOM-ENUM. If C is a class of graphs, we write #DOM-ENUM(C) to denote the problem #DOM-ENUM restricted to graphs in C. (Similarly for other enumeration or counting problems in hypergraphs.)

Theorem 2 ($\star$). #DOM-ENUM(*co-bipartite graphs*) *is #P-complete.*

It has been recently proved that for several graph classes C, DOM-ENUM(C) admits a positive answer [11,14]. In [11] the authors proved that #DOM-ENUM can be solved in polynomial time for several graph classes including interval graphs, permutation graphs, circular-arc graphs, trapezoid graphs, etc., and for fixed k, Dilworth-k graphs, complements of k-degenerate graphs, etc. In this paper we exhibit several graph classes C for which the problem #DOM-ENUM(C) is #P-complete, and examples are split graphs, chordal bipartite graphs, planar bipartite graphs and unit-disk graphs.

Main Tools. We introduce here the main known #P-complete problems from which we will do our reductions. Let G be a graph. A subset S of $V(G)$ is a *vertex-cover* if each edge of G has at least one of its endpoints in S. A subset F of $E(G)$ is a *perfect matching* if F induces a spanning subgraph of G that is 1-regular.

Problem. #VERTEX-COVER (or #INDEPENDENT-SET)
Input. A graph G.
Output. The number of vertex-covers (or independent sets) in G.

Since whenever S is a vertex-cover, the set $V(G) \setminus S$ is an independent set, both problems #VERTEX-COVER and #INDEPENDENT-SET are equivalent.

Theorem 3 ([21]). #VERTEX-COVER *and* #INDEPENDENT-SET *are #P-complete even with instances restricted to planar bipartite graphs of maximum degree* 4.

The following problem is the same as the computation of the permanent of a matrix [22].

Problem. #PERFECT-MATCHING
Input. A graph G.
Output. The number of perfect matchings in G.

Theorem 4 ([6]). #PERFECT-MATCHING *is #P-complete even with instances restricted to k-regular bipartite graphs for any $k \geq 3$.*

For a given set V and a set family $\mathcal{F} \subset 2^V$, a set covering is a subset X of $\mathcal{F}$ such that $\bigcup_{S \in X} S = V$.

Problem. #SET-COVER

Input. A set family $\mathcal{F} \subset 2^V$ for some finite set V.

Output. The number of set covers of $\mathcal{F}$.

Theorem 5 ([19]). #SET-COVER *is #P-complete.*

Our main reduction tool is the notion of *Vandermonde matrix*. An $m \times n$-matrix M is a *Vandermonde matrix* if it is of the following form

$$M := \begin{pmatrix} 1 & \alpha_1 & \alpha_1^2 & \cdots & \alpha_1^{n-1} \\ 1 & \alpha_2 & \alpha_2^2 & \cdots & \alpha_2^{n-1} \\ \vdots & \vdots & \vdots & \ddots & \vdots \\ 1 & \alpha_m & \alpha_m^2 & \cdots & \alpha_m^{n-1} \end{pmatrix}.$$

If all the α_i's are non-zero and $m = n$, then the determinant of M is non-zero and is equal to $\Pi_{1 \leq i \leq j \leq n}(\alpha_j - \alpha_i)$ [18]. Vandermonde matrices are used in the following setting. Consider that we want to prove that #DOM-ENUM($\mathcal{C}$) is #P-complete and assume that we dispose of a graph class $\mathcal{C}'$ such that #VERTEX-COVER($\mathcal{C}'$) is #P-complete. For each $G \in \mathcal{C}'$ and each $0 \leq i \leq m$ we construct a graph $G_i \in \mathcal{C}$ such that $|\mathcal{D}(G_i)| = \sum_{0 \leq j \leq m} f(i)^j x_j$ where x_j is the number of edge subsets of G with j edges that are not covered, m being the number of edges of G. Now, if all the $f(i)$'s are distinct, then the matrix

$$M := \begin{pmatrix} 1 & f(0) & f(0)^2 & \cdots & f(0)^m \\ 1 & f(1) & f(1)^2 & \cdots & f(1)^m \\ \vdots & \vdots & \vdots & \ddots & \vdots \\ 1 & f(m) & f(m)^2 & \cdots & f(m)^m \end{pmatrix}$$

is a Vandermonde matrix, and so if we assume that we can count in polynomial time the number of minimal dominating sets of any graph in $\mathcal{C}$, then we could resolve the following system of linear equations in polynomial time

$$M \cdot \begin{pmatrix} x_0 & x_1 & \cdots & x_m \end{pmatrix}^t = \begin{pmatrix} |\mathcal{D}(G_0)| & |\mathcal{D}(G_1)| & \cdots & |\mathcal{D}(G_m)| \end{pmatrix}^t$$

since the determinant of M is non null. Hence, we could solve in polynomial time #VERTEX-COVER($\mathcal{C}'$), contradicting its #P-completeness. Polynomial interpolations using Vandermonde matrices allow to obtain sub-exponential lower bounds. The EXPONENTIAL TIME HYPOTHESIS (ETH for short) says that given a SAT formula, with n variables, there is no algorithm with running time $2^{o(n)}$ checking whether it is satisfied. The counting part #ETH says similarly that there is no $2^{o(n)}$ counting algorithm for the positive assignments. Recently [5], it has been proved that, under #ETH, several counting problems in sparse graphs do not admit a $2^{o(n)}$ counting algorithm, *e.g.*, #PERFECT-MATCHING(sparse bipartite graphs), #VERTEX-COVER(sparse line graphs). We will explain in Sect. 4 how to apply the *block interpolation* introduced in [5] to get sub-exponential lower bounds.

3   #P-Completeness Results

The following is our set of #P-complete results. As in [21], if $\mathcal{C}$ is a class of graphs, we write $k\Delta$-$\mathcal{C}$ to denote the graphs in $\mathcal{C}$ of maximum degree k.

Theorem 6. *There is a Turing reduction using Vandermonde matrices from*

1. #SET-COVER *to* #DOM-ENUM(*split graphs*). ($\star$)
2. #VERTEX-COVER(*bipartite graphs*) *to* #DOM-ENUM(*bipartite graphs*).
3. #PERFECT-MATCHING(*sparse bipartite graphs*) *to* #DOM-ENUM(*line graphs*).
4. #INDEPENDENT-SET(*sparse graphs*) *to* #DOM-ENUM(*chordal bipartite graphs*).($\star$)
5. #VERTEX-COVER(4Δ-*planar bipartite graphs*) *to* #DOM-ENUM(*unit-disk graphs*).

We can derive the following lower bounds from [5], but we will improve them in Sect. 4.

Corollary 1. *Assuming #ETH, there is no $2^{o(\sqrt{n})}$ time algorithm for #DOM-ENUM even restricted to the following graph classes: line graphs, bipartite graphs, chordal bipartite graphs and split graphs.*

If S is a set, we denote by $S' := \{s' \mid s \in S\}$ a disjoint copy of S. Whenever we need to identify the two parts of a bipartite graph G, we denote it by (U_1, U_2, E) where U_1 and U_2 are its two independent set parts and E is the edge set.

If S is a vertex cover of a graph G, we denote by $uncov_G(S)$ the number of edges not covered by S. The length of a path is its number of edges. If v is a vertex of a graph G and H is a graph with a distinguished vertex v', called *vertex-gadget*, then the graph obtained from G by attaching H to v is the graph obtained from the disjoint union of G and H by identifying v and v'. And, similarly if H is a graph with two distinguished vertices u' and v', called *edge-gadget*, then the graph obtained from G by attaching H to uv is the graph obtained from the disjoint union of G and H by identifying u with u' and v with v'.

Proof (of Theorem 6(2)). Let $H := (U, F)$ be a (planar) graph. We construct the bipartite graph $G = (V, E)$ obtained from H by subdividing each edge once and adding a pendant vertex to each vertex of H. In other words, $V := U \cup U' \cup F$ and E is the set $\{uu' \mid u \in U\} \cup \{fu, fv \mid f = uv \in F\}$. For each positive integer i we denote by G_i the graph obtained from G by attaching i paths of length three to each vertex $f \in F$. It is straightforward to check that G_i is planar if H is planar.

Since for each vertex $u \in U$, the vertex $u' \in U'$ is a pendant vertex of u, then any minimal dominating set of G_i includes exactly one of u and u'. Therefore, for any minimal dominating set D of G_i, no vertex in U can be a private neighbour of $f \in F \cap D$.

Let $S \subseteq U$ and let D be a minimal dominating set of G_i such that $D \cap U = S$. Let $f \in F \cap N[S]$. Then for all paths $f - x - y - z$ attached to f, D contains either y or z, and if $f \in D$, then for at least one of the paths $f - x - y - z$ the vertex z is in D. So, for each $f \in N[S] \cap F$, we can extend S into at least $2^i + 2^i - 1$ minimal dominating sets. Let now $f \in F \setminus N[S]$. If $f \in D$, then for any path $f - x - y - z$ attached to f, D contains exactly either y or z, and if $f \notin D$, then for each path $f - x - y - z$ attached to f D contains either y or (x and z), or D contains x and y for exactly one path $f - x - y - z$ among the paths attached to f. So, for each $f \in F \setminus N[S]$ we can extend S into at least $2^{i+1} + 2^{i-1} + i - 1$ minimal dominating sets.

Moreover, since each $S \subseteq U$ is an irredundant set (each vertex has its copy as a private neighbour) and can be extended to a minimal dominating set excluding $U \setminus S$, the minimal dominating sets of G_i are characterised by their intersection with U.

Let x_j be the number of sets included in U such that $uncov_H(S) = j$ and let b_i be the number of minimal dominating sets of G_i. Observe that x_0 is the number of vertex covers of H. Then, from above

$$b_i = \sum_{j=0}^{|F|} \left((2^{i+1} - 1)^{|F|} \times \left(\frac{2^{i+1} + 2^{i-1} + i - 1}{2^{i+1} - 1} \right)^j \right) \times x_j$$

$$= (2^{i+1} - 1)^{|F|} \times \left(\sum_{j=0}^{|F|} \left(\frac{2^{i+1} + 2^{i-1} + i - 1}{2^{i+1} - 1} \right)^j \times x_j \right).$$

So, we have the following equality by letting $a_i := \frac{2^{i+1}+2^{i-1}+i-1}{2^{i+1}-1}$, $c_i := (2^{i+1} - 1)^{|F|}$

$$\begin{pmatrix} 1 & a_1 & a_1^2 & \cdots & a_1^{|F|} \\ 1 & a_2 & a_2^2 & \cdots & a_2^{|F|} \\ \vdots & \vdots & \vdots & \ddots & \vdots \\ 1 & a_{|F|+1} & a_{|F|+1}^2 & \cdots & a_{|F|+1}^{|F|} \end{pmatrix} \cdot \begin{pmatrix} x_0 \\ x_1 \\ \vdots \\ x_{|F|} \end{pmatrix} = \begin{pmatrix} \frac{b_1}{c_1} \\ \frac{b_2}{c_2} \\ \vdots \\ \frac{b_{|F|+1}}{c_{|F|+1}} \end{pmatrix}.$$

Since $a_i \neq a_j$ for any $i \neq j$, we can conclude the proof.

$\square$

Proof (of Theorem 6(3)). Since minimal dominating sets in line graphs and minimal edge dominating sets in graphs coincide, we will reduce #PERFECT-MATCHING to the enumeration of minimal edge dominating sets.

Let $H = (U, F)$ be a (bipartite) graph. We let $G = (V, E)$ be obtained from H by subdividing each edge once. For a pair of positive integers i and j, we let G_{ij} be the graph obtained from G by attaching i edges to each $f \in F$, and j edges to each $u \in U$. Of course, G_{ij} can be constructed in polynomial time in $|U| + |F| + i + j$.

Let $S \subseteq F$ and let D be a minimal edge dominating set of G_i such that $D \cap F = S$. Then, D does not include any of the i edges attached to $f \in F$ whenever S includes an edge incident to f; otherwise it contains exactly one of the i edges attached to f. Similarly, D does not include any of the j edges attached to $u \in U$ if S includes an edge incident to u; otherwise it includes exactly one the j edges attached to u. Let $p(S)$ be the number of vertices in U

incident to no edge of S, and $q(S)$ be the number of vertices in F adjacent to no edge of S. The number of such minimal edge dominating sets in G_{ij} is then $i^{q(S)} \cdot j^{p(S)}$.

Let $x_{i,j}$ be the number of irredundant sets S included in F such that $q(S) = i$ and $p(S) = j$, and let b_{ij} be the number of minimal edge dominating sets in G_{ij}. Then, we have the equation $b_{ij} = \sum_{0 \leq k \leq |F|, 0 \leq \ell \leq |U|} i^k \cdot j^\ell \cdot x_{k,\ell}$.

When $q(S) = 0$ and $p(S) = |F| - |U|/2$, the vertices in $S \cap F$ forms a perfect matching of H, thus $x_{0,(|F|-|U|/2)}$ is the number of perfect matchings in H. We now explain how to construct a solvable system of linear equations with variables the x_{ij}'s.

Let A and C be Vandermonde $(|F|+1) \times (|F|+1)$ and $(|U|+1) \times (|U|+1)$-matrices such that $a_{ij} := i^{j-1}$ and $c_{ij} := i^{j-1}$. We have the following equality (for convenience $n = |U| + 1$ and $m = |F| + 1$)

$$
\begin{pmatrix}
c_{11}A & c_{12}A & \cdots & c_{1n}A \\
c_{21}A & c_{22}A & \cdots & c_{2n}A \\
\vdots & \vdots & \ddots & \vdots \\
c_{n1}A & c_{n2}A & \cdots & c_{nn}A
\end{pmatrix}
\cdot
\begin{pmatrix}
x_{0,0} \\
\vdots \\
x_{0,n-1} \\
\vdots \\
x_{m-1,n-1}
\end{pmatrix}
=
\begin{pmatrix}
b_{11} \\
\vdots \\
b_{1n} \\
\vdots \\
b_{mn}
\end{pmatrix}.
$$

Because A and C are non-singular matrices, it is not hard to prove that the matrix

$$
\begin{pmatrix}
c_{11}A & c_{12}A & \cdots & c_{1n}A \\
c_{21}A & c_{22}A & \cdots & c_{2n}A \\
\vdots & \vdots & \ddots & \vdots \\
c_{n1}A & c_{n2}A & \cdots & c_{nn}A
\end{pmatrix}
$$

is non-singular, and hence one can solve the above system of equations in polynomial time, concluding the proof.   □

A *planar grid embedding* of a 4Δ-planar graph G is a planar embedding Q of G on a grid such that the vertices of G are mapped to the points of the grid and each edge uv of G is mapped into a path in the grid between $Q(u)$ and $Q(v)$. Notice that in a planar grid embedding two edges do not intersect, except possibly on their endpoints when they are adjacent. Before continuing let us consider the number of minimal dominating sets in paths.

Let $P_n := 1 - 2 - \cdots - n$ be the path with n vertices. For positive integers n and k, we let $P_{n,k}$ be the graph obtained from the path P_n by replacing the vertices 1 and n by cliques of size k, the vertices in each clique having the same neighbours as the initial end-point. We let $F_k(n)$ be the set of minimal dominating sets of $P_{n,k}$, $G_k(n)$ the set $\{D \subseteq V(P_{n,k}) \mid V(P_{n,k}) \setminus C(1) \subseteq N(D)$ and each vertex of D has a private neighbour in $V(P_{n,k}) \setminus C(1)\}$, and by $H_k(n)$ the set $\{D \subseteq V(P_{n,k}) \mid V(P_{n,k}) \setminus (C(1) \cup C(n)) \subseteq N(D)$ and each vertex of D has a private neighbour in $V(P_{n,k}) \setminus (C(1) \cup C(n))\}$ where $C(1)$ and $C(n)$ are the two cliques of size k. We also denote by $f_k(n)$, $g_k(n)$, $h_k(n)$ and $gh_k(n)$ the sizes of respectively $F_k(n)$, $G_k(n)$, $H_k(n)$ and $G_k(n) \cap H_k(n)$.

Lemma 1 ($\star$). *For all $n \geq 14$, all distinct $k, \ell \geq 0$*

$$\frac{f_k(n)}{gh_k(n)} \neq \frac{f_\ell(n)}{gh_\ell(n)}.$$

We are now ready to prove the last statement of Theorem 6.

Proof (of Theorem 6(5)). Let $H := (U, F)$ be a planar bipartite graph of maximum degree 4. From [20] H admits a planar embedding on a grid of size $O(n^2)$ and such a planar embedding can be found in polynomial time. Moreover, the length of every edge is bounded by $O(n)$. By stretching some sides of the grid we can assume that all edges have the same euclidean length $c = O(n)$. We remove from H all the non-necessary vertices and edges, and let G be the plane graph obtained from the planar grid embedding of H by adding a pendant edge to each vertex v of H; each such pendant edge is mapped into one of the faces incident with v, say the upper right one. We moreover assume that the euclidean distance between u and its new pendant vertex u' is 1. For each i, we let G_i be the graph obtained from G by replacing each grid path representing an edge of H by a copy of $P_{c,i}$. It is not hard to check that G_i is indeed a unit-disk: place in each point $Q(v)$ for $v \in U$ a unit disk, and now we can turn around $Q(v)$ and place a unit disk representing each of its at most $5 + i$ neighbours; the paths representing the edges of H can be replaced by paths of c unit disks. For convenience, for each edge uv of H, we denote by $Q(uv)$ the induced graph $P_{c,i}$ of G_i corresponding to the edge uv, excluding u and v.

First notice that a minimal dominating set of G_i cannot contain both u and its copy u', but because u' is a pendant vertex either u or u' should be included in any minimal dominating set. Let $S \subseteq U$ and let D be a minimal dominating set of G_i such that $D \cap U = S$. Let e be an edge of H. If none of the two end-points of $Q(e)$ are dominated by S, then the number of possible ways to extend S into D is the number of minimal dominating sets of $Q(e)$, *i.e.*, $f_i(c)$. Assume now that at least one of the end-points of $Q(e)$ is dominated by S, then the number of possible ways to extend S into D is the number of (inclusion-wise) minimal sets that dominate $Q(e)$, except possibly one or two of its end-points, *i.e.*, $gh_i(c)$. Similarly, any minimal dominating set D of G_i can be obtained in this way from $D \cap U$.

Now, let x_j be the number of vertex sets $S \subseteq U$ such that $uncov_H(S) = j$ and let b_i be the number of minimal dominating sets of G_i. Recall again that x_0 is the number of vertex covers of H. Then the following equation is obtained

$$b_i = \sum_{j=0}^{|F|} f_i(c)^j \cdot gh_i(c)^{m-j} \cdot x_j = gh_i(c)^m \times \left(\sum_{j=0}^{|F|} \left(\frac{f_i(c)}{gh_i(c)} \right)^j \cdot x_j \right).$$

If we let $b'_i := \frac{b_i}{gh_i(c)^m}$ and $\alpha_i := \frac{f_i(c)}{gh_i(c)}$, we have the following Vandermonde matrix

$$
\begin{pmatrix}
1 & \alpha_1 & \alpha_1^2 & \cdots & \alpha_1^{|F|} \\
1 & \alpha_2 & \alpha_2^2 & \cdots & \alpha_2^{|F|} \\
\vdots & \vdots & \vdots & \ddots & \vdots \\
1 & \alpha_{|F|+1} & \alpha_{|F|+1}^2 & \cdots & \alpha_{|F|+1}^{|F|}
\end{pmatrix}
\cdot
\begin{pmatrix}
x_0 \\
x_1 \\
\vdots \\
x_{|F|}
\end{pmatrix}
=
\begin{pmatrix}
b'_1 \\
b'_2 \\
\vdots \\
b'_{|F|+1}
\end{pmatrix}.
$$

By Lemma 1 $\alpha_i \neq \alpha_j$ for $i \neq j$, and hence we can conclude the statement. $\square$

4   Sub-exponential Lower Bounds Under #ETH

We use the *block interpolation* technique introduced in [5] to provide sub-exponential lower bounds for some of the considered graph classes. We will more or less follow the same notations as in [5]. A graph polynomial is a function p that maps a graph G to a polynomial $p(G) \in Q[\boldsymbol{x}]$ where $\boldsymbol{x}$ is some countable ordered set of indeterminates. If $p \in \mathbb{Q}[\boldsymbol{x}]$ has only one variable, then it is called *univariate*, otherwise it is called *multivariate*. For a graph polynomial p, we define the following computational problems $Coeff(p)$ and a family of problems $Eval_S(p)$ for $S \subseteq \mathbb{Q}$ such that

$Coeff(p)$. On input G, compute all coefficients of $p(G)$,

$Eval_S(p)$. On input G and $\mathbf{a} \in S^d$, evaluate $p(G)$ on $\mathbf{a}$, where d is the number of indeterminates. If p is univariate and $S = \{a\}$ is a singleton, we simply write $Eval_a(p)$ which asks for evaluating $p(G)$ on a.

As in [5], if $\boldsymbol{p}$ is a multivariate polynomial, and G' is a weighted graph (with weights on vertices and/or edges) with vertices/edges ordered, then $\boldsymbol{p}(G')$ denote the evaluation of $\boldsymbol{p}$ on input G', and with values of indeterminates the weights on vertices and/or edges.

Let $\mathscr{G}$ be the set of all finite undirected graphs and let $\mathcal{F} := \mathcal{V} \cup \mathcal{E}$ denote the countable set of all vertices and edges of such graphs. For a *sieve function* $\chi : \mathscr{G} \times 2^{\mathcal{F}} \to \mathbb{Q}$ and a *weight function* $\omega : \mathscr{G} \times 2^{\mathcal{F}} \to 2^{\mathcal{F}}$, we let $p_{\chi,\omega}$, called *subset-admissible*, be the univariate graph polynomial

$$
p_{\chi,\omega}(G; x) := \sum_{A \subseteq V(G) \cup E(G)} \chi(G, A) \cdot x^{|\omega(G,A)|}.
$$

The *multivariate generalisation polynomial* $\boldsymbol{p}_{\chi,\omega}$ on indeterminates $\mathbf{x} := \{x_a \mid x \in \mathcal{F}\}$ of $p_{\chi,\omega}$ is the multivariate polynomial given by

$$
\boldsymbol{p}_{\chi,\omega}(G; \mathbf{x}) := \sum_{A \subseteq V(G) \cup E(G)} \chi(G, A) \prod_{x \in \omega(G,x)} x_a.
$$

Definition 1 ([5]). *A* sub-exponential reduction family *from problem A to problem B is an algorithm $\mathbf{T}$ with oracle access to B. Its inputs are pairs (G, ϵ) where G is an input graph for A, ϵ with $0 < \epsilon < 1$ is a runtime parameter, such that*

1. **T** *computes $A(G)$, and it does so in time $f(\epsilon) \cdot 2^{\epsilon \cdot |V(G)|} \cdot |V(G)|^{O(1)}$,*
2. **T** *invokes the oracle B on graphs G' with at most $g(\epsilon).(|V(G)| + |E(G)|)$ vertices.*

*The functions f and g are computable functions that depend only on ϵ, and we write $A \leq_{serf} B$. If **T** invokes the oracle on instances with $O(g(\epsilon).(|V(G)| + |E(G)|))$ edges, then we say that it* preserves sparsity.

It is worth mentioning that R. Curticapean considered in [5] only sub-exponential reduction families that preserve sparsity as he was mostly interested in obtaining lower bounds for sparse graphs. However, the set of the following lemmas proved in [5] can be straightforwardly adapted to those that do not preserve sparsity as long as we are only interested in sub-exponential lower bounds depending on the number of vertices.

Lemma 2 ([5]). *If $A \leq_{serf} B$ and B can be solved in time $2^{o(n)} n^{O(1)}$ on graphs with n vertices, then A can be solved in time $2^{o(n)} n^{O(1)}$ on graphs with n vertices. If the reduction preserves sparsity, then if B can be solved in time $2^{o(n)} n^{O(1)}$ on graphs with n vertices and $O(n)$ edges, then so is A.*

Lemma 3 ([5]). *Let p be subset-admissible, with multivariate generalisation $\boldsymbol{p}$, and let $W := (w_0, w_1, \ldots)$ be an infinite countable ordered sequence of pairwise distinct numbers in $\mathbb{Q}$. Then $Coeff(p) \leq_{serf} Eval_W(\boldsymbol{p})$ by a reduction that preserves sparsity and satisfies the following for all inputs (G, ϵ): there is some d depending only on ϵ such that all queries $\boldsymbol{p}(G')$ are asked on graphs G' obtained from G by introducing edge-weights from $W_d := \{w_0, \ldots, w_d\}$.*

The following notion is a slight modification of a similar one in [5], where q equals p.

Definition 2. *Let p and q be subset-admissible, let $a \in \mathbb{Q}$, and*

1. *let $\mathcal{H} := (H_0, H_1, \ldots)$ be a countable sequence of vertex/edge-gadgets,*
2. *let $W := (w_0, w_1, \ldots,)$ be a countable sequence of pairwise distinct values in $\mathbb{Q}$,*
3. *let $F : \mathscr{G} \times \mathbb{Q} \to \mathbb{Q} \setminus \{0\}$ be a polynomial computable function, called* factor function

Let G be an admissible weighted graph with weights from W, i.e., if a vertex has weight w_i, then H_i is a vertex-gadget, and if the edge uv has weight w_j then H_j is an edge-gadget. Let $T(G)$ be obtained from G by attaching, for $i, j \in \mathbb{N}$, each vertex of weight w_i a copy of H_i, and similarly attach to each edge uv of weight w_j a copy of H_j.

The pair $(\mathcal{H}, W)$ allows to reduce $Eval_W(\boldsymbol{p})$ to $Eval_a(q)$ if for each admissible weighted graph with weights from W, $\boldsymbol{p}(G) = \frac{q(T(G),a)}{F(G,a)}$.

The proof of the following is the same as [5, Theorem 3.10] and uses Lemma 3.

Theorem 7 ([5]). *Let p and q be subset-admissible, and let $a \in \mathbb{Q}$ be fixed. Assuming #ETH, the problem $Eval_a(q)$ admits no $2^{o(n)}$ time algorithm on unweighted graphs with n vertices, provided that the following two conditions hold:*

1. *Assuming #ETH, the problem $Coeff(p)$ admits no $2^{o(n)}$ algorithm on graphs with n vertices.*
2. *There is a countable sequence $W := (w_0, w_1, \ldots)$ of pairwise distinct weights, a countable sequence of vertex/edge gadgets $\mathcal{H} := (H_0, H_1, \ldots)$ and a factor function F such that $(\mathcal{H}, F)$ allows to reduce $Eval_W(\boldsymbol{p})$ to $Eval_a(q)$.*

Sparsity is preserved if we use sub-exponential reduction families that preserve sparsity.

For a graph G, let $d_m(G) := \sum_{A \subseteq V(G)} [A \in \mathcal{D}(G)] \cdot x^{|A|}$. It is clear that d_m is subset admissible and $Eval_1(d_m)$ computes the number of minimal dominating sets of a graph.

Theorem 8 ($\star$). *Assuming #ETH, there is no $2^{o(n)}$ time algorithm for #DOM-ENUM in graphs with n vertices even restricted to the following graph classes: line graphs, bipartite graphs, chordal bipartite graphs, split graphs.*

Proof (Sketch). The idea consists in exhibiting from the proofs of Theorem 6 that there is a polynomial p such that $Coeff(p)$ does not admit a sub-exponential time algorithm, and that there is a pair $(\mathcal{H}, W)$ that allows to reduce $Eval_W(\boldsymbol{p})$ to $Eval_1(d_m)$. We exhibit the case of line graphs as a proof concept.

Let $W_v := (w_t^v)_{t \in \mathbb{N}}$ with $w_t^v := t$ and $\mathcal{H}_v := (H_t^v)_{t \in \mathbb{N}}$ with H_t^v, a vertex-gadget, a star with j leaves, its center being distinguished. Let $W_e := (w_t^e)_{t \in \mathbb{N}}$ with $w_t^e := t$ and $\mathcal{H}_e := (H_t^e)_{t \in \mathbb{N}}$ with H_t^e, an edge-gadget, a path $u_t - z_t - v_t$, (u_t, v_t) the distinguished vertices, with t edges attached to z_t. We let $W := W_v \cup W_e$ and $\mathcal{H} := \mathcal{H}_v \cup \mathcal{H}_e$.

Let $H := (U, F)$ be a bipartite graph. For an irredundant set $S \subseteq F$, i.e., each edge has an edge as a private neighbour, we let $q(S)$ be the set of vertices in U not incident with an edge of S, and let $p(S)$ be the set of edges in F not adjacent to an edge in S. We let

$$p_{pm}(H) := \sum_{S \subseteq F} [S \text{ irredundant set}] \cdot x^{|q(S)|} \cdot y^{|p(S)|}.$$

As observed in the proof of Theorem 6(3), the coefficient of $x^0 y^{|F|-|U|/2}$ gives the number of perfect matchings of H. Hence, from [5] $Coeff(p_{pm})$ does not admit a $2^{o(n)}$ time algorithm on bipartite graphs with n vertices and $O(n)$ edges under #ETH.

While p_{pm} is not univariate, its multivariate generalisation $\boldsymbol{p}_{pm}$ is well-defined, and the proof of Lemma 3 can be slightly modified so that $Coeff(\boldsymbol{p}_{pm})$ $\leq_{serf} Eval_W(\boldsymbol{p}_{pm})$. So, let H be a bipartite graph with vertex/edge-weights from W. From the proof of Theorem 6(3), $\boldsymbol{p}_{pm}(H) = d_m(T(G), 1)$. Hence, with

$F(H, 1) = 1$, $(\mathcal{H}, F)$ allows to reduce $Eval_W(\boldsymbol{p}_{pm})$ to $Eval_1(d_m)$. Therefore, we can deduce from Theorem 7 that #Dom-Enum(line graphs) does not admit a $2^{o(n)}$ time algorithm under #ETH on line graphs with n vertices and $O(n)$ edges.   □

5   Conclusion

We have proposed #P-completeness proofs for #Dom-Enum in some graph classes and give sub-exponential lower bounds, under #ETH. While it seems unlikely to have a dichotomy theorem between tractable and #P-complete classes, it would be nice to consider graph classes for which such a dichotomy can be obtained. We conjecture however that on chordal graphs a dichotomy theorem can be obtained. A k-sun is a graph obtained from a cycle of length $2k$ ($k \geq 3$) by adding edges to make the even-indexed vertices pairwise adjacent. Strongly chordal graphs are exactly chordal graphs without k-suns, for $k \geq 3$.

Conjecture 1. Let $\mathcal{C}$ be a class of chordal graphs. If $\mathcal{C}$ does not contain a k-sun as an induced subgraph, for $k \geq 4$, then #Dom-Enum($\mathcal{C}$) is in FP. Otherwise, #Dom-Enum(chordal graphs) can be reduced to #Dom-Enum($\mathcal{C}$) via Turing reductions.

The block interpolation used to obtain sub-exponential lower bounds fails in face of planar graphs. The reduction in [21] from #Vertex-Cover showing that #Vertex-Cover(planar bipartite graphs) is #P-complete uses a planarisation gadget that removes crossings. Since, even sparse graphs may have huge number of crossings in any embedding on the plane, the block interpolation technique fails. We ask whether sub-exponential time algorithms exist for #Dom-Enum(planar bipartite graphs) and #Dom-Enum(unit-disk graphs). We observe that fast exponential time algorithms for counting minimal dominating sets have been investigated, see for instance [2].

References

1. Berge, C.: Graphs. North-Holland Mathematical Library, vol. 6. North-Holland Publishing Co., Amsterdam (1985). Second revised edition of part 1 of the 1973 English version
2. Cochefert, M., Couturier, J.-F., Gaspers, S., Kratsch, D.: Faster algorithms to enumerate hypergraph transversals. In: Kranakis, E., Navarro, G., Chávez, E. (eds.) LATIN 2016. LNCS, vol. 9644, pp. 306–318. Springer, Heidelberg (2016). doi:10.1007/978-3-662-49529-2_23
3. Courcelle, B.: Linear delay enumeration and monadic second-order logic. Discrete Appl. Math. **157**(12), 2675–2700 (2009)
4. Creignou, N., Hermann, M.: Complexity of generalized satisfiability counting problems. Inf. Comput. **125**(1), 1–12 (1996)
5. Curticapean, R.: Block interpolation: a framework for tight exponential-time counting complexity. In: Halldórsson, M.M., Iwama, K., Kobayashi, N., Speckmann, B. (eds.) ICALP 2015. LNCS, vol. 9134, pp. 380–392. Springer, Heidelberg (2015). doi:10.1007/978-3-662-47672-7_31

6. Dagum, P., Luby, M.: Approximating the permanent of graphs with large factors. Theor. Comput. Sci. **102**(2), 283–305 (1992)

7. Diestel, R.: Graph Theory. Graduate Texts in Mathematics, vol. 173, 3rd edn. Springer, Heidelberg (2005)

8. Durand, A., Hermann, M.: On the counting complexity of propositional circumscription. Inf. Process. Lett. **106**(4), 164–170 (2008)

9. Eiter, T., Makino, K., Gottlob, G.: Computational aspects of monotone dualization: a brief survey. Discrete Appl. Math. **156**(11), 2035–2049 (2008)

10. Fredman, M.L., Khachiyan, L.: On the complexity of dualization of monotone disjunctive normal forms. J. Algorithms **21**(3), 618–628 (1996)

11. Golovach, P.A., Heggernes, P., Kanté, M.M., Kratsch, D., Sæther, S.H., Villanger, Y.: Output-polynomial enumeration on graphs of bounded (local) linear MIM-width. In: Elbassioni, K., Makino, K. (eds.) ISAAC 2015. LNCS, vol. 9472, pp. 248–258. Springer, Heidelberg (2015). doi:10.1007/978-3-662-48971-0_22

12. Haynes, T.W., Hedetniemi, S.T., Slater, P.J.: Fundamentals of domination in graphs. Monographs and Textbooks in Pure and Applied Mathematics, vol. 208. Marcel Dekker Inc., New York (1998)

13. Kanté, M.M., Limouzy, V., Mary, A., Nourine, L.: On the enumeration of minimal dominating sets and related notions. SIAM J. Discrete Math. **28**(4), 1916–1929 (2014)

14. Kanté, M.M., Limouzy, V., Mary, A., Nourine, L., Uno, T.: Polynomial delay algorithm for listing minimal edge dominating sets in graphs. In: Dehne, F., Sack, J.-R., Stege, U. (eds.) WADS 2015. LNCS, vol. 9214, pp. 446–457. Springer, Cham (2015). doi:10.1007/978-3-319-21840-3_37

15. Kijima, S., Okamoto, Y., Uno, T.: Dominating set counting in graph classes. In: Fu, B., Du, D.-Z. (eds.) COCOON 2011. LNCS, vol. 6842, pp. 13–24. Springer, Heidelberg (2011). doi:10.1007/978-3-642-22685-4_2

16. Kotek, T., Preen, J., Simon, F., Tittmann, P., Trinks. M.: Recurrence relations and splitting formulas for the domination polynomial. Electron. J. Comb. **19**(3), 27 (2012). Paper 47

17. Lawler, E.L., Lenstra, J.K., Rinnooy Kan, A.H.G.: Generating all maximal independent sets: NP-hardness and polynomial-time algorithms. SIAM J. Comput. **9**(3), 558–565 (1980)

18. Meyer, C.: Matrix analysis and applied linear algebra. Society for Industrial and Applied Mathematics (SIAM), Philadelphia (2000). With 1 CD-ROM (Windows, Macintosh and UNIX) and a solutions manual (iv+171 pp.)

19. Scott Provan, J., Ball, M.O.: The complexity of counting cuts and of computing the probability that a graph is connected. SIAM J. Comput. **12**(4), 777–788 (1983)

20. Tamassia, R., Tollis, I.G.: Planar grid embedding in linear time. IEEE Trans. Circ. Syst. **36**(9), 1230–1234 (1989)

21. Vadhan, S.P.: The complexity of counting in sparse, regular, and planar graphs. SIAM J. Comput. **31**(2), 398–427 (2001)

22. Valiant, L.G.: The complexity of computing the permanent. Theoret. Comput. Sci. **8**(2), 189–201 (1979)

On the Computational Complexity of Variants of Combinatorial Voter Control in Elections

Leon Kellerhals, Viatcheslav Korenwein, Philipp Zschoche, Robert Bredereck, and Jiehua Chen[✉]

TU Berlin, Berlin, Germany
{leon.kellerhals,viatcheslav.korenwein,zschoche}@campus.tu-berlin.de,
{robert.bredereck,jiehua.chen}@tu-berlin.de

Abstract. Voter control problems model situations in which an external agent tries to affect the result of an election by adding or deleting the fewest number of voters. The goal of the agent is to make a specific candidate either win (*constructive* control) or lose (*destructive* control) the election. We study the constructive and destructive voter control problems when adding and deleting voters have a *combinatorial flavor*: If we add (resp. delete) a voter v, we also add (resp. delete) a bundle $\kappa(v)$ of voters that are associated with v. While the bundle $\kappa(v)$ may have more than one voter, a voter may also be associated with more than one voter. We analyze the computational complexity of the four voter control problems for the Plurality rule.

We obtain that, in general, making a candidate lose is computationally easier than making her win. In particular, if the bundling relation is symmetric (i.e. $\forall w\colon w \in \kappa(v) \Leftrightarrow v \in \kappa(w)$), and if each voter has at most two voters associated with him, then destructive control is polynomial-time solvable while the constructive variant remains NP-hard. Even if the bundles are disjoint (i.e. $\forall w\colon w \in \kappa(v) \Leftrightarrow \kappa(v) = \kappa(w)$), the constructive problem variants remain intractable. Finally, the minimization variant of constructive control by adding voters does not admit an efficient approximation algorithm, unless P = NP.

1 Introduction

Since the seminal paper by Bartholdi III et al. [3] on controlling an election by adding or deleting the fewest number of voters or candidates with the goal of making a specific candidate to win (*constructive control*), a lot of research has been devoted to the study of control for different voting rules [4,13,15,20,22,23], on different control modes [16,17], or even on other controlling goals (e.g. aiming at several candidates' victory or a specific candidate's defeat) [19,25]. Recently, Bulteau et al. [8] introduced combinatorial structures to constructive control by

P. Zschoche—Supported by the Stiftung Begabtenförderung berufliche Bildung (SBB).

R. Bredereck—From September 2016 to September 2017 on postdoctoral leave at the University of Oxford (GB), supported by the DFG fellowship BR 5207/2.

© Springer International Publishing AG 2017
T.V. Gopal et al. (Eds.): TAMC 2017, LNCS 10185, pp. 348–361, 2017.
DOI: 10.1007/978-3-319-55911-7_25

adding voters: When a voter is added, a bundle of other voters is added as well. A combinatorial structure of the voter set allows us to model situations where an external agent hires speakers to convince whole groups of people to participate in (or abstain from) an election. In such a scenario, convincing a whole group of voters comes at the fixed cost of paying a speaker. Bulteau et al. [8] model this by defining a bundle of associated voters for each voter which will be convinced to vote "for free" when this voter is added or deleted. Moreover, the bundles of different voters could overlap. For instance, convincing two bundles of two voters each to participate in the election could result in adding a total of four, three or even just two voters.

We extend the work of Bulteau et al. [8] and investigate the cases where the agent wants to make a specific candidate win or lose by adding (resp. deleting) the fewest number of bundles. We study one of the simplest voting rules, the Plurality rule, where each voter gives one point to his favorite candidate, and the candidate with most points becomes a winner. Accordingly, an election consists of a set C of candidates and a set V of voters who each have a favorite candidate. Since real world elections typically contain only a small number of candidates, and a bundle of voters may correspond to a family with just a few members, we are especially interested in situations where the election has only few candidates and the bundle of each voter is small. Our goal is to ensure that a specific candidate p becomes a winner (or a loser) of a given election, by convincing as few voters from an unregistered voter set W as possible (or as few voters from V as possible), together with the voters in their bundles, to participate (or not to participate) in the election. We study the combinatorial voter control problems from both the classical and the parameterized complexity point of view. We confirm Bulteau et al.'s conjecture [8] that for the Plurality rule, the three problem variants: combinatorial constructive control by deleting voters and combinatorial destructive control by adding (resp. deleting) voters, behave similarly in complexity to the results of combinatorial constructive control by adding voters: They are NP-hard and intractable even for very restricted cases. We can also identify several special cases, where the complexity of the four problems behave differently. For instance, we find that constructive control tends to be computationally harder than destructive control. We summarize our results in Table 1.

Related Work. Bartholdi III et al. [3] introduced the complexity study of election control problems and showed that for the Plurality rule, the non-combinatorial variant of the voter control problems can be solved in linear time by a using simple greedy strategy. We refer the readers to the work by Faliszewski and Rothe [14], Rothe and Schend [26] for general expositions on election control problems.

In the original election control setting, a unit modification of the election concerns usually a single voter or candidate. The idea of adding combinatorial structure to election voter control was initiated by Bulteau et al. [8]: Instead of adding a voter at each time, one adds a "bundle" of voters to the election, and the bundles added to the election could intersect with each other.

They showed that combinatorial constructive control by adding the fewest number of bundles becomes notorious hard, even for the Plurality rule and for only two candidates. Chen [9] mentioned that even if each bundle has only two voters and the underlying bundling graph is acyclic, the problem still remains NP-hard. Bulteau et al. [8] and Chan [9] conjectured that

> "*the combinatorial addition of voters for destructive control, and combinatorial deletion of voters for either constructive or destructive control behave similarly to combinatorial addition of voters for constructive control.*"

The combinatorial structure notion for voter control has also been extended to candidate control [10] and electoral shift bribery [7].

Paper Outline. In Sect. 2, we introduce the notation used throughout the paper. In Sect. 3 we formally define the four problem variants, summarize our contributions, present results in which the four problem variants (constructive or destructive, adding voters or deleting voters) behave similarly, and provide reductions between the problem variants. Sections 4, 5 and 6 present our main results on three special cases

(1) when the bundles and the number of candidates are small,
(2) when the bundles are disjoint, and
(3) when the solution size could be unlimited.

We conclude in Sect. 7 with several future research directions. Due to space restrictions, some proofs are deferred to our technical report [21].

2   Preliminaries

The notation we use in this paper is based on Bulteau et al. [8]. We assume familiarity with standard notions regarding algorithms and complexity theory. For each $z \in \mathbb{N}$ we denote by $[z]$ the set $\{1, \ldots, z\}$.

Elections. An *election* $E = (C, V)$ consists of a set C of m *candidates* and a set V of *voters*. Each voter $v \in V$ has a favorite candidate c and we call voter v a c-voter. Note that since we focus on the Plurality rule, we simplify the notion of the preferences of voters in an election to the favorite candidate of each voter. For each candidate $c \in C$ and each subset $V' \subseteq V$ of voters, the *(Plurality) score* $s_c(V')$ of candidate c with respect to the voter set V' is defined as the number of voters from V' that have her as favorite candidate. We say that a candidate c is a *winner* of election (C, V) if c has the highest score $s_c(V)$. For the sake of convenience, for each $C' \subseteq C$, a C'-voter denotes a voter whose favorite candidate belongs to C'.

Combinatorial Bundling Functions. Given a voter set X, a *combinatorial bundling function* $\kappa \colon X \to 2^X$ (abbreviated as *bundling function*) is a function that assigns a set of voters to each voter; we require that $x \in \kappa(x)$. For the sake of convenience, for each subset $X' \subseteq X$, we define $\kappa(X') = \bigcup_{x \in X'} \kappa(x)$. For a voter $x \in X$, $\kappa(x)$ is named x's *bundle*; x is called the *leader* of the bundle. We let b denote the *maximum bundle size* of a given κ. Formally, $b = \max_{x \in X} |\kappa(x)|$. One can think of the bundling function as subsets of voters that can be added at a unit cost (e.g. $\kappa(x)$ is a group of voters influenced by x).

Bundling Graphs. The *bundling graph* of an election is a model of how the voters' bundles interact with each other. Let $\kappa \colon X \to 2^X$ be a bundling function. The *bundling graph* $G_\kappa = (V(G_\kappa), E(G_\kappa))$ of κ is a simple, directed graph, where for each voter $x \in X$ there is a vertex $x \in V(G_\kappa)$ with the same name, and for each two distinct voters $y, z \in X$ with $z \in \kappa(y)$, there is an arc $(y, z) \in E(G_\kappa)$.

We consider three special cases of bundling functions/graphs which we think are natural in real world. We say that a bundling function κ is *symmetric* if for each two distinct voters $x, y \in X$, it holds that $y \in \kappa(x)$ if and only if $x \in \kappa(y)$. The bundling graph for a symmetric bundling function always has two directed arcs connecting each two vertices. Thus, we can assume the graph to be undirected.

We say that κ is *disjoint* if for each two distinct voters $x, y \in X$, it holds that either $\kappa(x) = \kappa(y)$ or $\kappa(x) \cap \kappa(y) = \emptyset$. It is an easy exercise to verify that disjoint bundling functions are symmetric and the corresponding undirected bundling graphs consist only of disjoint complete subgraphs.

We say that κ is *anonymous* if for each two distinct voters x and y with the same favorite candidate, it holds that $\kappa(x) = \kappa(y)$, and that for all other voters z we have $x \in \kappa(z)$ if and only if $y \in \kappa(z)$.

Example 1. For an illustration, consider the following election $E := (C = \{a, b, c\}, V = \{v_1, v_2, \ldots, v_5\})$ in which the favorite candidate of voters v_1, v_2, v_3 is a, the favorite candidate of v_4 is b, and the favorite candidate of v_5 is c. The bundling graph corresponding to the bundling function of this election could be either the left or the right figure as depicted below. Note that the label above or below the circle (which represents the vertex) denotes the name of the voter and the label inside the circle indicates his favorite candidate. For instance, in the left figure below, the leftmost circle corresponds to voter v_1 and his favorite candidate is a.

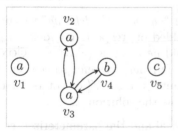 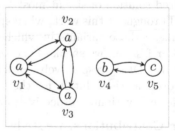

The bundling function corresponding to the left bundling graph is symmetric, but neither disjoint nor anonymous. The bundling function corresponding to the right bundling graph is symmetric, disjoint, and anonymous.

Parameterized Complexity. An instance (I, r) of a *parameterized problem* consists of the actual instance I and of an integer r referred to as the *parameter* [12, 18, 24]. A parameterized problem is called *fixed-parameter tractable* (in FPT) if there is an algorithm that solves each instance (I, r) in $f(r) \cdot |I|^{O(1)}$ time, where f is a computable function depending only on the parameter r.

There is also a hierarchy of hardness classes for parameterized problems, of which the most important ones are W[1] and W[2]. One can show that a parameterized problem L is (presumably) not in FPT by devising a *parameterized reduction* from a W[1]-hard or a W[2]-hard problem to L. A parameterized reduction from a parameterized problem L to another parameterized problem L' is a function that acts as follows: For two computable functions f and g, given an instance (I, r) of problem L, it computes in $f(r) \cdot |I|^{O(1)}$ time an instance (I', r') of problem L' so that $r' \leq g(r)$ and that $(I, r) \in L$ if and only if $(I', r') \in L'$. For a survey of research on parameterized complexity in computational social choice, we refer to Betzler et al. [5] and Bredereck et al. [6].

3   Central Problem

We consider the problem of *combinatorial voter control* in four variants. The variants differ in whether they are *constructive* or *destructive*, meaning that the goal is to make one selected candidate win or lose the election. This goal can be achieved by either *adding* voters to or *deleting* voters from the given election. Due to space constraints, we only provide the definition of constructive control. Destructive control is defined analogously.

COMBINATORIAL CONSTRUCTIVE CONTROL BY ADDING
(resp. **DELETING**) **VOTERS** [(resp. **C-CONS-DEL**)]
Input: An election $E = (C, V)$, a set W of unregistered voters with $V \cap W = \emptyset$, a bundling function $\kappa \colon W \to 2^W$ (resp. $\kappa \colon V \to 2^V$), a preferred winner $p \in C$, and an integer $k \in \mathbb{N}$
Quest.: Is there a size-at-most k subset $W' \subseteq W$ (resp. $V' \subseteq V$) of voters such that p wins the election $(C, V \cup \kappa(W'))$ (resp. $(C, V \setminus \kappa(V'))$)?

It is straight-forward to see that all four problem variants are contained in NP since we can check in polynomial time whether a given subset W' (or V') is a desired solution of size at most k.

Throughout this work, when speaking of the *"adding"* or *"deleting"* variants, we mean those variants in which voters are added or, respectively, deleted. In similar fashion, we speak of the *constructive* and *destructive* (abbr. by "CONS" and by "DES", respectively) problem variants. Further, we refer to the set W' of voters as the solution for the "adding" variants (the set V' of voters for the "deleting" variants, respectively) and denote k as the solution size.

Our Contributions. We study both the classical and the parameterized complexity of the four voter control variants. We are particularly interested in the real-world setting where the given election has a small number of candidates and where only a few voters are associated to a voter. On the one hand, we were able to confirm the conjecture given by Bulteau et al. [8] and Chan [9] that when parameterized by the solution size, C-CONS-DEL, C-DES-ADD, and C-DES-DEL are all intractable even for just two candidates or for bundle sizes of at most three, and that when parameterized by the number of candidates, they are fixed-parameter tractable for anonymous bundling functions. On the other hand, we identify interesting special cases where the four problems differ in their computational complexity. We conclude that in general, destructive control tends to be easier than constructive control: For symmetric bundles with at most three voters, C-CONS-ADD is known to be NP-hard, while both destructive problem variants are polynomial-time solvable. For disjoint bundles, constructive control is parameterized intractable (for the parameter "solution size k"), while destructive control is polynomial-time solvable. Unlike for C-CONS-DEL, a polynomial-time approximation algorithm for C-CONS-ADD does not exist, unless P = NP. Our results are gathered in Table 1.

Table 1. Computational complexity of the four combinatorial voter control variants with the Plurality rule. The parameters are "the solution size k", "the number m of candidates" and "the maximum bundle size b". We refer to $|I|$ as the instance size. The rows distinguish between different maximum bundle sizes b and the number m of candidates. All parameterized intractability results are for the parameter "solution size k". ILP-FPT means FPT based on a formulation as an integer linear program and the result is for the parameter "number m of candidates".

	C-CONS-ADD	C-CONS-DEL	C-DES-ADD	C-DES-DEL	References
κ symmetric					
$b = 2$	$O(|I|)$	P	$O(m \cdot |I|)$	$O(m \cdot |I|)$	Observation 2, Theorem 3 Theorem 5
$b = 3$					
$m = 2$	$O(|I|^5)$	$O(|I|^5)$	$O(|I|^5)$	$O(|I|^5)$	Theorem 2, Corollaries 1+2
m unbounded	NP-h	NP-h	$O(m \cdot |I|^5)$	$O(m \cdot |I|^5)$	Observation 1, Proposition 2, Corollary 2
b unbounded					
$m = 2$	W[2]-h	W[2]-h	W[2]-h	W[2]-h	[8], Theorem 1
m unbounded and κ disjoint	W[1]-h	W[2]-h	$O(m \cdot |I|)$	$O(m \cdot |I|)$	Theorems 4+5
κ anonymous	ILP-FPT	ILP-FPT	ILP-FPT	ILP-FPT	Theorem 1
κ arbitrary					
$b = 3, m = 2$	W[1]-h	W[1]-h	W[1]-h	W[1]-h	Theorem 1

The following theorem summarizes the conjecture given by Bulteau et al. [8] and Chan [9]. The corresponding proof can be found in our technical report [21].

Theorem 1. *All four combinatorial voter control problem variants are*

(i) *W[2]-hard with respect to the solution size k, even for only two candidates and for symmetric bundling functions κ.*

(ii) *W[1]-hard with respect to the solution size k, even for only two candidates and for bundle sizes of at most three.*

(iii) fixed-parameter tractable with respect to the number m of candidates if the bundling function κ is anonymous.

Relations Between the Four Problem Variants. We provide some reductions between the problem variants. They are used in several sections of this paper. The key idea for the reduction from destructive control to constructive control is to guess the candidate that will defeat the distinguished candidate and ask whether one can make this candidate win the election. The key idea for the reduction from the "deleting" to the "adding" problem variants is to build the "complement" of the registered voter set.

Proposition 1. *For each $X \in \{$ADD, DEL$\}$, C-DES-X with m candidates is Turing reducible to C-CONS-X with two candidates. For each $Y \in \{$CONS, DES$\}$, C-Y-DEL with two candidates is many-one reducible to C-Y-ADD with two candidates. All these reductions preserve the property of symmetry of the bundling functions.*

4 Controlling Voters with Symmetric and Small Bundles

In this section, we study combinatorial voter control when the voter bundles are symmetric and small. This could be the case when a voter's bundle models his close friends (including himself), close relatives, or office mates. Typically, this kind of relations is symmetric, and the number of friends, relatives, or office mates is small. We show that for symmetric bundles and for bundles size at most three, both destructive problem variants become polynomial-time solvable, while both constructive variants remain NP-hard. However, if there are only two candidates, then we can use dynamic programming to also solve the constructive control variants in polynomial time. If we restrict the bundle size to be at most two, then all four problem variants can be solved in polynomial time via simple greedy algorithms.

As already observed in Sect. 2, we only need to consider the undirected version of the bundling graph for symmetric bundles. Moreover, if the bundle size is at most two, then the resulting bundling graph consists of only cycles and trees. However, Bulteau et al. [8] already observed that C-CONS-ADD is NP-hard even if the resulting bundling graph solely consists of cycles, and Chan [9] observed that C-CONS-ADD remains NP-hard even if the resulting bundling graph consists of only directed trees of depth at most three.

Observation 1. C-CONS-ADD *is NP-hard even for symmetric bundling functions with maximum bundle size $b = 3$.*

It turns out that the reduction used by Bulteau et al. [8] to show the adding voters case (Observation 1) can be adapted to show NP-hardness for the deleting voters case.

Proposition 2. C-CONS-DEL *is NP-hard even for symmetric bundling functions with maximum bundle size $b = 3$.*

If, in addition to the bundles being symmetric and of size at most three, we have only two candidates, then we can solve C-Cons-Add in polynomial time. First of all, due to these constraints, we can assume that the bundling graph G_κ is undirected and consists of only cycles and paths. Then, it is easy to verify that we can consider each cycle and each path separately. Finally, we devise a dynamic program for the case when the bundling graph is a path or a cycle, maximizing the score difference between our preferred candidate p and the other candidate. The crucial idea behind the dynamic program is that the bundles of a minimum-size solution induce a subgraph where each connected component is small.

Lemma 1. *Let $(E = (C, V), W, \kappa, p, k)$ be a C-Cons-Add instance such that $C = \{p, g\}$, and κ is symmetric with G_κ being a path. Then, finding a size-at-most-k subset $W' \subseteq W$ of voters such that the score difference between p and g in $\kappa(W')$ is maximum can be solved in $O(|W|^5)$ time, where $|W|$ is the size of the unregistered voter set W.*

Proof. Since G_κ is a path, each bundle has at most three voters. We denote the path in G_κ by $(w_1, w_2, \ldots, w_{|W|})$ and introduce some definitions for this proof. The set $W(s, t) := \{w_i \in W \mid s \le i \le t\}$ contains all voters on a sequence from w_s to w_t. For every subset $W' \subseteq W$ we define $\mathsf{gap}(W') := s_p(\kappa(W')) - s_g(\kappa(W'))$ as the score difference between p and g. One can observe that if W' is a solution for $(E = (C, V), W, \kappa, p, k)$ then $\mathsf{gap}(W') \ge s_g(V) - s_p(V)$; note that we only have two candidates. An (s, t)-*proper-subset* W' is a subset of $W(s, t)$ such that $\kappa(W') \subseteq W(s, t)$. A *maximum* (s, t)-*proper-subset* W' additionally requires that each (s, t)-proper-subset $W'' \subseteq W$ with $|W''| = |W'|$ has $\mathsf{gap}(W'') \le \mathsf{gap}(W')$.

We provide a dynamic program in which a table entry $T[r, s, t]$ contains a maximum (s, t)-proper-subset W' of size r. We first initialize the table entries for the case where $t - s + 1 \le 9$ and $r \le 9$ in linear time.

For $t - s + 1 > 9$, we compute the table entry $T[r, s, t]$ by considering every possible partition of $W(s, t)$ into two disjoint parts.

$$T[r, s, t] := T[r - i, s, s + j] \cup T[i, s + j + 1, t],$$

$$\text{where } i, j = \underset{\substack{0 \le i \le r \\ 0 \le j \le t - s - 2}}{\arg\max} \ \mathsf{gap}(T[r - i, s, s + j]) + \mathsf{gap}(T[i, s + j + 1, t]).$$

Note that a maximum $(1, |W|)$-proper-subset W' of size $r - 1$ *could* have a higher gap than a maximum $(1, |W|)$-proper-subset W'' of size r.

To show the correctness of our program, we define the maximization and minimization function on a set of voters W', which return the largest and smallest index of all voters on the path induced by W', respectively:

$$\max(W') := \underset{i \in |W'|}{\arg\max}\{w_i \in (W')\} \ \text{ and } \ \min(W') := \underset{i \in |W'|}{\arg\min}\{w_i \in (W')\}.$$

First, we use the following claim to show that each maximum (s, t)-proper-subset W' can be partitioned into two (s, t)-proper-subsets W_1, W_2 such that the two sets $\kappa(W_1)$ and $\kappa(W_2)$ are disjoint. The correctness proof of the following claim can be found in our technical report [21].

Claim 1. *Let W' be a maximum (s,t)-proper-subset such that $(\max \kappa(W') - \min \kappa(W') + 1) > 9$. Then, there is a j with $s < j < t$ such that there is an (s,j)-proper-subset W_1 and a $(j+1,t)$-proper-subset W_2 with $|W_1| + |W_2| \leq |W'|$ and $\kappa(W_1 \cup W_2) = \kappa(W')$.*

Now, we show that the two subsets W_1 and W_2 from Claim 1 are indeed *optimal*: There is a j such that W_1 is a maximum (s,j)-proper-subset and W_2 is a maximum $(j + 1, t)$-proper-subset.

Assume towards a contradiction that W_2 is a $(j + 1, t)$-proper-subset but not a maximum $(j + 1, t)$-proper-subset. Therefore, there exists a maximum $(j+1,t)$-proper-subset W_2' where $|W_2| = |W_2'|$. This implies that $\mathsf{gap}(W_1 \cup W_2') > \mathsf{gap}(W_1 \cup W_2)$. This is a contradiction to $W' = W_1 \cup W_2$ being a maximum (s,t)-proper-subset. The case in which W_1 is not a maximum (s,j)-proper-subset is analogous.

Altogether, we have shown that we can compute $T[k, s, t]$ in constant time if $t - s + 1 \leq 9$, and that otherwise there exist an i and a j such that $T[k, s, t] = T[k - i, s, t - j] \cup T[i, t - j + 1, t]$. The dynamic program considers all possible i and j. The table entry $T[i, 1, |W|]$ contains a subset $W' \subseteq W$ of size i with maximum gap such that $\kappa(W') \subseteq W(1, |W|)$, which is identical to $\kappa(W') \subseteq W$.

This completes the correctness proof of our dynamic program. The table has $O(k \cdot |W|^2)$ entries. To compute one entry the dynamic program accesses $O(k \cdot |W|)$ other table entries. Note that the value $\mathsf{gap}(T[i, s, t])$ can be computed and stored after the entry $T[i, s, t]$ is computed. This takes at most $O(|W|)$ steps. Thus, the dynamic program runs in $O(|W|^5)$ time.   □
The dynamic program can be adapted to solve the same problem on cycles:

Lemma 2. *Let $(E = (C, V), W, \kappa, p, k)$ be a C-CONS-ADD instance such that $C = \{p, g\}$, and κ is symmetric with G_κ being a cycle. Then, finding a size-at-most-k subset $W' \subseteq W$ of voters such that the score difference between p and g in $\kappa(W')$ is maximum can be solved in $O(|W|^5)$ time, where $|W|$ is the size of the unregistered voter set W.*

Altogether, we obtain the following.

Theorem 2. *C-CONS-ADD with a symmetric bundling function, maximum bundle size of three, and for two candidates can be solved in $O(|W|^5)$ time, where $|W|$ is the size of the unregistered voter set.*

Proof. Let $(E = (C, V), W, \kappa, p, k)$ be a C-CONS-ADD instance, where the maximum bundle size b is three, κ is symmetric, and $C = \{p, g\}$. This means that all connected components $C_1, \ldots, C_\ell$ of G_κ are path or cycles. Furthermore, all bundles only contain voters from one connected component. We define a dynamic program in which each table entry $A[i, s, t]$ contains a solution $W' \subseteq W$ of size i, where $\kappa(W') \subseteq V(C_s) \cup \cdots \cup V(C_t)$ and $1 \leq s \leq t \leq \ell$:

(i) If $s = t = j$, then $A[i, s, t] = T[i, 1, |V(C_j)|]$, where T is the dynamic program of C_j, depending on whether C_j is a path or cycle.

(ii) Otherwise, we build the table as follows:

$$A[d, s, t] = A[d - i, s, s + j] \cup A[i, s + j + 1, t], \text{where}$$
$$i, j = \arg \max_{\substack{0 \le i \le d \\ 1 \le j \le t - s - 1}} \mathsf{gap}(A[d - i, s, s + j]) + \mathsf{gap}(A[i, s + j + 1, t]).$$

Each of the table entries $A[i, j, j]$ can be computed in $O(i^2 \cdot |V(C_j)|^3)$ time (see Lemmas 1 and 2) and each of the table entries $A[i, s, t]$ for $s < t$ can be computed in $O(k \cdot \ell)$ time. Since we have $k \cdot \ell^2$ entries, the total running time is

$$\sum_{i=1}^{\ell} O(k^2 \cdot |V(C_j)|^3) = O(k^2) \sum_{i=1}^{\ell} O(|V(C_i)|) = O(k^2 \cdot |W|^3). \qquad \square$$

From the polynomial-time solvability result of Theorem 2 and by Proposition 1, we obtain the following:

Corollary 1. C-CONS-DEL *with a symmetric bundling function, maximum bundle size of three, and two candidates can be solved in* $O(|V|^5)$ *time, where* $|V|$ *denotes the number of the voters.*

Corollary 2. C-DES-ADD *and* C-DES-DEL *with a symmetric bundling function and maximum bundle size three can be solved in time* $O(m \cdot |W|^5)$ *and* $O(m \cdot |V|^5)$, *respectively, where* m *is the number of candidates, and* $|W|$ *and* $|V|$ *are the numbers of unregistered and registered voters, respectively.*

5 Controlling Voters with Disjoint Bundles

We have seen in Sect. 4 that the interaction between the bundles influences the computational complexity of our combinatorial voter control problems. For instance, adding a voter v to the election may lead to adding another voter v' with $v \in \kappa(v)$. This is crucial for the reductions used to prove Theorem 1 and Observation 1. Thus, it would be interesting to know whether the problem becomes tractable if it is not necessary to add two bundles that share some voter(s). More specifically, we are interested in the case where the bundles are disjoint, meaning that we do not need to consider every single voter, but only the bundles as a whole, as it does not matter which voters of a bundle we select.

First, we consider disjoint bundles of size at most two. This is the case for voters who have a partner. If a voter is convinced to participate in or leaves the election, then the partner is convinced to do the same. Note that this is equivalent to having symmetric bundles of size at most two. Bulteau et al. [8, Theorem 6] constructed a linear-time algorithm for C-CONS-ADD if the maximum bundle size is two and κ is a full-d bundling function (which implies symmetry). We can verify that their algorithm actually works for disjoint bundles of size at most two. Thus, we obtain the following.

Observation 2. C-CONS-ADD *with a symmetric bundling function and with bundles of size at most two can be solved in* $O(|I|)$ *time, where* $|I|$ *is the input size.*

If we want to delete instead of add voter bundles, the problem reduces to finding a special variant of the f-FACTOR problem, which is a generalization of the well-known matching problem and can still be solved in polynomial time [1,2].

Theorem 3. C-CONS-DEL *with a symmetric bundling function and with bundles of size at most two can be solved in polynomial time.*

If we drop the restriction on the bundle sizes but still require the bundles to be disjoint, then C-CONS-ADD and C-CONS-DEL become parameterized intractable with respect to the solution size.

Theorem 4. *Parameterized by the solution size* k, C-CONS-ADD *and* C-CONS-DEL *are* W[1]*-hard and* W[2]*-hard respectively, even for disjoint bundles.*

Proof (with only the construction for the W[1]*-hardness proof of* C-CONS-ADD*).* We provide a parameterized reduction from the W[1]-complete problem INDEPENDENT SET (parameterized by the "solution size") which, given an undirected graph $G = (V(G), E(G))$ and a natural number $h \in \mathbb{N}$, asks whether G admits a size-h *independent set* $U \subseteq V(G)$, that is, all vertices in U are pairwise non-adjacent. Let (G, h) be an INDEPENDENT SET instance with $E(G) = \{e_1, \ldots, e_{m-1}\}$ and $V(G) = \{u_1, \ldots, u_n\}$. Without loss of generality, we assume that G is connected and $h \geq 3$. We construct an election $E = (C, V)$ with candidate set $C := \{p\} \cup \{g_j \mid e_j \in E(G)\}$. For each edge $e_j \in E$, we construct $h - 1$ registered voters that all have g_j as their favorite candidate. In total, V consists of $(h - 1) \cdot (m - 1)$ voters.

The unregistered voter set W is constructed as follows: For each vertex $u_i \in V(G)$, add a p-voter p_i, and for each edge e_j incident with u_i, add a g_j-voter $a_j^{(i)}$. The voters constructed for each vertex u_i are bundled by the bundling function κ. More formally, for each $u_i \in V(G)$ and each $e_j \in E(G)$ with $u_i \in e_j$, it holds that

$$\kappa(p_i) = \kappa(a_j^{(i)}) := \{p_i\} \cup \{a_{j'}^{(i)} \mid u_i \in e_{j'} \text{ for some } e_{j'} \in E(G)\}.$$

To finalize the construction, we set $k := h$. The construction is both a polynomial-time and a parameterized reduction, and all bundles are disjoint. To show the correctness, we note that p can only win if only if her score can be increased to at least h without giving any other candidate more than one more point. The solution corresponds to exactly to a subset of h vertices that are pairwise non-adjacent. The detailed correctness proof and the remaining proof for the W[2]-hardness result can be found in the our technical report [21].   □

For destructive control, it is sufficient to guess a potential defeater d out of $m - 1$ possible candidates that will have a higher score than p in the final election and use a greedy strategy similar to the one used for Observation 2 to obtain the following result.

Theorem 5. C-DES-ADD *and* C-DES-DEL *with a symmetric bundling function and disjoint bundles can be solved in* $O(m \cdot |I|)$ *time, where* $|I|$ *is the input size and* m *the number of candidates.*

6 Controlling Voters with Unlimited Budget

To analyze election control, it is interesting to know whether a solution exist at all, without bounding its size. Indeed, Bartholdi III et al. [3] already considered the case of unlimited solution size for the constructive candidate control problem. They showed that the problem is already NP-hard, even if the solution size is not bounded. (The non-combinatorial destructive control by adding unlimited amount of candidates is shown to be also NP-hard by Hemaspaandra et al. [19].) In contrast, the non-combinatorial voter control variants are linear-time solvable via simple greedy algorithms [3]. This leads to the question whether the combinatorial structure increases the complexity. To this end, we relax the four problem variants so that the solution can be of arbitrary size and call these problems C-CONS-ADD-UNLIM, C-DES-ADD-UNLIM, C-CONS-DEL-UNLIM and C-DES-DEL-UNLIM.

First of all, we observe that C-CONS-DEL-UNLIM becomes trivial if no unique winner is required.

Lemma 3. *Every* C-CONS-DEL-UNLIM *instance is a yes-instance.*

If we consider a voting rule $\mathcal{R}$ that only returns unique winners, then C-DES-DEL-UNLIM also becomes tractable since we only need to delete all voters.

For the constructive adding voters case, we obtain NP-hardness. The idea for the reduction derives from the W[1]-hardness proof of C-CONS-ADD shown by Bulteau et al. [8].

Lemma 4. C-CONS-ADD-UNLIM *is* NP-*hard.*

Lemma 4 immediately implies the following inapproximability result for the optimization variant of C-CONS-ADD (denoted as MIN-C-CONS-ADD), aiming at minimizing the solution size.

Theorem 6. *There is no polynomial-time approximation algorithm for* MIN-C-CONS-ADD, *unless* P = NP.

7 Conclusion

We extend the study of combinatorial voter control problems introduced by Bulteau et al. [8] and obtain that the destructive control variants tend to be computationally easier than their constructive cousins.

Our research leads to several open questions and further research opportunities. First, we have shown hardness results for the adding candidate case: if the bundling function consists of disjoint cliques, then parameterized by the solution size, C-CONS-ADD is W[1]-hard and C-DES-ADD is W[2]-hard. If one could also determine the complexity upper bound, that is, under the given restrictions, if C-CONS-ADD would be contained in W[1], then this would yield another difference in complexity between the destructive and the constructive variants. This also leads to the question whether the problem variants in their general setting are not only W[2]-hard, but W[2]-complete.

Second, we have only shown that MIN-C-CONS-ADD is inapproximable and MIN-C-DES-DEL is trivially polynomial-time solvable. For the other two problem variants, we do not know whether they can be approximated efficiently or not.

Another open question is whether there are FPT-results for any natural combined parameters. As a starting point, we conjecture that all problem variants can be formulated as a monadic second-order logic formula with length of at most $f(k, b, m)$ (where k is the solution size, b is the maximum bundle size, m is the number of candidates, and f is a computable function). Courcelle and Engelfriet [11] showed that every graph problem expressible as a monadic second-order logic formula ρ can be solved in $g(|\rho|, \omega) \cdot |I|$ time, where ω is the treewidth of the input graph and $|I|$ is the input size. Our conjecture would provide us with a fixed-parameter tractability result with respect to the solution size, the maximum bundle size, the number of candidates, and the treewidth of our bundling graph G_κ.

We have studied the Plurality rule exclusively. Thus it is still open which of our results also hold for other voting rules, especially for the Condorcet rule. Since with two candidates, the Condorcet rule is equivalent to the strict majority rule, we can easily adapt some of our results to work for the Condorcet rule as well. Other results (i.e., the Turing reductions) cannot be easily adapted to work for the Condorcet rule.

References

1. Anstee, R.P.: An algorithmic proof of Tutte's f-factor theorem. J. Algorithms **6**(1), 112–131 (1985)
2. Anstee, R.P.: Minimum vertex weighted deficiency of (g, f)-factors: a greedy algorithm. Discrete Appl. Math. **44**(1–3), 247–260 (1993)
3. Bartholdi, J.J., Tovey, C.A., Trick, M.A.: How hard is it to control an election? Math. Comput. Model. **16**(8–9), 27–40 (1992)
4. Betzler, N., Uhlmann, J.: Parameterized complexity of candidate control in elections and related digraph problems. Theor. Comput. Sci. **410**(52), 43–53 (2009)
5. Betzler, N., Bredereck, R., Chen, J., Niedermeier, R.: Studies in computational aspects of voting. In: Bodlaender, H.L., Downey, R., Fomin, F.V., Marx, D. (eds.) The Multivariate Algorithmic Revolution and Beyond. LNCS, vol. 7370, pp. 318–363. Springer, Heidelberg (2012). doi:10.1007/978-3-642-30891-8_16
6. Bredereck, R., Chen, J., Faliszewski, P., Guo, J., Niedermeier, R., Woeginger, G.J.: Parameterized algorithmics for computational social choice: nine research challenges. Tsinghua Sci. Technol. **19**(4), 358–373 (2014)
7. Bredereck, R., Faliszewski, P., Niedermeier, R., Talmon, N.: Large-scale election campaigns: combinatorial shift bribery. J. Artif. Intell. Res. **55**, 603–652 (2016)

8. Bulteau, L., Chen, J., Faliszewski, P., Niedermeier, R., Talmon, N.: Combinatorial voter control in elections. Theor. Comput. Sci. **589**, 99–120 (2015)
9. Chen, J.: Exploiting structure in computationally hard voting problems. Ph.D. thesis, Technische Universität Berlin (2016)
10. Chen, J., Faliszewski, P., Niedermeier, R., Talmon, N.: Elections with few voters: Candidate control can be easy. In: AAAI 2015, pp. 2045–2051 (2015)
11. Courcelle, B., Engelfriet, J.: Graph Structure and Monadic Second-Order Logic: A Language-Theoretic Approach, vol. 168. Cambridge University Press, New York (2012)
12. Downey, R.G., Fellows, M.R.: Fundamentals of Parameterized Complexity. Springer, London (2013)
13. Erdélyi, G., Fellows, M.R., Rothe, J., Schend, L.: Control complexity in Bucklin and fallback voting: a theoretical analysis. J. Comput. Syst. Sci. **81**(4), 632–660 (2015)
14. Faliszewski, P., Rothe, J.: Control and bribery in voting. In: Brandt, F., Conitzer, V., Endriss, U., Lang, J., Procaccia, A.D. (eds.) Handbook of Computational Social Choice, Chap. 7. Cambridge University Press, Cambridge (2016)
15. Faliszewski, P., Hemaspaandra, E., Hemaspaandra, L., Rothe, J.: Llull and Copeland voting computationally resist bribery and constructive control. J. Artif. Intell. Res. **35**, 275–341 (2009)
16. Faliszewski, P., Hemaspaandra, E., Hemaspaandra, L.: Multimode control attacks on elections. J. Artif. Intell. Res. **40**, 305–351 (2011)
17. Faliszewski, P., Hemaspaandra, E., Hemaspaandra, L.A.: Weighted electoral control. J. Artif. Intell. Res. **52**, 507–542 (2015)
18. Flum, J., Grohe, M.: Parameterized Complexity Theory. Springer, Heidelberg (2006)
19. Hemaspaandra, E., Hemaspaandra, L.A., Rothe, J.: Anyone but him: the complexity of precluding an alternative. Artif. Intell. **171**(5), 255–285 (2007)
20. Hemaspaandra, L.A., Lavaee, R., Menton, C.: Schulze and ranked-pairs voting are fixed-parameter tractable to bribe, manipulate, and control. Ann. Math. Artif. Intell. **77**(3–4), 191–223 (2016)
21. Kellerhals, L., Korenwein, V., Zschoche, P., Bredereck, R., Chen, J.: On the computational complexity of variants of combinatorial voter control in elections. Technical report arXiv:1701.05108 [cs.MA]. arXiv.org, January 2017
22. Liu, H., Zhu, D.: Parameterized complexity of control problems in Maximin election. Inf. Process. Lett. **110**(10), 383–388 (2010)
23. Liu, H., Feng, H., Zhu, D., Luan, J.: Parameterized computational complexity of control problems in voting systems. Theor. Comput. Sci. **410**(27–29), 2746–2753 (2009)
24. Niedermeier, R.: Invitation to Fixed-Parameter Algorithms. Oxford University Press, Oxford (2006)
25. Procaccia, A.D., Rosenschein, J.S., Zohar, A.: Multi-winner elections: Complexity of manipulation, control and winner-determination. In: IJCAI 2007, pp. 1476–1481 (2007)
26. Rothe, J., Schend, L.: Challenges to complexity shields that are supposed to protect elections against manipulation and control: A survey. Ann. Math. Artif. Intell. **68**(1–3), 161–193 (2013)

On the Computational Complexity of Read once Resolution Decidability in 2CNF Formulas

Hans Kleine Büning[1], Piotr Wojciechowski[2(✉)], and K. Subramani[2]

[1] Universität Paderborn, Paderborn, Germany
kbcsl@uni-paderborn.de
[2] LCSEE, West Virginia University, Morgantown, WV, USA
pwojciec@mix.wvu.edu, k.subramani@mail.wvu.edu

Abstract. In this paper, we analyze 2CNF formulas from the perspectives of Read-Once resolution (ROR) refutation schemes. We focus on two types of ROR refutations, viz., variable-once refutation and clause-once refutation. In the former, each variable may be used at most once in the derivation of a refutation, while in the latter, each clause may be used at most once. We show that the problem of checking whether a given 2CNF formula has an ROR refutation under both schemes is **NP-complete**. This is surprising in light of the fact that there exist polynomial refutation schemes (tree-resolution and DAG-resolution) for 2CNF formulas.

Keywords: 2SAT · Resolution · Read-once

1 Introduction

Resolution is a refutation procedure that was introduced in [11] to establish the unsatisfiability of clausal Boolean Formulas. Resolution is a sound and complete procedure, although it is not efficient in general. Resolution is one among many proof systems (refutation systems) that have been discussed in the literature [14]; indeed it is among the weaker proof systems, in that there exist propositional formulas for which short proofs exist (in powerful proof systems) but resolution proofs of unsatisfiability are exponentially long. Resolution remains an attractive option for studying the complexity of constraint classes on account of its simplicity and wide applicability; it is important to note that resolution is the backbone of a range of automated theorem provers.

There are a number of different types of resolution refutation that have been discussed in the literature [10,12]. The most important types of resolution refutation are *tree-like, dag-like* and *read-once*. One of the simplest types of resolution

This research was supported in part by the National Science Foundation through Award CCF-1305054.

This work was supported by the Air Force Research Laboratory under US Air Force contract FA8750-16-3–6003. The views expressed are those of the authors and do not reflect the official policy or position of the Department of Defense or the U.S. Government.

© Springer International Publishing AG 2017
T.V. Gopal et al. (Eds.): TAMC 2017, LNCS 10185, pp. 362–372, 2017.
DOI: 10.1007/978-3-319-55911-7_26

is Read-once Resolution (ROR). In an ROR refutation, each input clause and each derived clause may be used at most once. Iwama [5] showed that even in case of 3CNF formulas, the problem of checking ROR existence (henceforth, ROR decidability) is **NP-complete**.

It is well-known that 2CNF satisfiability is decidable in polynomial time. There are several algorithms for 2CNF satisfiability, most of which convert the clausal formula into a directed graph and then exploit the connection between the existence of labeled paths in the digraph and the satisfiability of the input formula. A natural progression of this research is to establish the ROR complexity of 2CNF formulas. We show that the problem of deciding whether an arbitrary 2CNF formula has a read-once refutation is **NP-complete**. Although ROR is an **incomplete** refutation technique, we may be able to find a refutation if clauses can be copied. We show that every 2CNF formula has an ROR refutation, if every clause can be copied once.

The rest of this paper is organized as follows: In Sect. 2, we discuss problem preliminaries and formally define the various types of refutations discussed in this paper. The minimal unsatisfiable subset problems is detailed in Sect. 3. In Sect. 4, the Variable Read Once Resolution (VAR-ROR) refutation problem is detailed. We also establish the computational complexity of this problem in 2CNF. We show that ROR decidability for 2CNF formulas is **NP-complete** in Sect. 5.

2   Preliminaries

In this section, we briefly discuss the terms used in this paper. We assume that the reader is familiar with elementary propositional logic. A literal is a variable x or its complement $\neg x$. x is termed a positive and $\neg x$ is termed a negative literal. A clause is a disjunction of literals. The empty clause, which is always false, is denoted as $\sqcup$.

A Boolean formula Φ is in CNF, if the formula is a conjunction of clauses. Note that a formula in CNF is a set of clauses and written as $\{\alpha_1, \ldots, \alpha_n\}$, $\alpha_1 \wedge \ldots \wedge \alpha_n$, or simply as $\Phi = \alpha_1, \ldots, \alpha_n$ for clauses α_i. A formula in CNF is in k-CNF, if it is of the form $\alpha_1 \wedge \alpha_2 \wedge \ldots \wedge \alpha_m$, where each α_i is a clause of at most k literals.

For a single resolution step with parent clauses $(\alpha \vee x)$ and $(\neg x \vee \beta)$ with resolvent $(\alpha \vee \beta)$, we write

$$(\alpha \vee x), (\neg x \vee \beta) \vdash_{RES}^{1} (\alpha \vee \beta).$$

The variable x is termed a matching or resolution variable. If for initial clauses $\alpha_1, \ldots, \alpha_n$, a clause π can be generated by a sequence of resolution steps we write

$$\alpha_1, \ldots, \alpha_n \vdash_{RES} \pi.$$

We now formally define the types of resolution refutation discussed in this paper.

Definition 1. *A formula is in* **Var-ROR** *(variable-read once resolution), if and only if there is a resolution refutation for which every variable is used at most once as a matching variable.*

A resolution derivation $\Phi \vdash_{RES} \pi$ is a Var-ROR derivation, if the matching variables are used at most once. We denote this as $\Phi \vdash_{Var\text{-}RO\text{-}Res} \pi$.

Definition 2. *A formula Φ is said to be* **minimally Var-ROR**, *if and only if $\Phi \in Var\text{-}ROR$ and every proper sub-formula is not in $Var\text{-}ROR$.*

Definition 3. *A* **Read-Once** *resolution refutation is a refutation in which each clause, π, can be used in only one resolution step. This applies to clauses present in the original formula and those derived as a result of previous resolution steps.*

ROR is the set of formulas in CNF, for which a read-once resolution refutation exists ($\Phi \in \text{ROR}$ if and only if $\Phi \vdash_{RO\text{-}Res} \sqcup$).

Definition 4. *A formula, Φ, is* **minimally ROR** *if and only if the formula is in ROR and every proper sub-formula is not in ROR.*

It is important to note that both types of Read-Once resolution (Var-ROR and ROR) are **incomplete** refutation procedures. Furthermore, ROR is a strictly more powerful refutation procedure than VAR-ROR.

3   Minimal Unsatisfiability

In the characterization of various read-once classes and in the proofs, we make use of minimal unsatisfiable formulas and the splittings of these formulas.

First, we recall some notions and results.

Definition 5. *A formula in CNF is* **minimal unsatisfiable**, *if and only if the formula is unsatisfiable and every proper sub-formula is satisfiable. The set of minimal unsatisfiable formulas is denoted as MU.*

Definition 6. *The* **deficiency** *of a formula Φ, written as $d(\Phi)$, is the number of clauses minus the number of variables. For fixed k, MU(k) is the set of MU-formulas with deficiency k.*

The problem of deciding whether a formula is minimal unsatisfiable is $\mathbf{D^P}$-**complete** [8]. $\mathbf{D^P}$ is the class of problems which can be represented as the difference of two **NP**-problems. Every minimal unsatisfiable formula has a deficiency greater or equal than 1 [1].

For fixed k, deciding if $\Phi \in \text{MU}(k)$ can be solved in polynomial time [3]. For formulas in 2CNF, there is no constant upper bound for the deficiency of minimal unsatisfiable formulas.

The proofs in this paper make use of so-called splitting formulas for MU-formulas.

Definition 7. *Let*

$$\Phi = (x \vee \pi_1), \ldots, (x \vee \pi_r), \sigma_1, \ldots, \sigma_t, (\neg x \vee \phi_1), \ldots, (\neg x \vee \phi_q)$$

be a minimal unsatisfiable formula, where neither the literal x nor the literal $\neg x$ occur in the clauses σ_i. A pair of formulas $(F_x, F_{\neg x})$ with

$$F_x = \pi_1, \ldots, \pi_r, \sigma_{i_1}, \ldots, \sigma_{i_s} \text{ and } F_{\neg x} = \sigma_{j_1}, \ldots, \sigma_{j_k}, \phi_1, \ldots, \phi_q$$

*is called a **splitting** of Φ over x, if F_x and $F_{\neg x}$ are minimal unsatisfiable.*

Definition 8. *A splitting $(F_x, F_{\neg x})$ is **disjunctive**, if*

$$\{\sigma_{i_1}, \ldots, \sigma_{i_s}\} \cap \{\sigma_{j_1}, \ldots, \sigma_{j_k}\} = \emptyset$$

*That is, F_x and $F_{\neg x}$ have no clause σ_i in common. If additionally F_x and $F_{\neg x}$ do not share any variables, then we say that the splitting is **variable-disjunctive**.*

We can continue to split both F_x and $F_{\neg x}$ to obtain a splitting tree. Splitting stops when the formula contains only one variable.

Definition 9. *A splitting tree is **complete** if every leaf of the tree is a formula that contains at most one variable.*

Note that, after splitting on the variable x, neither F_x nor $F_{\neg x}$ contains the variable x. Thus, the splitting tree can have depth at most $(n-1)$.

In case of disjunctive splittings, we speak about disjunctive splitting trees. It is known that formulas in MU(1) have variable-disjunctive splitting trees [2]. Moreover, every minimal unsatisfiable formula with a read-once resolution refutation has a disjunctive splitting tree and vice versa [6].

Let Φ be a minimal unsatisfiable 2CNF formula. We will now prove several properties of Φ.

Lemma 1. *If Φ contains a unit clause, then $\Phi \in MU(1)$.*

Proof. By induction on the number n of variables. Full proof in the journal version of the paper.

Lemma 2. *For every variable x, the splitting formulas $F_x, F_{\neg x}$ are in MU(1).*

Proof. We have that the splitting formulas F_x and $F_{\neg x}$ contain a unit clause and are minimal unsatisfiable. Thus, from Lemma 1, $F_x, F_{\neg x}$ are in MU(1). □

Lemma 3 [7]. *If for a variable x, there is a disjunctive splitting, then the splitting is unique.*

Lemma 4. *The problem of determining if Φ has a complete disjunctive splitting is in **P**. Furthermore, if Φ has a disjunctive splitting for some variable x, then there exists a complete splitting tree.*

Proof. In the journal version of the paper.

4   The Complexity of Var-ROR for 2CNF

In this section, we show that determining if a formula in 2CNF has a variable read-once refutation is **NP-complete**.

Let Φ be a formula in 2CNF.

Theorem 1. $\Phi \in$ *Var-ROR, if and only if there exists a sub-formula* $\Phi' \subseteq \Phi$ *such that* $\Phi' \in MU(1)$.

Theorem 1 follows immediately from Lemma 5.

Lemma 5. Φ *is minimally Var-ROR, if and only if* $\Phi \in MU(1)$.

Proof. Let Φ be minimally Var-ROR. We will show that $\Phi \in$ MU(1) by induction on the number of variables.

First, assume that Φ is a CNF formula over 1 variable. Thus, Φ has the form $(x) \wedge (\neg x)$. Obviously, $\Phi \in$ MU(1).

Now assume that Φ is a CNF formula over $(n + 1)$ variables. Let

$$(\alpha \vee x), (\neg x \vee \beta) \vdash_{\overline{RES}}^{1} (\alpha \vee \beta)$$

be the first resolution step in a Var-ROR refutation of Φ. Thus, the formula

$$\Phi' = (\Phi \setminus \{(\alpha \vee x), (\neg x \vee \beta)\} \cup \{(\alpha \vee \beta)\})$$

is minimally Var-ROR and contains no clause with x or $\neg x$. x has already been used as a matching variable. Thus, if any other clauses of Φ, or Φ', used x then Φ would not be minimally Var-ROR.

Φ' has n variables. Thus, by the induction hypothesis $\Phi' \in$ MU(1). Since Φ' is minimal unsatisfiable and consists of $(n + 1)$ clauses, Φ is minimal unsatisfiable and consists of $(n + 2)$ clauses. This means that $\Phi \in$ MU(1).

Now let Φ be a formula in MU(1). We will show that Φ is minimally Var-ROR by induction on the number of variables.

For every formula in MU(1), there exists a variable-disjunctive splitting tree [2]. Thus, we can easily construct a Var-ROR refutation for Φ.

First, assume that Φ is a CNF formula over 1 variable. Thus, $\Phi = (x) \wedge (\neg x)$ and is minimally Var-ROR.

Now assume that Φ is a CNF formula over $(n + 1)$ variables. Let $(F_x, F_{\neg x})$ be the first variable-disjunctive splitting in a variable-disjunctive splitting tree. Without loss of generality we assume that neither F_x nor $F_{\neg x}$ is the empty clause. By the induction hypothesis, both formulas are minimally Var-ROR, because, by Lemma 2, $F_x, F_{\neg x} \in$ MU(1).

Thus, there is a Var-ROR derivation

$$F_x^x \vdash_{Var\text{-}RO\text{-}Res} (x) \text{ and } F_{\neg x}^{\neg x} \vdash_{Var\text{-}RO\text{-}Res} (\neg x),$$

where F_x^x ($F_{\neg x}^{\neg x}$) is the formula we obtain by adding the removed literal x ($\neg x$) to the clauses in F_x ($F_{\neg x}$). The final step is to resolve (x) and $(\neg x)$. Note that the variable x has not been used as a matching variable in the derivations $F_x^x \vdash_{Var\text{-}RO\text{-}Res} (x)$ and $F_{\neg x}^{\neg x} \vdash_{Var\text{-}RO\text{-}Res} (\neg x)$. Hence, Φ is in Var-ROR and is minimally Var-ROR.   □

For a formula in MU(1), there always exists a complete variable-disjunctive splitting. Furthermore, a complete variable-disjunctive splitting tree can be computed in polynomial time. Thus, every formula in MU(1) has a Var-ROR refutation that can be computed in polynomial time.

Corollary 1. *Every formula in MU(1) has a Var-ROR refutation that can be computed in polynomial time.*

Next, we will show that determining if a 2CNF formula has a Var-ROR refutation is **NP-complete**. It can easily be seen that this problem is in **NP**. If the formula has n variables, then at most n resolution steps can be performed. The **NP-hardness** will be shown by a reduction from the vertex-disjoint path problem for directed graphs.

Definition 10. *Given a directed graph G and pairwise distinct vertexes s_1, t_1, s_2, and t_2, the* **vertex-disjoint path problem** *(2-DPP) consists of finding a pair of vertex-disjoint paths in G, one from s_1 to t_1 and the other from s_2 to t_2.*

The problem is known to be **NP-complete** [4]. Now we modify the problem as follows.

Definition 11. *Given a directed graph G and two distinct vertexes s and t, the* **vertex-disjoint cycle problem** *(C-DPP) consists of finding a pair of vertex-disjoint paths in G, one from s to t and the other from t to s.*

Note that the paths are vertex-disjoint, if the inner vertexes of the path from s to t are disjoint from the inner vertexes of the path from t to s.

Lemma 6. *C-DPP is* **NP-complete***.*

Proof. Obviously, the problem is in **NP**. We will show **NP-hardness** by a reduction from 2-DPP.

From $G = (V, E)$, s_1, t_1, s_2, and t_2 we construct the new graph

$$G' = (V \cup \{s, t\}, E \cup \{(s, s_1), (t_2, s), (t_1, t), (t, s_2)\}).$$

Assume that G has two vertex-disjoint paths, w_1 from s_1 to t_1, and w_2 from s_2 to t_2. Thus, the paths $(s, s_1), w_1, (t_1, t)$ and $(t, s_2), w_2, (t_2, s)$ in G' are vertex-disjoint. Note that s_1, s_2, t_1, t_2 are pairwise distinct. Thus, G' has the desired vertex-disjoint cycle.

Now assume that G' has two vertex-disjoint paths, w_1 from s to t, and w_2 from t to s. By construction, w_1 must contain a path from s_1 to t_1. Similarly, w_2 must contain a path from s_2 to t_2. Since w_1 and w_2 are vertex-disjoint these new paths must also be vertex-disjoint. Thus, G has the desired vertex-disjoint paths. □

Theorem 2. *Determining if a 2CNF formula has a Var-ROR refutation is* **NP-complete***.*

Proof. As previously stated, we only need to show **NP-hardness**. That will be done by a reduction from C-DPP.

From $G = (V, E)$, s, and t we construct a formula Φ in 2CNF as follows:

1. For each vertex $v_i \in V - \{s, t\}$, create the variable x_i.
2. Create the variable x_0.
3. Let $v_i, v_j \in V - \{s, t\}$.
   (a) If $(s, v_i) \in E$ add the clause $(x_0 \to x_i)$ to Φ.
   (b) If $(t, v_i) \in E$ add the clause $(\neg x_0 \to x_i)$ to Φ.
   (c) If $(v_i, s) \in E$ add the clause $(x_i \to x_0)$ to Φ.
   (d) If $(v_i, t) \in E$ add the clause $(x_i \to \neg x_0)$ to Φ.
   (e) If $(v_i, v_j) \in E$ add the clause $(x_i \to x_j)$ to Φ.

Assume that G has two vertex-disjoint paths,

$$w_1 = s, v_{i_1}, \ldots, v_{i_j}, t \text{ and } w_2 = t, v_{i_{j+1}}, \ldots, v_{i_k}, s.$$

Thus, there exist 2CNF formulas Φ_1 and Φ_2 such that:

$$\Phi_1 = \{(x_0 \to x_{i_1}), (x_{i_1} \to x_{i_2}), \ldots, (x_{i_j} \to \neg x_0)\}$$
$$\Phi_2 = \{(\neg x_0 \to x_{i_{j+1}}), (x_{i_{j+1}} \to x_{i_{j+2}}), \ldots, (x_{i_k} \to x_0)\}.$$

Clearly, $\Phi_1 \vdash_{\overline{Var\text{-}RO\text{-}Res}} (\neg x_0)$ and $\Phi_2 \vdash_{\overline{Var\text{-}RO\text{-}Res}} (x_0)$. Note that x_0 has not been used as a matching variable. Since w_1 and w_2 are vertex-disjoint, we have that

$$\{x_{i_1}, \ldots, x_{i_j}\} \cap \{x_{i_{j+1}}, \ldots, x_{i_k}\} = \emptyset.$$

Thus, $\Phi_1 \cup \Phi_2 \vdash_{\overline{Var\text{-}RO\text{-}Res}} \sqcup$. This means that $\Phi \supseteq \Phi_1 \cup \Phi_2$ is in Var-ROR.

Now assume that Φ is in Var-ROR. Let $\Phi' \subseteq \Phi$ be minimally Var-ROR. We have that Φ' contains clauses with x_0 and $\neg x_0$. Otherwise, the formula would be satisfiable by setting each x_i to **true**.

We proceed by induction on the number of clauses in Φ'.

The shortest formula is $\Phi' = (x_0 \to \neg x_0) \wedge (\neg x_0 \to x_0)$. This Φ' is generated when $(s, t), (t, s) \in E$. These edges form the desired vertex-disjoint paths.

Let y be the variable for which $(y) \wedge (\neg y) \vdash_{\overline{RES}}^{1} \sqcup$ is the last resolution step in $\Phi' \vdash_{\overline{Var\text{-}RO\text{-}Res}} \sqcup$. Thus, Φ', can be divided into two variable-disjoint sets of clauses, Φ'_1 and Φ'_2, such that $\Phi'_1 \vdash_{\overline{Var\text{-}RO\text{-}Res}} (y)$ and $\Phi'_2 \vdash_{\overline{Var\text{-}RO\text{-}Res}} (\neg y)$. Otherwise, a variable would be used twice in $\Phi' \vdash_{\overline{Var\text{-}RO\text{-}Res}} \sqcup$.

Let $(L \vee x_i) \wedge (\neg x_i \vee K) \vdash_{\overline{RES}}^{1} (L \vee K)$ a resolution step in $\Phi'_1 \vdash_{\overline{Var\text{-}RO\text{-}Res}} (y)$ such that $(L \vee x_i) \in \Phi'_1$ and $(\neg x_i \vee K) \in \Phi'_1$. Thus, no clause with x_i occurs in Φ'_2 or Φ_1 (except $(L \vee x_i)$ and $(\neg x_i \vee K)$). Moreover, we see that the formula

$$(\Phi' \setminus \{(L \vee a_i), (\neg a_i \vee K)\}) \cup \{(L \vee K)\}$$

is in Var-ROR. This formula represents the reduced graph where the edges $L \to a_i$ and $a_i \to K$ are replaced with the edge $L \to K$. By the induction hypothesis, there exists a vertex-disjoint cycle in this reduced graph. Thus, a vertex-disjoint cycle exists in G. ☐

.

For arbitrary formulas in CNF, the problem of deciding whether a formula Φ has a sub-formula Φ' such that $\Phi' \in MU(1)$ is known to be **NP-complete**. But it was only known for arbitrary CNF. Based on the Theorems above, we obtain as a corollary that the $MU(1)$ sub-formula problem is **NP-complete** for 2CNF, too. It follows that the problem of deciding whether a formula in 2CNF contains a minimal unsatisfiable formula with deficiency 1 is **NP-complete**.

5 The Complexity of ROR for 2CNF

In this section, we show that the ROR problem for 2CNF formulas is **NP-complete**. It was established in [13] that the Var-ROR problem for 2CNF can be reduced to the ROR problem for 2CNF. Unlike minimally Var-ROR formulas, minimally ROR formulas are not necessarily minimal unsatisfiable. They also can have deficiencies other than 1.

We now prove some properties of minimal unsatisfiable formulas in 2-CNF with one or two unit clauses. It can easily be seen that such formulas contain at most two unit clauses.

Lemma 7. *Let Φ be a minimal unsatisfiable 2CNF formula.*

1. *If Φ contains two unit clauses, then Φ has the form*

$$(L), (\neg L \vee L_1), \ldots, (\neg L_{t-1} \vee L_t), (\neg L_t \vee \neg K), (K)$$

   where $L, L_1, \ldots, L_t, K$ are pairwise distinct.
2. *If Φ contains exactly one unit clause, then Φ has the form*

$$(L), (\neg L \vee L_1), (\neg L_1 \vee L_2), \ldots, (\neg L_t \vee K),$$
$$(\neg K \vee S_1), (\neg S_1 \vee S_2), \ldots, (\neg S_q \vee R),$$
$$(\neg K \vee P_1), (\neg P_1 \vee P_2), \ldots, (\neg P_m \vee \neg R)$$

   where the literals are all pairwise distinct.
3. *If Φ contains at least one unit clause, then Φ has a read-once resolution refutation.*
4. *If Φ is in MU(1), then Φ has an ROR refutation.*

Proof. We prove each part of the lemma separately.

The proofs of part 1 and 2 are straightforward because no minimal unsatisfiable 2CNF formula contains more than two unit clauses.

3. If the formula has two unit clauses, then structure of the formula leads immediately to an ROR refutation.

   If the formula has one unit clause, then we can perform the desired resolution refutation as follows:

   (a) First, we resolve $(\neg K \vee S_1), (\neg S_1 \vee S_2), \ldots, (\neg S_q \vee R)$ to get $(\neg K \vee R)$.

   (b) Then, we resolve $(\neg K \vee P_1), (\neg P_1 \vee P_2), \ldots, (\neg P_m \vee \neg R)$ to get $(\neg K \vee \neg R)$.

(c) Next, we perform the resolution step

$$(\neg K \vee R) \wedge (\neg K \vee \neg R) \vdash_{\overline{RES}}^{1} (\neg K).$$

(d) Finally, the unit clause $(\neg K)$ together with the chain

$$(L), (\neg L \vee L_1), (\neg L_2 \vee L_3), \ldots, (\neg L_t \vee K)$$

resolve to finish the ROR refutation.

4. Every 2CNF formula in MU(1) has a complete disjunctive splitting tree. This guarantees the existence of an ROR refutation [6].   □

Theorem 3. *Let Φ be in 2CNF. Φ is in ROR, if and only if there exists a sub-formula $\Phi' \subseteq \Phi$ for which there exists a variable x and a disjunctive splitting $(F_x, F_{\neg x})$ over x, such that $F_x, F_{\neg x}$ are in MU(1).*

Proof. Suppose, there exists a sub-formula $\Phi' \subseteq \Phi$ with disjunctive splitting $(F_x, F_{\neg x})$, where $F_x, F_{\neg x} \in$ MU(1). We have that F_x and $F_{\neg x}$ each contain at least one unit clause. Now we reconstruct the clauses of F_x and $F_{\neg x}$ by adding the removed literal x (resp. $\neg x$) to the clauses in F_x (resp. $F_{\neg x}$). These new formulas are denoted as F_x^x and $F_{\neg x}^{\neg x}$.

From Lemma 7, every formula in MU(1) with a unit-clause has a read-once resolution refutation. We also have that (x) and $(\neg x)$ do not occur in the splitting formulas. Thus, we get $F_x \vdash_{\overline{RO\text{-}Res}} \sqcup$, $F_{\neg x} \vdash_{\overline{RO\text{-}Res}} \sqcup$, $F_x^x \vdash_{\overline{RO\text{-}Res}} (x)$, and $F_{\neg x}^{\neg x} \vdash_{\overline{RO\text{-}Res}} (\neg x)$. Now we have to guarantee that there is a read-once resolution for Φ. $(F_x, F_{\neg x})$ is disjunctive splitting. Thus, no clause of Φ occurs in both F_x^x and in $F_{\neg x}^{\neg x}$. We can combine the resolutions $F_x^x \vdash_{\overline{RO\text{-}Res}} (x)$ and $F_{\neg x}^{\neg x} \vdash_{\overline{RO\text{-}Res}} (\neg x)$, with the resolution step $(x) \wedge (\neg x) \vdash_{\overline{RES}}^{1} \sqcup$ to yield $\Phi \vdash_{\overline{RO\text{-}Res}} \sqcup$, since $F_x^x, F_{\neg x}^{\neg x} \subseteq \Phi$.

Now suppose that $\Phi \in$ ROR. Without loss of generality, we can also assume that Φ is minimally ROR. We will show that Φ contains the desired splitting. Let $x \wedge \neg x \vdash_{\overline{RES}} \sqcup$ the last resolution step in the read-once resolution refutation. Furthermore, let F_x^x ($F_{\neg x}^{\neg x}$ respectively) be the set of original clauses from Φ used in the derivation of x ($\neg x$ respectively). These sets have no clause in common because together they form a read-once resolution refutation for Φ.

The formulas have the form

$$F_x^x = (x \vee L_1) \wedge \ldots \wedge (x \vee L_t) \wedge \sigma_1 \text{ and } F_{\neg x}^{\neg x} = (\neg x \vee K_1) \wedge \ldots \wedge (\neg x \vee K_r) \wedge \sigma_2,$$

where $\sigma_1 \cap \sigma_2 = \emptyset$.

Thus, we can construct the formulas

$$F_x = (L_1) \wedge \ldots \wedge (L_t) \wedge \sigma_1 \text{ and } F_{\neg x} = (K_1) \wedge \ldots \wedge (K_r) \wedge \sigma_2,$$

where $\sigma_1 \cap \sigma_2 = \emptyset$.

By construction, $F_x \vdash_{\overline{RO\text{-}Res}} \sqcup$ and $F_{\neg x} \vdash_{\overline{RO\text{-}Res}} \sqcup$. Both F_x and $F_{\neg x}$ are minimal unsatisfiable. Otherwise, Φ would not be minimally ROR. This means that, $(F_x, F_{\neg x})$ is a disjunctive splitting. Both F_x and $F_{\neg x}$ contain unit clauses. Thus, by Lemma 1, F_x and $F_{\neg x}$ are MU(1).   □

We will now prove the **NP-completeness** or the ROR problem for 2CNF formulas. Instead of using the vertex-disjoint cycle problem, we will be reducing from the edge-disjoint cycle problem for directed graphs.

Definition 12. *Given a directed graph G and two distinct vertexes s and t, the* **edge-disjoint cycle problem** *(C-DEP) consists of finding a pair of edge-disjoint paths in G, one from s to t and the other from t to s.*

The problem is **NP-complete**. For two pairs of vertexes, the edge-disjoint path problem is **NP-complete** [9]. We can reduce the edge-disjoint path problem to C-DEP the same way we reduced 2-DPP to C-DPP.

Theorem 4. *The ROR problem for $2CNF$ formulas is* **NP-complete**.

Proof. ROR is in **NP** for arbitrary formulas in CNF [5]. Thus, we only need to show **NP-hardness**. That will be done by a reduction from C-DEP.

From $G = (V, E)$, s, and t we construct a formula Φ in 2CNF as follows:

1. For each vertex $v_i \in V - \{s, t\}$, create the variable x_i.
2. Create the variable x_0.
3. Let $v_i, v_j \in V - \{s, t\}$.
   (a) If $(s, v_i) \in E$ add the clause $(x_0 \to x_i)$ to Φ.
   (b) If $(t, v_i) \in E$ add the clause $(\neg x_0 \to x_i)$ to Φ.
   (c) If $(v_i, s) \in E$ add the clause $(x_i \to x_0)$ to Φ.
   (d) If $(v_i, t) \in E$ add the clause $(x_i \to \neg x_0)$ to Φ.
   (e) If $(v_i, v_j) \in E$ add the clause $(x_i \to x_j)$ to Φ.

Assume that G has two edge-disjoint paths,

$$w_1 = s, v_{i_1}, \ldots, v_{i_j}, t \text{ and } w_2 = t, v_{i_{j+1}}, \ldots, v_{i_k}, s.$$

Thus, there exist 2CNF formulas Φ_1 and Φ_2 such that:

$$\Phi_1 = \{(x_0 \to x_{i_1}), (x_{i_1} \to x_{i_2}), \ldots, (x_{i_j} \to \neg x_0)\}$$
$$\Phi_2 = \{(\neg x_0 \to x_{i_{j+1}}), (x_{i_{j+1}} \to x_{i_{j+2}}), \ldots, (x_{i_k} \to x_0)\}.$$

Obviously, $\Phi_1 \vdash_{RO\text{-}Res} (\neg x_0)$ and $\Phi_2 \vdash_{RO\text{-}Res} (x_0)$. Note that x_0 has not been used as a matching variable. Since w_1 and w_2 are edge-disjoint, we have that $\Phi_1 \cap \Phi_2 = \emptyset$. Thus, $\Phi_1 \cup \Phi_2 \vdash_{RO\text{-}Res} \sqcup$. This means that $\Phi \supseteq \Phi_1 \cup \Phi_2$ is in ROR.

Now assume that Φ is in ROR.

Let $\Phi' \subseteq \Phi$ be minimally ROR. We have that Φ' contains clauses with x_0 and $\neg x_0$. Otherwise, the formula would be satisfiable by setting each x_i to **true**.

We proceed by induction on the number of clauses in Φ'.

The shortest formula is $\Phi' = (x_0 \to \neg x_0) \wedge (\neg x_0 \to x_0)$. This Φ' is generated when $(s, t), (t, s) \in E$. These edges form the desired edge-disjoint paths.

Let $(L \to K) \wedge (K \to R) \vdash_{RES}^{1} (L \vee R)$ be a resolution step such that $(L \to K) \in \Phi'$ and $(K \to R) \in \Phi'$. Note that $(L \to R) \notin \Phi'$. Otherwise, Φ' would not be minimally ROR.

In a read-once refutation, we remove the parent clauses from Φ and add the resolvent $(L \to R)$. This new formula has a read-once resolution refutation and can be considered as obtained by a reduced graph without the edges $L \to K, K \to R$ but with the edge $L \to R$. By the induction hypothesis, this new graph contains the desired edge-disjoint cycle. If we replace the edge $L \to R$ in this cycle with $L \to K$ and $K \to R$, then we construct the desired edge-disjoint cycle in G. □

By Lemma 4, the problem of determining if an MU-formula in 2CNF has a disjunctive splitting whose splitting formulas are in MU(1) can be decided in polynomial time. Hence, based on Theorem 3, the ROR problem for minimal unsatisfiable formulas is solvable in polynomial time.

References

1. Aharoni, R., Linial, N.: Minimal non-two-colorable hypergraphs and minimal unsatisfiable formulas. J. Combin. Theor. Ser. A **43**, 196–204 (1986)
2. Davidov, G., Davydova, I., Kleine Büning, H.: An efficient algorithm for the minimal unsatisability problem for a subclass of CNF. Ann. Math. Artif. Intell. **23**, 229–245 (1998)
3. Fleischer, H., Kullmann, O., Szeider, S.: Polynomial-time recognition of minimal unsatisfiable formulas with fixed clause-variable difference. Theor. Comput. Sci. **289**(1), 503–516 (2002)
4. Fortune, S., Hopcroft, J.E., Wyllie, J.: The directed subgraph homeomorphism problem. Theor. Comput. Sci. **10**(2), 111–121 (1980)
5. Iwama, K., Miyano, E.: Intractability of read-once resolution. In: Proceedings of the 10th Annual Conference on Structure in Complexity Theory, pp. 29–36. IEEE (1995)
6. Kleine Büning, H., Zhao, X.: The complexity of read-once resolution. Ann. Math. Artif. Intell. **36**(4), 419–435 (2002)
7. Kleine Büning, H., Zhao, X.: On the structure of some classes of minimal unsatisfiable formulas. Discrete Appl. Math. **130**, 185–207 (2003)
8. Papadimitriou, C.H., Wolfe, D.: The complexity of facets resolved. J. Comput. Syst. Sci. **37**, 2–13 (1988)
9. Even, S., Itai, A., Shamir, A.: On the complexity of timetable and multicommodity flow problems. SIAM J. Comput. **5**, 691–703 (1976)
10. Beame, P., Pitassi, T.: Propositional proof complexity: past, present, future. Bull. EATCS **65**, 66–89 (1998)
11. Robinson, J.: A machine-oriented logic based on the resolution principle. J. ACM **12**(1), 23–41 (1965)
12. Bachmair, L., Ganzinger, H., McAllester, D., Lynch, C.: Bachmair, L., Ganzinger, H., McAllester, D., Lynch, C.: Chapter 2 - Resolution theorem proving. Handbook of Automated Reasoning, Issue. 1, pp. 19–99 (2001)
13. Szeider, S.: NP-Completeness of refutability by literal-once resolution. In: Automated Reasoning: First International Joint Conference, pp. 168–181 (2001)
14. Urquhart, A.: The complexity of propositional proofs. Bull. Symbolic Logic **1**(4), 425–467 (1995)

Vector Ambiguity and Freeness Problems in SL(2, ℤ)

Sang-Ki Ko$^{(\boxtimes)}$ and Igor Potapov

Department of Computer Science, University of Liverpool,
Ashton Street, Liverpool L69 3BX, UK
{sangkiko,potapov}@liverpool.ac.uk

Abstract. We study the vector ambiguity problem and the vector freeness problem in SL(2, ℤ). Given a finitely generated $n \times n$ matrix semigroup S and an n-dimensional vector $\mathbf{x}$, the vector ambiguity problem is to decide whether for every target vector $\mathbf{y} = M\mathbf{x}$, where $M \in S$, M is unique. We also consider the vector freeness problem which is to show that every matrix M which is transforming $\mathbf{x}$ to $M\mathbf{x}$ has a unique factorization with respect to the generator of S. We show that both problems are NP-complete in SL(2, ℤ), which is the set of 2×2 integer matrices with determinant 1. Moreover, we generalize the vector ambiguity problem and extend to the finite and k-vector ambiguity problems where we consider the degree of vector ambiguity of matrix semigroups.

Keywords: Matrix semigroup · SL(2, ℤ) · Vector ambiguity · Vector freeness · Decidability · NP-completeness

1 Introduction

Many computational problems for matrix semigroups and groups are proven to be undecidable starting from dimension three or four. On the other hand, a lot of questions for matrix semigroups in dimension two are open including the membership, vector reachability, scalar reachability problems and various problems on freeness. In this paper, we show decidaand reveal complexity of several questions for matrix semigroups in SL(2, ℤ), which is called the special linear group. The special linear group SL(2, ℤ) has been extensively exploited in hyperbolic geometry [10,13,30], dynamical systems [23], Lorenz/modular knots [20], braid groups [24], high energy physics [28], M/string theories [14], music theory [22], and so on.

Let $S = \langle G \rangle$ be a matrix semigroup finitely generated by a generating set G. The *membership problem* is to decide whether or not a given matrix M belongs to the matrix semigroup S. By restricting M to be the identity or zero matrix, we call the problems the *identity problem* or *mortality problem*, respectively.

This research was supported by EPSRC grant EP/M00077X/1.

T.V. Gopal et al. (Eds.): TAMC 2017, LNCS 10185, pp. 373–388, 2017.
DOI: 10.1007/978-3-319-55911-7_27

The *vector reachability problem*, which is a parameterized version of the membership problem, can be defined as follows: Given a finitely generated matrix semigroup S of $n \times n$ matrices and two vectors $\mathbf{x}, \mathbf{y}$ in dimension n, the vector reachability problem decides whether or not there exists a matrix M in S such that $M\mathbf{x} = \mathbf{y}$.

Due to its effective symbolic representation of matrices in $SL(2, \mathbb{Z})$, many decidability and complexity results have been established. For instance, it has been shown that the mortality, identity and vector reachability problems were at least NP-hard for $SL(2, \mathbb{Z})$ in [2,6], but for the finitely generated subgroups of the modular group, the membership was shown to be decidable in polynomial time by Gurevich and Schupp [15]. Choffrut and Karhumäki proved that the membership problem is decidable in $SL(2, \mathbb{Z})$, and the identity problem is decidable in $\mathbb{Z}^{2 \times 2}$ [12]. Moreover, Bell et al. [3] proved that the identity problem in $SL(2, \mathbb{Z})$ is NP-complete by developing a new effective technique to operate with compressed word representations of matrices. The decidability of the membership problem for matrix semigroups in dimension two over integers, rationals or complex numbers is an open question and the only known decidability result that is beyond $SL(2, \mathbb{Z})$ is the first algorithm for the membership problem for non-singular 2×2 integer matrices shown in [26].

Another fundamental problem for matrix semigroups is the *freeness problem*, where we want to know whether every matrix in the matrix semigroup has a unique factorization over G. Mandel and Simon [21] showed that the freeness problem is decidable in polynomial time for matrix semigroups with a single generator for any dimension over rational numbers.[1] Klarner et al. [16] proved that the freeness problem in dimension three over natural numbers is undecidable. Along with the membership problem, the freeness problem in dimension two is also an open problem for a long time [8,9] except certain special cases. For example Charlier and Honkala [11] showed that the freeness problem is decidable for upper-triangular matrices in dimension two over rationals when the products are restricted to certain bounded languages. Bell and Potapov [5] showed that the freeness problem is undecidable in dimension two for matrices over quaternions. Recently, the freeness problem in $SL(2, \mathbb{Z})$ is proven to be NP-complete where NP-hardness is shown in [17] by the reduction from the *equal subset sum problem* (ESSP) [29] and the NP algorithm is given in [3].

In case of vector (scalar) reachability, the question about uniqueness of transformations with respect to the given initial vector can be related to two different interpretations: the vector (scalar) ambiguity and vector (scalar) freeness. Let S be a matrix semigroup of $n \times n$ matrices and $\mathbf{x}$ be an n-dimensional vector. Bell and Potapov [4] showed that the problem of deciding whether S and $\mathbf{x}$ generate a non-repetitive set of vectors—the *vector ambiguity problem*—is undecidable in dimension four over integers and in dimension three over rationals. They used the fact that the problem of determining if a two-counter machine

---

[1] The freeness problem for matrix semigroups with a single generator is the complementary problem of the *matrix torsion problem* which asks whether there exist two integers $p, q \geq 1$ such that $M^p = M^{q+p}$.

has a periodic configuration is undecidable. Recently, the *scalar ambiguity and freeness problems* have been introduced [1]. In the scalar ambiguity and freeness problems, given a matrix semigroup S and two vectors $\mathbf{x}, \mathbf{y}$, we examine the set $\{\mathbf{x}^T M \mathbf{y} \mid M \in S\}$ of scalars and check whether there exists a unique matrix or a unique factorizations of matrices for each scalar. In 2016, Bell et al. [1] showed that both problems are undecidable over bounded languages.

In this paper, we study the *vector ambiguity problem* for matrix semigroups in SL(2, $\mathbb{Z}$) and show that the problem is decidable. Moreover, we prove that the vector ambiguity problem in SL(2, $\mathbb{Z}$) is NP-complete. We prove the NP-hardness of the vector ambiguity problem in SL(2, $\mathbb{Z}$) by the reduction from the *subset sum problem* and the membership in NP by the recent result that the identity problem in SL(2, $\mathbb{Z}$) is in NP [3]. We also examine the *vector freeness problem* in which we analyze the unique factorization of matrices leading to the same vector and show that the problem is also NP-complete in SL(2, $\mathbb{Z}$). Moreover, we generalize the vector ambiguity problem and extend to the finite and k-vector ambiguity problems where we consider the degree of vector ambiguity of matrix semigroups. In the table below, we are summarizing the results of this paper and position them in the context of currently known results in this area. Bold entries represent new results and dash line '—' means that ambiguity problems for matrix semigroups cannot be defined.

Problem	Domain	Matrix reachability	Vector reachability
Non-freeness	SL(2, $\mathbb{Z}$)	NP-complete [17]	**NP-complete**
	$\mathbb{N}^{3 \times 3}$	Undecidable [16]	**Undecidable**
k-non-freeness	SL(2, $\mathbb{Z}$)	EXPSPACE [17]	
	$\mathbb{N}^{3 \times 3}$	Undecidable [16]	**Undecidable**
Finite non-freeness	SL(2, $\mathbb{Z}$)	EXPSPACE [17]	
	$\mathbb{N}^{3 \times 3}$		**Undecidable**
Ambiguity	SL(2, $\mathbb{Z}$)	—	**NP-complete**
	$\mathbb{Z}^{4 \times 4}$	—	Undecidable [4]
k-ambiguity	SL(2, $\mathbb{Z}$)	—	**EXPSPACE**
	$\mathbb{Z}^{4 \times 4}$	—	Undecidable [4]
Finite ambiguity	SL(2, $\mathbb{Z}$)	—	**EXPSPACE**
	$\mathbb{Z}^{4 \times 4}$	—	Undecidable [4]

2   Preliminaries

In this section we formulate several problems, provide important definitions and notation as well as several technical lemmas used throughout the paper.

Basic definitions. A *semigroup* is a set equipped with an associative binary operation. Let S be a semigroup and X be a subset of S. Then, X is a *code* if

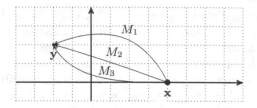

Fig. 1. Geometric interpretation of the vector ambiguity problem. If $M_1, M_2, M_3 \in S$, then the matrix semigroup S is ambiguous with respect to the vector $\mathbf{x}$ as there are already three matrices transforming $\mathbf{x}$ into $\mathbf{y}$.

and only if every element of S has a unique factorization over X. A semigroup S is *free* if there exists a subset $X \subseteq S$ which is a code and $S = X^+$.

Given an alphabet Σ, a word w is an element of Σ^*. For a letter $a \in \Sigma$, we denote by $\bar{a}$ the inverse letter of a such that $a\bar{a} = \varepsilon$ where ε is the empty word.

A *nondeterministic finite automaton* (NFA) is a tuple $A = (\Sigma, Q, \delta, q_0, F)$ where Σ is the input alphabet, Q is the finite set of states, $\delta \colon Q \times \Sigma \to 2^Q$ is the multivalued transition function, $q_0 \in Q$ is the initial state and $F \subseteq Q$ is the set of final states. In the usual way δ is extended as a function $Q \times \Sigma^* \to 2^Q$ and the language accepted by A is $L(A) = \{w \in \Sigma^* \mid \delta(q_0, w) \cap F \neq \emptyset\}$. The automaton A is a *deterministic finite automaton* (DFA) if δ is a single valued function $Q \times \Sigma \to Q$. It is well known that the deterministic and nondeterministic finite automata recognize the class of *regular languages* [27].

Vector ambiguity problem and freeness problems. Let S be an $n \times n$ matrix semigroup finitely generated by a set $G = \{M_1, M_2, \ldots, M_k\}$ of matrices (a generator) and $\mathbf{x}$ be an n-dimensional vector. Then, we assume that $M \cdot \mathbf{x} = \mathbf{y}$ for a matrix M in S. We can say that a vector $\mathbf{y}$ is reachable from $\mathbf{x}$ by S. If there is a unique matrix M in S that transforms $\mathbf{x}$ into $\mathbf{y}$, we say that $\mathbf{y}$ is *unambiguous* with respect to S and $\mathbf{x}$. Note that $\mathbf{y}$ is *ambiguous* with respect to S and $\mathbf{x}$ otherwise. Denote the set of reachable vectors from $\mathbf{x}$ by multiplying the elements of the matrix semigroup S on the left-hand side by V. Namely, $V = \{\mathbf{y} \mid \mathbf{y} = M\mathbf{x}, \ M \in S\}$. If every vector in V is unambiguous with respect to S and $\mathbf{x}$, then we say that the matrix semigroup S is unambiguous with respect to $\mathbf{x}$. In other words, if there is a unique matrix $M\mathbf{x} = \mathbf{y}$ for every target vector $\mathbf{y}$, then we say that the matrix semigroup S is *unambiguous* with respect to $\mathbf{x}$ (Fig. 1).

Similarly, we say that the matrix semigroup S is *free* with respect to $\mathbf{x}$ if every matrix M which transforms $\mathbf{x}$ into $M\mathbf{x}$ has a unique decomposition with respect to the generator G. Otherwise, S is said to be *non-free* with respect to $\mathbf{x}$. The problem of deciding whether or not a given matrix semigroup S is free (respectively, non-free) with respect to a given initial vector $\mathbf{x}$ is called the *vector freeness (respectively, non-freeness) problem*.

Here we consider the following problems for matrix semigroups in $SL(2, \mathbb{Z})$:

- The vector ambiguity problem: given a matrix semigroup S of $n \times n$ matrices and an n-dimensional vector $\mathbf{x}$, is S ambiguous with respect to $\mathbf{x}$?
- The vector non-freeness problem: given a semigroup S of $n \times n$ matrices and an n-dimensional vector $\mathbf{x}$, is S non-free with respect to $\mathbf{x}$?

Before tackling the problems, we establish relationships between the proposed problems and matrix semigroup freeness problem.

Lemma 1. *Given a semigroup S of $n \times n$ matrices and an n-dimensional vector $\mathbf{x}$, the following statements hold:*

1. *if S is free with respect to $\mathbf{x}$, then S is unambiguous with respect to $\mathbf{x}$,*
2. *if S is free with respect to $\mathbf{x}$, then S is free, and*
3. *if S is free and unambiguous with respect to $\mathbf{x}$, then S is free w.r.t. $\mathbf{x}$.*

(See the archive version for proof)

Group alphabet encodings. Let us introduce several technical lemmas that will be used in encodings for NP-hardness results. Our original encodings require the use of group alphabet and the following lemmas for showing the transformation from an arbitrary group alphabet into a binary group alphabet and later into matrix form that is computable in polynomial time.

It is well-known that $\{cd^i c^{-1} \mid i \geq 1\}$ freely generates a free subgroup of the free group $\langle c, d \rangle$ [7] and that the matrices $\begin{pmatrix} 1 & 2 \\ 0 & 1 \end{pmatrix}$ and $\begin{pmatrix} 1 & 0 \\ 2 & 1 \end{pmatrix}$ freely generates a free subgroup of $SL(2, \mathbb{Z})$ [19].

Let $\Sigma = \{z_1, z_2, \ldots, z_l\}$ be a group alphabet and $\Sigma_2 = \{c, d, \overline{c}, \overline{d}\}$ be a binary group alphabet. Define the mapping $\alpha : \Sigma \rightarrow \Sigma_2^*$ by: $\alpha(z_i) = c^i d\overline{c}^i, \alpha(\overline{z_i}) = c^i d\overline{c}^i$, where $1 \leq i \leq l$. It is easy to see that α is a monomorphism. Note that α can be extended to domain Σ^* in the usual way. We also define a monomorphism $f : \Sigma_2^* \rightarrow \mathbb{Z}^{2 \times 2}$ as follows:

$$f(c) = \begin{pmatrix} 1 & 2 \\ 0 & 1 \end{pmatrix}, \quad f(\overline{c}) = \begin{pmatrix} 1 & -2 \\ 0 & 1 \end{pmatrix}, \quad f(d) = \begin{pmatrix} 1 & 0 \\ 2 & 1 \end{pmatrix}, \quad f(\overline{d}) = \begin{pmatrix} 1 & 0 \\ -2 & 1 \end{pmatrix}.$$

The composition of two monomorphisms α and f gives us the following lemma that ensures that encoding the *subset sum problem* (SSP) and the *equal subset sum problem* (ESSP) [29] instances into matrix semigroups can be done in polynomial time.

Lemma 2 (Bell and Potapov [4]). *Let $z_j \in \Sigma$. For any $i \in \mathbb{N}$, $f(\alpha(z_j^i)) = f((c^j d\overline{c}^j)^i) = \begin{pmatrix} 1 + 4ij & -8ij^2 \\ 2i & 1 - 4ij \end{pmatrix}.$*

Lemma 3. *Let w and w' be any two distinct words in Σ^*. Then, for any non-zero integer t: $f(\alpha(w)) \cdot \begin{pmatrix} 1 & 1 \\ 0 & 1 \end{pmatrix}^t \neq f(\alpha(w')).$*

Symbolic representation of matrices from $\mathrm{SL}(2,\mathbb{Z})$. It is known that $\mathrm{SL}(2,\mathbb{Z})$ is generated by two matrices $\mathbf{S} = \begin{pmatrix} 0 & -1 \\ 1 & 0 \end{pmatrix}$ and $\mathbf{R} = \begin{pmatrix} 0 & -1 \\ 1 & 1 \end{pmatrix}$, which have respective orders 4 and 6. This implies that every matrix in $\mathrm{SL}(2,\mathbb{Z})$ is a product of $\mathbf{S}$ and $\mathbf{R}$. Since $\mathbf{S}^2 = \mathbf{R}^3 = -\mathbf{I}$, every matrix in $\mathrm{SL}(2,\mathbb{Z})$ can be uniquely brought to the following form:

$$(-\mathbf{I})^{i_0}\mathbf{R}^{i_1}\mathbf{SR}^{i_2}\mathbf{S}\cdots\mathbf{SR}^{i_{n-1}}\mathbf{SR}^{i_n}, \tag{1}$$

where $i_0 \in \{0,1\}$, $i_1, i_n \in \{0,1,2\}$, and $i_j \neq 0 \mod 3$ for $1 < j < n$.

Let $\Sigma = \{s, r\}$ be a binary alphabet. We define a mapping $\varphi : \Sigma \to \mathrm{SL}(2,\mathbb{Z})$ as follows: $\varphi(s) = \mathbf{S}$ and $\varphi(r) = \mathbf{R}$. Naturally, we can extend the mapping φ to the morphism $\varphi : \Sigma^* \to \mathrm{SL}(2,\mathbb{Z})$. Let $M \in \mathrm{SL}(2,\mathbb{Z})$ be a matrix of the form given in Eq. (1). Then, we say that the following word is the *canonical word* for M:

$$(ss)^{i_0} r^{i_1} s r^{i_2} s \cdots s r^{i_{n-1}} s r^{i_n}.$$

It is easy to see that every matrix in $\mathrm{SL}(2,\mathbb{Z})$ has a unique canonical word. We also call a word $w \in \Sigma^*$ *reduced* if there is no occurrence of substrings ss or rrr in w. Then, we have the following fact. For every matrix $M \in \mathrm{SL}(2,\mathbb{Z})$, there exists a unique reduced word $w \in \Sigma^*$ such that either $M = \varphi(w)$ or $M = -\varphi(w)$ [19].

Next we consider a language which contains all (s,r)-representations of a particular matrix in $\mathrm{SL}(2,\mathbb{Z})$.

Lemma 4. *Let M be a matrix in $SL(2,\mathbb{Z})$. Then, there exists a context-free language over $\Sigma = \{s, r\}$ which contains all unreduced representations $w \in \Sigma^*$ such that $\varphi(w) = M$. (See the archive version for proof)*

3   Vector Ambiguity and Freeness Problems in $\mathrm{SL}(2,\mathbb{Z})$

In this section, we prove that the vector ambiguity and freeness problems are NP-complete. Note that the vector ambiguity problem is undecidable over $\mathbb{Z}^{4\times 4}$ and over $\mathbb{Q}^{3\times 3}$ [4]. We later show that the vector freeness problem is undecidable over $\mathbb{N}^{3\times 3}$.

It was shown in [25] that if there is a matrix M from $\mathrm{SL}(2,\mathbb{Z})$ satisfying $M\mathbf{x} = \mathbf{y}$, where $\mathbf{x} = [x_1, x_2]^T$ and $\mathbf{y} = [y_1, y_2]^T$ are vectors from $\mathbb{Z} \times \mathbb{Z}$, then this equation either does not have a solution or all its solutions are given by the following formula

$$M = B\begin{pmatrix} 1 & k \\ 0 & 1 \end{pmatrix}^t C = B\begin{pmatrix} 1 & 1 \\ 0 & 1 \end{pmatrix}^{tk} C, \tag{2}$$

where $t \in \mathbb{Z}$ and B, C are matrices from $\mathrm{SL}(2,\mathbb{Z})$. Let us denote the matrix $\begin{pmatrix} 1 & 1 \\ 0 & 1 \end{pmatrix}$ by T from now on Moreover, if $M = BT^{kt}C$ and $M\mathbf{x} = \mathbf{y}$, then $\mathbf{y} = C\mathbf{x} = [d,0]^T$ and $\mathbf{v} = T^{kt}\mathbf{y} = [d,0]^T$, where $|d| = \gcd(x_1, x_2) = \gcd(y_1, y_2)$.

First, we state the following property of matrices in $SL(2, \mathbb{Z})$ and later exploit the property to establish the main results of the paper.

Lemma 5. *Let* $\mathbf{x} = [x_1, x_2]^T$ *be a vector from* $\mathbb{Z}^2$ *and* C *be a matrix from* $SL(2, \mathbb{Z})$ *such that* $C\mathbf{x} = [d, 0]^T$, *where* $|d| = \gcd(x_1, x_2)$. *Then, a matrix semi-group* S *of* 2×2 *integral matrices from* $SL(2, \mathbb{Z})$ *is unambiguous with respect to* $\mathbf{x}$ *if and only if for any matrix* B *in* $SL(2, \mathbb{Z})$, *there is at most one matrix* M *in* S *which is of the form of* $M = BT^tC$, *where* $t \in \mathbb{Z}$.

Proof. First we prove that if S is unambiguous with respect to $\mathbf{x}$, then there is at most one matrix $M \in S$ of the form $M = BT^tC$ for any $B \in S$, where $t \in \mathbb{Z}$.

Assume that S is ambiguous with respect to $\mathbf{x}$ and we have two different matrices $M = BT^tC$ and $M' = BT^{t'}C$, where $t \neq t'$. Let us denote the vector $[d, 0]^T$ by $\mathbf{d}$ for notational convenience. Since $C\mathbf{x} = \mathbf{d}$, $BT^t\mathbf{d}$ should not be equal to $BT^{t'}\mathbf{d}$ by the assumption. However, $BT^t\mathbf{d} = BT^{t'}\mathbf{d}$ always holds because $T\mathbf{d} = \mathbf{d}$. Therefore, this contradicts our assumption.

Let us consider the opposite direction. If there is a unique matrix $M \in S$ of form BT^tC, then S is unambiguous with respect to $\mathbf{x}$. Suppose that S is ambiguous with respect to $\mathbf{x}$. This implies that there are two matrices M and M' in S such that $M\mathbf{x} = M'\mathbf{x}$. From Eq. (2), we see that both M and M' can be represented in the form of BT^tC for some integer t.

Since $M \neq M'$, $M = BT^tC$ and $M' = BT^{t'}C$ such that $t \neq t'$, which contradicts our assumption. Therefore, we arrive at a contradiction and conclude the proof. □

Theorem 6. *The vector ambiguity problem for finitely generated matrix semi-groups in* $SL(2, \mathbb{Z})$ *is in NP.*

Proof. Suppose that we are given n matrices $M_1, M_2, \ldots, M_n \in SL(2, \mathbb{Z})$ as generators of the semigroup S. Namely, $S = \langle M_1, M_2, \ldots, M_n \rangle$. Let $w_1, w_2, \ldots, w_n \in \Sigma^*$ be words encoding the generators, such that $\varphi(w_i) = M_i$ for $1 \leq i \leq n$. Then, we can define a regular language L corresponding to S over $\Sigma = \{s, r\}$ as $L = \{w_1, w_2, \ldots, w_n\}^+$.

Recall that every matrix M that transforms a vector $\mathbf{x}$ into a vector $\mathbf{y}$ can be represented in the form of $M = BT^tC$ where $t \in \mathbb{Z}$ and B, C are matrices from $SL(2, \mathbb{Z})$. Moreover, we can compute two matrices B and C in polynomial time [25]. From Lemma 5, we can check whether or not S is ambiguous with respect to $\mathbf{x}$ by checking the existence of two different matrices in S which can be represented as BT^tC with different exponents for t.

We first compute a unique matrix C that transforms a given vector $\mathbf{x} = [x_1, x_2]^T$ into $[\gcd(x_1, x_2), 0]^T$. Then, take a inverse matrix C^{-1} of C and encode the matrix with a word w_C, namely, $\varphi(w_C) = C^{-1}$. Now we let $L' = L \cdot \{w_C\}$.

Then, $\varphi(L') = \{MC^{-1} \mid M \in S\}$. Moreover, we can obtain the following statement for $\varphi(L')$: $\varphi(L')$ has two matrices M and M' such that $M' = MT^t$ for some non-zero integer t if and only if $\varphi(L)$ has two matrices of form BT^tC with different exponents t.

Therefore, now it suffices to show that we can decide whether or not $\varphi(L')$ has two different matrices M and M' such that $M' = MT^t$. Because $\varphi(s^3 r) = T$ and $\varphi(r^5 s) = T^{-1}$, the following inequivalence implies that there are no such two matrices in S:

$$\varphi(L') \cap \varphi(L' \cdot (\{s^3 r\}^+ \cup \{r^5 s\}^+)) \neq \emptyset. \tag{3}$$

Note that the unary operator $+$ is called the Kleene plus, and for a set S, the Kleene plus on S, S^+, equals the concatenation of S with the Kleene plus on S, namely $S^+ = SS^*$. Since we know that there is an algorithm that decides whether or not the intersection of two regular subsets of $SL(2, \mathbb{Z})$ is empty [25], the vector ambiguity problem is decidable.

Here we go one step further to show that the vector ambiguity problem is in NP. We use the fact that the inequivalence given in Eq. (3) can be brought to the following form of inequivalence:

$$(M_1 + \cdots + M_n)^* C^{-1} \cap (M_1 + \cdots + M_n)^* C^{-1} (T^+ + (T^{-1})^+) \neq \emptyset.$$

This implies that the following regular subset of $SL(2, \mathbb{Z})$ contains the identity matrix: $(M_1 + \cdots + M_n)^* C^{-1} (T^+ + (T^{-1})^+) C (M_1^{-1} + \cdots + M_n^{-1})^* = I$.

Now the vector ambiguity problem reduces to the problem of determining whether the identity matrix is in a regular expression over matrices in $SL(2, \mathbb{Z})$, which is already proven to be in NP [3]. Therefore, we prove that the vector ambiguity problem for matrix semigroups in $SL(2, \mathbb{Z})$ is also in NP.   □

Theorem 7. *The vector ambiguity problem for finitely generated matrix semigroups in $SL(2, \mathbb{Z})$ is NP-complete. (See the archive version for full proof)*

Proof. The fact that the vector ambiguity problem in $SL(2, \mathbb{Z})$ is in NP is shown in Theorem 6. Now we show that it is NP-hard by using an encoding of the *subset sum problem* (SSP) into a set of two-dimensional integral matrices. The SSP is, given a set $U = \{s_1, s_2, \ldots, s_k\}$ of k integers, to decide whether or not there exists a subset $U' \subseteq U$ whose elements sum up to the given integer x. Namely, $\sum_{s \in U'} s = x$.

Define an alphabet $\Sigma = \{0, 1, \ldots, k-1, \ldots, \overline{1}, \overline{2}, \ldots, \overline{k}, a\}$. We define a set W of words which encodes the SSP instance as follows:

$$W = \{i \cdot a^{i+1} \cdot \overline{(i+1)}, \ i \cdot \varepsilon \cdot \overline{(i+1)}, \ 0 \cdot a^x \cdot \overline{k} \cdot \sigma \mid 0 \le i \le k-1\} \subseteq \{\Sigma \cup \{\sigma\}\}^*.$$

We define 'border letters' as letters from $\Sigma \setminus \{a, \overline{a}\}$ and the inner border letters of a word as all border letters excluding the first and last. We call a word a 'partial cycle' if all inner border letters in that word are inverse to a consecutive inner border letter. Note that any partial cycle $u \in W^+$ is of one of the following forms (i) $i \cdot a^m \cdot \overline{j}$ or (ii) $0 \cdot a^x \cdot \overline{k} \cdot \sigma$, where $i < j$ and m is any integer we can get as a subset sum of integers from s_{i+1} to s_j.

We introduce an additional letter σ which actually encodes the word $c^{2|\Sigma|}$, namely, $\alpha(\sigma) = c^{2|\Sigma|}$ and $f(\alpha(\sigma)) = T^{4|\Sigma|}$. We note that the introduction of the additional letter σ preserves the injectivity of α. For example, any word $w \in \Sigma^*$ of length l has the following image under the mapping α:

$$\alpha(w) = c^{i_1} d' c^{i_2} d' c^{i_3} d' \ldots d' c^{i_{l-1}} d' c^{i_l} d' c^{i_{l+1}},$$

where $-|\Sigma| + 1 \leq i_j \leq |\Sigma|$ for $1 \leq j \leq l+1$ and $d' \in \{d, \overline{d}\}$. Now we consider a word $w' \subset \{\Sigma \cup \{\sigma\}\}^*$. Then, the image of the word under α is

$$\alpha(w') = c^{i_1} d' c^{i_2} d' c^{i_3} d' \ldots d' c^{i_{l-1}} d' c^{i_l} d' c^{i_{l+1}}.$$

If any i_j for $1 \leq j \leq l+1$ is in a different range, for example, $[(2n-1)|\Sigma| + 1, 2n|\Sigma|]$, then we can immediately see that the substring σ^n is used.

Then, we prove that there is a solution to the SSP instance if and only if the matrix semigroup generated by a finite set of matrices corresponding to words in W is ambiguous with respect to a vector $\mathbf{x} = [1, 0]^T$. The full proof can be found in the archive version.   □

Now we consider the vector non-freeness problem in SL(2, $\mathbb{Z}$).

Theorem 8. *The vector non-freeness problem for finitely generated matrix semigroups in SL(2, $\mathbb{Z}$) is in NP.*

Proof. Let $S = \langle M_1, M_2, \ldots, M_k \rangle$ be a matrix semigroup and $\mathbf{x}$ be a vector. Let $V = \{\mathbf{v} \mid \mathbf{v} = M\mathbf{x}, M \in S\}$ be a set of target vectors transformed by a matrix in S from $\mathbf{x}$. Then, S is free with respect to $\mathbf{x}$ if there is a unique decomposition of $M \in S$ for every vector $\mathbf{v}$ such that $M\mathbf{x} = \mathbf{v}$.

Therefore, we can check whether or not S is free with respect to $\mathbf{x}$ by checking the existence of two factorizations of matrices that transform $\mathbf{x}$ into a vector in V. In other words, S is not free with respect to $\mathbf{x}$ if we find the following equivalence:

$$M_{s_1} M_{s_2} \ldots M_{s_n} \mathbf{x} = M_{p_1} M_{p_2} \ldots M_{p_m} \mathbf{x},$$

where $s_i, p_j \in [1, k]$ for $1 \leq i \leq n$ and $1 \leq j \leq m$ and $s_i \neq p_i$ for some i. By Eq. (2), we can represent these factorizations in the following way:

$$M_{s_1} M_{s_2} \ldots M_{s_n} = BT^i C \quad \text{and} \quad M_{p_1} M_{p_2} \ldots M_{p_m} = BT^j C,$$

where $i, j \in \mathbb{Z}$ and $B, C \in$ SL(2, $\mathbb{Z}$). Since C is in SL(2, $\mathbb{Z}$), we can multiply the inverse of C to the right and obtain the following equation:

$$M_{s_1} M_{s_2} \ldots M_{s_n} C^{-1} = M_{p_1} M_{p_2} \ldots M_{p_m} C^{-1} T^{i-j}.$$

Without loss of generality, we assume that $M_{s_1} \neq M_{p_1}$. Now we take the inverse of the right-hand side of the equation.

$$(M_{p_1} M_{p_2} \ldots M_{p_m} C^{-1} T^{i-j})^{-1} M_{s_1} M_{s_2} \ldots M_{s_n} C^{-1} = I$$
$$T^{j-i} C M_{p_m}^{-1} \ldots M_{p_2}^{-1} M_{p_1}^{-1} M_{s_1} M_{s_2} \ldots M_{s_n} C^{-1} = I \qquad (4)$$

From the above equation, we see that there exists such a multiplication sequence leading to the identity matrix if and only if the matrix semigroup S is not free with respect to the given vector $\mathbf{x}$.

Since the membership problem of a rational subset of matrices in $SL(2, \mathbb{Z})$ is known to be decidable [12], the vector freeness problem is also decidable, but in exponential space due to translations of matrices in $SL(2, \mathbb{Z})$ into words over a binary alphabet $\{s, r\}$. Recently, Bell et al. proved that the identity problem for matrix semigroups in $SL(2, \mathbb{Z})$ is NP-complete [3]. They also showed that the problem of deciding whether the identity matrix is in S, where S is an arbitrary regular subset of $SL(2, \mathbb{Z})$, is in NP. Since we decide whether a matrix semigroup S in $SL(2, \mathbb{Z})$ is non-free by checking whether the identity matrix exists in an arbitrary subset of $SL(2, \mathbb{Z})$ as presented in Eq. (4), we prove that the vector non-freeness problem for matrix semigroups in $SL(2, \mathbb{Z})$ is also in NP.   □

Moreover, the vector non-freeness problem is in fact, NP-complete in $SL(2, \mathbb{Z})$ and undecidable over $\mathbb{N}^{3 \times 3}$.

Theorem 9. *The vector non-freeness problem for finitely generated matrix semigroups in $SL(2, \mathbb{Z})$ is NP-complete. (See the archive version for proof)*

Lastly, we establish the following undecidability as a trivial corollary of Theorem 2 of [1].

Corollary 10. *The vector freeness problem for finitely generated matrix semigroups over $\mathbb{N}^{3 \times 3}$ is undecidable. (See the archive version for proof)*

4   On the Degree of Vector Ambiguity

Given a matrix semigroup S of $n \times n$ matrices and an n-dimensional vector $\mathbf{x}$, let V be a set of target vectors such that $V = \{\mathbf{y} \mid \mathbf{y} = M\mathbf{x}, \ M \in S\}$. Now we consider the problem of determining if there exists a vector $\mathbf{y} \in V$ such that there exists an infinite number of different matrices $M \in S$ such that $M\mathbf{x} = \mathbf{y}$. We call this problem the *finite vector ambiguity problem*.

First remark that if we restrict our attention to the specific target vector $\mathbf{y}$ and consider matrices transforming $\mathbf{x}$ into $\mathbf{y}$, then we can decide whether or not the number of such matrices is infinite.

Theorem 11. *Given two vectors $\mathbf{x}, \mathbf{y}$ and a finitely generated matrix semigroup S in $SL(2, \mathbb{Z})$, we can decide whether or not $\mathbf{y}$ is finitely ambiguous with respect to S and $\mathbf{x}$.*

Proof. We use a similar approach to the proof of Theorem 6. Recall that every matrix M that transforms $\mathbf{x}$ into $\mathbf{y}$ can be represented in the form of BT^tC where $t \in \mathbb{Z}$ and B, C are matrices from $SL(2, \mathbb{Z})$. We can also compute B and C in polynomial time [25]. Thus, it only remains to count the number of matrices in the form of BT^tC from the matrix semigroup S.

Suppose that we are given n matrices $M_1, M_2, \ldots, M_n$ from SL(2, ℤ) as generators of the semigroup S. Namely, $S = \langle M_1, M_2, \ldots, M_n \rangle$. Let $w_1, w_2, \ldots, w_n \in \Sigma^*$ be words encoding the generators, such that $\varphi(w_i) = M_i$ for $1 \le i \le n$. Then, we can define a regular language L corresponding to S over $\Sigma = \{s, r\}$ as $L = \{w_1, w_2, \ldots, w_n\}^+$. Let w_B and w_C be words over $\{s, r\}$ such that $\varphi(w_B) = B^{-1}$ and $\varphi(w_C) = C^{-1}$. Let $L' = \{w_B\} \cdot L \cdot \{w_C\}$. It is easy to see that $\varphi(L') = \{B^{-1} M C^{-1} \mid M \in S\}$.

We also define a regular language L_T corresponding to the set of matrices which are the powers of a matrix T or T^{-1} as follows: $L_T = \{s^3 r\}^* \cup \{r^5 s\}^*$. In other words, $\varphi(L_T) = \{T^m, T^{-m} \mid m \ge 0\}$. It is important to see that there exists only one word $w \in L_T$ which corresponds to the matrix T^m for any integer m and the word w is always reduced. For instance, ε is the only word in L_T for the matrix $T^0 = \mathbf{I}$ which is the identity matrix.

It remains to construct two signed automata $\mathcal{A}'$ and $\mathcal{A}_T$ for L' and L_T which recognize the set of words in L' and L_T, respectively, also with the set of reduced words corresponding to words in L' and L_T, respectively.[2] Then, we can see that the cardinality of the following set $L' \cap L_T$ implies the number of matrices in S of the form BT^tC with different exponents t.

Since we can decide the finiteness of any regular set of matrices, the problem of deciding whether there exists an infinite number of matrices in S transforming $\mathbf{x}$ into $\mathbf{y}$ is also decidable.   □

Theorem 12. *The finite vector ambiguity problem for finitely generated matrix semigroups in SL(2, ℤ) is decidable. (See the archive version for proof)*

In the context of semigroup freeness problem, we can also define the *finite vector non-freeness problem* as the problem of determining the existence of a target vector $\mathbf{v} \in V$ which is reachable from the initial vector $\mathbf{x}$ by an infinite number of different factorizations of matrices.

As in the finite vector ambiguity problem, we prove the case when the target vector $\mathbf{y}$ is fixed.

Theorem 13. *Given two vectors $\mathbf{x}, \mathbf{y}$ and a matrix semigroup S in SL(2, ℤ), we can decide whether or not $\mathbf{y}$ is finitely non-free with respect to S and $\mathbf{x}$. (See the archive version for full proof)*

Proof. First we mention that the problem is very similar to the problem of counting the number of matrices that transforms $\mathbf{x}$ into $\mathbf{y}$ considered in Theorem 11. The only difference is that we count the number of matrix factorizations instead of matrices from SL(2, ℤ). Since a matrix $M \in S = \langle M_1, M_2, \ldots, M_n \rangle$ can have multiple factorizations over the generating set $\{M_1, M_2, \ldots, M_n\}$, we cannot count the number of different factorizations of the form BT^tC by constructing singed automata and counting the number of matrices.

Since we are considering the number of factorizations of matrices, we need to keep the unreduced representations of matrices over $\{s, r\}$ instead of considering reduced representations.

---

[2] See the archive version for the formal definition of the signed automaton.

Let S be the matrix semigroup generated by the set $\{M_1, M_2, \ldots, M_n\}$. We compute the unique canonical word w_i for each matrix M_i for $1 \leq i \leq n$ and define $L_S = \{w_1, \ldots w_n\}^+$ be the regular language. We also compute two matrices B and C, and let w_B and w_C be the unique canonical words such that $\varphi(w_B) = B$ and $\varphi(w_C) = C$.

Recall that $\varphi(s^3 r) = T$ and $\varphi(r^5 s) = T^{-1}$. Based on Lemma 4, we define a context-free language L_{BT^*C} which is the set of all words corresponding to the set of matrices of the form $BT^m C$ where m is any integer. Then, we compute the intersection between L_S and L_{BT^*C} and check whether the cardinality is the context-free language is finite. Since we can decide whether or not a given context-free language is finite, we show that the problem is decidable. $\square$

We leave open the decidability of the finite vector non-freeness problem. We believe that the problem is also decidable but a little bit more complicated than the finite vector ambiguity problem because there is a possibility of losing some information about factorizations of matrices if we use signed automata in which we consider accepting computations on reduced words over $\{s, r\}$ for corresponding matrices.

Since we have considered the problem of determining finiteness of vector ambiguity of matrix semigroups, it is natural to compute the exact threshold of finitely vector ambiguous matrix semigroups. Given a matrix semigroup S, a vector $\mathbf{x}$, and a non-negative integer k (in unary representation), for every target vector $\mathbf{y}$, does there exist at most k different matrices in S which transform $\mathbf{x}$ into $\mathbf{y}$. We call the problem the k-vector ambiguity problem.

Interestingly, the k-vector ambiguity problem is PSPACE-hard by the reduction from the *DFA intersection emptiness problem* [18]. First we start with showing the decidability of the case when we are given both initial and target vectors as follows:

Corollary 14. *Given two vectors* $\mathbf{x}, \mathbf{y}$, *a finitely generated matrix semigroup* S *in* $SL(2, \mathbb{Z})$, *and a positive integer* $k \in \mathbb{N}$, *we can decide whether or not* $\mathbf{y}$ *is* k-*ambiguous with respect to* S *and* $\mathbf{x}$.

Proof. We use a similar approach to the proof of Theorem 11. The only difference is that here we need to count the number of matrices of the form $BT^t C$ from the matrix semigroup S instead of deciding the finiteness of the set of such matrices. Since the $L' \cap L_T$ is a regular set and we can simply enumerate every elements of the finite regular set, we can decide whether or not there exist at most k different matrices $M \in S$ such that $M\mathbf{x} = \mathbf{y}$. $\square$

Now we are ready to show that the k-vector ambiguity problem is decidable and PSPACE-hard.

Theorem 15. *The* k-*vector ambiguity problem for finitely generated matrix semigroups in* $SL(2, \mathbb{Z})$ *is decidable and PSPACE-hard.*

Proof. Recall that $BT^tCx = BT^{t'}Cx = y$ always holds from Lemma 5. Therefore, the finite vector ambiguity problem is equivalent to the problem of deciding whether or not there exists a finite number of different matrices M of the form $M = BT^tC$ in the matrix semigroup S such that $Mx = y$ for every target vector y. Remind that we can compute the matrix C in polynomial time based on the given vector $x = [x_1, x_2]^T$ such that $Cx = [d, 0]^T$ where $d - \gcd(x_1, x_2)$. In other words, the matrix semigroup S is k-*ambiguous* with respect to x if and only if the following condition holds:

$$\max\{x_B \mid B \in \text{SL}(2, \mathbb{Z}), \ x_B = |\{t \mid BT^tC \in S\}|\} \leq k,$$

where $Cx = [d, 0]^T$ and $d = \gcd(x_1, x_2)$.

Simply speaking, while enumerating all possible state subsets Q' satisfying $\bigcap_{q \in Q'}\{w \mid q \in \delta(I, w)\} \neq \emptyset$, we check the following condition: $|(L_{Q'} \cdot \{w_{C^{-1}}\}) \cap L_T| \leq k$. If there exists a state subset where the above inequality does not hold, we decide that S is not k-ambiguous with respect to x since there exist more than k matrices in S of the form BT^tC so that the matrices transform x into the same target vector. Otherwise, S is k-ambiguous with respect to x.

For the PSPACE-hardness of the problem, we reduce the DFA intersection problem [18] to the k-vector ambiguity problem. The DFA intersection problem is, given $k + 1$ DFAs $A_i, 1 \leq i \leq k + 1$, to decide whether or not the intersection of $k + 1$ DFAs is empty.

Let $A_i = (Q_i, \Sigma_A, \delta_i, s_i, F_i)$ be the ith DFA, where $Q_i = \{q_0, q_1, \ldots, q_n\}$ is a finite set of states, Σ_A is an alphabet, δ_i is the transition function, $s_i \in Q_i$ is the start state, and $F_i \subseteq Q_i$ is a finite set of final state. And let us define an alphabet

$$\Sigma = \Sigma_A \cup \{s\} \cup \bigcup_{i=1}^{k+1}(Q_i \cup \overline{Q_i}),$$

where $\overline{Q_i} = \cup_{q \in Q_i}\{\overline{q}\}$. We also define a set W of words which encodes the instance of the DFA intersection problem as follows. For each transition $p \in \delta_i(q, a)$ of A_i, we add a word $q \cdot a \cdot \overline{p}$ to the set W. For each start state s_i of A_i, we add $s \cdot s_i$. Then, it is very easy to see that $s \cdot w \cdot \overline{f_i} \in W^+$, where $f_i \in F_i$, if and only if $w \in L(A_i)$. Let us define an additional letter σ which encodes the word $c^{2|\Sigma|}$, namely, $\alpha(\sigma) = c^{2|\Sigma|}$ and $f(\alpha(\sigma)) = T^{4|\Sigma|}$. Now we additionally add the following words to the set W. For each final state f_i of A_i, $\{f_i \cdot \sigma^i \mid f_i \in F_i\} \subseteq W$. Then, we have $s \cdot w \cdot \sigma^i \in W^+$ if and only if $w \in L(A_i)$.

We claim that S_W is k-free with respect to the vector $[1, 0]^T$ if and only if the intersection of $k + 1$ DFAs is empty. We first prove that if S_W is k-free with respect to the vector $[1, 0]^T$, then the intersection of DFAs is empty. Assume that the intersection of $k + 1$ DFAs is not empty to prove by contradiction. This implies that there is a word that can be accepted by DFAs $A_1, \ldots A_{k+1}$. Therefore, there are $k+1$ words in W^+ as follows: $s \cdot w \cdot \sigma^i \in W^+$ for $1 \leq i \leq k+1$.

Since $f(\alpha(\sigma)) = T^{4|\Sigma|}$ and we have $k + 1$ matrices in S_W which can be represented as follows: $BT^{4|\Sigma|i} \in S_W$ for $1 \leq i \leq k + 1$, where B is a matrix from SL(2, $\mathbb{Z}$). By Lemma 5, we have a contradiction since S_W is not k-free with respect to $[1, 0]^T$.

Now we prove the opposite direction. Assume that S_W is not k-free with respect to $[1,0]^T$. This implies that there are at least $k+1$ matrices $M_1, \ldots, M_{k+1} \in S_W$ where each matrix can be decomposed into the form of BT^t. Since $M_i \in S_W, 1 \leq i \leq k+1$, we have a corresponding word $w_i \in W^+, 1 \leq i \leq k+1$ such that $f(\alpha(w_i)) = M_i$. By Lemma 3, each word w_i should be ending with a distinct number of special symbols σ. Moreover, since $k+1$ DFAs are not connected to each other, w_i should start with s which is an imaginary state connected to every state state of $k+1$ DFAs by ε-transitions. Since the symbol σ only appears after canceling a final state, the word w_i should be of the form $s \cdot w \cdot \sigma^i$, where $w \in L(A_i)$.

We reach the contradiction since there exists a word w that can be spelled out by all $k+1$ DFAs and thus, the intersection of DFAs is not empty. Note that the reduction process can be done in polynomial time. Hence, we prove that the k-vector ambiguity problem in $SL(2, \mathbb{Z})$ is PSPACE-hard.   □

Remark that the k-vector ambiguity problem is in fact, in EXPSPACE as the size of signed automata can be exponentially large in the size of representation of matrix semigroup S. Therefore, we have an EXPSPACE upper bound and PSPACE-hard lower bound for the k-vector ambiguity problem. However, the PSPACE-hardness still applies even if we are given all numeric values of the input in unary representation. Moreover, the size of signed automata stays polynomial if we assume that the input is given in unary representation. Hence, we have the following interesting corollary:

Corollary 16. *The k-vector ambiguity problem for finitely generated matrix semigroups in $SL(2, \mathbb{Z})$ is PSPACE-complete if we are given all numeric values of the input in unary representation.*

Lastly, we consider the k-vector non-freeness problem in which we consider the finite number of different factorizations transforming the given vector $\mathbf{x}$ into any target vector. As in the k-vector ambiguity case, the k-vector freeness problem is decidable if we are given a target vector $\mathbf{y}$ as follows:

Theorem 17. *Given two vectors $\mathbf{x}, \mathbf{y}$, a finitely generated matrix semigroup S in $SL(2, \mathbb{Z})$, and a positive integer $k \in \mathbb{N}$, we can decide whether or not $\mathbf{y}$ is k-non-free with respect to S and $\mathbf{x}$. (See the archive version for full proof)*

References

1. Bell, P.C., Chen, S., Jackson, L.: Scalar ambiguity and freeness in matrix semigroups over bounded languages. In: Dediu, A.-H., Janoušek, J., Martín-Vide, C., Truthe, B. (eds.) LATA 2016. LNCS, vol. 9618, pp. 493–505. Springer, Cham (2016). doi:10.1007/978-3-319-30000-9_38
2. Bell, P.C., Hirvensalo, M., Potapov, I.: Mortality for 2×2 matrices is NP-hard. In: Rovan, B., Sassone, V., Widmayer, P. (eds.) MFCS 2012. LNCS, vol. 7464, pp. 148–159. Springer, Heidelberg (2012). doi:10.1007/978-3-642-32589-2_16

3. Bell, P.C., Hirvensalo, M., Potapov, I.: The identity problem for matrix semigroups in SL₂(ℤ) is NP-complete. In: Proceedings of the Twenty-Eighth Annual ACM-SIAM Symposium on Discrete Algorithms, pp. 187–206 (2017)

4. Bell, P., Potapov, I.: Periodic and infinite traces in matrix semigroups. In: Geffert, V., Karhumäki, J., Bertoni, A., Preneel, B., Návrat, P., Bieliková, M. (eds.) SOFSEM 2008. LNCS, vol. 4910, pp. 148–161. Springer, Heidelberg (2008). doi:10.1007/978-3-540-77566-9_13

5. Bell, P.C., Potapov, I.: Reachability problems in quaternion matrix and rotation semigroups. Inf. Comput. 206(11), 1353–1361 (2008)

6. Bell, P.C., Potapov, I.: On the computational complexity of matrix semigroup problems. Fundam. Infomaticae 116(1–4), 1–13 (2012)

7. Birget, J.-C., Margolis, S.W.: Two-letter group codes that preserve aperiodicity of inverse finite automata. Semigroup Forum 76(1), 159–168 (2008)

8. Blondel, V.D., Cassaigne, J., Karhumäki, J.: Problem 10.3: freeness of multiplicative matrix semigroups. In: Unsolved Problems in Mathematical Systems and Control Theory, pp. 309–314. Princeton University Press (2004)

9. Cassaigne, J., Harju, T., Karhumäki, J.: On the undecidability of freeness of matrix semigroups. Int. J. Algebra Comput. 9(3–4), 295–305 (1999)

10. Chamizo, F.: Non-euclidean visibility problems. Proc. Indian Acad. Sci. - Math. Sci. 116(2), 147–160 (2006)

11. Charlier, E., Honkala, J.: The freeness problem over matrix semigroups and bounded languages. Inf. Comput. 237, 243–256 (2014)

12. Choffrut, C., Karhumäki, J.: Some decision problems on integer matrices. RAIRO - Theoret. Inf. Appl. 39(1), 125–131 (2010)

13. Elstrodt, J., Grunewald, F., Mennicke, J.: Arithmetic applications of the hyperbolic lattice point theorem. Proc. London Math. Soc. s3–57(2), 239–283 (1988)

14. García del Moral, M.P., Martín, I., Peña, J.M., Restuccia, A.: SL(2, ℤ) symmetries, supermembranes and symplectic torus bundles. J. High Energy Phys. 2011(9), 68 (2011)

15. Gurevich, Y., Schupp, P.: Membership problem for the modular group. SIAM J. Comput. 37(2), 425–459 (2007)

16. Klarner, D.A., Birget, J.-C., Satterfield, W.: On the undecidability of the freeness of integer matrix semigroups. Int. J. Algebra Comput. 01(02), 223–226 (1991)

17. Ko, S.-K., Potapov, I.: Matrix semigroup freeness problems in SL(2, ℤ). In: Steffen, B., Baier, C., Brand, M., Eder, J., Hinchey, M., Margaria, T. (eds.) SOFSEM 2017. LNCS, vol. 10139, pp. 268–279. Springer, Cham (2017). doi:10.1007/978-3-319-51963-0_21

18. Kozen, D.: Lower bounds for natural proof systems. In: Proceedings of the 18th Annual Symposium on Foundations of Computer Science, pp. 254–266 (1977)

19. Lyndon, R.C., Schupp, P.E.: Combinatorial Group Theory. Springer, Heidelberg (1977)

20. Mackenzie, D.: A new twist in knot theory. In: What's Happening in the Mathematical Sciences, vol. 7, pp. 3–17. American Mathematical Society (2009)

21. Mandel, A., Simon, I.: On finite semigroups of matrices. Theoret. Comput. Sci. 5(2), 101–111 (1977)

22. Noll, T.: Musical intervals and special linear transformations. J. Math. Music 1(2), 121–137 (2007)

23. Polterovich, L., Rudnick, Z.: Stable mixing for cat maps and quasi-morphisms of the modular group. Ergodic Theory Dyn. Syst. 24, 609–619 (2004)

24. Potapov, I.: Composition problems for braids. In: IARCS Annual Conference on Foundations of Software Technology and Theoretical Computer Science, vol. 24, pp. 175–187 (2013)
25. Potapov, I., Semukhin, P.: Vector reachability problem in SL(2, $\mathbb{Z}$). In: Proceedings of the 41st International Symposium on Mathematical Foundations of Computer Science, pp. 84:1–84:14 (2016)
26. Potapov, I., Semukhin, P.: Decidability of the membership problem for 2×2 integer matrices. In: Proceedings of the Twenty-Eighth Annual ACM-SIAM Symposium on Discrete Algorithms, pp. 170–186 (2017)
27. Shallit, J.: A Second Course in Formal Languages and Automata Theory, 1st edn. Cambridge University Press, New York (2008)
28. Witten, E.: SL(2, $\mathbb{Z}$) action on three-dimensional conformal field theories with abelian symmetry. In: From fields to strings: circumnavigating theoretical physics, vol. 2, pp. 1173–1200. World Scientific Publishing (2005)
29. Woeginger, G.J., Yu, Z.: On the equal-subset-sum problem. Inf. Process. Lett. **42**(6), 299–302 (1992)
30. Zagier, D.: Elliptic modular forms and their applications. In: Ranestad, K. (ed.) The 1-2-3 of Modular Forms. Universitext, pp. 1–103. Springer, Heidelberg (2008)

An $O(n^2)$ Algorithm for Computing Optimal Continuous Voltage Schedules

Minming Li[1]([✉]), Frances F. Yao[2], and Hao Yuan[3]

[1] Department of Computer Science, City University of Hong Kong,
Hong Kong SAR, Hong Kong
minming.li@cityu.edu.hk
[2] Institute for Interdisciplinary Information Sciences,
Tsinghua University, Beijing, China
csfyao@cityu.edu.hk
[3] Bopu Technologies, Shenzhen, China
hao@bopufund.com

Abstract. Dynamic Voltage Scaling techniques allow the processor to set its speed dynamically in order to reduce energy consumption. In the continuous model, the processor can run at any speed, while in the discrete model, the processor can only run at finite number of speeds given as input. The current best algorithm for computing the optimal schedules for the continuous model runs at $O(n^2 \log n)$ time for scheduling n jobs. In this paper, we improve the running time to $O(n^2)$ by speeding up the calculation of s-schedules using a more refined data structure. For the discrete model, we improve the computation of the optimal schedule from the current best $O(dn \log n)$ to $O(n \log \max\{d, n\})$ where d is the number of allowed speeds.

1 Introduction

Energy efficiency is always a primary concern for chip designers not only for the sake of prolonging the lifetime of batteries which are the major power supply of portable electronic devices but also for the environmental protection purpose when large facilities like data centers are involved. Currently, processors capable of operating at a range of frequencies are already available, such as Intel's SpeedStep technology and AMD's PowerNow technology. The capability of the processor to change voltages is often referred to in the literature as DVS (Dynamic Voltage Scaling) techniques. For DVS processors, since energy consumption is at least a quadratic function of the supply voltage (which is proportional to CPU speed), it saves energy to let the processor run at the

The work described in this paper was fully supported by a grant from Research Grants Council of the Hong Kong Special Administrative Region, China (Project No. CityU 117913).

H. Yuan—Part of this work was done while the author was working at City University of Hong Kong.

© Springer International Publishing AG 2017
T.V. Gopal et al. (Eds.): TAMC 2017, LNCS 10185, pp. 389–400, 2017.
DOI: 10.1007/978-3-319-55911-7_28

lowest possible speed while still satisfying all the timing constraints, rather than running at full speed and then switching to idle.

One of the earliest theoretical models for DVS was introduced by Yao, Demers and Shenker [26] in 1995. They assumed that the processor can run at any speed and each job has an arrival time and a deadline. They gave a characterization of the minimum-energy schedule (MES) and an $O(n^3)$ algorithm for computing it which is later improved to $O(n^2 \log n)$ by [21]. No special assumption was made on the power consumption function except convexity. Several online heuristics were also considered including the Average Rate Heuristic (AVR) and Optimal Available Heuristic (OPA). Under the common assumption of power function $P(s) = s^\alpha$, they showed that AVR has a competitive ratio of $2^{\alpha-1}\alpha^\alpha$ for all job sets. Thus its energy consumption is at most a constant times the minimum required. Later on, under various related models and assumptions, more algorithms for energy-efficient scheduling have been proposed.

Bansal et al. [7] further investigated the online heuristics for the model proposed by [26] and proved that the heuristic OPA has a tight competitive ratio of α^α for all job sets. For the temperature model where the temperature of the processor is not allowed to exceed a certain thermal threshold, they showed how to solve it within any error bound in polynomial time. Recently, Bansal et al. [5] showed that the competitive analysis of AVR heuristic given in [26] is essentially tight.

Pruhs et al. [23] studied the problem of minimizing the average flow time of a sequence of jobs when a fixed amount of energy is available and gave a polynomial time offline algorithm for unit-size jobs. Bunde [9] extended this problem to the multiprocessor scenario and gave some nice results for unit-size jobs. Chan et al. [10] investigated a slightly more realistic model where the maximum speed is bounded. They proposed an online algorithm which is $O(1)$-competitive in both energy consumption and throughput. More work on the speed bounded model can be found in [6, 11, 18].

Ishihara and Yasuura [16] initiated the research on discrete DVS problem where a CPU can only run at a set of given speeds. They solved the case when the processor is only allowed to run at two different speeds. Kwon and Kim [17] extended it to the general discrete DVS model where the processor is allowed to run at speeds chosen from a finite speed set. They gave an $O(n^3)$ algorithm for this problem based on the MES algorithm in [26], which is later improved in [19] to $O(dn \log n)$ where d is the allowed number of speeds.

When the CPU can only change speed gradually instead of instantly, [13] discussed about some special cases that can be solved optimally in polynomial time. Later, Wu et al. [25] extended the polynomial solvability to jobs with agreeable deadlines. Irani et al. [14] investigated an extended scenario where the processor can be put into a low-power sleep state when idle. A certain amount of energy is needed when the processor changes from the sleep state to the active state. The technique of switching processors from idle to sleep and back to idle is called Dynamic Power Management (DPM) which is the other major technique for energy efficiency. They gave an offline algorithm that achieves

2-approximation and online algorithms with constant competitive ratios. Recently, Albers and Antoniadis [3] proved the NP-hardness of the above problem and also showed some lower bounds of the approximation ratio. Pruhs et al. [22] introduced profit into DVS scheduling. They assume that the profit obtained from a job is a function on its finishing time and on the other hand money needs to be paid to buy energy to execute jobs. They give a lower bound on how good an online algorithm can be and also give a constant competitive ratio online algorithm in the resource augmentation setting. A survey on algorithmic problems in power management for DVS by Irani and Pruhs can be found in [15]. Most recent surveys by Albers can be found in [1,2].

In [20], the authors showed that the optimal schedule for tree structured jobs can be computed in $O(n^2)$ time. In this paper, we prove that the optimal schedule for general jobs can also be computed in $O(n^2)$ time, improving upon the previously best known $O(n^2 \log n)$ result [21]. The remaining paper is organized as follows. Section 2 will give the problem formulation. Section 3 will discuss the linear implementation of an important tool — the s-schedule used in the algorithm in [21]. Then we use the linear implementation to improve the calculation of the optimal schedule in Sect. 4. In Sect. 5, we give improvements in the computation complexity of the optimal schedule for the discrete model. Finally, we conclude the paper in Sect. 6.

2 Models and Preliminaries

We consider the single processor setting. A job set $J = \{j_1, j_2, \ldots, j_n\}$ over $[0, 1]$ is given where each job j_k is characterized by three parameters: arrival time a_k, deadline b_k, and workload R_k. Here workload means the required number of CPU cycles. We also refer to $[a_k, b_k] \subseteq [0, 1]$ as the interval of j_k. A schedule S for J is a pair of functions $(s(t), job(t))$ which defines the processor speed and the job being executed at time t respectively. Both functions are assumed to be piecewise continuous with finitely many discontinuities. A feasible schedule must give each job its required workload between its arrival time and deadline with perhaps intermittent execution. We assume that the power P, or energy consumed per unit time, is $P(s) = s^\alpha$ $(\alpha \geq 2)$ where s is the processor speed. The total energy consumed by a schedule S is $E(S) = \int_0^1 P(s(t))dt$. The goal of the min-energy feasibility scheduling problem is to find a feasible schedule that minimizes $E(S)$ for any given job set J. We refer to this problem as the continuous DVS scheduling problem. If the speed that can be chosen must come from a set $\{s_1, s_2, \ldots, s_d\}$, then the problem is referred to as the discrete DVS scheduling problem.

For the continuous DVS scheduling problem, the optimal schedule S_{opt} is characterized by using the notion of a critical interval for J, which is an interval I in which a group of jobs must be scheduled at maximum constant speed $g(I)$ in any optimal schedule for J. The algorithm MES in [26] proceeds by identifying such a critical interval I, scheduling those 'critical' jobs at speed $g(I)$ over I, then constructing a subproblem for the remaining jobs and solving it recursively. The details are given below.

Definition 1. *For any interval $I \subseteq [0,1]$, we use J_I to denote the subset of jobs in J whose intervals are completely contained in I. The intensity of an interval I is defined to be $g(I) = (\sum_{j_k \in J_I} R_k)/|I|$.*

An interval I^* achieving maximum $g(I)$ over all possible intervals I defines a critical interval for the current job set. It is known that the subset of jobs J_{I^*} can be feasibly scheduled at speed $g(I^*)$ over I^* by the earliest deadline first (EDF) principle. That is, at any time t, a job which is waiting to be executed and having earliest deadline will be executed during $[t, t + \epsilon]$. The interval I^* is then removed from $[0,1]$; all the remaining job intervals $[a_k, b_k]$ are updated to reflect the removal, and the algorithm recurses. We denote the optimal schedule which guarantees feasibility and consumes minimum energy in the continuous DVS model as OPT.

The authors in [21] later observed that in fact the critical intervals do not need to be located one after another. Instead, one can use a concept called s-schedule defined below to do bipartition on jobs which gradually approaches the optimal speed curve.

Definition 2. *For any constant s, the s-schedule for J is an EDF schedule which uses a constant speed s in executing any jobs of J. It will give up a job when the deadline of the job has passed. In general, s-schedules may have idle periods or unfinished jobs.*

Definition 3. *In a schedule S, a maximal subinterval of $[0,1]$ devoted to executing the same job j_k is called an execution interval for j_k (with respect to S). Denote by $I_k(S)$ the union of all execution intervals for j_k with respect to S. Execution intervals with respect to the s-schedule will be called s-execution intervals.*

It is easy to see that the s-schedule for n jobs contains at most $2n$ s-execution intervals, since the end of each execution interval (including an idle interval) corresponds to the moment when either a job is finished or a new job arrives. Also, the s-schedule can be computed in $O(n \log n)$ time by using a priority queue to keep all jobs currently available, prioritized by their deadlines. In the next section, we will show that the s-schedule can be computed in linear time.

By using s-schedule, one can divide the whole job set J into two subsets $J^{\geq s}$ and $J^{<s}$ where jobs in $J^{\geq s}$ will run at a speed at least s in the optimal schedule while jobs in $J^{<s}$ will run at a speed less than s in the optimal schedule [21]. For the continuous model, the algorithm proposed by [21] uses $g([0,1])$ as the speed s to do the first bipartition and then uses the intensity of the two subsets to do bipartition recursively. They showed that at most $2n$ bipartitions are needed since every bipartition will either split the jobs into two non-empty sets or finalize the optimal schedule within the concerned job set, therefore giving an algorithm with running time $O(nf(n))$ where $f(n)$ is the running time for carrying out one s-schedule.

3   Computing an s-Schedule in Linear Time

In this work, we assume that the underlying computational model is the unit-cost RAM model with word size $\Theta(\log n)$. This model is assumed only for the purpose of using a special union-find algorithm by Gabow and Tarjan [12].

Theorem 1. *If for each k, the rank of a_k in $\{a_1, a_2, \ldots, a_n\}$ and the rank of b_k in $\{b_1, b_2, \ldots, b_n\}$ are pre-computed, then the s-schedule can be computed in linear time in the unit-cost RAM model.*

We make the following two assumptions:

- the jobs are already sorted according to their deadlines;
- for each job j_k, we know the rank of a_k in the arrival time set $\{a_1, a_2, \ldots, a_n\}$.

Because of the first assumption and without loss of generality, we assume that $b_1 \le b_2 \le \ldots \le b_n$. Algorithm 1 schedules the jobs in the order of their deadlines. When scheduling job k, the algorithm tries to search for an earliest available time interval and schedule the job in it, and then repeat the process until all the workload of the job is scheduled or unable to find such a time interval before the deadline. A more detailed discussion of the algorithm is given below.

Let T be $\{a_1, a_2, \ldots, a_n, 1, 1+\epsilon\}$. Note that the times "1" and "1+ϵ" (where ϵ is any fixed positive constant) are included in T for simplifying the presentation

```
1  Initialize e_i ← t_i for 1 ≤ i < m. ;
2  for k=1 to n do
3  │   Let i be the rank of a_k in T, i.e., t_i = a_k. ;
4  │   Initialize r ← R_k, where r denotes the remaining workload to be scheduled.
   │     ;
5  │   while r > 0 do
6  │   │   Search for an earliest non-empty canonical time interval [e_p, t_{p+1}) such
   │   │     that e_p ≥ t_i. ;
7  │   │   if e_p ≥ b_k then
8  │   │   │   Break the while loop because the job cannot be finished.
9  │   │   end
10 │   │   Set u ← min{b_k, t_{p+1}}.
11 │   │   if r > s · (u − e_p) then
12 │   │   │   Schedule job k at [e_p, u). ;
13 │   │   │   Update e_p ← u. ;
14 │   │   │   Update r ← r − s · (u − e_p). ;
15 │   │   else
16 │   │   │   Schedule job k at [e_p, e_p + r/s). ;
17 │   │   │   Update e_p ← e_p + r/s. ;
18 │   │   │   Update r ← 0.
19 │   │   end
20 │   end
21 end
```

Algorithm 1. Algorithm for Computing an s-Schedule

of the algorithm. Denote the size of T by m. Denote t_i to be the i-th smallest element in T. Note that the rank of any a_k in T is known. During the running of the algorithm, we will maintain the following data structure:

Definition 4. *For each $1 \leq i < m$, the algorithm maintains a value e_i, whose value is in the range $[t_i, t_{i+1}]$. The meaning of e_i is that: the time interval $[t_i, e_i)$ is fully occupied by some jobs, and the time interval $[e_i, t_{i+1})$ is idle.*

If $[t_i, t_{i+1})$ is fully occupied, then e_i is t_{i+1}. Note that such a time e_i always exists during the running of the algorithm, which will be shown later when we discuss how to maintain e_i. At the beginning of the algorithm, we assume that the processor is idle for the whole time period. That means $e_i = t_i$ for $1 \leq i < m$ (see line 1 of Algorithm 1).

Example 1. An example for demonstrating the usage of the e_i data structure is given below: Assume that $T = \{0.1, 0.2, 0.3, 0.4, 0.5, 0.6, 0.7, 0.8, 0.9, 1, 1 + \epsilon\}$. At some point during the execution of the algorithm, if some jobs have been scheduled to run at time intervals $[0.2, 0.35), [0.6, 0.86), [0.9, 0.92)$, then we will have $e_1 = 0.1$, $e_2 = 0.3$, $e_3 = 0.35$, $e_4 = 0.4$, $e_5 = 0.5$, $e_6 = 0.7$, $e_7 = 0.8$, $e_8 = 0.86$, $e_9 = 0.92$, and $e_{10} = 1$.

Before we analyze the algorithm, we need to define an important concept called canonical time interval.

Definition 5. *During the running of the algorithm, a canonical time interval is a time interval of the form $[e_p, t_{p+1})$, where $1 \leq p < m$. When $e_p = t_{p+1}$, we call it an empty canonical time interval.*

Note that a non-empty canonical time interval is always idle based on the definition of e_p. Any arrival time a_k will not lie inside any canonical time interval but it is possible that a_k will touch any of the two ending points, i.e., for any $1 \leq p < m$, we have either $a_k \leq e_p$ or $a_k \geq t_{p+1}$. Therefore, if we want to search for a time interval to run a job at or after time a_k, then we should always look for the earliest non-empty canonical time interval $[e_p, t_{p+1})$ where $e_p \geq a_k$.

In Algorithm 1, a variable r is used to track the workload to be scheduled. Lines 5–20 try to schedule j_k as early as possible if $r > 0$. Line 6 tries to search for an earliest non-empty canonical time interval $[e_p, t_{p+1})$ no earlier than the arrival time of j_k (i.e., $e_p \geq a_k$). Such a p always exists because there is always a non-empty canonical time interval $[1, 1 + \epsilon)$. Lines 7–9 mean that, if e_p is not earlier than the deadline of j_k, then the job cannot be finished. Line 10 sets a value of u, whose meaning is that $[e_p, u)$ can be used to schedule the job. The value of u is no later than the deadline of j_k. Lines 12–14 process the case when the remaining workload of j_k cannot be finished in the time interval $[e_p, u)$. Lines 16–18 process the case when the remaining workload of j_k can be finished in the time interval $[e_p, u)$. In the first case, line 13 updates e_p to u because the time interval $[t_p, u)$ is occupied and $[u, t_{p+1})$ is idle. In the second case, a time of r/s is occupied by j_k after the time e_p, so e_p is increased by r/s.

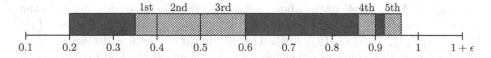

Fig. 1. An illustration for Example 2.

Example 2. Following the example provided in the previous example, assume that the speed is $s = 1$, if we are to schedule a job j_k, where $a_k = 0.3, b_k = 0.96, R_k = 0.35$, the algorithm will proceed as follows: At the beginning, r will be initialized to 0.35, and $i = 3$ (because $a_k = 0.3 = t_3$; see line 3). Line 6 will then get the interval $[e_3, t_4) = [0.35, 0.4)$ as an earliest non-empty canonical time interval, and a workload of $(0.4 - 0.35)s = 0.05$ is scheduled at that time interval. The values of e_3 will be updated to 0.4 accordingly. Now, r becomes $0.35 - 0.05 = 0.3$, and line 6 will get the time interval $[e_4, t_5) = [0.4, 0.5)$ to schedule the job. After that r becomes $0.3 - (0.5 - 0.4)s = 0.2$, and $e_4 = 0.5$. Line 6 then gets the time interval $[e_5, t_6) = [0.5, 0.6)$ to schedule the job, and r will be further reduced to 0.1. The values of e_5 will be updated to 0.6. The next time interval found will be $[e_8, t_9) = [0.86, 0.9)$, and r will become $0.1 - (0.9 - 0.86)s = 0.06$. The values of e_8 will be updated to 0.9. The remaining earliest non-empty canonical time interval is $[e_9, t_{10}) = [0.92, 1)$, but the deadline of the job is 0.96, so only $[0.92, 0.96)$ will be used to schedule the job, and r will be 0.02. The value of e_9 is then updated to 0.96. Finally, $[e_9, t_{10}) = [0.96, 1)$ is the remaining earliest non-empty canonical time interval, but $e_9 \geq b_k$, so line 7–9 will break the loop, and j_k will be an unfinished job. A graphical illustration is provided in Fig. 1. The solid rectangles represent the time intervals occupied by some jobs before scheduling j_k. The cross-hatched rectangles represent the time intervals that are used to schedule j_k. The q-th cross-hatched rectangle (where $1 \leq q \leq 5$) is the q-th time interval scheduled according to this example. Note that all the cross-hatched rectangles except the 5-th one are canonical time intervals right before scheduling j_k.

The most critical part of the algorithm is Line 6, which can be implemented efficiently by the following folklore method using a special union-find algorithm developed by Gabow and Tarjan [12] (see also the discussion of the decremental marked ancestor problem [4]). At the beginning, there is a set $\{i\}$ for each $1 \leq i < m$. The name of a set is the largest element of the set. Whenever e_p is updated to t_{p+1} (i.e., there is not any idle time in the interval $[t_p, t_{p+1})$), we make a union of the set containing p and the set containing $p + 1$, and set the name of this set to be the name of the set containing $p + 1$. After the union, the two old sets are destroyed. In this way, a set is always an interval of integers. For a set whose elements are $\{q, q + 1, \ldots, p\}$, the semantic meaning is that, $[t_q, e_p)$ is fully scheduled but $[e_p, t_{p+1})$ is idle. Therefore, to search for an earliest non-empty canonical time interval beginning at or after time t_i, we can find the set containing i, and let p be the name of the set, then $[e_p, t_{p+1})$ is the required time interval.

Example 3. An example of the above union-find process for scheduling j_k in the previous example is given below: Before scheduling j_k, we have the sets $\{1\}$, $\{2,3\}$, $\{4\}$, $\{5\}$, $\{6,7,8\}$, $\{9\}$, $\{10\}$. The execution of line 6 will always try to search for a set that contains the element $i = 3$. Therefore, the first execution will find the set $\{2,3\}$, so p will be 3. After that, e_3 becomes $t_4 = 0.4$, so the algorithm needs to make a union of the sets $\{2,3\}$ and $\{4\}$ to get $\{2,3,4\}$. Similarly, the next execution will find the set $\{2,3,4\}$, so $p = 4$. The algorithm will then make a union of $\{2,3,4\}$ and $\{5\}$ to get $\{2,3,4,5\}$. For the next execution, the set $\{2,3,4,5\}$ will be found, and it will be merged with $\{6,7,8\}$ to get $\{2,3,4,5,6,7,8\}$. In this case, $p = 8$, and the earliest non-empty canonical time interval is $[e_p, t_{p+1}) = [0.86, 0.9)$. After e_8 is updated to $t_9 = 0.9$, the algorithm will merge $\{2,3,4,5,6,7,8\}$ with $\{9\}$ and obtain $\{2,3,4,5,6,7,8,9\}$. Therefore, the next execution of line 6 will get $p = 9$. After the time interval $[0.92, 0.96)$ is scheduled and e_9 is updated to 0.96, so the algorithm will not do any union. The last execution finds $p = 9$ again, and a loop break is performed.

Now, we analyze the time complexity of the algorithm.

Lemma 1. *Each set always contains continuous integers.*

Proof. It can be proved by induction. At the beginning, each skeleton set is a continuous integer set. During the running of the algorithm, the union operation always merges two nearby continuous integer sets to form a larger continuous integer set.

Lemma 2. *There are at most $m - 2$ unions.*

Proof. It is because there are only $m - 1$ sets.

Lemma 3. *There are at most $2(m - 2) + n$ finds.*

Proof. Some $m - 2$ finds are from finding the set containing $p + 1$ during each union. Note that there is no need to perform a find operation to find the set containing p for union, because p is just the name of such a set, where the set contains continuous integers with p as the largest element. The other $(m-2)+n$ finds are from searching for earliest canonical time intervals beginning at or after time t_i. This can be analyzed in the following way: Let z_k be the number of times to search for an earliest non-empty canonical time interval when processing job j_k. Let w_k be the number of unions that are performed when processing job j_k. We have $z_k \leq w_k + 1$, because each of the first $z_k - 1$ finds must accompany a union. Therefore,

$$\sum_{1 \leq k \leq n} z_k \leq \sum_{1 \leq k \leq n} (w_k + 1) = \sum_{1 \leq k \leq n} w_k + n \leq (m - 2) + n.$$

Since these unions and finds are operated on the sets of integer intervals, such an interval union-find problem can be solved in $O(m + n)$ time in the unit-cost RAM model using Gabow and Tarjan's algorithm [12]. Note that $m = O(n)$, so the total time complexity is $O(n)$. Theorem 1 holds.

If the union-find algorithm is implemented in the pointer machine model [8] using the classical algorithm of Tarjan [24], the complexity of our s-schedule algorithm will become $O(n\alpha(n))$ where $\alpha(n)$ is the one-parameter inverse Ackermann function.

Note that, the number of finds can be further reduced with a more careful implementation of the algorithm as follows (but the asymptotic complexity will not change):

- Whenever the algorithm schedules a job j_k to run at a time interval $[e_p, b_k)$, the algorithm no longer needs to proceed to line 6 for the same job, because there will not be any idle time interval available before the deadline.
- For each job j_k, the first time to find a non-empty canonical time interval requires one find operation. In any of the later times to search for earliest non-empty canonical time intervals for the same job, there must be a union operation just performed. The p that determines the earliest non-empty canonical time interval $[e_p, t_{p+1})$ is just the name of that new set after that union, so a find operation is not necessary in this case. Note that the find operations that accompany the unions are still required.

Using the above implementation, the number of finds to search for earliest non-empty canonical time intervals can be reduced to n. Along with the $m - 2$ finds for unions, the total number of finds of this improved implementation is at most $(m - 2) + n$.

4   An $O(n^2)$ Continuous DVS Algorithm

We will first take a brief look at the previous best known DVS algorithm of Li, Yao and Yao [21]. As in [21], Define the "support" U of J to be the union of all job intervals in J. Define avr(J), the "average rate" of J to be the total workload of J divided by $|U|$. According to Lemma 9 in [21], using $s = $ avr(J) to do an s-schedule will generate two nonempty subsets of jobs requiring speed at least s or less than s respectively in the optimal schedule unless the optimal speed for J is a constant s. The algorithm will recursively do the scheduling based on the two subsets of jobs. Therefore, at most n calls of s-schedules on a job set with at most n jobs are needed before we obtain the optimal schedule for the whole job set. The most time-consuming part of their algorithm is the s-schedules.

To apply our improved s-schedule algorithm for solving the continuous DVS scheduling problem, we need to make sure that the ranks of the deadlines and arrival times are known before each s-schedule call. It can be done in the following way: Before the first call, sort the deadlines and arrival times and obtain the ranks. In each of the subsequent calls, in order to get the new ranks within the two subsets of jobs, a counting sort algorithm can be used to sort the old ranks in linear time. Therefore, the time to obtain the ranks is at most $O(n^2)$ for the whole algorithm. Based on the improved computation of s-schedules, the total time complexity of the DVS problem is now $O(n^2)$, improving the previous $O(n^2 \log n)$ algorithm of [21] by a factor of $O(\log n)$. We have the following theorem.

Theorem 2. *The continuous DVS scheduling problem can be solved in $O(n^2)$ time for n jobs in the unit-cost RAM model.*

5 Further Improvements

For the discrete DVS scheduling problem, we design an $O(n \log \max\{d, n\})$ algorithm to calculate the optimal schedule by doing binary testing on the given d speed levels, improving upon the previously best known $O(dn \log n)$ [19]. To be specific, given the input job set with size n and a set of speeds $\{s_1, s_2, \ldots, s_d\}$, we first choose the speed $s_{d/2}$ to bi-partition the job set into two subsets. Then within each subset, we again choose the middle speed level to do the bi-partition. We recursively do the bi-partition until all the speed levels are handled. In the recursion tree thus built, we claim that the re-sorting for subproblems on the same level can be done in $O(n)$ time which implies that the total time needed is $O(n \log d + n \log n) = O(n \log \max\{d, n\})$. The claim can be shown in the following way. Based on the initial sorting, we can assign a new label to each job specifying which subgroup it belongs to when doing bi-partitioning. Then a linear scan can produce the sorted list for each subgroup.

Hence we have

Theorem 3. *The discrete DVS scheduling problem can be solved in $O(n \log \max\{d, n\})$ time for n jobs and d speeds in the unit-cost RAM model.*

6 Conclusion

In this paper, we improve the time for computing the optimal continuous DVS schedule from $O(n^2 \log n)$ to $O(n^2)$. The major improvement happens in the computation of s-schedules. Originally, the s-schedule computation is done in an online fashion where the execution time is allocated from the beginning to the end sequentially and the time assigned to a certain job can be gradually decided. While in this work, we allocate execution time to jobs in an offline fashion. When jobs are sorted by deadlines, job j_i's execution time is totally decided before we go on to consider j_{i+1}. Then by using a suitable data structure and conducting a careful analysis, the computation time for s-schedules improves from $O(n \log n)$ to $O(n)$. We also design an algorithm to improve the computation of the optimal schedule for the discrete model from $O(dn \log n)$ to $O(n \log \max\{d, n\})$.

References

1. Albers, S.: Energy-efficient algorithms. Commun. ACM **53**(1), 86–96 (2010)
2. Albers, S.: Algorithms for dynamic speed scaling. In: STACS 2011, pp. 1–11 (2011)
3. Albers, S., Antoniadis, A.: Race to idle: new algorithms for speed scaling with a sleep state. In: SODA 2012, pp. 1266–1285 (2012)
4. Alstrup, S., Husfeldt, T., Rauhe, T.: Marked ancestor problems. In: FOCS 1998: Proceedings of the 39th Annual Symposium on Foundations of Computer Science, pp. 534–544 (1998)

5. Bansal, N., Bunde, D.P., Chan, H.-L., Pruhs, K.: Average rate speed scaling. In: Laber, E.S., Bornstein, C., Nogueira, L.T., Faria, L. (eds.) LATIN 2008. LNCS, vol. 4957, pp. 240–251. Springer, Heidelberg (2008). doi:10.1007/978-3-540-78773-0_21
6. Bansal, N., Chan, H.-L., Lam, T.-W., Lee, L.-K.: Scheduling for speed bounded processors. In: Aceto, L., Damgård, I., Goldberg, L.A., Halldórsson, M.M., Ingólfsdóttir, A., Walukiewicz, I. (eds.) ICALP 2008. LNCS, vol. 5125, pp. 409–420. Springer, Heidelberg (2008). doi:10.1007/978-3-540-70575-8_34
7. Bansal, N., Kimbrel, T., Pruhs, K.: Dynamic speed scaling to manage energy and temperature. In: Proceedings of the 45th Annual Symposium on Foundations of Computer Science, pp. 520–529 (2004)
8. Ben-Amram, A.M.: What is a "pointer machine"? SIGACT News **26**(2), 88–95 (1995)
9. Bunde, D.P.: Power-aware scheduling for makespan and flow. In: Proceedings of the 18th Annual ACM Symposium on Parallelism in Algorithms and Architectures, pp. 190–196 (2006)
10. Chan, H.L., Chan, W.T., Lam, T.W., Lee, L.K., Mak, K.S., Wong, P.W.H.: Energy efficient online deadline scheduling. In: Proceedings of the 18th Annual ACM-SIAM Symposium on Discrete Algorithms, pp. 795–804 (2007)
11. Chan, W.T., Lam, T.W., Mak, K.S., Wong, P.W.H.: Online deadline scheduling with bounded energy efficiency. In: Proceedings of the 4th Annual Conference on Theory and Applications of Models of Computation, pp. 416–427 (2007)
12. Gabow, H.N., Tarjan, R.E.: A linear-time algorithm for a special case of disjoint set union. In: STOC 1983: Proceedings of the Fifteenth Annual ACM Symposium on Theory of Computing, pp. 246–251. ACM, New York (1983)
13. Hong, I., Qu, G., Potkonjak, M., Srivastavas, M.B.: Synthesis techniques for low-power hard real-time systems on variable voltage processors. In: Proceedings of the IEEE Real-Time Systems Symposium, pp. 178–187 (1998)
14. Irani, S., Gupta, R.K., Shukla, S.: Algorithms for power savings. ACM Trans. Algorithms **3**(4), 41:1–41:23 (2007)
15. Irani, S., Pruhs, K.: Algorithmic problems in power management. ACM SIGACT News **36**(2), 63–76 (2005)
16. Ishihara, T., Yasuura, H.: Voltage scheduling problem for dynamically variable voltage processors. In: Proceedings of International Symposium on Low Power Electronics and Design, pp. 197–202 (1998)
17. Kwon, W., Kim, T.: Optimal voltage allocation techniques for dynamically variable voltage processors. In: Proceedings of the 40th Conference on Design Automation, pp. 125–130 (2003)
18. Lam, T.-W., Lee, L.-K., To, I.K.K., Wong, P.W.H.: Energy efficient deadline scheduling in two processor systems. In: Tokuyama, T. (ed.) ISAAC 2007. LNCS, vol. 4835, pp. 476–487. Springer, Heidelberg (2007). doi:10.1007/978-3-540-77120-3_42
19. Li, M., Yao, F.F.: An efficient algorithm for computing optimal discrete voltage schedules. SIAM J. Comput. **35**(3), 658–671 (2005)
20. Li, M., Liu, B.J., Yao, F.F.: Min-energy voltage allocation for tree-structured tasks. J. Comb. Optim. **11**(3), 305–319 (2006)
21. Li, M., Yao, A.C., Yao, F.F.: Discrete and continuous min-energy schedules for variable voltage processors. Proc. Nat. Acad. Sci. USA **103**(11), 3983–3987 (2006)
22. Pruhs, K., Stein, C.: How to schedule when you have to buy your energy. In: Serna, M., Shaltiel, R., Jansen, K., Rolim, J. (eds.) APPROX/RANDOM 2010. LNCS, vol. 6302, pp. 352–365. Springer, Heidelberg (2010). doi:10.1007/978-3-642-15369-3_27

23. Pruhs, K., Uthaisombut, P., Woeginger, G.: Getting the best response for your erg. In: Scandanavian Workshop on Algorithms and Theory, pp. 14–25 (2004)
24. Tarjan, R.E.: Efficiency of a good but not linear set union algorithm. J. ACM **22**(2), 215–225 (1975)
25. Wu, W., Li, M., Chen, E.: Min-energy scheduling for aligned jobs in accelerate model. Theor. Comput. Sci. **412**(12–14), 1122–1139 (2011)
26. Yao, F., Demers, A., Shenker, S.: A scheduling model for reduced CPU energy. In: Proceedings of the 36th Annual IEEE Symposium on Foundations of Computer Science, pp. 374–382 (1995)

Towards an Almost Quadratic Lower Bound on the Monotone Circuit Complexity of the Boolean Convolution

Andrzej Lingas[✉]

Department of Computer Science, Lund University, Lund, Sweden
Andrzej.Lingas@cs.lth.se

Abstract. We study the monotone circuit complexity of the so called semi-disjoint bilinear forms over the Boolean semi-ring, in particular the n-dimensional Boolean vector convolution. Besides the size of a monotone Boolean circuit, we consider also the and-depth of the circuit, i.e., the maximum number of and-gates on a path to an output gate, and the monom number of the circuit which is the number of distinct subsets of input variables induced by monoms at the output gates. We show that any monotone Boolean circuit of $\epsilon \log n$-bounded and-depth computing a Boolean semi-disjoint form with $2n$ input variables and q prime implicants has $\Omega(q/n^{2\epsilon})$ size. As a corollary, we obtain the $\Omega(n^{2-2\epsilon})$ lower bound on the size of any monotone Boolean circuit of so bounded and-depth computing the n-dimensional Boolean vector convolution. Furthermore, we show that any monotone Boolean circuit of $2^{n^{\epsilon}}$-bounded monom number, computing a Boolean semi-disjoint form on $2n$ variables, where each variable occurs in p prime implicants, has $\Omega(n^{1-2\epsilon}p)$ size. As a corollary, we obtain the $\Omega(n^{2-2\epsilon})$ lower bound on the size of any monotone Boolean circuit of $2^{n^{\epsilon}}$-bounded monom number computing the n-dimensional Boolean vector convolution. Finally, we demonstrate that in any monotone Boolean circuit for a semi-disjoint bilinear form with q prime implicants that has size substantially smaller than q, the majority of the terms at the output gates representing prime implicants have to have very large length (i.e., the number of variable occurrences). In particular, in any monotone circuit for the n-dimensional Boolean vector convolution of size $o(n^{2-4\epsilon}/\log n)$ almost all prime implicants of the convolution have to be represented by terms at the circuit output gates of length at least n^{ϵ}.

Keywords: Semi-disjoint bilinear form · Boolean vector convolution · Monotone Boolean circuit complexity

1 Introduction

A set f of quadratic polynomials over a semi-ring, defined on the set of variables $X \cup Y$ is a *semi-disjoint bilinear form* if the following properties hold.

© Springer International Publishing AG 2017
T.V. Gopal et al. (Eds.): TAMC 2017, LNCS 10185, pp. 401–411, 2017.
DOI: 10.1007/978-3-319-55911-7_29

1. For each polynomial Q in f and each variable $z \in X \cup Y$, there is at most one monomial (in the Boolean case, called a prime implicant [16]) of Q containing z.
2. The sets of monomials of polynomials in f are pairwise disjoint.
3. Each monomial of a polynomial in f consists of exactly one variable in X and one variable in Y.

The n-dimensional vector convolution and the $n \times n$ matrix product are important and popular examples of semi-disjoint bilinear forms (for the convolution, $|X| = |Y| = n$ and $|f| = 2n - 1$ while for the matrix product, $|X| = |Y| = |f| = n^2$). Both semi-disjoint bilinear forms in the arithmetic and Boolean case have a wide range of fundamental applications, for instance, in stringology (see, e.g., [5]) and graph algorithms (see, e.g., [20]).

Two $n \times n$ integer matrices can be arithmetically multiplied using $O(n^3)$ additions and multiplications following the definition of matrix product. This is optimal if neither other operations nor negative constants are allowed [8,11,14]. If additionally subtraction or negative constants are allowed then the so called fast matrix multiplication algorithms can be implemented using $O(n^\omega)$ operations [7,15,19], where $\omega < 3$. They rely on algebraic equations following from the possibility of term cancellation. Le Gall and Vassilevska Williams have recently shown the exponent ω of fast matrix multiplication to be smaller than 2.373 in [7,19]. The fast arithmetic algorithms run on $0 - 1$ matrices yield the same asymptotic upper time bounds for $n \times n$ Boolean matrix multiplication. On the other hand, Raz proved that if only addition, multiplication and products with constants of absolute value not exceeding one are allowed then $n \times n$ matrix multiplication requires $\Omega(n^2 \log n)$ operations [12].

Similarly, the arithmetic convolution of two n-dimensional vectors can be computed using $O(n^2)$ additions and multiplications. Next, the convolution of two n-dimensional vectors over a commutative ring with the so called principal n-th root of unity can be computed via Fast Fourier Transform using $O(n \log n)$ operations of the ring. The n-dimensional Boolean vector convolution also admits an $O(n \log n)$ algorithm by reduction to fast integer multiplication algorithm in turn relying on Fast Fourier Transform [5].

It is well known that for uniform problems, their Boolean circuit complexity corresponds up to logarithmic terms to their Turing complexity [16]. Unfortunately, up to today no super-linear lower bounds on the size of circuits using binary and unary Boolean operations forming a complete Boolean basis are known for natural problems [16]. On the other hand, such lower bounds are known in case of monotone Boolean circuits that use only the "or" and "and" binary operations [1,2,4,6,8–11,13,16–18]. There exist interesting connections between the general Boolean circuit complexity and the monotone one which motivate studying the latter for monotone functions [3].

In this paper, we study the complexity of monotone (i.e., using only and-gates and or-gates besides the input gates) Boolean circuits for Boolean semi-disjoint bilinear forms. In case of $n \times n$ Boolean matrix product, almost tight or even tight lower bounds of the form $\Omega(n^3)$ were presented in a series of

Table 1. Lower bounds on the monotone Boolean circuit complexity for n-dimensional Boolean vector convolution in a historical perspective.

Author	Year	Lower bound
N. Pippinger and L.G. Valiant [10]	1976	$\Omega(n \log n)$
E.A. Lamagna [6]	1979	$\Omega(n \log n)$
N. Blum [4]	1980	$n^{4/3}$
R. Weiss [18]	1981	$n^{3/2}$

papers [8,9,11] for more than three decades ago. As for the n-dimensional Boolean vector convolution, solely substantially sub-quadratic lower bounds on monotone Boolean circuit complexity are known in the literature in spite of the fact that one widely believes that n^2 and-gates and $n^2 - 2n + 1$ or-gates are required [3]. The best known lower bound on monotone Boolean circuit complexity for n-dimensional Boolean vector convolution is $n^{3/2}$ due to Weiss [18]. It improves on the previously best $n^{4/3}$ lower bound due to Blum [4], see also Table 1. Blum conjectures in [3] that one can derive an $\Omega((n/\log n)^2)$ lower bound for this problem mixing known techniques.

Besides the size of a monotone Boolean circuit, we shall also consider the and-depth of the circuit, i.e., the maximum number of and-gates on a path to an output gate, and the monom number of the circuit which is the number of distinct subsets of input variables induced by monoms at the output gates. First, we show that any monotone Boolean circuit of and-depth d (i.e., in which any directed path includes at most d and-gates) computing a Boolean semi-disjoint form with q prime implicants has to have at least $q/2^{2d}$ size (i.e., the number of non-input gates). As a corollary, we obtain the $\Omega(n^{2-2\epsilon})$ lower bound on the size of any monotone Boolean circuit of $\epsilon \log n$ and-depth computing the n-dimensional Boolean vector convolution. Our main result states that any monotone Boolean circuit of 2^{n^ϵ}-bounded monom number, computing a Boolean semi-disjoint form on $2n$ variables, where each variable occurs in p prime implicants, has $\Omega(n^{1-2\epsilon}p)$ size. As a corollary, we obtain the $\Omega(n^{2-2\epsilon})$ lower bound on the size of any monotone Boolean circuit of 2^{n^ϵ}-bounded monom number computing the n-dimensional Boolean vector convolution. Finally, we demonstrate that in any monotone Boolean circuit for a semi-disjoint bilinear form with q prime implicants that has size substantially smaller than q, the majority of the terms at the output gates representing prime implicants have to have very large length (i.e., the number of variable occurrences). In particular, in any monotone circuit for the n-dimensional Boolean vector convolution of size $o(n^{2-4\epsilon}/\log n)$, almost all prime implicants of the convolution have to be represented by terms at the circuit output gates of length at least n^ϵ.

2   Preliminaries

For two Boolean n-dimensional vectors $a = (a_0, ..., a_{n-1})$ and $b = (b_0, ..., b_{n-1})$ their convolution is a vector $c = (c_0, ..., c_{2n-2})$, where $c_i = \bigvee_{l=\max\{i-n+1,0\}}^{\min\{i,n-1\}} a_l \wedge b_{i-l}$ for $i = 0, ..., 2n - 2$.

A *monotone (Boolean) circuit* is a finite directed acyclic graph with the following properties:

1. The indegree of each vertex (termed gate) is either 0 or 2.
2. The source vertices (i.e., vertices with indegree 0 called input gates) are labeled by elements in some set of variables and the Boolean constants 0, 1.
3. The vertices of indegree 2 are labeled by elements of the set $\{and, or\}$ and termed and-gates and or-gates, respectively.

The *size* of a monotone Boolean circuit is the total number of gates of indegree two in the circuit, i.e., and-gates and or-gates. A monotone Boolean circuit is of *and-depth d* if the number of and-gates on any directed path in the circuit does no exceed d.

With each gate g of a monotone Boolean circuit, we associate a set $T(g)$ of terms in a natural way. Thus, with each input gate, we associate the singleton set consisting of the corresponding variable or constant. Next, with an or-gate, we associate the union of the sets associated with its direct predecessors. Finally, with an and-gate g, we associate the set of concatenations $t_1 t_2$ of all pairs of terms t_1, t_2, where $t_i \in T(g_i)$ and g_i stands for the i-th direct predecessor of g for $i = 1, 2$. The function computed at the gate is the disjunction of the functions (called monoms) represented by the terms in $T(g)$. A term in $T(g)$ is a zero-term if it contains the Boolean constant 0. Clearly, a zero-term represents the Boolean constant 0.

The *monom number* of a monotone circuit is the number of distinct subsets of variables induced by terms in $T(o)$ over all output gates o of the circuit. Note that the monom number never exceeds 2^m, where m is the number of input variables.

A Boolean bilinear form composed of k functions is computed by a monotone Boolean circuit if there are k distinguished gates (called output gates) computing the k functions.

An *implicant* of a set f of Boolean functions is the conjunction of some variables of f (monom) such that there is a function belonging to f which is true whenever the conjunction is true. An implicant of f that is minimal with respect to included variables is a *prime implicant* of f.

3   The Monotone Boolean Circuit Complexity

The following upper bound is straight-forward.

Lemma 1. *Each Boolean semi-disjoint bilinear form composed of k functions on $x_0, ..., x_{n-1}$ and $y_0, ..., y_{n-1}$ with q prime implicants in total can be computed by a monotone circuit of and-depth 1 and monom number q, with $q \leq n^2$ and-gates and $q - k$ or-gates.*

Proof. First, we use q and-gates to compute each prime implicant $x_i y_j$ separately. Then, we form k disjoint or-unions of the prime implicants corresponding to the k functions of the bilinear form using $q - k$ or-gates.   □

3.1   Warming up

In this subsection, we show how a restriction on the length (i.e., the number of variable occurrences in) of terms in $T(o)$ for output gates o of a monotone Boolean circuit computing a Boolean semi-disjoint form can be used to derive a non-trivial lower bound on the number of and-gates in the circuit.

Lemma 2. *Let S be a monotone circuit computing a semi-disjoint bilinear form f on the variables $x_0, ..., x_{n-1}$ and $y_0, ..., y_{n-1}$. Suppose that for each output gate o in S, each non-zero term in $T(o)$ contains at most k variables. Let h be a gate connected by directed paths with some output gates in S such that the function computed at h has prime implicants $z_{q_1}, ..., z_{q_{l(h)}}$ which are single variables and possibly some other non-single-variable prime implicants. The inequality $l(h) \leq k$ holds or h can be replaced by the Boolean constant 1.*

Proof. Consider a directed path P connecting h with some output gate o in S. At the output gate o, each z_{q_r}, $1 \leq r \leq l(h)$, has to appear in terms $t_1 z_{q_r} t_2$ in the associated set $T(o)$ (see Preliminaries) such that $t_1 t_2$ is a concatenation (i.e., conjunction) of some terms added by consecutive and-gates on P and $t_1 z_{q_r} t_2$ represents an implicant of the function f_o computed at o.

Suppose that there is such a $t_1 t_2$ that does not represent an implicant of f_o. It follows from the definition of $t_1 t_2$ that for any $z \in \{z_{q_r} | 1 \leq r \leq l(h)\}$, the term $t_1 z t_2$ also appears in the set $T(o)$ of terms associated with the output gate o and consequently it has to represent an implicant of f_o as well. Therefore, for each such a z, $t_1 t_2$ has to contain the unique variable z' for which zz' is a prime implicant of f_o. Note that if z is an x-variable then z' is an y-variable and *vice versa*. Since for different z the z' have to be different, $t_1 t_2$ has to contain at least $l(h)$ variables. We infer that $l(h) \leq k$.

On the contrary, if each such a term $t = t_1 t_2$ for each path P connecting h with any output gate o, represents an implicant of f_o then on each P we could connect the successor of the start vertex h with the Boolean constant 1 instead of h and the output gate o still would output f_o. To see this observe that then each $u \in T(h)$ is a part of the terms of the form $t_1 u t_2$ in $T(o)$, where $t_1 t_2$ represents an implicant of the function f_o.   □

We shall a call a class K of monotone Boolean circuits k-*nice* if (i) for each circuit $U \in K$, for each output gate o in U, each non-zero term in $T(o)$ contains at most k variables, and (ii) K is closed under the replacement of a gate in U by a Boolean constant.

Lemma 3. *Let S be a monotone circuit computing a semi-disjoint bilinear form f on the variables $x_0, ..., x_{n-1}$ and $y_0, ..., y_{n-1}$. Suppose that S belongs to a k-nice class K and achieves a minimum size among monotone circuits in K that compute f. Let g be an and-gate in S. Next, let S_g be the set of prime implicants s of f such that s is a prime implicant of the function computed at g, s is not a prime implicant of the function computed at any of the two direct predecessors of g, and there is a directed path connecting g with the output gate computing the function whose prime implicant is s. The inequality $|S_g| \leq k^2$ holds.*

Proof. We may assume w.l.o.g. $|S_g| \geq 1$. It follows that at least for one of the direct predecessor gates h of g, the function computed at h has at least $\sqrt{|S_g|}$ single variable prime implicants. By Lemma 2, we infer that either $\sqrt{|S_g|} \leq k$ or the gate h can be replaced by the constant 1. The latter possibility contradicts the minimality of S which yields the lemma.   □

Theorem 1. *Let S be a monotone circuit that computes a semi-disjoint bilinear form f on the variables $x_0, ..., x_{n-1}$ and $y_0, ..., y_{n-1}$, having q prime implicants in total. Suppose that S belongs to a k-nice class K and achieves a minimum size among monotone circuits in K that compute f. S has at least q/k^2 and-gates.*

Proof. For each prime implicant s of f there is at least one and-gate g having the properties described in the statement of Lemma 3, i.e., where $s \in S_g$. (To find such a gate g start from the output gate computing the function of f for which s is a prime implicant and iterate the following steps: check if the current gate g satisfies $s \in S_g$, if not go to the direct predecessor of g that computes a function having s as a prime implicant.) By the latter lemma, the same and-gate can have these properties for at most k^2 prime implicants of f.   □

Corollary 1. *Let S be a minimum-size monotone circuit of d-bounded and-depth that computes a semi-disjoint bilinear form f on the variables $x_0, ..., x_{n-1}$ and $y_0, ..., y_{n-1}$, having q prime implicants in total. S has at least $q/2^{2d}$ and-gates.*

Proof. By induction on the maximum number d of and-gates on a path from an input gate to a gate g in S, any term in $T(g)$ includes at most 2^d distinct variables. Hence, Theorem 1 yields the corollary.   □

Corollary 2. *Any minimum-size monotone circuit of $\epsilon \log n$-bounded and-depth that computes the n-dimensional Boolean vector convolution has at least $\Omega(n^{2-2\epsilon})$ and-gates.*

3.2 Large Monom Number

In this subsection, we combine an idea of elimination of long relevant terms in $T(o)$ for output gates o of a monotone Boolean circuit computing a Boolean semi-disjoint form with those from the preceding subsection in order to derive our lower bound on the monom number. The elimination idea consists in setting a specially chosen subset of input variables to the Boolean 0 using a probabilistic argument.

Theorem 2. *Let f be a semi-disjoint bilinear form on the variables $x_0, ..., x_{n-1}$ and $y_0, ..., y_{n-1}$, such that for each of the variables there are p prime implicants of f containing it. Any minimum-size monotone circuit S of $2^{n^{\epsilon}}$-bounded monom number that computes f has to include $\Omega(pn^{1-2\epsilon})$ and-gates.*

Proof. Let H be the set of all non-input gates h in S connected by directed paths with some output gates in S such that the function computed at h has prime implicants $z_{q_1}, ..., z_{q_{l(h)}}$ which are single variables and possibly some other non-single variable prime implicants. We let $Z_h = \{z_{q_r} | 1 \leq r \leq l(h)\}$. Note that by the optimality of S and Lemma 1, $|H| \leq 2n^2$ holds.

Let $h \in H$. Consider a directed path P connecting h with some output gate o in S. At the output gate o, each z_{q_r}, $1 \leq r \leq l(h)$, has to appear in terms $t_1 z_{q_r} t_2$ in the associated set $T(o)$ such that $t_1 t_2$ is a concatenation (i.e., conjunction) of some terms added by consecutive and-gates on P and $t_1 z_{q_r} t_2$ represents an implicant of the function f_o computed at o. Let $T(h, o)$ denote the set of all such terms $t = t_1 t_2$ for all possible directed paths P connecting h with o. Next, let $T_{im}(h, o)$ stand for the subset of all terms in $T(h, o)$ that represent implicants of f_o (in particular terms including the Boolean 0 are trivial implicants of f_o).

Suppose that $T(h, o) \setminus T_{im}(h, o) \neq \emptyset$, i.e., there are $t_1 t_2 \in T(h, o)$ that do not represent any implicant of f_o. It follows from the definition of $T(h, o)$ that for any $z \in \{z_{q_r} | 1 \leq r \leq l(h)\}$, the terms $t_1 z t_2$ also appear in the set $T(o)$ of terms associated with the output gate o and consequently they have to represent implicants of f_o as well. Therefore, for each $z \in \{z_{q_r} | 1 \leq r \leq l(h)\}$, each $t_1 t_2 \in T(h, o) \setminus T_{im}(h, o)$ has to contain the unique variable z' for which $z z'$ is a prime implicant of f_0. That is, if $z = x_i$ then $z' = y_j$, where y_j is the unique y-variable for which $x_i y_j$ is a prime implicant of f_o and *vice versa*. Since for different z the z' have to be different, each $t_1 t_2 \in T(h, o) \setminus T_{im}(h, o)$ has to contain all the $l(h)$ variables in the set $\{z'_{q_r} | 1 \leq r \leq l(h)\}$. We shall denote the latter set by $Z'_{h,o}$.

Pick uniformly at random a subset of $n/2$ variables from the set of the $2n$ input variables and set each variable in the subset to the Boolean 0. Note that each of the $2n$ input variables is set to Boolean 0 with probability $1/4$ and consequently it survives with probability $3/4$.

Let O stand for the set of output gates, and let $T = \bigcup_{o \in O} T(o)$. Then, for any $t \in T$ containing at least $(n^{\epsilon} + 2)/\log_2 \frac{4}{3}$ distinct variables, the probability that t becomes a zero-term representing trivial implicants of f_o, where $t \in T(o)$, is at least $1 - (3/4)^{(n^{\epsilon}+2)/\log_2 \frac{4}{3}} = 1 - \frac{1}{4} 2^{-n^{\epsilon}}$. Since the monom number of S is bounded by $2^{n^{\epsilon}}$, the probability that all terms $t \in T$ having at least $(n^{\epsilon}+2)/\log_2 \frac{4}{3}$ distinct variables become trivial implicants is at least $\frac{1}{2}$. Hence, there is a setting of $n/2$

input variables to Boolean 0 which turns the aforementioned terms to trivial implicants.

Observe that after the setting the resulting circuit S' computes a semi-disjoint bilinear form f' with at least $np - \frac{n}{2}p = np/2$ prime implicants. Consider any $h \in H$ in S' connected by directed paths with output gates o, where $l(h) \geq (n^\epsilon + 2)/\log_2 \frac{4}{3}$ after the zero-setting. Let z be any variable in the current Z_h. Suppose that there is a $t \in T(h,o) \setminus T_{im}(h,o)$ after the zero setting. Then, z jointly with t forms a term in the original $T(o)$ containing at least $(n^\epsilon + 2)/\log_2 \frac{4}{3}$ distinct variables complementing those in Z_h. By the choice of the zero setting, the aforementioned term has to include the Boolean 0. It follows that t is a trivial implicant of f_o, we obtain a contradiction. Thus, for all $h \in H$ in S' connected by directed paths with output gates o, where $l(h) \geq (n^\epsilon + 2)/\log_2 \frac{4}{3}$ after the zero-setting, we have $t \in T(h,o) = T_{im}(h,o)$. Consequently, we can eliminate any such a gate h by replacing with the Boolean 1 (cf. the last paragraph in the proof of Lemma 2).

In order to eliminate all such gates from the current circuit, while there is a gate $h \in H$ with $l(h) \geq (n^\epsilon + 2)/\log_2 \frac{4}{3}$ in the current circuit, we replace h with the Boolean 1. By our assumption on the size of S, the process has to stop latest after $2n^2$ iterations.

The proof that the resulting circuit still computes f' is by induction on the number of iterations, i.e., replacements of gates h by 1. By the preceding discussion, the current circuit computes f' after the first replacement. So suppose that the current circuit computes f' after i replacements of such gates with 1, and consider a candidate gate h with $l(h) \geq (n^\epsilon + 2)/\log_2 \frac{4}{3}$ in the current circuit for the $i + 1$st replacement. Similarly, as in the case of first replacement, consider $z \in Z_h$ and any $t \in T(h,o) \setminus T_{im}(h,o)$ for any output gate o reachable from h. The variable z jointly with t forms a term in the current $T(o)$ that originates from some term u in $T(o)$ in the circuit S' before the replacements by 1. Consequently, by the definition of the zero setting, u and also t have to contain a Boolean 0. We conclude again that $T(h,o) = T_{im}(h,o)$, so we can replace h with 1.

We infer that the circuit S'' resulting from all the replacements of gates h with $l(h) \geq (n^\epsilon + 2)/\log_2 \frac{4}{3}$ by 1 in the current circuit fulfills the following conditions:

1. the size of S'' does not exceed that of S';
2. S'' computes a semi-disjoint bilinear form f' on the same set of variables with at least $pn/2$ prime implicants in total;
3. for any gate h of S'', the cardinality of the set Z_h of single-variable implicants of the function f_h computed at h does not exceed $(n^\epsilon + 2)/\log_2 \frac{4}{3}$.

Let g be an arbitrary and-gate in S''. Next, let S_g be the set of prime implicants s of f' such that s is a prime implicant of the function f_g computed at g, s is not a prime implicant of the function computed at any of the two direct predecessors of g, and there is a directed path connecting g with the output gate computing the function belonging to f' whose prime implicant s is.

Suppose $|S_g| \geq 1$. It follows that at least for one of the direct predecessor gates h of g in S'', the function computed at h has at least $\sqrt{|S_g|}$ single-variable prime implicants. We infer from the construction of S'' that $\sqrt{|S_g|} \leq (n^\epsilon + 2)/\log_2 \frac{4}{3}$. The inequality $|S_g| \leq (n^\epsilon + 2)^2/(\log_2 \frac{4}{3})^2$ follows. Since for each prime implicant s of f' there must be at least one such a gate g where $s \in S_g$ (see the proof of Theorem 1 for how to find such a gate g), we conclude that S'' (and consequently S' and S) has at least $pn/(2 \times (n^\epsilon + 2)^2/(\log_2 \frac{4}{3})^2)$ and-gates. □

Corollary 3. *Any minimum-size monotone Boolean circuit of 2^{n^ϵ}-bounded monom number that computes the n-dimensional Boolean vector convolution has $\Omega(n^{2-2\epsilon})$ and-gates.*

4   Prime Implicant Terms of Large Length

Recall that for a term $t \in T(g)$, where g is a gate of a monotone Boolean circuit, by the length of t, we mean the total number of occurrences of variables in t. By using arguments from the preceding sections, we shall prove the following theorem.

Theorem 3. *Let S be a monotone circuit computing a semi-disjoint bilinear form f with $2n$ variables and q prime implicants. Suppose that an α fraction of the prime implicants of f is represented by terms in $T(o)$, where o ranges over output gates of S, having length smaller than r. Then, the size of the circuit S is $\Omega(\alpha q/(r^4 \log r))$.*

Proof. To begin with, we shall transform the circuit S into a monotone circuit S' such that each non-input gate g in S is represented by gates g_i, $i = 1, ..., r$, where for $i = 1, ..., r - 1$, $T(g_i)$ consists of all terms in $T(g)$ of length i, while $T(g_r)$ consists of all terms in $T(g)$ of length at least r. Such a transformation is folklore in arithmetic circuits.

One starts from input gates g, for which only g_1 is non-trivial, and then one proceeds bottom up. For an or-gate g in S, one needs solely to compute the disjunction of the outcomes of pairs of the gates in S' associated with the direct predecessors of g that correspond to the same length index. For an and-gate g in S, one creates auxiliary gates computing the and-product of pairs of outcomes of the gates associated with the direct descendants of g in S. Next, for $i = 1, ..., r-1$, g_i computes the disjunction of the outcomes of the auxiliary gates corresponding to the length i. This requires an introduction of $O(r^2 \log r)$ intermediate gates. Additionally, for each output gate o of S, one creates a corresponding output gate computing the disjunction of the outcomes of the gates o_1 through o_{r-1}, using $O(r \log r)$ intermediate gates. Note that in this way the terms in $T(o)$ of length at least r are disregarded.

Observe that in the worst case, one needs to introduce $O(r^2 \log r)$ gates in S' in order to simulate a single gate of S. Hence, the size of S' is at most $O(r^2 \log r)$ times larger than that of S. Also, it follows from our assumptions that S computes a semi-disjoint bilinear form f' with at least αq prime implicants.

Consider a gate h in S', where the function f_h computed at h has at least r single-variable implicants. Let o be any output gate in S' reachable from h, and let $t \in T(h, o) \setminus T_{im}(h, o)$. We know from the preceding sections, that t has to contain at least r distinct variables complementing those being implicants of f_h (see the proof of Lemma 2 or Theorem 2). This is however impossible, since we have disregarded all terms of length at least r constructing S'. It follows that we can eliminate h by replacing it with the Boolean 1 without affecting the bilinear form f' computed by S' (similarly as in the proof of Lemma 2 or Theorem 2). We keep replacing all such h with f_h currently having at least r single variable implicants, with Boolean 1, by using the aforementioned argument. The replacements may cause new gates h to get the number of single variable implicants over the $r - 1$ threshold. Nevertheless, the process has to stop because of the finiteness of S'.

Let S'' stand for the resulting circuit. S'' still computes f' and for each its gate h, f_h has at most $r - 1$ single variable implicants. Consequently, by considering and-gates g in S'' and the subsets S_g of prime implicants of f' that are prime implicants of f_g but not prime implicants of the functions computed at the direct predecessors of g, we infer that $|S_g| < r^2$ (similarly as in the proof of Theorems 1 or 2). Hence, S'' has to have at least $\alpha q / r^2$ and-gates. Since the size of S'' is $O(r^2 \log r)$ times larger than that of S, we conclude that the size of S is $\Omega(\alpha q / (r^4 \log r))$.   □

Corollary 4. *If the n-dimensional Boolean vector convolution can be computed by a monotone circuit of size $o(n^{2-4\epsilon} / \log n)$ then almost all (i.e., the fraction tends to 1 as n grows) prime implicants of the convolution have to be represented by terms at the circuit output gates of length at least n^ϵ.*

Acknowledgments. The author is grateful to Mia Persson for noting that the problem of deriving a quadratic or almost quadratic lower bound on the monotone cicuit complexity of Boolean vector convolution is still open, and to Norbert Blum and Mike Paterson for valuable comments on a very preliminary version of this paper. This research has been supported in part by Swedish Research Council grant 621-2011-6179.

References

1. Alon, N., Boppana, R.B.: The monotone circuit complexity of Boolean functions. Combinatorica **7**(1), 1–22 (1987)
2. Andreev, A.E.: On one method of obtaining constructive lower bounds for the monotone circuit size. Algebra Logics **26**(1), 3–26 (1987)
3. Blum, N.: On negations in boolean networks. In: Albers, S., Alt, H., Näher, S. (eds.) Efficient Algorithms. LNCS, vol. 5760, pp. 18–29. Springer, Heidelberg (2009). doi:10.1007/978-3-642-03456-5_2
4. Blum, N.: An $\Omega(n^{4/3})$ lower bound on the monotone network complexity of the n-th degree convolution. Theor. Comput. Sci. **36**, 59–69 (1985)
5. Fisher, M.J., Paterson, M.S.: String-matching and other products. In: Proceedings of the 7th SIAM-AMS Complexity of Computation, pp. 113–125 (1974)
6. Lamagna, E.A.: The complexity of monotone networks for certain bilinear forms, routing problems, sorting, and merging. IEEE Trans. Comput. **c-28**(10), 773–782 (1979)

7. Le Gall, F.: Powers of tensors and fast matrix multiplication. In: Proceedings of the 39th International Symposium on Symbolic and Algebraic Computation, pp. 296–303 (2014)
8. Mehlhorn, K., Galil, Z.: Monotone switching circuits and boolean matrix product. Computing **16**, 99–111 (1976)
9. Paterson, M.: Complexity of monotone networks for boolean matrix product. Theoret. Comput. Sci. **1**(1), 13–20 (1975)
10. Pippenger, N., Valiant, L.G.: Shifting graphs and their applications. J. ACM **23**(3), 423–432 (1976)
11. Pratt, R.: The power of negative thinking in multiplying boolean matrices. SIAM J. Comput. **4**(3), 326–330 (1975)
12. Raz, R.: On the complexity of matrix product. In: Proceedings of the STOC 2002, pp. 144–151 (2002)
13. Razborov, A.A.: Lower bounds on the monotone complexity of some boolean functions. Doklady Akademii Nauk **281**(4), 798–801 (1985)
14. Schnorr, C.-P.: A lower bound on the number of additions in monotone computations. Theor. Comput. Sci. **2**(3), 305–315 (1976)
15. Strassen, V.: Gaussian elimination is not optimal. Numer. Math. **13**, 354–356 (1969)
16. Wegener, I.: The Complexity of Boolean Functions. Wiley-Teubner Series in Computer Science, New York (1987)
17. Wegener, I.: Boolean functions whose monotone complexity is of size $n^2/\log n$. Theor. Comput. Sci. **21**, 213–224 (1982)
18. Weiss, J.: An $n^{3/2}$ lower bound on the monotone network complexity of the boolean convolution. Inf. Control **59**, 184–188 (1983)
19. Vassilevska Williams, V.: Multiplying matrices faster Coppersmith-Winograd. In: Proceedings of the 44th Annual ACM Symposium on Theory of Computing (STOC), pp. 887–898 (2012)
20. Zwick, U.: All pairs shortest paths using bridging sets and rectangular matrix multiplication. J. ACM **49**(3), 289–317 (2002)

Bounds for Semi-disjoint Bilinear Forms in a Unit-Cost Computational Model

Andrzej Lingas[1]([✉]), Mia Persson[2], and Dzmitry Sledneu[3]

[1] Department of Computer Science, Lund University, Lund, Sweden
Andrzej.Lingas@cs.lth.se
[2] Department of Computer Science, Malmö University, Malmö, Sweden
Mia.Persson@mah.se
[3] Centre for Mathematical Sciences, Lund University, Lund, Sweden
Dzmitry.Sledneu@math.lu.se

Abstract. We study the complexity of the so called semi-disjoint bilinear forms over different semi-rings, in particular the n-dimensional vector convolution and $n \times n$ matrix product. We consider a powerful unit-cost computational model over the ring of integers allowing for several additional operations and generation of large integers. We show the following dichotomy for such a powerful model: while almost all arithmetic semi-disjoint bilinear forms have the same asymptotic time complexity as that yielded by naive algorithms, matrix multiplication, the so called distance matrix product, and vector convolution can be solved in a linear number of steps. It follows in particular that in order to obtain a non-trivial lower bounds for these three basic problems one has to assume restrictions on the set of allowed operations and/or the size of used integers.

Keywords: Semi-disjoint bilinear form · Semi-ring · Vector convolution · Matrix multiplication · Distance product · Circuit complexity · Unit-cost ram · Time complexity

1 Introduction

A set F of quadratic polynomials over a semi-ring, defined on the set of variables $X \cup Y$ is a semi-disjoint bilinear form if the following properties hold.

1. For each polynomial P in F and each variable $z \in X \cup Y$, there is at most one monomial (in the Boolean case, called a prime implicant [17]) of P containing z.
2. The sets of monomials of polynomials in F are pairwise disjoint.
3. Each monomial of a polynomial in F consists of exactly one variable in X and one variable in Y.

The n-dimensional vector convolution and the $n \times n$ matrix product are important and popular examples of semi-disjoint bilinear forms (for the convolution, $|X| = |Y| = n$ and $|F| = 2n - 1$ while for the matrix product,

© Springer International Publishing AG 2017
T.V. Gopal et al. (Eds.): TAMC 2017, LNCS 10185, pp. 412–424, 2017.
DOI: 10.1007/978-3-319-55911-7_30

$|X| = |Y| = |F| = n^2$). Both semi-disjoint bilinear forms in the arithmetic and Boolean case have a wide range of fundamental applications, for instance, in stringology (see, e.g., [5]) and graph algorithms (see, e.g., [21]).

Two $n \times n$ integer matrices can be arithmetically multiplied using $O(n^3)$ additions and multiplications following the definition of matrix product. Similarly, the arithmetic convolution of two n-dimensional vectors can be computed using $O(n^2)$ additions and multiplications. Both are optimal if neither other operations nor negative constants are allowed [10,11,14]. If additionally subtraction or negative constants are allowed then the so called fast matrix multiplication algorithms can be implemented using $O(n^\omega)$ operations [4,9,16,20], where $\omega < 3$. They rely on algebraic equations following from the possibility of term cancellation. Le Gall and Vassilevska Williams have recently shown the exponent ω of fast matrix multiplication to be smaller than 2.373 in [9,20]. Next, the convolution of two n-dimensional vectors over a commutative ring with the so called principal n-th root of unity can be computed via Fast Fourier Transform using $O(n \log n)$ operations of the ring (Sect. 7 in [1]). On the other hand, Raz proved that if only addition, multiplication and products with constants of absolute value not exceeding one are allowed then $n \times n$ matrix multiplication requires $\Omega(n^2 \log n)$ operations [12].

Yuval was first to describe a reduction of the distance matrix product (equivalently, the $(\min, +)$ matrix product) of two $n \times n$ matrices to matrix multiplication of two $n \times n$ matrices over a ring, using $O(n^2)$ operations [19] (cf. [13]). If $A = (a_{i,j})$ and $B = (b_{i,j})$ are two $n \times n$ matrices then their distance product (equivalently, their $(\min, +)$ matrix product) $C = AB$ is an $n \times n$ matrix $C = (c_{i,j})$ such that $c_{ij} = \min\{a_{i,k} + b_{k,j} | 1 \leq k \leq n\}$ for $1 \leq i, j \leq n$.

The idea of the reduction is relatively simple [2,21]. Two input $n \times n$ matrices $A = (a_{i,j})$ and $B = (b_{i,j})$ with integer entries in $[-M, M]$ are transformed to two $n \times n$ matrices $A' = ((n+1)^{M-a_{i,j}})$ and $B' = ((n+1)^{M-b_{i,j}})$. It is not too difficult to see that if $C = (c_{i,j})$ is the distance product of A and B and $C' = (c'_{i,j})$ is the arithmetic matrix product of A' and B', then $c_{i,j} = 2M - \lfloor \log_{n+1} c'_{i,j} \rfloor$. Note that the reduction uses the exponentiation, logarithm and floor functions besides the arithmetic ring operations.

By combining the reduction with fast matrix multiplication, one obtains an algorithm for the distance matrix product, using $O(n^\omega)$ multiplications, additions and subtractions, and $O(n^2)$ exponentiation, logarithm and floor operations [13,19].

Since the entries in the transformed matrices A', B' are huge numbers that require $O(M)$ computer words of $\log n$ bits each, the matrix multiplication of A' and B' requires $O(Mn^\omega)$ algebraic operations on $O(\log n)$ bit numbers [2,21]. For this reason, the described algorithm for distance matrix product is interesting solely for smaller values of M and approximation purposes [21].

Recently, also a nondeterministic algorithm for $n \times n$ matrix multiplication using $O(n^2)$ arithmetic operations has been presented by Korec and Wiedermann in [8]. It results from a derandomization of Freivalds' randomized algorithm for matrix product verification [6]. Simply, the algorithm first guesses the product matrix and then verifies its correctness. Again, the derandomization involves huge numbers requiring $O(n)$ times more bits than the input numbers [8]. (Very recently, Wiedermann has presented two further, slightly slower nondeterministic algorithms for matrix multiplication, both running in $O(n^2 \log n)$ time and relying on the derandomization of Freivalds' algorithm. The first runs on a real RAM, the second on a unit-cost RAM using only integers of size proportional to that of the largest entry in the input matrices [18]).

In this paper, first, we consider the computational model of arithmetic circuits with non-input gates labeled by elements of a finite set of binary or unary arithmetic operations and input gates labeled by variables ranging over integers, and the arithmetic constants $0, 1$. We observe that for the arithmetic semi-disjoint bilinear forms the so called Shannon effect [17] holds in this computational model. In particular, we show that almost all such forms on $2n$ variables, where each variable occurs in n monomials like in the n-dimensional Boolean vector convolution, have $\Omega(n^2)$ arithmetic circuit complexity. Analogously, we show that almost all arithmetic semi-disjoint bilinear forms on $2n^2$ variables, where each variable occurs in n monomials like in the arithmetic matrix product, have $\Omega(n^3)$ arithmetic circuit complexity. Our results contrast with the aforementioned fast algorithms for the arithmetic vector convolution and matrix multiplication.

Next, we observe that if we allow for the use of division and the floor function (or exponentiation, logarithm and the floor function) besides multiplication, addition and subtraction in the unit-cost RAM model then we can compute the arithmetic convolution of two n-dimensional integer vectors in $O(n)$ steps and perform the arithmetic matrix multiplication of two integer $n \times n$ matrices in $O(n^2)$ steps. Similarly, as in the case of the reduction of distance matrix product to the arithmetic one, the idea is to use numbers requiring about $n \log n$ times more bits than any entry in the input matrices. If we combine the reduction with our algorithm for matrix multiplication, we obtain an algorithm for the distance matrix product using solely $O(n^2)$ operations on huge numbers requiring about $Mn \log n$ words of $\log n$ bits each. Analogously, we obtain an $O(n)$ algorithm for the (min, +) convolution of two n-dimensional integer vectors involving very large numbers.

Our deterministic method for matrix products in a way subsumes the aforementioned nondeterministic quadratic algorithm for matrix multiplication from [8] and the n^ω algorithms for distance product from [13, 19]. The power of large integers and the floor function in the case of computing matrix products appears even greater than that reported in the literature. Our upper time bounds yield analogous upper bounds on the size of the arithmetic circuit with an appropriate finite set of arithmetic operations for the aforementioned problems. It follows that any method for proving superlinear (in the input size) lower bounds for these problems has to assume a more restricted set of arithmetic operations and/or an upper bound on the size of allowed integers.

Finally, we show that the (min, +) integer vector convolution admits an arbitrarily close and fast approximation, e.g., in the logarithmic-cost RAM model, similar to that known for the distance matrix product [21].

2 Preliminaries

For two n-dimensional vectors $a = (a_0, ..., a_{n-1})$ and $b = (b_0, ..., b_{n-1})$ over a semi-ring $(\mathbb{U}, \oplus, \odot)$, their convolution over the semi-ring is a vector $c = (c_0, ..., c_{2n-2})$, where $c_i = \bigoplus_{l=\max\{i-n+1,0\}}^{\min\{i,n-1\}} a_l \odot b_{i-l}$ for $i = 0, ..., 2n-2$. Similarly, for a $p \times q$ matrix A and a $q \times r$ matrix B over the semi ring, their matrix product over the semi-ring is a $p \times r$ matrix C such that $C[i,j] = \bigoplus_{m=1}^{q} A[i,m] \odot B[m,j]$ for $1 \le i \le p$ and $1 \le j \le r$. In particular, for the semi-rings $(\mathbb{Z}, +, \times)$, $(\mathbb{Z}, \min, +)$, $(\mathbb{Z}, \max, +)$, and $(\{0,1\}, \vee, \wedge)$, we obtain the arithmetic, $(\min, +)$, $(\max, +)$, and the Boolean convolutions or matrix products, or semi-disjoint bilinear forms (see Introduction), respectively.

For two n-dimensional vectors $a = (a_0, ..., a_{n-1})$ and $b = (b_0, ..., b_{n-1})$ over a semi-ring $(\mathbb{U}, \oplus, \odot)$, their dot product $\bigoplus_{i=0}^{n-1} a_i \odot b_i$ is denoted by $a \circ b$. Note that the $(n-1)$-th coordinate c_{n-1} of the convolution of a and b is equal to $a \circ b^R$, where $b^R = (b_{n-1}, ..., b_0)$.

3 Lower Bounds on the Arithmetic Circuit Complexity

In this section, we show that for arithmetic semi-disjoint bilinear forms, where each variable occurs in the same given number p of monomials, the so called Shannon effect [17] holds. This means that the number of the aforementioned forms is large enough compared with the number of different arithmetic circuits of bounded size to yield a tight lower bound on the arithmetic circuit complexity for almost all of them. To start with, we need the following definition.

Suppose that we are given a priori a finite set (i.e., of size $O(1)$) of binary arithmetic operations, including multiplication and addition, and unary arithmetic operations (e.g., logarithm, exponentiation or the floor function). An *arithmetic circuit* is a finite directed acyclic graph with the following properties:

1. The indegree of each vertex (termed gate) is either 0, 1 or 2.
2. The source vertices (i.e., vertices with indegree 0 called input gates) are labeled by input variables, ranging over a subset of reals including 0, 1 (e.g., integers), and the arithmetic constants 0, 1.
3. The vertices of indegree 2 are labeled by elements in the aforementioned set of binary arithmetic operations.
4. The vertices of indegree 1 are labeled by elements in the aforementioned set of unary arithmetic operations.

Observe that all results in this section implicitly assume the $O(1)$-size of the aforementioned fixed set of binary and unary arithmetic operations that can be used in an arithmetic circuit.

The arithmetic functions computed at the gates of an arithmetic circuit are naturally defined by induction on the structure of the circuit starting from its input gates and constants. An arithmetic bilinear form composed of k functions is computed by an arithmetic circuit if there are k distinguished gates (called output gates) computing the k functions. The following upper bound is straightforward.

Lemma 1. *Each arithmetic semi-disjoint bilinear form composed of k functions on $x_0, ..., x_{n-1}$ and $y_0, ..., y_{n-1}$ with q monomials in total can be computed by an arithmetic circuit with $q \leq n^2$ multiplication gates and $q - k$ addition gates.*

Proof. First, we use q multiplication gates to compute each monomial $x_i y_j$ separately. Then, we form k disjoint sums of the monomials corresponding to the k functions of the bilinear form using $q - k$ addition gates.   □

The next lemma presents an upper bound on the number of different arithmetic bilinear forms composed of k functions on $2n$ variables that can be computed by arithmetic circuits of bounded size.

Lemma 2. *At most $(s + 2n + 1)^{2s} O(1)^s s(\sum_{l=1}^{k} \binom{2n+2+s}{l}) / s!$ arithmetic bilinear forms composed of at most k functions on $2n$ variables can be computed by arithmetic circuits with s non-input gates.*

Proof. We estimate the number of arithmetic circuits specified in the theorem as follows. For each non-input gate there are at most $2n + s - 1 + 2$ possibilities to choose each of its at most two predecessors among the $2n$ input gates, $s - 1$ remaining gates and the two constants. Each of the gates may be labeled by one of the $O(1)$ arithmetic operations and functions. The at most k output gates can be chosen in $\sum_{l=1}^{k} \binom{2n+2+s}{l}$ ways. On the other hand, each circuit can be counted $s!$ times since there are so many possible numberings of its gates.   □

Let $SBF(n, k, P, p)$ stand for the family of arithmetic semi-disjoint bilinear forms composed of at most k functions on the variables $x_0, ..., x_{n-1}$ and $y_0, ..., y_{n-1}$ with monomials in P such that each variable is contained in exactly p monomials in P. Note that the inequalities $|P| \leq n^2$ and $p \leq n$ hold.

The following lemma presents a lower bound on the cardinality of $SBF(n, k, P, p)$.

Lemma 3. *For $k = \Omega(n)$, and $p \geq 8$, the inequality $|SBF(n, k, P, p)| \geq p^{\Omega(np)} / k!$ holds.*

Proof. Form the regular bipartite graph $G = (X \cup Y, E)$, where $X = \{x_0, ..., x_{n-1}\}$ and $Y = \{y_0, ..., y_{n-1}\}$, and $\{x_i, y_j\}$ is an edge of G iff $x_i y_j$ is a monomial in P. G is regular since each vertex in G has degree p.

Observe that $F \in SBF(n, k, P, p)$ are in one-to-one correspondence with partial colorings of the edges of G with at most k colors. The edges of the same color in a partial edge coloring of G form a matching.

Set G^* to G, and iterate the following steps $\Omega(n)$ times:
Pick a set U of $\lceil p/2 \rceil$ vertices in G^* of smallest numbers that have degree at least $p/2$, and color $\lfloor p/8 \rfloor$ edges of G^* with a new color as follows. Pick a vertex u in U of smallest number that is not yet incident to a colored edge, pick an edge e incident to u whose other endpoint is not incident to a colored edge and color e. Note that there are at least $p/2 - 2(p/8)$ possible choices of e. After coloring $\lfloor p/8 \rfloor$ edges remove the edges from G^* (they define a new function of a bilinear form).

In each iteration, we have at least $(p/4)^{\lfloor p/8 \rfloor}$ possibilities for different sets of edges to be colored and the total degree of G^* drops solely at most by $p/4$. Hence, we can perform $\Omega(n)$ iterations, which leads to the $p^{\Omega(np)}/k!$ lower bound for $k = \Omega(n)$.   □

If all forms in $S(n,k,P,p)$, where $k = \Theta(n)$, $p = n^{\Omega(1)}$, admit arithmetic circuits of size s then the upper bound in terms of s from Lemma 2 should be at least as large as the lower bound on the number of such forms in Lemma 3. Hence, by straightforward calculations we infer that there are arithmetic bilinear forms in $SBF(n,k,P,p)$ whose circuit complexity is at least $\Omega(np)$ as long as $k = \Theta(n)$, $p = n^{\Omega(1)}$. Following the standard proof of the Shannon effect [17] for Boolean functions, we can strengthen this lower bound to include almost all members in $SBF(n,k,P,p)$.

Consider the subfamily $SBF^*(n,k,P,p)$ of $SBF(n,k,P,p)$ consisting of $|SBF(n,k,P,p)|/n$ forms of smallest arithmetic circuit complexity. Analogously, we obtain that for $k = \Theta(n)$, $p = n^{\Omega(1)}$, there are forms in $SBF^*(n,k,P,p)$ of circuit complexity $\Omega(np)$. On the other hand, Lemma 1 with $q = np$ provides the matching upper bound $O(np)$. Hence, we obtain:

Theorem 1. *For $k = \Theta(n)$ and $p = n^{\Omega(1)}$, the arithmetic circuit complexity of almost all bilinear forms in $SBF(n,k,P,p)$ is $\Theta(np)$.*

Note that the n-dimensional arithmetic convolution is in $SBF(2n, 2n - 1, P, n)$ while the $n \times n$ Boolean matrix product is in $SBF(2n^2, n^2, Q, n)$, for appropriate P, Q. By Theorem 1, almost all forms in $SBF(2n, 2n - 1, P, n)$ or $SBF(2n^2, n^2, Q, n)$ require arithmetic circuits of size $\Theta(n^2)$ or $\Theta(n^3)$, respectively.

4   The Arithmetic Algorithms

For an n-dimensional vector $a = (a_0, ..., a_{n-1})$ with integer coordinates let $a(x)$ denote the polynomial $\sum_{k=0}^{n-1} a_k x^k$. The following lemma is folklore (see Sect. 7.4 in [1]).

Lemma 4. *For $k = 0, ..., 2n - 2$, the k-th coordinate c_k of the convolution of the vectors $a = (a_0, ..., a_{n-1})$ and $b = (b_0, ..., b_{n-1})$ is the coefficient at x^k in the polynomial $a(x)b(x)$. Consequently, the coefficient at x^{n-1} is the dot product of a and the reversed vector $b^R = (b_{n-1}, ..., b_0)$, i.e., $\sum_{i=0}^{n-1} a_i b_{n-1-i}$.*

By Lemma 4, we obtain a linear algorithm for the convolution of integer vectors, see Fig. 1.

Theorem 2. *Let n, M, d be natural numbers such that $d \geq 2nM^2 + 1$. For $k = 0, ..., 2n - 2$, the k-th coordinate c_k of the convolution $c = (c_0, ..., c_{2n-2})$ of two integer vectors $a = (a_0, ..., a_{n-1})$ and $b = (b_0, ..., b_{n-1})$, each with n coordinates in $[-M, M]$, is equal to $\lfloor a(d)b(d)d^{-k} + \frac{1}{2} \rfloor - d \lfloor a(d)b(d)d^{-k-1} + \frac{1}{2} \rfloor$. Consequently, the convolution c of the n-dimensional vectors a and b can be computed using $O(n)$ additions, subtractions, multiplications, divisions and floor operations.*

Input: a natural number M and two n-dimensional vectors $a = (a_0, ..., a_{n-1})$ and $b = (b_0, ..., b_{n-1})$ with integer coordinates in $[-M, M]$.
Output: the convolution $c = (c_0, ..., c_{2n-2})$ of a and b.
1: $d \leftarrow 2nM^2 + 1$
2: $a(d) \leftarrow \sum_{l=0}^{n-1} a_l d^l$
3: $b(d) \leftarrow \sum_{l=0}^{n-1} b_l d^l$
4: $c(d) \leftarrow a(d)b(d)$
5: **for** $i = 0$ to $2n - 2$ **do**
6:     $c_i \leftarrow \lfloor c(d)/d^i + \frac{1}{2} \rfloor - d \lfloor c(d)/d^{i+1} + \frac{1}{2} \rfloor$
7: **end for**
8: $c \leftarrow (c_0, ..., c_{2n-2})$
9: **return** c

Fig. 1. A linear algorithm for computing the convolution c of two n-dimensional integer vectors a and b.

Proof. By Lemma 4, we have $a(d)b(d) = \sum_{l=0}^{2n-2} c_l d^l$. Hence, $\lfloor a(d)b(d)d^{-k} + \frac{1}{2} \rfloor = \sum_{l=k}^{2n-2} c_l d^{l-k} + \lfloor \sum_{l=0}^{k-1} c_l d^{l-k} + \frac{1}{2} \rfloor$. On the other hand, for $l = 0, ..., 2n - 2$, $|c_l| \leq nM^2$ and $d \geq 2nM^2 + 1$ hold. Let $p = d - 1$. For $k \geq 1$, we obtain

$$\left| \sum_{l=0}^{k-1} c_l d^l \right| \leq \frac{p}{2} \sum_{l=0}^{k-1} (p+1)^l = \frac{1}{2} p \frac{(p+1)^k - 1}{(p+1) - 1} = \frac{1}{2}(p+1)^k - \frac{1}{2} < \frac{1}{2}(p+1)^k \leq \frac{1}{2} d^k.$$

It follows that $\lfloor \sum_{l=0}^{k-1} c_l d^{l-k} + \frac{1}{2} \rfloor = 0$ and consequently $\lfloor a(d)b(d)d^{-k} + \frac{1}{2} \rfloor = \sum_{l=k}^{2n-2} c_l d^{l-k}$. Analogously, we have $\lfloor a(d)b(d)d^{-k-1} + \frac{1}{2} \rfloor = \sum_{l=k+1}^{2n-2} c_l d^{l-k-1}$. This and the previous inequality yield the equality $c_k = \lfloor a(d)b(d)d^{-k} + \frac{1}{2} \rfloor - d \lfloor a(d)b(d)d^{-k-1} + \frac{1}{2} \rfloor$. Hence, we obtain the algorithm for the convolution vector c depicted in Fig. 1. It uses a linear number of additions, subtractions, multiplications, divisions and floor operations. If M is not given as an input to the algorithm, we can upper bound M^2 by the sum of squares of the coordinates in the vectors a and b. $\square$

Assuming the notation of Theorem 2, we obtain the following corollary.

Corollary 1. *The dot product of two integer vectors $a = (a_0, ..., a_{n-1})$ and $b = (b_0, ..., b_{n-1})$, each with n coordinates in $[-M, M]$, is equal to $\lfloor a(d)b^R(d)d^{-n+1} + \frac{1}{2} \rfloor - d \lfloor a(d)b^R(d)d^{-n} + \frac{1}{2} \rfloor$, where $b^R = (b_{n-1}, ..., b_0)$.*

Corollary 1 yields in turn a quadratic algorithm for the matrix product of two $n \times n$ integer matrices, see Fig. 2.

Theorem 3. *The matrix product C of two $n \times n$ integer matrices A and B can be computed using $O(n^2)$ additions, subtractions, multiplications, divisions and floor operations.*

Proof. Our algorithm depicted in Fig. 2 is as follows. We set the constant d to $2nM^2 + 1$.

Input: a natural number M and two $n \times n$ matrices $A = (a_{i,j})$ and $B = (b_{i,j})$ with integer entries in $[-M, M]$.

Output: the matrix product $C = (c_{i,j})$ of A and B.

1: $d \leftarrow 2nM^2 + 1$
2: **for** $i = 1$ **to** n **do**
3:     $A_{i,*}(d) \leftarrow \sum_{l=1}^{n} a_{i,l} d^{l-1}$
4: **end for**
5: **for** $j = 1$ **to** n **do**
6:     $B_{*,j}(d) \leftarrow \sum_{l=1}^{n} b_{l,j} d^{n-l}$
7: **end for**
8: **for** $i = 1$ **to** n **do**
9:     **for** $j = 1$ **to** n **do**
10:        $C_{i,j}(d) \leftarrow A_{i,*}(d) B_{*,j}(d)$
11:        $c_{i,j} \leftarrow \lfloor C_{i,j}(d)/d^{n-1} + \frac{1}{2} \rfloor - d \lfloor C_{i,j}(d)/d^n + \frac{1}{2} \rfloor$
12:     **end for**
13: **end for**
14: $C \leftarrow (c_{i,j})$
15: **return** C

Fig. 2. A quadratic algorithm for computing the matrix product C of two integer $n \times n$ matrices A and B.

For $i = 1, ..., n$, for the i-th row $(a_{i,1}, ..., a_{i,n})$ of the matrix A, we consider the polynomial $A_{i,*}(x) = \sum_{k=1}^{n} a_{i,k} x^{k-1}$ and compute $A_{i,*}(d)$.

Symmetrically, for $j = 1, ..., n$, for the j-th column $(b_{1,j}, ..., b_{n,j})$ of the matrix B, we consider the polynomial $B_{*,j}(x) = \sum_{k=1}^{n} b_{k,j} x^{n-k}$ and compute $B_{*,j}(d)$.

The computation of $A_{i,*}(d)$ and $B_{*,j}(d)$, for $1 \leq i, \ j \leq n$, requires $O(n^2)$ multiplications and additions.

Finally, for $1 \leq i, \ j \leq n$, we compute the products $A_{i,*}(d) B_{*,j}(d)$, and then $\lfloor A_{i,*}(d) B_{*,j}(d)/d^{n-1} \rfloor - d \lfloor A_{i,*}(d) B_{*,j}(d)/d^n \rfloor$. It requires $O(n^2)$ multiplications, divisions, floor operations and subtractions.

By Corollary 1, in this way, we obtain the correct values of the entries $c_{i,j}$ of the product matrix C. $\qquad \square$

By combining the reduction of the distance matrix product to the arithmetic one outlined in the introduction [2,13,19,21] with Theorem 3, we obtain also the following corollary.

Corollary 2. *The matrix product C of two $n \times n$ matrices over the semi-ring $(\mathbb{Z}, \min, +)$ can be computed using $O(n^2)$ additions, subtractions, multiplications, divisions, and exponentiation, logarithm and floor operations.*

5   (min, +) Convolution

Recently, Bremner et al. have revived the interest in the problem of computing the convolution of two n-dimensional vectors over the semi-ring $(\mathbb{Z}, \min, +)$ [3]. They also provided the first slightly subquadratic algorithm for the $(\min, +)$

vector convolution by using $O(\sqrt{n})$ (min, +) matrix products (in a reasonable computational model).

The aforementioned reduction of the distance (i.e., (min, +)) matrix product to the arithmetic matrix product can be adapted to yield also an analogous reduction of the (min, +) vector convolution to the arithmetic one, see Fig. 3. Hence, we obtain the following corollary by Theorem 2.

Corollary 3. *The convolution of two n-dimensional vectors over the semi-ring $(\mathbb{Z}, \min, +)$ can be computed using $O(n)$ additions, subtractions, multiplications, divisions, and exponentiation, logarithm and floor operations.*

Further, we shall observe that the method of arbitrarily close approximation of the distance product due to Zwick [21] that is based on the reduction to the arithmetic matrix multiplication from [13,19] can be easily adopted to the (min, +) convolution.

Lemma 5. *The algorithm depicted in Fig. 3 computes the (min, +) convolution of two n-dimensional integer vectors whose coordinates are all of absolute value at most M or just $+\infty$ using $\tilde{O}(Mn)$ bit operations, where the $\tilde{O}()$ notation suppresses factors polylogarithmic in $M + n$.*

Proof. For $i = 0, ..., 2n - 2$, we have $c_i' = \sum_{l=\max\{i-n+1,0\}}^{\min\{i,n-1\}} (n+1)^{2M-(a_l+b_{i-l})}$ (note that if not all vector coordinates have their absolute value at most M then the terms $a_l + b_{i-l}$, where $|a_l| > M$ or $|b_{i-l}| > M$ are excluded from the summation). Since the number of the terms $(n+1)^{2M-(a_l+b_{i-l})}$ on the right-hand

---

Input: a natural number M and two n-dimensional vectors $a = (a_0, ..., a_{n-1})$ and
$\quad$ $b = (b_0, ..., b_{n-1})$ with integer coordinates.
Output: the (min, +) convolution $c = (c_0, ..., c_{2n-2})$ of a and b when the coordinates
$\quad$ of a and b are in $[-M, M] \cup \{+\infty\}$.
1: **for** $i = 0$ to $n - 1$ **do**
2: $\quad$ $a_i' \leftarrow$ **if** $|a_i| \leq M$ **then** $(n+1)^{M-a_i}$ **else** 0
3: **end for**
4: $a' \leftarrow (a_0', ..., a_{n-1}')$
5: **for** $i = 0$ to $n - 1$ **do**
6: $\quad$ $b_i' \leftarrow$ **if** $|b_i| \leq M$ **then** $(n+1)^{M-b_i}$ **else** 0
7: **end for**
8: $b' \leftarrow (b_0', ..., b_{n-1}')$
9: $c' \leftarrow$ **fast-arithmetic-convolution**(a', b')
10: **for** $l = 0$ to $2n - 2$ **do**
11: $\quad$ $c_l \leftarrow$ **if** $c_l' > 0$ **then** $2M - \lfloor \log_{n+1} c_l' \rfloor$ **else** $+\infty$
12: **end for**
13: $c \leftarrow (c_0, ..., c_{2n-2})$
14: **return** c

Fig. 3. A linear reduction of the (min, +) convolution of two n-dimensional integer vectors with coordinates in $[-M, M]$ to an arithmetic one.

side of the equality is at most n, we obtain $c_i = \min_{l=\max\{i-n+1,0\}}^{\min\{i,n-1\}} a_l + b_{i-l} = 2M - \lceil \log_{n+1} c_i' \rceil$.

A fast arithmetic convolution algorithm can be obtained for example by embedding the problem of computing the arithmetic vector convolution in a single large integer multiplication and using the classic Schönhage-Strassen integer multiplication algorithm [5]. For n-dimensional vectors with $O(M \log n)$ bit coordinates, the fast convolution algorithm will use $\tilde{O}(Mn)$ bit operations.

The computation of logarithms in the algorithm given in Fig. 3 can be easily implemented by binary search.   □

Lemma 6. *Let $\tilde{c}$ be the $(\min, +)$ convolution of the n-dimensional vectors obtained from a and b, respectively, by replacing each coordinate greater than M by $+\infty$. Let M and ϵ^{-1} be powers of two. Let c be the vector convolution returned by the algorithm depicted in Fig. 4. For $i = 0, \ldots, 2n - 2$, $\tilde{c}_i \leq c_i \leq (1 + \epsilon)\tilde{c}_i$ holds.*

Proof (sketch). The proof is analogous to that of Lemma 8.1 on an approximation algorithm for distance matrix product in [21].

The inequality $\tilde{c}_i \leq c_i$ follows from the fact the elements are always rounded up in the algorithm depicted in Fig. 4. Suppose that $\tilde{c}_i = a_k + b_{i-k}$. We may

Input: a natural number M, two n-dimensional vectors $a = (a_0, \ldots, a_{n-1})$ and $b = (b_0, \ldots, b_{n-1})$ with integer coordinates in $[0, M]$, an inverse of a power of 2 denoted by ϵ.
Output: an approximate convolution $c = (c_0, \ldots, c_{2n-2})$ of a and b.
1: **for** $i = 0$ to $2n - 2$ **do**
2:     $c_i \leftarrow +\infty$
3: **end for**
4: **for** $r = \lceil \log_2 \frac{4}{\epsilon} \rceil$ to $\lceil \log_2 M \rceil$ **do**
5:     **for** $i = 0$ to $n - 1$ **do**
6:         $a_i' \leftarrow$ **if** $0 \leq a_i \leq 2^r$ **then** $\lceil \frac{4a_i}{2^r \epsilon} \rceil$ **else** $+\infty$
7:     **end for**
8:     $a' \leftarrow (a_0', \ldots, a_{2n-2}')$
9:     **for** $i = 0$ to $n - 1$ **do**
10:         $b_i' \leftarrow$ **if** $0 \leq b_i \leq 2^r$ **then** $\lceil \frac{4b_i}{2^r \epsilon} \rceil$ **else** $+\infty$
11:     **end for**
12:     $b' \leftarrow (b_0', \ldots, b_{2n-2}')$
13:     compute the $(\min, +)$ convolution c' of a' and b' using the algorithm depicted in Fig. 3 with M set to $\frac{4}{\epsilon}$
14:     **for** $i = 0$ to $2n - 2$ **do**
15:         $c_i \leftarrow \min\{c_i, \frac{2^r \epsilon}{4} c_i'\}$
16:     **end for**
17: **end for**
18: $c \leftarrow (c_0, \ldots, c_{2n-2})$
19: **return** c

Fig. 4. A fast approximation algorithm for the convolution c of two n-dimensional integer vectors a and b with coordinates in $[0, M]$.

assume w.l.o.g. that $a_k \leq b_{i-k}$ and $2^{s-1} < b_{i-k} \leq 2^s$, where $1 \leq s \leq \log_2 M$. If $s \leq \log_2 4\epsilon^{-1}$ then the first iteration of the algorithm from Fig. 3 returns $c_i = \tilde{c}_i$. Otherwise, in the $r = s$ iteration, we have $\frac{2^r \epsilon}{4} a'_k \leq a_k + \frac{2^r \epsilon}{4}$ and $\frac{2^r \epsilon}{4} b'_{i-k} \leq b_{i-k} + \frac{2^r \epsilon}{4}$. Hence, after that iteration, we have $c_i \leq \frac{2^r \epsilon}{4} a'_k + \frac{2^r \epsilon}{4} b'_{i-k}$ $\leq a_k + b_{i-k} + \frac{2^{r+1} \epsilon}{4} \leq (1 + \epsilon) \tilde{c}_i$. □

By Lemmas 5 and 6 we obtain our main result in this section.

Theorem 4. *The algorithm depicted in Fig. 4 computes $1 + \epsilon$ approximations of the coordinates of the $(\min, +)$ convolution of two n-dimensional integer vectors whose coordinates are all of absolute value at most M using $\tilde{O}(\epsilon^{-1} n \log M)$ bit operations, where the $\tilde{O}()$ notation suppresses factors polylogarithmic in $\epsilon^{-1} + n + \log M$.*

6 Upper Bounds on the Arithmetic Circuit Complexity

In the exact arithmetic algorithms presented in the previous sections, we do not need to know the range M of the integer input variables. Instead, we can compute the sum of squares of the values of the input variables plus, say, 1. The computation of the sum requires a linear (in the number of input variables) number of multiplication and additions. When we fix the number of input variables, the aforementioned algorithms become oblivious, i.e., the type of operation performed in a given step does not depend on the input. Hence, they can be implemented by appropriate arithmetic circuits of sizes corresponding to their time performances in the unit-cost RAM model. We obtain the following theorem by Theorems 2, 3 and Corollaries 2, 3.

Theorem 5. *The convolution of two n-dimensional vectors over the semi-rings $(\mathbb{Z}, +, \times)$ and $(\mathbb{Z}, \min, +)$ can be computed by arithmetic circuits using $O(n)$ addition, subtraction, multiplication, division, exponentiation, logarithm and floor operation gates. Similarly, the matrix product C of two $n \times n$ matrices over the semi-rings $(\mathbb{Z}, +, \times)$ and $(\mathbb{Z}, \min, +)$ can be computed by arithmetic circuits using $O(n^2)$ addition, subtraction, multiplication, division, exponentiation, logarithm and floor operation gates.*

7 Final Remarks

For the majority of algorithmic community interested in matrix products and vector convolution, the main motivation is the design of faster algorithms for the aforementioned problems in a reasonable complexity model. However, in the case of this paper, our motivations, even for our exact algorithms, have been different.

The unrealistic power of unit-cost Random Access Machine using besides addition and subtraction also multiplication, division, and some additional operations (e.g., the floor function or/and bitwise operations) allowing for encoding long vectors or strings in huge numbers has been known since the 1970s [7,15].

In Sect. 4, we confirm the power of such a unrealistic, computational complexity model for the central problems of matrix multiplication, vector convolution, and the distance matrix product. Our study has been inspired by the known reductions of the distance matrix product to matrix multiplication in the aforementioned model [13,19]. It follows that in order to derive non-trivial lower bounds on the complexity of matrix multiplication or vector convolution one has to have some restriction on the size of used integers or the set of unit-cost operations. On the other hand, our lower bounds of Sect. 3 indicate that for almost all semi-disjoint bilinear forms even the use of the aforementioned unrealistically powerful model cannot asymptotically improve naive algorithms.

Acknowledgments. The authors are grateful to Christos Levcopoulos for valuable comments. This research has been supported in part by Swedish Research Council grant 621-2011-6179.

References

1. Aho, A.V., Hopcroft, J.E., Ullman, J.: The Design and Analysis of Computer Algorithms. Addison-Wesley Publishing Company, Reading (1974)
2. Alon, N., Galil, Z., Margalit, O.: On the exponent of all pairs shortest path problem. J. Comput. Syst. Sci. **54**, 25–51 (1997)
3. Bremner, D., Chan, T.M., Demaine, E.D., Erickson, J., Hurtado, F., Iacono, J., Langerman, S., Patrascu, M., Taslakian, P.: Necklaces, convolutions and X+Y. Algorithmica **69**, 294–314 (2014)
4. Coppersmith, D., Winograd, S.: Matrix multiplication via arithmetic progressions. J. Symbolic Comput. **9**, 251–280 (1990)
5. Fisher, M.J., Paterson, M.S.: String-matching and other products. In: Proceedings of the 7th SIAM-AMS Complexity of Computation, pp. 113–125 (1974)
6. Freivalds, R.: Probabilistic machines can use less running time. In: Proceedings of the IFIP Congress, pp. 839–842 (1977)
7. Hartmanis, J., Simon, J.: On the power of multiplication in random access machines. In: Proceedings of the SWAT (FOCS), pp. 13–23 (1974)
8. Korec, I., Wiedermann, J.: Deterministic verification of integer matrix multiplication in quadratic time. In: Geffert, V., Preneel, B., Rovan, B., Štuller, J., Tjoa, A.M. (eds.) SOFSEM 2014. LNCS, vol. 8327, pp. 375–382. Springer, Cham (2014). doi:10.1007/978-3-319-04298-5_33
9. Le Gall, F.: Powers of tensors and fast matrix multiplication. In: Proceedings of the 39th International Symposium on Symbolic and Algebraic Computation, pp. 296–303 (2014)
10. Mehlhorn, K., Galil, Z.: Monotone switching circuits and boolean matrix product. Computing **16**, 99–111 (1976)
11. Pratt, R.: The power of negative thinking in multiplying boolean matrices. SIAM J. Comput. **4**(3), 326–330 (1975)
12. Raz, R.: On the complexity of matrix product. In: Proceedings of the STOC, pp. 144–151 (2002)
13. Romani, F.: Shortest path problem is not harder than matrix multiplications. Inf. Process. Lett. **4**(6), 134–136 (1980)
14. Schnorr, C.-P.: A lower bound on the number of additions in monotone computations. Theoret. Comput. Sci. **2**(3), 305–315 (1976)

15. Schönhage, A.: On the power of random access machines. In: Maurer, H.A. (ed.) ICALP 1979. LNCS, vol. 71, pp. 520–529. Springer, Heidelberg (1979). doi:10. 1007/3-540-09510-1_42
16. Strassen, V.: Gaussian elimination is not optimal. Numer. Math. **13**, 354–356 (1969)
17. Wegener, I.: The Complexity of Boolean Functions. Wiley-Teubner Series in Computer Science, New York, Stuggart (1987)
18. Wiedermann, J.: Fast nondeterministic matrix multiplication via derandomization of freivalds' algorithm. In: Diaz, J., Lanese, I., Sangiorgi, D. (eds.) TCS 2014. LNCS, vol. 8705, pp. 123–135. Springer, Heidelberg (2014). doi:10.1007/ 978-3-662-44602-7_11
19. Yuval, G.: An algorithm for finding all shortest paths using $N^{2.81}$ infinite-precision multiplication. Inf. Process. Lett. **11**(3), 155–156 (1976)
20. Vassilevska Williams, V.: Multiplying matrices faster than coppersmith-winograd. In: Proceedings of the 44th Annual ACM Symposium on Theory of Computing (STOC), pp. 887–898 (2012)
21. Zwick, U.: All pairs shortest paths using bridging sets and rectangular matrix multiplication. J. ACM **49**(3), 289–317 (2002)

Bounding the Dimension of Points on a Line

Neil Lutz[1]([✉]) and D.M. Stull[2]

[1] Department of Computer Science, Rutgers University, Piscataway, NJ 08854, USA
njlutz@rutgers.edu
[2] Department of Computer Science, Iowa State University, Ames, IA 50011, USA
dstull@iastate.edu

Abstract. We use Kolmogorov complexity methods to give a lower bound on the effective Hausdorff dimension of the point $(x, ax + b)$, given real numbers a, b, and x. We apply our main theorem to a problem in fractal geometry, giving an improved lower bound on the (classical) Hausdorff dimension of generalized sets of Furstenberg type.

1 Introduction

In this paper we exploit fundamental connections between fractal geometry and information theory to derive both algorithmic and classical dimension bounds in the Euclidean plane.

Effective fractal dimensions, originally conceived by J. Lutz to analyze computational complexity classes [11,12], quantify the *density of algorithmic information* in individual infinite data objects. Although these dimensions were initially defined—and have primarily been studied—in Cantor space [4], they have been shown to be geometrically meaningful in Euclidean spaces and general metric spaces, and their behavior in these settings has been an active area of research (e.g., [3,7,20]).

This paper focuses on the *effective Hausdorff dimension*, $\dim(x)$, of individual points $x \in \mathbb{R}^n$, which is a potentially non-zero value that depends on the Kolmogorov complexity of increasingly precise approximations of x [18]. Given the pointwise nature of this quantity, it is natural to investigate the *dimension spectrum* of a set $E \subseteq \mathbb{R}^n$, i.e., the set $\{\dim(x) : x \in E\}$. Even for apparently simple sets, the structure of the dimension spectrum may not be obvious, as exemplified by a longstanding open question originally posed by J. Lutz [21]: *Is there a straight line $L \subseteq \mathbb{R}^2$ such that every point on L has effective Hausdorff dimension 1?*

J. Lutz and Weihrauch [16] have shown that the set of points in $\mathbb{R}^n$ with dimension less than 1 is totally disconnected, as is the set of points with dimension greater than $n - 1$. Turetsky has shown that the set of points in $\mathbb{R}^n$ of

---

N. Lutz—Research supported in part by National Science Foundation Grant 1445755.
D.M. Stull—Research supported in part by National Science Foundation Grants 1247051 and 1545028.

T.V. Gopal et al. (Eds.): TAMC 2017, LNCS 10185, pp. 425–439, 2017.
DOI: 10.1007/978-3-319-55911-7_31

dimension exactly 1 is connected [25], which implies that every line in $\mathbb{R}^2$ contains a point of dimension 1. J. Lutz and N. Lutz have shown that almost every point on any line with random slope has dimension 2 [14], despite the surprising fact that there are lines in every direction that contain no random points [13]. These results give insight into the dimension spectra of lines, but they also leave open the question of whether or not a line in $\mathbb{R}^2$ can have a singleton dimension spectrum.

We resolve this question in the negative with our main theorem, a general lower bound on the dimension of points on lines in $\mathbb{R}^2$. Our bound depends only on the dimension of the description (a, b) of the line (i.e., the ordered pair giving the line's slope and vertical intercept) and the dimension of the coordinate x relative to (a, b).

Theorem 1. *For all $a, b, x \in \mathbb{R}$,*

$$\dim(x, ax + b) \geq \dim^{a,b}(x) + \min \left\{ \dim(a, b), \dim^{a,b}(x) \right\}.$$

In particular, for almost every $x \in \mathbb{R}$, $\dim(x, ax + b) = 1 + \min\{\dim(a, b), 1\}$.

Since $\dim(0, b) \leq \min\{\dim(a, b), 1\}$, the second statement implies that every line contains two points whose dimensions differ by at least 1, and therefore that the dimension spectrum cannot be a singleton.

This theorem also implies a new result in classical geometric measure theory. It has been known for more than a decade [8] that for certain classes of sets,

$$\sup_{x \in E} \dim(x) = \dim_H(E), \tag{1}$$

where $\dim_H(E)$ is the (classical) Hausdorff dimension of E, i.e., the most standard notion of fractal dimension. Although (1) does not hold in general, this correspondence suggested that effective dimensions might provide new techniques for dimension bounds in classical fractal geometry.

A recent *point-to-set principle* of J. Lutz and N. Lutz [14] reinforces that prospect by characterizing the Hausdorff dimension of arbitrary sets in terms of effective dimension. This principle shows that for every set $E \subseteq \mathbb{R}^n$ there is an oracle relative to which (1) holds. In the same work, that principle is applied to give a new proof of an old result in fractal geometry. Namely, it gives an algorithmic information theoretic proof of Davies's 1971 theorem [2] stating that every *Kakeya set* in $\mathbb{R}^2$—i.e., every plane set that contains a unit segment in every direction— has Hausdorff dimension 2.

In this work, we apply the same point-to-set principle to derive a new result in classical fractal geometry from our main theorem. *Furstenberg sets* generalize Kakeya sets in $\mathbb{R}^2$; instead of containing segments in every direction, they contain α-(Hausdorff)-dimensional subsets of lines in every direction, for some parameter $\alpha \in (0, 1]$. While the theorem of Davies gives the minimum Hausdorff dimension of Kakeya sets in $\mathbb{R}^2$, the minimum Hausdorff dimension of Furstenberg sets is an important open question; the best known lower bound is $\alpha + \max\{1/2, \alpha\}$.[1]

---

[1] According to Wolff [27], this result is due, "in all probability," to Furstenberg and Katznelson. See [24] for a survey.

Molter and Rela [22] generalized this notion further by requiring α-dimensional subsets of lines in only a β-dimensional set of directions, for some second parameter $\beta \in (0, 1]$. They showed that any such set has Hausdorff dimension at least $\alpha + \max\{\beta/2, \alpha+\beta-1\}$. In Theorem 12, we give a lower bound of $\alpha + \min\{\beta, \alpha\}$, which constitutes an improvement whenever $\alpha, \beta < 1$ and $\beta/2 < \alpha$.

For the sake of self-containment, we begin in Sect. 2 with a short review of Kolmogorov complexity and effective Hausdorff dimension, along with some necessary technical lemmas. We discuss and prove our main theorem in Sect. 3, and we apply it to generalized Furstenberg sets in Sect. 4. We conclude with a brief comment on future directions.

2   Algorithmic Information Preliminaries

2.1   Kolmogorov Complexity in Discrete Domains

The *conditional Kolmogorov complexity* of $\sigma \in \{0,1\}^*$ given $\tau \in \{0,1\}^*$ is

$$K(\sigma|\tau) = \min_{\pi \in \{0,1\}^*} \{\ell(\pi) : U(\pi, \tau) = \sigma\},$$

where U is a fixed universal prefix-free Turing machine and $\ell(\pi)$ is the length of π. Any π that achieves this minimum is said to *testify* to, or be a *witness* to, the value $K(\sigma|\tau)$. The *Kolmogorov complexity* of σ is $K(\sigma) = K(\sigma|\lambda)$, where λ is the empty string. $K(\sigma)$ may also be called the *algorithmic information content* of σ. An important property of Kolmogorov complexity is the *symmetry of information* (attributed to Levin in [6]):

$$K(\sigma|\tau, K(\tau)) + K(\tau) = K(\tau|\sigma, K(\sigma)) + K(\sigma) + O(1).$$

These definitions and this symmetry extend naturally to other discrete domains (e.g., integers, rationals, and tuples thereof) via standard binary encodings. See [4, 10, 23] for detailed discussion of these topics.

2.2   Kolmogorov Complexity in Euclidean Spaces

The above definitions may also be lifted to Euclidean spaces by introducing variable precision parameters [14, 15]. Let $x \in \mathbb{R}^m$, and let $r, s \in \mathbb{N}$.[2] For $\varepsilon > 0$, $B_\varepsilon(x)$ denotes the open ball of radius ε centered on x.

The *Kolmogorov complexity of x at precision r* is

$$K_r(x) = \min\{K(p) : p \in B_{2^{-r}}(x) \cap \mathbb{Q}^m\}.$$

The *conditional Kolmogorov complexity of x at precision r given $q \in \mathbb{Q}^m$* is

$$\hat{K}_r(x|q) = \min\{K(p|q) : p \in B_{2^{-r}}(x) \cap \mathbb{Q}^m\}.$$

---

[2] As a matter of notational convenience, if we are given a nonintegral positive real as a precision parameter, we will always round up to the next integer. For example, $K_r(x)$ denotes $K_{\lceil r \rceil}(x)$ whenever $r \in (0, \infty)$.

The *conditional Kolmogorov complexity of x atprecision r given* $y \in \mathbb{R}^n$ *at precision s* is

$$K_{r,s}(x|y) = \max\left\{\hat{K}_r(x|q) : q \in B_{2^{-s}}(y) \cap \mathbb{Q}^n\right\}.$$

We abbreviate $K_{r,r}(x|y)$ by $K_r(x|y)$.

Although definitions based on K-minimizing rationals are better suited to computable analysis [26], it is sometimes useful to work instead with initial segments of infinite binary sequences. It has been informally observed that $K_r(x) = K(x{\upharpoonright}r) + o(r)$, where $x{\upharpoonright}r$ denotes the truncation of the binary expansion of each coordinate of x to r bits to the right of the binary point.

The following pair of lemmas show that the above definitions are only linearly sensitive to their precision parameters. Intuitively, making an estimate of a point slightly more precise only requires a small amount of information.

Lemma 2 (Case and J. Lutz [1]). *There is a constant* $c \in \mathbb{N}$ *such that for all* $n, r, s \in \mathbb{N}$ *and* $x \in \mathbb{R}^n$,

$$K_r(x) \le K_{r+s}(x) \le K_r(x) + K(r) + ns + a_s + c,$$

where $a_s = K(s) + 2\log(\lceil\frac{1}{2}\log n\rceil + s + 3) + (\lceil\frac{1}{2}\log n\rceil + 3)n + K(n) + 2\log n$.

Lemma 3 (J. Lutz and N. Lutz [14]). *For each* $m, n \in \mathbb{N}$, *there is a constant* $c \in \mathbb{N}$ *such that, for all* $x \in \mathbb{R}^m$, $y \in \mathbb{R}^n$, $q \in \mathbb{Q}^n$, *and* $r, s, t \in \mathbb{N}$,

(i) $\hat{K}_r(x|q) \le \hat{K}_{r+s}(x|q) \le \hat{K}_r(x|q) + ms + 2\log(1 + s) + K(r,s) + c$.
(ii) $K_{r,t}(x|y) \ge K_{r,t+s}(x|y) \ge K_{r,t}(x|y) - ns - 2\log(1 + s) + K(t,s) + c$.

In Euclidean spaces, we have a weaker version of symmetry of information, which we will use in the proof of Lemma 7.[3]

Lemma 4 *For every* $m, n \in \mathbb{N}$, $x \in \mathbb{R}^m$, $y \in \mathbb{R}^n$, *and* $r, s \in \mathbb{N}$ *with* $r \ge s$,

(i) $|K_r(x|y) + K_r(y) - K_r(x,y)| \le O(\log r) + O(\log\log\|y\|)$.
(ii) $|K_{r,s}(x|x) + K_s(x) - K_r(x)| \le O(\log r) + O(\log\log\|x\|)$.

Statement (i) is a minor refinement of Theorem 3 of [14], which treats x and y as constant and states that $K_r(x,y) = K_r(x|y) + K_r(y) + o(r)$. In fact, a precise sublinear term is implicit in earlier work by tracing back through several proofs in [1,14]. Our approach is more direct and is omitted here.

2.3   Effective Hausdorff Dimension

If $K_r(x)$ is the algorithmic information content of $x \in \mathbb{R}^n$ at precision r, then we may call $K_r(x)/r$ the *algorithmic information density* of x at precision r. This quantity need not converge as $r \to \infty$, but it does have finite asymptotes between

---

[3] Regarding asymptotic notation, we will treat dimensions of Euclidean spaces (i.e., m and n) as constant throughout this work but make other dependencies explicit, either as subscripts or in the text.

0 and n, inclusive [15]. Although effective Hausdorff dimension was initially developed by J. Lutz using generalized martingales [12], it was later shown by Mayordomo [18] that it may be equivalently defined as the lower asymptote of the density of algorithmic information. That is the characterization we use here. For more details on the history of connections between Hausdorff dimension and Kolmogorov complexity, see [4,19].

The *(effective Hausdorff) dimension* of $x \in \mathbb{R}^n$ is

$$\dim(x) = \liminf_{r \to \infty} \frac{K_r(x)}{r}.$$

This formulation has led to the development of other information theoretic apparatus for effective dimensions, namely mutual and conditional dimensions [1,14]. We use the latter in this work, including in the restatement of our main theorem in Sect. 3. The *conditional dimension* of $x \in \mathbb{R}^m$ given $y \in \mathbb{R}^n$ is

$$\dim(x|y) = \liminf_{r \to \infty} \frac{K_r(x|y)}{r}.$$

2.4 Algorithmic Information Relative to an Oracle

The above algorithmic information quantities may be defined relative to any oracle set $A \subseteq \mathbb{N}$. The *conditional Kolmogorov complexity relative to A* of $\sigma \in \{0,1\}^*$ *given* $\tau \in \{0,1\}^*$ is

$$K^A(\sigma|\tau) = \min_{\pi \in \{0,1\}^*} \{|\pi| : U^A(\pi,\tau) = \sigma\},$$

where U is now a universal prefix-free oracle machine and the computation $U^A(\pi,\tau)$ is performed with oracle access to A. This change to the underlying Turing machine also induces a relativized version of each other algorithmic information theoretic quantity we have defined.

Multiple oracle sets may be combined by simply interleaving them: given $A_1, \ldots, A_k \subseteq \mathbb{N}$, let $A = \bigcup_i \{kj - i + 1 : j \in A_i\}$. Then $K^{A_1,\ldots,A_k}(x)$ denotes $K^A(x)$. We will also consider algorithmic information relative to points in Euclidean spaces. For $y \in \mathbb{R}^n$, let $A_y \subseteq \mathbb{N}$ encode the interleaved binary expansions of y's coordinates in some standard way. Then $K_r^y(x)$ denotes $K_r^{A_y}(x)$. We will make repeated use of the following relationship between conditional and relative Kolmogorov complexity and dimension.

Lemma 5 (J. Lutz and N. Lutz [14]). *For each $m, n \in \mathbb{N}$, there is a constant $c \in \mathbb{N}$ such that, for all $x \in \mathbb{R}^m$, $y \in \mathbb{R}^n$, and $r, t \in \mathbb{N}$,*

$$K_r^y(x) \leq K_{r,t}(x|y) + K(t) + c.$$

In particular, $\dim^y(x) \leq \dim(x|y)$.

In pursuing a dimensional lower bound, we will use the fact that high-dimensional points are very common. Relative to any oracle $A \subseteq \mathbb{N}$, it follows from standard counting arguments that almost every point $x \in \mathbb{R}^n$ has

$\dim^A(x) = n$ and is furthermore *Martin-Löf random* relative to A, meaning there is some constant c such that, for all $r \in \mathbb{N}$, $K_r^A(x) \geq nr - c$ [4].

Finally, we note that all results in this paper, with unmodified proofs, relative to any given oracle. We present the unrelativized versions only to avoid notational clutter.

3   Bounding the Dimension of $(x, ax + b)$

In this section we prove Theorem 1, our main theorem. We first restate it in the form we will prove, which is slightly stronger than its statement in Sect. 1. The dimension of x in the first term is conditioned on—instead of relative to—(a, b), and even when working relative to an arbitrary oracle A, the last term $\dim^{a,b}(x)$ remains unchanged.

Theorem 1 (Restated). *For every $a, b, x \in \mathbb{R}$ and $A \subseteq \mathbb{N}$,*

$$\dim^A(x, ax + b) \geq \dim^A(x|a, b) + \min\left\{\dim^A(a, b),\ \dim^{a,b}(x)\right\}.$$

In particular, for almost every $x \in \mathbb{R}$, $\dim(x, ax + b) = 1 + \min\{\dim(a, b), 1\}$.

To prove this theorem, we proceed in three major steps, which we first sketch at a very high level here. In Sect. 3.1, we give sufficient conditions, at a given precision r, for a point $(x, ax + b)$ to have information content $K_r(x, ax + b)$ approaching $K_r(a, b, x)$. Notice that this is essentially the maximum possible value for $K_r(x, ax + b)$, since an estimate for (a, b, x) has enough information to estimate $(x, ax + b)$ to similar precision. Informally, the conditions are

(i) $K_r(a, b)$ is small.
(ii) If $ux + v = ax + b$, then either $K_r(u, v)$ is large or (u, v) is close to (a, b).

We show in Lemma 6 that when these conditions hold, we can algorithmically estimate (a, b, x) given an estimate for $(x, ax + b)$. In Sect. 3.2, we give a lower bound, Lemma 7, on $K_r(u, v)$ in terms of $\|(u, v) - (a, b)\|$, essentially showing that condition (ii) holds. Finally, we prove Theorem 1 in Sect. 3.3 by showing that there is an oracle which allows (a, b) to satisfy condition (i) without disrupting condition (ii) or too severely lowering $K_r(x, a, b)$.

3.1   Sufficient Conditions for a High-Complexity Point

Suppose that x, a, and b satisfy conditions (i) and (ii) above. Then, given an estimate q for the point $(x, ax + b)$, a machine can estimate (a, b) by simply running all short programs until some output approximates a pair (u, v) such that the line $L_{u,v} = \{(x, ux + v) : x \in \mathbb{R}\}$ passes near q. Since (u, v) was approximated by a short program, it has low information density and is therefore close to (a, b) by condition (ii). We formalize this intuition in the following lemma.

Lemma 6. *Suppose that $a, b, x \in \mathbb{R}$, $r \in \mathbb{N}$, $\delta \in \mathbb{R}_+$, and $\varepsilon, \eta \in \mathbb{Q}_+$ satisfy $r \geq \log(2|a| + |x| + 5) + 1$ and the following conditions.*

(i) $K_r(a, b) \leq (\eta + \varepsilon) r$.

(ii) For every $(u, v) \in B_1(a, b)$ such that $ux + v = ax + b$,

$$K_r(u, v) \geq (\eta - \varepsilon) r + \delta \cdot (r - t),$$

whenever $t = -\log \|(a, b) - (u, v)\| \in (0, r]$.

Then for every oracle set $A \subseteq \mathbb{N}$,

$$K_r^A(x, ax + b) \geq K_r^A(a, b, x) - \frac{4\varepsilon}{\delta} r - K(\varepsilon) - K(\eta) - O_{a,b,x}(\log r).$$

Proof. Let a, b, x, r, δ, ε, η, and A be as described in the lemma statement.

Define an oracle Turing machine M that does the following given oracle A and input $\pi = \pi_1 \pi_2 \pi_3 \pi_4 \pi_5$ such that $U^A(\pi_1) = (q_1, q_2) \in \mathbb{Q}^2$, $U(\pi_2) = h \in \mathbb{Q}^2$, $U(\pi_3) = s \in \mathbb{N}$, $U(\pi_4) = \zeta \in \mathbb{Q}$, and $U(\pi_5) = \iota \in \mathbb{Q}$.

For every program $\sigma \in \{0, 1\}^*$ with $\ell(\sigma) \leq (\iota + \zeta)s$, in parallel, M simulates $U(\sigma)$. If one of the simulations halts with some output $(p_1, p_2) \in \mathbb{Q}^2 \cap B_{2^{-1}}(h)$ such that $|p_1 q_1 + p_2 - q_2| < 2^{-s}(|p_1| + |q_1| + 3)$, then M halts with output (p_1, p_2, q_1). Let c_M be a constant for the description of M.

Now let π_1, π_2, π_3, π_4, and π_5 testify to $K_r^A(x, ax+b)$, $K_1(a, b)$, $K(r)$, $K(\varepsilon)$, and $K(\eta)$, respectively, and let $\pi = \pi_1 \pi_2 \pi_3 \pi_4 \pi_5$.

By condition (i), there is some $(\hat{p}_1, \hat{p}_2) \in B_{2^{-r}}(a, b)$ such that $K(\hat{p}_1, \hat{p}_2) \leq (\eta + \varepsilon) r$, meaning that there is some $\hat{\sigma} \in \{0, 1\}^*$ with $\ell(\hat{\sigma}) \leq (\eta + \varepsilon) r$ and $U(\hat{\sigma}) = (\hat{p}_1, \hat{p}_2)$. A routine calculation shows that

$$|\hat{p}_1 q_1 + \hat{p}_2 - q_2| < 2^{-r}(|\hat{p}_1| + |q_1| + 3),$$

for every $(q_1, q_2) \in B_{2^{-r}}(x, ax+b)$, so M is guaranteed to halt on input π. Hence, let $(p_1, p_2, q_1) = M(\pi)$. Another routine calculation shows that there is some

$$(u, v) \in B_{2^{\gamma-r}}(p_1, p_2) \subseteq B_{2^{-1}}(p_1, p_2) \subseteq B_{2^0}(a, b)$$

such that $ux + v = ax + b$, where $\gamma = \log(2|a| + |x| + 5)$.

We have $\|(p_1, p_2) - (u, v)\| < 2^{\gamma-r}$ and $|q_1 - x| < 2^{-r}$, so

$$(p_1, p_2, q_1) \in B_{2^{\gamma+1-r}}(u, v, x).$$

It follows that

$$
\begin{aligned}
K_{r-\gamma-1}^A(u, v, x) &\leq \ell(\pi_1 \pi_2 \pi_3 \pi_4 \pi_5) + c_M \\
&\leq K_r^A(x, ax + b) + K_1(a, b) + K(r) + K(\varepsilon) + K(\eta) + c_M \\
&= K_r^A(x, ax + b) + K(\varepsilon) + K(\eta) + O_{a,b}(\log r).
\end{aligned}
$$

Rearranging and applying Lemma 2,

$$K_r^A(x, ax + b) \geq K_r^A(u, v, x) - K(\varepsilon) - K(\eta) - O_{a,b,x}(\log r). \qquad (2)$$

By the definition of t, if $t > r$ then $B_{2^{-r}}(u, v, x) \subseteq B_{2^{1-r}}(a, b, x)$, which implies $K_r^A(u, v, x) \geq K_{r-1}^A(a, b, x)$. Applying Lemma 2 gives

$$K_r^A(u, v, x) \geq K_r^A(a, b, x) - O_{a,x}(\log r).$$

Otherwise, when $t \leq r$, we have $B_{2^{-r}}(u, v, x) \subseteq B_{2^{1-t}}(a, b, x)$, which implies $K_r^A(u, v, x) \geq K_{t-1}^A(a, b, x)$, so by Lemma 2,

$$K_r^A(u, v, x) \geq K_r^A(a, b, x) - 2(r - t) - O_{a,x}(\log r). \tag{3}$$

We now bound $r - t$. By our construction of M and Lemma 2,

$$\begin{aligned}
(\eta + \varepsilon)r &\geq K(p_1, p_2) \\
&\geq K_{r-\gamma}(u, v) \\
&\geq K_r(u, v) - O_{a,x}(\log r).
\end{aligned}$$

Combining this with condition (ii) in the lemma statement and simplifying yields

$$r - t \leq \frac{2\varepsilon}{\delta} r + O_{a,x}(\log r),$$

which, together with (2) and (3), gives the desired result.   □

3.2   Bounding the Complexity of Lines Through a Point

In this section we bound the information content of any pair (u, v) such that the line $L_{u,v}$ intersects $L_{a,b}$ at x. Intuitively, an estimate for (u, v) gives significant information about (a, b) whenever $L_{u,v}$ and $L_{a,b}$ are nearly coincident. On the other hand, estimates for (a, b) and (u, v) passing through x together give an estimate of x whose precision is greatest when $L_{a,b}$ and $L_{u,v}$ are nearly orthogonal. We make this dependence on $\|(a, b) - (u, v)\|$ precise in the following lemma.

Lemma 7. *Let* $a, b, x \in \mathbb{R}$. *For all* $u, v \in B_1(a, b)$ *such that* $ux + v = ax + b$, *and for all* $r \geq t := -\log \|(a, b) - (u, v)\|$,

$$K_r(u, v) \geq K_t(a, b) + K_{r-t,r}(x | a, b) - O_{a,b,x}(\log r).$$

Proof. Fix $a, b, x \in \mathbb{R}$. By Lemma 4(i), for all $(u, v) \in B_1(a, b)$ and every $r \in \mathbb{N}$,

$$K_r(u, v) \geq K_r(u, v | a, b) + K_r(a, b) - K_r(a, b | u, v) - O_{a,b}(\log r). \tag{4}$$

We bound $K_r(a, b) - K_r(a, b | u, v)$ first. Since $(u, v) \in B_{2^{-t}}(a, b)$, for every $r \geq t$ we have $B_r(u, v) \subseteq B_{2^{1-t}}(a, b)$, so

$$K_r(a, b | u, v) \leq K_{r,t-1}(a, b | a, b).$$

By Lemma 4(ii), then,

$$\begin{aligned}
K_r(a, b) - K_r(a, b | u, v) &\geq K_r(a, b) - K_{r,t-1}(a, b | a, b) \\
&\geq K_{t-1}(a, b) - O_{a,b}(\log r).
\end{aligned}$$

Lemma 2 tells us that

$$K_{t-1}(a, b) \geq K_t(a, b) - O(\log t).$$

Therefore we have, for every $u, v \in B_1(a, b)$ and every $r \geq t$,

$$K_r(a, b) - K_r(a, b|u, v) \geq K_t(a, b) - O_{a,b}(\log r). \tag{5}$$

We now bound the term $K_r(u, v|a, b)$. Let $(u, v) \in \mathbb{R}^2$ be such that $ux + v = ax + b$. If $t \leq r < t + |x| + 2$, then $r - t = O_x(1)$, so by Lemma 3(ii), $K_{r-t,r}(x|a, b) = O_x(1)$. In this case, $K_r(u, v|a, b) \geq K_{r-t,r}(x|a, b) - O_{a,b,x}(\log r)$ holds trivially. Hence, assume $r \geq t + |x| + 2$.

Let M be a Turing machine such that, whenever $q = (q_1, q_2) \in \mathbb{Q}^2$ and $U(\pi, q) = p = (p_1, p_2) \in \mathbb{Q}^2$, with $p_1 \neq q_1$,

$$M(\pi, q) = \frac{p_2 - q_2}{p_1 - q_1}.$$

For each $q \in B_{2-r}(a, b) \cap \mathbb{Q}^2$, let π_q testify to $\hat{K}_r(u, v|q)$. Then

$$U(\pi_q, q) \in B_{2-r}(u, v) \cap \mathbb{Q}^2.$$

It follows by a routine calculation that

$$|M(\pi_q, q) - x| = \left| \frac{p_2 - q_2}{p_1 - q_1} - \frac{b - v}{a - u} \right| < 2^{4+2|x|+t-r}.$$

Thus, $M(\pi_q, q) \in B_{2^{4+2|x|+t-r}}(x) \cap \mathbb{Q}^2$. For some constant c_M, then,

$$\hat{K}_{r-4-2|x|-t}(x|q) \leq \ell(\pi_q) + c_M$$
$$= \hat{K}_r(u, v|q) + c_M.$$

Taking the maximum of each side over $q \in B_{2-r}(a, b) \cap \mathbb{Q}^2$ and rearranging,

$$K_r(u, v|a, b) \geq K_{r-4-2|x|-t,r}(x|a, b) - c_M.$$

Then since Lemma 3(ii) implies that

$$K_{r-4-2|x|-t,r}(x|a, b) \geq K_{r-t,r}(x|a, b) - O_x(\log r),$$

we have shown, for every (u, v) satisfying $ux + v = ax + b$ and every $r \geq t$,

$$K_r(u, v|a, b) \geq K_{r-t,r}(x|a, b) - O_{a,b,x}(\log r). \tag{6}$$

The lemma follows immediately from (4), (5), and (6). $\qquad \square$

3.3   Proof of Main Theorem

To prove Theorem 1, we will show at every precision r that there is an oracle relative to which the hypotheses of Lemma 6 hold and $K_r(a, b, x)$ is still relatively large. These oracles will be based on the following lemma.

Lemma 8. *Let* $n, r \in \mathbb{N}$, $z \in \mathbb{R}^n$, *and* $\eta \in \mathbb{Q} \cap [0, \dim(z)]$. *Then there is an oracle* $D = D(n, r, z, \eta)$ *satisfying*

(i) For every $t \leq r$, $K_t^D(z) = \min\{\eta r, K_t(z)\} + O(\log r)$.
(ii) For every $m, t \in \mathbb{N}$ *and* $y \in \mathbb{R}^m$, $K_{t,r}^D(y|z) = K_{t,r}(y|z) + O(\log r)$ *and*
$K_t^{z,D}(y) = K_t^z(y) + O(\log r)$.

The proof of this lemma, which uses standard methods, is omitted here. Informally, for some $s \leq r$ such that $K_s(z)$ is near ηr, the oracle D encodes r bits of z conditioned on s bits of z. Unsurprisingly, access to this oracle lowers $K_t(z)$ to $K_s(z)$ whenever $t \geq s$ and has only a negligible effect when $t \leq s$, or when r bits of z are already known.

Theorem 1. *For every* $a, b, x \in \mathbb{R}$ *and* $A \subseteq \mathbb{N}$,

$$\dim^A(x, ax + b) \geq \dim^A(x|a, b) + \min\left\{\dim^A(a, b), \dim^{a,b}(x)\right\}.$$

In particular, for almost every $x \in \mathbb{R}$, $\dim(x, ax + b) = 1 + \min\{\dim(a, b), 1\}$.

Proof. Let $a, b, x \in \mathbb{R}$, and treat them as constant for the purposes of asymptotic notation here. Let $A \subseteq \mathbb{N}$,

$$H = \mathbb{Q} \cap \left[0, \dim^A(a, b)\right] \cap \left[0, \dim^{a,b}(x)\right),$$

and $\eta \in H$. Let $\delta = \dim^{a,b}(x) - \eta > 0$ and $\varepsilon \in \mathbb{Q}_+$. For each $r \in \mathbb{N}$, let $D_r = D(2, r, (a, b), \eta)$, as defined in Lemma 8. We claim that for every sufficiently large r, the conditions of Lemma 6, relativized to oracle D_r, are satisfied by these choices of a, b, x, r, δ, ε, η.

Property (i) of Lemma 8 guarantees that $K_r^{D_r}(a, b) \leq \eta r + O(\log r)$, so condition (i) of Lemma 6 is satisfied for every sufficiently large r.

To see that condition (ii) of Lemma 6 is also satisfied, let $(u, v) \in B_1(a, b)$ such that $ax + b = ux + v$ and $t = -\log \|(a, b) - (u, v)\| \leq r$. Then by Lemma 7, relativized to D_r, we have

$$K_r^{D_r}(u, v) \geq K_t^{D_r}(a, b) + K_{r-t,r}^{D_r}(x|a, b) - O(\log r).$$

Therefore, by Lemmas 5 and 8,

$$
\begin{aligned}
K_r^{D_r}(u, v) &\geq \min\{\eta r, K_t(a, b)\} + K_{r-t,r}(x|a, b) - O(\log r) \\
&\geq \min\{\eta r, K_t(a, b)\} + K_{r-t}^{a,b}(x) - O(\log r) \\
&\geq \min\{\eta r, \dim(a, b)t - o(t)\} + \dim^{a,b}(x)(r - t) - o(r) \\
&\geq \min\{\eta r, \eta t - o(t)\} + (\eta + \delta)(r - t) - o(r) \\
&= \eta t - o(t) + (\eta + \delta)(r - t) - o(r) \\
&= \eta r + \delta \cdot (r - t) - o(r) \\
&\geq (\eta - \varepsilon)r + \delta \cdot (r - t),
\end{aligned}
$$

whenever r is large enough.

For every sufficiently large r, then, the conclusion of Lemma 6 applies here. Thus, for constant a, b, ε, and η,

$$
\begin{aligned}
K_r^A(x, ax + b) &\geq K_r^{A, D_r}(x, ax + b) - O(1) \\
&\geq K_r^{A, D_r}(a, b, x) - 4\varepsilon r/\delta - O(\log r) \\
&= K_r^{A, D_r}(x|a, b) + K_r^{A, D_r}(a, b) - 4\varepsilon r/\delta - O(\log r) \\
&= K_r^A(x|a, b) + \eta r - 4\varepsilon r/\delta - O(\log r),
\end{aligned}
$$

where the last equality is due to the properties of D_r guaranteed by Lemma 8.

Dividing by r and taking limits inferior,

$$
\begin{aligned}
\dim^A(x, ax + b) &\geq \liminf_{r \to \infty} \frac{K_r^A(x|a, b) + \eta r - 4\varepsilon r/\delta - O(\log r)}{r} \\
&= \dim^A(x|a, b) + \eta - \frac{4\varepsilon}{\delta}.
\end{aligned}
$$

Since this holds for every $\eta \in H$ and $\varepsilon \in \mathbb{Q}_+$, we have

$$
\dim^A(x, ax + b) \geq \dim^A(x|a, b) + \min\left\{\dim^A(a, b), \dim^{a, b}(x)\right\}.
$$

The second part of the theorem statement follows easily, as relative to any given oracle for (a, b), almost every $x \in \mathbb{R}$ is Martin-Löf random and therefore has dimension 1. Applying Lemma 5, then, almost every $x \in \mathbb{R}$ has $\dim(x|a, b) \geq \dim^{a, b}(x) = 1$. □

We can now easily answer the motivating question of whether or not there is a line in $\mathbb{R}^2$ on which every point has effective Hausdorff dimension 1.

Corollary 9. *For every $a, b \in \mathbb{R}$, there exist $x, y \in \mathbb{R}$ such that*

$$
\dim(x, ax + b) - \dim(y, ay + b) \geq 1.
$$

In particular, there is no line in $\mathbb{R}^2$ on which every point has dimension 1.

Proof. Theorem 1 tells us that $\dim(x, ax + b) \geq 1 + \min\{\dim(a, b), 1\}$ for almost every $x \in \mathbb{R}$. For $y = 0$, we have $\dim(y, ay + b) = \dim(b) \leq \min\{\dim(a, b), 1\}$. □

There are lines for which the inequality in Corollary 9 is strict. Consider, for example, a line through the origin whose slope a is random. For every x that is random relative to a, the point (a, ax) has dimension $\dim(x) + \dim(a) = 2$, but the origin itself has dimension 0.

4   An Application to Classical Fractal Geometry

4.1   Hausdorff Dimension

As the name indicates, effective Hausdorff dimension was originally conceived as a constructive analogue to Hausdorff dimension, which is the most standard notion of dimension in fractal geometry. The properties and classical definition of Hausdorff dimension are beyond the scope of this paper; see [5,17] for discussion of those topics. Instead, we characterize it here according to a recent *point-to-set principle*:

Theorem 10 (J. Lutz and N. Lutz [14]). *For every* $n \in \mathbb{N}$ *and* $E \subseteq \mathbb{R}^n$, *the Hausdorff dimension of* E *is given by*

$$\dim_H(E) = \min_{A \subseteq \mathbb{N}} \sup_{x \in E} \dim^A(x).$$

4.2   Generalized Sets of Furstenberg Type

A *set of Furstenberg type* with parameter α is a set $E \subseteq \mathbb{R}^2$ such that, for every $e \in S^1$ (the unit circle in $\mathbb{R}^2$), there is a line ℓ_e in the direction e satisfying $\dim_H(E \cap \ell_e) \geq \alpha$. Finding the minimum possible dimension of such a set is an important open problem with connections to Falconer's distance set conjecture and to Kakeya sets [9,27].[4]

Molter and Rela [22] introduced a natural generalization of Furstenberg sets, in which the set of directions may itself have fractal dimension. Formally, a set $E \subseteq \mathbb{R}^2$ is in the class $F_{\alpha\beta}$ if there is some set $J \subseteq S^1$ such that $\dim_H(J) \geq \beta$ and for every $e \in J$, there is a line ℓ_e in the direction e satisfying $\dim_H(E \cap \ell_e) \geq \alpha$. They proved the following lower bound on the dimension of such sets.

Theorem 11 (Molter and Rela [22]). *For all* $\alpha, \beta \in (0,1]$ *and every set* $E \in F_{\alpha\beta}$,

$$\dim_H(E) \geq \alpha + \max\left\{\frac{\beta}{2}, \alpha + \beta - 1\right\}.$$

We now show that Theorem 1 yields an improvement on this bound whenever $\alpha, \beta < 1$ and $\beta/2 < \alpha$.

Theorem 12. *For all* $\alpha, \beta \in (0,1]$ *and every set* $E \in F_{\alpha\beta}$,

$$\dim_H(E) \geq \alpha + \min\{\beta, \alpha\}.$$

Proof. Let $\alpha, \beta \in (0,1]$, $\varepsilon \in (0, \beta)$, and $E \in F_{\alpha\beta}$. Using Theorem 10, let A satisfy

$$\sup_{z \in E} \dim^A(z) = \dim_H(E).$$

and $e \in S^1$ satisfy $\dim^A(e) = \beta - \varepsilon > 0$. Let ℓ_e be a line in direction e such that $\dim_H(\ell_e \cap E) \geq \alpha$. Since $\dim(e) > 0$, we know $e \notin \{(0,1), (0,-1)\}$, so we may let $a, b \in \mathbb{R}$ be such that $L_{a,b} = \ell_e$. Notice that $\dim^A(a) = \dim^A(e)$ because the

---

[4] Our main theorem also provides yet another alternative proof that every Kakeya set $E \subseteq \mathbb{R}^2$ has $\dim_H(E) = 2$. Briefly, let A be the minimizing oracle for E from Theorem 10, and let $a, b, x \in \mathbb{R}$ satisfy $\dim^A(a) = 1$, $\dim^{A,a,b}(x) = 1$, and $(x, ax+b) \in E$. Then Theorem 1 gives $\dim_H(E) \geq \dim^A(x, ax+b) \geq \dim^A(x|a,b) + \min\{\dim^A(a,b), \dim^{A,a,b}(x)\} \geq 2$.

mapping $e \mapsto a$ is computable and bi-Lipschitz in some neighborhood of e. Let $S = \{x : (x, ax + b) \in E\}$, which is similar to $\ell_e \cap E$, so $\dim_H(S) \geq \alpha$ also. We now have

$$\dim_H(E) = \sup_{z \in E} \dim^A(z)$$
$$\geq \sup_{z \in \ell_e \cap E} \dim^A(z)$$
$$= \sup_{x \in S} \dim^A(x, ax + b).$$

By Theorem 1 and Lemma 5, both relativized to A,

$$\sup_{x \in S} \dim^A(x, ax + b) \geq \sup_{x \in S} \left\{ \dim^{A,a,b}(x) + \min\{\dim^A(a, b), \dim^A(x|a, b)\} \right\}$$
$$\geq \sup_{x \in S} \left\{ \dim^{A,a,b}(x) + \min\{\dim^A(a, b), \dim^{A,a,b}(x)\} \right\}$$
$$\geq \sup_{x \in S} \dim^{A,a,b}(x) + \min\left\{ \dim^A(a), \sup_{x \in S} \dim^{A,a,b}(x) \right\}.$$

Theorem 10 gives

$$\sup_{x \in S} \dim^{A,a,b}(x) \geq \dim_H(S) \geq \alpha,$$

so we have shown, for every $\varepsilon \in (0, \beta)$, that $\dim_H(E) \geq \alpha + \min\{\beta - \varepsilon, \alpha\}$. □

5   Conclusion

With Theorem 1, we have taken a significant step in understanding the structure of algorithmic information in Euclidean spaces. Progress in that direction is especially consequential in light of Theorem 12, which, aside from its direct value as a mathematical result, demonstrates conclusively that algorithmic dimensional methods can provide new insights into classical fractal geometry. This motivates further investigation of algorithmic fractal geometry in general and of effective Hausdorff dimension on lines in particular; improvements on our lower bound or extensions to higher dimensions would have implications for important questions about Furstenberg or Kakeya sets. Our results also motivate a broader search for potential applications of algorithmic dimension to problems in classical fractal geometry.

References

1. Case, A., Lutz, J.H.: Mutual dimension. ACM Trans. Comput. Theory **7**(3), 12 (2015)
2. Davies, R.O.: Some remarks on the Kakeya problem. Proc. Cambridge Philos. Soc. **69**, 417–421 (1971)
3. Dougherty, R., Lutz, J.H., Mauldin, R.D., Teutsch, J.: Translating the Cantor set by a random real. Trans. Am. Math. Soc. **366**, 3027–3041 (2014)

4. Downey, R., Hirschfeldt, D.: Algorithmic Randomness and Complexity. Springer, New York (2010)
5. Falconer, K.J.: The Geometry of Fractal Sets. Cambridge University Press, Cambridge (1985)
6. Gács, P.: On the symmetry of algorithmic information. Sov. Math. Dokl. **15**, 1477–1480 (1974)
7. Gu, X., Lutz, J.H., Mayordomo, E., Moser, P.: Dimension spectra of random subfractals of self-similar fractals. Ann. Pure Appl. Logic **165**(11), 1707–1726 (2014)
8. Hitchcock, J.M.: Correspondence principles for effective dimensions. Theory Comput. Syst. **38**(5), 559–571 (2005)
9. Katz, N.H., Tao, T.: Some connections between Falconer's distance set conjecture and sets of Furstenburg type. N. Y. J. Math. **7**, 149–187 (2001)
10. Li, M., Vitányi, P.M.B.: An Introduction to Kolmogorov Complexity and Its Applications, 3rd edn. Springer, New York (2008)
11. Lutz, J.H.: Dimension in complexity classes. SIAM J. Comput. **32**(5), 1236–1259 (2003)
12. Lutz, J.H.: The dimensions of individual strings and sequences. Inf. Comput. **187**(1), 49–79 (2003)
13. Lutz, J.H., Lutz, N.: Lines missing every random point. Computability **4**(2), 85–102 (2015)
14. Lutz, J.H., Lutz, N.: Algorithmic information, plane Kakeya sets, and conditional dimension. In: Proceedings of the 34th International Symposium on Theoretical Aspects of Computer Science, STACS 2017, Hannover, Germany (to appear)
15. Lutz, J.H., Mayordomo, E.: Dimensions of points in self-similar fractals. SIAM J. Comput. **38**(3), 1080–1112 (2008)
16. Lutz, J.H., Weihrauch, K.: Connectivity properties of dimension level sets. Math. Logic Q. **54**, 483–491 (2008)
17. Mattila, P.: Geometry of Sets and Measures in Euclidean Spaces: Fractals and Rectifiability. Cambridge University Press, Cambridge (1995)
18. Mayordomo, E.: A Kolmogorov complexity characterization of constructive Hausdorff dimension. Inf. Process. Lett. **84**(1), 1–3 (2002)
19. Mayordomo, E.: Effective fractal dimension in algorithmic information theory. In: Cooper, S.B., Löwe, B., Sorbi, A. (eds.) New Computational Paradigms: Changing Conceptions of What is Computable, Observation of Strains, pp. 259–285. Springer, New York (2008). doi:10.1007/978-0-387-68546-5_12
20. Mayordomo, E.: Effective dimension in some general metric spaces. In: Proceedings of the 8th International Workshop on Developments in Computational Models, DCM 2012, Cambridge, United Kingdom, 17 June 2012, pp. 67–75 (2012)
21. Miller, J.S.: "Open Questions" section of personal webpage. http://www.math.wisc.edu/jmiller/open.html. Accessed 30 Oct 2016
22. Molter, U., Rela, E.: Furstenberg sets for a fractal set of directions. Proc. Am. Math. Soc. **140**, 2753–2765 (2012)
23. Nies, A.: Computability and Randomness. Oxford University Press Inc., New York (2009)
24. Rela, E.: Refined size estimates for Furstenberg sets via Hausdorff measures: a survey of some recent results. In: Cepedello Boiso, M., Hedenmalm, H., Kaashoek, M.A., Montes Rodríguez, A., Treil, S. (eds.) Concrete Operators, Spectral Theory, Operators in Harmonic Analysis and Approximation. OTAA, vol. 236, pp. 421–454. Springer, Basel (2014). doi:10.1007/978-3-0348-0648-0_27

25. Turetsky, D.: Connectedness properties of dimension level sets. Theor. Comput. Sci. **412**(29), 3598–3603 (2011)
26. Weihrauch, K., Analysis, C.: Computable Analysis: An Introduction. Springer, Heidelberg (2000)
27. Wolff, T.: Recent work connected with the Kakeya problem. In: Prospects in Mathematics, pp. 129–162 (1999)

Büchi Automata Recognizing Sets of Reals Definable in First-Order Logic with Addition and Order

Arthur Milchior[✉]

LACL, Université Paris-Est, Créteil, France
arthur.milchior@lacl.fr

Abstract. This work considers encodings of non-negative reals in a fixed base, and their encoding by weak deterministic Büchi automata. A Real Number Automaton is an automaton which recognizes all encodings of elements of a set of reals. We explain in this paper how to decide in linear time whether a set of reals recognized by a given minimal weak deterministic RNA is FO[$\mathbb{R}; +, <, 1$]-definable. Furthermore, it is explained how to compute in quasi-quadratic (respectively, quasi-linear) time an existential (respectively, existential-universal) FO[$\mathbb{R}; +, <, 1$]-formula which defines the set of reals recognized by the automaton.

As an additional contribution, the techniques used for obtaining our main result lead to a characterization of minimal deterministic Büchi automata accepting FO[$\mathbb{R}; +, <, 1$]-definable set.

1 Introduction

This paper deals with logically defined sets of numbers encoded by weak deterministic Büchi automata. The sets of tuples of integers whose encodings in base b are recognized by a finite automaton are called the b-recognizable sets. By [5], the b-recognizable sets of vectors of integers are exactly the sets which are FO[$\mathbb{N}; +, <, V_b$]-definable, where $V_b(n)$ is the greatest power of b dividing n. It was proven in [6,15] that the FO[$\mathbb{N}; +, <$]-definable sets are exactly the sets which are b- and b'-recognizable for every $b \geq 2$.

Those results naturally led to the following problem: deciding whether a finite automaton recognizes a FO[$\mathbb{N}; +, <$]-definable set of d-tuples of integers for some dimension $d \in \mathbb{N}^{>0}$. In the case of dimension $d = 1$, decidability was proven in [9]. For $d > 1$, decidability was proven in [14]. Another algorithm was given in [11], which solves this problem in polynomial time. For $d = 1$, a quasi linear time algorithm was given in [13].

The above-mentioned results about sets of tuples of natural numbers and finite automata have then been extended to sets of tuples of reals recognized by a Büchi automaton. The notion of Büchi automata is a formalism which describes languages of infinite words, also called ω-words. The Büchi automata are similar to the finite automata. The main difference is that finite automata accept finite words which admit runs ending on accepting state, while Büchi

© Springer International Publishing AG 2017
T.V. Gopal et al. (Eds.): TAMC 2017, LNCS 10185, pp. 440–454, 2017.
DOI: 10.1007/978-3-319-55911-7_32

automata accept infinite words which admit runs in which an accepting state appears infinitely often.

One of the main differences between finite automata and Büchi automata is that finite automata can be determinized while deterministic Büchi automata are less expressive than Büchi automata. For example, the language $L_{\text{fin}(a)}$ of words containing a finite number of times the letter a is recognized by a Büchi automaton, but is not recognized by any deterministic Büchi automaton. This statement implies, for example, that no deterministic Büchi automaton recognizes the set of reals of the form nb^p with $n \in \mathbb{N}$ and $p \in \mathbb{Z}$, that is, the set of reals whose encoding ends with 0 or $(b-1)$ repeated infinitely many times.

A main difference between the two classes of deterministic automata is that the class of languages recognized by deterministic finite automata is closed under complement while the class of languages recognized by deterministic Büchi automata is not. For example, $L_{\text{fin}(a)}$ is not recognized by any deterministic Büchi automaton while its complements $L_{\text{inf}(a)}$ is recognized by a deterministic Büchi automaton.

However, the set of weak deterministic Büchi automata is closed under complement. A weak deterministic Büchi automaton is a deterministic Büchi automaton whose set of accepting states is a union of strongly connected components. Handling weak Büchi automata is similar to manipulating finite automata. A set is said to be weakly b-recognizable if it is recognized by a weak automaton in base b. The class of weak deterministic Büchi automata is less expressive than the class of deterministic Büchi automata. For example, as mentionned above, the language $L_{\text{inf}(a)}$ is recognized by a deterministic Büchi automaton, but this language is not recognized by any weak deterministic Büchi automaton. This implies that, for example, no weak deterministic Büchi automaton recognizes the set of reals which are not of the form nb^p with $n \in \mathbb{N}$ and $p \in \mathbb{Z}$, since those reals are the ones whose encoding in base b contains an infinite number of non-0 digits. Furthermore, by [12], weak deterministic Büchi automata can be efficiently minimized.

A Real Vector Automaton (RVA, See e.g. [4]) of dimension d is a Büchi automaton $\mathcal{A}$ over alphabet $\{0, \ldots, b-1\}^d \cup \{\star\}$, which recognizes the set of encodings in base b of the elements of a set of vectors of reals. Equivalently, for w an infinite word encoding a vector of dimension d of real $(r_0, \ldots, r_{d-1})$, if w is accepted by $\mathcal{A}$, then all encodings w' of $(r_0, \ldots, r_{d-1})$ are accepted by $\mathcal{A}$. In the case where the dimension d is 1, those automata are called Real Number Automata (RNA, See e.g. [3]).

The sets of tuples of reals whose encoding in base b is recognized by a RVA are called the b-recognizable sets. By [18], they are exactly the FO $[\mathbb{R}, \mathbb{N}; +, <, X_b, 1]$-definable sets. The logic FO $[\mathbb{R}, \mathbb{N}; +, <, X_b, 1]$ is the first-order logic over reals with a unary predicate which holds over integers, addition, order, the constant one, and the function $X_b(x, u, k)$. The function $X_b(x, u, k)$ holds if and only if u is equal to some b^n with $n \in \mathbb{Z}$ and there exists an encoding in base b of x whose digit in position n is k. That is, u and x are of one of the two following forms:

$$u = 0 \ldots 0 \star 0 \ldots 0 \ 1 \ 0 \ldots$$
$$x = \quad \ldots \quad \star \quad \ldots \quad k \quad \ldots$$
or
$$u = 0 \ldots 0 \ 1 \ 0 \ldots 0 \star 0 \ldots$$
$$x = \quad \ldots \quad k \quad \ldots \quad \star \quad \ldots$$
.

By [4], a set is FO $[\mathbb{R}, \mathbb{N}; +, <]$-definable if and only if its set of encodings is weakly b-recognizable for all $b \geq 2$.

By [7], the logic FO $[\mathbb{R}; +, <, 1]$ admits quantifier elimination. By [17, Sect. 6], the set of reals which are FO $[\mathbb{R}; +, <, 1]$-definable are finite unions of intervals with rational bounds. Those sets are called the *simple sets*.

Standard definitions are recalled in Sect. 2. Sets of states of automata reading reals are studied in Sect. 3. Furthermore, a method to efficiently solve automaton problem is introduced. In Sect. 4, given a simple set, an automaton accepting it is constructed. A characterization of minimal deterministic Büchi automata accepting a FO $[\mathbb{R}; +, <, 1]$-definable set is given in Sect. 5. This characterization is similar to the insight given in [2] and leads to a linear time algorithm deciding whether a minimal RNA recognizes a FO $[\mathbb{R}; +, <, 1]$-definable set. This algorithm does not return any false positive on weak deterministic Büchi automata which are not RNA. A false negative is exhibited at the end of Sect. 5. Given a minimal weak RNA automaton accepting a simple set, it is shown in Sect. 6 that an existential (respectively, existential-universal) FO $[\mathbb{R}; +, <, 1]$-formula which defines R is computable in quasi-quadratic (respectively, quasi-linear) time.

2   Definitions

The definitions used in this paper are given in this section. Some basic lemmas are also given. Most definitions are standard.

Let $\mathbb{N}$, $\mathbb{Z}$, $\mathbb{Q}$ and $\mathbb{R}$ denote the set of non-negative integers, integers, rationals and reals, respectively. For $R \subseteq \mathbb{R}$, let $R^{\geq 0}$ and $R^{>0}$ denote the set of non-negative and of positive elements of R, respectively. For $n \in \mathbb{N}$, let $[n]$ represent $\{0, \ldots, n\}$. For $a, b \in \mathbb{R}$ with $a \leq b$, let $[a, b]$ denote the closed interval $\{r \in \mathbb{R} \mid a \leq r \leq b\}$, and let (a, b) denote the open interval $\{r \in \mathbb{R} \mid a < r < b\}$. Similarly, let $(a, b]$ (respectively, $[a, b)$) be the half-open interval equals to the union of (a, b) and of $\{b\}$ (respectively, $\{a\}$). For $r \in \mathbb{R}$ let $\lfloor r \rfloor$ be the greatest integer less than or equal to r.

2.1   Finite and Infinite Words

An alphabet is a finite set, its elements are called letters. A finite (respectively ω-) word over alphabet A is a finite (respectively infinite) sequence of letters of A. That is, a function from $[n]$ to A for some $n \in \mathbb{N}$ (respectively from $\mathbb{N}$ to A). A set of finite (respectively ω-) words over alphabet A is called a language (respectively, an ω-language) over alphabet A. The empty word is denoted ϵ.

Let w be a word, its length is denoted $|w|$, it is either a non-negative integer or the cardinality of $\mathbb{N}$. For $n < |w|$, let $w[n]$ denote the n-th letter of w. For v a finite word, let $u = vw$ be the concatenation of v and of w, that is, the word of length $|v| + |w|$ such that $u[i] = v[i]$ for $i < |v|$ and $u[|v| + i] = w[i]$ for $i < |w|$.

Let $w\,[< n]$ denote the *prefix* of w of length n, that is, the word u of length n such that $w[i] = u[i]$ for all $i \in [n-1]$. Similarly, let $w\,[\geq n]$ denote the suffix of w without its n-th first letters, that is, the word u of length $|w| - n$ such that $u[i] = w[i+n]$ for all $i \in [|w| - n]$. Note that $w = w\,[< i]\,w\,[\geq i]$ for all $i < |w|$.

Let L be a language of finite words and let L' be either a ω-language or a language of finite words. Let LL' be the set of concatenations of the words of L and of L'. For $i \in \mathbb{N}$, let L^i be the concatenations of i words of L. Let $L^* = \bigcup_{i \in \mathbb{N}} L^i$ and $L^+ = \bigcup_{i \in \mathbb{N}^{>0}} L^i$. If L is a language which does not contain the empty word, let L^ω be the set of infinite sequences of elements of L.

Encoding of Real Numbers. Let us now consider the encoding of numbers in an integer base $b \geq 2$. Let Σ_b be equal to $[b-1]$; it is the set of digits. The base b is fixed for the rest of this paper.

The function sending words to the number they encode are now introduced. Let w be an ω-word with exactly one $\star$. It is of the form $w = w_I \star w_F$, with $w_I \in \Sigma_b^*$ and $w_F \in \Sigma_b^\omega$. The word w_I is called the natural part of w and the ω-word w_F is called its fractional part. Let $[w_I]_b^I = \sum_{i=0}^{|w|-1} b^{|w|-1-i} w_I[i]$, let $[w_F]_b^F = \sum_{i \in \mathbb{N}} b^{-i-1} w_F[i]$ and finally, let $[w_I \star w_F]_b^R = [w_I]_b^I + [w_F]_b^F$. Some properties of concatenation and of encoding of reals are now stated.

Lemma 1. *Let $v \in \Sigma_b^*$, $v' \in \Sigma_b^+$, $w \in \Sigma_b^\omega$ and $a \in \Sigma_b$. Then:*

$$[w]_b^F = [0 \star w]_b^R, \qquad [aw]_b^F = \left(a + [w]_b^F\right)/b,$$
$$[va]_b^I = b\,[v]_b^I + a \text{ and } [v^\omega]_b^F = [v]_b^I / \left(b^{|v|} - 1\right).$$

Some basic facts about rationals are recalled (see e.g. [8]). The rationals are exactly the numbers which admit encodings in base b of the form $u \star vw^\omega$ with $u, v \in \Sigma_b^*$ and $w \in \Sigma_b^+$. Rationals of the form nb^p, with $n \in \mathbb{N}$ and $p \in \mathbb{Z}$, admit exactly two encodings in base b without leading 0 in the natural part. If $p < 0$, the two encodings are of the form $u \star va(b-1)^\omega$ and $u \star v(a+1)0^\omega$, with $u, v \in \Sigma_b^*$ and $a \in [b-2]$. Otherwise, if $p \geq 0$, the two encodings are of the form $ua(b-1)^q \star (b-1)^\omega$ and $u(a+1)0^q \star 0^\omega$ with $u \in \Sigma_b^*$, $a \in [b-2]$ and $q \in \mathbb{N}$. The rationals which are not of the form nb^p admit exactly one encoding in base b without leading 0 in the natural part.

Encoding of Sets of Reals. Relations between languages and sets of reals are now recalled. Given a language L which is a subset of $\Sigma_b^* \star \Sigma_b^\omega$, let $[L]_b^R$ be the set of reals admitting an encoding in base b in L. The language L is said to be an encoding in base b of the set of reals $[L]_b^R$. Reciprocally, given a set $R \subseteq \mathbb{R}^{\geq 0}$ of reals, $L_b(R)$ is the set of all encodings in base b of elements of R. For L a subset of Σ_b^ω, $[L]_b^F$ is the set of d-tuples of reals, belonging to $[0, 1]^d$, which admits an encoding in base b in L.

Following [11], a language L is said to be *saturated* if for any number r which admits an encoding in base b in L, all encodings in base b of r belong to L. The saturated languages are of the form $L_b(R)$ for $R \subseteq \mathbb{R}^{\geq 0}$. Note that $[L_b(R)]_b^R = R$ for all sets $R \subseteq \mathbb{R}^{\geq 0}$. Note also that $L \subseteq L_b([L]_b^R)$, and the subset relation is an equality if and only if L is saturated.

All non-empty sets of reals have infinitely many encodings in base b. For example, for $I \subseteq \mathbb{N}$ an arbitrary set, $0^* \star \{0,1\}^\omega \setminus \{0^i 1^\omega \mid i \in I\}$ is an encoding in base 2 of the simple set $[0,1]$. This language is saturated if and only if $I = \emptyset$.

2.2   Deterministic Büchi Automata

This paper deals with deterministic Büchi automata. This notion is now defined.

A *Deterministic Büchi automaton* $\mathcal{A}$ is a 5-tuple (Q, A, δ, q_0, F), where Q is a finite *set of states*, A is an alphabet, $\delta : Q \times A \to Q$ is the *transition function*, $q_0 \in Q$ is the *initial states* and $F \subseteq Q$ is the set of *accepting states*. A state belonging to $Q \setminus F$ is said to be a *rejecting state*.

From now on in this paper, all automata are assumed to be deterministic. The function δ is implicitly extended on $Q \times A^*$ by $\delta(q, \epsilon) = q$ and $\delta(q, wa) = \delta(\delta(q, w), a)$ for $a \in A$ and $w \in A^*$.

Let $\mathcal{A}$ be an automaton and w be an infinite word. A *run* π of $\mathcal{A}$ on w is a mapping $\pi : \mathbb{N} \mapsto Q$ such that $\pi(0) = q_0$ and $\delta(\pi(i), w[i]) = \pi(i+1)$ for all $i < |w|$. The run is accepting if there exists a state $q \in F$ such that there is an infinite number of $i \in \mathbb{N}$ such that $\pi(i) = q$. Let $\mathcal{A}$ be a finite automaton. Let $L_\omega(\mathcal{A})$ be the set of infinite words w such that a run of $\mathcal{A}$ on w is accepting.

Accessibility and Recurrent States. Some definitions related to the underlying labelled graph of Büchi automata are introduced in this section. A state q is said to be *accessible* from a state q' if there exists a finite non-empty word w such that $\delta(q', w) = q$. Following [12], a state q is said to be *recurrent* if it is accessible from itself and *transient* otherwise. Transient states are called *trivial* in [2]. The *strongly connected component* of a recurrent state q is the set of states q' such that q' is accessible from q and q is accessible from q'. A strongly connected component C is said to be a *leaf* if for all $a \in A$, for all $q \in C$, $\delta(q, a) \in C$. Let C be a strongly connected component. It is said to be a cycle if for each $q \in C$, there exists a unique $s_q \in A$ such that $\delta(q, s_q) \in C$.

The transient states of the automaton pictured in Fig. 2 are q_1, q_{10}, q_{11}, $q_{10\star}$ and $q_{11\star}$. All other states are recurrent. The cycles are $\{q_0\}$, $\{q_{0\star}, q_{0\star 0}\}$, $\{q_{10\star 0}\}$, $\{q_{10\star 1}, q_{10\star 10}\}$, $\{q_{11\star 0}\}$ and $\{q_{11\star 1}, q_{11\star 10}\}$. The strongly connected components which are not cycles are $\{q_{\emptyset, \mathcal{A}}\}$, $\{q_{\infty, \mathcal{A}}\}$ and $\{q_{[0,1], \mathcal{A}}\}$.

For $q \in Q$, let $\mathcal{A}_q$ be (Q_q, A, δ, q, F_q), where Q_q is the set of states of Q accessible from q, and $F_q = F \cap Q_q$. Note that, if there is no finite word w such that $\delta(q_0, w) = q_0$, then $Q_q \subsetneq Q$ for all $q \neq q_0$.

Quotients, Morphisms and Weak Büchi Automata. The Büchi automaton $\mathcal{A} = (Q, A, \delta, q_0, F)$ is said to be minimal if, for each distinct states q and q' of $\mathcal{A}$, $L_\omega(\mathcal{A}_q) \neq L_\omega(\mathcal{A}_{q'})$. Let $\mathcal{A} = (Q, A, \delta, q_0, F)$ be a Büchi automaton and $\mathcal{A}' = (Q', A, \delta', q_0', F')$ be a minimal Büchi automaton. A surjective function $\mu : Q \to Q'$ is a *morphism* of Büchi automata if and only if $\mu(q_0) = q_0'$ and, for all $q \in Q$, $L_\omega (\mathcal{A}_q) = L_\omega \left(\mathcal{A}'_{\mu(q)} \right)$.

The Büchi automaton $\mathcal{A}$ is said to be *weak* if for each recurrent accepting state q of $\mathcal{A}$, all states of the strongly connected components of q are accepting. An ω-language is said to be (weakly) *recognizable* if it is a set of word accepted by a (weak) Büchi automaton. An example of weak deterministic Büchi automaton is now given. This example is used through this paper to illustrate properties of Büchi automaton reading set of real numbers.

Example 1. Let $R = \left(\frac{1}{3}, 2\right] \cup \left(\frac{8}{3}, 3\right] \cup \left(\frac{11}{3}, \infty\right]$. The set of encodings in base 2 of reals of R is (weakly) recognized by the automaton pictured in Fig. 1. The run of $\mathcal{A}$ on the ω-word $011 \star (10)^\omega$ is $(q_0, q_0, q_1, q_3, q_{11\star}, q_{11\star1}, q_{11\star10}, \dots)$, with the two last states repeated infinitely often. The Büchi automaton $\mathcal{A}$ does not accept $011 \star (10)^\omega$ since this run does not contain any accepting state. The run of $\mathcal{A}$ on ω-word $\star 1^\omega$ is $\left(q_0, q_{0\star}, q_{[0,1],\mathcal{A}}, \dots\right)$ with the last state repeated infinitely often. The Büchi automaton $\mathcal{A}$ accepts $\star 1^\omega$ since the accepting state $q_{[0,1],\mathcal{A}}$ appears infinitely often in the run.

The main theorem concerning quotient of weak Büchi automata is now recalled.

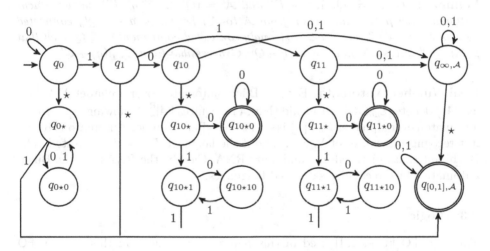

Fig. 1. Automaton $\mathcal{A}_R$ of Example 1

Theorem 1 ([12]). *Let $\mathcal{A} = (Q, A, \delta, q_0, F)$ be a weak Büchi automaton with n states such that all states of $\mathcal{A}$ are accessible from its initial state. Let c be the cardinality of A. There exists a minimal weak Büchi automaton $\mathcal{A}'$ such that there exists a morphism of automaton μ from $\mathcal{A}$ to $\mathcal{A}'$. The automaton $\mathcal{A}'$ and the morphism μ are computable in time $O\left(n \log(n) c\right)$ and space $O\left(nc\right)$.*

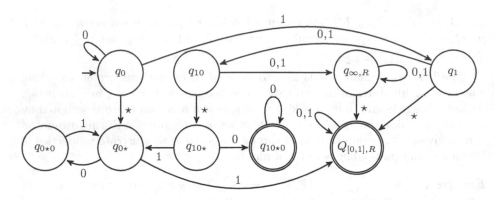

Fig. 2. Minimal quotient of automaton $\mathcal{A}_R$ of Fig. 1

The Büchi automaton $\mathcal{A}_R$ pictured in Fig. 1 is weak and is not minimal. Its minimal quotient is pictured in Fig. 2. The following lemma states that each strongly connected component of a quotient by a morphism μ from an automaton $\mathcal{A}$ is the image of a strongly connected component of $\mathcal{A}$. It allows to prove that some properties, such as being a cycle, is closed under taking quotient.

Lemma 2. Let $\mathcal{A} = (Q, A, \delta, q_0, F)$ and $\mathcal{A}' = (Q', \Sigma_b, \delta', q_0', F')$ be two Büchi automata. Let μ be a morphism from $\mathcal{A}$ to $\mathcal{A}'$. Let C' be a strongly connected component of $\mathcal{A}'$. There exists a strongly connected component $C \subseteq Q$ such that $\mu(C) = C'$ and such that, for all $q \in Q \setminus C$ accessible from C, $\mu(q) \notin C'$.

Real Number Automata. For $\mathcal{A}$ a Büchi automaton over alphabet $\Sigma_b \cup \{\star\}$, let $[\mathcal{A}]_b^R$ denote $[L_\omega(\mathcal{A})]_b^R$. It is said that $\mathcal{A}$ recognizes $[\mathcal{A}]_b^R$. Following [3], a Büchi automaton over alphabet $\Sigma_b \cup \{\star\}$ is said to be a Real Number Automaton (RNA) if it recognizes a subset of $\Sigma_b^* \star \Sigma_b^\omega$ and if the language $L_\omega(\mathcal{A})$ is saturated. The Büchi automata pictured in 1 and 2 are RNA. Clearly, the RNAs are the Büchi automata which recognize saturated languages.

2.3 Logic

The logic $\mathsf{FO}[\mathbb{R}; +, <, 1]$ used in this paper is introduced in this section. FO stands for first-order. The first parameter $\mathbb{R}$ means that the (free or quantified) variables are interpreted by non-negative real numbers. The $+$ and $<$ symbols mean that the addition function and the binary order relation over reals can be used in formulas. Finally, the last term, 1, means that the only constant is 1. The logic $\mathsf{FO}[\mathbb{R}; +, <, 1]$ is denoted by $\mathscr{L}$ in [7], where it is proven that this logic admits quantifier elimination. In this paper, all results deal with the quantifier-free, the existential fragment and the existential-universal fragment of $\mathsf{FO}[\mathbb{R}; +, <, 1]$ denoted by $\Sigma_0[\mathbb{R}; +, <, 1]$, $\Sigma_1[\mathbb{R}; +, <, 1]$ and $\Sigma_2[\mathbb{R}; +, <, 1]$ respectively.

In the rest of the paper, rationals are also used in the formulas. Admitting rationals does not change the expressivity since all rational constants are Σ_0 [$\mathbb{R}$; $+$, 1]-definable. The length of a formula ϕ is denoted by $|\phi|$. It is such that each symbol takes one bit of space, apart from integers n and rationals n/m which take $\log(1 + |n|)$ and $\log(1 + |n| + |m|)$ bits of space respectively.

First-Order Definable Sets of Reals. In this section, notations are introduced for the kind of sets studied in this paper: the FO [$\mathbb{R}$; $+$, $<$, 1]-definable sets. Following [17, Sect. 6], the FO [$\mathbb{R}$; $+$, $<$, 1]-definable sets are called the *simple sets*. By [17, Sect. 6], those sets are the finite unions of intervals with rational bounds. It implies that there exists an integer t_R such that for all $x, y \geq t_R$, x belongs to R if and only if y belongs to R. The least such integer t_R is called the *threshold of R*.

Note that every closed and half-closed intervals is the union of an open interval and of singletons, hence it can be assumed that any simple set R is of the form $R = \bigcup_{i=0}^{I-1}(\rho_{i,\mathfrak{L}}, \rho_{i,\mathfrak{R}}) \cup \bigcup_{i=0}^{J-1} \{\rho_{i,\mathfrak{S}}\}$, with $\rho_{i,\mathfrak{L}}, \rho_{i,\mathfrak{S}} \in \mathbb{Q}$ and $\rho_{i,\mathfrak{R}} \in \mathbb{Q} \cup \{\infty\}$. The $\rho_{i,\mathfrak{L}}$'s are the left bounds, the $\rho_{i,\mathfrak{R}}$'s are the right bounds and the $\rho_{i,\mathfrak{S}}$'s are the singletons.

For example, let $R = \left(\frac{1}{3}, 2\right] \cup \left(\frac{8}{3}, 3\right] \cup \left(\frac{11}{3}, \infty\right]$ as in Example 1. Then t_R is 4, $I = 3$, $J = 2$, $\rho_{1,\mathfrak{L}} = \frac{1}{3}$, $\rho_{2,\mathfrak{R}} = 2$, $\rho_{2,\mathfrak{L}} = \frac{8}{3}$, $\rho_{2,\mathfrak{R}} = 3$, $\rho_{3,\mathfrak{L}} = 11/3$, $\rho_{3,\mathfrak{R}} = \infty$, $\rho_{1,\mathfrak{S}} = 2$ and $\rho_{2,\mathfrak{S}} = 3$.

3 Some Sets of States of Automata Reading Reals

We now introduce five sets of states used in the algorithms of this paper.

Definition 1 ($Q_{\emptyset,\mathcal{A}}$, $Q_{[0,1],\mathcal{A}}$, $Q_{\infty,\mathcal{A}}$, $Q_{I,\mathcal{A}}$ and $Q_{F,\mathcal{A}}$). *Let $\mathcal{A} = (Q, A, \delta, q_0, F)$.*

- *Let $Q_{\emptyset,\mathcal{A}}$ be the set of states q such that $\mathcal{A}_q$ recognizes the empty language.*
- *Let $Q_{[0,1],\mathcal{A}}$ be the set of states q such that $\mathcal{A}_q$ recognizes Σ_b^ω.*
- *Let $Q_{\infty,\mathcal{A}}$ be the set of states q such that $\mathcal{A}_q$ recognizes the language $\Sigma_b^* \star \Sigma_b^\omega$.*
- *Let $Q_{I,\mathcal{A}}$ be the set of states q such that $\mathcal{A}_q$ recognizes a subset of $\Sigma_b^* \star \Sigma_b^\omega$.*
- *Let $Q_{F,\mathcal{A}}$ be the set of states q such that $\mathcal{A}_q$ recognizes a subset of Σ_b^ω.*

In [2], the strongly connected components included in $Q_{\emptyset,\mathcal{A}}$ are called empty and the ones included in $Q_{[0,1],\mathcal{A}}$ are called universal.

Intuitively the states belonging to $Q_{I,\mathcal{A}}$ and to $Q_{F,\mathcal{A}}$ are the states which can be visited while the automaton read the natural and the fractional part of the number respectively.

Let $\mathcal{A}$ be the automaton pictured in Fig. 2. Let $q_{\emptyset,\mathcal{A}}$ be the state $\delta(q_{10\star0}, 1)$, which is not pictured in Fig. 2. Then $Q_{[0,1],\mathcal{A}} = \{q_{[0,1],R}\}$, $Q_{\infty,\mathcal{A}} = \{q_{\infty,R}\}$ and $Q_{\emptyset,\mathcal{A}} = \{q_{\emptyset,\mathcal{A}}\}$. Furthermore, the states of $Q_{I,\mathcal{A}}$ are pictured in the top row of Fig. 2, they are q_0, q_1, q_{10}, $q_{\infty,R}$ and $q_{\emptyset,R}$. Finally, the states of $Q_{F,\mathcal{A}}$ are pictured in the second row of Fig. 2, they are $q_{10\star}$, $q_{10\star0}$, $q_{0\star}$, $q_{0\star0}$, $Q_{[0,1],R}$ and $q_{\emptyset,R}$.

In a minimal weak Büchi automaton $\mathcal{A}$, let $q_{\emptyset,\mathcal{A}}$, $q_{[0,1],\mathcal{A}}$ and $q_{\infty,\mathcal{A}}$ denote the only state q such that $\mathcal{A}_q$ recognizes the languages $\emptyset$, Σ_b^ω and $\Sigma_b^* \star \Sigma_b^\omega$ respectively. The following lemma states that the five sets introduced in Definition 1 are linear time computable.

Lemma 3. *Let $\mathcal{A}$ be a weak Büchi automaton with n states. Then the sets $Q_{\emptyset,\mathcal{A}}$, $Q_{[0,1],\mathcal{A}}$, $q_{\infty,\mathcal{A}}$, $Q_{I,\mathcal{A}}$ and $Q_{F,\mathcal{A}}$ are computable in time $O(nb)$.*

It is explained how to compute $Q_{\emptyset,\mathcal{A}}$. The other sets are computed similarly.

Proof. Tarjan's algorithm [16] can be used to compute the set of strongly connected component in time $O(nb)$, and therefore the set of recurrent states. By definition, $q \in Q_{\emptyset,\mathcal{A}}$ if and only if $\mathcal{A}_q$ accept no ω-word. It is equivalent to the fact that no recurrent state accessible from q are accepting. Equivalently, $Q_{\emptyset,\mathcal{A}}$ is the greatest set of states q such that, q is not a recurrent accepting state, and furthermore, for all $a \in \Sigma_b \cup \{\star\}$, $\delta(q,a) \in Q_{\emptyset,\mathcal{A}}$. This naturally leads to the following greatest fixed-point algorithm.

Two sets `PotentiallyEmpty` and `ToProcess` are used by the algorithm.

The algorithm initializes the set `PotentiallyEmpty` to Q and initializes the set `ToProcess` to the empty set. The algorithm runs on each recurrent state q. For each state q, if q is accepting, then q is removed from `PotentiallyEmpty` and added to `ToProcess`. The algorithm then runs on each element q of `ToProcess`. For each state q, the algorithms removes q from `ToProcess` and runs on each predecessors q' of q. For each q', if q' is in `PotentiallyEmpty`, then q' is removed from `PotentiallyEmpty` and added to `ToProcess`. Finally, when `ToProcess` is empty, the algorithm halts and $Q_{\emptyset,\mathcal{A}}$ is the value of `PotentiallyEmpty`.

4 From Simple Sets to Automata

Let us fix a simple non-empty set $R \subsetneq \mathbb{R}^{\geq 0}$. In this section a weak RNA $\mathcal{A}_R$ which recognize $L_\omega(R)$ is constructed. Since R is a simple set, there exists an integer $t_R \in \mathbb{N}^{\geq 0}$ such that $[t_R, \infty)$ is either a subset of R or is disjoint from R. Without loss of generality, it is assumed that $t_R \geq b$. As seen in Sect. 2.3, R can be expressed as $\bigcup_{i=0}^{I-1}(\rho_{i,\mathfrak{L}}, \rho_{i,\mathfrak{R}}) \cup \bigcup_{i=0}^{J-1}\{\rho_{i,\mathfrak{S}}\}$ with $\rho_{i,j} \in \mathbb{Q} \cap [0, t_R]$. Without loss of generality, it can be assumed that all integers n belonging to $[0, t_R]$ are of the form $\rho_{i,j}$ for some i, j. It suffices either to assume that there is some $i \in \mathbb{N}$ such that n is of the form $\rho_{i,\mathfrak{S}}$ if $n \in R$ and of the form $\rho_{i,\mathfrak{L}}$ and $\rho_{i,\mathfrak{R}}$ otherwise.

Since the $\rho_{i,j}$ are rationals, their encodings in base b are of the form $u_{i,j,k} v_{i,j,k}^\omega$ with $u_{i,j,k} \in \Sigma_b^* \star \Sigma_b^*$ such that $u_{i,j,k}[0] \neq 0$ and $v_{i,j,k} \in \Sigma_b^+$. Since there are at most two encodings, a third index, k, is also required. Up to replacing the words $u_{i,j,k}$ by $u_{i,j,k} v_{i,j,k}^n$, it can be assumed without loss of generality that, for all i, j, k, i', j', k', the word $u_{i,j,k}$ is not a strict prefix of $u_{i',j',k'}$ and if $u_{i,j,k} = u_{i',j',k'}$ then $v_{i,j,k} = v_{i',j',k'}$. The formal definition of $\mathcal{A}_R$ is now given.

Definition 2 ($\mathcal{A}_R$). *Let $R \subsetneq [0, \infty)$ be a simple non-empty set. Note that $t_R > 0$. Let $\mathcal{A}_R$ be the automaton $(Q, \Sigma_b \cup \{\star\}, \delta, q_0, F)$ where:*

- Q contains the states $q_{\emptyset,\mathcal{A}}$, $q_{[0,1],\mathcal{A}}$, $q_{\infty,\mathcal{A}}$, and a state q_w for each strict prefix w of a word $u_{i,j,k}v_{i,j,k}$.
- F contains $q_{[0,1],\mathcal{A}}$, and the q_w's, for $w \in \Sigma_b^* \star \Sigma_b^\omega$ some non-empty prefix of some $u_{i,\mathfrak{S},k}v_{i,\mathfrak{S},k}$.
- and the transition function is such that, for each word w and for each letter a:
  - $\delta(q_\epsilon, 0) = q_\epsilon$.
  - For wa a strict prefix of some $u_{i,j,k}v_{i,j,k}$, $\delta(q_w, a) = q_{wa}$.
  - For wa of the form $u_{i,j,k}v_{i,j,k}$, $\delta(q_w, a) = q_{u_{i,j,k}}$.
  It is now assumed that wa is not a prefix of or equal to any $u_{i,j,k}v_{i,j,k}$.
  - If $wa \in \Sigma_b^*$, then $\delta(q_w, a)$ is $q_{\infty,\mathcal{A}}$ if $[t_R, \infty) \subseteq R$ and $q_{\emptyset,\mathcal{A}}$ otherwise.
  - For $wa \in \Sigma_b^* \star \Sigma_b^*$, $\delta(q_w, a)$ is $q_{\emptyset,R}$ if $[wa0^\omega]_b^R \notin R$ and $q_{[0,1],\mathcal{A}}$ otherwise.
  - For q being $q_{[0,1],R}$, $q_{\infty,R}$ or $q_{\emptyset,R}$, $\delta(q, a) = q$.
  - $\delta(q_{\infty,R}, \star) = q_{[0,1],R}$.
  - For q being $q_{[0,1],\mathcal{A}}$ or $q_{\emptyset,\mathcal{A}}$ or q_w for $w \in \Sigma_b^* \star \Sigma_b^*$, $\delta(q, \star) = q_{\emptyset,\mathcal{A}}$.

It can be shown that $\mathcal{A}_R$ recognizes R. Let $R = \left(\frac{1}{3}, 2\right] \cup \left(\frac{8}{3}, 3\right] \cup \left(\frac{11}{3}, \infty\right]$ as in Example 1. The automaton $\mathcal{A}_R$ is pictured in Fig. 1, without the non accepting state $q_{\emptyset,\mathcal{A}}$. Its minimal quotient is pictured in Fig. 2.

A second example is now given, which shows that the minimal number of intervals of a simple set may be exponential in the number of state of the minimal Büchi automaton accepting this set. For every non-negative integer n, let R_n be $\left\{m2^{-(n-1)} \mid m \in [2^{n-1}]\right\}$. It is the set of reals which admit an encoding w in base 2 whose suffixes $w[\geq n]$ are either equal to 0^ω or to 1^ω. This set can not be described with less than 2^{n-2} intervals and is recognized by the automaton $\mathcal{A}_n$:

$$\mathcal{A}_n = \left(\{q_i \mid i \in [n]\} \cup \{q_{n+1,0}, q_{n+1,1}, q_{\emptyset,\mathcal{A}}\}, \Sigma_b, \delta, q_0, \{q_{n+1,0}, q_{n+1,1}\}\right),$$

where the transition function is such that, for $a \in \Sigma_2$, and $i \in [n-1] \setminus \{0\}$, $\delta(q_0, \star) = q_1$, $\delta(q_i, a) = q_{i+1}$, $\delta(q_n, a) = q_{n+1,a}$, $\delta(q_{n+1,a}, a) = q_{n+1,a}$. For each state q and letter a such that $\delta(q, a)$ is not defined above, $\delta(q, a) = q_{\emptyset,\mathcal{A}}$.

5 Deciding Whether an Automaton Recognizes a Simple Set

It is explained in this section how to decide whether a minimal weak RNA accepts a simple set. The first main theorem of this paper is now given.

Theorem 2. *It is decidable in time $O(nb)$ and space $O(n)$ whether a minimal weak Büchi RNA with n states recognizes a simple set.*

In order to prove this theorem, a proposition is is now given. This property is a general method used to efficiently decide properties of automata. This method is similar to the method used in [11] and in [13].

Proposition 1. *Let $\mathbb{A}'$ be a class of weak Büchi automata and let $\mathbb{L}'$ be the class of languages $\{L_\omega(\mathcal{A}) \mid \mathcal{A} \in \mathbb{A}'\}$. Let $\mathbb{L}$ be a class of languages over an alphabet such that there exists a class $\mathbb{A}$ of weak Büchi automata such that:*

1. *there exists an algorithm α which decides in time $t(n,b)$ and space $s(n,b)$ whether a Büchi automaton belongs to $\mathbb{A}$, for n the number of states and b the number of letters,*
2. *for each $L \in \mathbb{L} \cap \mathbb{L}'$, there exists an automaton $\mathcal{A} \in \mathbb{A}$ which recognizes L,*
3. *the minimal quotient of any automaton of $\mathbb{A}$ belongs to $\mathbb{A}$ and*
4. *every language recognized by an automaton belonging to $\mathbb{A}$ belongs to $\mathbb{L}$.*

The algorithm α decides in time $t(n,b)$ and space $s(n,b)$ whether a minimal automaton of $\mathbb{A}'$ recognizes a language of $\mathbb{L}$. Furthermore, the algorithm α applied to an automaton belonging to $\mathbb{A}' \setminus \mathbb{A}$ may not return a false positive.

Proof. Let $\mathcal{A}$ be an automaton which recognizes a language L. Let us assume that α accepts $\mathcal{A}$, by Proposition (1), $\mathcal{A} \in \mathbb{A}$, hence by Proposition (4), $L \in \mathbb{L}$.

Let us now assume that $\mathcal{A} \in \mathbb{A}'$ and that $L \in \mathbb{L}$. By definition of $\mathbb{L}'$, $L \in \mathbb{L}'$, hence $L \in \mathbb{L} \cap \mathbb{L}'$, thus by Proposition 2, there exists $\mathcal{A}' \in \mathbb{A}$ which recognizes L. Since $\mathcal{A}'$ and $\mathcal{A}$ recognize the same language, they have the same minimal quotient, which is $\mathcal{A}$. By Proposition 3, $\mathcal{A} \in \mathbb{A}$. Thus, by Proposition (1), α accepts $\mathcal{A}$.

In this paper, $\mathbb{A}'$ is the set of RNAs, hence $\mathcal{L}'$ is the class of saturated recognizable languages. The class of languages $\mathcal{L}$ is the class of base b encoding of non-empty sets $R \subsetneq \mathbb{R}^{\geq 0}$. The cases of $R = \mathbb{R}^{\geq 0}$ and of $R = \emptyset$ being special cases. The class $\mathbb{A}$ of automata is now introduced.

Definition 3 ($\mathbb{A}$). *Let $\mathbb{A}$ be the set of weak Büchi automata $\mathcal{A}$, of the form $(Q, \Sigma_b \cup \{\star\}, \delta, q_0, F)$, such that, for each strongly connected component $C \subseteq Q_{F,\mathcal{A}} \setminus (Q_{[0,1],\mathcal{A}} \cup Q_{\emptyset,\mathcal{A}})$, there exists $\beta_{<,C}$ and $\beta_{>,C}$, two states of $Q_{[0,1],\mathcal{A}} \cup Q_{\emptyset,\mathcal{A}}$, such that, for all $q \in C$:*

1. *C is a cycle. Recall that s_q is the only letter such that $\delta(q, s_q) \in C$.*
2. *For all $a > s_q$, $\delta(q,a)$ is $\beta_{>,C}$.*
3. *For all $a < s_q$, $\delta(q,a)$ is $\beta_{<,C}$.*
4. *There exists an accepting and a rejecting strongly connected component, accessible from the initial state, belonging to $Q_{F,\mathcal{A}}$.*
5. *The set $Q_{\emptyset,\mathcal{A}}$ contains exactly one recurrent state, denoted $q_{\emptyset,\mathcal{A}}$.*
6. *The set $Q_{\infty,\mathcal{A}}$ contains at most one recurrent state, denoted $q_{\infty,\mathcal{A}}$.*
7. *$\delta(q_0, 0) = q_0$.*
8. *$\delta(q_0, a) \neq q_0$ for all $0 < a < b$.*
9. *If $q_{\infty,\mathcal{A}}$ exists, then $\delta(q,a) \neq q_{\emptyset,\mathcal{A}}$ for all $q \in Q_{I,\mathcal{A}} \setminus \{q_{\emptyset,\mathcal{A}}\}$ and $a \in \Sigma_b$.*
10. *The recurrent states of $Q_{I,\mathcal{A}}$ are $q_{\emptyset,\mathcal{A}}$, q_0 and potentially $q_{\infty,\mathcal{A}}$.*

The automata of $\mathbb{A}$ admits the following property.

Lemma 4. *Let $\mathcal{A} \in \mathbb{A}$ be an automaton with n states recognizing a set R. If $\mathcal{A}$ contains a state $q_{\infty,\mathcal{A}}$, as in Definition 3, then $(b^{n-1}, \infty) \subseteq R$, otherwise $(b^{n-1}, \infty) \cap R = \emptyset$.*

Proof (Sketch of proof of Theorem 2). Using Lemma 3, the algorithms checks whether $\mathcal{A}$ accepts a subset L of $\Sigma_b^* \star \Sigma_b^\omega$, if it is not the case, the algorithm rejects. The algorithms also checks whether L is $\emptyset$ or $\Sigma_b^* \star \Sigma_b^\omega$. If it is the case, the algorithm accepts. It is now assumed that $\mathcal{A}$ accepts a non-empty language $L \subsetneq \Sigma_b^* \star \Sigma_b^\omega$. Let $\mathbb{L}'$ be the set of saturated languages and $\mathbb{A}'$ be the set of RNAs. Let $\mathbb{L}$ be the set of encoding of simple non-empty sets $R \subsetneq \mathbb{R}^{\geq 0}$. In order to prove this theorem, it suffices to show that $\mathbb{A}$ admits the four properties of Proposition 1.

Each property of Definition 3 is testable in time $O(nb)$ and space $O(n)$. Therefore, it is decidable in time $O(nb)$ and space $O(n)$ whether a weak Büchi automaton $\mathcal{A}$ with n states belongs to $\mathbb{A}$. Hence Property (1) of Proposition 1 holds.

For $R \subsetneq \mathbb{R}^{\geq 0}$ a non-empty simple set, the automaton $\mathcal{A}_R$ of Definition 2 belongs to $\mathbb{A}$. Therefore Property (2) of Proposition 1 holds.

Let $\mathcal{A} \in \mathbb{A}$ be a RNA. Let $\mathcal{A}'$ be its minimal quotient. It can be proven that $\mathcal{A}'$ satisfies the properties of Definition 3, hence $\mathcal{A}' \in \mathbb{A}$. Therefore Property (3) of Proposition 1 holds.

Property (4) of Proposition 1 is now considered. Automata satisfying Properties (1), (2) and (3) of Definition 3 are studied in [2]. It is shown that automata satisfying those properties accepts a set R such that $R \cap [i, i+1]$ is a finite union of intervals with rationals boundaries for all $i \in \mathbb{N}$. Lemma 4 ensures that furthermore, there is some $t \in \mathbb{N}$ such that $[t, \infty)$ is either a subset of R or is disjoint from R. Thus, an automaton of $\mathbb{A}$ recognize a finite union of interval with rational boundaries, i.e. a simple set. Therefore Property (4) of Proposition 1 holds.   □

The algorithm of Theorem 2 takes as input a minimal weak RNA and runs in time $O(nb)$. It should be noted that it is not known whether it is decidable in time $O(nb)$ whether a minimal Büchi automaton is a RNA. However, if the algorithm of Theorem 2 is applied to a weak Büchi automaton which is not a Real Number Automaton, the algorithm returns no false positive. An example of false negative is now given. The not-saturated language $L = (00)^* (01 + 2) \Sigma_3^* \star \Sigma_3^\omega$ encode the simple set of reals $\mathbb{R}^{>0}$. However, the minimal automaton recognizing L it is not accepted by the algorithm of Theorem 2.

6   From Automata to Simple Set

It is explained in this section how to compute a first-order formula which defines the simple set accepted by a weak RNA. The exact theorem is now stated.

Theorem 3. *Let $\mathcal{A} = (Q, \Sigma_b \cup \{\star\}, \delta, q_0, F)$ be a be a minimal weak RNA with n states which recognizes a simple set. There exists a $\Sigma_1 [\mathbb{R}; +, <, 1]$-formula computable in time $O(n^2 b \log(nb))$ which defines $[\mathcal{A}]_b^R$. There exists a $\Sigma_2 [\mathbb{R}; +, <, 1]$-formula computable in time $O(nb \log(nb))$ which defines $[\mathcal{A}]_b^R$.*

The proof of Theorem 3 consists mostly in encoding in a first-order formula $\phi(x)$ the run of A over an encoding w of x. The following lemma allows to consider two distinct part of the run on the fractional part of w. The first part of the run is of length at most n. The second part on the run begins on a state belonging to a restricted set of states.

Lemma 5. *Let $A \in \mathbb{A}$ be minimal with n states and $q \in Q_{F,A}$. Let $w_I \in \Sigma_b^*$ and $w_F \in \Sigma_b^\omega$. Let $Q \subseteq Q_{F,A}$ be a set containing exactly one state of each strongly connected component. Then, there exists $s \in [n]$ such that $\delta(q, w_I \star w_F [< s]) \in Q$.*

The following lemma allows to reduce the size of a formula by adding quantifications.

Lemma 6. *Let $\psi(x, x')$ be a formula of length l and $(x_i)_{i \in [n-1]}$ be n variables. Then $\bigwedge_{i \in [n-2]} \psi(x_i, x_{i+1})$ is equivalent to the following formula of length $O(n+l)$:*

$$\forall y, y'. \left\{ \bigvee_{i \in [n-2]} [y \doteq x_i \wedge y' \doteq x_{i+1}] \right\} \implies \psi(y, y').$$

A sketch of the proof of Theorem 3 can now be given.

Proof (Proof of Theorem 3). Let $R = [A]_b^R$. As shown in Sect. 5, it can be assumed that A belongs to $\mathbb{A}$. By Lemma 4, in order to construct a formula which defines R it suffices to construct a formula $\phi(x)$ which defines $R \cap [0, b^{n-1})$. The formula $\phi(x)$ is the conjunction of four subformulas of size $O(n^2 b \log(nb))$. Let $x \in [0, b^{n-1})$ and let w be an encoding of x without leading 0 in the natural part.

The first formula, $\psi(x, x_I, x_F)$, states that $x = x_I + x_F$ and that $x_I \in \mathbb{N}$. Since $x_I < b^{n-1}$, in order to state that $x_I \in \mathbb{N}$, it suffices to state that x_I is of the form $\sum_{i=0}^{n-2} a_i b^i$ for $a_i \in [b-1]$. More precisely, it suffices to state that x_I is of the form $(c_n + b(c_{n-1} + b(\cdots + b(c_0)\dots)))$ with the c_i belonging to $[b-1]$. This can be stated by existentially quantifying the $2n$ partial sums and products and taking disjunctions over each c_i. This can be done by a formula $\psi(x_I)$ of size $O(nb \log(b))$.

Let q be the state $\delta(q_0, w_I)$. The second formula, $\phi_I(x_I, q)$, states that the state $\delta(q_0, w_I \star)$ is equal to q. This formula existentially quantifies $2n$ variables. Those variables encode the n first steps of the runs and the values of $w_I [< i]$ for $i < n$. Each step of the computation can be encoded by a $\Sigma_0 [\mathbb{R}; +, <, 1]$-formula of length $O(nb \log(b))$, using the equalities of Lemma 1. Since $x_I < b^{n-1}$, $|w_I| < n$, the formula $\phi_I(x_I, q)$ have to consider at most n steps of the computation. The formula $\phi_I(x_I, q)$ is a conjunction of n formulas of size $O(n^2 b \log(b))$ and thus the size of $\phi_I(x_I, q)$ is $O(n^2 b \log(b))$.

Let Q be a set of states as in Lemma 5 and let q' be the first state of Q in the run of A on w. The third formula, $\phi_F(q, x_F, q', x_F')$ states that there exists $i \in [n]$ such that $\delta(q, w_F [< i]) = q'$, that $q' \in Q$ and that $x_F' = [w_F [\geq i]]_b^F$. By Lemma 5, i is at most n. Hence, similarly to $\phi_I(x_I, q)$, the size of the formula $\phi_F(q, x_F, q', x_F')$ is $O(n^2 b \log(b))$.

Finally, the fourth formula $\phi'_F(q', x'_F)$, states that $\mathcal{A}_{q'}$ accepts w_F [≥ i]. Let c be the number of strongly connected components in $\mathcal{A}$. For C a strongly connected components, let n_C be its number of state and q_C the only state of $C \cap \mathcal{Q}$. Let us assume that, for each strongly connected component C, there exists a formula $\phi'_C(q', x'_F)$ of length $O(n_C b \log(n_C b))$ which states that $\mathcal{A}_{q'}$ accepts w_F [≥ i]. Then the formula $\phi'_F(q', x'_F)$ is a disjunction of c formulas $q' \doteq q_C \wedge \phi'_C(q', x'_F)$ and its length is $O(\sum_C n_C b \log(n_C b)) = O(nb \log(nb))$.

It is now explained how to construct $\phi'_C(q', x'_F)$. Since $\mathcal{A} \in \mathbb{A}$, by Proposition 1 of Definition 3, strongly connected components of automata included in $Q_{F,\mathcal{A}}$ are either $\{q_{\emptyset,\mathcal{A}}\}$, $\{q_{[0,1],\mathcal{A}}\}$, or a cycle. In the first two cases, $\phi'_C(q', x'_F)$ is the formula False or True respectively. Let us consider the third cases. Let $v_{q'}$ be the word of size n_C such that $\delta(q', v_{q'}) = q'$. Since C is cycle, this word exists and is a unique. Then let $y = \left[v_{q'}^\omega\right]_b^F = \left[v_{q'}\right]_b^I / (b^{n_C} - 1)$. Recall that the notations $\beta_{<,C}$ and $\beta_{>,C}$ are introduced in Definition 3. Then the formula $\phi'_C(q', x'_F)$ states that $q' \in C'$ and that either $(x'_F < y$ and $\beta_{<,C} \in Q_{[0,1],\mathcal{A}})$, either $(x'_F = y$ and $q' \in F)$, or $(x'_F > y$ and $\beta_{>,C} \in Q_{[0,1],\mathcal{A}})$. It is indeed a formula of length $O(n_C b \log(n_C b))$.

Finally, in order to reduce the size of the formula to $O(nb \log(nb))$, it suffices to replace the conjunctions of $\phi_I(x_I, q)$ and of $\phi_F(q, x_F, q' x'_F)$ by a universal quantifications, as explained in Lemma 6.   □

7   Conclusion

In this paper, we proved that it is decidable in linear time whether a minimal weak Büchi Real Number Automaton $\mathcal{A}$ reading a set of real number R recognizes a finite union of intervals. It is proven that a quasi-linear sized existential-universal formula defining R exists. And that a quasi-quadratic sized existential formula defining R also exists.

The theorems of this paper lead us to consider two natural generalization. We intend to adapt the algorithm of this paper to similar problems for automata reading vectors of reals instead of automata reading reals. We also intend to solve the similar problem of deciding whether an RNA accepts a FO[ℝ, ℕ; +, <]-definable set of reals. Solving this problem requires to solve the problem of deciding whether an automaton reading natural number, beginning by the most-significant digit, recognizes an ultimately-periodic set. Similar problems has already been studied, see e.g. [1, 10] and seems to be difficult. Finally, we also intend to consider how to efficiently decide whether an automaton is a Real Number Automaton or a Real Vector Automaton.

The author thanks Bernard Boigelot, for a discussion about the algorithm of Theorem 3, which led to a decrease of the computation time. He also thanks the anonymous referees of for their remarks and suggestion to improve the paper.

References

1. Alexeev, B.: Minimal dfas for testing divisibility. CoRR cs.CC/0309052 (2003), http://arxiv.org/abs/cs.CC/0309052
2. Boigelot, B., Brusten, J.: A generalization of cobham's theorem to automata over real numbers. In: Arge, L., Cachin, C., Jurdziński, T., Tarlecki, A. (eds.) ICALP 2007. LNCS, vol. 4596, pp. 813–824. Springer, Heidelberg (2007). doi:10.1007/978-3-540-73420-8_70
3. Boigelot, B., Brusten, J., Bruyère, V.: On the sets of real numbers recognized by finite automata in multiple bases. Logical Methods Comput. Sci. 6(1) (2010)
4. Boigelot, B., Brusten, J., Leroux, J.: A generalization of semenov's theorem to automata over real numbers. In: Schmidt, R.A. (ed.) CADE 2009. LNCS (LNAI), vol. 5663, pp. 469–484. Springer, Heidelberg (2009). doi:10.1007/978-3-642-02959-2_34. https://hal.archives-ouvertes.fr/hal-00414753
5. Bruyre, V., Hansel, G., Michaux, C., Villemaire, R.: Logic and p-recognizable sets of integers. Bull. Belg. Math. Soc 1, 191–238 (1994)
6. Cobham, A.: On the base-dependence of sets of numbers recognizable by finite automata. Math. Syst. Theo. 3(2), 186–192 (1969). http://dx.doi.org/10.1007/BF01746527
7. Ferrante, J., Rackoff, C.: A decision procedure for the first order theory of real addition with order. SIAM J. Comput. 4(1), 69–76 (1975). http://dx.doi.org/10.1137/0204006
8. Hardy, G.H., Wright, E.M.: An introduction to the theory of numbers (1960). http://opac.inria.fr/record=b1133956, autres tirages avec corrections : 1962, 1965, 1968, 1971, 1975
9. Honkala, J.: A decision method for the recognizability of sets defined by number systems. ITA 20(4), 395–403 (1986). http://dblp.uni-trier.de/db/journals/ita/ita20.html#Honkala86
10. Lacroix, A., Rampersad, N., Rigo, M., Vandomme, É.: Syntactic complexity of ultimately periodic sets of integers and application to a decision procedure. Fundam. Inform. 116(1–4), 175–187 (2012). http://dx.doi.org/10.3233/FI-2012-677
11. Leroux, J.: Least significant digit first Presburger automata. CoRR abs/cs/0612037 (2006)
12. Löding, C.: Efficient minimization of deterministic weak omega automata (2001)
13. Marsault, V., Sakarovitch, J.: Ultimate periodicity of b-recognisable sets: a quasi-linear procedure. In: Béal, M.-P., Carton, O. (eds.) DLT 2013. LNCS, vol. 7907, pp. 362–373. Springer, Heidelberg (2013). doi:10.1007/978-3-642-38771-5_32
14. Muchnik, A.A.: The definable criterion for definability in Presburger arithmetic and its applications. Theor. Comput. Sci. 290(3), 1433–1444 (2003)
15. Semenov, A.L.: The presburger nature of predicates that are regular in two number systems. Siberian Math. J. 18, 289–299 (1977)
16. Tarjan, R.: Depth-first search and linear graph algorithms. SIAM J. Comput. 1(2), 146–160 (1972). http://dx.doi.org/10.1137/0201010
17. Weispfenning, V.: Mixed real-integer linear quantifier elimination (1999)
18. Wolper, P., Boigelot, B.: On the construction of automata from linear arithmetic constraints. In: Graf, S., Schwartzbach, M. (eds.) TACAS 2000. LNCS, vol. 1785, pp. 1–19. Springer, Heidelberg (2000). doi:10.1007/3-540-46419-0_1

qPCF: A Language for Quantum Circuit Computations

Luca Paolini[1](✉) and Margherita Zorzi[2](✉)

[1] Department of Computer Science, University of Torino, Turin, Italy
paolini@di.unito.it
[2] Department of Computer Science, University of Verona, Verona, Italy
margherita.zorzi@univr.it

Abstract. We propose qPCF, a functional language able to define and manipulate quantum circuits in an easy and intuitive way. qPCF follows the tradition of "quantum data & classical control" languages, inspired to the QRAM model. Ideally, qPCF computes finite circuit descriptions which are offloaded to a quantum co-processor (i.e. a quantum device) for the execution. qPCF extends PCF with a new kind of datatype: quantum circuits. The typing of qPCF is quite different from the mainstream of "quantum data & classical control" languages that involves linear/exponential modalities. qPCF uses a simple form of dependent types to manage circuits and an implicit form of monad to manage quantum states via a destructive-measurement operator.

1 Introduction

In the last fifteen years, the definition and the development of quantum programming languages catalyzed the attention of a part of the computer science research community. Quantum computers are a long term but concrete reality. Even if physicists and engineers have to continuously face tricky problems in the realization of quantum devices, the advance of these innovative technologies promises a noticeable speedup.

A calculus for quantum computable functions should present two different facets. On the first hand there is the unitary aspect of the calculus, that captures the essence of quantum computing as algebraic transformations of state vectors by means of unitary operators. On the other hand, it should be possible to *control* the quantum steps by means of classical computational steps, "embedding" the pure quantum evolution in a classical computation. Behind this second perspective we have the usual idea of computation as a sequence of discrete steps on (the mathematical description of) an abstract machine. The relationship between these different aspects gives rise to different approaches to quantum functional calculi (as observed in [1]). If we divide the two features, i.e. we separate data from control, we adopt the so called *quantum data & classical control (qd&cc)* approach. This means that classical computation is *hierarchical dependent* from the quantum part: a classical program (ideally in execution on a classical machine) computes some "directives": these directives are sent to a

© Springer International Publishing AG 2017
T.V. Gopal et al. (Eds.): TAMC 2017, LNCS 10185, pp. 455–469, 2017.
DOI: 10.1007/978-3-319-55911-7_33

hypothetical device which apply them to quantum data. Therefore classical program controls quantum data or, in other words, classical computational steps control the unitary part of the calculus. In general, the classical control acts on the quantum side of the computation in two way: by means of the selection of unitary transformations to be applied and by means of data observations, i.e. by means of measurements. A different approach based on the quantum control is the *superposition-of-programs paradigm*. See [34], Part III, for details. This idea is inspired to an architectural model called Quantum Random Access Machine (QRAM). The QRAM has been defined in [10] and can be viewed as a classically controlled machine enriched with a quantum device. On the grounds of the QRAM model, Selinger defined the first functional language based on the quantum data-classical control paradigm [28]. This work represents a milestone in the development of quantum functional calculi and inspired a number of different investigations. A key research line tried to retrace, in the quantum setting, foundational results about computability. In this direction, calculi for quantum computable functions have been defined, and equivalence results with other computational models, such as Quantum Turing Machine and Quantum Circuit Families, have been proved [4,11,12,35]. Moreover, interesting proposals to provide satisfactory denotational semantics for *qd&cc* functional calculi have been proposed in [8,19,29]. Many quantum programming languages [7,28,29] implementing the (*qd&cc*) approach have been proposed in literature. A recent and interesting proposal is Quipper which is an embedded, scalable functional programming language for quantum computing proposed in [28]. Quipper is essentially a high-level circuit description language: circuits can be created, manipulated, evaluated in a mixture of procedural and declarative programming styles. The most important quantum algorithms can be easily encoded thanks to a number of programming tools, macros, and extensive libraries of quantum functions. The idea of the separation between control and data is definitely reformulated in terms of *quantum-coprocessor* [31]. Quipper has been mainly developed as a *concrete* language. Authors are not interested in the foundational study of it. Quipper is based on the lambda calculus with classical control proposed in [28], and this relationship is discussed in [26], by means of a suitable calculus named Proto-quipper. In [14], the semantics of Proto-Quipper is further formalized by means of the linear specification logic SL. The type system is based on a linear logic approach that ensures the correct interaction of classical and quantum types. The "*qd&cc*" philosophy, in particular the circuit generation oriented approach, has been also adopted in the purely linear core-language QWire, introduced in [22]: a low-level quantum language developed to be a "quantum plugin" for a hosting classical language like Haskell. QWire and qPCF are based on some similar ideas. Differently from qPCF, QWire retains the focus on quantum states which is typical of the *qd&cc* tradition. In qPCF quantum states are not more atomic data, they are replaced by quantum circuits. In this paper we advance in the research on the languages for *qd&cc* paradigm by formalizing a flexible quantum language. We propose qPCF, a simple extension of PCF. We quickly list the main features of qPCF.

– *Absence of explicit linear constraints*: the management of linear resources is radically different from the ones proposed in languages inspired to Linear Logic such as [4,11,12,14,26,28,35]; so, we do not use linear/exponential modalities.
– *Use of dependent types*: we decouple the classic control from the quantum computation by adopting a simplified form of dependent types that provide a sort of linear interface.
– *Emphasis of the Standardization Theorem*: the Standardization Theorem, proved in [4,35], and largely used in circuit description languages such as Quipper, decomposes computations in different phases, according to the quantum circuit construction by classical instructions and the successive, independent, evaluation phase involving quantum operations.
– *Unique measurement at the end of the computation*: following the "principle of deferred measurement" which states that any quantum circuit is equivalent to one where all measurements are performed as the very last operations (see, e.g., [17]), we add an explicit measurement operator to qPCF syntax that models the *(von Neumann) Total Measurement* [17], a kind of measurement that reduces a quantum state to a classical one (a sequence of classical bit). Essentially we are using a monad-style programming, and we "embed" both, quantum evaluation and measurement into the operator dmeas (see Sect. 3 for dmeas operational behavior).
– *Implicit representation of quantum states*: differently from other proposals (e.g. [4,28,35]), we hide quantum states we are working on. This can be achieved thanks to the monadic-style approach we mentioned above.
– *Turing Completeness*: qPCF retains PCF expressive power. So, a qPCF term can represent an infinite class of circuits.

Synopsis: Section 2 introduces syntax and typing system of qPCF; the operational semantics of qPCF is in Sect. 3; Sect. 4 sketches some properties of qPCF; Sect. 5 contains some examples of qPCF circuit encodings; Sect. 6 is devoted to discuss conclusions and future work.

2   qPCF

In this section we describe qPCF, a programming language that pursue seriously the application of the standardization theorem of [4,35]: it states that, in the "quantum-data & classic control languages", the quantum evaluation can always be postponed after the classical execution. On the other hand, the classical evaluation designs a quantum circuit that can be evaluated in a second time. Ideally, qPCF computes a finite circuit description which is offloaded to a quantum co-processor for the execution. qPCF is definitively more flexible than the languages presented in [4,19,27–29]. It extends PCF with quantum circuits, viz. a new kind of classical data. Indeed, as observed in [22], quantum circuits can be freely duplicated and erased. We realized that the linearity of mainstream typing systems of "quantum-data & classic control" languages has been used to impose constraints on both the management of quantum-data and the management of classic control. qPCF neatly splits these linear facets by using two different solutions. On the one hand, qPCF shows that linearity for quantum control can be completely confined to atomic datatypes by using a simplified form of dependent types [23]. A dependent type picks up a family of types that bring in the type auxiliary information (just circuit arity in our case). On the other hand, the linearity needed to manage quantum state is hidden in a destructive measure operator (by means of an implicit form of monad) that model von Neumann Measurement [17] and allows us to avoid the explicit management of intermediate quantum states.

In the rest of the paper, we assume some familiarity with notions as quantum bits (the quantum equivalent of classical data), quantum states [15,16,32,33] (systems of n quantum bits), quantum circuit and quantum circuit families [18]. A quantum circuit generalizes the idea of classical circuit, replacing the elementary classical gates (AND, OR, NOT...) by elementary quantum gates [17], that enjoy matemathical descriptions in terms of *unitary operators* on suitable Hilbert spaces. A quantum circuit family can be (quite informally) considered as a function $\mathbf{K} : \mathbb{N} \to \mathcal{K}$ (denotable as $(K_i)_{i<\omega}$), where $\mathcal{K}$ is the set of circuit descriptions; $\mathbf{K}(n)$ returns K_n, i.e. the circuit of ariety n. See [35] for a friendly introduction to quantum computing. We remand to [17] for a complete overview about the topic. See also [9,18] for details about quantum circuit families and some crucial discussions about the universality of sets of quantum gates. Finally, see [25] for a rigorous algebraic characterisation of quantum computing.

2.1 Syntax

Dependent types have been widely used in strongly normalizing settings (usually, with logic goals) where the evaluation of expressions is always terminating. But in programming settings the strong normalization [21] is not realistic. Unfortunately, to allow types that embeds undefined terms (viz. not strongly normalizing ones) requires the management of "undefined types" [3]. We circumvent this issue by identifying a subclass of terms (always normalizing) that we use in our dependent types. qPCF extends PCF [5,6,24] to manage some additional atomic data structures: *indexes* (always normalizing number expressions) and *circuits*.

The row syntax of qPCF follows.

$$\texttt{M, N} ::= \texttt{x} \mid \lambda\texttt{x.M} \mid \texttt{MN} \mid \underline{\texttt{n}} \mid \texttt{pred} \mid \texttt{succ} \mid \texttt{if} \mid \texttt{Y}_\sigma \mid \texttt{set} \mid \texttt{get}$$
$$\mid \odot\texttt{EE}' \mid \underline{\texttt{s}} \mid \texttt{append} \mid \texttt{iter} \mid \texttt{reverse} \mid \texttt{size} \mid \texttt{dMeas} \ .$$

In the first row we extended PCF with syntactic sugar to facilitate the bitwise access to digit: $\texttt{get}$ allows us to read the i-th digit of the binary representation of a numeral, i.e. its i-th bit; $\texttt{set}$ allows us to modify the i-th bit of a numeral.

Index expressions (ranged over by $\mathbf{E}$) are completely formalized via the typing (cf. Table 1). They include numerals and some *total* operations on expressions: $\odot \in \{+, *\}$ (viz. sum, product).

We assume $\texttt{U}$ to range on a given set of selected gates (i.e. unitary operator, see [17]): if $\mathcal{U}$ is a fixed set of computable unitary operators then, we associate to each computable operator $\mathbf{U} \in \mathcal{U}$ a symbol $\texttt{U}$. We represent circuits by means

of *strings*, viz. the Kleene-closure of the following symbols: the parallel composition of circuits denoted ∥ (i.e. side-by-side placing of circuits), the sequential
composition of circuits ⦂ and gate-names U. append sequentializes two circuits
of the same arity. iter produces the parallel composition of a first circuit with a
given numbers of a second one. The goal of the operator reverse is to transform
a circuit in its adjoint one (in case the gate-base has been chosen closed under
adjunction, otherwise it will be meaningless).

size is an operator that applied to a circuit returns its arity: an index information. It is worth to notice that size do not add any expressivity to qPCF,
because the programmer can explicitly manage pairs of "circuit together with
arity": so that, a projection provides the arity of the circuit. We added size
to qPCF to emphasize the gain that dependent type can provide in a concrete
context, although this makes the proofs of the language properties more complex.

Last, but not least, we use dMeas (a.k.a. destructive measure) to evaluate
circuits initialized via a numeral (representing a binary classical state): dMeas
returns the classic state (encoded in the binary representation of a numeral)
obtained measuring the final quantum state of the considered circuit. Traditionally *qd&cc* languages focus on quantum states, while qPCF focuses on circuits
(hiding states in an monadic measure).

Typically, $\mathcal{U}$ will include a universal base for quantum circuits (see [9]). We
like to remark that we can instance $\mathcal{U}$ to interesting family of gate as reversible
ones: in these cases we are not properly building a quantum programming lan-
guage. If $k \in \mathbb{N}$ then we denote $\mathcal{U}(k)$ the gates in $\mathcal{U}$ having arity $k + 1$, so
$\mathcal{U} = \bigcup_0^\omega \mathcal{U}(k)$. Notice that we do not introduced explicit permutations, because
they can be provided by means of a convenient choice of quantum gates (see,
e.g. [30]). Thus, the choice $\mathcal{U}$ determines whose permutations our circuits can
use. We also notice that the gate identity is a particular permutation.

2.2 Typing System

Standard PCF types are extended to manage circuits and their dependencies.
We use types decorated by numerals to define a denumerable family of circuits.
Our approach is closely inspired to that mentioned in [23, Sect. 30.5] to manage
types of vectors (with dependencies): the decoration carry around some arity
information. We avoided general dependent types systems (see [3] for a survey)
because their great expressiveness is exceeding our need, we preferred to maintain
the qPCFtype system as simple as possible by aiming to show the feasibility of
the approach and its concrete benefits. Our approach to dependent types is based
on a special kind of numeric expressions that can be managed in a limited way:
index. Summing up, types of PCF (i.e. integers and arrows) are flanked by two
new types: circuits (viz. strings typed with dependent types that carry around
numeric information about arities) and indexes (that grasp a subset of numeric
expressions that express only terminating expressions).

Types of qPCF are formalized by the following grammar:

$$\sigma, \tau ::= \mathsf{Nat} \mid \mathsf{Idx} \mid \sigma \rightarrow \tau \mid \mathsf{circ(E)} \mid \Pi\mathsf{x}.\tau$$

where E is an index expressions (morally, a strongly normalizing numeric expression). As in standard dependent type system, we replace arrows involving dependencies by quantified types, namely an arrow $\sigma \to \tau$ is replaced by $\Pi x^{\sigma}.\tau$ whenever x^{σ} occurs in τ in order to emphasize that τ *depends* from x.

Index. Idx aims to pick up a subset of expressions on natural numbers being strongly normalizing, i.e. we want to cut out undefined PCF-expressions as $Y_{Nat}(\lambda x^{Nat}.x)$ (viz. a looping forever term). The leading use of Idx is to type terms M embodied in dependent types (i.e. used in types via circ(M)). The goal is to select numeric expressions that made the equivalence decidable (when such expressions are closed). We focus on a restricted, but revealing, syntax of index expressions is $E ::= x^{Idx} \mid \underline{n} \mid \odot E E'$ where $\odot \in \{+, *\}$ viz. operators denoting addition and multiplication. We are considering a very basic set of binary operators that can be conveniently extended in a concrete case, e.g. by adding the (positive subtraction) $\dot{-}$, or the %, or a selection if^x and so on.

Above expressions are typed Idx by the following rules:

$$\frac{}{B[x : \text{Idx}] \vdash x^{\text{Idx}} : \text{Idx}} \; (i_1) \qquad \frac{}{B \vdash \underline{n} : \text{Idx}} \; (i_2) \qquad \frac{B \vdash E_0 : \text{Idx} \quad B \vdash E_1 : \text{Idx}}{B \vdash \odot E_0 E_1 : \text{Idx}} \; (i_3)$$

where B denotes a standard typing base, i.e. sets of pairs (variable and type).

Index expression are closed when they do not contain any free variable. As usual for PCF, the evaluation is focused on closed expressions, and formalized by the following rules:

$$\frac{}{\underline{n} \Downarrow \underline{n}} \; (n) \qquad \frac{E_0 \Downarrow \underline{m} \quad E_1 \Downarrow \underline{n}}{\odot E_0 E_1 \Downarrow \underline{m} \odot \underline{n}} \; (\text{op})$$

where we use the $\odot$ to denote both its name and its straighforward semantics. We remark that we are considering a strict subset of the index expressions of qPCF in order to increase some intuition (e.g. by neglecting size).

It is immediate that the above index expressions are normalizing with the proposed evaluation strategy, when we focus on closed terms. Moreover, we can informally claim that they are strongly normalizing in the straightforward lambda-calculus behind our semantics, that can be obtained as usual by including some δ-rules for constants.

The most basic property of paradigmatic typing system is that well-typed terms do not "go wrong", i.e. types are preserved by the evaluation and, if the evaluation stops then the result is a value.

Theorem 1 (Preservation & Progress). *(i) If* $\vdash E : \text{Idx}$ *and* $E \Downarrow E'$ *then* $\vdash E' : \text{Idx}$. *(ii) If* $\vdash E : \text{Idx}$ *and* $E \Downarrow E'$ *then* E' *is a numeral.*

Remaining typing. We can now extend the typing to the whole qPCF: the typing system is given in Table 1 (be careful to implicit assumption remarked in the caption).

Table 1. Typing Rules. Each typing rule contains implicit premises: (i) all occurrences of circuits types (also those in bases) embody a term typed Idx with the given base; (ii) all free variables (occurring in terms and types) are typed in the base.

$$\frac{}{B[x:\sigma] \vdash x:\sigma} \; (p_1) \qquad \frac{}{B \vdash \underline{u}:\mathsf{Nat}} \; (p_2) \qquad \frac{}{B \vdash \underline{n}:\mathsf{Idx}} \; (i_2)$$

$$\frac{B[x:\sigma] \vdash N:\tau \quad x \notin FV(N)}{B \vdash \lambda x^\sigma.N:\sigma \to \tau} \; (p_3) \qquad \frac{B \vdash P:\sigma \to \tau \quad B \vdash Q:\sigma}{B \vdash PQ:\tau} \; (p_4)$$

$$\frac{B[x:\sigma] \vdash N:\tau \quad x \in FV(N)}{B \vdash \lambda x.N:\Pi x^\sigma.\tau} \; (x_1) \qquad \frac{B \vdash P:\Pi x^\sigma.\tau \quad B \vdash E:\sigma}{B \vdash PE:\tau[E/x]} \; (x_2)$$

$$\frac{}{B \vdash \mathsf{succ}:\mathsf{Nat} \to \mathsf{Nat}} \; (p_5) \qquad \frac{\sigma \in \{\mathsf{Nat},\mathsf{circ(E)}\}}{B \vdash \mathsf{if}:\mathsf{Nat} \to \sigma \to \sigma \to \sigma} \; (p_6)$$

$$\frac{}{B \vdash \mathsf{pred}:\mathsf{Nat} \to \mathsf{Nat}} \; (p_7) \qquad \frac{\sigma = \tau_1 \to \ldots \to \tau_n \to \tau \quad (n \geq 0) \quad \tau \in \{\mathsf{Nat},\mathsf{circ(E)}\}}{B \vdash Y_\sigma : (\sigma \to \sigma) \to \sigma} \; (p_8)$$

$$\frac{}{B \vdash \mathsf{get}:\mathsf{Nat} \to \mathsf{Nat} \to \mathsf{Nat}} \; (b_1) \qquad \frac{}{B \vdash \mathsf{set}:\mathsf{Nat} \to \mathsf{Nat} \to \mathsf{Nat}} \; (b_2)$$

$$\frac{B \vdash M:\mathsf{Idx}}{B \vdash M:\mathsf{Nat}} \; (x_3) \qquad \frac{B \vdash M:\mathsf{circ(E)}}{B \vdash \mathsf{size}\,M:\mathsf{Idx}} \; (x_4) \qquad \frac{B \vdash E_0:\mathsf{Idx} \quad B \vdash E_1:\mathsf{Idx}}{B \vdash \odot E_0\,E_1:\mathsf{Idx}} \; (i_3)$$

$$\frac{U \in \mathcal{U}(k)}{B \vdash U:\mathsf{circ}(\underline{k})} \; (c_1) \qquad \frac{}{B \vdash \mathsf{iter}:\Pi x^{\mathsf{Idx}}.\,\mathsf{circ}(E_0) \to \mathsf{circ}(E_1) \to \mathsf{circ}(E_0 + ((1+E_1) * x))} \; (c_2)$$

$$\frac{}{B \vdash \mathsf{?}:\mathsf{circ(E)} \to \mathsf{circ(E)} \to \mathsf{circ(E)}} \; (c_1') \qquad \frac{}{B \vdash \|:\mathsf{circ}(E_0) \to \mathsf{circ}(E_1) \to \mathsf{circ}(E_0 + E_1)} \; (c_1'')$$

$$\frac{}{B \vdash \mathsf{append}:\mathsf{circ(E)} \to \mathsf{circ(E)} \to \mathsf{circ(E)}} \; (c_3) \qquad \frac{}{B \vdash \mathsf{reverse}:\mathsf{circ(E)} \to \mathsf{circ(E)}} \; (c_4)$$

$$\frac{}{B \vdash \mathsf{dMeas}:\mathsf{Nat} \to \mathsf{circ(E)} \to \mathsf{Nat}} \; (c_5)$$

Finite sets of pairs *variable, type* are called bases whenever variable-names are disjoint: we use B to range on them. We write $B[x:\sigma]$ to denote the set where the pair $x:\sigma$ is added (possibly replacing an pair involving x. As usual, dependent type systems include a typing rule making explicit some type inter-convertibility. We consider types up to a congruence $\simeq$. We define $\simeq$ as the smaller equivalence that includes: (i) the α-conversion of bound variables and β-conversion, (ii) sum and product are associative and commutative; (iii) product distributes over the sum, i.e. $(*E(+E_0\,E_1)) \simeq (+(*EE_0)(*EE_1))$; (iv) $\underline{0}$ is the neuter element for the sum; and, (v) $\underline{1}$ is the neuter element for the product. We use the type equivalence often implicitly. In particular in the typing system (cf. Table 1) types (containing dependencies) are considered up to equivalence.

Rules $(p_1), (p_2), (p_3), (p_4), (p_5), (p_6), (p_7), (p_8)$ are directly inherited from PCF and do not require special care. We also note that (p_1) can be instantiated to (i_1) (which has not been included in the system). Rules $(p_6), (p_8)$ are restricted to excludes undefined index expressions. This restriction avoid types containing terms (i.e. index expressions) being not normalizing. The cases excluded by $(p_3), (p_4)$ are managed by rules $(x_1), (x_2)$. Rules $(x_1), (x_2)$ reflect the usual approach of dependent types.

Rule $(b_1), (b_2)$ type get and set that use the second numeral to select a bit in the binary representation of the first argument: get extract such bit, set modify it. The rule (x_3) allows us to transform an index in a numeral typed Nat.

The rule (x_4) is typing an operator that allows us to recover in the computation the index information carried around by the circuit type.

Rules (c_1), (c_1'), (c_1''),(c_2), (c_3), (c_4), (c_5) conclude our type-equipment. (c_1), (c_1'), (c_1'') type strings representing circuits. (c_2) allows to parallel compose circuits: a base circuit M and some copies of a circuit N. (c_3) allows to sequentialize circuits of the same arity. (c_4) transforms a circuit in its adjoint one.

Example 1. An interesting example of typed term that provides evidence of the circularity arising from dependent types is: $x : \Pi z^{\mathsf{Idx}}. \mathrm{circ}(z) \vdash x\,\mathrm{size}\,(M) :$ $\mathrm{circ}(\mathrm{size}\,(M))$ where M can be any term of qPCF typed by a circuit.

The above example shows that in types can occur undefined terms, maybe containing open variables not typed Idx. In particular, size can contain any term (that can be typed as a circuit of a given arity). Luckily, the evaluation of size does not need the normalization of its argument: it just requires the normalization of its type.

3   Operational Semantics

As standard for PCF, the evaluation focuses on closed terms of ground types, viz. Nat, Idx, $\mathrm{circ}(E)$. Because the inclusion of dependent types and the presence of the operator size, we assume that an evaluated terms brings implicitly in it, its whole typing information. We denote V the closed values of ground types, namely numerals (typed either Nat or Idx), and strings (typed as circuits of a given arity). The operational evaluation is formalized by means of the evaluation predicate $M \Downarrow V$: it holds whenever it is the conclusion of a derivation built on the rules presented in Table 2 (we included also the rule for the evaluation of index expressions). Table 2 includes the standard call-by-name semantics of PCF, namely the first two lines of rules. Since they are well-known, we do not insist further on them. The rules (sz), (op) compute some index expressions. In particular, (sz) recovers the numeral decorating the type circuit of a closed term. Since the involved expressions do not contain open variables, the evaluation does not pose any problem.

Let $\lceil \underline{m} \rceil^{\underline{n}}$ be notation for $\underbrace{((\underline{m} / \underline{2}) \ldots / \underline{2})}_{\underline{n}}\%\underline{2}$ where $/$ is the integer division and

$\%$ is the modulo. Thus, $\lceil \underline{m} \rceil^{\underline{0}}$ is the rightmost bit of the binary representation of $\underline{m}$. Moreover, if $\underline{k}$ is the logarithm (base 10) of $\underline{m}$ then $\lceil \underline{m} \rceil^{\underline{k-1}} = \underline{1}$ and, for all $\underline{h}$ greater than $\underline{k}$, $\lceil \underline{m} \rceil^{\underline{h}} = \underline{0}$. The rule (gt) and (st) get/set a bit of the first argument (the one selected by the second argument). Notice that set, get are syntactic sugar managing classical input states. In particular, the numeral $\mathrm{set}\,\underline{0}\,\underline{n+1}$ represents the state $1\underbrace{\underline{0} \ldots \underline{0}}_{n}$.

The rules (u), (u'), (u''), (r_0), (r_1), (r_2), (a), (d) build circuits, i.e. strings on $\S, \parallel$ and the gate-names U. The semantics of append is simply the sequential post-position of circuits. The semantics of iter is the parallel composition of circuits, driven by an argument of type Idx: thus the arity of the generated circuit is well

Table 2. Operational Semantics.

$$\frac{}{\underline{n} \Downarrow \underline{n}} \ (n) \qquad \frac{M \Downarrow \underline{n}}{s(M) \Downarrow \underline{n+1}} \ (s) \qquad \frac{M \Downarrow \underline{n+1}}{p(M) \Downarrow \underline{n}} \ (p) \qquad \frac{M[N/x]P_1 \cdots P_m \Downarrow V}{(\lambda x.M)NP_1 \cdots P_m \Downarrow V} \ (\beta)$$

$$\frac{M \Downarrow \underline{0} \quad L \Downarrow V}{\text{if } M \, L \, R \Downarrow V} \ (\text{if}_l) \qquad \frac{M \Downarrow \underline{n+1} \quad R \Downarrow V}{\text{if } M \, L \, R \Downarrow V} \ (\text{if}_r) \qquad \frac{M(YM)P_1 \cdots P_i \Downarrow V}{YMP_1 \cdots P_i \Downarrow V} \ (Y)$$

$$\frac{M : \text{circ}(E) \quad E \Downarrow \underline{n}}{\text{size} \, M \Downarrow \underline{n}} \ (\text{sz}) \qquad \frac{E_0 \Downarrow \underline{m} \quad E_1 \Downarrow \underline{n}}{\odot E_0 E_1 \Downarrow \underline{m} \odot \underline{n}} \ (\text{op}) \qquad \frac{M \Downarrow \underline{m} \quad N \Downarrow \underline{n}}{\text{get} \, M \, N \Downarrow \lceil \underline{m} \rceil^{\underline{n}}} \ (\text{gt})$$

$$\frac{M \Downarrow \underline{m} \quad N \Downarrow \underline{n} \quad \text{and } \underline{m}' \text{ is such that } \lceil \underline{m}' \rceil^{\underline{n}} = 1 \text{ and } \forall \underline{k} \neq \underline{n} \lceil \underline{m}' \rceil^{\underline{k}} = \lceil \underline{m} \rceil^{\underline{k}}}{\text{set} \, M \, N \Downarrow \underline{m}'} \ (\text{st})$$

$$\frac{}{U \Downarrow U} \ (\text{u}) \qquad \frac{M_0 \Downarrow C_0 \quad M_1 \Downarrow C_1}{M_0 \, \S \, M_1 \Downarrow C_0 \, \S \, C_1} \ (\text{u}') \qquad \frac{M_0 \Downarrow C_0 \quad M_1 \Downarrow C_1}{M_0 \parallel M_1 \Downarrow C_1 \parallel C_0} \ (\text{u}'')$$

$$\frac{M_0 \Downarrow C_0 \quad M_1 \Downarrow C_1}{\text{append} \, M_0 M_1 \Downarrow C_0 \, \S \, C_1} \ (\text{a}) \qquad \frac{E \Downarrow \underline{n} \quad M_0 \Downarrow C_0 \quad M_1 \Downarrow C_1}{\text{iter} \, E \, M_0 M_1 \Downarrow \underbrace{C_1 \parallel \cdots \parallel C_1}_{\underline{n}} \parallel C_0} \ (\text{it})$$

$$\frac{M \Downarrow U \quad (\ddagger U) = U'}{\text{reverse} \, M \Downarrow U'} \ (\text{r}_0) \qquad \frac{M \Downarrow C_0 \, \S \, C_1 \quad \text{reverse} \, C_0 \Downarrow C_0' \quad \text{reverse} \, C_1 \Downarrow C_1'}{\text{reverse} \, M \Downarrow C_1' \, \S \, C_0'} \ (\text{r}_1)$$

$$\frac{M \Downarrow C_0 \parallel C_1 \quad \text{reverse} \, C_0 \Downarrow C_0' \quad \text{reverse} \, C_1 \Downarrow C_1'}{\text{reverse} \, M \Downarrow C_0' \parallel C_1'} \ (\text{r}_2)$$

$$\frac{M \Downarrow \underline{m} \quad N \Downarrow C \quad N : \text{circ}(\underline{k}) \quad \text{circuitEval}(\underline{m}\lceil_{\underline{k}+1}, C) = \underline{n}}{\text{dMeas}(M, N) \Downarrow \underline{n}} \ (\text{m})$$

defined. The semantics of `reverse` is build to produce the adjoint circuits when a suitable endo-function $\ddagger$ is provided by the co-processor. If the co-processor do not provide it (for instance, because the set $\mathcal{U}$ of unitary gate is not closed under adjunction) then we let $\ddagger$ be the identity, so that `reverse` is well-defined but uninteresting.

Let $\underline{m}\lceil_{\underline{k}}$ to denote the numeral $\underline{k}$ such that the binary representation of $\underline{n}$ is the restriction of the binary representation of $\underline{m}$ on the first $\underline{k}$ bits. It is worth to recall that, conventionally, each classic state is represented via an integer having (implicitly) a number of relevant bits as the arity of the circuit. The rule (m) evaluates the `dMeas` arguments and uses the results of these evaluations to feed an external evaluation: morally, a quantum co-processor [31]. The co-processor call is done by using the auxiliary evaluation circuitEval. It executes the quantum circuits on the provided classical initialization, then it returns the measure of the whole final state. The rule explicitly restricts the evaluation of the first argument to the relevant number of bits (i.e. the arity of the circuit).

In order to define our co-processor we need two ingredients. The first one is the semantic for the evaluation of the circuit. We denote $Circ$ the valid strings of circuits, and $\mathcal{O}$ the set of unitary operators on finite dimensional Hilbert spaces (informally, we are mapping circuit descriptions into their mathematical

464 L. Paolini and M. Zorzi

denotation, i.e. into corresponding algebraic operators). So that we can define the circuit semantic by using the function Hilb : $Circ \rightarrow \mathcal{O}$ defined as follows: $\text{Hilb}(\mathtt{U}) ::= \mathbf{U}$; $\text{Hilb}(\mathtt{C_0} \parallel \mathtt{C_1}) ::= \text{Hilb}(\mathtt{C_0}) \otimes \text{Hilb}(\mathtt{C_1})$; $\text{Hilb}(\mathtt{C_0} \, \mathring{,} \, \mathtt{C_1}) ::= \text{Hilb}(\mathtt{C_0}) \circ \text{Hilb}(\mathtt{C_1})$. The second one is the semantic of the measure that rest on the von Neumann Measurement [9], here dubbed vMeas. We define $\text{circuitEval}(\underline{\mathtt{n}}, \mathtt{C})$ to be the measure of the application of the circuit to our initial state (in the assumed base), namely $\text{circuitEval}(\underline{\mathtt{n}}, \mathtt{C}) ::= \text{vMeas}(\text{Hilb}(\mathtt{C}), \underline{\mathtt{n}})$.

Equivalence. The operational equivalence can be defined by just considering closed terms of type Nat because the operational differences in the other types can be traced back to Nat (the reverse can be easily proved be false).

For many reasons, we are remarking the relevance of the notion of program of PCF, i.e. a closed term of type Nat. First, the result of a circuit-measure is a list of bits, viz. a natural number. Second, the circuits are represented by using strings on a finite alphabet, that still (in a Turing-complete setting) can be straightforwardly represented by numbers (at worst, paying some code-obfuscation). These remarks should make clear why the evaluation of qPCF is focused on natural numbers and the standard notion of program, as any PCF-like programming language.

4 Properties of qPCF

qPCF is, morally, a PCF-like language endowed with a quantum co-processor. This co-processor allows us to execute a quantum circuits that has been designed by executing a classical control. The co-processor returns a measure (total, in the sense that we measure the whole quantum state) of a run of the given circuit on a given input, to our classical processor.

A first property of a paradigmatic programming language as qPCF is some form of subject reduction. Moreover, we prove preservation, i.e. if a well-typed term takes a step of evaluation then the resulting term is also well typed. A second property expected for a programming language is progress [23]: well-typed terms evaluation do not stuck. Roughly, a term P is stuck whenever the evaluation of P ends in a normal form, which is not a ground value.

The main complexity in this proof comes from the fact that we have infinite (two plus a family) ground types (viz. Nat, Idx, $\text{circ}(\mathtt{E})$). Example 1 shows that each term can occurs in a type (in an index expression using size). To prove preservation and progress we must unravel the mutual relationship that holds between them.

Lemma 1. *If* $B, \mathtt{x} : \tau \vdash \mathtt{M} : \sigma$ *and* $B \vdash \mathtt{N} : \tau$ *then* $B \vdash \mathtt{M}[\mathtt{N}/\mathtt{x}] : \sigma[\mathtt{N}/\mathtt{x}]$ *and, moreover, if* $\sigma = \mathit{\Pi}\mathtt{z}^\tau.\sigma'$ *then* $B \vdash \mathtt{MN} : \sigma'[\mathtt{N}/\mathtt{x}]$.

Proof. The proof follows by induction on the derivation $B, \mathtt{x} : \tau \vdash \mathtt{M} : \sigma$.

Theorem 2 (Idx-safety). *If* $\vdash \mathtt{M} : \mathit{Idx}$ *then* $\mathtt{M} \Downarrow \underline{\mathtt{n}}$.

Proof. The proof is quite complex but it can be done by defining a suitable predicate of computability à la Tait.

Remark that Theorem 2 is stronger than both preservation and progress, in fact it immediately implies both: (i) if M is closed, ⊢ M : Idx and M ⇓ N then N is closed and ⊢ N : Idx, and (ii) if M is closed, ⊢ M : Idx and M ⇓ N then N is a numeral.

Theorem 3 (Preservation). *If* M *is a closed term such that* ⊢ M : σ *and* M ⇓ N *then* N *is a closed term such that* ⊢ N : σ.

Proof. The evaluation is applied only to terms typed by ground types, viz. Nat, Idx, circ(E). Indeed, the rule in Table 2 are applied to them. The proof follows by proving by mutual induction on the following statements: (i) if M is closed, ⊢ M : Nat and M ⇓ N then N is closed and ⊢ N : Nat; and, (ii) if M is closed, ⊢ M : circ(E) and M ⇓ N then N is closed and ⊢ N : circ(E). The fact that we are restricting our attention on closed terms typed by ground types simplify our proof by conveniently restricting the possible cases. The proof of (i) involves only the rules (n), (s), (p), (β), (if$_l$), (if$_r$), (Y), (gt), (st), (m), because the others are excluded by hypothesis. The proof of (ii) involves only the rules (β), (if$_l$), (if$_r$), (Y), (u), (u'), (u''), (r$_0$), (r$_1$), (r$_2$), (a), (it), because the others are excluded by hypothesis.

Likewise, a form of progress can be proved.

Theorem 4 (Progress)

- *If* M *is a closed term such that* ⊢ M : *Nat and* M ⇓ N *then* N *is a numeral.*
- *If* M *is a closed term such that* ⊢ M : *circ(E) and* M ⇓ N *then there is a numeral* k *such that* E ⇓ k *and* N *is a circuit of arity* k.

Proof. The proof is similar to the that of the Preservation Theorem.

Progress and preservation together tell us that a well-typed term can never reach a stuck state during evaluation.

We conclude this section with some preliminary comments about confluence. It is well-known that quantum-measures break the deterministic evolution of a quantum system. As a consequence, in presence of a measurement operator in a quantum language (equipped with an universal basis of quantum gates), one necessarily lost confluence. This loose of standard properties is typical in presence of "non classical" operators (this holds for examples also in languages including non deterministic or probabilistic choices [2,13]). Given an evaluation of a program P, a second evaluation can ends with a different result; in particular, the results of two evaluations of a same program can be different natural numbers. Clearly, the "measurement-free" fragment of qPCF, i.e. the whole calculus minus dMeas is patently deterministic. Finally, one can observe that the presence of the measure does not imply the loss of the determinism, if we limits the use of qPCF to deterministic circuits (by a suitable choice of unitary operators included in $\mathcal{U}$, e.g. only swaps).

5   Examples

In this section we propose some higher-order encoding of quantum circuit families. In the following examples, we exploit the full expressive power of the language. A qPCF term can be parametric, viz. it can represent an entire (infinite) quantum circuit family. In general, given an input numeral $\underline{n}$ we define a term that generates the description of the n-dimensional circuit of the family. Notice that, in some sense, circuits can be parameterized both in "horizontal" and in "vertical", that correspond to the two basic ways to built greater circuits from smaller ones, i.e. by sequential and parallel composition.

Example 2. The following term, applied to a circuit $C : \mathrm{circ}(\underline{k})$ and a numeral $\underline{n}$, concatenates $n + 1$ copies of C.

$\mathsf{M_{seq}} = \lambda u^{\mathrm{circ}(\underline{k})}.\lambda \underline{k}^{\mathrm{Nat}}.Y \mathrm{W} u \underline{k} : \mathrm{circ}(\underline{k}) \to \mathrm{Nat} \to \mathrm{circ}(\underline{k})$, where

$\mathrm{W} = \lambda w^{\sigma}.\lambda u^{\mathrm{circ}(\underline{k})}.\lambda \underline{k}^{\mathrm{Nat}}.\mathtt{if}\ x\ (u)\ \big(\mathtt{append}\,(u)\,((\lambda y^{\mathrm{circ}(\underline{k})}.\lambda z^{\mathrm{Nat}}.wyz)\,u\,\mathtt{pred}(x))\big)$

has type $\sigma \to \sigma$, with $\sigma = \mathrm{circ}(\underline{k}) \to \mathrm{Nat} \to \mathrm{circ}(\underline{k})$. It is easy to show that $\mathsf{M_{seq}} C \underline{0}$ generates the circuit built upon a single copy of the circuit C and so on.

It is straightforward to parameterize the above term in order to transform it in a template for a circuit-builder that can be used for any arity. It suffices to replace $\underline{k}$ with the variable k^{Idx} and to abstract it; so that, the resulting term is typed $\Pi k^{\mathrm{Idx}}.\,\mathrm{circ}(k) \to \mathrm{Nat} \to \mathrm{circ}(k)$.

Example 3. The following term, applied to a numeral $\underline{n}$ and two unitary gates U_1 and U_2 of arity $\underline{k}$ and $\underline{h}$ respectively, generates a simple circuit built upon a copy of gate U_1 in parallel with n copies of gate U_2:

$\mathsf{M_{par}} = \lambda x^{\mathrm{Idx}}.\lambda u^{\mathrm{circ}(\underline{k})}\lambda w^{\mathrm{circ}(\underline{h})}.\,\mathtt{iter}\ x\,u\,w : \Pi x^{\mathrm{Idx}}.\,\mathrm{circ}(\underline{k}) \to \mathrm{circ}(\underline{h}) \to \mathrm{circ}(x * \underline{h} + \underline{k})$

It is straighforward to parameterize the above example. It suffices to replace numerals $\underline{k}$ and $\underline{h}$ in the above example by variables and to abstract, by obtaining a term typed $\Pi k^{\mathrm{Idx}}.\Pi h^{\mathrm{Idx}}.\Pi x^{\mathrm{Idx}}.\,\mathrm{circ}(k) \to \mathrm{circ}(h) \to \mathrm{circ}(x * h + h)$.

Example 4 (Deutsch-Jozsa). We provide the qPCF encoding of the circuit that implements the generalised version of the Deutsch's problem [17].

The "simple case" of Deutsch's problem can be formulated as follows. Given a block box B_f implementing some function $f : \{0, 1\} \to \{0, 1\}$, determine whether f is constant or balanced. The classical computation to determine whether f is constant or balanced is very simple: one computes $f(0)$ and $f(1)$, and then check if $f(0) = f(1)$. This requires two different calls to B_f (i.e. one to compute $f(0)$ and one to compute $f(1)$) in the classical computing model. By means of the "quantum superpower", Deutsch showed how to achieve this result with a single call of B_f in the quantum case.

The problem can be generalised considering a function $f : \{0, 1\}^n \to \{0, 1\}$ which acts on many input bits. This yields the n-bit generalization of Deutsch's algorithm, known as the Deutsch-Josza algorithm. The following picture represents the circuit, up to the last, measurement phase, for the Deutsch-Josza problem.

When fed with a classical input state of the form $|0 \dots 01\rangle$, the output measurement of the first $n-1$ bits reveals if the function f is constant or not. If all $n-1$ measurement results are 0, we can conclude that the function was constant. Otherwise, if at least one of the measurement outcomes is 1, we conclude that the function was balanced. See [17] for details about Deutsch and Deutsch-Josza algorithm.

Consider now the following qPCF terms. They easily show how to encode different levels of the measurement-free parametric circuit for the Deutsch's problem. The last, evaluation-measurement phase, will be performed by our evaluation dMeas, suitably fed with the representation of the classical input state (i.e. set $\underline{0}\,\underline{0}$). Let H : circ($\underline{0}$) and I : circ($\underline{0}$) be the (unary) Hadamard and Identity gates respectively (so the index is 0). Suppose M^{B_f} : circ($\underline{n}$) is given for some n such that $\mathsf{M}^{B_f} \Downarrow \mathsf{U}_f$ where U_f : circ($\underline{n}$) is the qPCF§-circuit that represents the function f.

Observe that $\lambda \mathsf{x}^{\mathsf{Idx}}.\, \mathtt{iter}\ \mathsf{xHH} : \mathit{\Pi}\mathsf{x}^{\mathsf{Idx}}.\, \mathrm{circ}(\mathsf{x})$ clearly generates $x+1$ parallel copies of unary Hadamard gates H, and $\lambda \mathsf{x}^{\mathsf{Idx}}.\, \mathtt{iter}\ \mathsf{xIH} : \mathit{\Pi}\mathsf{x}^{\mathsf{Idx}}.\, \mathrm{circ}(\mathsf{x})$ concatenates in parallel x copies of unary Hadamard gates H and one copy of the unary identity gate I.

Thus the generator term of the parametric measurement-free Deutsch-Jozsa circuit, here dubbed DJ$^-$ can be defined as

DJ$^- = \lambda \mathsf{x}^{\mathsf{Idx}}.\lambda \mathsf{y}^{\mathrm{circ}(\mathsf{x})}.\, \mathtt{append}(\mathtt{append}(\mathtt{iter}\ \mathsf{x\,H\,H})\mathsf{y})(\mathtt{iter}\ \mathsf{x\,I\,H}) : \sigma$ where $\sigma = \mathit{\Pi}\mathsf{x}^{\mathsf{Idx}}.\, \mathrm{circ}(\mathsf{x}) \to \mathrm{circ}(\mathsf{x})$.

We finally evaluate DJ$^-$ by means of dMeas, providing the encoding M^{B_f} circ($\underline{n}$) of the black-box function f having arity $n+1$. Let us assume that the term dMeas(set$\underline{0}\underline{0}$, DJ$^-\underline{n}\mathsf{M}^{B_f}$) evaluated by means of $\Downarrow$, yields the numeral $\underline{m}$: the rightmost $\underline{n}$ digit of the binary representation of $\underline{m}$ are the result. Notice that DJ$^-\underline{0}\mathsf{M}^{B_f}$ returns the circuit description of Deutsch algorithm.

6   Conclusions and Future Work

We introduced qPCF, an extension of PCF for quantum circuit generation and evaluation. In this seminal work, we introduced qPCF syntax, typing rules and evaluation semantics. We started to study its properties and we provided some examples of parametric circuit families encoding. The presented research is the first step of some works in progress and for several short time investigations. First, we are further investigating qPCF properties sketched in Sect. 4. Second, we aim to deepen qPCF flexibility, e.g. studying *specialization* of qPCF: for example, we aim to focus on the (still "silent") reverse operator (of the calculus), also in different settings w.r.t quantum computing. We like to remark that gates can range on different interesting sets. Since *reversibility* is nowadays one of the most interesting trend in computer science [20], a reversible specialization of qPCF seems to be intriguing. Even if the use of total measurement does not represent a theoretical limitation, a partial measurement operator can represent an useful programming tool. Therefore, another interesting task will be to integrate in qPCF the possibility to perform partial measures of computation results.

References

1. Arrighi, P., Dowek, G.: Linear-algebraic lambda-calculus: higher-order, encodings and confluence. In: Proceedings of the 19th Annual Conference on Term Rewriting and Applications, pp. 17–31 (2008)
2. Aschieri, F., Zorzi, M.: Non-determinism, non-termination and the strong normalization of system T. In: Hasegawa, M. (ed.) TLCA 2013. LNCS, vol. 7941, pp. 31–47. Springer, Heidelberg (2013). doi:10.1007/978-3-642-38946-7_5
3. Aspinall, D., Hofmann, M.: Dependent types. In: Pierce, B. (ed.) Advanced Topics in Types and Programming Languages, Chap. 2, pp. 45–86. MIT Press (2005)
4. Dal Lago, U., Masini, A., Zorzi, M.: On a measurement-free quantum lambda calculus with classical control. Math. Struct. Comput. Sci. 19(2), 297–335 (2009)
5. Gaboardi, M., Paolini, L., Piccolo, M.: Linearity and PCF: a semantic insight! In: Proceeding of ICFP 2011, pp. 372–384 (2011)
6. Gaboardi, M., Paolini, L., Piccolo, M.: On the reification of semantic linearity. Math. Struct. Comput. Sci. 26(5), 829–867 (2016)
7. Grattage, J.: An overview of QML with a concrete implementation in haskell. Electron. Not. Theor. Comput. Sci. 270(1), 165–174 (2011)
8. Hasuo, I., Hoshino, N.: Semantics of higher-order quantum computation via geometry of interaction. In: LICS 2011, pp. 237–246 (2011)
9. Kaye, P., Laflamme, R., Mosca, M.: An Introduction to Quantum Computing. Oxford University Press, Oxford (2007)
10. Knill, E.: Conventions for quantum pseudocode. Technical Report, Los Alamos National Laboratory (1996)
11. Lago, U.D., Masini, A., Zorzi, M.: Quantum implicit computational complexity. Theor. Comput. Sci. 411(2), 377–409 (2010)
12. Lago, U.D., Masini, A., Zorzi, M.: Confluence results for a quantum lambda calculus with measurements. Electron. Not. Theor. Comput. Sci. 270(2), 251–261 (2011)
13. Lago, U.D., Zorzi, M.: Probabilistic operational semantics for the lambda calculus. RAIRO - Theor. Inf. Appl. 46(3), 413–450 (2012)
14. Mahmoud, M., Felty, A.P.: Formalization of Metatheory of the Quipper Programming Language in a Linear Logic. University of Ottawa, Canada (2016)
15. Masini, A., Viganò, L., Zorzi, M.: A qualitative modal representation of quantum register transformations. In: Proceedings of the 38th International Symposium on Multiple-Valued Logic (ISMVL), pp. 131–137 (2008)
16. Masini, A., Viganò, L., Zorzi, M.: Modal deduction systems for quantum state transformations. J. Multiple-Valued Logic Soft Comput. 17(5–6), 475–519 (2011)
17. Nielsen, M.A., Chuang, I.L.: Quantum Computation and Quantum Information. Cambridge University Press, Cambridge (2000)
18. Nishimura, H., Ozawa, M.: Perfect computational equivalence between quantum turing machines and finitely generated uniform quantum circuit families. Quant. Inf. Process. 8(1), 13–24 (2009)

19. Pagani, M., Selinger, P., Valiron, B.: Applying quantitative semantics to higher-order quantum computing. In: Proceedings of POPL 2014, pp. 647–658. ACM (2014)
20. Paolini, L., Piccolo, M., Roversi, L.: A class of reversible primitive recursive functions. Electron. Not. Theor. Comput. Sci. **322**(18605), 227–242 (2016)
21. Paolini, L., Pimentel, E., Rocca, S.R.: An operational characterization of strong normalization. In: Aceto, L., Ingólfsdóttir, A. (eds.) FoSSaCS 2006. LNCS, vol. 3921, pp. 367–381. Springer, Heidelberg (2006). doi:10.1007/11690634_25
22. Payikin, J., Rand, R., Zdancewic: QWire: A Core Language for Quantum Circuits. University of Pennsylvania, USA (2016)
23. Pierce, B.C.: Types and Programming Languages. The MIT Press, Cambridge (2002)
24. Plotkin, G.D.: LCF considered as a programming language. Theor. Comput. Sci. **5**, 223–255 (1977)
25. Roman, S.: Advanced Linear Algebra. Graduate Texts in Mathematics, vol. 135, 3rd edn. Springer, New York (2008)
26. Ross, N.: Algebraic and Logical Methods in Quantum Computation. Ph.D. thesis, Dalhousie University Halifax, Nova Scotia (2015)
27. Selinger, P.: Towards a quantum programming language. Math. Struct. Comput. Sci. **14**(4), 527–586 (2004)
28. Selinger, P., Valiron, B.: A lambda calculus for quantum computation with classical control. Math. Struct. Comput. Sci. **16**, 527–552 (2006)
29. Selinger, P., Valiron, B.: Quantum lambda calculus. In: Semantic Techniques in Quantum Computation, pp. 135–172. Cambridge University Press (2009)
30. Shende, V.V., Prasad, A.K., Markov, I.L., Hayes, J.P.: Synthesis of reversible logic circuits. IEEE Trans. Comput.-Aided Des. Integr. Circ. Syst. **22**(6), 710–722 (2003)
31. Valiron, B., Ross, N.J., Selinger, P., Alexander, D.S., Smith, J.M.: Programming the quantum future. Commun. ACM **58**(8), 52–61 (2015)
32. Viganò, L., Volpe, M., Zorzi, M.: Quantum state transformations and branching distributed temporal logic. In: Kohlenbach, U., Barceló, P., Queiroz, R. (eds.) WoLLIC 2014. LNCS, vol. 8652, pp. 1–19. Springer, Heidelberg (2014). doi:10.1007/978-3-662-44145-9_1
33. Viganò, L., Volpe, M., Zorzi, M.: A branching distributed temporal logic for reasoning about entanglement-free quantum state transformations. Inf. Comput., 1–24 (2017). http://dx.doi.org/10.1016/j.ic.2017.01.007
34. Ying, M.: Foundations of Quantum Programming. Morgan Kaufmann, Cambridge (2016)
35. Zorzi, M.: On quantum lambda calculi: a foundational perspective. Math. Struct. Comput. Sci. **26**(7), 1107–1195 (2016)

Blocking Independent Sets for H-Free Graphs via Edge Contractions and Vertex Deletions

Daniël Paulusma[1], Christophe Picouleau[2], and Bernard Ries[3($\boxtimes$)]

[1] Durham University, Durham, UK
daniel.paulusma@durham.ac.uk
[2] CNAM, Laboratoire CEDRIC, Paris, France
christophe.picouleau@cnam.fr
[3] University of Fribourg, Fribourg, Switzerland
bernard.ries@unifr.ch

Abstract. Let d and k be two given integers, and let G be a graph. Can we reduce the independence number of G by at least d via at most k graph operations from some fixed set S? This problem belongs to a class of so-called blocker problems. It is known to be co-NP-hard even if S consists of either an edge contraction or a vertex deletion. We further investigate its computational complexity under these two settings:

- we give a sufficient condition on a graph class for the vertex deletion variant to be co-NP-hard even if $d = k = 1$;
- in addition we prove that the vertex deletion variant is co-NP-hard for triangle-free graphs even if $d = k = 1$;
- we prove that the edge contraction variant is NP-hard for bipartite graphs but linear-time solvable for trees.

By combining our new results with known ones we are able to give full complexity classifications for both variants restricted to H-free graphs.

1 Introduction

A graph modification problem aims to modify a graph G, via a small number of operations, into some other graph H that has a certain desired property, which usually describes a certain graph class to which H must belong. In this way a variety of classical graph-theoretic problems is captured. For instance, if only k vertex deletions are allowed and H must be an independent set or a clique, one obtains the INDEPENDENT SET or CLIQUE problem, respectively.

Instead of specifying a graph class we can specify a graph parameter. That is, given a graph G, a set S of one or more graph operations and an integer k, we ask whether G can be transformed into a graph G' by using at most k operations from S such that $\pi(G') \leq \pi(G) - d$ for some *threshold* $d \geq 0$. Such problems are called *blocker problems*. This is because the set of vertices or edges involved can be viewed as "blocking" π. Identifying such sets may gives us some important information on the structure of the graph.

D. Paulusma received support from EPSRC (EP/K025090/1).

T.V. Gopal et al. (Eds.): TAMC 2017, LNCS 10185, pp. 470–483, 2017.
DOI: 10.1007/978-3-319-55911-7_34

Blocker problems have been well studied in the recent literature [1–3, 5, 7, 13, 14, 16, 18]; in particular, in [7, 14] several relations to other graph problems were identified, such as HADWIGER NUMBER, CLUB CONTRACTION and a number of graph transversal problems. So far, the graph parameters considered were the chromatic number, the independence number, the clique number, the matching number and the vertex cover number, whereas the set S consisted of a single graph operation, which was either the vertex deletion, edge contraction, edge deletion or the edge addition operation. In this paper we keep the restriction on the size of S, and we let S consist of either a single vertex deletion or a single edge contraction. We mainly consider the independence number α, but for the deletion variant we will also take the clique number ω into account (for reasons we explain later).

Before we can define our problems formally we first need to give some terminology. The *contraction* of an edge uv of a graph G removes the vertices u and v from G, and replaces them by a new vertex made adjacent to precisely those vertices that were adjacent to u or v in G (neither introducing self-loops nor multiple edges). We say that G can be *k-contracted* or *k-vertex-deleted* into a graph G' if G can be modified into G' by a sequence of at most k edge contractions or vertex deletions, respectively. We let π denote the (fixed) graph parameter; as mentioned, in this paper π belongs to $\{\alpha, \omega\}$.

CONTRACTION BLOCKER(π)

Instance: a graph G and two integers $d, k \geq 0$
Question: can G be k-contracted into a graph G' with $\pi(G') \leq \pi(G) - d$?

DELETION BLOCKER(π)

Instance: a graph G and two integers $d, k \geq 0$
Question: can G be k-vertex-deleted in a graph G' with $\pi(G') \leq \pi(G) - d$?

If we remove d from the input and fix it instead, we call the resulting problems d-CONTRACTION BLOCKER(π) and d-DELETION BLOCKER(π), respectively. Note that 1-DELETION BLOCKER(α) is equivalent to testing whether the input graph contains a set of S of size at most k that intersects every maximum independent set. If $k = 1$, this is equivalent to testing whether the input graph contains a vertex that is in every maximum independent set. The intersection of all maximum independent sets is known as the core of a graph. Properties of the core have been well studied (see for example [10–12]). In particular, Boros, Golumbic and Levit [4] proved that computing if the core of a graph has size at least ℓ is co-NP-hard for every fixed $\ell \geq 1$. Taking $\ell = 1$ gives co-NP-hardness of 1-DELETION BLOCKER(α), whereas 1-CONTRACTION BLOCKER(α) is known to be NP-hard [7].

Due to the above hardness results, it is natural to restrict the input to some special graph class. In a previous paper [7] we considered $\pi \in \{\alpha, \omega, \chi\}$, where

χ denotes the chromatic number of a graph, and we restricted the input to perfect graphs and subclasses of perfect graphs. We showed both new hardness results (e.g., for the class of perfect graphs itself) and tractable results (e.g., for cographs). In a follow-up paper [14] we extended the results of [7] by considering some more subclasses of perfect graphs for $\pi \in \{\omega, \chi\}$. Moreover, for every connected graph H and $\pi \in \{\omega, \chi\}$, we determined the computational complexity of CONTRACTION BLOCKER(π) and DELETION BLOCKER(π) for H-free graphs, that is, graphs that do not contain an induced subgraph isomorphic to H.

Our Results

We settle the computational complexity of CONTRACTION BLOCKER(α) and DELETION BLOCKER(α)restricted to H-free graphs for all graphs H (including those that are disconnected). We observe that DELETION BLOCKER(α) restricted to H-free graphs is equivalent to DELETION BLOCKER(ω) for $\overline{H}$-free graphs, where $\overline{H}$ denotes the complement of H. Hence, as a corollary, we obtain an extension of the aforementioned classification of [14] for DELETION BLOCKER(ω) for H-free graphs from connected graphs H to all graphs H.

To prove the above results we first show that CONTRACTION BLOCKER(α) is NP-hard for bipartite graphs in Sect. 3. In the same section we complement this result by showing that CONTRACTION BLOCKER(α) can be solved in linear time for trees. Then, in Sect. 4, we prove that DELETION BLOCKER(α) is co-NP-hard for triangle-free graphs even if $d = k = 1$ (in contrast the problem is polynomial-time solvable for bipartite graphs [2,5]). In Sect. 5 we extend our result for triangle-free graphs to other graph classes for which INDEPENDENT SET is NP-complete. That is, we give a sufficient condition on such a graph class $\mathcal{G}$, such that DELETION BLOCKER(α) is co-NP-hard for $\mathcal{G}$ even if $d = k = 1$. This condition is similar to a previous condition when $\pi \in \{\chi, \omega\}$ [14]. In Sect. 6 we combine our new results from Sects. 4 and 5 with known ones to obtain the classifications for H-free graphs. In Sect. 7 we compare our new results with the results of our previous paper [14] and list some open problems.

Recall that the deletion variant for $k = d = 1$ is equivalent to asking whether a graph has a vertex that is in every maximum independent set. As such, our hardness results in Sects. 4 and 5 strengthen the aforementioned result of Boros, Golumbic and Levit [4], who proved co-NP-hardness of the latter problem for general graphs. Note that for graph classes, for which INDEPENDENT SET is NP-complete, membership of our problems in NP is unknown. Contrary to those graph classes, for which INDEPENDENT SET is polynomial-time solvable and which are closed under the graph operation under consideration, a certificate consisting of a sequence of edge contractions or vertex deletions no longer suffices.

2   Preliminaries

We only consider finite, undirected graphs that have no self-loops and no multiple edges (we recall that when we contract an edge no self-loops or multiple edges are created). See [6] for undefined terminology and notation.

Let $G = (V, E)$ be a graph. For a family $\{H_1, \ldots, H_p\}$ of graphs, G is said to be $(H_1, \ldots, H_p)$-*free* if G has no induced subgraph isomorphic to a graph in $\{H_1, \ldots, H_p\}$; if $p = 1$ we may write H_1-free instead of (H_1)-free. The *complement* of G is the graph $\overline{G} = (V, \overline{E})$ with vertex set V and an edge between two vertices u and v if and only if $uv \notin E$. For a subset $S \subseteq V$, we let $G[S]$ denote the subgraph of G *induced* by S, which has vertex set S and edge set $\{uv \in E \mid u, v \in S\}$. We write $H \subseteq_i G$ if a graph H is an induced subgraph of G.

Let G be a graph. For a vertex $v \in V$, we write $G - v = G[V \setminus \{v\}]$ and for a subset $V' \subseteq V$ we write $G - V' = G[V \setminus V']$. Recall that the contraction of an edge $uv \in E$ removes the vertices u and v from G and replaces them by a new vertex that is made adjacent to precisely those vertices that were adjacent to u or v in G. In that case we may also say that u is *contracted onto* v, and we use v to denote the new vertex resulting from the edge contraction. The *subdivision* of an edge $uv \in E$ removes the edge uv from G and replaces it by a new vertex w and two edges uw and wv.

Let G_1 and G_2 be two vertex-disjoint graphs. The *disjoint union* $G_1 + G_2$ has vertex set $V(G_1) \cup V(G_2)$ and edge set $E(G_1) \cup E(G_2)$. The disjoint union of k copies of a graph G is denoted by kG. The *join* $G_1 \otimes G_2$ adds an edge between every vertex of G_1 and every vertex of G_2. For $r \geq 1$, the path, cycle and complete graph on r vertices are denoted by P_r, C_r and K_r respectively. The graph C_3 is also called the *triangle*. The *claw* $K_{1,3}$ is the 4-vertex star (that is, the graph with vertices u, v_1, v_2, v_3 and edges uv_1, uv_2, uv_3).

Let $G = (V, E)$ be a graph. A subset $K \subseteq V$ is called a *clique* of G if any two vertices in K are adjacent to each other. The *clique number* $\omega(G)$ is the number of vertices in a maximum clique of G. A subset $I \subseteq V$ is called an *independent set* of G if any two vertices in I are non-adjacent to each other. The *independence number* $\alpha(G)$ is the number of vertices in a maximum independent set of G. A subset of edges $M \subseteq E$ is called a *matching* if no two edges of M share a common end-vertex. The *matching number* $\mu(G)$ is the number of edges in a maximum matching of a graph G. A vertex v such that M contains an edge incident with v is *saturated* by M; otherwise v is *unsaturated* by M. A subset $S \subseteq V$ is a *vertex cover* of G if every edge of G is incident with at least one vertex of S.

The problems CLIQUE and INDEPENDENT SET are those of testing if a graph has a clique or independent set, respectively, of size at least k. The VERTEX COVER problem is that of testing if a graph has a vertex cover of size at most k.

A graph is *cobipartite* if it is the complement of a *bipartite* graph, that is, a graph whose vertex set can be partitioned into two sets that each form a (possibly empty) independent set. A graph is a *split graph* if it has a *split partition*, which is a partition of its vertex set into a clique K and an independent set I. Split graphs coincide with $(2P_2, C_4, C_5)$-free graphs [8]. A P_4-free graph is also called a *cograph*.

3   Bipartite Graphs and Trees

Our first lemma below follows directly from a result of Golovach, Heggernes, van't Hof and Paul [9] on the so-called s- CLUB CONTRACTION problem; see [7] for further details.

Lemma 1 (*[7]*). 1-CONTRACTION BLOCKER(α) *is* NP-*complete for cobipartite graphs.*

If $\pi \in \{\chi, \omega\}$, CONTRACTION BLOCKER(π) is trivial in bipartite graphs. To the contrary, for $\pi = \alpha$, we will show that CONTRACTION BLOCKER(π) is NP-hard for bipartite graphs. The complexity of d-CONTRACTION BLOCKER(α) remains open for bipartite graphs. Bipartite graphs are not closed under edge contraction. Therefore membership to NP cannot be established by taking a sequence of edge contractions as the certificate, even though due to König's Theorem (see, for example, [6]), INDEPENDENT SET is polynomial-time solvable for bipartite graphs.

Theorem 1. CONTRACTION BLOCKER(α) *is* NP-*hard on bipartite graphs.*

Proof. We know from Lemma 1 that 1-CONTRACTION BLOCKER(α) is NP-complete on cobipartite graphs, which have independence number 2. Consider a cobipartite graph G with m edges and an integer k, which together form an instance of 1-CONTRACTION BLOCKER(α). Subdivide each of the m edges of G in order to obtain a bipartite graph G'. We claim that (G, k) is a yes-instance of 1-CONTRACTION BLOCKER(α) if and only if $(G', \alpha(G') - 1, k + m)$ is a yes-instance of CONTRACTION BLOCKER(α).

First suppose that (G, k) is a yes-instance of 1-CONTRACTION BLOCKER(α). In G' we first perform m edge contractions to get G back. We then perform k edge contractions to get independence number $\alpha(G) - 1 = 1 = \alpha(G') - (\alpha(G') - 1)$. Hence, $(G', \alpha(G') - 1, k + m)$ is a yes-instance of CONTRACTION BLOCKER(α).

Now suppose that $(G', \alpha(G') - 1, k + m)$ is a yes-instance of CONTRACTION BLOCKER(α). Then there exists a sequence of $k + m$ edge contractions that transform G' into a complete graph K. We may assume that K has size at least 4 (as we could have added without loss of generality three dominating vertices to G without increasing k). As K has size at least 4, each subdivided edge must be contracted back to the original edge again. This operation costs m edge contractions, so we contract G to K using at most k edge operations. Hence, (G, k) is a yes-instance of 1-CONTRACTION BLOCKER(α). This proves the claim and hence the theorem.   □

We complement Theorem 1 by showing that CONTRACTION BLOCKER(α) is linear-time solvable on trees. In order to prove this result we make a connection to the matching number μ.

Theorem 2. CONTRACTION BLOCKER(α) *is linear-time solvable on trees.*

Proof. Let (T, d, k) be an instance of CONTRACTION BLOCKER(α), where T is a tree on n vertices. We first describe our algorithm and prove its correctness. Afterwards, we analyze its running time. Throughout the proof let M denote a maximum matching of T.

As $\alpha(T) + \mu(T) = n$ by König's Theorem (see, for example, [6]), we find that (T, d, k) is a no-instance if $d > n - \mu(T)$. Assume that $d \leq n - \mu(T)$. We observe that trees are closed under edge contraction. Hence, contracting an edge of T results in a new tree T'. Moreover, T' has $n - 1$ vertices and the edge contraction neither increased the independence number nor the matching number. As $\alpha(T) + \mu(T) = n$ and similarly $\alpha(T') + \mu(T') = n - 1$, this means that either $\alpha(T') = \alpha(T) - 1$ or $\mu(T') = \mu(T) - 1$.

First suppose that $d \leq n - 2\mu(T)$. There are exactly $\sigma(T) = n - 2\mu(T)$ vertices that are unsaturated by M. Let uv be an edge, such that u is unsaturated. As M is maximum, v must be saturated. Then, by contracting uv, we obtain a tree T' such that $\mu(T') = \mu(T)$. It follows from the above that $\alpha(T') = \alpha(T) - 1$. Say that we contracted u onto v. Then in T' we have that v is saturated by M, which is a maximum matching of T' as well. Thus, if $d \leq n - 2\mu(T)$, contracting d edges, one of the end-vertices of which is unsaturated by M, yields a tree T' with $\mu(T') = \mu(T)$ and $\alpha(T') = \alpha(T) - d$. Since an edge contraction reduces the independence number by at most 1, it follows that this is optimal. Hence, as $d \leq n - 2\mu(T)$, we find that (G, T, k) is a yes-instance if $k \geq d$ and a no-instance if $k < d$.

Now suppose that $d > n - 2\mu(T)$. Suppose that we first contract the $n - 2\mu(T)$ edges that have exactly one end-vertex that is unsaturated by M. It follows from the above that this yields a tree T' with $\mu(T') = \mu(T)$ and $\alpha(T') = \alpha(T) - (n - 2\mu(T))$. Since T' does not contain any unsaturated vertex, M is a perfect matching of T'. Then, contracting any edge in T' results in a tree T'' with $\mu(T'') = \mu(T') - 1$ and thus, $\alpha(T'') = \alpha(T')$. If we contract an edge $uv \in M$, the resulting vertex uv is unsaturated by $M' = M \setminus \{uv\}$ in T''. Hence, as explained above, if in addition we contract now an edge $(uv)w$, we obtain a tree T''' with $\alpha(T''') = \alpha(T'') - 1$ and $\mu(T''') = \mu(T'')$. Repeating this procedure, we may reduce the independence number of T by d with $n - 2\mu(T) + 2(d - n + 2\mu(T)) = 2(d + \mu(T)) - n$ edge contractions. Below we show that this is optimal.

Suppose that we contract p edges in T. Let T' be the resulting tree. We have $\alpha(T') + \mu(T') = n - p$. As $\mu(T') \leq \frac{1}{2}(n - p)$, this means that $\alpha(T') \geq \frac{1}{2}(n - p)$. If $p < 2(d + \mu(T)) - n$ we have $-\frac{p}{2} > -(d + \mu(T)) + \frac{n}{2}$, and thus

$$\alpha(T') \geq \tfrac{1}{2}(n - p)$$
$$> \tfrac{n}{2} - d - \mu(T) + \tfrac{n}{2}$$
$$= \alpha(T) - d.$$

So at least $2(d + \mu(T)) - n$ edge contractions are necessary to decrease the independence number by d. It remains to check if k is sufficiently high for us to allow this number of edge contractions.

As we can find a maximum matching of tree T (and thus compute $\mu(T)$) in $O(n)$ time by using the algorithm of Savage [17], our algorithm runs in $O(n)$ time. □

Remark 1. By König's Theorem, we have that $\alpha(G) + \mu(G) = |V(G)|$ for any bipartite graph G, but we can only use the proof of Theorem 2 to obtain a result for trees for the following reason: trees form the largest subclass of (connected) bipartite graphs that are closed under edge contraction, and this property plays a crucial role in our proof.

4   Triangle-Free Graphs

In this section we show that DELETION BLOCKER(α) is co-NP-hard for triangle-free graphs even if $d = k = 1$. We call a vertex *forced* if it is in every maximum independent set of a graph [5]. Recall that the set of all forced vertices is called the *core* of a graph and that Boros, Golumbic and Levit [4] proved that computing whether the core of a graph has size at least k is co-NP-hard for every fixed $k \geq 1$. As a special case of their result, the problem of testing the existence of a forced vertex is co-NP-hard. In this section we prove that the latter problem, or equivalently, DELETION BLOCKER(α) with $d = k = 1$, stays co-NP-hard even for triangle-free graphs.

We need some terminology and a well-known observation that follows from a construction of Poljak [15]. We say that we 2-*subdivide* an edge e of a graph G if we apply two consecutive edge subdivisions on e. It is readily seen that a graph G with m edges has an independent set of size k if and only if the graph obtained by 2-subdividing each edge of G has an independent set of size $k + m$ (see also [15]). Let $\mathcal{G}$ be a graph class. Then we let $\mathcal{G}^2$ be the graph class obtained from $\mathcal{G}$ after 2-subdividing each edge in every graph in $\mathcal{G}$.

Lemma 2. (*[15]*). *If* INDEPENDENT SET *is* NP-*complete for a graph class* $\mathcal{G}$, *then it is also* NP-*complete for* $\mathcal{G}^2$.

Two vertices in a graph G are *true twins* if they are adjacent to each other and apart from this have the same neighbours in G. The graph G^* obtained from a graph G by adding a new vertex u' for each vertex u of G that is a true twin of u is called the *twin graph* of G; see Fig. 1 for an example. We call u' the *copy* of u. Let $\mathcal{G}^*$ be the graph class obtained from a graph class $\mathcal{G}$ by replacing each graph in $\mathcal{G}$ by its twin graph. Note that $\alpha(G^*) = \alpha(G)$ for every graph G. Hence the following lemma holds.

Lemma 3. *If* INDEPENDENT SET *is* NP-*complete for a graph class* $\mathcal{G}$, *then it is also* NP-*complete for* $\mathcal{G}^*$.

Theorem 3. DELETION BLOCKER(α) *is co-*NP-*hard for triangle-free graphs even if* $d = k = 1$.

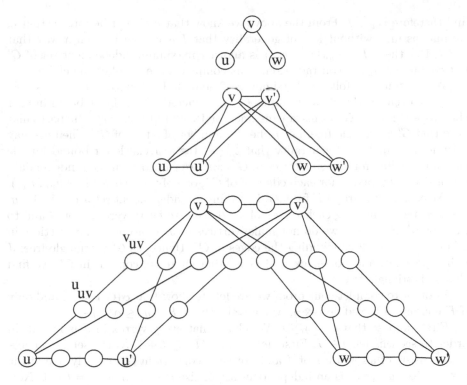

Fig. 1. An example of a graph G' constructed from a graph G via the graph G^*.

Proof. We prove that the equivalent problem of testing whether a triangle-free graph has a forced vertex is co-NP-hard via a reduction from INDEPENDENT SET. Let G be a graph with at least two vertices. From G we construct its twin graph G^*. We now subdivide each edge of G^* twice. We call the resulting graph G'. For an edge $e = uv$ in G^* (where $v = u'$ is possible), we call the two newly introduced vertices u_e and v_e, where u_e is the vertex adjacent to u and v_e the one adjacent to v. See Fig. 1 for an example of a graph G'.

We now show the following claim.

Claim. G' has no forced vertices.

We prove this claim as follows. For contradiction, suppose x is a forced vertex of G', that is, x belongs to every maximum independent set of G'. First suppose $x = u$ or $x = u'$ for some vertex u of G, say $x = u$. Then, by symmetry, its copy u' is also a forced vertex of G'. Let I be a maximum independent set of G'. Since u, u' are forced, we have $u, u' \in I$ and therefore $u_{uu'}, u'_{uu'} \notin I$. But then $(I \setminus \{u\}) \cup \{u_{uu'}\}$ is another maximum independent set of G' not containing u, a contradiction.

Now suppose $x = u_{uu'}$ for some vertex u of G. Then, by symmetry, $u'_{uu'}$ is a forced vertex as well. This is a contradiction, since $u_{uu'}$ and $u'_{uu'}$ are adjacent.

Finally suppose $x = u_{uv}$ for some vertices u, v of G^* with $v \neq u'$. Let I be a maximum independent set of G'. Since u_{uv} is a forced vertex, we have $u_{uv} \in I$

and therefore $v_{uv} \notin I$. From the above we know that v cannot be forced. Hence, we may assume without loss of generality that I is chosen in such a way that $v \notin I$. But then $(I \setminus \{u_{uv}\}) \cup \{v_{uv}\}$ is another maximum independent set of G' not containing u_{uv}, a contradiction. This completes the proof of the claim.

We continue as follows. By Lemmas 2 and 3, INDEPENDENT SET is NP-complete even for the class of graphs G' constructed above. Let ℓ be an integer that together with G' forms an instance of INDEPENDENT SET. In particular note that G' is triangle-free. Let m be the number of edges of G^*. Then we may assume without loss of generality that $\ell \geq m$ (as a trivial lower bound for the size of a maximum independent set in G' is m: we can construct an independent set of size m by taking for each edge uv of G^*, one of the two vertices u_{uv}, v_{uv}).

We construct a graph F from G' by taking an independent set J on $\ell + 1 - m$ vertices and by making each vertex of J adjacent to every vertex u of G and to its copy u'. Note that we do not make vertices of J adjacent to any vertices in G' obtained from 2-subdividing the edges of G^*. Hence, as G' is triangle-free, J is independent, and no vertex u of G is adjacent to its copy u' in G', we find that F is triangle-free.

In order to complete our proof we are left to show that $\alpha(G') \leq \ell$ if and only if F contains a forced vertex y, or equivalently, $\alpha(F - y) \leq \alpha(F) - 1$.

First suppose that $\alpha(G') \leq \ell$. We claim that every vertex in J is forced. In order to see this let $y \in J$. First note that $\alpha(F) \geq \ell + 1$, as the set of vertices that consists of all vertices of J and, for each edge uv in G^*, exactly one of the two vertices u_{uv}, v_{uv} is an independent set of size $\ell + 1 - m + m = \ell + 1$. Now let I be a maximum independent set of $F - y$. If I contains a vertex y' of J, then I must have size ℓ (since I cannot contain a vertex u of G or its copy u', as y' is adjacent to such vertices). If I does not contain a vertex of J, then I must be an independent set of G'. Then I has size at most $\alpha(G') \leq \ell$ by our assumption on $\alpha(G')$. In fact, as ℓ is a lower bound on the size of I (recall that $\alpha(F) \geq \ell + 1$), we have that I has size ℓ in this case as well. Hence, in both cases we find that $\alpha(F - y) = \ell \leq \alpha(F) - 1$ implying that y is a forced vertex of F.

Now suppose that F contains a forced vertex y, so $\alpha(F - y) \leq \alpha(F) - 1$. In fact we must have $\alpha(F - y) = \alpha(F) - 1$. We distinguish three cases.

First assume that y belongs to J. Let I be a maximum independent set of F. Then y must be in I, as y is forced. This means that I must have size $\ell + 1$, thus $\alpha(F) = \ell + 1$, as I cannot contain a vertex u of G or its copy u' (because $y \in I$) and I can contain, besides all vertices of J, exactly one of u_{uv}, v_{uv} for every edge uv of G^*. As y is forced, this implies that $\alpha(F - y) = \ell$. As G' is an induced subgraph of $F - y$, this means that $\alpha(G') \leq \ell$.

Now assume that $y = u$ or $y = u'$ for some u in G. Let I be a maximum independent set of F. As y is forced, y belongs to I. As y is adjacent to every vertex in J, we find that no vertex of J belongs to I. Then I is a maximum independent set of G'. However, in that case we can replace I by another maximum independent set of G', and thus of F, that does not contain y (by the above Claim). So we conclude that y is not a forced vertex of F, which is a contradiction.

Finally assume that $y = u_{uv}$ for some edge uv of G^* (where $v = u'$ is possible). If I shares no vertices with J, then we repeat the arguments of the previous case. Suppose I intersects with J. Then I does not contain v. Hence we may replace y by v_{uv} to get a maximum independent set of F that does not contain y. This implies that y is not forced, a contradiction. This completes the proof of Theorem 3.   ⊔

5   A Sufficient Condition for Hardness

In this section we give a sufficient condition for computational hardness of DELE-TION BLOCKER(α). Let $\mathcal{G}$ be a graph class with the following property: if $G \in \mathcal{G}$, then so are $G \otimes G$ and $G \otimes sP_1$ for any integer $s \geq 1$. We call such a graph class *stable-proof*. We show that determining the existence of a forced vertex is co-NP-hard on any stable-proof graph class, for which INDEPENDENT SET is NP-complete (note that we can only show co-NP-hardness for reasons discussed before).

Theorem 4. *If* INDEPENDENT SET *is* NP-*complete for a stable-proof graph class* $\mathcal{G}$, *then* DELETION BLOCKER*(α) is co-*NP-*hard for* $\mathcal{G}$, *even if* $d = k = 1$.

Proof. Let $\mathcal{G}$ be a graph class that is stable-proof. From a given graph $G \in \mathcal{G}$ and integer $\ell \geq 1$ we construct the graph $G' = G \otimes G \otimes (\ell+1)P_1$. Note that $G' \in \mathcal{G}$ by definition and that $\alpha(G') = \max\{\alpha(G), \ell+1\}$. We claim that $\alpha(G) \leq \ell$ if and only if G' can be 1-vertex-deleted into a graph G^* with $\alpha(G^*) \leq \alpha(G') - 1$.

First suppose that $\alpha(G) \leq \ell$. Then $\alpha(G') = \ell + 1$. In G' we delete a vertex v of the $(\ell+1)P_1$. This yields the graph $G^* = G \otimes G \otimes \ell P_1$. We have that $\alpha(G^*) = \max\{\alpha(G), \ell\} = \ell$. As $\alpha(G') = \ell + 1$, this means that $\alpha(G^*) \leq \alpha(G') - 1$.

Now suppose that G' can be 1-vertex-deleted into a graph G^* with $\alpha(G^*) \leq \alpha(G') - 1$. As deleting a vertex in one of the two copies of G does not lower the independence number of G', the deleted vertex must belong to the $(\ell + 1)P_1$. This means that $G^* = G \otimes G \otimes \ell P_1$. As $\alpha(G^*) = \max\{\alpha(G), \ell\} \leq \alpha(G') - 1 = \max\{\alpha(G), \ell + 1\} - 1$, we conclude that $\alpha(G) \leq \ell$. □

Remark 2. We cannot apply Theorem 4 on triangle-free graphs, as the class of triangle-free graphs is not stable-proof.

6   The Two Classifications

In this section we combine Theorems 3 and 4 with a number of known results for obtaining dichotomy results for our two blocker problems restricted to H-free graphs. Before we present these dichotomies we first state some known results that we need for their proofs.

Lemma 4. *([15]).* INDEPENDENT SET *is* NP-*complete for* C_5-*free graphs.*

Lemma 5. (*[15]*). VERTEX COVER *is* NP-*complete for* C_3-*free graphs.*

We also need two of our previous results.

Lemma 6. (*[14]*). *Let* G *be a triangle-free graph containing at least one edge and let* $k \geq 1$ *be an integer. Then* (G, k) *is a yes-instance of* 1-DELETION BLOCKER(ω) *if and only if* (G, k) *is a yes-instance of* VERTEX COVER.

Lemma 7. (*[7]*). *The problems* CONTRACTION BLOCKER(α) *and* DELETION BLOCKER(α) *are polynomial-time solvable for cographs but* NP-*complete on split graphs.*

We also use the following observation.

Lemma 8. *If* H *is a* $(3P_1, 2P_2)$-*free forest, then* $H \subseteq_i P_4$.

Proof. As H is $3P_1$-free, H contains at most two connected components. Suppose H contains exactly two connected components. Then, as H is $2P_2$-free, at least one of these components must be a P_1. As H is $3P_1$-free, this means that H is an induced subgraph of $P_2 + P_1$, so $H \subseteq_i P_4$. Suppose H is connected. As H is $3P_1$-free, H contains no claw and no path on more than five vertices. Hence, $H \subseteq_i P_4$. $\square$

We are now ready to present our first classification.

Theorem 5. *Let* H *be a graph. If* $H \subseteq_i P_4$, *then* CONTRACTION BLOCKER(α) *is polynomial-time solvable for* H-*free graphs, otherwise it is* NP-*hard for* H-*free graphs.*

Proof. Let H be a graph. Recall that a cograph is a P_4-free graph. Hence, if H is an induced subgraph of P_4, then we use Lemma 7 to obtain polynomial-time solvability.

Now suppose that H is not an induced subgraph of P_4. If H contains an induced cycle that is odd, then we use Theorem 1 to obtain NP-hardness. If H contains an induced cycle that is even, then H either contains an induced C_4 or, if the even cycle has at least six vertices, an induced $2P_2$. This means that we can use Lemma 7 to obtain NP-hardness after recalling that split graphs are $(2P_2, C_4)$-free. Assume H contains no cycle. Then H is a forest. If H contains an induced $3P_1$, then we use Lemma 1 to obtain NP-hardness, after observing that cobipartite graphs are $3P_1$-free. Assume H is $3P_1$-free. Then $2P_2 \subseteq_i H$ by Lemma 8, which means we can use Lemma 7 again to obtain NP-hardness. $\square$

Remark 3. In some cases of Theorem 5, such as when $H = C_5$, we could have applied Theorem 4 to obtain co-NP-hardness even if $d = k = 1$.

We now consider the vertex deletion variant and present our second classification.

Theorem 6. *Let* H *be a graph. If* $H \subseteq_i P_4$, *then* DELETION BLOCKER(α) *is polynomial-time solvable for* H-*free graphs, otherwise it is* NP-*hard or* co-NP-*hard for* H-*free graphs.*

Proof. Let H be a graph. If $H \subseteq_i P_4$, then we use Lemma 7 to obtain polynomial-time solvability. Suppose H is not an induced subgraph of P_4. First suppose H contains an induced cycle C_r. If $r = 3$, then we use Theorem 3 to find that the problem is co-NP-hard even if $d = k = 1$. If $r = 4$, then we use Lemma 7 (after recalling that split graphs are C_4-free) to find that the problem is NP-hard. If $r = 5$, then we combine Lemma 4 with Theorem 4 after observing that the class of C_5-free graphs is stable-proof. We then find that the problem is co-NP-hard even if $d = k = 1$. Note that we could have applied Lemma 7 to obtain NP-hardness, as split graphs are C_5-free, If $r \geq 6$, then H contains an induced $2P_2$ and we apply Lemma 7 (as split graphs are $2P_2$-free) to find that the problem is NP-hard.

Now assume that H is forest. By Lemma 8, either $2P_2 \subseteq_i H$ or $3P_1 \subseteq_i H$. If $2P_2 \subseteq_i H$, then we apply Lemma 7 again to obtain NP-hardness. If $3P_1 \subseteq_i H$, then we apply Lemmas 5 and 6 to obtain NP-hardness after observing that a graph is a yes-instance for 1-DELETION BLOCKER(α) if and only if its complement is a yes-instance for 1-DELETION BLOCKER(ω). $\qquad\square$

We are left to state our result for DELETION BLOCKER(ω), which follows immediately from Theorem 6 after making two observations. First, DELETION BLOCKER(ω) for H-free graphs is equivalent to DELETION BLOCKER(α) for $\overline{H}$-free graphs. Second, the graph P_4 is self-complementary, that is, $\overline{P_4} = P_4$.

Theorem 7. *Let H be a graph. If $H \subseteq_i P_4$, then* DELETION BLOCKER *(ω) is polynomial-time solvable for H-free graphs; otherwise it is* co-NP*-hard or* NP*-hard for H-free graphs.*

7   Conclusions

For every graph H we determined the computational complexities of CONTRACTION BLOCKER(α) and DELETION BLOCKER(π) ($\pi \in \{\alpha, \omega\}$) restricted to H-free graphs, and it would be interesting to generalize these results to families of more than one forbidden induced subgraph. In our previous paper [14] we obtained dichotomies for $\pi \in \{\omega, \chi\}$ but for three of the four classifications we needed to assume that H is connected. For comparing our new results with previous results we therefore need to restrict ourselves to connected graphs H. This leads to the following summary:

For a connected graph H, the following holds:

(i) *If $H \subseteq_i P_4$ or $H \subseteq_i \overline{P_1 + P_3}$ then* CONTRACTION BLOCKER(ω) *is polynomial time solvable for H-free graphs; otherwise it is* co-NP*-hard for H-free graphs.*

(ii) *For $\pi \in \{\alpha, \chi\}$, if $H \subseteq_i P_4$ then* CONTRACTION BLOCKER(π) *is polynomial time solvable for H-free graphs; otherwise it is* co-NP*-hard for H-free graphs.*

(iii) *For $\pi \in \{\alpha, \omega, \chi\}$, if $H \subseteq_i P_4$ then* DELETION BLOCKER(π) *is polynomial time solvable for H-free graphs; otherwise it is* co-NP *hard for H-free graphs.*

It is an open problem to generalize the results of the above summary from connected graphs H to arbitrary graphs H. For part (i) we need to settle one remaining case, namely $H = C_3 + P_1$ [14]. Part (ii) has been generalized to arbitrary graphs already; see [14] for the case when $\pi = \chi$ and see Sect. 6 for the case when $\pi \in \{\alpha, \omega\}$. Part (iii) has been settled for all graphs H already for $\pi \in \{\alpha, \omega\}$ (Sect. 6), whereas the situation for $\pi = \chi$ is less clear with a number of cases still being open; in particular polynomial-time results for disconnected graphs H exist incomparable to the case when $H \subseteq_i P_4$, e.g., if $H = 3P_1$ [14].

It is possible to construct graph classes for which a blocker problem is tractable, but the original problem is NP-complete. Take for instance the class of graphs G' from the proof of Theorem 3. The INDEPENDENT SET problem is NP-complete for this graph class, but its members are all no-instances of CONTRACTION BLOCKER(α) when $d = k = 1$. However, this class is not a hereditary graph class, that is, it is not closed under vertex deletion. In fact we do not know of such examples of hereditary graph classes. Hence, it would be interesting to prove for $\pi \in \{\alpha, \omega, \chi\}$ whether CONTRACTION BLOCKER(π) and DELETION BLOCKER(π) are computationally hard on every hereditary graph class $\mathcal{G}$, for which INDEPENDENT SET, CLIQUE or COLORING, respectively, is NP-complete.

Finally, we have shown that CONTRACTION BLOCKER(α) is NP-hard for bipartite graphs. We pose the question of determining the computational complexity of d-CONTRACTION BLOCKER(α) ($d \geq 1$) restricted to bipartite graphs as an open problem.

References

1. Bazgan, C., Bentz, C., Picouleau, C., Ries, B.: Blockers for the stability number and the chromatic number. Graphs Comb. **31**, 73–90 (2015)
2. Bazgan, C., Toubaline, S., Tuza, Z.: The most vital nodes with respect to independent set and vertex cover. Discrete Appl. Math. **159**, 1933–1946 (2011)
3. Bentz, C., Costa, M.-C., de Werra, D., Picouleau, C., Ries, B.: Weighted transversals and blockers for some optimization problems in graphs. In: Progress in Combinatorial Optimization, Wiley-ISTE (2012)
4. Boros, E., Golumbic, M.C., Levit, V.E.: On the number of vertices belonging to all maximum stable sets of a graph. Discrete Appl. Math. **124**, 17–25 (2002)
5. Costa, M.-C., de Werra, D., Picouleau, C.: Minimum d-blockers and d-transversals in graphs. J. Comb. Optim. **22**, 857–872 (2011)
6. Diestel, R.: Graph Theory. Springer, Heidelberg (2005)
7. Diner, Ö.Y., Paulusma, D., Picouleau, C., Ries, B.: Contraction blockers for graphs with forbidden induced paths. In: Paschos, V.T., Widmayer, P. (eds.) CIAC 2015. LNCS, vol. 9079, pp. 194–207. Springer, Cham (2015). doi:10.1007/978-3-319-18173-8_14
8. Földes, S., Hammer, P.L.: Split graphs. In: 8th South-Eastern Conference on Combinatorics, Graph Theory and Computing, Congressus Numerantium, vol. 19, pp. 311–315 (1977)
9. Golovach, P.A., Heggernes, P., Hof, P.V., Paul, C.: Hadwiger number of graphs with small chordality. In: Kratsch, D., Todinca, I. (eds.) WG 2014. LNCS, vol. 8747, pp. 201–213. Springer, Cham (2014). doi:10.1007/978-3-319-12340-0_17

10. Hammer, P.L., Hansen, P., Simeone, B.: Vertices belonging to all or to no maximum stable sets of a graph. SIAM J. Algebraic Discrete Methods **3**, 511–522 (1982)
11. Levit, V.E., Mandrescu, E.: Combinatorial properties of the family of maximum stable sets of a graph. Discrete Appl. Math. **117**, 149–161 (2002)
12. Levit, V.E., Mandrescu, E.: Vertices belonging to all critical sets of a graph. SIAM J. Discrete Math. **26**, 399–403 (2012)
13. Pajouh, F.M., Boginski, V., Pasiliao, E.L.: Minimum vertex blocker clique problem. Networks **64**, 48–64 (2014)
14. Paulusma, D., Picouleau, C., Ries, B.: Reducing the clique and chromatic number via edge contractions and vertex deletions. In: Cerulli, R., Fujishige, S., Mahjoub, A.R. (eds.) ISCO 2016. LNCS, vol. 9849, pp. 38–49. Springer, Cham (2016). doi:10. 1007/978-3-319-45587-7_4
15. Poljak, S.: A note on the stable sets and coloring of graphs. Comment. Math. Univ. Carol. **15**, 307–309 (1974)
16. Ries, B., Bentz, C., Picouleau, C., de Werra, D., Costa, M.-C., Zenklusen, R.: Blockers and transversals in some subclasses of bipartite graphs: when caterpillars are dancing on a grid. Discrete Math. **310**, 132–146 (2010)
17. Savage, C.: Maximum matchings and trees. Inf. Process. Lett. **10**, 202–205 (1980)
18. Toubaline, S.: Détermination des éléments les plus vitaux pour des problèmes de graphes, Ph. D thesis, Université Paris-Dauphine (2010)

A Density Theorem for Hierarchies of Limit Spaces over Separable Metric Spaces

Iosif Petrakis[✉]

University of Munich, Munich, Germany
petrakis@math.lmu.de

Abstract. In this paper, we show, almost constructively, a density theorem for hierarchies of limit spaces over separable metric spaces. Our proof is not fully constructive, since it relies on the constructively not acceptable fact that the limit relation induced by a metric space satisfies Urysohn's axiom for limit spaces. By adding the condition of strict positivity to Normann's notion of probabilistic projection we establish a relation between strictly positive general probabilistic selections on a sequential space and general approximation functions on a limit space. Showing that Normann's result, that a (general and strictly positive) probabilistic selection is definable on a separable metric space, admits a constructive proof, and based on the constructively shown in [18] cartesian closure property of the category of limit spaces with general approximations, our quite effective density theorem follows. This work, which is a continuation of [18], is within computability theory at higher types and Normann's Program of Internal Computability.

1 Introduction

Normann introduced the distinction between internal and external computability over a mathematical structure already in [11] and initiated, what can be called, a "Program of Internal Computability" (PIC) in [12–16] (see also [10]). As he mentions in [14], p. 300, "the internal concepts must grow out of the structure at hand, while external concepts may be inherited from computability over superstructures via, for example, enumerations, domain representations, or in other ways". Within PIC the characterization of functionals, like the Kleene-Kreisel functionals, is done without reference to any realizing objects, but via limit spaces. As Longley and Normann mention in [10], p. 374, the framework of limit spaces leads "in some cases to sharper results than other approaches; moreover, the limit space approach generalizes well to type structures over other base types such as $\mathbb{R}$".

Limit spaces were introduced in computability theory at higher types by Scarpellini in [19], while Hyland in [7] showed that Scarpellini's hierarchy is identical to Kleene's hierarchy of countable functionals over $\mathbb{N}$. In [12] Normann presented this hierarchy using limit spaces, and the corresponding density theorem using the notion of the nth approximation of a functional, for every $n \in \mathbb{N}$.

© Springer International Publishing AG 2017
T.V. Gopal et al. (Eds.): TAMC 2017, LNCS 10185, pp. 484–498, 2017.
DOI: 10.1007/978-3-319-55911-7_35

In [18] we generalized Normann's presentation by defining two new subcategories of limit spaces, the limit spaces with general approximations and the limit spaces with approximations. The constructively shown cartesian closure property for these subcategories enabled us to prove a constructive density theorem for hierarchies of limit spaces over $\mathbb{N}$ and the Cantor space $\mathcal{C}$. The corresponding density theorem for hierarchies of limit spaces over a compact metric space had an essentially classical proof.

In this paper we prove, almost constructively, a density theorem for hierarchies of limit spaces over an arbitrary separable metric space, generalizing and, in our view, computationally advancing the result of [18]. All main proofs included in this paper are within Bishop's informal system of constructive mathematics BISH (see [1–3]). Since the fact that the limit relation on $\mathbb{R}$ induced by its metric satisfies Urysohn's axiom of a limit space implies[1] the limited principle of omniscience (LPO), we cannot say now that our results are fully constructive. We discuss a constructive way out in the last section of this paper.

Nevertheless, our proof seems quite effective, since all the other parts of it are completely constructive. It uses again the cartesian closure property of the category of limit spaces with general approximations, and Normann's result on the existence of a probabilistic selection on a separable metric space. Adding the condition of strict positivity to Normann's notion of probabilistic selection a connection between strictly positive probabilistic selections and general approximation functions is established. This density theorem (Theorem 4) shows that limit spaces with general approximations provide a framework for characterizing hierarchies of functionals over base types maybe even more efficiently than general limit spaces.

2 Basic Notions and Facts

In order to be self-contained we include some basic definitions and facts necessary to the rest of the paper. For a classical treatment of limit spaces see [8,9], while for all general topological notions mentioned here see [6]. If X, Y are sets, $\mathbb{F}(X, Y)$ denotes the set of all functions from X to Y. The third condition of the definition of a limit space is known as Urysohn's axiom.

Definition 1. *A limit space is a structure* $\mathcal{L} = (X, \lim_X)$, *where X is a set, and* $\lim_X \subseteq X \times \mathbb{F}(\mathbb{N}, X)$ *is a relation satisfying the following conditions:*

(L_1) *If $x \in X$ and (x) denotes the constant sequence x, then* $\lim_X(x, (x))$.
(L_2) *If $\mathcal{S}$ denotes the set of all strictly monotone elements of the Baire space* $\mathbb{F}(\mathbb{N}, \mathbb{N})$, *then[2]* $\forall_{\alpha \in \mathcal{S}}(\lim_X(x, x_n) \to \lim_X(x, x_{\alpha(n)}))$.
(L_3) *If $x \in X$ and $(x_n)_{n \in \mathbb{N}} \in \mathbb{F}(\mathbb{N}, X)$, then* $\forall_{\alpha \in \mathcal{S}} \exists_{\beta \in \mathcal{S}}(\lim_X(x, x_{\alpha(\beta(n))})) \to \lim_X(x, x_n)$.

---

[1] This is a result of Hannes Diener (personal communication).
[2] If $(x_n)_{n \in \mathbb{N}} \subseteq X$, for simplicity we write $\lim_X(x, x_n)$ instead of $\lim_X(x, (x_n)_{n \in \mathbb{N}})$, and $\lim_X(x, x)$ instead of $\lim_X(x, (x))$.

If $\forall_{x,y\in X}\forall_{(x_n)_{n\in\mathbb{N}}\in\mathbb{F}(\mathbb{N},X)}(\lim(x,x_n) \rightarrow \lim(y,x_n) \rightarrow x = y)$, then the limit space has the uniqueness property. A subset D of X is called $\lim_X$-dense, if $\forall_{x\in X}\exists_{(d_n)_{n\in\mathbb{N}}\in\mathbb{F}(\mathbb{N},D)}(\lim_X(x,d_n))$, and $\mathcal{L}$ is called $\lim_X$-separable, if there is a countable $\lim_X$-dense subset of X. If $(X,\lim_X)$, $(Y,\lim_Y)$ are limit spaces, $f : X \rightarrow Y$ is called $\lim$-continuous, if $\forall_{x\in X}\forall_{(x_n)_{n\in\mathbb{N}}\in\mathbb{F}(\mathbb{N},X)}(\lim_X(x,x_n) \rightarrow \lim_Y(f(x),f(x_n)))$. The subset $\mathcal{O}$ of X is in the Birkhoff-Baer topology $\mathcal{T}_{\lim_X}$ on X, or is $\lim_X$-open, if $\forall_{x\in\mathcal{O}}\forall_{(x_n)_{n\in\mathbb{N}}\in\mathbb{F}(\mathbb{N},X)}(\lim_X(x,x_n) \rightarrow \mathrm{ev}(x_n,\mathcal{O}))$, where, if $A \subseteq X$, $\mathrm{ev}(x_n,A) :\leftrightarrow \exists_{n_0}\forall_{n\geq n_0}(x_n \in A)$. A topological space $(X,\mathcal{T})$ induces a limit space $(X,\lim_\mathcal{T})$, where $\lim_\mathcal{T}(x,x_n) :\leftrightarrow (x_n)_n \xrightarrow{\mathcal{T}} x$, and the symbol $(x_n)_n \xrightarrow{\mathcal{T}} x$ denotes the convergence of $(x_n)_{n\in\mathbb{N}}$ to x with respect to $\mathcal{T}$. If (X,d) is a metric space, $\lim_d$ denotes the limit relation on X induced by d. A limit space $(X,\lim_X)$ is called topological, if $\lim_X = \lim_{\mathcal{T}_{\lim_X}}$, and a topological space $(X,\mathcal{T})$ is called sequential, if $\mathcal{T} = \mathcal{T}_{\lim_\mathcal{T}}$.

It is easy to show constructively that D is dense in $(X,\mathcal{T}_{\lim_X})$, if D is lim-dense in $(X,\lim_X)$. Moreover, classically a metric space is a sequential space. The following proposition is folklore in the classical literature, but one can show that it holds constructively (see [17]).

Proposition 1. Let $\mathcal{L} = (X,\lim_X)$, $\mathcal{M} = (Y,\lim_Y)$ be limit spaces, and $A \subseteq X$. The relative limit space $\mathcal{L}_A := (A,\lim_A)$ is defined by $\lim_A = (\lim_X)_{|A\times\mathbb{F}(\mathbb{N},A)}$, and the product limit space $\mathcal{L} \times \mathcal{M} := (X \times Y, \lim_{X\times Y})$ is defined by the condition $\lim_{X\times Y}((x,y),(x_n,y_n)) :\leftrightarrow \lim_X(x,x_n) \wedge \lim_Y(y,y_n)$, for every $x \in X, y \in Y$, $(x_n)_{n\in\mathbb{N}} \in \mathbb{F}(\mathbb{N},X)$ and $(y_n)_{n\in\mathbb{N}} \in \mathbb{F}(\mathbb{N},Y)$. The exponential limit space $\mathcal{L} \rightarrow \mathcal{M} := (X \rightarrow Y, \lim_{X\rightarrow Y})$, where $X \rightarrow Y$ is the set of all lim-continuous functions from $\mathcal{L}$ to $\mathcal{M}$, is defined by the condition $\lim_{X\rightarrow Y}(f,f_n) :\leftrightarrow \forall_{x\in X}\forall_{(x_n)_{n\in\mathbb{N}}\in\mathbb{F}(\mathbb{N},X)}(\lim_X(x,x_n) \rightarrow \lim_Y(f(x),f_n(x_n)))$,

Definition 2. A limit space with general approximations is a structure $\mathcal{A} = (X,\lim_X,(\mathrm{XAppr}_n)_{n\in\mathbb{N}})$ such that $(X,\lim_X)$ is a limit space, and, for every $n \in \mathbb{N}$ the approximation functions $\mathrm{XAppr}_n : X \rightarrow X$ satisfy the following properties:
(A_1) If $x \in X$, then $\mathrm{XAppr}_n(\mathrm{XAppr}_n(x)) = \mathrm{XAppr}_n(x)$.
(A_2) $\mathrm{XAppr}_n(X) = \{\mathrm{XAppr}_n(x) \mid x \in X\}$ is an inhabited finite set.
(A_3) If $x \in X$ and $(x_n)_{n\in\mathbb{N}} \in \mathbb{F}(\mathbb{N},X)$, then

$$\lim_X(x,x_n) \rightarrow \lim_X(x,\mathrm{XAppr}_n(x_n)).$$

A limit space with general approximations is a limit spaces with approximations, if the following conditions are satisfied:
(A_1') If $x \in X$, then $\mathrm{XAppr}_n(\mathrm{XAppr}_m(x)) = \mathrm{XAppr}_{\min(n,m)}(x)$.
(A_4) XAppr_n is lim-continuous.
A limit space $(X,\lim_X)$ admits (general) approximations, if there are functions $(\mathrm{XAppr}_n)_{n\in\mathbb{N}}$ such that $(X,\lim_X,(\mathrm{XAppr}_n)_{n\in\mathbb{N}}))$ is a limit space with (general) approximations

The following two results were proved in [18] constructively.

Proposition 2. *If $\mathcal{A} = (X, \lim_X, (\mathrm{XAppr}_n)_{n \in \mathbb{N}})$ is a limit space with general approximations and $x \in X$, then $\lim_X (x, \mathrm{XAppr}_n(x))$. Moreover, the set $A = \bigcup_{n \in \mathbb{N}} \mathrm{XAppr}_n(X)$ is a countable $\lim_X$-dense subset of X, and therefore dense in $(X, \mathcal{T}_{\lim_X})$.*

Theorem 1. *If $\mathcal{A} = (X, \lim_X, (\mathrm{XAppr}_n)_{n \in \mathbb{N}})$, $\mathcal{B} = (Y, \lim_Y, (\mathrm{YAppr}_n)_{n \in \mathbb{N}})$ are limit spaces with (general) approximations, $n \in \mathbb{N}$, $x \in X, y \in Y$, and $f \in X \to Y$, we define*

$$(X \times Y)\mathrm{Appr}_n(x, y) := (\mathrm{XAppr}_n(x), \mathrm{YAppr}_n(y)),$$

$$f \mapsto (X \to Y)\mathrm{Appr}_n(f),$$

$$(X \to Y)\mathrm{Appr}_n(f)(x) := \mathrm{YAppr}_n(f(\mathrm{XAppr}_n(x))).$$

The structures $\mathcal{A} \times \mathcal{B} = (X \times Y, \lim_{X \times Y}, ((X \times Y)\mathrm{Appr}_n)_{n \in \mathbb{N}})$ and $\mathcal{A} \to \mathcal{B} = (X \to Y, \lim_{X \to Y}, ((X \to Y)\mathrm{Appr}_n)_{n \in \mathbb{N}})$ are limit spaces with (general) approximations.

From the last two results the following density theorem for a hierarchy of limit spaces over a compact metric space was shown in [18] classically.

Theorem 2. *Let (X, d) be a compact metric space. If $\iota = X \mid \rho \to \sigma$ is an inductively defined type system $\mathbb{T}$ over the base type X, then in the $\mathbb{T}$-typed hierarchy of limit spaces over X, defined by*

$$\mathcal{X}(\iota) := (X(\iota), \lim_\iota) := (X, \lim_d),$$

$$\mathcal{X}(\rho \to \sigma) := (X(\rho) \to X(\sigma), \lim_{\rho \to \sigma}),$$

the limit space $\mathcal{X}(\tau)$ admits general approximations $(\tau\mathrm{Appr}_n)_{n \in \mathbb{N}}$, for every type τ in $\mathbb{T}$. Moreover, there is a countable subset D_τ of $X(\tau)$, which is dense in $(X(\tau), \mathcal{T}_{\lim_\tau})$, for every type τ in $\mathbb{T}$.

A similar density theorem was shown constructively for $\iota = \mathbb{N}$ and $\iota = \mathcal{C}$, where $\mathcal{C}$ denotes the Cantor space. In Sect. 4 we show a density theorem for a hierarchy of limit spaces over an arbitrary separable space (Theorem 4), based again on Proposition 2 and Theorem 1. In this case though we use appropriately Normann's notion of a probabilistic selection on a sequential space to define general approximation functions on a separable metric space.

3   Positive and Strictly Positive Probabilistic Projections

The use of probability distributions in the study of hierarchies of functionals over $\mathbb{R}$ appeared first in Normann's work [12], following the work of DeJaeger in [5]. The next definition includes a slight variation[3] of Normann's definition of

---

[3] Namely, the continuity condition used by Normann is different from the condition (Γ_3) used here, but one can show that they are equivalent. Since no continuity condition affects the main density theorem, we do not include here the proof of their equivalence.

a probabilistic projection found in [14]. Moreover, the notions of general, positive and strictly positive probabilistic projections are introduced. Note that in the following definition we use Normann's starting point of a sequential space X, but what we only need for the proof of Theorem 4, and this is how one should read Definition 3 constructively, is that X is a metric space (recall that we need classical reasoning to show that a metric space is sequential).

Definition 3. *A structure* $\mathcal{P} = (X, \mathcal{T}, Y, (A_n)_{n \in \mathbb{N}}, (\mu_n)_{n \in \mathbb{N}})$ *is called a sequential space with a general probabilistic projection from* X *to* Y, *if* $(X, \mathcal{T})$ *is a sequential topological space,* $(A_n)_{n \in \mathbb{N}}$ *is a sequence of inhabited finite subsets of* X, *which is called the support of* $\mathcal{P}$, Y *is a subset of* X *such that*

$$A := \bigcup_{n \in \mathbb{N}} A_n \subseteq Y,$$

and $(\mu_n)_{n \in \mathbb{N}}$ *is a sequence of functions of type*

$$\mu_n : X \to \mathbb{F}(A_n, [0, 1])$$

$$x \mapsto \mu_n(x),$$

that satisfies the following properties:
(P_1) *For every* $n \in \mathbb{N}$ *the function* $\mu_n(x) : A_n \to [0, 1]$ *is a probability distribution on* A_n *i.e., it satisfies the condition*

$$\sum_{a \in A_n} \mu_n(x)(a) = 1.$$

(P_2) *If* $y \in Y$, $(y_n)_{n \in \mathbb{N}} \subseteq Y$ *such that* $\lim_{\mathcal{T}_{|Y}}(y, y_n)$, *where* $\lim_{\mathcal{T}_{|Y}}$ *is the limit relation on* Y *induced by the limit relation* $\lim_{\mathcal{T}}$ *on* X, *and if* $(a_n)_{n \in \mathbb{N}} \subseteq A$ *such that* $a_n \in A_n$, *for every* $n \in \mathbb{N}$, *the following implication holds:*

$$\forall_{n \in \mathbb{N}}(\mu_n(y_n)(a_n) > 0) \to \lim_{\mathcal{T}_{|Y}}(y, a_n).$$

The sequence of functions $(\mu_n)_{n \in \mathbb{N}}$ *is called a general probabilistic projection from* X *to* Y. *A sequential space* $(X, \mathcal{T})$ *admits a general probabilistic projection from* X *to* Y, *if there is a general probabilistic projection from* X *to* Y. *A structure* $\mathcal{P} = (X, \mathcal{T}, Y, (A_n)_{n \in \mathbb{N}}, (\mu_n)_{n \in \mathbb{N}})$ *is a sequential space with a probabilistic projection from* X *to* Y, *if* $(\mu_n)_{n \in \mathbb{N}}$ *satisfies also the following condition:*
(P_3) *If* $a \in A_n$, *for some* $n \in \mathbb{N}$, *the function* $\hat{a} : X \to [0, 1]$, *defined by*

$$x \mapsto \mu_n(x)(a),$$

for every $x \in X$, *is continuous.*
A general probabilistic projection $(\mu_n)_{n \in \mathbb{N}}$ *from* X *to* Y *is called positive, if the following conditions are satisfied:*
(P_4) *If* $a \in A_n$, *for some* $n \in \mathbb{N}$, *then*

$$\mu_n(a)(a) > 0,$$

$$\forall_{b \in A_n} (b \neq a \to \mu_n(a)(b) < \mu_n(a)(a)).$$

A positive probabilistic projection from X to Y is called strictly positive, if the following condition is satisfied:
(P_5) *If $a \in A_n$, for some $n \in \mathbb{N}$, then*

$$\mu_n(a)(a) = 1.$$

A (general) probabilistic projection $(\mu_n)_{n \in \mathbb{N}}$ from X to X is called a (general) probabilistic selection on X, and the structure $\mathcal{S} = (X, \mathcal{T}, X, (A_n)_{n \in \mathbb{N}}, (\mu_n)_{n \in \mathbb{N}})$, or simpler $\mathcal{S} = (X, \mathcal{T}, (A_n)_{n \in \mathbb{N}}, (\mu_n)_{n \in \mathbb{N}})$, is a sequential space with a (general) probabilistic selection.

By condition (P_1), if $(\mu_n)_{n \in \mathbb{N}}$ is a strictly positive probabilistic projection from X to Y, then

$$\forall_{b \in A_n} (b \neq a \to \mu_n(a)(b) = 0),$$

since, if $\mu_n(a)(b) > 0$, for some $b \in A_n$ such that $b \neq a$, then $\sum_{b \in A_n} \mu_n(a)(b) > 1$, which is a contradiction. Hence, $\mu_n(a)(b) \leq 0$, which together with the assumed condition $\mu_n(a)(b) \geq 0$ gives $\mu_n(a)(b) = 0$. A first constructive reading of condition (P_1) gives that $\neg\neg[\exists_{a \in A_n}(\mu_n(x)(a) > 0)]$; if $\neg[\exists_{a \in A_n}(\mu_n(x)(a) > 0)]$, then $\forall_{a \in A_n}(\mu_n(x)(a) \leq 0)$, since if $a \in A_n$ such that $\mu_n(x)(a) > 0$, then we get a contradiction, hence $\mu_n(x)(a) \leq 0$. Since $\forall_{a \in A_n}(\mu_n(x)(a) \geq 0)$, we get $\forall_{a \in A_n}(\mu_n(x)(a) = 0)$, hence $\sum_{a \in A_n} \mu_n(x)(a) = 0 = 1$. Next we show constructively how to shift double negation.

Proposition 3. *If $n \in \mathbb{N}$, $a_1, \ldots, a_n \geq 0$, and $l > 0$, then*

$$\sum_{i=1}^{n} a_i = l \to \exists_{j \in \{1, \ldots, n\}}(a_j > 0).$$

Proof. We show $\forall_{n \in \mathbb{N}} P(n)$, where

$$P(n) := \forall_{a_1, \ldots, a_n \geq 0} \forall_{l > 0} \left(\sum_{i=1}^{n} a_i = l \to \exists_{j \in \{1, \ldots, n\}}(a_j > 0) \right).$$

If $n = 1$, then $j = 1$. To show $P(n+1)$ from $P(n)$ let $a_1, \ldots, a_{n+1} \geq 0$, and $l > 0$ such that $\sum_{i=1}^{n+1} a_i = l$. If $b := \sum_{i=1}^{n} a_i \geq 0$, then $b + a_{n+1} = l$. By the constructive version of trichotomy of reals (see [2], p. 26) we have that $a_{n+1} > 0$ or $a_{n+1} < \frac{l}{2}$. In the first case we get that the required $j = n + 1$. If $a_{n+1} < \frac{l}{2}$, then $b = l - a_{n+1} > l - \frac{l}{2} = \frac{l}{2}$. Consequently, $\sum_{i=1}^{n} a_i = b > 0$, and by condition $P(n)$ on $a_1, \ldots, a_n$ and b we get some $j \in \{1, \ldots, n\}$ such that $a_j > 0$.

Hence, if $(\mu_n)_{n \in \mathbb{N}}$ is a general probabilistic projection from X to Y, the set

$$I_n(x) := \{a \in A_n | \mu_n(x)(a) > 0\}$$

is inhabited. The intuition behind the notion of a probabilistic projection from X to Y can be described as follows. The fact $\mu_n(x)(a) > 0$ expresses that a is "close" to x, and moreover, the closer to 1 the positive value $\mu_n(x)(a)$ is, the closer to x a is. The fact $\mu_n(x)(a) = 0$ expresses that a is "not close" to x. With this interpretation conditions (P_2) and (P_4) are quite natural. Note that the notion of a general probabilistic projection from X to Y corresponds to the notion of a limit space with general approximations, since in both cases a continuity condition is not necessarily satisfied. As in the case of limit spaces with (general) approximations, a dense subset is (classically) generated from a general probability projection.

Proposition 4. *(i) If $\mathcal{P} = (X, \mathcal{T}, Y, (A_n)_{n \in \mathbb{N}}, (\mu_n)_{n \in \mathbb{N}})$ is a sequential space with a general probability projection from X to Y, and Y is a closed, or open, subspace of X, then A is dense in Y.*
(ii) If $\mathcal{P} = (X, \mathcal{T}, (A_n)_{n \in \mathbb{N}}, (\mu_n)_{n \in \mathbb{N}})$ is a sequential space with a general probability selection, then A is dense in X.

Proof. We show (i), and (ii) follows immediately from (i). If $y \in Y$, let $(a_n)_{n \in \mathbb{N}} \subseteq A$ such that $a_n \in A_n$ and $\mu_n(y)(a_n) > 0$. The existence of such an element a_n of A_n follows from condition (P_1). Since $\lim_{\mathcal{T}}(y, y)$, by condition (P_2) we get $\lim_{\mathcal{T}_{|Y}}(y, a_n)$ i.e., A is $\lim_{\mathcal{T}_{|Y}}$-dense in X. Since a closed, or open, subspace of a sequential space is sequential, and since a $\lim_{\mathcal{S}}$-dense subset of a sequential space $(Z, \mathcal{S})$ is also dense in Z, we conclude that A is dense in Y.

Since A is countable, the relative space Y is separable. Consequently, if Y is not a separable subspace of X, there can be no probabilistic projection from X to Y. As in the density theorem for limit spaces with general approximations the continuity condition (P_3) plays no role in the above proof. Next follows the lim-version of Definition 3.

Definition 4. *A structure $\mathcal{N} = (X, \lim_X, Y, (A_n)_{n \in \mathbb{N}}, (\mu_n)_{n \in \mathbb{N}})$ is a limit space with a general lim-probabilistic projection from X to Y, if $(X, \lim_X)$ is a limit space and $Y, (A_n)_{n \in \mathbb{N}}, (\mu_n)_{n \in \mathbb{N}}$ are as in Definition 3, though the limit relation in (P_2) is the limit relation on Y inherited from $\lim_X$. A limit space with a lim-probabilistic projection from X to Y is a limit space with a general lim-probabilistic projection from X to Y such that the following condition is satisfied: (P_3') If $a \in A_n$, for some $n \in \mathbb{N}$, the function $\hat{a} : X \to [0,1]$, defined by $x \mapsto \mu_n(x)(a)$, for every $x \in X$, is lim-continuous i.e.,*

$$\lim_X(x, x_m) \to \lim_{[0,1]}(\mu_n(x)(a), \mu_n(x_m)(a)),$$

for every $x \in X$ and $(x_m)_{m \in \mathbb{N}} \subseteq X$, where $\lim_{[0,1]}$ is the limit relation on $[0,1]$ generated by its Euclidean metric. A limit space with a (general) lim-probabilistic selection, and the notions of a (strictly) positive (general) lim-probabilistic projection (selection) are defined as in Definition 3.

In the next classically shown proposition the hypothesis of positivity is used.

Proposition 5. *If* $(X, \lim_X, (A_n)_{n\in\mathbb{N}}, (\mu_n)_{n\in\mathbb{N}})$ *is a limit space with a positive, general* lim-*probabilistic selection, then there are approximation functions* $(\mathrm{XAppr}_n)_{n\in\mathbb{N}}$ *on* X *such that* $(X, \lim_X, (\mathrm{XAppr}_n)_{n\in\mathbb{N}})$ *is a limit space with general approximations, and* $\mathrm{XAppr}_n(X) = A_n$, *for every* $n \in \mathbb{N}$.

Proof. If $n \in \mathbb{N}$, suppose that $A_n = \{a_1^{(n)}, \ldots, a_{m(n)}^{(n)}\}$. If $x \in X$, let

$$i_{0,n}(x) := \left\{ i \in \{1, \ldots, m(n)\} \mid \mu_n(x)(a_i^{(n)}) > 0, \text{ and} \right.$$

$$\left. \forall_{j\in\{1,\ldots,m(n)\}}(\mu_n(x)(a_j^{(n)}) \leq \mu_n(x)(a_i^{(n)})) \right\}.$$

By the properties of the order on classical real numbers $i_{0,n}(x)$ is well-defined. For every $x \in X$ and every $n \in \mathbb{N}$ we define

$$\mathrm{XAppr}_n(x) := a_{i_{0,n}(x)}^{(n)}.$$

Since $(\mu_n)_{n\in\mathbb{N}}$ is positive, if $i \in \{1, \ldots, m(n)\}$, then

$$i_{0,n}(a_i^{(n)}) = \{i\},$$

and $\mathrm{XAppr}_n(a_i^{(n)}) = a_{i_{0,n}(a_i^{(n)})}^{(n)} = a_i^{(n)}$. The conditions $\mathrm{XAppr}_n(\mathrm{XAppr}_n(x)) = \mathrm{XAppr}_n(x)$ and $\mathrm{XAppr}_n(X) = A_n$ are then immediately satisfied. By the definition of $i_{0,n}(x)$ we have that

$$\mu_n(x)(\mathrm{XAppr}_n(x)) > 0.$$

If $(x_n)_{n\in\mathbb{N}} \subseteq X$ such that $\lim_X(x, x_n)$, then since $\mu_n(x_n)(\mathrm{XAppr}_n(x_n)) > 0$, for every $n \in \mathbb{N}$, by condition (P_2) of Definition 4 we get $\lim_X(x, \mathrm{XAppr}_n(x_n))$.

Note that constructively we can't find an algorithm providing an element of $i_{0,n}(x)$. We overcome this difficulty in Proposition 7, where the hypothesis of a strictly positive probabilistic selection is used. The next proposition is also shown classically.

Proposition 6. *(i) A limit space* $(X, \lim_X, (\mathrm{XAppr}_n)_{n\in\mathbb{N}})$ *with general approximations admits a strictly positive, general* lim-*probabilistic selection. (ii) A limit space* $(X, \lim_X, (\mathrm{XAppr}_n)_{n\in\mathbb{N}})$ *with approximations, where* $(X, \lim_X)$ *has the uniqueness property, admits a strictly positive general* lim-*probabilistic selection.*

Proof. (i) We define $A_n = \mathrm{XAppr}_n(X)$, and for every $x \in X$ the function $x \mapsto \mu_n(x)$ is defined by

$$\mu_n(x)(a) = \begin{cases} 1, & \text{if } a = \mathrm{XAppr}_n(x) \\ 0, & \text{ow.} \end{cases}$$

Clearly, $\mu_n(x)$ is a probability distribution on A_n. Since $\mu_n(x_n)(a_n) > 0 \leftrightarrow$ $a_n = \mathrm{XAppr}_n(x_n)$, we get $\lim_X(x, x_n) \to \lim_X(x, a_n)$. If $a \in \mathrm{XAppr}_n(X)$, there is some $x \in X$ such that $a = \mathrm{XAppr}_n(x)$, hence

$$\mu_n(a)(a) = \mu_n(\mathrm{XAppr}_n(x))(\mathrm{XAppr}_n(x)) = 1 > 0,$$

since $\mathrm{XAppr}_n(x) = \mathrm{XAppr}_n(\mathrm{XAppr}_n(x))$.
(ii) Suppose that $\lim_X(x, x_m)$ and that $\mu_n(x)(a) = 1 \leftrightarrow a = \mathrm{XAppr}_n(x)$. By the classical proof of Proposition 21(i) in [18], pp. 749–750, the sequence $(\mathrm{XAppr}_n(x_m))_{m \in \mathbb{N}}$ is eventually constant with value a. Thus, $(\mu_n(x_m)(a))_{m \in \mathbb{N}}$ is eventually constant 1. The case $a \neq \mathrm{XAppr}_n(x)$ is treated similarly.

The above proof corroborates the aforementioned intuition behind the existence of a probabilistic projection, that is $\mu_n(x)(a) > 0$ expresses a proximity of a to x, while $\mu_n(x)(a) = 0$ expresses a non-proximity of a to x. Regarding the proof of Proposition 6(ii), the lim-continuity of the approximation functions XAppr_n entails the lim-continuity of the function $\hat{a}$, where $a \in A$. Next follows the constructive version of Proposition 5, which is essential to the proof of Theorem 4. One needs to replace the condition of positivity by the condition of strict positivity.

Proposition 7. *If $(X, \lim_X, (A_n)_{n \in \mathbb{N}}, (\mu_n)_{n \in \mathbb{N}})$ is a limit space with a strictly positive, general lim-probabilistic selection, then there are approximation functions $(\mathrm{XAppr}_n)_{n \in \mathbb{N}}$ on X such that $(X, \lim_X, (\mathrm{XAppr}_n)_{n \in \mathbb{N}})$ is a limit space with general approximations, and $\mathrm{XAppr}_n(X) = A_n$, for every $n \in \mathbb{N}$.*

Proof. If $n \in \mathbb{N}$, suppose that $A_n = \{a_1^{(n)}, \dots, a_{m(n)}^{(n)}\}$. If $x \in X$, the set

$$I_{0,n}(x) := \{i \in \{1, \dots, m(n)\} \mid \mu_n(x)(a_i^{(n)}) > 0\}$$

is inhabited, i.e., for every $n \in \mathbb{N}$ there exists $i \in I_{0,n}(x)$. If $S_x \subseteq \mathbb{N} \times \bigcup_{n=1}^{\infty} I_{0,n}(x)$ is defined by $S_x(n, i) := i \in I_{0,n}(x)$, then by the principle of countable choice[4] there is a function $f_x : \mathbb{N} \to \bigcup_{n=1}^{\infty} I_{0,n}(x)$ such that $f_x(n) \in I_{0,n}(x)$, for every $n \in \mathbb{N}$. We define

$$\mathrm{XAppr}_n(x) := a_{f_x(n)}^{(n)},$$

for every $x \in X$ and every $n \in \mathbb{N}$. Since $(\mu_n)_{n \in \mathbb{N}}$ is a strictly positive probabilistic selection on X, if $i \in \{1, \dots, m(n)\}$, then $I_{0,n}(a_i^{(n)}) = \{i\}$, hence

$$f_{a_i^{(n)}}(n) = i,$$

and $\mathrm{XAppr}_n(a_i^{(n)}) = a_{f_{a_i^{(n)}}(n)}^{(n)} = a_i^{(n)}$. The conditions $\mathrm{XAppr}_n(\mathrm{XAppr}_n(x)) = \mathrm{XAppr}_n(x)$ and $\mathrm{XAppr}_n(X) = A_n$ are then immediately satisfied. By the definition of $I_{0,n}(x)$ we have that

$$\mu_n(x)(\mathrm{XAppr}_n(x)) > 0.$$

If $(x_n)_{n \in \mathbb{N}} \subseteq X$ such that $\lim_X(x, x_n)$, then since $\mu_n(x_n)(\mathrm{XAppr}_n(x_n)) > 0$, for every $n \in \mathbb{N}$, by condition (P_2) of Definition 4 we get $\lim_X(x, \mathrm{XAppr}_n(x_n))$.

---

[4] This principle is generally accepted within BISH (see [3], p. 12).

4   The Density Theorem

In [14] Normann proved that a complete and separable metric space X admits a probabilistic projection from X to a closed subspace Y of X of the form $Y = \overline{\bigcup_n \Lambda_n}$, where $A_n \subseteq A_{n+1} \subsetneq X$, for every $n \in \mathbb{N}$. In [13] Normann defined a probabilistic selection on a separable metric space. The proof is not given in [13], although it is actually in [14], which appeared later, but was written before [13]. In between Normann realized that completeness played no role in his original proof.

Here we show that the probabilistic selections defined by Normann differ in a crucial way. The one given in [14] is shown here to be positive, while the one given in [13] is shown to be strictly positive, a property crucial to the proof of Theorem 4. Next we give a new constructive treatment of Normann's result adding the properties of positivity and strict positivity, respectively. Note that Normann included his equivalent to (P_3) continuity condition to his results, but since the proof of continuity requires classical reasoning and does not play a role in our proof of Theorem 4, it is avoided here. The only non-effective element in the formulation of the following theorem (and not in its proof) is that $(X, \lim_d)$ is a limit space, hence that $\lim_d$ satisfies Urysohn's axiom.

Theorem 3 (Normann (BISH)). *Suppose that (X, d) is a separable metric space and $A = \{a_n \mid n \in \mathbb{N}\}$ is a countable dense subset of X, where $d(a_n, a_m) > 0$, if $n \neq m$. If $A_n = \{a_1, \ldots, a_n\}$, for every $n \in \mathbb{N}$ and, for every $1 \leq j \leq n$, we define[5]*

$$\mu_n(x)(a_j) := \frac{N_{n,x}(a_j)}{D_{n,x}},$$

$$N_{n,x}(a_j) := (d(x, A_n) + 2^{-n}) \mathbin{\dot{-}} d(x, a_j),$$

$$D_{n,x} := \sum_{i=1}^{n} [(d(x, A_n) + 2^{-n}) \mathbin{\dot{-}} d(x, a_i)],$$

$$\mu'_n(x)(a_j) := \frac{N'_{n,x}(a_j)}{D'_{n,x}},$$

$$N'_{n,x}(a_j) := (d(x, A_n) + \delta_n) \mathbin{\dot{-}} d(x, a_j),$$

$$D'_{n,x} := \sum_{i=1}^{n} [(d(x, A_n) + \delta_n) \mathbin{\dot{-}} d(x, a_i)],$$

where

$$d(x, A_n) := \min\{d(x, a_i) \mid 1 \leq i \leq n\},$$

$$\delta_n := \min\{2^{-n}, d(a_i, a_j) \mid i \neq j, i, j \in \{1, \ldots, n\}\},$$

$$a \mathbin{\dot{-}} b := (a - b) \vee 0.$$

---

[5] If $c, d \in \mathbb{R}$, we use the notations $c \vee d := \max\{c, d\}$, and $c \wedge d := \min\{c, d\}$.

(i) *The structure* $(X, \lim_d, (A_n)_{n\in\mathbb{N}}, (\mu_n)_{n\in\mathbb{N}})$ *is a limit space with a positive, general* lim-*probabilistic selection on* X.

(ii) *The structure* $(X, \lim_d, (A_n)_{n\in\mathbb{N}}, (\mu'_n)_{n\in\mathbb{N}})$ *is a limit space with a strictly positive, general* lim-*probabilistic selection on* X.

Proof. (i) The fact that $D_{n,x} > 0$ and conditions (P_1) and (P_2) are shown as in case (ii). For the positivity condition we have first that

$$\mu_n(a_j)(a_j) = \frac{2^{-n}}{D_{n,a_j}} > 0,$$

for every $j \in \{1, \ldots, n\}$. If $i \neq j$, then $N_{n,a_j}(a_i) = 2^{-n} \dotdiv d(a_j, a_i) = (2^{-n} - d(a_j, a_i)) \vee 0$. Since $2^{-n} - d(a_j, a_i) < 2^{-n}$ and $0 < 2^{-n}$, we get $(2^{-n} - d(a_j, a_i)) \vee 0 < 2^{-n}$ (here we used the following property of real numbers: $a \vee b < c \leftrightarrow a < c$ and $b < c$, see [4], p. 57, Ex. 3). Hence, $\mu_n(a_j)(a_i) < \mu_n(a_j)(a_j)$.

(ii) If $c_1, \ldots, c_n > 0$, then one shows[6] that their minimum $\bigwedge_{i=1}^n c_i > 0$, hence, since there are no repetitions in the sequence of A, we have that $\delta_n > 0$. Next we show[7] that $D'_{n,x} > 0$. The subspace A_n is totally bounded, since for every $\epsilon > 0$ it is an ϵ-approximation of itself, and since the distance d_x at x, defined by $a_j \mapsto d(x, a_j)$, is uniformly continuous on A_n, there exists $\inf d_x(A_n)$ (see [2], p. 94). It is immediate to see that $\inf d_x(A_n) = d(x, A_n)$ is the greatest lower bound of $\{d(x, a_j) \mid j \in \{1, \ldots, n\}\}$, and hence equal to $\bigwedge_{i=1}^n d(x, a_i)$, since $\bigwedge_{i=1}^n d(x, a_i)$ can be shown[8] to be the greatest lower bound of $\{d(x, a_j) \mid j \in \{1, \ldots, n\}\}$ too. By the definition of the infimum of a bounded below set of real numbers for $\frac{\delta_n}{2} > 0$ we get that the existence of some $j \in \{1, \ldots, n\}$ such that

$$d(x, a_j) < d(x, A_n) + \frac{\delta_n}{2} \rightarrow -\frac{\delta_n}{2} < d(x, A_n) - d(x, a_j) \rightarrow$$

$$0 < \delta_n - \frac{\delta_n}{2} < d(x, A_n) + \delta_n - d(x, a_j) \rightarrow$$

$$0 < \frac{\delta_n}{2} < (d(x, A_n) + \delta_n) \dotdiv d(x, a_j) \rightarrow$$

$$0 < D'_{n,x}.$$

Condition (P_1) is immediately satisfied. For the proof of condition (P_2) we fix $x \in X$, $(x_n)_{n\in\mathbb{N}} \subseteq X$, such that $\lim_d(x, x_n)$, and $(a_n)_{n\in\mathbb{N}}$ such that $a_n \in A_n$ and $\mu'_n(x_n)(a_n) > 0$, for every $n \in \mathbb{N}$. We need to show that $\lim_d(x, a_n) \leftrightarrow \forall_{\epsilon>0} \exists_{n_0} \forall_{n \geq n_0} (d(x, a_n) \leq \epsilon)$. Let $\epsilon > 0$. By the hypothesis $\lim_d(x, x_n)$ there

---

[6] The argument for the case of two positive numbers is the one used in the inductive step of the induction on n. If $c_1, c_2 > 0$, there are rationals q_1, q_2 such that $0 < q_1 < c_1$ and $0 < q_2 < c_2$ (see [2], p. 25). Since $q_1 \wedge q_2$ is either q_1 or q_2, we get that $q_1 \wedge q_2 < c_1$ and $q_1 \wedge q_2 < c_2$, hence $0 < q_1 \wedge q_2 \leq c_1 \wedge c_2$.

[7] Classically, this is trivial, since there is some $j \in \{1, \ldots, n\}$ such that $d(x, A_n) = d(x, a_j)$, hence $D'_{n,x} \geq (d(x, A_n) + \delta_n) \dotdiv d(x, a_i) = \delta_n \vee 0 = \delta_n > 0$.

[8] The proof is based on the fact that if $c \leq a$ and $c \leq b$, then $c \leq a \wedge b$, since if $c > a \wedge b$, then $c > a$ or $c > b$ (this is the dual of a property of the maximum of real numbers included in [4], p. 57, Ex. 3).

is $n_1 \in \mathbb{N}$ such that $\forall_{n \geq n_1}(d(x, x_n) \leq \frac{\epsilon}{4})$. By the density of A in X there exists $a \in A$ such that $d(x, a) \leq \frac{\epsilon}{4}$. If $a = a_{n_2}$, for some $n_2 \in \mathbb{N}$, we get that $\exists_{a \in A_{n_2}}(d(x, a) \leq \frac{\epsilon}{4})$. Clearly, there exists $n_3 \in \mathbb{N}$ such that $2^{-n_3} \leq \frac{\epsilon}{4}$. If we define $n_0 := \max(n_1, n_2, n_3)$, for every $n \in \mathbb{N}$ such that $n \geq n_0$ we get $A_n \supseteq A_{n_0}$ and $d(x, A_n) \leq d(x, a) \leq \frac{\epsilon}{4}$. Moreover, if $n \geq n_0$, then

$$d(x_n, A_n) \leq d(x_n, x) + d(x, A_n) \leq \frac{\epsilon}{4} + \frac{\epsilon}{4} = \frac{\epsilon}{2}.$$

The first inequality above is shown as follows: If $b \in A_n$, then using some basic properties of $\leq$ on $\mathbb{R}$ (see [2], p. 23) we get

$$d(x_n, A_n) \leq d(x_n, b) \leq d(x_n, x) + d(x, b) \rightarrow$$
$$d(x_n, A_n) - d(x_n, x) \leq d(x, b) \rightarrow$$
$$d(x_n, A_n) - d(x_n, x) \leq \min\{d(x, b) \mid b \in A_n\} = d(x, A_n) \rightarrow$$
$$d(x_n, A_n) \leq d(x_n, x) + d(x, A_n).$$

Moreover, if $n \geq n_0$, then

$$\mu'_n(x_n)(a_n) > 0 \rightarrow d(x_n, a_n) \leq \frac{3\epsilon}{4},$$

since, using the property[9] $\forall_{c \in \mathbb{R}}(c \vee 0 > 0 \rightarrow c \vee 0 = c)$ we have that

$$\mu'_n(x_n)(a_n) > 0 \rightarrow N'_{n, x_n}(a_n) > 0$$
$$\leftrightarrow (d(x_n, A_n) + \delta_n) \dot{-} d(x_n, a_n) > 0$$
$$\rightarrow (d(x_n, A_n) + \delta_n) - d(x_n, a_n) > 0$$
$$\rightarrow d(x_n, a_n) < d(x_n, A_n) + \delta_n \leq \frac{\epsilon}{2} + 2^{-n} \leq \frac{\epsilon}{2} + \frac{\epsilon}{4} = \frac{3\epsilon}{4}.$$

Hence, if $n \geq n_0$, we get

$$d(x, a_n) \leq d(x, x_n) + d(x_n, a_n) \leq \frac{\epsilon}{4} + \frac{3\epsilon}{4} = \epsilon.$$

Next we show the strict positivity of $(\mu'_n)_{n \in \mathbb{N}}$. If $n \in \mathbb{N}$ and $j \in \{1, \ldots, n\}$, then $N'_{n, a_j}(a_j) = \delta_n$, since $d(a_j, A_n) = d(a_j, a_j) = 0$. Moreover, $D'_{n, a_j}(a_j) = \sum_{i=1}^{n}[(d(a_j, A_n) + \delta_n) \dot{-} d(a_j, a_i)] = \sum_{i=1}^{n}(\delta_n \dot{-} d(a_j, a_i)) = \delta_n \dot{-} d(a_j, a_j) = \delta_n$, since for every $i \neq j$, we have that $\delta_n \dot{-} d(a_j, a_i) = 0$, since $\delta_n \leq d(a_j, a_i) \leftrightarrow \delta_n - d(a_j, a_i) \leq 0$. Consequently, $\mu'_n(a_j)(a_j) = 1$. Similarly, if $i \neq j$, we have that $N'_{n, a_j}(a_i) = \delta_n \dot{-} d(a_j, a_i) = 0$ i.e., $\mu'_n(a_j)(a_i) = 0$.

A "geometric" interpretation of the probabilistic selection $(\mu_n)_{n \in \mathbb{N}}$ of Theorem 3 goes as follows. By its definition $N_{n, x}(a_j) \geq 0$, while $\mu_n(x)(a_j) = 0 \leftrightarrow N_{n, x}(a_j) = 0 \leftrightarrow d(x, a_j) \geq d(x, A_n) + 2^{-n}$. If $x \notin A_n$ that can happen if a_j is sufficiently far from the point of A_n at which x attains its

---

[9] If $c \vee 0 > 0$, then $c > 0 \vee 0 > 0$ (see [4], p. 57). Hence, $c > 0$ in the case, and then we get immediately that $c \vee 0 = c$.

minimum distance from A_n, or if $x \in A_n$ and $d(x, a_j) \geq 2^{-n}$. Moreover, $\mu_n(x)(a_j) > 0 \leftrightarrow N_{n,x}(a_j) > 0 \leftrightarrow d(x, a_j) < d(x, A_n) + 2^{-n}$ i.e., either x attains its minimum distance from A_n at a_j or, otherwise, the distance $d(x, a_j)$ is less than 2^{-n}-close to the minimum distance $d(x, A_n)$ i.e., a_j is very close to the point of A_n at which x attains its minimum distance from A_n. A similar interpretation can be given for Normann's probabilistic selection $(\mu'_n)_{n \in \mathbb{N}}$. Note that a simpler definition, like

$$\nu_n(x)(a_j) = \frac{d(x, a_j)}{\sum_{i=1}^{n} d(x, a_i)}$$

gives rise to a probability distribution on A_n, which trivially satisfies the continuity condition, but it is not positive, and the hypothesis $\nu_n(x_n)(a_n) > 0$ is equivalent to $x_n \neq a_n$, which is far from satisfying condition (P_2) of a probabilistic selection.

Note that the constructive proof of Theorem 3 works for dense subsets A of X with a decidable equality, like $\mathbb{Q}$ in $\mathbb{R}$. Next follows a density theorem for hierarchies of limit spaces over separable metric spaces, the countable dense subsets of which are appropriately enumerated, or have a decidable equality.

Theorem 4 (density theorem). *Let (X, d) be a separable metric space, and let $A = \{a_n \mid n \in \mathbb{N}\}$ be a dense subset of X, where $d(a_n, a_m) > 0$, if $n \neq m$. If $\iota = X \mid \rho \to \sigma$ is an inductively defined type system $\mathbb{T}$ over the base type X, then in the $\mathbb{T}$-typed hierarchy of limit spaces over X, defined by*

$$\mathcal{X}(\iota) := (X(\iota), \lim_{\iota}) := (X, \lim_{d}),$$

$$\mathcal{X}(\rho \to \sigma) := (X(\rho) \to X(\sigma), \lim_{\rho \to \sigma}),$$

the limit space $\mathcal{X}(\tau)$ admits general approximations $(\tau \mathrm{Appr}_n)_{n \in \mathbb{N}}$, for every type τ in $\mathbb{T}$. Moreover, there is a countable subset D_τ of $X(\tau)$, which is $\lim_\tau$-dense in X_τ, therefore dense in $(X(\tau), \mathcal{T}_{\lim_\tau})$, for every type τ in $\mathbb{T}$.

Proof. If $\tau = \iota$, then by Theorem 3(ii) $(X, \lim_d, (A_n)_{n \in \mathbb{N}}, (\mu'_n)_{n \in \mathbb{N}})$ is a limit space with a strictly positive, general lim-probabilistic selection on X. By Proposition 7 there exist approximation functions $(\mathrm{XAppr}_n)_{n \in \mathbb{N}}$ on X such that the structure $(X, \lim_d, (\mathrm{XAppr}_n)_{n \in \mathbb{N}})$ is a limit space with general approximations, and $\mathrm{XAppr}_n(X) = A_n$, for every $n \in \mathbb{N}$. We define $\iota \mathrm{Appr}_n := \mathrm{XAppr}_n$, for every $n \in \mathbb{N}$. By Theorem 1, if $f \in X(\rho) \to X(\sigma)$, and $n \in \mathbb{N}$, then the function $(\rho \to \sigma)\mathrm{Appr}_n$, defined by

$$[(\rho \to \sigma)\mathrm{Appr}_n](f)(x) = \sigma \mathrm{Appr}_n(f(\rho \mathrm{Appr}_n(x))),$$

for every $x \in X(\rho)$, is the n-th approximation function that the limit space $\mathcal{X}(\rho \to \sigma)$ admits. The existence of the countable subset D_τ of $X(\tau)$ that is $\lim_\tau$-dense in X_τ, therefore dense in $(X(\tau), \mathcal{T}_{\lim_\tau})$, for every type τ in $\mathbb{T}$, follows from Proposition 2.

Note that constructively we only have that $\mathcal{T}_d \subseteq \mathcal{T}_{\lim_d}$, where $\mathcal{T}_d$ is the topology on X induced by its metric. Thus, what we determine through Theorem 4 are countable $\lim_\tau$-dense subsets of each limit space X_τ. Of course, classically, these are exactly the subsets one needs to find. Clearly, a density theorem for a hierarchy of limit spaces over more than one separable metric spaces can be shown similarly.

5   Concluding Remarks

The proof of the main density theorem presented in this paper reveals, in our view, the merits of the generalization of Normann's notion of the nth approximation of a functional in the typed hierarchy over $\mathbb{N}$ through the notion of a limit space with general approximations. The quite effective character of its proof is also worth noticing. As Normann writes in [14], p. 305,

[We would like to claim that an internal approach to computability in analysis will result in easy-to-use, high level, programming languages for computing in analysis, but the development cannot support this claim yet. The possibility of finding support for such a claim, together with basic curiosity, is nevertheless the motivation behind trying to find out what internally based algorithms might look like.]

The application of limit spaces with approximations to the (classical) study of limit spaces over other base types looks also promising. Moreover, it is interesting to see if the general idea behind the theory of limit spaces with approximations can be extended to other notions of space. Namely, to find a cartesian closed category $\mathcal{A}$, such that if X is an object of $\mathcal{A}$, general approximation functions XAppr_n of type $X \to X$ can be defined[10], for every $n \in \mathbb{N}$, such that the objects of $\mathcal{A}$ with general approximations form a cartesian closed subcategory of $\mathcal{A}$.

A plan to provide a fully constructive proof of Theorem 4 is the following. We expect that abstracting from the constructive properties of $\lim_d$ we can define a notion of a constructive limit space (X, clim_X) that preserves the cartesian closure property of limit spaces (with the same definition of the limit relation on the function space). In this case the proof of Theorem 4 goes through completely constructively, since the proof of Theorem 1 does not depend on Urysohn's axiom. We hope to realize this plan in future work.

Acknowledgments. We would like to thank Ulrich Berger for his insightful comments on an early draft of this paper and Hannes Diener for informing us on his result that relates Urysohn's axiom to LPO. We also thank the reviewers for their useful comments and suggestions and the Excellence Initiative of the LMU Munich for supporting our research.

---

[10] Where the notion of approximation, as it is expressed in condition (Λ_3) of Definition 2, will depend on the structure of X.

References

1. Bishop, E.: Foundations of Constructive Analysis. McGraw-Hill, New York (1967)
2. Bishop, E., Bridges, D.: Constructive Analysis. Springer, Heidelberg (1985)
3. Bridges, D., Richman, F.: Varieties of Constructive Mathematics. University Press, Cambridge (1987)
4. Bridges, D.S., Vîţă, L.S.: Techniques of Constructive Analysis. Universitext. Springer, New York (2006)
5. DeJaeger, F.: Calculabilité sur les réels. Thesis, Paris VII (2003)
6. Dugundji, J.: Topology. Universal Book Stall, New Delhi (1990)
7. Hyland, M.: Recursion theory on the countable functionals. Dissertation, Oxford (1975)
8. Kuratowski, K.: Topology, vol. I. Academic Press, New York (1966)
9. Kuratowski, K.: Topology, vol. II. Academic Press, New York (1968)
10. Longley, J., Normann, D.: Higher-Order Computability. Springer, Heidelberg (2015)
11. Normann, D.: External and internal algorithms on the continuous functionals. In: Metakides, G. (ed.) Patras Logic Symposion, pp. 137–144. North-Holland Publishing Company, Amsterdam (1982)
12. Normann, D.: Internal density theorems for hierarchies of continuous functionals. In: Beckmann, A., Dimitracopoulos, C., Löwe, B. (eds.) CiE 2008. LNCS, vol. 5028, pp. 467–475. Springer, Heidelberg (2008). doi:10.1007/978-3-540-69407-6_50
13. Normann, D.: A rich hierarchy of functionals of finite types. Logical Methods Comput. Sci. 5(3:11), 1–21 (2009)
14. Normann, D.: Experiments on an internal approach to typed algorithms in analysis. In: Cooper, S.B., Sorbi, A. (eds.) Computability in context: computation and logic in the real world, pp. 297–327. Imperial College Press, London (2011)
15. Normann, D.: Banach spaces as data types. Logical Methods Comput. Sci. 7(2:11), 1–20 (2011)
16. Normann, D.: The continuous functionals as limit spaces. In: Berger, U., Diener, H., Schuster, P., Seisenberger, M. (eds.) Logic, Construction, Computation Mathematical Logic, vol. 3, pp. 353–379. Ontos, Heusenstamm (2012)
17. Petrakis, I.: Limit spaces in computability at higher types, manuscript (2013)
18. Petrakis, I.: Limit spaces with approximations. Ann. Pure Appl. Logic 167(9), 737–752 (2016)
19. Scarpellini, B.: A model for barrecursion of higher types. Compositio Mathematica 23(1), 123–153 (1971)

On the Conjecture of the Smallest 3-Cop-Win Planar Graph

Photchchara Pisantechakool[1]($\boxtimes$) and Xuehou Tan[2]

[1] School of Science and Technology, Tokai University,
4-1-1 Kitakaname, Hiratsuka 259-1292, Japan
3btad008@mail.tokai-u.jp
[2] School of Information Science and Technology,
Tokai University, 4-1-1 Kitakaname, Hiratsuka 259-1292, Japan

Abstract. In the game of Cops and Robbers on a graph $G = (V, E)$, k cops try to catch a robber. The minimum number of cops required to win is called the cop number, denoted by $c(G)$. For a planar graph G, it is known that $c(G) \leq 3$. It is a conjecture that the regular dodecahedral graph of order 20 is the smallest planar graph whose cop number is three. As the very first attack on this conjecture, we provide the following evidences in this paper: (1) any planar graph of order at most 19 has the winning vertex at which two cops can capture the robber, and (2) a special planar graph of order 19 that is constructed from the regular dodecahedral graph has the cop number of two.

Keywords: Cops and Robbers game · Cop number · Capture strategy · Winning vertices

1 Introduction

In the pursuit-evasion games, the "evader" controlled by one player tries to avoid being captured by "pursuers" controlled by another player. The games have many versions, varied through many means such as whether the domain is discrete [7] or continuous [4], what information each player has [2], or how each player moves. The pursuers win the game if they can capture the evader, and the evader wins if he can avoid being captured indefinitely.

In a variant called Cops and Robbers game ([7,8,10]), both players have full knowledge of the terrain and other player's locations. Previous researches focused on how many cops it takes to successfully capture a robber on a given graph, or shortly, the *cop number* [1]. The minimum number of cops required to win on a graph G is denoted by $c(G)$. A graph G is said to be k-*cop-win* if and only if $c(G) = k$. When $k = 1$, it is simply called cop-win. Aigner and Fromme [1] proved that for any planar graph, $c(G) \leq 3$. A strategy using three cops for planar graphs was given in [9]. Although whether a planar graph is cop-win can be determined in polynomial time [5], there is still no known method of determining whether a planar graph is 2-cop-win or 3-cop-win.

© Springer International Publishing AG 2017
T.V. Gopal et al. (Eds.): TAMC 2017, LNCS 10185, pp. 499–514, 2017.
DOI: 10.1007/978-3-319-55911-7_36

From previous research, the smallest 3-cop-win graph is the Petersen graph, and thus any connected graph of order at most nine is 2-cop-win [3]. However, for planar graphs, the smallest 3-cop-win graph has not yet been studied.

It is a conjecture that the regular dodecahedral graph of order 20 is the smallest *planar* graph, whose cop number is three. In an attempt to prove this conjecture, we provide the following evidences: (1) any planar graph of order at most 19 has the winning vertices at which two cops can capture the robber, and (2) a special planar graph of order 19 that is constructed from the regular dodecahedral graph is 2-cop-win.

2   Basic Definitions

This paper refers to a standard graph theory notation written in [6]. For convenience, we introduce only new denotations and those which are heavily used in this paper. For a planar graph $G = (V(G), E(G))$, $F(G)$ is defined as a set of faces, which are also called the cycles of G. The number of faces is denoted by $|F(G)|$. For a face $f \in F(G)$, the number of its sides is denoted as $S(f)$, and also called the length of cycle f. The graph's *girth*, denoted by $g(G)$, is the size of its smallest cycle.

Let $N(v)$ denote the set of all neighbors of vertex v, and $\bar{N}(v) = N(v) + \{v\}$ the closed neighbors set of v. For any vertex v, if there exists a vertex $u \in N(v)$ such that $\bar{N}(v) \subseteq \bar{N}(u)$, then v is called a *dominated* vertex. For a vertex v, the *degree* of v, denoted by $d(v)$, is the number of the neighbors of v. The minimum degree of a graph G, denoted by $\delta(G)$, is the smallest degree of all vertices v in G.

In a graph G, when *contraction* of an edge $e \in E(G)$ with endpoints u, v is performed on G, the edge e is replaced by a single vertex such that the edges incident to the new vertex are those, other than e, which were incident to u or v (see Fig. 6). The contraction of e on G results in a graph with one edge and one vertex fewer than G. This graph operation is called an *edge contraction*. We called the resulting graph the *edge-contracted* version of the original graph, e.g., the edge-contracted regular dodecahedral graph is the result of performing an edge contraction on the regular dodecahedral graph.

In this paper, the game of Cops and Robbers is played by two players on a simple, undirected, connected planar graph G. The minimum number of cops required to win on a graph G is denoted by $c(G)$. A graph G is said to be *k-cop-win* if and only if $c(G) = k$. Denote by r the robber as well as the vertex he occupies, $C = \{c_1, c_2, \ldots, c_k\}$ the set of cops as well as the vertices they occupy. The set of vertices adjacent to or occupied by the cops is denoted by $\bar{N}(C) = \bar{N}(c_1) \cup \bar{N}(c_2) \cup \cdots \cup \bar{N}(c_k)$.

Let us consider the robber's final location on G before being captured by two cops. If the robber does not surrender, then the cops must *trap* him by restricting his movement just before he is captured [3]. The trapping condition using two cops against the robber r requires that $\bar{N}(r) \subseteq \bar{N}(C)$.

Definition 1. *For an vertex v, the vertices a and b are called the winning positions against v if $d(b, v) = 1$, $d(a, v) \leq 2$ and $\bar{N}(v) \subseteq \bar{N}(a) \cup \bar{N}(b)$. If the winning positions a and b against v exist, then v is called a winning vertex.*

3 Previous Results

We first review some known results on the cop numbers. Nowakowski and Winkler [7], and independently, Quilliot [10] provided the method of determining whether a graph is cop-win.

Theorem 1 ([7], Theorem 1, and [10]). *Graph G is a cop-win if and only if by successively removing dominated vertices, G can be reduced to a single vertex.*

Aigner and Fromme have shown that any planar graph requires no more than three cops, and the regular dodecahedral graph is 3-cop-win.

Theorem 2 ([1], Theorem 3). *If a graph G has $g(G) > 4$ then $c(G) \geq \delta(G)$.*

Theorem 3 ([1], Theorem 6). *For any planar graph G, the cop number of G is at most three.*

Baird et al. [3] use the results provided by Aigner and Fromme to prove that the Petersen graph is the smallest 3-cop-win. Since the Petersen graph is not planar, we conjecture that the regular dodecahedral graph is the smallest 3-cop-win planar graph. The regular dodecahedral graph is a planar graph of order 20, whose all vertices are of degree three and all faces are 5-cycles, as shown in Fig. 1. (It is easy to see that there are no wining vertices for two cops on the regular dodecahedral graph.) By Theorem 2, the following observation can be made.

Observation 1. *The regular dodecahedral graph is 3-cop-win.*

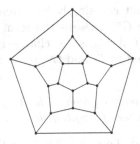

Fig. 1. The regular dodecahedral graph.

4   The Existence of Winning Vertices in Planar Graphs of Order at Most 19

A winning strategy using two cops requires two important parts: (1) a goal, and (2) a method to move each piece to its individual goal. In other words, two cops must force the robber to be at a winning vertex, and at the same time, two cops occupy the winning positions against that vertex. In this section, we prove that the winning vertices exist for any planar graph of order at most 19.

Baird et al. proved that the Petersen graph is the smallest 3-cop-win graph [3]. Since the Petersen graph is *not* planar, the next observation can also be made.

Observation 2. *For any planar graph of order 10 or smaller, the cop number is at most two.*

With Observation 2, we can focus on proving that a winning vertex exists for any planar graph of order between 11 and 19. Our proof consists of a series of lemmas, mainly on the relationships between $\delta(G)$ and the number of cycles of length at least five. For a graph G, denote by $S(F(G))$ the summation of sides of all faces in $F(G)$.

Lemma 1. *Suppose G is a planar graph with $11 \leq |V(G)| \leq 19$. If $\delta(G) = 3$, then $g(G) \leq 4$.*

Proof. It follows from Euler's Formula that $|V(G)| - |E(G)| + |F(G)| = 2$. Since $\delta(G) = 3$, we have this condition: (1) $|E(G)| \geq \frac{3|V(G)|}{2}$. Since an edge is shared by two faces, we have $S(F(G)) = 2|E(G)|$. Suppose (by contradiction) that $g(G) \geq 5$. Then, we have the second condition: (2) $S(F((G)))| \geq 5|F(G)|$. Let G^* be a planar graph with the minimum number of edges satisfying condition (1) for a given order n ($11 \leq n \leq 19$). To be precise, $G^* = (V(G^*), E(G^*))$ where $|V(G^*)| = n$ and $|E(G^*)| = \lceil \frac{3n}{2} \rceil$. We consider all of planar graphs G^* of order n and get the following results.

From Table 1, we know that for a given order n ($11 \leq n \leq 19$), the graph G^* has $g(G^*) \leq 4$. This result can simply be extended to any graph of order n, because adding i edges to G^* increases the term $|E(G)|$ by i, and the term $|F(G)|$ by i (as the term $|V(G)| = n$ is *not* changed). In other words, the term $5|F(G)|$ in Table 1 is increased by $5i$ and the term $S(F(G))$ is increased by $2i$. Hence, the lemma follows.   □

Lemma 2. *Suppose G is a planar graph with $11 \leq |V(G)| \leq 19$. If $\delta(G) = 4$, then there are at most six cycles of length at least five in G.*

Proof. Similar to the proof of Lemma 1, for G with $\delta(G) \doteq 4$, we have this condition: (1) $|E(G)| \geq 2|V(G)|$. Suppose there are seven cycles of length at least five in G. Even in the case where all the remaining faces are 3-cycles, we have the second condition: (2) $S(F((G)))| \geq 35 + 3(|F(G)| - 7) = 3|F(G)| + 14$. Again, let G^* be a planar graph with the minimum number of edges satisfying condition (1) for a given order n ($11 \leq n \leq 19$). That is, $|V(G^*)| = n$ and $|E(G^*)| = 2n$. We have the following results.

Table 1. Planar graphs G^* of order from 11 to 19 with $\delta(G^*) = 3$ do not have $g(G^*) \geq 5$.

n	$|E(G^*)|$	$|F(G^*)|$	$S(F(G^*))$	$5|F(G^*)|$	Is $S(F(G^*)) \geq 5|F(G^*)|$ (2)
11	17	8	34	40	False
12	18	8	36	40	False
13	20	9	40	45	False
14	21	9	42	45	False
15	23	10	46	50	False
16	24	10	48	50	False
17	26	11	52	55	False
18	27	11	54	55	False
19	29	12	58	60	False

Table 2. Planar graphs G of order 11 to 19 with $\delta(G) = 4$ do not have seven cycles of length at least five.

n	$|E(G^*)|$	$|F(G^*)|$	$S(F(G^*))$	$3|F(G^*)| + 14$	(2)
11	22	13	44	53	False
12	24	14	48	56	False
13	26	15	52	59	False
14	28	16	56	62	False
15	30	17	60	65	False
16	32	18	64	68	False
17	34	19	68	71	False
18	36	20	72	74	False
19	38	21	76	77	False

From Table 2, we know that for a given order n ($11 \leq n \leq 19$), the graph G^* has at most six cycles of length at least five. Similar to the proof of Lemma 1, it is sufficient to consider condition (2) on G^* (whose number of edges is the smallest to satisfy condition (1)), as for every i edges added to G^*, the term $S(F(G))$ is increased by $2i$ and the term $3|F(G)| + 14$ is increased by at least $3i$ (as $|V(G)| = n$ is *not* changed). So, the lemma follows.   □

Lemma 3. *Suppose G is a planar graph with $11 \leq |V(G)| \leq 19$. If $\delta(G) = 5$, then there are at most one cycle of length at least five in G.*[1]

Proof. Similar to the proof of Lemma 1, for G with $\delta(G) = 5$, we have this condition: (1) $|E(G)| \geq \frac{5|V(G)|}{2}$. Suppose there are two cycles of length at least five in G. Even in the case where all the remaining faces are 3-cycles, we have the

---

[1] Probably, the cycle of length at least five does not exist at all when $\delta(G) = 5$.

second condition: (2) $S(F((G)))| \geq 10+3(|F(G)|-2) = 3|F(G)|+4$. Again, let G^* be a planar graph with the minimum number of edges satisfying condition (1) for a given order n $(11 \leq n \leq 19)$. That is, $|V(G^*)| = n$ and $|E(G^*)| = \lceil \frac{5n}{2} \rceil$. Also, we have the following results.

Table 3. Planar graphs G of order 11 to 19 with $\delta(G) = 5$ do not have two cycles of length at least five.

n	$|E(G^*)|$	$|F(G^*)|$	$S(F(G^*))$	$3|F(G^*)|+4$	(2)
11	28	19	56	61	False
12	30	20	60	64	False
13	33	22	66	70	False
14	35	23	70	73	False
15	38	25	76	79	False
16	40	26	80	82	False
17	43	28	86	88	False
18	45	29	90	91	False
19	48	31	96	97	False

From Table 3, we know that for a given order n $(11 \leq n \leq 19)$, the graph G^* has at most one cycle of length at least five. Similar to the proof of Lemma 1, it is sufficient to consider condition (2) on G^* (whose number of edges is the smallest to satisfy condition (1)). Hence, the lemma follows.   □

Lemma 4. *Suppose that G is a planar graph of order between 11 and 19, with $\delta(G) = 4$. Then, there exists a vertex v of degree four such that v is the common vertex of either two 3-cycles or a 3-cycle and a 4-cycle.*

Proof. We claim that there are no graphs G of order at most 19, with $\delta(G) = 4$, such that every vertex of degree four belongs to four cycles of length at least four. We prove it by first constructing a graph of the smallest order such that there is only one vertex of degree four that belongs to four 4-cycles in the graph and all other vertices are of degree at least 5. Let us start a graph with vertex u of degree four at the center of a 3×3 grid, see Fig. 2(a). We then construct a graph such that all vertices, except for u, are of degree at least five, by adding the minimum number of vertices required to make the existing vertices be of degree five. This construction is shown step-by-step in Fig. 2, until our desired graph, shown in Fig. 2(d), is obtained. But, its order is 25 (this number cannot be further decreased), contradicting $|V(G)| \leq 19$. Furthermore, when the number of vertices of degree four is increased, the order of the resulting graph is increased. Hence, our claim holds.

By Lemma 2, there are at most six cycles of length at least five in G, and the remaining faces of G are 3- or 4-cycles. Since $\delta(G) = 4$, there are at most three

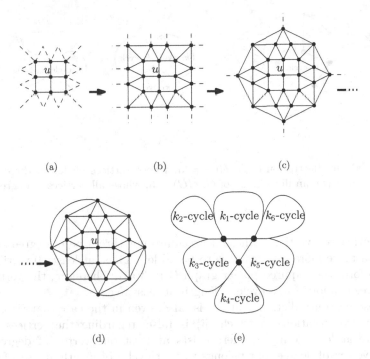

Fig. 2. From (a) to (d), an illustration of the proof of Lemma 4. In (e), an illustration of G, $\delta(G) = 4$, with six k_i-cycles, $k_i \geq 5$ for all $1 \leq i \leq 6$.

vertices of degree four, which are not the common vertices of either two 3-cycles or a 3-cycle and a 4-cycle, see Fig. 2(e). The lemma then follows from our claim and the inequalities $11 \leq |V(G)| \leq 19$. □

Lemma 5. *Suppose that G is a planar graph of order between 11 and 19. If $\delta(G) = 3$, then (i) there exists a vertex v of degree three such that v belongs to a 3- or 4-cycle, or (ii) there exists a vertex v of degree four such that v belongs to either two 3-cycles or a 3-cycle and a 4-cycle.*

Proof. This lemma has two statements: (i) there exists a vertex v of degree three such that v belongs to a 3- or 4-cycle, or (ii) there exists a vertex v of degree four such that v belongs to either two 3-cycles or a 3-cycle and a 4-cycle. For any G of order between 11 and 19 with $\delta(G) = 3$, at least one 3- or 4-cycle exists (Lemma 1). In the case that (i) is true, the lemma follows.

In the following, we prove that if (i) is false, then (ii) is true. First, we show that the vertices of degree three in G cannot form a cycle of length at least five. Note that the vertices of degree three do not belong to any 3- or 4-cycle (otherwise, (i) is true), and all other vertices are of degree at least four. Assume by contradiction that there are at least five vertices of degree three, which form a cycle of length, say, five. As can be seen in Fig. 3(a), six k_i-cycles ($k_i > 5$ for all $1 \leq i \leq 6$) are required to form this 5-cycle, and they use at least 15 vertices.

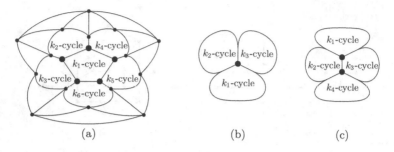

Fig. 3. In (a), an illustration of G, $\delta(G) = 3$, whose vertices of degree three form a cycle. In (b) and (c), an illustration of G, $\delta(G) = 3$, whose all vertices of degree three form a tree.

To make other ten vertices (whose degrees are not three) be of degree at least four, at least five more vertices have to be added. Thus, at least 20 vertices are needed to form the required planar graph (Fig. 3(a)). Therefore, the vertices of degree three cannot form a cycle of length at least five.

Suppose by contradiction that (ii) is false, even in the case that (i) is false. There are two situations in which (ii) is false, regarding the vertices whose degrees are at least four: (a) there exists at least one vertex of degree four and each vertex of degree four belongs to four cycles of length at least four, or (b) all the vertices, except for those of degree three, are of degree at least five.

We have shown in Lemma 4 that there are no graph G of order at most 19, with $\delta(G) = 4$, such that every vertex of degree four belongs to four cycles of length at least four. Since every vertex of degree three cannot belong to any 3- or 4-cycle (otherwise (a) is true), the argument given in the proof of Lemma 4 can also be made here. So, if there are the vertices of degree four, at least one of them belongs to either two 3-cycles or a 3-cycle and a 4-cycle. Hence, situation (a) never occurs.

Consider now the situation (b). From the discussion made above, the vertices of degree three cannot form a cycle. So, if the number of vertices of degree three is i, then there exist at least $i + 2$ cycles of length at least five (as can be observed from Fig. 3(b) and (c)). Since in G, all other vertices, except for those of degree three, are of degree at least five, we have this condition: (1) $\sum d(v) \geq (3 \times i) + 5(|V(G)| - i)$. Since $\sum d(v) = 2|E(G)|$, we can modify condition (1) to $|E(G)| \geq \frac{5|V(G)| - 2i}{2}$. By considering all other faces as 3-cycles, we have the second condition: (2) $S(F(G)) \geq 5(i + 2) + 3(|F(G)| - (i + 2)) = 3|F(G)| + 2i + 4$. Again, let G^* be a planar graph with the minimum number of edges satisfying condition (1) for a given order n ($11 \leq n \leq 19$). That is, $|V(G^*)| = n$ and $|E(G^*)| = \lceil \frac{5n - 2i}{2} \rceil$. We have the following results, for $i = 1$.

From Table 4, a planar graph G with $\delta(G) = 3$ and one vertex of degree three, which is common to three cycles of length at least five, cannot have all the remaining vertices of degree at least five. As discussed above, it is sufficient to consider condition (2) on G^* (whose number of edges is the smallest to satisfy condition (1)). Also, it is sufficient to consider the case of $i = 1$, since

Table 4. Planar graphs G of order 11 to 19 with $\delta(G) = 3$ and one vertex of degree three, which is common to three cycles of length at least five, cannot have all the remaining vertices of degree at least five.

n	$5n - 2i$	$|E(G^*)|$	$|F(G^*)|$	$S(F(G^*))$	$3|F(G^*| + 6$	(2)
11	53	27	18	54	60	False
12	58	29	19	58	63	False
13	63	32	21	64	69	False
14	68	34	22	68	72	False
15	73	37	24	74	78	False
16	78	39	25	78	81	False
17	83	42	27	84	87	False
18	88	44	28	88	90	False
19	93	47	30	94	96	False

when the term i is increased by j (without changing $|V(G^*)| = n$), the term $\frac{5n-2i}{2}$ is decreased by j. Thus, the term $|E(G^*)|$ is decreased by j (follows from $|E(G^*)| = \lceil \frac{5n-2i}{2} \rceil$) and so is the term $|F(G^*)|$ (Euler's Formula). That is, the term $S(F(G^*))$ is decreased by $2j$ and the term $3|F(G^*)| + 2i + 4$ is decreased by j. Hence, situation (b) never occurs either. The proof is complete. □

Theorem 4. *There exist winning vertices for any planar graph of order at most 19.*

Proof. Recall first that for any planar graph, $\delta(G) \le 5$ [6], and if G contains a dominated vertex v, then $c(G - \{v\}) = c(G)$ [3]. Hence, we only need to consider the four different situations in which $\delta(G) = 2, 3, 4$, or 5.

For the situation where $\delta(G) = 2$, any vertex v of degree 2 is a winning vertex, because two cops at two neighbors of v can trap the robber at v.

For the situation where $\delta(G) = 3$, by Lemma 5, there exist either vertices v of degree 3 belonging to a 3- or 4-cycle, or vertices v of degree 4, which are common to either two 3-cycles or a 3-cycle and a 4-cycle. In Fig. 4(a) and (b) (resp. Fig. 4(c) and (d)), we show the positions in which two cops can occupy to trap the robber at v of degree 3 (resp. degree 4).

For the situation where $\delta(G) = 4$, by Lemma 4, there exists a vertex v of degree 4, which is common to either two 3-cycles or a 3-cycle and a 4-cycle. From the discussion made for the case of $\delta(G) = 3$, the vertex v is the winning one.

Finally, consider the situation in which $\delta(G) = 5$. It follows from Lemma 3 that there exists a vertex v of degree 5 that does not belong to the k-cycle ($k \ge 5$), and is thus common to five 3- and/or 4-cycles. We show in Fig. 4(e)–(j) all the winning positions against v. Again, the winning vertex exists. The proof is complete. □

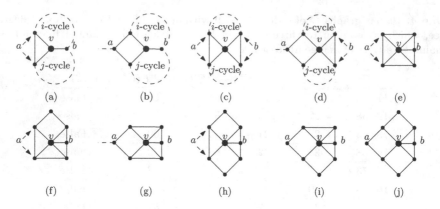

Fig. 4. Some instances of the winning vertex v in which $\bar{N}(v) \subseteq \bar{N}(a) \cup \bar{N}(b)$, and the cops' winning positions a and b.

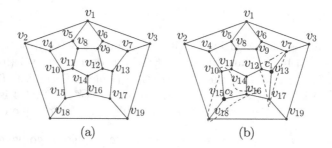

Fig. 5. The labeling of vertices in the edge-contracted dodecahedral graph (a), and the initial positions of two cops c_1 and c_2 in our strategy (b).

5   Two-Cops Strategy on the Edge-Contracted Regular Dodecahedral Graph

In this section, we show that the edge-contracted regular dodecahedral graph, which might be considered as the worst case of the planar graphs of order at most 19, has the cop number of two. Studying on the edge-contracted regular dodecahedral graph may give some insights on the proof that the regular dodecahedral graph is the smallest 3-cop-win planar graph and even on giving a method for determining whether a graph is 2-cop-win or 3-cop-win.

Note that since the regular dodecahedral graph is vertex-symmetry, performing an edge contraction on any edge results in the same graph with difference embeddings. As can be seen that a graph in Fig. 6(b) can be relabeled into a graph in Fig. 5(a). For simplicity, we will use the embedding shown in Fig. 5(a) to represent the edge-contracted regular dodecahedral graph.

Let the triple (a, b, v) denote the two winning positions a and b against v. For the instance shown in Fig. 5(b), we have the following triples: (v_5, v_{18}, v_2), (v_6, v_{19}, v_3), (v_1, v_{10}, v_4), (v_2, v_8, v_5), (v_3, v_9, v_6), and (v_1, v_{13}, v_7). The main part

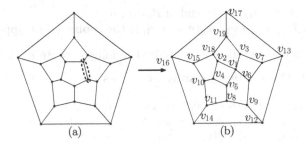

Fig. 6. Any graph resulted from performing an edge contraction on the regular dodec-ahedral graph is the same graph of different embeddings.

of a winning strategy is how to enforce the robber into such a vertex v, and at the same time two cops occupy the winning positions against v before the robber escapes. Since the edge-contracted regular dodecahedral graph is specific, we provide below a full strategy for it.

Initial placement of two cops and the robber: In our strategy, two cops c_1 and c_2 occupy v_{13} and v_{15} in the labeling shown in Fig. 5(b). If the robber initially occupies a vertex in $\bar{N}(C) = \bar{N}(v_{13}) \cup \bar{N}(v_{15})$, he will be captured by the cops in the very first round. So, the robber can initially occupy one of the following eleven vertices: $v_1, v_2, v_3, v_4, v_5, v_6, v_8, v_9, v_{11}, v_{14}$, and v_{19}.

In the strategy below, we provide the movements of the cops, as well as the choices of location the robber can move to without being caught. We use $c_1(v \to u)$ to denote the movement of c_1 from v to u, and the robber's choices of movements after the cops' turn (in the same round) are written after a semicolon. Note that our goal is to show the existence of a capture strategy, but not the attempt for giving an optimal strategy. We separate the vertices into three group; (1) the vertices whose distances to both v_{13} and v_{15} are two, (2) the vertices whose distances to v_{13} and v_{15} are two or three, and (3) the vertices whose distances to both v_{13} and v_{15} are at least three. We will give the strategy till the robber is trapped, since the robber can then be captured with one more round.

Let us first describe the scenarios in which the robber initially occupies a vertex in group (1). The vertices in group (1) are v_{14} and v_{19}.

Scenario R19: the robber initially occupies v_{19}.

> **Round 1** $c_1(v_{13} \to v_{13})$, $c_2(v_{15} \to v_{18})$; $r(v_{19} \to v_3)$.
> **Round 2** $c_1(v_{13} \to v_7)$, $c_2(v_{18} \to v_{18})$; $r(v_3 \to v_1)$.
> **Round 3** $c_1(v_7 \to v_7)$, $c_2(v_{18} \to v_2)$; $r(v_1 \to v_5)$.
> **Round 4** $c_1(v_7 \to v_6)$, $c_2(v_2 \to v_4)$; $r(v_5 \to v_8)$.
> **Round 5** $c_1(v_6 \to v_9)$, $c_2(v_4 \to v_{10})$; $r(v_8 \to v_5)$.
> **Round 6** $c_1(v_9 \to v_6)$, $c_2(v_{10} \to v_{11})$; (a) $r(v_5 \to v_5)$ or (b) $r(v_5 \to v_4)$.
> **Case (a):** $r(v_5 \to v_5)$ at the end of Round 6.
> > **Round 7** $c_1(v_6 \to v_1)$, $c_2(v_{11} \to v_{11})$; $r(v_5 \to v_4)$.
> > **Round 8** $c_1(v_1 \to v_1)$, $c_2(v_{11} \to v_{10})$; the robber r is trapped. (**Round 9** one cop moves to r.)

Case (b): $r(v_5 \rightarrow v_4)$ at the end of Round 6.
    Round 7 $c_1(v_6 \rightarrow v_1)$, $c_2(v_{11} \rightarrow v_{10})$; the robber r is trapped.

Scenario R19 ends within nine rounds (the longest route is Case (a)).

Scenario R14: the robber initially occupies v_{14}.

    Round 1 $c_1(v_{13} \rightarrow v_{12})$, $c_2(v_{15} \rightarrow v_{15})$; $r(v_{14} \rightarrow v_{11})$.
    Round 2 $c_1(v_{12} \rightarrow v_{12})$, $c_2(v_{15} \rightarrow v_{10})$; $r(v_{11} \rightarrow v_8)$.
    Round 3 $c_1(v_{12} \rightarrow v_9)$, $c_2(v_{10} \rightarrow v_{10})$; $r(v_8 \rightarrow v_5)$.
    Round 4 $c_1(v_9 \rightarrow v_6)$, $c_2(v_{10} \rightarrow v_{11})$; the same as the robber's turn at Round 6 in Scenario R19.

Scenario R14 ends within seven rounds (it uses the final three rounds from Scenario R19).

Next, we describe the scenarios in which the robber initially occupies a vertex in group (2). The vertices in group (2) are v_2, v_3, v_4, v_6, v_9, and v_{11}.

Scenario R2: the robber initially occupies v_2.

    Round 1 $c_1(v_{13} \rightarrow v_7)$, $c_2(v_{15} \rightarrow v_{18})$; (a) $r(v_2 \rightarrow v_1)$, or (b) $r(v_2 \rightarrow v_4)$.
    Case (a): $r(v_2 \rightarrow v_1)$ at the end of Round 1.
        Round 2 the same as Round 3 in Scenario R19.
    Case (b): $r(v_2 \rightarrow v_4)$ at the end of Round 1.
        Round 2 $c_1(v_7 \rightarrow v_6)$, $c_2(v_{18} \rightarrow v_{15})$; (b.1) $r(v_4 \rightarrow v_2)$, (b.2) $r(v_4 \rightarrow v_4)$, or (b.3) $r(v_4 \rightarrow v_5)$
        Case (b.1): $r(v_4 \rightarrow v_2)$ at the end of Round 2 in Case (b).
            Round 3 $c_1(v_6 \rightarrow v_1)$, $c_2(v_{15} \rightarrow v_{15})$; $r(v_2 \rightarrow v_4)$.
            Round 4 $c_1(v_1 \rightarrow v_1)$, $c_2(v_{15} \rightarrow v_{10})$; the robber is trapped.
        Case (b.2): $r(v_4 \rightarrow v_4)$ at the end of Round 2 in Case (b).
            Round 3 $c_1(v_6 \rightarrow v_1)$, $c_2(v_{15} \rightarrow v_{10})$; the robber is trapped.
        Case (b.3): $r(v_4 \rightarrow v_5)$ at the end of Round 2 in Case (b).
            Round 3 $c_1(v_6 \rightarrow v_1)$, $c_2(v_{15} \rightarrow v_{10})$; $r(v_5 \rightarrow v_8)$.
            Round 4 $c_1(v_1 \rightarrow v_6)$, $c_2(v_{10} \rightarrow v_{11})$; $r(v_8 \rightarrow v_5)$.
            Round 5 the same as Round 7 in Case (a) of Scenario R19.

In Scenario R2, the robber is captured within seven rounds (the longest route is Case (b.3), which uses final three rounds from Scenario R19).

Scenario R3: The robber initially occupies v_3.

    Round 1 $c_1(v_{13} \rightarrow v_7)$, $c_2(v_{15} \rightarrow v_{18})$; $r(v_3 \rightarrow v_1)$.
    Round 2 the same as Round 3 in Scenario R19.

Scenario R3 ends within eight rounds (one fewer round than Scenario R19).

Scenario R4: The robber initially occupies v_4.

    Round 1 $c_1(v_{13} \rightarrow v_7)$, $c_2(v_{15} \rightarrow v_{15})$; (a) $r(v_4 \rightarrow v_2)$, (b) $r(v_4 \rightarrow v_4)$, or (c) $r(v_4 \rightarrow v_5)$.
    Case (a): $r(v_4 \rightarrow v_2)$ at the end of Round 1.

Round 2 $c_1(v_7 \to v_3)$, $c_2(v_{15} \to v_{18})$; $r(v_2 \to v_4)$.
Round 3 $c_1(v_3 \to v_1)$, $c_2(v_{18} \to v_{15})$; the robber is trapped.
Case (b): $r(v_4 \to v_4)$ at the end of Round 1.
Round 2 $c_1(v_7 \to v_6)$, $c_2(v_{15} \to v_{15})$; the same as the robber's turn at Round 2 in Case (b) of Scenario R2.
Case (c): $r(v_4 \to v_5)$ at the end of Round 1.
Round 2 $c_1(v_7 \to v_6)$, $c_2(v_{15} \to v_{10})$; (c.1) $r(v_5 \to v_5)$, or (c.2) $r(v_5 \to v_8)$.
Case (c.1): $r(v_5 \to v_5)$ at the end of Round 2 in Case (c).
Round 3 $c_1(v_6 \to v_1)$, $c_2(v_{10} \to v_{11})$; $r(v_5 \to v_4)$.
Round 4 $c_1(v_1 \to v_1)$, $c_2(v_{11} \to v_{10})$; the robber is trapped.
Case (c.2): $r(v_5 \to v_8)$ at the end of Round 2 in Case (c).
Round 3 $c_1(v_6 \to v_6)$, $c_2(v_{10} \to v_{11})$; $r(v_8 \to v_5)$.
Round 4 the same as Case (a) at the end of Round 6 in Scenario R19.

Scenario R4 ends in at most seven rounds (the longest route is Case (b)).

Scenario R6: The robber initially occupies v_6.

Round 1 $c_1(v_{13} \to v_{13})$, $c_2(v_{15} \to v_{18})$; (a) $r(v_6 \to v_1)$, (b) $r(v_6 \to v_6)$ or (c) $r(v_6 \to v_9)$.
Case (a): $r(v_6 \to v_1)$ at the end of Round 1.
Round 2 $c_1(v_{13} \to v_7)$, $c_2(v_{18} \to v_2)$; $r(v_1 \to v_5)$.
Round 3 $c_1(v_7 \to v_6)$, $c_2(v_2 \to v_4)$; $r(v_5 \to v_8)$.
Round 4 $c_1(v_6 \to v_9)$, $c_2(v_4 \to v_{10})$; $r(v_8 \to v_5)$.
Round 5 the same as Round 4 in Scenario R14.
Case (b): $r(v_6 \to v_6)$ at the end of Round 1.
Round 2 $c_1(v_{13} \to v_{12})$, $c_2(v_{18} \to v_2)$; (b.1) $r(v_6 \to v_6)$, or (b.2) $r(v_6 \to v_7)$.
Case (b.1): $r(v_6 \to v_6)$ at the end of Round 2 in Case (b).
Round 3 $c_1(v_{12} \to v_{12})$, $c_2(v_2 \to v_1)$; $r(v_6 \to v_7)$.
Round 4 $c_1(v_{12} \to v_{13})$, $c_2(v_1 \to v_1)$; the robber is trapped.
Case (b.2): $r(v_6 \to v_7)$ at the end of Round 2 in Case (b).
Round 3 $c_1(v_{12} \to v_{13})$, $c_2(v_2 \to v_1)$; the robber is trapped.
Case (c): $r(v_6 \to v_9)$ at the end of Round 1.
Round 2 $c_1(v_{13} \to v_{12})$, $c_2(v_{18} \to v_2)$; (c.1) $r(v_9 \to v_6)$, or (c.2) $r(v_9 \to v_8)$.
Case (c.1): $r(v_9 \to v_6)$ at the end of Round 2 in Case (c).
Round 3 the same as Round 3 in Case (b.1) of Scenario R6.
Case (c.2): $r(v_9 \to v_8)$ at the end of Round 2 in Case (c).
Round 3 $c_1(v_{12} \to v_9)$, $c_2(v_2 \to v_4)$; $r(v_8 \to v_{11})$.
Round 4 $c_1(v_9 \to v_{12})$, $c_2(v_4 \to v_{10})$; $r(v_{11} \to v_8)$.
Round 5 the same as Round 3 in Scenario R14.

Scenario R6 ends within eight rounds (the longest route is Case (c.2)).

Scenario R9: The robber initially occupies v_9.

Round 1 $c_1(v_{13} \rightarrow v_{12})$, $c_2(v_{15} \rightarrow v_{18})$; (a) $r(v_9 \rightarrow v_6)$ or (b) $r(v_9 \rightarrow v_8)$.
Case (a): $r(v_9 \rightarrow v_6)$ at the end of Round 1.
   Round 2 $c_1(v_{12} \rightarrow v_{12})$, $c_2(v_{18} \rightarrow v_2)$; (a.1) $r(v_6 \rightarrow v_6)$, or (a.2) $r(v_6 \rightarrow v_7)$.
   Case (a.1): $r(v_6 \rightarrow v_6)$ at the end of Round 2 in Case (a).
      Round 3 $c_1(v_{12} \rightarrow v_{12})$, $c_2(v_2 \rightarrow v_1)$; $r(v_6 \rightarrow v_7)$.
      Round 4 $c_1(v_{12} \rightarrow v_{13})$, $c_2(v_1 \rightarrow v_1)$; the robber is trapped.
   Case (a.2): $r(v_6 \rightarrow v_7)$ at the end of Round 2 in Case (a).
      Round 3 $c_1(v_{12} \rightarrow v_{13})$, $c_2(v_2 \rightarrow v_1)$; the robber is trapped.
Case (b): $r(v_9 \rightarrow v_8)$ at the end of Round 1.
   Round 2 $c_1(v_{12} \rightarrow v_9)$, $c_2(v_{18} \rightarrow v_{15})$; (b.1) $r(v_8 \rightarrow v_5)$, or (b.2) $r(v_8 \rightarrow v_{11})$.
   Case (b.1): $r(v_8 \rightarrow v_5)$ at the end of Round 2 in Case (b).
      Round 3 $c_1(v_9 \rightarrow v_6)$, $c_2(v_{15} \rightarrow v_{10})$; the same as the robber's turn at Round 2 in Case (c) of Scenario R4.
   Case (b.2): $r(v_8 \rightarrow v_{11})$ at the end of Round 2 in Case (b).
      Round 3 $c_1(v_9 \rightarrow v_{12})$, $c_2(v_{15} \rightarrow v_{10})$; $r(v_{11} \rightarrow v_8)$.
      Round 4 the same as Round 3 in Scenario R14.

Scenario R9 ends within eight rounds (the longest route is Case (b.1)).

Scenario R11: The robber initially occupies v_{11}.

Round 1 $c_1(v_{13} \rightarrow v_{12})$, $c_2(v_{15} \rightarrow v_{10})$; $r(v_{11} \rightarrow v_8)$.
Round 2 the same as Round 3 in Scenario R14.

Scenario R11 ends within six rounds.
Lastly, we describe the scenarios in group (3), which consists of v_1, v_5, and v_8.

Scenario R1: The robber initially occupies v_1.

Round 1 $c_1(v_{13} \rightarrow v_7)$, $c_2(v_{15} \rightarrow v_{18})$; $r(v_1 \rightarrow v_5)$.
Round 2 $c_1(v_7 \rightarrow v_6)$, $c_2(v_{18} \rightarrow v_2)$; (a) $r(v_5 \rightarrow v_5)$, or (b) $r(v_5 \rightarrow v_8)$.
Case (a): $r(v_5 \rightarrow v_5)$ at the end of Round 2.
   Round 3 $c_1(v_6 \rightarrow v_9)$, $c_2(v_2 \rightarrow v_2)$; $r(v_5 \rightarrow v_5)$.
   Round 4 $c_1(v_9 \rightarrow v_8)$, $c_2(v_2 \rightarrow v_2)$; the robber is trapped.
Case (b): $r(v_5 \rightarrow v_8)$ at the end of Round 2.
   Round 3 $c_1(v_6 \rightarrow v_9)$, $c_2(v_2 \rightarrow v_4)$; $r(v_8 \rightarrow v_{11})$.
   Round 4 $c_1(v_9 \rightarrow v_{12})$, $c_2(v_4 \rightarrow v_{10})$; $r(v_{11} \rightarrow v_8)$.
   Round 5 the same as Round 4 in Case (b.2) of Scenario R9.

Scenario R1 ends within seven rounds (the longest route is Case (b)).

Scenario R5: The robber initially occupies v_5.

Round 1 $c_1(v_{13} \rightarrow v_{12})$, $c_2(v_{15} \rightarrow v_{10})$; (a) $r(v_5 \rightarrow v_1)$, (b) $r(v_5 \rightarrow v_5)$, or (c) $r(v_5 \rightarrow v_8)$.
Case (a): $r(v_5 \rightarrow v_1)$ at the end of Round 1.
   Round 2 $c_1(v_{12} \rightarrow v_9)$, $c_2(v_{10} \rightarrow v_4)$; (a.1) $r(v_1 \rightarrow v_1)$, or (a.2) $r(v_1 \rightarrow v_3)$.

Case (a.1): $r(v_1 \to v_1)$ at the end of Round 2 in Case (a).
    Round 3 $c_1(v_9 \to v_6)$, $c_2(v_4 \to v_4)$; $r(v_1 \to v_3)$.
    Round 4 $c_1(v_6 \to v_7)$, $c_2(v_4 \to v_2)$; $r(v_3 \to v_{19})$.
    Round 5 $c_1(v_7 \to v_{13})$, $c_2(v_2 \to v_{18})$; $r(v_{19} \to v_3)$.
    Round 6 the same as Round 2 in Scenario R19.
Case (a.2): $r(v_1 \to v_3)$ at the end of Round 2 in Case (a).
    Round 3 $c_1(v_9 \to v_6)$, $c_2(v_4 \to v_2)$; (a.2.1) $r(v_3 \to v_3)$, or (a.2.2) $r(v_3 \to v_{19})$.
    Case (a.2.1): $r(v_3 \to v_3)$ at the end of Round 3 in Case (a.2).
        Round 4 $c_1(v_6 \to v_6)$, $c_2(v_2 \to v_{18})$; $r(v_3 \to v_3)$.
        Round 5 $c_1(v_6 \to v_6)$, $c_2(v_{18} \to v_{19})$; the robber is trapped.
    Case (a.2.2): $r(v_3 \to v_{19})$ at the end of Round 3 in Case (a.2).
        Round 4 $c_1(v_6 \to v_7)$, $c_2(v_2 \to v_{18})$; $r(v_{19} \to v_{17})$.
        Round 5 $c_1(v_7 \to v_{13})$, $c_2(v_{18} \to v_{15})$; $r(v_{17} \to v_{19})$.
        Round 6 the same as Round 1 in Scenario R19.
Case (b): $r(v_5 \to v_5)$ at the end of Round 1.
    Round 2 $c_1(v_{12} \to v_9)$, $c_2(v_{10} \to v_4)$; $r(v_5 \to v_1)$.
    Round 3 the same as Round 3 in Case (a.1).
Case (c): $r(v_5 \to v_8)$ at the end of Round 1.
    Round 2 the same as Round 3 in Scenario R14.

Scenario R5 ends within fourteen rounds (the longest route is Case (a.1)).

Scenario R8: The robber initially occupies v_8.

    Round 1 $c_1(v_{13} \to v_{12})$, $c_2(v_{15} \to v_{10})$; (a) $r(v_8 \to v_5)$ or (b) $r(v_8 \to v_8)$.
Case (a): $r(v_8 \to v_5)$ at the end of Round 1.
    Round 2 the same as Round 2 in Case (b) of Scenario R5.
Case (b): $r(v_8 \to v_8)$ at the end of Round 1.
    Round 2 the same as Round 3 in Scenario R14.

Scenario R8 ends within seven rounds.

Theorem 5. *The edge-contracted dodecahedral graph of order 19 is 2-cop-win.*

Proof. The full capture strategy using two cops for the edge-contracted regular dodecahedral graph has been provided. As described above, the robber can be captured within fourteen rounds.   □

Finally, it is interesting to note that the distance between two cops is always kept to be at most three in our capture strategy.

6   Conclusions

We have proved in this paper that a winning vertex, at which two cops can trap and capture the robber, exists in any planar graph of order at most 19, and the planar graph resulted from performing a single edge contraction on the regular dodecahedral graph is 2-cop-win. Although the proof for the latter is very elementary, we hope a clever strategy can be developed in the future, not only for the edge-contracted regular dodecahedral graph, but also for any planar graph of order at most 19, so as to prove the conjecture that the regular dodecahedral graph is the smallest 3-cop-win planar graph.

References

1. Aigner, M., Fromme, M.: A game of cops and robbers. Discrete Appl. Math. **8**(1), 1–12 (1984)
2. Alspach, B.: Searching and sweeping graphs: a brief survey. Le Mathematiche **59**, 5–37 (2004)
3. Baird, W., Beveridge, A., Bonato, A., Codenotti, P., Maurer, A., McCauley, J., Valera, S.: On the minimum order of k-cop-win graphs. Contrib. Discrete Math. **9**, 70–84 (2014)
4. Bhaduaria, D., Klein, K., Isler, V., Suri, S.: Capturing an evader in polygonal environments with obstacles: the full visibility case. Int. J. Robot. Res. **31**(10), 1176–1189 (2012)
5. Bonato, A., Nowakowski, R.J.: The Game of Cops and Robbers on Graphs. American Mathematical Society, Providence (2011)
6. West, D.B.: Introduction to Graph Theory. Prentice Hall, Upper Saddle River (2001)
7. Nowakowski, R.J., Winkler, R.P.: Vertex-to-vertex pursuit in a graph. Discrete Math. **43**, 235–239 (1983)
8. Parsons, T.D.: Pursuit-evasion in a graph. In: Alavi, Y., Lick, D.R. (eds.) Theory and Applications of Graph. Lecture Notes in Mathematics, vol. 642, pp. 426–441. Springer, Heidelberg (1978). doi:10.1007/BFb0070400
9. Pisantechakool, P., Tan, X.: On the capture time of cops and robbers game on a planar graph. In: Chan, T.-H.H., Li, M., Wang, L. (eds.) COCOA 2016. LNCS, vol. 10043, pp. 3–17. Springer, Heidelberg (2016). doi:10.1007/978-3-319-48749-6_1
10. Quilliot, A.: Thèse de 3°cycle. In: Université de Paris VI, pp. 131–145 (1978)

On Complexity of Total Vertex Cover
on Subcubic Graphs

Sheung-Hung Poon[1]([✉]) and Wei-Lin Wang[2,3]

[1] School of Computing and Informatics,
Universiti Teknologi Brunei, Gadong, Brunei Darussalam
sheung.hung.poon@gmail.com
[2] Department of Computer Science,
National Chiao Tung University, Hsinchu, Taiwan
weblinkwang@gmail.com
[3] Institute of Information Systems and Applications,
National Tsing Hua University, Hsinchu, Taiwan

Abstract. A *total vertex cover* is a vertex cover whose induced subgraph consists of a set of connected components, each of which contains at least two vertices. A *t-total vertex cover* is a total vertex cover where each component of its induced subgraph contains at least t vertices. The *total vertex cover (TVC) problem* and the *t-total vertex cover (t-TVC) problem* ask for the corresponding cover set with minimum cardinality, respectively. In this paper, we first show that the t-TVC problem is NP-complete for connected subcubic grid graphs of arbitrarily large girth. Next, we show that the t-TVC problem is NP-complete for 3-connected cubic planar graphs. Moreover, we show that the t-TVC problem is APX-complete for connected subcubic graphs of arbitrarily large girth.

1 Introduction

A *vertex cover* C is a subset of vertices of a graph such that every edge of the graph is incident to at least one vertex in C. A vertex cover C is called *total*, if it is a vertex cover of a graph such that, each vertex is adjacent to at least one other vertex in C. In this paper, we consider *t-total vertex cover*, which is a total vertex cover where each component of its induced subgraph contains at least t vertices. By the definition of t-total vertex cover, a 2-total vertex cover is a total vertex cover, and a 1-total vertex cover is a vertex cover. The *vertex cover (VC) problem*, the *total vertex cover (TVC) problem*, and the *t-total vertex cover (t-TVC) problem*, each asks for a corresponding set with minimum cardinality, respectively. These formulated problems appear in various applications. For instance, in a real scenario that someone needs to place the security cameras at the cross junctions of streets so that the minimum number of cameras is required to monitor all streets, the problem can be formulated as the VC problem. Suppose that we make one further restriction that each camera has also be monitored by another camera. Then such a restricted scenario can be formulated as the TVC problem. If the security level is further increased up to

© Springer International Publishing AG 2017
T.V. Gopal et al. (Eds.): TAMC 2017, LNCS 10185, pp. 515–528, 2017.
DOI: 10.1007/978-3-319-55911-7_37

the level that each camera is connected to at least t other cameras through the monitoring relationship, then such a high security level scenario may be formulated as a t-TVC problem. As another example, the VC problem variants can also model another scenario that some workers have the duty to locate the optimal positions for placing ATMs at street corners of a city, such that every street needs to have at least one ATM, and the minimum number of ATMs are required.

Next, we define several technical terms used in this paper. A *cubic graph* (also called a *3-regular graph*) is a graph where every vertex has degree three. A *subcubic graph* is a graph where every vertex has degree at most three. A *grid graph* is an induced subgraph of a set of vertices from two dimensional grid. A *k-connected graph* is a graph which does not contain a set of $k-1$ vertices whose removal disconnects the graph. A ρ-*path* is a path of ρ vertices. A ρ-*cycle* is a cycle of ρ vertices. The *girth* of a graph is the shortest cycle length in the graph. The girth is considered to be infinity if the graph has no cycle. A connected component of a graph is a maximal connected subgraph of the graph. A *family* $\{a_i\}_{i \in I}$ is a set of elements, such that an element a_i in the family is associated with an index $i \in I$.

The VC problem is an important classical graph optimization problem. König's theorem [7,10] indicates that the VC problem on bipartite graphs is equivalent to the maximum matching problem on bipartite graphs, and therefore is solvable within polynomial time [8]. Karp [9] showed that the VC problem is NP-complete for general graphs. Garey and Johnson [5] showed that the VC problem is NP-complete for planar subcubic graphs. Murphy [11] showed that the VC problem is NP-complete for planar subcubic graphs of arbitrarily large girth. Uehara [12] later improved the result by showing that the VC problem is NP-complete for 3-connected cubic planar graphs. In this paper, we show that the t-TVC problem is also NP-complete for 3-connected cubic planar graphs.

The t-TVC problem is a close variation of the VC problem, and is investigated in depth in this paper. First, we present the related work on NP-completeness results for the TVC problem and the t-TVC problem. Both problems were first introduced by Fernau and Manlove [4]. They showed that the t-TVC problem for $t \geq 2$ is NP-complete for planar subcubic graphs. In this paper, we make a step forward to show that t-TVC problem for $t \geq 2$ is NP-complete for connected planar subcubic grid graphs of arbitrarily large girth.

There are also some existing work on approximability results for the t-TVC problem. Fernau [4] showed that the t-TVC problem for $t \geq 1$ is approximable within two for general graphs. This implies that the t-TVC problem is in APX for general graphs. In this paper, we further show that the t-TVC problem for $t \geq 2$ is APX-complete for subcubic graphs of arbitrarily large girth.

Outline. The rest of our paper is organized as follows. In Sect. 2, we first show that the t-TVC problem for $t \geq 2$ is NP-complete for connected subcubic grid graphs of arbitrarily large girth. Next, in Sect. 3, we show that the t-TVC problem for $t \geq 2$ is NP-complete for 3-connected cubic planar graphs. Furthermore, in Sect. 4, we show that the t-TVC problem for $t \geq 2$ is APX-complete for connected subcubic graphs of arbitrarily large girth.

2   NP-Completeness for Subcubic Grid Graphs of Arbitrarily Girth

This section starts by showing that the TVC problem is NP-complete for connected subcubic grid graphs of arbitrarily large girth in Subsect. 2.1. The NP-hardness proof is via reduction from the rectilinear planar monotone 3-SAT problem, which is known to be NP-complete [3], and uses very intricate gadgets in the reduction. Then in Subsect. 2.2, we extend our proof strategy to show that for any $t \geq 2$, the t-TVC problem for connected subcubic grid graphs of arbitrarily large girth is also NP-complete.

2.1   TVC Problem

Theorem 1. *The TVC problem is NP-complete for connected subcubic grid graphs of arbitrarily large girth.*

Proof. We first show that the TVC problem is in NP. The decision version of the TVC problem asks if there exists a total vertex cover of size of at most k, for a given fixed number k. We solve the decision problem through a polynomial number of guessing of the total vertex cover of size of at most k. In each guessing, each vertex set in the total vertex cover has its incident edges removed. After the edge removal, if there is no edge left, output "yes"; otherwise, output "no". Thus, the TVC problem is in NP. To show that the TVC problem is NP-hard, we reduce from the rectilinear planar monotone 3-SAT problem [3].

The input instance of the rectilinear planar monotone 3-SAT problem is a rectilinear representation drawing of a Boolean formula Φ. The Boolean formula $\Phi = c_0 \wedge \cdots \wedge c_{m-1}$ is a conjunction of m clauses. Each clause $c_h = l'_{\alpha,h} \vee l'_{\beta,h} \vee l'_{\gamma,h}$ of the Boolean formula Φ is a disjunction of three literals. The three literals of a clause are either any three variables in the family $\{x_i\}_{i=0}^{n-1}$, called *positive literals*, or the negation of the three variables, called *negative literals*. The rectilinear representation drawing consists of:

(a) A variable gadget X'_i for each variable x_i, represented by horizontal segments lying on the x-axis.
(b) A clause gadget D'_h for each clause c_h, represented by horizontal segments lying on the upper side the x-axis if it consists of only positive true literals, or lying on the lower side of the x-axis if it consists of only negative literals.
(c) A literal gadget $L'_{\alpha,h}$, $L'_{\beta,h}$, or $L'_{\gamma,h}$ for each of the three literals of clause c_h, represented by the vertical segment which links the clause gadget D'_h and the variable gadget X'_j together.

The outcome of the rectilinear planar monotone 3-SAT is "yes" if the Boolean formula is satisfiable, and "no" otherwise. In the following, we proceed to construct the TVC problem.

The input instance of the TVC problem is a connected subcubic grid graph G of arbitrarily large girth g. The graph G is constructed according to the

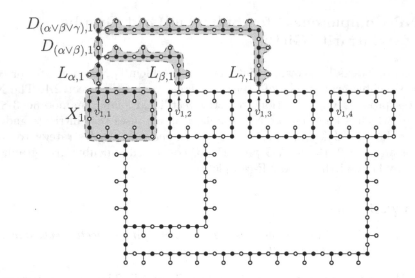

Fig. 1. A constructed graph G for the Boolean expression $\Phi = (x_1 \vee x_2 \vee x_3) \wedge (\overline{x_1} \vee \overline{x_2} \vee \overline{x_4})$ with a truth assignment $x_1 = \text{TRUE}$, $x_2 = \text{FALSE}$, $x_3 = \text{FALSE}$, and $x_4 = \text{TRUE}$. We let $p = 2$ here; thus, the girth of the graph is $6(m+p) = 24$. Solid bullets represent the 138 vertices in a minimum total vertex cover of G.

rectilinear representation drawing of the Boolean formula Φ. We modify the Boolean formula Φ, so that each of its clause $c = l'_{\alpha,h} \vee l'_{\beta,h} \vee l'_{\gamma,h} = \neg(l_{\alpha,h} \wedge l_{\beta,h} \wedge l_{\gamma,h})$, where $l_{\alpha,h} = \neg l'_{\alpha,h}$, $l_{\beta,h} = \neg l'_{\beta,h}$, $l_{\gamma,h} = \neg l'_{\gamma,h}$. However, the rectilinear representation drawing of the Boolean formula Φ remains the same. Graph G contains the following families of gadgets:

(1) The family of variable gadgets $\{X_i\}_{i=0}^{n-1}$: For any variable gadget X_i, its vertex set is $\{v_{j,i}\}_{j=1}^{6(m+p)} \cup \{d_{3j-1,i}\}_{j=1}^{m} \cup \{d_{3j-1,i}\}_{j=m+p+1}^{2m+p}$, for a specified number $p \in \mathbb{N} \setminus \{1\}$. (the second index of $v_{j,i}$ and $d_{3j+2,i}$ might be ignored for brevity). These vertices induce a $6(m+p)$-cycle $v_1 \ldots v_{6(m+p)} v_1$ and two families of edges $\{\{v_{3j-1}, d_{3j-1}\}\}_{j=1}^{m}$ and $\{\{v_{3j-1}, d_{3j-1}\}\}_{j=m+p+1}^{2m+p}$. Two subgraphs of the gadget, $v_{6(m+p)} v_1 \ldots v_{3m} v_{3m+1}$ and $v_{3(m+p)} \ldots v_{3(2m+p)+1}$, are $(3m+2)$-paths parallel to the x-axis; the other two subgraphs of the gadget, $v_{3m+1} \ldots v_{3(m+p)}$ and $v_{3(2m+p)+1} \ldots v_{6(m+p)}$, are $3p$-paths parallel to the y-axis. For a pair of variable gadgets in the family $\{(X_i, X_{i+1})\}_{i=0}^{n-2}$, they have the horizontal distance 2.

(2) The literal gadgets:
  – The literal gadget $L_{\alpha,h}$ for the first literal $l_{\alpha,h}$ of clause c_h: The vertex set of $L_{\alpha,h}$ is $\{u_{j,\alpha,h}\}_{j=1}^{3(2h+1)} \cup \{d_{3j-1,\alpha,h}\}_{j=1}^{2h+1}$ (the second and third indices of $u_{j,\beta,h}$ and $d_{3j-1,\alpha,h}$ might be ignored for brevity). These vertices induce a $(3(2h+1))$-path $u_1 \ldots u_{3(2h+1)}$ parallel to y-axis and the family of edges $\{\{v_{3j-1}, d_{3j-1}\}\}_{j=1}^{2h+1}$.

- The literal gadget $L_{\beta,h}$ for the second literal $l_{\beta,h}$ of clause c_h: The vertex set of $L_{\beta,h}$ is $\{u_{j,\beta,h}\}_{j=1}^{3(2h+1+(\beta-\alpha)(m+1))} \cup \{d_{3j-1,\beta,h}\}_{j=1}^{2h+1+(\beta-\alpha)(m+1)}$ (the second and third indices of $u_{j,\beta,h}$ and $d_{3j-1,\beta,h}$ might be ignored for brevity). These vertices induce a $(3(2h + 1 + (\beta - \alpha)(m + 1)))$-path $u_1 \ldots u_{(3(2h+1+(\beta-\alpha)(m+1)))}$ and the family of edges $\{\{v_{3j-1}, d_{3j-1}\}\}_{j=1}^{2h+1+(\beta-\alpha)(m+1)}$. Its subgraph $u_1 \ldots u_{3(2h+1)+1}$ is a $(3(2h + 1) + 1)$-path parallel to the y-axis. The other subgraph $u_{3(2h+1)+1} \ldots u_{3(2h+1+(\beta-\alpha)(m+1))}$ is a $(3(\beta - \alpha)(m + 1))$-path parallel to the x-axis.

- The literal gadget $L_{\gamma,h}$ for the third literal $l_{\gamma,h}$ of clause c_h: The vertex set of $L_{\gamma,h}$ is $\{u_{j,\gamma,h}\}_{j=1}^{3(2h+2+(\gamma-\alpha)(m+1))} \cup \{d_{3j-1,\gamma,h}\}_{j=1}^{2h+2+(\gamma-\alpha)(m+1)}$ (the second and third indices of $u_{j,\gamma,h}$ and $d_{3j-1,\gamma,h}$ might be ignored for brevity). These vertices induce a $(3(2h + 2 + (\gamma - \alpha)(m + 1)))$-path $u_1 \ldots u_{3(2h+2+(\gamma-\alpha)(m+1))}$ and the family of edges $\{\{v_{3j-1}, d_{3j-1}\}\}_{j=1}^{2h+2+(\gamma-\alpha)(m+1)}$. Its subgraph $u_1 \ldots u_{3(2h+2)+1}$ is a $(3(2h + 2) + 1)$-path parallel to the y-axis. The other subgraph $u_{3(2h+2)+1} \ldots u_{3(2h+2+(\gamma-\alpha)(m+1))}$ is a $(3(\gamma - \alpha)(m + 1))$-path parallel to the x-axis.

(3) The clause gadgets:
  - The clause gadget $D_{(\alpha\vee\beta),h}$ for the clause $c_{(\alpha\vee\beta),h} = l_{\alpha,h} \vee l_{\beta,h}$: The vertex set of $D_{(\alpha\vee\beta),h}$ is $\{u_{j,(\alpha\vee\beta),h}\}_{j=1}^{3} \cup \{d_{(\alpha\vee\beta),h}\}$. (the last two indices of $u_{j,(\alpha\vee\beta),h}$ and $d_{(\alpha\vee\beta),h}$ might be ignored for brevity). These vertices induce a 3-path $u_1 u_2 u_3$ parallel to the y-axis and the edge $\{u_2, d\}$.
  - The clause gadget $D_{(\alpha\vee\beta\vee\gamma),h}$ for the clause $c_{(\alpha\vee\beta\vee\gamma),h} = c_{(\alpha\vee\beta),h} \vee l_{\gamma,h} = l_{\alpha,h} \vee l_{\beta,h} \vee l_{\gamma,h}$: The vertex set of $D_{(\alpha\vee\beta\vee\gamma),h}$ is the family $\{u_{j,(\alpha\vee\beta\vee\gamma),h}\}_{j=1}^{2}$ of vertices (the second and third indices of $u_{j,(\alpha\vee\beta\vee\gamma),h}$ might be ignored for brevity). These vertices induce an edge $\{u_1, u_2\}$ parallel to the y-axis.

The gadgets are linked together as specified in the following. $D_{(\alpha\vee\beta\vee\gamma),h}$ links to $D_{(\alpha\vee\beta),h}$ and $L_{\gamma,h}$ through the edges $\{u_{1,(\alpha\vee\beta\vee\gamma),h}, u_{3,(\alpha\vee\beta),h}\}$ and $\{u_{1,(\alpha\vee\beta\vee\gamma),h}, u_{3(2h+2+(\gamma-\alpha)(m+1)),\gamma,h}\}$. $D_{(\alpha\vee\beta),h}$ links to $L_{\alpha,h}$ and $L_{\beta,h}$ through the edges $\{u_{1,(\alpha\vee\beta),h}, u_{3(2h+1),\alpha,h}\}$ and $\{u_{1,(\alpha\vee\beta),h}, u_{3(2h+1+(\beta-\alpha)(m+1)),\beta,h}\}$. If literal $l_{\alpha,h} = x_\alpha$, then $L_{1,\alpha,h}$ links to X_α through the edge $\{u_{1,\alpha,h}, v_{(1+3h),\alpha}\}$; otherwise, if literal $l_{\alpha,h} = \overline{x_\alpha}$, then $L_{1,\alpha,h}$ links to X_α through the edge $\{u_{1,\alpha,h}, v_{(3(2m+p)-3h),\alpha}\}$. In similar manners, $L_{\beta,h}$ links to X_β, and $L_{\gamma,h}$ links to X_γ.

The girth g of the constructed connected subcubic grid graph G is $6(m + p)$, for a specified number $p \in \mathbb{N} \setminus \{1\}$.

The outcome of the TVC problem is a minimum total vertex cover C, which is reduced from a truth assignment on the variables of Boolean formula Φ.

(i) We first consider the minimum total vertex cover C of graph G on variable gadgets. Every variable gadget $X_i = (V_i, E_i)$ has two alternative choices of $C \cap V_i$ of $4(m + p)$ vertices: One that is adjacent to the vertices of literal gadgets, consisting of the family of vertices $\{\{v_{3j-k}\}_{k=1}^{2}\}_{j=1}^{2(m+p)}$, chosen by us when variable $x_i = $ TRUE. The other that is not adjacent to the vertices of literal gadgets, consisting of the family of vertices $\{\{v_{3j-k}\}_{k=0}^1\}_{j=1}^{2(m+p)}$, chosen by us when variable $x_i = $ FALSE.

(ii) We then consider the minimum total vertex cover C of graph G on literal gadgets.

- Every literal gadget $L_{\alpha,h} = (V_{\alpha,h}, E_{\alpha,h})$ for the first literal $l_{\alpha,h}$ of clause c_h has two alternative categories of choices of $C \cap V_{\alpha,h}$ of $2(2h+1) = 2q_\alpha$ vertices. One that is adjacent to a vertex of variable gadget, consisting of either the family of vertices $\{\{v_{3j-s}\}_{s=0}^1\}_{j=1}^{q_\alpha}$, or $\{\{\{v_{3k-s}\}_{s=0}^1\}_{k=1}^j, \{\{v_{3k'-s'}\}_{s'=1}^2\}_{k'=j+1}^{q_\alpha}\}_{j=1}^{q_\alpha-1}$, or $\{\{v_{3j-s}\}_{s=0}^1, v_{3q_\alpha-1}, d_{3q_\alpha-1}\}_{j=1}^{q_\alpha-1}$, or $\{\{\{v_{3k-s}\}_{s=0}^1\}_{k=1}^j, v_{3(j+1)-1}, d_{3(j+1)-1}, \{\{v_{3k'-s'}\}\}_{s'=1}^2\}_{k'=j+2}^{q_\alpha}\}_{j=1}^{q_\alpha-2}$, chosen by us when either the literal $l_{\alpha,h} = x_\alpha = $ TRUE, or $l_{\alpha,h} = \overline{x_\alpha} = $ TRUE. The other that is adjacent to a vertex of variable gadget, consisting of the family of vertices $\{\{v_{3j-s}\}_{s=1}^2\}_{j=1}^{q_\alpha}$, chosen by us when either the literal $l_{\alpha,h} = x_\alpha = $ FALSE, or $l_{\alpha,h} = \overline{x_\alpha} = $ FALSE.

- Every literal gadget $L_{\beta,h} = (V_{\beta,h}, E_{\beta,h})$ for the second literal $l_{\beta,h}$ of clause c_h has two alternative categories of choices of $C \cap V_{\beta,h}$ of $2(2h+1+(\beta-\alpha)(m+1)) = 2q_\beta$ vertices. One that is adjacent to a vertex of variable gadget, consisting of either the family of vertices $\{\{v_{3j-s}\}_{s=0}^1\}_{j=1}^{q_\beta}$, or $\{\{\{v_{3k-s}\}_{s=0}^1\}_{k=1}^j, \{\{v_{3k'-s'}\}_{s'=1}^2\}_{k'=j+1}^{q_\beta}\}_{j=1}^{q_\beta-1}$, or $\{\{v_{3j-s}\}_{s=0}^1, v_{3q_\beta-1}, d_{3q_\beta-1}\}_{j=1}^{q_\beta-1}$, or $\{\{\{v_{3k-s}\}_{s=0}^1\}_{k=1}^j, v_{3(j+1)-1}, d_{3(j+1)-1}\{\{v_{3k'-s'}\}_{s'=1}^2\}_{k'=j+2}^{q_\beta}\}_{j=1}^{q_\beta-2}$, chosen by us when either the literal $l_{\beta,h} = x_\beta = $ TRUE, or $l_{\beta,h} = \overline{x_\beta} = $ TRUE. The other that is adjacent to a vertex of variable gadget, consisting of the family of vertices $\{\{v_{3j-s}\}_{s=1}^2\}_{j=1}^{q_\beta}$, chosen by us when either the literal $l_{\beta,h} = x_\beta = $ FALSE, or $l_{\beta,h} = \overline{x_\beta} = $ FALSE.

- Every literal gadget $L_{\gamma,h} = (V_{\gamma,h}, E_{\gamma,h})$ for the third literal $l_{\gamma,h}$ of clause c_h has two alternative categories of choices of $C \cap V_{\gamma,h}$ of $2(2h+2+(\gamma-\alpha)(m+1)) = 2q_\gamma$ vertices, similar to that of $L_{\beta,h}$.

(iii) We then consider the minimum total vertex cover C of graph G on clause gadgets.

- Every clause gadget $D_{(\alpha\vee\beta),h} = (V_{(\alpha\vee\beta),h}, E_{(\alpha\vee\beta),h})$ for the clause $c_{(\alpha\vee\beta),h}$ has two alternative categories of choices of $C \cap D_{(\alpha\vee\beta),h}$ of 2 vertices. One that is adjacent to the vertices of literal gadgets, consisting of either the set of vertices $\{u_1, u_3\}$, or $\{u_2, u_3\}$, chosen by us when the clause $c_{(\alpha\vee\beta),h} = $ TRUE. The other that is not adjacent to the vertices of literal gadgets, consisting of the set of vertices $\{u_1, u_2\}$, chosen by us when the clause $c_{(\alpha\vee\beta),h} = $ FALSE.

- Every clause gadget $D_{(\alpha\vee\beta\vee\gamma),h} = (V_{(\alpha\vee\beta\vee\gamma),h}, E_{(\alpha\vee\beta\vee\gamma),h})$ for the clause $c_{(\alpha\vee\beta\vee\gamma),h}$ has two alternative categories of choices of $C \cap D_{(\alpha\vee\beta\vee\gamma),h}$. One that consists of one vertex u_1, chosen by us when the clause $c_{(\alpha\vee\beta\vee\gamma),h} = $ TRUE. The other that consists of u_1 and one more vertex u_2, chosen by us when the clause $c_{(\alpha\vee\beta\vee\gamma),h} = $ FALSE, i.e. the Boolean formula Φ is not satisfied.

Thus, the Boolean formula Φ is satisfied by a truth assignment if and only if the constructed graph G has a total vertex cover C of size at most $n \cdot 4(m + p) + \sum_{h=0}^{m-1}(2(h+1)+2(2h+1+(\beta-\alpha)(m+1))+2(2h+2+(\gamma-\alpha)(m+1))+2+1) = 4n(m + p) + \sum_{h=0}^{m-1}(2(5h + 4 + (\beta + \gamma - 2\alpha)(m + 1)) + 3)$. This completes the proof. □

2.2 t-TVC Problem

We use the similar proof strategy as in Theorem 1, but with more involved gadgets, to prove the following theorem, whose proof is omitted here.

Theorem 2. *For any $t \geq 2$, the t-TVC problem is NP-complete for connected subcubic grid graphs of arbitrarily large girth.*

3   NP-Completeness for 3-Connected Cubic Planar Graphs

This section starts by showing that the TVC problem is NP-complete for 3-connected cubic planar graphs in Subsect. 3.1. The NP-hardness proof is via reduction from VC problem for any cubic planar graph, which is known to be NP-complete [6], and uses very intricate gadgets in the reduction. Then in Subsect. 3.2, we extend our proof strategy to show that for any $t \geq 2$, t-TVC problem for 3-connected cubic planar graphs is also NP-complete.

Lemma 1. *[12]   A cubic graph is 3-connected if and only if it is 3-edge-connected.*

To prove the NP-completeness for 3-connected cubic planar graphs, due to Lemma 1, we only need to prove the NP-completeness for 3-edge-connected cubic planar graphs, where a *k-edge-connected graph* is a graph that does not contain a set of $k - 1$ edges whose removal disconnects the graph.

3.1   TVC Problem

Theorem 3. *The TVC problem is NP-complete for 3-connected cubic planar graphs.*

Proof. Clearly, this problem is in NP. To show that this problem is NP-hard, we show a polynomial-time reduction from a known NP-complete problem, the VC problem for any cubic planar graph G_1 [6]. The reduction has three steps. First, the VC problem for any cubic planar graph G_1 is reduced to the first TVC problem for a 2-edge-connected cubic planar graph G_2. Second, the first TVC problem for a 2-edge-connected cubic planar graph G_2 is reduced to the second TVC problem for a 3-edge-connected cubic planar graph G_3. Third, we use the lemma given by Uehara [12] to show that G_3 is a 3-edge-connected cubic planar graph if and only if G_3 is a 3-connected cubic planar graph.

We show the first step of the reduction.

We formulate the VC problem in the following. The input instance of the VC problem is a cubic planar graph G_1. The outcome of the VC problem is the minimum vertex cover C_1 of graph G_1.

We formulate the first TVC problem in the following.

The input instance of the first TVC problem is a 2-edge-connected cubic planar graph G_2. G_2 is reduced from cubic planar graph G_1. For each edge $e_1 = \{u, v\}$ of G_1, we add a parallel edge $e'_1 = \{u, v\}$ to G_1. Hence, each vertex of G_1 becomes a vertex of degree 6. Then we map each vertex of degree 6 to the gadget H in graph G_2 shown in Fig. 2(a). We also map each of the 6 incident edges of the degree-6 vertex to an incident edge of each of the vertices t_1 to t_6, respectively.

The outcome of the first TVC problem is a minimum total vertex cover C_2 of graph G_2, which is reduced from the minimum vertex cover C_1 of graph G_1. For every vertex of G_1 set in C_1, it corresponds to the minimum total vertex cover of the gadget H shown in Fig. 2(b), which has size of 10; for every vertex of G_1 set not in C_1, it corresponds to the total vertex cover of size of 11 on the gadget H shown in Fig. 2(c).

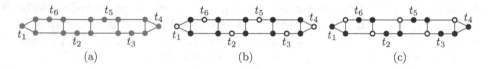

(a)   (b)   (c)

Fig. 2. (a) The gadget H for the TVC problem, which is reduced from a degree-6 vertex v, and has the six neighbors of vertex v adjacent to the vertices t_1 to t_6. (b) The minimum total vertex cover of the gadget H consisting of the 10 black vertices. (c) A total vertex cover of size of 11 on the gadget H represented by the black vertices.

We now show that cubic planar graph G_1 has a vertex cover C_1 of size k_1 if and only if 2-connected cubic planar graph G_2 has a total vertex cover C_2 of size $k_1 + 10n_1$, where n_1 is the number of vertices of graph G_1. The "only if" direction is straightforward. Now we prove the "if" direction. The 2-connected cubic planar graph G_2 has a number of n_1 gadgets H. If a gadget H has more than 11 vertices set in the minimum total vertex cover C_2 of graph G_2, then we can construct a new minimum total vertex cover C'_2 of graph G_2, which includes both the 11 black vertices of gadget H shown in Fig. 2(c) and the vertices of C_2 which are not in gadget H. The size of C'_2 is smaller than size of C_2; hence, C_2 is not the minimum total vertex cover of graph G_2. This results in a contradiction. Hence, for each gadget $H = (V_H, E_H)$ of graph G_2, $10 \leq |V_H \cap C_2| \leq 11$. Those gadgets H of graph G_2 where $|V_H \cap C_2| = 11$ can be mapped to vertices set in the vertex cover C_1 of graph G_1. Let x be the number of gadgets H where $|V_H \cap C_2| = 11$. $10(n_1 - x) + 11x = |C_2| = k_1 + 10n_1$. Thus, $x = k_1$. Hence, the graph G_1 has k_1 vertices set in C_1. This completes the first step of the reduction.

We show the second step of the reduction. The first TVC problem has been formulated in the first step. Now we formulate the second TVC problem in the following.

The input instance of the second TVC problem is a 3-edge-connected cubic planar graph G_3. G_3 is reduced from 2-connected cubic planar graph G_2. For each edge $e_2 = \{u, v\}$ of G_2, we add a parallel edge $e'_2 = \{u, v\}$ to G_2. Hence, each vertex of G_2 becomes a vertex of degree 6. Then we map each vertex of degree 6 to the gadget H in graph G_3 shown in Fig. 2(a). We also map each of the 6 incident edges of the degree-6 vertex to an incident edge of each of the vertices t_1 to t_6, respectively.

The outcome of the second TVC problem is a minimum total vertex cover C_3 of graph G_3, which is reduced from the minimum vertex cover C_2 of graph G_2. For every vertex of G_2 set in C_2, it corresponds to the minimum total vertex cover of the gadget H shown in Fig. 2(b), which has size of 10; for every vertex of G_2 set not in C_2, it corresponds to the total vertex cover of size of 11 on the gadget H shown in Fig. 2(c).

Then using similar arguments as in our first step of reduction, we can show that 2-connected cubic planar graph G_2 has a vertex cover C_2 of size k_2 if and only if 3-connected cubic planar graph G_3 has a total vertex cover C_3 of size $k_2 + 10n_2$, where n_2 is the number of vertices of graph G_2. This completes the second step of the reduction.

Lemma 1 says that G_3 is a 3-edge-connected cubic planar graph if and only if G_3 is a 3-connected cubic planar graph. Using this lemma, we complete the proof.   □

3.2 t-TVC Problem

We use the similar proof strategy as in Theorem 3, but with more involved gadgets, to prove the following theorem, whose proof is omitted here.

Theorem 4. *For any $t \geq 2$, the t-TVC problem is NP-complete for 3-connected cubic planar graphs.*

4   Approximation Hardness for Connected Subcubic Graphs of Arbitrarily Girth

Fernau and Manlove [4] showed that the t-TVC problem for general graphs is approximable within two. Thus, the t-TVC problem for general graphs, including subcubic graphs, is in APX. In this section, we further prove that the t-TVC problem for subcubic graphs is in fact APX-complete.

To prove the APX-completeness of a problem, we present an approximation-preserving reduction, called *L-reduction* ("linear reduction") [2], from the other APX-complete problem to this problem. An L-reduction preserves the approximability and the relative errors.

Next, we define the L-reduction formally. Let A be an optimization problem, so that given any instance x of problem A, we can find a solution through an approximation algorithm in polynomial time. The solutions of problem A are evaluated by the cost function c_A of problem A. We want to reduce optimization problem A to another optimization problem. Let B be another optimization problem, and c_B is the cost function of problem B. An L-reduction from problem A to problem B is a pair of polynomial-time-computable functions f and h. Thus, from an instance x of problem A, we derive the instance $f(x)$ of problem B in polynomial time; from the instance $f(x)$ of problem B, we derive the solution y' of problem B through an approximation algorithm in polynomial time; from the solution y' of problem B, we derive the solution $h(y')$ of problem A in polynomial time. L-reduction linearly preserves the relative error through two positive constant factors α and β, so that

1. $\text{OPT}_B(f(x)) \leq \alpha \text{OPT}_A(x)$, and
2. $|\text{OPT}_A(x) - c_A(h(y'))| \leq \beta |\text{OPT}_B(f(x)) - c_B(y')|$.

4.1   TVC Problem

In this subsection, we prove that the TVC problem for connected graphs of subcubic graphs of arbitrarily large girth is APX-complete.

We show an L-reduction from the VC problem for connected subcubic graphs, which is known to be APX-complete [1], to the TVC problem for connected subcubic graphs of arbitrarily large girth g. The instance of the VC problem is a connected subcubic graph $G_1 = (V_1, E_1)$. The instance of the TVC problem is a connected subcubic graph $G_2 = (V_2, E_2)$ of arbitrarily large girth g. The function f of the L-reduction is from the set of every G_1 to the set of every G_2. We then describe the function f in detail. We are given a graph $G_1 = (V_1, E_1)$. The function f then map each vertex $v \in V_1$ to a gadget $H = (V_H, E_H)$, where $V_H = \{v_i\}_{i=0}^{k(g+1)} \cup \{\{d_{3(j+1)+s+i(g+1)}\}_{s=0}^{1}\}_{j=0}^{k-1}\}_{i=0}^{k-1}$, and $E_H = \{\{v_{i(g+1)}v_{1+i(g+1)}, v_{(j \bmod g)+1+i(g+1)}v_{(j+1 \bmod g)+1+i(g+1)}, v_{-2+(i+1)(g+1)}$
$v_{(i+1)(g+1)}, \{v_{3(j+1)+s+i(g+1)}d_{3(j+1)+s+i(g+1)}\}_{s=0}^{1}\}_{j=0}^{k-1}\}_{i=0}^{k-1}$, $g = 3k + 1$, $k \in \mathbb{N}$. Moreover, for each vertex $v \in V_1$, the function f maps the set of its existing incident edges $\{\alpha v, \beta v, \gamma v \mid N(v) = \{\alpha, \beta, \gamma\}\}$ to the set of incident edges of some vertices of gadget H: $\{\alpha v_0, \beta v_0, \gamma v_{k(g+1)} \mid v_0, v_{k(g+1)} \in V_H\}$. See Fig. 3 for an example. The resulting graph is G_2.

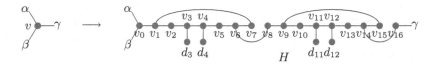

Fig. 3. The function f of the L-reduction replaces each vertex v of graph G by the gadget H in graph G_2 of girth $g = 3k + 1 = 7$, when we let $k = 2$.

For the gadget H, the following lemma holds.

Lemma 2. *A total vertex cover of gadget H has size $2k^2 + k$, $k \in \mathbb{N}$, if and only if it does not include the vertex set $\{v_0, v_{k(g+1)}\}$. Moreover, a total vertex cover of gadget H has size of $2k^2 + k + 1$ if and only if it includes the vertex set $\{v_0, v_{k(g+1)}\}$.*

Proof. First, we find the size of the minimum total vertex cover of gadget H. Hence, we need to find the size of the minimum total vertex cover of each g-cycle in gadget H. Each g-cycle in gadget H has $g = 3k + 1$ vertices, at least $2k + 1$ of them are placed in the minimum total vertex cover of gadget H. Since there are a number of k vertex-disjoint g-cycles in gadget H, the minimum total vertex cover of gadget H has size of $2k^2 + k$. Furthermore, we can let the minimum total vertex cover of gadget H to be $\{v_{1+i(g+1)}, \{v_{3j+1+s+i(g+1)}\}_{s=0}^1, v_{g-1+i(g+1)}, v_{g+i(g+1)}\}_{j=0}^{k-1}\}_{i=0}^{k-1}$, which includes none of the vertices v_0 and $v_{k(g+1)}$.

We then obtain a total vertex cover of gadget H of size equals to $2k^2 + k+1$: $\{\{v_{i(g+1)}, v_{1+i(g+1)}, \{v_{3j+1+s+i(g+1)}\}_{s=0}^1, v_{g-1+i(g+1)}, v_{g+i(g+1)}\}_{j=0}^{k-1}\}_{i=0}^{k-1}$, which includes both of the vertices v_0 and $v_{k(g+1)}$.   □

The solution of the TVC problem is a total vertex cover D on a connected subcubic graph $G_2 = (V_2, E_2)$ of arbitrarily large girth g. The solution of the VC problem is a vertex cover C on a connected subcubic graph $G_1 = (V_1, E_1)$ of arbitrarily large girth g. From the reduction, we have the function h from the set of every D to the set of every C. We define function h in detail in the following lemma.

Lemma 3. *There is a polynomial-time reduction h such that for each total vertex cover D in G_2, if $|D \cap V_H| = 2k^2 + k + 1$, then we replace $D \cap V_H$ by a vertex in $h(D)$; otherwise, if $|D \cap V_H| = 2k^2 + k$, then we replace $D \cap V_H$ by a vertex in $(V_1 \setminus h(D))$. Moreover, $h(D)$ is a vertex cover in G_1.*

Proof. To determine the function h, we have to confirm that its output is a vertex cover $h(D)$. By assumption, D is a total vertex cover of graph $G_2 = (V_2, E_2)$. Hence, for each edge $(u, v) \in E_2$, either $u \in D \wedge v \in D$, $u \in D \wedge v \notin D$, or $u \notin D \wedge v \in D$. Thus, for each edge (s_2, t_2) and two gadgets $H' = (V_{H'}, E_{H'})$, $H'' = (V_{H''}, E_{H''})$, so that $(s_2, t_2) \in E_2 \setminus (E_{H'} \cup E_{H''})$, $s_2 \in \{v_1', v_{k(g+1)}'\} \subset V_{H'}$, $t_2 \in \{v_1'', v_{k(g+1)}''\} \subset V_{H''}$, we have three cases:

(i) $s_2 \in D \wedge t_2 \in D$. Thus, $|D \cap V_{H'}| = 2k^2 + k + 1$, $|D \cap V_{H''}| = 2k^2 + k + 1$. Then function h maps $D \cap V_{H'}$ to vertex $s_1 \in h(D)$, and maps $D \cap V_{H''}$ to vertex $t_1 \in h(D)$. Hence, vertex cover $h(D)$ covers edge (s_1, t_1).

(ii) $s_2 \notin D \wedge t_2 \in D$. Thus, $|D \cap V_{H'}| = 2k^2 + k$, $|D \cap V_{H''}| = 2k^2 + k + 1$. Then function h maps $D \cap V_{H'}$ to vertex $s_1 \in (V_1 \setminus h(D))$, and maps $D \cap V_{H''}$ to vertex $t_1 \in h(D)$. Hence, vertex cover $h(D)$ covers edge (s_1, t_1).

(iii) $s_2 \in D \wedge t_2 \notin D$. Thus, $|D \cap V_{H'}| = 2k^2 + k + 1$, $|D \cap V_{H''}| = 2k^2 + k$. Then function h maps $D \cap V_{H'}$ to vertex $s_1 \in h(D)$, and maps $D \cap V_{H''}$ to vertex $t_1 \in (V_1 \setminus h(D))$. Hence, vertex cover $h(D)$ covers edge (s_1, t_1).

Therefore, $h(D)$ is the vertex cover of graph G_1.   □

We also introduce the inverse function of the function h, that is the function h^{-1} from the set of every C to the set of every D. We describe function h^{-1} in detail in the following lemma.

Lemma 4. *The inverse h^{-1} of the polynomial-time reduction h satisfies the following condition. For each vertex cover C in G_1, if v is in C, then we replace v by $(h^{-1}(C) \cap V_H)$ of size $2k^2 + k + 1$; otherwise, if v is in $(V_1 \setminus C)$, then we replace v by $(h^{-1}(C) \cap V_H)$ of size $2k^2 + k$ Moreover, $h^{-1}(C)$ is a total vertex cover in G_2.*

Proof. At the end, we need to confirm that $h^{-1}(C)$ for vertex cover C is a total vertex cover. By assumption, C is a vertex cover of graph $G_1 = (V_1, E_1)$. Hence, for each edge $(s_1, t_1) \in E_1$, we have three cases:

(i) $s_1 \in C \wedge t_1 \in C$. Then function h^{-1} maps vertex $s_1 \in C$ and $t_1 \in C$ to $h^{-1}(C) \cap V_{H'}$ and $h^{-1}(C) \cap V_{H''}$, so that $|h^{-1}(C) \cap V_{H'}| = 2k^2 + k + 1$ and $|h^{-1}(C) \cap V_{H''}| = 2k^2 + k + 1$. Thus, for each edge (s_2, t_2) and two gadgets $H' = (V_{H'}, E_{H'})$, $H'' = (V_{H''}, E_{H''})$, so that $(s_2, t_2) \in E_2 \setminus (E_{H'} \cup E_{H''})$, $s_2 \in \{v'_1, v'_{k(g+1)}\} \subset V_{H'}$, $t_2 \in \{v''_1, v''_{k(g+1)}\} \subset V_{H''}$, we can conclude that $s_2 \in h^{-1}(C)$ and $t_2 \in h^{-1}(C)$. Therefore, $h^{-1}(C)$ covers edge (s_2, t_2). Moreover, s_2 and t_2 are adjacent to vertices in $h^{-1}(C) \cap V_{H'}$ and $h^{-1}(C) \cap V_{H''}$.

(ii) $s_1 \notin C \wedge t_1 \in C$. Then function h^{-1} maps vertex $s_1 \notin C$ and $t_1 \in C$ to $h^{-1}(C) \cap V_{H'}$ and $h^{-1}(C) \cap V_{H''}$, so that $|h^{-1}(C) \cap V_{H'}| = 2k^2 + k$ and $|h^{-1}(C) \cap V_{H''}| = 2k^2 + k + 1$. Thus, for each edge (s_2, t_2) and two gadgets $H' = (V_{H'}, E_{H'})$, $H'' = (V_{H''}, E_{H''})$, so that $(s_2, t_2) \in E_2 \setminus (E_{H'} \cup E_{H''})$, $s_2 \in \{v'_1, v'_{k(g+1)}\} \subset V_{H'}$, $t_2 \in \{v''_1, v''_{k(g+1)}\} \subset V_{H''}$, we can conclude that $s_2 \notin h^{-1}(C)$ and $t_2 \in h^{-1}(C)$. Therefore, $h^{-1}(C)$ covers edge (s_2, t_2). Moreover, s_2 and t_2 are adjacent to vertices in $h^{-1}(C) \cap V_{H'}$ and $h^{-1}(C) \cap V_{H''}$.

(iii) $s_1 \in C \wedge t_1 \notin C$. Then function h^{-1} maps vertex $s_1 \in C$ and $t_1 \in C$ to $h^{-1}(C) \cap V_{H'}$ and $h^{-1}(C) \cap V_{H''}$, so that $|h^{-1}(C) \cap V_{H'}| = 2k^2 + k + 1$ and $|h^{-1}(C) \cap V_{H''}| = 2k^2 + k$. Thus, for each edge (s_2, t_2) and two gadgets $H' = (V_{H'}, E_{H'})$, $H'' = (V_{H''}, E_{H''})$, so that $(s_2, t_2) \in E_2 \setminus (E_{H'} \cup E_{H''})$, $s_2 \in \{v'_1, v'_{k(g+1)}\} \subset V_{H'}$, $t_2 \in \{v''_1, v''_{k(g+1)}\} \subset V_{H''}$, we can conclude that $s_2 \notin h^{-1}(C)$ and $t_2 \in h^{-1}(C)$. Therefore, $h^{-1}(C)$ covers edge (s_2, t_2). Moreover, s_2 and t_2 are adjacent to vertices in $h^{-1}(C) \cap V_{H'}$ and $h^{-1}(C) \cap V_{H''}$.

Therefore, $h^{-1}(C)$ is the total vertex cover of graph G_2.   □

By Lemma 3, for a total vertex cover D of G_2, we know that $|D| = |h(D)| + (2k^2 + k) \cdot |V_1|$. And by Lemma 4, for a vertex cover C of G_1, we know that $|C| = |h^{-1}(C)| - (2k^2 + k) \cdot |V_1|$. Thus we have the following Lemma.

Lemma 5. *Let h be the polynomial-time reduction from G_2 to G_1.*

(i) For a total vertex cover D of G_2, we have $|h(D)| = |D| - (2k^2 + k) \cdot |V_1|$.
(ii) For a vertex cover C of G_1, we have $|C| = |h^{-1}(C)| - (2k^2 + k) \cdot |V_1|$.

Using Lemma 5, we can derive the following lemma by performing simple calculations.

Lemma 6. *Let D^* be the minimum total vertex cover of a connected subcubic graph $G_2 = (V_2, E_2)$ of arbitrarily girth g, and C^* be the minimum vertex cover of a connected subcubic graph $G_1 = (V_1, E_1)$. Then $|C^*| = |D^*| - (2k^2 + k) \cdot |V_1|$.*

Proof. By Lemma 5, we have that $|C^*| \leq |h(D^*)| = |D^*| - (2k^2 + k) \cdot |V_1|$, and $|D^*| \leq |h^{-1}(C^*)| = |C^*| + (2k^2 + k) \cdot |V_1|$. Thus $|C^*| = |D^*| - (2k^2 + k) \cdot |V_1|$. $\square$

We first determine the positive factor β of the L-reduction. From Lemmas 5 and 6, we see that $|C| - |C^*| = |D| - |D^*|$. Thus, $\beta = 1$.

Next, we determine the positive factor α of the L-reduction. Recall that graph $G_1 = (V_1, E_1)$ is a connected subcubic graph. Thus, $|V_1| \leq |E_1| + 1 \leq \sum_{x \in C^*}(\deg(x)) + 1 \leq \sum_{x \in C^*}(\deg(x) + 1) = 4 \cdot |C^*|$. Hence, we can modify Lemma 6 as follows: $|D^*| = |C^*| + (2k^2 + k) \cdot |V_1| \leq |C^*| + 4(2k^2 + k) \cdot |C^*| = (8k^2 + 4k + 1) \cdot |C^*|$. Therefore, $\alpha = 8k^2 + 4k + 1$. Thus we obtain the L-reduction. Hence, we have the following theorem.

Theorem 5. *The TVC problem is APX-complete for connected subcubic graphs of arbitrarily large girth g.*

4.2 t-TVC Problem

We use the similar proof strategy as in Theorem 5, but with more involved gadgets, to prove the following theorem, whose proof is omitted here.

Theorem 6. *For any $t \geq 2$, the t-TVC problem is APX-complete for connected subcubic graphs of arbitrarily large girth g.*

References

1. Alimonti, P., Kann, V.: Some APX-completeness results for cubic graphs. Theoret. Comput. Sci. **237**(1–2), 123–134 (2000)
2. Crescenzi, P.: A short guide to approximation preserving reductions. In: Proceedings of the 12th Annual IEEE Conference on Computational Complexity, pp. 262–273 (1997)
3. de Berg, M., Khosravi, A.: Optimal binary space partitions in the plane. In: Proceedings of the 16th International Conference on Computing and Combinatorics, pp. 329–343 (2010)
4. Fernau, H., Manlove, D.F.: Vertex and edge covers with clustering properties: complexity and algorithms. J. Discrete Algorithms **7**(2), 149–167 (2009)
5. Garey, M.R., Johnson, D.S.: The rectilinear steiner tree problem is NP-complete. SIAM J. Appl. Math. **32**(4), 826–834 (1977)
6. Garey, M.R., Johnson, D.S.: Computers and Intractability: A Guide to the Theory of NP-Completeness. W. H. Freeman & Co., New York (1990)
7. Gross, J.L., Yellen, J.: Graph Theory and Its Applications. CRC Press, Boca Raton (2005)
8. Hopcroft, J.E., Karp, R.M.: A $n^{5/2}$ algorithm for maximum matchings in bipartite. In: Proceedings of 12th Annual IEEE Symposium on Switching and Automata Theory, pp. 225–231 (1971)

9. Karp, R.M.: Reducibility among combinatorial problems. In: Proceedings of a Symposium on the Complexity of Computer Computations, pp. 85–103 (1972)
10. König, D.: Vgráfok és mátrixok. Matematikai és Fizikai Lapok **38**, 116–119 (1931)
11. Murphy, O.J.: Computing independent sets in graphs with large girth. Discrete Appl. Math. **35**(2), 167–170 (1992)
12. Uehara, R.: NP-complete problems on a 3-connected cubic planar graph and their applications. Technical Report TWCU-M-0004, Tokyo Woman's Christian University (1996)

Nondeterministic Communication Complexity of Random Boolean Functions (Extended Abstract)

Mozhgan Pourmoradnasseri and Dirk Oliver Theis[(✉)]

Institute of Computer Science, University of Tartu, Ülikooli 17, 51014 Tartu, Estonia
{mozhgan,dotheis}@ut.ee

Abstract. We study nondeterministic communication complexity and related concepts (fooling sets, fractional covering number) of random functions $f \colon X \times Y \to \{0,1\}$ where each value is chosen to be 1 independently with probability $p = p(n)$, $n := |X| = |Y|$.

Keywords: Communication complexity · Random structures

1 Introduction

Communication Complexity lower bounds have found applications in areas as diverse as sublinear algorithms, space-time trade-offs in data structures, compressive sensing, and combinatorial optimization (cf., e.g., [11,27]). In combinatorial optimization especially, there is a need to lower bound *nondeterministic* communication complexity [18,30].

Let X, Y be sets and $f \colon X \times Y \to \{0,1\}$ a function. In nondeterministic communication, Alice gets an $x \in X$, Bob gets a $y \in Y$, and they both have access to a bit string supplied by a prover. In a protocol, Alice sends one bit to Bob; the decision whether to send 0 or 1 is based on her input x and the bit string z given by the prover. Then Bob decides based on his input y, the bit string z given by the prover, and the bit sent by Alice, whether to accept (output 1) or reject (output 0). The protocol is successful, if, (1) regardless of what the prover says, Bob never accepts if $f(x,y) = 0$, but (2) for every (x,y) with $f(x,y) = 1$, there is a proof z with which Bob accepts. The nondeterministic communication complexity is the smallest number ℓ of bits for which there is a successful protocol with ℓ-bit proofs.

Formally, the following basic definitions are common:

- The *support* is the set of all *1-entries:* supp $f := \{(x,y) \mid f(x,y) = 1\}$;
- a *1-rectangle* is a cartesian product of sets of inputs $R = A \times B \subseteq X \times Y$ all of which are 1-entries: $A \times B \subseteq$ supp f;
- a *cover* (or *1-cover*) is a set of 1-rectangles $\{R_1 = A_1 \times B_1, \dots, R_k = A_k \times B_k\}$ which together cover all 1-entries of f, i.e., $\bigcup_{j=1}^{k} R_j =$ supp f;
- the cover number $\mathsf{C}(f)$ of f is the smallest size of a 1-cover.

© Springer International Publishing AG 2017
T.V. Gopal et al. (Eds.): TAMC 2017, LNCS 10185, pp. 529–542, 2017.
DOI: 10.1007/978-3-319-55911-7_38

One can then define the *nondeterministic communication complexity* simply as $N(f) := \log_2 C(f)$ [21].

In combinatorial optimization, one wants to lower bound the nondeterministic communication complexity of functions which are defined based on relations between feasible points and inequality constraints of the optimization problem at hand: Alice has an inequality constraint, Bob has a feasible point, and they should reject (answer 0) if the point satisfies the inequality with equality.

Consider, the following example (it describes the so-called *permuthahedron*). Let $k \geq 3$ be a positive integer.

- Let Y denote the permutations π of $[k]$—the feasible points.
- Let X denote the set of non-empty subsets $U \subsetneq [k]$; such an U corresponds to an inequality constraint $\sum_{u \in U} \pi(u) \geq |U|(|U|+1)/2$.

Goemans [13] gave an $\Omega(\log k)$ lower bound for the nondeterministic communication complexity of the corresponding function:

$$f(\pi, U) = \begin{cases} 0, & \text{if } \sum_{u \in U} \pi(u) = |U|(|U|+1)/2; \\ 1, & \text{otherwise, i.e., } \sum_{u \in U} \pi(u) > |U|(|U|+1)/2. \end{cases}$$

For $k = 3$, see the following table. The rows are indexed by the set X, the columns by the set Y.

	123	132	213	231	312	321
{1}	0	0	1	1	1	1
{2}	1	1	0	1	0	1
{3}	1	1	1	0	1	0
{1,2}	0	1	0	1	1	1
{1,3}	1	0	1	0	1	1
{2,3}	1	1	1	1	0	0

In this situation, the nondeterministic communication complexity lower bounds the logarithm of the so-called *extension complexity:* the smallest number of linear inequalities which is needed to formulate the optimization problem. This relationship goes back to Yannakakis' 1991 paper [30], and has recently been the focus of renewed attention [2,20] and a source of some breakthrough results [9,10]. Other questions remain infamously open, e.g., the nondeterministic communication complexity of the minimum-spanning-tree function: For a fixed number k, Bob has a tree with vertex set $[k]$, Alice has one of a set of inequality constraints (see [28] for the details), and they are supposed to answer 1, if the tree does not satisfy the inequality constraint with equality.

In this paper, we focus on random functions, and we give tight upper and lower bounds for the nondeterministic communication complexity and its most important lower bounds: the fooling set bound; the ratio number of 1-entries over largest 1-rectangle; the fractional cover number. For that, we fix $|X| = |Y| = n$, and, we take $f(x,y)$, $(x,y) \in X \times Y$, to be independent Bernoulli random variables with parameter $p = p(n)$, i.e., $f(x,y) = 1$ with probability p and $f(x,y) = 0$ with probability $1 - p$.

In Communication Complexity, it is customary to determine these parameters up to within a constant factor of the number of bits, but in applications, this is often not accurate enough. E.g., the above question about the extension complexity of the minimum-spanning-tree polytope asks where in the range between $(1 + o(1))2\log n$ bits and $(1 + o(1))3\log n$ bits the nondeterministic communication complexity lies. (Here n should taken as $|Y| = 2^k - 2$). In our analyses, we focus on the constant factors in our communication complexity bounds.

1.1 Relationship to Related Work

In core (Communication) Complexity Theory, random functions are usually used for establishing that hard functions exist in the given model of computation. In this spirit, some easy results about the (nondeterministic) communication complexity of random functions and related parameters exist, with p a constant, mostly $p = 1/2$ (e.g., the fooling set bound is determined in this setting in [8]).

In contrast to this, in applications, the density of the matrices is typically close to 1, e.g., in combinatorial optimization, the number of 0s in a "row" $\{y \in Y \mid f(x, y) = 0\}$, is very often polylog of n. This makes necessary to look at these parameters in the spirit of the study of properties of random graph where $p = p(n) \to 1$ with $n \to \infty$. In an analogy to the fields of random graphs, the results become both considerably more interesting and also more difficult that way.

The random parameters we analyze have been studied in other fields beside Communication Complexity. Recently, Izhakian, Janson, and Rhodes [16] have determined asymptotically the triangular rank of random Boolean matrices with independent Bernoulli entries. (The triangular rank is itself important in Communication Complexity, and is a lower bound to the size of a fooling set). In that paper, determining the behavior for $p \to 0, 1$ is posed as an open problem.

The size of the largest monochromatic rectangle in a random Bernoulli matrix was determined in [26] when p is bounded away from 0 and 1, but their technique fails for $p \to 1$.

The nondeterministic communication complexity of a the clique-vs-stable set problem on random graphs was studied in [4].

The parameters we study in this paper are of importance beyond Communication Complexity and its direct applications. In combinatorics, e.g., the cover number coincides with strong isometric dimension of graphs [12], and has connections to extremal set theory and Coding Theory [14,15].

The size of the largest monochromatic rectangle is of interest in the analysis of gene expression data [26], and formal concept analysis [6].

Via a construction of Lovász and Saks [24], the 1-rectangles, covers, and fooling sets of a function f correspond to stable sets, colorings, and cliques, resp., in a graph constructed from the function. Consequently, determining these parameters could be thought of as analyzing a certain type of random graphs. This approach does not seem to be fruitful, as the probability distribution on the set of graphs seems to have little in common with those studied in random graph theory. Here is an important example for that. In the usual random graph models

(Erdős-Renyi, uniform regular), the chromatic number is within a constant factor of the independence ratio (i.e., the quotient independence number over number of vertices), and, in particular, of the fractional chromatic number (which lies between the two). The corresponding statement (replace "chromatic number" by "cover number"; "independence ratio" by "Hamming weight of f divided by the size of the largest 1-rectangle"; "fractional chromatic number" by "fractional cover number") is false for random Boolean functions, see Sect. 4.

This paper is organized as follows. We determine the size of the largest monochromatic rectangle in Sect. 2. Section 3 is dedicated to fooling sets: we give tight upper and lower bounds. Finally, in Sect. 4 we give bounds for both the covering number and the fractional covering number.

1.2 Definitions

A Boolean function $f\colon X \times Y \to \{0,1\}$ can be viewed as a matrix whose rows are indexed by X and the columns are indexed by Y. We will use the two concepts interchangeably. In particular, for convenience, we speak of "row" x and "column" y. We always take $n = |X| = |Y|$ without mentioning it. A *random Boolean function* $f\colon X \times Y \to \{0,1\}$ *with parameter p* is the same thing as a random $n \times n$ matrix with independent Bernoulli entries with parameter p.

We use the usual conventions for asymptotics: $g \ll h$ and $g = o(h)$ is the same thing. As usual, $g = \Omega(1)$ means that g is bounded away from 0. We are interested in asymptotic statements, usually for $n \to \infty$. A statement (i.e., a family of events E_n, $n \in \mathbb{N}$) holds *asymptotically almost surely, a.a.s.,* if its probability tends to 1 as $n \to \infty$ (more precisely, $\lim_{n\to\infty} \mathbf{P}(E_n) = 1$).

2   Largest 1-Rectangle

As mentioned in the introduction, driven by applications in bioinformatics, the size of the largest monochromatic rectangle in a matrix with independent (Bernoulli) entries, has been studied longer than one might expect. Analyzing computational data, Lonardi, Szpankowski, and Yang [22,23] conjectured the shape of the 1-rectangles. The conjecture was proven by Park and Szpankowski [26]. Their proof can be formulated as follows: Let $f\colon X \times Y \to \{0,1\}$ be a random Boolean function with parameter p.

- If $\Omega(1) = p \le 1/e$, then, a.a.s., the largest 1-rectangle consists of the 1-entries in a single row or column, and $\mathsf{R}^1(f) = (1 + o(1))pn$.
- If $p \ge 1/e$ but bounded away from 1, then with $a := \operatorname{argmax}_{b \in \{1,2,3,\dots\}} bp^b$, a.a.s. the largest 1-rectangle has a rows and $p^a n$ columns, or vice-versa.

The existence of these rectangles is fairly obvious. Proving that no larger ones exist requires some work. The problem with the union-bound based proof in [26] is that it breaks down if p tends to 1 moderately quickly. In our proofs, we work with strong tail bounds instead.

Our result extends the theorem in [26] for the case that p tends to 0 or 1 quickly.

For $K \subseteq X$, the *1-rectangle of f generated by K* is $R := K \times L$ with

$$L := \left\{ y \in Y \mid \forall\, x \in K\colon\ f(x,y) = 1 \right\}.$$

The 1-rectangle generated by a subset L of Y is defined similarly.

Theorem 2.1. *Let $f\colon X \times Y \to \{0,1\}$ be a random Boolean function with parameter $p = p(n)$.*

(a) If $5/n \le p \le 1/e$, then a.a.s., the largest 1-rectangle is generated by a single row or column, and if $p \gg (\ln n)/n$, its size is $(1 + o(1))pn$.

(b) Define

$$a_- := \lfloor \log_{1/p} e \rfloor,$$
$$a_+ := \lceil \log_{1/p} e \rceil, \text{ and} \tag{1}$$
$$a := \operatorname*{argmax}_{b \in \{a_-, a_+\}} b p^b \ = \ \operatorname*{argmax}_{b \in \{1,2,3,\dots\}} b p^b.$$

There exists a constant λ_0, such that if $1/e \le p \le 1 - \lambda_0/n$, then, a.a.s., a largest 1-rectangle is generated by a rows or columns and its size is $(1 + o(1))ap^a n$.

The proof requires us to upper bound the sizes of square 1-rectangles, i.e., $R = K \times L$ with $|K| = |L|$. Sizes of square 1-rectangles have been studied, too. Building on work in [6,7,26], it was settled in [29], for constant p. We need results for $p \to 0, 1$, but, fortunately, for our theorem, we only require weak upper bounds.

For the proof of (a), we say that a 1-rectangle is *bulky*, if it extends over at least 2 rows and also over at least 2 columns. We then proceed by considering three types of rectangles:

1. those consisting of exactly one row or column (they give the bound in the theorem);
2. square bulky rectangles;
3. bulky rectangles which are not square.

For the proof of (b), we also require an appropriate notion of "bulky": here, we say that a rectangle of dimensions $k \times \ell$ is bulky if $k \le \ell$. By again considering square rectangles, we prove that a bulky rectangle must have $k < n/\lambda^{2/3}$. (We always define λ through $p = 1 - \lambda/n$.) By exchanging the roles of rows and columns, and multiplying the final probability estimate by 2, we only need to consider 1-rectangles with at least as many columns as rows (i.e., bulky ones). Following that strategy yields the statement of the theorem.

The complete proof will be in the full version of the paper.

Remark 1.(a) If $p \ge 1/e$, then

$$1/e^2 \le \frac{p}{e} \le p \cdot p^{\log_{1/p} e} \le p^a \le \frac{1}{p} \cdot p^{\log_{1/p} e} \le \frac{1}{pe} \le 1/e, \tag{2}$$

i.e., $p^a \approx 1/e$, more accurately $p^a = (1 - o_{p \to 1}(1))/e$.

(b) With $p = 1 - \bar{p} = 1 - \lambda/n$, the following makes the range of $\mathsf{R}^1(f)$ clearer: Since $\bar{p} \leq \ln(1/(1-\bar{p})) \leq \bar{p} + \bar{p}^2$ holds when $\bar{p} \leq 1 - 1/e$, we have

$$\frac{1}{e\bar{p}} = \frac{n}{e\lambda} \leq p\frac{n}{\lambda} = \frac{p}{\bar{p}} \leq \frac{1}{1+\bar{p}} \cdot \frac{1}{\bar{p}} \leq \log_{1/p} e \leq \frac{1}{\bar{p}} = \frac{n}{\lambda} \tag{3}$$

Corollary 1. *For* $p = 1 - \frac{\lambda}{n}$ *with* $\lambda_0 \leq \lambda = o(n)$, *we have* $\mathsf{R}^1(f) = \frac{n^2}{e\lambda} + O(n)$.

See the full version of the paper for the proof.

3 Fooling Sets

A *fooling set* is a subset $F \subseteq X \times Y$ with the following two properties: (1) for all $(x,y) \in F$, $f(x,y) = 1$; and (2) and for all $(x,y), (x',y') \in F$, if $(x,y) \neq (x',y')$ then $f(x,y')f(x',y) = 0$. When f is viewed as a matrix, this means that, after permuting rows and columns, F identifies the diagonal entries of a submatrix which is 1 on the diagonal, and in every pair of opposite off-diagonal entries, at least one is 0. We denote by $\mathsf{F}(f)$ the size of the largest fooling set of f. The maximum size of a fooling set of a random Boolean function with $p = 1/2$ is easy to determine (e.g., [8]).

An obvious lower bound to the fooling set size is the *triangular rank*, i.e., the size of the largest triangular submatrix (again after permuting rows and columns). In a recent Proc. AMS paper, Izhakian, Janson, and Rhodes [16] determined the triangular rank of a random matrix with independent Bernoulli entries with constant parameter p. They left as an open problem to determine the triangular rank in the case when $p \to 0$ or 1, which is our setting.

Our constructions of fooling sets of random Boolean functions make use of ingredients from random graph theory. First of all, consider the bipartite H_f whose vertex set is the disjoint union of X and Y, and with $E(H_f) = \operatorname{supp} f \subseteq X$. For random f, this graph is an *Erdős-Renyi random bipartite graph:* each edge is picked independently with probability p. Based on the following obvious fact, we will use results about matchings in Erdős-Renyi random bipartite graphs:

Remark 2. Let $F \subseteq X \times Y$. The following are equivalent.

(a) F is a *fooling set*.
(b) F satisfies the following two conditions:
   - F is a matching, i.e., $F \subseteq E(H)$;
   - F is *cross-free*, i.e., for all $(x,y), (x',y') \in F$, if $(x,y) \neq (x',y')$ then $(x,y') \notin E$ or $(x',y) \notin E$.

Secondly, fooling sets can be obtained from stable sets in an auxiliary graph: For a random Boolean function f, this graph is an *Erdős-Renyi random graphs,* for which results are available yielding good lower bounds.

Figure 1 summarizes our upper and lower bounds: Upper bounds are above the dotted lines; lower bounds are below the dotted lines; the range for p is between the dotted lines. All upper bounds are by the 1st moment method.

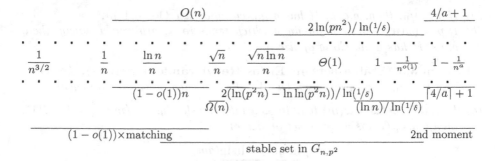

Fig. 1. Upper and lower bounds on fooling set sizes. $(\delta := 1 - p^2)$

We emphasize that the upper and lower bounds differ by at most a constant factor. If $p \to 1$ quickly enough, i.e., $\bar{p} = 1 - p = n^{-a}$ for a constant a, then the upper bounds and lower bounds are even the same except for rounding.

3.1 Statement of the Theorem, and a Glimpse of the Proof

Denote by $\nu(H)$ the size of the largest matching in a bipartite graph H. For $q = q(m)$, denote by $\mathbf{G}_{m,q}$ the graph with vertex set $\{1, \ldots, m\}$ in which each of the $\binom{m}{2}$ possible edges is chosen (independently) with probability q. Let $a(q) = a_m(q)$ be a function with the property that, a.a.s., every Erdős-Renyi random graph on m vertices with edge-probability q has an independent set of size at least $a_m(q)$.

Theorem 3.1. *Let $f : X \times Y \to \{0, 1\}$ be a random Boolean function with parameter $p = p(n)$. Define $\bar{p} := 1 - p$ and $\delta := 1 - p^2$.*

(a) For $n^{-3/2} \leq p = o(1/\sqrt{n})$, a.a.s., we have

$$\mathsf{F}(f) = (1 - o(1))\nu(H_f).$$

(b) If $pn - \ln n \to \infty$, then, a.a.s., $\mathsf{F}(f) \geq a(p^2)$.
(c) If $p \gg \sqrt{(\ln n)/n}$ and $\bar{p} \geq n^{-o(1)}$, then, a.a.s.,

$$\mathsf{F}(f) \leq 2\log_{1/\delta}(pn^2).$$

(d) If $a \in \,]0, 4[$ is a constant and $\bar{p} = n^{-a}$, then $\mathsf{F}(f) \leq 4/a + 1$. If, in addition, $a < 1$, then $\mathsf{F}(f) = \lfloor 4/a \rfloor + 1$

The proof is omitted due to space constraints (see full version).

To obtain the bounds in Fig. 1, the following facts from random graph theory are needed.

Theorem 3.2 (Matchings in Erdős-Renyi random bipartite graphs, cf., e.g., [17]). *Let $H = (X, Y, E)$ be a random bipartite graph with $|X| = |Y| = n$, and edge probability p.*

(a) If $p \gg 1/n$, then, a.a.s., H has a matching of size $(1 - o(1))n$.

(b) If $p = (\omega(n) + \ln n)/n$ for an ω which tends to ∞ arbitrarily slowly, then, a.a.s, H has a matching of size n.

Theorem 3.3 (Stable sets in Erdős-Renyi random graphs). *Let $G = ([m], E)$ be a random graph with $\{u, v\} \in E$ with edge probability $q = q(m)$.*

(a) *E.g., [17]: Let $\omega = \omega(m)$ tend to ∞ arbitrarily slowly. If $\omega/m \leq q = 1 - \Omega(1)$, then a.a.s., G has a stable set of size at least*

$$2\frac{\ln(qm) - \ln\ln(qm)}{\ln(1 - q)}.$$

(b) *Greedy stable set: If $q = \Omega(1)$, then, a.a.s., G has a stable set of size at least*

$$\frac{\ln(m)}{\ln(1 - q)}.$$

For the region $p = \Theta(1/\sqrt{n})$, there is a corresponding theorem (e.g., [5]). We give here an argument about the expectation based on Turán's theorem. Turán's theorem in the version for stable sets [1] states that in a graph with vertex set V, there exists a stable set of size at least

$$\sum_{v \in V} \frac{1}{\deg(v) + 1},$$

where $\deg(v)$ denotes the degree of vertex v. For random graphs on vertex set $V = [m]$ with edge probability $q = c/m$ for a constant c, using Jensen's inequality, we find that there expected size of the largest stable set is at least

$$\mathbf{E}\left(\sum_{v \in V} \frac{1}{\deg(v) + 1}\right) = \sum_{v \in V} \mathbf{E}\left(\frac{1}{\deg(v) + 1}\right)$$

$$\geq \sum_{v \in V} \frac{1}{\mathbf{E}\deg(v) + 1} = \frac{2m}{q(m - 1) + 1} \geq \frac{2m}{c + 1} = \Theta(m).$$

4   Fractional Cover Number and Cover Number

Armed with the fooling set and 1-rectangle-size lower bounds, we can now bound the cover number and the fractional cover number. We start with the easy case $p \leq 1/2$.

Let f be a random Boolean function $X \times Y \to \{0, 1\}$ with parameter p, as usual. If $1/n \ll p \leq 1/2$, we have $\mathsf{C}(f) = (1 - o(1))n$. Indeed, for $p = o(1/\sqrt{n})$, Theorem 3.1(a) gives the lower bound based on the fooling set lower bound. For $1/e \geq p \gg (\ln n)/n)$, Theorem 2.1(a) yields $\mathsf{R}^1(f) = (1 + o(1))pn$, a.a.s., and for $1/e \leq p \leq 1/2$, the value of a in Eq. (1) of Theorem 2.1(b) is 1, so that $\mathsf{R}^1(f) = (1 + o(1))pn$ there, too. We conclude that, a.a.s.,

$$\mathsf{C}(f) \geq \frac{|\operatorname{supp} f|}{\mathsf{R}^1(f)} = \frac{(1 - o(1))pn^2}{(1 - o(1))pn} = (1 - o(1))\,n.$$

As indicated in the introduction, the case $p > 1/2$ is more interesting, both from the application point of view and from the point of view of the proof techniques.

For the remainder of this section, we assume that $p > 1/2$. Define $\bar{p} := 1 - p$, and $\lambda := \bar{p}n$.

4.1   The Fractional Cover Number

We briefly review the definition of the fractional cover number. Let f be a fixed Boolean function, and let R be a random 1-rectangle of f, drawn according to a distribution π. Define

$$\gamma(\pi) := \min\Big\{ \mathop{\mathbf{P}}_{R \sim \pi} \big((x,y) \in R\big) \mid (x,y) \in \operatorname{supp} f \Big\}.$$

The *fractional cover number* is $\mathsf{C}^*(f) := \min_\pi 1/\gamma(\pi)$, where the minimum is taken over all distributions π on the set of 1-rectangles of f.

The following inequalities are well-known [21].

$$\left.\begin{array}{c} \dfrac{|\operatorname{supp} f|}{\mathsf{R}^1(f)} \\[4pt] \mathsf{F}(f) \end{array}\right\} \le \mathsf{C}^*(f) \le \mathsf{C}(f) \underset{(*)}{\le} \big(1 + \ln \mathsf{R}^1(f)\big)\, \mathsf{C}^*(f). \tag{5}$$

Lower Bound. Theorem 2.1(b) allows us to lower bound $\mathsf{C}^*(f)$. Let f be a random Boolean function $X \times Y \to \{0,1\}$ with parameter $p > 1/2$. With $\lambda/n = \bar{p} = 1 - p$, we have a.a.s.,

$$\frac{|\operatorname{supp} f|}{\mathsf{R}^1(f)} \ge \frac{(1 + o(1))pn^2}{(1 + o(1))n/e\ln(1/p)} = (1 + o(1))\, ep\ln(1/p)n \ge (1 - o(1))\, ep\lambda \tag{6}$$

where the last inequality follows from $\bar{p} \le \bar{p} + \bar{p}^2/2 + \bar{p}^3/3 + \cdots = \ln(1/(1-\bar{p}))$. For $\bar{p} = o(1)$, this is asymptotic to $e\lambda$. It is worth noting that the first inequality in (6) becomes an asymptotic equality if $\bar{p} = o(1)$.

Upper Bound. We now give upper bounds on $\mathsf{C}^*(f)$. To prove an upper bound b on the fractional cover number for a fixed function f, we have to give a distribution π on the 1-rectangles of f such that, if R is sampled according to π, we have, for all (x, y) with $f(x, y) = 1$,

$$\mathbf{P}\big((x,y) \in R\big) \ge 1/b.$$

To prove an "a.a.s." upper bound for a random f, we have to show that

$$\mathbf{P}\Big(\exists(x,y)\colon \ \mathbf{P}\big((x,y) \in R \mid f \ \& \ f(x,y) = 1\big) < 1/b\Big) = o(1). \tag{7}$$

Our random 1-rectangle R within the random Boolean function f is sampled as follows. Let K be a random subset of X, by taking each x into K independently, with probability q. Then let $R := K \times L$ be the 1-rectangle generated (see p. 5) by the row-set K, i.e., $L := \{y \mid \forall x \in K\colon f(x,y) = 1\}$.

For $y \in Y$, let the random variable Z_y count the number of $x \in X$ with $f(x, y) = 0$—in other words, the number of zeros in column y—and set $Z := \max_{y \in Y} Z$. For $(x, y) \in X \times Y$, conditioned on f and $f(x, y) = 1$, the probability that $(x, y) \in R$ equals

$$q(1 - q)^{Z_y} \geq q(1 - q)^Z,$$

so that for every positive integer z, using $1/b = q(1 - q)^z$ in (7),

$$\mathbf{P}\Big(\exists(x, y)\colon \ \mathbf{P}\big((x, y) \in R \mid f \ \& \ f(x, y) = 1\big) < q(1 - q)^z\Big) = \mathbf{P}(Z > z). \quad (8)$$

To obtain upper bounds on the fractional cover number, we give a.a.s. upper bounds on Z, and choose q accordingly.

Theorem 4.1. *Let $1/2 > p = 1 - \bar{p} = 1 - \lambda/n$.*

(a) If $\ln n \ll \lambda < n/2$, then, a.a.s., $(1 - o(1)) \, pe\lambda \leq \mathsf{C}^*(f) \leq (1 + o(1)) \, e\lambda$
(b) If $\lambda = \Theta(\ln n)$, then, a.a.s., $\mathsf{C}^*(f) = \Theta(\ln n)$.
(c) If $1 \ll \lambda = o(\ln n)$, then, a.a.s.,

$$(1 - o(1)) \, \lambda \leq \mathsf{C}^*(f) \leq (1 + o(1)) \, e \max\Big(2\lambda, \frac{\ln n}{\ln((\ln n)/\lambda)}\Big)$$

To summarize, we can determine the fractional cover number accurately in the region $\ln n \ll \lambda \ll n$. For $\lambda = \Theta(\ln n)$ and for $\lambda = \Theta(n)$, we can determine $\mathsf{C}^*(f)$ up to a constant. However, for $\lambda = o(\ln n)$, there is a large gap between our upper and lower bounds.

Proof. The lower bounds follow from the discussion above.
   Proof of the upper bound in (a). For every constant $t > 0$, let

$$\psi(t) := 1/\big((1 + t)\ln(1 + t) - t\big).$$

With

$$h(t) = h(t, n) := \frac{\lambda}{\psi(t)\ln n},$$

using the a standard Chernoff estimate (Theorem 2.1, Eq. (2.5) in [17]) we find that

$$\mathbf{P}\big(Z_1 \geq (1 + t)\lambda\big) \leq e^{-\lambda/\psi(t)} \leq e^{-h(t)}n,$$

so that, by the union bound,

$$\mathbf{P}\big(Z \geq (1 + t)\lambda\big) \leq e^{-h(t)}. \quad (10)$$

For every fixed $t > 0$, $h(t)$ tends to infinity with n, so that the RHS in (10) is $o(1)$. Using that in (8), we obtain

$$\mathbf{P}\Big(\exists(x, y)\colon \ \mathbf{P}\big((x, y) \in R \mid f \ \& \ f(x, y) = 1\big) < q(1 - q)^{(1+t)\lambda}\Big) = \mathbf{P}(Z > (1 + t)\lambda) = o(1),$$

and, taking $q := \frac{1}{(1+t)\lambda}$, we obtain, a.a.s.,

$$C^*(f) \le \frac{1}{q(1-q)^{(1+t)\lambda}} \le \frac{1+t}{1+\frac{1}{(1+t)\lambda}} e\lambda,$$

where we used $(1-\varepsilon)^k \ge (1-k\varepsilon^2)e^{-k\varepsilon}$ for $\varepsilon < 1$. Since this is true for every $t > 0$, we conclude that, a.a.s., $C^*(f) \le (1 - o(1))e\lambda$.

Proof of the upper bounds in (b), (c). Here we use a slightly different Chernoff bound: it is almost exactly Theorem 5.4 in [25], except that we allow $\lambda \to \infty$ slowly:

Lemma 1. *Let* $\bar{p} = \lambda/n$ *with* $1 < \lambda = o(n)$, *and* $2\lambda \le \alpha \le n/2$. *The probability that a* $\mathrm{Bin}(n, \bar{p})$ *random variable is at least* α *is at most*

$$O(1/\sqrt{\alpha}) \cdot e^{-\lambda} \left(\frac{e\lambda}{\alpha}\right)^{\alpha}. \tag{11}$$

No we can proceed with the main proof. For (b), suppose that $\lambda \le C \ln n$ for a constant $C > 1$. Using Lemma 1 with $\alpha = e^2 C \ln n$, we obtain

$$\mathbf{P}(Z_1 \ge e^2 C \ln n) = O(1/\sqrt{\ln n})e^{-\lambda}\left(\frac{eC \ln n}{e^2 C \ln n}\right)^{\alpha} = O(1/\sqrt{\ln n})e^{-\ln n}.$$

and thus

$$\mathbf{P}(Z \ge e^2 C \ln n) = o(1).$$

We conclude similarly as above: with $q := \frac{1}{e^2 C \ln n}$ we obtain, a.a.s., $C^*(f) \le e^3 C \ln n$.

Finally, for (c), if $\lambda = o(\ln n)$, let $\varepsilon > 0$ be a constant, and use Lemma 1 again, with

$$\alpha := \max\left(2\lambda, \frac{(1+\varepsilon)\ln n}{\ln\left(\frac{\ln n}{e\lambda}\right)}\right).$$

We find that

$$\mathbf{P}(Z_1 \ge \alpha) = o(e^{-\alpha \ln(\alpha/e\lambda)}),$$

and the usual calculation shows that $\alpha \ln(\alpha/e\lambda) \ge \ln n$, which implies

$$\mathbf{P}(Z \ge \alpha) = o(1).$$

Conclude similarly as above, with $q := \frac{1}{\alpha}$, we obtain, a.a.s.,

$$C^*(f) \le e\alpha = e \max\left(2\lambda, (1+\varepsilon)\frac{\ln n}{\ln\left(\frac{\ln n}{e\lambda}\right)}\right).$$

One obtains the statement in the theorem by letting ε tend to 0; the e-factor in the denominator of the ln of the denominator in α is irrelevant as $n \to \infty$.

The Cover Number. Inequality $(*)$ in (5) gives us corresponding upper bounds on the cover number.

Corollary 2. *We have $(1 - o(1))\lambda \leq C(f)$, and:*

(a) if $\ln n \ll \lambda = O(n/\ln n)$, then, a.a.s., $C(f) = O(\lambda \ln n)$;
(b) if $\lambda = \Theta(\ln n)$, then, a.a.s., $C(f) = O(\ln^2 n)$;

(c) if $1 \ll \lambda = o(\ln n)$, then, a.a.s., $C(f) = O\left(\max\left(\lambda \ln n, \dfrac{\ln^2 n}{\ln((\ln n)/\lambda)}\right)\right).$ ☐

4.2   Binary-Logarithm of the Number of Distinct Rows, and the Ratio C/C*

When we view f as a matrix, the binary logarithm of the number of distinct rows is a lower bound on the cover number of f [21]. We have the following.

Proposition 1

(a) If $1/2 \geq \bar{p} = \Omega(1/n)$, then, a.a.s., the 2-Log lower bound on $C(f)$ is $(1 - o(1))\log_2 n$.
(b) If $\bar{p} = n^{-\gamma}$ for $1 < \gamma \leq 3/2$, then a.a.s., the 2-Log lower bound on $C(f)$ is $(1 - o(1))(2 - \gamma)\log_2 n$.

Proof. Directly from the following Lemma 2 about the number of distinct rows, with $\lambda = n^{1-\gamma}$.

Lemma 2

(a) If $1/2 \geq \bar{p} = \Omega(1/n)$, then, a.a.s., f has $\Theta(n)$ distinct non-zero rows.
(b) With $\bar{p} = \lambda/n$, if $n^{-1/2} \leq \lambda \leq 1/2$, then, a.a.s., f has $\Omega(\lambda n)$ distinct non-zero rows.

(The constants in the big-Omegas are absolute)

Erdős-Renyi random graphs have the property that the chromatic number is within a small constant factor from the lower bound one obtains from the independence ratio. For the cover number of Boolean functions, this is not the case. Indeed, Theorem 4.1(c), together with Proposition 1, shows that, a.a.s.,

$$\frac{C(f)}{C^*(f)} \geq (1 + o(1))\frac{\log_2 n}{\frac{\ln n}{\ln\left(\frac{\ln n}{\lambda}\right)}} = \Omega\left(\ln\left(\frac{\ln n}{\lambda}\right)\right),$$

which is $\Omega(\ln \ln n)$ if $\lambda = \ln^{o(1)} n$.

This gap is more pronounced in the (not quite as interesting) situation when $\lambda = o(1)$. Consider, e.g., $\lambda = n^{-\varepsilon}$, for some $\varepsilon = \varepsilon(n) = o(1/\ln \ln n)$, say. Similarly to the proofs of Theorem 4.1, we obtain that $C^*(f) \leq e \max(10, 2/\varepsilon)$. (The max-term comes from the somewhat arbitrary upper bound $Z \leq \max(10, 2/\varepsilon)$.) For the Log-2 lower bound on the cover number, we have $(1 - \varepsilon)\log_2 n$, by Proposition 1, and thus

$$\frac{C(f)}{C^*(f)} = \Omega(\varepsilon \ln n).$$

Acknowledgments. The authors would like to thank the anonymous referees for their valuable comments.

Dirk Oliver Theis is supported by Estonian Research Council, ETAG (*Eesti Teadusagentuur*), through PUT Exploratory Grant #620. Mozhgan Pourmoradnasseri is recipient of the Estonian IT Academy Scholarship. This research is supported by the European Regional Fund through the Estonian Center of Excellence in Computer Science, EXCS.

References

1. Alon, N., Spencer, J.H.: The Probabilistic Method. Wiley, New York (2008)
2. Beasley, L.B., Klauck, H., Lee, T., Theis, D.O.: Communication complexity, linear optimization, and lower bounds for the nonnegative rank of matrices (dagstuhl seminar 13082). Dagstuhl Rep. **3**(2), 127–143 (2013)
3. Bollobás, B.: Random Graphs. Cambridge Studies in Advanced Mathematics, vol. 73, 2nd edn. Cambridge University Press, Cambridge (2001)
4. Braun, G., Fiorini, S., Pokutta, S.: Average case polyhedral complexity of the maximum stable set problem. In: Approximation, Randomization, and Combinatorial Optimization. Algorithms and Techniques, APPROX/RANDOM 2014, Barcelona, Spain, 4–6 September 2014, pp. 515–530 (2014). http://dx.doi.org/10.4230/LIPIcs. APPROX-RANDOM.2014.515
5. Dani, V., Moore, C.: Independent sets in random graphs from the weighted second moment method. In: Goldberg, L.A., Jansen, K., Ravi, R., Rolim, J.D.P. (eds.) APPROX/RANDOM 2011. LNCS, vol. 6845, pp. 472–482. Springer, Heidelberg (2011). doi:10.1007/978-3-642-22935-0_40
6. Dawande, M., Keskinocak, P., Swaminathan, J.M., Tayur, S.: On bipartite and multipartite clique problems. J. Algorithms **41**(2), 388–403 (2001). http://dx.doi. org/10.1006/jagm.2001.1199
7. Dawande, M., Keskinocak, P., Tayur, S.: On the biclique problem in bipartite graphs. Carnegie Mellon University (1996). GSIA Working Paper
8. Dietzfelbinger, M., Hromkovič, J., Schnitger, G.: A comparison of two lower-bound methods for communication complexity. Theoret. Comput. Sci. **168**(1), 39–51 (1996). http://dx.doi.org/10.1016/S0304-3975(96)00062-X, 19th International Symposium on Mathematical Foundations of Computer Science, Košice (1994)
9. Fiorini, S., Kaibel, V., Pashkovich, K., Theis, D.O.: Combinatorial bounds on nonnegative rank and extended formulations. Discrete Math. **313**(1), 67–83 (2013)
10. Fiorini, S., Massar, S., Pokutta, S., Tiwary, H.R., Wolf, R.: Linear vs. semidefinite extended formulations: exponential separation and strong lower bounds. In: STOC (2012)
11. Fiorini, S., Massar, S., Pokutta, S., Tiwary, H.R., Wolf, R.D.: Exponential lower bounds for polytopes in combinatorial optimization. J. ACM (JACM) **62**(2), 17 (2015)
12. Froncek, D., Jerebic, J., Klavžar, S., Kovár, P.: Strong isometric dimension, biclique coverings, and sperner's theorem. Comb. Probab. Comput. **16**(2), 271–275 (2007). http://dx.doi.org/10.1017/S0963548306007711
13. Goemans, M.X.: Smallest compact formulation for the permutahedron. Math. Program. **153**(1), 5–11 (2015)
14. Hajiabolhassan, H., Moazami, F.: Secure frameproof code through biclique cover. Discrete Math. Theor. Comput. Sci. **14**(2), 261–270 (2012). http://www. dmtcs.org/dmtcs-ojs/index.php/dmtcs/article/view/2131/4075

15. Hajiabolhassan, H., Moazami, F.: Some new bounds for cover-free families through biclique covers. Discrete Math. **312**(24), 3626–3635 (2012)
16. Izhakian, Z., Janson, S., Rhodes, J.: Superboolean rank and the size of the largest triangular submatrix of a random matrix. Proc. Am. Math. Soc. **143**(1), 407–418 (2015)
17. Janson, S., Łuczak, T., Rucinski, A.: Random Graphs. Wiley-Interscience Series in Discrete Mathematics and Optimization. Wiley-Interscience, New York (2000)
18. Kaibel, V.: Extended formulations in combinatorial optimization. Optima - Math. Optim. Soc. Newsl. **85**, 2–7 (2011). www.mathopt.org/Optima-Issues/optima85. pdf
19. Karp, R.M., Sipser, M.: Maximum matchings in sparse random graphs. In: FOCS, pp. 364–375 (1981)
20. Klauck, H., Lee, T., Theis, D.O., Thomas, R.R.: Limitations of convex programming: lower bounds on extended formulations and factorization ranks (dagstuhl seminar 15082). Dagstuhl Rep. **5**(2), 109–127 (2015)
21. Kushilevitz, E., Nisan, N.: Communication Complexity. Cambridge University Press, Cambridge (1997)
22. Lonardi, S., Szpankowski, W., Yang, Q.: Finding biclusters by random projections. In: Sahinalp, S.C., Muthukrishnan, S., Dogrusoz, U. (eds.) CPM 2004. LNCS, vol. 3109, pp. 102–116. Springer, Heidelberg (2004). doi:10.1007/978-3-540-27801-6_8
23. Lonardi, S., Szpankowski, W., Yang, Q.: Finding biclusters by random projections. Theor. Comput. Sci. **368**(3), 217–230 (2006)
24. Lovás, L., Saks, M.: Communication complexity and combinatorial lattice theory. J. Comput. Syst. Sci. **47**, 322–349 (1993)
25. Mitzenmacher, M., Upfal, E.: Probability and Computing – Randomized Algorithms and Probabilistic Analysis. Cambridge University Press, Cambridge (2006)
26. Park, G., Szpankowski, W.: Analysis of biclusters with applications to gene expression data. In: International Conference on Analysis of Algorithms. DMTCS Proc. AD, vol. 267, p. 274 (2005)
27. Roughgarden, T.: Communication complexity (for algorithm designers). arXiv preprint arXiv:1509.06257 (2015)
28. Schrijver, A.: Combinatorial Optimization. Polyhedra and Efficiency. Algorithms and Combinatorics, vol. 24. Springer, Berlin (2003)
29. Sun, X., Nobel, A.B.: On the size and recovery of submatrices of ones in a random binary matrix. J. Mach. Learn. Res **9**, 2431–2453 (2008)
30. Yannakakis, M.: Expressing combinatorial optimization problems by linear programs. J. Comput. Syst. Sci. **43**(3), 441–466 (1991). http://dx.doi.org/10.1016/0022-0000(91)90024-Y

The Smoothed Number of Pareto-Optimal Solutions in Non-integer Bicriteria Optimization

Heiko Röglin and Clemens Rösner[(⊠)]

Department of Computer Science, University of Bonn, Bonn, Germany
{roeglin,roesner}@cs.uni-bonn.de

Abstract. Pareto-optimal solutions are one of the most important and well-studied solution concepts in multi-objective optimization. Often the enumeration of all Pareto-optimal solutions is used to filter out unreasonable trade-offs between different criteria. While in practice, often only few Pareto-optimal solutions are observed, for almost every problem with at least two objectives there exist instances with an exponential number of Pareto-optimal solutions. To reconcile theory and practice, the number of Pareto-optimal solutions has been analyzed in the framework of smoothed analysis, and it has been shown that the expected value of this number is polynomially bounded for linear integer optimization problems. In this paper we make the first step towards extending the existing results to non-integer optimization problems. Furthermore, we improve the previously known analysis of the smoothed number of Pareto-optimal solutions in bicriteria integer optimization slightly to match its known lower bound.

1 Introduction

Optimization problems that arise from real-world applications often come with multiple objective functions. Since there is usually no solution that optimizes all objectives simultaneously, trade-offs have to be made. One of the most important solution concept in multi-objective optimization is that of *Pareto-optimal solutions*, where a solution is called Pareto-optimal if there does not exist another solution that is simultaneously better in all objectives. Intuitively Pareto-optimal solutions represent the reasonable trade-offs between the different objectives, and it is a common approach to compute the set of Pareto-optimal solutions to filter out all unreasonable trade-offs.

For many multi-objective optimization problems there exist algorithms that compute the set of Pareto-optimal solutions in polynomial time with respect to the input size and the number of Pareto-optimal solutions. These algorithms are not efficient in the worst case because for almost every problem with two or more objectives there exist instances with an exponential number of Pareto-optimal solutions. Since this does not reflect experimental results, where the number of Pareto-optimal solutions is usually small, there has been a significant interest in probabilistic analyses of multi-objective optimization problems in the last decade.

This research was supported by ERC Starting Grant 306465 (BeyondWorstCase).

T.V. Gopal et al. (Eds.): TAMC 2017, LNCS 10185, pp. 543–555, 2017.
DOI: 10.1007/978-3-319-55911-7_39

The analyses in the literature are restricted to *linear integer optimization problems*, in which the solutions can be encoded as integer vectors and there is a constant number of linear objective functions to be optimized. To be more precise, an instance of a linear integer optimization problem is given by a set $S \subseteq \{-k, -k+1, \ldots, k\}^n$ of feasible solutions for some $k \in \mathbb{N}$ and d linear objective functions $c^1, \ldots, c^d$ for some constant d. The function $c^i : S \to \mathbb{R}$ is of the form $c^i(x) = c_1^i x_1 + \ldots + c_n^i x_n$ for $x = (x_1, \ldots, x_n)$. Many well-known optimization problems can be formulated as a linear integer optimization problem. Consider, for example, the bicriteria shortest path problem in which one has to find a path in a graph $G = (V, E)$ from some source node s to some target node t and every edge has a certain length and induces certain costs. Then every s-t-path has a total length and total costs, and ideally one would like to minimize both simultaneously. A given instance of the bicriteria shortest path problem can easily be formulated as an instance of a linear bicriteria integer optimization problem by choosing $S \subseteq \{0, 1\}^{|E|}$ as the set of incidence vectors of s-t-paths. Then the coefficients in the two linear objective functions coincide with the edge lengths and costs.

A particular well-studied case are linear integer optimization problems with two objective functions. In the worst case it is very easy to come up with instances that have an exponential number of Pareto-optimal solutions. On the contrary, it has been proven that the expected number of Pareto-optimal solutions is polynomially bounded if the coefficients of one of the two objective functions are chosen at random, regardless of the choice of $S \subseteq \{-k, -k+1, \ldots, k\}^n$. This is not only true if the coefficients are chosen uniformly at random but also for more sophisticated probabilistic models like smoothed analysis, in which the coefficients can roughly be determined by an adversary and are only slightly perturbed at random. Furthermore, it suffices if only the coefficients of one of the objective functions are chosen at random; the other objective function can be adversarial and does not even have to be linear. This can be seen as a theoretical explanation for why in experiments usually only few Pareto-optimal solutions exist because already a small amount of random noise in the coefficients suffices to render it very unlikely to encounter an instance with many Pareto-optimal solutions.

The analyses in the literature are restricted to the case that the set of solutions is a subset of a discrete set $\{-k, -k+1, \ldots, k\}^n$. Consider, for example, the binary case $S \subseteq \{0, 1\}^n$, and assume that we allow a little bit more flexibility in choosing the set of feasible solutions as follows: every solution $x \in S \subseteq \{0, 1\}^n$ may be replaced by a solution $\bar{x}$ with $|x_i - \bar{x}_i| \leq \varepsilon$ for every component i for a small ε. This way a new set of feasible solutions $\bar{S} \subseteq [-\varepsilon, 1+\varepsilon]^n$ is obtained. Intuitively, if ε is very small, then the expected number of Pareto-optimal solutions with respect to S and with respect to $\bar{S}$ should be roughly the same. However, this is not covered by the previous analyses and indeed analyzing the expected number of Pareto-optimal solutions with respect to $\bar{S}$ seems to be a much harder problem.

In this paper we initiate the study of more general sets of feasible solutions. We do not solve the problem in full generality but we will make the first step towards understanding non-discrete sets of feasible solutions. The idea we use to obtain bounds for the expected number of Pareto-optimal solutions for the more general setting allows us also to improve slightly the best known bound for the bicriteria integer case, matching a known lower bound. In the following, we will first give a motivating example and then we will discuss our results and the previous work in more detail.

1.1   Knapsack Problem with Substitutes

In the knapsack problem, a set of n items with profits $p_1, \ldots, p_n$ and weights $w_1, \ldots, w_n$ is given. The goal is to find a vector $x \in \{0,1\}^n$ such that the total profit $p(x) = p_1 x_1 + \ldots + p_n x_n$ is maximized under the constraint that the total weight $w(x) = w_1 x_1 + \ldots + w_n x_n$ does not exceed a given capacity B. If one disregards the capacity, one can view the knapsack problem as a bicriteria optimization problem in which one seeks a solution from the set $\mathcal{S} = \{0,1\}^n$ with large profit and small weight. The assumption that profit and weight are linear functions is not always justified. If some items are substitute or complementary goods, then their joint profit can be smaller or larger than the sum of their single profits. Also if the weights represent costs and one gets a volume discount, the weight function is not linear.

In order to take this into account, we consider a more general version of the knapsack problem. We allow an arbitrary weight function $w : \{0,1\}^n \to \mathbb{R}$ that assigns a weight to every subset of items. Furthermore, we assume that some function $\alpha : \{0,1\}^n \to [0,1]^n$ is given and that the profit of a solution $x \in \{0,1\}^n$ is given as $p(x) = \alpha(x)_1 p_1 + \ldots + \alpha(x)_n p_n$. Hence, the function α determines for each item and each solution the fraction of the item's value that it contributes. One could, for example, encode rules like "if the second item is present, the first item counts only half, and if the second and third item are present, then the first item counts only a third".

The question we study in this paper is how many solutions from $\{0,1\}^n$ are Pareto-optimal with respect to the objective functions p and w. Formally, a solution $x \in \{0,1\}^n$ is Pareto-optimal if there does not exist a solution $y \in \{0,1\}^n$ that *dominates* x in the sense that y is at least as good as x in all criteria and strictly better than x in at least one criterion. Observe that we can reformulate the model in the following way so that it fits to our discussion above: We define $\bar{\mathcal{S}} = \{\alpha(x) \mid x \in \{0,1\}^n\} \subseteq [0,1]^n$, $\bar{w} : \bar{\mathcal{S}} \to \mathbb{R}$ by $\bar{w}(x) = w(\alpha^{-1}(x))$ for $x \in \bar{\mathcal{S}}$, assuming that α is injective, and $\bar{p}(x) = p_1 x_1 + \ldots + p_n x_n$. Now the goal is to minimize the arbitrary objective function $\bar{w}$ and to maximize the linear objective function $\bar{p}$ over the set $\bar{\mathcal{S}} \subseteq [0,1]^n$.

As even for the simple linear case one can easily find instances in which every solution from $\{0,1\}^n$ is Pareto-optimal, it does not make sense to study this question in a classical worst-case analysis. We will instead assume that the profits $p_1, \ldots, p_n$ are chosen at random and we will prove polynomial bounds for this case under the assumption that $\alpha(x)_i = 0$ for $x_i = 0$ and $\alpha(x)_i \geq \delta$

for $x_i = 1$ for some $\delta > 0$ for every $x \in \{0,1\}^n$ and every i (i.e., any item that is not part of a solution does not contribute any of its profit and any item that is part of a solution contributes at least some small fraction δ of its profit). In the literature only the case that α is the identity has been studied.

Since, we are only interested in the expected number of Pareto-optimal solutions, we do not care how the functions w and α are encoded. Our results are true for all functions, regardless of whether or not they can be encoded and evaluated efficiently.

1.2   Smoothed Analysis

Smoothed analysis has been introduced by Spielman and Teng [15] to explain why the simplex algorithm is efficient in practice despite its exponential worst-case behavior. We use the framework of smoothed analysis to study the number of Pareto-optimal solutions, and we will use the following model, which has already been used in the literature for the analysis of multi-objective linear optimization problems.

In our model, we assume that an arbitrary set $\mathcal{S} \subseteq [0,1]^n$ of feasible solutions that satisfies a certain property, which we define below, can be chosen by an adversary. Furthermore, the adversary can also choose an arbitrary objective function $w : \mathcal{S} \to \mathbb{R}$, which is to be minimized. Finally a second linear objective function $p : \mathcal{S} \to \mathbb{R}$ is given, which is to be maximized. This function is of the form $p(x) = p_1 x_1 + \ldots + p_n x_n$ and in contrast to a worst-case analysis we do not allow the adversary to choose the coefficients $p_1, \ldots, p_n$ exactly but we assume that they are chosen at random. For this, let $\phi \geq 1$ be some parameter and assume that the adversary can choose, for each coefficient p_i, a probability density function $f_i : [0,1] \to [0, \phi]$ according to which p_i is chosen independently of the other profits.

The smoothing parameter ϕ can be seen as a measure specifying how close the analysis is to a worst-case analysis. The larger ϕ, the more concentrated the probability mass can be: the adversary could for example define for each coefficient a uniform distribution on an interval of length $\frac{1}{\phi}$ from which it is chosen uniformly at random. This shows that for $\phi \to \infty$ our analysis approaches a worst-case analysis.

In the following, we will even allow a different parameter ϕ_i for each coefficient c_i, i.e., the density f_i is bounded from above by ϕ_i. Then $\phi = \max_{i \in [n]} \phi_i$, where $[n]$ denotes the set $\{1, \ldots, n\}$. We define the *smoothed number of Pareto-optimal solutions* as the largest expected number of Pareto-optimal solutions that the adversary can achieve by choosing the set $\mathcal{S}$, the objective function w, and the densities $f_1, \ldots, f_n$.

1.3   Previous Results

Multi-objective optimization is a well studied research area. There exist several algorithms to generate Pareto sets of various optimization problems like, e.g., the (bounded) knapsack problem [8,12], the bicriteria shortest path problem

[5,14], and the bicriteria network flow problem [6,11]. The running time of these algorithms depends crucially on the number of Pareto-optimal solutions and, hence, none of them runs in polynomial time in the worst case. In practice, however, generating the Pareto set is tractable in many situations [7,10].

Beier and Vöcking initiated the study of the expected number of Pareto-optimal solutions for binary optimization problems [2]. They consider the model described in Sect. 1.2 with $S \subseteq \{0,1\}^n$ and show that the expected number of Pareto-optimal solutions is bounded from above by $O(n^4\phi)$ and from below by $\Omega(n^2)$ even for $\phi = 1$. In [1] Beier, Röglin, and Vöcking analyze the smoothed complexity of bicriteria integer optimization problems. They show that the smoothed number of Pareto-optimal solutions is bounded from above by $O(n^2k^2\log(k)\phi)$ if $S \subseteq \{0,\ldots,k-1\}^n$. This improved the upper bound for the binary case to $O(n^2\phi)$. They also present a lower bound of $\Omega(n^2k^2)$ on the expected number of Pareto-optimal solutions for profits that are chosen uniformly from the interval $[-1,1]$.

Röglin and Teng generalized the binary setting $S \subseteq \{0,1\}^n$ to an arbitrary constant number $d \geq 1$ of linear objective functions with random coefficients plus one arbitrary objective function [13]. They showed that the smoothed number of Pareto-optimal solutions is in $O((n^2\phi)^{f(d)})$, for a function f that is roughly $f(d) = 2^d d!$. In [9] this bound was significantly improved to $O(n^{2d}\phi^{d(d+1)/2})$ by Moitra and O'Donnel. Brunsch et al. proved in [3] a lower bound of $\Omega(n^{d-1.5}\phi^d)$ for the same setting. Instead of binary optimization problems Brunsch and Röglin analyze the smoothed number of Pareto-optimal solutions for multi-objective integer optimization problems [4]. They consider $S \subseteq \{0,\ldots,k\}^n$ and show that the expected number of Pareto-optimal solutions is in $k^{2(d+1)^2} \cdot O(n^{2d}\phi^{d(d+1)})$.

None of these analyses applies to the case that the set S of feasible solutions is a non-integral subset of $[0,1]^n$.

1.4   Our Results

We study bicriteria optimization problems in which the set S of feasible solutions is a finite subset of $[0,1]^n$ and one wants to optimize one arbitrary objective function $w : S \to \mathbb{R}$ and one linear objective function $p : S \to \mathbb{R}$. We call w weight and p profit. We do not care about the exact values of w and will therefore assume that w is given as a ranking on S where solutions with a lower weight have a higher ranking. In order to obtain interesting results about the number of Pareto-optimal solutions, it is necessary to restrict the set S. We define the (k,δ)- property as follows.

Definition 1. *For given $k \in \mathbb{N}$ and $\delta \in (0,1]$, a set of solutions $S \subseteq [0,1]^n$ satisfies the (k,δ)-property if there exist finite sets $K_i \subseteq [0,1]$ with $|K_i| \leq k$ for $i \in [n]$, such that for each pair of solutions $s \neq s' \in S$ either $|\{i \in [n] \mid s_i \in K_i\}| \neq |\{i \in [n] \mid s_i' \in K_i\}|$ or there exist indices $i,j \in [n]$ such that $s_i \in K_i$, $|s_i - s_i'| \geq \delta$, and $s_j' \in K_j$, $|s_j - s_j'| \geq \delta$.*

Let $\mathcal{S} \subseteq [0,1]^n$ be an instance of the Knapsack Problem with Substitutes, as described in Sect. 1.1. Recall that different solutions $s \neq s' \in \mathcal{S}$ differ in the coordinates with a value of 0, i.e., there exists $i \in [n]$ such that either $s_i = 0 \neq s'_i$ or $s_i \neq 0 = s'_i$. Since the value of each coordinate has to be 0 or at least δ we can set $K_i = \{0\}$ for every $i \in [n]$ and see that $\mathcal{S}$ has the $(1, \delta)$-property.

For finite bicriteria integer optimization problems we have $\mathcal{S} \subseteq \{-k, -k + 1, \ldots, k-1, k\}^n$ for some $k \in \mathbb{N}$. For such sets the definition of the (k, δ)-property does not apply immediately. Instead we can first shift and then scale $\mathcal{S}$ to obtain $\hat{\mathcal{S}} \subseteq \{0 = \frac{0}{2k}, \frac{1}{2k}, \ldots, \frac{2k}{2k} = 1\}^n \subseteq [0,1]^n$ (First add k to every coordinate of every solution and then divide the result by $2k$). This shifting and scaling does not change the profit order and with the same ranking as before the shift, $\hat{\mathcal{S}}$ and $\mathcal{S}$ have the same number of Pareto-optimal solutions. With $K_i = \{0, \frac{1}{2k}, \ldots, 1\}$ it is easy to see that $\hat{\mathcal{S}}$ has the $(2k + 1, \frac{1}{2k})$-property.

In this paper, we present an approach for bounding the smoothed number of Pareto-optimal solutions for bicriteria optimization problems that have a finite set $\mathcal{S} \subseteq [0,1]^n$ of feasible solutions with the (k, δ)-property. The general idea underlying our analysis is similar to the one used by Beier, Röglin, and Vöcking [1] to analyze integer problems. The basic idea is to partition the Pareto-optimal solutions into different classes and to analyze the expected number in each class separately. Roughly the class of a Pareto-optimal solution x is determined by the indices in which it differs from the next Pareto-optimal solution, i.e., the Pareto-optimal solution with smallest weight among all Pareto-optimal solutions with larger profit than x.

To analyze the expected number of Pareto-optimal solutions in one class, we first partition the interval $[0, n]$ of possible profits of solutions from $\mathcal{S}$ uniformly into small subintervals. Then, by linearity of expectation, it suffices to bound for each subinterval I the expected number of Pareto-optimal solutions with a profit in I. Let $I = [t-\varepsilon, t)$ for some t and $\varepsilon > 0$ be such a subinterval. If ε is very small, then with high probability I contains either none or exactly one Pareto-optimal solution. Hence, the expected number of Pareto-optimal solutions in I equals almost exactly the probability that there exists a Pareto-optimal solution whose profit lies in I. In order to bound this probability, we use the principle of deferred decisions as follows. First we uncover all profits except for the profit p_i for one of the positions i in which x differs from its next Pareto-optimal solution. This information suffices to identify a set of candidates for a Pareto-optimal solution in I. That is, if there exists a Pareto-optimal solution in I, then it must come from this set of candidate solutions. Beier, Röglin and Vöcking [1] treated each of these candidates separately and used linearity of expectation to bound the probability that any of them becomes a Pareto-optimal solution with profit in I. This is not possible anymore in our more general setting because there could be an exponential number of candidates. We instead use a new method, in which we exploit dependencies between the different candidates. This allows us to treat the set of candidates as a whole and to obtain a better bound on the probability that one of them becomes a Pareto-optimal solution with profit in I.

Theorem 2. *Let $k \in \mathbb{N}$, $\delta \in (0, 1]$, and let $\mathcal{S} \subseteq [0, 1]^n$ be a set of feasible solutions with the (k, δ)-property and some arbitrary ranking w. Assume that each profit p_i is a random variable with density function $f_i : [0, 1] \to [0, \phi_i]$ and let $\phi = \max_{i \in [n]} \phi_i$. Let q denote the number of Pareto-optimal solutions in $\mathcal{S}$. Then*

$$E[q] = O\left(\frac{n^2}{\delta} \sum_{i=1}^{n} k_i \phi_i\right) = O\left(\frac{n^3 k \phi}{\delta}\right).$$

We will show that every set of solutions $\mathcal{S} \subseteq \{0, \ldots, k - 1\}^n$ can be scaled into a set of solutions $\mathcal{S}' \in [0, 1]^n$ with the $(k, \frac{1}{k})$-property. For bicriteria integer optimization problems we then further improve our analysis to improve the best previous result [1] and match the known lower bound $\Omega(n^2 k^2)$ [1] for constant ϕ.

Theorem 3. *Let $\mathcal{S} \subseteq \mathcal{D}^n$ be a set feasible solutions with a finite domain $\mathcal{D} = \{0, \ldots, k - 1\} \subseteq \mathbb{Z}$ and an arbitrary ranking w. Assume that each profit p_i is a random variable with density function $f_i : [0, 1] \to [0, \phi_i]$ and let $\phi = \max_{i \in [n]} \phi_i$. Let q denote the number of Pareto-optimal solutions in $\mathcal{S}$. Then*

$$E[q] = O\left(nk^2 \sum_{i=1}^{n} \phi_i\right) = O\left(n^2 k^2 \phi\right).$$

We also show a lower bound of $\Omega(\min\{(\frac{1}{\delta})^{\log_3(2)}, 2^n\})$ for the expected number of Pareto-optimal solutions in an instance with the $(1, \delta)$-property, where all profits are drawn according to a uniform distribution on the interval $[\frac{1}{2}, 1]$. This shows that the dependence on δ in Theorem 2 is necessary.

2 Upper Bound on the Expected Number of Pareto-Optimal Solutions

As discussed above, the methods and ideas we use in this chapter are inspired by the analysis of Beier, Röglin and Vöcking [1]. We adapt their approach to the non-integer setting and also improve their analysis of the integer case.

Lemma 4. *Let $k \in \mathbb{N}$, $\ell \in [n]$, $\delta \in (0, 1]$, and let $\mathcal{S} \subseteq [0, 1]^n$ be a set of solutions with the (k, δ)-property. Assume $K_i \subseteq [0, 1]$ with $|K_i| = k_i \leq k$ for $i \in [n]$ to be corresponding sets for the (k, δ)-property of $\mathcal{S}$. Also let $|\{i \in [n] \mid s_i \in K_i\}| = \ell$ for every solution $s \in \mathcal{S}$. Assume that each profit p_i is a random variable with density function $f_i : [0, 1] \to [0, \phi_i]$ and let $\phi = \max_{i \in [n]} \phi_i$. Let q denote the number of Pareto-optimal solutions in $\mathcal{S}$. Then*

$$E[q] \leq \left(\sum_{i=1}^{n} \frac{4nk_i \phi_i}{\delta}\right) + 1 \leq \frac{4n^2 k \phi}{\delta} + 1.$$

Proof. The idea of the proof is to partition the set of Pareto-optimal solutions into different classes and to compute the expected number of Pareto-optimal solutions in each of these classes separately. Let $\mathcal{P} \subseteq \mathcal{S}$ denote the set of all Pareto-optimal solutions. For each Pareto-optimal solution $s \in \mathcal{P}$, except for the one with largest profit, let $\text{next}(s):=\text{argmin}\{p(s') \mid s' \in \mathcal{P} \text{ and } p(s') > p(s)\}$ denote the Pareto-optimal solution with the next larger profit than s. Now let $s \in \mathcal{P}$ be an arbitrary Pareto-optimal solution that is not the one with the largest profit. By definition of the set $\mathcal{S}$, there has to be an $i \in [n]$ such that $\text{next}(s)_i = v$ for some $v \in K_i$ and $|s_i - \text{next}(s)_i| \geq \delta$. We then say that s belongs to the class (i, v). With the Pareto-optimal solution with the largest profit being a separate class by itself, every Pareto-optimal solutions is part of at least one of the classes. Note that a Pareto-optimal solution can belong to several different classes.

Let $i \in [n]$ and $v \in K_i$. We will now analyze the expected number of Pareto-optimal solutions in class (i, v). For this we consider the set

$$\mathcal{S}_{i,v} := \{s' \in [-1,1]^n \mid \exists s \in \mathcal{S} \text{ such that } s'_i = s_i - v \text{ and } \forall j \in [n] \setminus \{i\} : s_j = s'_j\}.$$

For each solution $s \in \mathcal{S}$ the set $\mathcal{S}_{i,v}$ contains a corresponding solution s', which is identical to s except for the i-th coordinate, where it is smaller by v. This does not change the profit order of the solutions because the profit of each solution in $\mathcal{S}_{i,v}$ is smaller by exactly $v \cdot p_i$ than the profit of its corresponding solution in $\mathcal{S}$. Given the same ranking (i.e., the same weight function) on $\mathcal{S}_{i,v}$ as on $\mathcal{S}$, there is a one-to-one correspondence between the Pareto-optimal solutions in $\mathcal{S}$ and $\mathcal{S}_{i,v}$. Hence, instead of analyzing the number of class (i, v) Pareto-optimal solution in $\mathcal{S}$, we can also analyze the number of class $(i, 0)$ Pareto-optimal solutions in $\mathcal{S}_{i,v}$. Instead of class $(i, 0)$ Pareto-optimal solutions, we will use the term *class i Pareto-optimal solutions* in the following. The following lemma concludes the proof by summing over the $\sum_{i=1}^{n} k_i$ different choices for the pair (i, v). Note that the term $+1$ in the lemma accounts for the Pareto-optimal solution with the largest profit.   □

Lemma 5. *Consider the setting described in Lemma 4 and let $i \in [n]$ and $v \in K_i$ be arbitrary. Let q' denote the number of class i Pareto-optimal solutions in $\mathcal{S}_{i,v}$. Then*

$$\boldsymbol{E}[q'] \leq \frac{4n\phi_i}{\delta}.$$

Proof. The key idea is to prove an upper bound on the probability that there exists a class i Pareto-optimal solution in $\mathcal{S}_{i,v}$, whose profit falls into a small interval $[t - \varepsilon, t)$, for arbitrary t and $\varepsilon > 0$. We use the principle of deferred decisions and will assume in the following that all profits p_j for $j \neq i$ are already fixed arbitrarily. We will only exploit the randomness of p_i.

We want to bound the probability that there exists a class i Pareto optimal solution, whose profit lies in the interval $[t-\varepsilon, t)$. Let $\mathcal{S}_{x_i=0} := \{s \in \mathcal{S}_{i,v} \mid s_i = 0\}$. Define x_1^* to be the highest ranked solution $x \in \mathcal{S}_{x_i=0}$ satisfying $p(x) \geq t$ and define $X^* := \{x_1^*, x_2^*, \ldots, x_{m_{t,\varepsilon}}^*\}$ to be the set containing x_1^* and all solutions

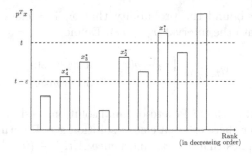

Fig. 1. Example of $\mathcal{S}_{x_i=0}$, with solutions in $X^* = \{x_1^*, x_2^*, x_3^*, x_4^*\}$ marked as such.

$x \in \mathcal{S}_{x_i=0}$ that are Pareto-optimal with respect to the set $\mathcal{S}_{x_i=0}$ and that satisfy $t - \varepsilon \le p(x) < t$. We assume X^* to be ordered such that for all $x_j^* \in X^*$ we have $p(x_j^*) < p(x_{j-1}^*)$ (see Fig. 1).

Note that the solutions in X^* do not have to be Pareto-optimal in $\mathcal{S}_{i,v}$ (they could be dominated by solutions outside of $\mathcal{S}_{x_i=0}$) and that X^* does not have to contain any solutions. If $\mathcal{S}_{i,v}$ contains a class i Pareto-optimal solution x, whose profit falls into the interval $[t - \varepsilon, t)$, we have next$(x) \in \mathcal{S}_{x_i=0}$. Since next$(x)$ is Pareto-optimal in $\mathcal{S}_{i,v}$, it has to be Pareto-optimal in $\mathcal{S}_{x_i=0}$ as well. We claim next$(x) \in X^*$. Assume next$(x) \notin X^*$, then $p(\text{next}(x))$ must be at least t and therefore next(x) must have a lower rank than x_1^*. Since next(x) is Pareto-optimal we must have $p(\text{next}(x)) > p(x_1^*)$. By the definition of next(x) there can be no Pareto-optimal solution x' in $\mathcal{S}_{i,v}$ with $p(x) < p(x') < p(\text{next}(x))$, which means that there can be no Pareto-optimal solution in $\mathcal{S}_{i,v}$ that dominates x_1^* but not next(x), which is a contradiction. Analogously it follows that next(x) is the solution with the highest rank among all $x_j^* \in X^*$ with $p(x_j^*) > p(x)$. If next$(x) = x_j^*$ for some $j < m_{t,\varepsilon}$, then $p(x) \in [p(x_{j+1}^*), p(x_j^*))$, and if next$(x) = x_{m_{t,\varepsilon}}^*$, then $p(x) \in [t - \varepsilon, p(x_{m,\varepsilon}^*))$.

In order to analyze the probability that there exists a class i Pareto-optimal solution, whose profit lies in the interval $[t - \varepsilon, t)$, we look at each $x_j^* \in X^*$. Let $r_1 = t$, $r_{m_{t,\varepsilon}+1} = t - \varepsilon$ and $r_j = p(x_j^*)$ for $j \in \{2, \ldots, m_{t,\varepsilon}\}$. We will, for each $j \in [m_{t,\varepsilon}]$, bound the probability that $\mathcal{S}_{i,v}$ contains a class i Pareto-optimal solution, whose profit lies in the interval $[r_{j+1}, r_j)$.

Let $j \in [m_{t,\varepsilon}]$ and let $\hat{x}_j$ denote the solution that has the largest profit among all solutions x with $|x_i| \ge \delta$ that are higher ranked than x_j^*. Assume that there exists a class i solution x with profit in the interval $[r_{j+1}, r_j)$. Then x_j^* has to be a Pareto-optimal solution and x has to be higher ranked than x_j^*, because otherwise x_j^* would dominate x. Let y denote the solution, among all solutions that are higher ranked than x_j^*, that has the largest profit. Since x_j^* is Pareto-optimal, y is Pareto-optimal as well and has less profit than x_j^*. Since we assume x to be a class i solution with profit in the interval $[r_{j+1}, r_j)$ we know next$(x) = x_j^*$ and therefore $x = y$ and $x = \hat{x}_j$.

We now aim to bound the probability that $\hat{x}_j$ is a class i Pareto-optimal solution and falls into the interval $[r_{j+1}, r_j)$. Define

$$\Lambda(t, j) = \begin{cases} r_j - p(\hat{x}_j) & \text{if } \hat{x}_j \text{ exists} \\ \bot & \text{otherwise.} \end{cases}$$

Let $\mathcal{P}$ denote the set of class i Pareto-optimal solutions and $\varepsilon_j = r_j - r_{j+1}$ for all $j \in [m_{t,\varepsilon}]$. Whenever there exists a class i solution $x \in \mathcal{P}$ with $p(x) \in [r_{j+1}, r_j)$, the choice of $\hat{x}_j$ implies that $x = \hat{x}_j$ and hence $\Lambda(t, j) \in (0, \varepsilon_j]$.
   Then

$$\mathbf{Pr}[\exists x \in \mathcal{P} : p(x) \in [r_{j+1}, r_j)] \leq \mathbf{Pr}[\Lambda(t, j) \in (0, \varepsilon_j]].$$

Since the expected number of class i Pareto-optimal solutions can be written as

$$\int_{-\infty}^{\infty} \lim_{\varepsilon \to 0} \frac{\mathbf{Pr}[\exists x \in \mathcal{P} : p(x) \in [t - \varepsilon, t)]}{\varepsilon} dt$$
$$\leq \int_{-\infty}^{\infty} \lim_{\varepsilon \to 0} \frac{\sum_{j=1}^{m_{t,\varepsilon}} \mathbf{Pr}[\exists x \in \mathcal{P} : p(x) \in [r_{j+1}, r_j)]}{\varepsilon} dt$$
$$\leq \int_{-\infty}^{\infty} \lim_{\varepsilon \to 0} \frac{\sum_{j=1}^{m_{t,\varepsilon}} \mathbf{Pr}[\Lambda(t, j) \in (0, \varepsilon_j]]}{\varepsilon} dt$$
$$= \int_{-n}^{n} \lim_{\varepsilon \to 0} \frac{\sum_{j=1}^{m_{t,\varepsilon}} \mathbf{Pr}[\Lambda(t, j) \in (0, \varepsilon_j]]}{\varepsilon} dt,$$

where the last equality comes from the fact that all solutions have a profit in $[-n, n]$, it remains to analyze the terms $\mathbf{Pr}[\Lambda(t, j) \in (0, \varepsilon_j]]$.
   Let $\mathcal{S}^{|x_i| \geq \delta} = \{x \in \mathcal{S}_{i,v} \mid |x_i| \geq \delta\}$ and $\mathcal{S}^{x_i = u} = \{x \in \mathcal{S}_{i,v} \mid x_i = u\}$. For all $j \in [m_{t,\varepsilon}]$ let $\mathcal{L}_j$ consist of all solutions from $\mathcal{S}^{|x_i| \geq \delta}$ that have a higher rank than x_j^* and let $\mathcal{L}_j^u$ consist of all solutions from $\mathcal{S}^{x_i = u}$ that have a higher rank than x_j^*. Let $\hat{x}_j^u$ denote the lowest ranked Pareto-optimal solution from the set $\mathcal{L}_j^u$, i.e., $\hat{x}_j^u$ has the largest profit among all solutions in $\mathcal{L}_j^u$.
   The identity of x_j^* is completely determined by the profits p_ℓ, $\ell \neq i$. For all $u \in [-1, 1]$ the set $\mathcal{L}_j^u$ and therefore the existence and identity of $\hat{x}_j^u$ are completely determined by those profits as well. Hence, if we fix all profits except for p_i, then $\hat{x}_j^u$ is fixed and its profit is $c_u + up_i$ for some constant c_u that depends only on the profits already fixed. The identity of $\hat{x}_j$ still depends on the exact value of p_i, but independent of p_i it has to be equal to $\hat{x}_j^u$ for some $u \in [-1, 1]$ with $|u| \geq \delta$. More specifically we have $\hat{x}_j = \operatorname{argmax}\{p(\hat{x}_j^u) \mid \hat{x}_j^u \text{ exists and } |u| \geq \delta\}$, which depends on the exact value of p_i. We can view $\{\hat{x}_j^u \mid \hat{x}_j^u \text{ exists and } |u| \geq \delta\}$ as the set of candidates, which could be a a class i Pareto-optimal solution with profit in the interval $[r_{j+1}, r_j)$.
   This means that $\hat{x}_j^u$ takes a profit in the interval $[r_{j+1}, r_j)$ if and only if p_i lies in the interval $[b_u, b_u + \frac{\varepsilon_j}{u}) := [\frac{r_{j+1} - c_u}{u}, \frac{r_j - c_u}{u})$ in case $u > 0$ and $(b_u, b_u + \frac{\varepsilon_j}{-u}] := (\frac{r_j - c_u}{u}, \frac{r_{j+1} - c_u}{u}]$ in case $u < 0$.

Let $b = \min\{b_u \mid u \in [\delta, 1]$ and $\hat{x}_j^u$ exists$\}$ and let $u' = \operatorname{argmin}\{b_u \mid u \in [\delta, 1]$ and $\hat{x}_j^u$ exists$\}$. Then for $p_i < b$ and all $u \in [\delta, 1]$ we have $p(\hat{x}_j^u) < r_{j+1}$ and for $p_i \geq b + \frac{\varepsilon_j}{u'}$ we have $p(\hat{x}_j^u) \geq r_j$. Let $b' = \max\{b_u \mid u \in [-1, -\delta]$ and $\hat{x}_j^u$ exists$\}$ and let $u'' = \operatorname{argmax}\{b_u \mid u \in [-1, -\delta]$ and $\hat{x}_j^u$ exists$\}$. Then for $p_i' \leq b'$ we have $p(\hat{x}_j^{u''}) \geq r_j$ and for $p_i > b' + \frac{\varepsilon_j}{-u''}$ and all $u \in [-1, -\delta]$ we have $p(\hat{x}_j^u) < r_{j+1}$. This implies that for all $p_i \notin [b, b + \frac{\varepsilon_j}{u'}] \cup (b', b' + \frac{\varepsilon_j}{-u''}]$ we have $p(\hat{x}_j) \notin [r_{j+1}, r_j)$. Hence we obtain $\mathbf{Pr}[\Lambda(t, j) \in (0, \varepsilon_j]] \leq \varepsilon_j \left(\frac{\phi_i}{u'} + \frac{\phi_i}{-u''} \right) \leq \varepsilon_j \frac{2\phi_i}{\delta}$.

Now we can bound the expected number of Pareto-optimal solutions:

$$\mathbf{E}[q'] \leq \int_{-n}^{n} \lim_{\varepsilon \to 0} \frac{\sum_{j=1}^{m_{t,\varepsilon}} \varepsilon_j \frac{2\phi_i}{\delta}}{\varepsilon} dt = \int_{-n}^{n} \lim_{\varepsilon \to 0} \frac{\varepsilon \frac{2\phi_i}{\delta}}{\varepsilon} dt = \int_{-n}^{n} \lim_{\varepsilon \to 0} \frac{2\phi_i}{\delta} dt = \frac{4n\phi_i}{\delta}.$$

□

We now show Theorem 2.

Proof (Proof (Theorem 2)). For $\ell \in [n]$ let $\mathcal{S}_\ell = \{s \in \mathcal{S} \mid |\{i \in [n] \mid s_i \in K_i\}| = \ell\}$, where the $K_i \subseteq [0, 1]$ with $|K_i| = k_i \leq k$ for $i \in [n]$ denote the corresponding sets for the (k, δ)-property of $\mathcal{S}$. Let $\mathcal{P}_\ell$ denote the set of Pareto-optimal solutions in $\mathcal{S}_\ell$ and let $\mathcal{P}$ be the set of Pareto-optimal solutions in $\mathcal{S}$. Let $s \in \mathcal{P}$ be a Pareto-optimal solution in $\mathcal{S}$. Then there exists no solution $s' \in \mathcal{S}$ that dominates s. For some $\ell \in [n]$ we have $s \in \mathcal{S}_\ell$. $\mathcal{S}_\ell \subseteq \mathcal{S}$ implies that no solution $s' \in \mathcal{S}_\ell$ dominates s. Therefore s is also Pareto-optimal in $\mathcal{S}_\ell$, i.e., $s \in \mathcal{P}_\ell$. This implies $\mathcal{P} \subseteq \bigcup_{\ell \in [n]} \mathcal{P}_\ell$. Let q_ℓ denote the number of Pareto-optimal solutions in $\mathcal{S}_\ell$, i.e., $q_\ell = |\mathcal{P}_\ell|$. Lemma 4 and linearity of expectation yield

$$\mathbf{E}[q] \leq \sum_{\ell \in [n]} \mathbf{E}[q_\ell] + 1 \leq \sum_{\ell \in [n]} \left(\frac{4n}{\delta} \sum_{i=1}^{n} k_i \phi_i + 1 \right) + 1$$

$$= \frac{4n^2}{\delta} \sum_{i=1}^{n} k_i \phi_i + n + 1 \leq \frac{4n^3 k\phi}{\delta} + n + 1.$$

Here the additional 1 comes from a possible solution $s \in \mathcal{S}$ with $|\{i \in [n] \mid s_i \in K_i\}| = 0$. The (k, δ)-property ensures that there can exist at most one such solution. This concludes the proof.   □

We now prove Theorem 3.

Proof (Proof (Theorem 3)). We take a look at the scaled version $\mathcal{S}' = \{\frac{1}{k}s \mid s \in \mathcal{S}\}$, where $\frac{1}{k}s$ denotes the solution s' with $s_i' = \frac{1}{k}s_i$ for all $i \in [n]$. Since this scaling operation changes the profit of every solution by a factor of $\frac{1}{k}$, the two sets $\mathcal{S}'$ and $\mathcal{S}$ have the same number of Pareto-optimal solutions. Setting $K = \{\frac{i}{k} \mid i \in \{0, \ldots, k-1\}\}$ and $\delta = \frac{1}{k}$ we can apply Lemma 4 to obtain

$$\mathbf{E}[q] \leq \sum_{i=1}^{n} \frac{4nk_i\phi_i}{1/k} + 1 \leq 4nk^2 \sum_{i=1}^{n} \phi_i + 1 = O\left(nk^2 \sum_{i=1}^{n} \phi_i \right)$$

□

3   A Lower Bound

In this section we will show a simple lower bound for the expected number of Pareto-optimal solutions. For a given $\delta \in (0,1]$ we will show how to find a set of solutions $\mathcal{S} \subseteq (\{0\} \cup [\delta,1])^n$ with the $(1,\delta)$-property such that the number of Pareto-optimal solutions in $\mathcal{S}$ is $\Omega(\min\{2^n, (\frac{1}{\delta})^{\log_3(2)}\})$, assuming that all profits are drawn uniformly at random from the interval $[\frac{1}{2},1]$. Furthermore, the coordinates of the solutions in $\mathcal{S}$ will take at most $n+1$ different values, showing that a bound on the number of different values alone is not sufficient to obtain a polynomial bound on the number of Pareto-optimal solutions.

Theorem 6. *Suppose profits are drawn according to a uniform distribution from the interval $[\frac{1}{2},1]$. Then for every $\delta \in (0,1]$ there exists a set $\mathcal{S} \subseteq (\{0\} \cup [\delta,1])^n$ with the $(1,\delta)$-property and a ranking on $\mathcal{S}$ such that the number of Pareto-optimal solutions in $\mathcal{S}$ is $\Omega(\min\{(\frac{1}{\delta})^{\log_3(2)}, 2^n\})$.*

Proof. For $i \in [n]$, let $x_i = \frac{1}{3^{i-1}}$ and let $\mathcal{S}' = \{0,x_1\} \times \{0,x_2\} \times \ldots \times \{0,x_n\}$. The choice of x_i guarantees that for all $i \in [n]$ we get $\frac{x_i}{2} \geq \sum_{j=i+1}^{n} x_j$. This implies that regardless of how the values of the profits p_i for $i \in [n]$ are chosen, the lexicographical order of the solutions is equal to their profit order. When we use the lexicographical order as our ranking as well, this implies that all solutions are Pareto-optimal. We will define $\mathcal{S} = \{s \in \mathcal{S}' \mid \forall i : s_i \in \{0\} \cup [\delta,1]\}$ to be the subset of $\mathcal{S}'$ that contains only the solutions, whose coordinates have values of 0 or at least δ. We get $\mathcal{S} = \{0,x_1\} \times \{0,x_2\} \times \ldots \times \{0,x_{\lfloor \log_3 \frac{1}{\delta}\rfloor+1}\} \times \{0\} \times \ldots \times \{0\}$ for the case $\lfloor \log_3 \frac{1}{\delta}\rfloor + 1 < n$ and $\mathcal{S} = \mathcal{S}'$ otherwise. With $K_i = \{0\}$ for all $i \in [n]$ we can see that $\mathcal{S}$ has the $(1,\delta)$-property. The set $\mathcal{S}$ contains $\min\{2^n, 2^{\lfloor \log_3 \frac{1}{\delta}\rfloor+1}\}$ different solutions, and as we have seen, all solutions are Pareto-optimal. The observation that $(\frac{1}{\delta})^{\log_3(2)} = 2^{\log_3(\frac{1}{\delta})}$ concludes the proof. $\square$

4   Conclusion and Open Problems

We defined for bicriteria optimization problems with a finite set of solutions $\mathcal{S} \subseteq [0,1]^n$ the (k,δ)-property and showed how to obtain an upper bound for the smoothed number of Pareto-optimal solutions in instances with the (k,δ)-property. It is easy to see that the (k,δ)-property can be applied to any finite set of solutions. However, in general δ can be arbitrarily small.

Lemma 7. *Let $\mathcal{S} \subseteq [0,1]^n$ be a finite set of solutions. There exist $k \in \mathbb{N}$ and $\delta \in (0,1]$, such that $\mathcal{S}$ has the (k,δ)-property.*

Proof. Let $E_{\mathcal{S}} = \{x \in [0,1] \mid \exists s \in \mathcal{S}, i \in [n] : s_i = x\}$ denote the set of values that the coordinates of the solutions in $\mathcal{S}$ take. Let $k = |E_{\mathcal{S}}|$ and $\delta = \min_{x \neq y \in E_{\mathcal{S}}} |x - y|$. Since $\mathcal{S}$ is a finite set, this is well defined and we get $k \in \mathbb{N}$ and $\delta \in (0,1]$. We choose $K_i = E_{\mathcal{S}}$ for all $i \in [n]$. Let $s \neq s' \in \mathcal{S}$ be two different solutions then there must exist $i \in [n]$ such that $s_i \neq s'_i$. By definition of K_i and δ we have $s_i \in K_i$, $s'_i \in K_i$ and $|s_i - s'_i| \geq \delta$, which yields that $\mathcal{S}$ has the (k,δ)-property. $\square$

As our upper bound on the expected number of Pareto-optimal solutions depends on both k and δ, one can ask if there exists an upper bound that depends only on k or only on δ. Theorem 6 shows there can be no polynomial upper bound only in n, ϕ, and k. On the other hand, we conjecture that there exists an upper bound for the smoothed number of Pareto-optimal solutions that depends polynomially on n, ϕ and the inverse of the minimum distance $\min_{s \neq s' \in \mathcal{S}} \|s - s'\|$ between solutions.

References

1. Beier, R., Röglin, H., Vöcking, B.: The smoothed number of pareto optimal solutions in bicriteria integer optimization. In: Fischetti, M., Williamson, D.P. (eds.) IPCO 2007. LNCS, vol. 4513, pp. 53–67. Springer, Heidelberg (2007). doi:10.1007/978-3-540-72792-7_5
2. Beier, R., Vöcking, B.: Random knapsack in expected polynomial time. J. Comput. Syst. Sci. **69**(3), 306–329 (2004)
3. Brunsch, T., Goyal, N., Rademacher, L., Röglin, H.: Lower bounds for the average and smoothed number of pareto-optima. Theory Comput. **10**(10), 237–256 (2014)
4. Brunsch, T., Röglin, H.: Improved smoothed analysis of multiobjective optimization. J. ACM **62**(1), 4 (2015)
5. Corley, H., Moon, I.: Shortest paths in networks with vector weights. J. Optim. Theor. Appl. **46**(1), 7986 (1985)
6. Ehrgott, M.: Integer solutions of multicriteria network flow problems. Investigacao Operacional **19**, 61–73 (1999)
7. Ehrgott, M., Gandibleux, X.: Multiple criteria optimization. In: Multiobjective Combinatorial Optimization, Lecture Notes in Economics and Mathematical Systems, vol. 491. Springer, Heidelberg (2000)
8. Klamroth, K., Wiecek, M.M.: Dynamic programming approaches to the multiple criteria knapsack problem. Naval Res. Logistics **47**(1), 57–76 (2000)
9. Moitra, A., O'Donnel, R.: Pareto optimal solutions for smoothed analysts. SIAM J. Comput. **41**(5), 1266–1284 (2012)
10. Müller-Hannemann, M., Weihe, K.: Pareto shortest paths is often feasible in practice. In: Brodal, G.S., Frigioni, D., Marchetti-Spaccamela, A. (eds.) WAE 2001. LNCS, vol. 2141, pp. 185–197. Springer, Heidelberg (2001). doi:10.1007/3-540-44688-5_15
11. Mustafa, A., Goh, M.: Finding integer efficient solutions for bicriteria and tricriteria network flow problems using DINAS. Comput. Oper. Res. **25**(2), 139–157 (1998)
12. Nemhauser, G.L., Ullmann, Z.: Discrete dynamic programming and capital allocation. Manage. Sci. **15**, 494–505 (1969)
13. Röglin, K., Teng, S.-H.: Smoothed analysis of multiobjective optimization. In: Proceedings of the 50th Annual IEEE Symposium on Foundations of Computer Science (FOCS), pp. 681–690 (2009)
14. Skriver, A.J.V., Andersen, K.A.: A label correcting approach for solving bicriterion shortest-path problems. Comput. Oper. Res. **27**(6), 507524 (2000)
15. Spielman, D.A., Teng, S.-H.: Smoothed analysis of algorithms: why the simplex algorithm usually takes polynomial time. J. ACM **51**(3), 385–463 (2004)

From Nonstandard Analysis
to Various Flavours of Computability Theory

Sam Sanders[1,2]([✉])

[1] Munich Center for Mathematical Philosophy, LMU Munich, Munich, Germany
sasander@me.com
[2] Department of Mathematics, Ghent University, Ghent, Belgium

Abstract. As suggested by the title, it has recently become clear that
theorems of *Nonstandard Analysis* (NSA) give rise to theorems in computability theory (no longer involving NSA). Now, the aforementioned
discipline divides into *classical* and *higher-order* computability theory,
where the former (resp. the latter) sub-discipline deals with objects of
type zero and one (resp. of all types). The aforementioned results regarding NSA deal exclusively with the *higher-order* case; we show in this
paper that theorems of NSA also give rise to theorems in *classical* computability theory by considering so-called *textbook proofs*.

1 Introduction

Computability theory naturally[1] includes two sub-disciplines: *classical* and
higher-order computability theory. The former deals with the computability of
objects of types zero and one (natural numbers and sets thereof) and the latter deals with the computability of *higher-order objects*, i.e. including objects
of type higher than zero and one. Friedman's closely related foundational program *Reverse Mathematics* (RM for short; see [22,23] for an overview) makes
use of *second-order arithmetic* which is also limited to type zero and one objects;
Kohlenbach has introduced *higher-order* RM in which all finite types are available [13].

As developed in [17,18,20,21], one can extract *higher-order* computability
results from theorems in Nonstandard Analysis. These results [18,20,21] involve
the 'Big Five' of RM and also the associated 'RM zoo' from [7], but all results
are part of *higher-order* RM. The following question thus naturally emerges:

(Q) *Is it possible to obtain* classical *computability theoretic results, including second-order Reverse Mathematics, from* NSA?

This research was supported by FWO Flanders, the John Templeton Foundation,
the Alexander von Humboldt Foundation, LMU Munich (via the Excellence Initiative), and the Japan Society for the Promotion of Science. The work was done partially while the author was visiting the Institute for Mathematical Sciences, National
University of Singapore in 2016. The visit was supported by the Institute.

[1] The distinction 'classical versus higher-order' is not binary as e.g. continuous functions on the reals may be represented by type one objects (See e.g. [23, II.6.1]).

T.V. Gopal et al. (Eds.): TAMC 2017, LNCS 10185, pp. 556–570, 2017.
DOI: 10.1007/978-3-319-55911-7_40

We will provide a positive answer to the question (Q) in this paper by studying an example based on the *monotone convergence theorem* in Sect. 3, after introducing Nonstandard Analysis and an essential fragment in Sect. 2. The notion *textbook proof* plays an important role. We also argue that our example generalises to many results in Nonstandard Analysis, as will be explored in [19].

Finally, we stress that our final results in (classical) computability theory are extracted *directly* from *existing theorems* of Nonstandard Analysis **without** modifications (involving computability theory or otherwise). In particular, no modification is made to the proofs or theorems in Nonstandard Analysis. We do consider special proofs in Nonstandard Analysis, which we christen *textbook proofs* due to their format. One could obtain the same results by mixing Nonstandard Analysis and computability theory, but one of the conceptual goals of our paper is to show that classical computability theory is *already implicit* in Nonstandard Analysis *pur sang*.

2   Internal Set Theory and Its Fragments

We discuss Nelson's axiomatic Nonstandard Analysis from [15], and the fragment P from [1]. The fragment P is essential to our enterprise due to Corollary 2.6.

2.1   Internal Set Theory 101

In Nelson's *syntactic* (or *axiomatic*) approach to Nonstandard Analysis [15], as opposed to Robinson's semantic one [16], a new predicate 'st(x)', read as 'x is standard' is added to the language of ZFC, the usual foundation of mathematics. The notations $(\forall^{st} x)$ and $(\exists^{st} y)$ are short for $(\forall x)(\text{st}(x) \to \dots)$ and $(\exists y)(\text{st}(y) \wedge \dots)$. A formula is called *internal* if it does not involve 'st', and *external* otherwise. The three external axioms *Idealisation*, *Standard Part*, and *Transfer* govern the new predicate 'st'; They are respectively defined[2] as:

(I) $(\forall^{st\ fin} x)(\exists y)(\forall z \in x)\varphi(z, y) \to (\exists y)(\forall^{st} x)\varphi(x, y)$, for any internal φ.

(S) $(\forall^{st} x)(\exists^{st} y)(\forall^{st} z)\big((z \in x \wedge \varphi(z)) \leftrightarrow z \in y\big)$, for any φ.

(T) $(\forall^{st} t)\big[(\forall^{st} x)\varphi(x, t) \to (\forall x)\varphi(x, t)\big]$, where $\varphi(x, t)$ is internal, and only has free variables t, x.

The system IST is (the internal system) ZFC extended with the aforementioned external axioms; The former is a conservative extension of ZFC for the internal language, as proved in [15].

In [1], the authors study fragments of IST based on Peano and Heyting arithmetic. In particular, they consider the systems H and P, introduced in the next section, which are conservative extensions of the (internal) logical systems E-HA$^\omega$ and E-PA$^\omega$, respectively *Heyting and Peano arithmetic in all finite types and the axiom of extensionality*. We refer to [12, Sect. 3.3] for the exact definitions of the

---

[2] The superscript 'fin' in (I) means that x is finite, i.e. its number of elements are bounded by a natural number.

(mainstream in mathematical logic) systems E-HA$^\omega$ and E-PA$^\omega$. Furthermore, E-PA$^{\omega*}$ and E-HA$^{\omega*}$ are the definitional extensions of E-PA$^\omega$ and E-HA$^\omega$ with types for finite sequences, as in [1, Sect. 2]. For the former, we require some notation.

Notation 2.1 (Finite sequences). The systems E-PA$^{\omega*}$ and E-HA$^{\omega*}$ have a dedicated type for 'finite sequences of objects of type ρ', namely ρ^*. Since the usual coding of pairs of numbers goes through in both, we shall not always distinguish between 0 and 0^*. Similarly, we do not always distinguish between 's^ρ' and '$\langle s^\rho \rangle$', where the former is 'the object s of type ρ', and the latter is 'the sequence of type ρ^* with only element s^ρ'. The empty sequence for the type ρ^* is denoted by '$\langle \rangle_\rho$', usually with the typing omitted.

Furthermore, we denote by '$|s| = n$' the length of the finite sequence $s^{\rho^*} = \langle s_0^\rho, s_1^\rho, \ldots, s_{n-1}^\rho \rangle$, where $|\langle \rangle| = 0$, i.e. the empty sequence has length zero. For sequences s^{ρ^*}, t^{ρ^*}, we denote by '$s*t$' the concatenation of s and t, i.e. $(s*t)(i) = s(i)$ for $i < |s|$ and $(s*t)(j) = t(|s| - j)$ for $|s| \le j < |s| + |t|$. For a sequence s^{ρ^*}, we define $\overline{s}N := \langle s(0), s(1), \ldots, s(N) \rangle$ for $N^0 < |s|$. For $\alpha^{0 \to \rho}$, we also write $\overline{\alpha}N = \langle \alpha(0), \alpha(1), \ldots, \alpha(N) \rangle$ for *any* N^0. By way of shorthand, $q^\rho \in Q^{\rho^*}$ abbreviates $(\exists i < |Q|)(Q(i) =_\rho q)$. Finally, we shall use $\underline{x}, \underline{y}, \underline{t}, \ldots$ as short for tuples $x_0^{\sigma_0}, \ldots x_k^{\sigma_k}$ of possibly different type σ_i.

2.2   The Classical System P

In this section, we introduce P, a conservative extension of E-PA$^\omega$ with fragments of Nelson's IST. We first introduce the system E-PA$_{\mathrm{st}}^{\omega*}$ using the definition from [1, Definition 6.1]. Recall that E-PA$^{\omega*}$ is the definitional extension of E-PA$^\omega$ with types for finite sequences as in [1, Sect. 2] and Notation 2.1. The language of E-PA$_{\mathrm{st}}^{\omega*}$ is that of E-PA$^{\omega*}$ extended with new symbols st$_\sigma$ for any finite type σ in E-PA$^{\omega*}$.

Notation 2.2. We write $(\forall^{\mathrm{st}} x^\tau)\Phi(x^\tau)$ and $(\exists^{\mathrm{st}} x^\sigma)\Psi(x^\sigma)$ for $(\forall x^\tau)[\mathrm{st}(x^\tau) \to \Phi(x^\tau)]$ and $(\exists x^\sigma)[\mathrm{st}(x^\sigma) \wedge \Psi(x^\sigma)]$. A formula A is 'internal' if it does not involve 'st', and external otherwise. The formula A^{st} is defined from internal A by appending 'st' to all quantifiers (except bounded number quantifiers).

The set $\mathcal{T}^*$ is defined as the collection of all the terms in the language of E-PA$^{\omega*}$.

Definition 2.3. The system E-PA$_{\mathrm{st}}^{\omega*}$ is defined as E-PA$^{\omega*} + \mathcal{T}_{\mathrm{st}}^* + \mathsf{IA}^{\mathrm{st}}$, where $\mathcal{T}_{\mathrm{st}}^*$ consists of the following axiom schemas.

1. The schema[3] $\mathrm{st}(x) \wedge x = y \to \mathrm{st}(y)$,
2. The schema providing for each closed[4] term $t \in \mathcal{T}^*$ the axiom $\mathrm{st}(t)$.
3. The schema $\mathrm{st}(f) \wedge \mathrm{st}(x) \to \mathrm{st}(f(x))$.

---

[3] The language of E-PA$_{\mathrm{st}}^{\omega*}$ contains a symbol st$_\sigma$ for each finite type σ, but the subscript is essentially always omitted. Hence $\mathcal{T}_{\mathrm{st}}^*$ is an *axiom schema* and not an axiom.

[4] A term is called *closed* in [1] (and in this paper) if all variables are bound via lambda abstraction. Thus, if $\underline{x}, \underline{y}$ are the only variables occurring in the term t, the term $(\lambda \underline{x})(\lambda \underline{y})t(\underline{x}, \underline{y})$ is closed while $(\lambda \underline{x})t(\underline{x}, \underline{y})$ is not. The second axiom in Definition 2.3 thus expresses that $\mathrm{st}_\tau((\lambda \underline{x})(\lambda \underline{y})t(\underline{x}, \underline{y}))$ if $(\lambda \underline{x})(\lambda \underline{y})t(\underline{x}, \underline{y})$ is of type τ.

The external induction axiom $\mathsf{IA}^{\mathrm{st}}$ states that for any (possibly external) Φ:

$$\Phi(0) \wedge (\forall^{\mathrm{st}} n^0)(\Phi(n) \to \Phi(n+1)) \to (\forall^{\mathrm{st}} n^0)\Phi(n). \qquad (\mathsf{IA}^{\mathrm{st}})$$

Secondly, we introduce some essential fragments of IST studied in [1].

Definition 2.4 (External axioms of P)

1. $\mathsf{HAC}_{\mathrm{int}}$: For any internal formula φ, we have

$$(\forall^{\mathrm{st}} x^\rho)(\exists^{\mathrm{st}} y^\tau)\varphi(x,y) \to (\exists^{\mathrm{st}} F^{\rho \to \tau^*})(\forall^{\mathrm{st}} x^\rho)(\exists y^\tau \in F(x))\varphi(x,y), \quad (2.1)$$

2. I: For any internal formula φ, we have

$$(\forall^{\mathrm{st}} x^{\sigma^*})(\exists y^\tau)(\forall z^\sigma \in x)\varphi(z,y) \to (\exists y^\tau)(\forall^{\mathrm{st}} x^\sigma)\varphi(x,y),$$

3. The system P is $\mathsf{E}\text{-}\mathsf{PA}^{\omega*}_{\mathrm{st}} + \mathsf{I} + \mathsf{HAC}_{\mathrm{int}}$.

Note that I and $\mathsf{HAC}_{\mathrm{int}}$ are fragments of Nelson's axioms *Idealisation* and *Standard part*. By definition, F in (2.1) only provides a *finite sequence* of witnesses to $(\exists^{\mathrm{st}} y)$, explaining its name *Herbrandized Axiom of Choice*.

The system P is connected to $\mathsf{E}\text{-}\mathsf{PA}^\omega$ by the following theorem. Here, the superscript 'S_{st}' is the syntactic translation defined in [1, Definition 7.1].

Theorem 2.5. *Let $\Phi(\underline{a})$ be a formula in the language of $\mathsf{E}\text{-}\mathsf{PA}^{\omega*}_{\mathrm{st}}$ and suppose $\Phi(\underline{a})^{S_{\mathrm{st}}} \equiv \forall^{\mathrm{st}} \underline{x} \exists^{\mathrm{st}} \underline{y}\, \varphi(\underline{x}, \underline{y}, \underline{a})$. If Δ_{int} is a collection of internal formulas and*

$$\mathsf{P} + \Delta_{\mathrm{int}} \vdash \Phi(\underline{a}), \qquad (2.2)$$

then one can extract from the proof a sequence of closed[5] terms t in $\mathcal{T}^*$ such that

$$\mathsf{E}\text{-}\mathsf{PA}^{\omega*} + \Delta_{\mathrm{int}} \vdash \forall \underline{x} \exists \underline{y} \in \underline{t}(\underline{x})\ \varphi(\underline{x}, \underline{y}, \underline{a}). \qquad (2.3)$$

Proof. Immediate by [1, Theorem 7.7].

The proofs of the soundness theorems in [1, Sects. 5, 6 and 7] provide an algorithm $\mathcal{A}$ to obtain the term t from the theorem. In particular, these terms can be 'read off' from the nonstandard proofs.

In light of [18], the following corollary (which is not present in [1]) is essential to our results. Indeed, the following corollary expresses that we may obtain effective results as in (2.5) from any theorem of Nonstandard Analysis which has the same form as in (2.4). It was shown in [18,20,21] that the scope of this corollary includes the Big Five systems of RM and the RM 'zoo' [7].

Corollary 2.6. *If Δ_{int} is a collection of internal formulas and ψ is internal, and*

$$\mathsf{P} + \Delta_{\mathrm{int}} \vdash (\forall^{\mathrm{st}} \underline{x})(\exists^{\mathrm{st}} \underline{y})\psi(\underline{x}, \underline{y}, \underline{a}), \qquad (2.4)$$

then one can extract from the proof a sequence of closed (See footnote 5) terms t in $\mathcal{T}^*$ such that

$$\mathsf{E}\text{-}\mathsf{PA}^{\omega*} + \mathsf{QF}\text{-}\mathsf{AC}^{1,0} + \Delta_{\mathrm{int}} \vdash (\forall \underline{x})(\exists \underline{y} \in \underline{t}(\underline{x}))\psi(\underline{x}, \underline{y}, \underline{a}). \qquad (2.5)$$

---

[5] Recall the definition of closed terms from [1] as sketched in Footnote 4.

Proof. Clearly, if for internal ψ and $\Phi(\underline{a}) \equiv (\forall^{st}\underline{x})(\exists^{st}\underline{y})\psi(x, y, a)$, we have $[\Phi(\underline{a})]^{S_{st}} \equiv \Phi(\underline{a})$, then the corollary follows immediately from the theorem. A tedious but straightforward verification using the clauses (i)–(v) in [1, Definition 7.1] establishes that indeed $\Phi(\underline{a})^{S_{st}} \equiv \Phi(\underline{a})$.

For the rest of this paper, the notion 'normal form' shall refer to a formula as in (2.4), i.e. of the form $(\forall^{st}x)(\exists^{st}y)\varphi(x, y)$ for φ internal.

Finally, we will use the usual notations for natural, rational and real numbers and functions as introduced in [13, p. 288–289]. (and [23, I.8.1] for the former). We only list the definition of real number and related notions in P.

Definition 2.7 (Real numbers and related notions in P)

1. A (standard) real number x is a (standard) fast-converging Cauchy sequence $q^1_{(\cdot)}$, i.e. $(\forall n^0, i^0)(|q_n - q_{n+i}| <_0 \frac{1}{2^n})$. We use Kohlenbach's 'hat function' from [13, p. 289] to guarantee that every sequence f^1 is a real.
2. We write $[x](k) := q_k$ for the k-th approximation of a real $x^1 = (q^1_{(\cdot)})$.
3. Two reals x, y represented by $q_{(\cdot)}$ and $r_{(\cdot)}$ are *equal*, denoted $x =_{\mathbb{R}} y$, if $(\forall n)(|q_n - r_n| \leq \frac{1}{2^n})$. Inequality $<_{\mathbb{R}}$ is defined similarly.
4. We write $x \approx y$ if $(\forall^{st}n)(|q_n - r_n| \leq \frac{1}{2^n})$ and $x \gg y$ if $x > y \wedge x \not\approx y$.
5. A function $F : \mathbb{R} \to \mathbb{R}$ mapping reals to reals is represented by $\Phi^{1 \to 1}$ mapping equal reals to equal reals as in $(\forall x, y)(x =_{\mathbb{R}} y \to \Phi(x) =_{\mathbb{R}} \Phi(y))$.
6. We write '$N \in \Omega$' as a *symbolic* abbreviation for $\neg st(N^0)$.

3 Main Results

In this section, we provide an answer to the question (Q) from Sect. 1 by studying the *monotone convergence theorem*. We first obtain the associated result in higher-order computability theory from NSA in Sect. 3.1. We then establish in Sect. 3.2 that the same proof in NSA also gives rise to *classical* computability theory.

3.1 An Example of the Computational Content of NSA

In this section, we provide an example of the *higher-order* computational content of NSA, involving the *monotone convergence theorem*, MCT for short, which is the statement *every monotone sequence in the unit interval converges*. In particular, we consider the equivalence between a nonstandard version of MCT and a fragment of Nelson's axiom *Transfer* from Sect. 2. From this nonstandard equivalence, an explicit RM equivalence involving higher-order versions of MCT and arithmetical comprehension is extracted as in (3.1).

Firstly, nonstandard MCT (involving *nonstandard* convergence) is:

$$(\forall^{st} c^{0 \to 1}_{(\cdot)})\left[(\forall n^0)(c_n \leq c_{n+1} \leq 1) \to (\forall N, M \in \Omega)[c_M \approx c_N]\right]. \qquad (MCT_{ns})$$

while the effective (or 'uniform') version of MCT, abbreviated $MCT_{ef}(t)$, is:

$$(\forall c^{0 \to 1}_{(\cdot)}, k^0)\left[(\forall n^0)(c_n \leq c_{n+1} \leq 1) \to (\forall N, M \geq t(c_{(\cdot)})(k))[|c_M - c_N| \leq \frac{1}{k}]\right].$$

We require two equivalent [13, Proposition 3.9] versions of arithmetical comprehension, respectively the *Turing jump functional* and *Feferman's mu-operator*, as follows

$$(\exists \varphi^2)\big[(\forall f^1)((\exists n^0) f(n) = 0 \leftrightarrow \varphi(f) = 0\big], \qquad (\exists^2)$$

$$(\exists \mu^2)\big[(\forall f^1)((\exists n^0) f(n) = 0 \rightarrow f(\mu(f)) = 0)\big], \qquad (\mu^2)$$

and also the restriction of Nelson's axiom *Transfer* as follows:

$$(\forall^{\mathrm{st}} f^1)\big[(\forall^{\mathrm{st}} n^0) f(n) \neq 0 \rightarrow (\forall m) f(m) \neq 0\big]. \qquad (\Pi_1^0\text{-TRANS})$$

Denote by $\mathsf{MU}(\mu)$ the formula in square brackets in (μ^2). We have the following nonstandard equivalence.

Theorem 3.1. *The system* P *proves that* $\Pi_1^0\text{-TRANS} \leftrightarrow \mathsf{MCT}_{\mathrm{ns}}$.

Proof. For the forward implication, assume $\Pi_1^0\text{-TRANS}$ and suppose $\mathsf{MCT}_{\mathrm{ns}}$ is false, i.e. there is a standard monotone sequence $c_{(\cdot)}$ such that $c_{N_0} \not\approx c_{M_0}$ for fixed nonstandard N_0, M_0. The latter is by definition $|c_{N_0} - c_{M_0}| \geq \frac{1}{k_0}$, where k_0^0 is a fixed standard number. Since N_0, M_0 are nonstandard in the latter, we have $(\forall^{\mathrm{st}} n)(\exists N, M \geq n)(|c_N - c_M| \geq \frac{1}{k_0})$. Fix standard n^0 in the latter and note that the resulting Σ_1^0-formula only involves *standard* parameters. Hence, applying the contraposition of $\Pi_1^0\text{-TRANS}$, we obtain $(\forall^{\mathrm{st}} n)(\exists^{\mathrm{st}} N, M \geq n)(|c_N - c_M| \geq \frac{1}{k_0})$. Applying[6] the previous formula $k_0 + 1$ times would make $c_{(\cdot)}$ escape the unit interval, a contradiction; $\mathsf{MCT}_{\mathrm{ns}}$ follows and the forward implication holds.

For the reverse implication, assume $\mathsf{MCT}_{\mathrm{ns}}$, fix standard f^1 such that $(\forall^{\mathrm{st}} n^0)(f(n) \neq 0)$ and define $c_{(\cdot)}^1$ as follows: c_k is 0 if $(\forall n \leq k)(f(n) \neq 0)$ and $\sum_{i=1}^{k} \frac{1}{2^i}$ otherwise. Note that $c_{(\cdot)}$ is standard (as f^1 is) and weakly increasing. Hence, $c_N \approx c_M$ for nonstandard N, M by $\mathsf{MCT}_{\mathrm{ns}}$. Now suppose m_0 is such that $f(m_0) = 0$ and also the least such number. By the definition of $c_{(\cdot)}$, we have that $0 = c_{m_0-1} \not\approx c_{m_0} = \sum_{i=1}^{m_0} \frac{1}{2^i} \approx 1$. This contradiction implies that $(\forall n^0)(f(n) \neq 0)$, and $\Pi_1^0\text{-TRANS}$ thus follows. $\square$

We refer to the previous proof as the 'textbook proof' of $\mathsf{MCT}_{\mathrm{ns}} \leftrightarrow \Pi_1^0\text{-TRANS}$. The reverse implication is indeed very similar to the proof of $\mathsf{MCT} \rightarrow \mathsf{ACA}_0$ in Simpson's textbook on RM, as found in [23, I.8.4]. This 'textbook proof' is special in a specific sense, as will become clear in the next section. Nonetheless, **any** nonstandard proof will yield higher-order computability results as in (3.1).

Theorem 3.2. *From* **any** *proof of* $\mathsf{MCT}_{\mathrm{ns}} \leftrightarrow \Pi_1^0\text{-TRANS}$ *in* P, *two terms* s, u *can be extracted such that* $\mathsf{E\text{-}PA}^{\omega *}$ *proves:*

$$(\forall \mu^2)\big[\mathsf{MU}(\mu) \rightarrow \mathsf{MCT}_{\mathrm{ef}}(s(\mu))\big] \wedge (\forall t^{1 \rightarrow 1})\big[\mathsf{MCT}_{\mathrm{ef}}(t) \rightarrow \mathsf{MU}(u(t))\big]. \qquad (3.1)$$

---

[6] To 'apply this formula k_0+1 times', apply $\mathsf{HAC}_{\mathrm{int}}$ to $(\forall^{\mathrm{st}} n)(\exists^{\mathrm{st}} N, M \geq n)(|c_N - c_M| \geq \frac{1}{k_0})$ to obtain standard $F^{0 \rightarrow 0^*}$ and define $G(n)$ as the maximum of $F(n)(i)$ for $i < |F(n)|$. Then $(\forall^{\mathrm{st}} n)(\exists N, M \geq n)(N, M \leq G(n) \wedge |c_N - c_M| \geq \frac{1}{k_0})$ and iterate the functional G at least $k_0 + 1$ times to obtain the desired contradiction.

Proof. We prove the second conjunct and leave the first one to the reader. Corollary 2.6 only applies to normal forms and we now bring $\mathsf{MCT_{ns}} \to \mathit{\Pi}_1^0$-$\mathsf{TRANS}$ into a suitable normal form to apply this corollary and obtain the second conjunct of (3.1). Clearly, $\mathit{\Pi}_1^0$-TRANS implies the following normal form:

$$(\forall^{\mathrm{st}} f^1)(\exists^{\mathrm{st}} n^0)\big[(\exists m)f(m) = 0) \to f(n) = 0\big]. \tag{3.2}$$

The nonstandard convergence of $c_{(\cdot)}$, namely $(\forall N, M \in \Omega)[c_M \approx c_N]$, implies

$$(\forall N, M)[(\forall^{\mathrm{st}} n^0)(M, N \geq n) \to (\forall^{\mathrm{st}} k)|c_M - c_N| < \tfrac{1}{k}],$$

in which we pull the standard quantifiers to the front as follows:

$$(\forall^{\mathrm{st}} k^0)\underline{(\forall N, M)(\exists^{\mathrm{st}} n^0)[M, N \geq n \to |c_M - c_N| < \tfrac{1}{k}]},$$

The contraposition of *idealisation* I applies to the underlined. We obtain:

$$(\forall^{\mathrm{st}} k^0)(\exists^{\mathrm{st}} z^{0^*})(\forall N, M)(\exists n^0 \in z)[M, N \geq n \to |c_M - c_N| < \tfrac{1}{k}],$$

and define K^0 as the maximum of $z(i)$ for $i < |z|$. We finally obtain:

$$(\forall^{\mathrm{st}} k^0)(\exists^{\mathrm{st}} K^0)(\forall N, M)[M, N \geq K \to |c_M - c_N| < \tfrac{1}{k}]. \tag{3.3}$$

and (3.3) is a normal form for nonstandard convergence. Hence, $\mathsf{MCT_{ns}}$ implies:

$$(\forall^{\mathrm{st}} c_{(\cdot)}^{0 \to 1}, k^0)(\exists^{\mathrm{st}} K^0)\big[(\forall n^0)(c_n \leq c_{n+1} \leq 1) \to (\forall N, M \geq K)[|c_M - c_N| \leq \tfrac{1}{k}]\big],$$

and let the formula in square brackets be $D(c_{(\cdot)}, k, K)$, while the formula in square brackets in (3.2) is $E(f, n)$. Then $\mathsf{MCT_{ns}} \to \mathit{\Pi}_1^0$-$\mathsf{TRANS}$ implies that

$$(\forall^{\mathrm{st}} c_{(\cdot)}^{0 \to 1}, k^0)(\exists^{\mathrm{st}} K^0)D(c_{(\cdot)}, k, K) \to (\forall^{\mathrm{st}} f^1)(\exists^{\mathrm{st}} n^0)E(f, n). \tag{3.4}$$

By the basic axioms in Definition 2.3, any *standard* functional Ψ produces standard output on standard input, which yields

$$(\forall^{\mathrm{st}} \Psi)\big[(\forall^{\mathrm{st}} c_{(\cdot)}^{0 \to 1}, k^0)D(c_{(\cdot)}, k, \Psi(k, c_{(\cdot)})) \to (\forall^{\mathrm{st}} f^1)(\exists^{\mathrm{st}} n^0)E(f, n)\big]. \tag{3.5}$$

We may drop the remaining 'st' in the antecedent of (3.5) to obtain:

$$(\forall^{\mathrm{st}} \Psi)\big[(\forall c_{(\cdot)}^{0 \to 1}, k^0)D(c_{(\cdot)}, k, \Psi(k, c_{(\cdot)})) \to (\forall^{\mathrm{st}} f^1)(\exists^{\mathrm{st}} n^0)E(f, n)\big],$$

and bringing all standard quantifiers to the front, we obtain a normal form:

$$(\forall^{\mathrm{st}} \Psi, f^1)(\exists^{\mathrm{st}} n^0)\big[(\forall c_{(\cdot)}^{0 \to 1}, k^0)D(c_{(\cdot)}, k, \Psi(k, c_{(\cdot)})) \to E(f, n)\big]. \tag{3.6}$$

Applying Corollary 2.6 to '$\mathsf{P} \vdash (3.6)$', we obtain a term t such that

$$(\forall \Psi, f^1)(\exists n^0 \in t(\Psi, f))\big[(\forall c_{(\cdot)}^{0 \to 1}, k^0)D(c_{(\cdot)}, k, \Psi(k, c_{(\cdot)})) \to E(f, n)\big]. \tag{3.7}$$

Define $s(f, \Psi)$ as the maximum of $t(\Psi, f)(i)$ for $i < |t(\Psi, f)|$. Then (3.6) implies

$$(\forall \Psi)\big[(\forall c_{(\cdot)}^{0 \to 1}, k^0)D(c_{(\cdot)}, k, \Psi(k, c_{(\cdot)})) \to (\forall f^1)(\exists n \leq s(f, \Psi))E(f, n)\big], \tag{3.8}$$

and we recognise the antecedent as the effective version of MCT; the consequent is (essentially) $\mathsf{MU}(s(f, \Psi))$. Hence, the second conjunct of (3.1) follows. $\qquad \square$

Note that the normal form (3.3) of nonstandard convergence is the 'epsilon-delta' definition of convergence with the 'epsilon' and 'delta' quantifiers enriched with 'st'. While the previous proof may seem somewhat magical upon first reading, one readily jumps from the nonstandard implication $\mathsf{MCT_{ns}} \to \Pi_1^0\text{-}\mathsf{TRANS}$ to (3.8) with some experience.

In conclusion, **any** proof of $\Pi_1^0\text{-}\mathsf{TRANS} \leftrightarrow \mathsf{MCT_{ns}}$ gives rise to the *higher-order* computability result (3.1). We may thus conclude the latter from the proof of Theorem 3.1. In the next section, we show that the latter theorem's 'textbook proof' is special in that it also gives rise to *classical* computability-theoretic results. The latter is non-trivial since both $\Pi_1^0\text{-}\mathsf{TRANS}$ and $\mathsf{MCT_{ns}}$ have a normal form starting with '$(\forall^{st} h^1)(\exists^{st} l^0)$' (up to coding). As a result, to convert the implication $\mathsf{MCT_{ns}} \to \Pi_1^0\text{-}\mathsf{TRANS}$ into a normal form, one has to introduce a *higher-order* functional like Ψ to go from (3.4) to (3.5). Note that replacing the sequence of reals $c_{(\cdot)}^{0\to1}$ in $\mathsf{MCT_{ns}}$ by a sequence of rationals $q_{(\cdot)}^1$ does not lower Ψ below type two. In a nutshell, the procedure in the previous proof (and hence most proofs in Nonstandard Analysis) *always* seems to produce higher-order computability results.

3.2 An Example of the Classical-Computational Content of NSA

In the previous section, we showed that **any** proof of $\Pi_1^0\text{-}\mathsf{TRANS} \leftrightarrow \mathsf{MCT_{ns}}$ gives rise to the *higher-order* equivalence (3.1). In this section, we show that the particular 'textbook proof' of $\Pi_1^0\text{-}\mathsf{TRANS} \leftrightarrow \mathsf{MCT_{ns}}$ in Theorem 3.1 gives rise to *classical* computability theoretic results as in (3.13) and (3.14).

First of all, we show that the 'textbook proof' of Theorem 3.1 is actually more uniform than the latter theorem suggests. To this end, let $\Pi_1^0\text{-}\mathsf{TRANS}(f)$ and $\mathsf{MCT_{ns}}(c_{(\cdot)})$ be respectively $\Pi_1^0\text{-}\mathsf{TRANS}$ and $\mathsf{MCT_{ns}}$ from Sect. 3.1 restricted to the function f^1 and sequence $c_{(\cdot)}$, i.e. the former principles are the latter with the quantifiers $(\forall f^1)$ and $(\forall c_{(\cdot)}^{0\to1})$ stripped off.

Theorem 3.3. *There are terms s, t such that the system P proves*

$$(\forall^{st} f^1)\left[\mathsf{MCT_{ns}}(t(f)) \to \Pi_1^0\text{-}\mathsf{TRANS}(f)\right], \tag{3.9}$$

$$(\forall^{st} c_{(\cdot)}^{0\to1})\left[(\forall^{st} n^0)\Pi_1^0\text{-}\mathsf{TRANS}(s(c_{(\cdot)}, n)) \to \mathsf{MCT_{ns}}(c_{(\cdot)})\right]. \tag{3.10}$$

All proofs are implicit in the 'textbook proof' of Theorem 3.1.

Proof. To establish (3.9), define the term $t^{1\to1}$ as follows for f^1, k^0:

$$t(f)(k) := \begin{cases} 0 & (\forall i \leq k)(f(i) \neq 0) \\ \sum_{i=0}^{k} \frac{1}{2^i} & \text{otherwise} \end{cases}. \tag{3.11}$$

The proof of Theorem 3.1 now yields (3.9). Indeed, fix a standard function f^1 such that $(\forall^{st} k^0)(f(k) \neq 0) \wedge (\exists n)(f(n) = 0)$ and $\mathsf{MCT_{ns}}(t(f))$. By the latter, the sequence $t(f)$ nonstandard convergences, while $0 = t(f)(n_0 - 1) \not\approx t(f)(n_0) \approx 1$

for n_0 the least (necessarily nonstandard) n such that $f(n) = 0$. From this contradiction, $\Pi_1^0\text{-TRANS}(f)$ follows, and thus also (3.9).

The remaining implication (3.10) is proved in exactly the same way. Indeed, the intuition behind the previous part of the proof is as follows: In the proof of the reverse implication of Theorem 3.1, to establish $\Pi_1^0\text{-TRANS}(f)$ for fixed standard f^1, we **only** used MCT_{ns} for **one** particular sequence, namely $t(f)$. Hence, we only need $\text{MCT}_{ns}(t(f))$, and not 'all of' MCT_{ns}, thus establishing (3.9). Similarly, in the proof of the forward implication of Theorem 3.1, to derive $\text{MCT}_{ns}(c_{(\cdot)})$ for fixed $c_{(\cdot)}$, we **only** applied $\Pi_1^0\text{-TRANS}$ to **one** specific Σ_1^0 formula with a standard parameters n^0 and $c_{(\cdot)}$. □

We are now ready to reveal the intended 'deeper' meaning of the term 'textbook proof': The latter refers to a proof (which may not exist) of an implication $(\forall^{st} f)A(f) \to (\forall^{st} g)B(g)$ which also establishes $(\forall^{st} g)[A(t(g)) \to B(g)]$, *and* in which the formula in square brackets is a formula in which all standard quantifiers involve variables of type zero. By Theorem 3.4, such a 'textbook proof' gives rise to results in **classical** computability theory.

We choose the term 'textbook proof' because proofs in Nonstandard Analysis (especially in textbooks) are quite explicit in nature, i.e. one often establishes $(\forall^{st} g)[A(t(g)) \to B(g)]$ in order to prove $(\forall^{st} f)A(f) \to (\forall^{st} g)B(g)$.

Before we can apply Corollary 2.6 to Theorem 3.3, we need some definitions, as follows. First, consider the following 'second-order' version of (μ^2):

$$(\forall e^0, n^0)\big[(\exists m, s)(\varphi_{e,s}^A(n) = m) \to (\exists m, s \leq \nu(e, n))(\varphi_{e,s}^A(n) = m)\big]. \quad (\text{MU}^A(\nu))$$

where '$\varphi_{e,s}^A(m) = n$' is the usual (primitive recursive) predicate expressing that the e-th Turing machine with input n and oracle A halts after s steps with output m; sets $A, B, C, \ldots$ are denoted by binary sequences. One easily defines the (second-order) Turing jump of A from ν^1 as in $\text{MU}^A(\nu)$ and vice versa.

Next, we introduce the 'computability-theoretic' version of $\text{MCT}_{ef}(t)$. To this end, let $\text{TOT}(e, A)$ be the formula '$(\forall n^0)(\exists m^0, s^0)(\varphi_{e,s}^A(n) = m)$', i.e. the formula expressing that the Turing machine with index e and oracle A halts for all inputs, also written '$(\forall n^0)\varphi_e^A(n) \downarrow$'. Assuming the latter formula to hold for e^0, A^1, the function φ_e^A is clearly well-defined, and will be used in P in the usual[7] sense of computability theory. We assume $\varphi_e^A(n)$ to code a rational number without mentioning the coding. We now introduce the 'second-order' version of $\text{MCT}_{ef}(t)$:

$$(\forall e^0)\big[\text{TOT}(e, A) \wedge (\forall n^0)(0 \leq \varphi_e^A(n) \leq \varphi_e^A(n+1) \leq 1) \qquad (\text{MCT}_{ef}^A(t))$$
$$\to (\forall k^0)(\forall N, M \geq t(e, k))[|\varphi_e^A(N) - \varphi_e^A(M)| \leq \tfrac{1}{k}]\big].$$

---

[7] For instance, written out in full '$0 \leq \varphi_e^A(n) \leq \varphi_e^A(n+1) \leq 1$' from $\text{MCT}_{ef}^A(t)$ is:

$$(\forall s^0, q^0, r^0)\big[(\varphi_{e,s}^A(n) = q \wedge \varphi_{e,s}^A(n+1) = r) \to 0 \leq_0 q \leq_0 r \leq_0 1\big], \qquad (3.12)$$

where we also omitted the coding of rationals.

Here, t has type $(0 \times 0) \to 0$ or $0 \to 1$, and we will usually treat the former as a type one object. Finally, let $\mathsf{MCT}^A_{\mathsf{ef}}(t, e)$ and $\mathsf{MU}^A(\nu, e, n)$ be the corresponding principles with the quantifiers outside the outermost square brackets removed.

Theorem 3.4. *From the textbook proof of* $\mathsf{MCT}_{\mathsf{ns}} \to \Pi^0_1\text{-}\mathsf{TRANS}$, *three terms* $s^{1 \to 1}, u^1, v^{1 \to 1}$ *can be extracted such that* $\mathsf{E\text{-}PA}^{\omega *}$ *proves:*

$$(\forall A^1, \psi^{0 \to 1})\left[\mathsf{MCT}^A_{\mathsf{ef}}(\psi) \to \mathsf{MU}^A(s(\psi, A))\right]. \tag{3.13}$$

$$(\forall e^0, n^0, A^1, \phi^1)\left[\mathsf{MCT}^A_{\mathsf{ef}}(\phi, u(e, n)) \to \mathsf{MU}^A(v(\phi, A, e, n), e, n)\right]. \tag{3.14}$$

Proof. Similar to the proof of Theorem 3.2, a normal form for $\Pi^0_1\text{-}\mathsf{TRANS}(f)$ is:

$$(\exists^{\mathsf{st}} n^0)\left[(\exists m)(f(m) = 0) \to (\exists i \le n)(f(i) = 0)\right], \tag{3.15}$$

while, for t as in (3.11), a normal from for $\mathsf{MCT}_{\mathsf{ns}}(t(f))$ is:

$$(\forall^{\mathsf{st}} k^0)(\exists^{\mathsf{st}} K^0)\left[(\forall n^0)(0 \le t(f)(n) \le t(f)(n+1) \le 1) \right. \tag{3.16}$$
$$\left. \to (\forall N^0, M^0 \ge K)(|t(f)(N) - t(f)(M)| \le \tfrac{1}{k})\right],$$

Let $C(n, f)$ (resp. $B(k, K, f)$) be the formula in (outermost) square brackets in (3.15) (resp. (3.16)). Then (3.9) is the formula

$$(\forall^{\mathsf{st}} f^1)[(\forall^{\mathsf{st}} k)(\exists^{\mathsf{st}} K)B(k, K, f) \to (\exists^{\mathsf{st}} n^0)C(n, f)],$$

which (following the proof of Theorem 3.2) readily implies the normal form:

$$(\forall^{\mathsf{st}} f^1, \psi^1)(\exists^{\mathsf{st}} n^0)[(\forall k)B(k, \psi(k), f) \to C(n, f)]. \tag{3.17}$$

Applying Corollary 2.6 to '$\mathsf{P}_0 \vdash$ (3.17)' yields a term z^2 such that

$$(\forall f^1, \psi^1)(\exists n \in z(f, \psi))\left[(\forall k)B(k, \psi(k), f) \to C(n, f)\right]$$

is provable in $\mathsf{E\text{-}PA}^{\omega *}$. Define the term $s(f, \psi)$ as the maximum of all $z(f, \psi)(i)$ for $i < |z(f, \psi)|$ and note that (by the monotonicity of C):

$$(\forall f^1, \psi^1)\left[(\forall k)B(k, \psi(k), f) \to C(s(f, \psi), f)\right]. \tag{3.18}$$

Now define f^2_0 as follows: $f_0(e, n, A, k) = 0$ if $(\exists m, s \le k)(\varphi^A_{e,s}(n) = m)$, and 1 otherwise. For this choice of function, namely taking $f^1 =_1 \lambda k.f_0$, the sentence (3.18) implies for all A^1, ψ^1, e^0, n^0 that

$$(\forall k')B(k', \psi(k'), \lambda k.f_0) \to C(s(\lambda k.f_0, \psi), \lambda k.f_0), \tag{3.19}$$

where we used the familiar lambda notation with some variables of f_0 suppressed to reduce notational complexity. Consider the term t from (3.11) and note that there are (primitive recursive) terms x^1, y^1 such that for all m we have $t(\lambda k.f_0(e, n, A, k))(m) = \varphi^A_{x(e,n),y(e,n)}(m)$; the definition of x^1, y^1 is implicit in the definition of t and f_0. Hence, with these terms, the antecedent and consequent of (3.19) are as required to yield (3.14).

To prove (3.13) from (3.19), suppose we have $(\forall k')B(k', \xi(e, n)(k'), \lambda k.f_0)$ for all e^0, n^0 and some $\xi^{0\to 1}$ and A^1, where $\xi(e, n)$ has type 1. By (3.19) we obtain

$$(\forall e^0, n^0)C(s(\lambda k.f_0, \xi(e, n)), \lambda k.f_0).$$

Putting the previous together, we obtain the sentence:

$$(\forall A^1, \xi^{0\to 1})\big[(\forall e^0, n^0, k')B(k', \xi(e, n)(k'), \lambda k.f_0) \tag{3.20}$$
$$\to (\forall e^0, n^0)C(s(\lambda k.f_0, \xi(e, n)), \lambda k.f_0)\big].$$

Clearly, the consequent of (3.20) implies that $s(\lambda k.f_0, \xi(e, n))$ provides the Turing jump of A as in $\mathsf{MU}^A(\lambda e \lambda n.s(\lambda k.f_0, \xi(e, n)))$. On the other hand, the antecedent of (3.20) expresses that the sequence $t(\lambda k.f_0(e, n, A, k))$ converges for all e, n as witnessed by the modulus $\xi(e, n)$. In light of the definitions of f_0 and t, the sequence $t(\lambda k.f_0)$ (considered as a type one object) is definitely computable from the oracle A (in the usual sense of Turing computability). Thus, the antecedent of (3.20) also follows from $\mathsf{MCT}^A_{\mathsf{ef}}(\xi)$. In other words, (3.20) yields

$$(\forall A^1, \xi^{0\to 1})\big[\mathsf{MCT}^A_{\mathsf{ef}}(\xi) \to \mathsf{MU}^A(\lambda e \lambda n.s(\lambda k.f_0, \xi(e, n)))\big], \tag{3.21}$$

which is as required for the theorem, with minor modifications to the term s. $\square$

Note that (3.14) expresses that in order to decide if the e-th Turing machine with oracle A and input n halts, it suffices to have the term s and a modulus of convergence for the sequence of rationals given by $\varphi^A_{u(e,n)}$. We do not claim these to be ground-breaking results in computability theory, but we do point out the surprising ease and elegance with which they fall out of *textbook proofs* in Nonstandard Analysis. Taking into account the claims[8] by Bishop and Connes that *Nonstandard Analysis be devoid of computational/constructive content*, we believe that the word 'surprise' is perhaps not misplaced to describe our results.

In a nutshell, to obtain the previous theorem, one first establishes the 'nonstandard uniform' version (3.9) of $\mathsf{MCT}_{\mathsf{ns}} \to \Pi^0_1\text{-}\mathsf{TRANS}$, which yields the 'superpointwise' version (3.18). The latter is then weakened into (3.14) and then weakened into (3.13); this modification should be almost identical for other similar implications. In particular, it should be straightforward, but unfortunately beyond the page limit, to obtain versions of Theorems 3.3 and 3.4 for *König's lemma* and *Ramsey's theorem* [23, III.7], or any theorem equivalent to ACA_0 in RM for that matter.

Furthermore, results related to *weak König's lemma*, the third Big Five system of RM [23, IV] and the RM zoo [7], can be obtained in the same way as

---

[8] Bishop (See [4, p. 513], [2, p. 1], and [3], which is the review of [11]) and Connes (See [6, p. 6207] and [5, p. 26]) have made rather strong claims regarding the nonconstructive nature of Nonstandard Analysis. Their arguments have been investigated in remarkable detail and were mostly refuted (See e.g. [8–10]).

above. For instance, one can easily obtain Π_1^0-TRANS $\to$ WKL$_{ns}$ where the latter is the nonstandard modification of WKL stating the existence of a *standard* path for every *standard* infinite binary tree. However, the existence of a 'textbook proof' (as discussed right below Theorem 3.3) for this implication (or the reverse implication) leads to a contradiction.

In conclusion, higher-order computability results can be obtained from arbitrary proofs of MCT$_{ns}$ $\to$ Π_1^0-TRANS, while the *textbook proof* as in the proof of Theorem 3.1 yields classical computability theory as in Theorem 3.4.

3.3 The Connection Between Higher-Order and Classical Computability Theory

This paper would not be complete without a discussion of the ECF-translation, which connects higher-order and second-order mathematics. In particular, we show that applying the ECF-translation to e.g. (3.1) does not yield e.g. (3.14).

We first define the central ECF-notion of 'associate' which some will know in an equivalent guise: Kohlenbach shows in [14, Proposition 4.4] that the existence of a 'RM code' for a continuous functional Φ^2 as in [23, II.6.1], is equivalent to the existence of an *associate* for Φ, and equivalent to the existence of a modulus of continuity for Φ, Simpson's claims from [23, I.8.9.5] notwithstanding.

Definition 3.5. The function α^1 is an *associate* of the continuous Φ^2 if:

  (i) $(\forall \beta^1)(\exists k^0)\alpha(\overline{\beta}k) > 0$,
  (ii) $(\forall \beta^1, k^0)(\alpha(\overline{\beta}k) > 0 \to \Phi(\beta) + 1 =_0 \alpha(\overline{\beta}k))$.

With regard to notation, it is common to write $\alpha(\beta)$, to be understood as $\alpha(\overline{\beta}k) - 1$ for large enough k^0 (See also Definition 3.8 below). Furthermore, we assume that every associate is a *neighbourhood function* as in [14], i.e. α also satisfies

$$(\forall \sigma^{0^*}, \tau^{0^*})[\alpha(\sigma) > 0 \wedge |\sigma| \leq |\tau| \wedge (\forall i < |\sigma|)(\sigma(i) = \tau(i)) \to \alpha(\sigma) = \alpha(\tau)].$$

We now sketch the ECF-translation; Note that RCA$_0^\omega$ is Kohlenbach's base theory for higher-order RM [13]; this system is essentially E-PA$^\omega$ weakened to one-quantifier-induction and with a fragment of the axiom of choice.

Remark 3.6 (ECF-translation). The translation '$[\cdot]_{ECF}$' is introduced in [24, Sect. 2.6.5] and we refer to the latter for the exact definition. Intuitively, applying the ECF-translation to a formula amounts to nothing more than *replacing all objects of type two or higher by associates*. Furthermore, Kohlenbach observes in [13, Sect. 2] that if RCA$_0^\omega$ $\vdash$ A then RCA$_0^2$ $\vdash$ $[A]_{ECF}$, i.e. $[\cdot]_{ECF}$ provides a translation from RCA$_0^\omega$ to (a system which is essentially) RCA$_0$, the base theory of RM.

Thus, we observe that the ECF-translation connects higher-order and second-order mathematics. We now show that the ECF-translation is not a 'magic bullet' in that $[A]_{ECF}$ may not always be very meaningful, as discussed next.

Example 3.7 (The ECF-translation of (μ^2)). The ECF-translation will interpret the *discontinuous*[9] functional μ^2 as in $\mathsf{MU}(\mu)$ as a *continuous* object satisfying the latter formula, which is of course impossible[10], and the same holds for theorems equivalent to (μ^2) as they involve discontinuous functionals as well. Hence, the ECF-translation reduces the implications in (3.1) to (correct) trivialities of the form '$0 = 1 \rightarrow 0 = 1$'.

By the previous example, we observe that the answer to question (Q) is not just 'apply ECF' in the case of theorems involving (μ^2). Nonetheless, we *could* apply the ECF-translation to (3.13) and (3.14) to replace the terms s, u, v by *associates*. To this end, we require definition of partial function application (See e.g. [24, 1.9.12] or [12, Definition 3.58]) for the final corollary.

Definition 3.8 (Partial function application). For α^1, β^1, '$\alpha(\beta)$' is defined as

$$\alpha(\beta) := \begin{cases} \alpha(\overline{\beta}k) - 1 & \text{If } k^0 \text{ is the least } n \text{ with } \alpha(\overline{\beta}n) > 0 \\ \text{undefined} & \text{otherwise} \end{cases},$$

and $\alpha|\beta := (\lambda n^0)\alpha(\langle n \rangle * \beta)$. We write $\alpha(\beta) \downarrow$ to denote that $\alpha(\beta)$ is defined, and $\alpha|\beta \downarrow$ to denote that $(\alpha|\beta)(n)$ is defined for all n^0. For β^1, γ^1, we define the paired sequence $\beta \oplus \gamma$ by putting $(\beta \oplus \gamma)(2k) = \beta(k)$ and $(\beta \oplus \gamma)(2k+1) = \gamma(k)$.

We now consider the following corollary to Theorem 3.4.

Corollary 3.9. *From the textbook proof of* $\mathsf{MCT_{ns}} \rightarrow \Pi_1^0\text{-}\mathsf{TRANS}$, *a term* z^1 *can be extracted such that* $\mathsf{E\text{-}PA}^{\omega*}$ *proves:*

$$(\forall \psi^1, A^1)\left[\mathsf{MCT}_{\mathsf{ef}}^A(\psi) \rightarrow [z|(\psi \oplus A) \downarrow \wedge \mathsf{MU}^A(z|(\psi \oplus A))]\right]. \tag{3.22}$$

Proof. Immediate from applying the ECF-translation to (3.13). □

Note that (3.22) is part of second-order arithmetic.

Acknowledgement. The author would like to thank Richard Shore, Anil Nerode, and Vasco Brattka for their valuable advice and encouragement.

References

1. van den Berg, B., Briseid, E., Safarik, P.: A functional interpretation for nonstandard arithmetic. Ann. Pure Appl. Logic **163**(12), 1962–1994 (2012)

---

[9] Suppose $f_1 = 11 \ldots$ and μ^2 from (μ^2) is continuous; then there is N_0^0 such that $(\forall g^1)(\overline{f_1}N_0 = \overline{g}N_0 \rightarrow \mu(f_1) =_0 \mu(g))$. Let N_1 be the maximum of N_0 and $\mu(f_1)$. Then $g_0 := \overline{f_1}N_1 * 00 \ldots$ satisfies $\overline{f_1}N_1 = \overline{g_0}N_1$, and hence $\mu(f_1) = \mu(g_0)$ and $f_1(\mu(f_1)) = g_0(\mu(g_0))$, but the latter is 0 by the definition of g_0 and μ, a contradiction..

[10] If a functional has an associate, it must be continuous on Baire space. We established in Footnote 9 that (μ^2) cannot be continuous, and thus cannot have an associate.

2. Bishop, E.: Aspects of constructivism. Notes on the Lectures Delivered at the Tenth Holiday Mathematics Symposium, New Mexico State University, Las Cruces, 27–31 December 1972
3. Bishop, E.: Review of [11]. Bull. Amer. Math. Soc **81**(2), 205–208 (1977)
4. Bishop, E.: The crisis in contemporary mathematics. In: Proceedings of the American Academy Workshop on the Evolution of Modern Mathematics, pp. 507–517 (1975)
5. Connes, A.: An interview with Alain Connes. EMS Newslett. **63**, 25–30 (2007). http://www.mathematics-in-europe.eu/maths-as-a-profession/interviews
6. Connes, A.: Noncommutative geometry and reality. J. Math. Phys. **36**(11), 6194–6231 (1995)
7. Dzhafarov, D.D.: Reverse Mathematics Zoo. http://rmzoo.uconn.edu/
8. Kanovei, V., Katz, M.G., Mormann, T.: Tools, objects, and chimeras: Connes on the role of hyperreals in mathematics. Found. Sci. **18**(2), 259–296 (2013)
9. Katz, M.G., Leichtnam, E.: Commuting and noncommuting infinitesimals. Am. Math. Mon. **120**(7), 631–641 (2013)
10. Keisler, H.J.: Letter to the editor. Not. Am. Math. Soc. **24**, 269 (1977)
11. Keisler, H.J.: Elementary Calculus. Prindle, Weber and Schmidt, Boston (1976)
12. Kohlenbach, U.: Applied Proof Theory: Proof Interpretations and Their Use in Mathematics. Springer Monographs in Mathematics. Springer, Berlin (2008)
13. Kohlenbach, U.: Higher order reverse Mathematics. In: Reverse Mathematics 2001. Lecture Notes in Logic, vol. 21, pp. 281–295. ASL (2005)
14. Kohlenbach, U.: Foundational and mathematical uses of higher types. In: Reflections on the Foundations of Mathematics (Stanford, CA, 1998). Lecture Notes in Logic, vol. 15, pp. 92–116. ASL (2002)
15. Nelson, E.: Internal set theory: a new approach to Nonstandard Analysis. Bull. Am. Math. Soc. **83**(6), 1165–1198 (1977)
16. Robinson, A.: Non-standard Analysis. North-Holland, Amsterdam (1966)
17. Sanders, S.: The Gandy-Hyland functional and a hitherto unknown computational aspect of Nonstandard Analysis (2015). Submitted. http://arxiv.org/abs/1502.03622
18. Sanders, S.: The unreasonable effectiveness of Nonstandard Analysis (2015). Submitted. arXiv: http://arxiv.org/abs/1508.07434
19. Sanders, S.: On the connection between Nonstandard Analysis and classical computability theory. In: Preparation (2016)
20. Sanders, S.: The taming of the reverse mathematics zoo. (2015). Submitted. http://arxiv.org/abs/1412.2022

21. Sanders, S.: The refining of the taming of the reverse mathematics zoo. Notre Dame J. Formal Logic (2016, to appear). http://arxiv.org/abs/1602.02270
22. Simpson, S.G. (ed.): Reverse Mathematics 2001. Lecture Notes in Logic, vol. 21. ASL, La Jolla (2005)
23. Simpson, S.G. (ed.): Subsystems of Second Order Arithmetic. Perspectives in Logic, 2nd edn. CUP, Cambridge (2009)
24. Troelstra, A.S.: Metamathematical Investigation of Intuitionistic Arithmetic and Analysis. Lecture Notes in Mathematics, vol. 344. Springer, Berlin (1973)

Hardness of Routing for Minimizing Superlinear Polynomial Cost in Directed Graphs

Yangguang Shi[1,2,4], Fa Zhang[1,3], and Zhiyong Liu[1,4(✉)]

[1] Institute of Computing Technology,
Chinese Academy of Sciences (ICT, CAS), Beijing, China
{shiyangguang,zhangfa,zyliu}@ict.ac.cn
[2] University of Chinese Academy of Sciences, Beijing, China
[3] Key Laboratory of Intelligent Information Processing, ICT, CAS, Beijing, China
[4] Beijing Key Laboratory of Mobile Computing and Pervasive Device,
ICT, CAS, Beijing, China

Abstract. We study the problem of routing in directed graphs with superlinear polynomial costs, which is significant for improving the energy efficiency of networks. In this problem, we are given a directed graph $G(V, E)$ and a set of traffic demands. Routing δ_e units of demands along an edge e will incur a cost of $f_e(\delta_e) = \mu_e(\delta_e)^\alpha$ with $\mu_e > 0$ and $\alpha > 1$. The objective is to find integral routing paths for minimizing $\sum_e f_e(\delta_e)$. Through developing a new labeling technique and applying it to a randomized reduction, we prove an $\Omega\left(\left(\frac{\log |E|}{\log \log |E|}\right)^\alpha \cdot |E|^{-\frac{1}{4}}\right)$-hardness factor for this problem under the assumption that $NP \not\subseteq$ ZPTIME($n^{\text{polylog}(n)}$).

Keywords: Hardness of approximation · Superlinear polynomial cost · Directed graphs · Network energy efficiency

1 Introduction

We investigate a routing problem whose objective is to minimize a superlinear polynomial cost of degree $\alpha > 1$ in directed graphs (abbrv. DirSPC). In this problem, we are given a directed graph $G(V, E)$ and a set of traffic demands, each of which specifies a source-target pair (s_i, t_i). Each directed edge $e \in E$ is associated with a superlinear polynomial cost function $f_e(x) = \mu_e \cdot x^\alpha$, where $\mu_e \geq 0$. Let δ_e be the traffic load on e, the objective is to find an integral path for each traffic demand to minimize the overall cost $\sum_e f_e(\delta_e) = \sum_e \mu_e(\delta_e)^\alpha$. A simplified version of this problem which sets all μ_e to 1 will be referred to as the homogeneous DirSPC problem (abbrv. Homo-DirSPC).

This work is supported by the National Natural Science Foundation of China (NSFC) Major International Collaboration Project 61520106005 and NSFC Project for Innovation Groups 61521092.

ⓒ Springer International Publishing AG 2017
T.V. Gopal et al. (Eds.): TAMC 2017, LNCS 10185, pp. 571–585, 2017.
DOI: 10.1007/978-3-319-55911-7_41

Theorem 1. *The general DirSPC problem is $\Omega(\beta')$-hard to approximate if the Homo-DirSPC problem has no $O(\beta')$-approximation.*

Our study on the DirSPC problem is motivated by the growing concern on the energy efficiency of networks. Related research works show that the modern communication infrastructures and network devices are responsible for considerable and increasing percentage of world wide electricity usage [10]. In such case, the energy conservation problem in communication networks is attracting more and more attention. Since the superlinear polynomial cost function of degree α can appropriately model the electric power consumption of the network devices employing the *speed scaling* mechanism [12,15], studies on DirSPC will be significant for improving the energy efficiency of networks globally [3,11].

Until now, many approximation algorithms have been developed for DirSPC (e.g., [6,13]), owing to the significance of this problem in the perspectives of both theory and economy. However, to the best of our knowledge, the hardness of this problem is still unknown. In this paper, we will bound the hardness of DirSPC through reducing from a *hard-to-approximate* general Constraint Sanctification Problem (CSP) [2,8] in a randomized manner. With this technique we prove that the DirSPC problem has no $\Omega\left(\left(\frac{\log|E|}{\log\log|E|}\right)^{\alpha}\cdot|E|^{-\frac{1}{4}}\right)$-approximation unless NP $\subseteq$ ZPTIME($n^{\text{polylog}(n)}$). Note that here n is not the size of the DirSPC problem instance but an abstract parameter for defining the complexity class [2].

1.1 Related Works

In order to improve the energy efficiency of networks, some approximation algorithms have been proposed for the DirSPC problem and its variations. In [6], Bampis et al. propose an $O(\tilde{B}_\alpha)$-approximation randomized routing algorithm for Homo-DirSPC, where $\tilde{B}_\alpha$ refers to the generalized Bell number for α. $\tilde{B}_\alpha$ follows Dobiński's formula [9]: $\tilde{B}_\alpha = \sum_{k=1}^{+\infty} \frac{k^\alpha \exp(-1)}{k!}$.

The best known result for the DirSPC problem is given by Makarychev and Sviridenko's randomized routing algorithm [13], whose approximation ratio is bounded by $(1+\epsilon)\tilde{B}_\alpha$. Note that this approximation ratio does not conflict with our hardness result, since for any fixed α, $\left(\frac{\log|E|}{\log\log|E|}\right)^{\alpha}|E|^{-\frac{1}{4}}$ can be bounded by a constant only depending on α when $|E|$ takes proper values.

Some research works (e.g., [1,7]) focus on a variation of the DirSPC problem, where the cost function $f_e(x)$ includes a *start-up cost* σ_e used to model the static power consumption of network devices. In such case, $f_e(x) = \sigma_e + \mu_e x^\alpha$ when $x > 0$, while $f_e(x) = 0$ when $x = 0$. [7] proves that this variation admits no polylogarithmic approximation ratio unless NP $\subseteq$ ZPTIME($n^{\text{polylog}(n)}$). However, their hardness results depend highly on the existence of a large enough start-up cost σ_e for some edge e, since the DirSPCwS problem instance in [7] admits a constant approximation ratio when $\sigma_e/\mu_e = o(1)$ [1,3].

In [14], Shi et al. prove that for a routing problem on an undirected graph where each edge is associated with a cost function $f_e(x) = x^\alpha$, no integral routing algorithm that is traffic-oblivious can guarantee an $O\left(|E|^{\frac{\alpha-1}{\alpha+1}}\right)$-competitive ratio. Nevertheless, this result only holds for the traffic-oblivious routing algorithms, but does not apply to the traffic-adaptive ones.

1.2 Overview

To analyze the inapproximability of the DirSPC problem, our reduction starts from a hard-to-approximate Constant Satisfaction Problem (CSP). An instance of CSP consists of a collection of constraints on k-combinations of a set of variables. We say a CSP instance is a Yes-instance if there exists some assignment to the variables such that all the constraints are satisfied, while a CSP instance is referred to as a No-instance if every assignment to the variables can only satisfy a tiny fraction sat_n of the constraints.

In our reduction, for a CSP instance Ψ we construct a directed graph $\mathbb{D}_\Psi$ and a set of traffic demands as the input of the DirSPC problem. The construction of $\mathbb{D}_\Psi$ guarantees that when Ψ is a Yes-instance, the demands can be routed with a low cost C_L, while if Ψ is a No-instance, any routings of the demands will incur a high cost C_H. It implies that if DirSPC problem has an $o\left(\frac{C_H-\varepsilon}{C_L}\right)$-approximation algorithm, it can be distinguished that whether Ψ is a Yes-instance or No-instance, which will conflict with complexity theories.

To obtain a well-bounded hardness factor, the key point is to establish a large enough lower bound on C_H. Owing to the superadditivity of $f_e(x)$, aggregating a large quantity of traffic into some edge e can incur a high cost. However, a large congestion does not imply a high overall cost directly. Consider the two directed graphs $\mathbb{D}_1$ and $\mathbb{D}_2$ in Fig. 1. In $\mathbb{D}_1$, the vertices s and t is connected by a directed edge from s to t and d paths with length d^α from t to s. The graph $\mathbb{D}_2$ is obtained by changing the direction of every edge in $\mathbb{D}_1$. In each of these two graphs, d traffic demands need to be routed from s to t. Although the demands in $\mathbb{D}_2$ can be routed on edge-disjoint paths with congestion 1 while routing in $\mathbb{D}_1$ will cause a congestion of d, it's easy to verify that the cost of any routings in $\mathbb{D}_2$ is $d = \Theta(|E|^{1/\alpha})$ times larger than the routing cost in $\mathbb{D}_1$.

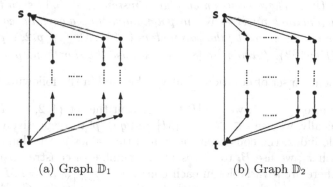

(a) Graph $\mathbb{D}_1$ (b) Graph $\mathbb{D}_2$

Fig. 1. Relationship between congestion and overall cost

The example above indicates that a high congestion c cannot guarantee a large enough overall cost if only a tiny fraction of edges have congestion c. To deduce a well-bounded inapproximability result, the construction of $\mathbb{D}_\Psi$ needs to make trade-off between the number of the heavy-loaded edges and the possible load on these edges. Specifically, we assign a collection of special paths called *canonical paths* [2,8] to each source-target pair. When Ψ is a No-instance, it is proved that any routings on the canonical paths will create high congestion $c = \Theta\left(\log|E|/\log\log|E|\right)$ on $\Omega(|E|^{3/4})$ edges with a high probability.

However, the traffic demands may be routed along the paths which deviate from the canonical paths to avoid the corresponding high cost. To eliminate such kind of routings, we adapt the *labeling technique* [8] into the construction of $\mathbb{D}_\Psi$. With this technique, we can guarantee that any paths deviating from the canonical paths cannot connect s to corresponding t. It implies that every traffic demand will be forced to go through one of its canonical paths. This completes our proof and establishes a hardness factor of $\Omega\left(\left(\frac{\log|E|}{\log\log|E|}\right)^\alpha |E|^{-\frac{1}{4}}\right)$.

2   Reduction and Graph Construction

Our reduction starts from the Constraint Satisfaction Problem. An instance of CSP consists of a collection of N variables $x_1, x_2, \cdots, x_N$ over a non-boolean domain $\mathcal{D} = \{1, 2, \cdots, p\}$ and a set $\mathcal{C}$ of M constraints, each of which is defined on a specific k-tuple of the variables. Each constraint C_j is satisfied by at most J local assignments to its variables. The objective of CSP is to distinguish between the following two cases:

- Yes-instance: There exists some assignments to the variables such that all the constraints can be satisfied.
- No-instance: Any assignments to the variables can satisfy at most a fraction sat_n of constraints.

We can characterize an instance of CSP with these parameters $(M, N, J, p, k, \text{sat}_n)$.

Theorem 2 ([8]). *There exists an absolute constant $\epsilon_0 \in (0, 1)$ such that for all large enough functions $p : \mathbb{N} \mapsto \mathbb{N}$, no polynomial-time algorithm can decide a CSP instance characterized by the parameters $\left(n^{\log p}, n^{\log p}, p, p, 2, 1/p^{\epsilon_0}\right)$ unless $NP \subseteq DTIME\left(n^{\log p}\right)$. (Similar to [8], here we use the shorthand $p = p(n)$.)*

Now we define a set of notations that will be used in the following reduction.

- $[I]$: For an integer $I \geq 1$, we use $[I]$ to represent the set $\{1, 2, \cdots, I\}$.
- i, j, q: Typically, we use $i \in [N]$, $j \in [M]$ and $q \in [p]$ to respectively represent the variable index, the constraint index, and the values in the domain $\mathcal{D}$.
- B_i: For each x_i, we use B_i to represent the number of constraints containing x_i. Since there are k variables in each constraint, $\sum_{i \in [N]} B_i = kM$ [2,8].
- ζ: It represents a local assignment that satisfies a constraint C_j. We say ζ is local for it only specifies the values of the variables in C_j.

- Γ_{ijq}: For each x_i, each C_j containing x_i and each q, we use Γ_{ijq} to represent the set of local assignments that satisfies C_j and assigns q to x_i.
- Γ_{iq}: It represents every pair (C_j, ζ) such that x_i is in C_j and $\zeta \in \Gamma_{ijq}$.

Based on a CSP instance Ψ characterized by $\left(n^{\log p}, n^{\log p}, p, p, 2, 1/p^{\epsilon_0}\right)$, next we will construct a directed graph $\mathbb{D}_\Psi$ in a randomized manner. The construction process is similar to [2]. The major difference between our graph $\mathbb{D}_\Psi$ and the graph constructed in [2] is that we adopt the 2-dimensional (2D) labeling techniques proposed in [8] to avoid the traffic demands deviating from the canonical paths. We use this technique to ensure that any routings of the traffic requests in the graph reduced from a No-instance will create a high congestion $c = c(n) = \log n$ on a large fraction of edges. Formally, a 2D label $\vec{\mu}(\tau) : \mathbb{N} \mapsto \mathbb{N}^2$ is a function that maps an integer τ to a 2-dimensional vector (τ, τ^2).

Before identifying the set $\mathcal{L}$ of labels, we need to declare two parameters Y, Z (their values will be specified later) and define *acceptance iteration* [2,8]. An acceptance iteration is a tuple (C_j, ζ, y) where ζ is a local assignment that satisfies C_j and $y \in [Y]$. It's easy to see that the number of acceptance iterations can be bounded by MJY. Consider a *acceptance function* $\tau = \tau(C_j, \zeta, y)$ that maps each acceptance iteration to a distinct integer in $[MJY]$ arbitrarily. For every acceptance iteration (C_j, ζ, y) and every $z \in [2Z]$, there will be a label $\vec{l}(\tau(C_j, \zeta, y), z) = \vec{\mu}(\tau(C_j, \zeta, y)) \cdot z$ in $\mathcal{L}$. Therefore, the size of $\mathcal{L}$ will be $(MJY)^3 \cdot (2Z)^2$. Two operators, '+' and '×', are defined over the set $\mathcal{L}$ as follows:

Definition 1 (Addition Operator [8]). *For any two vector $(a_1, a_2) \in \mathcal{L}$ and $(b_1, b_2) \in \mathcal{L}$, $(a_1, a_2) + (b_1, b_2)$ is defined to be $\left((a_1 + b_1) \bmod (2MJYZ), (a_2 + b_2) \bmod (2(MJY)^2 Z)\right)$.*

Definition 2 (Multiplication Operator). *For any vector $(a_1, a_2) \in \mathcal{L}$ and any integer $\lambda \in \mathbb{N}^+$, $(a_1, a_2) \times \lambda$ is defined to be $\left(a_1 \cdot \lambda \bmod (2MJYZ), a_2 \cdot \lambda \bmod (2(MJY)^2 Z)\right)$.*

The construction of $\mathbb{D}_\Psi$ starts from building certain variable gadgets. For each variable x_i, we will construct a gadget G_i in a randomized manner. Specifically, each G_i contains Z layers. For each $z \in [Z]$, there exists a matching M_z^i which consists of $X_i = YJB_i|\mathcal{L}|$ *special edges* $e_{s,\vec{l}}^{i,z} = \left(u_{s,\vec{l}}^{i,z}, v_{s,\vec{l}}^{i,z}\right)$ for $s \in [YJB_i]$ and $\vec{l} \in \mathcal{L}$. The vertices $u_{s,\vec{l}}^{i,z}$ and $v_{s,\vec{l}}^{i,z}$ will be called the tail and head of $e_{s,\vec{l}}^{i,z}$ respectively. We partition every matching M_z^i into $|\mathcal{L}|$ labeled matchings $\widetilde{M}_{i,z}^{\vec{l}}$, each of which consists of the special edges in M_z^i with the same label $\vec{l}$.

After constructing the special edges, the next step is to add $Z + 1$ layers of *connector vertices* into $\mathbb{D}_\Psi$. For each $z \in [Z+1]$, each $q \in [p]$ and each $\vec{l} \in \mathcal{L}$, we construct a set $W_{q,\vec{l}}^{i,z}$ of $Y|\Gamma_{iq}|$ connector vertices, each of which is corresponding to a tuple (C_j, ζ, y) where the pair $(C_j, \zeta) \in \Gamma_{iq}$ and $y \in [Y]$. Such a vertex in $W_{q,\vec{l}}^{i,z}$ will be represented by $\omega_{C_j,\zeta,y,\vec{l}}^{i,z}$. Recall that the acceptance function τ will map each acceptance iteration (C_j, ζ, y) to a distinct integer. Therefore, we will use the shorthand $\omega^{i,z}_{\tau,\vec{l}}$ to represent a connector vertex when the tuple (C_j, ζ, y) is obvious from the context. According to [2], for any $W^{i,z}_{q,\vec{l}}$:

$$W^{i,z}_{q,\vec{l}} \leq \sum_{q \in [p]} |W^{i,z}_{q,\vec{l}}| \leq YJB_i \tag{1}$$

The crucial part of the construction is to link the connector vertices to the special edges in a randomized manner. For each $W^{i,z}_{q,\vec{l}}$, a subset $S^{i,z}_{q,\vec{l}}$ of special edges of size $\left|W^{i,z}_{q,\vec{l}}\right|$ are **independently and uniformly** chosen from the group $\widetilde{M}^l_{i,z}$ **at random**. According to formulation (1), such a sub-matching always exists. Each $\omega^{i,z}_{C_j,\zeta,y,\vec{l}}$ will be connected to the *tail* of a special edge in $S^{i,z}_{q,\vec{l}}$ via a random matching. This special edge will be also called $\xi^{i,z}_{C_j,\zeta,y,\vec{l}}$. Then a directed edge will connect the *head* of $\xi^{i,z}_{C_j,\zeta,y,\vec{l}}$ to the connector vertex $\omega^{i,z+1}_{C_j,\zeta,y,\vec{l}+\vec{\mu}(\tau(C_j,\zeta,y))}$.

Consider two connector vertex ω_1 and ω_2. When $\omega_1 \in W^{i,z}_{q_1,\vec{l}}$ and $\omega_2 \in W^{i,z}_{q_2,\vec{l}}$ such that $q_1 \neq q_2$, they may be connected to the same special edge. Hence, the in-degree and out-degree of a special edge may be larger than 1. Let $\omega^{i,z+1}_{\tau,\vec{l'}}$ be an arbitrary connector vertex adjacent to the head of a special edge $e^{i,z}_{s,\vec{l}}$, then:

Lemma 1. *There exists an integer τ' such that $\vec{l'} = \vec{l} + \vec{\mu}(\tau')$.*

Proof. Corresponding to each connector vertex adjacent to the head of $e^{i,z}_{s,\vec{l}}$, there will be an $\omega^{i,z}_{\tau,\vec{l}}$ connected to tail of $e^{i,z}_{s,\vec{l}}$ such that $\vec{l'} = \vec{l} + \vec{\mu}(\tau)$. $\qquad\square$

We now add source-target pairs into $\mathbb{D}_\Psi$ and linking together the gadgets. Specifically, for each $C_j \in \mathcal{C}$, each $y \in [Y]$ and each $\vec{l} \in \mathcal{L}$, there will be a source-target pair $\left(s^{\vec{l}}_{j,y}, t^{\vec{l}}_{j,y}\right)$. For each $C_j \in \mathcal{C}$, let $x_{i_1}, x_{i_2}, \cdots, x_{i_k}$ be the k variables contained in C_j. We refer to $i_1, i_2, \cdots, i_k$ as the *local* indexes of the variables. For two variables $x_{i'}$ and $x_{i''}$ contained in C_j, **iff** $i' < i''$ they will be given local indexes i_{λ_1} and i_{λ_2} such that $\lambda_1 < \lambda_2$. For every local assignment ζ that satisfies C_j, we construct a path that connects $s^{\vec{l}}_{j,y}$ in the following way:

1. Connect the source $s^{\vec{l}}_{j,y}$ to $\omega^{i_1,1}_{C_j,\zeta,y,\vec{l}}$, and connect $\omega^{i_k,Z+1}_{C_j,\zeta,y,\vec{l}+\vec{\mu}(\tau(C_j,\zeta,y))\times Z}$ to the target $t^{\vec{l}}_{j,y}$.

2. For each $\lambda \in [k-1]$, connect $\omega^{i_\lambda,Z+1}_{C_j,\zeta,y,\vec{l}+\vec{\mu}(\tau(C_j,\zeta,y))\times Z}$ to $\omega^{i_{\lambda+1},1}_{C_j,\zeta,y,\vec{l}}$.

Since $\mathbb{D}_\Psi$ will be used as the input of the DirSPC problem, we assign a superlinear cost function $f(x) = x^\alpha$ to each edge to complete the construction.

Theorem 3. *The number of edges in $\mathbb{D}_\Psi$ can be bounded by $\Theta\left(k(MYJ)^4 Z^3\right)$.*

Proof. First, the number of the special edges can be bounded by

$$\sum_{i\in[N]} \sum_{z\in[Z]} |M_z^i| = Z \cdot \sum_{i\in[N]} X_i = ZYJ|\mathcal{L}| \cdot \sum_{i\in[N]} B_i = kMJYZ|\mathcal{L}|$$

Then we consider the bound on the number of the connector vertices:

$$\sum_{z\in[Z+1]} \sum_{\vec{i}\in\mathcal{L}} \sum_{i\in[N]} \sum_{q\in[p]} \left|W_{q,\vec{i}}^{i,z}\right| \leq (Z+1)|\mathcal{L}|YJ \sum_{i\in[N]} B_i \leq 2kMJYZ|\mathcal{L}|$$

Since the in-degree and out-degree of each connector vertices are at most 1, we can claim that the number of edges can be bounded by $\Theta(kMJYZ|\mathcal{L}|)$.   □

Corollary 1. *The time needed to construct $\mathbb{D}_\Psi$ can be bounded by $O\left(k(MJY)^4 Z^3\right)$.*

2.1   Parameters

Now we specify the values of the parameters related to the construction of $\mathbb{D}_\Psi$.

$$c = c(n) = \log n \qquad p = p(n) = (2c)^{k/\epsilon_0} \qquad r = \left(c^k \cdot \mathsf{sat}_n\right)^{-1}$$
$$\rho = 2pkJr \qquad\qquad Y = \left\lceil \frac{64\rho^c}{J} \right\rceil \qquad\qquad A_i = 2X_i/\rho$$
$$Z = 128c \cdot \rho^c$$

Note that the parameter ϵ_0 in the specification of p is the exponent of the reciprocal of sat_n.

Lemma 2

1. $\rho = \Theta(p^2)$.
2. $n^{\log p} = \Theta(\rho^{c/2})$.
3. $YJ = \Theta(\rho^c)$.
4. $|\mathcal{L}| = \Theta(c^2 \rho^{13c/2})$.
5. *The number of edges in $\mathbb{D}_\Psi$ can be bounded by $\Theta(c^3 M^4 \rho^{7c})$.*

Proof. The first three equalities are proved as follows:

1. According to Theorem 2, $J = p$, $k = 2$, $r = (\frac{c^k}{p^{\epsilon_0}})^{-1} = 4$. Thus, $\rho = 16p^2$.
2. Since $\log(\rho^c) = c \log \rho = 2\log n \cdot \log(4p)$ while $\log(n^{\log p}) = \log n \cdot \log p$, the second formulation follows.
3. Since $\frac{64\rho^c}{J} = \frac{64\rho^c}{p} > 1$, $\frac{64\rho^c}{J} \leq Y \leq 2 \cdot \frac{64\rho^c}{J}$, which implies that $YJ = \Theta(\rho^c)$.

According to Theorem 3, the last two equalities holds.   □

3 Inapproximability Analysis

3.1 Canonical Path

To analyze the routing cost, we start by investigating a set of special paths, which will be called *canonical paths* [5,8]. Before giving the formal definition of the canonical paths, first we need to consider the connectivity of the connector vertices. For any two connector vertices ω and ω', we say ω is *directed to* ω' if there exists at least one path connecting ω to ω'.

Lemma 3. *Consider a variable x_i and an acceptance iteration (C_j, ζ, y) such that C_j contains x_i and ζ satisfies C_j. For any $\vec{l} \in \mathcal{L}$ and any $z \in [Z]$, the connector vertex $\omega_{C_j,\zeta,y,\vec{l}}^{i,1}$ is directed to $\omega_{C_j,\zeta,y,\vec{l}+\vec{\mu}(\tau)\times z}^{i,z+1}$, where τ represents $\tau(C_j, \zeta, y)$.*

Proof. When $z = 1$, $\omega_{\tau,\vec{l}}^{i,1}$ will be connected to the tail of $\xi_{\tau,\vec{l}}^{i,1}$, whose head is connected to $\omega_{\tau,\vec{l}+\vec{\mu}(\tau)}^{i,2}$. Therefore the proposition follows in such case. Suppose it holds for any $z \in [z_0]$ where $z_0 \in [Z-1]$. In such case, $\omega_{\tau,\vec{l}}^{i,1}$ is directed to $\omega_{\tau,\vec{l}+\vec{\mu}(\tau)\times z_0}^{i,z_0+1}$ and $\omega_{\tau,\vec{l}+\vec{\mu}(\tau)\times z_0}^{i,z_0+1}$ is connected to $\omega_{\tau,\vec{l}+\vec{\mu}(\tau)\times(z_0+1)}^{i,z_0+2}$ via the special edge $\xi_{\tau,\vec{l}+\vec{\mu}(\tau)\times z_0}^{i,z_0+1}$. Thus this proposition follows when $z = z_0 + 1$. □

This lemma implies that there exists a path connecting $\omega_{C_j,\zeta,y,\vec{l}}^{i,1}$ to $\omega_{C_j,\zeta,y,\vec{l}+\vec{\mu}(\tau)\times Z}^{i,Z+1}$. Such a path will be denoted by $P[i,j,\zeta,y,\vec{l}]$. Recall that we denote the k variables in a constraint C_j by $x_{i_1}, x_{i_2}, \cdots, x_{i_k}$. For any $\lambda \in [k-1]$, the path $P[i_\lambda,j,\zeta,y,\vec{l}]$ is connected to $P[i_{\lambda+1},j,\zeta,y,\vec{l}]$ via the directed edge $\left(\omega_{\tau,\vec{l}+\vec{\mu}(\tau)\times Z}^{i_\lambda,Z+1}, \omega_{\tau,\vec{l}}^{i_{\lambda+1},1}\right)$. Thus, we have a path connecting $\omega_{\tau,\vec{l}}^{i_1,1}$ to $\omega_{\tau,\vec{l}+\vec{\mu}(\tau)\times Z}^{i_k,Z+1}$ which consists of every $P[i_\lambda,j,\zeta,y,\vec{l}]$. Such a path will be referred to as a **canonical path** $P[j,\zeta,y,\vec{l}]$. The construction of $\mathbb{D}_\Psi$ ensures that each source-target pair $\left(s_{j,y}^{\vec{l}}, t_{j,y}^{\vec{l}}\right)$ is connected by at most J canonical paths corresponding to distinct local assignment ζ that satisfies C_j. We will prove that the traffic demand between $\left(s_{j,y}^{\vec{l}}, t_{j,y}^{\vec{l}}\right)$ must go through one of these canonical paths.

We first consider a graph $\widetilde{\mathbb{D}}_\Psi$ formed by shrinking every gadget $G_i \in \mathbb{D}_\Psi$ to a single node g_i, which will be called a gadget vertex.

Lemma 4. *The graph $\widetilde{\mathbb{D}}_\Psi$ is acyclic.*

Proof. Here we construct an ordered sequence Q of vertices in $\widetilde{D}_\Psi$ as follows: first, add all the source vertices in $\mathcal{D}_\Psi$ into Q arbitrarily. The next step is to sequentially add the gadget vertices $g_1, g_2, \cdots, g_N$ into Q and finally we add the target vertices into it. Obviously, Q is a topological sort of $\widetilde{\mathbb{D}}_\Psi$. □

Remark 1. This lemma indicates that once a path P goes from a gadget $G_i \in \mathbb{D}_\Psi$ to a vertex out of G_i, it can never go back.

Lemma 5. *Let (C_j, ζ, y) and (C'_j, ζ', y') be two different acceptance iterations. For any $z \in [Z+1]$, let $\vec{l}' = \vec{l} + \vec{\mu}(\tau(C_j, \zeta, y)) \times (z-1)$. Then there will be no path that connects $\omega^{i,z}_{C'_j, \zeta', y', \vec{l}'}$ to $\omega^{i,Z+1}_{C_j, \zeta, y, \vec{l} + \vec{\mu}(\tau) \times Z}$, where τ represents $\tau(C_j, \zeta, y)$.*

Proof. Suppose that there exists P' that connects $\omega^{i,z}_{C'_j, \zeta', y', \vec{l}'}$ to $\omega^{i,Z+1}_{C_j, \zeta, y, \vec{l} + \vec{\mu}(\tau) \times Z}$. Lemma 4 indicates that here we only need to consider the paths inside the gadget G_i. According to the construction procedure, the layer indexes of the connector vertices going through by any path inside G_i will increase monotonically. Therefore, the path P' exists only when $z < Z+1$. In such case, it will contain exactly $(Z+1) - z + 1 = Z - z + 2$ connector vertices. Let the sequence of the connector vertices going through by P' be $\omega^{i,z_1}_{\tau_1, \vec{l}_1}, \omega^{i,z_2}_{\tau_2, \vec{l}_2}, \cdots, \omega^{i,Z+1}_{C_j, \zeta, y, \vec{l}_{Z+1}}$, where $\omega^{i,z_1}_{\tau_1, \vec{l}_1}$ represents $\omega^{i,z}_{C'_j, \zeta', y', \vec{l}'}$ and $\vec{l}_{Z+1} = \vec{l} + \vec{\mu}(\tau) \times Z$. According to Lemma 1, for any $z' \in [Z - z + 1]$, there exists $\tau_{z'} \in Z$ such that $\vec{l}_{z'+1} = \vec{l}_{z'} + \vec{\mu}(\tau_{z'})$. Since:

$$
\begin{aligned}
\vec{l}_{Z+1} - \vec{l}_1 &= \vec{l} + \vec{\mu}(\tau) \times Z - \vec{l}' \\
&= \vec{l} + \vec{\mu}(\tau) \times Z - \left[\vec{l} + \vec{\mu}(\tau) \times (z-1) \right] \\
&= \vec{\mu}(\tau) \times (Z - z + 1)
\end{aligned}
$$

then we have

$$
\sum_{z'=1}^{Z-z+1} \vec{\mu}(\tau_{z'}) = \vec{l}_{Z+1} - \vec{l}_z = \vec{\mu}(\tau) \times (Z - z) \tag{2}
$$

According to [8], the 2-dimensional labels are convexly independent, which implies that the formulation (2) does not hold unless $\tau_1 = \tau_2 = \cdots = \tau$. Note that the acceptance function maps different acceptance iterations to different integers in an injective manner. It indicates the equation $\tau_z = \tau$ conflicts with the assumption that $(C_j, \zeta, y) \neq (C'_j, \zeta', y')$. Thus, this proposition follows. $\square$

Lemma 6. *For an arbitrary integer $z \in [Z+1]$ and an arbitrary acceptance iteration (C_j, ζ, y), let $\vec{l}$ and $\vec{l}'$ be two labels such that $\vec{l}' \neq \vec{l} + \vec{\mu}(\tau(C_j, \zeta, y)) \times (z-1)$. Then there will be no path connecting $\omega^{i,z}_{\tau, \vec{l}'}$ to $\omega^{i,Z}_{\tau, \vec{l} + \vec{\mu}(\tau) \times Z+1}$.*

Proof. Suppose that $\omega^{i,z}_{C_j, \zeta, y, \vec{l}'}$ is directed to $\omega^{i,Z+1}_{C_j, \zeta, y, \vec{l} + \vec{\mu}(\tau) \times Z}$ via a path P'. According to Lemma 5, any connector vertices in P' is corresponding to (C_j, ζ, y). It implies that:

$$
\vec{l}' + \vec{\mu}(\tau(C_j, \zeta, y)) \times (Z + 1 - z) = \vec{l} + \vec{\mu}(\tau(C_j, \zeta, y)) \times Z
$$

It means $\vec{l}' = \vec{l} + \vec{\mu}(\tau(C_j, \zeta, y)) \times (z-1)$, which conflicts with the assumption. $\square$

Theorem 4. *Every traffic demand will be routed along a canonical path.*

Proof. Let's consider an arbitrary source-target pair $(s^{\vec{l}}_{j,y}, t^{\vec{l}}_{j,y})$. Suppose it is routed along a path $\mathcal{P}$. According to the construction of $\mathbb{D}_\Psi$, $\mathcal{P}$ must traverse a connector vertex $\omega^{i_k, Z+1}_{C_j, \zeta_0, \vec{l}+\vec{\mu}(\tau)\times Z}$ to reach the target $t^{\vec{l}}_{j,y}$, where ζ_0 is a local assignment that satisfies C_j. Next, we will show that $\mathcal{P} = P[j, \zeta_0, y, \vec{l}]$.

First, let G_{i_k} be the gadget corresponding to x_{i_k}. Any path that enters G_{i_k} must go though the first layers of connector vertices. According to Lemmas 5 and 6, we can infer that the vertex traversed by $\mathcal{P}$ in the first layer will be $\omega^{i_k, 1}_{C_j, \zeta_0, y, \vec{l}}$. Suppose that there exists a path $P' \neq P[i_k, j, \zeta_0, y, \vec{l}]$ that connects $\omega^{i_k, 1}_{C_j, \zeta_0, y, \vec{l}}$ to $\omega^{i_k, Z+1}_{C_j, \zeta_0, \vec{l}+\vec{\mu}(\tau)\times Z}$. Recall that the out-degree of any connector vertex and the tail of any special edge is 1, which implies that P' deviates from $P[i_k, j, \zeta_0, y, \vec{l}]$ at the head of some special edge. However, since the connector vertices adjacent to the head of the same special edge are corresponding to different acceptance iterations, Lemma 5 indicates that P' can never reach $\omega^{i_k, Z+1}_{C_j, \zeta_0, \vec{l}+\vec{\mu}(\tau)\times Z}$. Therefore, we can claim that $\mathcal{P} \cap G_{i_k} = P[i_k, j, \zeta_0, y, \vec{l}]$.

Note that the construction of $\mathbb{D}_\Psi$ ensures that the in-degree of $\omega^{i_k, 1}_{C_j, \zeta_0, y, \vec{l}}$ is 1. Therefore, to reach the target, $\mathcal{P}$ should traverse the edge from $\omega^{i_{k-1}, Z+1}_{C_j, \zeta_0, y, \vec{l}+\vec{\mu}(\tau)\times Z}$ to $\omega^{i_k, 1}_{C_j, \zeta_0, y, \vec{l}}$, which indicates that $\omega^{i_{k-1}, Z+1}_{C_j, \zeta_0, y, \vec{l}+\vec{\mu}(\tau)\times Z} \in \mathcal{P}$. With the techniques given in the last paragraph, now we can inductively prove that when $\lambda = k - 1, k - 2, \cdots, 1$, $\mathcal{P} \cap G_{i_\lambda} = P[i_\lambda, j, \zeta_0, \vec{l}]$. Thus, this proposition follows. $\square$

3.2 Routing Cost Corresponding to Yes-instance

Theorem 5. *Suppose there is an assignment η to the whole set of the N variables in Ψ such that all the M constraints are satisfied. In such case, the traffic demands can be routed along edge-disjoint paths.*

Proof. For each $C_j \in \mathcal{C}$, let ζ_j be the projection of η on C_j, i.e., for each variable x_i in C_j, ζ_j assigns $q \in [p]$ to x_i iff η assigns q to x_i. We will prove that the canonical paths $P[j, \zeta_j, y, \vec{l}]$ are edge-disjoint.

Consider two canonical path $P_1 = P[j_1, \zeta_{j_1}, y_1, \vec{l_1}]$ and $P_2 = P[j_2, \zeta_{j_2}, y_2, \vec{l_2}]$. Here we assume that the constraints C_{j_1} and C_{j_2} have a common variable x_i and there exists some $z \in [Z + 1]$ such that $\vec{l_1} + \vec{\mu}(\tau(C_{j_1}, \zeta_1, y_1)) \times (z - 1) = \vec{l_2} + \vec{\mu}(\tau(C_{j_2}, \zeta_2, y_2)) \times (z-1)$, otherwise P_1 and P_2 will be trivially edge-disjoint. Now let $\vec{l'} = \vec{l_1} + \vec{\mu}(\tau(C_{j_1}, \zeta_1, y_1)) \times (z-1)$. Since both ζ_{j_1} and ζ_{j_2} are projections of η, they will assign a same value q to x_i. Therefore, they will traverse the connector vertices belonging to the same $W^{i,z}_{q, \vec{l'}}$. The construction of $\mathbb{D}_\Psi$ ensures that P_1 and P_2 will go to distinct edges in the labeled matching $\widetilde{M}^{\vec{l'}}_{i,z}$. $\square$

Since every canonical path has a length of $\Theta(kZ)$, Theorem 5 implies:

Corollary 2. *Under the case that Ψ is a Yes-instance, we can route all the traffic demands in $\mathcal{D}_\Psi$ with the cost bounded by $O(MY|\mathcal{L}| \cdot kZ)$.*

3.3 Routing Cost Corresponding to No-instance

Definition 3 (Highlight Tuple). *For any $i \in [N]$, $q \in [p]$ and $\vec{l} \in \mathcal{L}$, we say a labeled variable-value tuple $(x_i, q, \vec{l})$ is highlighted by a source-target pair if it is routed along $P[j, \zeta, y, \vec{l}]$ such that $(C_j, \zeta) \in \Gamma_{iq}$.*

Definition 4 (Heavy and Light). *For any $i \in [N]$, $q \in [p]$ and $\vec{l} \in \mathcal{L}$, a tuple $(x_i, q, \vec{l})$ is said to be* **heavy** *if it is highlighted by more than A_i source-target pairs. A source-target pair $\left(s^{\vec{l}}_{j,y}, t^{\vec{l}}_{j,y}\right)$ is* **heavy** *if it is routed along a canonical path $P[j, \zeta, y, \vec{l}]$ such that for every $x_i \in C_j$, $(x_i, q, \vec{l})$ is heavy, where q is the value assigned to x_i by ζ. A source-target pair is said to be* **light** *if it is not heavy.*

Definition 5 (Over-ambiguous). *We say a label $\vec{l}$ is* **over-ambiguous** *if there exists an x_i such that more than c tuples $(x_i, q, \vec{l})$ corresponding to different $q \in [p]$ are heavy. A label which is not over-ambiguous is said to be* **unambiguous**.

Lemma 7. *For every label $\vec{l} \in \mathcal{L}$, the number of* **light** *source-target pairs corresponding to $\vec{l}$ is at most $p \sum_i A_i$.*

Proof. This proposition can be proved similarly to [2]. For a label $\vec{l}$, there will be at most A_i light source-target pairs corresponding to each tuple $(x_i, q, \vec{l})$. Thus, the number of light source-target pairs corresponding to $\vec{l}$ is at most $p \sum_i A_i$. □

Lemma 8. *For an* **unambiguous** *label $\vec{l}$, there will be at most $c^k \mathrm{sat}_n MY$ heavy source-target pairs corresponding to $\vec{l}$.*

Proof. Suppose that there are U heavy source-target pairs corresponding to $\vec{l}$ are heavy. Here we construct a set $\widehat{C}$ of distinct constraints C_j corresponding to the heavy source-target pairs. It's easy to see that $|\widehat{C}| \geq U/Y$. Denote the canonical path along which the source-target pair corresponding to $C_j \in \widehat{C}$ is routed by $P_j = P[j, \zeta_j, y_j, \vec{l}]$. Note that each ζ_j assigns a value q_j to a variable x_i that participates C_j. Then every variable x_i will be assigned at most c values by the canonical paths P_j, otherwise $\vec{l}$ will be over-ambiguous. For each variable x_i which participates a constraint $C_j \in \widehat{C}$, we can pick a value q' from the at most c possible values randomly and uniformly. In such case, each constraint $C_j \in \widehat{C}$ will be satisfied with probability at least $1/c^k$. Then the expectation of the constraints satisfied by this random assignment will be at least $\frac{U}{c^k \cdot Y}$. When Ψ is a No-instance, we have $\frac{U}{c^k \cdot Y} \leq \mathrm{sat}_n \cdot M$. □

Theorem 6. *When Ψ is a No-instance, every label $\vec{l}$ will be over-ambiguous.*

Proof. Suppose that there exists an unambiguous label $\vec{l}$. According to Lemmas 7 and 8, the number of source-target pairs corresponding to $\vec{l}$ is at most:

$$p\sum_i A_i + c^k \mathbf{sat}_n MY = \frac{2pYJ\sum_i B_i}{\rho} + c^k \mathbf{sat}_n MY$$

$$= \frac{2pYJkM}{2pkJr} + c^k \mathbf{sat}_n MY$$

$$= 2c^k \cdot \frac{1}{[(2c)^{k/\gamma}]^\gamma} MY < MY$$

which conflicts with the fact that there are MY source-target pair corresponding to each $\vec{l}$. $\square$

Given a variable x_i, let $q_1, q_2, \cdots, q_c$ be c distinct possible values of x_i. For $c' \in [c]$, we use $I_{c'}$ to represent a set of $A_i = 2X_i/\rho$ acceptance iterations (C_j, ζ, y) such that $(C_j, \zeta) \in \Gamma_{iq_{c'}}$. Given $z \in [Z]$ and $\vec{l} \in \mathcal{L}$, there will be A_i connector vertices in $W^{i,z}_{q_{c'}, \vec{l}}$ corresponding to the acceptance iterations in $I_{c'}$. A special edge e in $\widetilde{M}^l_{i,z}$ is said to be used by an acceptance (C_j, ζ, y) from $I_{c'}$ if the corresponding connector vertex $\omega^{i,z}_{(C_j,\zeta,y),\vec{l}} \in W^{i,z}_{q_{c'},\vec{l}}$ is mapped to e.

Definition 6 (Blobs and High Congestion). *Given a label $\vec{l} \in \mathcal{L}$ and an integer $z \in [Z]$, a blob $H^{\vec{l}}_z$ is formed by aggregating every labeled matching $\widetilde{M}^l_{i,z}$ corresponding to each $i \in [N]$. A series of c sets of acceptance iterations $\{I_1, I_2, \cdots, I_c\}$ is said to be highly congesting in a labeled sub-matching $\widetilde{M}^l_{i,z}$ if at least $(A_i/2X_i)^c X_i$ edges in $\widetilde{M}^l_{i,z}$ are used by an acceptance iteration from each $I_1, I_2, \cdots, I_c$. We say $I_1, I_2, \cdots, I_c$ is highly congesting in a blob $H^{\vec{l}}_z$ if they are highly congesting in a labeled sub-matching $\widetilde{M}^l_{i,z} \in H^{\vec{l}}_z$.*

Here we use $\text{Ev}\left(q_1, \cdots, q_c, I_1, \cdots, I_c, z, \vec{l}\right)$ to represent the *bad event* that $I_1, \cdots, I_c$ are not highly congesting in $H^{\vec{l}}_z$. Since we map the connector vertices in each $W^{i,z}_{q_{c'},\vec{l}}$ to $\widetilde{M}^l_{i,z}$ randomly and uniformly, the relationship between the sets $I_1, \cdots, I_c$ and $\widetilde{M}^l_{i,z}$ can be modeled by a *Balls-and-Bins* game [2], where c kinds of A_i balls are randomly thrown into X_i distinct bins. The results on the Balls-and-Bins games in [2] implies that:

Lemma 9. $\Pr\left[Ev\left(q_1, q_2, \cdots, q_c, I_1, I_2, \cdots, I_c, z, \vec{l}\right)\right] \leq 2\exp\left(-\frac{(A_i)^c}{16(2X_i)^{c-1}}\right).$

We use $\text{Ev}\left(q_1, \cdots, q_c, I_1, \cdots, I_c, \vec{l}\right)$ to represent the event that the bad event defined above happens at no less than $\frac{Z}{2}$ blobs $H^{\vec{l}}_z$ corresponding to the same label $\vec{l}$ and different layers $z \in [Z]$, and $\mathrm{Ev}\left(q_1, \cdots, q_c, \vec{l}\right)$ to represent there exist sets $I_1, \cdots, I_c$ such that $\mathrm{Ev}\left(q_1, \cdots, q_c, I_1, \cdots, I_c, \vec{l}\right)$ happens. According to [2]:

$$\Pr\left[\mathrm{Ev}\left(q_1, \cdots, q_c, \vec{l}\right)\right] \leq \exp\left(cA_i(\log X_i + 1 - \log A_i) + Z \ln 2 - \frac{ZX_i}{16}\left(\frac{A_i}{2X_i}\right)^c\right)$$

$$\leq \exp\left(-X_i\left(\frac{Z}{16\rho^c} - \frac{2c(\log \rho + 1)}{\rho}\right) + Z \ln 2\right) \quad (3)$$

According to Sect. 2.1, we can bound the exponent of formulation (3) by:

$$-X_i\left(\frac{Z}{16\rho^c} + \frac{2c(\log \rho + 1)}{\rho}\right) - Z \ln 2 = -X_i\left(\frac{128c\rho^c}{16\rho^c} - \frac{2c(\log \rho + 1)}{\rho}\right) + Z \ln 2$$
$$\leq -4c \cdot YJB_i + Z \ln 2$$
$$\leq -64c\rho^c$$

Thus, the probability can be bounded by $\exp(-\Omega(\rho^c))$.

Let $\mathrm{Ev}(\vec{l})$ be the event that there exists a variable x_i and c values $q_1, q_2, \cdots, q_c$ such that $\mathrm{Ev}(q_1, q_2, \cdots, q_c, \vec{l})$ happens. Then its probability will be:

$$\Pr\left[\mathrm{Ev}(\vec{l})\right] \leq Np^c \cdot \Pr\left[\mathrm{Ev}\left(q_1, q_2, \cdots, q_c, \vec{l}\right)\right] = n^{\log p} \cdot p^c \cdot \exp(-64c\rho^c)$$

The last equality follows from Theorem 2. According to Sect. 2.1, we have $n^{\log p} = \exp(\log n \cdot \log p) < \exp(c\rho)$. Thus, $\Pr\left[\mathrm{Ev}(\vec{l})\right] \leq \exp(-60c\rho^c)$.

Denote the event that there exists at least $|\mathcal{L}|/2$ labels $\vec{l}$ such that $\mathrm{Ev}(\vec{l})$ happens by $\mathcal{E}$. According to Lemma 2, we have:

$$\Pr[\mathcal{E}] \leq \binom{|\mathcal{L}|}{|\mathcal{L}|/2} \cdot \left(\Pr\left[\mathrm{Ev}(\vec{l})\right]\right)^{|\mathcal{L}|/2} \leq 2^{|\mathcal{L}|} \cdot \left(e^{-60c\rho^c}\right)^{|\mathcal{L}|/2} \in o\left(\frac{1}{\exp(|E|)}\right)$$

Theorem 7. *When Ψ is a No-instance, routing in $\mathcal{D}_\Psi$ will incur an $\Omega(c^\alpha \frac{YJ}{\rho^c} Z|\mathcal{L}|)$-cost with probability $1 - o(1/\exp(|E|))$.*

Proof. Suppose the event $\mathcal{E}$ does not happen. According to Theorem 6, every label $\vec{l}$ is over-ambiguous. For each label $\vec{l}$, let x_i be the variable such that there are c heavy tuples $(x_i, q, \vec{l})$ corresponding to c values $q_1, q_2, \cdots, q_c$. For each $c' \in [c]$, let $\mathcal{I}_{c'}$ be the set of A_i source-target pairs corresponding to the heavy tuple $(x_i, q_{c'}, \vec{l})$. When the event $\mathrm{Ev}(\vec{l})$ does not happen, at least $(\frac{A}{2X})^c X$ edges in at least $\frac{Z}{2}$ blobs corresponding to $\vec{l}$ will be used by c source-target pairs in $I_{q_1} \bigcup I_{q_2} \bigcup \cdots \bigcup I_{q_c}$, which will incur a cost at least $c^\alpha \cdot \left(\frac{A}{2X}\right)^c \frac{XZ}{2} = c^\alpha \frac{ZYJB_i}{2\rho^c} \geq c^\alpha \frac{YJZ}{2\rho^c}$. When $\mathcal{E}$ does not happen, there will be at least $\frac{|\mathcal{L}|}{2}$ labels $\vec{l}$ such that $\mathrm{Ev}(\vec{l})$ does not happen. Then the overall cost will be at least $c^\alpha \frac{YJZ|\mathcal{L}|}{4\rho^c}$. $\square$

3.4   Hardness Factors

Theorem 8. *The Homo-DirSPC problem cannot be approximated within a factor of* $o\left(\left(\frac{\log|E|}{\log\log|E|}\right)^{\alpha}|E|^{-\frac{1}{4}}\right)$ *unless* $NP \subseteq ZPTIME(n^{O(\log\log n)})$.

Proof. $\mathbb{D}_{\Psi}$ is a feasible input of the Homo-DirSPC problem since every edge e has the same cost function. According to Corollary 2 and Theorem 7, the gap between the routing costs is $\Omega\left(\frac{c^{\alpha}YJZ|\mathcal{L}|/\rho^{c}}{kMYZ|\mathcal{L}|}\right) \subseteq \Omega\left(\frac{c^{\alpha}}{M\rho^{c}}\right)$. Lemma 2 implies that $M\rho^{c} = O(|E|^{1/4})$. Thus, we can bound the gap by $\Omega\left(\left(\frac{\log|E|}{\log\log|E|}\right)^{\alpha} \cdot \frac{1}{|E|^{1/4}}\right)$.

According to Corollary 1 and Lemma 2, the time needed to construct the graph $\mathbb{D}_{\Psi}$ can be bounded by $\rho^{O(c)} = n^{O(\log p)} = n^{O(\log c)} = n^{O(\log\log n)}$. Therefore, the Homo-DirSPC problem cannot be approximated by $o\left(\left(\frac{\log|E|}{\log\log|E|}\right)^{\alpha}|E|^{-\frac{1}{4}}\right)$ unless $NP \subseteq coRTIME(n^{O(\log\log n)})$. According to complexity theories [2,4,5], this implies that our result holds if $NP \nsubseteq ZTIME(n^{O(\log\log n)})$.   □

Theorems 1 and 8 implies the following Corollary:

Corollary 3. *The general DirSPC problem cannot be approximated within a factor of* $o\left(\left(\frac{\log|E|}{\log\log|E|}\right)^{\alpha}|E|^{-\frac{1}{4}}\right)$ *unless* $NP \subseteq ZPTIME(n^{O(\log\log n)})$.

4   Conclusion

Motivated by the concern on the energy efficiency of communication networks, in this paper we study the inapproximability of the DirSPC problem. We start by reducing from a instance Ψ of a CSP problem to a directed graph $\mathbb{D}_{\Psi}$ as the input of DirSPC. $\mathbb{D}_{\Psi}$ is constructed in a randomized manner with $n^{O(\log\log n)}$-time.

To obtain a proper bound on the hardness factor, we adapt the labeling technique into the construction of $\mathbb{D}_{\Psi}$. Different from related works [5,8], the labeling technique in our paper can ensure that every traffic request follows a canonical path strictly, and can guarantee that every label is over-ambiguous. These two properties enables us to simultaneously guarantee proper bounds on the congestion and the number of edges with the congestion. In this way, we prove an $\Omega\left(\left(\frac{\log|E|}{\log\log|E|}\right)^{\alpha}|E|^{-\frac{1}{4}}\right)$-hardness factor for DirSPC.

When $\alpha \gg 3$, our hardness result is still not tight enough compared to the best known approximation ratio [13]. How to further improve our hardness result will be investigated in the future work.

References

1. Andrews, M., Antonakopoulos, S., Zhang, L.: Minimum-cost network design with (dis)economies of scale. In: 2010 51st Annual IEEE Symposium on Foundations of Computer Science (FOCS), FOCS 2010, pp. 585–592, October 2010

2. Andrews, M., Chuzhoy, J., Guruswami, V., Khanna, S., Talwar, K., Zhang, L.: Inapproximability of edge-disjoint paths and low congestion routing on undirected graphs. Combinatorica **30**(5), 485–520 (2010)
3. Andrews, M., Fernandez Anta, A., Zhang, L., Zhao, W.: Routing for power minimization in the speed scaling model. IEEE/ACM Trans. Networking **20**(1), 285–294 (2012)
4. Andrews, M., Zhang, L.: Hardness of the undirected congestion minimization problem. In: Proceedings of the Thirty-seventh Annual ACM Symposium on Theory of Computing, STOC 2005, pp. 284–293. ACM (2005)
5. Andrews, M., Zhang, L.: Logarithmic hardness of the directed congestion minimization problem. In: Proceedings of the Thirty-eighth Annual ACM Symposium on Theory of Computing, STOC 2006, pp. 517–526. ACM (2006)
6. Bampis, E., Kononov, A., Letsios, D., Lucarelli, G., Sviridenko, M.: Energy efficient scheduling and routing via randomized rounding. In: IARCS Annual Conference on Foundations of Software Technology and Theoretical Computer Science, FSTTCS 2013, Guwahati, India, 12–14 December 2013, pp. 449–460 (2013)
7. Bansal, N., Gupta, A., Krishnaswamy, R., Nagarajan, V., Pruhs, K., Stein, C.: Multicast routing for energy minimization using speed scaling. In: Even, G., Rawitz, D. (eds.) MedAlg 2012. LNCS, vol. 7659, pp. 37–51. Springer, Heidelberg (2012). doi:10.1007/978-3-642-34862-4_3
8. Chuzhoy, J., Guruswami, V., Khanna, S., Talwar, K.: Hardness of routing with congestion in directed graphs. In: Proceedings of the Thirty-ninth Annual ACM Symposium on Theory of Computing, STOC 2007, pp. 165–178. ACM (2007)
9. Dobiński, G.: Summirung der reihe $\sum n^m/n!$, für m= 1, 2, 3, 4, 5. Arch. für Mat. und Physik **61**, 333–336 (1877)
10. Fettweis, G., Zimmermann, E.: ICT energy consumption-trends and challenges. In: Proceedings of the 11th International Symposium on Wireless Personal Multimedia Communications, vol. 2, p. 6 (2008)
11. Gupta, A., Krishnaswamy, R., Pruhs, K.: Online primal-dual for non-linear optimization with applications to speed scaling. In: Erlebach, T., Persiano, G. (eds.) WAOA 2012. LNCS, vol. 7846, pp. 173–186. Springer, Heidelberg (2013). doi:10.1007/978-3-642-38016-7_15
12. Kaxiras, S., Martonosi, M.: Computer Architecture Techniques for Power-Efficiency, 1st edn. Morgan and Claypool Publishers (2008)
13. Makarychev, K., Sviridenko, M.: Solving optimization problems with diseconomies of scale via decoupling. In: 2014 IEEE 55th Annual Symposium on Foundations of Computer Science (FOCS), pp. 571–580, October 2014
14. Shi, Y., Zhang, F., Wu, J., Liu, Z.: Randomized oblivious integral routing for minimizing power cost. Theor. Comput. Sci. **607**(Part 2), 221–246 (2015)
15. Wierman, A., Andrew, L., Tang, A.: Power-aware speed scaling in processor sharing systems. In: INFOCOM 2009, pp. 2007–2015. IEEE, April 2009

A Cryptographic View of Regularity Lemmas: Simpler Unified Proofs and Refined Bounds

Maciej Skórski[✉]

IST Austria, Klosterneuburg, Austria
maciej.skorski@gmail.com

Abstract. In this work we present a short and unified proof for the Strong and Weak Regularity Lemma, based on the cryptographic technique called *low-complexity approximations*. In short, both problems reduce to a task of finding constructively an approximation for a certain target function under a class of distinguishers (test functions), where distinguishers are combinations of simple rectangle-indicators. In our case these approximations can be learned by a simple iterative procedure, which yields a unified and simple proof, achieving for any graph with density d and any approximation parameter ϵ the partition size
- a tower of 2's of height $O\left(d\epsilon^{-2}\right)$ for a variant of Strong Regularity
- a power of 2 with exponent $O\left(d\epsilon^{-2}\right)$ for Weak Regularity
The novelty in our proof is: (a) a simple approach which yields both strong and weaker variant, and (b) improvements when $d = o(1)$. At an abstract level, our proof can be seen a refinement and simplification of the "analytic" proof given by Lovasz and Szegedy.

Keywords: Regularity lemmas · Boosting · Low-complexity approximations · Convex optimization · Computational indistinguishability

1 Introduction

Szemeredi's Regularity Lemma was first used in his famous result on arithmetic progressions in dense sets of integers [Sze75]. Since then, it has emerged as an important tool in graph theory, with applications to extremal graph theory, property testing in computer science, combinatorial number theory, complexity theory and others. See for example [DLR95, FK99b, HMT88] to mention only few.

Roughly speaking, the lemma says that every graph can be partitioned into a finite number of parts such that the edges between these pairs behave randomly. There are two popular forms of this result, the original result referred to as the Strong Regularity Lemma and the weaker version developed by Frieze and Kannan [FK99b] for algorithmic applications.

This paper (with updates) is available on https://eprint.iacr.org/2016/965.pdf

M. Skórski—Supported by the European Research Council Consolidator Grant (682815-TOCNe).

T.V. Gopal et al. (Eds.): TAMC 2017, LNCS 10185, pp. 586–599, 2017.
DOI: 10.1007/978-3-319-55911-7_42

The purpose of this work is to give yet another proof of regularity lemmas, based on the cryptographic notion of *computational indistinguishability*. We don't revisit applications as it would be beyond the scope. For more about applications of regularity lemmas, we refer to surveys [KS96, RS, KR02]

From now, G is a fixed graph with a vertex set $V(G) = V$ and the edge set $E(G) = E \subset V^2$. By a partition of V we understand every family of disjoint subsets that cover V.

The rest of the paper is organized as follows: the remaining part of this section introduces necessary notions (Sect. 1.1), states regularity lemmas (Sect. 1.2), and summarizes our contribution (Sect. 1.3). In Sect. 2 we show how to obtain strong regularity and in Sect. 3 we deal with weak regularity. We conclude our work in Sect. 4.

1.1 Preliminaries

By the edge density of two vertex subsets we understand the fraction of pairs covered by graph edges.

Definition 1 (Edge density). *For two disjoint subsets T, S of a given graph G we define the edge density of the pair T, S as*

$$d_G(T, S) = \frac{E_G(T, S)}{|T||S|} \tag{1}$$

We slightly abuse the notation denoting $d_G = d_G(V, V)$ for the graph density.

Sets Regularity. The notion of *set irregularity* measures the difference between the number of actual edges and expected edges as if the graph was random. Note that for a random bipartite graph with a bipartition (T, S) we expect that for almost all subsets S', T' roughly the same fraction of vertex pairs is covered by graph edges. The deviation is precisely measured as follows

Definition 2 (Irregularity [LS07, FL14]). *The irregularity of a pair (S, T) of two vertex subsets is defined as*

$$\mathrm{irreg}_G(S, T) = \max_{S' \subset S, T' \subset T} |E(S', T') - d_G(S, T)|S'||T'||$$

If this quantity is a small fraction of $|S||T|$ then the edge distribution is "homogeneous" or, if we want, random-like.

In turn, two vertex subsets are called regular if the density is almost preserved on their (sufficiently big) subsets[1]

---

[1] The requirement of being "sufficiently big" is to make this notion equivalent with the irregularity above.

Definition 3 (Regularity). *A pair (S, T) of two disjoint subsets of vertices is said to be ϵ-regular, if*

$$|d_G(S', T') - d_G(S, T)| \leqslant \epsilon$$

for all $S' \subset S$, $T' \subset T$ such that $|S'| \geqslant \epsilon|S|$, $|T'| \geqslant \epsilon|T|$.

For completeness we mention that irregularity and regularity are pretty much equivalent (up to changing ϵ).

Remark 1 (Irregularity vs Regularity). It easy to see that $\text{irreg}_G(S, T) \leqslant \epsilon|S||T|$ is implied by ϵ-regularity, and it implies $\epsilon^{\frac{1}{3}}$-regularity.

Partition Regularity. The next important objects are regular partitions, for which almost all pairs of parts are regular. Note that irregular indexes are weighted by set sizes, to properly address partitions with parts of different size.

Definition 4 (Regular Partitions). *A partition $V_1, \ldots, V_k$ of the vertex set is said to be ϵ-regular if there is a set $I \subset V \times V$ such that*

$$\sum_{(i,j) \in I} |V_i||V_j| \leqslant \epsilon|V|^2$$

and for all $\forall(i, j) \notin I$ the pair (V_i, V_j) is ϵ-regular.

We say that a partition is equitable (or simply: is an equipartition) if any two parts differ in size by at most one. Note that for equitable partitions the above conditions simply means that all but ϵ-fraction of pairs are regular.

There is also a notion of partition irregularity based on sets irregularity

Definition 5 (Partition Irregularity). *The irregularity of a partition $\mathcal{V} = \{V_1, \ldots, V_k\}$ is defined to be $\text{irreg}(\mathcal{V}) = \sum_{i,j} \text{irreg}_G(V_i, V_j)$.*

Remark 2 (Partition Irregularity vs Partition Regularity). Again it it easy to see that both notions are equivalent up to a change in ϵ. Concretely, ϵ-regularity is implied by irregularity smaller than $\epsilon^4|V|^2$ and implies ϵ-irregularity [FL14].

The partition size in the Strong Regularity Lemma grows as fast as powers of twos. For completeness, we state the definition of the tower function.

Definition 6 (Power tower). *For any n we denote*

$$T(n) = \underbrace{2^{2^{\cdot^{\cdot^{2}}}}}_{n \; times} .$$

1.2   Regularity Lemmas

Summary of the State of the Art. Having introduced necessary notation, we are now in position to state regularity lemmas. There is a strong (original) and weak variant of the regularity lemma (developed later for algorithmic applications), which differ dramatically in the partition size. The strong variant has a few slightly relaxed statements, which are more convenient for applications and simpler to prove. These versions are equivalent up to a replacing ϵ by $\epsilon^{O(1)}$. The state of the art is that the variant of Strong Regularity Lemma (Theorem 2 below) and the Weak Regularity Lemma (Theorem 4 below) are tight in general, as shown recently[2] in [FL14]. For the sake of the completeness we note that there are more works offering the proofs for Regularity Lemmas, for example [FK99a] but they are not discussed here as they do not achieve optimal bounds.

Strong Regularity. The original variant of the Strong Regularity Lemma simply says that there is always an equipartition such that almost every pair of parts is regular, and the partition size is not dependent on the graph size.

Theorem 1 (Strong Regularity Lemma, original variant 1). *For any graph G there exists a partition $V_1, \ldots, V_k$ of vertices such that for all up to ϵ-fraction of pairs (i, j)*

$$|E(S,T) - d_G(V_i, V_j)|S||T|| \leqslant \epsilon |V_i||V_j|$$

for any $S \subset V_i, T \subset V_j$ such that $|S| \geqslant \epsilon|V_i|, |T| \geqslant \epsilon|V_j|$. Moreover, the size of partition is at most a power of twos of height $\mathrm{poly}(1/\epsilon)$.

It has been observed that proofs are much easier when one works with the total irregularity, rather than separate bounds for each pair. The following version is equivalent (up to changing ϵ).

Theorem 2 (Strong Regularity Lemma, variant 2 [FL14]). *For any graph G there exists a partition $V_1, \ldots, V_k$ of the vertices such that*

$$\sum_{i<j} \mathrm{irreg}_G(V_i, V_j) \leqslant \epsilon |V|^2. \tag{2}$$

Moreover, the partition size k is a power of twos of length $O(\epsilon^{-2})$.

The regularity lemma can be also formulated as an approximation by a weighted graph.

Theorem 3 (Strong Regularity Lemma, variant 3 [LS07]). *For any graph G there is a partition $V_1, \ldots, V_k$ of the vertices and real numbers $d_{i,j}$ such that*

$$\sum_{i<j} \max_{S\subset V_i, T\subset V_j} |E(S,T) - d_{i,j}|S||T|| \leqslant \epsilon|V|^2, \tag{3}$$

---

[2] Worse bounds were known before for example [Gow97].

and moreover the partition size k is at most a tower[3] of twos of height $O(\epsilon^{-2})$.

Weak Regularity. Finally, we state the weaker version obtained originally by Frieze and Kannan, with refined bounds due do Vadhan and Zheng [VZ13].

Theorem 4 (Weak Regularity Lemma [VZ13]). *For any graph G there exists a partition $V_1, \ldots, V_k$ of the vertices such that*

$$\left| \sum_{i,j} E(S \cap V_i, T \cap V_j) - \sum_{i,j} d_{i,j} |S \cap V_i||T \cap V_j| \right| \leqslant \epsilon |V|^2 \tag{4}$$

for all S, T. Moreover, the partition is generated[4] by $O\left(\epsilon^{-2} d_G \log(d_G^{-1})\right)$ subsets of V, where $d = |E|$. In particular, k is at most $2^{O(\epsilon^{-2} d_G \log d_G^{-1})}$.

1.3   Our Contribution and Related Works

We present a simple proof of both Regularity Lemmas, using the cryptographic framework of *low complexity approximations*. Our contribution is twofold: (a) conceptual, as we show how the Regularity Lemmas can be written and easy proved using the notion of indistinguishability, and (b) technical, as we improve known bounds by a factor equal to the graph density. We elaborate more on out techniques and results below.

A Simpler Proof. Our proof uses only a naive optimization algorithm, avoiding combinatoric calculations using energy arguments based on Cauchy-Schwarz inequalities, that appear in other proofs like [FL14].

Quantitative Improvements. For the Strong Regularity Lemma we bound the partition size by a tower of twos of height $O(\epsilon^{-2} d_G)$ which is an improvement by a factor of d_G over best results [FL14]. Similarly, for the Weak Regularity Lemma we prove that the partition is an overlay of $O(\epsilon^{-2} d_G)$ subsets (in particular has up to $2^{O(\epsilon^{-2} d_G)}$ members) which improves by a factor of $\log d_G$ the best bounds from [VZ13].

Note that for constant densities d_G, this matches both best upper and lower bounds [FL14]. Our improvements for smaller densities doesn't contradict the lower bounds as they depend on the density in a complicated and non-explicit way (and hence don't apply to all regimes of d_G).

---

[3] The original work [LS07] proves a bound being $O(\epsilon^{-2})$ iterations of the function $s(1) = 1$, $s(k+1) = 2^{s(1)^4 \cdots s(k)^4}$ starting at 1. It is easy to see that $s(k)$ can be bounded by a tower of height $k + O(1)$.

[4] The generated partition arises as intersections of the generating sets with their complements.

Regularity Lemmas as Low Complexity Approximations. We show that a variant of the Szemeredi Regularity Lemma, equivalent to the most often used statement, can be written in the following form

$$\forall f \in \mathcal{F}: \left| \underset{e \leftarrow \mathcal{X}}{\mathbb{E}} g(e)f(e) - \underset{e \leftarrow \mathcal{X}}{\mathbb{E}} h(e)f(e) \right| \leqslant \epsilon \qquad (5)$$

for some functions g, f and a class of functions $\mathcal{F}$ on a finite set $\mathcal{X}$, where h is "efficient" in terms of complexity. More precisely, the result states that a given function f (in our case related to the irregularity of the graph) can be efficiently approximated under a certain class of test functions (called also distinguishers). In cryptography results of this sort are known as *low complexity approximations* and are a powerful and elegant technique of proving complicated results [TTV09, VZ13, JP14]. The quantity in the absolute values in Eq. (5) is referred to as the advantage of f in distinguishing g and h, so the statement simply means that h is indistinguishable from g for small ϵ by all functions in $\mathcal{F}$. Depending on the class $\mathcal{F}$ it may be a good "replacement" for g in applications.

In our case the class of test functions changes depending on the problem. For weak regularity we use rectangle indicator functions, whereas for strong regularity we consider combinations of rectangle-indicator functions

$$\mathcal{F} = \{f : f = \pm 1_{T \times S}\} \qquad \text{(for Weak Regularity)}$$

$$\mathcal{F} = \left\{f : f = \sum_{i,j} \pm 1_{T_{i,j} \times S_{i,j}}\right\} \quad \text{(for Strong Regularity)}$$

The proof is in both cases very simple and can be viewed as a special case of the general subgradient descent algorithm well known in convex optimization[5]. The algorithm is given below in pseudocode (see Algorithm 1).

A similar result has been shown by Trevisan et al. [TTV09] with respect to the weak regularity lemma. It turns out that the weak regularity lemma can be directly translated to a form of Eq. (5). The case of the Strong Regularity Lemma is however a bit different, because the standard statement doesn't admit a direct translation to Eq. (5) so we need first to reduce the Regularity Lemma to a slightly relaxed form similar[6] to Theorem 2 and prove the relaxed statement by low complexity approximation tools. Also, the same class of functions appear in the analytic proof in [LS07] but in a different approximation technique.

Abstracting the Concept of Pseudo-regularity. In the Weak Regularity Lemma, we measure the irregularity of the partition as *average difference* between the actual number of edges and the expected number of edges across the pairs of

---

[5] If we consider the mapping $h \to \max_f \left| \underset{e \leftarrow \mathcal{X}}{\mathbb{E}} g(e)f(e) - \underset{e \leftarrow \mathcal{X}}{\mathbb{E}} h(e)f(e) \right|$ then its subgradient equals f for some $f \in \mathcal{F}$. Then the update is $h := h - t \cdot f$ precisely as in the proof of Sect. 2.1.

[6] The relaxed form we use is except that we allow any numbers $d_{i,j}$ in place of densities $d_G(V_i, V_j)$.

Algorithm 1. Low Complexity Approximations

    Input   : target function g to approximate,
                class of test functions $\mathcal{F}$,
                a starting point h^0,
                accuracy parameter ϵ,
                stepsize t
    Output: function h of low complexity w.r.t $\mathcal{F}$ and indistinguishable from g
                 (with respect to test functions $\mathcal{F}$)

  1.1 $n \leftarrow 0$
  1.2 while *can distinguish h^n and g by some $f \in \mathcal{F}$ with advantage ϵ* **do**
  1.3 | $n \leftarrow n + 1$
  1.4 | $h^n \leftarrow h^{n-1} - t \cdot f$

---

parts of the partition. Therefore, the Weak Regularity Lemma is obtained from the bound

$$\left| \sum_{i,j} E(T_i, S_j) - \sum_{i,j} d_{i,j} |T_i||S_j| \right| \ll |V|^2$$

(where T_i, S_j are subsets of V_i and V_j respectively; note that $\sum_{i,j} E(T_i, S_j) = E(T, S)$). In turn, to prove the Strong Regularity Lemma, we measure the *average absolute difference* between the actual number of edges and the expected number of edges. To prove our result we introduce the following condition (for some constants $d_{i,j}$)

$$\sum_{i,j} |E(T_{i,j}, S_{i,j}) - d_{i,j} |T_{i,j}||S_{i,j}|| \ll |V|^2.$$

($S_{i,j}, T_{i,j}$ being subsets of V_i and V_j respectively), and refer to this property as "pseudo-regularity"[7]. This condition extends slightly the notion of irregularity, where the true densities of pairs (V_i, V_j) appear in place of $d_{i,j}$. Note that pseudo-regularity can be understood as approximating the graph by a weighted graph, where we control the absolute deviation of number of edges across pairs of partition parts.

The approach with unrestricted constants is much easier to prove and is more flexible. In fact, the idea of relaxing restrictions on densities (equivalently: considering a weighted graph) goes back to [FK99b]. The concept of pseudoregularity is what allows us to connect the approximation lemma with the Strong Regularity Lemma.

1.4   Proof Techniques

The key ingredient of our proof is a descent algorithm, which translated back to the partition language is similar to the popular technique of proving regularity

---

[7] This property was also implicitly used in [LS07].

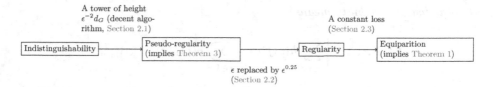

Fig. 1. An overview of our proof of the Strong Regularity Lemma

lemmas. As long as the current partition fails to satisfy the desired property, the algorithm uses sets being counterexamples to refine the partition. Moreover, we show that a certain quantity, called the energy function, decreases with every step by a constant (depending on ϵ). From this one concludes that the process of refining the partition halts after a number of step (the bound depends on concrete energy estimates).

Our proof is different with respect to the energy function, as we use simply the euclidean distance (second norm) between the candidate solution and the target. This allows us to decrease the number of rounds by the initial distance, which in our case equals d_G, as we start from $f = 1_E$ (where E is the edge set) and $g = 0$. An overview of the proof (of the Strong Regularity Lemma) is illustrated in Fig. 1.

The proof of the Weak Regularity Lemma is even simpler and consists of only first step (with the class of test functions changed accordingly).

1.5 Organization

In Sect. 2 we prove a variant of the Strong Regularity Lemma, in Sect. 3 we prove the Weak Regularity Lemma and conclude the work in Sect. 4.

2 Strong Regularity Lemma

2.1 Obtaining a Partition with Small Pseudo-irregularity

The key ingredient is the following approximation result, proved by the technique sketched in Algorithm 1.

Theorem 5 (Simulating against stepwise functions). *For any real function g on V^2 and any $\epsilon > 0$, there is a partition $V_1, \ldots, V_k$ of V and a piecewise function h constant on rectangles $V_i \times V_j$ such that h and g are ϵ-indistinguishable by functions piecewise constant on subrectangles of $V_i \times V_j$ where $i \leqslant j$*

$$\mathcal{F}_k = \left\{ f = \sum_{i \leqslant j} a_{i,j} 1_{S_{i,j} \times T_{i,j}} : \quad a_{i,j} = \pm 1, \quad S_{i,j} \subset V_i, T_{i,j} \subset V_j \right\}, \quad (6)$$

where indistinguishability means

$$\forall f \in \mathcal{F}_k : \quad \left| \sum_{e \in V^2} h(e)f(e) - \sum_{e \in V^2} g(e)f(e) \right| \leqslant \epsilon |V|^2, \tag{7}$$

and moreover k is not bigger than $O(d\epsilon^{-2})$ iterations of the function $k \to k \cdot 2^{k+1}$ at $k = 1$, where $d = \frac{1}{|V|^2} \sum_{e \in V^2} g(e)^2$. In particular, k is at most a tower of 2's of height $O\left(d\epsilon^{-2}\right)$.

Remark 3 (Symmetrizing the class $\mathcal{F}$). Note that ordering pairs $(i \leqslant j)$ in the definition of class $\mathcal{F}$ is crucial to obtain the complexity being a power of 2. Otherwise, we would obtain a (much worse) tower of 4's of the same height.

Remark 4. It is easy to see that the function is a power-tower of twos of height $O(d_G\delta^{-2})$ (a formal proof can obtained by induction as in [FL14].

As a corollary we obtain the following statement which is precisely the variant stated in Theorem 3.

Corollary 1 (Regularity Lemma in terms of pseudo-regularity (variant 3)). *For any graph G there is a partition $\{V_i\}_i$ of vertices V such that the absolute pseudo-irregularity is at most $\epsilon |V|^2$, that is for some numbers $d_{i,j}$*

$$\sum_{(i,j):i\leqslant j\leqslant k} \max_{S\subset V_i, T\subset V_j} |E(T,S) - d_{i,j} \cdot |T||S|| \leqslant \epsilon |V^2|, \tag{8}$$

where the number of partition parts is a tower of twos of height $O(d_G\delta^{-2})$.

Proof (Proof of Corollary 1). It suffices to apply Theorem 5 to $g = 1_E$ and $h = 0$. We have then $\sum_e g(e)t(e) = \sum_{i\leqslant j} a_{i,j}E(S_{i,j}, T_{i,j})$ and $\sum_e h(e)t(e) = \sum_{i\leqslant j} a_{i,j}d_{i,j}|S_{i,j}||T_{i,j}|$. The absolute values in Eq. (8) are achieved by fitting signs of the coefficients $a_{i,j} = \pm 1$.

Proof (of Theorem 5). Suppose we have a function h on a partition $V_1, \ldots, V_k$ which is $\delta|V|$-indistinguishable from g by a function f piecewise constant on squares $T_i \times S_j$, that is

$$\sum_e (g(e) - h(e))f(e) \geqslant \delta|V|^2 \tag{9}$$

Consider now $h' = h + t \cdot f$ and note that

$$\sum_e (h'(e) - g(e))^2 = \sum_e (h(e) - h(e))^2 - 2t \sum_e (g(e) - h(e))f(e) + t^2 \sum_e f(e)^2.$$

Setting $t = \delta$ in the above equation, by Eq. (9) we obtain

$$\sum_e (h'(e) - g(e))^2 \leqslant \sum_e (h(e) - h(e))^2 - \delta^2|V|^2,$$

which means that by replacing h by h' we decrease the distance to g by $\delta^2|V|^2$. Since our first choice for h is the zero function, the initial distance was equal to $\sum_e g(e)^2 = d|V|^2$ and the loop ends after at most $O(d\delta^{-2})$. Regarding the complexity of $h' = h + t\sum_{i \leqslant j} a_{i,j} 1_{S_{i,j} \times T_{i,j}}$ note that when adding step functions $1_{S_{i,j} \times T_{i,j}}$, any fixed partition member V_i is intersected by at most $k+1$ sets of the form $S_{i,j}$ or $T_{i,j}$ (because we consider only ordered pairs $i \leqslant j!$). Therefore, the function h' is piecewise constant on the partition $\mathcal{V}'$ generated by $\mathcal{V}$ and sets $S_{i,j}, T_{i,j}$ which has at most $k \cdot 2^{k+1}$ members.

2.2   Small Pseudo-irregularity Implies Regularity

In this section we show that pseudo-regularity implies regularity in the sense of Definition 3.

Proposition 1. *Suppose that for a partition $V_1, \ldots, V_k$ of V there exist numbers $d_{i,j}$ such that*

$$\sum_{i,j \leqslant k} |E(S_{i,j}, T_{i,j}) - d_{i,j} \cdot |T_{i,j}||S_{i,j}|| \leqslant \epsilon^4 |V|^2 \tag{10}$$

for all disjoint rectangles $T_{i,j} \times S_{i,j} \subset V_i \times V_j$. Then the partition is 2ϵ-regular.

Proof. Rewrite Eq. (10) as

$$\sum_{i,j \leqslant k} \frac{|S_{i,j}||T_{i,j}|}{|V|^2} |d_G(S_{i,j}, T_{i,j}) - d_{i,j}| \leqslant \epsilon^4$$

In particular, we get

$$\sum_{i,j \leqslant k} \frac{|V_i||V_j|}{|V|^2} |d_G(S_i, T_j) - d_{i,j}| \leqslant \epsilon^2 \tag{11}$$

when $|S_{i,j}|, |T_{i,j}| \geqslant \epsilon|V|$ for all i. Let $S'_{i,j}, T'_{i,j}$ (both bigger than $\epsilon|V|$) maximize $|d_G(S_{i,j}, T_{i,j}) - d_{i,j}|$. By the Markov inequality (applied to the probability weights $p_{i,j} = \frac{|V_i||V_j|}{|V|^2}$), there exists an "exceptional" set $I \subset \{1..k\}^2$ such that

$$\sum_{(i,j) \in I} |V_i||V_j| \leqslant \epsilon|V|^2$$

and

$$\forall(i,j) \notin I : \left| d_G(S'_{i,j}, T'_{i,j}) - d_{i,j} \right| \leqslant \epsilon.$$

By the choice of the pairs $(S'_{i,j}, T'_{i,j})$ this implies $|d_G(S_{i,j}, T_{i,j}) - d_{i,j}| \leqslant \epsilon$ for every pair $S_{i,j} \subset V_i, T_{i,j} \subset V_j$ (provided that $(i,j) \notin I$. In particular, this is true with $S_{i,j} = V_i$ and $T_{i,j} = V_j$ which gives $|d_G(V_i, V_j) - d_{i,j}| \leqslant \epsilon$. By the triangle inequality we have $|d_G(S_{i,j}, T_{i,j}) - d_G|V_i||V_j|| \leqslant 2\epsilon$ for $(i,j) \notin I$ which finishes the proof.

2.3 Enforcing Equipartition

To complete the last step of the proof we have to prove the following.

Lemma 1. *For any ϵ-regular partition $\mathcal{V}$ there exists a $O(\epsilon)$-regular equipartition $\mathcal{W}$ of size $|\mathcal{W}| = O\left(\epsilon^{-1}|\mathcal{V}|\right)$.*

The key observation is the following useful fact, which simply states that regularity is preserved under refinements. A simple proof is given in Appendix A.

Lemma 2 (Regularity preserved under refinements). *For any graph G, if (S, T) is ϵ-regular and $S' \subset S$, $T' \subset T$, then (S', T') is 2ϵ-regular.*

Consider now a coarser partition $\{V_{i,i'}\}_{i,i'}$ such that for every i the set V_i is partitioned into $k(i) \leqslant \frac{k}{\epsilon}$ parts $V_{i,i'}$ where $i' = 1, \ldots, k(i)$ which are all, up to one, of equal size

$$|V_{i,i'}| = \left\lceil \frac{|V|}{\ell} \right\rceil, \quad i' = 1, \ldots, k(i) - 1$$

$$|V_{i,i'}| < \left\lceil \frac{|V|}{\ell} \right\rceil, \quad i' = k(i)$$

Let $V' = \bigcup_i V_{k(i)}$. In other words, the set V' combines all "residual" parts into one component. We partition W again into equal (except one) parts $V'_1, \ldots, V'_r$ so that

$$|V'_i| = \left\lceil \frac{|V|}{\ell} \right\rceil, \quad i = 1, \ldots, r - 1$$

$$|V'_r| < \left\lceil \frac{|V|}{\ell} \right\rceil$$

Therefore, the family

$$\bigcup_{i=1,\ldots,k} \bigcup_{i'=1,\ldots,k(i)-1} \{V_{i,i'}\}_{i,i'} \cup \bigcup_{i=1,\ldots,r} \{V'_i\} \tag{12}$$

is a partition of V that has ℓ members, $\ell - 1$ of them being of size $\left\lceil \frac{|V|}{\ell} \right\rceil$ and one being a "remainder" of size smaller than $\left\lceil \frac{|V|}{\ell} \right\rceil$. It follows that the last term has to be of size at least $|V| - (l - 1)\left\lceil \frac{|V|}{\ell} \right\rceil$, that is between $\frac{|V|}{\ell}$ and $\frac{|V|}{\ell} - (l - 1)$. Now by moving up to one element from each of the other $\ell - 1$ components to the remaining component we arrive at an equipartition $W_1, \ldots, W_\ell$ where all members are of equal size up to one element, that is

$$||W_i| - |W_j|| \leqslant 1 \tag{13}$$

Note that we moved from sets V_i to V' at most $k \cdot \frac{|V|}{\ell} = O(\epsilon|V|)$ vertices, which by Eq. (12) belong to at most $O(\ell\epsilon)$ parts W_j. Therefore

Claim (Partition W_i is a refinement of V_i up to a small fraction of members).
For all up to a $O(\epsilon)$-fraction of pairs $(i,j) \in \{1,\dots,\ell\}^2$, the sets W_i, W_j are
subsets of some pair $V_{i'}, V_{j'}$.

Let I_W be the set of all pairs (i,j) such that the pair (W_i, W_j) is not ϵ-regular,
and let I_V be the set of pairs (i,j) such that (V_i, V_j) is not ϵ-regular.

$$\sum_{(i,j)\in I_W} |W_i||W_j| \leq \epsilon|V|^2 + \sum_{(i,j):W_i\subset V_{i'},W_j\subset V_{j'}} |W_i||W_j| \qquad (14)$$

$$\leq \sum_{(i,j)\in I_V} |V_i||V_j| \qquad (15)$$

$$\leq O\left(\epsilon|V|^2\right), \qquad (16)$$

where the first line follows by the last claim and the fact that W_i are disjoint,
the second line follows by the regularity of the partition V_i. Now Eq. (13) implies
$|I_W| = O(\epsilon\ell^2)$.

3   Weak Regularity Lemma

Theorem 6 (Simulating against rectangle-indicator functions). *For any
function $g : V^2 \to [-1,1]$, and any $\epsilon > 0$, there exists a partition $V_1,\dots,V_k$ and
a piece-wise function h constant on squares $V_i \times V_j$ such that f and g are ϵ-
indistinguishable by indicators of rectangles $V_i \times V_j$ where $i \leqslant j$*

$$\mathcal{F} = \{f = \pm 1_{S\times T} : \quad S \subset V_i, T \subset V_j\}, \qquad (17)$$

that is

$$\forall f \in \mathcal{F} : \quad \left|\sum_{e\in V^2} h(e)f(e) - \sum_{e\in V^2} g(e)f(e)\right| \leqslant \epsilon|V|^2. \qquad (18)$$

*Moreover, k is not bigger than $2^{O(d_G\epsilon^{-2})}$. In fact, the partition is an overlay of
$O(d_G\epsilon^{-2})$ subsets of vertices.*

By applying this result to the function 1_E on V^2 (being 1 for pairs $e = (v_1, v_2)$
which are connected and 0 otherwise) we reprove Theorem 4.

Corollary 2 (Deriving Weak Regularity Lemma). *The Weak Regularity
Lemma holds with $k = O(d_G\epsilon^{-2})$.*

This result, without the factor d_G was proved in [TTV09]. We skip the proof
of Theorem 6 as it merely repeats the argument from Theorem 5, noticing only
that the calculation of k is different because the class $\mathcal{F}$ is now simpler. Note
also that for this result the class $\mathcal{F}$ doesn't change with every round.

4 Conclusion

We have shown that both: weak and strong regularity lemmas can be written as indistinguishability statements, where the edge indicator function is approximated by a combination of rectangle-indicator functions.

This extends the result of Trevisan et al. for weak regularity to the case of Strong Regularity Lemma. Moreover, due to a different analysis of the underlying descent algorithm, our proof achieves quantitative improvements graphs with low edge densities.

A Proof of Lemma 2

Proof. Let d be the edge density of the pair (T, S) and d' be the edge density of the pair (T', S'). Denote $\epsilon = \mathrm{irreg}_G(T, S)$. For any two subsets $T'' \subset T', S'' \subset S'$, which are also subsets of T and S respectively, by the definition of d we have

$$\left| \frac{E(T', S')}{|T'||S'|} - d \right| \leqslant \epsilon.$$

which translates to

$$|d' - d| \leqslant \epsilon. \tag{19}$$

Therefore, by Eq. (19) and the triangle inequality

$$|E(T'', S'') - d' \cdot |T''||S''|| \leqslant |E(T'', S'') - d \cdot |T''||S''|| + \epsilon \cdot |T''||S''|. \tag{20}$$

Since the definition of d applied to $T'' \subset T, S'' \subset S$ implies

$$|E(T'', S'') - d' \cdot |T''||S''|| \leqslant \epsilon \cdot |T''||S''|,$$

from Eq. (20) we conclude that

$$|E(T'', S'') - d' \cdot |T''||S''|| \leqslant 2\epsilon \cdot |T''||S''|,$$

which finishes the proof.

References

[DLR95] Duke, R.A., Lefmann, H., Rdl, V.: A fast approximation algorithm for computing the frequencies of subgraphs in a given graph. SIAM J. Comput. **24**(3), 598–620 (1995)

[FK99a] Frieze, A., Kannan, R.: A simple algorithm for constructing szemeredi's regularity partition. Electron. J. Comb. [electronic only] **6**(1), Research paper R17, 7 p. (eng) (1999)

[FK99b] Frieze, A.M., Kannan, R.: Quick approximation to matrices and applications. Combinatorica **19**(2), 175–220 (1999)

[FL14]   Fox, J., Lovász, L.M.: A tight lower bound for szemerédi's regularity lemma. CoRR abs/1403.1768 (2014)

[Gow97]   Gowers, W.T.: Lower bounds of tower type for szemerédi's uniformity lemma. Geom. Funct. Anal. GAFA **7**(2), 322–337 (1997)

[HMT88]   Hajnal, A., Maass, W., Turán, G.: On the communication complexity of graph properties. In: Proceedings of the Twentieth Annual ACM Symposium on Theory of Computing, STOC 1988, New York, pp. 186–191. ACM (1988)

[JP14]   Jetchev, D., Pietrzak, K.: How to fake auxiliary input. In: Lindell, Y. (ed.) TCC 2014. LNCS, vol. 8349, pp. 566–590. Springer, Heidelberg (2014). doi:10.1007/978-3-642-54242-8_24

[KR02]   Kohayakawa, Y., Rdl, V.: Szemeredi's regularity lemma and quasi-randomness (2002)

[KS96]   Komls, J., Simonovits, M.: Szemeredi's regularity lemma and its applications in graph theory (1996)

[LS07]   Lovász, L., Szegedy, B.: Szemerédi's lemma for the analyst. Geom. Funct. Anal. **17**(1), 252–270 (2007). MR MR2306658 (2008a:05129)

[RS]   Rdl, V., Schacht, M.: Regularity lemmas for graphs

[Sze75]   Szemeredi, E.: On sets of integers containing no k elements in arithmetic progression (1975)

[TTV09]   Trevisan, L., Tulsiani, M., Vadhan, S.: Regularity, boosting, and efficiently simulating every high-entropy distribution. In: Proceedings of the 24th Annual IEEE Conference on Computational Complexity, CCC 2009, Washington, DC, USA, pp. 126–136. IEEE Computer Society (2009)

[VZ13]   Vadhan, S., Zheng, C.J.: A uniform min-max theorem with applications in cryptography. In: Canetti, R., Garay, J.A. (eds.) CRYPTO 2013. LNCS, vol. 8042, pp. 93–110. Springer, Heidelberg (2013). doi:10.1007/978-3-642-40041-4_6

On the Complexity of Breaking Pseudoentropy

Maciej Skorski[⊠]

IST Austria, Klosterneuburg, Austria
maciej.skorski@gmail.com

Abstract. Pseudoentropy has found a lot of important applications to cryptography and complexity theory. In this paper we focus on the foundational problem that has not been investigated so far, namely by how much pseudoentropy (the amount seen by computationally bounded attackers) differs from its information-theoretic counterpart (seen by unbounded observers), given certain limits on attacker's computational power?

We provide the following answer for HILL pseudoentropy, which exhibits a *threshold behavior* around the size exponential in the entropy amount:

- If the attacker size (s) and advantage (ϵ) satisfy $s \gg 2^k \epsilon^{-2}$ where k is the claimed amount of pseudoentropy, then the pseudoentropy boils down to the information-theoretic smooth entropy.
- If $s \ll 2^k \epsilon^2$ then pseudoentropy could be arbitrarily bigger than the information-theoretic smooth entropy.

Besides answering the posted question, we show an elegant application of our result to the complexity theory, namely that it implies the classical result on the existence of functions hard to approximate (due to Pippenger). In our approach we utilize non-constructive techniques: the duality of linear programming and the probabilistic method.

Keywords: Nonuniform attacks · Pseudoentropy · Smooth entropy · Hardness of boolean functions

1 Introduction

Pseudoentropy has recently attracted a lot of attention because of applications to complexity theory [RTTV08], leakage-resilient cryptography [DP08, Pie09], deterministic encryption [FOR15], memory delegation [CKLR11], randomness extraction [HLR07], key derivation, [SGP15] constructing pseudorandom number generators [VZ12, YLW13] or black-box separations [GW11].

What differs between pseudoentropy and information-theoretic entropy notions is the parametrization by adversarial resources. That is, pseudoentropy

---

The paper is available (with updates) at https://eprint.iacr.org/2016/1186.pdf.

M. Skorski—Supported by the European Research Council Consolidator Grant (682815-TOCNeT).

T.V. Gopal et al. (Eds.): TAMC 2017, LNCS 10185, pp. 600–613, 2017.
DOI: 10.1007/978-3-319-55911-7_43

not only has *quantity* k but also *quality*, which is typically described by the attacker size s and the advantage ϵ achieved in the security game.

Despite many works on applications of pseudoentropy, not much is known about relationships between k, s and ϵ for a given distribution X, in particular parameter settings that make pseudoentropy non-trivial (bigger than the information-theoretic entropy). Concrete numbers can be conjectured for some applications under assumptions about computational hardness, for example for outputs of pesudorandom generators, or keys obtained by the Diffie-Hellman protocol. Yet in many cases, like key derivation where pseudoentropy can model "weak" sources [SGP15], one simply assumes pseudoentropy of certain (strong enough) quality.

Without understanding relations between s, ϵ and k it is not clear how demanding are specific assumptions on pseudoentropy quality. This is precisely the issue we are going to address in this work.

1.1   Problem Statement

In this paper we are interested in separating pseudoentropy (entropy seen by computationally bounded attackers) from its information-theoretic counterpart (measured against unconstrained attackers).

An n-bit random variable X is said to have k bits of pseudoentropy[1] against attackers of size s and advantage ϵ if for some distribution Y of min-entropy k, no circuit of size s can distinguish it from Y with advantage bigger than ϵ (see Sect. 2.4)[2]. Note that the notion is parametrized by the adversarial specific size s and advantage ϵ. In particular the amount decreases when s gets bigger and ϵ gets smaller (it is harder to fool adversaries with bigger resources). When s is unbounded, pseudoentropy equals the information-theoretic smooth min-entropy (see Sect. 2.4).

To better understand possibilities and limitations of using pseudoentropy, it is natural to ask in what parameter regimes pseudoentropy provides non-trivial computational security, that is when we have a real gain in the entropy amount comparing to the information-theoretic case.

Q: How much computational power is needed to boil pseudoentropy down to information-theoretic smooth entropy?

1.2   Our Contribution

Nonuniform Attacks Against Pseudoentropy. Our result exhibit a *threshold phenomena*. Intuitively, with enough computational power (say size 2^n for n-bit random variables[3]) the notion of pseudoentropy is no more stronger than the corresponding information-theoretic entropy notion. We *estimate* the value of

---

[1] We consider here the most popular notion of HILL pseudoentropy.

[2] This matches the definition of pseudorandomness when k is the length of X.

[3] As this complexity is enough to compute every boolean function.

this threshold on the circuit size s, so that above there is no computational gain and below there exists non-trivial pseudoentropy. There result is somewhat surprising because: (a) the threshold doesn't depend on the length but the entropy amount and (b) the threshold depends also on the square of the advantage.

Theorem (Informal) (Breaking pseudoentropy with enough computational power). *For any k, and any s, ϵ satisfying*

$$s \gg 2^k \epsilon^{-2}$$

and for every distribution of min-entropy k, unbounded attackers and attackers of size s see the same entropy amount.

Theorem (Informal) (Lower bound). *For any k, and any s, ϵ satisfying*

$$s \ll 2^k \epsilon^2$$

there exists a distribution X such that

(a) (bounded attackers see k bits) pseudoentropy of X against circuits of size s and advantage ϵ is k.
(b) (k bits for unbounded attackers see less than k bits) information-theoretic entropy of X is k.

A short overview of our results is given in Table 1 below. We note that the result is tight with respect to k, but not with respect to ϵ.

Table 1. Overview of our results. The analyzed setting is k bits of pseudoentropy against size circuits of size s and advantage ϵ.

Regime	Result	Techniques	Reference
$s \gg 2^k \epsilon^{-2}$	Same entropy for attackers of size s as for $s = \infty$	LP duality distinguisher optimization	Theorem 1
$s \ll 2^k \epsilon^2$	Arbitrary gap in the amount for size s and $s = \infty$	Probabilistic method concentration bounds	Theorem 2

Proof Outline and Our Tools

Breaking pseudoentropy. We outline the proof of the first result below

1. We first consider somewhat weaker pseudoentropy notion, called Metric entropy, where the order of quantifiers is reversed. That is, for any distinguisher D there has to be some Y of min-entropy k which is close to X under that particular test D, that is $\mathbb{E}D(X) \approx \mathbb{E}D(Y)$.

2. We prove that this weaker pseudoentropy notion collapses when $s \gg 2^k$, by "compressing" distinguishers down to size 2^k. The intuitive reason for that is that we can always manipulate Y so that it has "small" support (only $O(2^k)$ elements), and if an attacker wants to maximize the advantage $|\mathbb{ED}(X) - \mathbb{ED}(Y)|$, the best strategy is to hardcode the elements x such that $\Pr[Y = x] > \Pr[X = x]$ which is a subset of the support of Y and can be implemented in size $\tilde{O}(2^k)$.

3. We use a generic transformation due to Barak et al. [BSW03, Sko15] to go back to our standard entropy notion. The transformation losses $\tilde{O}(\epsilon^2)$ in size and is based on the duality of linear programming.

This way we obtain that pseudoentropy with parameters (s, ϵ) becomes the same as the amount seen by unbounded attackers when $s = \tilde{O}\left(2^k \epsilon^{-2}\right)$. The details are explained in the proof of Theorem 1.

Matching lower bounds. The proof of the second result goes as follows

1. We take a random subset $\mathcal{X} \subset \{0,1\}^k$ of size $k - c$, where c will be the gap between what bounded and unbounded attackers can see. The distribution X is the uniform distribution over $\mathcal{X}$ plus a "random shift" of an ϵ-fraction of the probability mass.

2. We argue that the ϵ-smooth entropy is still roughly k, because we have shifted only that fraction of the total probability mass. This is handled by a result of independent interest, stating that "almost" smooth distributions cannot be further smoothened (see Corollary 2).

3. We argue that the distribution X is pseudorandom provided that the class of test functions is small enough. This fact is proved by applying concentration bounds twice, once to handle the random shift and for the second time to handle the choice of $\mathcal{X}$. Intuitively, the advantage of bounded attackers is much smaller than ϵ because they are "fooled" by the random shift of a part of the probability mass. In turn, the entropy amount seen by bounded attackers is much bigger than $k - c$ because $\mathcal{X}$ is a random subset of $\{0,1\}^k$.

Putting this all together we get a strict separation: not only the amount of entropy is bigger, but also the advantage is smaller. The necessary assumption to make it work is that the class of distinguishers is much smaller than $2^{2^{k-c}\epsilon^2}$ members. For the details see the proof of Theorem 2.

1.3   Related Works

Pseudorandomness exists unconditionally. The classical textbook results [Gol06] shows that pseudorandomness exists unconditionally, which can be seen as a separation between pseudorandomness and min-entropy.

Our Theorem 2 is stronger as we separate pseudoentropy from smooth min-entropy (and cannot derive it from the mentioned result). From a technical point of view, the main difference is the extra random mass fluctuation (Step 1 in the above explanation), which needs to be later handled by bit more subtle probability tools (we use concentration inequalities for random variables with local dependence due to Janson).

Complexity of non-uniform attacks against PRGs. De, Trevisan and Tulsani studied the complexity of nonuniform attacks against pseudorandom generators [DTT10]. Their results are specialized to outputs of PRGs and are constructive, whereas our results apply to any random variable (unfortunately don't offer non-trivial results for the case of PRGs). Also,

1.4 Applications

Hard-to-approximate boolean functions. Our Theorem 2 implies the classical result [Pip76] which states that for any n and $\delta \in (0,1)$, there exist δ-hard functions[4] for size $s = \tilde{\Omega}\left(2^n(1-\delta)^2\right)$. For details, see Sect. 5.1.

1.5 Organization

We start with explaining basic concepts and notions in Sect. 2. In Sect. 3 we prove useful auxiliary facts about smooth min-entropy. In Sect. 4 we give proofs of our main results. In Sect. 5 we discuss applications to the complexity of approximating boolean functions.

2 Preliminaries

2.1 Model of Computations

Our results hold in the non-uniform model. We consider general classes of distinguishers, denoted by $\mathcal{D}$, which are families of functions from n bits to real values. When discussing complexity applications, we restrict $\mathcal{D}$ to classes of circuits of certain size s, with boolean or real-valued outputs.

2.2 Basic Notions

Definition 1 (Statistical distance). *The statistical distance of two random variables X, Y taking values in the same finite set is defined as $\mathsf{SD}(X,Y) = \frac{1}{2}\sum_x |\Pr[X = x] - \Pr[Y = x]|$. Equivalently, $\mathsf{SD}(X,Y) = \max_\mathsf{D} |\mathbb{E}\mathsf{D}(X) - \mathbb{E}\mathsf{D}(Y)|$ where D runs over all boolean functions.*

2.3 Information-Theoretic Entropies

Definition 2 (Min-entropy). *We say that X has k bits of min-entropy if $\min_x \log \frac{1}{\Pr[X=x]} = k$.*

Definition 3 (Smooth min-entropy [RW05]). *We say that X has k bits of ϵ-smooth min-entropy, denoted by $\mathbf{H}_\infty^\epsilon(X) \geqslant k$, if X is ϵ-close in the statistical distance to some Y of min-entropy k.*

Remark 1. Smoothing allows for increasing the entropy amount by shifting a part of the probability mass, to make the distribution look "more flat" or "more smooth".

---

[4] f is δ hard for size s if every circuit of size s fails to predict f w.p. at least $\frac{1+\delta}{2}$.

2.4 Pseudoentropy

In what follows, X denotes an arbitrary n-bit random variable.

Definition 4 (HILL pseudoentropy [HILL88, BSW03]). *We say that X has k bits of HILL pseudoentropy against a distinguisher class $\mathcal{D}$ and advantage ϵ, denoted by*

$$\mathbf{H}_{s,\epsilon}^{\mathrm{HILL}}(X) \geqslant k$$

if there is a random variable Y of min-entropy at least k that ϵ-fools any $\mathsf{D} \in \mathcal{D}$, that is for every $\mathsf{D} \in \mathcal{D}$ we have such that $|\mathbb{E}\mathsf{D}(X) - \mathbb{E}\mathsf{D}(Y)| \leqslant \epsilon$.

Definition 5 (Metric Pseudoentropy [BSW03]). *We say that X has k bits of metric pseudoentropy against a distinguisher class $\mathcal{D}$ and advantage ϵ, denoted by*

$$\mathbf{H}_{s,\epsilon}^{\mathrm{Metric}}(X) \geqslant k$$

if for any $\mathsf{D} \in \mathcal{D}$ there is a random variable Y of min-entropy at least k that ϵ-fools this particular D that is such that $|\mathbb{E}\mathsf{D}(X) - \mathbb{E}\mathsf{D}(Y)| \leqslant \epsilon$.

Metric entropy is a convenient relaxation of HILL entropy, more suitable to work with in many cases. The important fact below shows that both notions are equivalent up to some loss in the circuit size.

Lemma 1 (Metric-to-HILL   Transformation   [BSW03, Sko15]). *If $\mathbf{H}_{s,\epsilon}^{\mathrm{Metric}}(X) \geqslant k$ then $\mathbf{H}_{s',\epsilon'}^{\mathrm{HILL}}(X) \geqslant k$ where $\epsilon' = 2\epsilon$ and $s' \approx s\epsilon^2/n$.*

Remark 2 (Abbreviations and equivalences for circuit classes). In the specific setting where $\mathcal{D}$ consists of deterministic boolean or deterministic real-valued circuits of size s we will slightly abuse the notation and write $\mathbf{H}_{s,\epsilon}^{\mathrm{Metric}}(X) = \mathbf{H}_{\mathcal{D},\epsilon}^{\mathrm{Metric}}(X)$. This is justified by the fact that for metric entropy deterministic real-valued circuits of size s give the same amount as deterministic boolean circuits of size $s' \approx s$ [FOR15]. In turn, for HILL entropy, deterministic boolean, deterministic randomized and deterministic real-valued circuits are equivalent with no entropy loss and with roughly same sizes [FOR15], so we also simply write $\mathbf{H}_{s,\epsilon}^{\mathrm{HILL}}(X) = \mathbf{H}_{\mathcal{D},\epsilon}^{\mathrm{HILL}}(X)$.

2.5 Relationships Between Entropy, Smooth Entropy, and Computational Entropy

The following proposition states that for extreme parameter regimes (unbounded attackers or zero advantage), pseudoentropy collapses to the information-theoretic notion of smooth-entropy (we skip the easy proof).

Proposition 1. *Let X be any n-bit random variable. Then we have*

(a) (Unbounded attackers) If $s = \infty^5$ then

$$\mathbf{H}^{\mathrm{Metric}}_{s,\epsilon}(X) = \mathbf{H}^{\mathrm{HILL}}_{s,\epsilon}(X) = \mathbf{H}^{\epsilon}_{\infty}(X) > \mathbf{H}_{\infty}(X).$$

(b) (No smoothing) If $\epsilon = 0$ then for any s

$$\mathbf{H}^{\mathrm{Metric}}_{s,\epsilon}(X) = \mathbf{H}^{\mathrm{HILL}}_{s,\epsilon}(X) = \mathbf{H}^{\epsilon}_{\infty}(X) = \mathbf{H}_{\infty}(X).$$

(c) (General) For any s, ϵ

$$\mathbf{H}^{\mathrm{Metric}}_{s,\epsilon}(X) \geqslant \mathbf{H}^{\mathrm{HILL}}_{s,\epsilon}(X) \geqslant \mathbf{H}^{\epsilon}_{\infty}(X) \geqslant \mathbf{H}_{\infty}(X).$$

2.6 Concentration Inequalities

The following lemma is a corollary from the famous concentration bound due to Jason, which exploits local dependencies.

Lemma 2 (Concentration bounds, local dependencies [Jan04]). *Let $X_1, \ldots, X_n$ be random variables with values in $[a, b]$, such that every X_i is not independent of at most Δ other variables $X_{i'}$. Let $\mu = n^{-1} \sum_{i=1}^{n} \mathbb{E} X_i$. Then*

$$\Pr\left[n^{-1} \sum_{i=1}^{n} X_i \geqslant \mu + \delta \right] \leqslant \exp\left(-\frac{2n\delta^2}{(a-b)^2(\Delta+1)} \right).$$

In particular, for $\Delta = 0$ we obtain the following bound.

Corollary 1 (Hoeffding's Inequality [Hoe63]). *Let $X_1, \ldots, X_n$ be independent random variables with values in $[a, b]$. Let $\mu = n^{-1} \sum_{i=1}^{n} \mathbb{E} X_i$. Then*

$$\Pr\left[n^{-1} \sum_{i=1}^{n} X_i \geqslant \mu + \delta \right] \leqslant \exp\left(-\frac{2n\delta^2}{(a-b)^2} \right).$$

Remark 3 (Hoeffding's Inequality for sampling without repetitions). The above inequality applies also the setting where X_i are random samples taken from the same distribution without repetitions [Ser74].

3 Auxiliary Facts

3.1 Auxiliary Results on Smooth Renyi Entropy

In the lemma below we show that smoothing doesn't help to increase entropy for flat distributions.

---

[5] If the domain consists of n-bit strings, it is enough to assume $s > 2^n$ as every function over n bits has complexity at most 2^n.

Lemma 3 (Flat distributions cannot be smoothened). *Let X be an n-bit random variable. Suppose that the distribution of X is flat and $\mathbf{H}_\infty(X) = k$. Then $\mathbf{H}_\infty^\epsilon(X) \leqslant k + \log\left(\frac{1}{1-\epsilon}\right)$ for every $\epsilon \in (0,1)$.*

Proof. Let X' be any distribution of min-entropy at least $k' > k + \log\left(\frac{1}{1-\epsilon}\right)$. Consider the distinguisher D which outputs $\mathsf{D}(x) = 1$ if $x \in \mathrm{supp}(X)$ and $\mathsf{D}(x) = 0$ otherwise. Note that $\mathbb{E}\mathsf{D}(X) = 1$ and $\mathbb{E}\mathsf{D}(X') = \frac{\mathrm{supp}(X)}{2^{k'}} < 1 - \epsilon$. Therefore $\mathbb{E}\mathsf{D}(X) - \mathbb{E}\mathsf{D}(X')$ and thus the statistical distance between X and X' is bigger ϵ.

Corollary 2 (Almost-flat distributions cannot be smoothened). *Suppose that X is ϵ_1-close to some X' being flat over 2^k elements. Then $\mathbf{H}_\infty^{\epsilon_2}(X) \leqslant k + \log\left(\frac{1}{1-\epsilon_1-\epsilon_2}\right)$ for any $\epsilon_1, \epsilon_2 > 0$ such that $\epsilon_1 + \epsilon_2 < 1$.*

Proof. Suppose not, then there exists X'' that is ϵ_2-close to X an has min-entropy at least $k' > k + \log\left(\frac{1}{1-\epsilon_1-\epsilon_2}\right)$. In particular, X'' is ϵ-close to X', where $\epsilon = \epsilon_1 + \epsilon_2$. Since X' is flat, Lemma 3 implies that the min-entropy of X'' is at most $k + \log\left(\frac{1}{1-\epsilon}\right)$, which is a contradiction.

4   Main Results

4.1   Complexity of Breaking Pseudoentropy

The following result specifies the attacker size for which pseudoentropy provides no computational security.

Theorem 1 (Breaking pseudoentropy is exponentially easy in the amount). *For any n bit random variable X, if $\mathbf{H}_\infty^\epsilon(X) = k$ then also $\mathbf{H}_{s,\epsilon}^{\mathrm{HILL}}(X) = k$ for $s > n^2 2^k \epsilon^{-2}$.*

The proof follows the steps explained in Sect. 1.2 and is given in Appendix A.

4.2   Matching Lower Bounds

Theorem 2 (Breaking pseudoentropy can be exponentially hard in the amount). *Let $\mathcal{S} \subset \{0,1\}^n$ be a set of cardinality 2^k, $\epsilon' \in (0,1)$ be arbitrary, and let $\mathcal{D}$ be a class of functions from $\mathcal{S}$ to $[0,1]$ such that*

$$|\mathcal{D}| < 2^{-2} \cdot 2^{2^{k-C-1}\epsilon'^2}.$$

Then for any $\epsilon < \frac{1}{4}$ there exists a random variable X on $\mathcal{S}$ such that

(a) $\mathbf{H}_{\mathcal{D},\epsilon'}^{\mathrm{HILL}}(X) = k$.

(b) $\mathbf{H}_\infty^\epsilon(X) = k - C + \log\left(\frac{1}{1-2\epsilon}\right)$.

Moreover, we have the following symmetry: *the probability mass function of X takes only two values on two subsets of S of equal size.*

Remark 4 (Doubly-strong separation: by the amount and the advantage). Note that the interesting setting of the parameters in the theorem above is when $\epsilon' \ll \epsilon$ so that not only we have a gap in the entropy amount, but even for much bigger advantage for unbounded distinguishers.

The proof follows the steps explained in Sect. 1.2 and appears in Appendix B.

5 Applications

5.1 Complexity of Hard Boolean Functions

For any function f, and a distribution μ on the domain of f we denote by $\mathsf{Guess}^D(f, \mu)$ the probability of guessing f by a function D when the input is sampled according to μ, that is $\mathsf{Guess}^D(f, \mu) = \Pr_{x \sim \mu}[D(x) = f(x)]$. We say that f on n bits is δ-hard[6] for size s if $\mathsf{Guess}^D(f, \mu) < 1 - \frac{\delta}{2}$ for every circuit D of size s and uniform μ (we also write $\mathsf{Guess}^D(f) < 1 - \frac{\delta}{2}$).

The corollary bellow is the classical result on the complexity of hard functions. Our result is optimal up to a factor linear in n (note that for large n, the value of n is negligible comparing to 2^n. Also, most interesting settings are with $\delta \approx 1$ with a negligible gap, and we get the optimal dependency on $1 - \delta$.).

Corollary 3 (Functions hard to approximate by circuits). *For any n and $\delta \in (0, 1)$, there exists an n-bit function which is δ-hard for all n-bit boolean circuits of size $s = \Omega\left(2^n(1 - \delta)^2\right)$.*

Proof (of Corollary 3). Let $D'(x) = 2D(x) - 1$. Denote for shortness $\mathsf{Adv}^D(X, Y) = \mathbb{E}D(X) - \mathbb{E}D(Y)$. Observe that for any X, Y we have

$$
\begin{aligned}
\mathsf{Adv}^D(X, Y) \\
&= \mathbb{E}D(X) - \mathbb{E}D(Y) \\
&= \frac{1}{2}\sum_x (2D(x) - 1)\left(\Pr[X = x] - \Pr[Y = x]\right) \\
&= \mathsf{SD}(X, Y)\mathbb{E}_{x \sim \mu}D'(x) \cdot \mathrm{sign}\left(\mathbf{P}_X(x) - \mathbf{P}_Y(x)\right) \\
&= \mathsf{SD}(X, Y)\left(\Pr_{x \sim \mu}[D'(x) = f(x)] - \Pr_{x \sim \mathbf{P}_X - \mathbf{P}_Y}[D'(x) \neq f(x)]\right) \\
&= \mathsf{SD}(X, Y)\left(2\Pr_{x \sim \mu}[D'(x) = f(x)] - 1\right) \\
&= \mathsf{SD}(X, Y) \cdot \left(2\mathsf{Guess}^D(f, \mu) - 1\right)
\end{aligned}
$$

---

[6] We use the convention for which $\delta = 1$ corresponds to completely unpredictable function. Some works substitute $1 - \delta$ in place of δ.

where $f(x) = \text{sign}(\mathbf{P}_X(x) - \mathbf{P}_Y(x))$ and $\mu(x) = \frac{|\mathbf{P}_X(x) - \mathbf{P}_Y(x)|}{2\text{SD}(X,Y)}$ (note that $\sum_x \mu(x) = 1$). Let us apply Theorem 2 to $k = n$, $\epsilon = \frac{1}{8}$, $\epsilon' = (1 - \delta)\epsilon$ and $\mathcal{D}$ being the class of deterministic circuits of size s. Let Y be the indistinguishable distribution from the definition of HILL entropy. Since in our case Y is uniform, the function f is well-defined and moreover $\text{SD}(X,Y) \geqslant \epsilon$ by (b). Thus

$$\text{Adv}^\mathcal{D}(X,Y) > \epsilon \cdot \left(2\text{Guess}^\mathcal{D}(f,\mu) - 1\right)$$

Moreover, $|\mathbf{P}_X(x) - \mathbf{P}_Y(x)|$ is constant by construction. Therefore μ is uniform and we obtain

$$\text{Adv}^\mathcal{D}(X,Y) > \epsilon \cdot \left(2\text{Guess}^\mathcal{D}(f) - 1\right)$$

Now $\text{Adv}^\mathcal{D}(X,Y) < \epsilon(1 - \delta)$ implies $\text{Guess}^\mathcal{D}(f) < 1 - \frac{\delta}{2}$ for any D, which means that f is $1 - \delta$-hard for size s (here we use the fact that there are exponentially many circuits of size s, so that $2^{O(s)} < 2^{2^{k-O(1)}(1-\delta)^2}$ and the assumption on the class size is satisfied).

A   Proof of Theorem 1

Proof. We start with proving a weaker result, namely that for Metric pseudoentropy (weaker notion) the threshold equals 2^k.

Lemma 4 (The complexity of breaking Metric pseudoentropy). *If $\mathbf{H}_\infty^\epsilon(X) = k$ then also $\mathbf{H}_{s,\epsilon}^{\text{Metric}}(X) = k$ for $s > n2^k$.*

Proof (Proof of Lemma 4). We will show the following claim which, by Proposition 1, implies the statement.

Claim. If $s > n2^k$ and $s' = \infty$ then $\mathbf{H}_{s,\epsilon}^{\text{Metric}}(X) = \mathbf{H}_{s',\epsilon}^{\text{Metric}}(X)$

Proof (Proof of Claim). It suffices to show only $\mathbf{H}_{s,\epsilon}^{\text{Metric}}(X) \leqslant \mathbf{H}_{s',\epsilon}^{\text{Metric}}(X)$ as the other implication is trivial. Our strategy is to show that any distinguisher D that negates the definition of Metric entropy can be implemented in size 2^k.

Suppose that $\mathbf{H}_{s',\epsilon}^{\text{Metric}}(X) < k$. This means that for some D of size s' and all Y of min-entropy at least k we have $|\mathbb{E}D(X) - \mathbb{E}D(Y)| \geqslant \epsilon$. Since the set of all Y of min-entropy at least k is convex, the range of the expression $|\mathbb{E}D(X) - \mathbb{E}D(Y)|$ is an interval, so we either have always $\mathbb{E}D(X) - \mathbb{E}D(Y) > \epsilon$ or $\mathbb{E}D(X) - \mathbb{E}D(Y) < -\epsilon$. Without loosing generality assume the first possibility (otherwise we proceed the same way with the negation $D'(x) = 1 - D(x)$). Thus

$$\mathbb{E}D(X) - \mathbb{E}D(Y) > \epsilon \quad \text{for all } n \text{ bit } Y \text{ of min-entropy } k$$

where by Remark 2 we can assume that D is boolean. In particular, the set $\{x : D(x) = 1\}$ cannot have more than 2^k elements, as otherwise we would put Y being uniform over x such that $D(x) = 1$ and get $\mathbb{E}D(X) - 1 > \epsilon > 0$

which contradicts the fact that D is boolean. But if D is boolean and outputs 1 at most 2^k times, can be implemented in size $n2^k$, by hardcoding this set and outputting 0 everywhere else. This means precisely that $\mathbf{H}^{\mathrm{Metric}}_{s,\epsilon}(X) < k$. Now by Proposition 1 we see that also $\mathbf{H}^\epsilon_\infty(X) < k$ which proves that $\mathbf{H}^{\mathrm{Metric}}_{s,\epsilon}(X) \leqslant \mathbf{H}^\epsilon_\infty(X)$ finishes the proof of the claim.

Having proven Lemma 4, we obtain the statement for HILL pseudoentropy by applying the transformation from Lemma 1.

B   Proof of Theorem 2

Proof (Proof of Theorem 2). Let $\mathcal{X}$ be a random subset of $\mathcal{S}$ of cardinality 2^{k-C}. Let $x_1, \ldots, x_{2^{k-c}}$ be the all elements of $\mathcal{X}$ enumerated according the lexicographic order. Define the following random variables $\xi(x)$

$$\xi(x) = \begin{cases} \text{random element from } \{-1, 1\}, & x = x_{2i-1} \text{ for some } i \\ -x_{2i-1}, & x = x_{2i} \text{ for some } i \end{cases} \tag{1}$$

for any x such that $x \in \mathcal{X}$. Once the choice of $\xi(x)$ is fixed, consider the distribution

$$\Pr[X = x] = \begin{cases} 2^{-k} + 2\epsilon \cdot 2^{-k} \cdot \xi(x) & x \in \mathcal{X} \\ 0, & x \notin \mathcal{X} \end{cases} \tag{2}$$

The rest of the proof splits into the following two claims:

Claim (X has small smooth min-entropy). For any choice of X and $\epsilon(x)$, we have $\mathbf{H}^\epsilon_\infty(X) \leqslant k - C + \log\left(\frac{1}{1-2\epsilon}\right)$.

Claim (X has large metric pseudo-entropy). We have $\mathbf{H}^{\mathrm{Metric}}_{D,\epsilon}(X) = k$.

Proof (Small smooth min-entropy). Note that by Eq. (1) the distribution of X is ϵ-close to the uniform distribution over $\mathcal{X}$. By Corollary 2 (note that k is replaced by $\log|\mathcal{X}| = k - C$), this means that the ϵ-smooth min-entropy of X is at most $k - C + \log\left(\frac{1}{1-2\epsilon}\right)$.

Proof (Large metric entropy). Note that for any D we have

$$\mathbb{E}D(X) = \sum_{x \in \mathcal{X}} D(x)\left(2^{-k} + \xi(x)2^{-k} \cdot 2\epsilon\right)$$

$$= \mathbb{E}D(U_\mathcal{X}) + 2^{-k} \cdot 2\epsilon \cdot \sum_{x \in \mathcal{X}} D(x)\xi(x)$$

In the next step we observe that the random variables $\xi(x)$ have the degree of dependence $\Delta = 1$. Indeed, by the construction in Eq. (1), for any fixed x the random variables $\xi(x')$ are independent of $\xi(x)$ except at most one value of x'. Now, by Lemma 2 applied to the random variables $D(x)\xi(x)$ we obtain

$$\Pr\left[2^{-k}\sum_{x \in \mathcal{X}} D(x)\xi(x) > \delta\right] \leqslant \exp\left(-2^{k-1}\delta^2\right)$$

for any $\delta > 0$, where the probability is over $\xi(x)$ after fixing the choice of the set $\mathcal{X}$ for $z \in \{0,1\}^m$. In other words, we have

$$\Pr_{\xi(x)} [\mathbb{ED}(X) \leqslant \mathbb{ED}(U_{\mathcal{X}}) + 2\delta\epsilon] \tag{3}$$

with probability $1 - \exp\left(2^{k-1}\delta^2\right)$ for any fixed choice of sets $\mathcal{X}$.

In the last step, we observe that since the choice of the sets $\mathcal{X}$ is random, we have $\mathbb{ED}(U_{\mathcal{X}}) \approx \mathbb{ED}(U_S)$ with high probability. Indeed, by the Hoeffding bound for samples taken without repetitions (see Remark 3)

$$\Pr_{\mathcal{X}} [\mathbb{ED}(U_{\mathcal{X}}) \leqslant \mathbb{ED}(U) + 2\delta\epsilon] \geqslant 1 - \exp(-2^{k-C+3}\delta^2\epsilon^2) \tag{4}$$

By combining Eqs. (3) and (4) for any D and any $\epsilon < \frac{1}{4}$ we obtain

$$\Pr_{\mathcal{X},\xi(x)} [\mathbb{ED}(X) \leqslant \mathbb{ED}(U_S) + 4\delta\epsilon] \geqslant 1 - 2\exp(-2^{k-C+3}\delta^2\epsilon^2). \tag{5}$$

Replacing δ with $\delta/4$ and applying the union bound over $\mathcal{D}$ we see that

$$\Pr_{\mathcal{X},\xi(x)} [\forall \mathsf{D} \in \mathcal{D} : \mathbb{ED}(X) \leqslant \mathbb{ED}(U_S) + \delta\epsilon] \geqslant 1 - 2|\mathcal{D}|\exp(-2^{k-C-1}\delta^2\epsilon^2).$$

and thus we have a distribution X such that

$$\forall \mathsf{D} \in \mathcal{D} : \mathbb{ED}(X) \leqslant \mathbb{ED}(U_S) + \delta\epsilon \tag{6}$$

as long as

$$2|\mathcal{D}| < 2^{2^{k-C-1}\delta^2\epsilon^2}. \tag{7}$$

Finally, note that by adding to the class $\mathcal{D}$ all negations (functions $\mathsf{D}'(x) = 1 - \mathsf{D}(x)$) we have $\mathbb{ED}(X) \leqslant \mathbb{ED}(U_S) + \delta\epsilon$ as well as $\mathbb{ED}(X) \geqslant \mathbb{ED}(U_S) - \delta\epsilon$, for every $\mathsf{D} \in \mathcal{D}$. In particular, we have

$$\forall \mathsf{D} \in \mathcal{D} : |\mathbb{ED}(X) - \mathbb{ED}(U_S)| < \delta\epsilon \tag{8}$$

as long as

$$4|\mathcal{D}| < 2^{2^{k-C-1}\delta^2\epsilon^2}. \tag{9}$$

It remains to observe that for every $\mathcal{X}$ the probability mass function of X takes two values on two halves of $\mathcal{X}$.

References

[BSW03] Barak, B., Shaltiel, R., Wigderson, A.: Computational analogues of entropy. In: Arora, S., Jansen, K., Rolim, J.D.P., Sahai, A. (eds.) APPROX/RANDOM -2003. LNCS, vol. 2764, pp. 200–215. Springer, Heidelberg (2003). doi:10.1007/978-3-540-45198-3_18

[CKLR11] Chung, K.-M., Kalai, Y.T., Liu, F.-H., Raz, R.: Memory delegation. In: Rogaway, P. (ed.) CRYPTO 2011. LNCS, vol. 6841, pp. 151–168. Springer, Heidelberg (2011). doi:10.1007/978-3-642-22792-9_9

[DP08] Dziembowski, S., Pietrzak, K.: Leakage-resilient cryptography in the standard model. IACR Cryptology ePrint Arch. 2008, 240 (2008)

[DTT10] De, A., Trevisan, L., Tulsiani, M.: Time space tradeoffs for attacks against one-way functions and PRGs. In: Rabin, T. (ed.) CRYPTO 2010. LNCS, vol. 6223, pp. 649–665. Springer, Heidelberg (2010). doi:10.1007/978-3-642-14623-7_35

[FOR15] Fuller, B., O'neill, A., Reyzin, L.: A unified approach to deterministic encryption: new constructions and a connection to computational entropy. J. Cryptol. 28(3), 671–717 (2015)

[Gol06] Goldreich, O.: Foundations of Cryptography: Volume 1. Cambridge University Press, New York (2006)

[GW11] Gentry, C., Wichs, D.: Separating succinct non-interactive arguments from all falsifiable assumptions. In: Proceedings of the 43rd ACM Symposium on Theory of Computing, STOC 2011, San Jose, CA, USA, 6–8 June 2011, pp. 99–108 (2011)

[HILL88] Håstad, J., Impagliazzo, R., Levin, L.A., Luby, M.: Pseudo-random generation from one-way functions. In: Proceedings of the 20th STOC, pp. 12–24 (1988)

[HLR07] Hsiao, C.-Y., Lu, C.-J., Reyzin, L.: Conditional computational entropy, or toward separating pseudoentropy from compressibility. In: Naor, M. (ed.) EUROCRYPT 2007. LNCS, vol. 4515, pp. 169–186. Springer, Heidelberg (2007). doi:10.1007/978-3-540-72540-4_10

[Hoe63] Hoeffding, W.: Probability inequalities for sums of bounded random variables. J. Am. Stat. Assoc. 58(301), 13–30 (1963)

[Jan04] Janson, S.: Large deviations for sums of partly dependent random variables. Random Struct. Algorithms 24(3), 234–248 (2004)

[Pie09] Pietrzak, K.: A leakage-resilient mode of operation. In: Joux, A. (ed.) EUROCRYPT 2009. LNCS, vol. 5479, pp. 462–482. Springer, Heidelberg (2009). doi:10.1007/978-3-642-01001-9_27

[Pip76] Pippenger, N.: Information theory and the complexity of boolean functions. Math. Syst. Theory 10(1), 129–167 (1976)

[RTTV08] Reingold, O., Trevisan, L., Tulsiani, M., Vadhan, S.: Dense subsets of pseudorandom sets. In: Proceedings of the 49th Annual IEEE Symposium on Foundations of Computer Science (Washington, DC, USA), FOCS 2008, pp. 76–85. IEEE Computer Society (2008)

[RW05] Renner, R., Wolf, S.: Simple and tight bounds for information reconciliation and privacy amplification. In: Roy, B. (ed.) ASIACRYPT 2005. LNCS, vol. 3788, pp. 199–216. Springer, Heidelberg (2005). doi:10.1007/11593447_11

[Ser74] Serfling, R.J.: Probability inequalities for the sum in sampling without replacement. Ann. Stat. 2(1), 39–48 (1974)

[SGP15] Skórski, M., Golovnev, A., Pietrzak, K.: Condensed unpredictability. In: Halldórsson, M.M., Iwama, K., Kobayashi, N., Speckmann, B. (eds.) ICALP 2015. LNCS, vol. 9134, pp. 1046–1057. Springer, Heidelberg (2015). doi:10.1007/978-3-662-47672-7_85

[Sko15] Skorski, M.: Metric pseudoentropy: characterizations, transformations and applications. In: Lehmann, A., Wolf, S. (eds.) ICITS 2015. LNCS, vol. 9063, pp. 105–122. Springer, Heidelberg (2015). doi:10.1007/978-3-319-17470-9_7

[VZ12] Vadhan, S., Zheng, C.J.: Characterizing pseudoentropy and simplifying pseudorandom generator constructions. In: Proceedings of the 44th symposium on Theory of Computing (New York, NY, USA), STOC 2012, pp. 817–836. ACM (2012)

[YLW13] Yu, Y., Li, X., Weng, J.: Pseudorandom generators from regular one-way functions: new constructions with improved parameters. In: Sako, K., Sarkar, P. (eds.) ASIACRYPT 2013. LNCS, vol. 8270, pp. 261–279. Springer, Heidelberg (2013). doi:10.1007/978-3-642-42045-0_14

Efficient Algorithms for Touring a Sequence of Convex Polygons and Related Problems

Xuehou Tan[2(✉)] and Bo Jiang[1]

[1] Dalian Maritime University, Linghai Road 1, Dalian, China
[2] Tokai University, 4-1-1 Kitakaname, Hiratsuka 259-1292, Japan
tan@wing.ncc.u-tokai.ac.jp

Abstract. Given a sequence of k convex polygons in the plane, a start point s, and a target point t, we seek a shortest path that starts at s, visits in order each of the polygons, and ends at t. This paper describes a simple method to compute the so-called *last step shortest path maps*, which were developed to solve this touring polygons problem by Dror et al. (STOC'2003). A major simplification is to avoid the (previous) use of point location algorithms. We obtain an $O(kn)$ time solution to the problem for a sequence of disjoint convex polygons and an $O(k^2n)$ time solution for possibly intersecting convex polygons, where n is the total number of vertices of all polygons. Our results improve upon the previous time bounds roughly by a factor of $\log n$.

Our new method can be used to improve the running times of two classic problems in computational geometry. We then describe an $O(n(k + \log n))$ time solution to the safari problem and an $O(n^3)$ time solution to the watchman route problem, respectively. The last step shortest path maps are further modified, so as to meet a new requirement that the shortest paths between a pair of consecutive convex polygons be contained in another bounding simple polygon.

1 Introduction

Shortest paths are of fundamental importance in computational geometry, robotics and autonomous navigation. The *touring polygons problem*, introduced by Dror et al. [6], is defined as follows. Suppose that we are given a start point $s = P_0$, a target point $t = P_{k+1}$, and a sequence of k possibly intersecting polygons P_1, P_2, ..., P_k in the plane. The objective is to compute a shortest path that begins at s, visits P_1, P_2, ..., P_k, in that order, and finally arrives at t. See Fig. 1.

The touring polygons problem has been well studied in the literature [8]. If the given polygons are disjoint and convex, the shortest touring polygons path can be computed in $O(kn \log(n/k))$ time [6], where n is the total number of vertices of all polygons. This simplest variant is related to the well-known *safari*

The work by Tan was partially supported by JSPS KAKENHI Grant Number 15K00023, and the work by Jiang was partially supported by National Natural Science Foundation of China under grant 61173034.

© Springer International Publishing AG 2017
T.V. Gopal et al. (Eds.): TAMC 2017, LNCS 10185, pp. 614–627, 2017.
DOI: 10.1007/978-3-319-55911-7_44

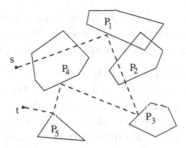

Fig. 1. Touring a sequence of convex polygons.

problem, in which all convex polygons are placed in the interior of a given polygon P and have a common edge with P [12]. In the case that the convex polygons are arbitrarily intersecting, the problem can be solved in $O(nk^2 \log n)$ time [6]. If the given polygons are not convex, the touring polygons problem is NP-hard [1].

The solution of the touring polygons problem finds applications in many fundamental geometric problems, including the parts cutting problem, the safari problem, the zoo-keeper problem and the watchman route problem; these problems involve finding a shortest path, within some constrained region, which visits a set of convex polygons in an order that is given, or can be computed. See [2,5,6,11–15]. The touring polygons problem is also related to the problem of finding the shortest path between two given points among polyhedral obstacles in three dimensions [6,8].

Previous work. A shortest touring path starting at s either goes through a polygon P_i ($1 \leq i \leq k$) or reflects on some edge of P_i, before it reaches t. In the case that all given polygons are convex, local optimality of a path implies global optimality [6]. To give an efficient solution, Dror et al. showed that the plane can be partitioned into regions such that the *last steps* of the shortest touring paths from s to all points in one region are combinatorially equivalent [6]. The location of the target point t in the subdivision of the plane (i.e., finding the region that contains t) then tells us the information on the last step of the shortest touring path. This data structure is termed the *last step shortest path map*.

The touring polygons problem is then solved using dynamic programming [6]. The last step shortest path maps are computed iteratively, one for each of the k polygons, in the order that they are given. The shortest touring path to a vertex v of P_{i+1} ($0 \leq i \leq k$), which visit $P_1, \ldots, P_i$ in order, can be computed by performing i point location queries in the last step shortest path maps for the previous polygons $P_i, \ldots, P_1$. To efficiently answer a point location query (roughly in $O(\log n)$ time), the last step shortest path map for a polygon P_j ($j \leq i$) is saved in a point location data structure [6].

Although the solution for disjoint convex polygons is simple, other three known algorithms (for possibly intersecting polygons, the safari problem and the watchman route problem) are quite complicated and not easy to be understood (e.g., a last step shortest path map consists of three data structures: S^R, S^F and S^A) [6].

It may be because they are obtained as by-products of a more general touring problem, in which intersecting polygons are allowed and the shortest paths between a pair of consecutive polygons are limited to a polygonal region.

Our work. A deep observation made in this paper is that instead of performing the locations of the vertices of P_{i+1} in the last step shortest path map for a polygon P_j ($j \leq i$) independently, we can do them together. Since all given polygons are convex, locating the vertices of P_{i+1} in a previous map can be handled by a constant number of traversals of that map. Hence, the point location data structure is not needed at all. This simple method *does* decrease the time complexities of all considered algorithms, roughly by a factor of $\log n$. Although this improvement is small, as only elementary computations are needed, our algorithms are quite simple and easy to implement. Also, we apply our new method to the safari and watchman route problems, and obtain the improved solutions to them.

The rest of this paper is organized as follows. Section 2 defines three types of contact of a shortest touring polygons path with an edge of a polygon. In Sect. 3, we describe how to construct the last step shortest path maps and give an $O(kn)$ time solution to the touring problem for disjoint convex polygons. To deal with the general case of possibly intersecting polygons, it requires to compute and distinguish the intersection points among the boundaries of the polygons. In Sect. 4, an $O(k^2n)$ time solution is further presented for the intersecting polygons. In Sect. 5, the applications of our new method to the watchman route problem and the safari problem are given. For these two problems, we further modify the shortest touring path maps so as to meet a new requirement that the shortest paths between a pair of consecutive convex polygons be contained in another bounding simple polygon. Our data structures are much simpler and essentially differ from that of Dror et al. (STOC'2003). Concluding remarks are given in Sect. 6.

2 Local Optimality of Shortest Touring Polygons Paths

Assume that all given polygons P_1, P_2, ..., P_k are convex, and the start point s and the target point t are outside of P_1 and P_k respectively; otherwise, P_1 or/and P_k needn't be considered. Also, we denote by $|P_i|$ the number of vertices of P_i, and let $n = \sum_1^k |P_i|$. We will simply call a touring polygons path, a *touring path*. Let $P_0 = s$ and $P_{k+1} = t$. Denote by T_{opt} a shortest touring path for the given polygons P_0, P_1, ..., P_k, P_{k+1}.

Suppose that T_{opt} visits in order the edges $e_1, e_2, \ldots, e_k$, $e_i \in P_i$ and $1 \leq i \leq k$. Note that local optimality of T_{opt} with respect to edge e_i ($1 \leq i \leq k$) is equivalent to global optimality [6,12]. Let us consider the contact point c_i of T_{opt} with the edge e_i, after T_{opt} visits $P_1, P_2, \ldots, P_{i-1}$. If T_{opt} goes across P_i, then we consider c_i as the *second* intersection point of the boundary of P_i with T_{opt}, and thus T_{opt} reaches the point c_i from the interior of P_i (see Fig. 2(c)). The point c_i occurs in one of the following situations:

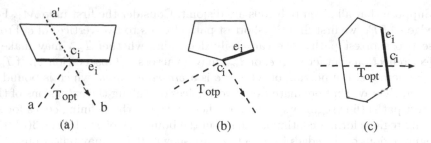

Fig. 2. Types of contact of T_{opt} with an edge e_i of polygon P_i.

1. *Perfect reflection contacts*: T_{opt} reflects on e_i at an *interior point* c_i of e_i such that the incoming angle of T_{opt} with the reflected edge e_i is equal to the outgoing angle. See Fig. 2(a). So, the *reflection* of the incoming segment of T_{opt}, with respect to the line through e_i, is collinear to its outgoing segment. In Fig. 2(a), the point a' is the reflection of a, which is obtained by reflecting a across the line (considered as a mirror) through e_i. This operation is called the *unfolding* of the point a with respect to edge e_i

2. *Bending contacts*: T_{opt} bends at a polygon vertex c_i, see Fig. 2(b).

3. *Crossing contacts*: A line segment of T_{opt} passes the edge e_i through an *interior point* c_i of e_i, see Fig. 2(c).

3   An $O(kn)$ Algorithm for Touring k Disjoint Polygons

In this section, we describe a new method to construct the last step shortest path maps, without invoking any point location algorithm.[1] We will also consider each polygon vertex as a target point, and call the path from s to a vertex of P_i, a *partial* touring path if it visits in order each of the polygons $P_1, \ldots, P_{i-1}$. The *last step shortest path map* for P_i, denoted by $\mathcal{M}_i$, is a subdivision of the plane, excluding the interior of P_i, into the regions such that the shortest partial touring paths to all the points in a region visit the same edge or vertex of P_i. In the following, if a path from s is said to be extended, it means that the last line segment of the path is extended to infinity, starting from the ending point of the path.

Observation 1. *A necessary condition for T_{opt} to make a perfect reflection (resp. crossing) contact with an edge e of P_i is that the shortest touring path from s to any point of e does not intersect (resp. intersects) the interior of P_i.*

---

[1] The authors have to point out an unhappy thing. The result (algorithm) obtained in this section was once given in *The Open Automation and Control Systems Journal*, 2015, 7, pp.1364–1368, mainly by a student of the second author, without permission from the authors of this paper. Needless to say more, that paper (titled "A new algorithm for the shortest path of touring disjoint convex polygons") had been RETRACTED.

Suppose that all given polygons are disjoint. Consider the first map $\mathcal{M}_1$. For an edge e of P_1, we first find the shortest paths from s to two vertices of e. From these two shortest paths, one can easily determine whether T_{opt} may make a perfect reflection contact with e, or T_{opt} may go across e (Observation 1). If T_{opt} may go across e, the portion of $\mathcal{M}_1$ for e is a *crossing region*, which is bounded by e and the rays that emanate from two vertices of e along the extensions of the shortest paths (line segments) from s to them. Clearly, the defining edges for all crossing regions form a continuous chain on the boundary of P_1. In Fig. 3(a), the crossing regions of the edges from e_3 to e_7 are shown. If T_{opt} may reflect on e, the portion of $\mathcal{M}_1$ for e consists of a *reflection region* and two *bending regions*, one per vertex. The reflection region is bounded by e and the two rays, which form perfect reflections on e with the incoming segments of the shortest paths from s to the vertices of e. Denote by v the common vertex of edge e with the other edge e'. The bending region of v is then the triangular region, with the apex at v, bounded by the two rays that emanate from v and are used in defining the reflection region of e and the reflection or crossing region of e'. (The bending regions of some vertices may be defined twice, but it can easily be handled.) In Fig. 3(a), the reflection regions and bending regions for e_1 and e_2 are shown.

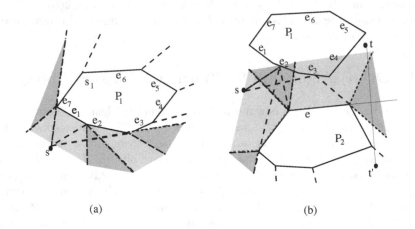

(a)   (b)

Fig. 3. The last step shortest path maps for disjoint convex polygons.

The portion of $\mathcal{M}_1$ for an edge e consists of at most three regions of constant size, and can thus be constructed in constant time. So, $\mathcal{M}_1$ is of size $O(|P_1|)$ and can be constructed in $O(|P_1|)$ time. All regions of $\mathcal{M}_1$ form a plane subdivision, excluding the interior of P_1, see Fig. 3(a). Moreover, the regions of the map $\mathcal{M}_1$ can be arranged into a *circular order* according to their defining edges/vertices on the boundary of P_1.

Assume now that the last step shortest path map $\mathcal{M}_i$ $(1 \le i < k)$ has been constructed. To construct the next map $\mathcal{M}_{i+1}$, we first describe how to compute the (unique) shortest partial touring path to a vertex v of P_{i+1} [6]. Suppose that the region of $\mathcal{M}_i$ containing v has been found. (The method of locating v in $\mathcal{M}_i$ will be given later.) We then distinguish the following three situations.

Case 1. v is contained in the bending region of a vertex u. Clearly, the last portion of the shortest partial touring path to v is the line segment with the endpoints u and v.

Case 2. v is contained in a crossing region. We recursively compute the shortest partial touring path to v by locating v in the map $\mathcal{M}_{i-1}$. In the case that v is located in the (empty) map $\mathcal{M}_0$, the shortest partial touring path to v is just the line segment having the endpoints v and s.

Case 3. v is in a reflection region. Let v' be the reflection of the point v with respect to the line through the reflected edge, say, e. We then recursively compute the shortest partial touring path to v' in $\mathcal{M}_{i-1}$: replacing the portion of the (currently) found path from the edge e to v' by the line segment from e to v gives the last segment of the shortest partial touring path to v. See Fig. 3(b) for an example, where the point t' is obtained by reflecting the endpoint t on edge e.

After the shortest partial touring paths to all vertices of P_{i+1} are computed, the reflection, crossing and bending regions of $\mathcal{M}_{i+1}$ can be defined analogously. Also, all the regions of $\mathcal{M}_{i+1}$ can be arranged into a circular order, and they form a plane subdivision, excluding the interior of P_{i+1}. See Fig. 3(b) for an instance of $\mathcal{M}_2$.

Let m denote the number of vertices of polygon P_{i+1}, and p_j $(1 \leq j \leq m)$ a vertex of P_{i+1}. Denote by $q_{h,j}$, $1 \leq h \leq i$, the (last) contact point of P_h with the shortest partial touring path to p_j in $\mathcal{M}_h$.

Lemma 1. *The last contact points $q_{h,1}, q_{h,2}, \ldots, q_{h,m}$ of P_h $(1 \leq h \leq i)$ with the shortest partial touring paths to all vertices of P_{i+1} form at most three x-monotone chains in $\mathcal{M}_h$.*

Proof. Let x and y be two adjacent vertices of P_{i+1}. Consider the shortest partial touring paths to x and y, which start at s and end at x and y, respectively. The last steps of the two subpaths of them in the map $\mathcal{M}_h$ do not cross, due to subpath optimality. Since P_h is convex, the points $q_{h,1}, q_{h,2}, \ldots, q_{h,m}$ then form two or three x-monotone chains, depending on whether or not the x-coordinate of $q_{h,1}$ is the minimum or maximum among those of $q_{h,1}, q_{h,2}, \ldots, q_{h,m}$. $\square$

Lemma 2. *Given $\mathcal{M}_1$, $\ldots$, $\mathcal{M}_i$, the map $\mathcal{M}_{i+1}$ can be constructed in time $O(i|P_{i+1}| + \sum_{h=1}^{i} |P_h|)$.*

Proof. The computation of $\mathcal{M}_{i+1}$ requires to visit, in this order, the maps $\mathcal{M}_i$, $\ldots$, $\mathcal{M}_1$. Locating in $\mathcal{M}_i$ the points $q_{i+1,j}$ (i.e., the vertices p_j of P), $1 \leq j \leq m$, can be done as follows. First, we locate a point, say, $q_{i+1,1}$ in $\mathcal{M}_i$. From the found region, locating all other points $q_{i+1,2}, \ldots, q_{i+1,m}$ in $\mathcal{M}_i$ can be done by a constant number of monotone scans (Lemma 1). The region containing $q_{i+1,1}$ can be found in $O(|P_i|)$ time. The total time required to locate $q_{i+1,1}, q_{i+1,2}, \ldots, q_{i+1,m}$ in $\mathcal{M}_i$ is clearly $O(|P_i| + |P_{i+1}|)$. This procedure is then recursively performed for $\mathcal{M}_h$, $1 \leq h < i$. Note that if a bending contact occurs in $\mathcal{M}_{h+1}$, the point

$q_{h+1,j}$ is then the bending vertex. If a crossing (reflection) contact occurs, then $q_{h+2,j}$ (the reflection of $q_{h+2,j}$) can be used as $q_{h+1,j}$, as they are in the same region of $\mathcal{M}_h$. Since $\mathcal{M}_i, \ldots, \mathcal{M}_1$ are all visited, the lemma thus follows.   □

Since $\sum_{i=1}^{k} i|P_i| = O(k(\sum_{i=1}^{k} |P_i|)) = O(kn)$, all maps $\mathcal{M}_1, \ldots, \mathcal{M}_k$ can be computed in $O(kn)$ time. After $\mathcal{M}_k$ is found, we can report in time $O(n)$ the shortest path from s to t that tours all polygons $P_1, \ldots, P_k$. On the other hand, following from the circluar property of the map $\mathcal{M}_i$ $(1 \le i \le k)$, a binary search can also be used to answer a point-in-region query in time $O(\log |P_i|)$. Thus, for a given query point t, we can report the shortest touring path to t in $O(k \log(n/k))$ time [6]. Hence, we obtain the first result of this paper.

Theorem 1. *For the problem of touring a sequence of k disjoint convex polygons, a data structure of $O(n)$ size can be built in time $O(kn)$ such that the shortest touring path to any given point t can be reported in time $O(n)$ or $O(k \log(n/k))$, where n is the total number of vertices of the given polygons.*

4 Touring a Sequence of Possibly Intersecting Polygons

Let us now extend the data structure developed in the previous section to the case of intersecting convex polygons. Without loss of generality, assume that no three (polygon) edges can intersect at a same point. Note that a vertex of P_i may be in the interior of P_{i+1}, and the shortest touring path map for P_{i+1} is affected by the intersection points among previous polygons. These new situations need be handled carefully.

First, we compute the intersection points among all convex polygons. Clearly, the total number of edge intersections among all *convex* polygons is bounded by $O(kn)$. All of these intersection points can simply be computed in time $O(k^2 n)$ or $O(kn \log n)$. In our algorithm, we consider the intersection points between P_i and P_h, $h < i$, as the *pseudo-vertices* of P_i, and the fragments of a polygon edge, which are introduced by the pseudo-vertices, as the *pseudo-edges*. Still, we denote by $|P_i|$ the total number of the vertices and pseudo-vertices of P_i, and $\mathcal{M}_i$ the shortest path touring map for P_i, $1 \le i \le k$. So, we have $\sum_{1}^{k} |P_i| = O(kn)$.

The first map $\mathcal{M}_1$ is the same as that constructed in Sect. 3, except that the interior of P_1 is now a crossing region of $\mathcal{M}_1$. This is because a portion of P_2 may be in the interior of P_1. Notice that the size of the crossing region P_1 is $|P_1|$, other than a constant. Clearly, the map $\mathcal{M}_1$ is a partition of the full plane.

Suppose that the shortest touring path maps $\mathcal{M}_1, \ldots, \mathcal{M}_i$ $(1 \le i < k)$ have been computed. Notice that the shortest partial touring path to a boundary point of P_i is unique, even in the case that the given polygons are arbitrarily intersecting [6]. We then compute the shortest partial touring paths to all vertices v of P_{i+1}, including the pseudo-vertices. Since Observation 1 also holds in this case, the reflection and crossing regions of $\mathcal{M}_{i+1}$ can analogously be computed for the pseudo-edges of P_{i+1}. Also, the interior of P_{i+1} is a crossing region of $\mathcal{M}_{i+1}$. For the instance given in Fig. 4(a), the map $\mathcal{M}_2$ is shown. (Clearly, the crossing regions of $\mathcal{M}_{i+1}$ for the pseudo-edges of a single edge can be merged.)

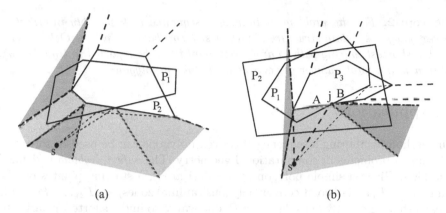

Fig. 4. Illustrating the maps for possibly intersecting polygons.

The bending regions of the vertices of P_{i+1} can be computed analogously, too. However, a new treatment needs for the pseudo-vertices. Let j be an intersection point of P_{i+1} and P_h, $1 \le h \le i$. The bending region of j exists only when (1) j is in the interiors of $P_{h+1}, \ldots, P_i$, and (2) the shortest partial touring path to j does not intersect the interior of P_h nor P_{i+1}. See Fig. 4(b). The second condition comes from the fact that if T_{opt} makes a bending contact at point j, then its outgoing segment (away from j) in the bounding region of j cannot intersect the interiors of both P_h and P_{i+1}; otherwise, T_{opt} can be shortened by slightly sliding its contact point on the edge of either P_h or P_{i+1}. Clearly, if the last segments of two shortest partial touring paths to j (in $\mathcal{M}_h$ and $\mathcal{M}_{i+1}$) overlap each other, the conditions (1) and (2) are true. The bending region of j (if it exists) is bounded by the two rays which form perfect reflections with the last segment of the shortest partial touring path to j, on the reflected edges of P_h and P_{i+1} respectively. By definitions of crossing and reflection regions, one of the regions adjacent to the bending region of j is the crossing region, and the other is the reflection region. See Fig. 4(b). In summary, at most two pseudo-vertices of a polygon may have their (extreme) bending regions, and two reflection regions may be adjacent on a polygon edge, because of the non-existence of bounding regions of some psuedo-vertices (see Fig. 4(a)).

The time required to construct the map $\mathcal{M}_{i+1}$ is still $O(i|P_{i+1}| + \sum_{h=1}^{i} |P_h|)$. This is because finding the P_{i+1}'s vertices that are in the interior (i.e., the crossing region) of a polygon P_h can be done in $O(|P_h| + |P_{i+1}|)$ time, as the intersection points between P_h and P_{i+1} have been precomputed. Also, since we have the last segments of two shortest partial touring paths to a pseudo-vertex at hand, whether the bending region of the pseudo-vertex exists can be determined in constant time, and such a bending region can be computed in constant time.

Since $\sum_{i=1}^{k} i|P_i| = O(k(\sum_{i=1}^{k} |P_i|)) = O(k^2 n)$, all last step shortest path maps can then be computed in time $O(k^2 n)$. After all maps are found, we can report in $O(kn)$ time the shortest touring path from s to t.

Theorem 2. *For the problem of touring a sequence of k possibly intersecting convex polygons, a data structure of $O(kn)$ size can be built in time $O(k^2n)$ such that the shortest touring path to any given point t can be reported in time $O(kn)$, where n is the total number of vertices of the given polygons.*

5 Applications

Our method for touring a sequence of convex polygons can be used to solve two well-known problems in computational geometry. The *safari problem* is defined as follows: Given a simple polygon P (a safari area), a starting point s on the boundary of P, and disjoint convex polygons (animal zones) $P_1, P_2, \ldots, P_k$ inside P, each sharing exactly one edge with P, one wants to find a shortest safari tour that starts at s, visits all polygons in order, and finally returns back to s. Assume that $P_1, P_2, \ldots, P_k$ appear on P clockwise, starting at s. The safari tour may enter the interior of a convex polygon P_i, $1 \le i \le k$. See Fig. 5(a)

In the *watchman route problem*, we are given a simple polygon P with a starting point s on its boundary. The objective is to find a shortest (closed) tour such that every point of P is visible from at least one point of the tour, starting and ending at s. See Fig. 5(b). This problem is quite interesting, because it deals with both the visibility and metric information [5,9].

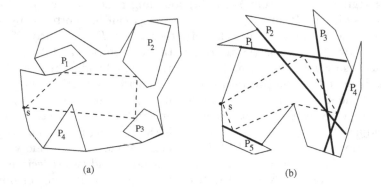

(a)   (b)

Fig. 5. Instances of the shortest safari and watchman routes.

5.1 New Last Step Shortest Path Maps for the Safari Problem

A vertex of the polygon P is *reflex* if its internal angle is strictly larger than π; otherwise, it is *convex*. For an instance of the safari problem, we denote by n the total number of the vertices of all the polygons P and P_i, $1 \le i \le k$. Let $P_{k+1} = s$. Again, we denote by $|P_i|$ the number of vertices of a convex polygon P_i, and thus $\sum_{i=1}^{k} |P_i| = O(n)$. The currently best result on the safari problem is an $O(kn \log n)$ solution [6].

The last step shortest path map $\mathcal{M}_i$ for a polygon P_i, $1 \leq i \leq k$, is almost the same as that defined in Sect. 3, except that the shortest paths between any two points (on convex polygons) have to be contained in the interior of P. With a careful treatment, we can construct the last step shortest path maps, which are independent of the enclosing polygon P. For simplicity, we will denote by l_i ($1 \leq i \leq k$) the vertex of the common edge of P_i and P, which is first encountered by a clockwise scan on P from s, and r_i the other vertex of that common edge.

To efficiently answer the shortest path queries inside P, we first preprocess P in $O(n)$ time such that the last segment of a shortest path between two query points in P (while ignoring all convex polygons) can be found in $O(\log n)$ time [7]. Also, we preprocess P in $O(n \log n)$ time such that a ray-shooting query inside P can be answered in $O(\log n)$ time [3]. A shortest path or a ray, as computed above, may go through the interior of a convex polygon P_i.

Consider how to construct the first map $\mathcal{M}_1$. We first compute the last segments of the shortest paths from s to all vertices of P_1 in $O(|P_1| \log n)$ time. For the P_1's edges that the shortest safari route may go across, we extend the last segments of the shortest paths to their vertices until the boundary of P is reached. For the P_1's edges that the shortest safari route may reflect on, we extend the *outgoing* line segments, which form perfect reflections with the last segments of the shortest paths to their vertices, until the boundary of P is reached. These first reached, or simply, *hit points* on P can be found by invoking the ray-shooting query algorithm [3]. We call the computed segments, which have the origins on P_1 and the hit points on P, the P_1-*rays* (see Fig. 6(a)).

A new situation here is that the shortest partial touring paths to some vertices of P_2 may turn, left or/and right, at reflex vertices of P. For left turns, we describe below a method to find the *first left turn point*, denoted by a_1. First, we find the first P_1-ray (as viewed from s), which does not intersect P_2 and whose hit point appears before the vertex l_2 on the boundary of P. Denote by u_1 (it always exists) the vertex of P_1 from which the found P_1-ray emanates, see Fig. 6(a). Whether or not a P_1-ray intersects P_2 can be determined in $O(\log |P_2|)$ time. Hence, the vertex u_1 can be computed in $O(|P_1| \log |P_2|)$, or simply $O(|P_1| \log n)$ time.

Observe that the point a_1 (if it exists) is contained in the bending region of u_1 or the reflection/crossing region of the edge having u_1 as its right vertex (as viewed from s). Although $\mathcal{M}_1$ has not yet been completely computed, the u_1's region is known at this moment. In the case that the u_1's region is a bending one, we compute the first turn point of the shortest path from u_1 to the vertex l_2. If that turn point is l_2 itself, then a_1 does not exist; otherwise, it is the point a_1. In the case that the u_1's region is a reflection or crossing one, a_1 is the first turn point of the shortest path from the other vertex of the edge defining the u_1's region to l_2. After a_1 (if it exists) is found, we compute using the u_1's region the last segment of the shortest touring path to a_1 and then extend the found segment till the boundary of P is reached. We also consider the extended segment (passing through a_1) as a special P_1-ray. See Fig. 6(a) for an example, where the P_1-ray through a_1 is drawn in bold dashed line. In this way, the vertex a_1 (if it exists) can be computed in $O(\log n)$ time.

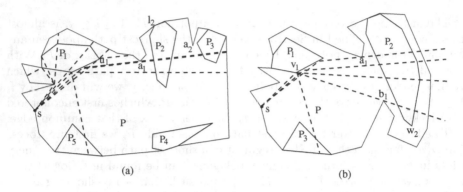

Fig. 6. Illustration of the last step shortest path maps for the safari problem.

Consider now the situation in which the shortest partial touring paths to some vertices of P_2 turn right. Again, we focus on the *first right turn point*, denoted by b_1. First, find the last P_1-ray (if it exists) that intersects P_2. It can simply be done in time $O(|P_1|\log|P_2|)$ or $O(|P_1|\log n)$. Denote by v_1 the vertex of P_1 from which the found P_1-ray emanates (Fig. 6(b)). Next, we find the vertex of P_2, which is to the right of the line containing the P_1-ray emanating from v_1 and whose distance to that line is maximum. This vertex, say, w_2, can be computed in $O(|P_2|)$ time. We then find in $O(\log n)$ time the first turn point of the shortest path from v_1 to w_2. If the found point differs from v_1, then it is the point b_1. Also, we find the special P_1-ray through b_1 (if it exists). See Fig. 6(b). The portion of P_2 that is to the right of the P_1-ray through b_1 can safely be ignored, because the shortest safari route needn't enter it.

The extensions of all P_1-rays between two P_1-rays through a_1 and b_1 into half-lines can be considered as a partition of the plane, excluding the interior of P_1; in the case that the point a_1 (b_1) does not exist, the restriction placed by the P_1-ray through a_1 (b_1) is completely released. See Fig. 6(b). We call the obtained partition, a *p-map* (short for "portion of the map"), and denote it by $\mathcal{N}_1$. In summary, $\mathcal{M}_1$ consists of a *p*-map $\mathcal{N}_1$, a left turn point a_1 and a right turn point b_1. At least one of $\mathcal{N}_1$, a_1 and b_1 exists. Note that $\mathcal{N}_1$ is completely independent of the enclosing polygon P.

Suppose now that the maps $\mathcal{M}_i$, for $1 \le i < k$, have been computed. A map $\mathcal{M}_i$ consists of at most three elements: a p-map $\mathcal{N}_i$, a left turn point a_i and a right turn point b_i. Again, let $m = |P_{i+1}|$, and denote by $q_{h,j}$, $1 \le h \le i$, the *last contact point* of P_h with the shortest partial touring path to a vertex p_j of P_{i+1}. (Some of $q_{h,1}, q_{h,2}, \ldots, q_{h,m}$ may be the same point.)

Lemma 3. *The last contact points $q_{h,1}, q_{h,2}, \ldots, q_{h,m}$ $(1 \le h \le i)$ of P_h with the shortest partial touring path to all vertices of P_{i+1} form at most three x-monotone chains.*

Proof. By an argument similar to the proof of Lemma 1, the lemma simply follows.   □

For the vertices of P_{i+1} that are to the left of the half-line through a_i (if it exists), we then compute the last steps of shortest partial touring paths to them by taking a_i as the local starting point. It clearly takes $O(|P_{i+1}|\log n)$ time. For the vertices of P_{i+1} that are to the left and right of the half-lines through a_i (if it exists) and b_i (if it exists) respectively, we compute the shortest partial touring paths to them using $\mathcal{N}_i$ and possibly the previous maps $\mathcal{M}_1, \ldots, \mathcal{M}_{i-1}$. It then follows from Lemma 3 that $\mathcal{N}_{i+1}$ can be computed in $O(i|P_{i+1}| + \sum_{h=1}^{i} |P_h|)$ time. Analogous to the operation for computing the map $\mathcal{M}_1$, we can find the points a_{i+1}, b_{i+1} (if they exist) in $O(|P_{i+1}|\log n + |P_{i+2}|)$ time.

Lemma 4. *Given $\mathcal{M}_1, \ldots, \mathcal{M}_i$, the map $\mathcal{M}_{i+1}$ can be constructed in time* $O((i + \log n)|P_{i+1}| + |P_{i+2}| + \sum_{h=1}^{i} |P_h|)$.

Proof. The lemma simply follows from the discussion made above. $\qquad\square$

Since $\sum_{i=1}^{k} i|P_i| = O(k(\sum_{i=1}^{k} |P_i|)) = O(kn)$, all maps $\mathcal{M}_1, \ldots, \mathcal{M}_k$ can be computed in $O(n(k+\log n))$ time. After $\mathcal{M}_k$ is found, we can report in time $O(n)$ the shortest path, starting and ending at s, for touring all polygons $P_1, \ldots, P_k$.

Theorem 3. *Suppose that a simple polygon P with a starting point on its boundary is given. Then, the safari problem can be solved in $O(n(k+\log n))$ time, where k is the number of convex polygons P_i inside P and n is the total number of the vertices of all polygons P and P_i, $1 \le i \le k$.*

5.2   An $O(n^3)$ Algorithm for the Watchman Route Problem

For the watchman route problem, Tan et al. [13,14] gave the first polynomial-time ($O(n^4)$) algorithm using dynamic programming, where n denotes the number of the vertices of the given polygon P. This result was later improved to $O(n^3 \log n)$ [6].

Let e be an edge incident to a reflex vertex v of P. Starting from v, one can extend e inside P, until the boundary of P is reached. This extension of e divides P into two pieces. It is called a *cut*, denoted by C, if the extension of e leads to a convex vertex at v in the piece containing s. The portion of P *not* containing s, is called a *pocket*. A cut C is *essential* if no other pocket is fully contained in the pocket of C. Also, a pocket is essential if its cut is essential. Then, a tour sees all points of P if and only if it visits all essential pockets, or equivalently, all essential cuts.

Let us denote by $P_1, \ldots, P_k$ ($C_1, \ldots, C_k$) the sequence of essential pockets (cuts), ordered by their appearances (i.e., their first points) on the boundary of P clockwise, starting from s. Clearly, some of $P_1, \ldots, P_k$ may intersect each other. Then, there exists a shortest watchman route that visits the pockets $P_1, \ldots, P_k$, in this order, see Fig. 5(b).[2]

---

[2] The shortest watchman route may *not* visit the essentail cuts $C_1, \ldots, C_k$, exactly in this order. But, the cuts on which the shortest watchman route reflects (e.g., the cuts of Γ_1, P_4 and P_5 in Fig. 5(b)) still follow that order.

Observe that the shortest watchman route needs not to visit any common boundary point between P and a pocket P_i, except for two endpoints of the cut C_i. So, the (convex) cut C_i can be considered as a simple representation of its (non-convex) pocket P_i. The last step shortest path maps for $P_1, \ldots, P_k$ can then be defined as those in Sect. 4, by considering the intersection points among $C_1, \ldots, C_k$ as the pseudo-vertices. By handling the enclosing polygon P as described in Sect. 5.1, we then compute the shortest partial watchman tours to the pseudo-vertices of C_i, in the increasing order of i. In this way, we can find the shortest watchman route through s. Since the total number of cut intersections is $O(n^2)$, we obtain a new result for the watchman route problem.

Theorem 4. *Suppose that a simple polygon P of n vertices and a starting point s on the boundary of P are given. Then, the watchman route problem can be solved in $O(n^3)$ time.*

6   Conclusions

We have presented an $O(kn)$ time solution to the problem of touring a sequence of disjoint convex polygons, and an $O(k^2 n)$ time solution for a sequence of possibly intersecting convex polygons, where k is the number of convex polygons and n is the total number of vertices of all polygons. Our results improve upon the previous time bounds roughly by a factor of $\log n$. We have also given an $O(n(k + \log n))$ time solution to the safari problem and an $O(n^3)$ time solution to the watchman route problem, improving upon the previous time bounds $O(kn \log n)$ and $O(n^3 \log n)$ respectively. The data structures presented in this paper are much simpler and essentially differ from that of Dror et al. (STOC'03). Whether or not our results can be improved is an interesting open problem.

References

1. Ahadi, A., Mozafari, A., Zarei, A.: Touring a sequence of disjoint polygons: complexity and extension. Theoret. Comput. Sci. **556**, 45–54 (2014)
2. Bespamyatnikh, S.: An $O(n \log n)$ algorithm for the zoo-keeper's problem. Comput. Geom. **24**, 63–74 (2002)
3. Chazelle, B., Guibas, L.: Visibility and intersection problem in plane geometry. Discrete Comput. Geom. **4**, 551–581 (1989)
4. Chazelle, B., Edelsbrunner, H.: An optimal algorithm for intersecting line segments in the plane. J. ACM **39**, 1–54 (1992)
5. Chin, W.P., Ntafos, S.: Optimum watchman routes. Inform. Process. Lett. **28**, 39–44 (1988)
6. Dror, M., Efrat, A., Lubiw, A., Mitchell, J.S.B.: Touring a sequence of polygons. In: Proceedings of the STOC 2003, pp. 473–482 (2003)
7. Guibas, L., Hershberger, J., Leven, D., Sharir, M., Tarjan, R.: Linear time algorithms for visibility and shortest path problems inside simple triangulated polygons. Algorithmica **2**, 209–233 (1987)

8. Mitchell, J.S.B.: Geometric shortest paths and network optimization. In: Sack, J.-R., Urrutia, J. (eds.) Handbook of Computational Geometry, pp. 633–701. Elsevier Science (2000)

9. Mitchell, J.S.B.: Approximating watchman routes. In: Proceedings of SODA 2013, pp. 844–855 (2013)

10. Ntafos, S.: Watchman routes under limited visibility. Comput. Geom. Theory Appl. **1**, 149–170 (1992)

11. Tan, X.: Fast computation of shortest watchman routes in simple polygons. Inform. Process. Lett. **77**, 27–33 (2001)

12. Tan, X., Hirata, T.: Finding shortest safari routes in simple polygons. Inform. Process. Lett. **87**, 179–186 (2003)

13. Tan, X., Hirata, T., Inagaki, Y.: An incremental algorithm for constructing shortest watchman routes. Int. J. Comput. Geom. Appl. **3**, 351–365 (1993)

14. Tan, X., Hirata, T., Inagaki, Y.: Corrigendum to "an incremental algorithm for constructing shortest watchman routes". Int. J. Comput. Geom. Appl. **9**, 319–323 (1999)

15. Tan, X., Wei, Q.: An improved on-line strategy for exploring unknown polygons. In: Lu, Z., Kim, D., Wu, W., Li, W., Du, D.-Z. (eds.) COCOA 2015. LNCS, vol. 9486, pp. 163–177. Springer, Cham (2015). doi:10.1007/978-3-319-26626-8_13

Parameterized Complexity of Fair Deletion Problems

Tomáš Masařík[1] and Tomáš Toufar[2(⊠)]

[1] Department of Applied Mathematics, Faculty of Mathematics and Physics,
Charles University, Prague, Czech Republic
masarik@kam.mff.cuni.cz
[2] Computer Science Institute of Charles University,
Faculty of Mathematics and Physics, Charles University, Prague, Czech Republic
toufi@iuuk.mff.cuni.cz

Abstract. Deletion problems are those where given a graph G and a graph property π, the goal is to find a subset of edges such that after its removal the graph G will satisfy the property π. Typically, we want to minimize the number of elements removed. In fair deletion problems we change the objective: we minimize the maximum number of deletions in a neighborhood of a single vertex.

We study the parameterized complexity of fair deletion problems with respect to the structural parameters of the tree-width, the path-width, the size of a minimum feedback vertex set, the neighborhood diversity, and the size of minimum vertex cover of graph G.

We prove the W[1]-hardness of the fair FO vertex-deletion problem with respect to the first three parameters combined. Moreover, we show that there is no algorithm for fair FO vertex-deletion problem running in time $n^{o(\sqrt[3]{k})}$, where n is the size of the graph and k is the sum of the first three mentioned parameters, provided that the Exponential Time Hypothesis holds.

On the other hand, we provide an FPT algorithm for the fair MSO edge-deletion problem parameterized by the size of minimum vertex cover and an FPT algorithm for the fair MSO vertex-deletion problem parameterized by the neighborhood diversity.

1 Introduction

We study the computational complexity of *fair deletion problems*. Deletion problems are a standard reformulation of some classical problems in combinatorial optimization examined by Yannakakis [20]. For a graph property π we can formulate an *edge deletion problem*. That means, given a graph $G = (V, E)$, find the minimum set of edges F that need to be deleted for graph $G' = (V, E \setminus F)$ to satisfy property π. A similar notion holds for the *vertex deletion problem*.

Research was supported by the project GAUK 338216 and by the project SVV-2016-260332.

T. Masařík—Supported by the project CE-ITI P202/12/G061.

© Springer International Publishing AG 2017
T.V. Gopal et al. (Eds.): TAMC 2017, LNCS 10185, pp. 628–642, 2017.
DOI: 10.1007/978-3-319-55911-7_45

Many classical problems can be formulated in this way such as MINIMUM VERTEX COVER, MAXIMUM MATCHING or MINIMUM FEEDBACK ARC SET. For example MINIMUM VERTEX COVER is formulated as a vertex deletion problem since we aim to find a minimum set of vertices such that the rest of the graph forms an independent set. An example of an edge deletion problem is PERFECT MATCHING: we would like to find a minimum edge set such that the resulting graph has all vertices being of degree exactly one. Many of such problems are NP-complete [1,13,19].

Fair deletion problems are such modifications where the cost of the solution should be split such that the cost is not too high for anyone. More formally, the FAIR EDGE DELETION PROBLEM for a given graph $G = (V, E)$ and a property π finds a set $F \subseteq E$ which minimizes the maximum degree of the graph $G^* = (V, F)$ where the graph $G' = (V, E \setminus F)$ satisfies the property π. Fair deletion problems were introduced by Lin and Sahni [17].

Minimizing the fair cost arises naturally in many situations, for example in defective coloring [5]. A graph is (k, d)-colorable if every vertex can be assigned a color from the set $\{1, \ldots, k\}$ in such a way that every vertex has at most d neighbors of the same color. This problem can be reformulated in terms of fair deletion; we aim to find a set of edges of maximum degree d such that after its removal the graph can be partitioned into k independent sets.

We focus on fair deletion problems with properties definable in either first order (FO) or monadic second order (MSO) logic. Our work extends the result of Kolman et al. [12]. They showed an XP algorithm for a generalization of fair deletion problems definable by MSO_2 formula on graphs of bounded tree-width.

We give formal definitions of the problems under consideration in this work.

FAIR FO EDGE-DELETION

Input: An undirected graph G, an FO sentence ψ, and a positive integer k.

Question: Is there a set $F \subseteq E(G)$ such that $G \setminus F \models \psi$ and for every vertex v of G, the number of edges in F incident with v is at most k?

Similarly, FAIR VERTEX DELETION PROBLEM finds, for a given graph $G = (V, E)$ and a property π, the solution which is the minimum of maximum degree of graph $G[W]$ where graph $G[V \setminus W]$ satisfy property π. Those problems are NP-complete for some formulas. For example Lin and Sahni [17] showed that deciding whether a graph G has a degree one subgraph H such that $G \setminus H$ is a spanning tree is NP-complete.

FAIR FO VERTEX-DELETION

Input: An undirected graph G, an FO sentence ψ, and a positive integer k.

Question: Is there a set $W \subseteq V(G)$ such that $G \setminus W \models \psi$ and for every vertex v of G, it holds that $|N(v) \cap W| \leq k$?

Both problems can be straightforwardly modified for MSO_1 or MSO_2.

The following notions are useful when discussing the fair deletion problems. The *fair cost of a set* $F \subseteq E$ is defined as $\max_{v \in V} |\{e \in F \mid v \in e\}|$. We refer to the function that assigns each set F its fair cost as the *fair objective function*. In case of vertex-deletion problems, the *fair cost of a set* $W \subseteq V$ is defined as $\max_{v \in V} |N(v) \cap W|$. The *fair objective function* is defined analogously. Whenever we refer to the fair cost or the fair objective function, it should be clear from context whether we mean the edge or the vertex version.

We now describe the generalization of fair deletion problems considered by Kolman et al. The main motivation is that sometimes we want to put additional constraints on the deleted set itself (e.g. CONNECTED VERTEX COVER, INDE-PENDENT DOMINATING SET). However, the framework of deletion problems does not allow that. To overcome this problem, we define the generalized problems as follows.

GENERALIZED FAIR MSO EDGE-DELETION

Input: An undirected graph G, an MSO formula ψ with one free edge-set variable, and a positive integer k.

Question: Is there a set $F \subseteq E(G)$ such that $G \models \psi(F)$ and for every vertex v of G, the number of edges in F incident with v is at most k?

GENERALIZED FAIR MSO VERTEX-DELETION

Input: An undirected graph G, an MSO formula ψ with one free vertex-set variable, and a positive integer k.

Question: Is there a set $W \subseteq V(G)$ such that $G \models \psi(W)$ and for every vertex v of G, it holds that $|N(v) \cap W| \le k$?

In this version, the formula ψ can force that G has the desired property after deletion as well as imposing additional constraints on the deleted set itself.

Courcelle and Mosbah [4] introduced a semiring homomorphism framework that can be used to minimize various functions over all sets satisfying a given MSO formula. A natural question is whether this framework can be used to minimize the fair objective function. The answer is no, as we exclude the possibility of an existence of an FPT algorithm for parameterization by tree-width under reasonable assumption. Note that there are semirings that capture the fair objective function, but their size is of order $O(n^{\text{tw}(G)})$, so this approach does not lead to an FPT algorithm.

1.1   Our Results

We prove that the XP algorithm given by Kolman et al. [12] is almost optimal under the exponential time hypothesis (ETH) for both the edge and the vertex version. Actually we proved something little bit stronger. We prove the hardness of the classical (weaker) formulation of FAIR DELETION PROBLEMS described in (weaker as well) FO logic.

Theorem 1. *If there is an* FPT *algorithm for* FAIR FO VERTEX-DELETION *parameterized by the size of the formula* ψ, *the pathwidth of* G, *and the size of minimum feedback vertex set of* G *combined, then* FPT $=$ W[1]. *Moreover, let* k *denote* $\mathrm{pw}(G) + \mathrm{fvs}(G)$. *If there is an algorithm for* FAIR FO VERTEX-DELETION *with running time* $f(|\psi|, k)n^{o(\sqrt[3]{k})}$, *then Exponential Time Hypothesis fails.*

Theorem 2. *If there is an* FPT *algorithm for* FAIR FO EDGE-DELETION *parameterized by the size of the formula* ψ, *the pathwidth of* G, *and the size of minimum feedback vertex set of* G *combined, then* FPT $=$ W[1]. *Moreover, let* k *denote* $\mathrm{pw}(G) + \mathrm{fvs}(G)$. *If there is an algorithm for* FAIR FO EDGE-DELETION *with running time* $f(|\psi|, k)n^{o(\sqrt[3]{k})}$, *then Exponential Time Hypothesis fails.*

By a small modification of our proofs we are able to derive tighter ($\sqrt{k}$ instead of $\sqrt[3]{k}$) results using MSO_2 logic or MSO_1 logic respectively. However, there is still a small gap that has been left open.

Theorem 3. *If there is an* FPT *algorithm for* FAIR MSO_1 VERTEX-DELETION *parameterized by the size of the formula* ψ, *the pathwidth of* G, *and the size of minimum feedback vertex set of* G *combined, then* FPT $=$ W[1]. *Moreover, let* k *denote* $\mathrm{pw}(G) + \mathrm{fvs}(G)$. *If there is an algorithm for* FAIR MSO_1 VERTEX-DELETION *with running time* $f(|\psi|, k)n^{o(\sqrt{k})}$, *then Exponential Time Hypothesis fails.*

Theorem 4. *If there is an* FPT *algorithm for* FAIR MSO_2 EDGE-DELETION *parameterized by the size of the formula* ψ, *the pathwidth of* G, *and the size of minimum feedback vertex set of* G *combined, then* FPT $=$ W[1]. *Moreover, let* k *denote* $\mathrm{pw}(G) + \mathrm{fvs}(G)$. *If there is an algorithm for* FAIR MSO_2 EDGE-DELETION *with running time* $f(|\psi|, k)n^{o(\sqrt{k})}$, *then Exponential Time Hypothesis fails.*

On the other hand we show some positive algorithmic results for the generalized version of the problems.

Theorem 5. GENERALIZED FAIR MSO_1 VERTEX-DELETION *is in* FPT *with respect to the neighborhood diversity* $\mathrm{nd}(G)$ *and the size of the formula* ψ.

We also provide an algorithm for the MSO_2 logic (strictly more powerful than MSO_1), however we need a more restrictive parameter because model checking of an MSO_2 formula is not even in XP for cliques unless E $=$ NE [3,15]. We consider the size of minimum vertex cover that allows us to attack the edge-deletion problem in FPT time.

Theorem 6. GENERALIZED FAIR MSO_2 EDGE-DELETION *is in* FPT *with respect to the size of minimum vertex cover* $\mathrm{vc}(G)$ *and the size of the formula* ψ.

2 Preliminaries

Throughout the paper we deal with simple undirected graphs. For further standard notation in graph theory, we refer to Diestel [6]. For terminology in parameterized computational complexity we refer to Downey and Fellows [7].

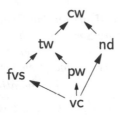

Fig. 1. Hierarchy of graph parameters. An arrow indicates that a graph parameter upper-bounds the other. Thus, hardness results are implied in direction of arrows and FPT algorithms are implied in the reverse direction.

2.1 Graph Parameters

We define several graph parameters being used throughout the paper (Fig. 1).

We start by definition of *vertex cover* being a set of vertices such that its complement forms an independent set. By vc (G) we denote the size of a smallest such set. This is the strongest of considered parameters and it is not bounded for any natural graph class.

A *feedback vertex set* is a set of vertices whose removal leaves an acyclic graph. Again, by fvs (G) we denote the size of a smallest such set.

Another famous graph parameter is *tree-width* introduced by Bertelé and Brioshi [2].

Definition 1 (Tree decomposition). *A tree decomposition of a graph G is a pair (T, X), where $T = (I, F)$ is a tree, and $X = \{X_i \mid i \in I\}$ is a family of subsets of $V(G)$ such that:*

- *the union of all X_i, $i \in I$ equals V,*
- *for all edges $\{v, w\} \in E$, there exists $i \in I$, such that $v, w \in X_i$ and*
- *for all $v \in V$ the set of nodes $\{i \in I \mid v \in X_i\}$ forms a subtree of T.*

The *width* of the tree decomposition is $\max(|X_i| - 1)$. The *tree-width* of a graph tw (G) is the minimum width over all possible tree decompositions of the graph G. The parameter of *path-width* (analogously pw (G)) is almost the same except the decomposition need to form a path instead of a general tree.

A less known graph parameter is the *neighborhood diversity* introduced by Lampis [14].

Definition 2 (Neighborhood diversity). *The neighborhood diversity of a graph G is denoted by nd (G) and it is the minimum size of a partition of vertices into classes such that all vertices in the same class have the same neighborhood, i.e. $N(v) \setminus \{v'\} = N(v') \setminus \{v\}$, whenever v, v' are in the same class.*

It can be easily verified that every class of neighborhood diversity is either a clique or an independent set. Moreover, for every two distinct classes C and C', either every vertex in C is adjacent to every vertex in C', or there is no edge between them. If classes C and C' are connected by edges, we refer to such classes as *adjacent*.

2.2 Parameterized Problems and Exponential Time Hypothesis

Definition 3 (Parameterized language). *Let Σ be a finite alphabet. A para-meterized language $L \subseteq \Sigma^* \times \mathbb{N}$ set of pairs (x, k) where x is a finite word over Σ and k is a nonnegative integer.*

We say that an algorithm for a parameterized problem L is an FPT *algorithm* if there exist a constant c and a computable function f such that the running time for input (x, k) is $f(k)|x|^c$ and the algorithm accepts (x, k) if and only if $(x, k) \in L$.

A standard tool for showing nonexistence of an FPT algorithm is W[1]-hardness (assuming FPT $\neq$ W[1]). For the definition of W[1] class and the notion of W[1]-hardness, we refer the reader to [7].

A stronger assumption than FPT $\neq$ W[1] that can be used to obtain hardness results is the Exponential Time Hypothesis (ETH for short). It is a complexity theoretic assumption introduced by Impagliazzo, Paturi and Zane [11]. We follow a survey on the topic of lower bounds obtained from ETH by Lokshtanov, Marx, and Saurabh [18], which contains more details on this topic.

The hypothesis states that there is no subexponential time algorithm for 3-SAT if we measure the time complexity by the number of variables in the input formula, denoted by n.

Exponential Time Hypothesis (ETH) [11]. There is a positive real s such that 3-SAT with parameter n cannot be solved in time $2^{sn}(n + m)^{O(1)}$.

Definition 4 (Standard parameterized reduction). *We say that parame-terized language L reduces to parameterized language L' by a standard parame-terized reduction if there are functions $f, g \colon \mathbb{N} \to \mathbb{N}$ and $h \colon \Sigma^* \times \mathbb{N} \to \Sigma^*$ such that function h is computable in time $g(k)|x|^c$ for a constant c, and $(x, k) \in L$ if and only if $(h(x, k), f(k)) \in L'$.*

For preserving bounds obtained from the ETH, the asymptotic growth of the function f need to be as slow as possible.

2.3 Logic Systems

We heavily use graph properties that can be expressed in certain types of logical systems. In the paper it is *Monadic second-order logic* (MSO) where monadic means that we allow quantification over sets (of vertices and/or edges). In *first order logic* (FO) there are no set variables at all.

We distinguish MSO_2 and MSO_1. In MSO_1 quantification only over sets of vertices is allowed and we can use the predicate of adjacency $\mathrm{adj}(u, v)$ returning true whenever there is an edge between vertices u and v. In MSO_2 we can addi-tionally quantify over sets of edges and we can use the predicate of incidence $\mathrm{inc}(v, e)$ returning true whenever a vertex v belongs to an edge e.

It is known that MSO_2 is strictly more powerful than MSO_1. For example, the property that a graph is Hamiltonian is expressible in MSO_2 but not in MSO_1 [16].

Note that in MSO_1 it is easy to describe several complex graph properties like being connected or having a vertex of a constant degree.

3 Hardness Results

In this section, we prove hardness of FAIR FO VERTEX-DELETION by exhibiting a reduction from EQUITABLE 3-COLORING.

Equitable 3-coloring

Input: An undirected graph G.

Question: Is there a proper coloring of vertices of G by at most 3 colors such that the size of any two color classes differ by at most one?

The following result was proven implicitly in [9].

Theorem 7. EQUITABLE 3-COLORING *is* W[1]-*hard with respect to* pw(G) *and* fvs(G) *combined. Moreover, if there exists an algorithm for* EQUITABLE 3-COLORING *running in time* $f(k)n^{o(\sqrt[3]{k})}$, *where k is* pw(G) + fvs(G), *then the Exponential Time Hypothesis fails.*

The proof in [9] relies on a reduction from MULTICOLORED CLIQUE [10] to EQUITABLE COLORING. The reduction transforms an instance of MULTI-COLORED CLIQUE of parameter k into an EQUITABLE COLORING instance of path-width and feedback vertex size at most $O(k)$ (though only tree-width is explicitly stated in the paper). Algorithm for EQUITABLE COLORING running in time $f(k)n^{o(\sqrt[3]{k})}$ would lead to an algorithm for MULTICOLORED CLIQUE running in time $f(k)n^{o(k)}$. It was shown by Lokshtanov, Marx, and Saurabh [18] that such algorithm does not exist unless ETH fails.

We now describe the idea behind the reduction from EQUITABLE 3-COLORING to FAIR FO VERTEX-DELETION. Let us denote by n the number of vertices of G and assume that 3 divides n. The vertices of G are referred to as *original vertices*. First, we add three vertices called *class vertices*, each of them corresponds to a particular color class. Then we add edge between every class vertex and every original vertex and subdivide each such edge. The vertices subdividing those edges are called *selector vertices*.

We can encode the partition of $V(G)$ by deleting vertices in the following way: if v is an original vertex and c is a class vertex, by deleting the selector vertex between v and c we say to the class represented by c. If we ensure that the set is deleted in such a way that every vertex belongs to exactly one class, we obtain a partition of $V(G)$.

The equitability of the partition will be handled by the fair objective function. Note that if we delete a subset W of selector vertices that encodes a partition then $|W| = n$. Those n vertices are adjacent to 3 class vertices, so the best possible fair cost is $n/3$ and thus a solution of the fair cost $n/3$ corresponds to an equitable partition.

Of course, not every subset W of vertices of our new graph encodes a partition. Therefore, the formula we are trying to satisfy must ensure that:

- every original vertex belongs to exactly one class,
- no original or class vertex was deleted,
- every class is an independent set.

However, the described reduction is too naive to achieve those goals; we need to slightly adjust the reduction. Let us now describe the reduction formally:

Proof (of Theorem 1). Let G be a graph on n vertices. We can assume without loss of generality (by addition of isolated vertices.) that 3 divides n and $n \geq 6$.

First we describe how to construct the reduction. All vertices of G will be referred to as *original vertices*. We add three vertices called *class vertices* and connect every original vertex with every class vertex by an edge. We subdivide each such edge once; the vertices subdividing those edges are called *selector vertices*. Finally, for every original vertex v, we add n new vertices called *dangling vertices* and connect each of them by an edge to v. We denote the graph obtained in this way as G'. For a schema of the reduction, see Fig. 2.

Now, we wish to find a set $W \subseteq V(G')$ such that it encodes an equitable 3-coloring of a graph G. The set is described by the following FO formula *eq_3_col* imposed on a graph $G \setminus W$. We claim that whenever this set satisfy following claims it encodes an equitable 3-coloring. A set W can contain only selector vertices and some dangling vertices (but those do not affect the coloring). For each vertex v of a graph there can be only one selector vertex in the set W and that vertex has only one class vertex as a neighbor. That vertex determine the color of v.

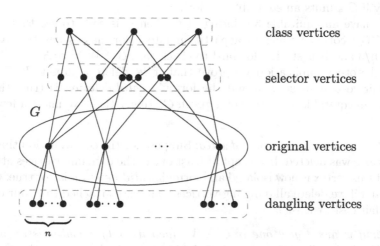

Fig. 2. The schema of the reduction

We use the following shorthand $\exists_{=k}$ meaning there are exactly k distinct elements satisfying a given predicate:

$$(\exists_{=k}w)(pred(w)) \equiv (\exists v_1, \ldots, v_k)\bigg(\bigwedge_{i=1}^{k} pred(v_i) \wedge \bigwedge_{1 \le i < j \le k}(v_i \ne v_j)$$

$$\wedge (\forall v')\Big(pred(v') \to \bigvee_{i=1}^{k}(v' = v_i)\Big)\bigg)$$

The building blocks for the formula are as follows:

$$isol(v) \equiv (\forall w)(\neg adj(v, w))$$
$$dangling(v) \equiv (\exists w)\big(adj(v, w) \wedge (\forall w')(adj(v, w') \to w = w')\big)$$
$$original(v) \equiv (\exists w)(dangling(w) \wedge adj(v, w))$$
$$selector(v) \equiv (\exists_{=2}w)(adj(v, w))$$
$$class(v) \equiv \neg orig(v) \wedge \neg selector(v) \wedge \neg dangling(v)$$
$$belongs\_to(v, a) \equiv original(v) \wedge class(a) \wedge \neg(\exists w)(adj(v, w) \wedge adj(w, a))$$
$$same\_class(v, w) \equiv original(v) \wedge original(w)$$
$$\wedge (\exists a)(class(a) \wedge belongs\_to(v, a) \wedge belongs\_to(w, a))$$
$$valid\_deletion \equiv (\forall v)(\neg isol(v))$$
$$\wedge (\forall v)\big(original(v) \to (\exists_{=1}c)(belongs\_to(v, c))\big)$$
$$eq\_3\_col \equiv valid\_deletion \wedge (\forall v, w)(same\_class(v, w) \to \neg adj(v, w))$$

The described reduction maps an instance G of an EQUITABLE COLORING into an instance $(G', eq\_3\_col, n/3)$ of FAIR FO VERTEX-DELETION.

We claim that there exists a set $W \subseteq V(G')$ of the fair cost at most $n/3$ if and only if G admits an equitable 3-coloring.

If we have an equitable 3-coloring of G then it is easy to see that the set $W \subseteq V(G')$ corresponding to a partition into color classes has the fair cost exactly $n/3$ and it is straightforward to check that $G' \setminus W \models eq\_3\_col$.

For the other implication we prove that if we delete a subset $W \subseteq V(G')$ of the fair cost at most $n/3$, and the formula $valid\_deletion$ is true, then we obtained an equitable 3-coloring of a graph G. To get there we made a few basic claims.

Claim 1: no original vertex was deleted: Suppose for the contradiction that original vertex v was deleted. If we kept at least one of the dangling vertices attached to v, but this vertex is now isolated and formula $valid\_deletion$ is not true. On the other hand if we delete all dangling vertices that were attached to v, our deleted set has fair cost at least n.

Claim 2: if w has degree one in $G' \setminus W$, then its only neighbor is an original vertex: If w is dangling, then its only neighbor is original vertex by the construction of G'. Suppose that w has degree one in $G' \setminus W$ but is not dangling. Since both class and original vertices have degree at least n in G', we cannot bring them down to degree one without exceeding the fair cost limit $n/3$. This leaves the only possibility that w is a selector and exactly one of its two neighbors is in the deleted set W. By Claim 1, the deleted neighbor must have been a class vertex so the only remaining neighbor of w in $G' \setminus W$ is an original vertex.

Claim 3: the formula original correctly recognizes original vertices: If v is original, then at least one of its dangling neighbors is not in W, otherwise we would exceed the fair cost. In this case the formula $original(v)$ is true. The other direction ($original(v)$ is true implies v is original) is proved by Claim 2.

Claim 4: if v is a dangling vertex such that $v \notin W$ then $dangling(v)$ is true: By Claim 1, we cannot delete the only neighbor of v, which means v has exactly one neighbor and so $dangling(v)$ is true.

Claim 5: the formula class(v) is true if and only if v is a class vertex that was not deleted: Suppose that $v \notin W$ is a class vertex. It cannot have neighbor of degree one in $G' \setminus W$, because that would mean that an original vertex was deleted which violates Claim 1. This means that $original(v)$ is false. Moreover, we cannot decrease the degree of v to two or less by deleting at most $n/3$ neighbors of v, so $dangling(v)$ and $selector(v)$ are false too. But then $class(v)$ is true.

For the other direction suppose that v is not a class vertex. If it is original or dangling, then $original(v)$ or $dangling(v)$ is true (by Claims 3 or 4) and hence $class(v)$ is false. If v is a selector then either none of its neighbors were deleted, v has degree two in $G' \setminus W$ and $selector(v)$ is true, or its class neighbor was deleted, v has degree one in $G' \setminus W$ and $dangling(v)$ is true. Either way, $class(v)$ is false as required.

Claim 6: no class vertex was deleted: since $valid\_deletion$ is true, we know that for every original vertex v there is exactly one class vertex c such that there is no path of length two between v and c (in other words, the selector vertex that was on the unique path of length two between v and c was deleted). Suppose for contradiction that one of the class vertices was deleted; then by Claim 5 we have at most two class vertices. But the $valid\_deletion$ formula implies that at least n selector vertices were deleted. By pigeonhole principle, one of the class vertices has at least $n/2$ deleted neighbors which means the fair cost is greater than $n/3$, a contradiction.

The chain of claims we just proved guarantees that the deleted set W indeed obeys the rules we required and corresponds to a partition (though we might have deleted a small number of dangling vertices, this does not affect the partition in any way). In order to meet the fair cost limit, each class of the partition must have at most $n/3$ vertices and since no original vertex was deleted, it has exactly $n/3$ vertices. Now it is easy to see that the formula eq_3_col forces that each class of the partition is independent and so the graph G has an equitable 3-coloring.

Let us now discuss the parameters and the size of the FAIR FO VERTEX DELETION instance. If G has a feedback vertex set S of size k, then the union of S with the set of class vertices is a feedback vertex set of G'. Therefore, $\text{fvs}(G') \le \text{fvs}(G) + 3$. To bound the path-width, observe that after deletion of the class vertices we are left with G with $O(n^2)$ added vertices of degree one; the addition of degree one vertices to the original vertices can increase the path-width by at most one and so we have $\text{pw}(G') \le \text{pw}(G) + 4$. Moreover it is clear that the size of instance is of size $O(n^2)$. It is obvious that the reduction can be carried out in polynomial time. $\square$

Let us mention that if we are allowed to use MSO formulas, we are actually able to reduce any equitable partition problem to fair vertex deletion. This allows us to reduce for example EQUITABLE CONNECTED PARTITION to FAIR MSO VERTEX-DELETION which in turn allows us to prove Theorem 3.

Equitable connected partition
Input: An undirected graph G, a positive integer r
Question: Is there a partition of $V(G)$ into r sets such that each of them
induces a connected graph and the sizes of every two sets differ
by at most one?

Enciso et al. [8] showed that EQUITABLE CONNECTED PARTITION is $W[1]$-hard for combined parameterization by $\text{fvs}(G)$, $\text{pw}(G)$, and the number of partitions r. The part that $f(k)n^{o(\sqrt{k})}$ algorithm would refute ETH is again contained only implicitly; the proof reduces an instance of MULTICOLORED CLIQUE of parameter k to an instance of EQUITABLE CONNECTED PARTITION of parameter $O(k^2)$.

Our reduction can be easily adapted to r parts (we just add r class vertices and we set the fair cost limit to n/r). We define the formula eq_conn as follows.

$$class\_set(W) \equiv (\exists v \in W) \land (\forall v, w \in W)(same\_class(v, w))$$
$$\land (\forall w \in W, z \notin W)(\neg same\_class(w, z))$$
$$eq\_conn \equiv (\forall W)(class\_set(W) \rightarrow connected(W))$$

By the same argument as in the proof of Theorem 1, we can show that there exists $W \subseteq V$ of fair cost at most n/r such that $G' \setminus W \models eq\_conn$ if and only if G admits an equitable connected partition.

Sketch of proof of Theorem 2. We do not present the complete proof, as the critical parts are the same as in proof of Theorem 1. The reduction follows the same idea as before: we add three class vertices and connect each class vertex to each original vertex by an edge. This time, we do not subdivide the edges, as the partition is encoded by deleting the edges.

The protection against tampering with the original graph has to be done in slightly different way: in this case, we add $n/3 + 1$ dangling vertices of degree one to each original vertex. Note that if we delete a set $F \subseteq E(G)$ of fair cost at most $n/3$, at least one of the added edges from every original vertex survives the deletion, so we can recognize the original vertices by having at least one neighbor of degree one. In our formula, we require that each vertex has at most two neighbors of degree one. This forces us to delete all of those added edges except two. Since at least one edge from the original vertex must be deleted to encode a partition, by deleting an edge of the original graph G we would exceed the fair cost limit $n/3$.

For the edge-deletion the formula eq_3_col is built as follows.

$$dangling(v) \equiv (\exists w)\big(adj(v, w) \wedge (\forall w')(adj(v, w') \rightarrow w = w')\big)$$
$$original(v) \equiv (\exists w)(dangling(w) \wedge adj(v, w))$$
$$class(v) \equiv \neg orig(v) \wedge \neg dangling(v)$$
$$belongs\_to(v, a) \equiv original(v) \wedge class(a) \wedge \neg adj(v, a)$$
$$same\_class(v, w) \equiv original(v) \wedge original(w)$$
$$\wedge\, (\exists a)(class(a) \wedge belongs\_to(v, a) \wedge belongs\_to(w, a))$$
$$valid\_deletion \equiv (\forall v)(\exists_{\leq 2}w)(adj(v, w) \wedge dangling(w))$$
$$\wedge\, (\forall v)\big(original(v) \rightarrow (\exists_{=1}c)(belongs\_to(v, c))\big)$$
$$eq\_3\_col \equiv valid\_deletion \wedge (\forall v, w)(same\_class(v, w) \rightarrow \neg adj(v, w))$$

The complete proof of correctness is omitted due to space considerations, however, it is almost exactly the same as in the proof of Theorem 1.   □

The transition between the FO case and the MSO case of edge-deletion (Theorem 4) is done in exactly the same way as before.

4   FPT Algorithms

We now turn our attention to FPT algorithms for fair deletion problems.

4.1   FPT Algorithm for Parameterization by Neighborhood Diversity

Definition 5. *Let $G = (V, E)$ be a graph of neighborhood diversity k and let $N_1, \ldots, N_k$ denote its classes of neighborhood diversity. A* shape *of a set $X \subseteq V$ in G is a k-tuple $s = (s_1, \ldots, s_k)$, where $s_i = |X \cap N_i|$.*

We denote by $\bar{s}$ the complementary shape *to s, which is defined as the shape of $V \setminus X$, i.e. $\bar{s} = (|N_1| - s_1, \ldots, |N_k| - s_k)$.*

Proposition 1. *Let $G = (V, E)$ be a graph, π a property of a set of vertices, and let $X, Y \subseteq V$ be two sets of the same shape in G. Then X satisfies π if and only if Y satisfies π.*

Proof. Clearly, we can construct an automorphism of G that maps X to Y.   □

Definition 6. *Let r be a non-negative integer and let $(s_1, \ldots, s_k)$, $(t_1, \ldots, t_k)$ be two shapes. The shapes are r-equivalent, if for every i:*

- $s_i = t_i$, or
- both s_i, t_i are strictly greater than r,

and the same condition hold for the complementary shapes $\bar{s}$, $\bar{t}$.

The following proposition gives a bound on the number of r-nonequivalent shapes.

Proposition 2. *For any graph G of neighborhood diversity k, the number of r-nonequivalent shapes is at most $(2r + 3)^k$.*

Proof. We show that for every i, there are at most $(2r + 3)$ choices of s_i. This holds trivially if $|N_i| \leq 2r + 3$. Otherwise we have following $2r + 3$ choices:

- $s_i = k$ and $\bar{s_i} > r$ for $k = 0, 1, \ldots, r$, or
- both $s_i, \bar{s_i} > r$, or
- $s_i > r$ and $\bar{s_i} = k$ for $k = 0, 1, \ldots, r$. $\qquad\square$

The next lemma states that the fair cost of a set can be computed from its shape in a straightforward manner. Before we state it, let us introduce some auxiliary notation.

If a graph G of neighborhood diversity k has classes of neighborhood diversity $N_1, \ldots, N_k$, we write $i \sim j$ if the classes N_i and N_j are adjacent. If the class N_i is a clique, we set $i \sim i$. Moreover, we set $\eta_i = 1$ if the class N_i is a clique and $\eta_i = 0$ if it is an independent set. The classes of size one are treated as cliques for this purpose.

Lemma 1. *Let $G = (V, E)$ be a graph of neighborhood diversity k and let N_i be its classes of neighborhood diversity. Moreover, let $X \subseteq V$ be a set of shape s. Then the fair vertex cost of X is*

$$\max_i \left(\left(\sum_{j : i \sim j} s_j \right) - \eta_i \right).$$

Proof. It is straightforward to check that vertex $v \in N_i$ has exactly $\sum_{j : i \sim j} s_j - \eta_i$ neighbors in X. $\qquad\square$

Our main tool is a reformulation of Lemma 5 from [14]:

Lemma 2. *Let ψ be an MSO_1 formula with one free vertex-set variable, q_E vertex element quantifiers, and q_S vertex set quantifiers. Let $r = 2^{q_S} q_E$. If $G = (V, E)$ is a graph of neighborhood diversity k and $X, Y \subseteq V$ are two sets such that their shapes are r-equivalent, then $G \models \psi(X)$ if and only if $G \models \psi(Y)$.*

The last result required is the MSO_1 model checking for graphs of bounded neighborhood diversity [14]:

Theorem 8. *Let ψ be an MSO_1 formula with one free vertex-set variable. There exists an FPT algorithm that given a graph $G = (V, E)$ of neighborhood diversity k and a set $X \subseteq V$ decides whether $G \models \psi(X)$. The running time of the algorithm is $f(k, |\psi|) n^{O(1)}$.*

We now have all the tools required to prove Theorem 5.

Proof (Proof of Theorem 5). Let ψ be an MSO_1 formula in the input of FAIR MSO_1 VERTEX-DELETION. Denote by q_S the number of vertex-set quantifiers in ψ, by q_E the number of vertex-element quantifiers in ψ, and set $r = 2^{q_S} q_E$.

By Proposition 1, the validity of $\psi(X)$ depends only on the shape of X. Let us abuse notation slightly and write $G \models \psi(s)$ when "X has shape s" implies $G \models \psi(X)$. Similarly, Lemma 1 allows us to refer to the fair cost of a shape s.

From Lemma 2 it follows that the validity of $\psi(s)$ does not depend on the choice of an r-equivalence class representative. The fair cost is not same for all r-equivalent shapes, but since the fair cost is monotone in s, we can easily find the representative of the minimal fair cost.

Suppose we have to decide if there is a set of a fair cost at most ℓ. The algorithm will proceed as follows: For each class of r-equivalent shapes, pick a shape s of the minimal cost, if the fair cost is at most ℓ and $G \models \psi(s)$, output `true`, if no such shape is found throughout the run, output `false`.

By the previous claims, the algorithm is correct. Let us turn our attention to the running time. The number of shapes is at most $(2r + 3)^k$ by Proposition 2, and so it is bounded by $f(|\psi|, k)$ for some function f. The MSO_1 model checking runs in time $f'(|\psi|, k)n^{O(1)}$ by Theorem 8, so the total running time is $f(|\psi|, k)f'(|\psi|, k)n^{O(1)}$, so the described algorithm is in FPT.   □

4.2   FPT Algorithm for Parameterization by Vertex Cover

The FPT algorithm for parameterization by the size of minimum vertex cover uses the same idea. We use the fact that every MSO_2 formula can be translated to MSO_1 formula — roughly speaking, every edge-set variable is replaced by vc (G) vertex-set variables.

We only sketch translation from MSO_2 to MSO_1, for the proof we refer the reader to Lemma 6 in [14]. Let $G = (V, E)$ be a graph with vertex cover $C = \{v_1, \ldots, v_k\}$ and $F \subseteq E$ a set of edges. We construct vertex sets $U_1, \ldots, U_k$ in the following way: if w is a vertex such that an edge in F connects w with v_i, we put w into U_i. It is easy to see that the sets $U_1, \ldots, U_k$ together with the vertex cover $v_1, \ldots, v_k$ describe the set F.

In this way, we reduce the problem of finding a set F to finding k-tuple of sets $(U_1, \ldots, U_k)$. We can define shapes and classes of r-equivalence in an analogous way as we did in previous section. Since the number of r-equivalence classes defined in this way is still bounded, we can use essentially the same algorithm: for each class of r-equivalence, run a model checking on a representative of this class. From those representatives that satisfy ψ, we choose the one with best fair cost.

Due to the space limitation we omit more detailed treatment.

5   Open Problems

The main open problem is whether the bound in Theorems 1 and 2 can be improved to $f(|\psi|, k)n^{o(k/\log k)}$ or even to $f(|\psi|, k)n^{o(k)}$.

Acknowledgements. The authors would like to thank Martin Koutecký and Petr Hliněný for helpful discussions.

References

1. Ae, T., Watanabe, T., Nakamura, A.: On the NP-hardness of edge-deletion and -contraction problems. Discrete Appl. Math. **6**, 63–78 (1983)
2. Bertelè, U., Brioschi, F.: Nonserial Dynamic Programming. Mathematics in Science and Engineering. Academic Press, Orlando (1972)
3. Courcelle, B., Makowsky, J.A., Rotics, U.: Linear time solvable optimization problems on graphs of bounded clique-width. Theor. Comput. Syst. **33**, 125–150 (2000)
4. Courcelle, B., Mosbah, M.: Monadic second-order evaluations on tree-decomposable graphs. Theor. Comput. Sci. **109**, 49–82 (1993)
5. Cowen, L.J., Cowen, R., Woodall, D.R.: Defective colorings of graphs in surfaces: partitions into subgraphs of bounded valency. J. Graph Theor. **10**, 187–195 (1986)
6. Diestel, R.: Graph Theory. Graduate Texts in Mathematics, vol. 173, 4th edn. Springer, Heidelberg (2012)
7. Downey, R.G., Fellows, M.R.: Fundamentals of Parameterized Complexity. Texts in Computer Science. Springer, London (2013)
8. Enciso, R., Fellows, M.R., Guo, J., Kanj, I., Rosamond, F., Suchý, O.: What makes equitable connected partition easy. In: Chen, J., Fomin, F.V. (eds.) IWPEC 2009. LNCS, vol. 5917, pp. 122–133. Springer, Heidelberg (2009). doi:10.1007/978-3-642-11269-0_10
9. Fellows, M., Fomin, F.V., Lokshtanov, D., Rosamond, F., Saurabh, S., Szeider, S., Thomassen, C.: On the complexity of some colorful problems parameterized by treewidth. In: Dress, A., Xu, Y., Zhu, B. (eds.) COCOA 2007. LNCS, vol. 4616, pp. 366–377. Springer, Heidelberg (2007). doi:10.1007/978-3-540-73556-4_38
10. Fellows, M.R., Hermelin, D., Rosamond, F.A., Vialette, S.: On the parameterized complexity of multiple-interval graph problems. Theor. Comput. Sci. **410**, 53–61 (2009)
11. Impagliazzo, R., Paturi, R., Zane, F.: Which problems have strongly exponential complexity? J. Comput. Syst. Sci. **63**, 512–530 (2001)
12. Kolman, P., Lidický, B., Sereni, J.-S.: Fair edge deletion problems on treedecomposable graphs and improper colorings (2010)
13. Krishnamoorthy, M.S., Deo, N.: Node-deletion NP-complete problems. SIAM J. Comput. **8**, 619–625 (1979)
14. Lampis, M.: Algorithmic meta-theorems for restrictions of treewidth. Algorithmica **64**, 19–37 (2011)
15. Lampis, M.: Model checking lower bounds for simple graphs. Logical Methods Comput. Sci. **10**, 1–21 (2014)
16. Libkin, L.: Elements of Finite Model Theory. Texts in Theoretical Computer Science. An EATCS Series. Springer, Heidelberg (2004)
17. Lin, L., Sahni, S.: Fair edge deletion problems. IEEE Trans. Comput. **38**, 756–761 (1989)
18. Lokshtanov, D., Marx, D., Saurabh, S.: Lower bounds based on the exponential time hypothesis. Bull. EATCS **105**, 41–72 (2011)
19. Yannakakis, M.: Node- and edge-deletion NP-complete problems. In: ACM STOC, pp. 253–264 (1978)
20. Yannakakis, M.: Edge-deletion problems. SIAM J. Comput. **10**, 297–309 (1981)

Degrees of Word Problem for Algebras Without Finitely Presented Expansions

Guohua Wu$^{(\boxtimes)}$ and Huishan Wu

Division of Mathematical Sciences, School of Physical and Mathematical Sciences,
Nanyang Technological University, Singapore 637371, Singapore
guohua@ntu.edu.sg

Abstract. Bergstra and Tucker [1,2] proved that computable universal algebras have finitely presented expansions. Bergstra and Tucker, and Goncharov, independently, asked whether all finitely generated computably enumerable algebras have finitely presented expansions. Khoussainov and Hirschfeldt [3] constructed finitely generated, infinite c.e. semigroups without finitely presented expansions; furthermore, Khoussainov and Miasnikov [6] found such examples in class of groups and algebras over finite fields. In this paper, we consider Turing degrees of the word problem for semigroups constructed in [3] and for algebras over finite fields constructed in [6], and prove that the word problem for such semigroups and algebras appears in all nonzero c.e. degrees respectively.

Keywords: Computably enumerable universal algebra · Finitely presented expansion · Word problem

1 Introduction

Given a signature $\sigma = (f_1, \cdots, f_n, c_1, \cdots, c_m)$ with k_i-ary function symbols f_i, $1 \le i \le n$ and constant symbols c_j, $1 \le j \le m$, a universal algebra (or simply, an algebra) for σ is a structure $\mathfrak{A} = (A, f_1^{\mathfrak{A}}, \cdots, f_n^{\mathfrak{A}}, c_1^{\mathfrak{A}}, \cdots, c_m^{\mathfrak{A}})$ with domain A, k_i-ary functions $f_i^{\mathfrak{A}} : A^{k_i} \to A$, and constants $c_j^{\mathfrak{A}} \in A$. An algebra is *computable* if its domain and all its functions are computable. An example of computable algebras is the term algebra. Here a term for a signature σ is defined by induction as usual: all variables and constants are terms, and if t_i, $1 \le i \le k$, are terms and f is a k-ary function symbol in σ, then $f(t_1, \cdots, t_k)$ is also a term.

Definition 1. *A ground term of σ is a term containing no variables. We use T_G to denote the set of all ground terms of σ. Based on T_G, we can define the following interpretation: the interpretation of constant symbol is itself, and for a k-ary function symbol f in σ, the interpretation of f is f^{T_G}, which is a function on T_G^k such that for any $(t_1, \cdots, t_k) \in T_G^k$, $f^{T_G}(t_1, \cdots, t_k)$ is the ground term $f(t_1, \cdots, t_k)$. We denote this algebra by T_G.*

Both authors are partially supported by MOE2011-T2-1-071 (ARC 17/11, M45110030) from Ministry of Education of Singapore, and by AcRF grants RG29/14, M4011274 from Nanyang Technological University.

T.V. Gopal et al. (Eds.): TAMC 2017, LNCS 10185, pp. 643–653, 2017.
DOI: 10.1007/978-3-319-55911-7_46

For an algebra $\mathfrak{A} = (A, f_1^{\mathfrak{A}}, \cdots, f_n^{\mathfrak{A}}, c_1^{\mathfrak{A}}, \cdots, c_m^{\mathfrak{A}})$, and an equivalence relation E on domain A, say that E is a congruence relation on $\mathfrak{A}$ if each k_i-ary function $f_i^{\mathfrak{A}} : A^{k_i} \to A$ satisfies the condition: for any $(x_1, \cdots, x_{k_i}), (y_1, \cdots, y_{k_i}) \in A^{k_i}$, if $x_j \, E \, y_j$ for all $1 \le j \le k_i$, then $f_i^{\mathfrak{A}}(x_1, \cdots, x_{k_i}) E f_i^{\mathfrak{A}}(y_1, \cdots, y_{k_i})$, where the symbol $x \, E \, y$ means x and y are equivalent under relation E.

Let E be a congruence relation on an algebra $\mathfrak{A} = (A, f_1^{\mathfrak{A}}, \cdots, f_n^{\mathfrak{A}}, c_1^{\mathfrak{A}}, \cdots, c_m^{\mathfrak{A}})$, the set of all equivalence classes A/E of E can be turned into an algebra $\mathfrak{A}/E = (A/E, f_1^{\mathfrak{A}/E}, \cdots, f_n^{\mathfrak{A}/E}, c_1^{\mathfrak{A}/E}, \cdots, c_m^{\mathfrak{A}/E})$ such that for each i,

$$f_i^{\mathfrak{A}/E}([x_1], \cdots, [x_{k_i}]) = [f_i^{\mathfrak{A}}(x_1, \cdots, x_{k_i})]$$

and for each j, $c_j^{\mathfrak{A}/E} = [c_j^{\mathfrak{A}}]$, where for $x \in A$, the symbol $[x]$ stands for the equivalence class of E containing x. We will call $\mathfrak{A}/E$ *the quotient algebra of* $\mathfrak{A}$ *modulo the congruence relation* E.

Definition 2. *An algebra is computably enumerable (c.e. for short) if it is the quotient algebra $\mathfrak{A}/E$ of a computable algebra $\mathfrak{A}$ modulo a c.e. congruence relation E on $\mathfrak{A}$.*

A finitely presented algebra $\mathfrak{A}$ is given by a finite set G of generators (G is the interpretations of all constants of the signature of $\mathfrak{A}$) and a finite set R of relations among generators, then $\mathfrak{A}$ is the quotient algebra of term algebra modulo the congruence relation generated by R. Finitely presented algebras are finitely generated and computably enumerable, but the converse is false, as there are finitely generated and c.e. algebras which are not finitely presented, for example, $(\omega, 0, S, 2^x)$, where S is the successor function on ω, is generated by 0 but not finitely presented, note the expansion $(\omega, 0, S, 2^x, +, \times)$ by two binary operations, the usual addition and multiplication on natural numbers, is finitely presented.

A well known result of Bergstra and Tucker is that for any computable universal algebra, there is an expansion by signatures (that is, by adding finitely many function symbols or constant symbols) which is finitely presented. After proving this, Bergstra and Tucker, and Goncharov, independently, asked whether we can extend this result to finitely generated c.e. universal algebras? A negative answer is first provided by Kasymov in 1987 [4], and Khoussainov also gave a negative answer in 1998 [5] by an alternative approach. A recent work of Hirschfeldt and Khoussainov [3] show the existence of finitely generated and c.e. semigroups with no finitely presented expansions; furthermore, Khoussainov and Miasnikov [6] proved the existence of finitely generated and c.e. groups and algebras over finite fields having no finitely presented expansions.

For a c.e. algebra $\mathfrak{A}/E$ of a finitely generated computable algebra $\mathfrak{A}$ modulo the c.e. congruence relation E, the word problem for $\mathfrak{A}/E$ is the problem of deciding whether any two elements in $\mathfrak{A}$ are equal in $\mathfrak{A}/E$, so the word problem for $\mathfrak{A}/E$ is Turing equivalent to the c.e. congruence relation E.

In this paper, we will prove that the word problem for c.e. semigroups constructed by Hirschfeldt and Khoussainov in [3] occurs in each nonzero c.e. degree:

Theorem 1. *For each nonzero c.e. degree* **c**, *there is a finitely generated, c.e. semigroup without any finitely presented expansions whose word problem has degree* **c**.

Khoussainov and Miasnikov [6] proved the existence of finitely generated c.e. algebras over finite fields without finitely presented expansions. We will prove that the word problem for such algebras exists in any nonzero c.e. degree.

Theorem 2. *For each nonzero c.e. degree* **c**, *there is a finitely generated, c.e. algebra over finite fields without any finitely presented expansions whose word problem has degree* **c**.

We will follow basic ideas of Hirschfeldt and Khoussainov [3] and Khoussainov and Miasnikov [6]. The following properties are needed:

Definition 3. *A c.e. algebra* $\mathfrak{A}/E$ *is algorithmically finite if there is no infinite c.e. sequence of mutually nonequivalent elements of* $\mathfrak{A}$ *under relation* E. *Otherwise,* $\mathfrak{A}/E$ *is called effectively infinite.*

Definition 4. *A universal algebra* $\mathfrak{A}$ *is residually finite if for any two different elements* a, b *of* $\mathfrak{A}$, *there is a homomorphism* φ *(depending on* a, b*) from* $\mathfrak{A}$ *onto some finite algebra keeping* $\varphi(a) \neq \varphi(b)$.

A crucial point of both constructions is the following sufficient condition for a universal algebra with no finitely presented expansions:

Theorem 3 *(NFP Theorem [6]). For a signature* σ, *let* E *be a c.e. congruence relation on the term algebra* $\mathcal{T}_G$. *If the c.e. algebra* $\mathcal{T}_G/E$ *is infinite, finitely generated, algorithmically finite and residually finite, then* $\mathcal{T}_G/E$ *has no finitely presented expansions.*

2   Semigroup $\mathcal{S}_A$ and Basic Idea of Constructions

Consider the set S of all finite strings of 0's and 1's. It forms a semigroup $\mathcal{S} = (S, \frown, 0, 1)$ under the usual concatenation $\frown$, and with two constants 0, 1. We first define a computable linear order $\leq_{\mathcal{S}}$ on domain S of $\mathcal{S}$, and then deal with permitting and coding constructions within $\mathcal{S}$ under $\leq_{\mathcal{S}}$. Let $\mathcal{S}_i$ be the set of all strings of length i in $\mathcal{S}$ for all i, and λ the empty string with length $|\lambda| = 0$. We can assign a linear order $\leq_{\mathcal{S}}$ on domain S of $\mathcal{S}$ according to the length of strings as follows: on each $\mathcal{S}_i$, $\leq_{\mathcal{S}}$ is the usual lexicographical linear order with $0 <_{\mathcal{S}} 1$; for each i, j with $i < j$, set $\max\mathcal{S}_i <_{\mathcal{S}} \min\mathcal{S}_j$. For example, $\lambda <_{\mathcal{S}} 0 <_{\mathcal{S}} 1 <_{\mathcal{S}} 00 <_{\mathcal{S}} 01 <_{\mathcal{S}} 10 <_{\mathcal{S}} 11 <_{\mathcal{S}} 000$. Assume $\mathcal{S} := \{v_0^0 <_{\mathcal{S}} v_1^0 <_{\mathcal{S}} \cdots \}$. By definition, $\leq_{\mathcal{S}}$ is a computable linear order on $\mathcal{S}$ such that $|u| < |v|$ implies $u <_{\mathcal{S}} v$ for all $u, v \in \mathcal{S}$, this property is important for our permitting and coding constructions.

Let A be a subset of $\mathcal{S}$, and define a congruence relation η_A on $\mathcal{S}$ as follows:

$$u \; \eta_A \; v \text{ if } u = v \text{ or } u, v \in \mathrm{R}(A),$$

where $R(A) := \{u_1^\frown v^\frown u_2 : u_1, u_2 \in \mathcal{S}, v \in A\}$, the realization set of A, the elements in $R(A)$ are just binary strings containing a substring in A. We often write $u^\frown v$ as uv for convenience. It is easy to see that the equivalence classes of η_A are $R(A)$ and all singletons $\{x\}$ with $x \notin R(A)$. Let $\mathcal{S}_A$ be the quotient semigroup $\mathcal{S}/\eta_A$ of $\mathcal{S}$ modulo the c.e. congruence relation η_A. As the word problem for $\mathcal{S}_A$ is Turing equivalent to η_A and $\eta_A \equiv_T R(A)$, the word problem for $\mathcal{S}_A$ is Turing equivalent to $R(A)$.

Definition 5. *A c.e. subset A of $\mathcal{S}$ is simple if its complement is infinite and contains no infinite c.e. subsets.*

The desired semigroup in [3] is the quotient semigroup $\mathcal{S}_A = \mathcal{S}/\eta_A$ with A satisfying:

- A is a simple set;
- for all e, A has at most e elements of length $< e + 5$.

if these requirements hold, $\mathcal{S}_A$ is c.e., infinite and algorithmically finite, $\mathcal{S}_A$ is obvious finitely generated, according to Kasymov [4], $\mathcal{S}_A$ is residually finite. By above NFP Theorem, $\mathcal{S}_A$ has no finitely presented expansions.

Let $\mathbf{c}$ be a nonzero c.e. degree, and D a c.e. subset of natural numbers in $\mathbf{c}$. Above linear order $\leq_{\mathcal{S}}$ provides an effective bijection between $\mathcal{S}$ and natural numbers in that the i-th element v_i^0 of $\mathcal{S}$ under $\leq_{\mathcal{S}}$ maps to i for all $i \in \omega$, let $\delta : \mathcal{S} \to \omega$ be the corresponding effective bijection. Choose $C = \delta^{-1}(D) := \{v \in \mathcal{S} : \delta(v) \in D\}$, then C is a c.e. subset of $\mathcal{S}$ and the degree of C is just $\mathbf{c}$. We will construct a semigroup of the form $\mathcal{S}_A = \mathcal{S}/\eta_A$ with $R(A) \equiv_T C$, then the word problem for $\mathcal{S}_A$ has degree $\mathbf{c}$, this implies Theorem 1.

Theorem 4. *For any noncomputable c.e. subset C of $\mathcal{S}$, there is a simple set A such that the word problem for semigroup $\mathcal{S}_A = \mathcal{S}/\eta_A$ has the same degree as C and that $\mathcal{S}_A$ satisfies all the conditions of NFP Theorem, and hence has no finitely presented expansions.*

2.1   Requirements

As the degree of the word problem for $\mathcal{S}_A = \mathcal{S}/\eta_A$ is the same as the degree of $R(A)$, we need to show that $R(A) \equiv_T C$. By $R(A) \leq_T A$, we just need to show that $A \leq_T C \leq_T R(A)$. We adopt usual permitting strategy to ensure $A \leq_T C$, so we enumerate a string v into A only when some string $u \leq_{\mathcal{S}} v$ is enumerated into C; we adopt usual coding strategy to ensure $C \leq_T R(A)$, for each string v and stage s, we will define a coding marker $\sigma(v, s)$ which is an element in the complement of $R(A_s)$ at stage s, and if v is enumerated into C at stage s for first time, we enumerate the coding marker $\sigma(v, s)$ into A_s which is a subset of $R(A_s)$, recall $R(A_s)$ is the set of all strings which have a substring in A_s.

The desired A will be constructed to meet following requirements. As in [3], let $\{W_e : e \in \omega\}$ be an enumeration of all c.e. subsets of $\mathcal{S}$.

$\mathcal{Q}: C = \Sigma^{R(A)}$, $A = \Gamma^C$, with Σ, Γ total computable functionals built by us.
$\mathcal{P}_e$: If W_e is infinite, then $A \cap W_e \neq \emptyset$.
$\mathcal{R}_e$: A has at most e elements of length less than $e + 5$.

If $\mathcal{Q}$ is satisfied, as $A \leq_T C \leq_T R(A) \leq_T A$, we have $R(A) \equiv_T C$. If all the $\mathcal{R}_e$ requirements are satisfied, then the complement of $R(A)$ is infinite, as $A \subseteq R(A)$, this also implies $\overline{A}$, the complement of A, is infinite. If all the $\mathcal{P}_e$ requirements are satisfied, then $\overline{A}$ contains no infinite c.e. subsets.

Recall $\mathcal{S} = \{v_0^0 <_\mathcal{S} v_1^0 <_\mathcal{S} \cdots\}$. Fix a computable enumeration $\{C_s\}_{s \in \omega}$ of C such that for each stage s, $C_s \subseteq \{v_0^0 <_\mathcal{S} \cdots <_\mathcal{S} v_s^0\}$ and $|C_{s+1} - C_s| \leq 1$. There are two types of elements in A, one type is enumerated by positive requirements $\{\mathcal{P}_e\}_{e \in \omega}$ and we say such elements are permitted elements as they must be permitted by C, the other type is enumerated as coding markers to make $C = \Sigma^{R(A)}$.

2.2 A $\mathcal{R}_e$-strategy

For each e, our construction will ensure at most e permitted elements of A with length $< 2e + 5$, and those elements are only provided by requirements $\mathcal{P}_j$ with $j < e$. We also want the number of coding markers of length $< 2e + 5$ in A is bounded by a number $g(e)$ with $g(e) \leq e$ for each e. Then for all e,

$$|\{v \in A : |v| < 2e + 5\}| \leq e + g(e) \leq 2e.$$

So all $\mathcal{R}_{2e}$ hold. Function g is also properly chosen such that all $\mathcal{R}_{2e+1}$ hold, we choose $g(e) = \frac{e}{2}$, then $g(e+1) = \frac{e+1}{2} \leq e$ for all $e \geq 1$, so $\mathcal{R}_{2e+1}$ holds for all $e \geq 1$, in fact,

$$|\{v \in A : |v| < (2e+1) + 5\}| \leq |\{v \in A : |v| < 2(e+1) + 5\}|$$
$$\leq (e+1) + g(e+1) = \frac{3(e+1)}{2} \leq 2e + 1.$$

During the construction, we will make $|\{v \in A : |v| < 2e + 5\}| \leq \frac{3e}{2}$ for all e. Then $\mathcal{R}_{2e}$ holds for all e and $\mathcal{R}_{2e+1}$ holds for all $e \geq 1$, it is easy to check $\mathcal{R}_1$ also holds.

2.3 The Coding Strategy

At stage s, for $v \in \mathcal{S}$ with $v \leq_\mathcal{S} v_s^0$, define $\Sigma^{R(A_s)}(v) = C_s(v)$ with use $\sigma(v, s)$, where $\sigma(v, s)$ is the coding marker defined below, and $\Sigma^{R(A)}(v) = \lim_{s \to \infty} \Sigma^{R(A_s)}(v)$ with use $\sigma(v) = \lim_{s \to \infty} \sigma(v, s)$. To maintain $\Sigma^{R(A)}(v) = C(v)$, we need to effectively find a coding marker $\sigma(v, s)$ from $\overline{R(A_s)}$, the complement of the realization set $R(A_s)$ of A_s, to put it into A_{s+1} when v is first enumerated into C at stage $s + 1$, and satisfying the overall requirement: the number of all coding markers of length $< 2e + 5$ in A is $\leq \frac{e}{2}$.

Before defining coding marker $\sigma(v,s)$, we first define a computable bijection ϕ from $\mathcal{S}$ to a set $\mathcal{O}$ containing exactly one element of each length and $|\phi(v)| \geq |v|$ for all $v \in \mathcal{S}$ as follows: for $i \in \omega$, let $o_i = \min \mathcal{S}_i$, where $\mathcal{S}_i$ is the set of binary strings of length i, then $o_0 = \lambda$ and o_i is the string with i many 0's if $i \geq 1$, the set $\mathcal{O} - \{o_i : i \in \omega\}$ has exactly one element of length i, define $\phi : \mathcal{S} \to \mathcal{O}; v_i^0 \mapsto o_i$, recall $\mathcal{S} = \{v_0^0 <_{\mathcal{S}} v_1^0 <_{\mathcal{S}} \cdots\}$. We now start to define $\sigma(v,s)$, notice that if $|\sigma(v,s)| \geq 4|\phi(v)| + 7$, then the coding markers of length $< 2e + 5$ in A can only be provided by those $\phi(v)$ with $|\phi(v)| < \frac{e-1}{2}$, moreover,

$$|\{v \in \mathcal{S} : |\phi(v)| < \frac{e-1}{2}\}| \leq \frac{e}{2}.$$

So there are at most $\frac{e}{2}$ many strings in $\mathcal{S}$ whose coding markers are of length $< 2e + 5$, and thus the number of all coding markers of length $< 2e + 5$ in A is $\leq \frac{e}{2}$. For each v and s, we choose $\sigma(v,s) \in \overline{R(A_s)} \cap [\bigcup_{i \geq 4|\phi(v)|+7} \mathcal{S}_i]$, and define

$\sigma(v,s) := \min(\overline{R(A_s)} \cap [\bigcup_{i \geq 4|\phi(v)|+7} \mathcal{S}_i])$, that is, $\sigma(v,s)$ is the minimal element of

set $\overline{R(A_s)} \cap [\bigcup_{i \geq 4|\phi(v)|+7} \mathcal{S}_i]$, where minimal is taking under the linear order $\leq_{\mathcal{S}}$ on domain of $\mathcal{S}$.

We now show how the functional $\Sigma^{R(A)}$ computes C. For $v \in \mathcal{S}$, let $v = v_{s_0}^0$, then at stage s_0, $\Sigma^{R(A_{s_0})}(v)$ is first defined as $C_{s_0}(v)$ with use $\sigma(v, s_0)$. Suppose at stage $s \geq s_0$, we have $\Sigma^{R(A_s)}(v) = C_s(v)$ with use $\sigma(v,s)$, positive requirements or coding requirements may enumerate elements $\leq_{\mathcal{S}} \sigma(v,s)$ into A at later stages to undefine $\Sigma^{R(A_s)}(v)$, say $t > s$ is the least stage such that a new $x \leq_{\mathcal{S}} \sigma(v,s)$ is put into A_t, then $\Sigma^{R(A_t)}(v)$ is undefined after enumerating x into A_t, and then $\Sigma^{R(A_t)}(v)$ is redefined as $C_t(v)$ with use $\sigma(v,t) \geq_{\mathcal{S}} \sigma(v,s)$ at the end of stage t. If there is a stage $t > s_0$ such that $v \in C_t - C_{t-1}$, we have $\Sigma^{R(A_{t-1})}(v) = C_{t-1}(v) = 0$ with use $\sigma(v, t-1)$ at the end of stage $t - 1$, then we enumerate the use $\sigma(v, t-1)$ into A_t to undefine $\Sigma^{R(A_t)}(v)$, and then redefine $\Sigma^{R(A_t)}(v) = 1$ with new use $\sigma(v,t) >_{\mathcal{S}} \sigma(v, t-1)$ at the end of stage t. So for the definition of $\Sigma^{R(A)}(v)$, its approximation values $\{\Sigma^{R(A_s)}(v) : s \geq s_0\}$ will change from 0 to 1 for at most once, and its final use will be $\sigma(v) = \min(\overline{R(A)} \cap [\bigcup_{i \geq 4|\phi(v)|+7} \mathcal{S}_i])$, as we will ensure all $\mathcal{R}_e$ requirements hold, $\overline{R(A)}$ is infinite, then $\overline{R(A)} \cap [\bigcup_{i \geq 4|\phi(v)|+7} \mathcal{S}_i]$ is a nonempty subset of $\mathcal{S}$, when the least element in this set appears in $\overline{R(A_s)}$ at some stage s, this least element is never enumerated into A as it is in $\overline{R(A)}$ which is a subset of the complement of A, so $\sigma(v,t) = \sigma(v,s)$ for all $t \geq s$, and $\Sigma^{R(A)}(v) = C_s(v) = C(v)$ with use $\sigma(v,s)$.

2.4   A $\mathcal{P}_e$-strategy with Permitting

At a stage s, for $v \in \mathcal{S}$ with $v \leq_{\mathcal{S}} v_s^0$, define $\Gamma^{C_s}(v) = A_s(v)$ with use v, and $\Gamma^C(v) = \lim_{s \to \infty} \Gamma^{C_s}(v)$ with use v. The number of permitted elements of length $< 2e + 5$ is at most e, so we only put elements with length $\geq 2e + 5$ of W_e into

A. For a subset B of $\mathcal{S}$ and an element v of $\mathcal{S}$, the symbol $B \upharpoonright_v$ stands for the set $\{u \in B : u \leq_{\mathcal{S}} v\}$.

To satisfy $\mathcal{P}_e$, a $\mathcal{P}_e$ strategy proceeds with cycles: cycle $k(k \in \omega)$ proceed as follows.

(1) Wait for a witness $x_k \in W_{e,s_k}$ at some stage s_k satisfying $x_k \notin A_{s_k}$, $|x_k| \geq 2e + 5$ and $x_k >_{\mathcal{S}} x_{k-1}$.
(2) Open cycle $k + 1$. Wait for $C_{s+1} \upharpoonright_{x_k} \neq C_s \upharpoonright_{x_k}$ at some stage $s + 1$.
(3) Now $\Gamma^{C_{s+1}}(x_k)$ is undefined, redefine $\Gamma^{C_{s+1}}(x_k) = 1$ and put x_k into A_{s+1}. Close all other cycles.

For $v \in \mathcal{S}$, let $v = v_{s_0}^0$, $\Gamma^{C_{s_0}}(v)$ is first defined as $A_{s_0}(v)$ with use v at stage s_0, and at stage $s \geq s_0$, the use of $\Gamma^{C_s}(v)$ is always v to make sure $\Gamma^C(v)$ totally defined. Assume we have $\Gamma^{C_s}(v) = A_s(v)$, if $t > s$ is the least stage such that a new element $\leq_{\mathcal{S}} v$ goes into C_t, then $\Gamma^{C_t}(v)$ is undefined and we redefine it as $A_t(v)$ with same use v at the end of stage t. For $s \geq s_0$, on the one hand, we enumerate v into A_{s+1} for positive requirements only when $\Gamma^{C_{s+1}}(v)$ is undefined by $C_{s+1} \upharpoonright_v \neq C_s \upharpoonright_v$, that is, some new element $v' \leq_{\mathcal{S}} v$ goes into C at stage $s + 1$, and then redefine $\Gamma^{C_{s+1}}(v) = 1$; on the other hand, v maybe enumerated into A_{s+1} as a coding element $\sigma(u, s)$ for some $u \in C_{s+1} - C_s$, as our $\sigma(u, s)$ is selected such that $|u| < |\phi(u)| < |\sigma(u, s)|$, then $u <_{\mathcal{S}} \sigma(u, s) = v$, we also have $C_{s+1} \upharpoonright_v \neq C_s \upharpoonright_v$, and $\Gamma^{C_{s+1}}(v)$ is undefined, so we have a chance to redefine $\Gamma^{C_{s+1}}(v) = 1$.

3 Construction

We say $\mathcal{P}_e$ *requires attention* at stage s if

(1) $\mathcal{P}_e$-strategy has already opened cycle 0 at some previous stage t_e.
(2) There is a cycle k of a basic $\mathcal{P}_e$-strategy which has already reached (2) at some previous stage $s_k > t_e$, that is, there is a witness $x_k \in W_{e,s_k}$ we found at stage s_k, under a basic $\mathcal{P}_e$-strategy, cycle $k + 1$ is opened at stage s_k and we started to wait for $C_{s_k} \upharpoonright_{x_k}$ to change; moreover, stage s is the least stage larger than t_e such that there is a cycle of $\mathcal{P}_e$-strategy arriving at (3), and this cycle is cycle k.

We now state the construction of A.

Stage 0: Let $A_0 := \emptyset$, $\overline{A_0} := \mathcal{S}$, and fix $\delta : \mathcal{S} \to \omega; v_i^0 \mapsto i$. Start $\mathcal{P}_0$-strategy: open cycle 0 for requirement $\mathcal{P}_0$, that is, we start to wait for witness x_0 at some stage s_0 satisfying $x_0 \in W_{0,s_0}$ with $x_0 \notin A_{s_0}$ and $|x_0| \geq 5$, and then proceed according to a basic $\mathcal{P}_0$-strategy unless $\mathcal{P}_0$ is satisfied.

Assume at the end of stage s, we have set A_s, and for v with $\delta(v) \leq s$, we have $\Sigma^{\mathrm{R}(A_s)}(v) = C_s(v)$ with use $\sigma(v, s) = \min(\overline{\mathrm{R}(A_s)} \cap [\bigcup_{i \geq 4|\phi(v)|+7} S_i])$ and $\Gamma^{C_s}(v) = A_s(v)$ with use v. We say requirement $\mathcal{P}_e(e \leq s)$ is satisfied at stage s if $W_{e,s} \cap A_s \neq \emptyset$, otherwise, $\mathcal{P}_e$ is unsatisfied.

Stage $s + 1$:

(i) *Coding.* If $C_{s+1}(v) \neq C_s(v)$, and $\delta(v) \leq s$, in order to change $\Sigma^{R(A_{s+1})}(v)$, we first put the use $\sigma(v, s)$ into A_{s+1} to undefine $\Sigma^{R(A_{s+1})}(v)$. Let $A'_{s+1} = A_s \cup \{\sigma(v, s)\}$ if $C_{s+1}(v) \neq C_s(v)$, and $A'_{s+1} = A_s$ otherwise.

(ii) *Permitting.* For those unsatisfied $\mathcal{P}_e$ with $e \leq s$ at stage s, if $W_{e,s+1} \cap A'_{s+1} \neq \emptyset$, $\mathcal{P}_e$ is satisfied now, close all existed cycles of $\mathcal{P}_e$-strategy. For those unsatisfied $\mathcal{P}_e$ with $e \leq s$ at stage s and $W_{e,s+1} \cap A'_{s+1} = \emptyset$, there are two cases:

    (1) If there is a least $e \leq s$ such that $\mathcal{P}_e$ requires attention at stage $s + 1$, assume cycle k is the first cycles of $\mathcal{P}_e$-strategy reaching (3), we have $C_{s_k} \restriction_{x_k} = C_s \restriction_{x_k} \neq C_{s+1} \restriction_{x_k}$, where stage s_k is the stage at which cycle k reached (2) by finding witness x_k, and we started to wait for $C_{s_k} \restriction_{x_k}$ to change. Now $\Gamma^{C_{s+1}}(x_k)$ is undefined, we enumerate x_k into A_{s+1} and redefine $\Gamma^{C_{s+1}}(x_k) = 1$ with use x_k, $\mathcal{P}_e$ is satisfied now and close all cycles for $\mathcal{P}_e$. Now $A_{s+1} = A'_{s+1} \cup \{x_k\}$, for each unsatisfied requirement $\mathcal{P}_j$ with $j \leq s$ at stage s, if $W_{j,s+1} \cap A_{s+1} \neq \emptyset$, it is satisfied now, close all its cycles if such cycles are still opened.

    (2) Otherwise, no $\mathcal{P}_e$ requires attention at stage $s + 1$ for $e \leq s$, let $A_{s+1} = A'_{s+1}$. If there is a least $j \leq s + 1$ such that $\mathcal{P}_j$ has no cycles opened, and $\mathcal{P}_j$ is not satisfied so far, that is, $W_{j,s+1} \cap A_{s+1} = \emptyset$, then we start a basic $\mathcal{P}_j$-strategy by opening cycle 0: we first start to wait for witness x_0 at some stage s_0 such that $x_0 \in W_{j,s_0}$ with $x_0 \notin A_{s_0}$ and $|x_0| \geq 2j + 5$, and then proceed as in a basic $\mathcal{P}_j$-strategy unless $\mathcal{P}_j$ is satisfied.

(iii) *Defining Σ and Γ.* For v with $\delta(v) \leq s + 1$, let $\Sigma^{R(A_{s+1})}(v) = C_{s+1}(v)$ with use $\sigma(v, s + 1) = \min(\overline{R(A_{s+1})} \cap [\bigcup_{i \geq 4|\phi(v)|+7} S_i])$, and $\Gamma^{C_{s+1}}(v) = A_{s+1}(v)$ with use v.

This completes the stage $s + 1$ of the construction and hence the construction of A.

4 Verification

Lemma 1. *For all e, (1) $\mathcal{R}_e$ holds; (2) $\mathcal{P}_e$ holds.*

Proof. (1) There are two types of elements in A, one is coding markers, the other is permitted elements. The number of coding markers of length $< 2e + 5$ in A is at most $\frac{e}{2}$, and the number of permitted elements of length $< 2e + 5$ in A is at most e, so $|\{v \in A : |v| < 2e + 5\}| \leq \frac{3e}{2}$, this implies $|\{v \in A : |v| < e + 5\}| \leq e$ according to $\mathcal{R}_e$-strategy.

(2) Suppose $\mathcal{P}_e$ fails, then W_e is infinite and $A \cap W_e = \emptyset$, we will show C is computable. Imagine how the construction looks like under $\mathcal{P}_e$ strategies: as W_e is infinite, there are infinitely many witnesses of length $\geq 2e + 5$ in W_e, so infinitely many cycles are opened for $\mathcal{P}_e$; each witness $x_k \in W_{e,s_k}$ in cycle k arrives at (2) but never reaches (3), that is, for each stage $t > s_k$, we have $C_t \restriction_{x_k} = C_{s_k} \restriction_{x_k}$, then for all k, $C \restriction_{x_k} = C_{s_k} \restriction_{x_k}$. Now for a given $v \in S$, choose such a stage s_k with witness $x_k >_S v$, then $C(v) = C_{s_k}(v)$. □

Lemma 2. $C = \Sigma^{\mathrm{R}(A)}$.

Proof. For each v, as $\overline{\mathrm{R}(A)} \cap [\bigcup_{i \geq 4|\phi(v)|+7} S_i] \neq \emptyset$, there is a least stage s such that $\sigma(v) = \min(\overline{\mathrm{R}(A)} \cap [\bigcup_{i \geq 4|\phi(v)|+7} S_i]) = \min(\overline{\mathrm{R}(A_s)} \cap [\bigcup_{i \geq 4|\phi(v)|+7} S_i]) = \sigma(v, s)$, then $\sigma(v) = \sigma(v, t)$ for all $t \geq s$. By $\sigma(v) \in \overline{\mathrm{R}(A)} \subseteq \overline{A}$, $C_t(v) = C_s(v)$ for all $t \geq s$, so $\Sigma^{\mathrm{R}(A)}(v) = \Sigma^{\mathrm{R}(A_s)}(v) = C_s(v) = C(v)$ with use $\sigma(v)$. $\square$

Lemma 3. $A = \Gamma^C$.

Proof. For $v \in S$, find the least stage s with $C \upharpoonright_v = C_s \upharpoonright_v$. Then $A(v) = A_s(v)$, in fact, if $A_s(v) = 1$, of course $A(v) = 1$; if $A_s(v) = 0$, on the one hand, v can not be put into A by permitting strategies at later stages, on the other hand, for all $u \in C_{t+1} - C_t$ with $t \geq s$, $\sigma(u, t) >_S u \geq_S v$, then $A_t(v) = 0$ for all $t \geq s$. So $A(v) = A_s(v) = \Gamma^{C_s}(v) = \Gamma^C(v)$ with use v. $\square$

As mentioned at the beginning of Sect. 2.1, $\mathrm{R}(A) \leq_T A$ [to see whether τ is in $\mathrm{R}(A)$ or not, we check whether there exists some substring τ' of τ in A, and the reduction follows as τ has only finitely many substrings]. Thus, by Lemma 3, $\mathrm{R}(A) \leq_T C$. Together with Lemma 2, we have $\mathrm{R}(A) \equiv_T C$. This completes the proof of Theorem 4, and hence the proof of Theorem 1.

5   Basic Ideas of Proving Theorem 2

For a field K, an algebra $\mathfrak{A}$ over K is a ring $(A, +_A, \cdot_A, 0_A, 1_A)$ with identity 1_A together with a K-vector space structure under abelian group $(A, +_A, 0_A)$ and a scalar multiplication $\cdot$ such that $\cdot$ is compatible with the multiplication of rings, that is, for all $k \in K$, $a, b \in A$, $k \cdot (a \cdot_A b) = (a \cdot k) \cdot_A b = a \cdot_A (k \cdot b) = (a \cdot_A b) \cdot k$.

Let $\mathcal{F} = K\{x_1, \cdots, x_m\}$, $m \geq 2$, be the free algebra on $\{x_1, \cdots, x_m\}$, where K is a finite field, then $\mathcal{F} = \bigoplus_{i \in \omega} \mathcal{F}_i$, where $\mathcal{F}_i$ is the subspace of $\mathcal{F}$ generated by words on $\{x_1, \cdots, x_m\}$ of length i. For each i, a nonzero polynomial in $\mathcal{F}_i$ is called a homogeneous polynomial of degree i. A subset of $\mathcal{F}$ is called homogeneous if any its nonzero polynomial is homogeneous. Note a set is homogeneous iff it is a subset of $\bigcup_{i \in \omega} \mathcal{F}_i$. An ideal (two-sided) of $\mathcal{F}$ is homogeneous if it is generated by a homogeneous set.

As the field K is finite, we have the following Proposition 1 which is an analogue of $\mathrm{R}(A) \leq_T A$ in the proof of Theorem 1.

Proposition 1. *Let H be a homogeneous set of $\mathcal{F}$ and (H) be the ideal generated by H, then $(H) \leq_T H$.*

In this section, we provide some basic idea of constructing a computable linear order $\leq_\mathcal{F}$ on $\mathcal{F}$ such that for each i, $\max(\mathcal{G}_i) <_\mathcal{F} \min(\mathcal{F}_{i+1})$, where $\mathcal{G}_i = \bigoplus_{0 \leq k \leq i} \mathcal{F}_k$. This property is crucial for us to deal with permitting and coding in our construction.

We first partition $\mathcal{F}$ into a computable sequence of disjoint finite subsets $Q_{(i,j)}$ with $i \leq j$, and then build the order level by level under this partition.

For $i \leq j$, define

$$\mathcal{G}_{(i,j)} := \bigoplus_{i \leq k \leq j} \mathcal{F}_k,$$

where $\mathcal{F}_k$ is the subspace of $\mathcal{F}$ generated by words on $\{x_1, \cdots, x_m\}$ of length k, $Q_{(i,j)}$ will be a subset of $\mathcal{G}_{(i,j)}$. As K is finite, $\mathcal{G}_{(i,j)}$ is finite. So we can fix a linear order $\leq_{\mathcal{G}_{(i,j)}}$ on $\mathcal{G}_{(i,j)}$ first for each (i,j).

Now let

$$Q_{(i,i)} := \mathcal{F}_i;$$

and for $i < j$, let

$$Q_{(i,j)} := \mathcal{G}_{(i,j)} - (\mathcal{G}_{(i,j-1)} \cup \mathcal{G}_{(i+1,j)}).$$

Then $\mathcal{F} = \bigcup_{i \leq j} Q_{(i,j)}$, and $Q_{(i,j)} \cap Q_{(k,l)} = \emptyset$ if $(i,j) \neq (k,l)$, where $(i,j) = (k,l)$ means $i = k$ and $j = l$.

Define an one to one map φ on $\{(i,j) : i \leq j\}$ by letting $\varphi(0,0) = 1$, and for $i \leq j$,

$$\varphi(i,j) = \begin{cases} (t+1)^2 - i, & \text{if } i+j = 2t; \\ (t+1)^2 + i + 1, & \text{if } i+j = 2t+1. \end{cases}$$

Then we can have an order of $\{Q_{(i,j)} : i \leq j\}$, where $\varphi(i,j)$ is the location number of $Q_{(i,j)}$ in this order. This provides an order $\leq_{\mathcal{F}}$ on $\mathcal{F}$: for $f, g \in \mathcal{F}$, if f, g are in the same $Q_{(i,j)}$, then $f \leq_{\mathcal{F}} g$ iff $f \leq_{\mathcal{G}_{(i,j)}} g$; if $f \in Q_{(i,j)}$ and $g \in Q_{(i',j')}$ with $(i,j) \neq (i',j')$, then $f <_{\mathcal{F}} g$ iff $\varphi(i,j) < \varphi(i',j')$. We can check $\max(\mathcal{G}_i) <_{\mathcal{F}} \min(\mathcal{F}_{i+1})$ for each i by using

$$\mathcal{G}_i \subseteq \bigcup_{\varphi(i',j') < \varphi(i+1,i+1)} Q_{(i',j')}.$$

This completes the construction of wanted computable linear ordering $\leq_{\mathcal{F}}$.

The following form of Golod-Shafarevich Theorem for $\mathcal{F}$ is used in [6] to build infinite dimensional algebras.

Theorem 5 (*Golod-Shafarevich Theorem*). *If a homogeneous set B of $\mathcal{F} = K\{x_1, \cdots, x_m\}$ satisfies the condition: there is an ε with $0 < \varepsilon \leq \frac{m}{2}$ such that for all $i \geq 2$, the number d_i of homogeneous polynomials of degree i in B is less than or equal to $\varepsilon^2 (m - 2\varepsilon)^{i-2}$, then the quotient algebra $\mathcal{F}/(B)$ has an infinite dimension, where (B) is the ideal generated by B.*

If a homogeneous c.e. subset B of $\mathcal{F}$ satisfies the condition of Golod-Shafarevich Theorem, and $\mathcal{F}/(B)$ is also algorithmically finite, then as in [6], $\mathcal{F}/(B)$ satisfies all conditions in the NFP Theorem, and hence $\mathcal{F}/(B)$ has no finitely presented expansions. As the word problem for $\mathcal{F}/(B)$ is Turing equivalent to (B), the ideal generated by B, to prove Theorem 2, we just need to prove that such (B) occurs in each nonzero c.e. degree, where a standard permitting and coding argument will be used within $\mathcal{F}$ under above computable linear order $\leq_{\mathcal{F}}$, as we did in the proof of Theorem 1.

Acknowledgement. G. Wu is partially supported by AcRF grants MOE2016-T2-1-083 from Ministry of Education of Singapore, RG29/14, M4011274 and RG32/16, M4011672 from Nanyang Technological University and Ministry of Education of Singapore.

References

1. Bergstra, J.A., Tucker, J.V.: Initial and final algebra semantics for data type specifications: two characterization theorems. SIAM J. Comput. **12**, 366–387 (1983)
2. Bergstra, J.A., Tucker, J.V.: Algebraic specifications of computable and semicomputable data types. Theor. Comput. Sci. **50**, 137–181 (1987)
3. Hirschfeldt, D.R., Khoussainov, B.: Finitely presented expansions of computably enumerable semigroups. Algebra Logic **51**, 435–444 (2012)
4. Kasymov, N.Kh.: Algebras with finitely approximable positively representable enrichments. Algebra Logic **26**, 441–450 (1987)
5. Khoussainov, B.: Randomness, computability, and algebraic specifications. Ann. Pure Appl. Logic **91**, 1–15 (1998)
6. Khoussainov, B., Miasnikov, A.: Finitely presented expansions of semigroups, groups, and algebras. Trans. Am. Math. Soc. **366**, 1455–1474 (2014)
7. Soare, R.I.: Recursively Enumerable Sets and Degrees. Springer, Heidelberg (1987)

Kernelization and Parameterized Algorithms for 3-Path Vertex Cover

Mingyu Xiao[(⊠)] and Shaowei Kou

School of Computer Science and Engineering,
University of Electronic Science and Technology of China, Chengdu 611731, China
myxiao@gmail.com, kou_sw@163.com

Abstract. A 3-path vertex cover in a graph is a vertex subset C such that every path of three vertices contains at least one vertex from C. The parameterized 3-path vertex cover problem asks whether a graph has a 3-path vertex cover of size at most k. In this paper, we give a kernel of $5k$ vertices and an $O^*(1.7485^k)$-time algorithm for this problem, both new results improve previous known bounds.

1 Introduction

A vertex subset C in a graph is called an *ℓ-path vertex cover* if every path of ℓ vertices in the graph contains at least one vertex from C. The ℓ-path vertex cover problem, to find an ℓ-path vertex cover of minimum size, has been studied in the literature [5,6]. When $\ell = 2$, this problem becomes the famous vertex cover problem and it has been well studied. In this paper we study the 3-path vertex cover problem. A 3-path vertex cover is also known as a 1-*degree-bounded deletion set*. The d-degree-bounded deletion problem [11,25,26] is to delete a minimum number of vertices from a graph such that the remaining graph has degree at most d. The 3-path vertex cover problem is exactly the 1-degree-bounded deletion problem. Several applications of 3-path vertex covers have been proposed in [6,16,27].

It is not hard to establish the NP-hardness of the 3-path vertex cover problem by reduction from the vertex cover problem. In fact, it remains NP-hard even in planar graphs [28] and in C_4-free bipartite graphs with vertex degree at most 3 [4]. There are several graph classes, in which the problem can be solved in polynomial time [2–4,6,7,14,15,18–20].

The 3-path vertex cover problem has been studied from approximation algorithms, exact algorithms and parameterized algorithms. There is a randomized approximation algorithm with an expected approximation ratio of $\frac{23}{11}$ [16]. In terms of exact algorithms, Kardoš et al. [16] gave an $O^*(1.5171^n)$-time algorithm to compute a maximum dissociation set in an n-vertex graph.

This work is supported by National Natural Science Foundation of China, under the grant 61370071, and the Fundamental Research Funds for the Central Universities, under the grant ZYGX2015J057.

T.V. Gopal et al. (Eds.): TAMC 2017, LNCS 10185, pp. 654–668, 2017.
DOI: 10.1007/978-3-319-55911-7_47

Chang et al. [8] gave an $O^*(1.4658^n)$-time algorithm and the result was further improved to $O^*(1.3659^n)$ later [27].

In parameterized complexity, this problem is fixed-parameter tractable by taking the size k of the 3-path vertex cover as the parameter. The running time bound of parameterized algorithm for this problem has been improved at least three times during the last one year. Tu [22] showed that the problem can be solved in $O^*(2^k)$ time. Wu [24] improved the result to $O^*(1.882^k)$ by using the measure-and-conquer method. Katrenič designed an $O^*(1.8172^k)$-time algorithm [17]. Very recently, Chang et. al. [29] gave an $O^*(1.7964^k)$-time polynomial-space algorithm and an $O^*(1.7485^k)$-time exponential-space algorithm. In this paper, we show that this problem can be solved in $O^*(1.7485^k)$ time and polynomial space Another important issue in parameterized complexity is kernelization. A kernelization algorithm is a polynomial-time algorithm which, for an input graph with a parameter (G, k) either concludes that G has no 3-path vertex cover of size k or returns an equivalent instance (G', k'), called a *kernel*, such that $k' \leq k$ and the size of G' is bounded by a function of k. Kernelization for the d-degree-bounded deletion problem has been studied in the literature [11,25]. For $d = 1$, Fellows et al.'s algorithm [11] implies a kernel of 15 k vertices for the 3-path vertex cover problem, and Xiao's algorithm [25] implies a kernel of 13 k vertices. There is another closed related problem, called the *3-path packing* problem. In this problem, we are going to check if a graph has a set of at least k vertex-disjoint 3-paths. When we discuss kernelization algorithms, most structural properties of the 3-path vertex cover problem and the 3-path packing problem are similar. Several previous kernelization algorithms for the 3-path packing problem are possible to be modified for the 3-path vertex cover problem. The bound of the kernel size of the 3-path packing problem has been improved for several times from the first bound of 15 k [21] to 7 k [23] and then to 6 k [10]. Recently, there is a paper claiming a bound of 5 k vertices for the 3-path packing problem in net-free graphs [9]. Although the paper [9] provides some useful ideas, the proof in it is incomplete and the algorithm may not stop. Several techniques for the 3-path packing problem in [23] and [9] will be used in our kernelization algorithm. We will give a kernel of 5 k vertices for the 3-path vertex cover problem.

Omitted proofs in this extended abstract can be found in the full version of this paper.

2   Preliminaries

We let $G = (V, E)$ denote a simple and undirected graph with $n = |V|$ vertices and $m = |E|$ edges. A singleton $\{v\}$ may be simply denoted by v. The vertex set and edge set of a graph G' are denoted by $V(G')$ and $E(G')$, respectively. For a subgraph (resp., a vertex subset) X, the subgraph induced by $V(X)$ (resp., X) is simply denoted by $G[X]$, and $G[V \setminus V(X)]$ (resp., $G[V \setminus X]$) is also written as $G \setminus X$. A vertex in a subgraph or a vertex subset X is also called a X *vertex*. For a vertex subset X, let $N(X)$ denote the set of *open neighbors* of X, i.e., the vertices in $V \setminus X$ adjacent to some vertex in X, and $N[X]$ denote the set of *closed neighbors* of X, i.e., $N(X) \cup X$. The *degree* of a vertex v in a graph G, denoted by $d(v)$, is defined to be the number of vertices adjacent to v in G. Two vertex-disjoint subgraphs X_1 and X_2 are *adjacent* if there is an edge with one endpoint in X_1 and the other in X_2. The number of connected components in a graph G is denoted by $Comp(G)$ and the number of connected components of size i in a graph G is denoted by $Comp_i(G)$. Thus, $Comp(G) = \sum_i Comp_i(G)$.

A *3-path*, denoted by P_3, is a simple path with three vertices and two edges. A vertex subset C is called a *3-path vertex cover* or a *P_3VC-set* if there is no 3-path in $G \setminus C$. Given a graph $G = (V, E)$, a *P_3-packing* $\mathcal{P} = \{L_1, L_2, ..., L_t\}$ of size t is a collection of vertex-disjoint P_3 in G, i.e., each element $L_i \in \mathcal{P}$ is a 3-path in G and $V(L_{i_1}) \cap V(L_{i_2}) = \emptyset$ for any two different 3-paths $L_{i_1}, L_{i_2} \in \mathcal{P}$. A P_3-packing is *maximal* if it is not properly contained in any strictly larger P_3-packing in G. The set of vertices in 3-paths in $\mathcal{P}$ is denoted by $V(\mathcal{P})$.

Let $\mathcal{P}$ be a P_3-packing and A be a vertex set such that $A \cap V(\mathcal{P}) = \emptyset$ and A induces a graph of maximum degree 1. We use A_i to denote the set of degree-i vertices in the induced graph $G[A]$ for $i = 0, 1$. A component of two vertices in $G[A]$ is called an A_1-edge. For each $L_i \in \mathcal{P}$, we use $A(L_i)$ to denote the set of A-vertices that are in the components of $G[A]$ adjacent to L_i. For a 3-path $L_i \in \mathcal{P}$, the degree-2 vertex in it is called the *middle vertex* of it and the two degree-1 vertices in it are call the *ending vertices* of it.

3   A Parameterized Algorithm

In this section we will design a parameterized algorithm for the 3-path vertex cover problem. Our algorithm is a branch-and-reduce algorithm that runs in $O^*(1.7485^k)$ time and polynomial space, improving all previous results. In branch-and-reduce algorithms, the exponential part of the running time is determined by the branching operations in the algorithm. In a branching operation, the algorithm solves the current instance I by solving several smaller instances. We will use the parameter k as the measure of the instance and use $T(k)$ to denote the maximum size of the search tree generated by the algorithm running on any instance with parameter at most k. A branching operation, which generates l small branches with measure decrease in the i-th branch being at least c_i, creates a recurrence relation $T(k) \leq T(k-c_1) + T(k-c_2) + \cdots + T(k-c_l) + 1$. The largest root of the function $f(x) = 1 - \sum_{i=1}^{l} x^{-c_i}$ is called the *branching factor* of the recurrence. Let γ be the maximum branching factor among all branching factors in the algorithm. The running time of the algorithm is bounded by $O^*(\gamma^k)$. More details about the analysis and how to solve recurrences can be found in the monograph [13]. Next, we first introduce our branching rules and then present our algorithm.

3.1   Branching Rules

We have four branching rules. The first branching rule is simple and easy to observe.

Branching rule (B1): *Branch on a vertex v to generate $|N[v]|+1$ branches by either*

(i) deleting v from the graph, including it to the solution set, and decreasing k by 1, or

(ii) deleting $N[v]$ from the graph, including $N(v)$ to the solution set, and decreasing k by $|N(v)|$, or

(iii) for each neighbor u of v, deleting $N[\{u,v\}]$ from the graph, including $N(\{u,v\})$ to the solution set, and decreasing k by $|N(\{u,v\})|$.

A vertex v is *dominated* by a neighbor u of it if v is adjacent to all neighbors of u. The following property of dominated vertices has been proved and used in [27].

Lemma 1. *Let v be a vertex dominated by u. If there is a minimum 3-path vertex cover C not containing v, then there is a minimum 3-path vertex cover C' of G such that $v, u \notin C'$ and $N(\{u,v\}) \subseteq C'$.*

Based on this lemma, we design the following branching rule.

Branching rule (B2): *Branch on a vertex v dominated by another vertex u to generate two instances by either*

(i) deleting v from the graph, including it to the solution set, and decreasing k by 1, or

(ii) deleting $N[\{u,v\}]$ from the graph, including $N(\{u,v\})$ to the solution set, and decreasing k by $|N(\{u,v\})| = |N(v)| - 1$.

For a vertex v, a vertex $s \in N_2(v)$ is called a *satellite* of v if there is a neighbor p of v such that $N[p] - N[v] = \{s\}$. The vertex p is also called the *parent* of the satellite s at v.

Lemma 2. *Let v be a vertex that is not dominated by any other vertex. If v has a satellite, then there is a minimum 3-path vertex cover C such that either $v \in C$ or $v, u \notin C$ for a neighbor u of v.*

Branching rule (B3): *Let v be a vertex that has a satellite but is not dominated by any other vertex. Branch on v to generate $|N[v]|$ instances by either*

(i) deleting v from the graph, including it to the solution set, and decreasing k by 1, or

(ii) for each neighbor u of v, deleting $N[\{u,v\}]$ from the graph, including $N(\{u,v\})$ to the solution set, and decreasing k by $|N(\{u,v\})|$.

Lemma 3. *Let v be a degree-3 vertex with a degree-1 neighbor u_1 and two adjacent neighbors u_2 and u_3. There is a minimum 3-path vertex cover C such that either $C \cup \{u_1, v\} = \emptyset$ or $C \cup \{u_1, u_2, u_3\} = \emptyset$.*

Branching rule (B4): *Let v be a degree-3 vertex with a degree-1 neighbor u_1 and two adjacent neighbors u_2 and u_3. Branch on v to generate two instances by either*

(i) deleting $N[\{u_1, v\}]$ from the graph, including $\{u_2, u_3\}$ to the solution set, and decreasing k by 2, or

(ii) deleting $N[\{u_2, u_3\}] \cup \{u_1\}$ from the graph, including $N(\{u_2, u_3\})$ to the solution set, and decreasing k by $|N(\{u_2, u_3\})|$.

3.2 The Algorithm

We will use P3VC(G, k) to denote our parameterized algorithm. The algorithm contains 7 steps. When we execute one step, we assume that all previous steps are not applicable anymore on the current graph. We will analyze each step after describing it.

Step 1 (Trivial cases). *If $k \leq 0$ or the graph is an empty graph, then return the result directly. If the graph has a component of maximum degree 2, find a minimum 3-path vertex cover S of it directly, delete this component from the graph, and decrease k by the size of S.*

After Step 1, each component of the graph contains at least four vertices. A degree-1 vertex v is called a *tail* if its neighbor u is a degree-2 vertex. Let v be a tail, u be the degree-2 neighbor of v, and w be the other neighbor of u. We show that there is a minimum 3-path vertex cover containing w but not containing any of u and v. At most one of u and v is contained in any minimum 3-path vertex cover C, otherwise $C \cup \{w\} \setminus \{u, v\}$ would be a smaller 3-path vertex cover. If none of u and v is in a minimum 3-path vertex cover C, then w must be in C to cover the 3-path uvw and then C is a claimed minimum 3-path vertex cover. If exactly one of u and v is contained in a minimum 3-path vertex cover C, then $C' = C \cup \{w\} \setminus \{u, v\}$ is a claimed minimum 3-path vertex cover.

Step 2 (Tails). *If there is a degree-1 vertex v with a degree-2 neighbor u, then return* p3vc$(G \setminus N[\{v, u\}], k - 1)$.

Step 3 (Dominated vertices of degree $\geq$ 3). *If there is a vertex v of degree ≥ 3 dominated by u, then branch on v with Rule (B2) to generate two branches*

$$\text{p3vc}(G \setminus \{v\}, k - 1) \quad and \quad \text{p3vc}(G \setminus N[\{v, u\}], k - |N(\{v, u\})|).$$

Lemma 1 guarantees the correctness of this step. Note that $|N(\{v, u\})| = d(v) - 1$. This step gives a recurrence

$$T(k) \leq T(k-1) + T(k - (d(v) - 1)) + 1, \tag{1}$$

where $d(v) \geq 3$. For the worst case that $d(v) = 3$, the branching factor of it is 1.6181.

A degree-1 vertex with a degree-1 neighbor will be handled in Step 1, a degree-1 vertex with a degree-2 neighbor will be handled in Step 2, and a degree-1 vertex with a neighbor of degree ≥ 3 will be handled in Step 3. So after Step 3, the graph has no vertex of degree ≤ 1. Next we consider degree≥ 4 vertices.

Step 4 (Vertices of degree ≥ 4 with satellites). *If there is a vertex v of $d(v) \geq 4$ having a satellite, then branch on v with Rule (B3) to generate $d(v)+1$ branches*

$$\text{p3vc}(G\backslash\{v\}, k-1) \quad and \quad \text{p3vc}(G\backslash N[\{v, u\}], k-|N(\{v, u\})|) \text{ for each } u \in N(v).$$

The correctness of this step is guaranteed by lemma 2. Note that there is no dominated vertex after Step 3. Each neighbor u of v is adjacent to at least one vertex in $N_2(v)$ and then $|N(\{v, u\})| \geq d(v)$.

This step gives a recurrence

$$T(k) \leq T(k-1) + d(v) \cdot T(k - d(v)) + 1, \tag{2}$$

where $d(v) \geq 4$. For the worst case that $d(v) = 4$, the branching factor of it is 1.7485.

After Step 4, if there is still a vertex of degree ≥ 4, we use the following branching rule. Note that now each neighbor u of v is adjacent to at least two vertices in $N_2(v)$ and then $|N(\{v, u\})| \geq d(v) + 1$.

Step 5 (Normal vertices of degree ≥ 4). *If there is a vertex v of $d(v) \geq 4$, then branch on v with Rule (B1) to generate $d(v) + 2$ branches*

$$\text{p3vc}(G \backslash \{v\}, k-1), \quad \text{p3vc}(G \backslash N[v], k - |N(v)|)$$
$$and \quad \text{p3vc}(G \backslash N[\{v, u\}], k - |N(\{v, u\})|) \text{ for each } u \in N(v).$$

Since $|N(\{v, u\})| \geq d(v) + 1$, this step gives a recurrence

$$T(k) \leq T(k-1) + T(k - d(v)) + d(v) \cdot T(k - (d(v) + 1)) + 1, \tag{3}$$

which $d(v) \geq 4$. For the worst case that $d(v) = 4$, the branching factor of it is 1.6930.

After Step 5, the graph has only degree-2 and degree-3 vertices. We first consider degree-2 vertices.

A path $u_0 u_1 u_2 u_3$ of four vertices is called a *chain* if the first vertex u_0 is of degree ≥ 3 and the two middle vertices are of degree 2. Note that there is no chain with $u_0 = u_3$ after Step 3. So when we discuss a chain we always assume that $u_0 \neq u_3$. A chain can be found in linear time if it exists. In a chain $u_0 u_1 u_2 u_3$, u_2 is a satellite of u_0 with a parent u_1.

Step 6 (Chains). *If there is a chain $u_0 u_1 u_2 u_3$, then branch on u_0 with Rule (B3). In the branch where u_0 is deleted and included to the solution set, u_1 becomes a tail and we further handle the tail as we do in Step 2.*

We get the following branches

$$\text{p3vc}(G \backslash N[\{u_1, u_2\}], k - 2)$$
$$and \quad \text{p3vc}(G \backslash N[\{u_0, u\}], k - |N(\{u_0, u\})|) \text{ for each } u \in N(u_0).$$

Note that $|N(\{u_0, u\})| \geq d(u_0)$ since there is no dominated vertex. We get a recurrence

$$T(k) \leq T(k-2) + d(u_0) \cdot T(k - d(u_0)) + 1,$$

where $d(u_0) \geq 3$. For the worst case that $d(u_0) = 3$, the branching factor of it is 1.6717.

After Step 6, each degree-2 vertex must have two nonadjacent degree-3 vertices. Note that no degree-2 is in a triangle if there is no dominated vertex.

Step 7 (Degree-2 vertices with a neighbor in a triangle). *If there is a degree-2 vertex v with $N(v) = \{u, w\}$ such that a neighbor u of it is in a triangle $u u_1 u_2$, then branch on w with Rule (B1) and then in the branch w is deleted and included in the solution set further branch on u with Rule (B4). We get the following branches*

$$\mathbf{p3vc}(G \setminus N[\{u, v\}], k - |N(\{u, v\})|),$$
$$\mathbf{p3vc}(G \setminus N[\{u_1, u_2\}] \cup \{u, w\}, k - |N(\{u_1, u_2\}) \cup \{w\}|), \text{ and}$$
$$\mathbf{p3vc}(G \setminus N[\{w, u'\}], k - |N(\{w, u'\})|) \text{ for each } u' \in N(w).$$

There two neighbors u and w of v are degree-3 vertices. Since there is no dominated vertex, for any edge $v_1 v_2$ it holds $|N(\{v_1, v_2\})| \geq \min\{d(v_1), d(v_2)\}$. We know that $|N(\{u, v\})| \geq d(u) = 3$, $|N(\{u_1, u_2\}) \cup \{w\}| \geq |N(\{u_1, u_2\})| \geq 3$ (since no degree-2 vertex is in a triangle) and $|N(\{w, u'\})| \geq d(w)$ for each $u' \in N(w)$. We get the following recurrence

$$T(k) \leq T(k-3) + T(k-3) + 3 \cdot T(k-3) + 1.$$

The branching factor of it is 1.7100.

After Step 7, no degree-3 vertex in a triangle is adjacent to a degree-2 vertex.

Step 8 (Degree-2 vertices v with a degree-3 vertex in $N_2(v)$). *If there is a degree-2 vertex v such that at least one of its neighbors u and w, say u, has a degree-3 neighbor u_1, then branch on u with Rule (B1) and in the branch where u is deleted and included to the solution set, branch on w with Rule (B2). We get the branches*

$$\mathbf{p3vc}(G \setminus \{u, v, w\}, k - 2), \; \mathbf{p3vc}(G \setminus N[\{w, v\}], k - |N(\{w, v\})|),$$
$$\text{and } \mathbf{p3vc}(G \setminus N[\{u, u'\}], k - |N(\{u, u'\})|) \text{ for each } u' \in N(u).$$

Note that $d(u) = d(w) = 3$. It holds $|N(\{w, v\})| \geq d(w) = 3$ and $|N(\{u, u'\})| \geq d(u) = 3$ for $u' \in N(u)$. Furthermore, we have that $|N(\{u, u_1\})| \geq 4$ because u and u_1 are degree-3 vertices not in any triangle. We get the following recurrence

$$T(k) \leq T(k-2) + T(k-3) + 2 \cdot T(k-3) + T(k-4).$$

The branching factor of it is 1.7456.

Lemma 4. *After Step 8, if the graph is not an empty graph, then each component of the graph is either a 3-regular graph or a bipartite graph with one side of degree-2 vertices and one side of degree-3 vertices.*

Lemma 5. *Let $G = (V_1 \cup V_2, E)$ be a bipartite graph such that all vertices in V_1 are of degree 2 and all vertices in V_2 are of degree 3. The set V_1 is a minimum 3-path vertex cover of G.*

Step 9 (Bipartite graphs). *If the graph has a component H being a bipartite graph with one side V_1 of degree-2 vertices and one side V_2 of degree-3 vertices, then return* p3vc$(G \setminus H, k - |V_1|)$.

Step 10 (3-regular graphs). *If the graph is a 3-regular graph, pick up an arbitrary vertex v and branch on it with Rule (B1).*

Lemma 4 shows that the above steps cover all the cases, which implies the correctness of the algorithm. Note that all the branching operations except Step 10 in the algorithm have a branching factor at most 1.7485. We do not analyze the branching factor for Step 10, because this step will not exponentially increase the running time bound of our algorithm. Any proper subgraph of a connected 3-regular graph is not a 3-regular graph. For each connected component of a 3-regular graph, Step 10 can be applied for at most one time and all other branching operations have a branching factor at most 1.7485. Thus each connected component of a 3-regular graph can be solved in $O^*(1.7485^k)$ time. Before getting a connected component of a 3-regular graph, the algorithm always branches with branching factors of at most 1.7485. Therefore,

Theorem 1. *The 3-path vertex cover problem can be solved in $O^*(1.7485^k)$ time and polynomial space.*

4   Kernelization

In this section, we show that the parameterized 3-path vertex cover problem allows a kernel with at most $5k$ vertices.

4.1   Graph Decompositions

The kernelization algorithm is based on a vertex decomposition of the graph, called *good decomposition*, which can be regarded as an extension of the crown decomposition [1]. Based on a good decomposition we show that an optimal solution to a special local part of the graph is contained in an optimal solution to the whole graph. Thus, once we find a good decomposition, we may be able to reduce the graph by adding some vertices to the solution set directly. We only need to find good decompositions in polynomial time in graphs with a large size to get problem kernels. Some previous rules to kernels for the parameterized 3-path packing problem [10,12,23] are adopted here to find good decompositions in an effective way.

Definition 1. *A good decomposition of a graph $G = (V, E)$ is a decomposition (I, C, R) of the vertex set V such that*

1. the induced subgraph $G[I]$ has maximum degree at most 1;
2. the induced subgraph $G[I \cup C]$ has a P_3-packing of size $|C|$;
3. no vertex in I is adjacent to a vertex in R.

Lemma 6. *A graph G that admits a good decomposition (I, C, R) has a P_3-vertex cover (resp., P_3-packing) of size k if and only if $G[R]$ has a P_3-vertex cover (resp., P_3-packing) of size $k - |C|$.*

Lemma 6 provides a way to reduce instances of the parameterized 3-path vertex cover problem based on a good decomposition (I, C, R) of the graph: deleting $I \cup C$ from the graph and adding C to the solution set. Here arise a question: how to effectively find good decompositions? It is strongly related to the quality of our kernelization algorithm. The kernel size will be smaller if we can polynomially compute a good decomposition in a smaller graph. Recall that we use $Comp(G')$ and $Comp_i(G')$ to denote the number of components and number of components with i vertices in a graph G', respectively. For a vertex subset A that induces a graph of maximum degree at most 1 and $j = \{1, 2\}$, we use $N_j(A) \subseteq N(A)$ to denote the set of vertices in $N(A)$ adjacent to at least one component of size j in $G[A]$, and $N_2'(A) \subseteq N_2(A)$ be the set of vertices in $N(A)$ adjacent to at least one component of size 2 but no component of size 1 in $G[A]$. We will use the following lemma to find good decompositions, which was also used in [9] to design kernel algorithms for the 3-path packing problem.

Lemma 7. *Let A be a vertex subset of a graph G such that each connected component of the induced graph $G[A]$ has at most 2 vertices. If*

$$Comp(G[A]) > 2|N(A)| - |N_2'(A)|, \tag{4}$$

then there is a good decomposition (I, C, R) of G such that $\emptyset \neq I \subseteq A$ and $C \subseteq N(A)$. Furthermore, the good decomposition (I, C, R) together with a P_3-packing of size $|C|$ in $G[I \cup C]$ can be computed in $O(\sqrt{n}m)$ time.

By using Lemma 7, we can get a linear kernel for the parameterized 3-path vertex cover problem quickly. We find an arbitrary maximal P_3-packing S and let $A = V \setminus V(S)$. We assume that S contains less than k 3-paths and then $|V(S)| < 3k$, otherwise the problem is solved directly. Note that $|N(A)| \subseteq |V(S)|$. If $|A| > 12k$, then $Comp(G[A]) \geq \frac{|A|}{2} > 6k > 2|V(S)| \geq 2|N(A)|$ and we reduce the instance by Lemma 7. So we can get a kernel of $15k$ vertices. This bound can be improved by using a special case of Lemma 7.

For a vertex subset A such that $G[A]$ has maximum degree at most 1. Let A_0 be the set of degree-1 vertices in $G[A]$. Note that $Comp(G[A_0]) = Comp_2(G[A])$ and $|N(A_0)| = |N_2(A_0)| = |N_2(A)|$. By applying Lemma 7 on A_0, we can get

Corollary 1. *Let A be a vertex subset of a graph G such that each connected component of the induced graph $G[A]$ has at most 2 vertices. Let $N_2(A) \subseteq N(A)$ be the set of vertices in $N(A)$ adjacent to at least one vertex in a component of size 2 in $G[A]$. If*

$$Comp_2(G[A]) > |N_2(A)|, \tag{5}$$

then there is a good decomposition (I, C, R) of G such that $\emptyset \neq I \subseteq A$ and $C \subseteq N(A)$. Furthermore, the good decomposition (I, C, R) together with a P_3-packing of size $|C|$ in $G[I \cup C]$ can be computed in $O(\sqrt{n}m)$ time.

Note that $|A| = Comp_1(G[A]) + 2 \cdot Comp_2(G[A])$. If $|A| > 9k$, then $Comp_1(G[A]) + 2 \cdot Comp_2(G[A]) = |A| > 9k > 3|V(S)| \geq 3|N(A)| \geq (2|N(A)| - |N'_2(A)|) + |N_2(A)|$ and at least one of (4) and (5) holds. Then by using Lemma 7 and Corollary 1, we can get a kernel of size $9k + 3k = 12k$. It is possible to bound $|N(A)|$ by k and then to get a kernel of size $3k + 3k = 6k$. To further improve the kernel size to $5\,k$, we need some sophisticated techniques and deep analyses on the graph structure.

4.2   A 5 k Kernel

In this section, we use "crucial partitions" to find good partitions. A vertex partition (A, B, Z) of a graph is called a *crucial partition* if it satisfies *Basic Conditions* and *Extended Conditions*. Basic Conditions include the following four items:

(B1) A induces a graph of degree at most 1;
(B2) B is the vertex set of a P_3-packing $\mathcal{P}$;
(B3) No vertex in A is adjacent to a vertex in Z;
(B4) $|Z| \leq 5 \cdot \gamma(G[Z])$, where $\gamma(G[Z])$ is the size of a minimum P_3VC-set in the induced subgraph $G[Z]$.

Before presenting the definition of Extended Conditions, we give some used definitions. We use $\mathcal{P}_j$ to denote the collection of 3-paths in $\mathcal{P}$ having j vertices adjacent to A-vertices ($j = 0, 1, 2, 3$). Then $\mathcal{P} = \mathcal{P}_0 \cup \mathcal{P}_1 \cup \mathcal{P}_2 \cup \mathcal{P}_3$. We use $\mathcal{P}^1$ to denote the collection of 3-paths $L \in \mathcal{P}$ such that $|A(L)| = 1$. We also partition $\mathcal{P}_1 \setminus \mathcal{P}^1$ into two parts:
let $\mathcal{P}_M \subseteq \mathcal{P}_1 \setminus \mathcal{P}^1$ be the collection of 3-paths with the middle vertex adjacent to some A-vertices;
let $\mathcal{P}_L \subseteq \mathcal{P}_1 \setminus \mathcal{P}^1$ be the collection of 3-paths L_i such that $|A(L_i)| \geq 2$ and one ending vertex of L_i is adjacent to some A-vertices.
    A vertex in a 3-path in $\mathcal{P}$ is *free* if it is not adjacent to any A-vertex. A 3-path in $\mathcal{P}_0$ is *bad* if it has at least two vertices adjacent to some free-vertex in a 3-path in $\mathcal{P}_L$ and *good* otherwise. A 3-path in $\mathcal{P}_L$ is *bad* if it is adjacent to a bad 3-path in $\mathcal{P}_0$ and *good* otherwise.
    Extended Conditions include the following seven items:

(E1) For each 3-path $L_i \in \mathcal{P} \setminus \mathcal{P}^1$, at most one vertex in L_i is adjacent to some vertex in A, i.e., $\mathcal{P} \setminus \mathcal{P}^1 = \mathcal{P}_0 \cup \mathcal{P}_1$;
(E2) No 3-path in $\mathcal{P}_M$ is adjacent to both of A_0-vertices and A_1-vertices;
(E3) No free-vertex in a 3-path in $\mathcal{P}_L$ is adjacent to a free-vertex in another 3-path in $\mathcal{P}_L$;
(E4) No free-vertex in a 3-path in $\mathcal{P}_L$ is adjacent to a free-vertex in a 3-path in $\mathcal{P}_M$;

(E5) Each 3-path in $\mathcal{P}^1$ has at most one vertex adjacent to a free-vertex in a 3-path in $\mathcal{P}_L$;

(E6) If a 3-path in $\mathcal{P}^0$ has at least two vertices adjacent to some free-vertex in a 3-path in $\mathcal{P}_L$, then all those free-vertices are from one 3-path in $\mathcal{P}_L$, i.e., each bad 3-path in $\mathcal{P}^0$ is adjacent to free-vertices in only one bad 3-path in $\mathcal{P}_L$;

(E7) No free-vertex in a 3-path in $\mathcal{P}_L$ is adjacent to a vertex in Z.

Lemma 8. *A crucial partition of the vertex set of any given graph can be found in polynomial time.*

After obtaining a crucial partition (A, B, Z), we use the following three reduction rules to reduce the graph. In fact, Extended Conditions are mainly used for the third reduction rule and the analysis of the kernel size.

Refinement Rule 1. *If the number of 3-paths in $\mathcal{P}$ is greater than $k - |Z|/5$, halt and report it as a no-instance.*

Note that each P_3VC-set of the graph G must contain at least $|Z|/5$ vertices in Z by Basic Condition (B4) and each P_3VC-set must contain one vertex from each 3-path in $\mathcal{P}$. If the number of 3-paths in $\mathcal{P}$ is greater than $k - |Z|/5$, then any P_3VC-set of the graph has a size greater than k.

Refinement Rule 2. *If $Comp_2(G[A]) > |N_2(A)|$ (the condition in Corollary 1) holds, then find a good decomposition by Corollary 1 and reduce the instance based on the good decomposition.*

Reduction Rule 2 is easy to observe. Next, we consider the last reduction rule. Let B^* be the set of free-vertices in good 3-paths in $\mathcal{P}_L$ and let A^* be the set of A_0-vertices adjacent to 3-paths in $\mathcal{P}^1$. Let $A' = A \cup B^* \setminus A^*$. By the definition of crucial decompositions, we can get that

Lemma 9. *The set A' still induces of a graph of maximum degree 1.*

Proof. Vertices in B^* are free-vertices and then any vertex in B^* is not adjacent to a vertex in A. Furthermore, no two free-vertices in B^* from two different 3-paths in $\mathcal{P}_L$ are adjacent by Extended Condition (E3). Since A induces a graph of maximum degree 1, we know that $A \cup B^*$ induces a graph of maximum degree 1. The set $A' = A \cup B^* \setminus A^*$ is a subset of $A \cup B^*$ and then A' induces of a graph of maximum degree 1. □

Based on Lemma 9, we can apply the following reduction rule.

Refinement Rule 3. *If $Comp(G[A']) > 2|N(A')| - |N'_2(A')|$ (the condition in Lemma 6 on set A') holds, then find a good decomposition by Lemma 6 and reduce the instance based on the good decomposition.*

Next, we assume that none of the three reduction rules can be applied and prove that the graph has at most $5k$ vertices.

We consider a crucial partition (A, B, Z) of the graph. Let k_1 be the number of 3-paths in $\mathcal{P}$. Since Reduction Rule 1 cannot be applied, we know that

$$k_1 \leq k - |Z|/5. \tag{6}$$

Since Reduction Rule 2 and Reduction Rule 3 cannot be applied, we also have the following two relations

$$Comp_2(G[A]) \leq |N_2(A)|, \tag{7}$$

and

$$Comp(G[A']) \leq 2|N(A')| - |N_2'(A')|. \tag{8}$$

By Extended Condition (E1), we know that $\mathcal{P} = \mathcal{P}_0 \cup \mathcal{P}_1 \cup \mathcal{P}^1 = \mathcal{P}_0 \cup \mathcal{P}_L \cup \mathcal{P}_M \cup \mathcal{P}^1$. Let x_1 and x_2 be the numbers of good and bad 3-paths in $\mathcal{P}_L$, respectively. Let y_i $(i = 0, 1)$ be the number of 3-paths in $\mathcal{P}_0$ with i vertices adjacent to some free-vertex in a 3-path in $\mathcal{P}_L$, and y_2 be the number of 3-paths in $\mathcal{P}_0$ with at least two vertices adjacent to some free-vertex in a 3-path in $\mathcal{P}_L$, i.e., the number of bad 3-paths in $\mathcal{P}_0$. Let z_1 and z_2 be the numbers of 3-paths in $\mathcal{P}_M$ adjacent to only A_0-vertices and only A_1-vertices, respectively. Let w_1 be the number of 3-paths in $\mathcal{P}^1$ adjacent to some free-vertex in a 3-path in $\mathcal{P}_L$ and w_2 be the number of 3-paths in $\mathcal{P}^1$ not adjacent to any free-vertex in a 3-path in $\mathcal{P}_L$. We get that

$$k_1 = x_1 + x_2 + y_0 + y_1 + y_2 + z_1 + z_2 + w_1 + w_2. \tag{9}$$

By Extended Conditions (E1) and (E2), we know that

$$|N(A)_2| \leq x_1 + x_2 + z_2. \tag{10}$$

Extended Condition (E6) implies the number of bad 3-paths in $\mathcal{P}_L$ is at most the number of bad 3-paths in $\mathcal{P}_0$, i.e.,

$$x_2 \leq y_2. \tag{11}$$

Each 3-path in $\mathcal{P}^1$ is adjacent to only one A_0-vertex. Since A^* is the set of A_0-vertices adjacent to 3-paths in $\mathcal{P}^1$, we know that $|A^*|$ is not greater than $w_1 + w_2$, i.e., the number of 3-paths in $\mathcal{P}^1$. By the definition of A', we know that

$$Comp(G[A']) \geq Comp(G[A]) + x_1 - (w_1 + w_2). \tag{12}$$

Next, we consider $|N(A')|$ and $|N_2'(A')|$. Note that each 3-path has at most one vertex adjacent to vertices in $A \setminus A^*$ by Extended Condition (E1). This property will also hold for the vertex set $A' = (A \setminus A^*) \cup B^*$. We prove the following two relations

$$|N(A')| \leq x_1 + x_2 + y_0 + y_1 + z_1 + z_2 + w_1, \tag{13}$$

and

$$|N_2'(A')| \geq y_1 + z_2 + w_1. \tag{14}$$

By Extended Conditions (E1) and (E3), we know that each 3-path in $\mathcal{P}_L$ has at most one vertex in $N((A \setminus A^*) \cup B^*) = N(A')$. By the definition of good 3-paths in $\mathcal{P}_0$, we know that each good 3-path in $\mathcal{P}_0$ has no vertex adjacent to vertices in A and has at most one vertex adjacent to vertices in B^* (which will be in a component of size 2 in $G[A']$). There are exactly y_1 vertices in good 3-paths in $\mathcal{P}_0$ adjacent to vertices in B^*. No vertex in a bad 3-path in $\mathcal{P}_0$ is adjacent to a vertex in $A \cup B^*$ by the definitions of bad 3-paths and B^*. Each 3-path in $\mathcal{P}_M$ has at most one vertex adjacent to A' by Extended Conditions (E1) and (E4). Only z_2 vertices in 3-paths in $\mathcal{P}_M$ are adjacent to vertices in A', all of which are vertices of degree-1 in $G[A']$. No vertex in a 3-path in $\mathcal{P}^1$ is adjacent to a vertex in $A \setminus A^* \supseteq A'$ by the definition of A^*. Furthermore, each 3-path in $\mathcal{P}^1$ has at most one vertex adjacent to vertices in B^* (which will be in a component of size 2 in $G[A']$) by Extended Condition (E5) and there are exactly w_1 vertices in 3-paths in $\mathcal{P}^1$ adjacent to vertices in B^*. No vertex in Z is adjacent to a vertex in $A \cup B^*$ by Basic Condition (B3) and Extended Condition (E7). Summing all above up, we can get (13) and (14).

Relations (8), (12), (13) and (14) imply

$$Comp(G[A]) \leq 2(x_2 + y_0 + z_1 + w_1) + x_1 + y_1 + z_2 + w_2. \tag{15}$$

According to (7) and (10), we know that

$$Comp_2(G[A]) \leq x_1 + x_2 + z_2. \tag{16}$$

Note that $|A| = Comp(G[A]) + Comp_2(G[A])$, we get

$$
\begin{aligned}
|A| &= Comp(G[A]) + Comp_2(G[A]) \\
&\leq 2(x_1 + x_2 + y_0 + z_1 + z_2 + w_1) + x_2 + y_1 + w_2 && \text{by (15) and (16)} \\
&\leq 2(x_1 + x_2 + y_0 + z_1 + z_2 + w_1) + y_2 + y_1 + w_2 && \text{by (11)} \\
&\leq 2k_1 && \text{by (9).}
\end{aligned}
$$

Note that $|B| = 3k_1$ and $k_1 \leq k - |Z|/5$ by (6). We get that

$$
\begin{aligned}
|V| &= |A| + |B| + |Z| \\
&\leq 5k_1 + |Z| \leq 5k.
\end{aligned}
$$

Theorem 2. *The parameterized 3-path vertex cover problem allows a kernel of at most $5k$ vertices.*

References

1. Abu-Khzam, F.N., Collins, R.L., Fellows, M.R., Langston, M.A.: Kernelization algorithms for the vertex cover problem: theory and experiments. In: ALENEX/ANALC, pp. 62–69 (2004)

2. Alekseev, V.E., Boliac, R., Korobitsyn, D.V., Lozin, V.V.: NP-hard graph problems and boundary classes of graphs. Theor. Comput. Sci. **389**(1–2), 219–236 (2007)
3. Asdre, K., Nikolopoulos, S.D., Papadopoulos, C.: An optimal parallel solution for the path cover problem on P_4-sparse graphs. J. Parallel Distrib. Comput. **67**(1), 63–76 (2007)
4. Boliac, R., Cameron, K., Lozin, V.V.: On computing the dissociation number and the induced matching number of bipartite graphs. Ars Combin. **72**, 241–253 (2004)
5. Brešar, B., Jakovac, M., Katrenič, J., Semanišin, G., Taranenko, A.: On the vertex k-path cover. Discrete Appl. Math. **161**(13–14), 1943–1949 (2013)
6. Brešar, B., Kardoš, F., Katrenič, J., Semanišin, G.: Minimum k-path vertex cover. Discrete Appl. Math. **159**(12), 1189–1195 (2011)
7. Cameron, K., Hell, P.: Independent packings in structured graphs. Math. Program. **105**(2–3), 201–213 (2006)
8. Chang, M.-S., Chen, L.-H., Hung, L.-J., Liu, Y.-Z., Rossmanith, P., Sikdar, S.: An $O^*(1.4658^n)$-time exact algorithm for the maximum bounded-degree-1 set problem. In: The 31st Workshop on Combinatorial Mathematics and Computation Theory, pp. 9–18 (2014)
9. Chang, M.-S., Chen, L.-H., Huang, L.-J.: A $5k$ kernel for P_2-packing in net-free graphs. In: International Computer Science and Engineering Conference, pp. 12–17. IEEE (2014)
10. Chen, J., Fernau, H., Shaw, P., Wang, J., Yang, Z.: Kernels for packing and covering problems. In: Snoeyink, J., Lu, P., Su, K., Wang, L. (eds.) AAIM/FAW -2012. LNCS, vol. 7285, pp. 199–211. Springer, Heidelberg (2012). doi:10.1007/978-3-642-29700-7_19
11. Fellows, M.R., Guo, J., Moser, H., Niedermeier, R.: A generalization of Nemhauser and Trotter's local optimization theorem. JCSS **77**(6), 1141–1158 (2011)
12. Fermau, H., Raible, D.: A parameterized perspective on packing paths of length two. J. Comb. Optim. **18**(4), 319–341 (2009)
13. Fomin, F.V., Kratsch, D.: Exact Exponential Algorithms. Springer, Berlin (2010)
14. Göring, F., Harant, J., Rautenbach, D., Schiermeyer, I.: On F-independence in graphs. Discussiones Math. Graph Theor. **29**(2), 377–383 (2009)
15. Hung, R.-W., Chang, M.-S.: Finding a minimum path cover of a distance-hereditary graph in polynomial time. Discrete Appl. Math. **155**(17), 2242–2256 (2007)
16. Kardoš, F., Katrenič, J.: On computing the minimum 3-path vertex cover and dissociation number of graphs. Theoret. Comput. Sci. **412**(50), 7009–7017 (2011)
17. Katrenič, J.: A faster FPT algorithm for 3-path vertex cover. Inf. Process. Lett. **116**(4), 273–278 (2016)
18. Lozin, V.V., Rautenbach, D.: Some results on graphs without long induced paths. Inf. Process. Lett. **88**(4), 167–171 (2003)
19. Orlovich, Y., Dolgui, A., Finke, G., Gordon, V., Werner, F.: The complexity of dissociation set problems in graphs. Disc. Appl. Math. **159**(13), 1352–1366 (2011)
20. Papadimitriou, C.H., Yannakakis, M.: The complexity of restricted spanning tree problems. J. ACM **29**(2), 285–309 (1982)
21. Prieto, E., Sloper, C.: Looking at the stars. Theor. Comp. Sci. **351**(3), 437–445 (2006)
22. Tu, J.: A fixed-parameter algorithm for the vertex cover P3 problem. Inf. Process. Lett. **115**(2), 96–99 (2015)
23. Wang, J., Ning, D., Feng, Q., Chen, J.: An improved kernelization for P_2-packing. Inf. Process. Lett. **110**(5), 188–192 (2010)

24. Wu, B.Y.: A measure and conquer approach for the parameterized bounded degree-one vertex deletion. In: Xu, D., Du, D., Du, D. (eds.) COCOON 2015. LNCS, vol. 9198, pp. 469–480. Springer, Heidelberg (2015). doi:10.1007/978-3-319-21398-9_37

25. Xiao, M.: On a generalization of Nemhauser and Trotter's local optimization theorem. J. Comput. Syst. Sci. **84**, 97–106 (2017)

26. Xiao, M.: A parameterized algorithm for bounded-degree vertex deletion. In: Dinh, T.N., Thai, M.T. (eds.) COCOON 2016. LNCS, vol. 9797, pp. 79–91. Springer, Heidelberg (2016). doi:10.1007/978-3-319-42634-1_7

27. Xiao, M., Kou, S.: Exact algorithms for the maximum dissociation set and minimum 3-path vertex cover problems. Theoret. Comput. Sci. **657**, 86–97 (2017)

28. Yannakakis, M.: Node-deletion problems on bipartite graphs. SIAM J. Comput. **10**(2), 310–327 (1981)

29. Chang, M-S., Chen, L-H., Hung, L-J., Rossmanith, p., Su, P-C.: Fixed-parameter algorithms for vertex cover P3. Discrete Optim. 19, 12–22 (2016)

Fast Searching on Cartesian Products of Graphs

Yuan Xue and Boting Yang[✉]

Department of Computer Science, University of Regina, Regina, SK, Canada
boting.yang@uregina.ca

Abstract. Given a graph that contains an invisible fugitive, the fast searching problem is to find the fast search number, i.e., the minimum number of searchers to capture the fugitive in the fast search model. In this paper, we give a new lower bound on the fast search number. Using the new lower bound, we prove an explicit formula for the fast search number of the Cartesian product of an Eulerian graph and a path. We also give formulas for the fast search number of variants of the Cartesian product. We present an upper bound and a lower bound on the fast search number of hypercubes, and extend the results to a broader class of graphs including toroidal grids.

1 Introduction

Motivated by applied problems in the real world and theoretical issues in computer science and mathematics, graph searching has become a hot topic. It has many models, such as edge searching, node searching, mixed searching, fast searching, etc. These models are basically defined by the class of graphs, the actions of searchers and fugitives, visibility of fugitives, and conditions of captures [1,3,4,8,9].

Megiddo et al. [12] first introduced the edge search problem. In the edge search model, there are three actions for searchers: placing a searcher on a vertex, removing a searcher from a vertex and sliding a searcher along an edge from one endpoint to the other. An edge is cleared if both of its endpoints are occupied by searchers or cleared by a sliding action. Kirousis and Papadimitriou [10] introduced the node search problem, in which there are two actions for searchers: placing and removing. An edge uv becomes cleared if both u and v are guarded by searchers. Bienstock and Seymour [2] introduced the mixed search problem, which is a combination of the edge searching and node searching.

Throughout this paper, we only consider finite undirected graphs with no loops or multiple edges. Let $G = (V, E)$ be a graph with vertex set V and edge set E. We also use $V(G)$ and $E(G)$ to denote the vertex set and edge set of G respectively. We use uv to denote an edge with endpoints u and v. For a vertex $v \in V$, the *degree* of v is the number of edges incident on v, denoted $\deg_G(v)$. A *leaf* is a vertex that has degree one. A vertex is *odd* when its degree is odd.

B. Yang—Research supported in part by an NSERC Discovery Research Grant, Application No.: RGPIN-2013-261290.

© Springer International Publishing AG 2017
T.V. Gopal et al. (Eds.): TAMC 2017, LNCS 10185, pp. 669–683, 2017.
DOI: 10.1007/978-3-319-55911-7_48

An *odd graph* is a graph with vertex degrees all odd. Similarly, a vertex is *even* when its degree is even. Define $V_{\text{odd}}(G) = \{v \in V : v \text{ is odd}\}$.

For a subset $V' \subseteq V$, we use $G[V']$ to denote the subgraph induced by V', which consists of all vertices of V' and all the edges between vertices in V'. We use $G - V'$ to denote the induced subgraph $G[V \setminus V']$. For a subset $E' \subseteq E$, we use $G - E'$ to denote the subgraph $(V, E \setminus E')$. Let $G_1 = (V_1, E_1)$ and $G_2 = (V_2, E_2)$ be two subgraphs of G. The *union* of two graphs G_1 and G_2 is the graph $G_1 \cup G_2 = (V_1 \cup V_2, E_1 \cup E_2)$. We use $G_1 + V_2$ to denote the induced subgraph $G[V_1 \cup V_2]$ and we also use $G_1 + E_2$ to denote the subgraph $(V_1 \cup V(E_2), E_1 \cup E_2)$, where $V(E_2)$ is the vertex set of E_2.

Given two graphs H_1 and H_2, the *Cartesian product* of H_1 and H_2, denoted by $H_1 \square H_2$, is the graph whose vertex set is the Cartesian product $V(H_1) \times V(H_2)$ of the two vertex sets $V(H_1)$ and $V(H_2)$, and in which two vertices $(u, v), (u', v') \in V(H_1) \times V(H_2)$ are adjacent in $H_1 \square H_2$ if and only if $u = u'$ and v is adjacent to v' in H_2, or $v = v'$ and u is adjacent to u' in H_1.

A *walk* is a list $v_0, e_1, v_1, \ldots, e_k, v_k$ of vertices and edges such that each edge e_i, $1 \leq i \leq k$, has endpoints v_{i-1} and v_i. A *path* is a walk in which every vertex appears once, except that its first vertex might be the same as its last. We use $v_0 v_1 \ldots v_k$ to denote a path with ends v_0 and v_k. A *cycle* is a path in which its first vertex is the same as its last vertex. We use $v_0 v_1 \ldots v_k v_0$ to denote a cycle with $k + 1$ vertices. We will also use P_n to denote a path with n vertices and C_n to denote a cycle with n vertices, respectively. A *trail* is a walk that does not contain the same edge twice. For a connected subgraph G' with at least one edge, an *Eulerian trail* of G' is a trail that traverses every edge of G' exactly once. A *circuit* is a trail that begins and ends on the same vertex. An *Eulerian circuit* is an Eulerian trail that begins and ends on the same vertex. A graph is called *Eulerian* if it contains an Eulerian circuit that traverses all its edges. We will use B_m to denote an Eulerian graph with m vertices. Note that we only consider finite graphs with no loops or multiple edges. So an Eulerian circuit or Eulerian subgraph contains at least three edges throughout this paper. A *trail cover* of a graph G is a family of edge-disjoint trails in G that contain every edge of G. The minimum number of such trails is called the *trail cover number* of G and is denoted by $\tau(G)$.

Dyer et al. [7] introduced the fast search model, in which the fugitive hides either on a vertex or along an edge. The fugitive can move at a high speed at any moment from a vertex to another vertex along a path that contains no searchers. We call an edge uv *contaminated* if uv may contain the fugitive. An edge uv that does not contain the fugitive is called *cleared*. In the fast search model, all edges are contaminated initially. One of the two actions can happen in each step: *placing* a searcher on a vertex or *sliding* a searcher along a contaminated edge from one endpoint to the other. An edge uv can be cleared in one of the following two ways: if u is occupied by at least two searchers, one of them slides along uv from u to v; or if u is occupied by only one searcher and uv is the only contaminated edge incident on u, the searcher on u slides to v along uv. Since searchers are allowed to slide only on contaminated edges, every edge can only be traversed exactly once. A *fast search strategy* of a graph is a sequence of placing and sliding actions that clear all edges of the graph. The *fast search number* of G, denoted by $\mathrm{fs}(G)$, is the smallest number of searchers needed to capture the fugitive in G.

Stanley and Yang [13] gave a linear time algorithm for computing the fast search number of Harlin graphs and their extensions. They also presented a quadratic time algorithm for computing the fast search number of cubic graphs, while the problem of finding the node search number of cubic graphs is NP-complete [11]. Yang [14] proved that the problem of finding the fast search number of a graph is NP-complete; and it remains NP-complete for Eulerian graphs. He also proved that the problem of determining whether the fast search number of G is a half of the number of odd vertices in G is NP-complete; and it remains NP-complete for planar graphs with maximum degree 4. Dereniowski et al. [6] characterized graphs for which 2 or 3 searchers are sufficient in the fast search model. They also proved an NP-completeness result.

This paper is organized as follows. In Sect. 2, we first consider the trail covering problem, and show its relations to the fast searching problem. We then give a new lower bound on the fast search number. Using the new lower bound, we give an explicit formula for the fast search number of the Cartesian product of an Eulerian graph and a path. In Sect. 3, we investigate the fast search number of hypercubes. We prove an upper bound and a lower bound respectively on the fast search number of hypercubes. Section 4 concludes the paper with some open problems.

2   Lower Bounds and Cartesian Products

In this section, we first give lower bounds on the fast search number. We then apply the lower bounds to proving a formula for the fast search number of the Cartesian product of an Eulerian graph and a path.

2.1   Lower Bounds

We first consider two lower bounds on $\mathrm{fs}(G)$.

Lemma 1 ([7]).   *For any connected graph G, $\mathrm{fs}(G) \geq \frac{1}{2}|V_{\mathrm{odd}}(G)|$.*

Lemma 2 ([13]).   *For any connected graph G with no leaves, $\mathrm{fs}(G) \geq \frac{1}{2}|V_{\mathrm{odd}}(G)| + 2$.*

We now establish relations between a graph and its subgraph under some constraints.

Lemma 3. *Given a graph G, let W be a subset of $V(G)$. If G' is a graph obtained from G by adding a pendant edge to each vertex in W, then $\mathrm{fs}(G) \leq \mathrm{fs}(G')$.*

Lemma 4. *Given a graph G, let W be a subset of $V_{\mathrm{odd}}(G)$. If H is a graph obtained from G by adding a pendant edge to every vertex in W, then $\mathrm{fs}(G) - \mathrm{fs}(H)$.*

The following lemma shows a relation between the trail cover number and the number of odd vertices.

Lemma 5. *Let G be a graph that contains at least one edge. Then $\tau(G) = \mu(G) + |V_{\text{odd}}(G)|/2$, where $\mu(G)$ is the number of connected components in G that are Eulerian.*

Proof. If $V_{\text{odd}}(G) = \emptyset$, then each connected component of G is Eulerian. Thus $\tau(G) = \mu(G)$. Suppose that $V_{\text{odd}}(G) \neq \emptyset$. Let G' be a graph obtained from G by deleting all connected components of G that are Eulerian. Note that the number of odd vertices in a graph is even. Since each odd vertex must be an end vertex of a trail in any trail cover of G', we know that $\tau(G') \geq |V_{\text{odd}}(G')|/2$.

We now show that there is a trail cover of G' whose cardinality is at most $|V_{\text{odd}}(G')|/2$. Let a and b be any two odd vertices in a connected component of G' and let P be a trail between them. After deleting all edges of P from G', the remaining graph $G' - E(P)$ has two less odd vertices than G', that is, $V_{\text{odd}}(G' - E(P)) = V_{\text{odd}}(G') \setminus \{a, b\}$. We repeat this process until the remaining graph H contains no odd vertices. Thus H is a graph with vertex degrees all even. Let S be the set of all trails deleted from G' and H' be the graph obtained from H by deleting all isolated vertices from H. For each connected component H'' in H', it must have a vertex that is contained in some trail of S, say P. Since H'' is Eulerian, we can merge H'' and P into a new trail P', and replace P of S by P'. In this way, every edge of G' is contained in a trail of S. Thus S is a trail cover of G' whose cardinality is $|V_{\text{odd}}(G')|/2$. Hence $\tau(G') \leq |V_{\text{odd}}(G')|/2$. Therefore, we have

$$\tau(G) = \mu(G) + \tau(G') = \mu(G) + |V_{\text{odd}}(G')|/2 = \mu(G) + |V_{\text{odd}}(G)|/2.$$

Theorem 1. *Let G be a connected graph and H be a graph obtained from G by adding two pendant edges on each vertex of G. If G has at least one odd vertex, then $\text{fs}(H) = \tau(H) = |V(G)| + \tau(G)$; otherwise, $\text{fs}(H) = \tau(H) = |V(G)|$.*

Corollary 1. *Let G be a connected graph that contains at least one edge and P be a path that contains at least two edges. Then $\text{fs}(G \square P) \leq |V(G)| + \tau(G)|P|$.*

A subset E' of the edge set of a connected graph G is an *edge cut* of G, if $G - E'$ is disconnected. We now use an edge cut to give a lower bound on the fast search number.

Theorem 2. *Let G be a connected graph and E_χ be an edge cut of G such that the graph $G - E_\chi$ consists of two connected components G_1 and G_2. If each edge of E_χ connects a vertex of $V(G_1)$ to a vertex of $V(G_2)$, then $\text{fs}(G) \geq \text{fs}(G_1 + E_\chi) + \text{fs}(G_2 + E_\chi) - |E_\chi|$.*

Proof. Let S be an optimal fast search strategy of G. We first consider the graph $G_1 + E_\chi$. We modify S to obtain a fast search strategy S_1 that can clear $G_1 + E_\chi$ in the following way: We first delete all actions from S that are related only to G_2, i.e., "placing a searcher on a vertex of G_2" or "sliding a searcher along an edge of G_2"; for each edge v_1v_2 of E_χ with $v_1 \in V(G_1)$ and $v_2 \in V(G_2)$, if it is cleared by the action of "sliding a searcher from v_2 to v_1" in S, then immediately before this sliding action, we insert a new placing action, i.e., "placing a searcher on v_2". Let m_1 be the total number of new placing actions added to S_1.

We now show how to use S_1 to clear $G_1 + E_\chi$. Considering an edge $v_1v_2 \in E_\chi$ with $v_1 \in V(G_1)$ and $v_2 \in V(G_2)$, if it is cleared by sliding a searcher from v_2 to v_1 in S, since we have inserted the action of placing a searcher on v_2 in S_1, then the searcher can be used to clear v_1v_2 by sliding from v_2 to v_1. Further, in S_1, since we keep all sliding actions of S on edges of $G_1 + E_\chi$, we know $G_1 + E_\chi$ can be cleared in the same way as in S. Therefore, S_1 can clear $G_1 + E_\chi$.

Similarly, we can also modify the fast search strategy S to obtain a fast search strategy S_2 that can clear $G_2 + E_\chi$. Let m_2 be the total number of new placing actions added to S_2.

From the above, we know that the total number of searchers used to clear $G_1 + E_\chi$ and $G_2 + E_\chi$ is $\text{fs}(G) + m_1 + m_2$. Thus, $\text{fs}(G_1 + E_\chi) + \text{fs}(G_2 + E_\chi) \leq \text{fs}(G) + m_1 + m_2$. It is easy to see that m_1 is the number of edges in E_χ that are cleared by sliding actions in S from G_2 to G_1, and m_2 is the number of edges in E_χ that are cleared by sliding actions in S from G_1 to G_2. Since every edge of E_χ can be traversed exactly once, we have $|E_\chi| = m_1 + m_2$. Therefore, $\text{fs}(G_1 + E_\chi) + \text{fs}(G_2 + E_\chi) \leq \text{fs}(G) + |E_\chi|$.

2.2 Cartesian Products of Eulerian Graphs and Paths

Recall that B_m is an Eulerian graph with $m \geq 3$ vertices and P_n is a path with n vertices. We will use $G_{m \times n}$ to denote $B_m \square P_n$. Let $v_{i,j}$ denote a vertex of $G_{m \times n}$, which corresponds to the vertex v_i on B_m and the vertex v_j on P_n. So we use B_m^j, $1 \leq j \leq n$, to denote the Eulerian graph in $G_{m \times n}$ with vertex set $\{v_{1,j}, v_{2,j}, \ldots, v_{m,j}\}$ (the j-th copy of B_m). Similarly, we use P_n^i, $1 \leq i \leq m$, to denote the path in $G_{m \times n}$ with vertex set $\{v_{i,1}, v_{i,2}, \ldots, v_{i,n}\}$ (the i-th copy of P_n). We first give an upper bound on $\text{fs}(G_{m \times n})$.

Lemma 6. *For $m \geq 3$ and $n \geq 2$, $\text{fs}(G_{m \times n}) \leq m + n$.*

Proof. Here is a fast search strategy that clears all edges of $G_{m \times n}$ using $m + n$ searchers.

1. Place a searcher λ_i on each vertex $v_{i,1} \in V(G_{m \times n})$, $1 \leq i \leq m$, and place a searcher γ_j on each vertex $v_{1,j} \in V(G_{m \times n})$, $1 \leq j \leq n$.
2. Slide γ_1 along the Eulerian circuit of B_m^1 to clear all its edges. Let $j = 1$.
3. Slide each λ_i to B_m^{j+1}, $1 \leq i \leq m$, and then slide γ_{j+1} along the Eulerian circuit of B_m^{j+1} to clear all its edges. If $j + 1 = n$, then stop; otherwise, $j \leftarrow j + 1$ and repeat step 3.

Since the above fast search strategy has $m + n$ placing actions, we know that $\text{fs}(G_{m \times n}) \leq m + n$.

Lemma 7. *For $m \geq 3$, $\text{fs}(G_{m \times 2}) = m + 2$.*

In order to obtain a lower bound for fs($G_{m\times 3}$), we need to consider a graph obtained from $G_{m\times n}$ by adding a pendant edge to every vertex of B_m^n (see Figs. 1 and 2). This new graph is denoted by $G'_{m\times n}$. Since $G'_{m\times 1}$ has $2m$ odd vertices, it follows from Lemma 1 that fs($G'_{m\times 1}$) $\geq m$. In Lemma 9, we will give a fast search strategy for $G'_{m\times 1}$ that uses m searchers. Before that, we need a structure property of the Eulerian subgraph in $G'_{m\times 1}$, which is described in Lemma 8.

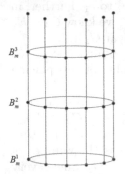

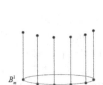

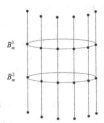

Fig. 1. An instance of $G'_{6\times 3}$.   **Fig. 2.** An instance of $G'_{6\times 1}$.   **Fig. 3.** An instance of $G''_{6\times 2}$.

Let B be an Eulerian graph and a be a vertex of B. Let $d = \deg_B(a)/2$. We repeat the following process d times until the degree of a is dropped to 0: select a cycle containing a, which has the shortest length among all the cycles containing a, and then remove all the edges of the cycle from B. Let $C_a^1, C_a^2, \ldots, C_a^d$ be the cycles selected in each iteration. The cycle C_a^d has the following property.

Lemma 8. *Let a be a vertex of an Eulerian graph B. Let $C_a^1, C_a^2, \ldots, C_a^d$ be the cycles described above. Then C_a^d contains two neighbors of the vertex a which are not contained in any C_a^j, $j < d$.*

From Lemma 8, we can prove the following results.

Lemma 9. fs($G'_{m\times 1}$) $= m$.

Proof. It follows from Lemma 1 that fs($G'_{m\times 1}$) $\geq m$. So we only need to describe a fast search strategy that uses m searchers to clear $G'_{m\times 1}$. Let B be the Eulerian graph obtained from $G'_{m\times 1}$ by deleting all its leaves. Let a be a vertex that has the minimum degree among all vertices of B, and let $d = \deg_B(a)/2$. Let $u \in V(B)$ be a neighbor of a and $C_a^1, C_a^2, \ldots, C_a^d$ be the cycles in Lemma 8 such that $u \in V(C_a^d)$ and $u \notin V(C_a^j)$ for $j < d$. If $d \geq 2$, then let H_u be a connected component that contains u after all edges of $\cup_{1\leq i\leq d-1}E(C_a^i)$ are deleted from B, and let $\overline{H}_u$ be a subgraph of B obtained from B by deleting all edges of H_u from B. Note that both H_u and $\overline{H}_u$ are Eulerian, and $E(H_u)$ and $E(\overline{H}_u)$ form a partition of $E(B)$. If $d = 1$, then let $H_u = B$. Now we give a fast search strategy for $G'_{m\times 1}$ that uses m searchers.

1. Place a searcher on every leaf of $G'_{m\times1}$, except the leaf neighbor of u; then slide these searchers to their non-leaf neighbors. Place another searcher on a. If $H_u = B$, go to Step 3.
2. Note that $\overline{H}_u$ is an Eulerian subgraph, in which a is occupied by two searchers and each other vertex is occupied by one searcher. Slide one of the two searchers on a from a to itself along all edges of $\overline{H}_u$.
3. Note that H_u is an Eulerian subgraph with $\deg_{H_u}(a) = 2$, in which a is occupied by two searchers, u is not occupied, and each other vertex is occupied by one searcher. Slide one of the two searchers on a from a to u along the edge au, and slide the other searcher from a to u along all edges of H_u except the edge au. After H_u is cleared, slide one searcher from u to its leaf neighbor on $G'_{m\times1}$.

Since only m searchers are placed on $G'_{m\times1}$ in Step 1, the above strategy clears $G'_{m\times1}$ using m searchers.

Lemma 10. *Each optimal fast search strategy of $G'_{m\times1}$ has the following properties: (1) the first cleared edge is cleared by sliding a searcher from a leaf to its neighbor; and (2) the last cleared edge is cleared by sliding a searcher from a non-leaf vertex to its leaf neighbor.*

From Theorem 2, Lemmas 4, 7, 9 and 10, we can prove the following result.

Lemma 11. *For $m \geq 3$, $\mathrm{fs}(G'_{m\times3}) = m + 3$.*

Proof. Recall that $G'_{m\times3}$ is a graph obtained from $B_m \square P_3$ by adding a pendant edge to every vertex of B_m^3 (see Fig. 1). Let $G'_{m\times1}$ be the subgraph of $G'_{m\times3}$ which contains B_m^1 (see Fig. 2) and $G'''_{m\times2}$ be the subgraph of $G'_{m\times3}$ after deleting all edges on B_m^1 (see Fig. 3). Recall that B_m^j, $1 \leq j \leq 3$, is an Eulerian graph with vertex set $\{v_{1,j}, v_{2,j}, \ldots, v_{m,j}\}$. For convenience, we will use $v_{i,4}$, $1 \leq i \leq m$, to denote the leaf neighbor of vertex $v_{i,3}$.

Note that $G''_{m\times2}$ is the graph obtained from $G_{m\times2}$ by adding a pendant edge to every odd vertex of $G_{m\times2}$. From Lemmas 4 and 7, we have $\mathrm{fs}(G''_{m\times2}) = \mathrm{fs}(G_{m\times2}) = m+2$. So it follows from Theorem 2 and Lemma 9 that $\mathrm{fs}(G'_{m\times3}) \geq \mathrm{fs}(G''_{m\times2}) + \mathrm{fs}(G'_{m\times1}) - m = m + 2$.

For the sake of contradiction, we assume that $\mathrm{fs}(G'_{m\times3}) = m + 2$. Then let $\mathcal{S}_{G'_{m\times3}}$ be an optimal fast search strategy of $G'_{m\times3}$ that uses $m + 2$ searchers. Note that all placing actions in $\mathcal{S}_{G'_{m\times3}}$ take place before all sliding actions. So each odd vertex of $G'_{m\times3}$ contains at least one searcher either before the first sliding action or at the end of $\mathcal{S}_{G'_{m\times3}}$. Thus, $V_{\mathrm{odd}}(G'_{m\times3})$ should contain at least m searchers at the beginning or at the end. If $V_{\mathrm{odd}}(G'_{m\times3})$ contains at least m searchers at the end of $\mathcal{S}_{G'_{m\times3}}$, then, using the method in the proof of Theorem 4.4 in [13], we can reverse the strategy $\mathcal{S}_{G'_{m\times3}}$ so that $V_{\mathrm{odd}}(G'_{m\times3})$ contains at least m searchers initially. So we can assume that $V_{\mathrm{odd}}(G'_{m\times3})$ contains m searchers before the first sliding action in $\mathcal{S}_{G'_{m\times3}}$.

Note that the edge set $\{v_{i,1}v_{i,2} \mid 1 \leq i \leq m\}$ is an edge cut of $G'_{m\times3}$. We can modify the strategy $\mathcal{S}_{G'_{m\times3}}$, using the method described in the first paragraph of the proof of Theorem 2, to obtain two separate fast search strategies $\mathcal{S}_{G'_{m\times1}}$ and $\mathcal{S}_{G''_{m\times2}}$ for $G'_{m\times1}$ and $G''_{m\times2}$, respectively. Since $\{v_{i,1}v_{i,2} \mid 1 \leq i \leq m\}$ contains m edges, we need to add m additional placing actions to obtain $\mathcal{S}_{G'_{m\times1}}$ and $\mathcal{S}_{G''_{m\times2}}$. Totally, $2m+2$ searchers are used in $\mathcal{S}_{G'_{m\times1}}$ and $\mathcal{S}_{G''_{m\times2}}$. Because $\mathrm{fs}(G'_{m\times1}) = m$ and $\mathrm{fs}(G''_{m\times2}) = m+2$, we know $\mathcal{S}_{G'_{m\times1}}$ uses at least m searchers and $\mathcal{S}_{G''_{m\times2}}$ uses at least $m+2$ searchers. Therefore, $\mathcal{S}_{G'_{m\times1}}$ uses exactly m searchers and $\mathcal{S}_{G''_{m\times2}}$ uses exactly $m+2$ searchers, which also means that $\mathcal{S}_{G'_{m\times1}}$ and $\mathcal{S}_{G''_{m\times2}}$ are both optimal. We can assume that all placing actions in $\mathcal{S}_{G'_{m\times1}}$ and $\mathcal{S}_{G''_{m\times2}}$ are carried out before all sliding actions.

From Lemma 10, the optimal fast search strategy $\mathcal{S}_{G'_{m\times1}}$ must have the following properties: The first cleared edge of $G'_{m\times1}$ must be cleared by sliding a searcher from a leaf, say $v_{i_1,2}$, to its neighbor $v_{i_1,1}$; and the last cleared edge of $G'_{m\times1}$ must be cleared by sliding a searcher from a non-leaf vertex, say $v_{i_2,1}$, to its leaf neighbor $v_{i_2,2}$.

Since $G'_{m\times1}$ and $G''_{m\times2}$ share each edge $v_{i,1}v_{i,2}$, $1 \leq i \leq m$, we know $\mathcal{S}_{G'_{m\times1}}$ and $\mathcal{S}_{G''_{m\times2}}$ must share the sliding action on each edge $v_{i,1}v_{i,2}$, $1 \leq i \leq m$. Thus $\mathcal{S}_{G''_{m\times2}}$ must satisfy the following conditions: The first cleared edge in the set $\{v_{i,1}v_{i,2} \mid 1 \leq i \leq m\}$ must be cleared by sliding a searcher from $v_{i_1,2}$ to $v_{i_1,1}$; and the last cleared edge in the set $\{v_{i,1}v_{i,2} \mid 1 \leq i \leq m\}$ must be cleared by sliding a searcher from $v_{i_2,1}$ to $v_{i_2,2}$.

From the assumption for $\mathcal{S}_{G''_{m\times3}}$, we know that $V_{\mathrm{odd}}(G''_{m\times2})$ contains at least m searchers before the first sliding action in $\mathcal{S}_{G''_{m\times2}}$. Thus, B_m^2 and B_m^3 contain at most two searchers before the first sliding action of $\mathcal{S}_{G''_{m\times2}}$. Moreover, B_m^2 should contain at least one searcher before the first sliding action of $\mathcal{S}_{G''_{m\times2}}$. This is because the first cleared edge in the set $\{v_{i,1}v_{i,2} \mid 1 \leq i \leq m\}$ is cleared by sliding a searcher from $v_{i_1,2}$ to $v_{i_1,1}$. If B_m^2 does not contain any searcher before the first sliding action, then we know no searcher can slide from $v_{i_1,2}$ to $v_{i_1,1}$. Suppose that B_m^2 contains two searchers before the first sliding action of $\mathcal{S}_{G''_{m\times2}}$. One of the following two cases must happen.

Case 1. There are two vertices of B_m^2, each of which contains one searcher before the first sliding action of $\mathcal{S}_{G''_{m\times2}}$. It is easy to see that the two searchers cannot move until another searcher slides to them. We also know that each searcher placed on $v_{j,1}$, $1 \leq j \leq m$, cannot move until a searcher slides from $v_{i_1,2}$ to $v_{i_1,1}$. Searchers that are placed on a subset of $\{v_{j,4} \mid 1 \leq j \leq m\}$ can slide to B_m^3. However, they would be stuck on B_m^3 since B_m^3 contains no searchers. Thus, no edges between B_m^2 and B_m^3 can be cleared, which is a contradiction.

Case 2. There is a vertex $v_{k,2}$, $1 \leq k \leq m$, of B_m^2 that contains two searchers before the first sliding action of $\mathcal{S}_{G''_{m\times2}}$. Then we have three subcases.

Case 2.1. The first cleared edge incident on $v_{k,2}$ is cleared by sliding one of the two searchers from $v_{k,2}$ to $v_{k,1}$ (in this case, $k = i_1$). After this action, the other searcher contained in $v_{k,2}$ cannot move until another searcher slides to $v_{k,2}$. Further, since all the vertices of B_m^2 and B_m^3 except $v_{k,2}$ contain no searcher initially, we know searchers sliding from $v_{i,1}$ to $v_{i,2}$, $1 \leq i \leq m$, or from $v_{j,4}$ to

$v_{j,3}$, $1 \leq j \leq m$, are stuck on B_m^2 or B_m^3 respectively. Thus, no edges between B_m^2 and B_m^3 can be cleared. This is a contradiction.

Case 2.2. The first cleared edge incident on $v_{k,2}$ is cleared by sliding one of the two searchers from $v_{k,2}$ to $v_{k,3}$. After this action, the other searcher contained in $v_{k,2}$ cannot move until another searcher slides to $v_{k,2}$. If a searcher slides from $v_{j,3}$ to $v_{j,2}$, $1 \leq j \leq m$ and $j \neq k$, then we know the searcher must be stuck on $v_{j,2}$ since then. Note that the first cleared edge between B_m^2 and B_m^1 is cleared by sliding a searcher from $v_{i_1,2}$ to $v_{i_1,1}$. But no searcher can slide from $v_{i_1,2}$ to $v_{i_1,1}$. This is a contradiction.

Case 2.3. The first cleared edge incident on $v_{k,2}$ is cleared by sliding one of the two searchers from $v_{k,2}$ to $v_{k',2}$, which is a neighbor of $v_{k,2}$ on B_m^2. Similar to Case 1, no edges between B_m^2 and B_m^3 can be cleared, which brings a contradiction.

Therefore, B_m^2 must contain exactly one searcher before the first sliding action of $\mathcal{S}_{G''_{m \times 2}}$. Further, B_m^3 also contains exactly one searcher before the first sliding action of $\mathcal{S}_{G''_{m \times 2}}$; otherwise, no searcher can slide from B_m^3 to B_m^2. Note that the remaining m searchers are placed on m odd vertices of $G''_{m \times 2}$ respectively. Since each leaf contains at most one searcher all the time in an optimal strategy, we know these m searchers are placed on m distinct odd vertices of $G''_{m \times 2}$.

Before the first sliding action of $\mathcal{S}_{G''_{m \times 2}}$, let $v_{\ell_1,2}$ be a vertex of B_m^2 that contains a searcher, and let $v_{\ell_2,3}$ be a vertex of B_m^3 that contains a searcher. Since $\deg(v_{\ell_1,2}) = \deg(v_{\ell_2,3}) = 4$, both $v_{\ell_1,2}$ and $v_{\ell_2,3}$ will be occupied by at least one searcher throughout $\mathcal{S}_{G''_{m \times 2}}$. Note that all other m searchers will stop on the m leaves of $G''_{m \times 2}$, on which no searchers are placed before the first sliding action of $\mathcal{S}_{G''_{m \times 2}}$. Hence, at the end of $\mathcal{S}_{G''_{m \times 2}}$, B_m^2 contains exactly one searcher that occupies $v_{\ell_1,2}$, and B_m^3 contains exactly one searcher that occupies $v_{\ell_2,3}$. If no searcher slides from $v_{\ell_1,3}$ to $v_{\ell_1,2}$, then searchers sliding from B_m^3 to B_m^2 would be stuck on B_m^2 and no searcher can clear the edge $v_{i_1,2}v_{i_1,1}$. Thus the edge $v_{\ell_1,2}v_{\ell_1,3}$ is cleared by sliding a searcher from $v_{\ell_1,3}$ to $v_{\ell_1,2}$. Similarly, the edge $v_{\ell_2,3}v_{\ell_2,4}$ is cleared by sliding a searcher from $v_{\ell_2,4}$ to $v_{\ell_2,3}$; otherwise, searchers sliding from a subset of $\{v_{j,4} \mid 1 \leq j \leq m\}$ to B_m^3 would be stuck on B_m^3.

Let t denote the moment just after the last contaminated edge, i.e., $v_{i_2,1}v_{i_2,2}$, in $\{v_{i,1}v_{i,2} \mid 1 \leq i \leq m\}$ is cleared. Since the edge $v_{i_2,1}v_{i_2,2}$ is cleared by sliding a searcher from $v_{i_2,1}$ to $v_{i_2,2}$, we know that at the moment t, $v_{i_2,2}$ should contain at least one searcher.

If all edges between B_m^2 and B_m^3 are cleared at the moment t, B_m^2 would contain at least two searchers at the end because $v_{\ell_1,2}$ always contains a searcher throughout $\mathcal{S}_{G''_{m \times 2}}$. This is a contradiction. If there are edges between B_m^2 and B_m^3 that are not cleared at the moment t, we have two cases.

Case 1. All edges of B_m^2 are cleared at t. Then $v_{i_2,2}$ contains two searchers at t. We have two subcases:

Case 1.1. $i_2 = \ell_1$. Because the edge $v_{\ell_1,2}v_{\ell_1,3}$ is cleared by sliding a searcher from $v_{\ell_1,3}$ to $v_{\ell_1,2}$, the searchers contained in $v_{i_2,2}$ cannot slide to B_m^3 along $v_{i_2,2}v_{i_2,3}$. Since all edges of B_m^2 are cleared, we know $v_{i_2,2}$ would contain at least two searchers at the end of $\mathcal{S}_{G''_{m \times 2}}$, which is a contradiction.

Case 1.2. $i_2 \neq \ell_1$. Since all edges of B_m^2 are cleared, there is at most one contaminated edge incident on $v_{i_2,2}$. Thus, $v_{i_2,2}$ must contain at least one searcher at the end of $\mathcal{S}_{G''_{m \times 2}}$. This is also a contradiction.

Case 2. There are edges of B_m^2 that are not cleared at t. Consider a sliding action that leaves all edges of B_m^2 cleared. Let $v_{j_1,2}v_{j_2,2}$ denote the last cleared edge of B_m^2 which is cleared by sliding a searcher from $v_{j_1,2}$ to $v_{j_2,2}$. Then $v_{j_2,2}$ contains two searchers at the moment when $v_{j_1,2}v_{j_2,2}$ becomes cleared. If $j_2 = \ell_1$, $v_{j_2,2}$ would contain two searchers at the end since the edge $v_{j_2,2}v_{j_2,3}$ must be cleared by sliding a searcher from $v_{j_2,3}$ to $v_{j_2,2}$. This is a contradiction. If $j_2 \neq \ell_1$, since the edge $v_{j_2,2}v_{j_2,1}$ has been cleared, we know that $v_{j_2,2}$ would contain a searcher at the end, which is a contradiction.

From the above, we know $\mathrm{fs}(G'_{m \times 3}) \geq m+3$. Therefore, from Lemmas 6 and 4, we have $\mathrm{fs}(G'_{m \times 3}) = m + 3$.

Lemma 12. *For $m \geq 3$ and $n \geq 2$, $\mathrm{fs}(G_{m \times n}) \geq m + n$.*

Proof. If $n = 2$, it follows from Lemma 7 that $\mathrm{fs}(G_{m \times 2}) = m + 2$. If $n = 3$, from Lemmas 4 and 11, we have $\mathrm{fs}(G_{m \times 3}) = \mathrm{fs}(G'_{m \times 3}) = m + 3$. We now suppose that $n \geq 4$.

If n is odd, then we can decompose $G_{m \times n}$ into one $G_{m \times 3}$ and $(n-3)/2$ copies of $G_{m \times 2}$ (see Fig. 4). From Theorem 2, we have

$$
\begin{aligned}
\mathrm{fs}(G_{m \times n}) &\geq \mathrm{fs}(G'_{m \times 3}) + \mathrm{fs}(G'_{m \times (n-3)}) - m \\
&\geq m + 3 + \mathrm{fs}(G''_{m \times 2}) + \mathrm{fs}(G'_{m \times (n-5)}) - 2m \\
&\geq m + 3 + \frac{1}{2}(n-3)(m+2) - \frac{1}{2}(n-3)m \\
&= m + n.
\end{aligned}
$$

If n is even, then we decompose $G_{m \times n}$ into $n/2$ copies of $G_{m \times 2}$ (see Fig. 5). Similar to the above case, we have

$$
\mathrm{fs}(G_{m \times n}) \geq \frac{1}{2}n(m+2) - (\frac{1}{2}n - 1)m = m + n.
$$

Therefore, $\mathrm{fs}(G_{m \times n}) \geq m + n$ when $n \geq 2$.

From Lemmas 6 and 12, we have the main result of this section.

Theorem 3. *For $m \geq 3$ and $n \geq 2$, $\mathrm{fs}(B_m \square P_n) = m + n$.*

2.3 Variants of $B_m \square P_n$

From the proofs in Sect. 2.2, we know that even if each Eulerian graph B_m^j, $1 \leq j \leq n$, is replaced by an arbitrary Eulerian graph with m vertices (all these Eulerian graphs may be different), we can still prove the same results. This means Theorem 3 holds for a larger class of graphs including all $B_m \square P_n$.

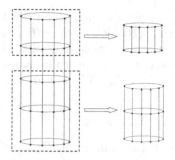

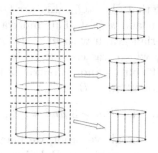

Fig. 4. Decomposition of $G_{6\times5}$ into $G_{6\times2}$ and $G_{6\times3}$.

Fig. 5. Decomposition of $G_{6\times6}$ into three copies of $G_{6\times2}$.

Theorem 4. *Let $W_{m,n}$ be a graph obtained from $B_m \square P_n$ ($m \geq 3$, $n \geq 2$) by replacing each Eulerian graph B_m^j ($1 \leq j \leq n$) by an arbitrary Eulerian graph with m vertices. Then $\mathrm{fs}(W_{m,n}) = m + n$.*

We now consider another variant of the Cartesian product $B_m \square P_n$. For every B_m^i on $B_m \square P_n$, $1 \leq i \leq n$, we select a vertex $v_{x_i,i}$, $1 \leq x_i \leq m$. We then add an edge between $v_{x_i,i}$ and $v_{x_{i+1},i+1}$ for every $i = 1, \ldots, n-1$. Let the new graph denoted by $Z_{m\times n}$.

Lemma 13. *For $m \geq 3$ and $n \geq 2$, $\mathrm{fs}(Z_{m\times n}) = m + 1$.*

It is easy to see that the lower bound and the fast search strategy in the proof of Lemma 13 can also be applied to the graph $W'_{m,n}$ defined in the following corollary.

Corollary 2. *Let $W'_{m,n}$ be a graph obtained from $Z_{m\times n}$ ($m \geq 3$, $n \geq 2$) by replacing each Eulerian graph B_m^j ($1 \leq j \leq n$) by an arbitrary Eulerian graph with m vertices. Then $\mathrm{fs}(W'_{m,n}) = m + 1$.*

Note that by adding a path with n vertices to $B_m \square P_n$, the fast search number of the new graph $Z_{m\times n}$ can be arbitrarily smaller than $\mathrm{fs}(B_m \square P_n)$. This demonstrates that the fast searching problem is not subgraph-closed.

3   Hypercubes and Toroidal Grids

In this section, we investigate the fast search number of hypercubes and the fast search number of toroidal grids. Let Q_k, $k \geq 0$, denote a k-dimensional hypercube.

Theorem 5. *If k is odd and $k \geq 3$, then $\mathrm{fs}(Q_k) = 2^{k-1} + 2$.*

Proof. Note that Q_k has 2^k vertices and every vertex has degree k. Since k is odd and $k \geq 3$, we know that Q_{k-1} is an Eulerian graph with 2^{k-1} vertices. It follows from Theorem 3 that $\mathrm{fs}(Q_k) = \mathrm{fs}(Q_{k-1} \square P_2) = 2^{k-1} + 2$.

Observe that $Q_k = Q_{k-2} \square Q_2 = Q_{k-2} \square C_4$ and Q_{k-2} is an Eulerian graph when k is even and $k \geq 4$. This motivates us to consider $\mathrm{fs}(B_m \square C_n)$, where B_m is an Eulerian graph with m vertices. Although $B_m \square C_n$ is a simple extension of $B_m \square P_n$ which was considered in the previous section, it turns out to be much more difficult to find a nontrivial lower bound on $\mathrm{fs}(B_m \square C_n)$. We first give an upper bound on $\mathrm{fs}(B_m \square C_n)$.

Lemma 14. *If $m \geq 3$ and $n \geq 3$, then $\mathrm{fs}(B_m \square C_n) \leq 2m + n - 2$.*

From the above proof, we know that even if each Eulerian graph B_m^j, $1 \leq j \leq n$, is any arbitrary Eulerian graph with m vertices, we can still prove the same results.

Corollary 3. *Let $D_{m,n}$ be a graph obtained from $B_m \square C_n$ $(m \geq 3, n \geq 3)$ by replacing each Eulerian graph B_m^j $(1 \leq j \leq n)$ by an arbitrary Eulerian graph with m vertices. Then $\mathrm{fs}(D_{m,n}) \leq 2m + n - 2$.*

From the structure of $B_m \square C_n$, the fast search strategy described in the proof is essential and the upper bound in Lemma 14 seems hard to beat. But to our surprise, we can improve this upper bound when B_m is a cycle with at least four vertices.

Theorem 6. *If $n \geq m \geq 4$, then $\mathrm{fs}(C_m \square C_n) \leq 2m + n - 3$.*

We now consider the upper bound of $\mathrm{fs}(Q_k)$ when k is even. Since $Q_k = Q_{k-2} \square C_4$ and Q_{k-2} is an Eulerian graph when k is even and $k \geq 4$, it follows from Lemma 14 that $\mathrm{fs}(Q_k) \leq 2^{k-1} + 2$. In the following theorem, we use a new technique to improve this upper bound.

Theorem 7. *If k is even and $k \geq 4$, then $\mathrm{fs}(Q_k) \leq 2^{k-1} + 1$.*

Proof. If $k = 4$, then from Theorem 6 we have $\mathrm{fs}(Q_4) \leq 9$, and so the theorem is proved. Suppose that $k \geq 6$ and k is even. We observe that $Q_k = Q_{k-4} \square Q_4 = Q_{k-4} \square C_4 \square C_4$. Let $Q_{k-4}^{(i,j)}$, $1 \leq i, j \leq 4$, denote a copy of Q_{k-4} in $C_4 \square C_4$ (see Fig. 6). Let $Q_{k-2}^{(j)}$, $1 \leq j \leq 4$, denote a copy of Q_{k-2} in Q_k which is induced by the vertices of $Q_{k-4}^{(1,j)}$, $Q_{k-4}^{(2,j)}$, $Q_{k-4}^{(3,j)}$, and $Q_{k-4}^{(4,j)}$. We now describe a fast search strategy that clears Q_k using $2^{k-1} + 1$ searchers.

(1) Place two searchers on each vertex of $Q_{k-4}^{(1,1)}$; place one searcher on each vertex of $Q_{k-4}^{(2,1)}$, place three searchers on each vertex of $Q_{k-4}^{(4,1)}$; place two searchers on each vertex of $Q_{k-4}^{(3,2)}$; and place an additional searcher on just one vertex of $Q_{k-2}^{(3)}$.

(2) For $Q_{k-4}^{(1,1)}$, slide a searcher from one of its vertices along all its edges and back to this vertex. At the end of this step, all edges of $Q_{k-4}^{(1,1)}$ are cleared.

(3) Slide a searcher from each vertex of $Q_{k-4}^{(4,1)}$ to its neighbor on $Q_{k-4}^{(1,1)}$, and then slide further to a new neighbor on $Q_{k-4}^{(2,1)}$. At the end of this step, each vertex of $Q_{k-4}^{(1,1)}$, $Q_{k-4}^{(2,1)}$ and $Q_{k-4}^{(4,1)}$ contains two searchers.

(4) Slide a searcher from each vertex of $Q_{k-4}^{(1,1)}$, $Q_{k-4}^{(2,1)}$ and $Q_{k-4}^{(4,1)}$ to its neighbor on $Q_{k-2}^{(2)}$ along the edge between them.

(5) Slide a searcher from a vertex of $Q_{k-4}^{(3,2)}$ along all the edges of $Q_{k-2}^{(2)}$ and back to this vertex. Then slide a searcher from each vertex of $Q_{k-4}^{(3,2)}$ to its neighbor on $Q_{k-4}^{(3,1)}$ along the edge between them.

(6) After Step (5), each vertex of $Q_{k-2}^{(1)}$ and $Q_{k-2}^{(2)}$ contains exactly one searcher. Since all edges of $Q_{k-2}^{(2)}$ are cleared, slide a searcher from each vertex of $Q_{k-2}^{(2)}$ to its neighbor on $Q_{k-2}^{(3)}$ along the edge between them.

(7) First slide the additional searcher placed in Step (1) along all edges of $Q_{k-2}^{(3)}$ to clear them, and then slide a searcher from each vertex of $Q_{k-2}^{(3)}$ to its neighbor on $Q_{k-2}^{(4)}$ along the edge between them.

(8) After Step (7), each vertex of $Q_{k-2}^{(1)}$ and $Q_{k-2}^{(4)}$ contains exactly one searcher. From Steps (2) and (3), we know that each vertex of $Q_{k-4}^{(1,1)}$ has only one contaminated edge incident on it. So, slide a searcher from each vertex of $Q_{k-4}^{(1,1)}$ to its neighbor on $Q_{k-4}^{(1,4)}$ along the edge between them.

(9) For $Q_{k-2}^{(4)}$, slide a searcher from one of the vertices on $Q_{k-4}^{(1,4)}$ along all edges of $Q_{k-2}^{(4)}$. Then slide a searcher from each vertex of $Q_{k-4}^{(i,4)}$, $2 \le i \le 4$, to its neighbor on $Q_{k-2}^{(1)}$ along the edge between them.

(10) For each of $Q_{k-4}^{(i,1)}$, $2 \le i \le 4$, slide one searcher from one of its vertices along all its edges and back to this vertex. At the end of this step, all edges of $Q_{k-4}^{(i,1)}$, $2 \le i \le 4$, are cleared.

(11) Slide a searcher from each vertex of $Q_{k-4}^{(2,1)}$ to its neighbor on $Q_{k-4}^{(3,1)}$, and then slide further to a new neighbor on $Q_{k-4}^{(4,1)}$. At the end of this step, every edge of Q_k is cleared.

From Step (1), we know that $3 \cdot 2^{k-3}$ searchers are placed on $Q_{k-2}^{(1)}$, 2^{k-3} searchers are placed on $Q_{k-2}^{(2)}$, and one additional searcher is placed on $Q_{k-2}^{(3)}$. In total, $2^{k-1} + 1$ searchers are placed on Q_k. Therefore, $\mathrm{fs}(Q_k) \le 2^{k-1} + 1$.

Applying the same idea as that used in the proof of Theorem 7, we can show the following.

Corollary 4. *For $m \ge 3$, $\mathrm{fs}(B_m \square Q_4) \le 8m + 1$.*

Similar to Corollary 3, we can extend Corollary 4 to a broader class of graphs.

Corollary 5. *Let $Q_{m,n}$ be a graph obtained from $B_m \square Q_4$ ($m \ge 3$) by replacing each copy of B_m by an arbitrary Eulerian graph with m vertices. Then $\mathrm{fs}(Q_{m,n}) \le 8m + 1$.*

A graph is *even* if every vertex has an even degree.

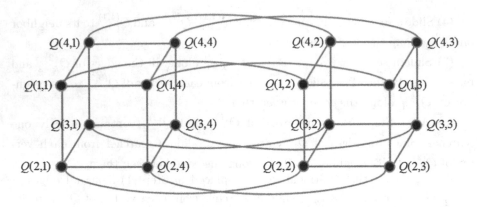

Fig. 6. $Q_k = Q_{k-4} \square Q_4$: the vertex $\mathcal{Q}(i,j)$, $1 \leq i, j \leq 4$, represents $Q_{k-4}^{(i,j)}$.

Corollary 6. *Let $n \geq 6$ and H be a graph with m vertices. If n and H are even, or n and H are odd, then $\mathrm{fs}(H \square Q_n) \leq m2^{n-1} + 1$. If one of n and H is even and the other is odd, then $\mathrm{fs}(H \square Q_n) = m2^{n-1} + 2$.*

Before giving a lower bound for $\mathrm{fs}(Q_k)$ when k is even, we first consider the treewidth of Q_k, denoted by $tw(Q_k)$. From [5], we have the following lower bound for $tw(Q_k)$.

Lemma 15. ([5]) *There is a constant k_0 such that, for any $k \geq k_0$, $tw(Q_k) \geq \frac{12 \cdot 2^k}{25\sqrt{k}}$.*

Since $\mathrm{fs}(G) \geq pw(G) \geq tw(G)$, where $pw(G)$ is the pathwidth of G, we have the following result (we rewrite the lower bound in a power of 2 so that it can be easily compared with the upper bound of $\mathrm{fs}(Q_k)$ when k is even).

Theorem 8. *There is a constant k_0 such that $\mathrm{fs}(Q_k) \geq \frac{3}{25}2^{k+2-\log\sqrt{k}}$ for any $k \geq k_0$.*

4   Open Problems

We conclude this paper by listing some open problems that we consider worth to investigate.

(1) For toroidal grids $C_m \square C_n$ ($n \geq m \geq 4$), we proved that $\mathrm{fs}(C_m \square C_n) \leq 2m + n - 3$. We conjecture that $2m + n - 3$ is also a lower bound when $n \geq m \geq 4$. We also conjecture that $\mathrm{fs}(B_m \square C_n) \geq 2m + n - 3$, $n \geq m \geq 4$.

(2) We proved that $\mathrm{fs}(Q_k) \leq 2^{k-1} + 1$ when k is even and $\mathrm{fs}(Q_k) \geq \frac{3}{25}2^{k+2-\log\sqrt{k}}$ for large k. We conjecture that $\mathrm{fs}(Q_k) = 2^{k-1} + 1$ when k is even.

(3) In [13], Stanley and Yang showed that $\mathrm{fs}(P_m \square P_n) = m + n - 2$ ($m \geq 2$, $n \geq 2$). We also believe that it would be interesting to consider algorithms for computing $\mathrm{fs}(T_m \square P_n)$, where T_m is a tree with m vertices and P_n is a path with n vertices.

References

1. Bienstock, D.: Graph searching, path-width, tree-width and related problems (a survey). DIMACS Ser. Discrete Math. Theoret. Comput. Sci. **5**, 33–49 (1991)
2. Bienstock, D., Seymour, P.: Monotonicity in graph searching. J. Algorithms **12**(2), 239–245 (1991)
3. Bonato, A., Nowakowski, R.J.: The Game of Cops and Robbers on Graphs. American Mathematical Soc., Providence (2011)
4. Bonato, A., Yang, B.: Graph searching and related problems. In: Pardalos, P., Du, D.-Z., Graham, R. (eds.) Handbook of Combinatorial Optimization, 2nd edn., pp. 1511–1558. Springer, New York (2013)
5. Chandran, S., Kavitha, T.: The treewidth and pathwidth of hypercubes. Discrete Math. **306**(3), 359–365 (2006)
6. Dereniowski, D., Diner, Ö., Dyer, D.: Three-fast-searchable graphs. Discrete Appl. Math. **161**, 1950–1958 (2013)
7. Dyer, D., Yang, B., Yaşar, Ö.: On the fast searching problem. In: Fleischer, R., Xu, J. (eds.) AAIM 2008. LNCS, vol. 5034, pp. 143–154. Springer, Heidelberg (2008). doi:10.1007/978-3-540-68880-8_15
8. Fomin, F.V., Thilikos, D.M.: An annotated bibliography on guaranteed graph searching. Theoret. Comput. Sci. **399**(3), 236–245 (2008)
9. Hahn, G.: Cops, robbers and graphs. Tatra Mt. Math. Publ. **36**(163), 163–176 (2007)
10. Kirousis, L.M., Papadimitriou, C.H.: Searching and pebbling. Theoret. Comput. Sci. **47**, 205–218 (1986)
11. Makedon, F.S., Papadimitriou, C.H., Sudborough, I.H.: Topological bandwidth. SIAM J. Algebraic Discrete Methods **6**(3), 418–444 (1985)
12. Megiddo, N., LouisHakimi, S., Garey, M.R., Johnson, D.S., Papadimitriou, C.H.: The complexity of searching a graph. J. ACM **35**(1), 18–44 (1988)
13. Stanley, D., Yang, B.: Fast searching games on graphs. J. Comb. Optim. **22**(4), 763–777 (2011)
14. Yang, B.: Fast edge-searching and fast searching on graphs. Theoret. Comput. Sci. **412**, 1208–1219 (2011)

Sequentialization Using Timestamps

Anand Yeolekar$^{(\boxtimes)}$, Kumar Madhukar, Dipali Bhutada, and R. Venkatesh

Tata Consultancy Services Ltd, Pune, India
anand.yeolekar@tcs.com

Abstract. Given a run of a concurrent program and the underlying memory model, we can view the shared memory accesses as a chronological sequence of read and write operations. This chronological sequence of shared memory accesses exactly characterizes the run. We present an approach to sequentialization that captures these sequences by assigning timestamps to the memory accesses. The axioms of the underlying memory model can be encoded as constraints on the timestamps, within the sequentialized program, to generate precisely the set of traces permissible by the original concurrent program. Experimental evaluation shows that the encoding can be efficiently checked by the backend model checker.

1 Introduction

As multi-core processors gain widespread adoption, multi-threaded software is being increasingly designed, developed and deployed. The exponential number of interleavings exhibited by concurrent software poses a challenge during the validation phase of the software development lifecycle. Further, many architectures support weak memory models allowing concurrent program behaviours that need not conform to sequential consistency. Consequently, model checking is necessary to exhaustively check the concurrent program for *heisenbugs* - bugs that lie deep inside an interleaving and are near-impossible to detect or reproduce using testing. In this work, we present an approach to sequentialization of concurrent C programs to enable efficient checking of program assertions.

Consider two threads T_1:x++; and T_2:x--;, incrementing and decrementing a shared variable, respectively. The threads issue a read operation of the shared variable from memory and then a write to store the updated value. Let r^i and w^i denote the read and write operation, resp., for thread i. Under Sequential Consistency (SC) memory model [1], any run must satisfy the following conditions: (i) program order be maintained among operations within a thread, and (ii) the execution appears to be the result of a single sequential order across threads (atomicity). The set of sequences of the shared memory read r and write w operations, corresponding to all possible runs of the threads, can be listed as: $\{\langle r^1, r^2, w^1, w^2 \rangle, \langle r^1, r^2, w^2, w^1 \rangle, \langle r^1, w^1, r^2, w^2 \rangle, \langle r^2, w^2, r^1, w^1 \rangle\}$. Note that we ignore sequences that differ only in a read sub-sequence permutation, for e.g. $\langle r^2, r^1, w^1, w^2 \rangle$.

It is well-known [3,16] that the set of all *correct* sequences can be captured as solutions to constraints derived from the program and the underlying memory model. For the example shown above, we obtain two relations from the

© Springer International Publishing AG 2017
T.V. Gopal et al. (Eds.): TAMC 2017, LNCS 10185, pp. 684–696, 2017.
DOI: 10.1007/978-3-319-55911-7_49

threads, namely $po(r^1, w^1)$ and $po(r^2, w^2)$, where po denotes the per-thread program order between memory accesses. Additionally, as per condition (i) of SC stated earlier, $po(m_1, m_2) \Leftrightarrow hb(m_1, m_2)$ must hold for all m_1, m_2 belonging to the same thread, in every correct sequence. We say that $hb(m_1, m_2)$ holds in a sequence if m_1 precedes m_2 in that sequence. Thus we obtain the constraint $hb(r^1, w^1) \wedge hb(r^2, w^2)$, which can be solved to get precisely the sequences listed above.

The hb relation is commonly referred to as the *happens before* relation in the literature, relating memory accesses in an execution trace. The second condition of SC specifies how a given sequence or trace can be interpreted: a read must return the value of the *freshest* write, i.e. if a read r links to a write w then $\nexists w'$: $hb(w, w') \wedge hb(w', r)$, unless w, w' write to different memory locations. Observe that both the SC conditions can be expressed using the happens-before relation, which can be naturally modeled if we could refer to the *time* of occurrence of the shared memory operations. For example, if t_m and $t_{m'}$ denote the time of occurrence of memory access m and m' respectively, then $hb(m, m')$ is simply the constraint $t_m < t_{m'}$.

In this paper, we propose the use of *timestamps* for sequentialization. A timestamp is a natural number that encodes the logical time of occurrence of a shared memory access. We assign timestamps to shared memory accesses to map reads with writes as permitted by the underlying memory model. The set of permitted read-write maps is defined by the axiomatic specification of the memory model. It is this specification that we encode as constraints on timestamps, at the source level.

We construct a sequentialized program, that encodes the constraints on timestamps described above, as follows. We introduce global arrays to store timestamps of writes along with their values. Timestamps are assigned non-deterministically and constrained to be monotonically increasing. We encode the requirements for SC by rewriting instructions accessing shared memory i.e., read and write operations. The program is instrumented as follows: (i) a *write* access is redirected to an array location whose timestamp is larger than a *locally* tracked current time, (ii) a *read* reads from a location with a timestamp *closest* to the current time i.e. the timestamp of the *successive* write (in the array) must be larger than current time. The sequentialization is completed by issuing calls to the thread function bodies directly. Locks are modeled as shared variables. Locking sets the variable provided the latest access to the variable was a reset, and unlocking resets the variable. We show in Sect. 2.3 that the sequentialized program exhibits precisely the set of behaviours of the concurrent programs.

We propose that using timestamps can *naturally* describe the runs of a concurrent program under any memory consistency model and yield a simple yet efficient sequentialized program. We make the following contributions through this work.

- A sequentialization approach based on timestamps that encodes axioms of SC.
- A prototype tool, ConSequence, that implements the proposed encoding.
- An experimental evaluation demonstrating the usability of this approach.

The rest of the paper is organized as follows. In Sect. 2, we illustrate our encoding under SC with an example, formalize the encoding and present an argument for its correctness. Our experimental results are presented in Sect. 3. We discuss the related work in Sect. 4 before concluding in Sect. 5 and listing some immediate directions of future work.

2   Sequential Consistency Memory Model

2.1   Illustrative Example

We illustrate our approach with an example, shown in Fig. 1(a), that computes the Fibonacci sequence with two threads, under SC. The assertion can be violated only when context switches between the threads follow a certain order (the reads and writes occur hand-in-hand). This makes the analysis challenging for tools that rely on under-approximations such as write-bounding or context-bounding.

The sequentialized code is shown in Fig. 1(b). We use two procedures, write_var() and read_var(), for every shared variable var, to instrument its memory writes and reads, respectively. Thread creation in main function

```
#include <pthread.h>
#define N 3
int i=1, j=1;
void* f1(void* arg){
 int x;
 for(x=0;x<N;x++) {
  i = i+j; }
}
void* f2(void* arg){
 int x;
 for(x=0;x<N;x++) {
  j = j+i; }
}
int main() {
 pthread_t t1,t2;
 pthread_create(&t1,0,f1,0);
 pthread_create(&t2,0,f2,0);
 pthread_join(t1,0);
 pthread_join(t2,0);
 assert(i<21);
}
             (a)
```

```
#define N 3
int i=1, j=1;

f1(){
 int x;
 for(x=0;x<N;x++) {
  write_i(read_i()+read_j());}
}
f2(){
 int x;
 for(x=0;x<N;x++) {
  write_j(read_j()+read_i());}
}
int main() {
 sysinit();
 procinit(); t1(); procend();
 procinit(); t2(); procend();
 sysend();

 assert(i<21);
}
             (b)
```

Fig. 1. Example concurrent program fib.c (a) and its sequentialization (b)

```
#define MAXW_i (N+1)
#define MAXW_j (N+1)
#define MAXT ((MAXW_i-1)+(MAXW_j-1)+1)

int *value_i,*value_j;
unsigned short
        *ts_i,loc_i=0,count_i=0,last_i=0,
        *ts_j,loc_j=0,count_j=0,last_j=0,
        ct=0;
_Bool *free_i,*free_j;

void sysinit(){
 value_i=(int *)malloc(
        sizeof(int) * MAXW_i);
 value_i[0]=i;

 ts_i=(unsigned short*)malloc(
        sizeof(unsigned short)*(MAXW_i+1));
 ts_i[0]=0; ts_i[MAXW_i]=MAXT;
 for (int k=1; k<MAXW_i+1; k++)
   assume(ts_i[k-1] < ts_i[k]);

 free_i=(_Bool *)malloc(
        sizeof(_Bool)*MAXW_i);
 free_i[0]=false;
 /* similarly for j */ }

void procinit() {
 ct=0; loc_i=0; loc_j=0;}
```

```
void procend() {
 if (last_i<loc_i) last_i=loc_i;
 if (last_j<loc_j) last_j=loc_j;}

void sysend() {
 assume(last_i==count_i);
 i=value_i[last_i];
 /* similarly for j */ }

void write_i(int value) {
 unsigned short loc=*;
 assume(loc_i < loc < MAXW_i &&
        free_i[loc] &&
        ts_i[loc] > ct &&
        value_i[loc] == value);
 loc_i = loc;
 free_i[loc] = false;
 icount++;
 ct = ts_i[loc];}

int read_i() {
 unsigned short loc=*;
 assume(loc_i <= loc < MAXW_i &&
        ts_i[loc+1] > ct);
 loc_i = loc;
 if (ct<ts_i[loc]) ct=ts_i[loc];
 return value_i[loc];}

 /* similarly for write_j, read_j */
```

Fig. 2. Datastructures and auxiliary code for sequentialization

is replaced with direct calls to the thread function body, augmented with pre-
and post-processing code.

The auxiliary datastructures and code used for sequentialization is shown in
Fig. 2. We assume fib.c is *structurally bounded* with loops executing N number
of times. Let MAXW_var denote the maximum number of writes to shared memory
variable var that may occur along *any* program path, across all threads of the
program. In the example of Fig. 1(a), MAXW_i = MAXW_j = N+1, as each thread
writes to a variable exactly once in a loop iteration, apart from the initialization.
Additionally, we define MAXT as the total number of unique timestamps needed
for write accesses across all shared variables, where initializations of all shared
variables get the same timestamp.

For each shared variable, we use additional memory as explained here. Arrays
value_var,ts_var store the value of a write access and its timestamp, respectively, and free_var tracks if an index in value and timestamp arrays is *available*.
An index ceases to be available once it is written to. We refer to the three arrays
as a *timestore* for the shared variable. Auxiliary variable count_var records the
number of writes and last_var tracks the largest index accessed by a read or
write in each timestore. The variable ct, common to all shared memory variables
of the concurrent program, tracks the time of the latest memory access issued
by a thread procedure. Note that ct is updated locally by each thread, though
declared as a global variable.

Procedure `sysinit()` initializes each timestore to non-deterministic values (timestamps are bounded by the respective maximum number of writes along any path) through `malloc` calls. It further adds the constraint that timestamps in `ts_var` increase strictly monotonically.

We explain the write and read instrumentation scheme wrt. shared variable i of Fig. 1(a), presented in the procedures `write_i()` and `read_i()` of Fig. 2. Intuitively, in the sequentialized program, a write advances the *local* clock to allow for interfering writes from other threads to happen. We select an empty location by non-deterministically advancing from the current location `loc_i` in the timestamp array and store the value (`value_i[loc]==value`). We also add a constraint to ensure that this write occurs *after* the current time `ct` (`ts_i[loc]>ct`). A read in the sequentialized program, may return the value at any index of the timestore, provided this is the most *recent* write relative to the read. We first select a write in the timestore array by non-deterministically advancing from the last accessed (read or write) location by this thread. Next, to ensure that this is the most *recent* write relative to the read, we add a constraint that the *successive* write occurs after the current time (`ts_i[loc+1]>ct`), which is intuitively the time at which the read happens. Recall that the timestores are sorted on timestamps *a-priori*; this guarantees the succession of writes. Note that an explicit assignment of timestamps to read accesses is not required to encode the aforementioned constraint; we thus do not assign timestamps to read accesses.

The procedure `procinit()` resets variables `ct` and `loc_i`, `loc_j`. The procedure `procend()` tracks the last write location updated by thread procedures. Finally, procedure `sysend()` ensures that writes are stored contiguously in the timestore by adding the constraint `last==count` for each shared variable and finally reinstates the shared variables.

	value_i[]	ts_i[]	free_i[]
0	1	0	false
1	3	2	false
2	8	4	false
3	21	6	false
4	-	7	-

	value_j[]	ts_j[]	free_j[]
0	1	0	false
1	2	1	false
2	5	3	false
3	13	5	false
4	-	7	-

Fig. 3. Datastructures populated by the counterexample produced by CBMC

The resulting sequentialized program can be analyzed by any sequential model checker such as CBMC [6]. Figure 3 shows how the datastructures are populated by the counterexample returned by CBMC, violating the assertion, for N=3.

2.2 Formalization

Let P_C be a *structurally bounded* concurrent C program consisting of threads $T_1, .., T_n$, invoking procedures $f_1, .., f_n$, respectively, using the `pthreads` API.

Let V denote the set of variables shared by the threads. We assume that procedure main invokes the threads and waits for the threads to join, followed by an assertion ϕ to be checked. Let G^k denote the unfolded control flow graph of thread id k, with each statement containing at most one read r or write w access to a shared variable. We denote a memory access by m when we do not distinguish between a read or write. We use the notation m^v to represent the memory access m operates on the shared variable v.

Definition 1. *The per-thread program order po is a relation that statically orders memory accesses.*

$$\forall m, m', path(m, m') \Leftrightarrow (m, m') \in po \qquad (1)$$

where $path(i, j)$ holds iff there is a path from i to j in G^k.

We encode po in terms of *happens-before* relation $\hat{hb}$, i.e. hb (stated in Sect. 1) restricted to same-thread memory accesses, as follows.

$$\forall m, m', po(m, m') \Leftrightarrow \hat{hb}(m, m') \qquad (2)$$

Any interleaving or trace of P_C is a sequence τ of memory accesses that is a solution to the po constraints encoded as the $\hat{hb}$ relation (Eq. 2). The interpretation of τ, in terms of the values of the memory accesses, comes from the underlying memory model as a *read-from* relation.

Definition 2. *The read-from relation rf maps every read to a write in τ.*

Under SC, rf enforces the condition that in a trace, a read returns the value of the *most recent* write to the same variable. We refer to this condition as *atomicity*.

We encode rf in terms of the happens-before relation:

$$\forall r \in \tau, \exists w \mid val(r) = val(w) \wedge hb(w, r) \wedge \nexists w' : hb(w, w') \wedge hb(w', r) \qquad (3)$$

where $val(.)$ returns the value of the memory access and we interpret $hb(i, j)$ over the trace as i precedes j in τ.

Timestamps allow us to model the hb relation naturally and succinctly. In fact,

$$\forall m, m', hb(m, m') \Leftrightarrow t_m < t_{m'} \qquad (4)$$

Sequentialization. The sequentialization is presented in example Fig. 1(b) and 2. The listing in Fig. 2 encodes the SC conditions stated in Eqs. 2 and 3, in procedures read and write.

In the next subsection, we show that the encoding correctly captures these conditions in terms of timestamps (Eq. 4).

Note that in Eq. 4, the inequality need not be strict when m' is a read access. It suffices to have distinct timestamps for writes. As an optimization, wherever possible, we allow a read access to implicitly acquire the timestamp of the preceding memory access, whether read or write.

2.3   Correctness of the Sequentialization

In this section we take a closer look at the auxiliary code for sequentialization shown in Fig. 2. Recall that the timestore records writes from all threads in a single global order, obtained by sorting the timestamps a priori. Also, procedures read and write are the ones that encode the axioms of the underlying memory model, when accessing timestore.

The intra-thread program order, encoded as constraints on timestamps, is preserved through (i) the constraint ts[loc]>ct when writing and ct getting updated to ts[loc], and (ii) advancing ct when reading in the future.

Lemma 1. *The conditions listed above guarantee program order between per-thread memory accesses.*

Proof. Consider $m, m' \mid (m, m') \in po$. When m' is a write access, then condition (i) ensures that $t_m < t_{m'}$ as a write always updates ct to a larger value. Note that ct, intuitively the current time, is at least as large as t_m when the memory access m has occurred. When m' is a read, the timestamp of m' is either the same as ct (when reading in the past) or advanced (when reading in the future) as per condition (ii). This ensures $t_m <= t_{m'}$ (note that the equality on timestamps is an optimization not violating SC, as described earlier). Thus the proposed sequentialization guarantees the per-thread program order.

The atomicity condition of SC is preserved by (i) the constraints iloc<loc and iloc<=loc when writing and reading, respectively, and (ii) the constraint ts_i[loc+1]>ct when reading.

Lemma 2. *The conditions (i) and (ii), above, guarantee atomicity under SC.*

Proof. The variable ct also accounts for thread interference. Whenever a read maps to a write of a different thread, the timestamps of the read and write are compared. If the write has occurred in the past relative to the read, we ensure the atomicity condition of SC by constraining the *successive write in global order* to occur *after* this read, through the constraint ts_i[loc+1]>ct. If the write is located in the timestore in the future wrt. this read, then this implies that the read should have occurred *after* this write and *before* the subsequent write (in global order); thus the current time, which is the time this read happens, is updated to match the write's time, again guaranteeing atomicity.

The existential quantifier stated in the SC atomicity condition is handled implicitly in our implementation by sorting the timestore during the initialization. The timestore reflects the write serialization to main memory i.e. the sequence of writes to main memory as agreed upon by all threads.

To summarize, shared memory read and write operations are constrained exactly according to the axioms of the memory model. The runs of the sequentialized program can be obtained by solving the read and write constraints in addition to the program logic. This set of runs exactly corresponds to the set of runs of the parallel program, since the runs of the sequentialized program are exactly the solutions to the system of constraints of the axiomatic memory model, encoded at source level.

Since our sequentialization approach does not impose (or relax) any other constraints, the only constraints present in the sequentialized program are the memory model axioms which define the correct behavior. Therefore, the transformation precisely captures correct behavior. The approach easily generalizes to a framework where one can plug the memory model axioms, akin to a library call, to obtain a transformation corresponding to a different memory model.

Corollary 1. *The proposed encoding correctly captures the constraints of SC at the source level i.e. the sequentialization exactly encodes the behaviours of the concurrent program.*

3 Experiments

We have implemented the timestamp-based sequentialization approach in a prototype tool ConSequence. Given a structural or unwind bound u, ConSequence transforms the multi-threaded C code by instantiating a header file from a template, which can be viewed as a library that replaces thread API calls. Read and write accesses to shared memory variables are identified using PRISM[1]and replaced with calls to respective procedures. ConSequence automatically computes the maximum number of writes for every shared variable. Thread creation and joining calls are replaced with calls to thread procedures with pre and post processing code. ConSequence uses CBMC to check assertions in the resulting sequentialized program (other model checkers can be used). For efficient performance with CBMC, ConSequence implements several tweaks/optimizations during the sequentialization.

Table 1. Results for Sequential Consistency

S. No.	Benchmark	Unwind	Assertion	ConSequence	MU-CSeq 0.3	CBMC5.5	Corral	CIVL
1	stateful01.c	1	safe	0.17	0.53	1.18	53.71	1.72
2	stateful01.c	1	unsafe	0.17	0.52	1.07	47.64	1.72
3	fib_bench_longer.c	7	safe	2.8	17.12	17	timeout	timeout
4	3Var-nolock-2threads.c	4	safe	0.65	38	7.21	16.5	18.4
5	27_Boop.c	10	unsafe	1.1	52.7	2.75	error	timeout
6	fib_bench_longest.c	12	safe	130	timeout	timeout	timeout	timeout
7	fib_bench_longest.c	12	unsafe	101	543.6	749	error	timeout
8	peterson.c	60	safe	1.5	42.9	15.3	2.07	1.61
9	szymanski.c	21	safe	2.30	45.02	7.5	1.91	1.65
10	inc-dec-lock-2threads.c	9	safe	8.9	44.5	timeout	timeout	4.8
11	dekker.c	9	safe	6	37	4.94	error	timeout

Table 1 shows the comparison between ConSequence, MU-CSeq 0.3 [18], CBMC 5.5 [6], Corral [2] and CIVL [15]. Though MU-CSeq 0.3 comes with CBMC

---

[1] A static analysis framework developed at TRDDC, Pune [5,11].

4.9 by default, in order to have an unbiased evaluation, we have run both Con-Sequence and MU-CSeq with CBMC 5.5 as the backend model checking tool. The benchmarks have been selected from the concurrency category of verification tasks at SV-COMP 2016 [4], except the ones numbered 4 and 10 that are synthetic. inc-dec has 2 threads and one shared variable incremented and decremented under a lock, in a loop. 3var has 3 shared variables with two threads incrementing and decrementing the variables in a loop. For every benchmark, we report the unwinding depth chosen, state whether the benchmark was safe or unsafe for the chosen unwinding, and list the time taken (in seconds) by the tools to analyze the benchmark correctly. We set a timeout of 60 s for these experiments, except for fib_bench_longest program (rows 6 and 7) where we set a timeout of 900 s. All the benchmarks, except fib_bench_longest, were run with the largest possible unwind such that at least 3 out of 5 tools run to completion, i.e. we allowed no more than two tools to time out. The term error indicates that the tool either crashed, or terminated without producing the correct result. For the comparison to be unambiguous, the reported values of time are in fact the time taken by the underlying decision procedure in the analysis, for all the tools except CIVL. In case of CIVL we could not determine the time taken by the decision procedure separately, and hence we have reported the total time taken. We conducted our experiments on an Intel Xeon 2.2 GHz 32-core machine with 20 GB RAM. As seen from Table 1, ConSequence outperforms all other tools on both safe and unsafe instances. The benchmark programs, their sequentialized version (as produced by ConSequence), the exact commands used to invoke the tools, and the corresponding log files are available at http://www.cmi.ac.in/~madhukar/ConSequence/.

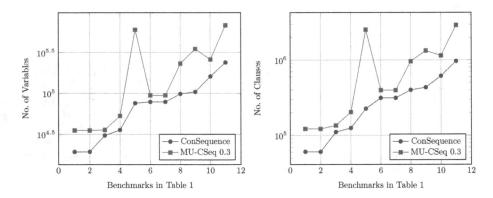

Fig. 4. Plots comparing the number of variables (left), and the number of clauses (right), as reported by the backend decision procedure in ConSequence and MU-CSeq 0.3, for the benchmarks shown in Table 1.

The graphs shown in Fig. 4 compare the number of variables and clauses generated by ConSequence and MU-CSeq (by the backend decision procedure of

CBMC, in both the cases), during the analysis of benchmarks shown in Table 1. Our encoding consistently generates fewer variables and clauses, explaining the order of magnitude reduction in the model checking time with ConSequence.

4   Related Work

The idea of sequentialization was proposed by Qadeer and Wu [14] with the motivation to leverage analysis techniques developed for sequential programs. Lal and Reps [13] proposed a sequentialization for a given bound on context switches. Their scheme implemented a non-deterministic scheduler by instrumenting the code and storing the program state at each switch. An improved version of this algorithm was implemented in [7]. Inverso et al. proposed a further enhancement [8,9] combining [12] and bounded model checking. Recently, Tomasco et al. [19] proposed a new technique for sequentialization that bounds the number of shared memory write accesses. This approach explores an orthogonal set of interleavings compared to the earlier context-bounding approaches.

Tomasco et al. [19] propose memory unwinding, i.e. a sequence of write operations into the shared memory. The technique guesses the sequence and simulates the executions of a multi-threaded program according to any scheduling that respects it. However, it bounds the total number of write operations into the shared memory. To track writes, they use arrays and duplicate the shared memory state (of unmodified variables) each time a write occurs in a thread (read-explicit scheme), or use pointers to track the last relevant write at each memory location (read-implicit scheme). The main difference is how time is stored and used to form constraints: in [19], the array location implicitly represents time whereas we use an auxiliary array to explicitly store timestamps. The advantage of our encoding is that the successive write is actually stored in the next location. This avoids costly duplication of the read-explicit scheme and maintaining pointers in the read-implicit scheme, scaling better as the number of memory accesses and/or threads increases. Further, linking constraints can be naturally and compactly expressed with respect to timestamps instead of locations of (arrays representing) shared memory.

The memory unwinding technique has been extended further [19] to use timestamps and for the analysis of TSO and PSO [20]. The authors view a concurrent program as two independent subsystems, separating computation (individual threads) from communication (shared memory). This is similar in spirit to our approach but we differ in the encoding and implementation details. This reflects as an order of magnitude reduction in both the number of clauses and the number of variables for the benchmarks (see Fig. 4).

Partial orders employing memory model axioms to link read and write events are presented in [3,10,16]. A two-stage approach involving intra-thread summarization and composition under sequential consistency is presented in [16,17]. The method constructs a concurrent control flow graph (as part of an *inter-ference skeleton*) to discover intra-thread causal ordering of events. The linking phase gives rise to redundant pairing of *read-write* events, requiring pruning using dataflow analysis. In contrast, our approach works at source-level, provides a syntactic transformation under SC, and does not require any pruning of constraints. Several sequential program analysis techniques can be applied on the transformed program including procedure summarization.

5   Conclusion and Future Work

We have presented an approach to sequentialization of concurrent programs that uses timestamps to map reads with appropriate writes. The possible read-write maps, defined by the axiomatic memory model, are encoded as constraints over the timestamps. The solutions to these constraints yield traces that precisely capture all valid interleavings of the concurrent program. Our encoding, based on timestamps, naturally captures the semantics of memory models expressed as axiomatic composition rules on reads and writes. Further, the encoding is compact, simple and efficiently analyzable by a bounded model checker like CBMC.

The use of model checkers to explore interleavings encoded as constraints on timestamps has the potential to face state space explosion problem. In particular, the technique may explore redundant orders of writes whenever the number of writes produced some choice of paths is less than the (statically) allocated space in the timestore. Further, when the size of shared memory is large, such as when entire arrays are shared, the timestores may become prohibitively large to analyze. Another limitation is that the programs need to be structurally bounded, which effectively bounds the number of threads by bounding loops and recursive function calls. However, given that bounded model checking is the most popular form of model checking used in the industry, this limitation may be acceptable in practice.

The future steps for this work are:

- Improve tool support for more thread operations from posix-API.
- Optimize the timestamps space by assigning the same timestamps to independent writes (or fixing their order of exploration). Note that the size of timestamps directly affects the model checker's state space.
- Use invariants to reduce the timestore size.
- Extend the work to support relaxed memory models i.e. program order and atomicity relaxation.

References

1. Adve, S.V., Gharachorloo, K.: Shared memory consistency models: a tutorial. IEEE Comput. **29**, 66–76 (1995)
2. Lal, A., Qadeer, S., Lahiri, S.: Corral: a solver for reachability modulo theories. Technical report (January 2012). https://www.microsoft.com/en-us/research/publication/corral-a-solver-for-reachability-modulo-theories/
3. Alglave, J., Kroening, D., Tautschnig, M.: Partial orders for efficient bounded model checking of concurrent software. In: Sharygina, N., Veith, H. (eds.) CAV 2013. LNCS, vol. 8044, pp. 141–157. Springer, Heidelberg (2013). doi:10.1007/978-3-642-39799-8_9

4. Beyer, D.: Reliable and reproducible competition results with benchexec and witnesses (report on SV-COMP 2016). In: Chechik, M., Raskin, J.-F. (eds.) TACAS 2016. LNCS, vol. 9636, pp. 887–904. Springer, Heidelberg (2016). doi:10.1007/978-3-662-49674-9_55

5. Chimdyalwar, B., Kumar, S.: Effective false positive filtering for evolving software. In: Proceedings of the 4th India Software Engineering Conference, ISEC 2011, pp. 103–106. ACM, New York (2011). http://doi.acm.org/10.1145/1953355.1953369

6. Clarke, E., Kroening, D., Lerda, F.: A tool for checking ANSI-C programs. In: Jensen, K., Podelski, A. (eds.) TACAS 2004. LNCS, vol. 2988, pp. 168–176. Springer, Heidelberg (2004). doi:10.1007/978-3-540-24730-2_15

7. Fischer, B., Inverso, O., Parlato, G.: CSeq: A concurrency pre-processor for sequential C verification tools. In: 2013 IEEE/ACM 28th International Conference on Automated Software Engineering (ASE), pp. 710–713. IEEE (2013)

8. Inverso, O., Tomasco, E., Fischer, B., La Torre, S., Parlato, G.: Bounded model checking of multi-threaded C programs via lazy sequentialization. In: Biere, A., Bloem, R. (eds.) CAV 2014. LNCS, vol. 8559, pp. 585–602. Springer, Cham (2014). doi:10.1007/978-3-319-08867-9_39

9. Inverso, O., Tomasco, E., Fischer, B., Torre, S., Parlato, G.: Lazy-CSeq: a lazy sequentialization tool for C. In: Ábrahám, E., Havelund, K. (eds.) TACAS 2014. LNCS, vol. 8413, pp. 398–401. Springer, Heidelberg (2014). doi:10.1007/978-3-642-54862-8_29

10. Kahlon, V., Gupta, A., Sinha, N.: Symbolic model checking of concurrent programs using partial orders and on-the-fly transactions. In: Ball, T., Jones, R.B. (eds.) CAV 2006. LNCS, vol. 4144, pp. 286–299. Springer, Heidelberg (2006). doi:10.1007/11817963_28

11. Khare, S., Saraswat, S., Kumar, S.: Static program analysis of large embedded code base: an experience. In: Proceedings of the 4th India Software Engineering Conference, ISEC 2011, pp. 99–102. ACM, New York (2011). http://doi.acm.org/10.1145/1953355.1953368

12. Torre, S., Madhusudan, P., Parlato, G.: Reducing context-bounded concurrent reachability to sequential reachability. In: Bouajjani, A., Maler, O. (eds.) CAV 2009. LNCS, vol. 5643, pp. 477–492. Springer, Heidelberg (2009). doi:10.1007/978-3-642-02658-4_36

13. Lal, A., Reps, T.: Reducing concurrent analysis under a context bound to sequential analysis. Formal Methods Syst. Des. 35(1), 73–97 (2009)

14. Qadeer, S., Wu, D.: Kiss: keep it simple and sequential. In: ACM SIGPLAN Notices, vol. 39, pp. 14–24. ACM, New York (2004)

15. Siegel, S.F., Dwyer, M.B., Gopalakrishnan, G., Luo, Z., Rakamaric, Z., Thakur, R., Zheng, M., Zirkel, T.K.: Civl: The concurrency intermediate verification language. Technical report UD-CIS-2014/001, Department of Computer and Information Sciences, University of Delaware (2014)

16. Sinha, N., Wang, C.: Staged concurrent program analysis. In: Proceedings of the Eighteenth ACM SIGSOFT International Symposium on Foundations of Software Engineering, pp. 47–56. ACM, New York (2010)

17. Sinha, N., Wang, C.: On interference abstractions. In: ACM SIGPLAN Notices, vol. 46, pp. 423–434. ACM, New York (2011)

18. Tomasco, E., Inverso, O., Fischer, B., Torre, S., Parlato, G.: MU-CSeq: sequentialization of C programs by shared memory unwindings. In: Ábrahám, E., Havelund, K. (eds.) TACAS 2014. LNCS, vol. 8413, pp. 402–404. Springer, Heidelberg (2014). doi:10.1007/978-3-642-54862-8_30

19. Tomasco, E., Inverso, O., Fischer, B., Torre, S., Parlato, G.: Verifying concurrent programs by memory unwinding. In: Baier, C., Tinelli, C. (eds.) TACAS 2015. LNCS, vol. 9035, pp. 551–565. Springer, Heidelberg (2015). doi:10.1007/978-3-662-46681-0_52
20. Tomasco, E., Nguyen Lam, T., Fischer, B., La Torre, S., Parlato, G.: Separating computation from communication: a design approach for concurrent program verification (2016)

Author Index

Printed in the United States
By Bookmasters